Conversion factors

		$= 32.174\ \text{lbm·ft/s}^2$
	1 N	$= 10^5$ dyne
	1 kg (force)	$= 9.80665$ N
Pressure (and stress):	1 lbf/in^2	$= 6.895$ kPa (kPa $\equiv$ kN/m^2)
	1 lbf/ft^2	$= 47.88$ Pa
	1 atm	$= 1.01 \times 10^5$ Pa
		$= 760$ mm Hg
		$= 2116.2$ lbf/ft^2
	1 bar	$= 10^5$ Pa
	1 mmHg	$= 13.3322$ Pa
Density:	1 lbm/ft^3	$= 16.0166$ kg/m^3
	1 kg/m^3	$= 6.243 \times 10^{-2}$ lbm/ft^3
		$= 1.9404 \times 10^{-3}$ slug/ft^3
Thermal conductivity:	1 BTU/ft.h.°R	$= 1.73073$ W/m·K
Velocity:	1 mi/h	$= 1.467$ ft/s $= 1.609$ km/h $= 0.4470$ m/s
	1 knot	$= 1.15155$ mile/h
	1 km/h	$= 0.2278$ m/s $= 0.6214$ mile/h
		$= 0.9113$ ft/s
	1 ft/s	$= 0.3048$ m/s $= 12.0$ in/s
Temperature:	1 K	$= 1.8$°R
		$= $°C $+ 273.15$
	1 °R	$= $°F $+ 459.67$
	T(°F)	$= 1.8$T(°C) $+ 32$
	T(°C)	$= 5/9[$T(°F) $- 32]$
	0°C	$= 273.15$ K
	0°F	$= 459.67$ °R
Viscosity:	1 kg/m·s	$= 0.67197$ lbm/ft·s
	1 centipoise	$= 0.001$ Pa·s $= 0.001$ kg/m·s
		$= 6.7197 \times 10^{-4}$ lbm/ft·s
	1 stoke	$= 1.0 \times 10^{-4}$ m^2/s (kinematic viscosity)
	1 lbf-s/ft^2	$= 47.88025$ kg/m·s
	1 lbm/ft.s	$= 1.48817$ N·s/m^2
Thrust specific fuel consumption (TSFC):	1 lbm/h/lbf	$= 28.33$ mg/N·s
Specific Thrust:	1 lbf/lbm/s	$= 9.807$ N·s/kg
Gas constant R:	1 ft^2/s^2.°R	$= 0.1672$ m^2/s^2·K
Specific Enthalpy:	1 BTU/lbm	$= 2.326$ kJ/kg

Aircraft
Propulsion

Aircraft Propulsion

SAEED FAROKHI

University of Kansas

WILEY

John Wiley & Sons, Inc.

PUBLISHER	Don Fowley
ASSOCIATE PUBLISHER	Dan Sayre
ACQUISITIONS EDITOR	Jennifer Welter
SENIOR PRODUCTION EDITOR	Nicole Repasky
MARKETING MANAGER	Christopher Ruel
CREATIVE DIRECTOR	Harry Nolan
DESIGNER	Michael St. Martine
PRODUCTION MANAGEMENT SERVICES	Thomson Digital Limited
EDITORIAL ASSISTANT	Mark Owens
MEDIA EDITOR	Lauren Sapira

On the Cover: Rolls-Royce's latest engine, the Trent 1000, which is the lead engine to power the Boeing 787. Photo courtesy of Rolls-Royce plc.

This book was set in Times Roman by Thomson Digital Limited and printed and bound by Hamilton Printing. The cover was printed by Phoenix Color.

This book is printed on acid free paper. ∞

To order books or for customer service please, call 1-800-CALL WILEY (225-5945).

Library of Congress Cataloging in Publication Data:

Farokhi, Saeed.
    Aircraft propulsion / Saeed Farokhi.
        p. cm.
    Includes index.
    ISBN 978-0-470-03906-9 (pbk.)
1. Airplanes—Jet propulsion. 2. Airplanes—Motors—Design and construction. I. Title.
    TL709.F34 2008
    629.134′35—dc22   2007028896

Printed in the United States of America

10 9 8 7 6 5 4 3 2

This book is dedicated to the memory of my parents who gave me the opportunity to pursue my dreams.

Preface

Intended Audience

This book is intended to provide a foundation for the analysis and design of aircraft engines. The target audience for this book is upperclassmen, undergraduates, and first-year graduate students in aerospace and mechanical engineering. The practicing engineers in the gas turbine and aircraft industry will also benefit from the integration and system discussions in the book. Background in thermodynamics and fluid mechanics at a fundamental level is assumed.

Motivation

In teaching undergraduate and graduate propulsion courses for the past 23 years, I accumulated supplemental notes on topics that were not covered in most of our adopted textbooks. The supplemental materials ranged from issues related to the propulsion system integration into aircraft to the technological advances that were spawned by research centers around the world. I could have continued handing out supplemental materials to the textbooks to my classes, except that I learned that the presentation style to undergraduate students had to be (pedagogically) different than for the graduate students. For example, leaving out many steps in derivations of engineering principles can lead to confusion for most undergraduate students. Although it is more important to grasp the underlying principles than the *mechanics* of some derivations, but if we lose the students in the derivation phase, they may lose sight of the underlying principles as well. Another motivation for attention to details in analysis is my conviction that going back to basics and showing how the end results are obtained demystifies the subject and promotes students' confidence in their own abilities.

Mathematical Level

The mathematics in the present book is intentionally kept at the calculus and basic differential equations level, which makes the book readily accessible to undergraduate engineering students. Physical interpretations of mathematical relations are always offered in the text to help students grasp the physics that is hidden and inherent in the formulas. This approach will take the mystery out of formulas and let engineering students go beyond symbols and into understanding concepts.

Chapter Organization and Topical Coverage

The first chapter is an introduction to airbreathing aircraft engines and is divided in two parts. The first part reviews the history of gas turbine engine development, and the second part highlights modern concepts in aircraft engine and vehicle design. Young engineering students are excited to learn about the new opportunities and directions in aircraft engine design that are afforded by advances in materials, manufacturing, cooling technology, computational methods, sensors, actuators, and controls. Renewed interest in hypersonic airbreathing engines in general and supersonic combustion ramjets in particular as well as a sprawling interest in Uninhabited Aerial Vehicles (UAVs) has revitalized the ever-popular X-planes. The goal of Chapter 1 is first to inform students about the history, but more importantly to excite them about the future of aerospace engineering.

Chapter 2 is a review of compressible flow with heat and friction. The conservation principles are reviewed and then applied to normal and oblique shocks, conical shocks, and expansion waves, quasi-one-dimensional flows in ducts as well as Rayleigh and Fanno flows. At the closing of Chapter 2, the impulse concept and its application to gas turbine engine components are introduced.

Chapter 3 is on engine thrust and performance parameters. Here, we introduce internal and external performance of aircraft engines and their installation effect.

Chapter 4 describes aircraft gas turbine engine cycles. The real and ideal behaviors of engine components are described *simultaneously* in this chapter. Efficiencies, losses, and figures of merit are defined both physically and mathematically for each engine component in Chapter 4. Once we define the real behavior of all components in a cycle, we then proceed to calculate engine performance parameters, such as specific thrust, specific fuel consumption and thermal and propulsive efficiencies. The ideal cycle thus becomes a special case of a real cycle when all of its component efficiencies are equal to one.

The next five chapters treat aircraft engine components. Chapter 5 deals with aircraft inlets and nozzles. Although the emphasis throughout the book is on internal performance of engine components, the impact of external or installation effects is always presented for a balanced view on aircraft propulsion. As a building block of aircraft inlet aerodynamics, we have thoroughly reviewed two-dimensional and conical diffuser performance. Some design guidelines, both internal and external to inlet cowl, are presented. Transition duct aero-dynamics also plays an important role in design and understanding of aircraft inlets and is thus included in the treatment. Supersonic and hypersonic inlets with their attendant shock losses, boundary layer management, and instabilities such as buzz and starting problem are included in the inlet section of Chapter 5. The study of aircraft exhaust systems comprises the latter part of Chapter 5. Besides figures of merit, the performance of a convergent nozzle is compared with the de Laval or a convergent–divergent nozzle. The requirements of reverse- and vector thrust are studied in the context of thrust reversers and modern thrust vectoring nozzles. In the hypersonic limit, the exhaust nozzle is fully integrated with the vehicle and introductory design concepts and off-design issues are presented. Nozzle cooling is introduced for high-performance military aircraft engine exhaust systems and the attendant performance penalties and limitations are considered. Plug nozzle and its on- and off-design performances are introduced. Since mixers are an integral part of long-duct turbofan engines, their effect on gross thrust enhancement is formulated and presented in the nozzle section in Chapter 5.

Chemical reaction is studied on a fundamental basis in Chapter 6. The principles of chemical equilibrium and kinetics are used to calculate the composition of the products of combustion in a chemical reaction. These principles allow the calculation of flame temperature and pollutant formations that drive the design of modern aircraft gas turbine combustors. Further details of flame speed, stability, and flameholding are presented in the context of combustion chamber and afterburner design. Pollutant formation and its harmful impact on ozone layer as well as the greenhouse gases in the exhaust are presented to give students an appreciation for the design issues in modern combustors. Aviation fuels and their properties and a brief discussion of combustion instability known as screech are included in Chapter 6.

Turbomachinery is introduced in three chapters. *Chapter 7 deals with axial-flow compressors in two and three dimensions.* The aerodynamics of axial-flow compressors and stage performance parameters are derived. The role of cascade data in two-dimensional design is presented. Emphasis throughout this chapter is in describing the physical phenomena that lead to losses in compressors. Shock losses and transonic fans are introduced. The physics of compressor instability in stall and surge is described. A simple model by Greitzer that teaches the value of characteristic timescales and their relation to compressor instability is outlined. *Chapter 8 discusses the aerodynamics and performance of centrifugal compressors.* Distinctive characters of centrifugal compressors are highlighted and compared with axial-flow compressors. Turbine aerodynamics and cooling are presented in Chapter 9. Component matching and engine parametric study is discussed in Chapter 10. Finally, chemical rocket and hypersonic propulsion is presented in Chapter 11.

Instructor Resources

The following resources are available to instructors who adopt this book for their course. Please visit the website at www.wiley.com/college/farokhi to request a password and access these resources.

- Solutions Manual
- Image Gallery

Acknowledgments

I express my sincere appreciation and gratitude to all those who have contributed to my understanding of fluid mechanics and propulsion. Notable among these are my professors in Illinois and MIT. Hermann Krier, Jack Kerrebrock, James McCune, William Hawthorne, and Ed Greitzer contributed the most. The fellow graduate students in the Gas Turbine Lab were also instrumental in my education. Choon Tan, Maher El-Masri, Allan Epstein, Arun Sehra, Mohammad Durali, Wai Cheng, Segun Adebayo, James Fabunmi, and Anthony Nebo discussed their dissertations with me and helped me understand my own. In the Gas Turbine Division of Brown, Boveri and Co. in Baden, Switzerland, I learned the value of hardware engineering and testing, advanced product development, and component research. My colleagues, Meinhard Schobeiri, Konrad Voegeler, Hans Jakob Graf, Peter Boenzli, and Horst Stoff, helped me understand how industry works and how it engineers new products. At the

University of Kansas, my graduate students were my partners in research and we jointly advanced our understanding of fluid mechanics and propulsion. My doctoral students, Ray Taghavi, Gary Cheng, Charley Wu, Ron Barrett, and Kyle Wetzel, taught me the most. I appreciate the contributions of 30 M.S. students whom I chaired their theses to our ongoing research. The colleagues at NASA-Lewis (now Glenn) who sponsored my research and provided insightful discussions and hospitality over the summer months in Cleveland are Ed Rice, Khairul Zaman, Ganesh Raman, Bernie Anderson, Reda Mankbadi, James Scott, and Charlie Towne who welcomed me into their laboratory (and their homes), and we enjoyed some fruitful research together. The faculty and staff in the Aerospace Engineering Department of the University of Kansas have been very supportive for the past 23 years, and I would like to express my sincere appreciation to all of them. Vince Muirhead, Jan Roskam, Eddie Lan, Dave Downing, Howard Smith, Dave Ellis, Tae Lim, John Ogg, James Locke, Mark Ewing, Rick Hale, and Trevor Sorenson taught me an appreciation for their disciplines in aerospace engineering. I joined my colleagues in GE-Aircraft Engines in teaching propulsion system design and integration short courses to engineers in industry, FAA, and NASA for many years. I learned from Don Dusa and Jim Younghans from GE and Bill Schweikhard of KSR some intricate aspects of propulsion engineering and flight-testing.

I would like to thank the following colleagues who reviewed the draft manuscript:

David Benson, Kettering University
Kirby S. Chapman, Kansas State University
Mohamed Gad-el-Hak, Virginia Commonwealth University
Knox Millsaps, Naval Postgraduate School
Alex Moutsoglou, South Dakota State University
Norbert Mueller, Michigan State University
Meinhard T. Schobeiri, Texas A&M University
Ali R. Ahmadi, California State University and Polytechnic—Pomona
Ganesh Raman, Illinois Institute of Technology

Finally, I express my special appreciation to my wife of 36 years, Mariam, and our three lovely daughters, Kamelia, Parisa, and Farima (Fallon) who were the real inspiration behind this effort. I could not have contemplated such a huge project without their love, understanding, encouragement, and support. I owe it all to them.

SAEED FAROKHI
LAWRENCE, KANSAS
MARCH 16, 2007

Table of Contents

5

Aircraft Engine Inlets and Nozzles   225

6

Combustion Chambers and Afterburners 308

7

Axial Compressor Aerodynamics   389

Nomenclature

Latin	Definition	Unit
a	Local speed of sound	m/s, ft/s
a	Semimajor axis of inlet elliptic lip (internal)	m, ft
a	Swirl profile parameter	—
a_t	Speed sound based on total temperature	m/s, ft/s
A	Area	m^2, ft^2
A_n	Projection of area in the normal direction	m^2, ft^2
A_9	Nozzle exit flow area	m^2, ft^2
A_{ref}	Reference area	m^2, ft^2
A_0	Inlet (freestream) capture area	m^2, ft^2
A_1	Inlet capture area	m^2, ft^2
A_8, A_{8geo}	Nozzle throat area (geometrical area)	m^2, ft^2
A_{8eff}	Effective nozzle throat area	m^2, ft^2
A_B	Blocked area (due to boundary layer)	m^2, ft^2
A_b	Burning area of grain in solid rocket motors	m^2, ft^2
A_E	Effective area	m^2, ft^2
A_{HL}	Inlet highlight area	m^2, ft^2
A_M	Maximum nacelle area	m^2, ft^2
A_{th}	Inlet throat area	m^2, ft^2
A^*	Sonic throat, choked area	m^2, ft^2
b	Semiminor axis of inlet elliptic lip (internal)	m, ft
b	Swirl profile parameter	—
B	Blockage	—
B	Compressor instability parameter due to Greitzer	—
$\vec{C}$	Absolute velocity vector	m/s, ft/s
C	Absolute flow speed, i.e., $\sqrt{C_r^2 + C_\theta^2 + C_z^2}$	m/s, ft/s
c	Chord length	m, ft
c	Effective exhaust velocity in rockets	m/s, ft/s
c^*	Characteristic velocity in rockets	m/s, ft/s
C_r, C_θ, C_z	Radial, tangential, axial velocity components in the absolute frame of reference	m/s, ft/s

C_D	drag coefficient	—
C_f	Friction drag coefficient	—
c_f	Local skin friction coefficient	—
C_F	Force coefficient	—
C_p	Pressure coefficient	—
C_{PR}	Diffuser static pressure recovery coefficient	—
C_A	Nozzle flow angularity loss coefficient	—
C_{D8}	Nozzle (throat) discharge coefficient	—
C_{fg}	Nozzle gross thrust coefficient	—
C_V	Nozzle exit velocity coefficient	—
C_d	Sectional profile drag coefficient	—
C_{Di}	Induced drag coefficient	—
C_l	Sectional lift coefficient	—
C_h	Enthalpy-equivalent of the static pressure rise coefficient due to Koch	—
c_p	Specific heat at constant pressure	J/kg·K
c_v	Specific heat at constant volume	J/kg·K
$\bar{c}_p$	Molar specific heat at constant pressure	J/kmol·K
d	Flameholder width	m, ft
D	Diameter, drag	m, N
D	Liquid fuel droplet diameter	micron
$D_{flameholder}$	Flameholder drag	N, lbf
D_{add}	Additive drag	N, lbf
$D_{nacelle}$	Nacelle drag	N, lbf
D_{pylon}	Pylon drag	N, lbf
D_r	Ram drag	N, lbf
$D_{spillage}$	Spillage drag	N, lbf
$D_{aft\text{-}end}$	Nozzle aft-end drag	N, lbf
$D_{boattail}$	Nozzle boattail drag	N, lbf
$D_{plug\text{-}friction}$	Friction drag on the plug nozzle	N, lbf
D	Diffusion factor	—
D'	Two-dimensional or sectional profile drag	N/m
$\hat{e}$	Unit vector	—
e	Specific internal energy	J/kg
e_c, e_t	Polytropic efficiency of compressor or turbine	—
E	Internal energy	J
E_a	Activation energy	kcal/mol
f	Fuel-to-air ratio	—
f_{stoich}	Stoichiometric fuel-to-air ratio	—
F_g	Gross thrust	N, lbf
F_{lip}	Lip suction force	N, lbf
F_{plug}	Axial force on the nozzle plug	N, lbf
F_n	Net thrust	N, lbf
F	Force	N, lbf
F_θ, F_z	Tangential force, axial force	N, lbf
f_D	D'Arcy (pipe) friction factor	—
g	Staggered spacing (s.cos β in a rotor and s.cos α in a stator)	m
g_0	Gravitational acceleration on earth	m/s², ft/s²
h	Specific enthalpy	J/kg
h_t	Specific total enthalpy	J/kg
h	Heat transfer rate per unit area per unit temp. difference	W/m²K
h	Altitude above a planet	km, kft

h_t	Specific total (or stagnation) enthalpy in the absolute frame; $h + C^2/2$	J/kg
h_{tr}	Specific total enthalpy in relative frame of reference; $h + W^2/2$	J/kg
h_{lg}	Latent heat of vaporization	J/kg
HHV	Higher heating value	J/kg, BTU/lbm
H	Enthalpy	J, ft-lbf
H	Afterburner duct height	m, ft
i	Blade section incidence angle	deg
i_{opt}	Optimum incidence angle	deg
I_s	Specific impulse	s
I_t	Total impulse	N · s, lbf · s
I	Impulse	N, lbf
K_p	Equilibrium constant based on partial pressure	$(bar)^x$
K_n	Equilibrium constant based on molar concentration	—
L	Length	m, ft
L	Lift	N, lbf
L	Flameholder length of recirculation zone	m, ft
L	Diffuser wall length	m, ft
L	Diffusion length scale in a blade row	m, ft
LHV	Lower heating value	J/kg, BTU/lbm
L/D	Aircraft lift-to-drag ratio	—
M_b	Blowing parameter in film cooling, $\rho_c u_c / \rho_g u_g$	—
M_T	Blade tangential Mach number U/a	—
M_z	Axial Mach number, C_z/a	—
M_r	Relative Mach number; $(M_z^2 + M_T^2)^{1/2}$	—
M	Mach number	—
M^*	Characteristic Mach number	—
M_s	Gas Mach number upstream of a shock in nozzle	—
m	Parameter in Carter's rule for deviation angle	—
m	Mass	kg, lbm
$\dot{m}$	Mass flow rate	kg/s, lbm/s
$\dot{m}_c$	Corrected mass flow rate	kg/s, lbm/s
$\dot{m}_0$	Air mass flow rate	kg/s, lbm/s
$\dot{m}_f$	Fuel mass flow rate	kg/s, lbm/s
$\dot{m}_p$	Propellant (oxidizer and fuel) mass flow rate	kg/s, lbm/s
$\dot{m}_s$	Mass flow rate through the side of the control volume	kg/s, lbm/s
$\dot{m}_c$	Coolant flow rate	kg/s, lbm/s
MW	Molecular weight	kg/kmol
n	Exponent of superellipse	—
n	Polytropic exponent; parameter in general swirl distribution	—
N	Number of blades; shaft rotational frequency; number of stages	—
N	Number of bluff bodies in a flameholder	—
N	Diffuser axial length	m, ft
N_a	Avagadro's number (6.023×10^{23} molecules per gmole)	—
N_B	Inlet lip bluntness parameter	—
N_c	Corrected shaft speed	rad/s, rpm
$\hat{n}$	Unit normal vector (pointing out of a surface)	—
Nu	Nusselt number	—
Pr	Prandtl number	—
p	Static pressure (absolute)	bar, Pa, psia
p_t	Total pressure	bar, Pa, psia
p_s	Static pressure upstream of a shock (in nozzle)	bar, Pa, psia

$\wp$	Power	W, hp
$\wp_s$	Shaft power	W, BTU/s
PF	Pattern factor	—
P_f	Profile factor	—
Q	Heat exchange	J, BTU
q	Dynamic pressure	bar, atm
q	Heat transfer per unit area (heat flux)	W/m^2, $BTU/s.ft^2$
q	Heat transfer rate per unit mass flow rate	J/kg, BTU/lbm
Q_R	Fuel heating value	kJ/kg, BTU/lbm
$\dot{Q}$	Heat transfer rate	W, BTU/s
$\Re$	Aircraft range	nm
Re	Reynolds number	—
R	Gas constant	$J/kg \cdot K$, BTU/lbm°R
$R_{l.e.}$	Blade leading-edge radius	m, in
$\bar{R}$	Universal gas constant	$J/kmol \cdot K$
r	Mixture ratio (oxidizer to fuel) in liquid propellant rockets	—
r	Burning rate in solid propellant rockets	cm/s, in/s
r	Radius	m, ft
r	Cylindrical or spherical coordinate	—
r_h	Hub radius	m, ft
r_t	Tip radius	m, ft
r_m	Pitchline or mean radius $(r_h+r_t)/2$	m, ft
°R	Stage degree of reaction	—
S	Entropy	J/K
S_L	Laminar flame speed	m/s, ft/s
S_T	Turbulent flame speed	m/s, ft/s
St	Stanton number	—
SN	Smoke number	—
s	Specific entropy	$J/kg \cdot K$
s	Blade spacing	m, ft
t	Blade thickness	m, ft
t	Time	s
t_{max}	Maximum blade thickness	m, ft
$t_{reaction}$	Reaction time scale	ms
t_i	Ignition delay time	ms
t_e	Evaporation time scale	ms
T	Static temperature	K, °R, °C, °F
T_t	Total temperature	K, °R, °C, °F
T_f	Reference temperature, 298.16 K	K, °R, °C, °F
T_g	Gas temperature	K, °R, °C, °F
T_c	Coolant temperature	K, °R, °C, °F
T_{af}	Adiabatic flame temperature	K, °R, °C, °F
T_{aw}	Adiabatic wall temperature	K, °R, °C, °F
Tu	Turbulence intensity, $[(u'^2 + v'^2 + w'^2)/3]^{1/2}/V_m$	—
u	Speed, velocity normal to a shock	m/s, ft/s
u	Gas speed	m/s, ft/s
u'_{rms}	Turbulent fluctuating speed (root mean square)	m/s, ft/s
$\bar{U}$	Rotational velocity vector of rotor; $\omega \cdot r\hat{e}_\theta$	m/s, ft/s
U_T	Blade tip rotational speed, ωr_t	m/s, ft/s
u', v', w'	Root mean square of fluctuating velocities in 3 spatial directions	m/s, ft/s
$\bar{v}$	Average gas speed in the mixing layer	m/s, ft/s

V	Volume	m^3, ft^3
V	Speed	m/s, ft/s
V_m	Mean speed (used in stall margin)	m/s, ft/s
V'	Relative speed used in the stall margin analysis	m/s, ft/s
V_c	Compressor or chamber volume	m^3, ft^3
V_p	Plenum volume	m^3, ft^3
W	Weight	N, lbf
W	Flame width	m, ft
W	Width	m, ft
$\vec{W}, W$	Relative velocity vector, relative flow speed	m/s, ft/s
W_r, W_θ, W_z	Radial, tangential and axial velocity components in relative frame of reference	m/s, ft/s
w_c	Rotor specific work (rotor power per unit mass flow rate; $\wp/\dot{m}$)	J/kg, BTU/lbm
w	Specific work	J/kg, BTU/lbm
w	Tangential speed to an oblique shock	m/s, ft/s
w_p	Propellant weight	N, lbf
$W_{visc.}$	Rate of work done by the viscous force	W, BTU/s
X	Solid flow fraction in a rocket nozzle	—
X	Semimajor axis of an elliptic external cowl	m, ft
Y	Semiminor axis of an elliptic external cowl	m, ft
z	Axial coordinate in the cylindrical coordinate system	—
z	Airfoil camber	m, ft
z_{max}	Maximum airfoil camber	m, ft
x, y, z	Cartesian coordinates	—

Greek	**Definition**	**Unit**
α	Bypass ratio in a turbofan engine	—
α	Angle of attack	deg
α	Absolute flow angle with respect to the axial direction	deg
$\Delta\alpha$	Flow turning angle across a stator blade section	deg
Δp	Pressure drop	Pa, psi
β	Plane oblique shock wave angle	degree
β	Relative flow angle with respect to the axial direction	deg
β_m	Mean flow angle corresponding to an average swirl across a blade row	deg
$\Delta\beta$	Flow turning angle across a rotor blade section	deg
δ	Boundary layer thickness	m, ft
δ^*	Boundary layer displacement thickness	m, ft
δ	Ratio of total pressure to reference (standard sea level) pressure; p/p_{ref}	—
δ_T	Thermal boundary layer thickness	m, ft
δ^*	Deviation angle defined at the blade trailing edge, a cascade parameter	deg
$\Delta\bar{h}_f^0$	(Standard) molar heat of formation	J/kmol
Δh_f^0	(Standard) specific heat of formation	J/kg
ε	Tip clearance; slip factor	—
ε	A small quantity ($\ll 1$)	—
ε_g	Emissivity of gas	—
κ	Coefficient of thermal conductivity	W/m·K
κ_1	Blade leading-edge angle	deg
κ_2	Blade trailing-edge angle	deg
π	Total pressure ratio	—
ω	Angular speed	rad/s, rpm
ϖ	Total pressure loss parameter in a cascade; $\Delta p_t/q_r$	—

ϕ	Spherical coordinate	—
ϕ	Equivalence ratio	—
ϕ	Diffuser wall divergence angle	deg
ϕ	Flow coefficient; C_z/U	—
φ	Camber angle, $\kappa_1 - \kappa_2$	deg
Φ	Cooling effectiveness parameter	—
γ	Ratio of specific heats	—
Γ	Circulation (of a vortex filament)	m^2/s, ft^2/s
Υ	Cascade stagger angle or blade setting angle	deg
ρ	Fluid density	kg/m^3, lbm/ft^3
μ	Coefficient of viscosity	$N \cdot s/m^2$
μ	Mach angle	degree
ν	Kinematic viscosity $\equiv \mu/\rho$	m^2/s, ft^2/s
ν	Prandtl–Meyer angle	degree
π_c	Compressor total pressure ratio	—
π_b	Burner total pressure ratio	—
π_d	Inlet total pressure recovery	—
π_n	Nozzle total pressure ratio	—
π_K	Temperature sensitivity of chamber pressure in solid rockets	%/K, %/F
Π_M	Mach index $\equiv U_T/a_{t1}$	—
θ	Flow angle, cylindrical or spherical coordinate	degree
θ	Nozzle exit flow angle (from axial direction)	deg
θ	Ratio of total temperature to the reference (standard sea level) temperature; T/T_{ref}	—
θ	Circumferential extent of the inlet spoiled or distortion sector	deg
θ^*	Momentum deficit thickness in the boundary layer	m
σ	Cascade or blade solidity; c/s	—
σ	Stefan–Boltzmann constant	W/m^2K^4
σ_p	Temperature sensitivity of burning rate in solid propellant grain	%/K, %/F
τ	Shear stress	Pa, lbf/ft^2
τ	Total temperature ratio	—
τ	Characteristic timescale	s
τ_r, τ_s	Rotor torque, stator torque	N·m, ft-lbf
τ_t	Turbine total temperature ratio, T_{t5}/T_{t4}	—
τ_λ	Cycle limit enthalpy ratio, $c_{pt}T_{t4}/c_{pc}T_0$	—
$\tau_{\lambda AB}$	Limit enthalpy ratio with afterburner, $c_{p,AB}T_{t7}/c_{pc}T_0$	—
$\tau_{resident}$	Resident timescale	ms
η_b	Burner efficiency	—
η_o	Overall efficiency	—
η_p	Propulsive efficiency	—
η_{th}	Thermal efficiency	—
η_d	Adiabatic efficiency of a diffuser	—
η_n	Adiabatic efficiency of a nozzle	—
ξ	Coordinate along the vortex sheet	—
ψ	Stage loading parameter; $\Delta h_t/U^2$	—
ψ	stream function	m^2/s, ft^2/s
∇	Vector operator, Del	m^{-1}
∇p	Pressure gradient	bar/m
χ	Mole fraction	—
ζ	Propellant mass fraction	—

Subscripts

1, 2	Stations up- and downstream of a shock, or inlet and exit of a duct
C.S.	Control surface
C.V.	Control volume
e	Boundary layer edge
h	Hydraulic (in hydraulic diameter)
max	Maximum
n	Normal to an oblique shock
net	Net
r	Rotor, relative
rev	Reversible
s	Shock, shaft, stator
s	Isentropic
t	Total or stagnation
w	Wall
∞	Free stream

Superscripts

*	Sonic or critical state

Abbreviations and acronyms

AGARD	Advisory Group for Aeronautical Research and Development
AIAA	American Institute of Aeronautics and Astronautics
AR	Blade aspect ratio
AR	Diffuser area ratio
ASME	American Society of Mechanical Engineers
BLING	Bladed ring
BLISK	Bladed disk
BPR	Bypass ratio
C–D	Convergent–divergent
CEV	Crew Exploration Vehicle
C.G.	Center of gravity
CFD	Computational fluid dynamics
CO	Carbon monoxide
C.V.	Control volume
C.S.	Control surface
CDA	Controlled-diffusion airfoil
DCA	Double-circular arc blade
E^3	Energy efficient engine
EPA	Environmental Protection Agency
ET	External tank
GE	General Electric company
GNC	Guidance–navigation–control
HPC	High-pressure compressor
HPT	High-pressure turbine
IGV	Inlet guide vane
ICAO	International Civil Aviation Organization
K–D	Kantrowitz–Donaldson inlet
LE	Leading edge
LHS	Left-hand side
LOX	Liquid oxygen

LPC	Low-pressure compressor
LPT	Low-pressure turbine
MFR	Inlet mass flow ratio
MCA	Multiple-circular arc blade
MEMS	Micro-electro-mechanical systems
MIT	Massachusetts Institute of Technology
MR	Mass ratio
NACA	National Advisory Committee on Aeronautics
NASA	National Aeronautics ands Space Administration
NPR	Nozzle pressure ratio
NO_x	Nitric oxide(s)
N.S.	Normal shock
OPR	Overall pressure ratio
O.S.	Oblique shock
PDE	Pulse detonation engine
PR	Pressure ratio
PS	Pressure surface
P&W	Pratt & Whitney
RBCC	Rocket-based combined cycle
RJ	Ramjet
RMS	Root mean square (of a fluctuating signal, e.g. turbulence or total pressure distortion)
RR	Rolls-Royce
RAE	Royal Aeronautical Establishment
RHS	Right-hand side
SCRJ	Scramjet
Sfc	Specific fuel consumption (same as TSFC)
SM	Stall margin
SS	Suction surface
SSME	Space Shuttle Main Engine
TE	Trailing edge
TF	Turbofan
TJ	Turbojet
TP	Turboprop
TPC	Thermal protection coating
TO	Takeoff
TSFC	Thrust specific fuel consumption
UAV	Uninhabited aerial vehicles
UCAV	Uninhabited combat air vehicle
UDF	Unducted fan
UHB	Ultra-high bypass
UHC	Unburned hydrocarbons
VTOL	Vertical takeoff and landing

Aircraft
Propulsion

CHAPTER 1

Introduction

1.1   History of Airbreathing Jet Engine, a Twentieth-Century Invention—The Beginning

Powered flight is a twentieth-century invention. The era of powered flight began on December 17, 1903 with the Wright brothers who designed, fabricated, and flew "The Flyer" in Kitty Hawk, North Carolina. The power onboard The Flyer was a gas powered, 12-hp reciprocating intermittent combustion engine. This type of engine, with a propeller, provided power to all (manned) aircraft until late 1930s. The history of aircraft gas turbine engine started in January 1930 with a patent issued to Frank Whittle in Great Britain. Figure 1.1 shows a p–v diagram and components of the Whittle engine as they appeared in the patent application. The flow pattern and engine assembly are shown in Fig. 1.2. The performance of W1 engine and the aircraft that flew it are shown in Fig. 1.3. An engineer at work, Sir Frank Whittle, the inventor of jet engine, with a slide rule is shown in Fig. 1.4. For more details on Whittle turbojet see Meher-Homji (1997).

The gas turbine engine of Fig. 1.1 is based on Brayton cycle. The compression in Whittle engine is achieved via a double-sided centrifugal compressor. The axial compressor had not been developed due to aerodynamic stability complications. The combustion takes place in a reverse-flow burner that is very large relative to other engine components. The straight through-flow burner had posed problems with stable combustion and thus a reverse-flow combustor provided the needed flame stability in the burner. The compressor shaft power is delivered from a single-stage axial flow turbine.

In an independent effort, Hans-Joachim Pabst von Ohain invented a turbojet engine in Germany that was granted a patent in 1936. In 1937, von Ohain's engine designated as the He S-1 turbojet engine with hydrogen fuel was tested and produced a thrust of 250 pounds at 10,000 rpm. Von Ohain's engine was the first to be developed ahead of the Whittle engine and flew on the first jet-powered aircraft, Heinkel 178, in 1939. Both Whittle and von Ohain are

1

■ **FIGURE 1.1**
Patent drawings of Sir Frank Whittle jet engine

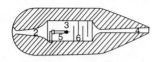

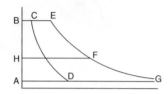

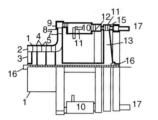

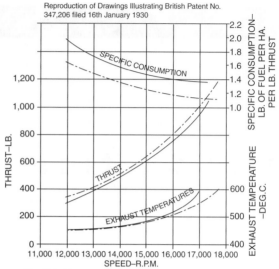

Reproduction of Drawings Illustrating British Patent No. 347,206 filed 16th January 1930

The W1 Engine: Curves of Thrust, Specific Fuel Consumption, and Exhaust Temperatures plotted against speed.

―――――――― Test results
―·―·―·―·―·― "Desing" performance

■ **FIGURE 1.2**
The assembly and flow pattern in Whittle jet engine

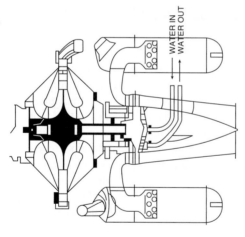

Assembly of W1 Engine. (Combustion chamber details not shown)

PROGRESS IN JET PROPULSION

The company formed by Whittle, known as Power Jets Ltd. produced the W.2.B. engine which was a classic of its type. It had the reverse flow combustion system which was typical of the Whittle designs. It was eventually developed to give nearly three times the thrust of the W.1 without occupying more space.

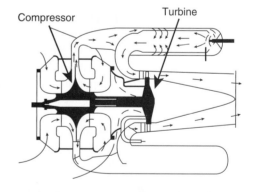

Compressor   Turbine

■ **FIGURE 1.3**
Performance testing of Whittle jet engine, known as W1, and the experimental aircraft, Gloster E28/39 that flew it in 1941

Gloster Experimental Aeroplane E28/39 at Takeoff. (Royal Aerospace Establishment, Crown Copyright.)

■ **FIGURE 1.4**
Sir Frank Whittle with a slide rule

■ **FIGURE 1.5**   The first historic meeting between the two inventors of the jet engine took place in WPAFB on May 3, 1978 (Courtesy AFRL/AFMC)

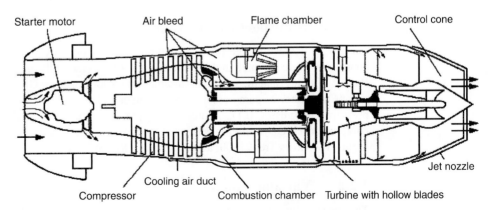

■ **FIGURE 1.6**   The first production jet aircraft, Me 262

credited as the coinventors of airbreathing gas turbine engine. Figure 1.5 shows the two inventors of the jet engine, a historical meeting on May 3, 1978.

The first production jet aircraft was Messerschmitt Me 262, shown in Fig. 1.6. Two Jumo 004B turbojet engines powered the Messerschmitt Me 262 jet fighter. The Me 262 first-flight was on July 18, 1942. Dr. Anselm Franz of the Junkers Engine Company designed the Jumo 004, which was based on von Ohain's patent. The Jumo 004B engine cutaway is shown in Fig. 1.7. This engine has many modern gas turbine features such as axial-flow compressor and a straight throughflow combustor with air-cooling of the turbine and the nozzle. For more details see Meher-Homji (1996).

■ **FIGURE 1.7**   Jumo 004B engine cutaway features an axial-flow compressor, a straight throughflow combustor, an air-cooled axial turbine, and an exhaust nozzle

■ **FIGURE 1.8**
The first U.S. produced aircraft gas turbine engine (Courtesy of Air Force Museum)

Drawing of Jumo 004B turbojet engine in Fig. 1.7 shows an air-cooling system that bleeds air from the compressor and cools the turbine and the exhaust nozzle. The engine produces ~2000 lb of thrust at airflow of 46.6 lb/s. The engine pressure ratio is 3.14, turbine inlet temperature is 1427°F, and the specific fuel consumption is 1.4 lbm/h/lbf-thrust. The engine dry weight is ~1650 lb, its diameter and length are ~30 and 152 in., respectively. Engine component efficiencies are reported to be 78% compressor, 95% combustor, and 79.5% turbine. We will put these numbers in perspective when we compare them with their modern counterparts.

The jet engine comes from Great Britain to the United States in 1941. The J-31 (also known by its company designation, I-16) was the first turbojet engine produced in quantity in the United States. It was developed from the General Electric I-A, which was a copy of the highly secret British "Whittle" engine. Figure 1.8 shows J-31 gas turbine engine (courtesy of Air Force Museum).

1.2   Innovations in Aircraft Gas Turbine Engines

In this section, we introduce the most significant innovations in gas turbine industry since the introduction of aircraft jet engine by Whittle and von Ohain. Dawson (1991), Wallace (1996) and NASA web sites (references 5 and 7) as well as NASA publication (reference 8) should be consulted for further reading/information.

1.2.1   Multispool Configuration

In order to achieve a high-pressure compression system, two distinct and complementary approaches were invented in the United States. One is the multispool concept (developed by Pratt & Whitney) and the second is variable stator (developed by GE). The multispool concept groups a number of compressor stages together in two or three groups, known as the low-pressure compressor (LPC), intermediate-pressure compressor (IPC), and high-pressure compressor (HPC). A different shaft that spins at different rotational speed drives each group. Figure 1.9 shows Trent 1000, a modern Rolls-Royce engine that employs three spools.

1.2.2   Variable Stator

The need to adjust the flow direction in a multistage high-pressure ratio compressor (in starting and off-design) prompted Gerhard Neumann of GE to invent variable stator. By allowing the

■ FIGURE 1.9 Three-spool gas turbine engine is developed by Rolls-Royce (Courtesy of Rolls-Royce, plc)

stators to rotate in pitch, compressors can operate at higher pressure ratios and away from stall. Modern gas turbine engines use variable stators in their LPC and IPC. The high-temperature environment of HPC has not been hospitable to variable stators.

1.2.3   Transonic Compressor

Better understanding of supersonic flow and the development of high strength-to-weight ratio, titanium alloy allowed the development of supersonic tip fan blades. The transonic fan is born at a high shaft speed that creates a relative supersonic flow at the tip and a subsonic flow at the hub. A modern transonic fan stage produces a stage pressure ratio of ∼1.6. The Jumo 004B produced a cycle pressure ratio of 3.14 with eight stages, which means an average stage pressure ratio of ∼1.15. Therefore to achieve a pressure ratio of 3.14, we need only two transonic fan stages instead of eight. The higher compression per stage has allowed a reduction in engine weight, size,

■ **FIGURE 1.10**
(a) CFD in transonic compressor rotor flowfield (Courtesy of NASA); (b) advanced transonic fan (Courtesy of Rolls-Royce, plc)

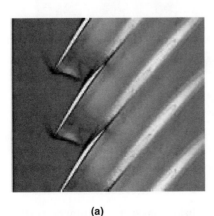

(a)   (b)

and part-count and has improved reliability. The advances in computational fluid dynamics (CFD) and nonintrusive testing techniques have paved the way for a better understanding of supersonic flow in compressors. A compressor flow simulation is shown in Fig. 1.10(a). The rotor passage shock, boundary layer interaction, and flow separation are clearly visualized in Fig. 1.10(a). An advanced transonic fan is shown in Fig. 1.10(b) from Rolls-Royce.

1.2.4   Low-Emission Combustor

The gas turbine combustor has perhaps seen the most dramatic innovations/changes since the Whittle reverse-flow burner. The better understanding of combustion process, from atomization and vaporization of the fuel to mixing with air and chemical reaction, has allowed efficient combustion to take place in small spaces. For example, compare the relative length and volume of the combustor in GP7000, shown in Fig. 1.11, to the Whittle engine or Jumo 004B.

In the textbox of Fig. 1.11, we note that the combustor emissions are characterized by their nitric oxide formation, the so-called NO_x, the unburned-hydrocarbon (UHC) emission, and finally carbon monoxide formation in the exhaust nozzle flow. In order to achieve low levels of pollutant emissions, different concepts in "staged combustion" are developed by aircraft engine manufacturers (as shown in Fig. 1.12).

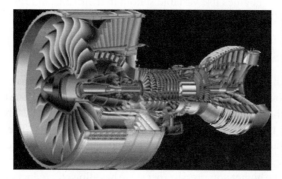

Engine Alliance engine: GP7000
T.O. Thrust: 76,500 lbs/340 kN
OPR (max. climb): 45.6
BPR (cruise): 8.7
Fan Diameter: 116.7 in.
Emissions:
NO_x: 59.7 g/kN
UHC: 3.9 g/kN
CO: 33.8 g/kN
Noise: 22.9 dB Margin to Stage 3

■ **FIGURE 1.11   Engine Alliance engine GP7000 (Courtesy of Engine Alliance) [Note: Engine Alliance is a 50/50 joint venture between GE Aviation and Pratt & Whitney]**

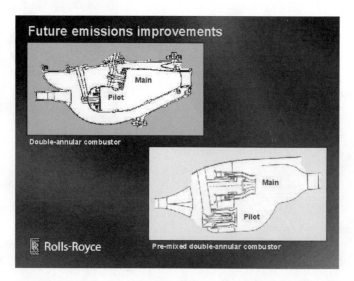

■ **FIGURE 1.12   Concepts in low-emission combustor design (Courtesy of Rolls-Royce, plc)**

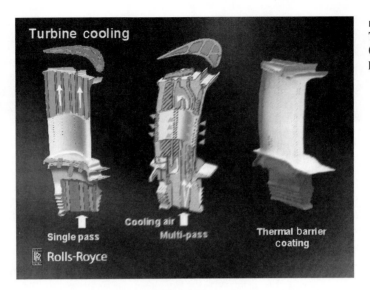

■ **FIGURE 1.13**
**Turbine blade cooling
(Courtesy of Rolls-Royce,
plc)**

1.2.5 Turbine Cooling

The need to cool the turbine stems from being able to operate the combustor at higher temperature (to produce more thrust) and to achieve turbine durability, that is, an improved component life. The first production turbojet engine, Jumo 004B, utilized internal cooling for the turbine blades. So, the concept is as old as the turbojet engine itself. Improved manufacturing techniques and better understanding of the flow physics involved in coolant ejection, mixing with hot gas, and three-dimensional flow in turbines have allowed for a rationed approach to coolant usage as well as component life enhancement. Figure 1.13 shows a single- and a multipass internal cooling of a turbine blade that incorporates film cooling as well as the thermal protection (or barrier) coating (TPC) to reduce the heat transfer to turbine blades.

1.2.6 Exhaust Nozzles

The concept of an exhaust nozzle for aircraft jet engine has changed from a simple convergent duct that was used to propel the hot exhaust gases to a variable-geometry and multitasked component in modern designs. The new tasks involve thrust reversing, thrust vectoring, noise suppression, and dynamic stability enhancement of maneuvering aircraft. To achieve these goals, advancements in nozzle cooling, actuation, and manufacturing had to be realized. Figure 1.14 shows a sophisticated propulsion layout (and nozzle system) in F-35 aircraft that

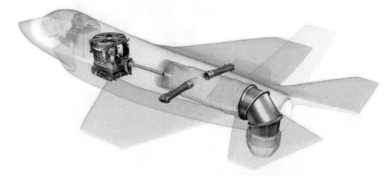

■ **FIGURE 1.14** **Propulsion layout for vertical landing and stability of F-35 Joint Strike Fighter (Courtesy of Rolls-Royce, plc)**

■ **FIGURE 1.15** **F119 engine that powers F-22 Raptor is shown in vector thrust (Courtesy of Pratt & Whitney)**

has vertical takeoff/landing capability as well as roll control in hover. Figure 1.15 shows a $\pm 20°$ vector thrust in F119 engine developed by Pratt & Whitney for F-22 "supercruise" aircraft.

1.2.7   Modern Materials and Manufacturing Techniques

Nonmetallics and composite materials represent a sizable change in modern material usage in aircraft and jet engines. Metal matrix composites technology offers high strength to weight ratio relative to titanium and nickel superalloys suitable for fan blades. Single crystal turbine blades offer more resistance to vibration and thus fatigue failure. Manufacturing technique that utilizes honeycomb core with a composite skin offers weight and stress reductions in fan blades. Compressor weight savings are derived from bladed disk "Blisk" and bladed ring "Bling" manufacturing technology. All these are shown in Fig. 1.16.

An example of a modern engine is EJ200, which powers the "Eurofighter" Typhoon (shown in Fig. 1.17). Its design features are tabulated in Table 1.1.

■ **FIGURE 1.16**
**Advanced materials
and manufacturing
techniques (Courtesy
of Rolls-Royce, plc)**

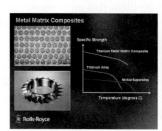
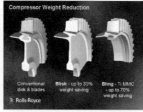
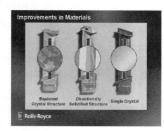
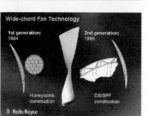

■ **FIGURE 1.17**
**Cutaway of EJ200,
an afterburning
turbofan engine
designed for the
Eurofighter
(Courtesy of
Rolls-Royce, plc)**

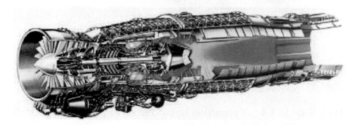

■ **TABLE 1.1**
EJ200 Specifications

		Fan/compressor stages	1/3/5
		LPT/HPT	1/1
		Max. diameter	29 in.
Two-spool configuration		OPR	26:1
Fan technology		BPR	0.4
Wide chord		Length	157 in.
Single-crystal "Blisk" (Bladed Disk)		Dry weight	2,286 lbf
No IGV		Sfc (max. power)	0.81 lbm/h/lbf
Three-stage LPC:	4.2 PR	Sfc w. AB	1.75 lbm/h/lbf
Mass flow:	77 kg/s or 170 lbm/s	Thrust (SL)	13,500 lbf
HPC:	Single crystal Blisk	Thrust w. AB	20,250–22,250 lbf
T_{t4}:	1800 K (or 2780°F)	Thrust/weight (Dry)	5.92
HPT:	Air-cooled + TPC (two-layers)	Thrust/weight (AB)	9.1
C-D nozzle:	Titanium alloy	Thrust vectoring:	23° any direction
		Engine management	FADEC + monitoring unit

The modern materials and the manufacturing techniques that we have discussed are described in Table 1.1. Compare the turbine inlet temperature (T_{14}) in EJ200 and Jumo 004B, or thrust-to-weight ratio.

1.3   New Engine Concepts

In this section, we examine new concepts in aircraft propulsion. The first two concepts harness unsteadiness as a means of thrust production. The third is a triumph of micro-electro-mechanical (MEM) device manufacturing. The rest are combined cycles.

1.3.1   Wave Rotor Topping Cycle

Wave rotor creates a pressure gain in the combustor, instead of the baseline pressure drop, thereby enhances cycle efficiency. As a simple example of a higher efficiency cycle that takes advantage of constant-volume combustion, we may examine the Humphrey cycle. Schematics of the wave rotor topping cycle concept, a wave rotor hardware, and a test rig at NASA-Glenn Research Center are shown in Fig. 1.18. A performance chart of the wave rotor topping cycle for small turboshaft engines, also in Fig. 1.18, shows nearly 10% fuel savings compared with a baseline engine.

1.3.1.1   Humphrey Cycle versus Brayton Cycle. An ideal Humphrey cycle is shown in Fig. 1.19 in a pressure–volume and temperature–entropy diagrams. Combustion takes place at constant volume in a Humphrey cycle, whereas it takes place at constant pressure in an ideal Brayton cycle. We utilize the definition of cycle efficiency and thermodynamic principles to get Brayton and Humphrey cycle efficiencies.

Cycle efficiency of a constant-pressure combustion (Brayton) cycle: 1–2–5–6–1, is

$$\eta_{th} = 1 - \frac{T_1}{T_2}$$

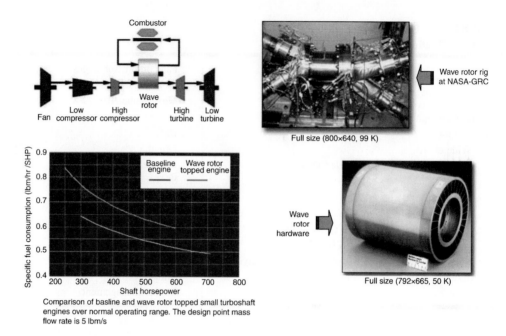

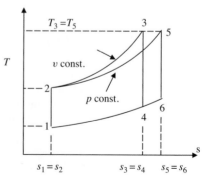

Wave rotor rig at NASA-GRC

Full size (800×640, 99 K)

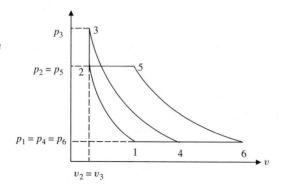

Wave rotor hardware

Full size (792×665, 50 K)

Comparison of basline and wave rotor topped small turboshaft engines over normal operating range. The design point mass flow rate is 5 lbm/s

■ **FIGURE 1.18** **Schematics of the wave rotor toping cycle wave rotor hardware and a test rig at NASA (Courtesy of NASA)**

Constant-volume combustion (Humphrey cycle: 1–2–3–4–1),

$$\eta_{\text{th}} = \frac{1 - \gamma \dfrac{T_1}{T_2} \left[\left(\dfrac{T_3}{T_2} \right)^{\frac{1}{\gamma}} - 1 \right]}{\left[\dfrac{T_3}{T_2} - 1 \right]}$$

where γ is the ratio of specific heats.

Cycle efficiency in Humphrey cycle depends on T_1/T_2 and on the temperature ratio T_3/T_2 (in effect p_3/p_2). Figure 1.20 shows the ideal cycle thermal efficiency of a Brayton and a Humphrey cycle for $T_1 = 288$ K, $T_2 = 800$ K, and T_3 that varies between 1600 and 2500 K, for $\gamma = 1.4$.

■ **FIGURE 1.19**
Constant-volume and constant-pressure combustion cycles

■ FIGURE 1.20
Ideal thermal efficiency
of Humphrey and
Brayton cycles for
$\gamma = 1.4$, and $T_1 = 288$ K,
$T_2 = 800$ K, and T_3
that varies between
1600 and 2500 K

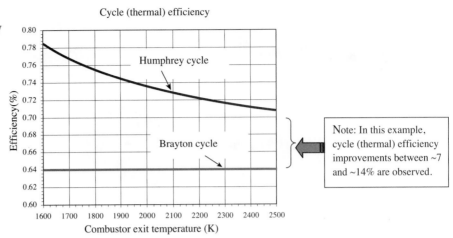

Note: In this example, cycle (thermal) efficiency improvements between ~7 and ~14% are observed.

■ FIGURE 1.21
The Pulse Detonation
engine with a trigger
chamber (Courtesy
of NASA)

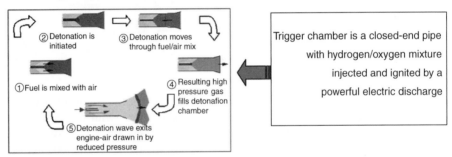

PDE wave cycle

Trigger chamber is a closed-end pipe with hydrogen/oxygen mixture injected and ignited by a powerful electric discharge

1.3.2   Pulse Detonation Engine (PDE)

Pulse Detonation Engine (PDE) is a constant-volume combustion ramjet that is capable of producing static thrust. The operation of a PDE is similar to a pulsejet except combustion in a pulsejet is based on the principle of deflagration that is a slow wave front with low-pressure ratio. The PDE creates a detonation wave, which is akin to an explosion that creates high-pressure shock waves. To get a feel for how often these explosions occur, we note the frequency of these explosions that is ~60 detonations per second. The PDE wave cycle is shown in Fig. 1.21.

1.3.3   Millimeter-Scale Gas Turbine Engines: Triumph of MEMS

Microchip manufacturing techniques and some vivid imaginations have given birth to millimeter-scale gas turbine engines. Figure 1.22 shows a "button" size gas turbine engine that is designed, manufactured, and tested at MIT. At these scales, the rotor has to spin at ~1,000,000 rpm to achieve the needed compression for the cycle. The process of fuel injection, atomization, vaporization, and combustion is a challenge among the myriad of other mechanical challenges in the manufacturing of millimeter-scale gas turbine engine.

1.3.4   Combined Cycle Propulsion: Engines from Takeoff to Space

There are several developments that address combined cycles as a means of producing efficient propulsion over a wide range of flight speeds, typically from takeoff to hypersonic Mach numbers. An example of this approach is found in airbreathing rocket engine, which is a

■ **FIGURE 1.22**
Millimeter-scale gas
turbine engine with
the rotor and external
shell (Courtesy of MIT)

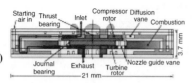

■ **FIGURE 1.23**
An RBCC air-
augmented rocket with
an ejector nozzle
(with Mach contours
computed) (Courtesy
of NASA)

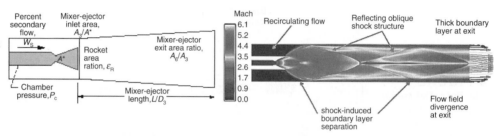

Rocket-Based Combined Cycle (RBCC) engine. At takeoff where conventional ramjets are incapable of producing thrust, a rocket is fired (with an ejector nozzle configuration to get a thrust boost) that accelerates the vehicle to say Mach 2. At Mach 2, the rocket is turned off and air intakes are opened to start a subsonic ramjet engine operation. The airbreathing engine switches from the subsonic to supersonic combustion ramjet (scramjet) near Mach 5. Scramjet will accelerate the vehicle to say Mach 15. The air intakes close at Mach 15 and rocket operation resumes accelerating the vehicle to orbital speeds (~Mach 25 or higher). The rocket with the ejector nozzle and computational results of Mach contours are shown in Fig. 1.23. An RBCC engine is capable of reducing launch costs by two orders of magnitude. Artist's concept of the vehicle is shown in Fig. 1.24. An RBCC flight weight engine system test was conducted in 2006. Figure 1.25 shows the test firing of the airbreathing rocket.

■ **FIGURE 1.24** Artist's drawing of
an advanced launch vehicle using RBCC
propulsion (Courtesy of NASA)

■ **FIGURE 1.25** Testing of an
airbreathing rocket at NASA (Courtesy of
NASA)

1.4   New Vehicles

There are exciting new vehicles on the drawing board for many different missions at many different speeds. The interest in uninhabited aerial vehicles (UAVs) has prompted new configurations such as Northrop–Grumman X-47 "Pegasus," or the tailless agility aircraft, X-36 from Boeing, or X-45A Unmanned Combat Air Vehicle (UCAV). NASA's interest in hypersonic flight and scramjet propulsion has prompted the X43 series of technology demonstrator vehicles. These aircraft are shown in Fig. 1.26.

Northrop–Grumman X–47 Pegasus

Boeing X-36

Boeing X-45A UCAV

NASA X43-A Mach 10 airbreathing
technology demonstrator

NASA X43-B technology demonstrator

NASA X43-C technology demonstrator

■ **FIGURE 1.26**   **Uninhabited aerial vehicles and NASA X43 technology demonstrators (Courtesy of NASA)**

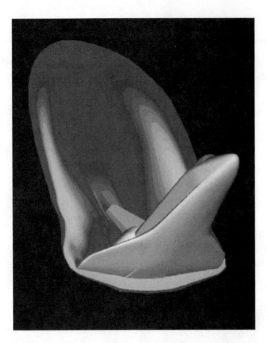

■ **FIGURE 1.27** **Flowfield simulations around the Space Shuttle in reentry using computational fluid dynamics (Courtesy of NASA)**

1.5   Summary

There are exciting developments in aerospace propulsion and vehicle design:

- Physics-based computer simulation/design
- Advanced composite materials
- Manufacturing will take unprecedented precision from nano-scale up
- Exciting new vehicles on the horizon
- NASA X-planes are back!
- Harnessing unsteadiness as a means of propulsion
- Rocket-based combined cycle propulsion: from takeoff to orbit!
- New lunch vehicles and missions for hypersonic aircraft and space exploration
- *"There's a lot of room at the bottom"* Richard Feynman said. Enter MEMS-GT engines!
- Manned-mission to Mars

An additional example of computational flow simulation is shown in Fig. 1.27.

1.6   Roadmap for the Book

We begin our studies in propulsion with a review of compressible flow that involves friction and heat transfer in Chapter 2. Engine thrust and performance parameters are discussed in Chapter 3 where rigorous derivation of uninstalled thrust and installation effects are presented. Gas turbine engine cycle analysis both for ideal and real components are studied in Chapter 4. Aircraft engine inlets and nozzles, over a wide speed range, are analyzed in Chapter 5. The principles of combustion are detailed in Chapter 6. The specific characteristics of the primary and afterburners, as in flameholding, are discussed in the same chapter. The turbomachinery principles and their application to axial-flow compressor, centrifugal compressor, and the axial-flow turbine are extensively derived and discussed in Chapters 7 through 9. Chapter 10 aims to integrate all the gas turbine engine components into a unified system, from component matching to engine off-design analysis. Chapter 11 is dedicated to chemical rocket and hypersonic propulsion where rockets, ramjets, scramjet, and combined cycles are discussed. An overview of available computational and online resources and links, related to propulsion, is also assembled in a separate appendix.

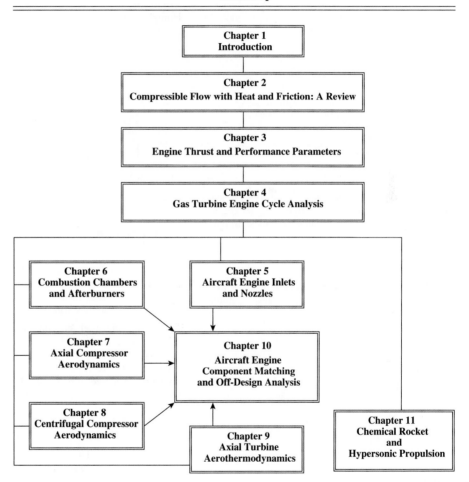

Aircraft propulsion
Roadmap

References

1. Dawson, V.P., *Engines and Innovation: Lewis Laboratory and American Propulsion Technology*, NASA SP-4306, 1991.

2. Meher-Homji, C.B., "The Development of Junkers Jumo 004B—The World's First Production Turbojet," ASME Paper No. 96-GT-457, 1996.

3. Meher-Homji, C.B., "The Development of Whittle Turbojet," ASME Paper No. 97-GT-528, 1997.

4. Meher-Homji, C.B., "Pioneering Turbojet Developments of Dr. Hans von Ohain—From HeS 01 to HeS 011," ASME Paper No. 99-GT-228, 1999.

5. NASA History Division's Web site: http://history.nasa.gov/

6. Wallace, L.E., *Flights of Discovery: Fifty Years at the NASA Dryden Flight Research Center*, National Aeronautics and Space Administration, Washington, DC, 1996.

7. http://www.nasa.gov/centers/dryden/news/FactSheets

8. "Celebrating a Century of Flight," NASA Publication SP-2002-09-511-HQ.

Problems

1.1 The Carnot cycle sets the limit on thermal efficiency of a heat engine operating between two temperature limits. Show that ideal Carnot efficiency is

$$\eta_{th} = 1 - \frac{T_1}{T_2}$$

What is the thermal efficiency if $T_1 = 288$ K and $T_2 = 2000$ K?

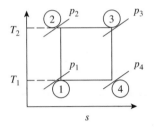

■ **FIGURE P1.1**

1.2 The ideal Brayton cycle operates between two pressure limits as shown. It is the model of an airbreathing jet engine, such as a turbojet or ramjet engine. Show that ideal Brayton cycle efficiency is

$$\eta_{th} = 1 - \frac{T_1}{T_2}$$

What is the thermal efficiency of the Brayton that has $T_1 = 288$ K and $T_2 = 864$ K? Note that maximum cycle temperature T_3 has no effect on cycle thermal efficiency.

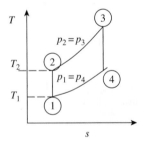

■ **FIGURE P1.2**

1.3 Humphrey cycle operates a constant-volume combustor instead of a constant-pressure cycle like Brayton. Show that

$$\eta_{th} = 1 - \gamma \frac{T_1}{T_2} \left[\left(\frac{T_3}{T_2} \right)^{\frac{1}{\gamma}} - 1 \right] \bigg/ \left[\frac{T_3}{T_2} - 1 \right]$$

is the thermal efficiency of an ideal Humphrey cycle.

Let us use the same T_1 as in Problems 1.1 and 1.2, that is, $T_1 = 288$ K. Let us use the same temperature T_2 as in Problem 1.2, that is, $T_2 = 864$ K.

Finally, let us use the same maximum cycle temperature as in Carnot (Problem 1.1), that is, $T_{max} = 2000$ K. With the ratio of specific heats $\gamma = 1.4$, calculate the thermal efficiency of Humphrey cycle. Compare the answer with Brayton cycle efficiency.

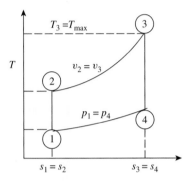

■ **FIGURE P1.3**

1.4 The rotor of a millimeter-scale gas turbine engine has a radius of 1 mm. It has to reach a tip, or rim speed of nearly the speed of sound for an effective compression. Assuming that the speed of sound is 340 m/s, calculate the rotor rotational speed in revolutions per minute (rpm).

1.5 Specific fuel consumption (sfc) projects the fuel economy of an engine, that is, it measures the fuel flow rate (say in pound-mass per hour or g/s) that leads to a production of a unit thrust (say 1 pound-force or 1 Newton). Two sets of numbers are copied from Table 1.1 (from EJ200 specification), which are

Sfc (max. power)	0.81 lbm/h/lbf
Sfc w. AB	1.75 lbm/h/lbf
Thrust (SL)	13,500 lbf
Thrust w. AB	20,250–22,250 lbf

First note that afterburner (AB) use more than doubles the fuel consumption while boosting the thrust by only ~50%. This explains the sparse use of an afterburner in aircraft mission. Now to quantify, calculate the amount of additional fuel burned in 30 min of afterburner use (producing 21,000 lbf thrust) as compared with 30 min of no afterburner use (producing 13,500 lbf thrust).

CHAPTER 2

Compressible Flow with Friction and Heat: *A Review*

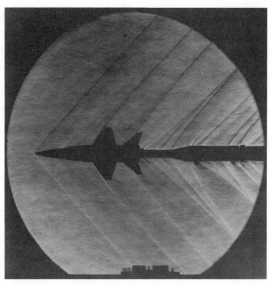

Schlieren visualization of waves around an X-15 model in a supersonic Wind tunnel (Courtesy of NASA)

2.1   Introduction

The study of propulsion is intimately linked to understanding of internal fluid mechanics. This means that duct flows are of prime interest to propulsion. The physical phenomena in a real duct include the following effects:

- Friction
- Heat transfer through the walls
- Chemical reaction within the duct
- Area variation of the duct
- Compressibility effects, e.g. appearance of shocks

The laws of thermodynamics govern the relationship between the state variables of the gas, namely density ρ, pressure p, absolute temperature T, entropy s, internal energy e, and derived properties such as enthalpy h and specific heats at constant pressure and volume, c_p and c_v, respectively. In addition to the laws of thermodynamics, the fluid flow problems need to obey other conservation principles that were introduced in Newtonian mechanics. These are conservation of mass and momentum as described in classical mechanics. Since, the study of gas turbine engines and propulsion in undergraduate curricula in mechanical and aerospace engineering follows the introductory courses in thermo-fluid dynamics, we shall review only the principles that have a direct impact on our study of jet engines.

The purpose of this chapter is thus to provide a review of the working principles in aerothermodynamics, which serve as the foundation of propulsion. The reader should consult textbooks on thermodynamics, such as the classical work of Sonntag, Borgnakke, and Van

17

Wylen (2003), and modern fluid mechanics books, such as Munson, Young, and Okiishi (2006) or John Anderson's books on aerodynamics (2005). For detailed exposition of compressible flow, the two volumes written by Shapiro (1953) should be consulted.

2.2   A Brief Review of Thermodynamics

To get started, we need to characterize the medium that flows through the duct. In gas turbine engines the medium is a perfect gas, often air. The perfect gas law that relates the pressure, density, and the absolute temperature of the gas may be derived rigorously from the kinetic theory of gases. Prominent assumptions in its formulation are (1) intermolecular forces between the molecules is negligibly small and (2) the volume of molecules that occupy a space is negligibly small and may be ignored. These two assumptions lead us to

$$p = \rho R T \tag{2.1}$$

where R is known as the gas constant, which is inversely proportional to the molecular weight of the gas, i.e.,

$$R \equiv \frac{\overline{R}}{MW} \tag{2.2-a}$$

where $\overline{R}$ is the universal gas constant expressed in two systems of units as

$$\overline{R} = 8314 \frac{\text{J}}{\text{kmol} \cdot \text{K}} \tag{2.2-b}$$

$$\overline{R} = 4.97 \times 10^4 \frac{\text{ft} \cdot \text{lbf}}{\text{slug} \cdot \text{mol} \cdot {}^{\circ}\text{R}} \tag{2.2-c}$$

The thermodynamic relations for a perfect gas in terms of specific heats at constant pressure and volume are

$$dh \equiv c_p dT \tag{2.3}$$

$$de \equiv c_v dT \tag{2.4}$$

In general, the specific heats at constant pressure and volume are functions of gas temperature,

$$c_p = c_p(T) \tag{2.5}$$

$$c_v = c_v(T) \tag{2.6}$$

The gas is then called a *thermally perfect* gas. There is often a simplifying assumption of constant specific heats, which is a valid approximation to gas behavior in a narrow temperature range. In this case,

$$c_p = \text{Constant} \tag{2.5-a}$$

$$c_v = \text{Constant} \tag{2.6-a}$$

The gas is referred to as a *calorically perfect* gas.

EXAMPLE 2.1

Calculate the density of air where its static pressure and temperature are 3.0 MPa and 25°C, respectively.

First express the static temperature in absolute scale, $T = 25 + 273 = 298$ K.

The gas constant for air is calculated from Eq. 2.2-a based on its (average) molecular weight of 29 kg/kmol, i.e.,

$$R = \frac{\bar{R}}{MW} = \frac{8,314 \text{ J/kmol·K}}{29 \text{ kg/kmol}}$$

$$= 286.69 \text{ J/kg·K} \cong 287 \text{ J/kg·K}$$

We use the perfect gas law (Eq. 2.1) to get the density, i.e.,

$$\rho = \frac{p}{RT} = \frac{3 \times 10^6 \text{ N/m}^2}{287 \dfrac{\text{J}}{\text{kg·K}} (298 \text{ K})} \cong 35.08 \frac{\text{kg}}{\text{m}^3} \qquad \boxed{\rho \cong 35.08 \frac{\text{kg}}{\text{m}^3}}$$

The first law of thermodynamics is the statement of conservation of energy for a system of fixed mass m, namely,

$$\delta q = de + \delta w \qquad (2.7)$$

where the element of heat transferred to the system from the surrounding is considered positive and on a per-unit-mass basis is δq with a unit of energy/mass, e.g., J/kg. The element of work done by the gas on the surrounding is considered positive and per-unit-mass of the gas is depicted by δw. The difference in the convention for positive heat and work interaction with the system explains the opposite sides of equation where the two energy exchange terms are located in Eq. 2.7, otherwise they both represent energy exchange with the system. The net energy interaction with the system results in a change of energy of the system; where again on a per-unit-mass basis is referred to as de. The three terms of the first law of thermodynamics have dimensions of energy per mass. The elemental heat and work exchange are shown by a delta "δ" symbol instead of an exact differential "d" as in "de." This is in recognition of *path-dependent* nature of heat and work exchange, which differ from the thermodynamic property of the gas "e," which is independent of the path, i.e.,

$$\int_1^2 \delta q \neq q_2 - q_1 \qquad (2.8)$$

rather

$$\int_1^2 \delta q =\, _1q_2 \qquad (2.9)$$

Whereas in the case of internal energy (or any other thermodynamic property),

$$\int_1^2 de = e_2 - e_1 \qquad (2.10)$$

Note that in the eyes of the first law of thermodynamics, there is no distinction between heat and mechanical work exchange with the system. It is their "net" interaction with the system that needs to be accounted for in the energy balance. The application of the first law to a closed cycle is of importance to engineering and represents a balance between the heat and work exchange in a cyclic process, i.e.,

$$\oint \delta q = \oint \delta w \qquad (2.11)$$

Also, for an adiabatic process, i.e., $\delta q = 0$, with no mechanical exchange of work, i.e., $\delta w = 0$, the energy of a system remains constant, namely $e_1 = e_2 = \text{constant}$. We are going to use this principle in conjunction with a control volume approach in the study of inlet and exhaust systems of an aircraft engine.

The second law of thermodynamics introduces the absolute temperature scale and a new thermodynamic variable s, the entropy. It is a statement of impossibility of a heat engine exchanging heat with a single reservoir and producing mechanical work continuously. It calls for a second reservoir at a lower temperature where heat is rejected to by the heat engine. In this sense, the second law of thermodynamics distinguishes between heat and work. It asserts that all mechanical work may be converted into system energy whereas not all heat transfer to a system may be converted into system energy continuously. A corollary to the second law incorporates the new thermodynamic variable s and the absolute temperature T into an inequality, known as the Clausius inequality,

$$Tds \geq \delta q \tag{2.12}$$

where the equal sign holds for a reversible process. The concept of irreversibility ties in closely with frictional losses, viscous dissipation, and the appearance of shock waves in supersonic flow. The pressure forces within the fluid perform reversible work, and the viscous stresses account for dissipated energy of the system (into heat). Hence the reversible work done by a system per unit mass is

$$\delta w_{\text{rev.}} = pdv \tag{2.13}$$

where v is the specific volume, which is the inverse of fluid density ρ. A combined first and second law of thermodynamics is known as the Gibbs equation, which relates entropy to other thermodynamic properties, namely

$$Tds = de + pdv \tag{2.14}$$

Although it looks as if we have substituted the reversible forms of heat and work into the first law to obtain the Gibbs equation, it is applicable to irreversible processes as well. Note that in an irreversible process, all the frictional forces that contribute to *lost work* are dissipated into heat. Now, we introduce a derived thermodynamic property known as enthalpy h as

$$h \equiv e + pv \tag{2.15}$$

This derived property, i.e., h combines two forms of fluid energy, namely internal energy (or thermal energy) and what is known as the flow work, pv, or the pressure energy. The other forms of energy such as kinetic energy and potential energy are still unaccounted by the enthalpy h. We shall account for the other forms of energy by a new variable called the *total enthalpy* later in this chapter.

Now, let us differentiate Eq. 2.15 and substitute it in the Gibbs equation, to get

$$Tds = dh - vdp \tag{2.16}$$

By expressing enthalpy in terms of specific heat at constant pressure, via Eq. 2.3, and dividing both sides of Eq. 2.16 by temperature T, we get

$$ds = c_p \frac{dT}{T} - \frac{v}{T} dp = c_p \frac{dT}{T} - R \frac{dp}{p} \tag{2.17}$$

We incorporated the perfect gas law in the last term of Eq. 2.17. We may now integrate this equation between states 1 and 2 to arrive at

$$s_2 - s_1 \equiv \Delta s = \int_1^2 c_p \frac{dT}{T} - R \ln \frac{p_2}{p_1} \tag{2.18}$$

An assumption of a calorically perfect gas will enable us to integrate the first term on the right-hand side of Eq. 2.18, i.e.,

$$\Delta s = c_p \ln \frac{T_2}{T_1} - R \ln \frac{p_2}{p_1} \tag{2.19}$$

Otherwise, we need to refer to a tabulated thermodynamic function ϕ defined as

$$\int_1^2 c_p \frac{dT}{T} \equiv \phi_2 - \phi_1 \tag{2.20}$$

From the definition of enthalpy, let us replace the flow work, pv, term by its equivalent from the perfect gas law, i.e., RT, and then differentiate the equation as

$$dh \equiv c_p dT = de + R dT = c_v dT + R dT \tag{2.21}$$

Dividing through by the temperature differential dT, we get

$$c_p = c_v + R \tag{2.22-a}$$

$$\frac{c_p}{R} = \frac{c_v}{R} + 1 \tag{2.22-b}$$

This provides valuable relations among the gas constant and the specific heats at constant pressure and volume. The ratio of specific heats is given by a special symbol γ due to its frequency of appearance in compressible flow analysis, i.e.,

$$\gamma \equiv \frac{c_p}{c_v} = \frac{c_v + R}{c_v} = 1 + \frac{1}{c_v/R} \tag{2.23}$$

In terms of the ratio of specific heats γ and R, we express c_p and c_v as

$$c_p = \frac{\gamma}{\gamma - 1} R \tag{2.24-a}$$

$$c_v = \frac{1}{\gamma - 1} R \tag{2.24-b}$$

The ratio of specific heats is related to the degrees of freedom of the gas molecules, n, via

$$\gamma = \frac{n + 2}{n} \tag{2.25}$$

The degrees of freedom of a molecule are represented by the sum of the energy states that a molecule possesses. For example, atoms or molecules possess kinetic energy in three spatial directions. If they rotate as well, they have kinetic energy associated with their rotation. In a molecule, the atoms may vibrate with respect to each other, which then create kinetic energy of vibration as well as the potential energy of intermolecular forces. Finally, the electrons in an atom or molecule are described by their own energy levels (both kinetic energy and potential) that depend on their position around the nucleus. As the temperature of the gas increases, the successively higher energy states are excited; thus the degrees of freedom increases. A monatomic gas, which may be modeled as a sphere, has at least three degrees of freedom, which represent translational motion in three spatial directions. Hence, for a monatomic gas, under "normal" temperatures, the ratio of specific heats is

$$\gamma = \frac{5}{3} \cong 1.667 \quad \text{Monatomic gas at "normal" temperatures} \tag{2.26}$$

A monatomic gas has negligible rotational energy about the axes that pass through the atom due to its negligible moment of inertia. A monatomic gas will not experience a vibrational energy, as vibrational mode requires at least two atoms. At higher temperatures, the electronic energy state of the gas is affected, which eventually leads to ionization of the gas. For a diatomic gas, which may be modeled as a dumbbell, there are five degrees of freedom, under "normal" temperature conditions, three of which are in translational motion and two are in rotational direction. The third rotational motion along the intermolecular axis of the dumbbell is negligibly small. Hence for a diatomic gas such as air (near room temperature), hydrogen, nitrogen, etc, the ratio of specific heats is

$$\gamma = \frac{7}{5} = 1.4 \quad \text{Diatomic gas at ''normal'' temperatures} \tag{2.27-a}$$

At high temperatures, molecular vibrational modes and the excitation of electrons add to the degrees of freedom and that lowers γ. For example, at ~600 K vibrational modes in air are excited, thus the degrees of freedom of diatomic gasses are initially increased by 1, i.e., it becomes $5 + 1 = 6$, when the vibrational mode is excited. Therefore, the ratio of specific heats for diatomic gases at elevated temperatures becomes

$$\gamma = \frac{8}{6} \approx 1.33 \quad \text{Diatomic gas at elevated temperatures} \tag{2.27-b}$$

The vibrational mode represents two energy states corresponding to the kinetic energy of vibration and the potential energy associated with the intermolecular forces. When fully excited, the vibrational mode in a diatomic gas, such as air, adds 2 to the degrees of freedom, i.e., it becomes 7. Therefore, the ratio of specific heats becomes

$$\gamma = \frac{9}{7} \approx 1.29 \quad \text{Diatomic gas at higher temperatures} \tag{2.27-c}$$

For example, air at 2000 K has its translational, rotational, and vibrational energy states fully excited. This temperature level describes the combustor or afterburner environment. Gases with a more complex structure than a diatomic gas have higher degrees of freedom and thus their ratio of specific heats is less than 1.4. Figure 2.1 (from Anderson, 2003) shows the

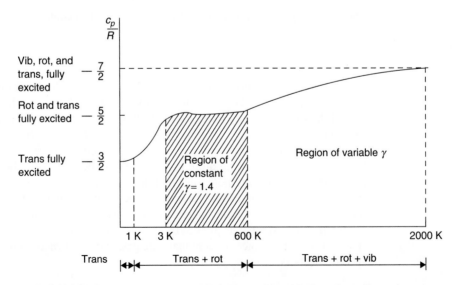

■ **FIGURE 2.1**   **Temperature dependence of specific heat for a diatomic gas (from Anderson, 2003)**

behavior of a diatomic gas from 0 to 2000 K. The nearly constant specific heat ratio between 3 and 600 K represents the calorically perfect gas behavior of a diatomic gas such as air with $\gamma = 1.4$. Note that near absolute zero (0 K), $c_v/R \to 3/2$, therefore a diatomic gas ceases to rotate and thus behaves like a monatomic gas, i.e., it exhibits the same degrees of freedom as a monatomic gas, i.e., $n = 3, \gamma = 5/3$.

2.3   Isentropic Process and Isentropic Flow

For an isentropic process, where entropy remains constant, the Gibbs equation relates the pressure and temperature ratios by an isentropic exponent:

$$\frac{p_2}{p_1} = \left(\frac{T_2}{T_1}\right)^{\frac{c_p}{R}} = \left(\frac{T_2}{T_1}\right)^{\frac{\gamma}{\gamma-1}} \quad \text{(valid for a calorically perfect gas)} \tag{2.28}$$

Now, using the perfect gas law by replacing the temperature ratio by pressure and density ratios in Eq. 2.28 and simplifying the exponents, we get

$$\frac{p_2}{p_1} = \left(\frac{\rho_2}{\rho_1}\right)^{\gamma} \quad \text{(valid for a calorically perfect gas)} \tag{2.29}$$

EXAMPLE 2.2

Air is compressed from $p_1 = 100\,\text{kPa}, T_1 = 15°\text{C}$ to $p_2 = 1.0\,\text{MPa}$ isentropically. Calculate the exit temperature and density of air. Assume $\gamma = 1.4, R = 287\,\text{J/Kg·K}$, and the gas is calorically perfect.

First convert the temperature to absolute scale, i.e., $T_1 = 15 + 273 = 288$ K.

From isentropic relation (2.28), we get the exit temperature according to

$$T_2 = T_1 \left(\frac{p_2}{p_1}\right)^{\frac{\gamma-1}{\gamma}} = (288K)(10)^{0.4/1.4} \cong 556\,\text{K}$$

There are two ways that we can calculate the exit density. The first method calls for the application of the perfect gas law at the exit, that is,

$$\rho_2 = \frac{p_2}{RT_2} = \frac{10^6\,\text{Pa}}{(287\,\text{J/kg·K})(556\,\text{K})} \cong 6.267\,\text{kg/m}^3$$

The second method uses the isentropic relation between the inlet and exit density, i.e., Eq. 2.29. We need to calculate the inlet density first,

$$\rho_1 = \frac{p_1}{RT_1} = \frac{100\,\text{kPa}}{(287\,\text{J/kg·K})(288\,\text{K})} \cong 1.2098\,\text{kg/m}^3$$

Now, we may use the isentropic relation 2.29 to get the exit density ρ_2

$$\rho_2 = \rho_1 \left(\frac{p_2}{p_1}\right)^{\frac{1}{\gamma}} = (1.2098\,\text{kg/m}^3)(10)^{1/1.4} \cong 6.267\,\text{kg/m}^3$$

2.4   Conservation Principles for Systems and Control Volumes

A system is a collection of matter of fixed identity, hence fixed mass, whereas a control volume is a fixed region in space with fluid crossing its boundaries. A control volume approach seems to be a more practical method of treating the fluid flow problems in gas turbine engines. However, all the classical laws of Newtonian physics are written for a matter of fixed mass, i.e., the system approach. For example, the mass of an object (in our case a collection of matter described by a system) does not change with time, is a Newtonian mechanics principle. The counterpart of that expressed for a control volume is known as the continuity equation and is derived as follows. The starting point is the system expression, namely

$$\left.\frac{dm}{dt}\right|_{\text{system}} = 0 \qquad (2.30)$$

We now express mass as the integral of density over volume, as

$$\frac{d}{dt}\iiint\limits_{V(t)} \rho dV = 0 \qquad (2.31)$$

where $V(t)$ is the volume of the system at any time t. Note that to encompass the entire original mass of the gas in the system, the boundaries of the system become a function of time. Expression 2.31 is a derivative with respect to time of an integral with time-dependent limits. The Leibnitz' rule of differentiating such integrals is the bridge between the system and the control volume approach, namely

$$\frac{d}{dt}\iiint\limits_{V(t)} \rho dV = \iiint\limits_{V(t_0)} \frac{\partial \rho}{\partial t} dV + \iint\limits_{S(t_0)} \rho \vec{V} \cdot \hat{n} dS \qquad (2.32)$$

The limits of the integrals on the RHS of Eq. 2.32 describe a volume $V(t_0)$ and a closed surface $S(t_0)$, which represents the boundaries of the system at time t_0. We choose our control volume to coincide with the boundaries of the system at time t_0. Since the time t_0 is arbitrary, the above equation represents the control volume formulation for the medium, namely our continuity equation or law of conservation of mass, for a control volume is written as

$$\iiint\limits_{C.V.} \frac{\partial \rho}{\partial t} dV + \iint\limits_{C.S.} \rho \vec{V} \cdot \hat{n} dS = 0 \qquad (2.33)$$

The fluid velocity vector is $\vec{V}$ and $\hat{n}$ represents a unit vector normal to the control surface and pointing outward. The first integral in Eq. 2.33 accounts for unsteady accumulation (or depletion) of mass within the control volume and of course vanishes for steady flows. We may demonstrate this by switching the order of integration and differentiation for the first integral, i.e.,

$$\iiint\limits_{C.V.} \frac{\partial \rho}{\partial t} = \frac{\partial}{\partial t}\iiint\limits_{C.V.} \rho dV = \frac{\partial}{\partial t}(m)_{C.V.} \qquad (2.34)$$

A positive value depicts mass accumulation in the control volume and a negative value shows mass depletion. The second integral takes the scalar product of the fluid velocity vector and the unit normal to the control surface. Therefore, it represents the net crossing of the mass through the control surface per unit time. Flow enters a control volume through one or more inlets and exits the control volume through one or more exits. There is no mass crossing the boundaries of a control volume at any other sections besides the "inlets" and "outlets." Since the dot product of the velocity vector and the unit normal vector vanishes on all surfaces of the control volume where mass does not cross the boundaries, the second integral makes a contribution to the continuity equation through its inlets and outlets, namely

$$\iint\limits_{C.S.} \rho \vec{V} \cdot \hat{n} dS = \iint\limits_{\text{Outlets}} \rho \vec{V} \cdot \hat{n} dS + \iint\limits_{\text{Inlets}} \rho \vec{V} \cdot \hat{n} dS \qquad (2.35)$$

We also note that a unit normal at an inlet points in the opposite direction to the incoming velocity vector, therefore the mass flux at an inlet contributes a negative value to the mass balance across the control volume (through a negative dot product). An outlet has a unit normal pointing in the same direction as the velocity vector, hence contributes positively to the mass balance in the continuity equation. In general, the inlet and exit faces of a control surface are not normal to the flow, but still the angle between the normal and the velocity vector is obtuse for inlets and acute for exits, hence a negative and a positive dot product over the inlets and outlets, respectively. Therefore, the net flux of mass crossing the boundaries of a control volume, in a steady flow, is zero. We may assume a uniform flow over the inlet and exits of the control volume, to simplify the integrals of Eq. 2.35, to

$$\Sigma(\rho VA)_{\text{out}} = \Sigma(\rho VA)_{\text{in}} \qquad (2.36)$$

The product VA in the above equation is more accurately written as $V_n A$ or VA_n where a normal component of velocity through an area A contributes to the mass flow through the area, or equivalently, the product of velocity and a normal projection of the area to the flow contributes to the mass flow through the boundary. Often the subscript "n" is omitted in the continuity equation for convenience; however, it is always implied in writing the mass flow through a boundary. To write the conservation of mass for a control volume, let us combine the unsteady term and the net flux terms as

$$\frac{d}{dt}(m)_{C.V.} + \dot{m}_{\text{out}} - \dot{m}_{\text{in}} = 0 \qquad (2.37)$$

where the mass flow rate is now given the symbol $\dot{m}$ (with units of kg/s, lbm/s, or slugs/s). We note that the difference between a positive outgoing mass and a negative incoming mass appears as a mass accumulation or depletion in the control volume. If the exit mass flow rate is higher than the inlet mass flow rate, then mass depletes within the control volume and vice versa.

EXAMPLE 2.3

Consider the control volume shown. There are two inlets and one-exit boundaries where flow crosses the control volume. Assuming the flow is steady and uniform, calculate the exit flow rate from the known inlet conditions.

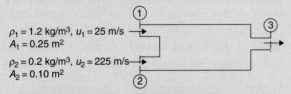

$\rho_1 = 1.2$ kg/m³, $u_1 = 25$ m/s
$A_1 = 0.25$ m²

$\rho_2 = 0.2$ kg/m³, $u_2 = 225$ m/s
$A_2 = 0.10$ m²

The mass flow rate entering boundary 1, is

$$\dot{m}_1 = \rho_1 A_1 u_1 = (1.2\,\text{kg/m}^3)(0.25\,\text{m}^2)(25\,\text{m/s})$$
$$= 7.5\,\text{kg/s}$$

The mass flow rate entering boundary 2, is

$$\dot{m}_2 = \rho_2 A_2 u_2 = (0.2\,\text{kg/m}^3)(0.10\,\text{m}^2)(225\,\text{m/s})$$
$$= 4.5\,\text{kg/s}$$

Therefore, the mass flow rate out, of station 3, is the sum of the mass flow rate into the box, i.e.,

$$\dot{m}_3 = \dot{m}_1 + \dot{m}_2 = (7.5 + 4.5)\,\text{kg/s} = 12\,\text{kg/s}$$

The momentum equation, according the Newtonian mechanics, relates the time rate of change of linear momentum of an object of fixed mass to the net external forces that act on the object. We may write this law as

$$\frac{d}{dt}(m\vec{V}) = \vec{F}_{\text{net}} \qquad (2.38)$$

Again, we propose to write the mass as the volume integral of the density within the system as

$$\frac{d}{dt}\iiint\limits_{V(t)} \rho\vec{V}dV = \vec{F}_{\text{net}} \tag{2.39}$$

The Eq. 2.39 is suitable for a system. We apply Leibnitz' rule to the momentum equation to arrive at the control volume formulation, namely

$$\iiint\limits_{C.V.}\frac{\partial}{\partial t}(\rho\vec{V})dV + \iint\limits_{C.S.} \rho\vec{V}(\vec{V}\cdot\hat{n})dS = \vec{F}_{\text{net}} \tag{2.40}$$

The first integral measures the unsteady momentum within the control volume and vanishes identically for a steady flow. The second integral is the net flux of momentum in and out of the control surface. Assuming uniform flow at the boundaries of the control surface inlets and outlets, we may simplify the momentum equation to a very useful engineering form, namely

$$(\dot{m}\vec{V})_{\text{out}} - (\dot{m}\vec{V})_{\text{in}} = \vec{F}_{\text{net}} \tag{2.40-a}$$

Note that momentum equation (2.40-a) is a vector equation and a shorthand notation for the momentum balance in three spatial directions. For example, in Cartesian coordinates, we have

$$(\dot{m}V_x)_{\text{out}} - (\dot{m}V_x)_{\text{in}} = F_{\text{net},x} \tag{2.41-a}$$

$$(\dot{m}V_y)_{\text{out}} - (\dot{m}V_y)_{\text{in}} = F_{\text{net},y} \tag{2.41-b}$$

$$(\dot{m}V_z)_{\text{out}} - (\dot{m}V_z)_{\text{in}} = F_{\text{net},z} \tag{2.41-c}$$

In cylindrical coordinates (r, θ, z), we may write the momentum equation as

$$(\dot{m}V_r)_{\text{out}} - (\dot{m}V_r)_{\text{in}} = F_{\text{net},r} \tag{2.42-a}$$

$$(\dot{m}V_\theta)_{\text{out}} - (\dot{m}V_\theta)_{\text{in}} = F_{\text{net},\theta} \tag{2.42-b}$$

$$(\dot{m}V_z)_{\text{out}} - (\dot{m}V_z)_{\text{in}} = F_{\text{net},z} \tag{2.42-c}$$

We may write these equations in spherical coordinates (r, θ, ϕ) as well. The force terms in the momentum equation represent the net external forces exerted on the fluid at its boundaries and any volume forces, known as body forces, such as gravitational force. If the control volume contains/envelopes an object, then the force acting on the fluid is equal and opposite to the force experienced by the body. For example a body that experiences a drag force D imparts on the fluid a force equal to $-D$.

EXAMPLE 2.4

In placing a curved plate in front of a garden hose with a water flow rate of 0.1 kg/s, we have to exert an axial force on the plate, as shown, to hold it in place. Estimate the axial and lateral forces, F_x and F_y, respectively, that are needed to support the plate.

   (note: u and v are x- and y-component of velocity, respectively).

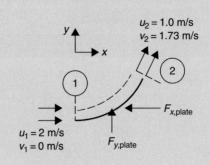

For a quick estimation of the axial force, we assume the flow is uniform and steady. Then, we may apply Eq. 2.41-a to get

$$F_{x,\text{fluid}} = \dot{m}_2 u_2 - \dot{m}_1 u_1 = \dot{m}(u_2 - u_1) = 0.1 \, \text{kg/s}(-1 \, \text{m/s})$$

$$= -0.1 \, \text{N}$$

The axial force acting on the curved plate is equal and opposite to $F_{x,\text{fluid}}$. Therefore, the external force needed to hold the plate horizontally is -0.1 N (that is in negative x-

direction, as shown). Also, the lateral force on the fluid is calculated from the lateral momentum balance, i.e.,

$$F_{y,\text{fluid}} = \dot{m}_2 v_2 - \dot{m}_1 v_1 = \dot{m}(v_2 - v_1)$$

$$= 0.1 \, \text{kg/s} \, (1.73 \, \text{m/s}) = 0.173 \, \text{N}$$

The lateral force on the plate is in opposite direction (i.e., in $-y$ direction) and thus the external force to hold the plate laterally is $+0.173$ N, as shown.

The law of conservation of energy for a control volume starts with the first law of thermodynamics applied to a system. Let us divide the differential form of the first law by an element of time dt to get the rate of energy transfer, namely

$$\dot{Q} = \frac{dE}{dt} + \dot{W} \tag{2.43}$$

where the equation is written for the entire mass of the system. The energy E is now represented by the internal energy e times mass, as well as the kinetic energy of the gas in the system and the potential energy of the system. The contribution of changing potential energy in gas turbine engines or most other aerodynamic applications is negligibly small and often ignored. We write the energy as the mass integral of specific energy over the volume of the system, and apply Leibnitz' rule of integration to get the control volume version, namely

$$\frac{dE}{dt} = \frac{d}{dt} \iiint_{V(t)} \rho\left(e + \frac{V^2}{2}\right) dV = \iiint_{C.V.} \frac{\partial}{\partial t}\left[\rho\left(e + \frac{V^2}{2}\right)\right] dV + \iint_{C.S.} \rho\left(e + \frac{V^2}{2}\right) \vec{V} \cdot \hat{n} dS \tag{2.44}$$

The first term on the RHS of Eq. 2.44 is the time rate of change of energy within the control volume, which identically vanishes for a steady flow. The second integral represents the net flux of fluid power (i.e., the *rate* of energy) crossing the boundaries of the control surface. The rates of heat transfer to the control volume and the rate of mechanical energy transfer by the gases inside the control volume on the surrounding are represented by $\dot{Q}$ and $\dot{W}$ terms in the energy equation 2.43. Now, let us examine the forces at the boundary that contribute to the rate of energy transfer. These surface forces are due to pressure and shear acting on the boundary. The pressure forces act normal to the boundary and point inward, i.e., opposite to $\hat{n}$,

$$- p\hat{n} dS \tag{2.45}$$

To calculate the rate of work done by a force, we take the scalar product of the force and the velocity vector, namely

$$- p\vec{V} \cdot \hat{n} dS \tag{2.46}$$

Now, we need to sum this elemental rate of energy transfer by pressure forces over the surface, via a surface integral, that is,

$$\iint_{C.S.} - p\vec{V} \cdot \hat{n} dS \tag{2.47}$$

Since the convention on the rate of work done in the first law is positive when it is performed "on" the surroundings, and Eq. 2.46 represents the rate of work done by the surroundings on the control volume, we need to incorporate an additional negative factor for this term in the energy equation. The rate of energy transfer by the shear forces is divided into a shaft power $\wp_s$

that crosses the control surface in the form of shaft shear and the viscous shear stresses on the boundary of the control volume. Hence,

$$\dot{W} = \oiint_{C.S.} p\vec{V} \cdot \hat{n}\, dS + \wp_s + \dot{W}_{\text{viscous-shear}} \tag{2.48}$$

Let us combine the expression 2.48 with Eqs. 2.44 and 2.43 to arrive at a useful form of energy equation for a control volume, i.e.,

$$\iiint_{C.V.} \frac{\partial}{\partial t} \left[\rho \left(e + \frac{V^2}{2} \right) \right] dV + \oiint_{C.S.} \rho \left(e + \frac{V^2}{2} \right) \vec{V} \cdot \hat{n}\, dS + \oiint_{C.S.} p\vec{V} \cdot \hat{n}\, dS = \dot{Q} - \wp_s - \dot{W}_{\text{visc}} \tag{2.49}$$

The closed surface integrals on the LHS may be combined and simplified to

$$\iiint_{C.V.} \frac{\partial}{\partial t} \left[\rho \left(e + \frac{V^2}{2} \right) \right] dV + \oiint_{C.S.} \rho \left(e + \frac{V^2}{2} + \frac{p}{\rho} \right) \vec{V} \cdot \hat{n}\, dS = \dot{Q} - \wp_s - \dot{W}_{\text{visc}} \tag{2.50}$$

We may replace the internal energy and the flow work terms in Eq. 2.50 by enthalpy h and define the sum of the enthalpy and the kinetic energy as the total or stagnation enthalpy, to get

$$\iiint_{C.V.} \frac{\partial}{\partial t} \left[\rho \left(e + \frac{V^2}{2} \right) \right] dV + \oiint_{C.S.} \rho h_t \vec{V} \cdot \hat{n}\, dS = \dot{Q} - \wp_s - \dot{W}_{\text{visc.}} \tag{2.51}$$

where the total enthalpy h_t is defined as

$$h_t \equiv h + \frac{V^2}{2} \tag{2.52}$$

In steady flows the volume integral that involves a time derivative vanishes. The second integral represents the net flux of fluid power across the control volume. The terms on the RHS are the external energy interaction terms, which serve as the *drivers* of the energy flow through the control volume. In adiabatic flows, the rate of heat transfer through the walls of the control volume vanishes. In the absence of shaft work, as in inlets and nozzles of a jet engine, the second term on the RHS vanishes. The rate of energy transfer via the viscous shear stresses is zero on solid boundaries (since velocity on solid walls obeys the no slip boundary condition) and nonzero at the inlet and exit planes. The contribution of this term over the inlet and exit planes is, however, small compared with the net energy flow in the fluid, hence neglected.

The integrated form of the energy equation for a control volume, assuming uniform flow over the inlets and outlets, yields a practical solution for quick engineering calculations

$$\Sigma(\dot{m}h_t)_{\text{out}} - \Sigma(\dot{m}h_t)_{\text{in}} = \dot{Q} - \wp_s \tag{2.53}$$

The summations in Eq. 2.53 account for multiple inlets and outlets of a general control volume. In flows that are adiabatic and involve no shaft work, the energy equation simplifies to

$$\Sigma(\dot{m}h_t)_{\text{out}} = \Sigma(\dot{m}h_t)_{\text{in}} \tag{2.54}$$

For a single inlet and a single outlet, the energy equation is even further simplified, as the mass flow rate also cancels out, to yield

$$h_{t-\text{exit}} = h_{t-\text{inlet}} \tag{2.55}$$

Total or stagnation enthalpy then remains constant for adiabatic flows with no shaft power, such as inlets and nozzles or across shock waves.

EXAMPLE 2.5

Let us consider a control volume with heat transfer rate and mechanical power (i.e., shaft power) exchange specified at its boundaries. There is a single inlet and a single outlet where mass crosses the boundary in a steady and uniform flow, as shown. Calculate the exit total and static temperatures T_{t2} and T_2, respectively (note that the gas is not calorically perfect).

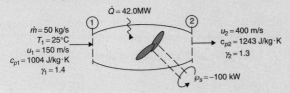

First, we conclude that the exit mass flow rate is 50 kg/s, to satisfy the continuity equation for steady flow. Second, we cast the inlet static temperature in the absolute scale, i.e.,

$$T_1 = 25 + 273 = 298 \text{ K}$$

Third, we apply the conservation of energy equation 2.53 to this problem, which requires the knowledge of total

enthalpy at the inlet. We use the definition of total enthalpy:

$$h_{t1} = h_1 + u_1^2/2 = c_{p1}T_1 + u_1^2/2 = (1004 \text{ J/kg·K})(298 \text{ K})$$
$$+ (150)^2/2 \text{ J/kg}$$
$$= 310,442 \text{ J/Kg}$$

Now, Eq. 2.53 gives the exit total enthalpy as

$$h_{t2} = h_{t1} + \left(\frac{\dot{Q} - \wp_s}{\dot{m}}\right) = 310.442 \text{ kJ/kg}$$
$$+ \left(\frac{(42,000 + 100) \times 1000 \text{ W}}{50 \text{ kg/s}}\right)$$
$$\cong 1,152.4 \text{ kJ/kg}$$

$$T_{t2} = \frac{h_{t2}}{c_{p2}} = \frac{1,152.4 \text{ kJ/kg}}{1.243 \text{ kJ/kg−K}} \cong 927 \text{ K}$$

$$T_2 = T_{t2} - u_2^2/2c_{p2}$$
$$= 927 \text{ K} - (400)^2/(2 \times 1243 \text{ J/kg·K}) \cong 863 \text{ K}$$

2.5   Speed of Sound & Mach Number

Sound waves are infinitesimal pressure waves propagating in a medium. The propagation of sound waves, or acoustic waves, is reversible and adiabatic, hence isentropic. Since sound propagates through collision of fluid molecules, the speed of sound is higher in liquids than gas. The derivation of the speed of sound is very simple and instructive. Assume a plane sound wave propagates in a medium at rest with speed u. The fluid behind the wave is infinitesimally set in motion at the speed du with an infinitesimal change of pressure, temperature, and density. The fluid ahead of the wave is at rest and yet unaffected by the approaching wave. A schematic drawing of this wave and fluid properties are shown in Fig. 2.2.

By switching observers from the gas at rest to an observer that moves with the wave at the wave speed u, we change the unsteady wave propagation problem into a steady one. This is known as the Lorentz' transformation. The transformed problem is shown in Fig. 2.3.

The fluid static properties, pressure, density, and temperature are independent of the motion of an observer; hence they remain unaffected by the observer transformation. We may

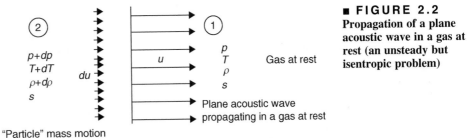

■ **FIGURE 2.2**
Propagation of a plane acoustic wave in a gas at rest (an unsteady but isentropic problem)

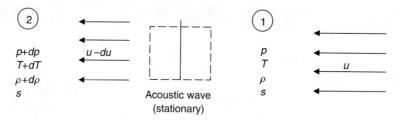

■ **FIGURE 2.3**   **Flow as seen by an observer fixed at the wave (a steady problem)**

now apply steady conservation principles to a control volume, as shown in Fig. 2.3. The control volume is a box with its sides parallel to the flow, an inlet and an exit area normal to the flow. We may choose the entrance and exit areas of the box to be unity. The continuity demands

$$\rho u = (\rho + d\rho)(u + du) \tag{2.56}$$

which simplifies to

$$\rho du = -u d\rho \tag{2.57}$$

The momentum equation for this one-dimensional wave problem is

$$(\rho u)(u + du) - (\rho u)u = p - (p + dp) \tag{2.58}$$

The change of momentum from inlet to exit is shown on the LHS of Eq. 2.58. The first parenthesis of each momentum term is the mass flow rate, which is multiplied by the respective flow speeds at the exit and inlet. The driving forces on the RHS of Eq. 2.58 are the pressure–area terms in the direction and opposite to the fluid motion, acting on areas that were chosen to be unity. The momentum equation simplifies to

$$\rho u du = dp \tag{2.59}$$

Now, let us substitute the continuity equation 2.57 in the momentum equation 2.59, to get an expression for the square of the acoustic wave propagation in terms of pressure and density changes that occur as a result of wave propagation in a medium at rest.

$$u^2 d\rho = dp \tag{2.60}$$

or

$$u^2 = \frac{dp}{d\rho} = \left(\frac{\partial p}{\partial \rho}\right)_s \tag{2.61}$$

Since for isentropic flow, $p = \text{const} \cdot \rho^\gamma$, Eq. 2.61 reduces to

$$u^2 = \frac{\gamma p}{\rho} = \gamma R T \tag{2.62}$$

We replaced the ratio of pressure to density by RT from perfect gas law in Eq. 2.62. The symbol we use in this book for the speed of sound is "a," hence, local speed of sound in a gas is

$$a = \sqrt{\frac{\gamma p}{\rho}} = \sqrt{\gamma R T} = \sqrt{(\gamma - 1)c_p T} \tag{2.63}$$

The speed of sound is a *local* parameter, which depends on local absolute temperature of the gas. Its value changes with gas temperature, hence it drops when fluid accelerates (or *expands*) and increases when the gas decelerates (or *compresses*). The speed of sound in air at standard sea level conditions is ~340 m/s or ~1100 ft/s. The type of gas also affects the speed of

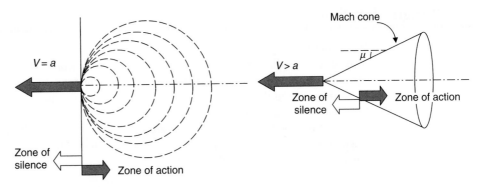

■ **FIGURE 2.4**   **Acoustic wave propagation in sonic and supersonic flows (or the case of a moving source)**

propagation of sound through its molecular weight. We may observe this behavior by the following substitution:

$$a = \sqrt{\gamma R T} = \sqrt{\gamma \left(\frac{\overline{R}}{MW}\right) T} \qquad (2.64)$$

A light gas, like hydrogen (H_2) with a molecular weight of 2, causes an acoustic wave to propagate *faster* than a heavier gas, such as air with (a mean) molecular weight of 29. If we substitute these molecular weights in Eq. 2.64, we note that sound propagates in gaseous hydrogen *nearly four times faster* than air. Since both hydrogen and air are diatomic gases, the ratio of specific heats remains (nearly) the same for both gases at the same temperature.

The equation for the speed of propagation of sound that we derived is for a gas at rest. Let us superimpose a uniform collective gas speed in a particular direction to the wave front, then the wave propagates as the vector sum of the two, namely, $\vec{V} + \vec{a}$. For waves propagating normal to a gas flow, we get either $(V + a)$ or $(V - a)$ as the propagation speed of sound. It is the $(V - a)$ behavior that is of interest here. In case the flow is sonic, then $(a - a = 0)$, which will not allow the sound to travel upstream and hence creates a zone of silence upstream of the disturbance. In case the flow speed is even faster than the local speed of sound, i.e., known as supersonic flow, the acoustic wave will be confined to a cone. These two behaviors for small disturbances are shown in Fig. 2.4.

The ratio of local gas speed to the speed of sound is called Mach number, M

$$M \equiv \frac{V}{a} \qquad (2.65)$$

The envelope of the waves that create the zones of action and silence is the Mach wave. It makes a local wave angle with respect to the flow μ, which from the geometry of wave propagation, as shown in Fig. 2.5, is

$$\mu = \sin^{-1}\left(\frac{a \cdot t}{V \cdot t}\right) = \sin^{-1}\left(\frac{1}{M}\right) \qquad (2.66)$$

■ **FIGURE 2.5**   **Wave front created by a *small disturbance* moving at a supersonic speed**

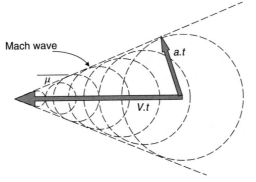

2.6   Stagnation State

We define the stagnation state of a gas as the state reached in decelerating a flow to rest reversibly and adiabatically and without any external work. Thus, the stagnation state is reached isentropically. This state is also referred to as the *total* state of the gas. The symbols for stagnation state in this book use a subscript "t" for total. The total pressure is p_t, the total temperature is T_t, and the total density is ρ_t. Since the stagnation state is reached isentropically, the static and total entropy of the gas are the same, i.e., $s_t = s$. Based on the definition of stagnation state, the total energy of the gas does not change in the deceleration process, hence the stagnation enthalpy h_t takes on the form

$$h_t \equiv h + \frac{V^2}{2} \tag{2.52}$$

which we defined earlier in this chapter. Assuming a *calorically perfect gas*, we may simplify the total enthalpy relation 2.52 by dividing through by c_p to get an expression for total temperature according to

$$T_t = T + \frac{V^2}{2c_p} \quad \text{(valid for a calorically perfect gas)} \tag{2.67}$$

This equation is very useful in converting the local static temperature and gas speed into the local stagnation temperature. To nondimensionalize Eq. 2.67, we divide both sides by the static temperature, that is,

$$\frac{T_t}{T} = 1 + \frac{V^2}{2c_p T} \tag{2.68}$$

The denominator of the kinetic energy term on the RHS is proportional to the square of the local speed of sound a^2 according to Eq 2.63, which simplifies to

$$\frac{T_t}{T} = 1 + \left(\frac{\gamma - 1}{2}\right)\frac{V^2}{a^2} = 1 + \left(\frac{\gamma - 1}{2}\right)M^2 \tag{2.69}$$

Therefore to arrive at the stagnation temperature of the gas, we need to have its static temperature as well as either the gas speed V or the local Mach number of the gas M to substitute in Eq. 2.68 or 2.69, respectively. The ratio of stagnation to static temperature of a (calorically perfect) gas is a unique function of local Mach number, according to Eq. 2.69. We may also use a spreadsheet tabulation of this function for later use for any gas characterized by its ratio of specific heats γ. From isentropic relations between the pressure and temperature ratio that we derived earlier based on Gibbs equation of thermodynamics, we may now relate the ratio of stagnation to static pressure and the local Mach number via

$$\frac{p_t}{p} = \left(\frac{T_t}{T}\right)^{\frac{\gamma}{\gamma-1}} = \left[1 + \left(\frac{\gamma - 1}{2}\right)M^2\right]^{\frac{\gamma}{\gamma-1}} \quad \text{(valid for a calorically perfect gas)} \tag{2.70}$$

We used the isentropic relation between the stagnation and static states based on the definition of the stagnation state that is reached isentropically. Also the stagnation density is higher than the static density according to

$$\frac{\rho_t}{\rho} = \left(\frac{p_t}{P}\right)^{\frac{1}{\gamma}} = \left[1 + \left(\frac{\gamma - 1}{2}\right)M^2\right]^{\frac{1}{\gamma-1}} \quad \text{(valid for a calorically perfect gas)} \tag{2.71}$$

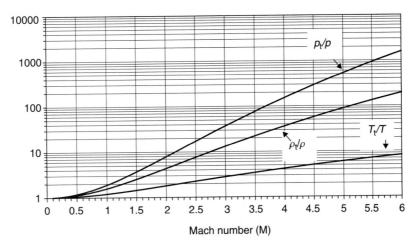

■ FIGURE 2.6   **Variation of the stagnation state with Mach number for a diatomic gas, $\gamma = 1.4$ (For a calorically perfect gas, i.e., c_p, c_v, and γ are constants)**

We note that the ratio of stagnation-to-static state of a gas is totally described by the local Mach number and the type of gas γ. The tabulation of these functions, i.e., p_t/p, T_t/T, and ρ_t/ρ are made in isentropic tables for a gas described by its ratio of specific heats γ. Figure 2.6 shows the variation of the stagnation state with Mach number for a diatomic gas, such as air ($\gamma = 1.4$). The least variation is noted for the stagnation temperature, followed by the density and pressure. To graph these variations (up to Mach 6), we had to use a logarithmic scale, as the total pressure rise with Mach number reaches above one thousand (times the static pressure) while the stagnation temperature stays below ten (times the static temperature). We may examine higher Mach numbers in Fig. 2.7. The assumption in arriving at the stagnation state properties relative to the static properties was *a calorically perfect gas*. The impact of high-speed flight on the atmosphere is to cause molecular dissociation, followed by ionization of the oxygen and nitrogen molecules. A host of other chemical reactions takes place, which result in a violation of our initial assumption. In reality, both the specific heats (c_p and c_v) are functions of temperature as well as the pressure at hypersonic Mach numbers, i.e., $M > 5$. There are two lessons to be learned here. One, always examine the validity of your assumptions and the other be cautious about data extrapolations!

Although the pressure ratio (p_t/p) is relieved through real gas effects due to a high-speed flight, it still remains high. Consequently, the high-speed flight has to be scheduled at a high altitude where the static pressure is low. For example, the static pressure at 25 km is only ~2.5% and at 50 km altitude is 0.07% of the sea level pressure.

To identify the local static and stagnation states of a gas in a Mollier (h–s) diagram is important in studying propulsion. We use the stagnation enthalpy definition to identify the local stagnation properties of a gas, for example, we start with the static state and then build up a kinetic energy, as in Fig. 2.8.

We may also apply the Gibbs equation to the stagnation state of the gas to get

$$\Delta s = c_p \ell n(T_{t2}/T_{t1}) - R\ell n(p_{t2}/p_{t1}) \tag{2.72}$$

For adiabatic flows the first term on the RHS identically vanishes, since the total temperature remains constant. Hence, the total pressure for adiabatic flows with losses, e.g., due to frictional losses, always decreases, since the entropy change has to be positive. The exponential

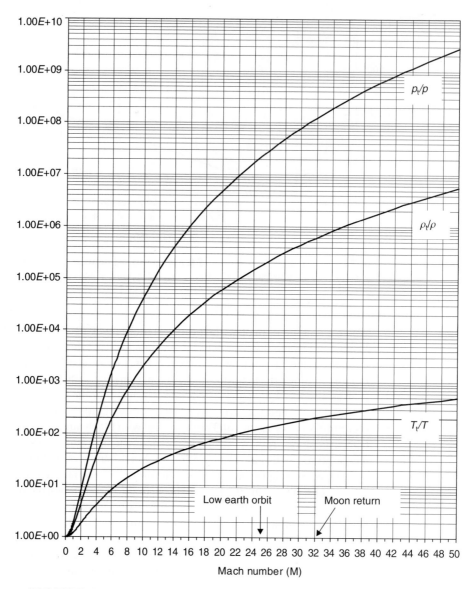

■ **FIGURE 2.7**   **Variation of stagnation state of gas with flight Mach number for** $\gamma = 1.4$ **(for a *calorically perfect gas*)**

relationship between the total pressure ratio and the entropy rise in an adiabatic flow is very useful to propulsion studies and we write it in Eq. 2.73.

$$p_{t2} / p_{t1} = e^{-\frac{\Delta s}{R}} \qquad (2.73)$$

All adiabatic flows with loss result in a drop of total pressure. For example, in a supersonic inlet with shocks and frictional losses in the boundary layer, we encounter a total pressure loss. Since, the total pressure is now cast as a measure of loss via Eq. 2.73, it serves as a commodity that propulsion engineers try to preserve as much as possible. The T–s diagram of an adiabatic flow with loss is shown in Fig. 2.9.

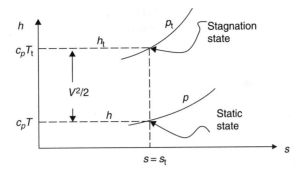

■ **FIGURE 2.8**
Stagnation and static states of a gas in motion shown on an h–s diagram

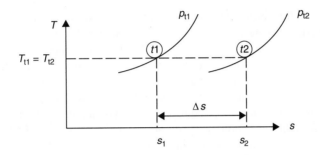

■ **FIGURE 2.9** Adiabatic flow that shows total pressure loss Δp_t, e.g., flows containing shocks

2.7 Quasi-One-Dimensional Flow

In duct flows with area variations, such as a diffuser or a nozzle, where the boundary layers are completely attached, the streamwise variation of flow variables dominate the lateral variations. In essence the flow behaves as a uniform flow with pressure gradient in a variable-area duct. A schematic drawing of a duct with area variation is shown in Fig. 2.10. The primary flow direction is labeled as "x" and the lateral direction is "y." Although the flow next to the wall assumes the same slope as the wall, the cross-sectional flow properties are based on uniform parallel flow at the cross section. The flow area at a cross section enters the conservation laws, but the variation in the cross section from the centerline to the wall is neglected. Such flows are called quasi-one-dimensional flows.

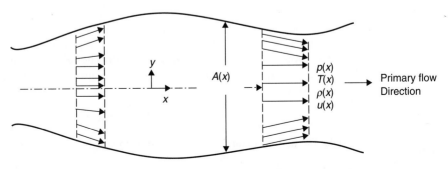

■ **FIGURE 2.10** Schematic drawing of a variable-area duct with attached boundary layers

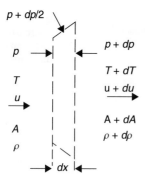

■ **FIGURE 2.11** *Slab* of inviscid fluid in a variable-area duct (walls exert pressure on the fluid)

For high Reynolds number flows where the boundary layer is thin compared with lateral dimensions of a duct and assuming an attached boundary layer, the assumption of *inviscid* fluid leads to a reasonable approximation of the duct flow. This means that an inviscid fluid model accurately estimates the pressure, temperature, and flow development along the duct axis. Attached boundary layers require a slow variation in the duct area, which is almost invariably the case in propulsion engineering. Inviscid flow analysis also produces a limit performance capability of a component when boundary layers are infinitely thin. In a practical sense, ducts with boundary layer suction approach an inviscid flow model. To develop the quasi-one-dimensional flow equations in a variable-area duct with inviscid fluid, we examine a slice, known as a *slab*, of the flow and apply the Taylor series approximation to the two sides of the slab. Figure 2.11 shows a thin slice of a duct flow with streamwise length dx separating the two sides of the slab.

We assume general parameters on one side and step in the x-direction using Taylor series approximation of an analytical function expanded in the neighborhood of a point. Since the step size dx is small, we may truncate the infinite Taylor series after the first derivative. Note that the walls exert one-half of the incremental pressure $(dp/2)$ over the upstream pressure p, since we take only one-half the step size $(dx/2)$ to reach its center. Now, let us apply the conservation principles to this thin slice of the flow. Continuity demands

$$\rho u A = (\rho + d\rho)(u + du)(A + dA) \tag{2.74}$$

Ignoring higher order terms involving the products of two small parameters, such as $du \cdot dA$, we get

$$\rho u \, dA + u A \, d\rho + \rho A \, du = 0 \tag{2.75}$$

Now dividing both sides by a nonzero constant namely $\rho u A$, we get the continuity equation in differential form

$$\frac{dA}{A} + \frac{d\rho}{\rho} + \frac{du}{u} = 0 \tag{2.76}$$

This is also known as the *logarithmic derivative* of the local mass flow rate, $\rho u A = $ constant. Since we use the logarithmic derivative often in our analysis, let us examine it now. From the continuity equation

$$\rho u A = \text{Constant} \tag{2.77}$$

We take the natural logarithm of both sides, to get the sum

$$\ell n(\rho) + \ell n(u) + \ell n(A) = \text{Constant} \tag{2.78}$$

Now, if we take the differential of both sides we get Eq. 2.76, which is called logarithmic derivative of the continuity equation.

The momentum balance in the streamwise direction gives

$$(\rho u A)(u + du) - (\rho u A)u = pA - (p + dp)(A + dA) + (p + dp/2)dA \tag{2.79}$$

The first term on the LHS is the rate of momentum out of the slab, which is the mass flow rate times the velocity out of the box. The second term on the LHS is the rate of momentum into the box. The first term on the LHS is the pressure force pushing the fluid out. The second term on the LHS is the pressure force in the opposite direction to the flow, hence negative. The last term on the LHS is the pressure force contribution of the walls. First, we note that its direction is in the flow direction, hence positive. Second, we note that the projection of the sidewalls in the flow direction is dA, which serves as the *effective* area for the wall pressure to push the fluid out of the box. We may simplify this equation by canceling terms and neglecting higher order quantities to get

$$\rho u A du = -A dp \tag{2.80}$$

The flow area term is cancelled from Eq. 2.80 (as expected) to yield

$$\rho u du = -dp \tag{2.81}$$

The energy equation for an adiabatic (non heat-conducting) flow with no shaft work is the statement of conservation of total enthalpy, which differentiates into

$$dh_t = dh + udu = 0 \tag{2.82}$$

The equation of state for a perfect gas may be written in logarithmic derivative form as

$$dp/p = d\rho/\rho + dT/T \tag{2.83}$$

Also, as we stipulated an isentropic flow through the duct, i.e., $ds = 0$, the pressure–density relationship follow the isentropic rule, namely $p/\rho^\gamma = $ constant, which has the following logarithmic derivative:

$$\frac{dp}{p} - \gamma \frac{d\rho}{\rho} = 0 \tag{2.84}$$

The set of governing equations for an isentropic flow through a duct of variable area are summarized below:

$$d\rho/\rho + du/u + dA/A = 0$$
$$\rho u du = -dp$$
$$dh + udu = 0$$
$$dp/p = d\rho/\rho + dT/T$$

The unknowns are the pressure p, density ρ, temperature T, and velocity u. The enthalpy for a perfect gas follows $dh = c_p dT$, hence we need to know the specific heat at constant pressure of the gas as well. The entropy remains constant, hence the static pressure is related to the speed of sound and density according to:

$$p = \rho a^2/\gamma \tag{2.85}$$

throughout the flow. We shall apply these governing equations in the following sections to fluid flow problems of interest to propulsion.

2.8   Area-Mach Number Relationship

Let us divide the momentum equation by $d\rho$, to get

$$\frac{\rho}{d\rho}udu = -\frac{dp}{d\rho} = -a^2 \tag{2.86}$$

which yields

$$\frac{d\rho}{\rho} = -\frac{udu}{a^2} \tag{2.87}$$

Now, let us substitute the density ratio in the continuity equation to relate the area variation to Mach number according to

$$-\frac{udu}{a^2} + \frac{du}{u} + \frac{dA}{A} = 0 \tag{2.88-a}$$

This simplifies to

$$(M^2 - 1)\frac{du}{u} = \frac{dA}{A} \tag{2.88-b}$$

This equation relates area variation to speed variation for the subsonic and supersonic flows. Assuming a subsonic flow, the parenthesis involving Mach number becomes a negative quantity. Hence an area-increase in a duct, i.e., $dA > 0$, results in $du < 0$, to make the signs of both sides of 2.88-b consistent. A negative du means flow deceleration. Hence a subsonic diffuser requires a duct with $dA/dx > 0$. A subsonic nozzle with $du > 0$ demands $dA < 0$. These two duct geometries for a subsonic flow are shown in Fig. 2.12.

For a supersonic flow, the parenthesis that involves Mach number in Eq. 2.88-b is positive. Hence, an area increase results in flow acceleration and vice versa. These geometric relationships between area variation and flow speed for supersonic flow are shown in Fig. 2.13.

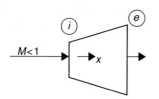
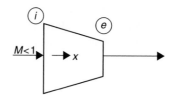

(a) A subsonic diffuser ($dA/dx > 0$)   (b) A subsonic nozzle ($dA/dx < 0$)

■ **FIGURE 2.12**   **Area variations for a subsonic diffuser and a nozzle**

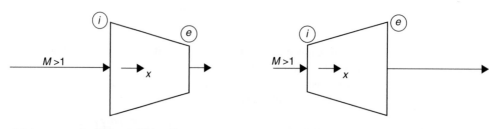

(a) A supersonic diffuser ($dA/dx < 0$)   (b)A supersonic nozzle ($dA/dx > 0$)

■ **FIGURE 2.13**   **Area variations for a supersonic diffuser and nozzle**

We note that the geometric requirements for a desired fluid acceleration or deceleration in a duct are opposite to each other in subsonic and supersonic flow regimes. This dual behavior is seen repeatedly in aerodynamics. We shall encounter several examples of this in the present chapter.

2.9   Sonic Throat

A unique flow condition separates the subsonic from supersonic flow. It is the sonic flow that marks the boundary between the two flows. In a subsonic flow, we learned that area reduction accelerates the gas and ultimately could reach a sonic state at the exit of the subsonic nozzle. A sonic flow may accelerate to supersonic Mach numbers if the area of the duct increases in the streamwise direction. This initial contraction and later expansion of the duct is capable of accelerating a subsonic flow to supersonic exit flow. This is called a convergent–divergent duct, with an internal throat, i.e., the sonic throat. Conversely, a convergent–divergent duct with a supersonic entrance condition has the capability of decelerating the flow to subsonic exit conditions through a sonic throat. These duct geometries are shown in Fig. 2.14.

The sonic state where the gas speed and the local speed of sound are equal is distinguished with a star. For example, the pressure at the sonic point is p^* and the temperature is T^*, the speed of sound is a^*, etc. Since the flow is isentropic throughout the duct the stagnation pressure remains constant and as a consequence of adiabatic duct flow the stagnation temperature remains constant. By writing the stagnation temperature in terms of the local static temperature and local Mach number, we may relate the local static temperature to sonic temperature according to

$$T_t = T\left[1 + \left(\frac{\gamma - 1}{2}\right)M^2\right] = T^*\left[1 + \left(\frac{\gamma - 1}{2}\right)(1)\right] = T^*\left(\frac{\gamma + 1}{2}\right) \qquad \text{(2.89)}$$

Taken in ratio, the above equation yields

$$\frac{T}{T^*} = \frac{\dfrac{\gamma + 1}{2}}{1 + \left(\dfrac{\gamma - 1}{2}\right)M^2} \qquad \text{(2.90)}$$

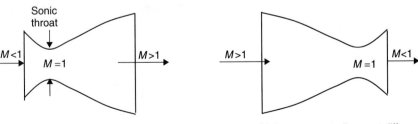

(a) A convergent–divergent nozzle   (b) A convergent–divergent diffuser

■ **FIGURE 2.14**   **Convergent–divergent duct with an internal sonic throat**

We may apply the isentropic relationship between temperature ratio and pressure ratio (Eq. 2.24) to get an expression for p/p^*.

$$\frac{p}{p^*} = \left[\frac{\frac{\gamma + 1}{2}}{1 + \left(\frac{\gamma - 1}{2}\right)M^2} \right]^{\frac{\gamma}{\gamma - 1}} \tag{2.91}$$

From perfect gas law and the two expressions for temperature and pressure ratio (2.90) and (2.91), we get an expression for the density ratio

$$\frac{\rho}{\rho^*} = \frac{p}{p^*} \Big/ \frac{T}{T^*} = \left[\frac{\frac{\gamma + 1}{2}}{1 + \left(\frac{\gamma - 1}{2}\right)M^2} \right]^{\frac{1}{\gamma - 1}} \tag{2.92}$$

The above expressions for the sonic ratios describe the properties of a section of the duct where flow Mach number is M to the flow properties at the sonic point. The mass flow rate balance between any section of the duct and the sonic throat yields an expression for the area ratio A/A^*. First, let us write the continuity equation for a uniform flow in terms of the total gas properties, p_t and T_t.

$$\dot{m} = \rho A V = \frac{\gamma p}{a^2} A V = \frac{\gamma p A M}{a} = \sqrt{\frac{\gamma}{R}} \frac{p}{\sqrt{T}} A M \tag{2.93-a}$$

We replaced the density with speed of sound and pressure to get Eq. 2.93-a. Now, we may substitute the total pressure and temperature for the static pressure and temperature of Eq. 2.93-a and their respective functions of Mach number to get Eq. 2.93-b.

$$\dot{m} = \sqrt{\frac{\gamma}{R}} \frac{p_t}{\sqrt{T_t}} A \cdot M \left(\frac{1}{1 + \frac{\gamma - 1}{2}M^2} \right)^{\frac{\gamma + 1}{2(\gamma - 1)}} \tag{2.93-b}$$

According to Eq. 2.93-b, we need the local flow area A, the local Mach number M, the local total pressure and temperature p_t and T_t, respectively, and the type of gas, γ, R, to calculate the mass flow rate. The total temperature remains constant in adiabatic flows with no shaft work and the total pressure remains constant for isentropic flows with no shaft work, hence they are convenient parameters in Eq. 2.93-b. Since the mass flow rate between any two sections of a duct in steady flow has to be equal, Eq. 2.93-b is a Mach number-area relationship. We use the sonic flow area A^*, with $M = 1$, to express a nondimensional area ratio.

$$\frac{A}{A^*} = \frac{1}{M} \left(\frac{1 + \frac{\gamma - 1}{2}M^2}{\frac{\gamma + 1}{2}} \right)^{\frac{\gamma + 1}{2(\gamma - 1)}} \tag{2.94}$$

We may tabulate the functions p/p^*, T/T^*, and A/A^* in addition to p_t/p, T_t/T in an isentropic table for a specific gas, i.e., $\gamma = $ constant. Figure 2.15 shows the isentropic flow parameters as a function of Mach number for a diatomic gas, such as air.

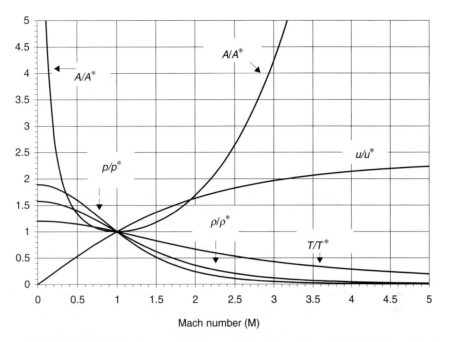

■ **FIGURE 2.15**   **Isentropic flow parameters for a diatomic (calorically perfect) gas, $\gamma = 1.4$**

At the minimum-area section, the ratio of mass flow rate to the cross-sectional area becomes maximum. Therefore to demonstrate that $M = 1$ maximizes the mass flow per unit area, we differentiate the following equation and set it equal to zero,

$$\frac{\dot{m}}{A} = const \cdot M \cdot \left(\frac{1}{1 + \frac{\gamma - 1}{2}M^2}\right)^{\frac{\gamma+1}{2(\gamma-1)}}$$

$$\frac{d}{dM}\left(\frac{\dot{m}}{A}\right) = const \cdot \left[\left(\frac{1}{1 + \frac{\gamma - 1}{2}M^2}\right)^{\frac{\gamma+1}{2(\gamma-1)}}\right. \tag{2.95}$$

$$\left. + M\left(\frac{-(\gamma + 1)}{2(\gamma - 1)}\right)\left((\gamma - 1)M\right)\left(\frac{1}{1 + \frac{\gamma - 1}{2}M^2}\right)^{\frac{\gamma+1}{2(\gamma-1)} + 1}\right] = 0$$

After some minor simplifications we get

$$1 - \left(\frac{\gamma + 1}{2}M^2\right) \Big/ \left(1 + \frac{\gamma - 1}{2}M^2\right) = 0 \tag{2.96}$$

Equation 2.96 is identically satisfied by $M = 1$. We may also demonstrate this principle graphically by plotting the function $f(M)$ (from Eq. 2.93-b):

$$f(M) = M \Big/ \left(1 + \frac{\gamma - 1}{2}M^2\right)^{\frac{\gamma+1}{2(\gamma-1)}} \tag{2.97}$$

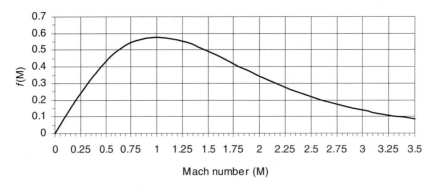

■ **FIGURE 2.16** Nondimensional mass flow parameter variation with Mach number shows a sonic condition at the minimum-area section, i.e., the throat

Figure 2.16 is a graph of the nondimensional mass flow rate per unit area, i.e., Eq. 2.97.

2.10 Waves in Supersonic Flow

In a subsonic flow, a disturbance travels upstream to alert the flow of an upcoming object, say an airfoil. The streamlines in the flow make the necessary adjustments prior to reaching the obstacle to "make room" for the body that has a thickness. Figure 2.17 shows the phenomenon of "upwash" ahead of a subsonic wing, as evidence of subsonic flow adjustment before reaching the body.

In contrast to a subsonic flow, the flowfield in a supersonic flow has to make adjustment for a body abruptly and within the zone of action. For small disturbances, the zone of action is the Mach cone, as shown in Fig. 2.4. Larger disturbances, as in thick bodies or bodies at an angle of attack, create a system of compression and expansion waves that turn the supersonic flow around the body. The finite compression waves in aerodynamics are called shock waves due to their abrupt nature. A diamond airfoil in a supersonic flow shows the wave pattern about the body in Fig. 2.18.

■ **FIGURE 2.17**
"Upwash" ahead of a
subsonic wing section

■ **FIGURE 2.18**
Sketch of waves about a
diamond airfoil at an
angle of attack
in supersonic flow

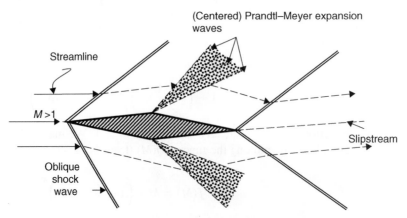

The attached oblique shocks at the leading edge turn the flow parallel to the surface and the expansion waves, known as the Prandtl–Meyer waves, cause the flow to turn at the airfoil shoulders. The trailing-edge waves turn the flow parallel to each other and the slipstream. The flow downstream of the tail waves, in general, is not parallel to the upstream flow. This is also in contrast to the subsonic flow that recovers the upstream flow direction further downstream of the trailing edge. The slipstream in supersonic flow allows for parallel flows and a continuous static pressure across the upper and lower parts of the trailing flowfield. It is the latter boundary condition that determines the slipstream inclination angle.

A special case of an oblique shock wave is the normal shock. This wave is normal to the flow and causes a sudden deceleration; static pressure, temperature, density, and entropy rise; Mach number and total pressure drop across the wave. We derive the *jump conditions* across a normal shock in the following section.

2.11   Normal Shocks

We apply the conservation principles to a normal shock flow, as shown in Fig. 2.19. The shape of the control volume is chosen to simplify the applications of the conservation laws. The frame of reference is placed (with an observer) at the shock wave for a steady flow.

To avoid writing the flow area of our arbitrary control volume, we choose it to be unity. The continuity equation of a steady flow through the control volume demands

$$\rho_1 u_1 = \rho_2 u_2 \tag{2.98}$$

The momentum equation in the *x*-direction is

$$\rho_2 u_2^2 - \rho_1 u_1^2 = p_1 - p_2 \tag{2.99-a}$$

Now separating the flow conditions on the two sides of the shock, we get

$$p_1 + \rho_1 u_1^2 = p_2 + \rho_2 u_2^2 \tag{2.99-b}$$

Each side of Eq. 2.99-b is called fluid impulse per unit area, I/A. Since the flow area is constant across a normal shock, we conclude that *impulse* is conserved across the shock. We will return to impulse later in this chapter. By factoring the static pressure and introducing the speed of sound in Eq. 2.99-b, we get an alternative form of the conservation of momentum across the shock that involves Mach number and pressure, namely

$$p_1(1 + \gamma M_1^2) = p_2(1 + \gamma M_2^2) \tag{2.100}$$

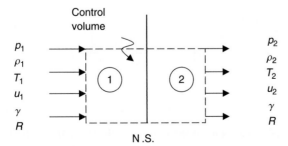

Control volume

p_1
ρ_1
T_1
u_1
γ
R

1

2

p_2
ρ_2
T_2
u_2
γ
R

N.S.

■ **FIGURE 2.19**   **Control volume for a normal shock (for an observer fixed at the shock)**

The energy equation for an adiabatic flow with no shaft work demands the conservation of the total enthalpy, that is,

$$c_p T_1 + \frac{u_1^2}{2} = c_p T_2 + \frac{u_2^2}{2} \quad (\text{assuming } c_p = \text{constant}) \tag{2.101}$$

The essential unknowns in a normal shock flow are p_2, ρ_2, T_2, and u_2. The other unknowns are derivable from these basic parameters. For example, from the static temperature, T_2 we calculate the local speed of sound a_2 and then the Mach number downstream of the shock, M_2. Also, with the Mach number known, we may calculate the total pressure p_{t2} and temperature T_{t2} downstream of a shock. The three conservation laws and an equation of state for the gas provide the four equations needed to solve for the four basic unknowns, that is, gas properties and the speed. These governing equations are coupled, however, and not immediately separable. We introduce the steps to the derivation of a normal shock flow in the following section.

We may relate the enthalpy to speed of sound, via Eq. 2.61-a, and derive alternative forms of the energy equation

$$\frac{a^2}{\gamma - 1} + \frac{u^2}{2} = \frac{a^{*2}}{\gamma - 1} + \frac{a^{*2}}{2} = \frac{\gamma + 1}{2(\gamma - 1)} a^{*2} = \frac{a_t^2}{\gamma - 1} = \text{Constant} \tag{2.102-a}$$

Divide through by u^2, and multiply by $(\gamma - 1)$, to get

$$\frac{1}{M^2} + \frac{\gamma - 1}{2} = \frac{\gamma + 1}{2} \cdot \frac{1}{M^{*2}} \tag{2.102-b}$$

Here, we have introduced a new parameter M^*, which is called the characteristic Mach number and has a definition

$$M^* \equiv \frac{u}{a^*} \tag{2.103}$$

The characteristic Mach number is the ratio of local gas speed to the speed of sound at the sonic point. Hence it is the ratio of two speeds at two different points in the flow. It is related to the local Mach number M through Eq. 2.102-b. If we solve for M^{*2} in terms of M^2, we get the expression 2.104, which is plotted in Fig. 2.20.

$$M^{*2} = \frac{(\gamma + 1)M^2}{2 + (\gamma - 1)M^2} \tag{2.104}$$

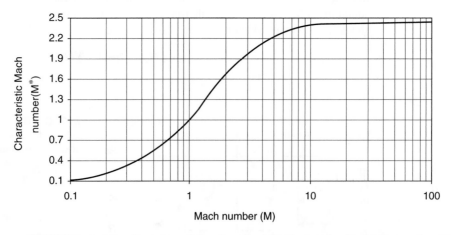

■ **FIGURE 2.20**   Variation of the characteristic Mach M^*, with Mach number M, $\gamma = 1.4$

We note that for $M < 1$, M^* is less than 1, for $M = 1$, M^* is equal to 1, and for $M > 1$, M^* is greater than 1. The characteristic Mach number is finite, however, as the Mach number M approaches infinity, M^* approaches a finite value, i.e.,

$$M^* \to \sqrt{\frac{\gamma + 1}{\gamma - 1}} \quad \text{as } M \to \infty. \tag{2.105}$$

For $\gamma = 1.4$, the limiting value of M^* is $\sqrt{6} = 2.449$.

The momentum balance across the shock may be written as

$$\rho_1 u_1 \left(\frac{p_1}{\rho_1 u_1} + u_1 \right) = \rho_2 u_2 \left(\frac{p_2}{\rho_2 u_2} + u_2 \right) \tag{2.106}$$

From continuity equation, we note that the product of density–velocity is a constant of motion, i.e., $\rho_1 u_1 = \rho_2 u_2$, hence it may be cancelled from Eq. 2.106. Also, we recognize that the ratio of static pressure to density is proportional to the speed of sound squared according to

$$p/\rho = a^2/\gamma$$

Hence, the modified form of the momentum equation is

$$\frac{a_1^2}{\gamma u_1} + u_1 = \frac{a_2^2}{\gamma u_2} + u_2 \tag{2.107}$$

From energy Equation 2.101, we may replace the local speed of sound a by the local speed u and the speed of sound at the sonic point a^* to get an equation in terms of the gas speeds across the shock and the speed of sound at the sonic point.

$$\frac{1}{\gamma u_1} \left[\frac{\gamma + 1}{2} a^{*2} - \frac{\gamma - 1}{2} u_1^2 \right] + u_1 = \frac{1}{\gamma u_2} \left[\frac{\gamma + 1}{2} a^{*2} - \frac{\gamma - 1}{2} u_2^2 \right] + u_2 \tag{2.108}$$

This equation simplifies to the Prandtl relation for a normal shock, i.e.,

$$u_1 u_2 = a^{*2} \tag{2.109}$$

Prandtl relation states that the product of gas speeds on two sides of a normal shock is a constant and that is the square of the speed of sound at the sonic point. This powerful expression relates the characteristic Mach number upstream and downstream of a normal shock via an inverse relationship, namely

$$M_2^* = 1/M_1^* \tag{2.110}$$

We may also conclude from this equation that the flow downstream of a normal shock is subsonic, since the characteristic Mach number upstream of the shock is greater than 1. From Eq. 2.104, we replace the characteristic Mach number by the local Mach number to get

$$M_2^{*2} = \frac{1}{M_1^{*2}} = \frac{2 + (\gamma - 1)M_1^2}{(\gamma + 1)M_1^2} = \frac{(\gamma + 1)M_2^2}{2 + (\gamma - 1)M_2^2} \tag{2.111-a}$$

This is a one equation, one unknown expression for M_2^2 that results in

$$M_2^2 = \frac{2 + (\gamma - 1)M_1^2}{2\gamma M_1^2 - (\gamma - 1)}$$

(2.111-b)

The downstream Mach number of a normal shock is finite even for $M_1 \to \infty$, which results in

$$M_2 \to \sqrt{\frac{\gamma - 1}{2\gamma}} \qquad \text{as } M_1 \to \infty$$

(2.111-c)

The density ratio across the shock is inversely proportional to the gas speed ratio via the continuity equation, which is related to upstream Mach number according to

$$\frac{\rho_2}{\rho_1} = \frac{u_1}{u_2} = \frac{M_1^*}{M_2^*} = M_1^{*2} = \frac{(\gamma + 1)M_1^2}{2 + (\gamma - 1)M_1^2}$$

(2.112-a)

We note that the density ratio is also finite as $M_1 \to \infty$, i.e.,

$$\frac{\rho_2}{\rho_1} \to \frac{\gamma + 1}{\gamma - 1} \qquad \text{as } M_1 \to \infty$$

(2.112-b)

To derive the static pressure ratio across a shock, we start with the gas momentum equation. We may regroup the pressure and momentum terms in 2.99-b according to

$$p_2 - p_1 = \rho_1 u_1^2 \left(1 - \frac{\rho_2}{\rho_1} \cdot \frac{u_2^2}{u_1^2} \right) = \rho_1 u_1^2 \left(1 - \frac{\rho_1}{\rho_2} \right)$$

(2.113)

Now, we may divide both sides by p_1 and substitute for the density ratio from Eq. 2.112, and simplify to get

$$\frac{p_2}{p_1} = 1 + \frac{2\gamma}{\gamma + 1} (M_1^2 - 1)$$

(2.114)

From the perfect gas law, we establish the static temperature ratio across the shock in terms of the static pressure and density ratios in terms M_1:

$$\frac{T_2}{T_1} = \frac{p_2}{p_1} \frac{\rho_1}{\rho_2} = \left[1 + \frac{2\gamma}{\gamma + 1} (M_1^2 - 1) \right] \left[\frac{2 + (\gamma - 1)M_1^2}{(\gamma + 1)M_1^2} \right]$$

(2.115)

The stagnation pressure ratio across a normal shock may be written as the static pressure ratio and the function of local Mach numbers according to

$$\frac{p_{t2}}{p_{t1}} = \frac{p_2}{p_1} \left[\frac{1 + \dfrac{\gamma - 1}{2} M_2^2}{1 + \dfrac{\gamma - 1}{2} M_1^2} \right]^{\frac{\gamma}{\gamma - 1}}$$

$$= \left[1 + \frac{2\gamma}{\gamma + 1} (M_1^2 - 1) \right] \left[\frac{1 + \dfrac{\gamma - 1}{2} \left(\dfrac{2 + (\gamma - 1)M_1^2}{2\gamma M_1^2 - (\gamma - 1)} \right)}{1 + \dfrac{\gamma - 1}{2} M_1^2} \right]^{\frac{\gamma}{\gamma - 1}}$$

(2.116)

The entropy rise across a shock is related to the stagnation pressure loss following Eq. 2.73,

$$\Delta s / R = -\ell n (p_{t2} / p_{t1})$$

(2.117)

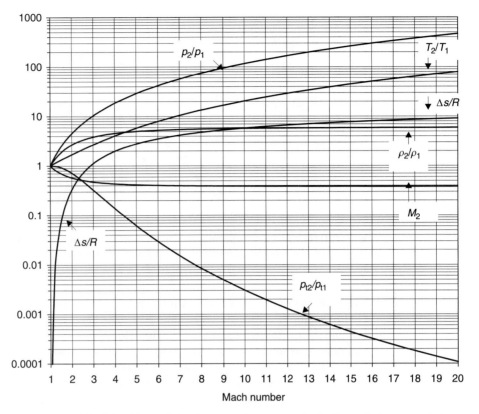

■ **FIGURE 2.21**   Normal shock parameters for a calorically perfect gas, $\gamma = 1.4$

In summary, we established the downstream Mach number in terms of upstream Mach number, all the jump conditions across a normal shock, i.e., $p_2/p_1, T_2/T_1, \rho_2/\rho_1, p_{t2}/p_{t1}, \Delta s/R$, as well as identified the constants of motion, namely:

$$\rho u = \text{Constant}$$

$$p + \rho u^2 = p(1 + \gamma M^2) = \text{Constant}$$

$$h + \frac{u^2}{2} = \frac{a^2}{\gamma - 1} + \frac{u^2}{2} = \frac{(\gamma + 1)}{2(\gamma - 1)} a^{*2} = \frac{a_t^2}{\gamma - 1} = \text{Constant}$$

$$u_1 u_2 = a^{*2} = \text{Constant}$$

Figure 2.21 shows the normal shock parameters for $\gamma = 1.4$ on a log-linear scale. Note that the density ratio as well as the Mach number downstream of the shock approach constant values as Mach number increases.

2.12   Oblique Shocks

Compression Mach waves may coalesce to form an oblique shock wave. A normal shock is a special case of an oblique shock with a wave angle of $90°$. Figure 2.22 shows a schematic drawing of an oblique shock flow with a representative streamline that abruptly changes direction across the shock. The shock wave angle with respect to upstream flow is called β

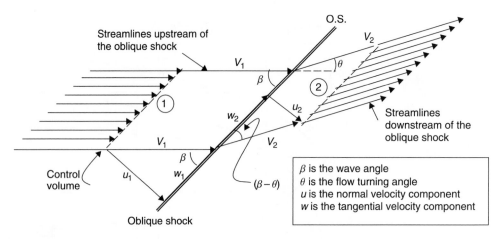

■ **FIGURE 2.22**   **Definition sketch of velocity components normal and parallel to an oblique shock, with the wave angle and flow turning angle definitions**

and the flow-turning angle (again with respect to the upstream flow) is θ. A representative control volume is also shown in Fig. 2.22. The flow is resolved into a normal and a tangential direction to the shock wave. The velocity components are u and w that are normal and tangential to the shock wave front, respectively. Two velocity triangles are shown upstream and downstream of the shock with vertex angles β and $(\beta - \theta)$, respectively. The control volume is confined between a pair of streamlines and the entrance, and exit planes are parallel to the shock. We choose the area of the entrance and exit to be one for simplicity.

To arrive at the shock jump conditions across the shock, we satisfy the conservation of mass, momentum normal and tangential to the shock, and the energy across the oblique shock. The continuity equation demands

$$\rho_1 u_1 = \rho_2 u_2 \tag{2.118}$$

The conservation of normal momentum requires the momentum change normal to the shock to be balanced by the fluid forces acting in the normal direction on the control volume. The only forces acting normal to the shock are the pressure forces acting on the entrance and the exit planes of the control surface.

$$\rho_2 u_2^2 - \rho_1 u_1^2 = p_1 - p_2 \tag{2.119}$$

The conservation of momentum along the shock front, i.e., tangential to the shock, requires

$$\rho_2 u_2 w_2 - \rho_1 u_1 w_1 = 0 \tag{2.120}$$

The net pressure forces in the tangential direction is zero, hence the tangential momentum is conserved. We may cancel the continuity portion of Eq. 2.120 to arrive at the simple result of the conservation of tangential velocity across an oblique shock, namely

$$w_1 = w_2 \tag{2.121}$$

The energy equation demands the conservation of total enthalpy, therefore

$$h_1 + \frac{u_1^2 + w_1^2}{2} = h_2 + \frac{u_2^2 + w_2^2}{2} \tag{2.122}$$

Since the tangential velocity is conserved across the shock the energy equation is simplified to

$$h_1 + \frac{u_1^2}{2} = h_2 + \frac{u_2^2}{2} \tag{2.123}$$

Now, we note that the conservation equations normal to the shock take on the exact form of the normal shock equations that we solved earlier to arrive at the shock jump conditions. Consequently, all the jump conditions across an oblique shock are established uniquely by the normal component of the flow to the oblique shock, i.e., M_{1n}.

From the wave angle β, we get

$$M_{1n} = M_1 \sin\beta \tag{2.124}$$

The normal shock relation 2.111-b establishes the normal component of the Mach number downstream of an oblique shock, according to

$$M_{2n}^2 = \frac{2 + (\gamma - 1)M_{1n}^2}{2\gamma M_{1n}^2 - (\gamma - 1)} = \frac{2 + (\gamma - 1)M_1^2 \sin^2\beta}{2\gamma M_1^2 \sin^2\beta - (\gamma - 1)} \tag{2.125}$$

From the velocity triangle downstream of the oblique shock, we have

$$M_2 = \frac{M_{2n}}{\sin(\beta - \theta)} \tag{2.126}$$

To summarize an oblique shock jump conditions are established by the normal component of the flow and the downstream Mach number follows Eq. 2.126. To establish a relationship between the flow turning angle and the oblique shock wave angle, we use the velocity triangles upstream and downstream of the shock,

$$\tan\beta = \frac{u_1}{w_1} \tag{2.127}$$

$$\tan(\beta - \theta) = \frac{u_2}{w_2} \tag{2.128}$$

The ratio of these two equations relates the angles and the normal velocity ratio according to

$$\frac{\tan\beta}{\tan(\beta - \theta)} = \frac{u_1}{u_2} = \frac{\rho_2}{\rho_1}$$

Now substituting for the density ratio from the normal shock relations, we get

$$\frac{\tan\beta}{\tan(\beta - \theta)} = \frac{(\gamma + 1)M_1^2 \sin^2\beta}{2 + (\gamma - 1)M_1^2 \sin^2\beta} \tag{2.129}$$

Expression 2.129 is called "θ–β–M" equation and a graph of the oblique shock angle versus the turning angle is plotted for a constant upstream Mach number to create oblique shock charts. In the limit of zero turning angle, i.e., as $\theta \to 0$,

$$(\gamma + 1)M_1^2 \sin^2\beta = 2 + (\gamma - 1)M_1^2 \sin^2\beta \tag{2.130}$$

There are two solutions to Eq. 2.130. The first is

$$\beta = \sin^{-1}\left(\frac{1}{M}\right) \equiv \mu \tag{2.131}$$

which identifies an infinitesimal strength shock wave as a Mach wave and the second is, $\beta = 90°$, which identifies a normal shock as the strongest oblique shock. Thus, the weakest oblique shock is a Mach wave and the strongest oblique shock is a normal shock. Since the

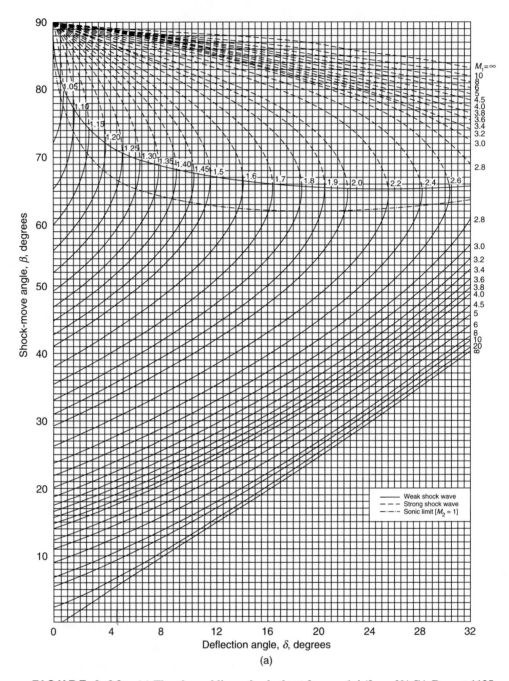

■ **FIGURE 2.23** (a) The plane oblique shock chart for $\gamma = 1.4$ (from NACA Report 1135, 1953); (b) The plane oblique shock chart-continued (from NACA Report 1135, 1953)

strength of the wave depends on $M \cdot \sin\beta$, an oblique shock wave angle is larger than the Mach angle, i.e.,

$$\mu \leq \beta \leq 90° \tag{2.132}$$

Figure 2.23 is a graph of Eq. 2.129 with Mach number as a running parameter (from NACA Report 1135, 1953). There are two solutions for the wave angle β for each turning angle θ. The

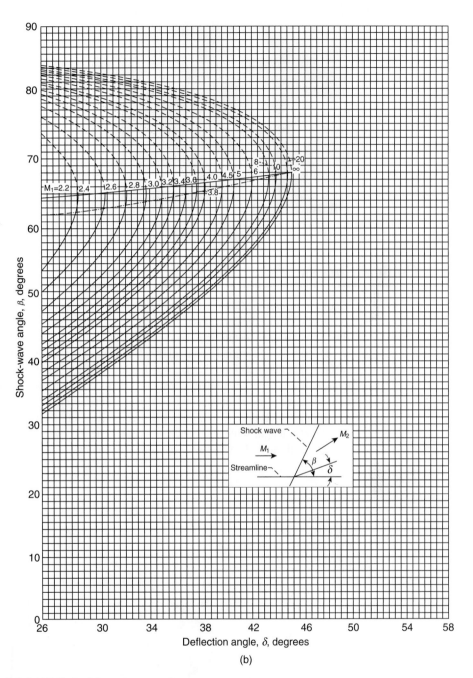

■ **FIGURE 2.23** (*Continued*)

high wave angle is referred to as the strong solution, and the lower wave angle is called the weak solution. A continuous line and a broken line distinguish the weak and strong solutions in Fig. 2.23, respectively. Also, there is a maximum turning angle θ_{max} for any supersonic Mach number. For example, a Mach 2 flow can turn only $\sim 23°$ via an attached plane oblique shock wave. For higher turning angles than θ_{max}, a detached shock will form upstream of the body. This shock, due to its shape, is referred to as the bow shock.

SR-71 in flight (Courtesy of NASA)

2.13   Conical Shocks

A cone in supersonic flow creates a conical shock. In case of zero angle of attack, the flowfield is axisymmetric and the cone axis and the conical shock axis coincide. The jump conditions at the shock are, however, established by the (local) normal component of Mach number to the shock front. The flowfield behind a conical shock continues to vary from the shock surface to the cone surface. In a plane oblique shock case, the flowfield undergoes a single jump at the shock and then its flow properties remain constant. The mechanism for a continuous change behind a conical shock is the compression Mach waves emanating from the cone vertex. These are the rays along which flow properties remain constant. Note that an axisymmetric conical shock flow lacks a length scale and hence θ becomes the only variable of the problem. Figure 2.24 shows a definition sketch of an axisymmetric conical shock flow.

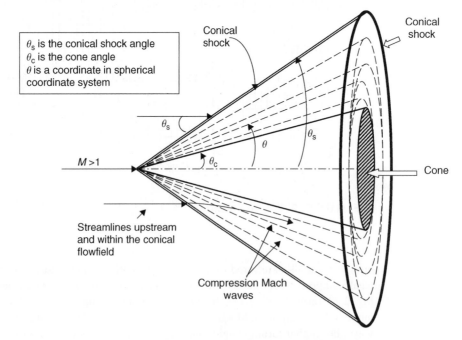

■ **FIGURE 2.24**   Definition sketch of a conical shock flowfield

All the jump conditions across the conical shock depend on $M \cdot \sin\theta_s$ as we derived earlier in this section. The flow properties downstream of the shock, e.g., p, T, ρ, V, are a function of the conical angle θ, i.e., $p(\theta)$, etc. Therefore all the rays emanating from the cone vertex have different flow properties. The cone surface is a ray itself and thus experiences constant flow properties such as pressure, temperature, density, velocity, and Mach number. The streamlines downstream of a conical shock continue to turn in their interaction with the compression Mach waves. The flowfield downstream of the conical shock remains reversible, however, as the compression Mach waves are each of infinitesimal strength. The flowfield downstream of an axisymmetric conical shock is irrotational since the shock is straight, hence of constant strength. We apply conservation principles to arrive at the detail flowfield downstream of a conical shock. In addition, we utilize the irrotationality condition to numerically integrate the governing conical flowfield equation, known as the Taylor–Maccoll equation. A cone of semivertex angle θ_c would create a conical shock wave of angle θ_s, which is smaller than a corresponding plane oblique shock on a two-dimensional ramp of the same vertex angle. This results in a weaker shock in three-dimensional space as compared with a corresponding two-dimensional space. This important flow behavior is referred to as *3D relieving effect* in aerodynamics. Figure 2.25 shows a two-dimensional compression ramp and a conical ramp of equal vertex angle in supersonic flow.

A conical shock chart for three Mach numbers is shown in Fig. 2.26 (from Anderson, 2003). This figure contains only the weak solution. Note that the maximum turning angle for a given Mach number is larger for a cone than a two-dimensional ramp of the same nose angle. This too is related to the well-known *3-D relieving effect*.

Finally, we graph a shock wave static and stagnation states on a T–s diagram, in Fig. 2.27. All shocks are irreversible and hence suffer an entropy rise, as graphically demonstrated in Fig. 2.27. The static path connects points 1 and 2 across the shock wave. The stagnation path

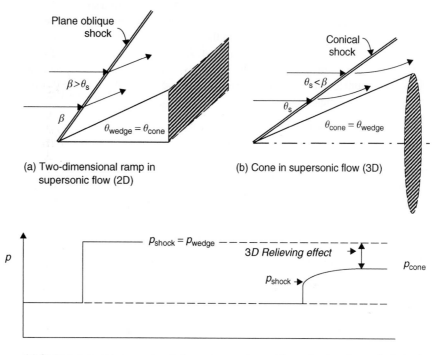

(a) Two-dimensional ramp in supersonic flow (2D)

(b) Cone in supersonic flow (3D)

(c) Static pressure jump and variation across a plane oblique shock and a conical shock

■ **FIGURE 2.25**   **Three-dimensional relieving effect on a cone**

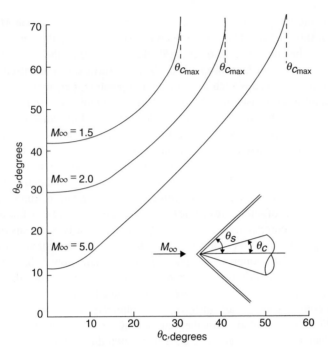

■ **FIGURE 2.26**   Conical shock chart for a calorically perfect gas, $\gamma = 1.4$ (from Anderson, 2003)

connects "$t1$" to "$t2$." The constant-pressure lines indicate a static pressure rise and a stagnation pressure drop between the two sides of a shock. The static temperature rises and the stagnation temperature remains constant (assuming c_p is constant as in a calorically perfect gas). We also note a drop in the kinetic energy of the gas, which is converted into heat through viscous dissipation within the shock. Shocks are generally thin and are of the order of the mean free path of the molecules in the gas. Thus, shocks are often modeled as having a zero thickness, unless the gas is in near vacuum of upper atmosphere where the mean free path is of the order of centimeters.

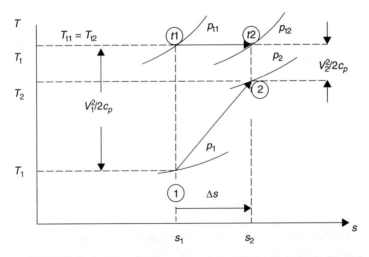

■ **FIGURE 2.27**   Static and stagnation states of a (calorically perfect) gas across a shock wave

2.14   Expansion Waves

The oblique shocks are compression waves in a supersonic flow that abruptly turn the flow and compress the gas in the process. An expansion wave causes a supersonic flow to turn and the static pressure to drop. The expansion waves are Mach waves that make a Mach angle with respect to the local flow. The flow through expansion waves is inherently isentropic, as the Mach waves are of infinitesimal strength and hence reversible. An expansion Mach wave is capable of only turning the flow incrementally and accelerates the flow incrementally. Such a flow turning and flow acceleration is shown in Fig. 2.28 with exaggeration of the turning angle and the flow acceleration increment. The local wave angle with respect to upstream flow is at the Mach angle μ. The incremental turning angle is labeled as $d\theta$. The change of velocity is shown by a magnitude of dV and a direction $d\theta$ with respect to upstream flow. Since the velocity downstream of the Mach wave has to have the same tangential component as the upstream flow, the right triangles that share the same tangential velocity are labeled with V and $V + dV$ vectors in Fig. 2.28.

We may apply the law of sines to the triangle with sides V and $V + dV$ to relate the turning angle and incremental speed change,

$$\frac{V + dV}{\sin\left(\frac{\pi}{2} - \mu\right)} = \frac{V}{\sin\left(\frac{\pi}{2} - \mu - d\theta\right)} \tag{2.133}$$

Let us simplify the sines and isolate the velocity terms, such as

$$1 + \frac{dV}{V} = \frac{\cos\mu}{\cos(\mu + d\theta)} \cong \frac{\cos\mu}{\cos\mu - d\theta \cdot \sin\mu} = \frac{1}{1 - d\theta \cdot \tan\mu} \approx 1 + d\theta \cdot \tan\mu \tag{2.134}$$

We applied a small angle approximation to sine and cosine $d\theta$, and in the last approximation on the RHS, we used a binomial expansion and truncation at the linear term for a small ε:

$$\frac{1}{1 \pm \varepsilon} \approx 1 \mp \varepsilon \tag{2.135}$$

Now from the Mach triangle shown in Fig. 2.29, we deduce that

$$\tan\mu = \frac{1}{\sqrt{M^2 - 1}} \tag{2.136}$$

Substitute this expression in Eq. 2.134 to get a simple relationship between the flow turning and the incremental acceleration across a Mach wave in *a* supersonic flow, i.e.,

$$d\theta = \sqrt{M^2 - 1} \frac{dV}{V} \tag{2.137}$$

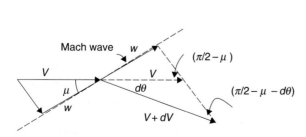

■ **FIGURE 2.28**   **Incremental turning across a Mach wave**

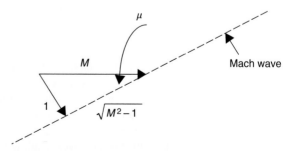

■ **FIGURE 2.29**   **Sketch of a Mach wave and its right triangle**

We first express the incremental velocity ratio in terms of Mach number and then proceed to integrate Eq. 2.137. Now, we employ logarithmic derivative as a tool to derive the functions of Mach number. We start with a basic definition and then build on that, for example:

$$V = M \cdot a = \text{constant} \cdot M \cdot T^{1/2} \tag{2.138}$$

The logarithmic derivative of Eq. 2.138 is

$$\frac{dV}{V} = \frac{dM}{M} + \frac{1}{2}\frac{dT}{T} \tag{2.139}$$

To express the temperature ratio in terms of Mach number, we use

$$T = \text{constant} \cdot \left(1 + \frac{\gamma - 1}{2}M^2\right)^{-1} \tag{2.140}$$

Note that the constant in Eq. 2.140 is the total temperature, which remains constant across the supersonic waves. Now, we may logarithmically differentiate Eq. 2.140 to get

$$\frac{dT}{T} = \frac{-(\gamma - 1)MdM}{1 + \dfrac{\gamma - 1}{2}M^2} \tag{2.141}$$

We substitute Eq. 2.141 into 2.139 and the result into 2.137 to get

$$d\theta = \frac{\sqrt{M^2 - 1}}{1 + \dfrac{\gamma - 1}{2}M^2}\frac{dM}{M} \tag{2.142}$$

Now, this equation may be integrated, as we have isolated a single variable M on the RHS of Eq. 2.142. The limits of integral are set at zero angle for $M = 1$ and $v(M)$ at Mach M. The angle $v(M)$ is called the Prandtl–Meyer function.

$$\int_0^{v(M)} d\theta = v(M) = \int_1^M \frac{\sqrt{M^2 - 1}}{1 + \dfrac{\gamma - 1}{2}M^2}\frac{dM}{M} \tag{2.143-a}$$

The definite integral on the RHS of Eq. 2.143-a is integrable in closed form by partial fraction and simple change of variables to cast the integrand in the form of the derivative of an inverse tangent function. The result of integration is

$$v(M) = \sqrt{\frac{\gamma + 1}{\gamma - 1}}\ \tan^{-1}\sqrt{\frac{\gamma - 1}{\gamma + 1}(M^2 - 1)} - \tan^{-1}\sqrt{M^2 - 1} \tag{2.143-b}$$

Figure 2.30 shows the Prandtl–Meyer function and the Mach angle as a function of Mach number for $\gamma = 1.4$.

We may tabulate this function in terms of Mach number for any gas, described by its specific heat ratio γ. A physical interpretation of the Prandtl–Meyer angle at Mach M is the needed turning angle to turn a sonic flow into a supersonic flow at Mach M. A schematic drawing of an expanding flow around a sharp corner is shown in Fig. 2.31. The wall turning angle is $v(M)$ and the downstream flow is at M. Note that the streamlines expand through Prandtl–Meyer waves to allow for a higher Mach number (thus lower density) downstream. Since the flow in an expansion wave is isentropic, the area between streamlines follows the A/A^* rule, derived earlier.

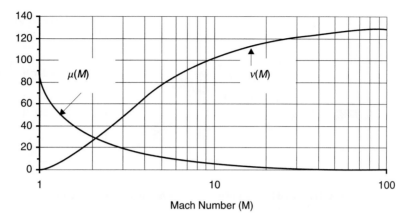

■ **FIGURE 2.30** Prandtl–Meyer function and Mach angle (in degrees) for a diatomic gas, $\gamma = 1.4$

The limiting values of the Prandtl–Meyer angle are zero at Mach 1 and in the limit of Mach number approaching infinity, the turning angle is

$$v(M_2 \to \infty) = \left(\sqrt{\frac{\gamma + 1}{\gamma - 1}} - 1 \right) \frac{\pi}{2} \tag{2.144}$$

For a ratio of specific heats of $\gamma = 1.4$, the maximum wall angle leading to infinite Mach number is $\sim 130.5°$, as also shown in Fig. 2.31. The lead Mach wave in Fig. 2.31 is normal to the sonic flow and the tail Mach wave makes an angle $\mu(M)$ with the downstream velocity vector. In general, the wall-turning angle for any upstream flow $(M_1 > 1)$ is the difference between the Prandtl–Meyer angles downstream and upstream of the corner,

$$\Delta\theta = \theta_{\text{wall}} = v_2 - v_1 \tag{2.145}$$

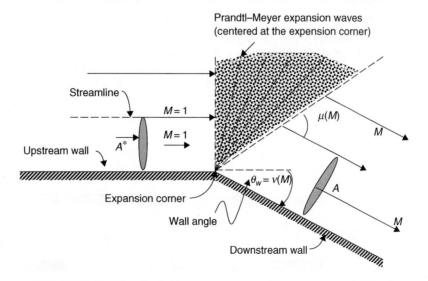

■ **FIGURE 2.31** Sonic flow turns to expand into a supersonic flow downstream of a sharp corner

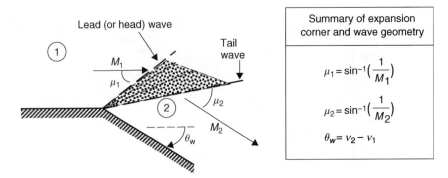

■ **FIGURE 2.32** An expansion corner in a supersonic flow

The wall-turning angle in a supersonic flow uniquely determines the flow Mach number downstream of the corner. The wave structure at the sharp corner in a supersonic flow is the same as that shown in Fig. 2.31, except the lead Mach wave is now at the local Mach angle $\sin^{-1}(1/M_1)$ instead of $90°$ for the sonic flow, and the wave envelope is rotated in the direction of wall-turning angle (see Fig. 2.32). The ratio of flow parameters follows the isentropic relation, namely,

$$\frac{T_2}{T_1} = \frac{1 + \dfrac{\gamma - 1}{2}M_1^2}{1 + \dfrac{\gamma - 1}{2}M_2^2} \tag{2.146}$$

$$\frac{p_2}{p_1} = \left(\frac{1 + \dfrac{\gamma - 1}{2}M_1^2}{1 + \dfrac{\gamma - 1}{2}M_2^2}\right)^{\frac{\gamma}{\gamma - 1}} \tag{2.147}$$

Rayleigh flow

2.15 Frictionless, Constant-Area Duct Flow with Heat Transfer

The class of one-dimensional compressible flow problems includes the effects of heat transfer and friction. These two effects have been traditionally separated to make the analysis simple and the effects of friction and heat transfer individually and separately studied. We first examine the effect of heat transfer on a compressible flow in a constant-area duct without friction. This fluid problem is called the Rayleigh flow.

We first define the geometry and the boundary conditions of the problem in a schematic drawing (see Fig. 2.33) to assist with the application of conservation principles.

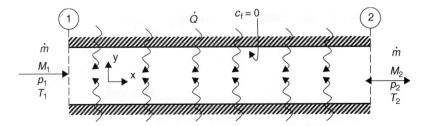

■ **FIGURE 2.33** Uniform frictionless flow in a constant-area duct with heat transfer

The boundary conditions are uniform flow, zero wall friction, a known inlet flow condition, i.e., $\dot{m}$, p_1, T_1, and heat transfer rate $\dot{Q}$. The unknowns are the flow parameters at the exit of the duct, namely, M_2, p_2, T_2, etc. The medium is a perfect gas.

The continuity equation demands

$$\rho_1 u_1 = \rho_2 u_2 \tag{2.148}$$

The momentum equation along the flow direction is

$$\rho_2 u_2^2 - \rho_1 u_1^2 = p_1 - p_2 \tag{2.149-a}$$

$$p_1 + \rho_1 u_1^2 = p_2 + \rho_2 u_2^2 \quad \Longrightarrow \quad I_1 = I_2 \tag{2.149-b}$$

Again, we note that due to our frictionless flow assumption in Rayleigh flow, the fluid impulse I remains constant.

The energy equation that accounts for the heat transfer through the walls is

$$\dot{m}(h_{t2} - h_{t1}) = \dot{Q} \tag{2.150-a}$$

The mass flow rate is the size parameter and may be divided out of the energy equation and the heat transfer rate per unit mass flow rate is called q, to get

$$h_{t2} = h_{t1} + q \tag{2.150-b}$$

The exit total enthalpy is immediately calculated based on the inlet condition h_{t1} and the known heat transfer per unit mass flow rate q. The governing equations, i.e., the mass flow rate and the momentum equation, are identical to a normal shock flow, except for the energy equation of the Rayleigh flow that includes heat transfer.

The momentum equation is written in terms of pressure and Mach number as in the normal shock flow,

$$p_2/p_1 = (1 + \gamma M_1^2)/(1 + \gamma M_2^2) \tag{2.151}$$

The continuity equation leads us to the density ratio in terms of the Mach numbers and temperature ratio following a simple manipulation,

$$\frac{\rho_2}{\rho_1} = \frac{u_1}{u_2} = \frac{M_1 a_1}{M_2 a_2} = \frac{M_1}{M_2}\sqrt{\frac{T_1}{T_2}} \tag{2.152}$$

Now, let us use the perfect gas law for the density ratio in favor of the pressure and temperature ratio to simplify Eq. 2.152 according to

$$\frac{p_2}{p_1} \cdot \frac{T_1}{T_2} = \frac{1 + \gamma M_1^2}{1 + \gamma M_2^2} \cdot \frac{T_1}{T_2} = \frac{M_1}{M_2}\sqrt{\frac{T_1}{T_2}} \tag{2.153-a}$$

This equation simplifies to

$$\frac{T_2}{T_1} = \frac{M_2^2}{M_1^2}\left(\frac{1 + \gamma M_1^2}{1 + \gamma M_2^2}\right)^2 \tag{2.153-b}$$

The density ratio is the ratio of pressure ratio to the temperature ratio, therefore

$$\frac{\rho_2}{\rho_1} = \frac{\dfrac{1 + \gamma M_1^2}{1 + \gamma M_2^2}}{\dfrac{M_2^2}{M_1^2}\left(\dfrac{1 + \gamma M_1^2}{1 + \gamma M_2^2}\right)^2} = \frac{M_1^2}{M_2^2}\left(\frac{1 + \gamma M_2^2}{1 + \gamma M_1^2}\right) \tag{2.154}$$

All the exit conditions are expressed in terms of the inlet and the exit Mach numbers. The exit Mach number is determined from the energy equation according to Eq. 2.155. The only unknown in this equation is the exit Mach number M_2. To facilitate numerical calculations, we take the exit flow condition to be sonic and tabulate functions, $p/p^*, T/T^*, \rho/\rho^*, p_t/p_t^*, T_t/T_t^*$.

$$\frac{T_{t2}}{T_{t1}} = \frac{T_2}{T_1}\left(\frac{1+\frac{\gamma-1}{2}M_2^2}{1+\frac{\gamma-1}{2}M_1^2}\right) = \frac{M_2^2}{M_1^2}\left(\frac{1+\gamma M_1^2}{1+\gamma M_2^2}\right)^2\left(\frac{1+\frac{\gamma-1}{2}M_2^2}{1+\frac{\gamma-1}{2}M_1^2}\right) = 1 + \frac{q}{c_p T_t} \qquad (2.155)$$

These functions are written for a general inlet condition M and an exit Mach number of 1. A flow that reaches a sonic state is said to be *choked*, and the effect of heat transfer may lead to an exit choking condition, which is referred to as a *thermally choked* flow. The sonic, i.e., choked condition is used as a reference state in compressible flow problems. The Rayleigh flow functions are

$$\frac{p}{p^*} = \frac{\gamma+1}{1+\gamma M^2} \qquad (2.156)$$

$$\frac{T}{T^*} = M^2\left(\frac{\gamma+1}{1+\gamma M^2}\right)^2 \qquad (2.157)$$

$$\frac{\rho}{\rho^*} = \frac{1}{M^2}\left(\frac{1+\gamma M^2}{\gamma+1}\right) \qquad (2.158)$$

$$\frac{T_t}{T_t^*} = M^2\left(\frac{\gamma+1}{1+\gamma M^2}\right)^2\left(\frac{1+\frac{\gamma-1}{2}M^2}{\frac{\gamma+1}{2}}\right) \qquad (2.159)$$

$$\frac{p_t}{p_t^*} = \left(\frac{\gamma+1}{1+\gamma M^2}\right)\left(\frac{1+\frac{\gamma-1}{2}M^2}{\frac{\gamma+1}{2}}\right)^{\frac{\gamma}{\gamma-1}} \qquad (2.160)$$

Now, let us examine the thermodynamics of this flow on a T–s diagram, which is called a Rayleigh line. The Gibbs equation is

$$ds = c_p dT/T - R dp/p \qquad (2.161)$$

The momentum equation in terms of pressure and Mach number is

$$p(1 + \gamma M^2) = \text{constant} \qquad (2.162)$$

The logarithmic derivative of the momentum equation relates incremental pressure change to the Mach number variation, that is,

$$\frac{dp}{p} + \frac{2\gamma M dM}{1+\gamma M^2} = 0 \qquad (2.163)$$

We may substitute the incremental pressure ratio in terms of dM/M in Eq. 2.161,

$$ds = c_p dT/T + R[2\gamma M^2/(1+\gamma M^2)]dM/M \qquad (2.164)$$

Now, we use continuity equation

$$d\rho/\rho + du/u = 0 \tag{2.165}$$

We may now substitute for the density ratio in terms of the pressure and temperature ratio following the perfect gas law,

$$dp/p - dT/T + dM/M + dT/2dT = dp/p - dT/2dT + dM/M = 0 \tag{2.166}$$

Let us substitute for the incremental pressure from the momentum equation 2.163 to relate the Mach increment to static temperature change dT,

$$dM/M = \frac{1}{2}dT/T + [2\gamma M^2/(1 + \gamma M^2)]dM/M \tag{2.167-a}$$

This equation simplifies to

$$dM/M = (dT/2T)[(1 + \gamma M^2)/(1 - \gamma M^2)] \tag{2.167-b}$$

Substitute the incremental Mach number expression in the Gibbs equation 2.164 to get

$$ds = c_p\frac{dT}{T} + R\frac{2\gamma M^2}{1 + \gamma M^2}\left(\frac{1}{2}\frac{dT}{T}\frac{(1 + \gamma M^2)}{(1 - \gamma M^2)}\right) = c_p\frac{dT}{T}\left[1 + (\gamma - 1)\left(\frac{M^2}{1 - \gamma M^2}\right)\right] \tag{2.168}$$

This equation simplifies to $s(T, M)$ for a Rayleigh line that may be graphed on a T–s diagram where the parameter along the T–s curve is the local Mach number M.

$$ds = c_p[(1 - M^2)/(1 - \gamma M^2)]dT/T \tag{2.169}$$

A graph of this function is shown in Fig. 2.34. Critical points of this function are at $ds/dT = 0$ and $dT/ds = 0$. These points represent the maximum entropy and the maximum temperature on the Rayleigh line, respectively. The point of maximum entropy occurs at $ds/dT = 0$, which Eq. 2.169 indicates the sonic condition at this point, i.e.,

$$ds/dT = 0 \text{ at } M = 1 \tag{2.170}$$

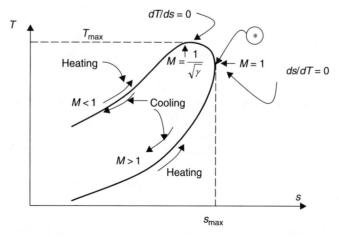

■ **FIGURE 2.34**   Static states of a gas in a frictionless, one-dimensional flow in constant-area duct with heat transfer

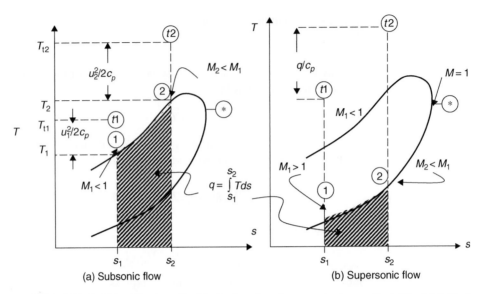

■ **FIGURE 2.35**   Heating of a frictionless flow in a constant-area duct (Rayleigh line)

and the maximum temperature point is at $dT/ds = 0$, which again Eq. 2.169 requires

$$dT/ds = 0 \text{ at } M = 1/\sqrt{\gamma} \tag{2.171}$$

There are two branches of the Rayleigh line, namely a subsonic and a supersonic branch. These two branches are separated at the sonic point, labeled (*) in Fig. 2.34. Also the path of heating tends to choke both the subsonic and supersonic flows and the cooling path drives the flow Mach number down in subsonic and up in supersonic flows. Interestingly, the point of maximum temperature occurs at a subsonic Mach number of $(1/\gamma)^{1/2}$ and further heating actually reduces the static temperature toward the exit sonic condition. The stagnation temperature continues to increase, however, with heating. The heating that leads to a choked exit condition is called the critical heating q^*. Since, the point of s_{max} occurs at the sonic point, the sonic condition has to occur at the exit of the duct.

The Rayleigh line, as shown in Fig. 2.34, is the loci of all the points with a constant mass flow rate $\rho u A$. Since the area of the duct is constant, the product of density and velocity remains constant, i.e., mass flow rate per unit area. The momentum equation of the Rayleigh line and a normal shock is the same. In deriving the functional form of the Rayleigh line in the T–s diagram, we did not use the energy equation. Therefore, this flow at least at one point represents a normal shock. We will establish this point when we study an adiabatic flow in a constant-area duct with friction in the next section.

The Rayleigh line does not involve any frictional dissipation hence the heating/cooling of the gas is reversible. Therefore the area under the T–s curve represents the heat transfer per unit mass of the fluid, as in $\delta q_{rev} = Tds$. This is graphically shown in Fig. 2.35 for a subsonic inlet condition and a supersonic case. The gas at point 1 is in a subsonic flow (in part (a) of Fig. 2.35) and upon heating it reaches a higher Mach number represented by point 2 and a higher static temperature. For a supersonic case, we refer to part (b) of Fig. 2.35. The heating decelerates the gas from Mach number M_1 to a lower supersonic flow Mach number with $M_2 < M_1$. The stagnation enthalpy difference between points 1 and 2 is equal to q following the energy equation.

The critical heating q^* will cause the exit of the duct to be choked. The heat transfer to a supersonic flow decelerates the gas and we deduce that in a supersonic combustion (as in a scramjet) there must be a critical fuel flow rate that causes the combustion chamber flow to

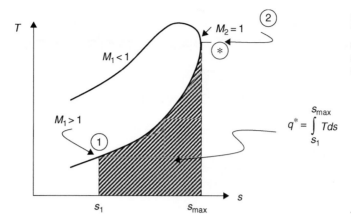

■ **FIGURE 2.36**
Thermally choking a
supersonic flow in a constant-
area frictionless duct (model
of a scramjet combustor)

choke at its exit. This is shown in Fig. 2.36. We shall examine supersonic combustion further in this book.

The critical heating q^* is a function of the inlet condition, as evident from Fig. 2.36. Any additional heating beyond q^* can be accommodated only by a reset inlet condition, through a reduced mass flow rate. A family of Rayleigh lines is shown in Fig. 2.37 with successively reduced mass flow rate.

The inlet flow condition shifts from state 1 to state 1′, as shown in Fig. 2.37, for a reduced flow rate in the duct. The physical inlet flow may be depicted as an inlet with flow spillage, as shown in Fig. 2.38. A subsonic inlet flow is shown with a curved streamline that allows for mass spillage. The diverging area of the streamtube in subsonic flow decelerates the gas to a lower speed as it increases the gas static pressure and temperature. These features are shown on the lower half of Fig. 2.39. For a supersonic throughflow with $q^* > q_1^*$, the inlet is in the zone of silence and hence it may not be reset. However, the boundary layers that form on solid surfaces are partially subsonic and serve as the communication line, or a wave-guide, between the exit and the inlet planes and hence the inlet may be reset to accommodate a higher heating rate. Since our assumption of one-dimensional flow could not allow for a boundary layer formation, in the strictest sense of this theory, a reset supersonic Rayleigh flow is not possible.

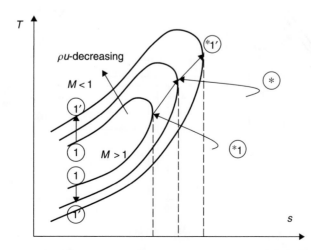

■ **FIGURE 2.37**   Family of
Rayleigh lines with a decreasing
flow rate in the duct caused
by $q > q_1^*$

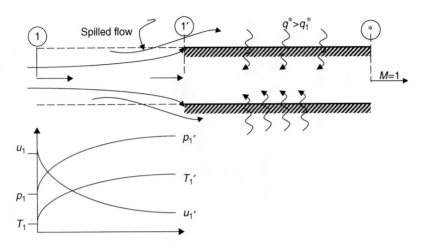

■ **FIGURE 2.38** Excessive heating of a gas in constant-area frictionless duct causes the inlet flow to spill and a new inlet state 1′ to be established

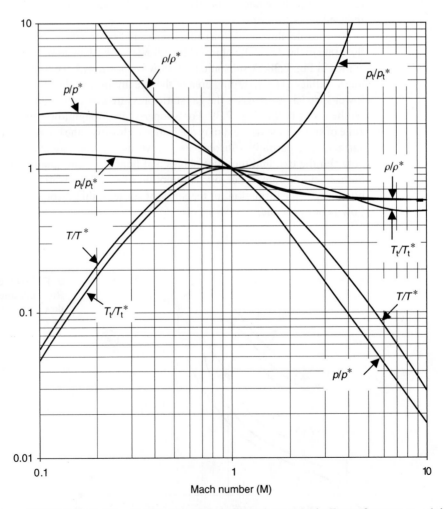

Mach number (M)

■ **FIGURE 2.39** Rayleigh flow parameters for a calorically perfect gas, $\gamma = 1.4$

EXAMPLE 2.6

Air enters a constant-area duct of circular cross section with diameter $D = 20$ cm. The inlet conditions are Mach number $M_1 = 0.2$, $p_{t1} = 100$ kPa, and $T_{t1} = 288$ K. There is a heat transfer to the fluid at the rate of $\dot{Q} = 100$ kW. Assuming the fluid is inviscid that allows us to neglect the frictional drag force on the fluid, calculate

(a) Mass flow rate through the duct

(b) The critical heat flux that would choke the duct for the given M_1

(c) The exit Mach number M_2

(d) The percent total pressure loss

(e) The entropy rise $\Delta s/R$

(f) The static pressure drop Δp

(g) Label the inlet and exit states on a Rayleigh line on a T–s diagram

Assume $R = 287$ J/kg · K and $\gamma = 1.4$.

SOLUTION

The inlet mass flow rate, assuming the inlet conditions prevail, is

$$\dot{m} = \sqrt{\frac{\gamma}{R}} \frac{p_{t1}}{\sqrt{T_{t1}}} \cdot A \cdot \frac{M_1}{\left(1 + \frac{\gamma - 1}{2}M_1^2\right)^{\frac{\gamma+1}{2(\gamma-1)}}}$$

$$= \sqrt{\frac{1.4}{287(\text{J/kg} \cdot \text{K})}} \frac{100 \times 1000(\text{N/m}^2)}{\sqrt{288(\text{K})}}$$

$$\times (0.1)^2 \cdot \pi(\text{m}^2) \cdot \frac{(0.2)}{(1 + 0.2 \times 0.04)^3} \approx 2.525 \text{ kg/s}$$

The heat per unit mass q is the ratio of heat flow rate to the fluid mass flow rate

$$q = \frac{\dot{Q}}{\dot{m}} = \frac{100 \text{ kW}}{2.525 \text{ kg/s}} \approx 39.61 \text{ kJ/kg}$$

Now, let us compare the actual heat flux q to the critical heat flux q^* by going through the Rayleigh tables,

$$M_1 = 0.2 \xrightarrow[\gamma = 1.4]{\text{Rayleigh table}} T_{t1}/T_t^* = 0.1736$$

$$p_{t1}/p_{t1}^* = 1.2346$$
$$(\Delta s/R)_1 = 6.3402$$
$$p_1/p^* = 2.2727$$

$T_t^* = 288 \text{ K}/0.1736 \cong 1,659 \text{ K}$
$p^* = 100 \text{ kPa}/2.2727 = 44 \text{ kPa}$

From energy equation, we have

$$T_{t2} = T_{t1} + q/c_p$$
$$= 288 \text{ K} + (39.61 \text{ kJ/kg})/(1.004 \text{ kJ/kg} \cdot \text{K})$$
$$= 327.4 \text{ K}$$

Where c_p was calculated from γ and R via

$$c_p = \frac{\gamma}{\gamma - 1}R = 1004 \text{ J/kg} \cdot \text{K}$$

Since the exit total temperature is less than T_t^*, the heat flux is less than the critical value and thus the exit is unchoked.

The critical heat flux is calculated from energy equation, for an exit total temperature of T_t^*, namely,

$$q_1^* = c_p(T_t^* - T_{t1})$$
$$\cong 1.004(1659 - 288) \text{ kJ/kg} \cong 1,376.5 \text{ kJ/kg}$$

For the exit Mach number, we form the ratio T_{t2}/T_t^*

$$T_{t2}/T_t^* = (327.4 \text{ K})/(1659 \text{ K}) = 0.1973 \xrightarrow{\text{Rayleigh Table}}$$

$$M_2 \cong 0.22$$

$$p_{t2}/p_t^* \cong 1.2281$$

$$(\Delta s/R)_2 \cong 5.7395$$
$$p_2/p^* \cong 2.2477$$

The percent total pressure drop is

$$\frac{(p_{t1}/p_t^*) - (p_{t2}/p_t^*)}{(p_{t1}/p_t)} \times 100$$

$$\cong \frac{1.2346 - 1.2281}{1.2346} \times 100 \approx 0.526\%$$

The entropy rise is the difference between $(\Delta s/R)_1$ and $(\Delta s/R)_2$

$$\Delta s/R \approx 6.3402 - 5.7395 = 0.6007$$

The static pressure drop in the duct due to heat transfer is

$$\Delta p = [(p_1/p^*) - (p_2/p^*)] \times p^*$$
$$= [2.2727 - 2.2477](44 \text{ kPa}) = 1.1 \text{ kPa}$$

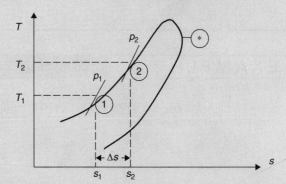

EXAMPLE 2.7

Air enters a constant-area combustion chamber at Mach 3.0 at a pressure and temperature of $p_{t1} = 45\,\text{kPa}$ and $T_{t1} = 1,800\,\text{K}$, respectively. Hydrogen is the fuel with a lower heating value of $120,000\,\text{kJ/kg}$. Calculate

(a) Exit total temperature T_{t2} (K) if the exit is choked

(b) The maximum heat release per unit mass of air, q_*

(c) The fuel-to-air ratio f to thermally choke the combustor exit

(d) The total pressure loss in the supersonic combustor

Assume the gas mixture in the combustor is a perfect gas and has the following properties: $\gamma = 1.3$ and $R = 0.287\,\text{kJ/kg}\cdot\text{K}$.

SOLUTION

As a first approximation, we may neglect the wall friction and thus treat the fluid as inviscid. In addition, we will make a uniform flow approximation to enable us to use the Rayleigh flow theory. The mass addition in the combustor due to fuel flow rate is assumed to be small compared with the inlet airflow rate. Finally, we assume a 100% burner efficiency that allows all of the fuel heating value to be released as thermal energy to the fluid.

From Rayleigh flow table (for $\gamma = 1.3$), we get

$$M_1 = 3.0 \quad \xrightarrow[\gamma = 1.3]{\text{Rayleigh table}} \quad T_{t1}/T_t^* = 0.6032$$

$$p_{t1}/p_t^* = 4.0073$$

Therefore the exit total temperature is $T_{t2}/T_t^* = 1800\,\text{K}/0.6032 \approx 2,984\,\text{K}$

$$c_p = \gamma R/(\gamma - 1)$$

$$= 1.3(0.287\,\text{kJ/kg}\cdot\text{K})/0.3 \cong 1.2437\,\text{kJ/kg}\cdot\text{K}$$

From energy balance across the burner, we get

$$q_1^* = c_p(T_t^* - T_{t1}) = c_p T_{t1}\left(\frac{T_t^*}{T_{t1}} - 1\right)$$

$$= 1.2437(1800)\left(\frac{1}{0.6032} - 1\right)\text{kJ/kg}$$

$$q_1^* \cong 1472.6\,\text{kJ/kg(of air)}$$

Therefore the fuel-to-air ratio f is derived from

$$f(120,000\,\text{kJ/kg}) = q_1^* = 1472.6\,\text{kJ/kg} \quad \rightarrow \quad f \approx 1.23\%$$

The total pressure loss is $\Delta p_t/p_{t1}$ or $1 - p_t^*/p_{t1}$

Therefore, Total pressure loss $= 1 - (1/4.0073) \approx 75\%$

REMARKS Solving a scramjet problem using a simple Rayleigh flow approximation produces a quick estimation of the *ballpark* values of the fuel flow rate and the stagnation pressure losses. We may add wall friction to our analysis without a major difficulty, as we shall see in the next section. The effect of duct area variation may also be modeled. Some of the most complex elements of the analysis deal with the chemical reaction in the burner. We face the issues of flame stability, reaction rates, and, finally, the combustion efficiency. These fundamental issues are addressed in an introductory manner in the chapter called combustion chambers and afterburners in this book.

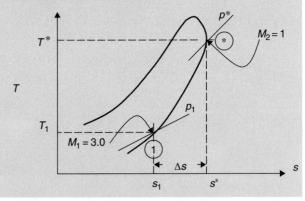

EXAMPLE 2.8

A subsonic flow in a constant-area duct is subjected to heat transfer. For an inlet flow parameters

$$M_1 = 0.5, \ p_{t1} = 14.7\,\text{psia},$$

$$T_{t1} = 50°\text{F}, \gamma = 1.4, \text{ and } R = 53.34\,\text{ft}\cdot\text{lbf/lbm}\cdot°\text{R}$$

Neglecting the wall friction, calculate

(a) The new inlet Mach number $M_{1'}$, if $q = 1.2q_1^*$

(b) The percent spilled flow at the inlet

SOLUTION

We first convert the inlet total temperature to the absolute scale, i.e.,

$$T_{t1} = (50 + 460)°R = 510°R$$

Next, we refer to the Rayleigh table (for $\gamma = 1.4$) to calculate the critical heat flux q_1^*

$$M_1 = 0.5 \xrightarrow[\gamma = 1.4]{\text{Rayleigh table}} T_{t1}/T_t^* = 0.69136$$

The energy equation establishes q_1^* according to

$$q_1^* = c_p(T_t^* - T_{t1}^*)$$

From γ and R, we calculate the specific heat at constant pressure c_p

$$c_p = \gamma R/(\gamma - 1) = 1.4(53.34)/0.4 \text{ ft·lbf/lbm · °R}$$

$$= 186.69 \text{ ft · lbf/lbm · °R}$$

Therefore,

$$q_1^* = (186.69)(510°R)(1/0.69136 - 1) \text{ ft·lbf/lbm · °R}$$
$$\cong 42,505 \text{ ft · lbf/lbm}$$

The heat flux is specified to be $1.2\, q_1^* = 1.2(42,505) = 51,006 \text{ ft·lbf/lbm}$

From this value of heat flux and the energy equation, we calculate the new exit total temperature $T_{t1'}^*$, which reflects the additional q, according to

$$q_{1'}^* = c_p(T_{t1'}^* - T_{t1})$$

Therefore the new exit total temperature is

$$T_{t1'}^* = T_{t1} + \frac{q_{1'}^*}{c_p} = 510 + \frac{51,006}{186.69} \approx 783\, °R$$

The ratio of the $T_{t1}/T_{t1'}^*$ may now be calculated as

$$T_{t1}/T_{t1'}^* = 510/783$$
$$= 0.65116 \xrightarrow[\gamma = 1.4]{\text{Rayleigh table}} M_{1'} \cong 0.473 \text{ (via interpolation)}$$

The percent spilled mass flow rate is

$$\frac{\dot{m}_1 - \dot{m}_{1'}}{\dot{m}_1} \times 100$$

Since the total pressure and temperature remain constant between the states 1 and 1', the mass flow in both cases is proportional to the mass flow parameter $f(M)$,

$$f(M) = M \Big/ \left(1 + \frac{\gamma - 1}{2}M^2\right)^{\frac{\gamma+1}{2(\gamma-1)}}$$

Hence, $f(M_1) = 0.4319$ and $f(M_{1'}) = 0.4148$
Therefore, % inlet flow spillage $= 3.966\% \approx 4\%$

2.16   Adiabatic Flow of a Calorically Perfect Gas in a Constant-Area Duct with Friction

In this section, a compressible flow in a constant-area duct is subjected to the effect of friction. The walls are insulated to heat transfer and thus the flow is assumed to be adiabatic. The medium is perfect gas and the flow is steady. This flow is called the *Fanno* flow in fluid mechanics. The governing equations are the conservation principles applied to a control volume with specified boundary conditions. Let us first define the geometry of the problem/control volume and identify its boundary conditions. Figure 2.40 shows the schematic drawing of this problem.

Fanno flow

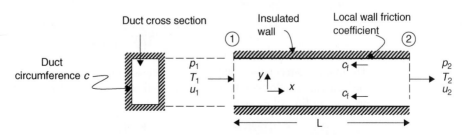

■ **FIGURE 2.40**   Geometry of an adiabatic flow in a duct with friction

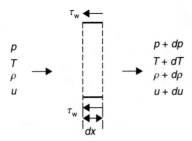

■ **FIGURE 2.41**   **Slab of fluid in a constant-area insulated duct with friction**

To apply the conservation principles to this problem, we choose an elemental slab of fluid, of incremental length, dx. The derivation involves differential equations of motion and is very useful as a practice drill in fundamentals. Figure 2.41 shows a representative slab of a fluid in an adiabatic constant-area duct flow with friction.

The continuity equation demands

$$\rho u = (\rho + d\rho)(u + du) \tag{2.172}$$

It simplifies to the following differential forms:

$$\rho \, du = -u \, d\rho \tag{2.173-a}$$

or

$$\frac{d\rho}{\rho} = -\frac{du}{u} \tag{2.173-b}$$

The momentum equation per unit area of the flow, in the x-direction, is

$$\rho u(u + du) - \rho u^2 = p - (p + dp) - \tau_w \cdot c \cdot dx / A \tag{2.174}$$

We may introduce the hydraulic diameter $(D_h \equiv 4A/c)$ in Eq. 2.174 and simplify it to

$$\rho u \, du = -dp - 4\tau_w dx / D_h \tag{2.175}$$

Note that the fluid impulse is not conserved in Fanno flow due to wall friction. The fluid impulse is reduced along the flow direction (note the minus sign in front of wall shear in Eq. 2.175).

The energy equation for the slab is

$$dh + u \, du = c_p dT + u \, du = 0 \tag{2.176}$$

The energy equation may be divided by u^2 to get

$$c_p \frac{dT}{u^2} + \frac{du}{u} = \frac{\gamma}{\gamma - 1} \frac{RT}{u^2} \frac{dT}{T} + \frac{du}{u} = \frac{1}{(\gamma - 1)M^2} \frac{dT}{T} + \frac{du}{u} = 0 \tag{2.177-a}$$

We may isolate du/u and express it in terms of dT/T, as

$$\frac{du}{u} = -\frac{1}{(\gamma - 1)M^2} \frac{dT}{T} \tag{2.177-b}$$

The momentum equation may be nondimensionalized by dividing through by ρu^2 and writing the local skin friction coefficient c_f as the ratio of wall shear stress divided by the local dynamic pressure in Eq. 2.174 yields

$$\frac{du}{u} = -\frac{p}{\rho u^2} \frac{dp}{p} - 2c_f \frac{dx}{D_h} \tag{2.178}$$

where

$$c_f \equiv \frac{\tau_w}{\rho u^2 / 2} \tag{2.179}$$

The skin friction coefficient is a function of Reynolds number based on duct diameter and the surface roughness of the wall. In supersonic flow, the skin friction coefficient is a function of local Mach number as well. To simplify the Fanno flow problem, we introduce an average wall friction coefficient C_f (also known as the friction drag coefficient), which is defined as

$$C_f \equiv \frac{1}{L} \int_0^L c_f dx \tag{2.180}$$

We use Eq. 2.177-b to replace the incremental speed ratio du/u in the momentum equation. The ratio of static pressure to density on the RHS of Eq. 2.178 is the product of $1/\gamma$ and the square of the local speed of sound a^2, and we may replace the incremental pressure ratio with density and temperature ratio following the perfect gas law, according to

$$\frac{1}{(\gamma - 1)M^2} \frac{dT}{T} = \frac{1}{\gamma M^2} \left(\frac{d\rho}{\rho} + \frac{dT}{T} \right) + 2c_f \frac{dx}{D_h} \tag{2.181}$$

We replace the density ratio from continuity equation and then substitute for the Mach temperature via the logarithmic derivatives to get

$$\frac{1}{(\gamma - 1)M^2} \frac{dT}{T} = \frac{1}{\gamma M^2} \left(\frac{1}{(\gamma - 1)M^2} \frac{dT}{T} + \frac{dT}{T} \right) + 2c_f \frac{dx}{D_h} \tag{2.182-a}$$

$$\frac{M^2 - 1}{\gamma M^2 \cdot (\gamma - 1)M^2} \frac{dT}{T} = 2c_f \frac{dx}{D_h} \tag{2.182-b}$$

Since the stagnation temperature T_t remains constant, we may use the logarithmic derivative to represent dT/T in terms of dM/M following:

$$T_t = T \left(1 + \frac{\gamma - 1}{2} M^2 \right) \tag{2.183-a}$$

$$\frac{dT_t}{T_t} = \frac{dT}{T} + \frac{(\gamma - 1)MdM}{1 + \frac{\gamma - 1}{2}M^2} = \frac{dT}{T} + \frac{(\gamma - 1)M^2}{1 + \frac{\gamma - 1}{2}M^2} \frac{dM}{M} = 0 \tag{2.183-b}$$

Now, let us combine Eqs. 2.183-b and 2.182-b to get

$$\left[\frac{1 - M^2}{\gamma M^2 \left(1 + \frac{\gamma - 1}{2}M^2 \right)} \right] \frac{dM}{M} = 2c_f \frac{dx}{D_h} \tag{2.184-a}$$

Since the terms in the bracket involve M^2, we may express dM/M as $(1/2)dM^2/M^2$, which produces

$$4c_f \frac{dx}{D_h} = \left[\frac{1 - M^2}{\gamma M^2 \left(1 + \frac{\gamma - 1}{2}M^2 \right)} \right] \frac{dM^2}{M^2} \tag{2.184-b}$$

The LHS of this equation is always positive. The RHS should also be positive for any M. Therefore, for a subsonic flow, i.e., $M < 1$, dM^2 has to be positive, and for supersonic flow where $M > 1$, then dM^2 has to be negative. This suggests that the effect of friction on a subsonic flow is to accelerate it toward the sonic state. Also, a supersonic flow is decelerated toward the sonic point by friction. This impact of friction on compressible flow is similar to the Rayleigh flow where heating causes the flow to approach a choking condition. In both cases, friction and heating act like a *valve*, or in effect a flow area reduction. Now, we may integrate this equation from $x = 0$ to L^* while M^2 varies from M^2 to 1, to get an expression for the Fanno parameter $4C_f L^*/D_h$,

$$\frac{4}{D_h}\int_0^{L^*} c_f dx = 4C_f \frac{L^*}{D_h} = \int_{M^2}^{1} f(M^2, \gamma)dM^2 = \frac{1 - M^2}{\gamma M^2} + \frac{\gamma + 1}{2\gamma}\ell n\left[\frac{(\gamma + 1)M^2}{2\left(1 + \frac{\gamma - 1}{2}M^2\right)}\right] \quad \textbf{(2.185)}$$

We have selected the choking condition to represent a reference state, as in Rayleigh flow. The energy equation for Fanno flow maintains a constant total temperature, therefore,

$$\frac{T}{T^*} = \frac{\gamma + 1}{2\left(1 + \frac{\gamma - 1}{2}M^2\right)} \quad \textbf{(2.186)}$$

The density ratio from the continuity equation may be written as

$$\frac{\rho}{\rho^*} = \frac{u^*}{u} = \frac{1}{M}\sqrt{\frac{T^*}{T}} = \frac{1}{M}\sqrt{\frac{2\left(1 + \frac{\gamma - 1}{2}M^2\right)}{\gamma + 1}} \quad \textbf{(2.187)}$$

The pressure ratio is the product of the density and temperature ratio, therefore,

$$\frac{p}{p^*} = \frac{1}{M}\sqrt{\frac{\gamma + 1}{2\left(1 + \frac{\gamma - 1}{2}M^2\right)}} \quad \textbf{(2.188-a)}$$

The fluid impulse ratio I/I^* is related to static pressure ratio and Mach number following

$$\frac{I}{I^*} = \frac{p(1 + \gamma M^2)}{p^*(1 + \gamma)} = \left(\frac{p}{p^*}\right)\frac{1 + \gamma M^2}{\gamma + 1} = \frac{1}{M}\sqrt{\frac{\gamma + 1}{2[1 + (\gamma - 1)M^2/2]}}\left(\frac{1 + \gamma M^2}{\gamma + 1}\right) \quad \textbf{(2.188-b)}$$

The stagnation pressure ratio is expressed in terms of static pressure ratio and Mach number as

$$\frac{p_t}{p_t^*} = \frac{p}{p^*}\left(\frac{1 + \frac{\gamma - 1}{2}M^2}{\frac{\gamma + 1}{2}}\right)^{\frac{\gamma}{\gamma - 1}} = \frac{1}{M}\left(\frac{1 + \frac{\gamma - 1}{2}M^2}{\frac{\gamma + 1}{2}}\right)^{\frac{\gamma + 1}{2(\gamma - 1)}} \quad \textbf{(2.189)}$$

Figure 2.42 shows a log–log graph of the Fanno flow parameters for a diatomic gas ($\gamma = 1.4$).

To interpret the parameter variations properly in Fig. 2.42, we need to remember that the movement is toward the sonic point on both sides of $M = 1$ axis. For example, the static pressure, temperature, and density drop in subsonic flow and all increase in a supersonic flow with friction. The total pressure drops in both cases.

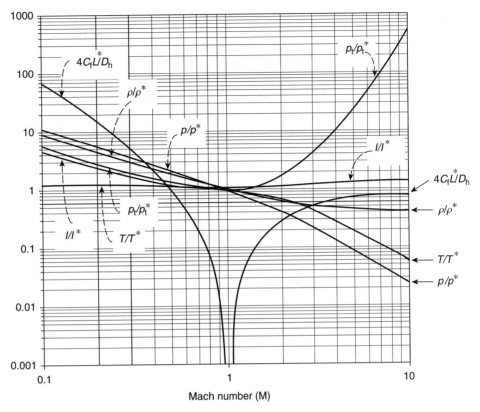

■ **FIGURE 2.42**   Fanno flow parameters for a calorically-perfect diatomic gas with $\gamma = 1.4$

To graph the thermodynamic state variation of this flow on a T–s diagram, which is referred to as the Fanno line, we need to derive the fluid entropy in terms of temperature and local Mach number. We start with the Gibbs equation,

$$ds = c_p dT/T - Rdp/p \tag{2.190}$$

We replace the logarithmic derivative of pressure by the logarithmic derivatives of density and temperature using the perfect gas law, namely

$$dp/p = d\rho/\rho + dT/T \tag{2.191-a}$$

The density ratio is related to incremental speed ratio via the continuity equation, therefore

$$dp/p = -du/u + dT/T \tag{2.191-b}$$

Now, replace the logarithmic derivative of gas speed by the logarithmic derivatives of Mach number and temperature via the definition of Mach number, i.e.,

$$\frac{dp}{p} = -\left(\frac{dM}{M} + \frac{1}{2}\frac{dT}{T}\right) + \frac{dT}{T} = \frac{1}{2}\frac{dT}{T} - \frac{dM}{M} \tag{2.191-c}$$

Substituting Eq. 2.191-c in the Gibbs equation, we get

$$ds = c_p dT/T - R(dT/2dT - dM/M) \tag{2.192}$$

From the energy equation for an adiabatic flow, we may write the following logarithmic derivative:

$$dT/T + [(\gamma - 1)M dM]\bigg/\left[1 + \frac{\gamma - 1}{2}M^2\right] = 0 \qquad \textbf{(2.193-a)}$$

Now, we isolate dM/M in terms of dT/T according to

$$\frac{dM}{M} = -\left(\frac{1 + \dfrac{\gamma - 1}{2}M^2}{(\gamma - 1)M^2}\right)\frac{dT}{T} \qquad \textbf{(2.193-b)}$$

Substituting 2.193-b into 2.192 gives

$$\frac{ds}{R} = \left[\frac{\gamma}{\gamma - 1} - \frac{1}{2} - \left(\frac{1 + \dfrac{\gamma - 1}{2}M^2}{(\gamma - 1)M^2}\right)\right]\frac{dT}{T} \qquad \textbf{(2.194)}$$

The bracket may be simplified to get the final form of $s(T, M)$ as

$$ds/R = \left[(M^2 - 1)/((\gamma - 1)M^2)\right]dT/T \qquad \textbf{(2.195)}$$

The critical point of this function is at $ds/dT = 0$, which occurs at the sonic condition, i.e., when the numerator of the bracket in 2.195 vanishes. A graph of this function in a T–s diagram is shown in Fig. 2.43. There is a subsonic and a supersonic branch separated by the sonic point (designated by *). An arrow on these branches shows the direction of the thermodynamic states in a compressible, adiabatic one-dimensional flow with friction on a T–s diagram. The effect of friction is seen to increase the entropy of the gas as expected, and hence to push a subsonic or supersonic flow toward the sonic point. This phenomenon is referred to as the *frictional choking* effect in a duct with compressible flow.

In arriving at the $s(T, M)$ for the Fanno flow, we did not use the momentum equation, which contained the friction term. The continuity and the energy equations that we used in the derivation of $s(T, M)$ are identical to a normal shock flow equations. Hence a transition from a supersonic flow to a subsonic flow, with an associated entropy rise, may depict a normal shock

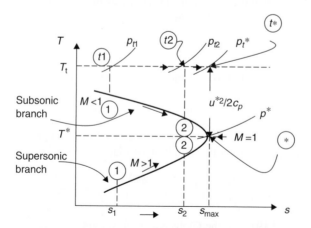

■ **FIGURE 2.43** **The path of a one-dimensional adiabatic flow with friction on a** T–s **diagram**

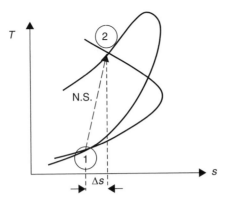

■ **FIGURE 2.44** The intersection of a Rayleigh and a Fanno lines with the same mass flow rate satisfying the conservation principles across a normal shock

on a Fanno line. This is similar to a Rayleigh flow that shares the momentum and mass conservation with a normal shock. Fanno line shares the mass and energy conservation with a normal shock flow. Hence the intersection of a Rayleigh line and a Fanno flow with the same mass flow rate per unit area, i.e., ρu, satisfies the mass, momentum, and the energy conservation principles across a normal shock. Figure 2.44 shows a possible normal shock at the intersection of a Fanno and a Rayleigh line.

For a duct longer than the choking length of the inlet Mach number, i.e., $L > L_1^*$, a subsonic flow will reset itself with a lower mass flow rate at its inlet, similar to a Rayleigh line flow spillage at the inlet due to excessive heating, $q > q^*$. In case of a supersonic flow in the duct with friction, a normal shock appears in the duct that transitions the flow to subsonic. The schematic drawing of these two cases is shown in Fig. 2.45. We introduce a family of Fanno lines, each representing a mass flow rate (per unit area) on a T–s diagram, similar to the Rayleigh line family. The inlet flow is reset to a lower mass flow rate, hence a reduced inlet Mach number when the throughflow is subsonic. The exit remains choked. Figure 2.46 shows a Fanno line family.

To show the case of supersonic inlet flow and a shock in the duct, we use a schematic drawing on a T–s diagram. Figure 2.47 shows the shock formation for $L > L_1^*$ in a supersonic Fanno flow.

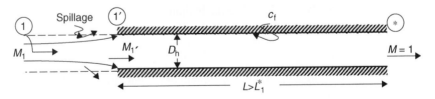

a) Subsonic inlet flow spills to lower $\dot{m}$ in a Fanno duct with $L > L_1^*$

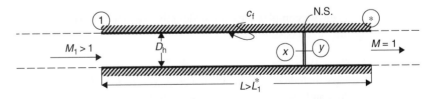

b) supersonic flow in a Fanno duct creates a shock (inside the duct) for $L > L_1^*$

■ **FIGURE 2.45** Adjustment mechanisms for a Fanno flow with $L > L_1^*$ for subsonic and supersonic throughflow cases

■ **FIGURE 2.46**
**Family of Fanno lines on
a *T–s* diagram with
decreasing mass flow rate**

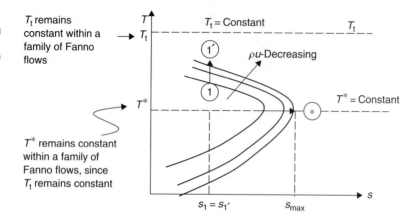

■ **FIGURE 2.47**
**Frictional choking with a
normal shock in the duct**

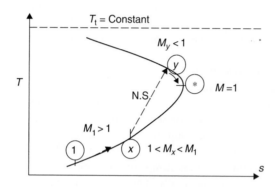

To establish the exit flow conditions in a Fanno line problem, we may use the tabulated functions according to the following steps:

Given: M_1, L, Cross-sectional shape of the duct, C_f and γ
Calculate: M_2, p_2/p_1, T_2/T_1, ρ_2/ρ_1, p_{t2}/p_{t1}, $\Delta s/R$, u_2/u_1, etc.

Solution: We first calculate the hydraulic diameter of the duct from the cross-sectional shape according to

$$D_h \equiv 4A/c$$

Next, we look up the choking length of the duct for its inlet Mach number from the tabulated functions, i.e.,

$$M_1, \gamma \rightarrow 4C_f L_1^*/D_h$$

Then, we compare the choking length L_1^* and the physical length of the duct, L. Let us examine the three possible cases:

Case 1: $L < L_1^*$
We conclude that the exit is not choked, the inlet flow remains unaffected by the length of the duct and there is no normal shock in the duct. Figure 2.48 shows this case.

The length of the fictitious duct is $L_1^* - L$, which is the choking length for the exit Mach number, M_2. Hence, we calculate L_2^* according to

$$L_2^* = L_1^* - L$$

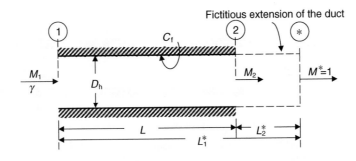

■ **FIGURE 2.48**
Length of a duct that
does not choke the exit

We form the Fanno parameter at the exit of the duct, which results in M_2 from the tabulated functions:

$$4C_f L_2^* / D_h \rightarrow M_2$$

We read the parameters $p_1/p^*, T_1/T^*$, etc. from M_1 value of the table and $p_2/p^*, T_1/T^*$, etc. from the M_2 value of the Fanno table. Since the choking condition is unique, the ratio of these parameters reveals the duct ratios, namely

$$\frac{p_2}{p_1} = \frac{p_2/p^*}{p_1/p^*}$$

Case 2: $L = L_1^*$

This case is the easiest of the three, since the exit is choked and all exit conditions are sonic, namely, $M_2 = 1$, $p_2 = p^*$, etc.

Case 3: $L > L_1^*$

The length of the duct is longer than the choking length based on the inlet Mach number M_1. In case of subsonic flow, then the inlet flow is reset to a lower mass flow rate (i.e., Mach number), such that the physical length of the duct is now the choking length for the new inlet condition (1'). We immediately conclude that the exit is choked, i.e.,

$$M_2 = 1, p_2 = p^*, T_2 = T^*, \text{etc.}$$

Since the length of the duct is the choking length for the new inlet condition, the Fanno parameter $4C_f L_{1'}^*/D$ is actually, $4C_f L/D$, from the Fanno table, we get $M_{1'}$ according to

$$4C_f L_{1'}^* / D_h = 4C_f L/D_h \rightarrow M_{1'}$$

The inlet static pressure and temperature is now increased due to spillage, while the stagnation pressure and temperature at the inlet is preserved, i.e.,

$$p_{t1'} = p_{t1} \text{ and } T_{t1'} = T_{t1}$$

With the new Mach number at the inlet, $M_{1'}$, we may calculate the new inlet static pressure and temperature $p_{1'}$ and $T_{1'}$ using the fundamental relations we derived earlier, namely

$$p_{1'} = \frac{p_{t1'}}{\left(1 + \dfrac{\gamma - 1}{2} M_{1'}^2\right)^{\frac{\gamma}{\gamma - 1}}} = \frac{p_{t1}}{\left(1 + \dfrac{\gamma - 1}{2} M_{1'}^2\right)^{\frac{\gamma}{\gamma - 1}}}$$

$$T_{1'} = \frac{T_{t1'}}{\left(1 + \dfrac{\gamma - 1}{2} M_{1'}^2\right)} = \frac{T_{t1}}{\left(1 + \dfrac{\gamma - 1}{2} M_{1'}^2\right)}$$

From $M_{1'}$ and the Fanno table, we read p/p^*, etc. and the ($*$) represents the exit state and the p, T, etc. are the newly calculated $p_{1'}, T_{1'}$, etc.

■ **FIGURE 2.49**
Normal shock in an
insulated duct with
friction

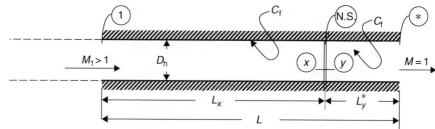

Now, we address the case of supersonic flow with $L > L_1^*$. The inlet flow is unchanged and the exit flow Mach number is also known to be sonic. The key unknown is the location of a normal shock that splits the length of the duct into a supersonic Fanno and a subsonic Fanno flows. The physical constraint is the length of the duct and the governing equations on both sides of the shock.

We may solve this problem iteratively by guessing the normal shock Mach number M_x and check if the length constraint of the duct is satisfied. Figure 2.49 shows the relationship between the shock position and its associated length scales along the duct.

The solution procedure is as follows:

Guess an M_x (less than M_1 obviously) then use the table of Fanno functions,

$$M_x \rightarrow 4C_f L_x^*/D_h$$

Then, we establish the length, L_x according to

$$L_x = L_1^* - L_x^*$$

Now, we are ready to tackle the shock jump and the remaining portion of the duct, first by using the normal shock table to get M_y, then the Fanno table to get $4C_f L_y^*/D_h$, i.e.,

$$M_x \rightarrow M_y$$
$$M_y \rightarrow 4C_f L_y^*/D_h$$

Does the length L_y^* and the L_x add up to the physical duct length L? If yes, then our initial guess of M_x was correct and if not, then we need to guess again. To know whether to guess higher or lower depends on the relative magnitude of the sum $(L_x + L_y^*)$ and L.

- If $(L_x + L_y^*) > L$, then our shock was too strong, and we need to lower the shock Mach number, M_x.
- If $(L_x + L_y^*) < L$, then our initial guess of the shock Mach number was not strong enough, hence we need to increase M_x.

An iteration process has a *target* value that needs to be met. The target value is a *constraint* imposed by the geometry and/or physics of the problem. We need to specify a level of accuracy that we demand of our final iterated solution. For example, we may expect/specify that the target value to be met should be within 0.1% of the last iteration value.

EXAMPLE 2.9

Air enters a constant-area duct of circular cross section with diameter $D = 10$ cm. The length of the duct is $L = 20$ m. The duct is insulated hence the flow inside the duct is assumed to be adiabatic. The average wall friction coefficient is $C_f = 0.005$. The inlet Mach number is $M_1 = 0.24$. Calculate

(a) The choking length of the duct, L_1^*

(b) The exit Mach number M_2

(c) The percent total pressure loss

(d) The static pressure drop

(e) Loss of fluid impulse due to friction

Assume $\gamma = 1.4$.

SOLUTION

From Fanno table, we get the Fanno friction parameter based on the inlet Mach number, i.e.,

$$M_1 = 0.24 \xrightarrow[\text{Fanno table}]{} 4C_f L_1^*/D_h = 9.3866$$

$$p_{t1}/p_t^* = 2.4956$$
$$p_1/p^* = 4.5383 \text{ and}$$
$$I_1/I^* = 2.043$$

Now, we calculate the choking length of the duct, L_1^* (note that $D = 0.1$ m)

$$L_1^* = [9.3866(0.1 \text{ m})]/[4(0.005)] = 46.933 \text{ m}$$

$$\therefore \boxed{L_1^* \cong 46.933 \text{ m}}$$

We note that the physical length of the duct is less than the choking length; therefore the exit Mach number will be less than 1. To arrive at the exit Mach number, we form the Fanno parameter $4C_f L_2^*/D_h$

$$L_2^* = L_1^* - L = 46.933 \text{ m} - 20 \text{ m} = 26.933 \text{ m}$$

$$4C_f L_2^*/D_h = [4(0.005)(26.933\text{m})]/0.1\text{m} = 5.3866$$

$$4C_f L_2^*/D_h = 5.3866 \xrightarrow{\text{Fanno table}} \boxed{M_2 \approx 0.3}$$

$$p_{t2}/p_t^* = 2.0351$$
$$p_2/p^* = 3.6191$$
$$\text{and } I_2/I^* = 1.698.$$

Therefore the percent total pressure drop due to friction is

$$\frac{p_{t1} - p_{t2}}{p_{t1}} \times 100 = \frac{(p_{t1}/p_t^*) - (p_{t2}/p_t^*)}{(p_{t1}/p_t^*)} \times 100$$

$$= \frac{2.4956 - 2.0351}{2.4956} \times 100$$

$$\frac{\Delta p_t}{p_{t1}} \times 100 \cong 18.45\%$$

Similarly, the static pressure drop is $(4.5383 - 3.6191)/4.5383 = 20.25\%$

The loss of fluid impulse due to friction is
$$I_2 - I_1 \approx (1.698 - 2.043)I^* \approx -0.345I^*$$

EXAMPLE 2.10

Air flows in a duct with a rectangular cross section of 1 cm $\times$ 2 cm dimensions. The average skin friction coefficient is 0.005. Assuming the walls of this duct are insulated and the entrance Mach number is 0.5, calculate

(a) The maximum length of the duct that will support its inlet condition

(b) The new inlet condition $M_{1'}$, if $L = 2.16 L_{\max}$

(c) The percent inlet mass flow drop due to the longer length of the duct

SOLUTION

The maximum length with the inlet flow remaining intact is L_1^*. From the Fanno tables we have

$$M_1 = 0.5 \xrightarrow[\text{Fanno table}]{} 4C_f L_1^*/D_h = 1.0691$$

The hydraulic diameter is
$$D_h = 4A/c = 4(2 \times 1)/[2(2 + 1)] \text{ cm} = 4/3 \text{ cm}$$

$$\boxed{L_{\max} = L_1^* \cong 71.27 \text{ cm}}$$

Therefore, $L_1^* = 1.069(4/3 \text{ cm})/[4(0.005)] = 71.27 \text{ cm}$

The longer duct is still choked, however, with a reduced flow rate. Since the duct is choked, its physical length is the choking length for its new inlet flow condition, namely the inlet Mach number $M_{1'}$.

$$4C_f L_{1'}^*/D_h = 4(0.005)[2.16(71.27 \text{ cm})]/(4/3 \text{ cm})$$

$$= 2.309$$

$$4C_f L_{1'}^*/D_h = 2.309 \xrightarrow{\text{Fanno table}} M_{1'} = 0.4 \quad \boxed{M_{1'} = 0.4}$$

The inlet total pressure and temperature remain the same, therefore the mass flow rate in the duct is proportional to

$$f(M) = \frac{M}{\left(1 + \dfrac{\gamma - 1}{2}M^2\right)^{\frac{\gamma+1}{2(\gamma-1)}}}$$

via continuity equation. We evaluate this mass flow parameter at two inlet Mach numbers M_1 and $M_{1'}$ to get

$$f(M_1) = \frac{0.5}{(1 + 0.2 \times 0.25)^3} = 0.4319$$

$$f(M_{1'}) = \frac{0.4}{(1 + 0.2 \times 0.16)^3} = 0.3639$$

Therefore the mass flow drop is:

$$\frac{\Delta \dot{m}}{\dot{m}_1} = \frac{0.4319 - 0.3639}{0.4319} = 0.1574 \quad \therefore \quad \boxed{\frac{\Delta \dot{m}}{\dot{m}_1} \cong 15.74\%}$$

EXAMPLE 2.11

Consider a supersonic flow of air in a long duct with friction but with negligible heat transfer. The duct length-to-diameter ratio is 40. The inlet Mach number is $M_1 = 2.6$ and an average wall skin friction coefficient is 0.005. First demonstrate that a normal shock will appear in the duct and then calculate

(a) The location of the normal shock in the duct

(b) The T–s diagram for this flow

SOLUTION

From the Fanno table, we find the Fanno friction parameter $4C_f L_1^*/D_h$, i.e.,

$$M_1 = 2.6 \xrightarrow{\text{Fanno table}} 4C_f L_1^*/D_h = 0.45253$$

Therefore, $L_1^*/D_h = 0.45253/[4(0.005)] = 22.62$. Since $L_1^*/D_h > L/D_h$, a normal shock will appear in the duct to support a choked exit flow, as shown in Fig. 2.45(b) or Fig. 2.46.

Guess M_x to be 1.5

$$M_x = 1.5 \xrightarrow{\text{Fanno table}} 4C_f L_x^*/D_h = 0.13602$$

$$4C_f L_x/D_h = (4C_f L_1^*/D_h) - (4C_f L_x^*/D_h)$$

$$= 0.45253 - 0.13602 = 0.31651$$

Then, we establish the shock jump conditions, namely

$$M_x = 1.5 \xrightarrow{\text{Normal shock table}} M_y = 0.7011$$

$$M_y = 0.70 \xrightarrow{\text{Fanno table}} 4C_f L_y^*/D_h \cong 0.2081$$

Now, we check the criterion: $L_x + L_y^* = L$ or equivalently

$$4C_f L_x/D_h + 4C_f L_y^*/D_h = 0.31651 + 0.2081 = 0.5246$$

$4C_f L/D_h = 4(0.005)(40) = 0.8 \neq 0.5246$, therefore our initial guess on M_x was incorrect. Do we need to guess a higher or a lower M_x? Since the sum of L_x and L_y^* was less than the physical length of the duct, L, our shock was not strong enough to choke the exit flow. Hence, we need to guess a higher M_x, namely, let

$$M_x = 2.0 \xrightarrow{\text{Fanno table}} 4C_f L_x^*/D_h = 0.30495$$

$$4C_f L_x/D_h = (4C_f L_1^*/D_h) - (4C_f L_x^*/D_h)$$

$$= 0.45253 - 0.30495 = 0.14758$$

$$M_x = 2.0 \xrightarrow{\text{Normal shock table}} M_y = 0.5774$$

$$M_y = 0.5774 \approx 0.58 \xrightarrow{\text{Fanno table}} 4C_f L_y^*/D_h \cong 0.57572$$

$$4C_f L_x/D_h + 4C_f L_y^*/D_h = 0.14758 + 0.57572 = 0.7233$$

$4C_f L/D_h = 0.8 \neq 0.7233$

Our shock Mach number of 2.0 was slightly weaker than it should be. Hence, we choose a new shock Mach number, namely,

$$M_x = 2.1 \xrightarrow{\text{Fanno table}} 4C_f L_x^*/D_h = 0.33381$$

$$4C_f L_x/D_h = (4C_f L_1^*/D_h) - (4C_f L_x^*/D_h)$$

$$= 0.45253 - 0.33381 = 0.11872$$

$$M_x = 2.1 \xrightarrow{\text{Normal stock table}} M_y = 0.5613$$

$$M_y = 0.5613 \approx 0.56 \xrightarrow{\text{Fanno table}} 4C_f L_y^*/D_h \cong 0.67362$$

$$4C_f L_x/D_h + 4C_f L_y^*/D_h = 0.11872 + 0.67362 = 0.7923$$

Although we are close to our target value of 0.8, we need to slightly increase the shock Mach number (to perhaps 2.12). Also, instead of using the nearest values from the tables, we may use the Fanno flow equations to improve accuracy. However, we have demonstrated the principles and we stop at this value of $M_x \sim 2.1$. Figure 2.50 shows the T–s diagram for this problem.

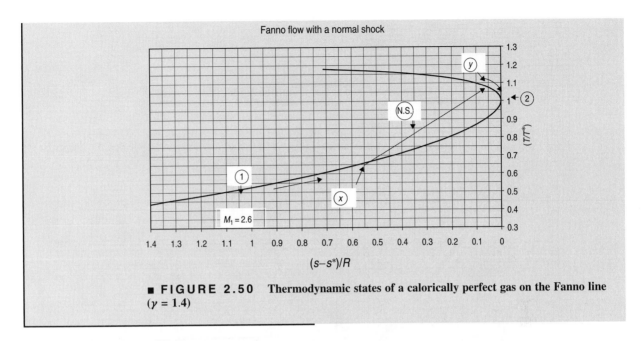

■ **FIGURE 2.50**   **Thermodynamic states of a calorically perfect gas on the Fanno line** $(\gamma = 1.4)$

2.17   Friction (Drag) Coefficient C_f and D'Arcy Friction Factor f_D

Let us consider a fully developed pipe flow, where the velocity profile remains constant, i.e., preserved, along the length of the pipe, as shown in Fig. 2.51.

Since the momentum of the fluid remains constant along the length of the pipe, the frictional drag on the fluid has to be in balance with the pressure forces acting on the fluid. Therefore,

$$\int_0^L \tau_w \cdot c \cdot dx = (p_1 - p_2)A = \Delta p \cdot A \qquad (2.196)$$

where c is the circumference of the pipe. Dividing both sides by the dynamic pressure and introducing the friction drag coefficient C_f, we get

$$C_f \cdot L \cdot c = \frac{\Delta p}{\rho V^2/2} \cdot A \qquad (2.197)$$

Rearranging Eq. 2.197 and using the definition of hydraulic diameter D_h, we get

$$4 \cdot C_f = \left(\frac{\Delta p}{(\rho V^2/2)(L/D_h)}\right) \qquad (2.198\text{-a})$$

The nondimensional term in the bracket on the RHS of Eq. 2.198-a is referred to as the D'Arcy friction factor f_D. Moody (1944) developed the functional dependence of the friction factor on pipe Reynolds number for incompressible flow. A graphic presentation of the friction factor f_D

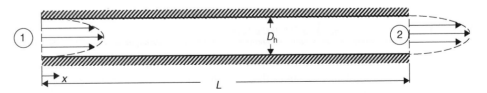

■ **FIGURE 2.51**   **Schematic drawing of a fully developed pipe flow**

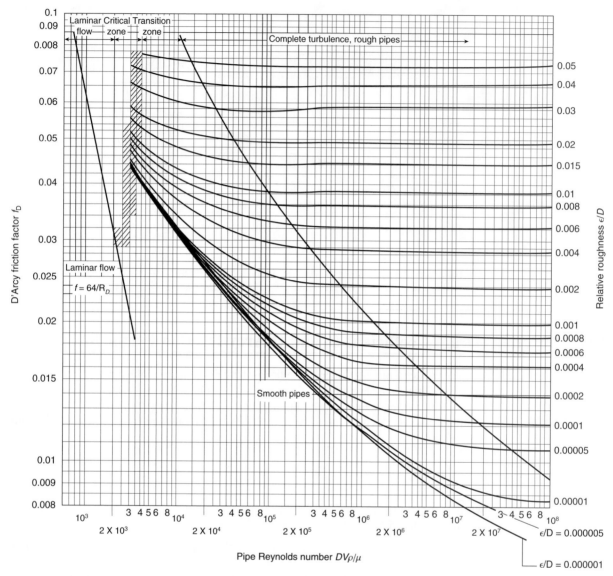

■ **FIGURE 2.52**   Moody diagram for the friction factor f_C, in terms of pipe Reynolds number and the relative roughness of a pipe with an incompressible fluid flow (Data from Moody, 1944)

due to Moody is shown in Fig. 2.51. The Fanno flow calculations that need the friction coefficient C_f need to take 1/4 of the friction factor data presented in Moody diagram (Fig. 2.52).

$$C_f = f_D/4 \qquad (2.198\text{-b})$$

2.18   Dimensionless Parameters

Following Buckingham Π theorem, a group of independent dimensionless parameters appear in fluid mechanics that govern its behavior and hence introduce dynamic similarity between flows that share these parameters. The major dimensionless groups that are relevant to our studies are

- Mach number $M \equiv V/a$
- Reynolds number $Re_l \equiv \rho Vl/\mu$
- Prandtl number $Pr \equiv \mu \cdot c_p/k$
- Nusselt number $Nu_l \equiv h \cdot l/k$, where h is the rate of heat transfer per unit area per unit temperature difference

$$h \equiv \frac{\dot{Q}/A}{T_w - T_e}$$

where T_w and T_e are the wall and thermal boundary layer edge temperatures, respectively.

- Stanton number, $St \equiv Nu/Pr \cdot Re$
- Pressure coefficient $C_p \equiv (p - p_\infty)/(\rho_\infty V_\infty^2/2)$
- Force coefficient $C_F \equiv F \left/ \left[\left(\frac{\rho_\infty V_\infty^2}{2} \right) A \right] \right.$
- Knudsen number $Kn \equiv \lambda/l$, where λ is the mean-free path of the molecules in the gas

The pressure coefficient, C_p, may be written in terms of Mach number and static pressure ratio, such as

$$C_p \equiv \frac{p - p_\infty}{\rho_\infty V_\infty^2/2} = \frac{2 p_\infty}{\rho_\infty V_\infty^2} \left(\frac{p}{p_\infty} - 1 \right) = \frac{2}{\gamma V_\infty^2} \cdot \frac{\gamma p_\infty}{\rho_\infty} \left(\frac{p}{p_\infty} - 1 \right) = \frac{2}{\gamma M_\infty^2} \left(\frac{p}{p_\infty} - 1 \right) \quad \textbf{(2.199)}$$

There are no approximations in expressing the pressure coefficient in terms of the Mach number and the pressure ratio, as shown in Eq. 2.199. We may use the Bernoulli equation

$$p_t = p + \rho V^2/2 \quad \textbf{(2.200)}$$

which is valid for a low speed flow, to get the low-speed approximation of the pressure coefficient, such as

$$C_p \approx 1 - V^2/V_\infty^2 \quad \textbf{(2.201)}$$

The low speed range of validity for the Bernoulli equation or the pressure coefficient representation, as in Eq. 2.201, is for flows with $M < 0.3$. From the compressible flow equation for the total and static densities we have

$$\frac{\rho_t}{\rho} = \left(1 + \frac{\gamma - 1}{2} M^2 \right)^{\frac{1}{\gamma - 1}} \quad \textbf{(2.202)}$$

For a Mach number of 0.3 or below, the maximum density variation, i.e., between the total and static densities, is less than $\sim 5\%$. A density variation with Mach number is shown in Fig. 2.53.

Now, let us compare the ratio of total-to-static pressure at low speed and at compressible levels. The low-speed version, which is based on the Bernoulli equation, is a special case of the general expression,

$$p_t/p = [1 + (\gamma - 1)M^2/2]^{\frac{\gamma}{\gamma - 1}} \quad \textbf{(2.203)}$$

We may use binomial expansion of Eq. 2.203 for a low Mach number and keep the first term in the expansion, such as

$$\frac{p_t}{p} \cong 1 + \left(\frac{\gamma - 1}{2} \right) \left(\frac{\gamma}{\gamma - 1} \right) M^2 = 1 + \frac{\gamma}{2} M^2 \quad \textbf{(2.204)}$$

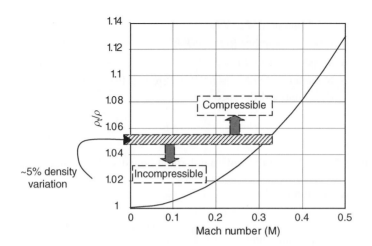

■ FIGURE 2.53 Maximum density ratio at low subsonic Mach numbers for $\gamma = 1.4$

By replacing the Mach number in Eq. 2.204 with the gas speed and the speed of sound, we show that Eq. 2.204 is Bernoulli equation,

$$\frac{p_t}{p} \cong 1 + \frac{\gamma}{2}\frac{V^2}{a^2} = 1 + \frac{1}{RT} \cdot \frac{V^2}{2} = 1 + \frac{\rho}{p} \cdot \frac{V^2}{2} \tag{2.205}$$

We have demonstrated that the Bernoulli equation is only a special case of the general equation relating stagnation and static pressure via Mach number. Hence, the Bernoulli equation should be used only at low speeds. Let us graph the general and the special case of total pressure equation as a function of Mach number in Fig. 2.54 to appreciate the shortcomings of Bernoulli at compressible speeds. To capture the discrepancy between the Bernoulli and the general expression at high Mach numbers, we use a logarithmic scale on the ordinate of Fig. 2.54.

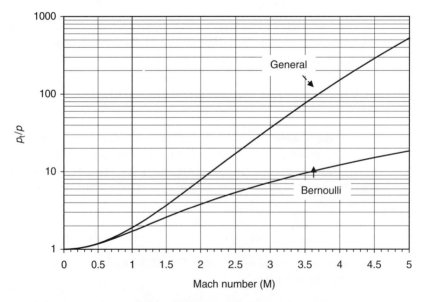

■ FIGURE 2.54 Comparison between the Bernoulli and the general equation for total pressure as a function of local Mach number ($\gamma = 1.4$)

2.19   Fluid Impulse

Let us consider a duct, or device, where fluid enters at certain inlet conditions, e.g., mass flow rate, speed, static pressure, and temperature, and exits the duct or device at a different state. A definition sketch is shown in Fig. 2.55.

Note that in part (b), the p_w represents the pressure exerted *on the fluid* from the wall, and the term τ_w is the shear stress acting *on the fluid*, i.e., exerted by the wall on the fluid. The pressure and shear felt at the wall is the opposite in direction and equal in magnitude to the ones acting on the fluid. Assuming that the flow is uniform at the inlet and exit planes, we may write the momentum equation for the fluid (in the x-direction) as

$$\dot{m}V_2 - \dot{m}V_1 = p_1 A_1 - p_2 A_2 + F_x|_{\text{fluid}} \tag{2.206}$$

where $F_{x,\text{fluid}}$ is the integral of pressure and shear forces exerted on the fluid by the wall, in the $+x$-direction. Now, we can isolate this force exerted by the wall on the fluid and express it in terms of fluid impulse I, which is defined as

$$I \equiv pA + \dot{m}V = pA(1 + \gamma M^2) \tag{2.207}$$

at the inlet and exit planes of the duct, namely,

$$F_x|_{\text{fluid}} = (\dot{m}V_2 + p_2 A_2) - (\dot{m}V_1 + p_1 A_1) = I_2 - I_1 \tag{2.208}$$

In turn the force felt by the duct (in the x-direction), in reaction to fluid flow through it, is equal and opposite to $F_{x,\text{fluid}}$, i.e.,

$$F_x|_{\text{walls}} = -F_x|_{\text{fluid}} = I_1 - I_2 \tag{2.209}$$

There are three lessons here. One, we may arrive at the integrated wall force, as in $F_{x,\text{wall}}$ above, by simply balancing the fluid impulse at the inlet and exit of a duct. Two, depending on the change of fluid impulse from inlet to exit, $I_1 - I_2$, the force acting on the wall will be in the thrust or drag direction, i.e., in the $+x$ or $-x$-direction. Finally, the force that we call $F_{x,\text{wall}}$ is force on the *inner wall* and thus accounts for no external drag contribution.

In a diffuser where $I_2 > I_1$, the duct will experience a thrust (i.e., $-x$) force. In air-breathing jet engines, the inlet is responsible for a percentage of net thrust production. The

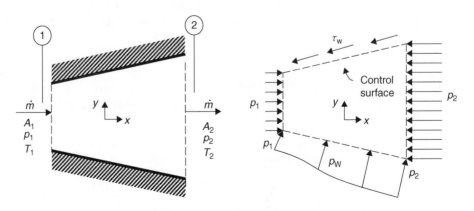

(a) Duct geometry and flow parameters   (b) Forces acting on the *fluid* (control surface)

■ **FIGURE 2.55**   **Definition sketch for a duct flow and its corresponding (closed) control surface**

percentage of inlet contribution to thrust increases with flight Mach number. In hypersonic flight, the inlet is responsible for more than 50% of thrust production of the engine. Note that we are addressing the *internal performance* of an inlet. The external aerodynamic drag plays a dominant role in inlet selection and airframe integration and contributes to the net *installed thrust*.

EXAMPLE 2.12 (Subsonic Diffuser)

Consider a diffusing duct of the following geometrical and flow characteristics:

A_1 is the inlet area

$M_1 = 0.7$

Total pressure loss in the diffuser is 1% of the inlet total pressure, i.e., $p_{t2}/p_{t1} = 0.99$

The exit area is $A_2 = 1.237 A_1$

Assume that the flow in the diffuser is adiabatic and unseparated and the exit flow is uniform, calculate

(a) the exit Mach number M_2

(b) the static pressure recovery in the diffuser C_{PR}

(c) the force acting on the diffuser inner wall, i.e., $F_{x,\text{wall}}$, nondimensionalized by the inlet static pressure and area, i.e., $p_1 A_1$

SOLUTION

Expressing continuity equation in terms of fluid total or stagnation state and Mach number, we can calculate the exit Mach number M_2 via

$$\sqrt{\frac{\gamma}{R}}\frac{p_{t1}}{\sqrt{T_{t1}}}A_1 M_1 \left(\frac{1}{1+\dfrac{\gamma-1}{2}M_1^2}\right)^{\frac{\gamma+1}{2(\gamma-1)}}$$

$$=\sqrt{\frac{\gamma}{R}}\frac{p_{t2}}{\sqrt{T_{t2}}}A_2 M_2 \left(\frac{1}{1+\dfrac{\gamma-1}{2}M_2^2}\right)^{\frac{\gamma+1}{2(\gamma-1)}}$$

which simplifies to:

$$\frac{M_2}{(1+0.2M_2^2)^3}=\left(\frac{p_{t1}}{p_{t2}}\right)\frac{A_1}{A_2}\frac{M_1}{(1+0.2M_1^2)^3}$$

$$=\left(\frac{1}{0.99}\right)\left(\frac{1}{1.237}\right)\frac{0.7}{1.3237}=0.4318$$

The (subsonic) solution to the above equation for M_2 is

$$\boxed{M_2 \approx 0.50}$$

The static pressure recovery in a diffuser is defined as

$$C_{\text{pr}} \equiv \frac{p_2 - p_1}{q_1}=\frac{1}{\gamma M_1^2}\left(\frac{p_2}{p_1}-1\right)$$

We may write the static pressure ratio in terms of total pressure ratio and a function of Mach number as

$$\frac{p_2}{p_1}=\frac{p_{t2}}{p_{t1}}\left(\frac{1+\dfrac{\gamma-1}{2}M_1^2}{1+\dfrac{\gamma-1}{2}M_2^2}\right)^{\frac{\gamma}{\gamma-1}}=(0.99)\left(\frac{1.098}{1.05}\right)^{3.5}$$

$$=1.169353$$

Therefore, the static pressure recovery is $\boxed{C_{PR} \approx 0.4937}$

$$C_{PR}=\frac{2}{(1.4)(0.7)^2}(0.169353)=0.4937$$

The physical interpretation of the value of C_{PR} that we calculated is that the diffuser has converted ~49.4% of its inlet dynamic pressure into static pressure rise. We note that in the flow deceleration in the diffuser, from Mach 0.7 at the inlet to 0.5 at the exit.

What is the internal force that the diffuser feels? To answer this question, we look at the change in fluid impulse, $I_1 - I_2$, according to Eq. 2.209:

$$F_x|_{\text{walls}}=I_1-I_2=A_1 p_1(1+\gamma M_1^2)-A_2 p_2(1+\gamma M_2^2)$$

As suggested in the problem statement, we may use inlet static pressure and area as the nondimensionalizing force (although we could have used inlet dynamic pressure and the area, instead), therefore,

$$\frac{F_x|_{\text{walls}}}{p_1 A_1}=1+\gamma M_1^2-\left(\frac{A_2}{A_1}\right)\left(\frac{p_2}{p_1}\right)(1+\gamma M_2^2)$$

Upon substitution of the values that we calculated and were given in the problem statement, in the above expression, we get

$$\boxed{\frac{F_x|_{\text{walls}}}{p_1 A_1}=-0.26676}$$

COMMENTS The negative sign on the force acting on the diffuser inner wall identifies the force as acting in the

–*x*-direction, i.e., in *the thrust* direction. Note that we calculated the force acting on the inner wall, including the viscous force (that caused the total pressure drop in the diffuser), without the knowledge of either the wall static pressure distribution or the viscous shear stress distribution on the wall. The balance of fluid impulse is therefore a powerful tool that we use to calculate forces on the engine components.

Also, note that the shape of the duct (cross section or in longitudinal direction) did not enter the problem, in other words, for the same flow conditions at the inlet, area ratio, and the total pressure loss, the duct shape could have been as drawn in Fig. 2.56.

However, one manifestation of different cross sections and longitudinal shapes is in the extent of total pressure loss that is created in the duct. Boundary layer development is sensitive to both streamwise and *transverse pressure gradients* and therefore potential flow separation as well as extensive secondary flow losses will be different in ducts of different geometries.

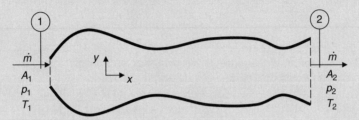

■ **FIGURE 2.56** **Diffuser with different cross section and longitudinal shapes as that shown in Fig. 2.55 (with the same inlet and exit flow conditions)**

EXAMPLE 2.13 (Supersonic Nozzle)

Consider a convergent–divergent nozzle as shown in Fig. 2.57. The flow conditions are

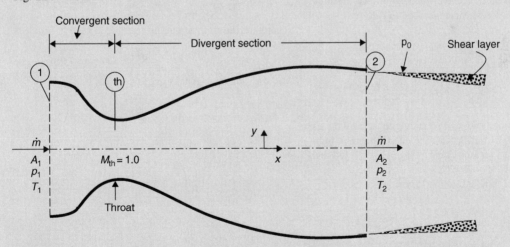

■ **FIGURE 2.57** **Sketch of a convergent–divergent nozzle with choked throat and the free shear layer**

(a) inlet Mach number $M_1 = 0.5$

(b) inlet nozzle total pressure is $p_{t1} = 10\,p_0$, where p_0 is the ambient pressure

(c) total pressure loss in the convergent section of the nozzle is 1%, i.e. $(p_{t1} - p_{t,th})/p_{t1} = 0.01$

(d) total pressure loss in the divergent section of the nozzle is 2%, i.e. $(p_{t,th} - p_{t2})/p_{t,th} = 0.02$

(e) nozzle area expansion ratio is $A_2/A_{th} = 2.0$

In addition, we assume that the gas is perfect and its properties remain unchanged throughout the nozzle. The gas is characterized by

(a) $\gamma = 1.4$

(b) $R = 287\,\text{J/kg·K}$

The flow in the nozzle is assumed to be steady and adiabatic, therefore the total enthalpy remains constant, i.e.,

(a) $h_{t2} = h_{t1}$

Calculate

(a) the exit Mach number M_2

(b) the exit static pressure in terms of ambient pressure p_2/p_0

(c) the nondimensional axial force acting on the convergent nozzle $F_{x,\text{con.}-\text{wall}}/A_{\text{th}}\,p_{t1}$

(d) the nondimensional axial force acting on the divergent nozzle $F_{x,\text{div.}-\text{wall}}/A_{\text{th}}\,p_{t1}$

(e) the total (nondimensional) axial force acting on the nozzle $F_{x,\text{nozzle}}/A_{\text{th}}\,p_{t1}$

SOLUTION

By setting the mass flow rate at the exit and throat equal to each other, we calculate the exit Mach number M_2. We used the same approach in the previous example on subsonic diffusers.

$$\sqrt{\frac{\gamma}{R}}\frac{p_{t,\text{th}}}{\sqrt{T_{t,\text{th}}}}A_{\text{th}}M_{\text{th}}\left(\frac{1}{1+\dfrac{\gamma-1}{2}M_{\text{th}}^2}\right)^{\frac{\gamma+1}{2(\gamma-1)}}$$

$$=\sqrt{\frac{\gamma}{R}}\frac{p_{t2}}{\sqrt{T_{t2}}}A_2 M_2\left(\frac{1}{1+\dfrac{\gamma-1}{2}M_2^2}\right)^{\frac{\gamma+1}{2(\gamma-1)}}$$

"Frozen" gas properties along the nozzle will cause the elimination of similar terms on both sides of the continuity equation. We will cancel the total temperature and substitute $M_{\text{th}} = 1.0$ for the throat Mach number, to get

$$\frac{M_2}{(1+0.2M_2^2)^3}=\left(\frac{p_{t,\text{th}}}{p_{t2}}\right)\frac{A_{\text{th}}}{A_2}\frac{M_{\text{th}}}{(1+0.2M_{\text{th}}^2)^3}$$

$$=\left(\frac{1}{0.98}\right)\left(\frac{1}{2}\right)\frac{1.0}{1.728}=0.295257$$

There are two solutions to the above equation. One is subsonic, i.e., $M_2 = 0.313$, which is unacceptable in light of a large nozzle pressure ratio $(p_{t1}/p_0 = 10)$ and the second solution is $M_2 = 2.174$.

Therefore the acceptable solution for exit Mach number is

$$\boxed{M_2 = 2.174}$$

The nozzle exit static pressure is related to the exit total pressure and the exit Mach number via

$$p_2=\frac{p_{t2}}{\left(1+\dfrac{\gamma-1}{2}M_2^2\right)^{\frac{\gamma}{\gamma-1}}}$$

The nozzle exit total pressure is calculated based on nozzle losses, e.g.,

$$p_{t2}=\frac{p_{t2}}{p_{t,\text{th}}}\frac{p_{t,\text{th}}}{p_{t1}}\frac{p_{t1}}{p_0}p_0$$

Therefore, the nozzle exit static pressure is related to ambient static pressure by

$$\frac{p_2}{p_0}=\frac{(0.98)(0.99)(10)}{[1+0.2(2.174)^2]^{3.5}}=0.9450$$

$$\boxed{p_2/p_0 \cong 0.945}$$

Since $p_2 < p_0$, the (nozzle) flow is overexpanded (but just slightly). A weak oblique shock at the nozzle lip will appear to balance the static pressure across the jet shear layer. Let us apply Eq. 2.209 to calculate the force acting on the wall, i.e.,

$$F_x|_{\text{wall}}=-F_x|_{\text{fluid}}=I_1-I_2$$

Therefore, the force on the convergent section of the nozzle is

$$F_{x,\text{con}-\text{wall}}=p_1 A_1(1+\gamma M_1^2)-p_{\text{th}}A_{\text{th}}(1+\gamma M_{\text{th}}^2)$$

and the nondimensional force is

$$\frac{F_{x,\text{con.}-\text{wall}}}{A_{\text{th}}\,p_{t1}}=\left(\frac{p_1}{p_{t1}}\right)\left(\frac{A_1}{A_{\text{th}}}\right)\left(1+1.4(0.5)^2\right)-\left(\frac{p_{\text{th}}}{p_{t1}}\right)(2.4)$$

We may calculate the inlet area ratio A_1/A_{th} using the continuity equation,

$$\sqrt{\frac{\gamma}{R}}\frac{p_{t1}}{\sqrt{T_{t1}}}A_1 M_1\left(\frac{1}{1+\dfrac{\gamma-1}{2}M_1^2}\right)^{\frac{\gamma+1}{2(\gamma-1)}}$$

$$=\sqrt{\frac{\gamma}{R}}\frac{p_{t,\text{th}}}{\sqrt{T_{t,\text{th}}}}A_{\text{th}}M_{\text{th}}\left(\frac{1}{1+\dfrac{\gamma-1}{2}M_{\text{th}}^2}\right)^{\frac{\gamma+1}{2(\gamma-1)}}$$

Which simplifies to

$$\frac{A_1}{A_{\text{th}}}=\frac{p_{t,\text{th}}}{p_{t1}}\frac{1}{M_1}\left(\frac{1+\dfrac{\gamma-1}{2}M_1^2}{\dfrac{\gamma+1}{2}}\right)^{\frac{\gamma+1}{2(\gamma-1)}}$$

$$=(0.99)(1/0.5)\left(\frac{1+0.2(0.25)}{1.2}\right)^3=1.326445$$

We can calculate the throat static pressure, p_{th} in terms of the inlet total pressure p_{t1} according to

$$p_{th} = \frac{p_{t,th}}{\left(\dfrac{\gamma+1}{2}\right)^{\frac{\gamma}{\gamma-1}}} = \frac{0.99\,p_{t1}}{(1.2)^{3.5}} = 0.523\,p_{t1}$$

The static pressure at the inlet is written in terms of the inlet total pressure and the Mach number,

$$p_1 = \frac{p_{t1}}{\left(1+\dfrac{\gamma-1}{2}M_1^2\right)^{\frac{\gamma}{\gamma-1}}} = \frac{p_{t1}}{[1+0.2(0.25)]^{3.5}} = 0.84302$$

Now, we are ready to calculate the nondimensional force on the convergent section of the nozzle wall,

$$\frac{F_{x,con.-wall}}{A_{th}\,p_{t1}} = (0.84302)(1.326445)(1.35) - (0.523)$$

$$(2.4) \cong 0.2544$$

$$\boxed{F_{x,\,con.\,wall}/A_{th}\cdot p_{t1} \cong 0.2544}$$

The force on the divergent section of the nozzle is

$$F_{x,div.-wall} = p_{th}A_{th}(1+\gamma M_{th}^2) - p_2 A_2(1+\gamma M_2^2)$$

The nondimensional force on the nozzle divergent section is

$$\frac{F_{x,div.wall}}{A_{th}\,p_{t1}} = \left(\frac{p_{th}}{p_{t1}}\right)(2.4) - \left(\frac{p_2}{p_{t1}}\right)\left(\frac{A_2}{A_{th}}\right)(1+\gamma M_2^2)$$

The ratio of exit static pressure to inlet total pressure is

$$\frac{p_2}{p_{t1}} = \frac{p_2}{p_0}\frac{p_0}{p_{t1}} = (0.945)(1/10) = 0.0945$$

Now, we may substitute for all the parameters in the nondimensional axial force acting on the divergent section of the nozzle

$$\frac{F_{x,div.wall}}{A_{th}\,p_{t1}} = (0.523)(2.4) - (0.0945)(2)[1+1.4(2.174)^2]$$

$$= -0.18437$$

$$\boxed{F_{x,div.\,wall}/A_{th}\,p_{t1} \cong -0.1844}$$

The total axial force acting on the nozzle wall is the sum of the convergent and divergent sections, namely

$$\frac{F_{x,nozzle}}{A_{th}\,p_{t1}} \cong 0.2544 - 0.1844 = 0.07$$

$$\boxed{F_{x,\,nozzle}/A_{th.}\,p_{t1} \cong 0.07}$$

The net positive sign on the axial force acting on the convergent–divergent nozzle wall shows the C–D nozzle as a drag-producing component of an airbreathing engine.

EXAMPLE 2.14 (Axial-Flow Compressor)

Let us apply the fluid impulse principle to an axial-flow compressor, as shown in Fig. 2.58.

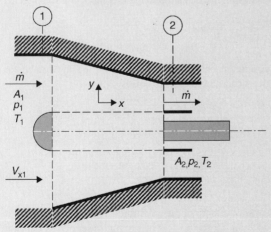

■ **FIGURE 2.58** Axial-flow compressor with its inlet and exit flow parameters identified

For a constant-axial velocity through the compressor, as a design choice, we note that flow cross-sectional area shrinks inversely as the density rise, i.e. (steady-flow) continuity demands

$$\frac{A_2}{A_1} = \frac{\rho_1}{\rho_2}$$

The density ratio across the compressor may be linked to static pressure ratio, which for simplicity, we may assume the compression is achieved reversibly and adiabatically. Therefore, for a compressor pressure ratio of say, 20,

$$\frac{p_2}{p_1} = 20$$

the density ratio is

$$\frac{\rho_2}{\rho_1} = \left(\frac{p_2}{p_1}\right)^{1/\gamma} = (20)^{1/1.4} \cong 8.4978$$

Therefore, the area ratio is the inverse of 8.4978, i.e.,

$$\frac{A_2}{A_1} \cong \frac{1}{8.4978} \approx 0.1177$$

Guided by these basic principles, we choose the following parameters for our compressor:

$$p_2/p_1 = 20$$
$$V_{x1} = V_{x2}$$
$$A_2/A_1 = 0.12$$

The application of impulse principle, Eq. 2.209, produces the net axial force experienced by all the surfaces interacting with the flow between stations 1 and 2, i.e.,

$$F_x|_{\text{walls}} = -F_x|_{\text{fluid}} = I_1 - I_2$$
$$= p_1 A_1 + \dot{m} V_{x1} - (p_2 A_2 + \dot{m} V_{x2}) = p_1 A_1 - p_2 A_2$$

Note that by our (constant-axial velocity) design choice, the axial momentum across the compressor remained constant. The non-dimensional axial force $F_{x,\text{walls}}/p_1 A_1$ is therefore

$$\frac{F_{x,\text{walls}}}{p_1 A_1} = 1 - \left(\frac{p_2}{p_1}\right)\left(\frac{A_2}{A_1}\right) = 1 - (20)(0.12) = -1.4$$

$$\boxed{F_{x,\text{walls}}/p_1 A_1 = -1.4}$$

The negative sign on the axial force experienced by the compressor structure signifies a thrust production by this component.

EXAMPLE 2.15 (Combustor)

Combustor represents a complex problem for a simple application of the impulse principle. A typical combustor has several ports that carry two different fluids, i.e., fuel and the compressed air, into a reaction chamber. The compressed air itself is introduced at different levels along the combustor length. Let us simplify the problem by assuming a constant-area duct, which takes all the fluid at its entrance (in gaseous form) at certain pressure and temperature conditions and allow a heated fluid (as a result of chemical reaction) to leave the combustor at nearly the same pressure as the inlet, i.e., $p_1 \cong p_2$. Since the pressure is assumed constant, the density drops inversely proportional to the rising temperature. Also with a constant area duct assumption, the axial velocity rises inversely proportional to the fluid density drop, which in turn is inversely proportional to the static temperature rise. All of these assumptions and trends are simplifications to the real and complex process; however, by making these simplifications, we can apply the simple impulse principle and examine the direction of axial force on the combustor. Let us assume that Fig. 2.59 is a simple model of a combustor.

Therefore, let the combustor be modeled as a duct with

$$A_2 = A_1$$
$$p_2 \approx p_1$$
$$T_2 = 1.8 T_1$$

The density ratio will be the inverse of 1.8, i.e., $\rho_2/\rho_1 \approx 1/1.8 = 0.5556$

The axial velocity ratio will be equal to the temperature ratio (or inverse of the density ratio),
$$V_{x2}/V_{x1} = 1.8$$

Now the impulse equation written for this simple duct yields

$$F_x|_{\text{wall}} = -F_x|_{\text{fluid}} = I_1 - I_2$$
$$= p_1 A_1 + \dot{m} V_{x1} - (p_2 A_2 + \dot{m} V_{x2}) = \dot{m}(V_{x1} - V_{x2})$$

The nondimensional axial force acting on the combustor walls (according to our model) is

$$\frac{F_x|_{\text{wall}}}{\dot{m} V_{x1}} = 1 - \frac{V_{x2}}{V_{x1}} = -0.8$$

$$\boxed{F_x|_{\text{wall}}/\dot{m} V_{x1} = -0.8}$$

The negative sign of the axial force again signifies a thrust production by the device, i.e., in this case, our model of a combustor.

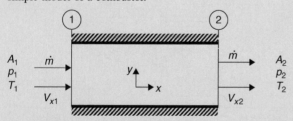

■ **FIGURE 2.59**   Simple model of a combustor flow

EXAMPLE 2.16 (Axial-Flow-Turbine)

We anticipate an opposite behavior from the turbine as compared with a compressor. To quantify that, we model an uncooled turbine flow as a reversible and adiabatic expansion of a perfect gas. Figure 2.60 shows the schematics of an axial-flow turbine.

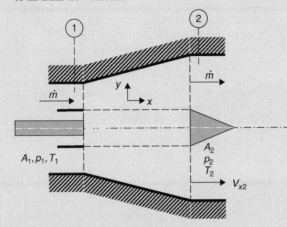

■ **FIGURE 2.60**   Schematic drawing of an uncooled axial-flow turbine

Assuming the turbine is designed for a constant axial velocity $V_{x1} = V_{x2}$ similar to the compressor, the area ratio will be inversely proportional to the density ratio, i.e.,

$$A_2/A_1 = \rho_1/\rho_2$$

The density ratio is then related to the temperature ratio isentropically, namely

$$\rho_2/\rho_1 = (T_2/T_1)^{\frac{1}{\gamma-1}}$$

The turbine temperature ratio is calculated from the power balance between the compressor/fan and the turbine, as discussed in the cycle analysis. For now, we assume the turbine expansion is known to be

$$T_2/T_1 \cong 0.79$$

Therefore, the density ratio will be $\rho_2/\rho_1 = (0.79)^{2.5} = 0.5547$

The area ratio is the inverse of the density ratio, i.e., $A_2/A_1 = 1.803$

And the pressure ratio is related to density ratio according to

$$p_2/p_1 = (\rho_2/\rho_1)^\gamma = (0.5547)^{1.4} = 0.4382$$

The impulse equation will again simplify to

$$F_Y|_{\text{walls}} = -F_x|_{\text{fluid}} = I_1 - I_2$$
$$= p_1A_1 + \dot{m}V_{x1} - (p_2A_2 + \dot{m}V_{x2}) = p_1A_1 - p_2A_2$$

and the nondimensional axial force is

$$\frac{F_{x,\text{walls}}}{p_1A_1} = 1 - \left(\frac{p_2}{p_1}\right)\left(\frac{A_2}{A_1}\right)$$
$$= 1 - (0.4382)(1.803) \approx 0.210$$

$$\boxed{F_{x,\text{walls}}/p_1A_1 \approx 0.210}$$

As expected, the turbine behavior is opposite to the compressor, i.e., the turbine walls experience an axial force in the drag direction.

2.20   Summary of Fluid Impulse

We have applied fluid impulse principle to various aircraft engine components. Figure 2.61 represents a schematic drawing of a turbojet engine and the (axial) forces internally developed

■ **FIGURE 2.61**
Turbojet engine is shown with its component's (internal) axial force (not-to-scale) contributing to engine force

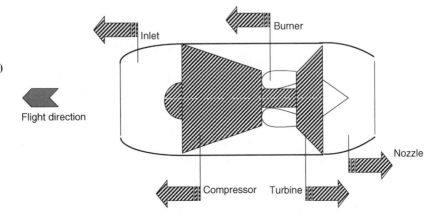

by each component. The vectors indicate the direction and are not drawn to scale. The vector sum of these component forces is transmitted to the aircraft as engine (internal) force. There are external forces (both in the drag and thrust directions) that act on the nacelle that we have not accounted for through our treatment of *internal flow* in engine components. We shall discuss *external flow* effects in Chapter 5.

References

1. Anderson, J.D., Jr., *Modern Compressible Flow*, 3rd edition, McGraw-Hill, New York, 2003.
2. Anderson, J.D., Jr., *Fundamentals of Aerodynamics*, 4th edition, McGraw-Hill, New York, 2005.
3. Moody, L.F., "Friction Factors for Pipe Flow," Transactions of ASME, November 1944.
4. Munson, B.R., Young, D.F., and Okiishi, T.H., *Fundamentals of Fluid Mechanics*, 5th edition, John Wiley & Sons, Inc., New York, 2006.
5. NACA Report 1135, Ames Research Staff, "Equations, Tables and Charts for Compressible Flow,"1953.
6. Shapiro, A.H., *The Dynamics and Thermodynamics of Compressible Fluid Flow*, Vol. I and II, The Ronald Press, New York, 1953.
7. Sonntag, R.E., Borgnakke, C., and Van Wylen, G.J., *Fundamentals of Thermodynamics*, 6th Ed., John Wiley & Sons, Inc., New York, 2003.

Problems

ASSUME ALL GASES ARE CALORICALLY PERFECT, UNLESS OTHERWISE SPECIFIED.

2.1   A normal shock flow is characterized by stagnation speed of sound a_t and speed of sound a as shown.
Calculate:

(a) M_1

(b) M_2^*

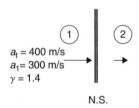

N.S.

■ **FIGURE P2.1**

2.2   A two-dimensional projectile with a sharp nose is exposed to a Mach 3 flow, as shown. Assuming the base pressure is p_1, calculate

(a) p_2/p_1

(b) p_3/p_1

(c) $C_d \equiv D'/q_1 \cdot c$

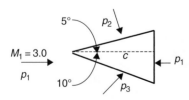

■ **FIGURE P2.2**

2.3   Calculate the stagnation pressure measured by a Pitot tube on an inclined ramp in a supersonic flow, as shown.

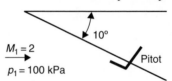

■ **FIGURE P2.3**

2.4   A reflected oblique shock has the following geometry. Calculate

(a) M_1

(b) M_2

(c) M_3

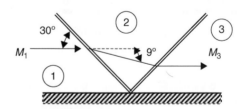

■ **FIGURE P2.4**

2.5   A symmetrical half-diamond airfoil has a leading-edge angle of 5°. This airfoil is set at 5° angle of attack, as shown, and is placed in a wind tunnel with $M_{T.S.} = 2.0$, $p_{t,T.S.} = 100\,\text{kPa}$, $T_{t,T.S.} = 25°\text{C}$. Assume $\gamma = 1.4$, $c_p = 1.004\,\text{kJ/kg·K}$.
Calculate

(a) p_2 (kPa)

(b) p_3 (kPa)

(c) p_4 (kPa)

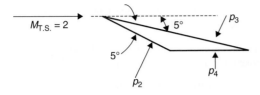

■ **FIGURE P2.5**

2.6 A supersonic flow expands around a sharp corner, as shown. Calculate the following parameters:

(a) M_2

(b) A_2/A_1

(c) The angle of Prandtl-Meyer fan envelope δ

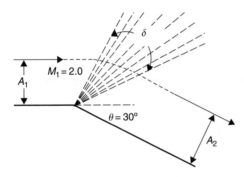

■ **FIGURE P2.6**

2.7 A symmetrical half-diamond airfoil has a nose angle $\theta_{nose} = 30°$ and is exposed to a supersonic flow. A pitot tube is installed on each of the three surfaces, as shown. Calculate the pitot tube readings on the airfoil surfaces at zero angle of attack.

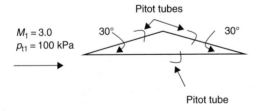

■ **FIGURE P2.7**

2.8 A flat plate is in a Mach 5 flow at $30°$ angle-of-attack. Calculate

(a) lift-to-drag ratio, L'/D'

(b) pitching moment coefficient about the L.E.

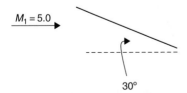

■ **FIGURE P2.8**

2.9 A normal shock is in a Mach 2.0 flow. Upstream gas temperature is $T_1 = 15°C$, the gas constant is $R = 287\,\text{J/kg}\cdot$K and $\gamma = 1.4$. Calculate

(a) a^* in m/s

(b) u_2 in m/s (use Prandtl's relation)

(c) a_t in m/s

(d) h_2 in kJ/kg

■ **FIGURE P2.9**

2.10 A supersonic tunnel has a test section (T.S.) Mach number of $M_{T.S.} = 2.0$. The reservoir for this tunnel is the room with $T_{room} = 15°C$ and $p_{room} = 100\,\text{kPa}$. The test section has two windows ($10\,\text{cm} \times 20\,\text{cm}$ each). Calculate

(a) the speed of sound in the test section

(b) the force on each glass window

Assume $R = 287\,\text{J/kg}\cdot$K, $\gamma = 1.4$.

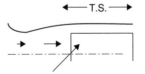

Glass window

■ **FIGURE P2.10**

2.11 Calculate the wave drag coefficient of a 2D sharp-nosed projectile, as shown, assuming the base pressure is ambient, i.e., $p_{base} = p_\infty$. The 2D wave drag coefficient is defined based on b. Assume the angle of attack iz zero.

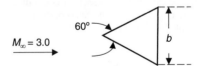

■ **FIGURE P2.11**

2.12 A Prandtl–Meyer centered expansion wave is visualized with the wave angles as shown. Calculate

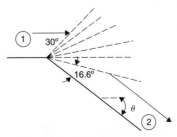

■ **FIGURE P2.12**

(a) the flow turning angle, i.e., wall angle θ

(b) velocity ratio across the expansion wave V_2/V_1

2.13   A diamond airfoil with a nose angle of $15°$ and a thickness-to-chord ratio of 10% is in a supersonic flow, as shown. Calculate

(a) the chordwise location of the max thickness point (in % c)

(b) the trailing-edge angle

(c) the nondimensional pressure on the three surfaces, p_1/p_∞, p_2/p_∞, p_3/p_∞

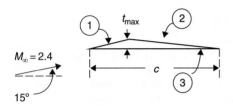

■ **FIGURE P2.13**

2.14   A blunt-nosed vehicle is in supersonic flight at an altitude where static pressure is $p_\infty = 20\,\text{kPa}$. Calculate the stagnation pressure on the vehicle.

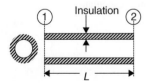

■ **FIGURE P2.14**

2.15   An adiabatic constant-area duct has an inlet flow of $M_1 = 0.5$, $T_1 = 260°\text{C}$, $p_1 = 1\,\text{MPa}$. The average skin friction coefficient, $C_f = 0.005$ and the duct cross section is circular with inner diameter $d = 10\,\text{cm}$. Assuming the gas is perfect and has properties $\gamma = 1.4$ and $R = 286.8\,\text{J/kg·K}$, calculate

(a) mass flow rate $\dot{m}$

(b) L_{max} to choke this duct at the exit

(c) p_t loss in the duct at $L = L_{\text{max}}$

■ **FIGURE P2.15**

2.16   Calculate the critical heat flux q^* to a frictionless, constant-area duct with an inlet Mach number $M_1 = 2.0$ and an

inlet temperature $T_1 = -15°\text{C}$. Assume $\gamma = 1.4$ and $c_p = 1{,}004\,\text{J/kg·K}$.

2.17   A constant-area duct has an L/D of 100 and is frictionally choked. For an entrance Mach number of 0.5, calculate

(a) average wall friction coefficient C_f

(b) percent static pressure drop in the flow

Assume the fluid is air, with $\gamma = 1.4$.

2.18   A frictionless, constant-area duct flow of a perfect gas is heated to a choking condition. The rate of heat transfer per unit mass flow rate is $500\,\text{kJ/kg}$. Assuming inlet Mach number is $M_1 = 3.0$ and $c_p = 1.004\,\text{kJ/kg·K}$, calculate

(a) inlet total temperature T_{t1}

(b) percent static pressure rise in the flow

2.19   In a Rayleigh flow, as shown, the maximum temperature T_{max} is reached on the subsonic branch. The sonic condition is shown as *, on the T–s diagram. Assuming the medium is air, with $\gamma = 1.4$, calculate

(a) T_{max}/T^*

(b) $p_{@T-\text{max}}/p^*$

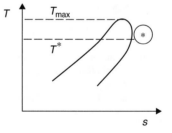

■ **FIGURE P2.19**

2.20   In an adiabatic, constant area flow of a perfect gas, the inlet conditions are $p_1 = 100\,\text{kPa}$, $\rho_1 = 1\,\text{kg/m}^3$, and $u_1 = 100\,\text{m/s}$. At a downstream station the gas is at $200\,\text{m/s}$. Assuming the medium is air, calculate the corresponding static pressure and density, p_2 and ρ_2, respectively.
$[R_{\text{air}} = 287\,\text{J/kg·K}, \gamma_{\text{air}} = 1.4]$

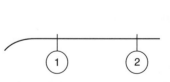

■ **FIGURE P2.20**

2.21   In a frictionless, constant area flow of a perfect gas, the inlet conditions are $p_1 = 100\,\text{kPa}$, $\rho_1 = 1\,\text{kg/m}^3$, and

$u_1 = 100\,\text{m/s}$. At a downstream station the gas is at $200\,\text{m/s}$. Assuming the medium is air, calculate the corresponding static pressure and density, p_2 and ρ_2, respectively.

[$R_{\text{air}} = 287\,\text{J/kg·K}$, $\gamma_{\text{air}} = 1.4$]

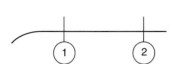

■ **FIGURE P2.21**

2.22 Consider a one-dimensional adiabatic flow of a perfect gas with friction. Air enters a circular duct of 20-cm diameter at Mach 0.5 and an average wall friction coefficient of $C_f = 0.005$. If the exit total pressure drops to 85% of the inlet value, i.e., $p_{t2} = 0.85\,P_{t1}$, calculate the length of the duct, L.

Fanno

2.23 A perfect gas flows in a well-insulated, constant area pipe with friction at $M_1 = 2.0$. For an average wall friction coefficient of $C_f = 0.004$, calculate

(a) the maximum length of the pipe that can transmit the flow

(b) total pressure loss at this length

Assume the pipe cross section is rectangular and has the dimensions of $10\,\text{cm} \times 20\,\text{cm}$.

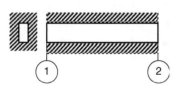

■ **FIGURE P2.23**

Rayleigh!

2.24 Air enters a frictionless, constant-area pipe at $p_1 = 60$ psia., $T_1 = 500\,°R/$ and $M_1 = 0.6$. If heat is transferred to the air in the pipe at $q = 300\,\text{BTU/lbm}$ of air, calculate

(a) the exit Mach number

(b) static and total pressure and temperature at the exit, p_2, T_2, P_{t2}, T_{t2}

(c) the critical heat flux q^* that will thermally choke the pipe.

$c_p = 0.24\,\text{BTU/lbm.}°R$, $\gamma = 1.4$

2.25 A constant-area frictionless pipe is thermally choked. The inlet total temperature and the heat flux are $T_{t1} = 650\,°R$ and $q^* = 500\,\text{BTU/lbm}$, respectively. Calculate

(a) the inlet Mach number M_1

(b) exit total temperature T_{t2}

$c_p = 0.24\,\text{BTU/lbm.}°R$, $\gamma = 1.4$

2.26 If we neglect the friction in a jet engine combustion chamber and assume the flow through the burner may be

modeled as a Rayleigh line flow, calculate the combustor exit Mach number if $M_3 = 0.2$, $T_{t3} = 900\,°R$. The fuel heating value is $18,400\,\text{BTU/lbm}$ (of fuel) and the fuel-to-air ratio is 2%. Assume $\gamma = 1.33$ and $c_p = 0.27\,\text{BTU/lbm.}°R$ and neglect the added mass of the fuel.

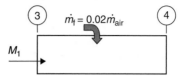

■ **FIGURE P2.26**

2.27 A perfect gas flows in a constant-area tube with no heat interaction with the surroundings. The inlet Mach number is $M_1 = 0.5$. The average wall friction coefficient is $C_f = 0.005$. Calculate

(a) the choking L/D of the pipe

(b) the new inlet Mach number $M_{1'}$ if the pipe is 20% longer than part (a)

Assume $\gamma = 1.4$.

2.28 Air enters an insulated duct with a constant area with an average wall friction coefficient of $C_f = 0.004$. The inlet Mach number is $M_1 = 2.0$. There is a choking length L_1^*, for this duct that corresponds to the inlet Mach number M_1. For any duct longer than the choking length L_1^*, a normal shock appears in the duct. Assuming the length of the duct is 10% longer than L_1^*, i.e., $L = 1.1\,L_1^*$, calculate

(a) the shock location along the duct x_s/D

(b) percent total pressure loss in the longer duct

(c) percent loss of fluid impulse

Assume $\gamma = 1.4$.

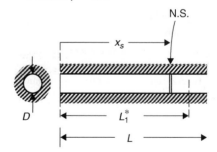

■ **FIGURE P2.28**

2.29 Consider an adiabatic flow of a perfect gas in a duct with friction. The geometry of the duct is shown. Assuming $\gamma = 1.4$ and the exit flow is choked, calculate

(a) two possible inlet Mach numbers M_1, one subsonic and the second supersonic

(b) the static pressure drop in the subsonic case between the inlet and exit, i.e., $\Delta p/p_1$

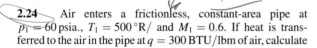

(c) the static pressure rise in the supersonic case, i.e., $\Delta p/p_1$

(d) the loss of fluid impulse due to friction

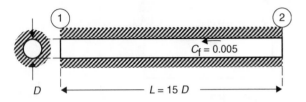

■ **FIGURE P2.29**

2.30 A frictionless flow in a constant-area duct encounters heat transfer through the sidewall as shown. Calculate

(a) the exit Mach number

(b) the static temperature ratio T_2/T_1

(c) the percent total pressure loss in the duct, $\Delta p_t/p_{t1}(\times 100)$

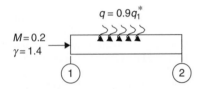

■ **FIGURE P2.30**

2.31 The inlet and exit flow conditions of a subsonic diffuser are shown. The diffuser has an area ratio of $A_2/A_1 = 1.25$. Assuming the fluid is air and is treated as a calorically perfect gas with $\gamma = 1.4$ and $R = 287\,\mathrm{J/Kg\cdot K}$, calculate

(a) The mass flow rate in kg/s

(b) The exit Mach number M_2

(c) the diffuser static pressure recovery $C_{PR} \equiv (p_2 - p_1)/q_1$

(d) the axial force exerted on the diffuser wall by the fluid in kN

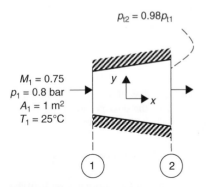

■ **FIGURE P2.31**

2.32 Consider the flow of perfect gas (air) on a vane, as shown. Apply momentum principles to the fluid to calculate the components of force F_x and F_y that act on the vane.

Assume the flow is uniform, and due to low speeds, $p_1 = p_2 = p_0$, where p_0 is the ambient pressure.

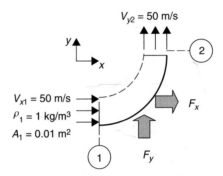

■ **FIGURE P2.32**

2.33 A subsonic nozzle is choked, i.e., $M_2 = 1.0$. Also the exit static pressure is equal to the ambient pressure p_0. Calculate the axial force that is experienced by the nozzle, i.e., $F_{x,\,noz}$, nondimensionalized by p_1A_1.

Assume $\gamma = 1.4$.

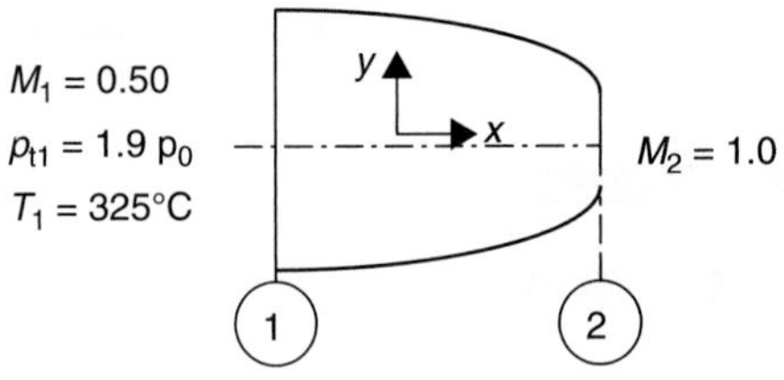

$$M_1 = 0.50$$
$$p_{t1} = 1.9\,p_0$$
$$T_1 = 325°C$$

■ **FIGURE P2.33**

2.34 A scramjet combustor has a supersonic inlet condition and a choked exit. The combustor flow area increases linearly in the flow direction, as shown.

The inlet and exit flow conditions are

$$M_1 = 3.0$$
$$p_1 = 1\,\mathrm{bar}$$
$$T_1 = 1000\,\mathrm{K}$$
$$A_1 = 1\mathrm{m}^2$$
$$M_2 = 1.0$$
$$A_2 = 1.4\,\mathrm{m}^2$$
$$\gamma = 1.4,\ R = 287\,\mathrm{J/kg\cdot K}$$

The total heat release due to combustion, per unit flow rate in the duct, is initially *assumed* to be 15 MJ/kg. If we divide the combustor into three constant-area sections, with stepwise jumps in the duct area, we may apply Rayleigh flow principles to each segment, as shown. The heat release per segment is then 1/3 of the total heat release in the duct, i.e., 5000 kJ/kg. As the exit condition of a segment needs to be matched to the inlet condition of the following segment, we propose to satisfy continuity equation at the boundary through an isentropic step area expansion, i.e., p_t, T_t remain the same and only the Mach number jumps isentropically through area expansion.

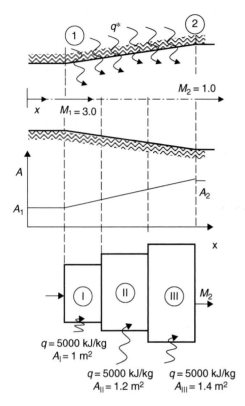

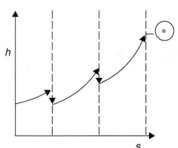

■ **FIGURE P2.34**

If we march from the inlet condition toward the exit with the assumed heat release rates, we calculate the exit Mach number M_2. Since the exit flow is specified to be choked, then we need to adjust the total heat release in order to get a choked exit. Calculate the critical heat release in the above duct that leads to thermal choking of the flow.

2.35 The center spike of an external-compression inlet is a cone. Use conical shock charts in appendix to calculate

(a) surface Mach number M_{cone}

(b) surface pressure coefficient $C_{p,\,cone}$

(c) cone pressure drag (also known as wave drag) coefficient $C_{D,\,cone}$.

Assume the cone base pressure is p_∞.

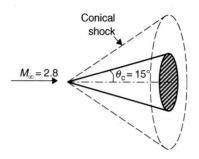

■ **FIGURE P2.35**

2.36 A supersonic inlet is shown. Based on the flow conditions specified, calculate

(a) the inlet mass flow rate $\dot{m}$ in kg/s

(b) exit area A_2 in m^2

(c) exit static pressure p_2 in bar

(d) fluid impulse at the inlet I_1

(e) fluid impulse at the exit I_2

(f) the (internal) axial force on the inlet F_x (in kN)

Assume the flow in the inlet is adiabatic as well as the inlet and exit flows are uniform.

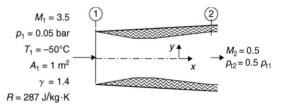

■ **FIGURE P2.36**

2.37 A steady-adiabatic flow in a convergent-divergent nozzle (as shown in Fig. 2.57) is characterized by

(a) inlet Mach number $M_1 = 0.55$

(b) inlet nozzle total pressure is $p_{t1} = 6\,p_0$, where p_0 is the ambient pressure

(c) total pressure loss in the convergent section of the nozzle is 1%, i.e. $(p_{t1} - p_{t,th})/p_{t1} = 0.01$

(d) total pressure loss in the divergent section of the nozzle is 2%, i.e. $(p_{t,th} - p_{t2})/p_{t,th} = 0.02$

(e) nozzle area expansion ratio is $A_2/A_{th} = 1.50$

Assuming that the gas is perfect and its properties remain unchanged throughout the nozzle with

$\gamma = 1.4$

$R = 287 \, \text{J/kg·K}$

Calculate

(a) the exit Mach number M_2

(b) the exit static pressure in terms of ambient pressure p_2/p_0

(c) the nondimensional (internal) axial force acting on the convergent nozzle, $F_{x,\text{con.}-\text{wall}}/A_{\text{th}} \, p_{t1}$

(d) the nondimensional (internal) axial force acting on the divergent nozzle, $F_{x,\text{div.}-\text{wall}}/A_{\text{th}} \, p_{t1}$

(e) the total (nondimensional) axial force acting on the nozzle, $F_{x,\text{nozzle}}/A_{\text{th}} \, p_{t1}$

2.38   A subsonic diffuser flow is steady and adiabatic with inlet Mach number $M_1 = 0.6$. Total pressure loss in the diffuser is 2% of the inlet total pressure, i.e., $p_{t2}/p_{t1} = 0.98$. The diffuser area ratio is $A_2/A_1 = 1.50$

Assuming that the flow in the diffuser has a uniform exit flow, calculate

(a) the exit Mach number M_2

(b) the static pressure recovery in the diffuser C_{PR}

(c) the (internal) force acting on the diffuser, i.e., $F_{x,\text{wall}}$, nondimensionalized by the inlet static pressure and area, i.e., $p_1 A_1$

2.39   An axial-flow compressor (as shown in Fig. 2.58) has a pressure ratio $p_2/p_1 = 15$. In addition, we have

$$V_{x1} = V_{x2}$$
$$A_2/A_1 = 0.155$$

Calculate the nondimensional internal force that acts in the axial direction on the compressor, $F_{x,\text{wall}}/A_1 \, p_1$ if the flow is assumed to be steady and uniform at the compressor inlet and exit.

2.40   An axial-flow turbine (as shown in Fig. 2.60) is designed for constant axial velocity, $V_{x1} = V_{x2}$. Turbine pressure ratio is measured to be $p_2/p_1 = 0.36$. The turbine area ratio is $A_2/A_1 = 2.1$. Assuming the flow is steady and uniform at the inlet and exit of the turbine, calculate the nondimensional internal force that acts in the axial direction on the turbine, $F_{x,\text{wall}}/A_1 \, p_1$.

CHAPTER 3

Engine Thrust and Performance Parameters

Trent 1000 Turbofan Engine (Courtesy of Rolls-Royce, plc)

3.1   Introduction

An aircraft engine is designed to produce thrust F (or sometimes lift in VTOL/STOL aircraft, e.g., the lift fan in the Joint Strike Fighter, F-35). In an airbreathing engine, a mass flow rate of air $\dot{m}_0$ and fuel $\dot{m}_f$ are responsible for creating that thrust. In a liquid rocket engine, the air is replaced with an onboard oxidizer $\dot{m}_{ox}$, which then reacts with an onboard fuel $\dot{m}_f$ to produce thrust. Although we will discuss the internal characteristics of a gas turbine engine in Chapter 4, it is instructive to show the station numbers in a two-spool turbojet engine with an afterburner. Figure 3.1 is schematic drawing of such engine. The air is brought in through the air intake, or inlet, system, where station 0 designates the unperturbed flight condition, station 1 is at the inlet (or cowl) lip, and station 2 is at the exit of the air intake system, which corresponds to the inlet of the compressor (or fan). The compression process from stations 2 to 3 is divided into a low-pressure compressor (LPC) spool and a high-pressure compressor (HPC) spool. The exit of LPC is designated by station 2.5 and the exit of the HPC is station 3. The HPC is designed to operate at a higher shaft rotational speed than the LPC spool. The compressed gas enters the main or primary burner at 3 and is combusted with the fuel to produce hot high-pressure gas at 4 to enter the high-pressure turbine (HPT). Flow expansion through the HPT and the low-pressure turbine (LPT) produces the shaft power for the HPC and LPC, respectively. An afterburner is designated between stations 5 and 7 where an additional fuel is combusted with the turbine discharge flow before it expands in the exhaust nozzle. The station 8 is at the throat of the nozzle and station 9 designates the nozzle exit.

To derive an expression for the engine thrust, it is most convenient to describe a control volume surrounding the engine and apply momentum principles to the fluid flow crossing the

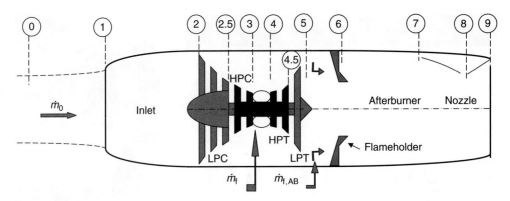

■ **FIGURE 3.1** Station numbers for an afterburning turbojet engine

boundaries of the control volume. Let us first consider an airbreathing engine. From a variety of choices that we have in describing the control volume, we may choose one that shares the same exit plane as the engine nozzle, and its inlet is far removed from the engine inlet so not to be disturbed by the nacelle lip. These choices are made for convenience. As for the sides of the control volume, we may choose either stream surfaces, with the advantage of no flow crossing the sides, or a constant-area box, which has a simple geometry but fluid flow crosses the sides. Regardless of our choice of the control volume, however, the physical expression derived for the engine thrust has to produce the same force in either method. Figure 3.2 depicts a control volume in the shape of a box around the engine.

The pylon by necessity is cut by the control volume, which is enclosing the engine. The thrust force F and its reaction are shown in Fig. 3.2. We may assume the sides of the control volume are not affected by the flowfield around the nacelle, i.e., the static pressure distribution on the sides is nearly the same as the ambient static pressure p_0. In the same spirit, we may assume that the exit plane also sees an ambient pressure of p_0 with an exception of the plane of the jet exhaust, where a static pressure p_9 may prevail. To help us balance momentum flux and net forces acting on the fluid crossing the control surface, we show the external pressure distributions as in Fig. 3.3.

From our study of fluid mechanics, we know that the closed surface integral of a constant pressure results in a zero net force acting on the closed surface, i.e.,

$$\oiint - p\hat{n}ds = 0 \quad \text{(for constant } p\text{)} \tag{3.1}$$

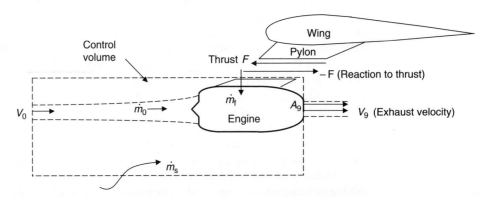

■ **FIGURE 3.2** Schematic drawing of an airbreathing engine with a box-like control volume positioned around the engine (Note the flow of air to the engine, through the sides, and the fuel flow rate)

■ FIGURE 3.3
Simplified model of static pressure distribution on the surface of the control volume

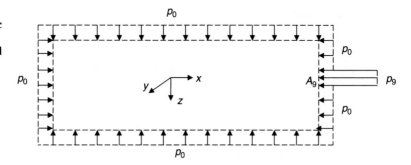

where $\hat{n}$ is the unit normal vector pointing out of the control surface, ds is the elemental surface area, and the negative sign shows that the direction of the pressure is in the inward normal direction to the closed surface. The equivalence of the closed surface integral and the volume integral via the gradient theorem may be used to prove the above principle, namely;

$$\oiint p\hat{n}\,ds = \iiint \nabla p\,dv \tag{3.2}$$

where dv is the element of volume and ∇p is the pressure gradient vector. Note that $\nabla p \equiv 0$ for a constant p. Now, let us subtract the constant ambient pressure p_0 from all sides of the closed surface of the control volume shown in Fig. 3.4, to get a net pressure force acting on the control surface.

As we are interested in deriving an expression for a force, namely thrust, we have to use the momentum conservation principles, namely, we need to balance the fluid momentum in the x-direction and the resultant external forces *acting on the fluid* in the x-direction. Before doing that, we must establish the fluid flow rates in and out of the control volume. This is achieved by applying the law of conservation of mass to the control volume. In its simplest form, i.e., steady flow case, we have

$$\sum \dot{m}_{\text{in}} = \sum \dot{m}_{\text{out}} \tag{3.3}$$

Applying Eq. 3.3 to the control volume surrounding our engine (Fig. 3.5), gives the following bookkeeping expression on the mass flow rates in and out of the box, namely,

$$\rho_0 V_0 A + \dot{m}_{\text{s}} + \dot{m}_{\text{f}} = (\dot{m}_0 + \dot{m}_{\text{f}}) + \rho_0 V_0 (A - A_9) \tag{3.4}$$

■ FIGURE 3.4
Simplified model of the static pressure force acting on the control surface

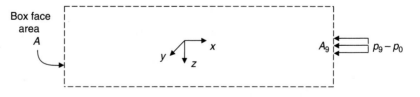

■ FIGURE 3.5
Control volume and the mass flow rates in and out of its boundaries

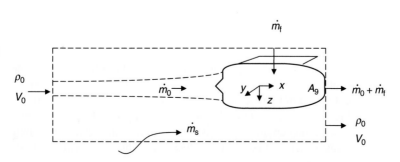

which simplifies to:

$$\dot{m}_s = \dot{m}_0 - \rho_0 V_0 A_9 \tag{3.5}$$

Now, we are ready to apply the momentum balance to the fluid entering and leaving our control volume. Again assuming steady, uniform flow, we may use the simplest form of the momentum equation, i.e.,

$$\Sigma(\dot{m}V_x)_{\text{out}} - \Sigma(\dot{m}V_x)_{\text{in}} = \Sigma(F_x)_{\text{fluid}} \tag{3.6}$$

which states that the difference between the fluid (time rate of change of) momentum out of the box and into the box is equal to the net forces acting on the fluid in the x-direction on the boundaries and within the box. Now, let us spell out each of the contributions to the momentum balance of Eq. 3.6, i.e.,

$$\Sigma(\dot{m}V_x)_{\text{out}} = (\dot{m}_0 + \dot{m}_f)V_9 + [\rho_0 V_0(A - A_9)]V_0 \tag{3.7}$$

$$\Sigma(\dot{m}V_x)_{\text{in}} = (\rho_0 V_0 A)V_0 + \dot{m}_s V_0 \tag{3.8}$$

$$\Sigma(F_x)_{\text{fluid}} = (-F)_{\text{fluid}} - (p_9 - p_0)A_9 \tag{3.9}$$

Substituting Eqs. 3.7, 3.8, and 3.9 into 3.6 and for $\dot{m}_s$ from Eq. 3.5, we get

$$\begin{aligned}(\dot{m}_0 + \dot{m}_f)V_9 &+ \rho_0 V_0(A - A_9)V_0 - \rho_0 V_0 A V_0 - (\dot{m}_0 - \rho_0 V_0 A_9)V_0 \\ &= (-F_x)_{\text{fluid}} - (p_9 - p_0)A_9\end{aligned} \tag{3.10}$$

which simplifies to

$$(\dot{m}_0 + \dot{m}_f)V_9 - \dot{m}_0 V_0 = (-F_x)_{\text{fluid}} - (p_9 - p_0)A_9 \tag{3.11}$$

Now, through the action–reaction principle of the Newtonian mechanics, we know that an equal and opposite axial force is exerted on the engine, by the fluid, i.e.,

$$-F_x)_{\text{fluid}} = F_x)_{\text{pylon}} = F_x)_{\text{engine}} \tag{3.12}$$

Therefore, calling the axial force of the engine "thrust," or simply F, we get the following expression for the engine thrust

$$F = (\dot{m}_0 + \dot{m}_f)V_9 - \dot{m}_0 V_0 + (p_9 - p_0)A_9 \tag{3.13}$$

This expression for the thrust is referred to as the "net uninstalled thrust" and sometimes a subscript "n" is placed on F to signify the "net" thrust. Therefore, the thrust expression of Eq. 3.13 is better written as

$$\boxed{F_n)_{\text{uninstalled}} = (\dot{m}_0 + \dot{m}_f)V_9 - \dot{m}_0 V_0 + (p_9 - p_0)A_9} \tag{3.14}$$

Now, we attribute a physical meaning to the three terms on the RHS of Eq. 3.14, which contribute to the engine net (uninstalled) thrust. We first note that the RHS of Eq. 3.14 is composed of two momentum terms and one pressure–area term. The first momentum term is the exhaust momentum through the nozzle contributing positively to the engine thrust. The second momentum term is the inlet momentum, which contributes negatively to the engine thrust in effect it represents a drag term. This drag term is called "ram drag," or simply the penalty of bringing air in the engine with a finite momentum. It is often given the symbol of D_{ram} and expressed as

$$D_{ram} = \dot{m}_0 V_0 \qquad (3.15)$$

The last term in Eq. 3.14 is a pressure–area term, which acts over the nozzle exit plane, i.e., area A_9, and will contribute to the engine thrust only if there is an imbalance of static pressure between the ambient and the exhaust jet. As we remember from our aerodynamic studies, a nozzle with a subsonic jet will always expand the gasses to the same static pressure as the ambient condition, and a sonic or supersonic exhaust jet may or may not have the same static pressure in their exit plane as the ambient static pressure. Depending on the "mismatch" of the static pressures, we categorized the nozzle flow as

> if $p_9 < p_0$, the nozzle is *overexpanded*
> which can happen in supersonic jets only (i.e., in convergent–divergent nozzles with area ratio *larger* than needed for perfect expansion)

> if $p_9 = p_0$, the nozzle is *perfectly expanded*
> which is the case for *all subsonic jets* and sometimes in sonic or supersonic jets (i.e., with the "right" nozzle area ratio)

> if $p_9 > p_0$, the nozzle is *underexpanded*
> which can happen in sonic or supersonic jets only (i.e., with inadequate nozzle area ratio)

EXAMPLE 3.1

An airbreathing engine has a flight Mach number of $M_0 = 0.85$ at an altitude where the speed of sound is 300 m/s. The air mass flow rate into the engine is 50 kg/s. Calculate the ram drag (in kN) for this engine.

SOLUTION

The flight speed is $V_0 = M_0 \cdot a_0 = (0.85)(300 \, \text{m/s}) = 255 \, \text{m/s}$

The ram drag is given in Eq. 3.15, therefore,

$$D_{ram} = (50 \, \text{kg/s})(255 \, \text{m/s}) = 12.75 \, \text{kN}$$

Examining the various contributions from Eq. 3.14 to the net engine (uninstalled) thrust, we note that the thrust is the difference between the nozzle contributions (both momentum and pressure-area terms) and the inlet contribution (the momentum term). The nozzle contribution to thrust is called *gross thrust* and is given a symbol F_g, i.e.,

$$F_g \equiv (\dot{m}_0 + \dot{m}_f)V_9 + (p_9 - p_0)A_9 \qquad (3.16)$$

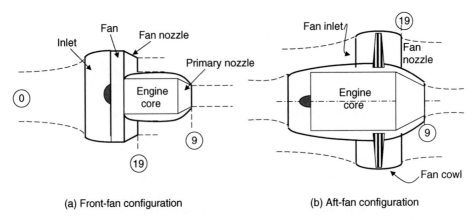

(a) Front-fan configuration   (b) Aft-fan configuration

■ **FIGURE 3.6**   Schematic drawing of a turbofan engine with separate exhausts

and as explained earlier, the inlet contribution was negative and was called "ram drag" D_{ram}, therefore

$$F_n)_{uninstalled} = F_g - D_{ram} \qquad (3.17)$$

Now, we can generalize the result expressed in Eq. 3.17 to aircraft engines with more than a single stream, as, for example, the turbofan engine with separate exhausts. This task is very simple as we account for all gross thrusts produced by all the exhaust nozzles and subtract all the ram drag produced by all the air inlets to arrive at the engine uninstalled thrust, i.e.,

$$F_n)_{uninstalled} = \Sigma(F_g)_{nozzles} - \Sigma(D_{ram})_{inlets} \qquad (3.18)$$

Another way of looking at this is to balance the momentum of the exhaust stream and the inlet momentum with the pressure thrust at the nozzle exit planes and the net uninstalled thrust of the engine. In a turbofan engine, the captured airflow is typically divided into a "core" flow, where the combustion takes place, and a fan flow, where the so-called "bypass" stream of air is compressed through a fan and later expelled through a fan exhaust nozzle. This type of arrangement, i.e., bypass configuration, leads to a higher overall efficiency of the engine and lower fuel consumption. A schematic drawing of the engine is shown in Fig. 3.6.

In this example, the inlet consists of a single (or dual) stream, and the exhaust streams are split into a primary and a fan nozzle. We may readily write the uninstalled thrust produced by the engine following the momentum principle, namely,

$$F_n)_{uninstalled} = \dot{m}_9 V_9 + \dot{m}_{19} V_{19} + (p_9 - p_0)A_9 + (p_{19} - p_0)A_{19} - \dot{m}_0 V_0 \qquad (3.19)$$

The first four terms account for the momentum and pressure thrusts of the two nozzles (what is known as gross thrust) and the last term represents the inlet ram drag.

EXAMPLE 3.2

A separate flow turbofan engine develops exhaust velocities of 450 m/s and 350 m/s in the core and fan nozzles, respectively. The nozzles are perfectly expanded. The mass flow rate through the core nozzle is 50 kg/s and through the fan nozzle is 350 kg/s. Calculate the gross thrust of the core and the fan nozzles.

SOLUTION

Since the core and fan nozzles are both perfectly expanded, the gross thrust is simply the momentum thrust at the nozzle exit, i.e.,

$$F_{g,core} = \dot{m}_9 V_9 = (50\,kg/s)(450\,m/s) = 22.5\,kN$$

$$F_{g,fan} = \dot{m}_{19} V_{19} = (350\,kg/s)(350\,m/s) = 122.5\,kN$$

3.1.1   Takeoff Thrust

At takeoff, the air speed V_0 ("flight" speed) is often ignored in the thrust calculation, therefore the ram drag contribution to engine thrust is neglected, i.e.,

$$F_{takeoff} \approx F_g = (\dot{m}_0 + \dot{m}_f)V_9 + (p_9 - p_0)A_9 \qquad (3.20)$$

For a perfectly expanded nozzle, the pressure thrust term vanishes to give

$$F_{takeoff} \approx (\dot{m}_0 + \dot{m}_f)V_9 \approx \dot{m}_0 V_9 \qquad (3.21)$$

Therefore the takeoff thrust is proportional to the captured airflow.

3.2   Installed Thrust—Some Bookkeeping Issues on Thrust and Drag

As indicated by the description of the terms "installed thrust" and "uninstalled thrust," they refer to the *actual propulsive force* transmitted to the aircraft by the engine and the thrust produced by the engine if it had zero external losses, respectively. Therefore, for the installed thrust, we need to account for the installation losses to the thrust such as the nacelle skin friction and pressure drags that are to be included in the propulsion side of the drag bookkeeping. On the contrary, the pylon and the engine installation that affects the wing aerodynamics (in podded nacelle, wing-mounted configurations), namely by altering its "clean" drag polar characteristics, causes an "interference" drag that is accounted for in the aircraft drag polar. In the study of propulsion, often times, we concentrate on the engine "internal" performance, i.e., the uninstalled characteristics, rather than the installed performance because the external drag of the engine installation depends not only on the engine nacelle geometry but also on the engine–airframe integration. Therefore accurate installation drag accounting will require CFD analysis and wind tunnel testing at various flight Mach numbers and engine *throttle settings*. In its simplest form, we can relate the installed and uninstalled thrust according to

$$F_{installed} = F_{uninstalled} - D_{nacelle} \qquad (3.22)$$

In our choice of the control volume as depicted in Fig. 3.2, we made certain assumptions about the exit boundary condition imposed on the aft surface of the control volume. We made assumptions about the pressure boundary condition as well as the velocity boundary condition. About the pressure boundary condition, we stipulated that the static pressure of flight p_0 imposed on the exit plane, except at the nozzle exit area of A_9, therefore we allowed for an under- or overexpanded nozzle. With regard to the velocity boundary condition at the exit plane, we stipulated that the flight velocity V_0 is prevailed, except at the nozzle exit where the jet velocity of V_9 prevails. In reality, the aft surface of the control volume is by necessity downstream of the nacelle and pylon, and therefore it is in the middle of the wake generated by the nacelle and the pylon. This implies that there would be a momentum deficit in the wake and

■ **FIGURE 3.7**
Velocity profile in the
nacelle and pylon wake
showing a momentum
deficit region

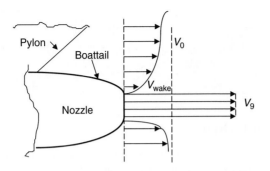

the static pressure nearly equals the free stream pressure. The velocity profile in the exit plane, i.e., the wake of the nacelle and pylon, is more likely to be represented by the following schematic drawing (Fig. 3.7).

However, the force transmitted through the pylon to the aircraft is not the "uninstalled" thrust, as we called it in our earlier derivation, rather the *installed* thrust and pylon drag. But the integral of momentum deficit and the pressure imbalance in the wake is exactly equal to the nacelle and pylon drag contributions, i.e., we have

$$D_{\text{nacelle}} + D_{\text{pylon}} = \iint \rho V(V_0 - V)dA + \iint (p_0 - p)dA \qquad \textbf{(3.23)}$$

where the surface integral is taken over the exit plane downstream of the nacelle and pylon.

Now, let us go back and correct for the wake contributions noted in Eq. 3.21 as well as the *actual* force transmitted through the pylon, namely,

$$F_{\text{uninstalled}} - D_{\text{nacelle}} - D_{\text{pylon}} = (\dot{m}_0 + \dot{m}_f)V_9 - \dot{m}_0 V_0 + (p_9 - p_0)A_9$$
$$- \iint \rho V(V_0 - V)dA - \iint (p_0 - p)dA \qquad \textbf{(3.24)}$$

After canceling the integrals on the RHS with the drag terms of the LHS in Eq. 3.24, we recover our Eq. 3.14 that we derived earlier for the *uninstalled thrust*, i.e.

$$F_{\text{n}})_{\text{uninstalled}} = (\dot{m}_0 + \dot{m}_f)V_9 - \dot{m}_0 V_0 + (p_9 - p_0)A_9$$

As noted earlier, we are not limited to the choice of the control volume that we made in the form of a box wrapped around the engine and cutting through the pylon (see Fig. 3.2). Let us examine other logical choices that we could have made. For example, we could use the inlet-captured streamtube as the upstream portion of the control volume and allow the nacelle external surface to serve as the remaining portion of the control volume, and then truncate the control volume at the nozzle exit plane. This choice of control volume is schematically shown in Fig. 3.8.

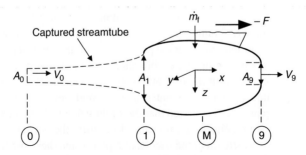

■ **FIGURE 3.8**   Schematic drawing of an alternative control volume

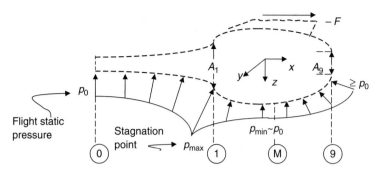

■ **FIGURE 3.9**   Sketch of the static pressure distribution on the sides of the control surface, including the capture streamtube and the nacelle

The choice of the control volume depicted in Fig. 3.8 offers the advantage of no airflow crossing the sides of the control surface. There is of course the fuel flow rate through the pylon and the airflow rates through the inlet-captured streamtube (station 0) and the nozzle exit plane (station 9), as expected. The penalty of using this control volume is in the more complicated force balance terms, which appear in the momentum equation. The pressure distribution on the captured streamtube, for example, will correspond to a diffusing stream pattern, from station 0 to 1, therefore, the static pressure increases along the captured streamtube. Then, the static pressure is decreased to the point of maximum nacelle diameter (station labeled M) as the flow accelerates from the inlet lip stagnation point. From the position of minimum static pressure at M, the flow is diffused on the aft end of the nacelle and will recover most of the static pressure and almost reach the flight ambient pressure p_0. The static pressure distribution on the control surface is schematically shown in Fig. 3.9.

Following Newton's second law of motion, we stipulate that the change of momentum of fluid crossing the boundaries of the control volume (Fig. 3.8) in the x-direction, say, is balanced by the net forces acting on the fluid in the x-direction. Therefore,

$$(\dot{m}_0 + \dot{m}_f)V_9 - \dot{m}_0V_0 = \Sigma(F_{x-\text{external}})_{\text{control-surface}} \tag{3.25}$$

where the LHS is the net gain in x-momentum, which is created by the net forces in the positive x-direction acting on the fluid, i.e., the RHS terms of Eq. 3.25.

The external forces are due to pressure and frictional stresses acting on the control surface in the x-direction. In the x-balance of momentum, the reader notes that the gravitational force does not enter the calculation at any case and in general we do not include the effect of gravitational force, as it represents a negligible contribution to the force balance involving aircraft jet propulsion. Now, let us breakdown the elements of external force, which act on the control surface. First, the pressure integral over the captured streamtube acting on the fluid in the x-direction can be written

$$(F_x)_{\text{pressure},0-1} = \iint(p - p_0)dA_n \tag{3.26}$$

where the element of area dA_n is the change of the area of the captured streamtube normal to the x-direction, i.e., varying from A_0 at the inlet to A_1 at the exit, therefore experiencing $(A_1 - A_0)$ change over the length of the captured streamtube. We can use the following graph (Fig. 3.10) as an aid in demonstrating what is meant by dA_n.

On the captured streamtube depicted in Fig. 3.8 (or 3.9), the local pressure p is greater than the ambient pressure p_0, and dA_n is also positive, as the captured streamtube widens $(dA > 0)$, hence the pressure integral of Eq. 3.26 is positive. This means that the flow deceleration outside the inlet will contribute a positive force in the x-direction on the control surface, which represents a drag for the engine installation. This drag term is called *pre-entry drag* or *additive drag*. It is

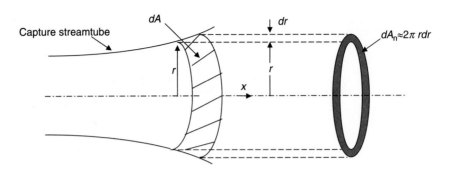

■ **FIGURE 3.10** Definition sketch for dA and its projection normal to the x-axis dA_n

interesting to note that if the captured streamtube were inverted, i.e., it caused external flow acceleration instead of deceleration, a typical scenario for takeoff and climb or the engine static testing, then the local static pressure would drop below ambient pressure, which causes the integrand $(p - p_0)$ to be negative. Despite this, we still encounter a pre-entry drag, as we note that dA_n is also negative for a shrinking streamtube, hence the pre-entry momentum contribution remains as a penalty, i.e., a drag term. This is schematically shown in Fig. 3.11.

The axial force due to static pressure distribution over the cowl lip from station 1 to M (Fig. 3.9) can be written as the integral:

$$(F_x)_{\text{pressure},1-M} = \iint (p - p_0)dA_n \tag{3.27}$$

where the element of area is again represented by the change of area normal to the x-direction. The integral of pressure over the inlet cowl (i.e., from 1 to M) represents a thrust contribution, as the wall static pressure on the forward region of the cowl is below the ambient pressure, i.e., $p - p_0 < 0$, and dA_n is positive, therefore the pressure integral of Eq. 3.27 is negative. A negative force in the x-direction represents a thrust term. This force may be thought of as the suction force on a *circular wing* (to represent the inlet cowl lip) projected in the x-direction. To demonstrate this point, we examine the following schematic drawing (Fig. 3.12).

For an axisymmetric inlet, the aerodynamic side force created on the nacelle integrates to zero, but as most engine nacelles are asymmetrical for either ground clearance or accessibility of the engine accessories reasons, a net side force is created on the nacelle, as well.

The inlet additive drag is thus for the most part balanced by the cowl lip thrust and the difference is called the *spillage drag*, i.e.,

$$D_{\text{spillage}} = \iint_{0-1} (p - p_0)dA_n - \iint_{1-M} (p - p_0)dA_n \tag{3.28}$$

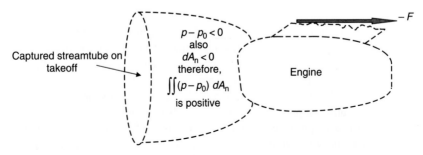

■ **FIGURE 3.11** Captured streamtube at takeoff

■ **FIGURE 3.12**
Flow detail near a blunt cowl lip showing a lip thrust component and a side force

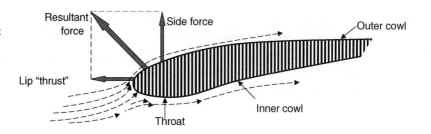

In general, for well-rounded cowl lips of subsonic inlets, the spillage drag is rather small and it becomes significant for only supersonic, sharp-lipped inlets. The nacelle frictional drag also contributes to the external force as well as the aft-end pressure drag of the boat tail, which may be written as

$$\sum(F_{x\text{-external}})_{\text{control-surface}} = D_{\text{spillage}} + \iint_{M-9}(p - p_0)dA_n + \iint_{1-9}\tau_w\,dA_x + D_{\text{pylon}} - F_{x,\text{fluid}}$$

$$\text{(I)} \qquad\qquad \text{(II)} \qquad\qquad \text{(III)} \qquad \text{(IV)} \qquad \text{(V)}$$

$$- (p_9 - p_0)A_9$$

$$\text{(VI)} \tag{3.29}$$

The first term on the RHS, i.e., (I), is the spillage drag that we discussed earlier, the second term (II) is the pressure drag on the nacelle aft end or boat tail pressure drag, the third term (III) is the nacelle viscous drag, the fourth term (IV) is the pylon drag, the fifth term (V) is the reaction to the installed thrust force and pylon drag acting on the fluid, and the last term (VI) is the pressure thrust due to imperfect nozzle expansion. To summarize

- Terms I + II + III are propulsion system installation drag losses
- Term IV is accounted for in the aircraft drag polar
- Terms V and VI combine with I, II, III, and IV to produce the uninstalled thrust

Therefore, we recover the expression for the uninstalled thrust that we had derived using another control volume for an airbreathing engine, namely

$$F_n)_{\text{uninstalled}} = (\dot{m}_0 + \dot{m}_f)V_9 - \dot{m}_0V_0 + (p_9 - p_0)A_9$$

Figure 3.13 shows the elements of force that contribute to installed net thrust.

■ **FIGURE 3.13**
Definition of *installed net thrust* in terms of internal and external parameters of an engine nacelle (adopted from Lotter, 1977)

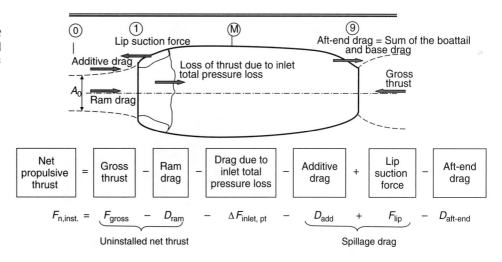

3.3   Engine Thrust Based on the Sum of Component Impulse

We have derived an expression for a component internal force based on the fluid impulse at its inlet and exit in Chapter 2. We wish to demonstrate the relationship between the sum of the component forces and the engine thrust.

The axial component of force felt by the inner walls of the diffuser (from 1 to 2), see Fig. 3.14, is

$$F_{x,\text{diffuser}} = I_1 - I_2 \tag{3.30}$$

The axial component force felt by the inner walls of the compressor (from 2 to 3) is

$$F_{x,\text{compressor}} = I_2 - I_3 \tag{3.31}$$

The axial component force felt by the inner walls of the combustor (from 3 to 4) is

$$F_{x,\text{burner}} = I_3 - I_4 \tag{3.32}$$

The axial component force felt by the inner walls of the turbine (from 4 to 5) is

$$F_{x,\text{turbine}} = I_4 - I_5 \tag{3.33}$$

The axial component force felt by the inner walls of the nozzle (from 5 to 9) is

$$F_{x,\text{nozzle}} = I_5 - I_9 \tag{3.34}$$

Therefore, the total internal axial force acting on the inner walls of the engine is

$$F_{x,\text{engine}} = F_{x,\text{diffuser}} + F_{x,\text{compressor}} + F_{x,\text{burner}} + F_{x,\text{turbine}} + F_{x,\text{nozzle}}$$
$$= I_1 - I_2 + I_2 - I_3 + I_3 - I_4 + I_4 - I_5 + I_5 - I_9 = I_1 - I_9$$
$$= \dot{m}_1 V_1 + p_1 A_1 - (\dot{m}_9 V_9 + p_9 A_9) \tag{3.35}$$

The thrust acts in the $-x$-direction, therefore the engine internal thrust is (minus the above)

$$\text{Internal} - \text{Thrust} = \dot{m}_9 V_9 - \dot{m}_1 V_1 + p_9 A_9 - p_1 A_1 \tag{3.36}$$

However, in flight we neither know V_1 nor p_1. To connect the inlet flow to the flight condition, we examine the captured streamtube, as shown in Fig. 3.15.

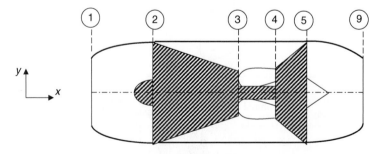

■ FIGURE 3.14   Definition sketch used in engine thrust calculation based on component impulse

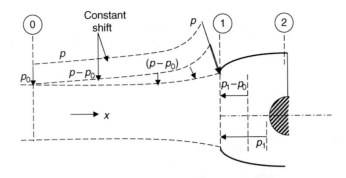

■ **FIGURE 3.15**   **Pressure distribution on the captured streamtube**

We have subtracted a constant pressure p_0 from all sides, as shown in Fig. 3.15. The momentum balance in the x-direction gives

$$\dot{m}_1 V_1 - \dot{m}_0 V_0 = \int_0^1 (p - p_0)dA_n - (p_1 - p_0)A_1 \tag{3.37}$$

The integral on the right-hand side is called the *additive drag* D_{add}, therefore,

$$\dot{m}_1 V_1 + p_1 A_1 = \dot{m}_0 V_0 + p_0 A_1 + D_{add} \tag{3.38}$$

$$\text{Internal} - \text{Thrust} = \dot{m}_9 V_9 - \dot{m}_0 V_0 + p_9 A_9 - p_0 A_1 + D_{add} \tag{3.39}$$

The integral of constant pressure distribution p_0 around the nacelle, as in Fig. 3.16, leads to

$$\oiint p_0 \, dA_n = p_0 A_1 - p_0 A_9 + \int_{sides} p_0 \, dA_n = 0 \tag{3.40}$$

Therefore, we may replace $p_0 A_1$ by $p_0 A_9$ and the integral on the sides to get

$$\text{Internal} - \text{Thrust} = \dot{m}_9 V_9 - \dot{m}_0 V_0 + p_9 A_9 - p_0 A_9 + D_{add} - \int_{sides} p_0 dA_n \tag{3.41}$$

The external forces acting on the nacelle (in the x-direction) arise from pressure and shear integrals, namely

$$F_{x-nacelle} = \int_{sides} p dA_n + F_{x-viscous} \tag{3.42}$$

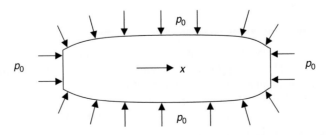

■ **FIGURE 3.16**   **Nacelle under a constant pressure distribution**

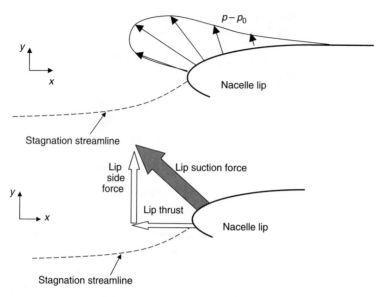

The integral of pressure on the nacelle lip produces a lip thrust force, which almost cancels the inlet additive drag, see Fig. 3.17, i.e.,

$$\int_{\text{nacelle-lip}} (p - p_0)dA_{\text{n}} = \text{lip} - \text{thrust} \approx -D_{\text{add}} \tag{3.43}$$

Therefore the *net installed thrust*, after canceling inlet additive drag with the lip suction force, is

$$\text{Installed} - \text{Thrust} = \underbrace{\dot{m}_9 V_9 - \dot{m}_0 V_0 + p_9 A_9 - p_0 A_9}_{\text{Uninstalled thrust}} - \underbrace{\int_{\text{sides-lip}} (p - p_0)dA_n - F_{x-\text{viscous}}}_{\text{Installation drag}} \tag{3.44}$$

The uninstalled thrust thus connects the free stream to the nozzle exit, after we account for the additive drag and the nacelle lip suction force to cancel each other. The leftover external pressure and shear integral are then attributed to installation drag.

$$\textit{Uninstalled thrust} = \dot{m}_9 V_9 - \dot{m}_0 V_0 + (p_9 - p_0)A_9 \tag{3.45}$$

3.4   Rocket Thrust

Although our primary effort in this book is to study airbreathing propulsion, the subject of rocket propulsion has been presented in Chapter 11. For now, we may think of the rocket thrust as a special case of an airbreathing engine with a "sealed inlet"! It is simply too tempting to ignore this special case of jet propulsion with no inlet penalty. Figure 3.18 shows a schematic drawing of a rocket engine where all the propellant, i.e., fuel and oxidizer, is stored in specialized tanks onboard the rocket.

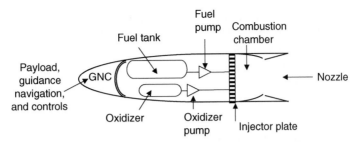

■ FIGURE 3.18   Schematic drawing of liquid propellant rocket engine internal components

The thrust produced by a rocket engine amounts to only the *gross thrust* produced by an airbreathing engine. We remember that the gross thrust was the *nozzle* contribution to thrust, which a rocket also possesses to accelerate the combustion gas. The air intake system is missing in a rocket, and therefore the inlet ram drag contribution is also absent from the rocket thrust equation. Consequently, we can express the rocket thrust as

$$F_{\text{rocket}} = (\dot{m}_{\text{ox}} + \dot{m}_{\text{f}})V_9 + (p_9 - p_0)A_9 \tag{3.46}$$

It is interesting to note that the rocket thrust is independent of flight speed V_0, as Eq. 3.46 does not explicitly involve V_0. However, if the flight trajectory has a vertical component to it, then the ambient pressure p_0 changes with altitude, and therefore the pressure thrust term of Eq. 3.46 will then affect the thrust magnitude. For example, space shuttle main engine produces about 25% more thrust in "vacuum" than at sea level, mainly due to the pressure thrust difference between sea level p_0 of 14.7 psia (or 1 atm) and p_0 of zero corresponding to "vacuum" in near earth orbit.

Again, for the net propulsive force acting on the rocket, we need to account for the vehicle external drag. Although, the Eq. 3.46 is not explicitly referred to as "uninstalled" thrust in rocket propulsion literature, the reader should view it as the *internal performance* of a rocket engine. The external aerodynamic analysis will produce control forces and vehicle external drag contributions.

EXAMPLE 3.3

A liquid propellant rocket engine consumes 200 kg/s of oxidizer and 50 kg/s of fuel. After the combustion, the gas is accelerated in a convergent–divergent nozzle where it attains $V_9 = 4000$ m/s and $p_9 = 200$ kPa. The rocket engine is at sea level where $p_0 = 100$ kPa. If the exit diameter of the nozzle is 2 m, calculate

(a) the pressure thrust

(b) the rocket gross thrust

SOLUTION

The pressure thrust is $(p_e - p_0)A_e$. The nozzle exit area is

$$A_9 = \pi D_e^2/4 \approx 3.1416\,\text{m}^2$$

The pressure thrust is

$(p_9 - p_0)A_9 = (200 - 100)\text{kPa} \cdot (3.1416\,\text{m}^2) \approx 314.16\,\text{kN}$

The momentum thrust is $\dot{m}_9 V_9 = (250 \text{ kg/s})(4000 \text{ m/s}) = 1$ MN

$$F_{\text{g}} = 1000\,\text{kN} + 314.16\,\text{kN} \approx 1314\,\text{kN}$$

3.5   Airbreathing Engine Performance Parameters

The engine thrust, mass flow rates of air and fuel, the rate of kinetic energy production across the engine or the mechanical power/shaft output, and engine dry weight, among other parameters, are combined to form a series of important performance parameters, known as the propulsion system figures of merit.

3.5.1   Specific Thrust

The size of the air intake system is a design parameter that establishes the flow rate of air, $\dot{m}_0$. Accordingly, the fuel pump is responsible for setting the fuel flow rate in the engine, $\dot{m}_f$. Therefore, in producing thrust in a "macroengine," the engine size seems to be a "scaleable" parameter. The only exception in scaling the jet engines is the "microengines" where the component losses do not scale. In general, the magnitude of the thrust produced is directly proportional to the mass flow rates of the fluid flow through the engine. Then, it is logical to study thrust per unit mass flow rate as a figure of merit of a candidate propulsion system. In case of an airbreathing engine, the ratio of thrust to air mass flow rate is called *specific thrust* and is considered to be an engine performance parameter, i.e.,

---

Airbreathing engine performance parameter #1:

$\dfrac{F}{\dot{m}_0}$ "specific thrust" with a metric unit of $[\text{N} \cdot \text{s/kg}]$ and a British unit of $[\text{lbf} \cdot \text{s/lbm}]$

The target for this parameter, i.e., specific thrust, in a cycle analysis is usually to be maximized, i.e., to produce thrust with the least quantity of air flow rate, or equivalently to produce thrust with a minimum of engine frontal area. However, with subsonic cruise Mach numbers, the drag penalty for engine frontal area is far less severe than their counterparts in supersonic flight. Consequently, the specific thrust as a figure of merit in a commercial transport aircraft (e.g., Boeing 777 or Airbus A-340) takes a back seat to the lower fuel consumption achieved in a very large bypass ratio turbofan engine at subsonic speeds. As noted, specific thrust is a dimensional quantity with the unit of force per unit mass flow rate. A nondimensional form of the specific thrust, which is useful for graphing purposes and engine comparisons, is (following Kerrebrock, 1992)

$$\text{Nondimensional specific thrust} \equiv \frac{F}{\dot{m}_0 a_0} \tag{3.47}$$

where a_0 is the ambient speed of sound taken as the reference velocity.

3.5.2   Specific Fuel Consumption and Specific Impulse

The ability to produce thrust with a minimum of fuel expenditure is another parameter, which is considered to be a performance parameter in an engine. In the commercial world, e.g., the airline business, specific fuel consumption represents perhaps the most important parameter of the engine. After all, the money spent on fuel is a major expenditure in operating an airline, for example. However, the reader is quickly reminded of the unspoken parameters of *reliability and maintainability* that have a direct impact on the cost of operating commercial engines and therefore they are at least as important, if not more important, as the engine-specific fuel consumption. In the military world, the engine fuel consumption parameter takes a decidedly second role to other aircraft performance parameters, such as stealth, agility, maneuverability, and survivability. For an airbreathing engine, the ratio of fuel flow rate per unit thrust force produced is called thrust-specific fuel consumption (TSFC), or sometimes just the specific fuel consumption, i.e., sfc, and is defined as

> Airbreathing engine performance parameter #2:
> $$\frac{\dot{m}_f}{F} \text{ ``thrust-specific fuel consumption''}$$
> (TSFC) with a metric unit of $[\text{mg/s/N}]$ and a British unit of: $[\text{lbm/h/lbf}]$

The target for this parameter, i.e., TSFC, in a cycle analysis is to be minimized, i.e., to produce thrust with a minimum of fuel expenditure. This parameter, too, is dimensional. For a rocket, on the contrary, the oxidizer as well as the fuel both contribute to the "expenditure" in the engine to produce thrust, and as such the oxidizer flow rate $\dot{m}_{ox}$ needs to be accounted for as well. The word "propellant" is used to reflect the combination of oxidizer and fuel in a liquid propellant rocket engine or a solid propellant rocket motor. It is customary to define a corresponding performance parameter in a rocket as thrust per unit propellant weight flow rate. This parameter is called *specific impulse* I_s, i.e.,

$$I_s \equiv \frac{F}{\dot{m}_p g_0} \tag{3.48}$$

where

$$\dot{m}_p \equiv \dot{m}_f + \dot{m}_{ox} \tag{3.49}$$

and g_0 is the gravitational acceleration on the surface of earth, i.e. 9.8 m/s^2 or 32.2 ft/s^2. The dimension of $\dot{m}_p g_0$ in denominator of Eq. 3.48 is then the *weight flow rate* of the propellant based on earth's gravity, or force per unit time. Consequently, the dimension of specific impulse is "Force/Force/second" which simplifies to just the "second." All propulsors, rockets, and airbreathers, then could be compared using a unifying figure of merit, namely their specific impulse in seconds. An added benefit of the specific impulse is that regardless of the units of measurement used in the analysis, i.e., either metric or the British on both sides of the Atlantic, specific impulse comes out as seconds in both systems. The use of specific impulse as a unifying figure of merit is further justified in the twenty-first century as we attempt to commercialize the space with potentially reusable rocket-based combined cycle (RBCC) power plants in propelling a variety of single-stage-to-orbit (SSTO) vehicles. To study advanced propulsion concepts where the mission calls for multimode propulsion units, as in air-ducted rockets, ramjets, and scramjets, all combined into a single "package," the use of specific impulse becomes even more obvious. In summary,

> Specific impulse: $I_s \equiv \dfrac{F}{\dot{m}_f g_0} [\text{s}]$ for an airbreathing engine
>
> Specific impulse: $I_s \equiv \dfrac{F}{\dot{m}_p g_0} [\text{s}]$ for a rocket

An important goal in an engine cycle design is to maximize this parameter, i.e., the ability to produce thrust with the least amount of fuel or propellant consumption in the engine.

3.5.3 Thermal Efficiency

The ability of an engine to convert the thermal energy inherent in the fuel (which is unleashed in a chemical reaction) to a net kinetic energy gain of the working medium is called the engine thermal efficiency, η_{th}. Symbolically, thermal efficiency is expressed as

$$\eta_{th} \equiv \frac{\Delta K\dot{E}}{\wp_{thermal}} = \frac{\dot{m}_9 \dfrac{V_9^2}{2} - \dot{m}_0 \dfrac{V_0^2}{2}}{\dot{m}_f Q_R} = \frac{(\dot{m}_0 + \dot{m}_f)V_9^2 - \dot{m}_0 V_0^2}{2\dot{m}_f Q_R} \tag{3.50}$$

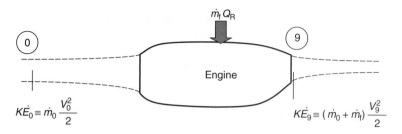

■ **FIGURE 3.19**   Thermal power input and the mechanical power production (output)

where $\dot{m}$ signifies mass flow rate corresponding to stations 0 and 9 and subscript 'f' signifies fuel, and Q_R is the fuel heating value. The unit of Q_R is energy per unit mass of the fuel (e.g., kJ/kg or BTU/lbm) and is tabulated as a fuel property. Equation 3.50 compares the *mechanical power* production in the engine to the *thermal power* investment in the engine. Figure 3.19 is a definition sketch, which is a useful tool to help remember thermal efficiency definition as it graphically depicts the energy sources in an airbreathing engine. The rate of thermal energy consumption in the engine and the rate of mechanical power production by the engine are not equal. Yet, we are not violating the law of conservation of energy. The thermal energy production in an engine is not actually "lost," as it shows up in the hot jet exhaust stream, rather this energy is "wasted" and we were unable to convert it to a "useful" power. It is important to know, i.e., to quantify, this inefficiency in our engine.

EXAMPLE 3.4

Consider an aircraft engine (a turbojet) in takeoff condition with the following parameters:

$\dot{m}_0 = 100\,\text{kg/s}$ (this is equivalent to about 220 lbm/s of air flow rate)

$V_0 \approx 0\,\text{m/s}$ (this is a typical takeoff assumption that ignores low takeoff speeds)

$\dot{m}_f = 2\,\text{kg/s}$ (this represents a 2% fuel-to-air ratio)

$Q_R = 42{,}000\,\text{kJ/kg}$ (this is the heating value of a typical hydrocarbon fuel)

$V_9 = 900\,\text{m/s}$ (this represents a high speed exhaust jet)

If we substitute these numbers in Eq. 3.34, we get

$$\eta_{\text{th}} = \frac{(102)(900)^2}{2(2)(42000)(1000)} \cong 49.2\%$$

The engine thrust at takeoff is

$$F_n \approx F_g \approx \dot{m}_9 V_9 = 102(900)\,\text{N} = 91.8\,\text{kN}$$
$$= (91{,}800/4.448)\,\text{lbf} \approx 20{,}640\,\text{lbf}$$

Equation 3.50, which defines thermal efficiency, is simply the ratio of "net mechanical output" to the "thermal input," as we had learned in thermodynamics.

Example 3.4 shows a turbojet engine (with the specified parameters) is only about 49.2% thermally efficient! What happened to the rest of the fuel (thermal) energy? We know that $\sim 49.2\%$ of it was converted to a net mechanical output and the remaining 51.8% then must have been untapped and left in the exhaust gas as *thermal energy*. The thermal energy in the exhaust gas is of no use to the vehicle and in fact, in many applications, costs an additional weight that needs to be considered for cooling/thermal protection of the exhaust nozzle and nearby structures. The thermal energy in the exhaust gas of an aircraft engine basically goes to waste. Therefore, the lower the exhaust gas temperature, the more useful energy is extracted from the combustion gases and hence the cycle is more efficient in the thermal context. Now, you may wonder what other context is there, besides the thermal efficiency context in a heat engine? The answer is found in the *application* of the engine. The purpose of an aircraft engine is to provide propulsive power to the aircraft. In simple terms, the engine has to produce a thrust force that can accelerate the vehicle to a desired speed, e.g., from takeoff to cruise condition, or maintain the vehicle speed, i.e., just to overcome vehicle drag in cruise. Of course, one may add the hover and lifting applications to an aircraft engine purpose as, for example, in helicopters and short takeoff and landing (STOL) aircraft, respectively. Therefore, there is another context other than the thermal efficiency for an aircraft engine, namely the *propulsive efficiency*. Before leaving our treatment of the thermal efficiency of the jet engines and investigating their propulsive efficiency, let us explore various ways of improving the thermal efficiency for an aircraft engine.

To lower the exhaust gas temperature, we can place an additional turbine wheel in the high pressure, hot gas stream and produce shaft power. This shaft power can then be used to power a propeller, a fan, or a helicopter rotor, for example. The concept of additional turbine stages to extract thermal energy from the combustion gases and powering a fan in a jet engine has led to the development of more efficient two- or three-spool turbofan engines. Therefore, the mechanical output of these engines is enhanced by the additional shaft power. The thermal efficiency of a cycle that produces a shaft power can therefore be written as

$$\eta_{th} = \frac{\wp_{shaft} + \Delta K\dot{E}}{\dot{m}_f Q_R} \tag{3.51}$$

A schematic drawing of an aircraft gas turbine engine, which is configured to produce shaft power, is shown in Fig. 3.20. In part (a), the "power turbine" provides shaft power to a propeller, whereas in part (b), the power turbine provides the shaft power to a helicopter main rotor.

The gas generator in Fig. 3.20 refers to the compressor, burner, and the turbine combination, which are detailed in the next chapter. In turboprops and turboshaft engines, the mechanical output of the engine is dominated by the shaft power that in the definition of the thermal efficiency of such cycles the rate of kinetic energy increase is neglected, i.e.,

$$\eta_{th} \equiv \frac{\wp_s}{\dot{m}_f Q_R} \quad \text{in turboprop and turboshaft engines} \tag{3.52}$$

In addition to a shaft-power turbine concept, we can lower the exhaust gas temperatures by placing a heat exchanger in the exhaust stream to preheat the compressor air prior to combustion. The exhaust gas stream is cooled as it heats the cooler compressor gas and the less fuel is needed to burn to achieve a desired turbine entry temperature. This scheme is referred to as *regenerative cycle* and is shown in Fig. 3.21.

All of the cycles shown in Figs. 3.20 and 3.21 produce less wasted heat in the exhaust nozzle; consequently they achieve a higher thermal efficiency than their counterparts without the extra shaft power or the heat exchanger. However, we engineers need to examine a *bottom*

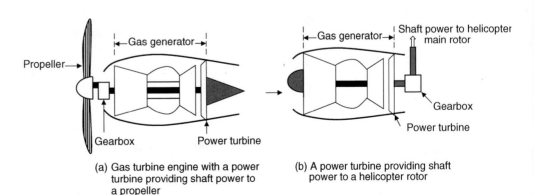

(a) Gas turbine engine with a power turbine providing shaft power to a propeller

(b) A power turbine providing shaft power to a helicopter rotor

■ **FIGURE 3.20** Schematic drawing of a power turbine placed in the exhaust of a gas turbine engine

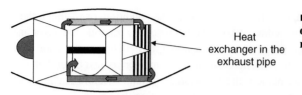

■ **FIGURE 3.21**   Schematic drawing of a gas turbine engine with a regenerative scheme

line question all the time, i.e., "will our gains outweigh our losses?" Obviously we gain in thermal efficiency but our systems in all instances require more complexity and weight. System complexity ties in with the issues of reliability and maintenance and the system weight ties in with the added cost and thus market acceptability. Therefore, we note that a successful engine does not necessarily have the highest thermal efficiency, but rather its *overall system performance and cost* is designed to meet the *customer's requirements* in an optimum manner. And in fact, its thermal efficiency is definitely compromised by the engine designers within the propulsion system design optimization loop/process. We will return to the issues surrounding thermal efficiency in every cycle we study in the next chapter.

Finally, one last question before we leave this subject is that "is it possible to achieve a 100% thermal efficiency in an aircraft engine?" This was actually a trick question to see who remembered his/her thermo! The answer is obviously no! We cannot violate the second law of thermodynamics. We remember that the highest thermal efficiency attainable in a heat engine operating between two temperature limits was that of a Carnot cycle operating between those temperatures. Figure 3.22 shows the Carnot cycle on a *T–s* diagram.

As noted in Fig. 3.22, both heat rejection at absolute zero $(T_1 = 0)$ and heating to infinite temperatures $(T_2 = \infty)$ are impossibilities, therefore we are thermodynamically bound by the Carnot thermal efficiency as the maximum (<100%). A Brayton cycle, which gas turbine engines are represented by, experiences a lower thermal efficiency than the Carnot cycle and this subject is presented in the cycle analysis (Chapter 4) in more detail.

3.5.4   Propulsive Efficiency

The fraction of the net mechanical output of the engine which is converted into thrust power, is called the propulsive efficiency. The net mechanical output of the engine is $\Delta K\dot{E}$ for perfectly expanded nozzle and the thrust power is $F \cdot V_0$, therefore, the propulsive efficiency is defined as their ratio:

$$\eta_{\mathrm{p}} \equiv \frac{F \cdot V_0}{\Delta K\dot{E}} \tag{3.53}$$

■ **FIGURE 3.22**   *T–s* diagram of a Carnot cycle

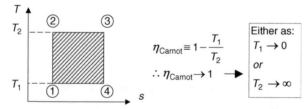

■ **FIGURE 3.23**   Schematic drawing of an engine installation showing the mechanical and thrust power

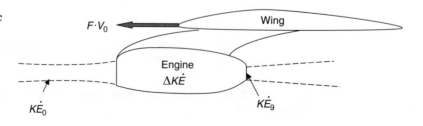

A graphical depiction of propulsive efficiency is shown in Fig. 3.23 to help the reader to remember the definition of propulsive efficiency.

Although the thrust power represented by $F \cdot V_0$ in Eq. 3.53 is based on the *installed* thrust, for simplicity, it is often taken as the uninstalled thrust power to highlight a very important, and at first astonishing, result about the propulsive efficiency. Now, let us substitute the uninstalled thrust of a perfectly expanded jet in the above definition to get

$$\eta_p \approx \frac{[(\dot{m}_0 + \dot{m}_f)V_9 - \dot{m}_0 V_0]V_0}{(\dot{m}_0 + \dot{m}_f)\dfrac{V_9^2}{2} - \dot{m}_0 \dfrac{V_0^2}{2}} \tag{3.54}$$

We recognize that the fuel flow rate is but a small fraction ($\sim 2\%$) of the air flow rate and thus can be ignored relative to the air flow rate and thus Eq. 3.54 can be simplified to

$$\eta_p \approx \frac{(V_9 - V_0)V_0}{\dfrac{1}{2}(V_9^2 - V_0^2)} = \frac{2V_0}{V_9 + V_0} = \frac{2}{1 + \dfrac{V_9}{V_0}} \tag{3.55}$$

Equation 3.55 as an approximate expression for the propulsive efficiency of a jet engine is cast in terms of a single parameter, namely the jet-to-flight velocity ratio V_9/V_0. We further note that 100% propulsive efficiency (within the context of approximation presented in the derivation) is mathematically possible and will be achieved by engines whose exhaust velocity is as fast as the flight velocity, i.e., $V_9 = V_0$. But a practical question poses itself that how can we produce thrust in an airbreathing jet engine if the jet velocity is not *faster* than the flight velocity? The short answer is that we cannot! Therefore, we conclude that some overspeeding in the jet, compared with flight speed, is definitely needed to produce reaction thrust in an airbreathing jet engine. Therefore for practical reasons, a 100% propulsive efficiency is not possible, just as the thermal efficiency of 100% was not possible. However, the smaller the increment of velocity rises across the engine, the higher its propulsive efficiency will be. In order to achieve a small velocity increment across a gas turbine engine for a given fuel flow rate, we need to drain the thermal energy in the combustion gas further and convert it to additional shaft power that in turn can act on a larger mass flow rate of air, in a secondary or a bypass stream, e.g., a fan or a propeller, to produce the desired level of thrust.

EXAMPLE 3.5

A turbojet engine is flying at 200 m/s. The products of combustion achieve an exhaust velocity of 900 m/s. Estimate the engine propulsive efficiency.

SOLUTION

Equation 3.55 is used to estimate the propulsive efficiency of $\sim 36.3\%$ (see Fig. 3.24).

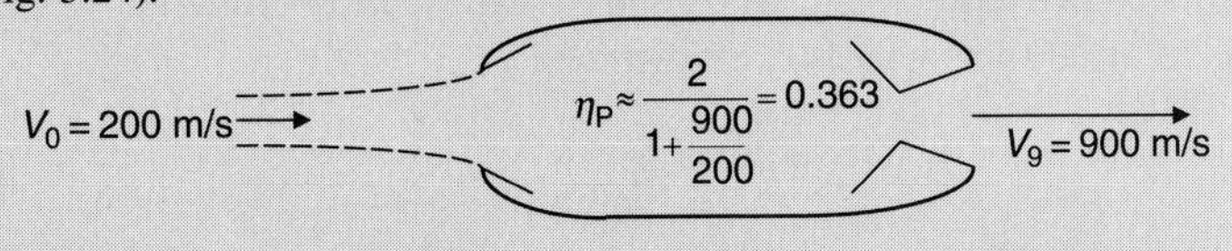

$$V_0 = 200 \text{ m/s} \longrightarrow \qquad \eta_P \approx \frac{2}{1 + \dfrac{900}{200}} = 0.363 \qquad V_9 = 900 \text{ m/s} \longrightarrow$$

■ **FIGURE 3.24** Schematic drawing of a turbojet engine in flight with a velocity ratio of 4.5

EXAMPLE 3.6

Consider a turbofan engine flying at the same speed as the previous example turbojet engine, see Fig. 3.25, i.e., 200 m/s. Now, the exhaust velocity is significantly reduced as more airflow rate is accelerated. For simplicity, assume that the exhaust velocity in the primary and fan exhaust nozzles is 250 m/s. Estimate the engine's propulsive efficiency.

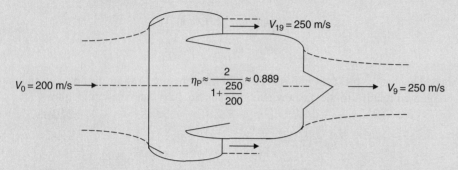

■ **FIGURE 3.25**   Schematic drawing of a turbofan engine in flight with a velocity ratio of 1.25

SOLUTION

As the exhaust velocities are the same in the primary and fan nozzles, this engine becomes equivalent to a single stream engine with a reduced velocity ratio of $250/200 = 1.25$. Therefore using Eq. 3.55, we arrive at a propulsive efficiency of 88.9%. In comparison to the previous turbojet engine with a velocity ratio of 4.5, our turbofan engine has achieved a 52.6% increase in propulsive efficiency.

Propulsive efficiency of a turboprop engine is defined as the fraction of mechanical power, which is converted to the total thrust (i.e., the sum of propeller and engine nozzle thrust) power, namely

$$\eta_p \equiv \frac{F \cdot V_0}{\wp_s + \Delta \dot{KE}} \approx \frac{F \cdot V_0}{\wp_s} \quad \text{[turboprop]} \tag{3.56}$$

Again, this definition compares the propulsive *output*$(F \cdot V_0)$ to the mechanical power *input* (shaft and jet kinetic energy change) in an aircraft engine. The fraction of shaft power delivered to the propeller, which is converted to the propeller thrust, is called *propeller efficiency* η_{pr}

$$\eta_{pr} \equiv \frac{F_{prop} \cdot V_0}{\wp_{s,prop}} \tag{3.57}$$

Due to large diameters of propellers, it is necessary to reduce their rotational speed to avoid severe tip shock-induced losses. Consequently, the power turbine rotational speed is mechanically reduced in a reduction gearbox and a small fraction of shaft power is lost in the gearbox, which is referred to as gear box efficiency, i.e.,

$$\eta_{gb} \equiv \frac{\wp_{s,prop}}{\wp_{s,turbine}} \tag{3.58}$$

We will address turboprop performance in Chapter 4.

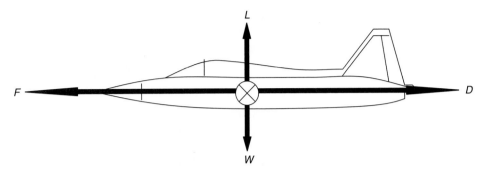

■ **FIGURE 3.26**   Aircraft in an un-accelerated level flight

3.5.5   Engine Overall Efficiency and Its Impact on Aircraft Range and Endurance

The product of the engine thermal and propulsive efficiency is called the engine overall efficiency

$$\eta_0 \equiv \eta_{th} \cdot \eta_p = \frac{\Delta \dot{KE}}{\dot{m}_f Q_R} \frac{F \cdot V_0}{\Delta \dot{KE}} = \frac{F \cdot V_0}{\dot{m}_f Q_R} \tag{3.59}$$

The overall efficiency of an aircraft engine is therefore the fraction of the fuel thermal power, which is converted into the thrust power of the aircraft. Again, a useful output is compared with the input investment in this efficiency definition. In an aircraft performance course that typically precedes the aircraft propulsion class, the engine overall efficiency is tied with the aircraft range, through the Breguet range equation. The derivation of the Breguet range equation is both fundamental to our studies and surprisingly simple enough for us to repeat it here for review purposes. An aircraft in level flight cruising at the speed of V_0, see Fig. 3.26, experiences a drag force that is entirely balanced by the engine thrust (for no acceleration), i.e.,

$$F_{engine} = D_{aircraft} \tag{3.60}$$

The aircraft lift L is also balanced by the aircraft weight to maintain level flight, i.e.,

$$L = W \tag{3.61}$$

We can multiply Eq. 3.60 by the flight speed V_0 and then replace the resulting thrust power, $F \cdot V_0$, by $\eta_0 \dot{m}_f Q_R$, via the definition of the engine overall efficiency, to get

$$F \cdot V_0 = \eta_0 \dot{m}_f Q_R = D \cdot V_0 \tag{3.62}$$

Now, let us divide the RHS term in Eq. 3.62 by lift L and the middle term of Eq. 3.62 by aircraft weight (which is the same as lift), to get

$$\frac{\eta_0 \dot{m}_f Q_R}{W} = \frac{D}{L} \cdot V_0 \tag{3.63}$$

Noting that the fuel flow rate, $\dot{m}_f = \frac{-1}{g_0} \frac{dW}{dt}$, i.e., the rate at which the aircraft is losing mass (thus the negative sign), we can substitute this expression in Eq. 3.63 and rearrange to get

$$-\frac{\eta_0 Q_R}{g_0} \frac{dW}{W} = \frac{D}{L} \cdot V_0 dt = \frac{dR}{L/D} \tag{3.64}$$

where g_0 is Earth's gravitational acceleration and $V_0 dt$ is interpreted as the aircraft elemental range dR, which is the distance traveled in time dt by the aircraft at speed V_0. Now, we can proceed to integrate Eq. 3.64 by making the assumptions of constant lift-to-drag ratio and constant engine overall efficiency, over the cruise period, to derive the Breguet range equation as

$$R = \eta_0 \cdot \frac{Q_R}{g_0} \cdot \frac{L}{D} \cdot \ell n \frac{W_i}{W_f} \tag{3.65}$$

where W_i is the aircraft initial weight and W_f is the aircraft final weight (note that the initial weight is larger than the final weight by the weight of the fuel burned in flight). Equation 3.65 is known as the Breguet range equation, which owes its simplicity and elegance to our assumptions of (1) unaccelerated level flight and (2) constant lift-to-drag ratio and engine overall efficiency. Also note that the range segment contributed by the takeoff and landing distances is not accounted for in the Breguet range equation. The direct proportionality of the engine overall efficiency and the aircraft range is demonstrated by the Breguet range equation, i.e.,

$$\text{Aircraft range } R \propto \eta_0$$

Now, let us replace the overall efficiency of the engine in the range equation by the ratio of thrust power to the thermal power in the fuel according to

$$R = \frac{F_n V_0}{\dot{m}_f Q_R} \frac{Q_R}{g_0} \frac{L}{D} \ell n \frac{W_i}{W_f} \tag{3.66}$$

We may express the flight speed in terms of a product of flight Mach number and the speed of sound, in addition, we may substitute the thrust specific fuel consumption for the ratio of fuel flow rate to the engine net thrust, to get

$$R = \left(M_0 \frac{L}{D} \right) \frac{a_0/g_0}{\text{TSFC}} \ell n \frac{W_i}{W_f} \tag{3.67}$$

The result of this representation of the aircraft range is the emergence of (ML/D) as the aerodynamic figure of merit for aircraft range optimization, known as the *range factor,*

$$\text{Aircraft range } R \propto M_0 \frac{L}{D} \tag{3.68}$$

$$\text{Aircraft range } R \propto \frac{1}{\text{TSFC}} \tag{3.69}$$

Now, if we use a more energetic fuel than the current jet aviation fuel, e.g. hydrogen, we will be able to reduce the thrust specific fuel consumption, or we can see the effect of fuel energy content on the range following Eq. 3.65, which shows

$$\text{Aircraft range } R \propto Q_R \tag{3.70}$$

Equivalently, we may seek out the effect of engine overall efficiency, or the specific fuel consumption on aircraft endurance, which for our purposes is the ratio of aircraft range to the flight speed,

$$\text{Aircraft endurance} = \frac{R}{V_0} = \frac{\eta_0}{V_0} \cdot \frac{Q_R}{g_0} \cdot \frac{L}{D} \cdot \ell n \frac{W_i}{W_f} \tag{3.71}$$

This again points out the importance of engine overall efficiency on aircraft performance parameters such as endurance. Now, let us substitute for engine overall efficiency and recast this equation in terms of TSFC, to get

$$\text{Aircraft endurance} = \left(\frac{L}{D}\right) \frac{1/g_0}{\text{TSFC}} \ell n \frac{W_i}{W_f} \tag{3.72}$$

The engine thrust-specific-fuel consumption appears in the denominator as in the engine impact on the range equation, and this time the aerodynamic figure of merit is L/D, instead of ML/D, as expected for the aircraft endurance.

$$\text{Aircraft endurance} \propto \frac{1}{\text{TSFC}} \tag{3.73}$$

$$\text{Aircraft endurance} \propto \frac{L}{D} \tag{3.74}$$

For additional reading on the subject, Anderson (1999), Newman (2002) and Pratt & Whitney operations manual 200 (1988) are recommended.

3.6   Summary

In this chapter, we defined engine thrust and the factors outside the engine that influenced its "installed" performance. We noted that the specific fuel consumption is a "fuel economy" parameter and is thus very critical to the direct operating expenses of an aircraft. We also used component impulse formulation to demonstrate the thrust equation for an airbreathing engine. Thermal and propulsive efficiencies each measured the internal performance and thrust production efficiency of an engine in flight, respectively. The overall efficiency of the engine, as the product of thermal and propulsive efficiencies, is tied to both the fuel economy as well as aircraft range. A unifying figure of merit for airbreathing engines and rockets is "specific impulse" with units of seconds.

Airbreathing engines incur *ram drag*, which is the product of air mass flow rate and the flight speed. The exhaust nozzle produces *gross thrust*, which is the sum of the momentum thrust and a pressure force contribution that occurs for a nozzle with imperfect expansion. The gross thrust is maximized for a *perfectly expanded nozzle* where $p_9 = p_0$. The sum of the gross thrust and the ram drag is the engine uninstalled thrust. The installation effects are primarily due to inlet and nacelle aft-end drag contributions.

Propulsive efficiency improves when the exhaust and flight speeds are closer to each other in magnitude, as in a turbofan engine. The parameter that brings the exhaust and flight speeds closer to each other in a turbofan engine is the *bypass ratio*. The trend in subsonic engine development/manufacturing is in developing ultrahigh bypass (UHB) engines, with bypass ratio in 10–15 range. Current turbofan engines offer a bypass ratio of ~ 8. Since the turboprop affects a much larger airflow at an incrementally smaller speed increase, it offers the highest propulsive efficiency for a low-speed (i.e., subsonic) aircraft. Advanced turboprops may utilize counterrotating propellers and may include a *slimline nacelle*, as in a ducted fan configuration.

Thermal efficiency is a cycle-dependent parameter. The highest thermal efficiency of a heat engine operating between two temperature limits corresponds to the Carnot cycle. Consequently, a higher compression (Brayton) cycle, still bound by two temperature limits, offers a higher thermal efficiency. The trend in improving engine thermal efficiency is in developing high-pressure ratio compressors. The current maximum (compressor pressure ratio) is ~ 45–50.

References

1. Kerrebrock, J.L., *Aircraft Engines and Gas Turbines*, 2nd edition, MIT Press, Cambridge, Massachusetts, 1992.

2. Lotter, K., "Aerodynamische Probleme der Integration von Triebwerk und Zelle beim Kampfflugzeugen," *Proceedings of the 85th Wehrtechnischen Symposium*, Mannheim, Germany, 1977.

3. Anderson, J.D., *A, Aircraft Performance and Design*, McGraw-Hill, New York, 1999.

4. Newman, D., *Interactive Aerospace Engineering and Design*, McGraw-Hill, New York, 2002.

5. Pratt & Whitney *Aircraft Gas Turbine Engine and Its Operation*, P&W Operations Manual 200, 1988.

Problems

3.1   The total pressures and temperatures of the gas in an afterburning turbojet engine are shown (J57 "B" from Pratt & Whitney, 1988). The mass flow rates for the air and fuel are also indicated at two engine settings, the Maximum Power and the Military Power. Use the numbers specified in this engine to calculate

(a) the fuel-to-air ratio f in the primary burner and the afterburner, at both power settings

(b) the low- and high-pressure spool compressor pressure ratios and the turbine pressure ratio (note that these remain constant with the two power settings)

(c) the exhaust velocity V_9 for both power settings by assuming the specified thrust is based on the nozzle gross thrust (because of sea level static) and *neglecting any pressure thrust* at the nozzle exit

(d) the thermal efficiency of this engine for both power settings (at the sea level static operation), assuming

the fuel heating value is $18,600\,\text{BTU/lbm}$ and $c_p = 0.24\,\text{BTU/lbm}\cdot{}^\circ\text{R}$. Explain the lower thermal efficiency of the Maximum power setting

(e) the thrust specific fuel consumption in lbm/h/lbf in both power settings

(f) the Carnot efficiency of a corresponding engine, i.e., operating at the same temperature limits, in both settings

(g) the comparision of percent thrust increase to percent fuel flow rate increase when we turn the afterburner on

(h) that why is it that we don't get proportional thrust increase with fuel flow increase (when it is introduced in the afterburner), i.e., doubling the fuel flow in the engine (through afterburner use) does not double the thrust

3.2   The total pressures and temperatures of the gas are specified for a turbofan engine with separate exhaust streams (JT3D-3B from Pratt & Whitney, 1974). The mass flow rates

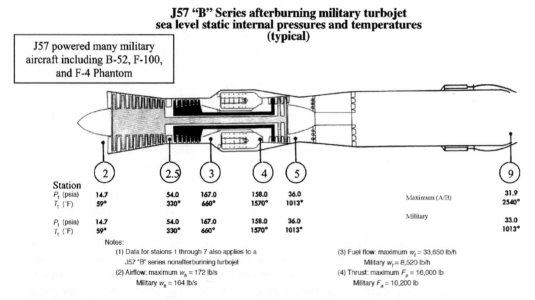

J57 "B" Series afterburning military turbojet
sea level static internal pressures and temperatures
(typical)

J57 powered many military aircraft including B-52, F-100, and F-4 Phantom

Station	2		2.5	3		4	5		9
P_t (psia)	14.7		54.0	167.0		158.0	36.0	Maximum (A/B)	31.9
T_t (°F)	59°		330°	660°		1570°	1013°		2540°
P_t (psia)	14.7		54.0	167.0		158.0	36.0	Military	33.0
T_t (°F)	59°		330°	660°		1570°	1013°		1013°

Notes:
(1) Data for staions 1 through 7 also applies to a J57 "B" series nonafterburining turbojet
(2) Airflow: maximum w_a = 172 lb/s
    Military w_a = 164 lb/s

(3) Fuel flow: maximum w_f = 33,650 lb/h
    Military w_f = 8,520 lb/h
(4) Thrust: maximum F_a = 16,000 lb
    Military F_a = 10,200 lb

■ **FIGURE P3.1**   **(from Pratt & Whitney Operation Manual 200)**

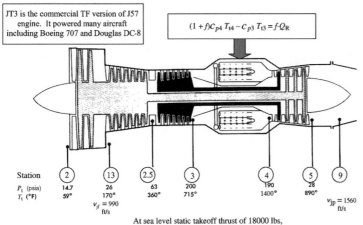

**JT3D-3B Turbofan
Internal pressures and temperatures**

■ **FIGURE P3.2** **(from Pratt & Whitney
Operation Manual 200)**

JT3 is the commercial TF version of J57 engine. It powered many aircraft including Boeing 707 and Douglas DC-8

$(1 + f)c_{p4} T_{t4} - c_{p3} T_{t3} = f \cdot Q_R$

Station	2	13	2.5	3	4	5	9
P_t (psia)	14.7	26	63	200	190	28	
T_t (°F)	59°	170°	360°	715°	1400°	890°	

$v_{jf} = 990$ ft/s

$v_{jp} = 1560$ ft/s

At sea level static takeoff thrust of 18000 lbs,
$W_{af} = 265$ lbs/s $W_{ap} = 195$ lbs/s

in the engine core (or primary) and the engine fan are also specified for the sea level static operation. Calculate

(a) the engine bypass ratio α defined as the ratio of fan-to-core flow rate

(b) from the total temperature rise across the burner, estimate the fuel-to-air ratio and the fuel flow rate in lbm/h, assuming the fuel heating value is $Q_R \sim 18,600$ BTU/lbm and the specific heat at constant pressure is 0.24 and 0.26 BTU/lbm ·°R at the entrance and exit of the burner, respectively

(c) the engine static thrust based on the exhaust velocities and the mass flow rates *assuming perfectly expanded nozzles* and compare your answer to the specified thrust of 18,000 lbs

(d) the engine thermal efficiency η_{th}

(e) the thermal efficiency of this engine compared to the afterburning turbojet of Problem 1. Explain the major contributors to the differences in η_{th} in these two engines

(f) the engine thrust specific fuel consumption in lbm/h/lbf

(g) the nondimensional engine specific thrust

(h) the Carnot efficiency corresponding to this engine

(i) the engine overall pressure ratio p_{t3}/p_{t2}

(j) fan nozzle exit Mach number [use $T_t = T + V^2/2c_p$ to calculate local static temperature at the nozzle exit, then local speed of sound $a = (\gamma RT)^{1/2}$]

3.3 A mixed exhaust turbofan engine (JT8D from Pratt & Whitney, 1988) is described by its internal pressures and temperature, as well as air mass flow rates and the mixed jet (exhaust) velocity. Let us examine a few parameters for this engine, for a ballpark approximation.

(a) Estimate the fuel flow rate from the total temperature rise across the burner assuming the fuel heating value is ~18,600 BTU/lbm and the specific heat at constant pressure is 0.24 and 0.26 BTU/lbm ·°R at the entrance and exit of the burner, respectively

(b) Calculate the momentum thrust at the exhaust nozzle and compare it to the specified thrust of 14,000 lbs

(c) Estimate the thermal efficiency of this engine and compare it to Problems 3.1 and 3.2 as well as a Carnot cycle operating between the temperature extremes of this engine. Explain the differences

(d) Estimate the specific fuel consumption for this engine in lbm/h/lbf

(e) The overall pressure ratio (of the fan–compressor section) p_{t3}/p_{t2}

(f) What is the bypass ratio α for this engine at takeoff

(g) What is the Carnot efficiency corresponding to this engine

(h) Estimate nozzle exit Mach number [look at part (j) in Problem 3.2]

(i) What is the low-pressure compressor (LPC) pressure ratio $p_{t2.5}/p_{t2}$

(j) What is the high-pressure compressor (HPC) pressure ratio $p_{t3}/p_{t2.5}$

3.4 A large bypass ratio turbofan engine (JT9D engine from Pratt & Whitney, 1988) is described by its fan and core engine gas flow properties.

(a) What is the overall pressure ratio (OPR) of this engine

(b) Estimate the fan gross thrust $F_{g,fan}$ in lbf

(c) Estimate the fuel-to-air ratio based on the energy balance across the burner, assuming the fuel heating value is ~18,600 BTU/lbm and the specific heat at

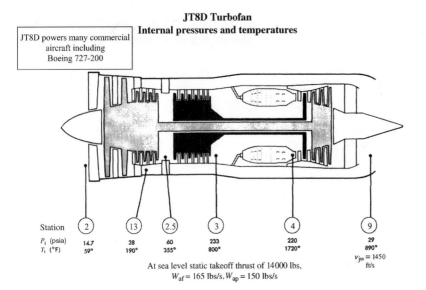

JT8D Turbofan
Internal pressures and temperatures

JT8D powers many commercial
aircraft including
Boeing 727-200

Station	②	⑬	②⑤	③	④	⑨
P_t (psia)	14.7	28	60	233	220	29
T_t (°F)	59°	190°	355°	800°	1720°	890°
						$v_{jm} = 1450$ ft/s

At sea level static takeoff thrust of 14000 lbs,
$W_{af} = 165$ lbs/s, $W_{ap} = 150$ lbs/s

■ **FIGURE P3.3**   (from Pratt &
Whitney Operation Manual 200)

constant pressure is 0.24 and 0.26 BTU/lbm · °R at
the entrance and exit of the burner, respectively

(d) Calculate the core gross thrust and compare the sum of
the fan and the core thrusts to the specified engine thrust
of 43,500 lbf

(e) Calculate the engine thermal efficiency and compare it
to Problems 3.1–3.3. Explain the differences

(f) Estimate the thrust-specific fuel consumption (TSFC),
in lbm/h/lbf

(g) What is the bypass ratio of this turbofan engine

(h) What is the Carnot efficiency η_{Carnot} corresponding to
this engine

(i) What is the LPC pressure ratio $p_{t2.5}/p_{t2}$

(j) What is the HPC pressure ratio $p_{t3}/p_{t2.5}$

(k) Estimate the fan nozzle exit Mach number [see part (j)
in Problem 3.2]

(l) Estimate the primary nozzle exit Mach number

3.5   An airbreathing engine flies at Mach $M_0 = 2.0$ at an
altitude where the ambient temperature is $T_0 = -50°C$ and
ambient pressure is $p_0 = 10\,kPa$. The airflow rate to the engine
is 25 kg/s. The fuel flow rate is 3% of airflow rate and has a
heating value of 42,800 kJ/kg. Assuming the exhaust speed is
$V_9 = 1050\,m/s$, and the nozzle is perfectly expanded, i.e.,
$p_9 = p_0$, calculate

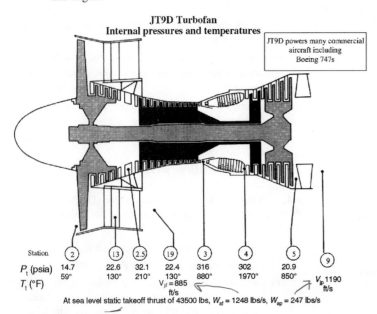

JT9D Turbofan
Internal pressures and temperatures

JT9D powers many commercial
aircraft including
Boeing 747s

Station	②	⑬	②⑤	⑲	③	④	⑤	⑨
P_t (psia)	14.7	22.6	32.1	22.4	316	302	20.9	
T_t (°F)	59°	130°	210°	130°	880°	1970°	850°	

$V_{jf} = 885$ ft/s V_{jp} 1190 ft/s

At sea level static takeoff thrust of 43500 lbs, $W_{af} = 1248$ lbs/s, $W_{ap} = 247$ lbs/s

■ **FIGURE P3.4**   (from Pratt &
Whitney Operation Manual 200)

(a) ram drag in kN

(b) gross thrust in kN

(c) net (uninstalled) thrust in kN

(d) thrust-specific fuel consumption in kg/h/N

(e) engine thermal efficiency η_{th}

(f) propulsive efficiency η_p

(g) engine overall efficiency η_o

3.6 A turbo-propeller-driven aircraft is flying at $V_0 = 150\,\text{m/s}$ and has a propeller efficiency of $\eta_{pr} = 0.75$. The propeller thrust is $F_{prop} = 5000\,\text{N}$ and the airflow rate through the engine is $5\,\text{kg/s}$. The nozzle is perfectly expanded and produces 1000 N of gross thrust. Calculate

(a) the shaft power delivered to the propeller in kW

(b) the nozzle exit velocity in m/s (neglect fuel flow rate in comparison to the air flow rate)

(c) in using Eq. 3.56, $\eta_p \equiv \dfrac{F \cdot V_0}{\wp_s + \Delta \dot{K}\dot{E}}$, first show that the contribution of the net kinetic power produced by the engine $\Delta \dot{K}\dot{E}$ is small compared to the shaft power $\wp_s$ in denominator of Eq. 3.56. Second, estimate the propulsive efficiency η_p for the turboprop engine from this equation.

3.7 Let us consider the control volume shown to represent the capture streamtube for an airbreathing engine at takeoff. The air speed is 10 m/s in area A_0 and 100 m/s in A_1.

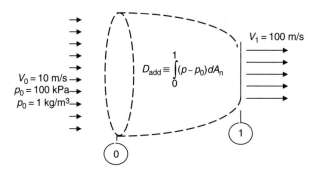

■ FIGURE P3.7

(a) Use incompressible flow assumption to estimate capture area ratio A_0/A_1

(b) Use Bernoulli equation to estimate p_1/p_0

(c) Use momentum balance to estimate nondimensional additive drag $D_{add}/A_1 p_0$

3.8 A rocket motor burns propellant at a rate of 50 kg/s. The exhaust speed is 3500 m/s and the nozzle is perfectly expanded. Calculate

(a) the rocket thrust in kN

(b) the rocket motor specific impulse I_s (s).

3.9 For the turbofan engine shown, calculate

(a) ram drag D_{ram} in kN

(b) primary nozzle gross thrust F_{g9}, in kN

(c) fan nozzle gross thrust F_{g19}, in kN

(d) the engine net thrust F_n, in kN

(e) the propulsive efficiency η_p $(-)$

Hint: To calculate the pressure thrust for the primary and fan nozzles, you may calculate the flow areas at A_9 and A_{19} using the mass flow rate information as well as the density that you may calculate from pressure and temperature (via the speed of sound) using perfect gas law.

3.10 A ramjet is flying at Mach 2.0 at an altitude where $T_0 = -50°\text{C}$ and the engine airflow rate is 10 kg/s. If the exhaust Mach number of the ramjet is equal to the flight Mach number, i.e., $M_9 = M_0$, with perfectly expanded nozzle and $T_{t9} = 2500\,\text{K}$, calculate

(a) the engine ram drag D_{ram} in kN

(b) the nozzle gross thrust F_g in kN

(c) the engine net thrust F_n in kN

(d) the engine propulsive efficiency η_p

Assume gas properties remain the same throughout the engine, i.e., assume $\gamma = 1.4$ and $c_p = 1004\,\text{J/kg·K}$. Also, assume that the fuel flow rate is 4% of airflow rate.

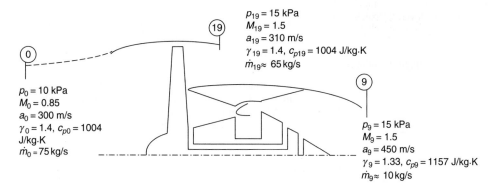

■ FIGURE P3.9

3.11   A turbojet-powered aircraft cruises at $V_0 = 300\,\text{m/s}$ while the engine produces an exhaust speed of $600\,\text{m/s}$. The air mass flow rate is $100\,\text{kg/s}$ and the fuel mass flow rate is $2.5\,\text{kg/s}$. The fuel heating value is $Q_R = 42{,}000\,\text{kJ/kg}$. Assuming that the nozzle is perfectly expanded, calculate:

(a) engine ram drag in kN

(b) engine gross thrust in kN

(c) engine net thrust in kN

(d) engine thrust-specific fuel consumption (TSFC) in mg/s/kN

(e) engine thermal efficiency

(f) engine propulsive efficiency

(g) aircraft range R for L/D of 10 and the W_i/W_f of 1.25

(h) if this aircraft make it across the Atlantic Ocean?

3.12   We wish to investigate the range of a slender supersonic aircraft where its lift-to-drag ratio as a function of flight Mach number is described by

$$\frac{L}{D} \approx 3\frac{M_0 + 3}{M_0}$$

Using Eq. 3.67, i.e.,

$$R = \left(M_0 \frac{L}{D}\right)\frac{a_0/g_0}{\text{TSFC}}\ell n\frac{W_i}{W_f}$$

for range equation, vary the thrust specific fuel consumption TSFC between 1.0 and 2.0 lbm/h/lbf to graph R for flight Mach number ranging between 2.0 and 4.0. You may assume a_0 is 1000 ft/s and the aircraft initial-to-final weight ratio is $W_i/W_f = 2.0$.

3.13   A turboshaft engine consumes fuel with a heating value of $42{,}000\,\text{kJ/kg}$ at the rate of $1\,\text{kg/s}$. Assuming the thermal efficiency is 0.333, calculate the shaft power that this engine produces.

3.14   A rocket engine consumes propellants at the rate of $1000\,\text{kg/s}$ and achieves a specific impulse of $I_s = 400\,\text{s}$. Assuming the nozzle is perfectly expanded, calculate

(a) the rocket exhaust speed V_9 in m/s

(b) the rocket thrust in MN

3.15   A rocket engine has a nozzle exit diameter of $D_9 = 2\,\text{m}$. It is perfectly expanded at sea level. Calculate the rocket pressure thrust in vacuum.

3.16   A ramjet engine is in supersonic flight. Its inlet flow parameters are shown.

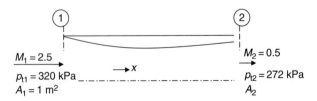

$M_1 = 2.5$

$p_{t1} = 320$ kPa

$A_1 = 1$ m²

$M_2 = 0.5$

$p_{t2} = 272$ kPa

A_2

■ **FIGURE P3.16**

Assuming the flow is adiabatic and $\gamma = 1.4$, calculate

(a) the diffuser exit area A_2 in m²

(b) impulse (in kN) in stations 1 and 2, I_1 and I_2

(c) internal force exerted (by the fluid) on the inlet in flight (or $-x$) direction

3.17   A convergent nozzle is perfectly expanded with exit Mach number $M_8 = 1.0$. The exit total pressure is 98% of the inlet total pressure. The nozzle inlet Mach number is $M_7 = 0.5$ and the nozzle area at the exit is $A_8 = 0.5\,\text{m}^2$.

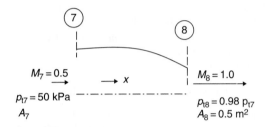

$M_7 = 0.5$

$p_{t7} = 50$ kPa

A_7

$M_8 = 1.0$

$p_{t8} = 0.98\,p_{t7}$

$A_8 = 0.5$ m²

■ **FIGURE P3.17**

Assuming the gas ratio of specific heats is $\gamma = 1.33$, and the flow is adiabatic, calculate

(a) nozzle inlet area A_7 in m²

(b) nozzle inlet impulse I_7 in kN

(c) nozzle exit impulse I_8 in kN

(d) the axial force (i.e., in the x-direction) exerted by the fluid on the nozzle

3.18   A turboprop engine flies at $V_0 = 200\,\text{m/s}$ and produces a propeller thrust of $F_{\text{prop}} = 40\,\text{kN}$ and a core thrust of $F_{\text{core}} = 10\,\text{kN}$. Engine propulsive efficiency η_p is 85%. Calculate

(a) total thrust produced by the turboprop in kN

(b) thrust power in MW

(c) shaft power produced by the engine $\wp_s$ in MW

3.19   A turbojet engine produces a net thrust of 40,000 N at the flight speed of V_0 of $300\,\text{m/s}$. For a propulsive efficiency of $\eta_p = 0.40$, estimate the turbojet exhaust speed V_9 in m/s.

3.20   Calculate the engine specific impulse in seconds for Problem 3.19. Also, assuming the fuel heating value is $42000\,\text{kJ/kg}$ and the thermal efficiency is 45%, estimate the fuel-to-air ratio consumed in the burner.

Courtesy of Rolls-Royce plc

CHAPTER 4

Gas Turbine Engine Cycle Analysis

4.1   Introduction

In this chapter, we examine the aerothermodynamics of aircraft gas turbine engines. We first start our analysis with the basic building block or the common features of any gas turbine engine, namely the gas generator. We will then *construct* a variety of aircraft engines employing the gas generator. These include the turbojet, the afterburning turbojet, the separate and mixed-exhaust turbofan, and the turboprop engines.

4.2   The Gas Generator

At the heart of an aircraft gas turbine engine *is a gas generator.* It is composed of three major components, a compressor, a burner (sometimes referred to as combustor or combustion chamber), followed by a turbine. The schematics of a gas generator and station numbers are shown in Fig. 4.1. The parameters that define the physical characteristics of a gas generator are noted in Table 4.1.

Compressor total pressure ratio π_c is a design parameter. An aircraft engine designer has the design choice of the compressor staging, i.e., the number and the type of compressor stages. That choice is a strong function of flight Mach number, or what we will refer to as *ram pressure ratio.* As a rule of thumb, the higher the flight Mach number, the lower the compressor pressure ratio the cycle requires to operate efficiently. In fact at the upper supersonic Mach numbers, i.e., $M_0 \geq 3$, an airbreathing engine will not even require *any* mechanical compression, i.e., the compressor is totally unneeded. Such engines work on the principle of ram compression and are called *ramjets.* We will refer to this later on in our analysis. The compressor air mass flow rate is the sizing parameter, which basically scales the engine face diameter. The takeoff gross weight of the aircraft is the primary parameter that most often *sizes* the engine. Other parameters,

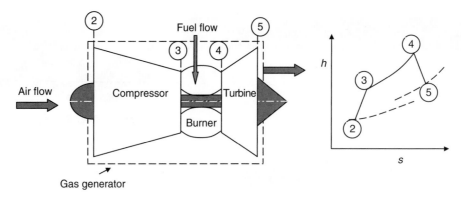

■ **FIGURE 4.1** Schematic drawing of a gas generator

■ **TABLE 4.1**

The Parameters in a Gas Generator

1. Compressor pressure ratio $\pi_c \equiv \dfrac{p_{t3}}{p_{t2}}$ · · · · · · · · · · ·("design" parameter)

2. Compressor air mass flow rate $\dot{m}_0$ · · · · · · · · · · ·("size" parameter)

3. Combustor fuel flow rate $\dot{m}_f$ or turbine entry temperature T_{t4} · · · · · ·("temperature limit" parameter)

4. Fuel heating value Q_R · · · · · · · · · · ·("ideal fuel energy" parameter)

5. Component efficiency · · · · · · · · · · ·("irreversibility" or loss parameter)

which contribute to engine sizing, are the engine-out rate of climb requirements, the transonic acceleration, and the allowable use of afterburner, among other mission specification parameters. The burner fuel flow rate is the fuel energy release rate parameter, which may be replaced by the turbine entry temperature T_{t4}. Both of these parameters establish a *thermal limit* identity for the engine, which dictate the material and the cooling technologies to be employed in the engine *hot section* (i.e., the turbine and nozzle) at the design stage. Fuel heating value, or heat of reaction Q_R, represents the (ideal) fuel energy density, i.e., the fuel thermal energy per unit mass of fuel. Finally, the component efficiencies are needed to describe the extent of losses or stated in thermodynamic language "irreversibility" in each component.

4.3   Aircraft Gas Turbine Engines

In this section, we examine the variety of aircraft gas turbine engines that are possible around the basic theme of a gas generator. The simplest example is the turbojet engine.

4.3.1   The Turbojet Engine

An aircraft turbojet (TJ) engine is basically a gas generator fitted with an inlet and exhaust system. A schematic drawing of a TJ-engine is shown in Fig. 4.2.

The additional parameters, which are needed to calculate the performance of a turbojet engine, are the inlet and exhaust component efficiencies. These parameters will be defined in

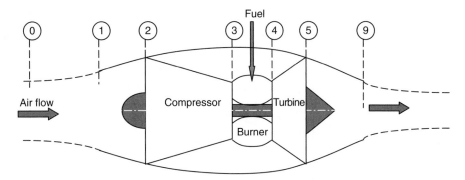

■ **FIGURE 4.2**   Schematic drawing of a turbojet engine

the next section. The station numbers in a turbojet engine are defined at the unperturbed flight condition, 0, inlet lip, 1, compressor face, 2, compressor exit, 3, burner exit, 4, turbine exit, 5, and the nozzle exit plane, 9.

4.3.1.1   *The Inlet.*

The *basic* function of the inlet is to deliver the air to the compressor at the *right* Mach number M_2 and the *right* quality, i.e., low distortion. The subsonic compressors are designed for an axial Mach number of $M_2 = 0.5-0.6$. Therefore, if the flight Mach number is higher than 0.5 or 0.6, which includes all commercial (fixed wing) transports and military (fixed wing) aircraft, then the inlet is required to *decelerate* the air efficiently. Therefore the main function of an inlet is to *diffuse or decelerate* the flow, and hence it is also called a diffuser. Flow deceleration is accompanied by the static pressure rise or what is known as the *adverse pressure gradient* in fluid dynamics. As one of the first principles of fluid mechanics, we learned that the boundary layers, being of a low-energy and momentum-deficit zone, facing an adverse pressure gradient environment tend to separate. Therefore one of the challenges facing an inlet designer is to prevent inlet boundary layer separation. One can achieve this by tailoring the geometry of the inlet to avoid rapid diffusion or possibly through variable geometry inlet design. Now, it becomes obvious why an aircraft inlet designer faces a bigger challenge if the inlet has to decelerate a Mach 2 or 3 stream to the compressor face Mach number of 0.5 than an aircraft that flies at Mach 0.8 or 0.9. In the present section, we will examine the *thermodynamics* of an aircraft inlet. This exciting area of propulsion, i.e., inlet aerodynamics, will be treated in more detail in the next chapter.

An ideal inlet is considered to provide a *reversible and adiabatic,* i.e., isentropic, compression of the captured flow to the engine. The adiabatic aspect is actually met in real inlets as well. This requirement says that there is no heat exchange between the captured stream and the ambient air, through the diffuser walls. We remember (from the Fourier law of heat conduction) that for heat transfer to take place through the inlet wall, we need to set up a *temperature gradient* across the wall, such that

$$q_{\mathrm{n}} \equiv \frac{\dot{Q}_{\mathrm{n}}}{A} = -k\frac{\partial T}{\partial n} \tag{4.1}$$

where n denotes the direction of heat conduction, k the thermal conductivity of the wall, q_{n} the heat flux, which is defined as the heat transfer rate $\dot{Q}_{\mathrm{n}}$ per unit area A. Equation 4.1 signifies the importance of a *gradient* to set up the heat transfer. The temperature gradient across the inlet wall (i.e., nacelle) is negligibly small, and consequently the inlet aerodynamics is considered to be *adiabatic*, even in a real flow. Therefore, it is only the *reversible* aspect of our ideal inlet flow assumption that negates the realities of wall friction and any shocks, which are invariably present in *real* supersonic flows. The process of compression in a real inlet can be shown on the *h–s* diagram of Fig. 4.3.

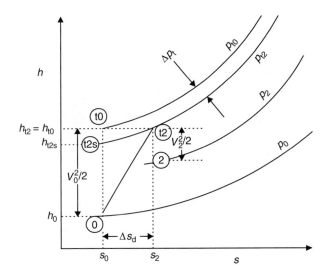

■ **FIGURE 4.3** *h–s* diagram of an aircraft engine inlet flow under real and ideal conditions

Now, let us describe the information we observe in Fig. 4.3. First, four isobars are shown on the *h–s* diagram in Fig. 4.3. From the lowest to the highest value, these are p_0, the static pressure at the flight altitude, p_2, the static pressure at the engine face, p_{t2}, the total pressure at the inlet discharge (or compressor face), and p_{t0}, the flight total pressure. We also note five thermodynamic states identified, as the flight static (0), the flight total (t0), the compressor face or inlet discharge static (2), and total (t2) and a stagnation state (t2s) that does not actually exist! Note, that the first four states, i.e., 0, t0, 2, and t2 do all in reality exist and are measured. The states 0 and t0 are measured by a pitot-static tube on an aircraft, and the static and total pressures at the engine face are measured by a pressure rake or inlet pitot tubes. We note that the state (t2s) is at the intersection of s_0 and p_{t2}, which is arrived at isentropically from state (0) to a state that shares the same total pressure as that of a real inlet (t2). We note an entropy rise Δs_d across the inlet as well. The total enthalpy of the real inlet h_{t2} is identified to be the same as the total enthalpy of flight h_{t0}. This is based on the earlier assertion that a *real* inlet flow may be considered to be *adiabatic*. It is to be noted that although the total energy of the fluid in a real inlet is not changing, there is an *energy conversion* that takes place in the inlet. The inlet kinetic energy is partially converted to the static pressure (rise) and partially it is dissipated into heat. It is the latter portion, i.e., the dissipation, which renders the process irreversible and cause the entropy to rise. The entropy rise in an adiabatic process leads to a total pressure loss Δp_t following the combined first and second law of thermodynamics, i.e., Gibbs equation, according to

$$\frac{p_{t2}}{p_{t0}} = e^{-\frac{s_2 - s_0}{R}} = e^{-\frac{\Delta s}{R}} \qquad (4.2)$$

From Fig. 4.3, we also note that the states (t0) and (0) are separated *isentropically* by the amount of $V_0^2/2$, as expected from the definition of stagnation state and its relation to the static state. The states (t2s) and (0) are separated isentropically by a kinetic energy amount, which produces p_{t2} as its stagnation state and obviously the rest of the kinetic energy, i.e., the gap between the states (t0) and (t2s), represents the amount of dissipated kinetic energy into heat. Therefore, the smaller the gap between the states (t2s) and (t0), the more efficient will the diffuser flow process be. Consequently, we use the fictitious state (t2s) in a definition of inlet efficiency (or a figure of merit) known as the *inlet adiabatic efficiency*. Symbolically, the inlet adiabatic efficiency is defined as

$$\eta_d \equiv \frac{h_{t2s} - h_0}{h_{t2} - h_0} = \frac{(V^2/2)_{\text{ideal}}}{V_0^2/2} \qquad (4.3)$$

The practical form of the above definition is derived when we divide the numerator and denominator by h_0 to get

$$\eta_d = \frac{\dfrac{h_{t2s}}{h_0} - 1}{\dfrac{h_{t2}}{h_0} - 1} = \frac{\dfrac{T_{t2s}}{T_0} - 1}{\dfrac{h_{t0}}{h_0} - 1} = \frac{\left(\dfrac{p_{t2}}{p_0}\right)^{\frac{\gamma-1}{\gamma}} - 1}{\dfrac{\gamma-1}{2}M_0^2} \qquad (4.4)$$

where we have used the isentropic relation between the states (t2s) and (0). Note that the only unknown in Eq. 4.4 is p_{t2} for a given flight altitude p_0, flight Mach number M_0, and an inlet adiabatic efficiency η_d. We can separate the unknown term p_{t2} and write the following expression:

$$\frac{p_{t2}}{p_0} = \left\{ 1 + \eta_d \frac{\gamma-1}{2} M_0^2 \right\}^{\frac{\gamma}{\gamma-1}} \qquad (4.5)$$

It is interesting to note that Eq. 4.5 recovers the isentropic relation for a 100% efficient inlet or $\eta_d = 1.0$. Another parameter, or a *figure of merit*, that describes the inlet performance is the total pressure ratio between the compressor face and the (total) flight condition. This is given a symbol π_d and is often referred to as the *inlet total pressure recovery*:

$$\pi_d \equiv \frac{p_{t2}}{p_{t0}} \qquad (4.6)$$

As expected, the two figures of merit for an inlet, i.e., η_d or π_d are not independent from each other and we can derive a relationship between η_d and π_d working the left-hand side of Eq. 4.5, as follows:

$$\frac{p_{t2}}{p_0} = \frac{p_{t2}}{p_{t0}} \frac{p_{t0}}{p_0} = \left\{ 1 + \eta_d \frac{\gamma-1}{2} M_0^2 \right\}^{\frac{\gamma}{\gamma-1}} \qquad (4.6\text{-a})$$

$$\pi_d = \frac{\left\{ 1 + \eta_d \dfrac{\gamma-1}{2} M_0^2 \right\}^{\frac{\gamma}{\gamma-1}}}{\dfrac{p_{t0}}{p_0}} = \left\{ \frac{1 + \eta_d \dfrac{\gamma-1}{2} M_0^2}{1 + \dfrac{\gamma-1}{2} M_0^2} \right\}^{\frac{\gamma}{\gamma-1}} \qquad (4.6\text{-b})$$

Therefore, Eq. 4.6-b relates the inlet total pressure recovery π_d to the inlet adiabatic efficiency π_d at any flight Mach number M_0. We note that Eq. 4.6-b as $\eta_d \to 1$, then $\pi_d \to 1$ as well, as expected. Figure 4.3 also shows the static state 2, which shares the same entropy as the total state t2 and lies below it by the amount of kinetic energy at 2, namely, $V_2^2/2$. Compare the kinetic energy at 2 and at 0, shown in Fig. 4.3, and then justify the static pressure rise achieved in the inlet (diffuser), $p_2 - p_0$.

So far, we have considered the cruise condition or the high-speed end of the flight envelope and have thought of inlets as diffusers. Under low-speed or takeoff conditions, the captured stream tube will instead undergo acceleration to the engine face and as such the inlet acts like a nozzle! In Fig. 4.4, the schematics of a captured stream tube, under a low-speed flight condition, is shown. Note that the stagnation point is on the outer cowl and flow accelerates to the engine face ($A_0 > A_2$). The aerodynamics of the outer nacelle geometry affects the drag divergence of the inlet and plays a major role in the propulsion system integration studies of engine installation. However, the cycle analysis phase usually disregards the external performance and concentrates on the internal evaluation of the propulsion system.

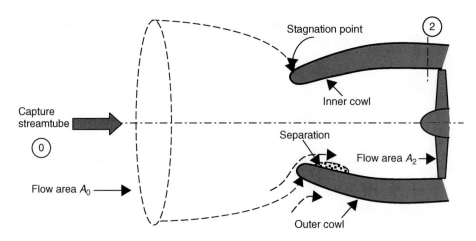

■ **FIGURE 4.4** Schematic drawing of an inlet flow at low-speed and/or takeoff flight condition (note: $A_0 > A_2$)

The nature of the flow path into the inlet at low forward speed or under static engine testing conditions require the inlet to act as a nozzle; therefore, it explains the use of a bell mouth in static test rigs. Figure 4.5 shows the schematic drawing of an inlet configuration in a static test rig. In Fig. 4.5, we note that the static pressure along the stream lines drop from the ambient p_0 to the engine face pressure of p_2. Also, we note that the gas speed starts at almost zero, since the captured flow area A_0 is fairly large, and continues to grow to the engine face speed of V_2.

In summary, we have learned that

■ the inlet flow may be considered to be adiabatic, i.e., $h_{t2} = h_{t0}$
■ the inlet flow is always irreversible, i.e., $p_{t2} < p_{t0}$, with viscous dissipation in the boundary layer and in a shock as the sources of irreversibility $(s_2 > s_0)$
■ there are two figures of merit that describe the extent of *losses* in the inlet and these are η_d and π_d
■ the two figures of merit are related (via Eq. 4.6-b)
■ in cruise, $A_0 < A_2$ and hence a diffusing passage and at low speed or takeoff, $A_0 > A_2$, i.e., a nozzle
■ the outer nacelle geometry of an inlet dictates the drag divergence and high angle of attack characteristics of the inlet and is crucial for the installed performance

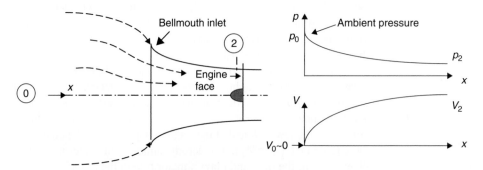

■ **FIGURE 4.5** Bellmouth inlet guides the flow smoothly into the engine on a static test rig

EXAMPLE 4.1

An aircraft is flying at an altitude where the ambient static pressure is $p_0 = 10\,\text{kPa}$ and the flight Mach number is $M_0 = 0.85$. The total pressure at the engine face is measured to be $p_{t2} = 15.88\,\text{kPa}$. Assuming the inlet is adiabatic and $\gamma = 1.4$, calculate

(a) the inlet total pressure recovery π_d

(b) the inlet adiabatic efficiency η_d

(c) the nondimensional entropy rise caused by the inlet $\Delta s_\text{d}/R$

SOLUTION

We first calculate the flight total pressure p_{t0}, and from definition of π_d (i.e., Eq. 4.6), the inlet total pressure recovery.

$$p_{t0} = p_0[1 + (\gamma - 1)M_0^2/2]^{\frac{\gamma}{\gamma-1}} = 10\,\text{kPa}[1 + 0.2(0.85)^2]^{3.5}$$

$$= 16.04\,\text{kPa}$$

$$\pi_\text{d} \equiv p_{t2}/p_{t0} = 15.88/16.04 = 0.990$$

Inlet adiabatic efficiency η_d is calculated from Eq. 4.4

$$\eta_\text{d} = \left[\left(\frac{p_{t2}}{p_0}\right)^{\frac{\gamma-1}{\gamma}} - 1\right] \bigg/ \left(\frac{\gamma - 1}{2}M_0^2\right)$$

$$= [(15.88/10)^{0.2857} - 1]/[0.2(0.85)^2] \cong 0.9775$$

The entropy rise is linked to the inlet total pressure loss parameter π_d via Eq. 4.2,

$$\Delta s_\text{d}/R = -\ln(\pi_\text{d}) = -\ln(0.99) \cong 0.010$$

4.3.1.2   ***The Compressor.*** The thermodynamic process in a gas generator begins with the mechanical compression of air in the compressor. As the compressor discharge contains higher energy gas, i.e., the compressed air, it requires *external power* to operate. The power comes from the turbine via a shaft, as shown in Fig. 4.1 in case of an operating gas turbine engine. Other sources of external power may be used to *start* the engine, which are in the form of electric motor, air-turbine, and hydraulic starters. The flow of air in a compressor is considered to be an essentially adiabatic process, which suggests that only a *negligible* amount of heat transfer takes place between the air inside and the ambient air outside the engine. Therefore, even in a real compressor analysis, we will still treat the flow as adiabatic. Perhaps a more physical argument in favor of neglecting the heat transfer in a compressor can be made by examining the order of magnitude of the energy transfer sources in a compressor. The power delivered to the medium in a compressor is achieved by one or more rows of rotating blades (called rotors) attached to one or more spinning shafts (typically referred to as *spools*). Each rotor blade, which changes the *spin* (or swirl) of a medium, will experience a countertorque as a reaction to its own action on the fluid. If we denote the rotor torque as τ, then the power delivered to the medium by the rotor spinning at the angular speed ω follows Newtonian mechanics, i.e.,

$$\wp = \tau \cdot \omega \tag{4.7}$$

A typical axial-flow compressor contains hundreds of rotor blades (distributed over several stages), which interact with the medium according to the above equation. Therefore, the rate of mechanical energy transfer in a typical modern compressor is usually measured in mega-Watt (MW) and is several orders of magnitude larger than heat transfer through the compressor wall. Symbolically, we may present this as

$$\wp_\text{c} \ggg \dot{Q}_\text{w} \tag{4.8}$$

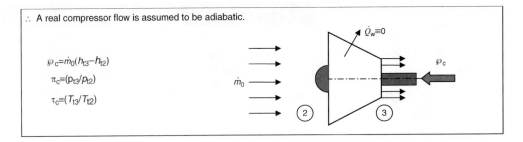

∴ A real compressor flow is assumed to be adiabatic.

$\wp_c = \dot{m}_0 (h_{t3} - h_{t2})$

$\pi_c \equiv (p_{t3}/p_{t2})$

$\tau_c \equiv (T_{t3}/T_{t2})$

As a real process, however, the presence of wall friction acting on the medium through the boundary layer and shock waves caused by the relative supersonic flow through compressor blades will render the process irreversible. Therefore, the sources of irreversibility in a compressor are due to viscosity of the medium and its consequences (boundary layer formation, wakes, vortex shedding) and the shock formation in relative supersonic passages.

The measure of irreversibility in a compressor may be thermodynamically defined through some form of compressor efficiency. There are two methods of compressor efficiency definitions. These are

- compressor adiabatic efficiency η_c
- compressor polytropic efficiency e_c

To define the compressor adiabatic efficiency η_c, we depict a "real" compression process on an h–s diagram, as shown in Fig. 4.6 and compare it to an ideal, i.e., isentropic process. The state "t2" represents the *total* (or stagnation) state of the gas entering the compressor, typically designated by p_{t2} and T_{t2}. An actual flow in a compressor will follow the solid line from "t2" to "t3" thereby experiencing an entropy rise in the process, Δs_c. The actual total state of the gas is designated by "t3" in Fig. 4.6. The ratio p_{t3}/p_{t2} is known as the compressor pressure ratio, with a shorthand notation π_c. The compressor total temperature ratio is depicted by, the shorthand notation, $\tau_c \equiv T_{t3}/T_{t2}$. Since the state "t3" is the actual state of the gas at the exit of the compressor and is not achieved via an isentropic process, we cannot expect the isentropic relation between τ_c and π_c, to hold, i.e.,

$$\tau_c \neq \pi_c^{\frac{\gamma-1}{\gamma}} \tag{4.9}$$

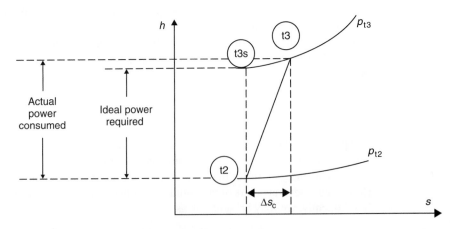

■ **FIGURE 4.6** Enthalpy–entropy (h–s) diagram of an actual and ideal compression process (note Δs_c)

It can be seen from Fig. 4.6 that the actual τ_c is larger than the ideal, i.e., isentropic τ_c, which is denoted by the end state T_{t3s}. This fact actually helps with the exponent memorization in a real compression process. The compressor adiabatic efficiency is the ratio of the ideal power required to the power consumed by the compressor, i.e.,

$$\eta_c \equiv \frac{h_{t3s} - h_{t2}}{h_{t3} - h_{t2}} = \frac{\Delta h_{t,\text{isentropic}}}{\Delta h_{t,\text{actual}}} \tag{4.10}$$

The numerator in Eq. 4.10 is the power-per-unit mass flow rate in an *ideal compressor* and the denominator is the power-per-unit mass flow rate in the actual compressor. If we divide the numerator and denominator of Eq. 4.10 by h_{t2}, we get

$$\eta_c = \frac{T_{t3s}/T_{t2} - 1}{T_{t3}/T_{t2} - 1} \tag{4.11}$$

Since the thermodynamic states "t3s" and "t2" are on the same isentrope, the temperature and pressure ratios are then related via the isentropic formula, i.e.,

$$\frac{T_{t3s}}{T_{t2}} = \left(\frac{p_{t3s}}{p_{t2}}\right)^{\frac{\gamma-1}{\gamma}} = \left(\frac{p_{t3}}{p_{t2}}\right)^{\frac{\gamma-1}{\gamma}} = \pi_c^{\frac{\gamma-1}{\gamma}} \tag{4.12}$$

Therefore, compressor adiabatic efficiency may be expressed in terms of compressor pressure and temperature ratios as

$$\eta_c = \frac{\pi_c^{\frac{\gamma-1}{\gamma}} - 1}{\tau_c - 1} \tag{4.13}$$

Equation 4.13 involves three parameters, η_c, π_c, and τ_c. It can be used to calculate τ_c for a given compressor pressure ratio and adiabatic efficiency. As compressor pressure ratio and efficiency are typically known and assumed quantities in a gas turbine cycle analysis, the only unknown in Eq. 4.13 is τ_c.

A second efficiency parameter in a compressor is *polytropic efficiency* e_c. As might be expected, compressor adiabatic and polytropic efficiencies are related. The definition of compressor polytropic efficiency is

$$e_c \equiv \frac{dh_{ts}}{dh_t} \tag{4.14}$$

It is interesting to compare the definition of compressor adiabatic efficiency, involving finite jumps (Δh_t), and the polytropic efficiency, which takes infinitesimal steps (dh_t). The conclusion can be reached that the polytropic efficiency is actually the adiabatic efficiency of a compressor with *small* pressure ratio. Consequently, compressor polytropic efficiency is also called *small stage efficiency*. From the combined first and second law of thermodynamics, we have

$$T_t ds = dh_t - \frac{dp_t}{\rho_t} \tag{4.15}$$

we deduce that for an isentropic process, i.e., $ds = 0$, $dh_t = dh_{ts}$ and therefore,

$$dh_{ts} = \frac{dp_t}{\rho_t} \tag{4.16}$$

If we substitute Eq. 4.16 in 4.14 and replace density with pressure and temperature from the perfect gas law, we get

$$e_c = \frac{\dfrac{dp_t}{p_t}}{\dfrac{dh_t}{RT_t}} = \frac{\dfrac{dp_t}{p_t}}{\dfrac{C_p dT_t}{RT_t}} = \frac{\dfrac{dp_t}{p_t}}{\dfrac{\gamma}{\gamma-1}\dfrac{dT_t}{T_t}} \tag{4.17}$$

$$\frac{dp_t}{p_t} = \frac{\gamma e_c}{\gamma-1}\frac{dT_t}{T_t} \tag{4.18}$$

which can now be integrated between the inlet and exit of the compressor to yield

$$\frac{p_{t3}}{p_{t2}} = \pi_c = \left(\frac{T_{t3}}{T_{t2}}\right)^{\frac{\gamma e_c}{\gamma-1}} = (\tau_c)^{\frac{\gamma e_c}{\gamma-1}} \tag{4.19}$$

To express the compressor total temperature ratio in terms of compressor pressure ratio and polytropic efficiency, Eq. 4.19 can be rewritten as

$$\tau_c = \pi_c^{\frac{\gamma-1}{\gamma e_c}} \tag{4.20}$$

The presence of e_c in the denominator of the above exponent (Eq. 4.20) causes the exponent of π_c to be greater than its isentropic exponent (which is $(\gamma-1)/\gamma$), therefore,

$$\tau_{c,real} \succ \tau_{c,isentropic} \quad \text{or} \quad T_{t3} > T_{t3s} \tag{4.21}$$

as noted earlier (see Fig. 4.6). The physical argument for higher actual T_t than the isentropic T_t (to achieve the same compressor pressure ratio) can be made on the ground that *lost work* to overcome the irreversibility in the real process (friction, shock) is converted into heat, 'a higher exit T_t is reached in a real machine due to dissipation. On the contrary, for a given compressor pressure ratio π_c, an ideal compressor consumes less power than an actual compressor (the factor being η_c, as defined earlier). Again, the absence of dissipative mechanisms, leading to lost work, is cited as the reason for a reversible flow machine to require less power to run.

Now, to relate the two types of compressor efficiency description, e_c and η_c, we may substitute Eq. 4.20 into 4.13, to get

$$\eta_c = \frac{\pi_c^{\frac{\gamma-1}{\gamma}} - 1}{\tau_c - 1} = \frac{\pi_c^{\frac{\gamma-1}{\gamma}} - 1}{\pi_c^{\frac{\gamma-1}{\gamma \cdot e_c}} - 1} \tag{4.22}$$

Equation 4.22 is plotted in Fig. 4.7.

The compressor adiabatic efficiency η_c is a function of compressor pressure ratio, while the polytropic efficiency is independent of it. Consequently, in a cycle analysis, we usually assume the polytropic efficiency e_c as the figure of merit for a compressor (and turbine) and then we can maintain e_c as constant in our engine off-design analysis. Typical values for the polytropic efficiency in modern compressors range in the 88–92%.

So far, we have studied the thermodynamics of the mechanical compression through the stagnation states of the working medium. It is also instructive to examine the static states of the gas in a compressor as well. The mental connection between these states will help us with the fluid dynamics of the compressors. Figure 4.8 is basically an elaborate version of an earlier Fig. 4.6, to which we have added the static states of the gas at the inlet and exit of the compressor. These states, i.e., the local stagnation and static states, when plotted on the *h–s*

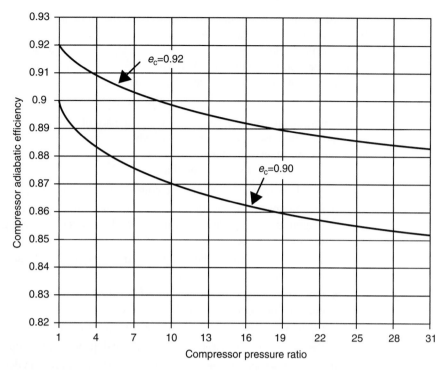

■ **FIGURE 4.7** Variation of compressor adiabatic efficiency with the pressure ratio and polytropic efficiency

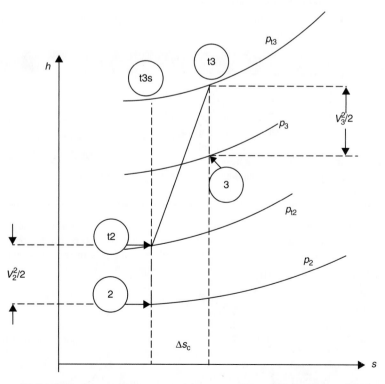

■ **FIGURE 4.8** Static and stagnation states at the compressor inlet and exit, on an h–s diagram

diagram, are distanced from each other only by the amount of local (specific) kinetic energy $V^2/2$ and on the same isentrope.

The flow velocity at the compressor inlet, V_2, is nearly the same as the fluid velocity at the compressor exit, V_3, by design. This explains the vertical gap shown in Fig. 4.8, between the static and stagnation states at 2 and 3 is nearly the same, i.e.,

$$V_2^2/2 \approx V_3^2/2 \tag{4.23}$$

This design philosophy, i.e., $V_2 = V_3$, is a part of the so-called constant throughflow assumption often used in compressor aerodynamics. We will return to this topic, in detail, in the turbomachinery chapter.

In summary, we have learned that

- a real compressor flow may be considered adiabatic, i.e., $\dot{Q}_{\text{compressor}} \approx 0$
- the energy transfer to the fluid due to shaft, in a compressor, is several orders of magnitude higher than any heat transfer that takes place through the casing, thus heat transfer is neglected
- viscous dissipation in the wall boundary layer and shocks account for the sources of irreversibility in a compressor
- there are two figures of merit that describe the compressor efficiency, one is the adiabatic compressor efficiency η_c and the second is the polytropic or small-stage efficiency e_c
- the two compressor efficiencies are interrelated, i.e., $\eta_c = \eta_c(\pi_c, e_c)$
- the compressor polytropic efficiency is independent of compressor pressure ratio π_c
- the compressor adiabatic efficiency is a function of π_c and decreases with increasing pressure ratio
- to achieve a high-pressure ratio in a compressor, multistaging and multispool configurations are needed
- in a gas turbine engine, the compressor power is derived from a shaft that is connected to a turbine

EXAMPLE 4.2

A multistage axial-flow compressor has a mass flow rate of 50 kg/s and a total pressure ratio of 35. The compressor polytropic efficiency is $e_c = 0.90$. The inlet flow condition to the compressor is described by $T_{t2} = 288$ K and $p_{t2} = 100$ kPa. Assuming the flow in the compressor is adiabatic, and constant gas properties throughout the compressor are assumed, i.e., $\gamma = 1.4$ and $c_p = 1004$ J/kg · K, calculate

(a) compressor exit total temperature T_{t3} in K

(b) compressor adiabatic efficiency η_c

(c) compressor shaft power $\wp_c$ in MW

SOLUTION

Following Eq. 4.20, we relate compressor total temperature and pressure ratio via polytropic efficiency,

$$\tau_c = \pi_c^{\frac{\gamma-1}{\gamma \cdot e_c}} = (35)^{0.31746} \cong 3.0916$$

Therefore, the exit total temperature is $T_{t3} = 3.0916\,T_{t2} = 3.0916(288\,\text{K}) \cong 890.4$ K.

Compressor adiabatic efficiency is related to the polytropic efficiency and compressor pressure ratio, via Eq. 4.13

$$\eta_c = \frac{\pi_c^{\frac{\gamma-1}{\gamma}} - 1}{\tau_c - 1} = \frac{35^{0.2857} - 1}{3.0916 - 1} \approx 0.8422$$

Therefore, compressor adiabatic efficiency is $\eta_c \cong 84.22\%$.

Compressor shaft power is proportional to the mass flow rate (i.e., the size of the compressor) as well as the total enthalpy rise across the compressor, according to

$$\wp_c = \dot{m}(h_{t3} - h_{t2}) = \dot{m}c_p(T_{t3} - T_{t2})$$
$$= (50\,\text{kg/s})(1004\,\text{J/kg}\cdot\text{K})(890.4 - 288)\text{K} \approx 30.24\,\text{MW}$$

Therefore the shaft power delivered to the compressor is $\wp_c \approx 30.24$ MW

4.3.1.3 *The Burner.*

In the combustor, the air is mixed with the fuel and a chemical reaction ensues which is *exothermic*, i.e., it results in a heat release. The ideal burner is considered to behave like a reversible heater, which in the combustion context, means very slow burning, $M_b \approx 0$, and with no friction acting on its walls. Under such circumstances, the total pressure remains conserved.

In a real combustor, due to wall friction, turbulent mixing and chemical reaction at finite Mach number, the total pressure drops, i.e.,

$$\pi_b = \frac{p_{t4}}{p_{t3}} \prec 1 \quad \text{``\textit{real} combustion chamber''} \tag{4.24}$$

$$\pi_b = 1 \qquad \text{``\textit{ideal} combustion chamber''} \tag{4.25}$$

Kerrebrock gives an approximate expression for π_b in terms of the average Mach number of the gas in the burner, M_b, as

$$\pi_b \approx 1 - \varepsilon \frac{\gamma}{2} M_b^2 \qquad \text{where,} \qquad 1 \prec \varepsilon \prec 2 \tag{4.26}$$

The total pressure loss in a burner is then proportional to the average dynamic pressure of the gases inside the burner, i.e., $\propto (\gamma/2)M_b^2$, where the proportionality coefficient is ε. Assuming an average Mach number of gases of 0.2 and $\varepsilon = 2$, we get $\pi_b \approx 0.95$ (for a $\gamma \approx 1.33$). This formula is obviously not valid for supersonic combustion throughflows, rather it points out the merits of *slow* combustion in a conventional burner. A schematic diagram of a combustion chamber, and its essential components, is shown in Fig. 4.9.

A preliminary discussion of the components of the combustion chamber is useful at this time. The inlet diffuser decelerates the compressor discharge flow to a Mach number of about 0.2–0.3. The low-speed flow will provide an efficient burning environment in the combustor. Mixing improvement with the fuel in the combustor primary zone is achieved via the air swirler. A recirculation zone is created that provides the necessary stability in the primary combustion zone. To create a fuel-rich environment in the primary zone to sustain combustion, a large percentage of air is diverted around a dome-like structure. The airflow that has bypassed the burner primary zone will enter the combustor as the cooling flow through a series of cooling

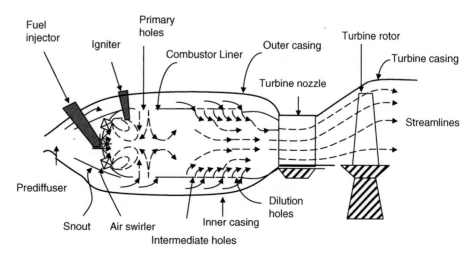

■ **FIGURE 4.9** Components of a conventional combustion chamber and the first turbine stage

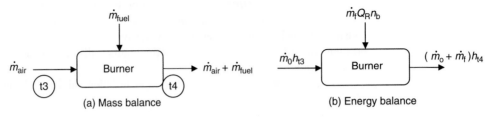

(a) Mass balance   (b) Energy balance

■ **FIGURE 4.10**   Block diagram of a burner with mass and energy balance

holes (as shown). The fuel-air mixture is ignited in the combustor primary zone via an igniter properly positioned in the dome area.

For the purposes of *cycle analysis*, a combustor flow is analyzed only at its inlet and outlet. Thus, we will not consider the details of combustion processes such as atomization, vaporization, mixing, chemical reaction, and dilution in the cycle analysis phase. Also we do not consider pollutant formation and the means of reducing them in this chapter. We will address these issues in a later chapter dealing with combustor flow and design considerations. Obviously, with the details of flow and reaction omitted, we need to make assumptions regarding the loss of total pressure and burner efficiency in the cycle analysis only. A block diagram representation of a combustor, useful in cycle analysis, is shown in Fig. 4.10. Figures 4.10 (a) and (b) that are the steady-state mass and the energy balance applied to the combustion chamber, i.e.,

$$\dot{m}_4 = \dot{m}_0 + \dot{m}_f = \dot{m}_0(1 + f) \tag{4.27}$$

where, f is the fuel-to-air ratio, $f \equiv \dfrac{\dot{m}_f}{\dot{m}_0}$, and energy balance in Fig. 4.9 (b) gives

$$\dot{m}_0 h_{t3} + \dot{m}_f Q_R \eta_b = (\dot{m}_0 + \dot{m}_f)h_{t4} = \dot{m}_0(1 + f)h_{t4} \tag{4.28}$$

The fuel is characterized by its *energy content per unit mass*, i.e., the amount of thermal energy inherent in the fuel, capable of being released in a chemical reaction. This parameter is heat of reaction and is given the symbol Q_R. The unit for this parameter is energy/mass, which in the metric system is kJ/kg and in the British system of units is BTU/lbm. In an actual combustion chamber, primarily due to volume limitations, the entirety of the Q_R cannot be realized. The fraction that can be realized is called *burner efficiency* and is given the symbol η_b. Therefore,

$$\eta_b \equiv \frac{Q_{R,\text{Actual}}}{Q_{R,\text{Ideal}}} \tag{4.29}$$

The ideal heat of reaction or heating value of typical hydrocarbon fuels, to be used in our cycle analysis, is

$$Q_R = 42,000 \, \text{kJ/kg} \tag{4.30-a}$$

or

$$Q_R = 18,000 \, \text{BTU/lbm} \tag{4.30-b}$$

However, the most energetic fuel is the hydrogen, which is capable of releasing roughly three times the energy of typical hydrocarbon fuels per unit mass, i.e.,

$$Q_R = 127,500 \, \text{kJ/kg} \tag{4.30-c}$$

or

$$Q_R = 55,400 \, \text{BTU/lbm} \tag{4.30-d}$$

Consistent with the theory of *no free lunch,* we need to note the drawbacks of a fuel such as hydrogen. The low-molecular weight of hydrogen makes it the lightest fuel (with a density ratio of about 1/10 of typical hydrocarbon fuels, such as octane). This implies a comparatively very large volume requirement for the hydrogen fuel. We may want to think of this as *volumetric efficiency* of the hydrogen is the lowest of all fuels. Secondly, hydrogen in liquid form is cryogenic, which means a very low boiling point temperature, i.e., $-423°F$ or 20 K at ambient pressure. The cryogenic aspect of hydrogen requires thermally insulated fuel tanks, fuel lines, valves, and the associated weight penalty. Therefore, due to space limitations and system requirement weight penalties on board conventional atmospheric-flight aircraft, conventional hydrocarbon fuels are preferred over hydrogen. An exception to this is found in rocketry and hypersonic airbreathing propulsion where the regenerative cooling of the engine components requires a cryogenic fuel such as hydrogen to withstand the thermal loads of aerodynamic heating.

Typically in modern gas turbine engines, the burner efficiency can be as high as 98–99%. In a cycle analysis, we need to make assumptions about the loss parameters in every component, which in a combustion chamber are $\pi_b \prec 1$ and $\eta_b \prec 1$. The real and ideal combustion process can be depicted on a *T–s* diagram, which later will be used to perform cycle analysis. Figure 4.11 shows the burner thermodynamic process on a *T–s* diagram.

The isobars p_{t3} and p_{t4} drawn at the entrance and exit of the combustor in Fig. 4.11, clearly show a total pressure drop in the burner, $\Delta p_{t, \text{burner}}$. The maximum temperature limit T_{t4} is governed by the level of cooling technology, material selection, and the thermal protective coating used in the turbine. Typical current values for the maximum T_{t4} is about 3200– 3,600 °R or 1775–2,000 K. Another burner parameter is the temperature rise ΔT_t across the combustion chamber, as shown in Fig. 4.11. The thermal power invested in the engine (by the fuel) is proportional to the temperature rise across the combustor, i.e., it is nearly equal to $\dot{m}_0 c_p (\Delta T_t)_{\text{burner}}$.

The application of energy balance across the burner, i.e., Eq. 4.28, will yield the fuel-to-air ratio f as the only unknown parameter. To derive an expression for the fuel-to-air ratio, we will divide Eq. 4.28 by $\dot{m}_0$, the air mass flow rate, to get

$$h_{t3} + f Q_R \eta_b = (1 + f) h_{t4} \qquad (4.31)$$

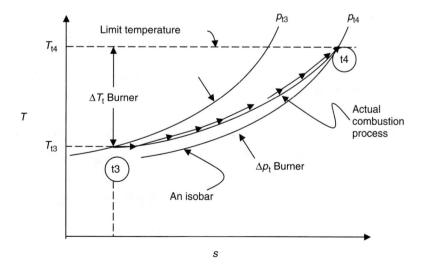

■ **FIGURE 4.11** **Actual flow process in a burner (note total pressure loss Δp_t across the burner)**

The unknown parameter f can be isolated and expressed as

$$f = \frac{h_{t4} - h_{t3}}{Q_R \eta_b - h_{t4}} \qquad \text{(4.32)}$$

Knowing the fuel property Q_R, assuming burner efficiency η_b, and having specified a turbine inlet temperature T_{t4}, the denominator of Eq. 4.32 is fully known. The compressor discharge temperature T_{t3} is established via compressor pressure ratio, efficiency, and inlet condition, as described in the compressor section and in Eq. 4.19, which renders the numerator of Eq. 4.32 fully known as well. Therefore, application of energy balance to a burner usually results in the establishment of fuel-to-air ratio parameter f. It is customary to express Eq. 4.32 in terms of nondimensional parameters, by dividing each term in the numerator and denominator by the flight static enthalpy h_0 to get

$$f = \frac{\dfrac{h_{t4}}{h_0} - \dfrac{h_{t3}}{h_0}}{\dfrac{Q_R \eta_b}{h_0} - \dfrac{h_{t4}}{h_0}} = \frac{\tau_\lambda - \tau_r \tau_c}{\dfrac{Q_R \eta_b}{h_0} - \tau_\lambda} \qquad \text{(4.33)}$$

where we recognize the product $\tau_r \tau_c$ as h_{t3}/h_0, and τ_λ as the cycle thermal limit parameter h_{t4}/h_0.

In summary, we learned that

- the fuel is characterized by its heating value Q_R (maximum releasable thermal energy per unit mass)
- the burner is characterized by its efficiency η_b, and its total pressure ratio π_b
- burning at finite Mach number, frictional losses on the walls and turbulent mixing are identified as the sources of irreversibility, i.e., losses, in a burner
- the fuel-to-air ratio f and the burner exit temperature T_{t4} are the thrust control/engine design parameters
- the application of the energy balance across the burner yields either f or T_{t4}

EXAMPLE 4.3

A gas turbine combustor has inlet condition $T_{t3} = 800\,\text{K}$, $p_{t3} = 2\,\text{Mpa}$, air mass flow rate of $50\,\text{kg/s}$, $\gamma_3 = 1.4$, $c_{p3} = 1004\,\text{J/kg·K}$.

A hydrocarbon fuel with ideal heating value $Q_R = 42000\,\text{kJ/kg}$ is injected in the combustor at a rate of $1\,\text{kg/s}$. The burner efficiency is $\eta_b = 0.995$ and the total pressure at the combustor exit is 96% of the inlet total pressure, i.e., combustion causes a 4% loss in total pressure. The gas properties at the combustor exit are $\gamma_4 = 1.33$ and $c_{p4} = 1156\,\text{J/kg·K}$. Calculate

(a) fuel-to-air ratio f

(b) combustor exit temperature T_{t4} in K and p_{t4} in MPa

SOLUTION

The air and fuel flow rates are specified at 50 and 1 kg/s, respectively, in the problem, therefore, $f = 1/50 = 0.02$ or 2%.

We calculate combustor exit temperature by energy balance,

$$\dot{m}_0 h_{t3} + \dot{m}_f Q_R \eta_b = (\dot{m}_0 + \dot{m}_f) h_{t4} = \dot{m}_0 (1 + f) h_{t4}$$

Therefore

$$h_{t3} + f Q_R \eta_b = (1 + f) h_{t4}$$

$$T_{t4} = \frac{(c_{p3}/c_{p4}) T_{t3} + f Q_R \eta_b / c_{p4}}{1 + f}$$

$$= \frac{(1004/1156) 800\,\text{K} + 0.02 (42000\,\text{kJ/kg})(0.995)/1.156\,\text{kJ/kg·K}}{1.02}$$

Therefore, combustor exit temperature is $T_{t4} \cong 1390\,\text{K}$ and $p_{t4} = 0.96(2\,\text{MPa}) = 1.92\,\text{MPa}$

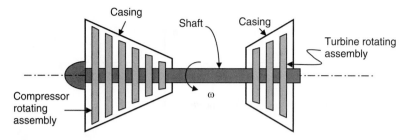

■ **FIGURE 4.12** **Common shaft in a gas generator connects the compressor and turbine (ω is the shaft rotational speed)**

4.3.1.4 ***The Turbine.*** The high pressure and temperature gas that leaves the combustor is directed into a turbine. The turbine may be thought of as a *valve,* because on one side, it has a high-pressure gas and on the other side, it has a low-pressure gas of the exhaust nozzle or the tailpipe. Therefore, the first *valve,* i.e., the throttle station, in a gas turbine engine is at the turbine. The throat of an exhaust nozzle in a supersonic aircraft is the second and final throttle station in an engine. Thus, the flow process in a turbine (and exhaust nozzle) involves significant (static) pressure drop, and in harmony with it, the (static) temperature drop, which is called *flow expansion.* The flow expansion produces the necessary power for the compressor and the propulsive power for the aircraft. The turbine is connected to the compressor via a common shaft, which provides the shaft power to the compressor (see Fig. 4.12). In drawing an analogy, we can think of the expansion process in a gas turbine engine as the counterpart of *power stroke* in an intermittent combustion engine. However, in a turbine, the power transmittal is continuous.

Due to high temperatures of the combustor exit flow, the first few stages of the turbine, i.e., the high-pressure turbine (HPT), need to be cooled. The coolant is the air bleed from the compressor, which may be bled from different compression stages, e.g., between the low- and high-pressure compressor and at the compressor exit. It has been customary, however, to analyze an *uncooled* turbine in the preliminary cycle analysis, then followed by an analysis of the cycle with cooling effects in the turbine and the exhaust nozzle. Other cooling media, such as water, have been used in stationary gas turbine power plants. However, in a design-to-weight environment of aircraft carrying extra water to cool, the turbine blades are not feasible. A cooling solution, which uses the engine cryogenic fuel, such as hydrogen or methane, as the coolant to cool the engine and aircraft components is called *regenerative cooling* and has proven its effectiveness in liquid propellant chemical rocket engines for decades.

A real flow process in an uncooled turbine involves irreversibilities such as frictional losses in the boundary layer, tip clearance flows, and shock losses in transonic turbine stages. The viscous dominated losses, i.e., boundary layer separation, reattachment, and tip vortex flows, are concentrated near the end walls, thus a special attention in turbine flow optimization is made on the *end-wall* regions. We will address the end wall losses and treatments in the turbomachinery chapters. Another source of irreversibility in a real turbine flow is related to the cooling losses. Coolant is typically injected from the blade attachment (to the hub or casing) into the blade, which provides internal convective cooling and usually external film cooling on the blades. The turbulent mixing associated with the coolant stream and the hot gases is the primary mechanism for (the turbine stage) cooling losses. We will examine the question of turbine cooling and cycle thermal efficiency later in this chapter.

The thermodynamic process for an uncooled turbine flow may be shown in an *h–s* diagram (see in Fig. 4.13). The actual expansion process in the turbine is depicted by the solid line connecting the total (or stagnation) states t4 and t5 (see Fig. 4.13). The isentropically reached exit state t5s represents the ideal, loss-free flow expansion in the turbine to the same

■ **FIGURE 4.13**
Expansion process in an
uncooled turbine (note
the entropy rise across
the turbine, Δs_t)

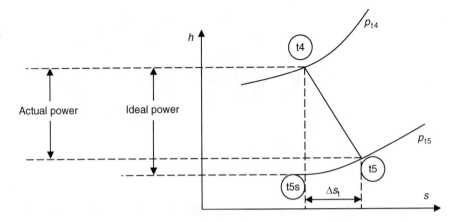

backpressure p_{t5}. The relative height, on the enthalpy scale in Fig. 4.13, between the inlet total and outlet total condition of the turbine represents the power production potential (i.e., ideal) and the actual power produced in a turbine. The ratio of these two heights is called the turbine adiabatic efficiency η_t, i.e.,

$$\wp_{t,\,actual} = \dot{m}_t(h_{t4} - h_{t5}) = \dot{m}_t \Delta h_{t,\,actual} \tag{4.34}$$

$$\wp_{t,\,ideal} = \dot{m}_t(h_{t4} - h_{t5s}) = \dot{m}_t \Delta h_{t,\,isentropic} \tag{4.35}$$

$$\eta_t \equiv \frac{h_{t4} - h_{t5}}{h_{t4} - h_{t5s}} = \frac{\Delta h_{t,\,actual}}{\Delta h_{t,\,isentropic}} \tag{4.36}$$

In Eqs. 4.34 and 4.35, the turbine mass flow rate is identified as $\dot{m}_t$ which accounts for the air and fuel mass flow rate that emerge from the combustor and expand through the turbine, i.e.,

$$\dot{m}_t = \dot{m}_0 + \dot{m}_f = (1 + f)\dot{m}_0 \tag{4.37}$$

The numerator in Eq. 4.36 is the actual power produced in a *real uncooled turbine,* and the denominator is the *ideal* power that in a reversible and adiabatic turbine could be produced. If we divide the numerator and denominator of Eq. 4.36 by h_{t4}, we get

$$\eta_t = \frac{1 - T_{t5}/T_{t4}}{1 - T_{t5s}/T_{t4}} \tag{4.38}$$

Since the thermodynamic states "t5s" and "t4" are on the same isentrope, the temperature and pressure ratios are then related via the isentropic formula, i.e.,

$$\frac{T_{t5s}}{T_{t4}} = \left(\frac{p_{t5s}}{p_{t4}}\right)^{\frac{\gamma-1}{\gamma}} = \left(\frac{p_{t5}}{p_{t4}}\right)^{\frac{\gamma-1}{\gamma}} = \pi_t^{\frac{\gamma-1}{\gamma}} \tag{4.39}$$

Therefore, turbine adiabatic efficiency may be expressed in terms of the turbine total pressure and temperature ratios as

$$\eta_t = \frac{1 - \tau_t}{1 - \pi_t^{\frac{\gamma-1}{\gamma}}} \tag{4.40}$$

We may also define a *small-stage* efficiency for a turbine, as we did in a compressor, and call it the turbine polytropic efficiency e_t. For a small expansion, representing a small stage, we can replace the finite jumps, i.e., Δs, with incremental step d in Eq. 4.36 and write

$$e_t \equiv \frac{dh_t}{dh_{ts}} = \frac{dh_t}{\dfrac{dp_t}{\rho_t}} \tag{4.41}$$

Expressing enthalpy in terms of temperature and specific heat at constant pressure as $dh_t = c_p dT_t$, which is suitable for a perfect gas, and simplify Eq. 4.41 similar to the compressor section, we get

$$\tau_t = \pi_t^{\frac{(\gamma-1)e_t}{\gamma}} \quad \text{or} \quad \pi_t = \tau_t^{\frac{\gamma}{(\gamma-1)e_t}} \tag{4.42}$$

$$\tau_t \equiv T_{t5}/T_{t4} \tag{4.43}$$

$$\pi_t \equiv p_{t5}/p_{t4} \tag{4.44}$$

We note that in Eq. 4.42, in the limit of e_t approaching 1, i.e., isentropic expansion, we will recover the isentropic relationship between the temperature and pressure ratio, as expected. Replacing π_t in Eq. 4.39 by its equivalent expression (from Eq. 4.42), we derive a relation between the two types of turbine efficiencies, η_t and e_t, as

$$\eta_t = \frac{1 - \tau_t}{1 - \tau_t^{1/e_t}} \tag{4.45}$$

Equation 4.45 is plotted in Fig. 4.14. For a small-stage efficiency of 90%, i.e., $e_t = 0.90$, the turbine adiabatic efficiency η_t grows with the inverse of turbine expansion parameter $1/\tau_t$. This may initially defy logic that how is it possible to become more efficient if we *add* more stages to the turbine? Wouldn't it add to losses? The opposite of this trend occurred in the compressor, i.e., compressor adiabatic efficiency for a finite-size compressor *was* lower than the efficiency of a small-stage compressor (see Fig. 4.7). The explanation is that the energy transfer occurs from the fluid to the rotor in the turbine; therefore, more stages offer more opportunities to convert fluid energy into shaft power. The opposite happens in a compressor, i.e., the compressor stages *consume* power, which means that the additional stages add losses to the process. In simple terms, the turbine stages may be thought of as *opportunities* and compressor stages as the *burden*.

The temperature ratio parameter τ_t across the turbine is established via a power balance between the turbine, compressor, and other shaft power extraction (e.g., electric generator) on the gas generator. Let us first consider the power balance between the turbine and compressor in its simplest form and then try to build on the added parameters. Ideally, the compressor absorbs all the turbine shaft power, i.e.,

$$\wp_t = \wp_c \tag{4.46-a}$$

$$\dot{m}_0(1 + f)(h_{t4} - h_{t5}) = \dot{m}_0(h_{t3} - h_{t2}) \tag{4.46-b}$$

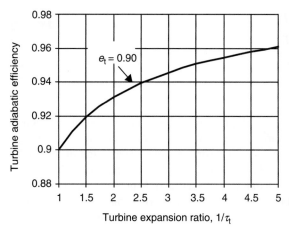

■ **FIGURE 4.14** Variation of turbine adiabatic efficiency η_t with the inverse of turbine expansion ratio $1/\tau_t$

which simplifies to the following nondimensional form

$$(1 + f)\tau_\lambda(1 - \tau_t) = \tau_r(\tau_c - 1) \tag{4.46-c}$$

The only unknown in the above equation is τ_t, as all other parameters either flow from upstream components, e.g., combustor will provide f, the compressor produces τ_c, etc., or are design parameters such as τ_λ or τ_r. Hence, the turbine expansion parameter τ_t can be written as

$$\tau_t = 1 - \frac{\tau_r(\tau_c - 1)}{(1 + f)\tau_\lambda} \tag{4.46-d}$$

Next, we consider the practical issue of hydrodynamic (frictional) losses in (radial) bearings holding the shaft in place and provide dynamic stability under operating conditions to the rotating assemblies of turbine and compressor. Therefore, a small fraction of the turbine power output is dissipated through viscous losses in the bearings, i.e.,

$$\wp_t = \wp_c + \Delta\wp_{bearings} \tag{4.47}$$

where $\Delta\wp_{bearings}$ is the power loss due to bearings. In addition, an aircraft has electrical power needs for its flight control system and other aircraft subsystems, which requires tapping into the turbine shaft power. Consequently, the power balance between the compressor and turbine should account for the electrical power extraction, which usually accompanies the gas generator, i.e.,

$$\wp_t = \wp_c + \Delta\wp_{bearing} + \Delta\wp_{electric\ generator} \tag{4.48-a}$$

$$\wp_c = \wp_t - \Delta\wp_{bearing} - \Delta\wp_{electric\ generator} \tag{4.48-b}$$

In a simple cycle analysis, it is customary to lump all power dissipation and power extraction terms into a single *mechanical efficiency* parameter η_m that is multiplied by the turbine shaft power, to derive the compressor shaft power, i.e.,

$$\wp_c = \eta_m \wp_t \tag{4.49}$$

where η_m is the mechanical efficiency parameter that needs to be specified a'priori, e.g., $\eta_m = 0.95$.

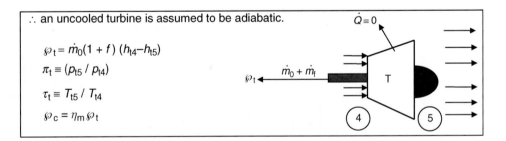

> ∴ an uncooled turbine is assumed to be adiabatic. $\dot{Q} \equiv 0$
>
> $\wp_t = \dot{m}_0(1 + f)(h_{t4} - h_{t5})$
>
> $\pi_t \equiv (p_{t5} / p_{t4})$
>
> $\tau_t \equiv T_{t5} / T_{t4}$
>
> $\wp_c = \eta_m \wp_t$
>
> $\wp_t \leftarrow$ $\dot{m}_0 + \dot{m}_f$ T
>
> 4   5

To cool the high-pressure turbine stages, a small fraction of compressor air can be diverted from various stages of the compression. Engine cooling is essentially a *pressure-driven* process, which calls for a pressure scheduling of the coolant to achieve the highest cooling efficiency. For example, to cool the first nozzle and the first rotor in a turbine, we need to tap the compressor exit air, as it has the *right* pressure. Earlier stages of the compressor have not yet developed the necessary pressure to overcome downstream pressure in the high-pressure end of a turbine. To cool the medium-pressure turbine stages, we need to divert compressor air

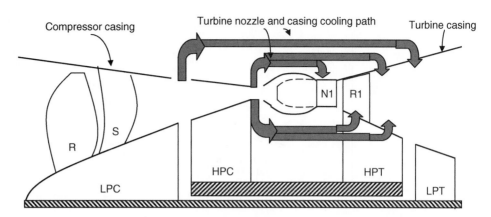

■ **FIGURE 4.15** **Two-spool gas turbine engine with two stations of compressor bleed for cooling purposes (LPC: Low-pressure compressor, HPC: High-pressure compressor, HPT: High-pressure turbine, LPT: Low-pressure turbine, R: Rotor, S: Stator, N: Nozzle)**

from the medium-pressure compressor. We could use higher pressure compressor exit air as well but the extra compression work we spent compressing the gas to the exit condition is essentially a wasted effort, i.e., energy loss for the cycle. Therefore, the problem of *pressure matching* between the coolant stream and the hot gas in a turbine is real and significant. The consequence of a mismatch is either inadequate cooling flow in a turbine blade with a possibility of even a *reverse flow* of hot gases into the cooling channels or the need for excessive throttling of the coolant stream and associated cycle losses. Figure 4.15 shows possible coolant extraction in the compressor for turbine cooling.

We can simplify the investigation of the thermodynamics of cooled turbine stages significantly by assuming that each blade row coolant is discharged at the blade row exit, i.e., trailing edge ejection. This simplifies the thermodynamics as we consider *dumping* the entire coolant used in a given blade row at the trailing edge of the blades and then apply conservation laws to the coolant-hot gas mixing process. This approach avoids the discrete coolant-hot gas mixing, which may be distributed over the suction and pressure surfaces of the blade row, as in film-cooled blades. Application of the energy equation to the coolant-hot gas mixing process will result in a lower mixed-out temperature for the hot gas. Consequently, the hot-gas stream will experience a reduced entropy state after mixing with the coolant. In return, the coolant stream, which is heated by the hot gas, will experience an entropy rise, which according to the second law of thermodynamics, is greater than the entropy reduction in the hot-gas stream. The expansion process in a turbine with two cooled stages followed by an uncooled low-pressure turbine is depicted in an *h*–*s* diagram in Fig. 4.16. The mixing process at the exit of

■ **FIGURE 4.16**
***h*–*s* diagram for a turbine expansion with two cooled stages followed by an uncooled LPT (blade row exit mixing with the coolant is indicated by a negative Δ*s* and a temperature drop)**

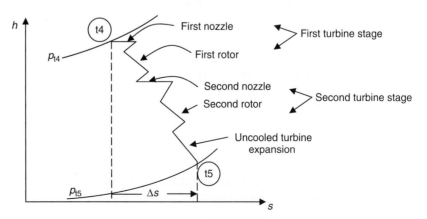

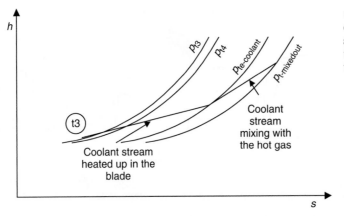

■ **FIGURE 4.17** *h–s* diagram of the coolant stream, first inside the blade then mixing with the hot turbine gas

each blade row is identified with negative entropy production, and the cooling path is roughly drawn in as a nearly constant pressure process. In contrast, the coolant stream will undergo a heating process from the blade and then through mixing with the hot gases in the turbine. The thermodynamic state of the coolant is shown in Fig. 4.16.

The *h–s* diagram for the hot gas (shown in Fig. 4.16) shows a constant h_t process for the first and second turbine nozzle. Constant total enthalpy indicates an adiabatic process with no mechanical energy exchange. These two criteria are met in an uncooled turbine nozzle, which is stationary and consequently exchanges no mechanical energy with the fluid. Across the nozzle, due to frictional and shock losses, a total pressure drop is indicated in Fig. 4.16. Regarding the blade cooling, however, we will introduce the entire blade row coolant at the blade row exit, as suggested earlier. The cooling process results in a lower total temperature of the mixture of hot and cold gas, as well as a total pressure drop due to turbulent mixing of the two streams. This process can be seen in Fig. 4.17. The drop of total enthalpy across the rotor is proportional to the shaft power production by the rotor blade row. This shaft power may be written as

$$\wp_{rotor} = \tau_{rotor}\,\omega = \dot{m}_{rotor}\Delta h_t \tag{4.50}$$

The product of rotor torque τ and the angular speed ω of the rotor is the shaft power a' la Newton. The irreversibility in the flow through the rotor is shown via a process involving entropy rise across the rotor in Fig. 4.16. Again the flow expansion after the rotor is cooled off by the blade row coolant, which is injected at the blade row trailing edge and the process continues. To establish the mixed-out state of the coolant and the hot gas, we will apply the conservation principles, as noted earlier. The law of conservation of energy applied to a coolant stream of $\dot{m}_c$ mass flow rate carrying a total enthalpy of $h_{t,c}$ mixing with a hot gas stream of $\dot{m}_g$ with a total enthalpy of $h_{t,g}$ is written as

$$\dot{m}_c h_{t,c} + \dot{m}_g h_{t,g} = (\dot{m}_c + \dot{m}_g)h_{t,mixed-out} \tag{4.51}$$

where the only unknown is the mixed-out enthalpy state of the coolant-hot gas mixture, $h_{t,mixed-out}$. We can express the above equation in nondimensional form by dividing both sides by $\dot{m}_g h_{t,g}$, to get

$$\frac{h_{t,mixed-out}}{h_{t,g}} = \frac{1 + \dfrac{\dot{m}_c}{\dot{m}_g}\dfrac{h_{t,c}}{h_{t,g}}}{1 + \dfrac{\dot{m}_c}{\dot{m}_g}} \approx \left(1 - \frac{\dot{m}_c}{\dot{m}_g}\right)\left(1 + \frac{\dot{m}_c}{\dot{m}_g}\frac{h_{t,c}}{h_{t,g}}\right) \tag{4.52}$$

In Eq. 4.52, two nondimensional parameters emerge that basically govern the energetics of a hot and a cold mixture. These parameters are

- $\dfrac{\dot{m}_c}{\dot{m}_g}$ the coolant mass fraction (typically $\ll 1$)

- $\dfrac{h_{t,c}}{h_{t,g}}$ cold-to-hot total enthalpy ratio

We have also used a mathematical approximation in Eq. 4.52, in the form of binomial expansion:

$$\frac{1}{1+\varepsilon} \approx 1 - \varepsilon \qquad (4.53)$$

where $\varepsilon \prec\prec 1$ and the largest neglected term in the series is $O(\varepsilon^2)$.

The coolant stream experiences a total pressure drop in the cooling channels of the blade as a result of frictional losses and a total pressure drop as a result of turbulent mixing with the hot gas. The enthalpy–entropy diagram representing a coolant stream is depicted in Fig. 4.17.

In summary we learned that

- the flow expansion in the turbine produces the needed shaft power for the compressor and other propulsion system or aircraft needs, e.g., an electric generator
- there are two figures of merit in a turbine, which measure the extent of irreversibility in a turbine, η_t and e_t and they are related
- the gas expansion in an uncooled turbine is treated as adiabatic
- the frictional losses on the blades and the casing as well as any shock losses, in relative supersonic passages, are the sources of irreversibility in an uncooled turbine
- turbulent mixing losses between the coolant and the hot gas is an added source of irreversibility in a cooled turbine
- the turbine entry temperature T_{t4} is a design parameter that sets the stage for the turbine material and cooling requirements
- the power balance between the turbine, the compressor, and other known power drainage, establishes the turbine expansion ratio τ_t
- cooling of the high-pressure turbine is achieved through compressor air bleed that is injected through the blade root in the rotor and the casing for the turbine nozzle
- the turbine nozzle is choked (i.e., the throat Mach number is 1), over a wide operating range of the engine, and as such is the first *throttle station* of the engine.

EXAMPLE 4.4

Consider an uncooled gas turbine with its inlet condition the same as the exit condition of the combustor described in Example 4.3. The turbine adiabatic efficiency is 88%. The turbine produces a shaft power to drive the compressor and other accessories at $\wp_t = 45$ MW. Assuming that the gas properties in the turbine are the same as the burner exit in Example 4.3, calculate

(a) turbine exit total temperature T_{t5} in K
(b) turbine polytropic efficiency, e_t
(c) turbine exit total pressure p_{t5} in kPa
(d) turbine shaft power $\wp_t$ based on turbine expansion ΔT_t

SOLUTION

The turbine shaft power is proportional to the mass flow rate through the turbine, which from Example 4.3 is 51 kg/s (50 for air and 1 for fuel flow rate), as well the total enthalpy drop, i.e.,

$$\wp_t = \dot{m}_t(h_{t4} - h_{t5})$$

Therefore, we isolate h_{t5} from above equation to get

$$h_{t5} = h_{t4} - \wp_t/\dot{m}_t = c_{p4}T_{t4} - 35 \times 10^6/51$$

$$= 1.156\,\text{kJ/kg·K}(1390\,\text{K})$$

$$- 45000\,\text{kW}/51\,\text{kg/s} \approx 724.5\,\text{kJ/kg}$$

$$T_{t5} = h_{t5}/c_{p5}$$

$$= (724.5\,\text{kJ/kg})/(1.156\,\text{kJ/kg·K}) \approx 626.7\,\text{K}$$

Turbine polytropic efficiency e_t may be related to its adiabatic efficiency and τ_t via equation

$$\eta_t = \frac{1 - \tau_t}{1 - \tau_t^{1/e_t}}$$

The turbine expansion parameter $\tau_t = 626.7/1390 = 0.4509$, and if we isolate e_t from above equation we get

$$e_t = \ln(\tau_t)/\ln[1 - (1 - \tau_t)/\eta_t]$$

$$= \ln(0.4509)/\ln[1 - 0.5491/0.88] \cong 0.8144$$

We know that turbine pressure and temperature ratios are related by the polytropic efficiency via

$$\pi_t = \tau_t^{\frac{\gamma}{(\gamma-1)e_t}} = (0.4509)^{\frac{1.33}{0.33(0.8144)}} \cong 0.01941$$

The turbine exit total pressure is therefore $p_{t4} \cdot \pi_t$. We had found $p_{t4} = 1.92\,\text{MPa}$ in Example 4.3, therefore,

$$p_{t5} = 0.01941(1.92\,\text{MPa}) = 37.26\,\text{kPa}$$

The turbine shaft power is the product of the turbine mass flow rate and the total enthalpy drop across the turbine, i.e.,

$$\wp_t = \dot{m}_t c_{pt}(T_{t4} - T_{t5})$$

$$= 51\,\text{kg/s}(1156\,\text{J/kg·K})(1390 - 626.7)\text{K} \cong 45\,\text{MW}$$

EXAMPLE 4.5

Consider a turbine nozzle blade row with a hot gas mass flow rate of 50 kg/s and $h_{tg} = 1850\,\text{kJ/kg}$. The nozzle blades are internally cooled with a coolant mass flow rate of 0.5 kg/s and $h_{tc} = 904\,\text{kJ/kg}$ as the coolant is ejected through nozzle blades trailing edge. The coolant mixes with the hot gas and causes a reduction in the mixed-out enthalpy of the gas. Calculate the mixed-out total enthalpy after the nozzle. Also for the $c_{p,\text{mixed-out}} = 1594\,\text{J/kg·K}$, calculate the mixed-out total temperature.

SOLUTION

A simple energy balance between the mixed-out state and the hot and cold streams solves this problem, namely,

$$\dot{m}_c h_{t,c} + \dot{m}_g h_{t,g} = (\dot{m}_c + \dot{m}_g)h_{t,\text{mixed-out}}$$

$$50.5\,\text{kg/s}\,(h_{t,\text{mixed-out}}) = 0.5\,\text{kg/s}\,(904\,\text{kJ/kg})$$

$$+ 50\,\text{kg/s}\,(1850\,\text{kJ/kg})$$

$$= 92952\,\text{kW}$$

$$h_{t,\text{mixed-out}} = 92952\,\text{kW}/50.5\,\text{kg/s} = 1840.6\,\text{kJ/kg}$$

$$T_{t,\text{mixed-out}} = 1840.6\,\text{kJ/kg}/1.594\,\text{kJ/kg·K} = 1154.7\,\text{K}$$

EXAMPLE 4.6

Consider the internally cooled turbine nozzle blade row of Example 4.5. The hot gas total pressure at the entrance of the nozzle blade is $p_{t4} = 1.92\,\text{MPa}$, $c_{pg} = 1156\,\text{J/kg·K}$, and $\gamma_g = 1.33$. The mixed-out total pressure at the exit of the nozzle has suffered 2% loss due to both mixing and frictional losses in the blade row boundary layers. Calculate the entropy change $\Delta s/R$ across the turbine nozzle blade row.

SOLUTION

The hot gas total temperature is h_{tg}/c_{pg}, which for a $c_{pg} = 1156\,\text{J/kg·K}$ is calculated to be $T_{tg} = [1850\,\text{kJ/kg}]/[1.156\,\text{kJ/kg·K}] = 1600\,\text{K}$. This is the same temperature as T_{t4} at the entrance to the nozzle. The coolant total temperature is $T_{tc} = h_{tc}/c_{pc} = 904\,\text{kJ/kg}/1.04\,\text{kJ/kg·K} = 900.4\,\text{K}$. Also, the mixed-out total pressure is 98% of the incoming (hot gas) total pressure, i.e., $p_{t,\text{mixed-out}} = 0.98(1.92\,\text{MPa}) = 1.176\,\text{MPa}$. We may use the Gibbs equation (assuming constant gas specific heats) to calculate the entropy change, namely,

$$\Delta s/R = \frac{\gamma_g}{\gamma_g - 1}\ln(T_{t,\text{mixed-out}}/T_{t4}) - \ln(p_{t,\text{mixed-out}}/p_{t4})$$

Upon substitution, we get

$$\Delta s/R \cong -0.824$$

The negative sign of entropy change is due to cooling.

4.3.1.5 *The Nozzle.*

The primary function of an aircraft engine exhaust system is to accelerate the gas efficiently. The nozzle parameter that is of utmost importance in *propulsion* is the gross thrust F_g. The expression we had earlier derived for the gross thrust was

$$F_g = \dot{m}_9 V_9 + (p_9 - p_0)A_9 \tag{4.54}$$

In this equation, the first term on the right-hand-side is called the momentum thrust and the second term is called the pressure thrust. It is interesting to note that the nozzles produce a "signature," typically composed of infrared radiation, thermal plume, smoke, and acoustic signatures, which are key design features of a stealth aircraft exhaust system *in addition* to the main propulsion requirement of the gross thrust.

As the fluid accelerates in a nozzle, the static pressure drops and hence a *favorable pressure gradient* environment is produced in the nozzle. This is in contrast to diffuser flows where an *adverse pressure gradient* environment prevails. Therefore, boundary layers are, by and large, well behaved in the nozzle and less cumbersome to treat than the inlet. For a subsonic exit Mach number, i.e., $M_9 < 1$, the nozzle expansion process will continue all the way to the ambient pressure p_0. This important result means that in subsonic streams, the static pressure inside and outside of the jet are the same. In fact, there is no mechanism for a pressure jump in a subsonic flow, which is in contrast to the supersonic flows where shock waves and expansion fans allow for static pressure discontinuity. We have depicted a convergent nozzle in Fig. 4.18 with its exhaust stream (i.e., a jet) emerging in the ambient gas of static state (0). The outer shape of the nozzle is called a boat tail, which affects the *installed performance* of the exhaust system. The external aerodynamics of the nozzle installation belongs to the propulsion system integration studies and does not usually enter the discussions of the *internal performance*, i.e., the cycle analysis. However, we need to be aware that our decisions for the internal flow path optimization, e.g., the

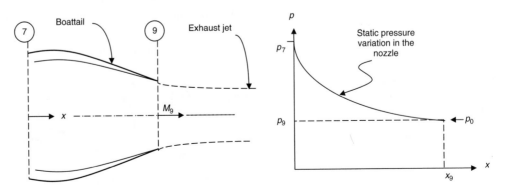

■ **FIGURE 4.18** Schematic drawing of a subsonic nozzle with its static pressure distribution

nozzle exit-to-throat area ratio, could have adverse effects on the installed performance that may offset any gains that may have been accrued as a result of the internal optimization.

We have learned in aerodynamics that a convergent duct, as shown in Fig. 4.18, causes flow acceleration in a subsonic stream to a maximum Mach number of 1, which can only be reached at the minimum area of the duct, namely at its exit. So we stipulate that for all subsonic jets, i.e., $M_9 < 1$, there is a static pressure equilibrium in the exhaust stream and the ambient fluid, i.e., $p_9 = p_0$. Let us think of this as the *Rule 1* in nozzle flow.

> Rule 1 :   If $M_{jet} < 1$,   then $p_{jet} = p_{ambient}$

The expression "jet" in the above rule should not be confusing to the reader, as it relates to the flow that emerges from the nozzle. In that context, M_{jet} is the same as M_9, and $p_{ambient}$ is the same as p_0. We recognize that not only it is entirely possible but also desirable for the sonic and supersonic jets to expand to the ambient static pressure as well. Such nozzle flows are called *perfectly expanded*. Actually, the nozzle gross thrust can be maximized if the nozzle flow is perfectly expanded. We state this principle here without proof, but we will address it again, and actually prove it, in the next chapters. Let us think of this as the *Rule 2* in nozzle flow.

> Rule 2 :   If $p_{jet} = p_{ambient}$
>
> then we have a perfectly expanded nozzle which results in $F_{g,max}$

Here, the stipulation is only on the *static pressure match* between the jet exit and the ambient static pressure and not *perfect flow* inside the nozzle. A real nozzle flow experiences total pressure loss due to viscous dissipation in the boundary layer as well as shock waves, and yet it is possible for it to *perfectly expand* the gas to the ambient condition. Remembering compressible duct flows in aerodynamics, we are reminded that the exit pressure p_9 is a direct function of the nozzle area ratio A_{exit}/A_{throat}, which in our notation, it becomes A_9/A_8, and the nozzle pressure ratio (NPR). Recalling the definition of NPR,

$$\text{NPR} \equiv \frac{p_{t7}}{p_0} \tag{4.55}$$

we will demonstrate a critical value of NPR will result in the choking condition at the nozzle throat, i.e., $M_8 = 1.0$, when $\text{NPR} \geq (\text{NPR})_{critical}$. Let us think of this as *Rule 3* in convergent or convergent–divergent nozzle aerodynamics. These rules do not apply if the divergent section of a C–D nozzle becomes a subsonic a diffuser.

> Rule 3 :   If $\text{NPR} \geq (\text{NPR})_{critical}$
>
> then the nozzle throat velocity is sonic (i.e., choked), $M_8 = 1.0$

A schematic drawing of a convergent–divergent supersonic nozzle is shown in Fig. 4.19.

■ **FIGURE 4.19**
A choked convergent–divergent nozzle and the static pressure distribution along the nozzle axis

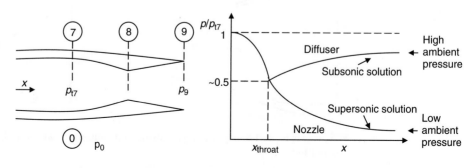

The nozzle throat becomes choked at a static pressure of about 50% of the nozzle total pressure, i.e., $p_{\text{throat}} \sim 1/2\, p_{t7}$. This fact provides for an important *rule of thumb* in nozzle flows, which is definitely worth remembering. Let us consider this, as *Rule 4*.

Rule 4 : If $\dfrac{p_{t7}}{p_0} \geq \sim 2$ then the nozzle throat can be choked, i.e., $M_8 = 1.0$

For example, a passage (or a duct) connecting a pressure vessel of about 30 psia pressure to a room of about 15 psia pressure will experience sonic speed at its minimum (orifice) area. As with any rules of thumb, the Rule 4 is stated as an *approximation* and the exact value of the critical nozzle pressure ratio depends on the nozzle expansion efficiency, i.e., the extent of losses in the nozzle.

The two possible solutions in static pressure distribution along the nozzle axis are shown in Fig. 4.19. One is a subsonic solution downstream of the nozzle throat and the second is a supersonic flow solution downstream of the throat. The subsonic solution is clearly a result of high backpressure, i.e., p_0, and causes a flow deceleration in the divergent duct downstream of the throat. Therefore, the divergent portion of the duct is actually a *diffuser* and not a *nozzle*. The supersonic solution is a result of low backpressure and therefore the flow continues to expand (i.e., accelerates) beyond the throat to supersonic speeds at the exit. Also note that only a perfectly expanded nozzle flow is shown as the supersonic branch of the nozzle flow in Fig. 4.19. For ambient pressures in between the two pressures shown in Fig. 4.19, there are a host of shock solutions, which occur as an oblique shock at the lip or normal shock inside the nozzle. The range of backpressures for a given nozzle area ratio, which lead to a normal shock solution inside the nozzle is shown in Fig. 4.20. The jump in static pressure across a normal shock is shown as an abrupt (i.e., vertical) pressure rise in Fig. 4.20, whereas in reality shocks have finite thickness, and therefore, all the *jump* conditions are then *diffused* or *spread* over the scale of shock thickness. We note the diffusing nature of the flow, i.e., the static pressure increasing, after the shock occurs inside the nozzle, which signifies a *diffuser environment* after the shock. For a lower range of backpressures, the oblique shock solution is possible, which indicate an oblique shock is hanging on the nozzle lip. Then we reach a unique nozzle pressure ratio, which leads to perfect expansion, as shown in Fig. 4.19 and repeated in Fig. 4.20. For all backpressures lower than the perfect expansion, we get the isentropic centered expansion fan solution. All these waves are the necessary mechanisms to satisfy the physical requirement of continuous static pressure across the slipstream in the jet exit flow. Let us call this important gas dynamic *boundary condition* as the rule number 5 in nozzle flows. Note that this rule is not unlike the

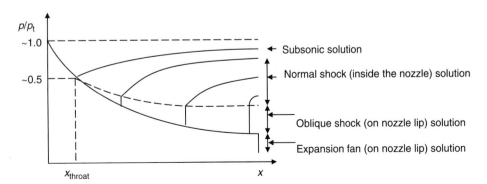

■ **FIGURE 4.20** Schematic drawing of possible static pressure distributions inside a choked supersonic nozzle

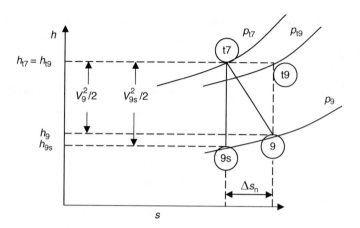

■ **FIGURE 4.21**   Enthalpy–entropy diagram of nozzle flow expansion

familiar Kutta condition in airfoil aerodynamics, where we prevented a static pressure jump at the sharp tailing edge of an airfoil.

To examine the efficiency of a nozzle in expanding the gas to an exit (static) pressure p_9, we create an enthalpy–entropy diagram, very similar to an inlet. In the inlet studies, we took the ambient static condition 0 and compressed it to the total intake exit condition t2. But since a nozzle may be treated as a reverse-flow diffuser (and vice versa), we take the nozzle inlet gas, at the total state t7, and expand it to the exit static condition 9. This process is shown in Fig. 4.21. The analogy in the thermodynamic analysis of a nozzle and a diffuser is helpful in better learning both components. The actual nozzle expansion process is shown by a solid line connecting the gas total state t7 to the exit static state 9. As the real nozzle flows may still be treated as adiabatic, the total enthalpy h_t, remains constant in a nozzle.

> Rule 5 :   Static pressure must be continuous across the slipstream of a jet exhaust plume.

$$h_{t7} = h_{t9} \tag{4.56}$$

We remember that the inlet flow was also considered to be adiabatic. An ideal exit state 9s is reached isentropically from the total state t7 to the same exit pressure p_9, in Fig. 4.21. The vertical gap between a total and static enthalpy states, in an h–s diagram, represents the kinetic energy of the gas, $V^2/2$, which is also shown in Fig. 4.21. Due to frictional and shock losses in a nozzle, the flow will suffer a total pressure drop, i.e.,

$$p_{t9} \prec p_{t7} \tag{4.57}$$

as depicted in Fig. 4.21. Again, this behavior is quite analogous to a diffuser. We can immediately define the total pressure ratio across a nozzle as its figure of merit, similar to the total pressure recovery of an inlet, namely,

$$\pi_n \equiv \frac{p_{t9}}{p_{t7}} \tag{4.58}$$

Now we can quantify the entropy rise in an adiabatic nozzle, as a result of total pressure losses, using the combined first and second law of thermodynamics,

$$\frac{\Delta s_n}{R} = -\ln \pi_n \tag{4.59}$$

We may also define a nozzle adiabatic efficiency η_n very similar to the inlet adiabatic efficiency, as

$$\eta_n \equiv \frac{h_{t7} - h_9}{h_{t7} - h_{9s}} = \frac{V_9^2/2}{V_{9s}^2/2} \tag{4.60}$$

Let us interpret the above definition in physical terms. The fraction of an ideal nozzle exit kinetic energy $V_{9s}^2/2$, which is realized in a real nozzle $V_9^2/2$, is called the nozzle adiabatic efficiency. The loss in total pressure in a nozzle manifests itself as a loss of kinetic energy. Consequently, the adiabatic efficiency η_n, which deals with the loss of kinetic energy in a nozzle, has to be related to the nozzle total pressure ratio π_n. Again, in a very similar manner as in the inlet, let us divide the numerator and denominator of Eq. 4.60 by h_{t7}, to get

$$\eta_n = \frac{1 - \dfrac{h_9}{h_{t7}}}{1 - \dfrac{h_{9s}}{h_{t7}}} = \frac{1 - \dfrac{h_9}{h_{t7}}}{1 - \left(\dfrac{p_9}{p_{t7}}\right)^{\frac{\gamma-1}{\gamma}}} = \frac{1 - \left(\dfrac{p_9}{p_{t9}}\right)^{\frac{\gamma-1}{\gamma}}}{1 - \left(\dfrac{p_9}{p_{t7}}\right)^{\frac{\gamma-1}{\gamma}}} \tag{4.61}$$

where we had used $h_{t7} = h_{t9}$ in the numerator and the isentropic relation between the temperature and pressure ratios in both numerator and denominator. It is interesting to note that as the nozzle exit total pressure p_{t9} approaches the value of the nozzle inlet total pressure p_{t7}, the nozzle adiabatic efficiency will approach 1. We may treat Eq. 4.61 as containing only one unknown and that is p_{t9}. Therefore, for a given nozzle exit static pressure p_{t9} and the nozzle inlet total pressure p_{t7}, the nozzle adiabatic efficiency η_n will result in a knowledge of the total pressure at the exit of the nozzle, p_{t9}. Now, if we multiply the numerator and denominator of the right-hand side of Eq. 4.61 by $(p_{t7}/p_9)^{\frac{\gamma-1}{\gamma}}$, we will reach our goal of relating the two figures of merit in a nozzle, i.e.,

$$\eta_n = \frac{\left(\dfrac{p_{t7}}{p_9}\right)^{\frac{\gamma-1}{\gamma}} - \pi_n^{-\frac{\gamma-1}{\gamma}}}{\left(\dfrac{p_{t7}}{p_9}\right)^{\frac{\gamma-1}{\gamma}} - 1} \tag{4.62}$$

There are three parameters in Eq. 4.62. The two figures of merit, η_n and π_n, and the ratio of nozzle inlet total pressure to the nozzle exit static pressure, p_{t7}/p_9. The last parameter is a known quantity, as we know the nozzle inlet total pressure p_{t7} from the upstream component analysis (in a turbojet, it is the turbine, $p_{t7} = p_{t5}$), and the nozzle exit pressure p_9 is a direct function of the nozzle area ratio A_9/A_8. To express Eq. 4.62 in terms of the NPR, we may split p_{t7}/p_9 into

$$\frac{p_{t7}}{p_9} = \frac{p_{t7}}{p_0} \frac{p_0}{p_9} = \text{NPR} \cdot \left(\frac{p_0}{p_9}\right) \tag{4.63}$$

Now, substituting the above expression in Eq. 4.62, we get

$$\eta_n = \frac{\left\{\text{NPR}\left(\dfrac{p_0}{p_9}\right)\right\}^{\frac{\gamma-1}{\gamma}} - \pi_n^{-\frac{\gamma-1}{\gamma}}}{\left\{\text{NPR}\left(\dfrac{p_0}{p_9}\right)\right\}^{\frac{\gamma-1}{\gamma}} - 1} \tag{4.64}$$

A plot of Eq. 4.64 is shown in Fig. 4.22 for a perfectly expanded nozzle, i.e., $p_9 = p_0$.

It is comforting to note that in Eq. 4.64, the nozzle adiabatic efficiency will approach 1, as the nozzle total pressure ratio approaches 1. The parameter p_0/p_9 represents a measure of mismatch between the nozzle exit static pressure and the ambient static pressure. For $p_9 > p_0$, the flow is considered to be *underexpanded*. For $p_9 < p_0$, the flow is defined as *overexpanded*.

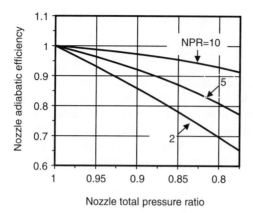

■ **FIGURE 4.22** Two figures of merit in a nozzle plotted as a function of NPR ($\gamma = 1.33$)

In the under-expanded scenario, the nozzle area ratio is not adequate, i.e., not large enough, to expand the gas to the desired ambient static pressure. In the overexpanded nozzle flow case, the nozzle area ratio is too large for perfect expansion. As noted earlier, a perfectly expanded nozzle will have $p_0 = p_9$, therefore, the above equation is further simplified to

$$\eta_n = \frac{\{NPR\}^{\frac{\gamma-1}{\gamma}} - \pi_n^{-\frac{\gamma-1}{\gamma}}}{\{NPR\}^{\frac{\gamma-1}{\gamma}} - 1} \tag{4.65}$$

for a perfectly expanded nozzle. It is instructive to show the three cases of nozzle expansion, i.e., under-, over-, and perfectly expanded cases, on an h–s diagram (see Fig. 4.23).

In Fig. 4.23 (a), the nozzle area ratio is smaller than necessary for perfect expansion and that explains the high exit pressure p_9. However, an expansion fan at the nozzle exit plane will adjust the pressure down to the ambient level, after the jet leaves the nozzle, in accordance to Rule 5. After all, the slipstream cannot take a (static) pressure jump. In Fig. 4.23 (b), the nozzle area ratio is too large for this altitude, which explains a lower-than-ambient static pressure at the nozzle exit. To adjust the low exit pressure up to the ambient level, a shock wave is formed on the nozzle exit lip, which for mild overexpansion will be an oblique shock. Higher levels of over expansion will strengthen the shock to a normal position and eventually cause the shock to enter the nozzle. Figure 4.23 (c) shows a perfect expansion scenario, which implies the nozzle area ratio is perfectly matched to the altitude requirements of ambient pressure. Again, as noted in Fig. 4.23, all nozzle expansions are depicted as irreversible processes with an associated entropy rise; therefore, *perfect expansion* is not to be mistaken as *perfect (isentropic) flow*.

In summary, we learned that:

- the primary function of a nozzle is to accelerate the gas efficiently
- the gross thrust parameter F_g signifies nozzle's contribution to the thrust production

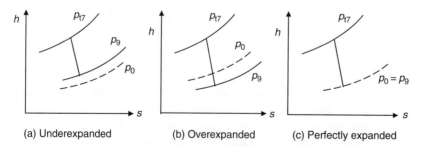

(a) Underexpanded   (b) Overexpanded   (c) Perfectly expanded

■ **FIGURE 4.23** The h–s diagram for the three possible nozzle expansions

- the gross thrust reaches a maximum when the nozzle is perfectly expanded; i.e., $p_9 = p_0$
- real nozzle flows may still be considered as adiabatic
- a nozzle pressure ratio (NPR) that causes a Mach-1 flow at the throat (i.e., choking condition) is called the *critical nozzle pressure ratio*, and as a rule of thumb, we may remember an $(NPR)_{crit}$ of ~ 2
- there are two efficiency parameters that quantify losses or the degree of irreversibility in a nozzle and they are related
- nozzle losses manifest themselves as the total pressure loss
- all subsonic exhaust streams have $p_{jet} = p_{ambient}$
- a perfect nozzle expansion means that the nozzle exit (static) pressure and the ambient pressure are equal
- an imperfect nozzle expansion is caused by a mismatch between the nozzle area ratio and the altitude of operation
- underexpansion is caused by smaller-than-necessary nozzle area ratio, leading to $p_9 > p_0$
- overexpansion is caused by larger-than-necessary nozzle area ratio, leading to $p_9 < p_0$

EXAMPLE 4.7

Consider a convergent–divergent nozzle with a pressure ratio NPR = 10. The gas properties are $\gamma = 1.33$ and $c_p = 1{,}156\,\text{J/kg·K}$ and remain constant in the nozzle. The nozzle adiabatic efficiency is $\eta_n = 0.94$. Calculate

(a) nozzle total pressure ratio π_n

(b) nozzle area ratio A_9/A_8 for a perfectly expanded nozzle

(c) nozzle exit Mach number M_9 (perfectly expanded)

SOLUTION

We may use Eq. 4.64 that relates the figures of merit of a nozzle (η_n and π_n) with *NPR*,

$$\eta_n = \frac{\left\{NPR\left(\dfrac{p_0}{p_9}\right)\right\}^{\frac{\gamma-1}{\gamma}} - \pi_n^{-\frac{\gamma-1}{\gamma}}}{\left\{NPR\left(\dfrac{p_0}{p_9}\right)\right\}^{\frac{\gamma-1}{\gamma}} - 1}$$

For a perfectly expanded nozzle, $p_0 = p_9$, therefore, we get the following expression for π_n

$$\pi_n = \left[NPR^{\frac{\gamma-1}{\gamma}} - \eta_n\left(NPR^{\frac{\gamma-1}{\gamma}} - 1\right)\right]^{\frac{-\gamma}{\gamma-1}} \cong 0.8335$$

The entropy rise in an adiabatic nozzle is a function of the total pressure ratio π_n, according to (Eq. 4.59)

$$\Delta s_n/R = -\ell n\pi_n \cong -\ell n(0.8335) \approx 0.1822$$

We can calculate the local Mach number M_9 if we know the total and static pressures p_{t9} and p_9 simultaneously. From the NPR and the total pressure ratio, we may write

$$p_{t9}/p_9 = (p_{t9}/p_{t7})(p_{t7}/p_0)(p_0/p_9) = 0.8335(10)(1)$$
$$= 8.335$$

From the general expression for total pressure and Mach number

$$p_t/p = (1 + (\gamma - 1)M^2/2)^{\frac{\gamma}{\gamma-1}}$$

We can isolate Mach number as follows

$$M_9 = \sqrt{\frac{2}{\gamma - 1}\left[(p_{t9}/p_9)^{\frac{\gamma-1}{\gamma}} - 1\right]}$$

$$= \sqrt{\frac{2}{0.33}\left[(8.335)^{0.3/1.33} - 1\right]} \cong 2.05$$

4.3.1.6   *Thermal Efficiency of a Turbojet Engine.*

An ideal turbojet engine (with the components of inlet, gas generator, and the nozzle) is the same as an ideal Brayton cycle that we learned in thermodynamics. We remember the Brayton cycle efficiency as

$$\eta_{th} = 1 - \frac{T_0}{T_3} \tag{4.66}$$

where T_3 is the compressor discharge temperature and T_0 is the inlet temperature, we note that the Brayton cycle thermal efficiency improves with increasing T_3. We can recast Eq. 4.66 in terms of the compressor pressure ratio and the ram temperature ratio, as follows:

$$\eta_{th} = 1 - \frac{1}{\dfrac{T_3}{T_0}} = 1 - \frac{1}{\dfrac{T_{t3}}{T_{t2}} \dfrac{T_{t2}}{T_0}} = 1 - \frac{1}{\tau_c \tau_r} = 1 - \frac{1}{\pi_c^{\frac{\gamma-1}{\gamma}}\left(1 + \dfrac{\gamma-1}{2}M_0^2\right)} \tag{4.67}$$

From Eq. 4.67, we note that for a given flight Mach number M_0, i.e., for a constant *ram temperature ratio* τ_r, the cycle thermal efficiency increases with the compressor temperature ratio τ_c. However, the higher the cycle pressure ratio, the higher the compressor discharge temperature will be, which implies the following attributes for the high-pressure compressor:

- using turbine material in HPC
- cooling needs of the HPC (potentially regenerative fuel cooling)
- narrow passage (vanishing flow annulus) in HPC with large secondary flow losses

Furthermore, for a constant turbine inlet temperature T_{t4}, the cycle work output w_{cycle} will eventually vanish in the limit of $T_{t3} = T_{t4}$. It is also interesting to note the emergence of Carnot-looking cycles with successively higher cycle pressure ratios (see Fig. 4.24). We remember from basic thermodynamics that a Carnot cycle operating between the temperature limits T_0 and T_4 will attain a thermal efficiency of

$$\eta_{th-Carnot} = 1 - \frac{T_0}{T_4} \tag{4.68}$$

which is the highest thermal efficiency of any heat engine operating between the temperature limits T_0 and T_4. The technology limit on compressor discharge temperature is shown in Fig. 4.24 for the standard sea level condition at takeoff, i.e., $M_0 \approx 0$.

The information presented in Fig. 4.25 identifies a limitation on the compressor pressure ratio π_c. Based on the practical limits of the compressor materials and potential cooling requirements, a compressor pressure ratio of $\sim 40 - 45$ seems to represent the upper limit of the

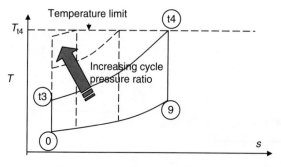

■ **FIGURE 4.24**   Simple Brayton cycle with increasing cycle pressure ratio

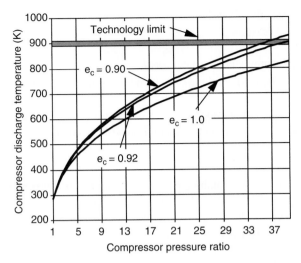

■ **FIGURE 4.25**   Influence of compressor pressure ratio on the discharge temperature T_{t3} ($M_0 = 0$, standard sea level condition)

current technology for the subsonic commercial aircraft gas turbine engines. The useful portion of Eq. 4.67 is repeated here for the purpose of discussion.

$$\eta_{\text{th}} = 1 - \frac{1}{\pi_{\text{c}}^{\frac{\gamma-1}{\gamma}}\left(1 + \frac{\gamma - 1}{2}M_0^2\right)} \tag{4.67}$$

The effect of flight Mach number on the thermal efficiency of an ideal turbojet engine can be discerned from Eq. 4.67. For a given compressor pressure ratio π_{c}, the thermal efficiency of an ideal turbojet engine increases with an increasing flight Mach number. The rationale for this behavior is that the higher flight stagnation temperature T_{t2}, combined with a constant compressor pressure ratio, leads to higher T_{t3}s. Going back to Eq. 4.66 or 4.67, we observe that a high compressor discharge temperature T_{t3} leads to a high thermal efficiency. On the contrary, aircraft engines experience their lowest thermal efficiency at takeoff. Although the fact that thermal efficiency improves with flight Mach number is important to know, it is of little practical value, because the flight Mach number (e.g., at cruise) is established by the mission requirements and not the engine thermal efficiency requirement. A family of thermal efficiency curves with the flight Mach number is plotted in Fig. 4.26, for different compressor pressure ratios.

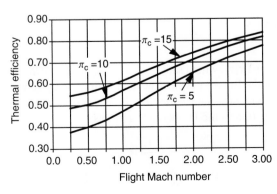

■ **FIGURE 4.26**   The effect of flight Mach number on the thermal efficiency of an *ideal* turbojet engine

The thermal efficiency shown above, for the most part, is not realized in a turbojet engine for the following reasons

1. the engine components are not *ideal*
2. for a high flight Mach number, e.g., $M_0 = 3$, there is no need for a compressor!

To appreciate item number 2, we need to examine the cycle output w_c more closely. For an ideal turbojet engine, neglecting a small mass flow contribution due to the fuel, the cycle specific work can be written as

$$w_c \cong (h_{t4} - h_9) - (h_{t3} - h_0) \tag{4.68}$$

where the first parenthesis contains the turbine and nozzle output (per unit mass flow rate) and the second parenthesis contains the inlet and compressor input (per unit mass flow rate). Therefore, the difference between the *expansion leg* and the *compression leg* in a Brayton cycle is the net output of the cycle. To put it differently, in an ideal cycle, the difference between the power production and consumption can be entirely converted into useful cycle power output. Now, let us manipulate Eq. 4.68 in order to cast it in terms of useful nondimensional quantities, i.e.,

$$
\begin{aligned}
w_c &\cong h_{t4}\left(1 - \frac{T_9}{T_{t4}}\right) - h_{t3}\left(1 - \frac{T_0}{T_{t3}}\right) \\
&= h_{t4}\left\{1 - \left(\frac{p_9}{p_{t4}}\right)^{\frac{\gamma-1}{\gamma}}\right\} - h_{t3}\left\{1 - \left(\frac{p_0}{p_{t3}}\right)^{\frac{\gamma-1}{\gamma}}\right\}
\end{aligned}
\tag{4.69}
$$

But since $p_9 = p_0$ and $p_{t4} = p_{t3}$ in an ideal cycle with a perfectly expanded nozzle, we can rewrite Eq. 4.69 in the following form;

$$
\begin{aligned}
w_c &\cong (h_{t4} - h_{t3})\left(1 - \frac{T_0}{T_{t3}}\right) = h_0(\tau_\lambda - \tau_c\tau_r)\left(1 - \frac{1}{\tau_c\tau_r}\right) \\
&= h_0\left(\tau_\lambda - \frac{\tau_\lambda}{\tau_r\tau_c} - \tau_r\tau_c + 1\right)
\end{aligned}
\tag{4.70}
$$

where $\tau_\lambda \equiv h_{t4}/h_0$ is the engine thermal limit parameter, $\tau_r \equiv T_{t0}/T_0 = 1 + \gamma - 1/2\,M_0^2$ is the ram temperature ratio, and $\tau_c \equiv T_{t3}/T_{t2}$ is the compressor temperature ratio. To find a maximum value for the specific work of the cycle, for a constant turbine inlet temperature, i.e., τ_λ, we may treat the product $\tau_r\tau_c$ as one and only variable in Eq. 4.70. Therefore, we can differentiate w_c/h_0 with respect to $\tau_r\tau_c$ and set it equal to zero,

$$\frac{d\left(\dfrac{w_c}{h_0}\right)}{d(\tau_r\tau_c)} = \frac{\tau_\lambda}{(\tau_r\tau_c)^2} - 1 = 0 \tag{4.71}$$

which implies that

$$\tau_r\tau_c = \sqrt{\tau_\lambda} \tag{4.72}$$

Therefore, the product $\tau_r\tau_c$, which satisfies Eq. 4.72, yields the *largest area under the curve* in the *T–s* diagram of the Brayton cycle. The area within the cycle walls, on a *T–s* diagram, we learned in thermodynamics that represents the net cycle specific work. We can isolate the compressor contribution to the temperature ratio as

$$\tau_c = \frac{\sqrt{\tau_\lambda}}{\tau_r} = \frac{\sqrt{\tau_\lambda}}{1 + \dfrac{\gamma - 1}{2}M_0^2} \tag{4.73}$$

Equation 4.73 is useful in telling us that for a given burner temperature, the compressor pressure ratio requirement for the maximum cycle output (i.e., the *optimum* compressor pressure ratio) goes down with the flight Mach number. We can recast Eq. 4.73 in terms of the (optimum) compressor pressure ratio as

$$\pi_{c\,Optimum} = \tau_c^{\frac{\gamma}{\gamma-1}} = \left(\frac{\sqrt{\tau_\lambda}}{1 + \frac{\gamma-1}{2}M_0^2}\right)^{\frac{\gamma}{\gamma-1}} \quad \text{For maximum thrust} \qquad (4.74)$$

Before plotting Eq. 4.74, we recognize that the lowest compressor pressure ratio is 1, which basically states that there is no need for a compressor, i.e., in this limit we reach a *ramjet*. Setting $\pi_c = 1$, in Eq. 4.74 and isolating the flight Mach number M_0, we arrive at the following expression for the maximum flight Mach number, which renders a compressor in an aircraft engine as useless, i.e.,

$$M_{0\,max} = \sqrt{\frac{2}{\gamma-1}\left(\sqrt{\tau_\lambda} - 1\right)} \qquad (4.75)$$

Now, let us plot Eq. 4.75 for a gas specific heat ratio of 1.4 ($\gamma = 1.4$).

Although we chose a turbine inlet temperature corresponding to a $\tau_\lambda = 6.5$ or 7 in plotting Eq. 4.74, the physical principle remains the same for other turbine inlet temperatures. The optimum compressor pressure ratio drops nonlinearly with the flight Mach number. In this case, substituting $\tau_\lambda = 6.5$, in Eq. 4.75 yields a maximum flight Mach number of $M_{0,max} = \sqrt{5(\sqrt{6.5} - 1)} = 2.783$, which corresponds to a compressor pressure ratio of 1, in Fig. 4.27, and $\tau_\lambda = 7.0$ yields a maximum M_0 of 2.868. Reexamining Eq. 4.72 can reveal an interesting relationship between the states t3 and 9, i.e., the compressor discharge and the nozzle exit, for a cycle with maximum (specific) work output capability. Figure 4.28 shows an ideal turbojet cycle with a fixed limit temperature T_{t4} but varying compressor discharge temperatures T_{t3}. We remember that the state t3 is reached through ram compression, as well as the mechanical compression. The shaded area represents the cycle (specific energy) output, which is maximized by Eq. 4.72.

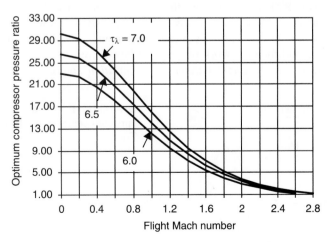

■ **FIGURE 4.27** Optimum compressor pressure ratio variation with the flight Mach number (for $\tau_\lambda = 6.0-7.0$)

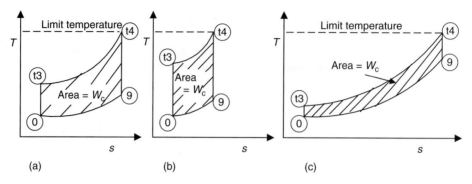

■ **FIGURE 4.28** *T–s* diagram of an ideal turbojet cycle with varying T_{t3}, or cycle pressure ratio

We will proceed to demonstrate that the maximum cycle area, which is guaranteed by the following equation, requires T_{t3} to be equal to the nozzle exit temperature T_9.

$$\tau_r \tau_c = \sqrt{\tau_\lambda} \qquad (4.72)$$

but

$$\tau_r \tau_c = \frac{T_{t0}}{T_0} \frac{T_{t3}}{T_{t2}} = \frac{T_{t3}}{T_0} \qquad (4.76)$$

which is set equal to $\sqrt{\tau_\lambda}$ and manipulated as follows:

$$\frac{T_{t3}}{T_0} = \sqrt{\frac{T_{t4}}{T_0}} = \sqrt{\frac{T_{t4}}{T_9} \frac{T_9}{T_0}} = \sqrt{\frac{T_{t3}}{T_0} \frac{T_9}{T_0}} = \frac{\sqrt{T_{t3}T_9}}{T_0} \qquad (4.77)$$

where we used the fact that $T_{t4}/T_9 = T_{t3}/T_0$, as they are isentropically bound by a pair of isobars, i.e., $p_9 = p_0$ and $p_{t4} = p_{t3}$. Now, we can simplify Eq. 4.77, to get the desired result:

$$\boxed{T_{t3} = T_9 \qquad Q.E.D.} \qquad (4.78)$$

None of the cycles drawn in Fig. 4.28 satisfy the above requirement and, as such, have not created the largest area in the *T–s* diagram. Figure 4.29 shows the optimum compression for a Brayton cycle. We need to remember that the optimum work cycle, as shown in Fig. 4.29, is not achieved at the highest cycle thermal efficiency. The highest cycle thermal efficiency is achieved in the limit of $T_{t3} = T_{t4}$, which corresponds to Carnot cycle efficiency ($\eta_{th,max} = 1 - T_0/T_{t4}$), and then the work output vanishes! Here, we learned an important lesson in propulsion, namely, *a single parameter optimization* approach in an engine may lead to

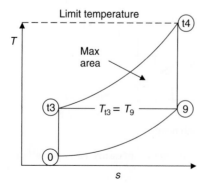

■ **FIGURE 4.29** *T–s* diagram of an optimum compression Brayton cycle shows $T_{t3} = T_9$

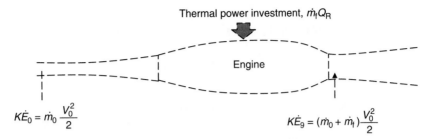

Thermal power investment, $\dot{m}_f Q_R$

Engine

$$KE_0 = \dot{m}_0 \frac{V_0^2}{2}$$

$$K\dot{E}_9 = (\dot{m}_0 + \dot{m}_f)\frac{V_0^2}{2}$$

■ **FIGURE 4.30**   Schematic drawing of an aircraft engine with fuel thermal power as input and $\Delta K\dot{E}$ as output

impractical results for the engine performance (e.g., the highest thermal efficiency in a Brayton cycle will lead to no work output!). An engine, in reality, is a multiparameter system and an *overall optimum* is, by necessity, a mix of compromises between the component efficiencies, performance output, and the issues of reliability, maintainability, operating cost, and other market-driven practical limitations based on *customer* input/demands.

We may examine the description of aircraft engine thermal efficiency from another perspective. From the energy standpoint, the function of an airbreathing engine is to increase the kinetic energy of the medium per unit time, i.e., $\Delta K\dot{E}$. This may be thought as the *mechanical output* of the engine with a perfectly expanded nozzle. The thermal investment in the engine is through the fuel flow rate, liberating a thermal energy equal to the heating value of the fuel per unit mass, per unit time. Graphically, we may present the change in kinetic power of the stream and the thermal investment through the fuel as in Fig. 4.30.

Now, it is entirely reasonable to think that the thermal efficiency of this engine is the fraction of the thermal power investment that is converted into useful mechanical power output, i.e.,

$$\eta_{th} \equiv \frac{\Delta K\dot{E}}{\dot{m}_f Q_R} = \frac{\frac{\dot{m}_0}{2}\{(1 + f)V_9^2 - V_0^2\}}{\dot{m}_f Q_R} = \frac{(1 + f)V_9^2 - V_0^2}{2 f Q_R} \tag{4.79}$$

Is this expression for thermal efficiency the same as $\eta_{th} = 1 - \dfrac{T_0}{T_{t3}}$? The answer is obviously affirmative as we will demonstrate that Eq. 4.79 can, under some simplifying assumptions, reduce to the thermodynamic efficiency of Brayton cycle, described earlier. First, let us ignore the fuel-to-air ratio f in favor of 1 in the numerator of Eq. 4.79, i.e., since $f \ll 1$, $1 + f \approx 1$

$$\eta_{th} \approx \frac{V_9^2 - V_0^2}{2 f Q_R} \approx \frac{(h_{t9} - h_9) - (h_{t0} - h_0)}{h_{t4} - h_{t3}} \tag{4.80}$$

We had replaced the kinetic energy term by the difference between the total and static enthalpy in Eq. 4.80, as well as the burner input by the total enthalpy rise across it. In the approximation made to the denominator of Eq. 4.80, we again neglected the small contribution of fuel-to-air ratio, as compared to one (see Eq. 4.31, for example). Second, let us replace h_{t9} by h_{t5} and h_{t0} by h_{t2}, as the nozzle and inlets are treated as adiabatic ducts, therefore, total enthalpy is conserved, i.e.,

$$\eta_{th} \approx \frac{(h_{t5} - h_9) - (h_{t2} - h_0)}{h_{t4} - h_{t3}} \tag{4.81}$$

From the power balance between the compressor and turbine, again neglecting f in favor of one, we can replace h_{t5} by $[h_{t4} - (h_{t3} - h_{t2})]$, in Eq. 4.81, to get

$$\eta_{th} \approx \frac{h_{t4} - h_{t3} + h_{t2} - h_9 - h_{t2} + h_0}{h_{t4} - h_{t3}} = \frac{(h_{t4} - h_9) - (h_{t3} - h_0)}{h_{t4} - h_{t3}} \tag{4.82}$$

The thermodynamic states t4 and 9 and t3 and 0 are grouped together in Eq. 4.82, as they are on the same isentrope, respectively. The temperature ratio T_{t4}/T_9 and the pressure ratio p_{t4}/p_9 are then linked via isentropic exponent. Using these simple steps in Eq. 4.82 yield

$$\eta_{th} \approx \frac{h_{t4}\left(1 - \dfrac{T_9}{T_{t4}}\right) - h_{t3}\left(1 - \dfrac{T_0}{T_{t3}}\right)}{h_{t4} - h_{t3}} = \frac{h_{t4}\left\{1 - \left(\dfrac{p_9}{p_{t4}}\right)^{\frac{\gamma-1}{\gamma}}\right\} - h_{t3}\left\{1 - \left(\dfrac{p_0}{p_{t3}}\right)^{\frac{\gamma-1}{\gamma}}\right\}}{h_{t4} - h_{t3}} \tag{4.83}$$

We realize that the two parenthesis in Eq. 4.83 are the same, as $p_9 = p_0$, i.e., a perfectly expanded nozzle, and $p_{t4} = p_{t3}$, as a perfect combustion, consequently, we can further simplify the above equation to get the desired result, i.e.,

$$\boxed{\eta_{th} \approx \frac{(h_{t4} - h_{t3})\left(1 - \dfrac{T_0}{T_{t3}}\right)}{h_{t4} - h_{t3}} = 1 - \frac{T_0}{T_{t3}} \qquad Q.E.D.} \tag{4.84}$$

Therefore, we demonstrated that our general definition of an airbreathing engine thermal efficiency (Eq. 4.79), under the assumptions of perfect nozzle expansion, ideal components, and negligible fuel flow rate as compared to the air flow rate, reduces to the ideal Brayton cycle efficiency that we learned in thermodynamics.

EXAMPLE 4.8

Prove the ideal thermal efficiency for the Brayton cycle, as shown, is $\eta_{th} = 1 - T_0/T_{t3}$

SOLUTION

From the first (and second) law of thermodynamics, we have

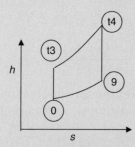

$$\oint \delta w = \oint \delta q = \oint T ds$$

$Tds = dh - vdp = dh$, for constant pressure processes, therefore, the closed cycle integral gives

$$\oint \delta w = (h_{t4} - h_{t3}) + (h_0 - h_9) = (h_{t4} - h_9) - (h_{t3} - h_0)$$

The (specific) thermal power investment in the cycle comes from a constant-pressure heating between t3 and t4, i.e.,

$$\int_{t3}^{t4} T ds = \int_{t3}^{t4} dh = h_{t4} - h_{t3}$$

The cycle thermal efficiency is the ratio of net work produced by the cycle to the net thermal investment in the cycle, i.e.,

$$\eta_{th} = \frac{h_{t4} - h_{t3} + h_0 - h_9}{h_{t4} - h_{t3}} = 1 - \frac{h_9 - h_0}{h_{t4} - h_{t3}}$$

For a constant c_p, we may replace the enthalpies with temperatures,

$$\eta_{th} = 1 - \frac{T_9 - T_0}{T_{t4} - T_{t3}} = 1 - \frac{T_0}{T_{t3}}[(T_9/T_0 - 1)/(T_{t4}/T_{t3} - 1)]$$

If we show that the last bracket is unity, then we have proven what we set out to demonstrate. Is $T_9/T_0 = T_{t4}/T_{t3}$? The answer becomes obvious when we switch the terms in the equality, to get.

Is $T_9/T_{t4} = T_0/T_{t3}$? Now, since the temperature ratios, T_9/T_{t4} and T_0/T_{t3}, are represented by isentropic processes,

they may be expressed in terms of the pressure ratios raised to isentropic exponent, i.e.,

$$T_9/T_{t4} = (p_9/p_{t4})^{(\gamma-1)/\gamma}$$
$$T_0/T_{t3} = (p_0/p_{t3})^{(\gamma-1)/\gamma}$$

But since $p_0 = p_9$ and $p_{t3} = p_{t4}$, the temperature ratios are equal to each other and we have proven the last bracket in the thermal efficiency expression is unity.

Therefore, $\eta_{th} = 1 - \dfrac{T_0}{T_{t3}}$

EXAMPLE 4.9

Prove the thermal efficiency of an ideal Carnot cycle is $\eta_{th} = 1 - T_0/T_{t4}$.

SOLUTION

The net specific work of the cycle is

$$\oint \delta w = \oint \delta q = \oint T ds$$

Therefore the net specific work is

$$w = (T_{t3} - T_0)\Delta s$$

The constant temperature heating from t3 to t4 represents the specific thermal investment in the cycle, i.e.,

$$\int_{t3}^{t4} T ds = T_{t3} \cdot \Delta s$$

The thermal efficiency of the Carnot cycle is the fraction of the thermal investment that is converted into mechanical work, namely,

$$\eta_{th} = \frac{(T_{t3} - T_0)\Delta s}{T_{t3}\Delta s} = 1 - \frac{T_0}{T_{t3}}$$

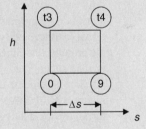

4.3.1.7 *Propulsive Efficiency of a Turbojet Engine.* The fraction of the net kinetic power created by the engine, which is converted into thrust power delivered to the vehicle, is called the propulsive efficiency of the engine. Symbolically, we may represent this as

$$\eta_p \equiv \frac{F \cdot V_0}{\Delta K\dot{E}} = \frac{F \cdot V_0}{\dot{m}_9 \dfrac{V_9^2}{2} - \dot{m}_0 \dfrac{V_0^2}{2}} \tag{4.85}$$

For the thrust power expression in the numerator, we have an option of using the net *uninstalled* thrust, or include the effect of installations on the thrust and calculate the *installed* thrust power. The latter is more difficult, as it includes the throttle-dependent inlet and exhaust system drag

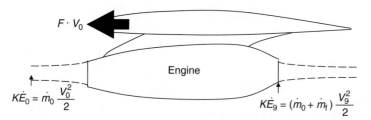

■ **FIGURE 4.31**   **Definition sketch used in defining propulsive efficiency (nozzle is perfectly expanded)**

and the engine-airframe integration, and in order to simplify the calculations, the general form of the net uninstalled thrust is

$$F_{n_{\text{uninstalled}}} = \dot{m}_9 V_9 - \dot{m}_0 V_0 + (p_9 - p_0)A_9 \tag{4.86}$$

However, if we assume a perfectly expanded nozzle, then the pressure thrust, i.e., the last term in Eq. 4.86, vanishes identically. Furthermore, if we neglect small contribution of the fuel flow rate in the exhaust momentum term, the mass flow rate in the exit plane is nearly equated to the air mass flow rate, therefore,

$$F_{n_{\text{uninstalled}}} \approx \dot{m}_0(V_9 - V_0) \tag{4.87}$$

Now, if we substitute Eq. 4.87 into Eq. 4.85 and simplify, we get

$$\eta_p \cong \frac{2(V_9 - V_0)V_0}{V_9^2 - V_0^2} \cong \frac{2V_0}{V_9 + V_0} \cong \frac{2}{1 + (V_9/V_0)} \tag{4.88}$$

Figure 4.31 is a useful tool in the definition of the propulsive efficiency.

The propulsive efficiency expressed in Eq. 4.88 is an approximation. It is cast in terms of only one variable, V_9/V_0, which is the ratio of the (exhaust) jet speed to the flight speed of the aircraft. The propulsive efficiency of an airbreathing aircraft engine drops inversely proportional to this velocity ratio. Also, we note that, in the limit of this velocity ratio approaching 1, the propulsive efficiency reaches 100%, i.e., for $V_9/V_0 = 1$, then $\eta_p \approx 100\%$. The dilemma here is that an engine, which operates near 100% propulsive efficiency, i.e., $V_9 = V_0$, cannot produce significant momentum thrust! The solution seems to be an engine that captures, i.e., interacts, with large quantities of air and only imparts a small velocity jump to the flow. A turboprop or a large bypass ratio turbofan engine fit this description. The point is made, however, that for improved propulsive efficiency, the ratio of exhaust stream to flight speed needs to be slightly larger than 1. A plot of Eq. 4.88 is a useful reminder of this fact and hence is shown in Fig. 4.32.

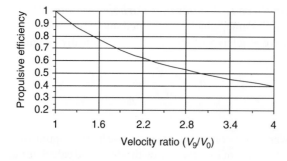

■ **FIGURE 4.32**   **Propulsive efficiency variation with the velocity ratio V_9/V_0**

EXAMPLE 4.10

Calculate the propulsive efficiency of a turbojet engine under the following two flight conditions that represent takeoff and cruise, namely:

1. $V_0 = 160\,\text{km/h}$ and $V_9 = 1000\,\text{km/h}$

2. $V_0 = 800\,\text{km/h}$ and $V_9 = 1000\,\text{km/h}$

SOLUTION

We may use the approximation $\eta_\text{p} \cong 2/(1 + V_9/V_0)$ to estimate engine propulsive efficiencies,

$$\eta_\text{p} \cong 2/(1 + 1000/160) \cong 0.2759$$
$$\eta_\text{p} \cong 2/(1 + 1000/800) = 0.8889$$

4.3.1.8 ***The Overall Efficiency of a Turbojet Engine.*** The overall efficiency of a turbojet engine is defined as the product of the thermal and the propulsive efficiency of the engine, namely,

$$\eta_\text{o} \equiv \eta_\text{th} \cdot \eta_\text{p} = \frac{F \cdot V_0}{\dot{m}_\text{f} Q_\text{R}} \tag{4.89}$$

which states that the fraction of thermal power invested in the engine by the fuel, which is converted into thrust power, is called the engine overall efficiency. The above equation for the overall efficiency, Eq. 4.89, can be expressed in terms of specific thrust by dividing the numerator and denominator by the air mass flow rate through the engine, i.e.,

$$\eta_\text{o} = \frac{(F/\dot{m}_0) V_0}{f Q_\text{R}} \tag{4.90}$$

4.3.1.9 ***Performance Evaluation of a Turbojet Engine.*** As outlined in the engine performance chapter, the performance parameters for an aircraft engine are

(a) specific thrust $F/\dot{m}$

(b) thrust specific fuel consumption (TSFC) or specific impulse I_s

(c) thermal, propulsive, and overall efficiencies

Our approach to calculating the performance parameters of an aircraft engine is to *march through* an engine, component by component, until we calculate the target parameters, which are

(a) fuel-to-air ratio f

(b) exhaust velocity V_9

Once we established these unknowns, i.e., V_0, f, and V_9, the specific thrust, as a figure of merit for an aircraft engine, may be written as

$$\frac{F_\text{n}}{\dot{m}_0} = (1 + f)V_9 - V_0 + \frac{(p_9 - p_0)A_9}{\dot{m}_0} \tag{4.91}$$

which can be expressed in terms of the calculated parameters by recognizing that the air mass flow rate is only a factor $(1 + f)$ away from the exhaust mass flow rate, i.e.,

$$\dot{m}_0 = \frac{\dot{m}_9}{1 + f} = \frac{\rho_9 A_9 V_9}{1 + f} = \frac{p_9 A_9 V_9}{R T_9 (1 + f)} \tag{4.92}$$

Now, substituting Eq. 4.92 into 4.90, we can get

$$\frac{F_n}{\dot{m}_0} = (1 + f)V_9 - V_0 + \frac{(p_9 - p_0)A_9}{\dfrac{p_9 A_9 V_9}{RT_9(1 + f)}} = (1 + f)V_9 - V_0 + \frac{RT_9(1 + f)}{V_9}\left(1 - \frac{p_0}{p_9}\right) \quad \textbf{(4.93-a)}$$

Note that we have calculated all the terms on the right-hand side of the Eq. 4.93-a, namely, f, V_9, V_0, and T_9. Therefore,

$$\frac{F_n}{\dot{m}_0} = (1 + f)V_9 - V_0 + \frac{RT_9(1 + f)}{V_9}\left(1 - \frac{p_0}{p_9}\right) \quad \textbf{(4.93-b)}$$

The primary contribution to the specific thrust in an airbreathing engine comes from the first two terms in Eq. 4.93-b, i.e., the momentum contribution. The last term would identically vanish if the nozzle is perfectly expanded, i.e., $p_9 = p_0$. Otherwise, its contribution is small compared with the momentum thrust. It is very tempting to insert a γ in the numerator and denominator of the last term in Eq. 4.93-b, and identify the γRT_9 as the a_9^2 to recast the above equation in a more elegant form, as

$$\frac{F_n}{\dot{m}_0} = (1 + f)V_9\left(1 + \frac{1}{\gamma M_9^2}\left(1 - \frac{p_0}{p_9}\right)\right) - V_0 \quad \textbf{(4.93-c)}$$

Thrust specific fuel consumption was defined as

$$\text{TSFC} \equiv \frac{\dot{m}_f}{F_n} = \frac{f}{F_n/\dot{m}_0} \quad \textbf{(4.94)}$$

and we have calculated the fuel-to-air ratio and the specific thrust that are combined as in Eq. 4.94 to create the fuel efficiency parameter, or figure of merit, of the engine.

The overall efficiency, which was derived earlier in Eq. 4.90, can be recast in terms of the TSFC as

$$\eta_0 = \frac{(F_n/\dot{m}_0)V_0}{fQ_R} = \frac{V_0/Q_R}{\text{TSFC}} \quad \textbf{(4.95-a)}$$

The inverse proportionality between the engine overall efficiency and the thrust specific fuel consumption, i.e., the lower the specific fuel consumption, the higher the engine overall efficiency, is noted in Eq. 4.95, i.e.,

$$\eta_0 \propto \frac{1}{\text{TSFC}} \quad \textbf{(4.95-b)}$$

In concluding the section on turbojet engines, it is appropriate to show a *real cycle* depicted on a *T–s* diagram. This is shown in Fig. 4.33.

4.3.2   The Turbojet Engine with an Afterburner

4.3.2.1   Introduction. To augment the thrust of an aircraft gas turbine engine, an afterburner can be used. This solution is often sought in military aircraft, primarily due to its simplicity and effectiveness. It is interesting to note that an afterburner has the potential of nearly doubling the thrust produced in an aircraft gas turbine engine while in the process more than quadruple the engine fuel consumption rate. Figure 4.34 shows the schematics of an afterburning turbojet (AB-TJ) engine. The new station numbers 7 and 8 refer to the exit of the afterburner and the nozzle throat, respectively. Suitable nozzle geometry for an afterburning turbojet (or turbofan) engine is a *supersonic nozzle* with a convergent–divergent (C–D) geometry, i.e., a nozzle with a well-defined

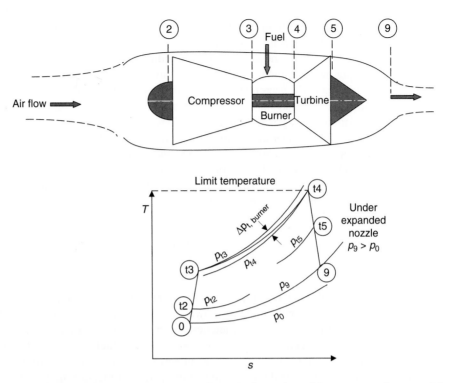

■ FIGURE 4.33 *T–s* diagram of a turbojet engine with component losses and imperfect expansion in the nozzle

throat before a divergent cone or ramp. The throat Mach number is 1, over a wide range of operating conditions, which is referred to as the *choked* throat. As the afterburner operation causes the gas temperature to rise, the gas density shall decrease in harmony with the temperature rise. Consequently, to accommodate the lower density gas at the sonic condition prevailing at the nozzle throat and satisfy continuity equation, the throat area needs to be opened. This is referred to as a *variable-geometry* nozzle requirement of the exhaust system. The continuity equation for a steady flow of a perfect gas is rewritten here to demonstrate the extent of the nozzle throat-opening requirement in a variable–geometry, convergent–divergent nozzle.

$$\dot{m} = \sqrt{\frac{\gamma}{R}} \cdot \frac{p_t}{\sqrt{T_t}} A.M. \left(\frac{1}{1 + \frac{\gamma - 1}{2} M^2} \right)^{\frac{\gamma+1}{2(\gamma-1)}} \quad \textbf{(4.96)}$$

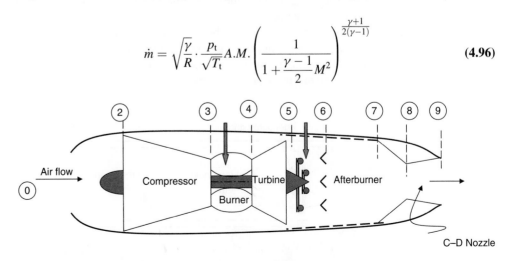

■ FIGURE 4.34 Schematic drawing of an afterburning turbojet engine

Let us examine the various terms in Eq. 4.96, as a result of the afterburner operation. First, the mass flow rate increases slightly by the amount of $\dot{m}_{f,AB}$, which is but a small percentage of the gas flow with the afterburner-off condition, say 3–4%. The total pressure p_t drops with the afterburner operation, but this too is a small percentage, say 5–8%. The throat Mach number will remain as 1; therefore, except for the gas property variation, the last parentheses in Eq. 4.96 will remain unchanged, as a result of afterburner operation. Now, again neglecting small variations in the gas properties γ and R, which are in reality a function of the gas temperature, the first term under the square root in Eq. 4.96 shall be unchanged. This brings us to the main driver for the nozzle throat area increase requirement when the afterburner is in operation, namely the gas total temperature T_t. Therefore the nozzle throat area needs to be opened directly proportional to the square root of the gas total temperature in the nozzle throat. This is symbolically expressed as

$$\frac{A_{8,\text{AB-ON}}}{A_{8,\text{AB-OFF}}} \approx \sqrt{\frac{T_{t8,\text{AB-ON}}}{T_{t8,\text{AB-OFF}}}} \qquad (4.97)$$

The afterburner is composed of an inlet diffuser, a fuel spray bar, and a flameholder to stabilize the combustion, as shown in Figs. 4.34 and 4.35. The inlet diffuser decelerates the gas to allow for higher efficiency combustion in the afterburner. A similar prediffuser is found at the entrance to the primary burner. The fuel spray bar is typically composed of one or more rings with distinct fuel injection heads circumferentially distributed around the ring. The V-shaped flame holder ring(s) create a fuel–air mixture recirculation region in its turbulent wake, which allows for a stable flame front to be established. There is also a perforated liner, which serves two functions. One, it serves as a cooling conduit that blankets, i.e., protects, the outer casing from the hot combustion gases. Two, it serves as an acoustic liner that dampens the high frequency noise, i.e., screech, which is generated as a result of combustion instability.

An afterburner is considered to be an adiabatic duct with insulated walls, which neglects the small heat transfer through the casing, i.e.,

$$_5(\dot{Q}_{\text{wall}})_7 \cong 0 \qquad (4.98)$$

We notice in Fig. 4.34 that the basic building block of an afterburning turbojet engine is again a simple gas generator.

Additional parameters (beyond the simple turbojet engine) are needed to analyze the performance of an afterburning turbojet engine. These are related to the physical characteristics of the afterburning system, e.g., the type of fuel used in the afterburner (which is typically the same as the fuel in the primary burner), either the fuel flow setting of the afterburner fuel pump,

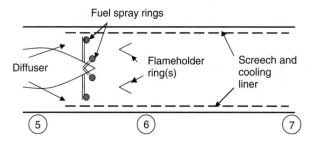

■ **FIGURE 4.35**   **Schematic drawing of an afterburner**

or the exit total temperature of the afterburner, the afterburner efficiency and the afterburner total pressure loss.

Symbolically, we represent the new parameters as

New parameters in an afterburner

1. $Q_{R,AB}$ — (ideal) heating value of the fuel in the afterburner
2. $\dot{m}_{f,AB}$ or T_{t7} or $\tau_{\lambda-AB} \equiv \dfrac{c_{pAB}T_{t7}}{c_{pc}T_0}$ — fuel flow rate or the exit temperature
3. $\eta_{AB} \equiv \dfrac{(Q_{R,AB})_{actual}}{(Q_{R,AB})_{ideal}}$ — afterburner efficiency (<1)
4. $\pi_{AB} \equiv \dfrac{p_{t7}}{p_{t5}}$ — total pressure ratio across the afterburner (<1)

4.3.2.2   Analysis. A nondimensional *T–s* diagram for an ideal afterburning turbojet engine is shown in Fig. 4.36. It is interesting to note the appearance of τ parameters on the nondimensional temperature axis. Also, we note that the thermodynamic process from the flight static 0 to the turbine exit, i.e., t5, is unaffected by the afterburner. Although, the *T–s* diagram depicts an ideal afterburning turbojet engine in Fig. 4.36, the real engine afterburner is to operate with no *back influence* on the upstream components as well. We will continue using the *marching technique* in establishing the stagnation flow properties throughout the engine as we used in the turbojet section. The above stipulation, that the operation of an afterburner does not affect the upstream components, will bring our analysis, unchanged, to station t5. We will take up the analysis from station t5 and establish exit conditions at t7.

To allow marching through the afterburner, we need to know or estimate its losses in total pressure and its inefficiency in heat release in the confines of the finite volume of the afterburner. These are π_{AB} and η_{AB}, respectively. So far the afterburner loss depiction through the total pressure and heat release capability is the same as the primary or main burner. However, unlike the primary burner, an afterburner may or may not be in operation and this has a large influence on the total pressure loss characteristics of the afterburner. It is only obvious to expect that the operating total pressure loss in the afterburner to be larger

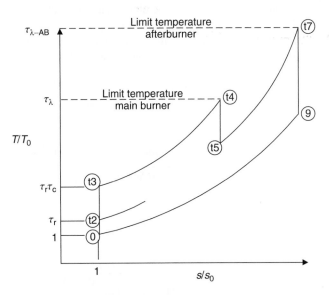

■ **FIGURE 4.36**   Nondimensional *T–s* diagram for an *ideal* afterburning turbojet engine

than the afterburner-off total pressure loss. One may think of it as the *compounding* of the burning losses as well as the frictional losses on the walls and the flameholder, which add up to the total pressure loss in an operating afterburner. Thus, we distinguish between $\pi_{\text{AB-OFF}}$ and $\pi_{\text{AB-ON}}$ by

$$\pi_{\text{AB-OFF}} \succ \pi_{\text{AB-ON}} \tag{4.99}$$

We perform an energy balance across the afterburner to establish the afterburner fuel-to-air ratio f_{AB}.

$$(\dot{m}_0 + \dot{m}_f + \dot{m}_{f,\text{AB}})h_{t7} - (\dot{m}_0 + \dot{m}_f)h_{t5} = \dot{m}_{f,\text{AB}}Q_{R,\text{AB}}\eta_{\text{AB}} \tag{4.100}$$

Divide both sides by the air mass flow rate $\dot{m}_0$ to get

$$(1 + f + f_{\text{AB}})h_{t7} - (1 + f)h_{t5} = f_{\text{AB}}Q_{R,\text{AB}}\eta_{\text{AB}} \tag{4.101}$$

Now, we can isolate f_{AB}, as

$$f_{\text{AB}} = \frac{(1 + f)(h_{t7} - h_{t5})}{Q_{R,\text{AB}}\eta_{\text{AB}} - h_{t7}} \approx \frac{(1 + f)(T_{t7} - T_{t5})}{\dfrac{Q_{R,\text{AB}}\eta_{\text{AB}}}{c_{p\text{AB}}} - T_{t7}} \tag{4.102}$$

where $c_{p\text{AB}}$ represents an average specific heat at constant pressure between the temperatures T_{t5} and T_{t7}.

The only unknown in Eq. 4.102 is the fuel-to-air ratio in the afterburner, as we had calculated all the upstream parameters, e.g., f, T_{t5}, values and the afterburner fuel heating value and the efficiency as well as the exit temperature are all specified. The total pressure at the afterburner exit, p_{t7}, is

$$p_{t7} = \pi_{\text{AB}} \cdot p_{t5} \tag{4.103}$$

The operation of the afterburner changes the inlet conditions to the nozzle in so far as the mass flow rate, total pressure, and temperature are concerned. The mass flow rate due to the fuel in the afterburner is calculated via Eq. 4.102, the total pressure is calculated via Eq. 4.103, and the total temperature at the exit of the afterburner constitutes another thermal limit, which is often specified. The analysis of the nozzle, therefore, remains unchanged, except the nozzle inlet values are different as a result of the afterburner operation.

The fundamental definitions of the cycle thermal and propulsive efficiency as given earlier in the turbojet section remain the same for an afterburning turbojet engine as well, i.e.,

$$\eta_{\text{th}} \equiv \frac{\Delta K\dot{E}}{\wp_{\text{thermal}}} = \frac{\Delta K\dot{E}}{\dot{m}_f \cdot Q_R + \dot{m}_{f,\text{AB}} \cdot Q_{R,\text{AB}}} \tag{4.104-a}$$

$$\eta_{\text{th}} = \frac{(\dot{m}_0 + \dot{m}_f + \dot{m}_{f,\text{AB}})\dfrac{V_9^2}{2} - \dot{m}_0 \dfrac{V_0^2}{2}}{\dot{m}_f \cdot Q_R + \dot{m}_{f,\text{AB}} \cdot Q_{R,\text{AB}}} \tag{4.104-b}$$

$$\eta_p \equiv \frac{F \cdot V_0}{\Delta K\dot{E}} \tag{4.105-a}$$

$$\eta_p = \frac{(F_n/\dot{m}_0)V_0}{(1 + f + f_{\text{AB}})\dfrac{V_9^2}{2} - \dfrac{V_0^2}{2}} \tag{4.105-b}$$

The only term that has been added to the denominator of Eq. 4.104 (as compared with the turbojet) is the afterburner contribution to the thermal power investment in the engine. We can expand on the nondimensional form of the nozzle total pressure ratio for a turbojet, i.e.,

$$\pi_n = \left\{ \left(\pi_t \pi_b \pi_c \pi_d \pi_r \frac{p_0}{p_9} \right)^{\frac{\gamma-1}{\gamma}} - \eta_n \left[\left(\pi_t \pi_b \pi_c \pi_d \pi_r \frac{p_0}{p_9} \right)^{\frac{\gamma-1}{\gamma}} - 1 \right] \right\}^{\frac{-\gamma}{\gamma-1}} \quad \text{(Turbojet)} \quad \textbf{(4.106)}$$

by inserting the π_{AB} in the total pressure chain, to get the afterburning version of the equation, namely,

$$\pi_n = \left\{ \left(\pi_{AB} \pi_t \pi_b \pi_c \pi_d \pi_r \frac{p_0}{p_9} \right)^{\frac{\gamma-1}{\gamma}} - \eta_n \left[\left(\pi_{AB} \pi_t \pi_b \pi_c \pi_d \pi_r \frac{p_0}{p_9} \right)^{\frac{\gamma-1}{\gamma}} - 1 \right] \right\}^{\frac{-\gamma}{\gamma-1}} \quad \text{(AB-TJ)}$$

$$\textbf{(4.107)}$$

Now, we insert π_{AB} in the expression for the nozzle exit Mach number M_9 that we had derived for a turbojet, i.e.,

$$M_9 = \sqrt{\frac{2}{\gamma-1} \left[\left(\pi_n \pi_{AB} \pi_t \pi_b \pi_c \pi_d \pi_r \frac{p_0}{p_9} \right)^{\frac{\gamma-1}{\gamma}} - 1 \right]} \quad \text{(AB-TJ)} \quad \textbf{(4.108)}$$

The exhaust velocity will remain as the product of the exit Mach number and the speed of sound in station 9,

$$
\begin{aligned}
V_9 &= M_9 \cdot a_9 = M_9 \sqrt{\gamma R \frac{T_{t9}}{1 + \frac{\gamma-1}{2} M_9^2}} \\
&= M_9 \sqrt{\gamma R \frac{T_0 \tau_{\lambda,AB}}{1 + \frac{\gamma-1}{2} M_9^2}} = a_0 M_9 \sqrt{\frac{\tau_{\lambda,AB}}{1 + \frac{\gamma-1}{2} M_9^2}} \quad \text{(AB-TJ)} \quad \textbf{(4.109)}
\end{aligned}
$$

The specific thrust for an afterburning turbojet engine is defined as

$$
\begin{aligned}
\frac{F_n}{\dot{m}_0} &\equiv \frac{(\dot{m}_0 + \dot{m}_f + \dot{m}_{f,AB}) V_9 - \dot{m}_0 V_0 + (p_9 - p_0) A_9}{\dot{m}_0} \\
&= (1 + f + f_{AB}) V_9 - V_0 + \frac{(1 + f + f_{AB}) p_9 \left(1 - \dfrac{p_0}{p_9} \right) A_9}{\rho_9 V_9 A_9} \\
&= (1 + f + f_{AB}) V_9 - V_0 + (1 + f + f_{AB}) \frac{R T_9 \left(1 - \dfrac{p_0}{p_9} \right)}{V_9} \quad \text{(AB} - \text{TJ)} \quad \textbf{(4.110)} \\
&= (1 + f + f_{AB}) V_9 \left(1 + \frac{1 - \dfrac{p_0}{p_9}}{\gamma M_9^2} \right) - V_0
\end{aligned}
$$

The thrust specific fuel consumption for an afterburning turbojet engine is defined as

$$\text{TSFC} \equiv \frac{\dot{m}_f + \dot{m}_{f,AB}}{F_n} = \frac{f + f_{AB}}{F_n / \dot{m}_0} \quad \textbf{(4.111)}$$

4.3.2.3 Optimum Compressor Pressure Ratio for Maximum (Ideal) Thrust Turbojet Engine with Afterburner.

In an ideal turbojet engine with an afterburner where the thermal limit parameters τ_λ and $\tau_{\lambda,AB}$ are specified and remain fixed, we search for a compressor pressure ratio, which would maximize the performance of the engine. The desired performance parameter to be maximized is the specific thrust, $F_n/\dot{m}_0$. To simplify the analysis, let us assume the nozzle is perfectly expanded and thus there is no pressure thrust contribution. Thus, the specific thrust becomes

$$\frac{F_n}{\dot{m}_0} \approx (1 + f + f_{AB})V_9 - V_0 \approx V_9 - V_0 \qquad (4.112)$$

Now, if we multiply both sides by the *average* flight and exhaust speed, i.e., $1/2\,(V_0 + V_9)$, we get the cycle specific work, namely,

$$w_c \approx \frac{V_9^2}{2} - \frac{V_0^2}{2} \qquad (4.113)$$

Therefore, to maximize the specific work of the cycle at a given flight speed, we need to maximize the exhaust kinetic energy. The exhaust kinetic energy is the difference between the stagnation and static enthalpies at the nozzle exit, i.e.,

$$\frac{V_9^2}{2} \equiv h_{t9} - h_9 = h_{t9}\left(1 - \frac{T_9}{T_{t9}}\right) = h_0 \cdot \tau_{\lambda,AB}\left(1 - \frac{1}{1 + \dfrac{\gamma - 1}{2}M_9^2}\right) \qquad (4.114)$$

Therefore, to maximize the exit kinetic energy for a given $\tau_{\lambda,AB}$ and h_0, we need to maximize the exit Mach number M_9, or M_9^2. We can express the exit Mach number in terms of the total and static pressures at the nozzle exit, as in the turbojet section, as

$$M_9^2 = \frac{2}{\gamma - 1}\left[\left(\frac{p_{t9}}{p_9}\right)^{\frac{\gamma-1}{\gamma}} - 1\right] \qquad (4.115)$$

To maximize the exit Mach number then would be satisfied if the ratio of total-to-static pressure at the exit is maximized, namely,

$$\frac{p_{t9}}{p_9} = \frac{p_{t9}}{p_{t7}}\frac{p_{t7}}{p_{t5}}\frac{p_{t5}}{p_{t4}}\frac{p_{t4}}{p_{t3}}\frac{p_{t3}}{p_{t2}}\frac{p_{t2}}{p_{t0}}\frac{p_{t0}}{p_0}\frac{p_0}{p_9}$$
$$= \pi_n \pi_{AB}\pi_t\pi_b\pi_c\pi_d\pi_r\frac{p_0}{p_9} \qquad (4.116)$$

Assuming ideal components and perfect expansion and substitute for the turbine pressure ratio, we get

$$\frac{p_{t9}}{p_9} = 1.1.\tau_t^{\frac{\gamma}{\gamma-1}}.1.\tau_c^{\frac{\gamma}{\gamma-1}}.1.\tau_r^{\frac{\gamma}{\gamma-1}}.1 = \left[\tau_r\tau_c\left(1 - \frac{\tau_r(\tau_c - 1)}{\tau_\lambda}\right)\right]^{\frac{\gamma}{\gamma-1}} \qquad (4.117)$$

Note the exponent in Eq. 4.117 is the inverse of the pressure exponent in Eq. 4.115, therefore, we differentiate the terms in the bracket with respect to τ_c and set it equal to zero, namely,

$$\frac{d}{d\tau_c}\left[\tau_r\tau_c\left(1 - \frac{\tau_r(\tau_c - 1)}{\tau_\lambda}\right)\right] = \tau_r - \frac{2\tau_r^2\tau_c}{\tau_\lambda} + \frac{\tau_r^2}{\tau_\lambda} = 0 \qquad (4.118)$$

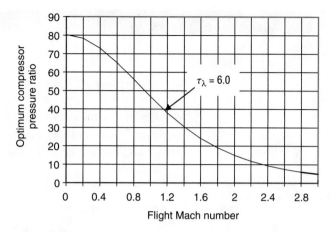

■ **FIGURE 4.37**   Optimum compressor pressure ratio for a maximum-thrust turbojet engine with afterburner

which simplifies to

$$\tau_\lambda - 2\tau_r\tau_c + \tau_r = 0 \tag{4.119-a}$$

$$\tau_c = \frac{\tau_r + \tau_\lambda}{2\tau_r} \tag{4.119-b}$$

$$\pi_{c,\text{Optimum}} = \left(\frac{\tau_r + \tau_\lambda}{2\tau_r}\right)^{\frac{\gamma}{\gamma-1}} \quad \text{For maximum (ideal) thrust} \tag{4.120}$$

We have plotted Eq. 4.120 in Fig. 4.37 for a $\tau_\lambda = 6.0$, in terms of the flight Mach number M_0. It is interesting to note that the optimum compressor pressure ratio in an afterburning turbojet engine is independent of the afterburner temperature limit $\tau_{\lambda,\text{AB}}$. Also by comparing the compressor pressure ratio for a maximum thrust nonafterburning turbojet (Fig. 4.27) with the afterburning engine (Fig. 4.37), we note that at low speed, very large compressor pressure ratios are required for the afterburning engine. In fact, at subsonic flight Mach numbers, the optimum compressor pressure ratios depicted in Fig. 4.37 are unreasonably high. Figure 4.38 shows a comparison between the optimum compressor pressure ratio for a maximum-thrust turbojet with and without afterburning.

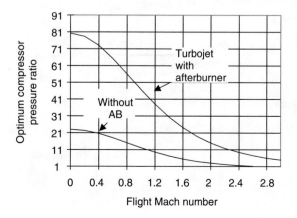

■ **FIGURE 4.38**   Compressor pressure ratio for a maximum-thrust turbojet engine with and without afterburner ($\tau_\lambda = 6.0$)

EXAMPLE 4.11

An afterburning turbojet engine operates at

$M_0 = 2.0$, $p_0 = 10\,\text{kPa}$, $T_0 = -45°C$, $\gamma_c = 1.4$,

$c_{pc} = 1004\,\text{J/kg·K}, \pi_d = 0.88$

$\pi_c = 12, e_c = 0.90$

$\tau_\lambda = 8.0, Q_R = 42000\,\text{kJ/kg}$,

$\eta_b = 0.98, \pi_b = 0.95$

$\gamma_t = 1.33, c_{pt} = 1156\,\text{J/kg·K}$,

$e_t = 0.82, \eta_m = 0.995$

$\tau_{\lambda AB} = 11, Q_{RAB} = 42,000\,\text{kJ/kg}$,

$\eta_{AB} = 0.98, \pi_{AB} = 0.93$

$\gamma_{AB} = 1.30, c_{pAB} = 1,243\,\text{J/kg·K}$

$\pi_n = 0.93, p_9 = p_0$

Calculate

(a) total pressure and temperature throughout the engine as well as fuel-to-air ratio f and f_{AB}

(b) nondimensional specific thrust $F_n/\dot{m}a_0$

(c) thrust specific fuel consumption (TSFC) in mg/s/N

(d) thermal and propulsive efficiencies η_{th} and η_p, respectively

SOLUTION

We first covert the ambient temperature into absolute scale, $T_0 = -45 + 273 = 228\,\text{K}$. The corresponding speed of sound and flight speeds are

$$a_0 = \sqrt{(\gamma_c - 1)c_{pc}T_0} = \sqrt{0.4(1004)(228)}\,\text{m/s} \cong 302.6\,\text{m/s}$$

$$V_0 = M_0 \cdot a_0 = (2)(302.6\,\text{m/s}) \cong 605.2\,\text{m/s}$$

Flight total pressure and temperature are

$$p_{t0} = p_0\left[1 + (\gamma_c - 1)M_0^2/2\right]^{\gamma_c/(\gamma_c-1)} = 10\,\text{kPa}(1.8)^{3.5} \cong 78.24\,\text{kPa}$$

$$T_{t0} = T_0\left[1 + (\gamma_c - 1)M_0^2/2\right] = 228\,\text{K}(1.8) \cong 410.4\,\text{K}$$

⎱ Flight total pressure and temperature ⎰ **⓪**

Since inlets are adiabatic, $T_{t2} = T_{t0} = 410.4\,\text{K}$

$$p_{t2} = p_{t0} \cdot \pi_d (78.24\,\text{kPa})(0.88) \cong 68.85\,\text{kPa}$$

⎱ total pressure and temperature at ⎰ **②**

Compressor exit total pressure is $p_{t3} = p_{t2} \cdot \pi_c = 68.85\,\text{kPa} \cdot (12) \cong 826.26\,\text{kPa}$

Compressor temperature ratio is related to pressure ratio via

$$\tau_c = \pi_c^{\frac{\gamma-1}{\gamma e_c}} = (12)^{1.4/0.4/0.9} \cong 2.201 \qquad \tau_c = \pi_c^{\frac{\gamma-1}{\gamma e_c}}$$

⎱ Compressor exit ⎰ **③**

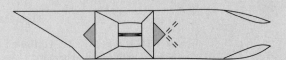

Therefore, $T_{t3} = T_{t2} \cdot \tau_c = (410.4\,\text{K})(2.201) \cong 903.24\,\text{K}$

Combustor exit temperature is $\tau_\lambda \equiv c_{pt}T_{t4}/c_{pc}T_0$, therefore,

$$T_{t4} = c_{pc}T_0\tau_\lambda/c_{pt} = (1004)(228\,\text{K})(8)/1156 \cong 1,584.2\,\text{K}$$

$$p_{t4} = p_{t3} \cdot \pi_b = 826.26\,\text{kPa}\,(0.95) \cong 784.95\,\text{kPa}$$

⎱ Burner exit ⎰ **④**

For fuel-to-air ratio in the burner, we use the energy balance $c_{pc}T_{t3} + fQ_R\eta_b = (1 + f)c_{pt}T_{t4}$

Where $f = \dfrac{c_{pt}T_{t4} - c_{pc}T_{t3}}{Q_R\eta_b - c_{pt}T_{t4}} = \dfrac{(1156)(1584.2) - (1004)(903.24)}{42000(1000)(0.98) - 1156(1584.2)} \cong 0.02351$

For the turbine exit temperature, we use the power balance between the compressor and turbine:

$$c_{pc}(T_{t3} - T_{t2}) = \eta_m(1 + f)c_{pt}(T_{t4} - T_{t5})$$

Therefore

$$T_{t5} = T_{t4} - c_{pc}(T_{t3} - T_{t2})/c_{pt}\eta_m(1 + f)$$

$$= 1584.2 - 1004(903.24 - 410.4)/1156(0.995)(1 + 0.02351) \cong 1,163.85\,\text{K}$$

⎱ Turbine exit ⎰ **⑤**

$$\pi_t = \tau_t^{\frac{\gamma_t}{(\gamma_t - 1)e_t}} = (1163.85/1584.2)^{1.33/0.33/0.82} \cong 0.2197$$

$$p_{t5} = p_{t4} \cdot \pi_t = 784.95 \, \text{kPa}(0.2197) \cong 172.47 \, \text{kPa}$$

The afterburner exit total pressure is $p_{t7} = p_{t5} \cdot \pi_{AB} = (172.47 \, \text{kPa})(0.93) = 160.4 \, \text{kPa}$

The afterburner exit total temperature is $T_{t7} = c_{pc} T_0 \tau_{\lambda AB} / c_{pAB} = (1004)(228)(11)/(1243) \cong 2025.8 \, \text{K}$

$$f_{AB} = \frac{(1 + f)(h_{t7} - h_{t5})}{Q_{R,AB}\eta_{AB} - h_{t7}} = \frac{(1 + 0.02351)[1243(2025.8) - 1156(1163.85)]}{42000(1000)(0.98) - 1243(2025.8)} \cong 0.0311$$

$\left.\right\}$ Afterburner exit ⑦

The nozzle exit total pressure is $p_{t9} = p_{t7} \cdot \pi_n = (160.4 \, \text{kPa})(0.93) \cong 149.17 \, \text{kPa}$

Since the flow in the nozzle is adiabatic, $T_{t9} = T_{t7} = 2025.8 \, \text{K}$

Since $p_9 = p_0 = 10 \, \text{kPa}$, we can calculate M_9 from

$$M_9 = \sqrt{\frac{2}{\gamma_{AB} - 1}\left[\left(\frac{p_{t9}}{p_9}\right)^{\frac{\gamma_{AB} - 1}{\gamma_{AB}}} - 1\right]} = \sqrt{\frac{2}{0.3}[(14.917)^{0.3/1.3} - 1]} \cong 2.402$$

$$T_9 = \frac{T_{t9}}{1 + (\gamma_{AB} - 1)M_9^2/2} = \frac{2025.8 \, \text{K}}{1 + 0.15(2.402)^2} \cong 1{,}085.8 \, \text{K}$$

$$a_9 = \sqrt{(\gamma_{AB} - 1)c_{pAB}T_9} = \sqrt{0.3(1243)(1{,}085.8)} \cong 636.3 \, \text{m/s}$$

$$V_9 = M_9 \cdot a_9 = 2.402(636.3 \, \text{m/s}) = 1{,}528.67 \, \text{m/s}$$

$\left.\right\}$ Nozzle exit ⑨

The specific thrust when the nozzle is perfectly expanded as in this problem is

$$F_n/\dot{m}_0 = (1 + f + f_{AB})V_9 - V_0$$

The nondimensional specific thrust is

$$F_n/\dot{m}_0 a_0 = (1 + f + f_{AB})V_9/a_0 - M_0 = (1 + 0.02351 + 0.0311)(1528.67)/302.6 - 2.0 \cong 3.3275$$

Also, the thrust specific fuel consumption is

$$\text{TSFC} = \frac{f + f_{AB}}{F_n/\dot{m}_0} = \frac{0.02351 + 0.0311}{3.3275(302.6)}(\text{kg/s/N})(10^6 \, \text{mg/kg}) \cong 54.2 \, \text{mg/s/N}$$

The cycle thermal efficiency is

$$\eta_{th} = \frac{(1 + f + f_{AB})\dfrac{V_9^2}{2} - \dfrac{V_0^2}{2}}{f \cdot Q_R + f_{AB} \cdot Q_{R,AB}} = \frac{(1 + 0.02351 + 0.0311)(1528.67)^2 - (605.2)^2}{2[0.02351(42000)(1000) + 0.0311(42000)(1000)]} \cong 0.4578$$

$\left.\right\}$ Performance parameters

The cycle propulsive efficiency in the exact form is

$$\eta_p = \frac{(F_n/\dot{m}_0)V_0}{(1 + f + f_{AB})\dfrac{V_9^2}{2} - \dfrac{V_0^2}{2}} = \frac{3.3275(302.6)(605.2)}{(1 + 0.02351 + 0.0311)(1528.67)^2/2 - (605.2)^2/2} \cong 0.5809$$

The approximate form of propulsive efficiency is

$$\eta_p \approx \frac{2}{1 + V_9/V_0} = 2/(1 + 1528.67/605.2) \approx 0.5672$$

EXAMPLE 4.12

Use a spreadsheet or write a computer program (e.g., MATLAB) to study the effect of the compressor pressure ratio on an afterburning turbojet engine. Use compressor pressure ratio range from 1 to 24, and the remaining engine parameters from Example 4.11. Our choice of the starting compressor pressure ratio of 1 allows us to simulate a ramjet. We are interested in TSFC in mg/s/N, nondimensional specific thrust, thermal, propulsive, and overall efficiency. Offer physical explanations for the behavior observed.

SOLUTION

Spreadsheet calculations are performed, and tables of performance parameters are produced. The graph of TSFC as a function of compressor pressure ratio is shown first.

We observe that the engine consumes the highest amount of fuel if the compressor is eliminated, i.e., in the ramjet mode, $\pi_c = 1.0$. That corresponds to a very low cycle thermal efficiency (of $\sim 30\%$), as observed in the following graph. Also, we observe that there is an optimum pressure ratio (of ~ 12) that minimizes the fuel consumption. This behavior is also observed in specific thrust figure, which shows a maximum thrust is produced for an optimum compressor pressure ratio. Propulsive efficiency is less profoundly affected than the cycle thermal efficiency. A decrease in propulsive efficiency with increasing cycle pressure ratio is due to an increase in exhaust velocity. The minimum exhaust velocity that occurs in a ramjet mode is $V_9 \approx 1288\,\text{m/s}$. The maximum exhaust velocity occurs at the optimum compressor pressure ratio (since the cycle produces maximum work) and is $V_9 \approx 1,588\,\text{m/s}$.

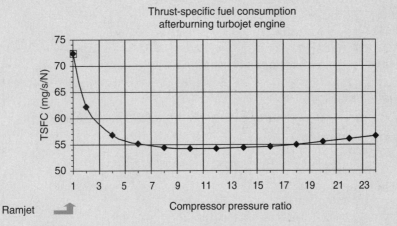

Thrust-specific fuel consumption afterburning turbojet engine

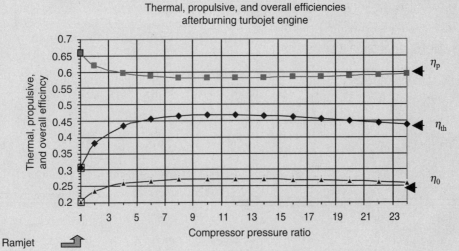

Thermal, propulsive, and overall efficiencies afterburning turbojet engine

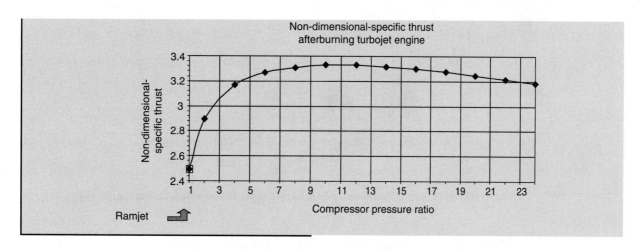

4.3.3   The Turbofan Engine

4.3.3.1   ***Introduction.*** To create a turbofan engine, a basic gas generator is followed by an additional turbine stage(s), which tap into the exhaust stream thermal energy to provide shaft power to a fan. This arrangement of multiple loading demands on the turbine stages lead to a multiple shaft arrangement, referred to as *spools*. The fan stages and potentially several (low-pressure) compressor stages may be driven by a shaft, which is connected to the low-pressure turbine. The high-pressure compressor stages are driven by the high-pressure turbine stage(s). The rotational speeds of the two shafts are called N_1 and N_2, respectively, for the low- and high-pressure spools. An additional two new parameters enter our gas turbine vocabulary when we consider turbofan engines. The first is the *bypass ratio* and the second is the *fan pressure ratio*. The ratio of the flow rate in the fan *bypass duct* to that of the gas generator (i.e., the *core*) is called the bypass ratio α. The fan pressure ratio is the ratio of total pressure at the fan exit to that of the fan inlet. It is given a symbol π_f.

$$\alpha \equiv \dot{m}_{\text{fan-bypass}} / \dot{m}_{\text{core}} \tag{4.121}$$

$$\pi_f \equiv p_{t13} / p_{t2} \tag{4.122}$$

The principle behind the turbofan concept comes from sharing the power with a larger mass flow rate of air at a smaller velocity increment pays dividend at low-speed flight. As we have seen earlier, the smaller the velocity increment across the engine, the higher the propulsive efficiency will be. This principle is Mach number independent; however, for supersonic flight the *installation drag* of large bypass ratio turbofan engines become excessive and consequently *small-frontal-area engines* are more suitable to high-speed flight. The fan exhaust stream may be separate from the so-called *core* stream, which is then referred to as *separate exhaust turbofan engine*. Figure 4.39 shows the schematic drawing of a two-spool, separate-exhaust turbofan engine. Note that the gas generator is still the heart of this engine.

    It is possible to extend the fan duct and provide mixing of the fan and the core stream in a common exhaust nozzle. The weight penalty of the long-duct turbofan engine needs to be assessed against the thrust enhancement potential of mixing a cold and hot stream in a mixer before the exhaust nozzle. The schematic drawing of this configuration is shown in Fig. 4.40.

4.3.3.2   ***Analysis of a Separate-Exhaust Turbofan Engine.*** In applying our marching technique developed in the analysis of a turbojet engine, to a turbofan engine,

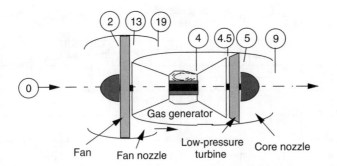

■ **FIGURE 4.39** Schematic drawing of a separate-exhaust turbofan engine with two spools

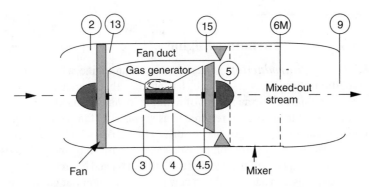

■ **FIGURE 4.40** Schematic of a long-duct turbofan engine with a mixer

we are confronted with the fan. Therefore the logical question is "how do we treat the thermodynamics of a fan?" The short answer is "it is treated the same way as a compressor." This short answer should suffice and it enables us to proceed with the analysis of a turbofan engine for now. The long answer is postponed to the chapter on turbomachinery. As in a compressor, a fan is characterized by its pressure ratio and efficiency. We may define, in an analogous manner to the compressor, an adiabatic fan efficiency η_f or a fan polytropic efficiency e_f. Let us depict the fan compression process on an h–s diagram, as in Fig. 4.41.

The fan adiabatic efficiency is defined as

$$\eta_f \equiv \frac{h_{t13s} - h_{t2}}{h_{t13} - h_{t2}} = \frac{\Delta h_{t,\,\text{isentropic}}}{\Delta h_{t,\,\text{actual}}} = \frac{\text{ideal power}}{\text{actual power}} \tag{4.123}$$

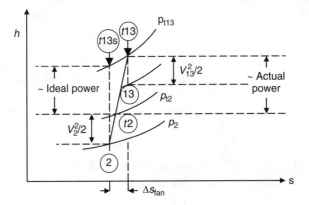

■ **FIGURE 4.41** h–s diagram representing the flow process in an (adiabatic) fan

Upon dividing the numerator and denominator of Eq. 4.123 by h_{t2}, we get

$$\eta_f = \frac{T_{t13s}/T_{t2} - 1}{T_{t13}/T_{t2} - 1} = \frac{\pi_f^{\frac{\gamma-1}{\gamma}} - 1}{\tau_f - 1} \tag{4.124}$$

which relates the fan pressure and temperature ratio through adiabatic fan efficiency, i.e.,

$$\tau_f = 1 + \frac{1}{\eta_f}\left(\pi_f^{\frac{\gamma-1}{\gamma}} - 1\right) \tag{4.125}$$

$$\pi_f = \{1 + \eta_f(\tau_f - 1)\}^{\frac{\gamma}{\gamma-1}} \tag{4.126}$$

Except for the subscript "f," the above expressions for the fan are identical to those we derived for the compressor (see Eq. 4.13, for example).

The power balance between the turbine and compressor now includes the fan, i.e., we have to write

$$\eta_m \dot{m}_0(1 + f)(h_{t4} - h_{t5}) = \dot{m}_0(h_{t3} - h_{t2}) + \alpha \dot{m}_0(h_{t13} - h_{t2}) \tag{4.127}$$

Note that the mass flow rates in a turbofan engine are

$\dot{m}_0$ air mass flow rate through the engine core
$\alpha\dot{m}_0$ air mass flow rate through the fan duct
$(1 + \alpha)\dot{m}_0$ air mass flow rate through the inlet (which later splits into core and the fan duct flow rates)
$\dot{m}_f$ the fuel flow rate in the main burner and the fuel-to-air ratio f is defined as $\dot{m}_f/\dot{m}_0$, as logically expected

This is depicted graphically in Fig. 4.42.

From the power balance Equation 4.127, the core mass flow rate $\dot{m}_0$ cancels out and if we divide both sides by flight static enthalpy h_0 to nondimensionalize the equation, we get

$$\eta_m(1 + f)\frac{h_{t4}}{h_0}(1 - \tau_t) = \frac{h_{t2}}{h_0}[(\tau_c - 1) + \alpha(\tau_f - 1)] \tag{4.128}$$

We recognize the ratio of total enthalpy at the turbine inlet to the flight static enthalpy as τ_λ and the ratio h_{t2}/h_0 as τ_r, and we proceed to isolate τ_t and express it in terms of all the known quantities, namely

$$\tau_t = 1 - \frac{\tau_r[(\tau_c - 1) + \alpha(\tau_f - 1)]}{\eta_m(1 + f)\tau_\lambda} \tag{4.129}$$

In the above equation, all the terms on the RHS are either directly specified, e.g., α, η_m, or τ_λ, or easily calculated from the given engine design parameters. For example, flight Mach number M_0 will produce τ_r, or compressor pressure ratio π_c and efficiency e_c will produce τ_c and similarly, fan

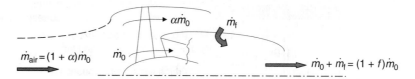

■ **FIGURE 4.42**   **Definition sketch for various mass flow rates in a separate-exhaust turbofan engine**

pressure ratio π_f and efficiency e_f produce τ_f. The parameter fuel-to-air ratio f in Eq. 4.129 is calculated by applying a power balance *across the burner*, which makes it independent of the bypass ratio and dependent only on the flight condition, compressor exit temperature, the specified turbine inlet temperature, and the fuel heating value and burner efficiency. The fuel-to-air ratio in a turbofan engine is thus expressed the same way as the fuel-to-air ratio in a turbojet engine, i.e.,

$$f = \frac{\tau_\lambda - \tau_r \tau_c}{\dfrac{Q_R \eta_b}{h_0} - \tau_\lambda} \tag{4.130}$$

The only point of caution about the above expression is that the fuel flow rate is referenced to the airflow rate *in the core* of a turbofan engine and not the entire airflow rate through the inlet that partially goes through the fan.

Now, let us examine Eq. 4.129 more closely. We immediately note that the turbine expansion parameter τ_t is a strong function of fan bypass ratio and the compressor pressure ratio, as fully expected. After all the power to compress the air in the fan and compressor comes directly from the turbine. However, we note that under certain conditions of, say, large fan pressure ratio or the engine bypass ratio, we may end up in a negative turbine expansion!! To see this, let us split Eq. 4.129 into its rational constituents, namely,

$$\tau_t = 1 - \frac{\tau_r(\tau_c - 1)}{\eta_m(1 + f)\tau_\lambda} - \frac{\tau_r \alpha(\tau_f - 1)}{\eta_m(1 + f)\tau_\lambda} = 1 - A - B \tag{4.131}$$

where A and B are the second and third terms on the RHS of Eq. 4.131. Let us further rearrange Eq. 4.131 as follows:

$$1 - \tau_t = A + B \tag{4.132}$$

We note that $(1 - \tau_t)$ term that appears on the LHS of Eq. 4.132 is a nondimensional expression for the turbine power output, as may be seen in Eq. 4.133,

$$1 - \tau_t = \frac{T_{t4} - T_{t5}}{T_{t4}} = \frac{\wp_t}{\dot{m}_t c_{pt} T_{t4}} \tag{4.133}$$

The quantity "A" represents the turbine expansion needed to power the compressor, as in the turbojet analysis (see, for example, Eq. 4.46-d). The magnitude of A is therefore independent of the fan bypass and pressure ratio. The last term in Eq. 4.131, namely the B term, represents the additional turbine expansion of gases needed to run the fan. Now, this is the term that could render Eq. 4.131 physically meaningless, i.e., lead to a negative τ_t! We further note that the maximum expansion in an aircraft gas turbine engine is produced when the turbine exit pressure is equal to the ambient pressure, i.e., $p_{t5} = p_0$. This assumes that there is no *exhaust diffuser* downstream of the turbine (in place of the exhaust nozzle) as it is customary in stationary gas turbine power plants. Therefore, not only a negative τ_t is impossible but also a $1 - \tau_t \succ \wp_{t,max}/\dot{m}_t c_{pt} T_{t4}$, based on the expansion beyond the ambient pressure, is impossible. All of these arguments are presented graphically in Fig. 4.43.

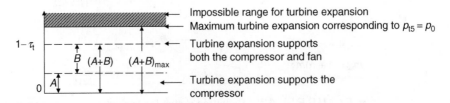

■ **FIGURE 4.43** **Various turbine expansion possibilities including the impossible range**

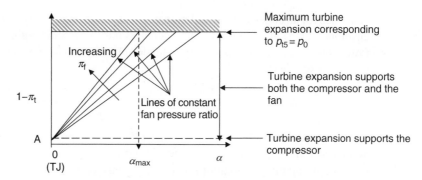

■ **FIGURE 4.44** **Variation of turbine power parameter $(1-\tau_t)$ with the fan bypass and pressure ratios**

We can also present the impact of the engine bypass ratio as well as the fan pressure ratio on the power balance equation 4.132 graphically, for example, by noting the linear dependence of turbine expansion and bypass ratio and plotting that in Fig. 4.44.

We learn from Fig. 4.44 that for a desired fan pressure ratio π_f, the bypass ratio is limited by the maximum turbine expansion corresponding to $p_{t5} = p_0$. Also, we note that with an increasing fan pressure ratio, the bypass ratio needs to be reduced for a given turbine power output. In essence, what is graphically depicted in Fig. 4.44 is intuitive.

We are now ready to develop the thrust and performance parameters for a separate-exhaust turbofan engine. The question of thrust developed in a turbofan engine is best answered if we apply the momentum principles learned in the turbojet section to two separate streams and write the gross thrust for two nozzles and account for the entire ram drag through the inlet, namely,

$$F_n = \dot{m}_0(1 + f)V_9 + \alpha \dot{m}_0 V_{19} - (1+\alpha)\dot{m}_0 V_0 + (p_9 - p_0)A_9 + (p_{19} - p_0)A_{19} \quad (4.133)$$

We recognize the terms on the RHS of Eq. 4.133 as the two nozzle momentum thrusts followed by the ram drag and then the pressure thrust of the primary and the fan nozzles. Once we learn the principles of thrust development of a single-stream engine, as in a turbojet, the generalization to a two-stream engine, as in the separate-exhaust turbofan engine, is very simple and straightforward. The specific thrust for a turbofan engine is the ratio of the net (uninstalled) thrust divided by the entire airflow rate through the inlet, i.e., specific thrust is

$$\frac{F_n}{(1+\alpha)\dot{m}_0} = \frac{1+f}{1+\alpha}V_9 + \frac{\alpha}{1+\alpha}V_{19} - V_0 + \frac{(p_9 - p_0)A_9}{(1+\alpha)\dot{m}_0} + \frac{(p_{19} - p_0)A_{19}}{(1+\alpha)\dot{m}_0} \quad (4.134)$$

We immediately note that the specific thrust drops for a turbofan engine (see $1 + \alpha$ in the denominator of the LHS of Eq. 4.134), in comparison to a turbojet engine that has zero bypass ratio, as more air is handled to produce the thrust in a bypass configuration. Therefore, this observation of lower specific thrust for a turbofan engine leads to the conclusion that bypass engines have by necessity larger frontal areas, again in comparison to a turbojet engine producing the same thrust. Now, we know that the larger frontal areas will result in a higher level of nacelle and installation drag. Therefore, the benefit of higher propulsive efficiency promised in a bypass engine should be carefully weighed against the penalties of higher installation drag for such configurations. The general conclusions are that for subsonic applications, the external drag penalty of a turbofan engine installation is significantly smaller than the benefits of higher engine propulsive efficiency. As a result, in a subsonic cruise civil transport, for example, the trend has been and continues to be in developing higher bypass ratio engines in order to save on fuel consumption. Ground clearance for an under wing installation poses the main constraint on the bypass ratio/engine envelope. As the flight Mach number increases into supersonic regime, the optimum bypass ratio is then reduced. For a perfectly expanded primary and fan nozzles, the expression 4.134 simplifies to

$$\frac{F_n}{(1+\alpha)\dot{m}_0} = \frac{1+f}{1+\alpha}V_9 + \frac{\alpha}{1+\alpha}V_{19} - V_0 \tag{4.135}$$

We may divide both sides of Eq. 4.134 or 4.135 by the ambient speed of sound a_0 to arrive at a nondimensional expression for the specific thrust.

Applying the definition of specific fuel consumption to a turbofan engine, we get

$$\text{TSFC} = \frac{\dot{m}_f}{F_n} = \frac{f}{F_n/\dot{m}_0} = \frac{f/(1+\alpha)}{F_n/(1+\alpha)\dot{m}_0} \tag{4.136}$$

4.3.3.3   Thermal Efficiency of a Turbofan Engine.
The ideal cycle thermal efficiency η_{th} is unaffected by the placement of a fan stage in a gas generator where the fan discharge is then expanded in a nozzle. The reason is that the power utilized to compress the gas in the fan is balanced by the power turbine and therefore the net energy exchange in the bypass duct comes from the balance of the fan inlet and the fan nozzle. A perfectly expanded nozzle will recover the kinetic energy exchange of the inlet diffuser, and thus the ideal thermal efficiency of a turbofan engine is unaffected. We can see this, for an ideal fan inlet and a nozzle, in the following T–s diagram (see Fig. 4.45).

Therefore, the thermal efficiency of an ideal turbofan engine is identical to a turbojet engine with the same overall, i.e., cycle, compression ratio, namely,

$$\eta_{th} = 1 - \frac{1}{\tau_r\tau_c} = 1 - \frac{1}{\left(1+\frac{\gamma-1}{2}M_0^2\right)\pi_c^{\frac{\gamma-1}{\gamma}}} \tag{4.137}$$

For the real, i.e., nonideal, turbofan engine, we start from the definition of thermal efficiency:

$$\eta_{th} \equiv \frac{\Delta K\dot{E}}{\dot{m}_f Q_R} = \frac{\alpha\dot{m}_0\frac{V_{19}^2}{2}+(1+f)\dot{m}_0\frac{V_9^2}{2}-(1+\alpha)\dot{m}_0\frac{V_0^2}{2}}{\dot{m}_f Q_R}$$

$$= \frac{\alpha V_{19}^2 + (1+f)V_9^2 - (1+\alpha)V_0^2}{2 f Q_R} = \frac{(1+f)V_9^2 - V_0^2}{2 f Q_R} + \alpha\frac{V_{19}^2 - V_0^2}{2 f Q_R} \tag{4.138-a}$$

The right-hand side of Eq. 4.138-a identifies the bypass ratio, the two exhaust velocities, the fuel-to-air ratio, the fuel heating value, and the burner efficiency as the parameters in a turbofan engine that affect the thermal efficiency of the engine. The typical unknowns, however, for the design point analysis of a turbofan engine with a given bypass ratio and a fuel type, are the two

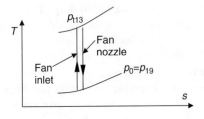

■ **FIGURE 4.45**   *T–s* diagram of a compression (fan) inlet stream followed by the expansion in a nozzle

nozzle exhaust velocities and the combustor fuel-to-air ratio. We note from the last equality in Eq. 4.138 that all the bypass contribution to engine cycle thermal efficiency is lumped in the last term and the core contribution is in the first term. Ideally, the lower kinetic energy of the core due to power drainage of the fan, i.e., the first term, appears as the kinetic energy rise in the bypass stream, i.e., the second term. Therefore, the thermal efficiency of a turbofan engine is ideally unaffected by the magnitude of the bypass stream.

The Eq. 4.138-a is correct for a turbofan engine with perfectly expanded nozzles. In case the exhaust nozzles are not perfectly expanded, we have to use *effective exhaust speeds* $V_{19\text{eff}}$ and $V_{9\text{eff}}$ in place of actual exit velocities V_{19} and V_9. We define the nozzle effective exhaust speed as

$$V_{19\text{eff}} \equiv \frac{F_{\text{g-fan}}}{\dot{m}_{19}} \qquad \textbf{(4.138-b)}$$

$$V_{9\text{eff}} \equiv \frac{F_{\text{g-core}}}{\dot{m}_9} \qquad \textbf{(4.138-c)}$$

4.3.3.4   Propulsive Efficiency of a Turbofan Engine.

The propulsive efficiency of a turbofan engine, in contrast to the thermal efficiency of such engines, is a strong function of the engine bypass ratio. Starting from the definition of propulsive efficiency

$$\eta_{\text{p}} \equiv \frac{F_{\text{n}}V_0}{\Delta K\dot{E}} = \frac{[(1+f)V_9 + \alpha V_{19} - (1+\alpha)V_0 + (p_9 - p_0)A_9 + (p_{19} - p_0)A_{19}]V_0}{(1+f)\dfrac{V_{9\text{eff}}^2}{2} + \alpha\dfrac{V_{19\text{eff}}^2}{2} - (1+\alpha)\dfrac{V_0^2}{2}} \qquad \textbf{(4.139)}$$

and assuming the nozzles are perfectly expanded and the fuel-to-air ratio f is negligible compared to 1, the above equation may be further simplified to

$$\eta_{\text{p}} \approx \frac{2V_0[V_9 + \alpha V_{19} - (1+\alpha)V_0]}{V_9^2 + \alpha V_{19}^2 - (1+\alpha)V_0^2} \qquad \textbf{(4.140)}$$

Let us recast Eq. 4.140 in terms of nondimensional velocity ratios V_9/V_0 and V_{19}/V_0, by dividing the numerator and denominator of Eq. 4.140 by the square of the flight velocity V_0^2 to get

$$\eta_{\text{p}} \approx \frac{2\left[\left(\dfrac{V_9}{V_0}\right) + \alpha\left(\dfrac{V_{19}}{V_0}\right) - (1+\alpha)\right]}{\left(\dfrac{V_9}{V_0}\right)^2 + \alpha\left(\dfrac{V_{19}}{V_0}\right)^2 - (1+\alpha)} \qquad \textbf{(4.141)}$$

In a turbofan engine, the propulsive efficiency seems to be influenced by three parameters, as shown in Eq. 4.141, the bypass ratio α, the primary jet-to-flight velocity ratio V_9/V_0, and the fan jet-to-flight velocity ratio V_{19}/V_0. In case the fan nozzle velocity and the primary nozzle velocity are equal, Eq. 4.141 reduces to the familiar expression we derived earlier for a turbojet, namely,

$$\eta_{\text{p}} \approx 2/(1 + V_9/V_0) \qquad \textbf{(4.142)}$$

which interestingly does not explicitly depend on the bypass ratio α! The effect of bypass ratio is implicit, however, in the exhaust velocity reduction that appears in the denominator of Eq. 4.142. Again as noted earlier, if the nozzles are not perfectly expanded, we have to use the effective exhaust speeds instead of the "actual" velocities. We will examine these issues in the next two examples.

EXAMPLE 4.13

A high-bypass ratio (separate-exhaust) turbofan engine powers a commercial transport. At the cruise condition, the flight and engine operating conditions are

$M_0 = 0.88$, $p_0 = 15\,\text{kPa}$, $T_0 = -40°C$, $\gamma_c = 1.4$,
$c_{pc} = 1004\,\text{J/kg·K}$
$\pi_d = 0.995$
$\pi_f = 1.6$, $e_f = 0.90$, $\alpha = 8.0$

Fan nozzle is convergent with $\pi_{fn} = 0.95$

$\pi_c = 40$, $e_c = 0.90$
$\tau_\lambda = 8.0$, $c_{pt} = 1,152\,\text{J/kg·K}$,
$\gamma_t = 1.33$, $Q_R = 42,000\,\text{kJ/kg}$, $\pi_b = 0.95$,
$\eta_b = 0.992$
$\eta_m = 0.95$, $e_t = 0.85$

Primary nozzle is of convergent design and operates at $\pi_n = 0.98$

CALCULATE

(a) total pressures and temperatures throughout the engine and the fuel-to-air ratio

(b) nozzle exit static pressure p_{19} and p_9

(c) actual and *effective* (fan and core) nozzle exit velocities V_{19}, V_9 and $V_{19\text{eff}}$, $V_{9\text{eff}}$

(d) ratio of fan-to-core thrust

(e) nondimensional specific thrust

(f) TSFC in mg/s/N

(g) all engine efficiencies η_{th}, η_p, and η_o

Note: For (convergent) exhaust nozzles that are not perfectly expanded, the thermal and propulsive efficiencies are defined based on effective exhaust velocities:

$$\eta_{th} \equiv \frac{\Delta \dot{KE}}{\dot{m}_f Q_R} = \frac{\dot{m}_{19} V_{19\text{eff}}^2 + \dot{m}_9 V_{19\text{eff}}^2 - (\dot{m}_{19} + \dot{m}_0) V_0^2}{2\dot{m}_f Q_R}$$

$$\eta_p = \frac{2F_\eta V_0}{\dot{m}_{19} V_{19\text{eff}}^2 + \dot{m}_9 V_{9\text{eff}}^2 - (\dot{m}_{19} + \dot{m}_0) V_0^2}$$

where the effective exhaust velocity $V_{19\text{eff}}$ is defined as

$$V_{19\text{eff}} \equiv \frac{F_{g-\text{jet}}}{\dot{m}_{19}}$$

$$V_{9\text{eff}} \equiv \frac{F_{g-\text{carg}}}{\dot{m}_9}$$

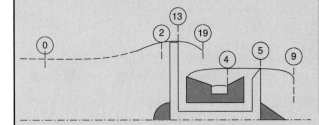

SOLUTION

To practice the principles that we have learned, we march component-by-component through the engine. First, we convert ambient static temperature to absolute scale, i.e.,

$$T_0 = (-40 + 273)\,K = 233\,K$$

Speed of sound at the flight altitude is $a_0 = \sqrt{(\gamma_c - 1)c_{pc}T_0} = \sqrt{0.4(1004)(233)}$ m/s $\cong 305.9\,\text{m/s}$

The flight speed is $V_0 = M_0$, $a_0 = (0.88)(305.9\,\text{m/s}) \cong 269.2\,\text{m/s}$

The total pressure of flight is $p_{t0} = p_0[1 + (\gamma_c - 1)M_0^2/2]^{\gamma/(\gamma-1)} = 15\,\text{kPa}[1 + 0.2(0.88)^2]^{3.5} \cong 24.83\,\text{kPa}$

The total temperature of flight is $T_{t0} = T_0[1 + (\gamma_c - 1)M_0^2/2] = 233K[1 + 0.2(0.88)^2] \cong 269.1\,\text{K}$

The total pressure at the engine face is $p_{t2} = p_{t0} \cdot \pi_d = 24.83\,\text{kPa}(0.995) \cong 24.7\,\text{kPa}$

The total temperature at the engine face is equal to flight total temperature, since inlets are adiabatic,

$T_{t2} = T_{t0} = 269.1\,\text{K}$

At this point, air is divided into two streams, one that goes through the fan and the second that enters the engine core. First, let us analyze the fan stream.

The fan exit total pressure is $p_{t13} = p_{t2} \cdot \pi_f = 24.7 \, \text{kPa}(1.6) \cong 39.53 \, \text{kPa}$

We calculate fan exit total temperature from its efficiency, namely $\tau_f = \pi_f^{(\gamma_c - 1)/e_f\gamma_c} = 1.6^{0.4/0.90/1.4} \cong 1.1609$

The total temperature at the fan exit is $T_{t13} = T_{t2} \cdot \tau_f = (269.1 \, \text{K})(1.1437) \cong 312.4 \, \text{K}$

The fan nozzle exit total pressure is $p_{t19} = p_{t13} \cdot \pi_{fn} = 39.53 \, \text{kPa}(0.95) \cong 37.55 \, \text{kPa}$

Since the fan nozzle is of convergent design, the highest exit Mach number that it can attain is Mach 1. Let us check whether the nozzle exit reaches Mach 1. The nozzle exit static pressure for a Mach 1 exit is

$$p_{19} = p_{t19}/[1 + (\gamma_c - 1)/2]^{\gamma_c/(\gamma_c - 1)} = 37.55 \, \text{kPa}/(1.2)^{3.5} \cong 19.84 \, \text{kPa}$$

Since $p_{19} > p_0$, the nozzle flow is underexpanded, therefore the convergent fan nozzle has reached its peak exit velocity, i.e., sonic. Therefore, $M_{19} = 1.0$.

The fan nozzle exit static temperature is $T_{19} = T_{t13}/(1.2) = 312.4 \, \text{K}/1.2 \cong 260.3 \, \text{K}$

The speed of sound at the fan nozzle exit is $a_{19} = \sqrt{(\gamma_c - 1)c_{pc}T_{19}} = \sqrt{0.4(1004)(260.3)} \, \text{m/s} \cong 323.3 \, \text{m/s}$

Since the fan nozzle exit Mach number is 1, $V_{19} = a_{19} = 323.3 \, \text{m/s}$

The fan gross thrust is $F_{g-fan} = \dot{m}_{19}V_{19} + (p_{19} - p_0)A_{19}$

The effective velocity at the fan nozzle exit is defined as $V_{19eff} \equiv F_{g-fan}/\dot{m}_{19}$

Therefore the effective fan nozzle exit velocity is

$$V_{19eff} = V_{19} + \frac{(p_{19} - p_0)A_{19}}{\dot{m}_{19}} = V_{19} + \frac{p_{19}(1 - p_0/p_{19})A_{19}}{\rho_{19}V_{19}A_{19}} = V_{19} + \left(\frac{\gamma_c p_{19}}{\rho_{19}}\right)\frac{(1 - p_0/p_{19})}{\gamma_c V_{19}}$$

The terms in the parenthesis on the RHS of the above equation represent the speed of sound at the fan nozzle exit $(a_{19})^2$. We have calculated all these terms, so upon substitution, we get $V_{19eff} \cong 379.7 \, \text{m/s}$

Now, we are ready to calculate the core stream.

The compressor exit total pressure is $p_{t3} = p_{t2} \cdot \pi_c = 24.7 \, \text{kPa}(40) \cong 988.2 \, \text{kPa}$

The compressor exit total temperature is calculated via $\tau_c = \pi_c^{(\gamma_c - 1)/e_c\gamma_c} = 40^{0.4/1.4/0.90} \cong 3.225$

The exit total temperature is $T_{t3} = T_{t2} \cdot \tau_c = 269.1 \, \text{K}(3.225) \approx 868 \, \text{K}$

The burner exit total pressure is $p_{t4} = p_{t3} \cdot \pi_b = 988.2 \, \text{kPa}(0.95) \approx 938.8 \, \text{kPa}$

The burner exit total temperature is $T_{t4} = c_{pc}T_0\tau_\lambda/c_{pt} = 1004(233 \, \text{K})(8)/1152 \approx 1,624.5 \, \text{K}$

The fuel-to-air ratio is calculated from the energy balance across the burner,

$$f = (c_{pt}T_{t4} - c_{pc}T_{t3})/(Q_R\eta_b - c_{pt}T_{t4})$$

We substitute the values in the above equation to get $f = 0.0251$

To get the total temperature at the turbine exit, we apply the power balance between the turbine and the compressor and fan, i.e., $(1 + f)c_{pt}(T_{t4} - T_{t5}) \cdot \eta_m = c_{pc}(T_{t3} - T_{t2}) + \alpha c_{pc}(T_{t13} - T_{t2})$

We have calculated all the parameters in the above equation, except T_{t5}. Upon substitution, we get $T_{t5} \approx 778.6 \, \text{K}$

To calculate the turbine exit total pressure, we use the polytropic efficiency e_t in $\pi_t = \tau_t^{\gamma_t/e_t(\gamma_t - 1)}$

The turbine $\tau_t = T_{t5}/T_{t4} = 778.6/1624.5 \approx 0.4793$

Therefore, $\pi_t = (0.4793)^{1.33/0.33/0.85} \cong 0.0306$. The turbine exit total pressure is $p_{t5} = p_{t4} \cdot \pi_t = 938.8 \, \text{kPa}(0.0306)$ $\approx 28.72 \, \text{kPa}$

The nozzle exit total pressure is $p_{t9} = p_{t5} \cdot \pi_n = 28.72 \, \text{kPa}(0.98) \approx 28.14 \, \text{kPa}$

Since the maximum Mach number at the convergent nozzle exit is $M_9 = 1.0$, we examine the static pressure p_9 and compare it to p_0. If $p_9 \geq p_0$, then the exit Mach number is 1. Otherwise, the exit Mach number is less than 1 and the exit static pressure is equal to the ambient static pressure. We calculate p_9 based on p_{t9} and the choked exit condition according to

$$p_9 = p_{t9}/[(\gamma_t + 1)/2]^{\gamma_t/(\gamma_t - 1)} = 28.14 \, \text{kPa}(1.165)^{1.33/0.33} \cong 15.21 \, \text{kPa}$$

Since $p_9 > p_0$, the exit Mach number is $M_9 = 1.0$ and $p_9 = 15.21 \, \text{kPa}$

The nozzle exit static temperature is $T_9 = T_{t5}/(\gamma_t + 1)/2 = 778.6 \, \text{K}/(1.165) = 668.3 \, \text{K}$

The speed of sound at the nozzle exit is $a_9 = \sqrt{(\gamma_t - 1)c_{pt}T_9} = \sqrt{0.33(1152)(668.39)} \, \text{m/s} \cong 504.1 \, \text{m/s}$

Since $M_9 = 1$, $V_9 = a_9 = 504.1 \, \text{m/s}$

The effective exhaust speed is $V_{9eff} = F_{g-core}/\dot{m}_9 = V_9 + \frac{(p_9 - p_0)A_9}{\dot{m}_9} = V_9 + \frac{a_9^2(1 - p_0/p_9)}{\gamma_t \cdot V_9} \cong 509.2 \, \text{m/s}$

The fan net thrust is $F_{n-fan} = F_{g-fan} - \alpha\dot{m}_0V_0 = \alpha\dot{m}_0V_{19eff} - \alpha\dot{m}_0V_0$

The nondimensional specific fan thrust (based on the inlet mass flow rate) is $\dfrac{F_{n-fan}}{(1+\alpha)\dot{m}_0 \cdot a_0} = \dfrac{\alpha(V_{19eff} - V_0)}{(1+\alpha)\cdot a_0}$

Upon substitution, we get $\dfrac{F_{n-fan}}{(1+\alpha)\dot{m}_0 \cdot a_0} \cong 0.3210$

The net core thrust is $F_{n-core} = F_{g-core} - \dot{m}_0 V_0 = (1+f)\dot{m}_0 V_{9eff} - \dot{m}_0 V_0$

The nondimensional specific core thrust (based on the inlet mass flow rate) is $\dfrac{F_{n-core}}{(1+\alpha)\dot{m}_0 \cdot a_0} = \dfrac{(1+f)V_{9eff} - V_0}{(1+\alpha)\cdot a_0}$

Upon substitution, we get $\dfrac{F_{n-core}}{(1+\alpha)\dot{m}_0 \cdot a_0} \cong 0.09184$

The ratio of fan-to-core thrust is $\dfrac{F_{n-fan}}{F_{n-core}} = \dfrac{0.3210}{0.09184} \cong 3.495$

The fan contribution to the engine net thrust is $F_{n-fan}/F_{n-total} = 0.3210/(0.3210 + 0.09184) \approx 77.75\%$

The core contribution to the total net thrust is $F_{n-core}/F_{n-total} = 0.09184/(0.3210 + 0.09184) \approx 22.25\%$

The thrust specific fuel consumption is TSFC $= \dot{m}_f/F_n$. In terms of the nondimensional specific thrust, TSFC may be written

as TSFC $= \dfrac{f}{(1+\alpha)a_0 F_n/[(1+\alpha)\dot{m}_0 a_0]} = \dfrac{(0.0251)\cdot10^6}{(9)(305.9)(0.4128)}$ mg/s/N $= 22\cdot11$ mg/s/N

Note that without the 10^6 in the numerator, we would get kg/s/N. Therefore, TSFC ≈ 22.11 mg/s/N.

Since the fan and core nozzles were not perfectly expanded, we use the effective exhaust speeds in calculating thermal and propulsive efficiencies.

$$\eta_{th} \equiv \frac{\Delta K\dot{E}}{\dot{m}_f Q_R} = \frac{\dot{m}_{19}V_{19eff}^2 + \dot{m}_9 V_{9eff}^2 - (\dot{m}_{19} + \dot{m}_0)V_0^2}{2\dot{m}_f Q_R} = \frac{\alpha V_{19eff}^2 + (1+f)V_{9eff}^2 - (1+\alpha)V_0^2}{2fQ_R}$$

Upon substitution, we get $\eta_{th} \approx 36.32\%$

The propulsive efficiency is

$$\eta_p \equiv \frac{2F_n V_0}{\dot{m}_{19}V_{19eff}^2 + \dot{m}_9 V_{9eff}^2 - (\dot{m}_{19} + \dot{m}_0)V_0^2} = \frac{2\left[\dfrac{F_n}{(1+\alpha)\dot{m}_0 a_0}\right](1+\alpha)a_0 V_0}{\alpha V_{19eff}^2 + (1+f)V_{9eff}^2 - (1+\alpha)V_0^2}$$

We have calculated the terms on the RHS, upon substitution, we get $\eta_p \approx 79.80\%$

The overall efficiency is the product of the thermal and propulsive efficiency, $\eta_0 \approx (0\cdot3632)(0\cdot798) \approx 28\cdot99\%$

EXAMPLE 4.14

Write a computer program or use the spreadsheet to calculate and graph the performance of the separate-exhaust turbofan engine described in Example 4.13 for a range of bypass ratios from 0 to 8. The zero bypass ratio simulates a turbojet. Explain the results that you graphically depict for TSFC, non-dimensional-specific thrust, and thermal, propulsive, and overall efficiencies.

SOLUTION

Spreadsheet calculations are performed in the range of 0–8 bypass ratio turbofan of Example 4.13. A plot of TSFC in (mg/s/N) shows continuous improvement with increasing bypass ratio. Indeed, the turbojet case of $\alpha = 0$ consumes the most fuel. The modern ultrahigh bypass ratio turbofan engines in commercial aviation with a bypass ratio of ~8 use ~40% less fuel than a turbojet or ~10% less fuel than a bypass ratio 5 turbofan engine.

The specific thrust of the turbofan engine as a function of bypass ratio is shown in the following graph. Since turbofan produces its thrust by acting over a large mass flow of air, its specific thrust is lower than a turbojet. We note that the engine thrust per unit *inlet mass flow rate*, nondimensionalized by the ambient speed of sound a_0 is reduced with increasing bypass ratio. However, if we graph the turbofan engine thrust per unit *core mass flow rate* (again nondimensionalized by a_0), we note that engine thrust continuously increases with bypass ratio.

The thermal efficiency of the turbofan engine as a function of bypass ratio is graphed. As indicated in the text, the ideal thermal efficiency of a turbofan engine doesn't change with bypass ratio and as seen, this point is borne by data for the "real" turbofan engine as well. The propulsive efficiency, which was the target of improvement in a turbofan engine, shows continuous increase with bypass ratio. The lowest propulsive efficiency corresponds to the turbojet engine ($\alpha = 0$) at $\sim 44\%$ in this example. The overall efficiency improves with bypass ratio similar to the propulsive efficiency.

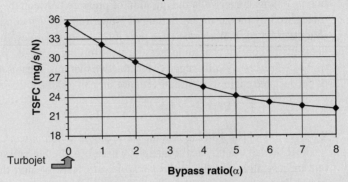

Turbofan-specific fuel consumption

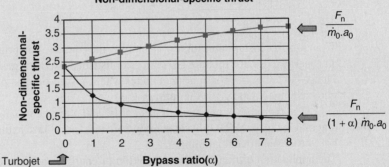

Non-dimensional-specific thrust

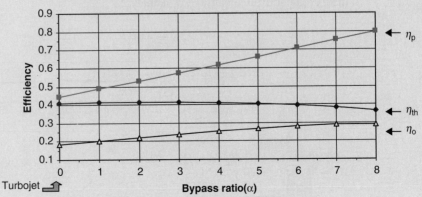

Turbofan thermal, propulsive, and overall efficiencies

4.4 Analysis of a Mixed-Exhaust Turbofan Engine with an Afterburner

The schematic of Fig. 4.46 of a mixed-exhaust turbofan engine shows the station numbers for this engine and identifies a new component, namely the mixer. The engine core discharges a hot gas into the mixer, whereas the fan duct injects a "cold" gas into the mixer. Upon mixing of the two streams, the gas will attain a "mixed-out" state at station 6M. Now, what about the pressure of the two incoming streams? On the question of pressure between the two streams, as they enter the mixer, we rely on our understanding of Kutta trailing-edge condition applied to airfoils with a sharp trailing edge. We learned this principle in aerodynamics. In simple terms, Kutta condition demands the continuity of static pressure at the trailing edge, i.e., $p_{upper} = p_{lower}$ on an airfoil at its trailing edge. It physically suggests that the two merging streams at the trailing edge of an airfoil, or fan duct and engine core streams cannot support a *static pressure jump*. Therefore, we demand

$$p_{15} = p_5 \qquad (4.143)$$

We can think of Eq. 4.143 as that the static pressure is *communicated* between the core and the fan duct, which is a physical principle. Therefore, the fan pressure ratio and hence mass flow rate are basically set by the *engine backpressure*. Here, we note that the two parameters of a fan, namely its pressure ratio and the bypass ratio, cannot be independently set in this configuration. Here, we can stipulate only one parameter and the second parameter *falls out* of the common backpressure requirement.

4.4.1 Mixer

To analyze a constant-area mixer, we employ the conservation principles of mass, momentum, and energy. Also, the mixture gas laws establish the mixed-out gas properties at the mixer exit. Figure 4.47 shows a constant-area mixer where mixed-out gas properties c_{p6M} and c_{v6M} are mass averaged and γ_{6M} is the ratio of the two.

The thermodynamic state of gas before and after the mixing process is shown on a T–s diagram in Fig. 4.48. The physical mixing process takes place in the shear layer where cold and hot vortices interject and mingle.

The application of the law of conservation of energy to an insulated mixer gives

$$\dot{m}_5 h_{t5} + \dot{m}_{15} h_{t15} = \dot{m}_{6M} h_{t6M} \qquad (4.144)$$

Assuming the inlet conditions are known for the mixer, the only unknown in the above equation in the mixed-out total enthalpy h_{t6M}, namely,

$$h_{t6M} = \frac{(1 + f)h_{t5} + \alpha h_{t15}}{1 + \alpha + f} \qquad (4.145)$$

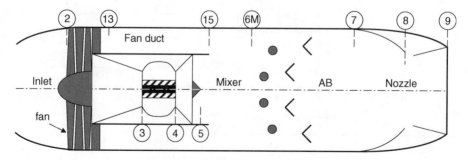

■ **FIGURE 4.46** Long-duct turbofan engine with a mixer and an afterburner

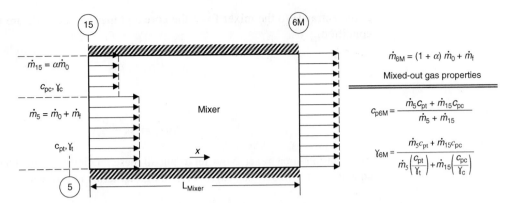

■ **FIGURE 4.47** **Mixer control volume showing a nonuniform inlet and *mixed-out* exit velocity profiles**

Here we can introduce the component nondimensional parameters in the above equation as follows:

$$h_{t6M} = h_0 \cdot \frac{(1 + f)\tau_t \cdot \tau_\lambda + \alpha\tau_r \cdot \tau_f}{1 + \alpha + f} \qquad (4.146)$$

In order to establish the mixed-out pressure p_{6M} at the exit of the mixer, we apply the conservation of momentum in the streamwise direction, i.e., x as shown in Fig. 4.47. A simple expression for the x-momentum is just $\dot{m}V_x$, which holds for a steady uniform flow and may be applied to the mixer control volume as follows:

$$\dot{m}_{6M}V_{6M} - [\dot{m}_5V_5 + \dot{m}_{15}V_{15}] = \Sigma F_x \qquad (4.147)$$

In the above expression, the right-hand side is the sum of all external forces acting on the control volume in the x-direction. The external forces are nothing more than the integrals of pressure and wall shear over the flow and wetted areas, respectively, namely,

$$\Sigma F_x = p_5A_5 + p_{15}A_{15} - p_{6M}A_{6M} - \int \tau_w dA_{\text{wetted}} \qquad (4.148)$$

where the last term is the wall friction drag force acting on the mixer wall, $D_{f_{\text{mixer}}}$, associated with the mixer in the above flow condition. We also note that the static pressures

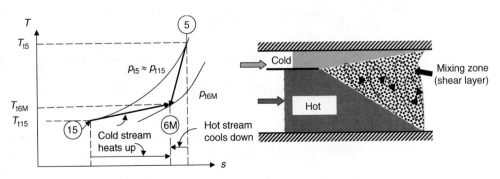

■ **FIGURE 4.48** **Mixing of two streams at different temperatures on a *T–s* diagram and in physical domain**

at the entrance to the mixer from the core and the fan duct side are the same a' la Kutta condition,

$$p_5 = p_{15} \tag{4.149}$$

and

$$A_{6M} = A_5 + A_{15} \tag{4.150}$$

for a constant-area mixer. Now, let us put all these relationships together in the x-momentum equation of the mixer, i.e., Eq. 4.147, and group terms to arrive at

$$(\dot{m}_{6M}V_{6M} + p_{6M}A_{6M}) - (\dot{m}_5 V_5 + p_5 A_5) - (\dot{m}_{15}V_{15} + p_{15}A_{15}) = -D_f \tag{4.151}$$

The grouped terms in the above equation are the *impulse* (given a symbol I, as we discussed in Chapter 2) in corresponding stations, namely 6M, 5, and 15, respectively. We note that the fluid impulse out of the mixer is somewhat less than the (sum of the) fluid impulse entering the mixer and this difference or loss is caused by the friction drag on the mixer wall. Therefore, if we had assumed a frictionless flow in the mixer, we would have concluded that the impulse is conserved, i.e.,

$$I_{6M} = I_5 + I_{15} \tag{4.152}$$

We proceed with our analysis using the assumption of zero boundary layer frictional losses on the walls of the mixer in order to derive a closed form solution for the unknown mixer exit conditions. We then suggest a correction factor due to friction that accounts for the viscous flow losses on the total pressure.

The impulse function "*I*" attains a very simple form when we express it in terms of flow Mach number. The steps are outlined in the following equation:

$$I \equiv \dot{m}V + pA = A(\rho V^2 + p) = Ap\left(\frac{\rho V^2}{p} + 1\right) = Ap(1 + \gamma M^2) \tag{4.153}$$

We had used the equation for the speed of sound $a^2 = \gamma p/\rho$ in the above derivation. Hence, the flow impulse at any station is the product of pressure–area term and $(1 + \gamma M^2)$ at that station. Now, in light of this expression for impulse, let us recast Eq. 4.152 in terms of flow Mach number and pressure area terms as

$$p_{6M}A_{6M}\left(1 + \gamma_{6M}M_{6M}^2\right) - p_5 A_5(1 + \gamma_5 M_5^2) - p_{15}A_{15}(1 + \gamma_c M_{15}^2) = 0 \tag{4.154}$$

Now, if we divide the above equation by a reference area, namely turbine exit area A_5, we will eliminate the effect of size from the above equation. Therefore, we can express Eq. 4.154 as

$$p_{6M}\lfloor 1 + \gamma_{6M}M_{6M}^2 \rfloor (1 + A_{15}/A_5) - p_5 \lfloor (1 + \gamma_5 M_5^2) + (A_{15}/A_5)(1 + \gamma_c M_{15}^2) \rfloor = 0 \tag{4.155}$$

Equation 4.155 has two unknowns, namely p_{6M} and M_{6M} that need to be determined. The continuity equation can be expressed in terms of pressure and Mach number according to

$$\dot{m} \equiv \rho AV = AVp/RT = \frac{\gamma\, pAM}{\sqrt{\gamma RT}} = \frac{\gamma\, pAM}{\sqrt{\gamma RT_t}}\sqrt{1 + \frac{\gamma - 1}{2}M^2} \tag{4.156}$$

Therefore, continuity equation as expressed in terms of pressure and Mach number, applied to the mixer, can be written as

$$
\frac{p_{6M} M_{6M} \sqrt{1 + (\gamma_{6M} - 1)M_{6M}^2/2}}{\sqrt{\gamma_{6M} R_{6M} T_{t6M}}} (1 + A_{15}/A_5)
$$

$$
= \frac{p_5 M_5 \sqrt{1 + (\gamma_t - 1)M_5^2/2}}{\sqrt{\gamma_t R_t T_{t5}}} + \frac{p_{15} M_{15} \sqrt{1 + (\gamma_c - 1)M_{15}^2/2}}{\sqrt{\gamma_c R_c T_{t15}}} (A_{15}/A_5) \quad \textbf{(4.157)}
$$

The above equation and the impulse equation, 4.155 and 4.157, form a two equation, two unknown system in terms of the unknowns p_{6M} and M_{6M}. We can solve the equations in closed form as they reduce to a quadratic equation in M_{6M}^2.

4.4.2 Cycle Analysis

To reinforce the powerful *marching technique* that we have learned so far in cycle analysis, let us apply it to a mixed-flow turbofan engine at this time.

First, we will identify a *mix of parameters* that we need to specify in order to start a cycle analysis in a mixed-flow turbofan engine.

Flight condition: We need the flight Mach number (or speed), static temperature, and pressure at the flight altitude, i.e., M_0 (or V_0), T_0, and p_0.

Inlet: We need the inlet adiabatic efficiency or its total pressure recovery parameter, i.e., η_d or π_d, only one figure-of-merit is needed.

Fan: We need fan's pressure ratio $\pi_f \equiv (p_{t13}/p_{t2})$ *or* the fan bypass ratio α and its polytropic efficiency e_f or its adiabatic efficiency η_f.

Compressor: We need compressor's total pressure ratio $\pi_c \equiv p_{t3}/p_{t2}$ and its polytropic efficiency e_c or its adiabatic efficiency η_c.

Combustor: We need combustor's total pressure ratio π_b, combustion efficiency η_b, and fuel heating value Q_R and either turbine inlet temperature T_{t4} or the fuel-to-air ratio f.

Turbine: We need turbine's polytropic efficiency e_t or adiabatic efficiency η_t, mechanical efficiency η_m exit Mach number M_5, and an extra assumption about the pressure matching condition between the core and fan ducts at their merging boundary, typically $p_{t5} = p_{t15}$ or $p_5 = p_{15}$.

Mixer: We need a viscous loss parameter that accounts for the mixer wall friction on the total pressure loss across the mixer, namely, $\pi_{M,f}$.

Afterburner: We need the afterburner efficiency η_{AB}, the maximum temperature T_{t7} (or f_{AB}), afterburner total pressure ratio π_{AB}, and the heating value of the fuel $Q_{R,AB}$ in the afterburner.

Nozzle: We need adiabatic efficiency of the nozzle η_n or its total pressure loss parameter π_n, and the nozzle exit pressure p_9 or instead of exit pressure, we may specify nozzle's (exit-to-throat) area ratio A_9/A_8.

4.4.2.1 *Solution Procedure.* Let us first map out the total temperature and pressure distribution throughout the engine. Since our analysis is limited to a (steady) one-dimensional flow study of components, i.e., we treat the fluid and flow properties to be uniform at each and every stations of the engine. Therefore, we propose to follow a candidate streamline that goes through the core and a candidate streamline that goes through the fan. Once we establish the total fluid properties after each interaction along the streamline path, we have the 1D information for the core and the fan streams. Then, we allow the Kutta condition to establish the communication between the two streams before the mixer takes over and achieves a mixed-out state of the gas prior to discharge into the nozzle or afterburner.

Station 0

$$p_{t0} = p_0 \left(1 + \frac{\gamma_c - 1}{2} M_0^2 \right)^{\frac{\gamma_c}{\gamma_c - 1}}$$

$$\pi_r = \left(1 + \frac{\gamma_c - 1}{2} M_0^2 \right)^{\frac{\gamma_c}{\gamma_c - 1}} \tag{4.158}$$

$$T_{t0} = T_0 \left(1 + \frac{\gamma_c - 1}{2} M_0^2 \right)$$

$$\tau_r = \left(1 + \frac{\gamma_c - 1}{2} M_0^2 \right) \tag{4.159}$$

$$a_0 = \sqrt{\gamma_c R_c T_0}$$

$$V_0 = a_0 M_0$$

Station 2

$p_{t2} = \pi_d \cdot p_{t0}$ or in terms of adiabatic efficiency,

$$p_{t2} = p_0 \left(1 + \eta_d \frac{\gamma_c - 1}{2} M_0^2 \right)^{\frac{\gamma_c}{\gamma_c - 1}} \tag{4.160}$$

$$T_{t2} = T_{t0}$$

Now, following a streamline that enters the core, we proceed to station 3.

Station 3

$$
\begin{aligned}
p_{t3} &= p_{t2} \cdot \pi_c \\
T_{t3} &= T_{t2} \cdot \pi_c^{\frac{\gamma - 1}{\gamma \cdot e_c}} \\
\tau_c &= \pi_c^{\frac{\gamma_c - 1}{\gamma_c e_c}}
\end{aligned} \tag{4.161}
$$

Station 4

$$p_{t4} = p_{t3} \cdot \pi_b$$

T_{t4} is either explicitly given as the engine maximum temperature or can be calculated from the fuel-to-air ratio f according to

$$h_{t4} = C_{pt} T_{t4} = \frac{1}{1 + f} (h_{t3} + f Q_R \eta_b)$$

We also calculate the fuel-to-air ratio in the burner, if only T_{t4} is specified, i.e.,

$$f = \frac{h_{t4} - h_{t3}}{Q_R \eta_b - h_{t4}} = \frac{\tau_\lambda - \tau_r(\tau_c - 1)}{\dfrac{Q_R \eta_b}{h_0} - \tau_\lambda} \tag{4.162}$$

where $\tau_\lambda \equiv \dfrac{h_{t4}}{h_0}$

Before marching to station 5, let us follow a streamline that goes through the fan.

Station 13

$$
\begin{aligned}
p_{t13} &= p_{t2} \cdot \pi_f \\
T_{t13} &= T_{t2} \cdot \pi_f^{\frac{\gamma_c - 1}{\gamma_c e_f}}
\end{aligned}
$$

or

$$\tau_f = \pi_f^{\frac{\gamma_c - 1}{\gamma_c} e_f} \tag{4.163}$$

Station 15

Here we assume the fan duct to be frictional but adiabatic and we characterize the frictional aspect of it by a total pressure ratio π_{fd}.

$$p_{t15} = p_{t13} \cdot \pi_{fd} \tag{4.164}$$

where π_{fd} is given. Also our adiabatic flow assumption in the fan duct earns the following condition:

$$T_{t15} = T_{t13} \quad [\text{for a calorically-perfect gas, otherwise, } h_{t15} = h_{t13}]$$

or

$$\tau_{fd} = 1$$

Now, let us proceed through the turbine by a representative "core" streamline.

Station 5

$$\eta_m(1 + f)(h_{t4} - h_{t5}) = (h_{t3} - h_{t2}) + \alpha(h_{t13} - h_{t2}) \tag{4.165}$$

We note that in the above equation there are two unknowns, namely the bypass ratio α and the turbine exit (total) enthalpy h_{t5}. Let us nondimensionalize the above equation by dividing both sides by the flight static enthalpy h_0 to get

$$\eta_m(1 + f)\tau_\lambda(1 - \tau_t) = \tau_r(\tau_c - 1) + \alpha\tau_r(\tau_f - 1) \tag{4.166}$$

Here, we still have the two unknowns in terms of the bypass ratio and the turbine expansion ratio. We can make the assumption that the turbine expansion parameter as depicted by τ_t is linked to compressor and fan pressure ratios such that

$$p_{t5} \approx p_{t15} \tag{4.167}$$

This immediately ties the turbine pressure ratio to the fan pressure ratio if we divide both sides by the engine face total pressure p_{t2} to get

$$\frac{p_{t5}}{p_{t2}} = \pi_t \cdot \pi_b \cdot \pi_c = \frac{p_{t15}}{p_{t2}} = \pi_{fd} \cdot \pi_f \tag{4.168}$$

Consequently, turbine total pressure ratio may be expressed in terms of all known parameters of the cycle as

$$\pi_t = \frac{\pi_{fd} \cdot \pi_f}{\pi_b \cdot \pi_c} \tag{4.169}$$

The turbine temperature expansion ratio τ_t may be linked to π_t via the polytropic exponent, namely,

$$\tau_t = \pi_t^{\frac{\gamma_t - 1}{\gamma_t} e_t} = \left(\frac{\pi_{fd} \cdot \pi_f}{\pi_b \cdot \pi_c}\right)^{\frac{\gamma_t - 1}{\gamma_t} e_t} \tag{4.170}$$

Now, we see that the power balance equation written above for the turbine (Eq. 4.166), fan, and compressor contains only one unknown and that is the bypass ratio α, which is expressed as

$$\alpha = \frac{\eta_m(1 + f)\tau_\lambda(1 - \tau_t) - \tau_r(\tau_c - 1)}{\tau_r(\tau_f - 1)} \tag{4.171}$$

The assumption of constant total pressure at the exit of the turbine and the fan duct is a reasonable assumption for a design-point analysis. This has enabled us to calculate the turbine pressure ratio and hence the temperature ratio linked via the polytropic exponent. The power balance then gave us the bypass ratio according to Eq. 4.171.

Station 6M

From the energy balance across an adiabatic mixer, we obtained

$$h_{t6M} = h_0 \cdot \frac{(1+f)\tau_t \cdot \tau_\lambda + \alpha\tau_f \cdot \tau_r}{1+\alpha+f} \tag{4.172}$$

All parameters in the above equation are either given as design/limit parameters or have been calculated in the previous steps, thus enabling us to march through the gas turbine. Now, we are ready to calculate the flow Mach number at the fan duct exit M_{15} in terms of turbine exit Mach number M_5 according to

$$p_{t15} = p_{15} \cdot \left(1+\frac{\gamma_c-1}{2}M_{15}^2\right)^{\frac{\gamma_c}{\gamma_c-1}} = p_5 \cdot \left(1+\frac{\gamma_t-1}{2}M_5^2\right)^{\frac{\gamma_t}{\gamma_t-1}} \tag{4.173}$$

which would reduce to the trivial solution, $M_{15} = M_5$, except when we account for a different ratio of specific heats in the core and the fan duct exit, namely, γ_c and γ_t. Now assuming a different ratio of specific heats, we can express the fan duct exit Mach number as

$$M_{15}^2 = \frac{2}{\gamma_c-1}\left\{\left[\left(1+\frac{\gamma_t-1}{2}M_5^2\right)^{\frac{\gamma_t}{\gamma_t-1}}\right]^{\frac{\gamma_c-1}{\gamma_c}}-1\right\} \tag{4.174}$$

We can use the mass flow rate expression that utilizes the static pressure and Mach number as its variables to calculate the ratio of flow areas between the fan duct and the core flow, namely,

$$\dot{m}_{15} = \frac{\gamma_c\, p_{15}A_{15}M_{15}}{a_{15}} = \alpha\dot{m}_0 \tag{4.175}$$

$$\dot{m}_5 = \frac{\gamma_t\, p_5A_5M_5}{a_5} = (1+f)\dot{m}_0 \tag{4.176}$$

Now, if we take the ratio of the above two expressions and note that static pressures are identical (a' la Kutta condition), we get an expression for the unknown area ratio, i.e.,

$$\frac{A_{15}}{A_5} = \frac{\alpha}{1+f}\cdot\left(\frac{\gamma_t}{\gamma_c}\right)\frac{a_{15}}{a_5}\frac{M_5}{M_{15}} \tag{4.177}$$

So far, we have established the mixed-out total enthalpy h_{t6M} and all the upstream parameters to the mixer, such as M_{15} and the area ratio A_{15}/A_5. In the mixer analysis, we had derived two equations and two unknowns based on the continuity and the (inviscid) momentum balance across the mixer, Eqs. 4.155 and 4.157.

$$p_{6M}\left\lfloor 1+\gamma_{6M}M_{6M}^2\right\rfloor(1+A_{15}/A_5) - p_5\left\lfloor(1+\gamma_t M_5^2)+(A_{15}/A_5)(1+\gamma_c M_{15}^2)\right\rfloor = 0 \tag{4.178}$$

$$\frac{\gamma_{6M}\, p_{6M}M_{6M}\sqrt{1+(\gamma_{6M}-1)M_{6M}^2/2}}{\sqrt{(\gamma_{6M}-1)c_{p6M}T_{t6M}}}(1+A_{15}/A_5)$$
$$= \frac{\gamma_t\, p_5M_5}{a_5} + \frac{\gamma_c\, p_{15}M_{15}}{a_{15}}(A_{15}/A_5) \tag{4.179}$$

Let us nondimensionalize Eq. 4.178 by dividing through by p_5 and rewriting it in the following form:

$$\frac{p_{6M}}{p_5}[1 + \gamma_{6M}M_{6M}^2] = [(1 + \gamma_t M_5^2) + (A_{15}/A_5)(1 + \gamma_c M_{15}^2)]/(1 + A_{15}/A_5) = C_1 \quad (4.180)$$

The right-hand side of the above equation is known and we have given it a shorthand notation of C_1. Now, we can isolate the same unknowns in Eq. 4.179 to get

$$\frac{p_{6M}}{P_5}\left(M_{6M}\sqrt{1 + (\gamma_{6M} - 1)M_{6M}^2/2}\right)$$
$$= \left[\left(\frac{\gamma_t}{\gamma_{6M}}\right)\frac{M_5}{a_5} + \left(\frac{\gamma_c}{\gamma_{6M}}\right)\frac{M_{15}(A_{15}/A_5)}{a_{15}}\right]\frac{\sqrt{(\gamma_{6M} - 1)c_{p6M}T_{t6M}}}{(1 + A_{15}/A_5)} = C_2 \quad (4.181)$$

Again the RHS of the above equation is known and it is labeled C_2. By dividing Eq. 4.180 by 4.181, we can eliminate the mixer static pressure ratio and arrive at the mixer exit Mach number M_{6M} via

$$\frac{1 + \gamma_{6M}M_{6M}^2}{M_{6M}\sqrt{1 + (\gamma_{6M} - 1)M_{6M}^2/2}} = \frac{C_1}{C_2} = \sqrt{C} \quad (4.182)$$

Let us cross multiply and square both sides of Eq. 4.182 to arrive at a quadratic equation in M_{6M}^2, as follows:

$$(1 + \gamma_{6M}M_{6M}^2)^2 = C \cdot M_{6M}^2(1 + (\gamma_{6M} - 1)M_{6M}^2/2) \quad (4.183)$$

$$\lfloor \gamma_{6M}^2 - C(\gamma_{6M} - 1)/2 \rfloor M_{6M}^4 + (2\gamma_{6M} - C)M_{6M}^2 + 1 = 0 \quad (4.184)$$

$$M_{6M}^2 = \frac{C - 2\gamma_{6M} - \sqrt{(C - 2\gamma_{6M})^2 - 4[\gamma_{6M}^2 - C(\gamma_{6M} - 1)/2]}}{2\gamma_{6M}^2 - C(\gamma_{6M} - 1)} \quad (4.185)$$

where C is defined in Eq. 4.182 and is expressible in terms of earlier established parameters C_1 and C_2. Now, the mixer static pressure ratio can be determined from

$$\frac{p_{6M}}{p_5} = \frac{C_1}{1 + \gamma_{6M}M_{6M}^2} \quad (4.186)$$

The *ideal* (i.e., inviscid) total pressure ratio across the mixer is p_{t6M}/p_{t5}, which is expressible in terms of the static pressure ratio of Eq. 4.186 and a function of Mach number according to

$$\pi_{M_i} \equiv \frac{p_{t6M}}{p_{t5}} = \frac{p_{6M}}{p_5}\frac{[1 + (\gamma_{6M} - 1)M_{6M}^2/2]^{\frac{\gamma_{6M}}{\gamma_{6M}-1}}}{[1 + (\gamma_t - 1)M_5^2/2]^{\frac{\gamma_t}{\gamma_t-1}}} \quad (4.187)$$

The *actual* total pressure ratio across the mixer should also account for the mixer wall frictional losses, which may be expressed as a multiplicative factor of the ideal parameter, namely,

$$\pi_M \equiv \pi_{M_i} \cdot \pi_{M_f} \quad (4.188)$$

where the frictional loss parameter π_{M_f} needs to be specified.

Station 7

Application of the energy equation to the afterburner yields an expression for the fuel-to-air ratio in the afterburner in terms of known parameters as follows:

$$(\dot{m}_{6M} + \dot{m}_{fAB})h_{t7} - \dot{m}_{6M}h_{t6M} = \dot{m}_{fAB}Q_{R,AB}\eta_{AB} \tag{4.189}$$

$$f_{AB} \equiv \frac{\dot{m}_{fAB}}{\dot{m}_{6M}} = \frac{h_{t7} - h_{t6M}}{Q_{R,AB}\eta_{AB} - h_{t7}} = \frac{\tau_{\lambda AB} - h_{t6M}/h_0}{\dfrac{Q_{R,AB}\eta_{AB}}{h_0} - \tau_{\lambda AB}} \tag{4.190}$$

The total pressure at the exit of the afterburner is known via the loss parameter according to

$$p_{t7} = p_{t6M} \cdot \pi_{AB} \tag{4.191}$$

Note that we need two afterburner total pressure loss parameters, one for the afterburner on and the second for the afterburner off, i.e., $\pi_{AB\text{-}On}$ and $\pi_{AB\text{-}Off}$.

Station 9

From adiabatic flow assumption in the nozzle, we maintain the total enthalpy in the nozzle, namely,

$$T_{t9} = T_{t7} \tag{4.192}$$

and the total pressure at the nozzle exit p_{t9} is expressible in terms of the nozzle total pressure ratio parameter, according to

$$p_{t9} = p_{t7} \cdot \pi_n \tag{4.193}$$

or in case of an adiabatic nozzle efficiency η_n, we can relate the total pressure ratio across the nozzle to this parameter according to

$$\eta_n = \left[\left(\frac{p_{t7}}{p_9}\right)^{\frac{\gamma_{AB}-1}{\gamma_{AB}}} - \pi_n^{-\gamma_{AB}}\frac{\gamma_{AB}-1}{}\right] \bigg/ \left[\left(\frac{p_{t7}}{p_9}\right)^{\frac{\gamma_{AB}-1}{\gamma_{AB}}} - 1\right] \tag{4.194}$$

where the only unknown is π_n.

Now, with the total pressure calculated at the nozzle exit and the static pressure p_9, as a given in the cycle analysis, we can establish the exit Mach number, according to

$$M_9^2 = \frac{2}{\gamma_{AB}-1}\left[\left(\frac{p_{t9}}{p_9}\right)^{\frac{\gamma_{AB}-1}{\gamma_{AB}}} - 1\right] \tag{4.195}$$

and the exit speed of sound from the exit Mach number and total temperature, according to

$$a_9^2 = \frac{\gamma_{AB}R_{AB}T_{t9}}{1 + (\gamma_{AB}-1)M_9^2/2} \tag{4.196}$$

and exit velocity, as the product of $M_9 \cdot a_9$, i.e.,

$$V_9 = M_9 \cdot a_9 \tag{4.197}$$

Now, we have completely established the desired parameters inside the engine, albeit a one-dimensional or an *average* value per station, the fuel-to-air ratios in the primary and the afterburner and the nozzle exhaust velocity. We can insert these parameters in the performance equations for the specific thrust, propulsive and thermal efficiencies, and specific fuel consumption parameter.

EXAMPLE 4.15

A mixed-exhaust turbofan engine with afterburner is flying at

$$M_0 = 2.0, \ p_0 = 10 \text{ kPa}, \ T_0 = 223 \text{ K}$$

The engine inlet total pressure loss is characterized by $\pi_d = 0.90$

The fan pressure ratio is $\pi_f = 1.9$ and polytropic efficiency of the fan is $e_f = 0.90$

The flow in the fan duct suffers 1% total pressure loss, i.e., $\pi_{fd} = 0.99$

The compressor pressure ratio and polytropic efficiency are $\pi_c = 13$ and $e_c = 0.90$, respectively

The combustor exit temperature is $T_{t4} = 1600 \text{ K}$, fuel heating value is $Q_R = 42{,}000 \text{ kJ/kg}$, total pressure ratio is $\pi_b = 0.95$, and the burner efficiency is $\eta_b = 0.98$

The turbine polytropic efficiency is $e_t = 0.80$, its mechanical efficiency is $\eta_m = 0.95$, and $M_5 = 0.5$

The constant-area mixer suffers a total pressure loss due to friction, which is characterized by $\pi_{M,f} = 0.98$

The afterburner is on with $T_{t7} = 2{,}000 \text{ K}$, $Q_{R,AB} = 42{,}000 \text{ kJ/kg}$, $\pi_{AB\text{-}On} = 0.92$, $\eta_{AB} = 0.98$

The nozzle has a total pressure ratio of $\pi_n = 0.95$ and $p_9/p_0 = 3.8$

The gas behavior in the engine is dominated by temperature (in a thermally perfect gas), thus we consider four distinct temperature zones:

Inlet, fan, and compressor section :	$\gamma_c = 1.4$, $c_{pc} = 1004 \text{ J/kg·K}$
Turbine section :	$\gamma_t = 1.33$, $c_{pt} = 1152 \text{ J/kg·K}$
Mixer exit :	γ_{6M}, c_{p6M} (to be calculated based on mixture of gases)
Afterburner and nozzle section :	$\gamma_{AB} = 1.30$, $c_{p,AB} = 1241 \text{ J/kg·K}$

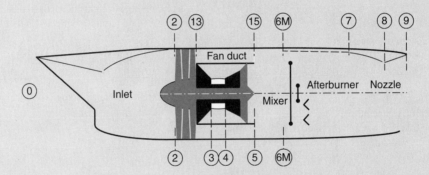

CALCULATE

(a) total pressure and temperature throughout the engine, the fan bypass ratio α, and include the contributions of fuel-to-air ratio in the primary and afterburner f and f_{AB}

(b) engine performance parameters, i.e., TSFC in mg/s/N, specific thrust, and cycle efficiencies

SOLUTION

Marching through the engine, we start with the flight condition. The remaining parameters to know in station "0" are p_{t0} and T_{t0}, the speed of sound a_0, and flight speed V_0.

$$p_{t0} = p_0 \left(1 + \frac{\gamma_c - 1}{2} M_0^2 \right)^{\frac{\gamma_c}{\gamma_c - 1}} = 78.24 \text{ kPa},$$

$$T_{t0} = T_0 \left(1 + \frac{\gamma_c - 1}{2} M_0^2 \right) = 401.4 \text{ K}$$

$$\pi_r = p_{t0}/p_0 = 7.824, \ \tau_r = T_{t0}/T_0 = 1.8$$

$$a_0 = \sqrt{(\gamma_c - 1)c_{pc}T_0} \cong 299.3 \quad \text{m/s}$$

$$V_0 = a_0 \cdot M_0 = 598.5 \text{ m/s}$$

The inlet is adiabatic and thus $T_{t2} = T_{t0}$. The total pressure in the inlet is reduced due to shocks and boundary layer friction, $p_{t2} = p_{t0} \cdot \pi_d \cong 70.42 \text{ kPa}$.

The fan exit pressure is $p_{t13} = p_{t2} \cdot \pi_f \approx 133.8$ kPa. The fan temperature and pressure ratios are related via $\tau_f = \pi_f^{\frac{\gamma_c - 1}{\gamma_c e_f}} = 1.226$. The fan exit temperature is $T_{t13} = T_{t0} \cdot \tau_f \approx 492.12$ K. The fan duct is also adiabatic, which gives $T_{t15} = T_{t13}$. The flow in the fan duct is frictional; therefore it suffers a total pressure loss, $p_{t15} = p_{t13} \cdot \pi_{fd} \approx 132.46$ kPa

The compressor calculations are similar to fan, which results in $p_{t3} = p_{t2} \cdot \pi_c = 915.46$ kPa, the temperature ratio $\tau_c = 2.2575$, and the compressor exit temperature is $T_{t3} = 906.2$ K. This temperature represents an upper limit for an uncooled compressor.

The burner exit total pressure is $p_{t4} = p_{t3} \cdot \pi_b \approx 869.69$ kPa. We can calculate fuel-to-air ratio in the burner from

$$f = \frac{h_{t4} - h_{t3}}{Q_R \eta_b - h_{t4}} \approx 0.0237$$

We impose the same total pressure at the turbine exit as the fan duct exit (namely, we assume $p_{t5} = p_{t15}$) to get the turbine pressure ratio π_t

$$\pi_t = \frac{\pi_{fd} \cdot \pi_f}{\pi_b \cdot \pi_c} \approx 0.1523$$

The turbine temperature expansion ratio is related to its pressure ratio following

$$\tau_t = \pi_t^{\frac{\gamma_t - 1}{\gamma_t} e_t} \approx 0.6883$$

The turbine exit total temperature is $T_{t5} = T_{t4} \cdot \tau_t \approx 1101$ K. Knowing turbine exit Mach number (given as 0.5), we can calculate T_5 and then a_5. These are 1057.6 K and 634.1 m/s, respectively.

We can calculate the fan bypass ratio α by the energy balance between the turbine and fan/compressor,

$$\alpha = \frac{\eta_m (1 + f) \tau_\lambda (1 - \tau_t) - \tau_r (\tau_c - 1)}{\tau_r (\tau_f - 1)} \approx 0.5707$$

The mixer energy balance produces the mixed-out total enthalpy at the exit of the mixer,

$$h_{t6M} = h_0 \cdot \frac{(1 + f) \tau_t \cdot \tau_\lambda + \alpha \tau_f \cdot \tau_r}{1 + \alpha + f}$$

where $\tau_\lambda \equiv c_{pt} T_{t4} / c_{pc} T_0 \approx 8.2325$. Upon substitution, we get $h_{t6M} \approx 991.423$ kJ/kg. The gas properties c_p and c_v at the mixer exit are related to the cold and hot streams entering the mixer according to

$$c_{p6M} = \frac{\dot{m}_5 c_{pt} + \dot{m}_{15} c_{pc}}{\dot{m}_5 + \dot{m}_{15}} \text{ and}$$

$$\gamma_{6M} = \frac{\dot{m}_5 c_{pt} + \dot{m}_{15} c_{pc}}{\dot{m}_5 \left(\dfrac{c_{pt}}{\gamma_t} \right) + \dot{m}_{15} \left(\dfrac{c_{pc}}{\gamma_c} \right)}$$

Upon substitution in the above expressions, we get, $c_{p6M} = 1099$ J/kg·K and $\gamma_{6M} \approx 1.352$, respectively.

To get the mixer exit static pressure and Mach number, we balance the mass flow and fluid impulse across the ideal mixer. Since the gas properties in the fan duct are different than the turbine exit, the gas Mach number in station "15" is derived from Eq. 4.174.

$$M_{15}^2 = \frac{2}{\gamma_c - 1} \left\{ \left[\left(1 + \frac{\gamma_t - 1}{2} M_5^2 \right)^{\frac{\gamma_t}{\gamma_t - 1}} \right]^{\frac{\gamma_c - 1}{\gamma_c}} - 1 \right\}$$

which results in $M_{15} \approx 0.4881$

The gas temperature and static pressure at the fan duct exit are calculated from the total temperature and pressure and the flow Mach number M_{15}, $T_{15} = 469.74$ K and $p_{15} = 112.55$ kPa.

The gas static temperature and pressure at the turbine exit may similarly be calculated, $T_5 = 1057.64$ K and $p_5 = 112.55$ kPa. There is no discrepancy in static pressure between stations 5 and 15. Different gas properties in stations 5 and 15 resulted in a slightly lower M_{15} than M_5. The area ratio between the fan duct and the turbine exit flow is expressed via continuity equation in 4.177.

$$\frac{A_{15}}{A_5} = \frac{\alpha}{1 + f} \cdot \left(\frac{\gamma_t}{\gamma_c} \right) \frac{a_{15}}{a_5} \frac{M_5}{M_{15}}$$

Upon substitution, we get $A_{15}/A_5 = 0.3716$.

From impulse balance across the ideal mixer, we get (Eq. 4.180).

$$\frac{p_{6M}}{p_5} \left[1 + \gamma_{6M} M_{6M}^2 \right] = \left[(1 + \gamma_t M_5^2) + (A_{15}/A_5) \right.$$
$$\left. (1 + \gamma_c M_{15}^2) \right] / (1 + A_{15}/A_5)$$
$$= C_1 = 1.33278$$

From mass balance, we get

$$\frac{p_{6M}}{P_5} \left(M_{6M} \sqrt{1 + (\gamma_{6M} - 1) M_{6M}^2 / 2} \right)$$
$$= \left[\left(\frac{\gamma_t}{\gamma_{6M}} \right) \frac{M_5}{a_5} + \left(\frac{\gamma_c}{\gamma_{6M}} \right) \frac{M_{15} (A_{15}/A_5)}{a_{15}} \right]$$
$$\frac{\sqrt{(\gamma_{6M} - 1) c_{p6M} T_{t6M}}}{(1 + A_{15}/A_5)} = C_2$$

Upon substitution, we get $C_2 = 0.52037$.

Taking the ratio of the above two equations eliminates the pressure and leaves a quadratic in M_{6M}^2, as in

$$M_{6M}^2 =$$
$$\frac{C - 2\gamma_{6M} - \sqrt{(C - 2\gamma_{6M})^2 - 4[\gamma_{6M}^2 - C(\gamma_{6M} - 1)/2]}}{2\gamma_{6M}^2 - C(\gamma_{6M} - 1)}$$

where $C = (C_1/C_2)^2$

The answer to mixer exit Mach number is $M_{6M} \approx 0.5218$ and from

$$\frac{p_{6M}}{p_5} = \frac{C_1}{1 + \gamma_{6M} M_{6M}^2}$$

we get the exit static pressure $p_{6M} \approx 109.63$ kPa. Note that there is a slight static pressure drop in the mixer, i.e, from ~ 112.5 kPa to ~ 109.6 kPa. The (ideal) total pressure at the mixer exit is $p_{t6Mi} = 131.23$ kPa. This results in an ideal mixer total pressure ratio of $\pi_{M,i} \approx 0.9907$ and the frictional contribution is specified in $\pi_{M,f}$ in the problem (as 0.98), therefore, $\pi_M \approx 0.9709$. The actual mixer exit total pressure is $p_{t6M} = p_{t6Mi} \cdot \pi_{Mf} = 128.6$ kPa

The afterburner is on with $T_{t7} = 2000$ K and the total pressure loss is given by $\pi_{AB-On} = 0.92$ Therefore, $p_{t7} = 118.32$ kPa. The energy balance across the afterburner yields the fuel-to-air ratio f_{AB} according to

$$f_{AB} \equiv \frac{\dot{m}_{fAB}}{\dot{m}_{6M}} = \frac{h_{t7} - h_{t6M}}{Q_{R,AB}\eta_{AB} - h_{t7}} = 0.03854$$

The nozzle exit total pressure is $p_{t9} = p_{t7} \cdot \pi_n$, which gives $p_{t9} = 112.40$ kPa. The exit static pressure is specified to be $p_9/p_0 = 3.8$, which gives $p_9 = 38$ kPa. Knowing total and static pressure at the nozzle exit, we calculate the local Mach number $M_9 = 1.377$. The static temperature at the nozzle exit is calculated from the total temperature and the Mach number information at the exit, i.e., we get $T_9 = 1557.2$ K. The speed of sound at "9" is $a_9 = 761.4$ m/s and the exhaust velocity at the nozzle exit is

$V_9 = M_9 \cdot a_9 = 1048.4$ m/s. Since the nozzle did not achieve perfect expansion, the effective exhaust speed is

$$V_{9eff} = F_g/\dot{m}_9 = V_9 + \frac{(p_9 - p_0)A_9}{\dot{m}_9}$$

$$= V_9 + \frac{a_9^2(1 - p_0/p_9)}{\gamma_{AB} \cdot V_9} \cong 1361.8 \text{ m/s}$$

The nondimensional specific thrust is

$$\frac{F_n}{\dot{m}_0(1 + \alpha)a_0} = \left(\frac{1 + \alpha + f + f_{AB}}{1 + \alpha}\right)\frac{V_{9eff}}{a_0} - M_0 \cong 2.731$$

The thrust-specific fuel consumption is

$$\text{TSFC} = \frac{(f + f_{AB})/(1 + \alpha)a_0}{F_n/\dot{m}_0(1 + \alpha)a_0} \times 10^6 \cong 48.515 \text{ mg/s/N}$$

The thermal efficiency is calculated from

$$\eta_{th} \equiv \frac{\Delta \dot{KE}}{\dot{m}_f Q_R + \dot{m}_{f,AB}Q_{R,AB}}$$

$$= \frac{(1 + \alpha + f + f_{AB})V_{9eff}^2 - (1 + \alpha)V_0^2}{2(fQ_R + f_{AB}Q_{R,AB})} \cong 47.13\%$$

The propulsive efficiency is calculated from

$$\eta_p \equiv \frac{2F_n V_0}{\dot{m}_9 V_{9eff}^2 - (1 + \alpha)\dot{m}_0 V_0^2}$$

to be $\eta_p \approx 62.3\%$ Therefore the overall efficiency is the product of the thermal and propulsive efficiencies, namely $\eta_o = 29.37\%$.

EXAMPLE 4.16

A mixed exhaust turbofan engine with afterburner has the same parameters as those specified in Example 4.15, except for the compressor pressure ratio. Write a computer program, e.g., MATLAB, or use spreadsheet to calculate engine performance for a range of compressor pressure ratios from 6 to 16.

SOLUTION

Using a spreadsheet, we calculate the engine performance parameters, summarized in the following table.

The graphical representations of these parameters are shown. In the following figure, TSFC shows the familiar bucket curve behavior as a function of compressor pressure ratio. The optimum compressor pressure ratio, in this engine and flight condition, is ~ 10.

In addition, the variations of engine bypass ratio and total fuel to air ratio are calculated and graphed.

π_c	TSFC (mg/s/N)	Specific thrust/a_0	η_{th}	η_p	η_o
6	50.858	2.7386	0.4491	0.6240	0.2802
7	48.902	2.7305	0.4674	0.6235	0.2914
8	47.823	2.7263	0.4782	0.6232	0.2980
9	47.316	2.7246	0.4834	0.6230	0.3012
10	47.214	2.7245	0.4845	0.6230	0.3018
11	47.416	2.7257	0.4824	0.6230	0.3005
12	47.863	2.7279	0.4779	0.6231	0.2977
13	48.515	2.7310	0.4713	0.6232	0.2937
14	49.350	2.7348	0.4632	0.6234	0.2888
15	50.352	2.7394	0.4538	0.6236	0.2830
16	51.513	2.7446	0.4434	0.6239	0.2766

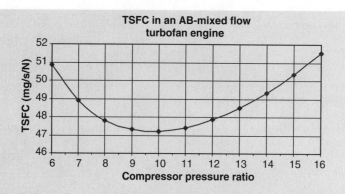

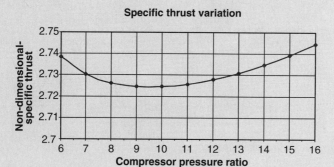

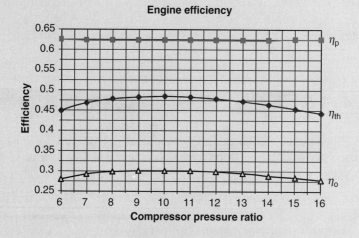

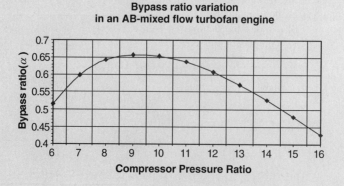

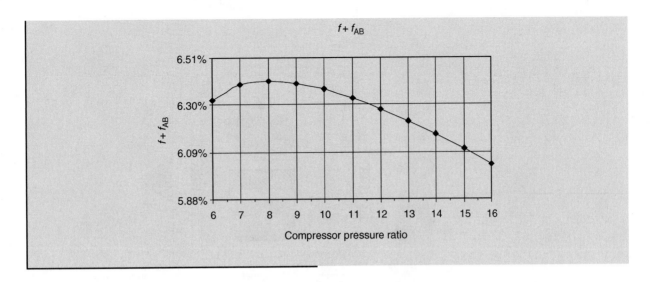

4.5   The Turboprop Engine

4.5.1   Introduction

To construct a turboprop engine, we start with a gas generator.

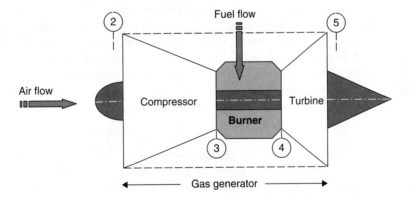

The turbine in the gas generator provides the shaft power to the compressor. However, we recognize that the gas in station 5 is still highly energetic (i.e., high p_t and T_t) and capable of producing shaft power, similar to a turbofan engine. Once this shaft power is produced in a follow-on turbine stage that is called "power" or "free" turbine, we can supply the shaft power to a propeller. A schematic drawing of this arrangement is shown in Fig. 4.49.

The attractiveness of a turbopropeller engine as compared with a turbo-fan engine lies in its ability to offer a very large bypass ratio, which may be between 30 and 100. The large bypass ratio, by necessity, will then cut back on the exhaust velocities of the propulsor, thereby attaining higher propulsive efficiencies for the engine. The high propulsive efficiency, however, comes at a price. The limitation on the tip Mach number of a rotating propeller, say to less than 1.3, leads to a cruise Mach number in the 0.7–0.8 range for advanced turboprops and to 0.4–0.6 for conventional propellers. We do not have *this limitation* on cruise Mach number in a turbofan engine. Also, the large diameter of the propeller often requires a reduction gearbox that adds to the engine weight and system complexity with its attendant reliability and maintainability issues.

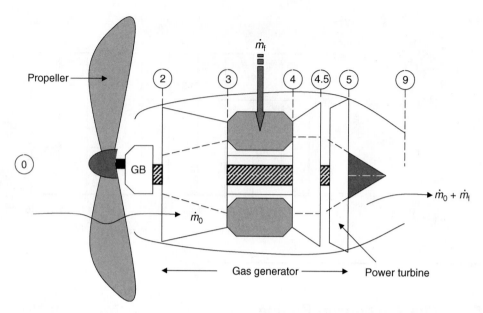

■ **FIGURE 4.49** Schematic drawing of a turboprop engine with station numbers identified

4.5.2 Cycle Analysis

4.5.2.1 The New Parameters. Let us identify the new parameters that we have introduced by inserting a propeller in the gas turbine engine. We will examine the turboprop from the power distribution point of view as well as its thrust producing capabilities namely the propeller thrust contribution to the overall thrust, which includes the core thrust.

From the standpoint of power, the low-pressure turbine power is supplied to a gearbox, which somewhat diminishes it in its frictional loss mechanism in the gearing and then delivers the remaining power to the propeller. We will call the fractional delivery of shaft power through the gearbox, the *gearbox efficiency* $\eta_{gb,}$ which symbolically is defined as

$$\eta_{gb} \equiv \wp_{prop}/\wp_{LPT} \tag{4.198}$$

where the numerator is the power supplied to the propeller and the denominator is the shaft power provided by the power turbine to the gearbox. Also, we define the fraction of *propeller shaft power* that is converted in the *propeller thrust power* as the *propeller efficiency* η_{prop} as

$$\eta_{prop} \equiv F_{prop} \cdot V_0/\wp_{prop} \tag{4.199}$$

Now, let us examine the overall thrust picture of a turboprop engine. We recognize that the propeller and the engine core both contribute to thrust production. We can express this fact as

$$F_{total} = F_{prop} + F_{core} \tag{4.200}$$

The contribution of the engine core to the overall thrust, which we have called as the *core thrust*, takes on the familiar form of the gross thrust of the nozzle minus the ram drag of the air flow rate that enters the engine, i.e.,

$$F_{core} = (\dot{m}_0 + \dot{m}_f)V_9 - \dot{m}_0V_0 + (p_9 - p_0)A_9 \tag{4.201}$$

The pressure thrust contribution of the nozzle, i.e., the last term in Eq. 4.201, for a turboprop engine is often zero due to perfectly expended exhaust, i.e., $p_9 = p_0$. So, for all practical purposes, the engine core of a turboprop produces a thrust based solely on the momentum balance between the exhaust and the intake of the engine, namely,

$$F_{\text{core}} \cong (\dot{m}_0 + \dot{m}_f) \ V_9 - \dot{m}_0 V_0 \qquad (4.202)$$

4.5.2.2 Design Point Analysis.
We require the following set of input parameters in order to estimate the performance of a turboprop engine. The following list, which sequentially proceeds from the flight condition to the nozzle exit, summarizes the input parameters per component. In this section, we will practice the powerful marching technique that we have learned so far in this book.

Station 0

The flight Mach number M_0, the ambient pressure and temperature p_0 and T_0, and air properties γ and R are needed to characterize the flight environment. We can calculate the flight total pressure and temperature p_{t0} and T_{t0}, the speed of sound at the flight altitude a_0, and the flight speed V_0, based on the input.

Station 2

At the engine face, we need to establish the total pressure and temperature p_{t2} and T_{t2}. From adiabatic flow assumption in the inlet, we conclude that

$$T_{t2} = T_{t0}$$

To account for the inlet frictional losses and its impact on the total pressure recovery of the inlet, we need to define an inlet total pressure ratio parameter π_d or adiabatic diffuser efficiency η_d. This results in establishing p_{t2}, similar to our earlier cycle analysis, for example,

$$p_{t2} = \pi_d \cdot p_{t0}$$

Station 3

To continue our march through the engine, we need to know the compressor pressure ratio π_c, which again is treated as a design choice, and the compressor polytropic efficiency e_c. This allows us to calculate the compressor temperature ratio in terms of compressor pressure ratio using the polytropic exponent, i.e.,

$$\tau_c = \pi_c^{\frac{\gamma_c-1}{\gamma_c e_c}}$$

Now, we have established the compressor discharge total pressure and temperature p_{t3} and T_{t3}.

Station 4

To establish the burner exit conditions, similar to earlier analysis, we need to know the loss parameters η_b and π_b as well as the limiting temperature T_{t4}. The fuel type with its energy content, that we had called the *heating value* of the fuel, Q_R, needs to be specified. Again, we establish the fuel-to-air ratio f by energy balance across the burner and the total pressure at the exit, p_{t4}, by loss parameter π_b.

Station 4.5

For the upstream turbine, or the so-called the HPT, we need to know the mechanical efficiency η_{mHPT}, which is a power transmission efficiency, and the turbine polytropic efficiency e_{tHPT}, which measures the internal efficiency of the turbine. The power balance between the compressor and high-pressure turbine is

$$\eta_{mHPT}(1+f)(h_{t4}-h_{t4.5})=h_{t3}-h_{t2} \qquad \textbf{(4.203-a)}$$

$$h_{t4.5}=h_{t4}-\frac{h_{t3}-h_{t2}}{\eta_{mHPT}(1+f)} \qquad \textbf{(4.203-b)}$$

leads to the only unknown in the above equation, which is $h_{t4.5}$. The total pressure at station 4.5 may be linked to the turbine total temperature ratio according to

$$\frac{p_{t4.5}}{p_{t4}}=\left(\frac{T_{t4.5}}{T_{t4}}\right)^{\frac{\gamma_t}{(\gamma_t-1)e_{tHPT}}} \qquad \textbf{(4.204)}$$

$$\pi_{HPT}=\tau_{HPT}^{\frac{\gamma_t}{(\gamma_t-1)e_{tHPT}}} \qquad \textbf{(4.205)}$$

Stations 5 and 9

Since the power turbine drives a *load*, i.e., the propeller, we need to specify the turbine expansion ratio that supports this *load*. In this sense, we consider the propeller as an external load to the cycle and hence as an input parameter to the turboprop analysis. It serves a purpose to put this and the following station, i.e., 9, together, as both are responsible for the thrust production. Another view of stations 5 and 9 downstream of 4.5 points to the *power split*, decision made by the designer, between the propeller and the exhaust jet. The following T–s diagram best demonstrates this principle.

In the T–s diagram (Fig. 4.50), both the actual and the ideal expansion processes are shown. We will use this diagram to define the component efficiencies as well as the power split choice. For example, we define the power turbine (i.e., LPT) adiabatic efficiency as

$$\eta_{LPT}\equiv\frac{h_{t4.5}-h_{t5}}{h_{t4.5}-h_{t5s}}=\frac{h_{t4.5}(1-\tau_{LPT})}{h_{t4.5}\left[1-\pi_{LPT}^{\frac{\gamma_t-1}{\gamma_t}}\right]}=\frac{1-\tau_{LPT}}{1-\pi_{LPT}^{\frac{\gamma_t-1}{\gamma_t}}} \qquad \textbf{(4.206)}$$

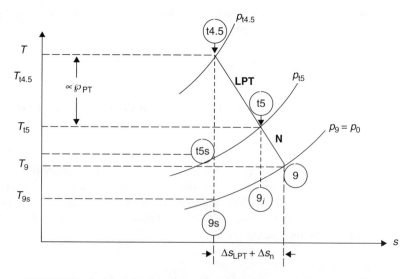

■ **FIGURE 4.50** Thermodynamic states of an expansion process in free turbine and nozzle of a turboprop engine

Also, we may define the nozzle adiabatic efficiency η_n as

$$\eta_n \equiv \frac{h_{t5} - h_9}{h_{t5s} - h_{9s}} \tag{4.207}$$

We note that the above definition for the nozzle adiabatic efficiency deviates slightly from our earlier definition in that we have assumed

$$h_{t5} - h_{9i} \approx h_{t5s} - h_{9s} \tag{4.208}$$

which in light of small expansions in the nozzle and hence near parallel isobars, this approximation is considered reasonable. The total ideal power available at station 4.5, per unit mass flow rate, may be written as

$$\frac{\wp}{\dot{m}_0(1+f)} = h_{t4.5} - h_{9s} = h_{t4.5}\left[1 - \left(\frac{p_9}{p_{t4.5}}\right)^{\frac{\gamma_t-1}{\gamma_t}}\right] = \wp_{i,\text{total}}/\dot{m}_9 \tag{4.209}$$

If we examine the RHS of the above equation, we note that all terms on the RHS are known. Therefore, the total ideal power is known to us a' priori. Now, let us assume that the power split between the free turbine and the nozzle is, say α and $1 - \alpha$, respectively, as shown in Fig. 4.51.

We can define the power split as

$$\alpha \equiv \frac{h_{t4.5} - h_{t5s}}{h_{t4.5} - h_{9s}} = \frac{\dfrac{\wp_{\text{LPT}}/\dot{m}_9}{\eta_{\text{LPT}}}}{\wp_{i,\text{total}}/\dot{m}_9} \tag{4.210}$$

which renders the following expression for the free turbine (LPT) power in terms of a given α,

$$\wp_{\text{LPT}} = \dot{m}_9 \eta_{\text{LPT}} \cdot \alpha \cdot h_{t4.5}\left[1 - \left(\frac{p_9}{p_{t4.5}}\right)^{\frac{\gamma_t-1}{\gamma_t}}\right] = \dot{m}_9(h_{t4.5} - h_{t5}) \tag{4.211}$$

This expression for the power turbine (LPT) can be applied to the propeller through gearbox and propeller efficiency in order to arrive at the thrust power produced by the propeller, namely,

$$F_{\text{prop}} \cdot V_0 = \eta_{\text{prop}} \cdot \eta_{\text{gb}} \cdot \eta_{\text{mLPT}} \cdot \wp_{\text{LPT}}$$

$$= \dot{m}_0(1+f)[\alpha\eta_{\text{LPT}}\eta_{\text{mLPT}}\eta_{\text{gb}}\eta_{\text{prop}}]h_{t4.5}\left[1 - \left(\frac{p_9}{p_{t4.5}}\right)^{\frac{\gamma_t-1}{\gamma_t}}\right] \tag{4.212}$$

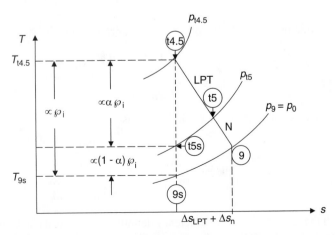

■ **FIGURE 4.51**　Definition of power split α in a turboprop engine

We note that the RHS of the above equation, per unit mass flow rate, is known. Now, let us examine the nozzle thrust. The kinetic energy per unit mass at the nozzle exit may be linked to

$$V_9^2/2 \cong \eta_n(h_{t5s} - h_{9s}) = \eta_n(1 - \alpha)\wp_{i,\text{total}}/\dot{m}_9 = (1 - \alpha)\eta_n h_{t4.5}\left[1 - \left(\frac{p_9}{p_{t4.5}}\right)^{\frac{\gamma_t - 1}{\gamma_t}}\right] \tag{4.213}$$

Therefore, the exhaust velocity is now approximated by the power split parameter α and the total ideal power available after the gas generator, i.e., station 4.5, according to

$$V_9 \approx \sqrt{2(1 - \alpha)\eta_n h_{t4.5}\left[1 - \left(\frac{p_9}{p_{t4.5}}\right)^{\frac{\gamma_t - 1}{\gamma_t}}\right]} \tag{4.214}$$

A more accurate expression for exhaust velocity is derived based on nozzle adiabatic efficiency that is defined based on the states t5, 9 and 9i. Example 4.17 calculates the nozzle exhaust velocity using the more accurate method.

Now, the turboprop thrust per unit air mass flow rate (through the engine) can be expressed in terms of the propeller thrust of expression 4.212 and the core thrust expression 4.202 with Eq. 4.214 incorporated for the exhaust velocity.

$$\frac{F_{\text{total}}}{\dot{m}_0} = (1 + f)V_9 - V_0 + \frac{(1 + f)[\alpha\eta_{\text{PT}}\eta_{\text{mLPT}}\eta_{\text{gb}}\eta_{\text{prop}}]h_{t4.5}\left[1 - \left(\frac{p_9}{p_{t4.5}}\right)^{\frac{\gamma_t - 1}{\gamma_t}}\right]}{V_0} \tag{4.215}$$

The fuel efficiency of a turboprop engine is often expressed in terms of the fraction of the fuel consumption in the engine to produce a unit shaft/mechanical power, according to

$$\text{PSFC} \equiv \frac{\dot{m}_f}{\wp_{\text{prop}} + \wp_{\text{core}}} = \frac{f}{\left(\dfrac{\wp_{\text{prop}}}{\dot{m}_0} + \dfrac{\wp_{\text{core}}}{\dot{m}_0}\right)} \tag{4.216}$$

$$\wp_{\text{prop}} = \eta_{\text{gb}}\eta_{\text{mLPT}}\wp_{\text{LPT}} = \dot{m}_0(1 + f)\eta_{\text{gb}}\eta_{\text{mLPT}}\eta_{\text{LPT}} \cdot \alpha \cdot h_{t4.5}\left[1 - \left(\frac{p_9}{p_{t4.5}}\right)^{\frac{\gamma_t - 1}{\gamma_t}}\right] \tag{4.217}$$

$$\wp_{\text{core}} = \frac{\dot{m}_0}{2}[(1 + f)V_9^2 - V_0^2] \quad \text{for a perfectly expanded nozzle, otherwise use } V_{9\text{eff}} \tag{4.218}$$

We can define the thermal efficiency of a turboprop engine as

$$\eta_{\text{th}} \equiv \frac{\wp_{\text{prop}} + \wp_{\text{core}}}{\dot{m}_f Q_R} = \frac{\left(\dfrac{\wp_{\text{prop}}}{\dot{m}_0} + \dfrac{\wp_{\text{core}}}{\dot{m}_0}\right)}{fQ_R} \tag{4.219}$$

The propulsive efficiency η_p may be defined as

$$\eta_p \equiv \frac{F_{\text{total}} \cdot V_0}{\wp_{\text{prop}} + \wp_{\text{core}}} = \frac{\dfrac{F_{\text{total}}}{\dot{m}_0} \cdot V_0}{\left(\dfrac{\wp_{\text{prop}}}{\dot{m}_0} + \dfrac{\wp_{\text{core}}}{\dot{m}_0}\right)} \tag{4.220}$$

where all the terms in the above efficiency definitions have been calculated in earlier steps and the overall efficiency is again the product of the thermal and propulsive efficiencies.

4.5.2.3 Optimum Power Split Between the Propeller and the Jet.

For a given fuel flow rate, flight speed, compressor pressure ratio, and all internal component efficiencies, we may ask a very important question, which is "at what power split α would the total thrust be maximized?" This is a simple mathematics question. What we need first, is to express the total thrust in terms of all independent parameters, i.e., f, V_0, π_c, etc., and then differentiate it with respect to α and set the derivative equal to zero. From that equation, obtain the solution(s) for α that satisfies the equation. We go to Eq. 4.215 for an expression for the total thrust.

$$\frac{F_{\text{total}}}{\dot{m}_0} = (1 + f)V_9 - V_0 + \frac{(1 + f)[\alpha \eta_{\text{LPT}} \eta_{\text{mLPT}} \eta_{\text{gb}} \eta_{\text{prop}}]h_{t4.5}\left[1 - \left(\frac{p_9}{p_{t4.5}}\right)^{\frac{\gamma_t - 1}{\gamma_t}}\right]}{V_0}$$

We express the exhaust velocity V_9 as (Eq. 4.214)

$$V_9 \approx \sqrt{2(1 - \alpha)\eta_n h_{t4.5}\left[1 - \left(\frac{p_9}{p_{t4.5}}\right)^{\frac{\gamma_t - 1}{\gamma_t}}\right]}$$

We note that the bracketed term in the above equations, i.e.,

$$1 - \left(\frac{p_9}{p_{t4.5}}\right)^{\frac{\gamma_t - 1}{\gamma_t}} = 1 - \left(\frac{p_9}{p_0}\frac{p_0}{p_{t0}}\frac{p_{t0}}{p_{t2}}\frac{p_{t2}}{p_{t3}}\frac{p_{t3}}{p_{t4}}\frac{p_{t4}}{p_{t4.5}}\right)^{\frac{\gamma_t - 1}{\gamma_t}}$$

$$= 1 - \left(\frac{p_9}{p_0}\left(\pi_r \pi_d \pi_c \pi_b \pi_{\text{HPT}}\right)^{-1}\right)^{\frac{\gamma_t - 1}{\gamma_t}} \tag{4.221}$$

which is a constant. Also let us examine the total enthalpy at station 4.5, $h_{t4.5}$,

$$h_{t4.5} = \frac{h_{t4.5}}{h_{t4}}\frac{h_{t4}}{h_0}h_0 = \tau_{\text{HPT}} \cdot \tau_\lambda \cdot h_0 \tag{4.222}$$

which is a constant, as well. Therefore, the expression for the total thrust of the engine (per unit mass flow rate in the engine nozzle) is essentially composed of a series of constants and the dependence on α takes on the following simplified form:

$$\frac{F_{\text{total}}}{(1 + f)\dot{m}_0} = C_1\sqrt{1 - \alpha} + C_2\alpha + C_3 \tag{4.223}$$

where C_1, C_2, and C_3 are all constants. Now, let us differentiate the above function with respect to α and set the derivative equal to zero, i.e.,

$$\frac{d}{d\alpha}\left[\frac{F_{\text{total}}}{(1 + f)\dot{m}_0}\right] = \frac{-C_1}{2\sqrt{1 - \alpha}} + C_2 = 0 \tag{4.224}$$

which produces a solution for the power split parameter α that maximizes the total thrust of a turboprop engine. Hence, we may call this special value of α, the "optimum" α, namely

$$\alpha_{\text{opt}} = 1 - \left(\frac{C_1}{2C_2}\right)^2 \tag{4.225}$$

Now, upon substitution for the constants C_1 and C_2 and some simplification, we get

$$\alpha_{\text{opt}} = 1 - \frac{\eta_n}{(\eta_{\text{PT}} \eta_{\text{mLPT}} \eta_{\text{gb}} \eta_{\text{prop}})^2} \cdot \frac{\gamma_c - 1}{2} \frac{M_0^2}{\tau_{\text{HPT}}\tau_\lambda\left[1 - \left(\frac{p_9/p_0}{\pi_r \pi_d \pi_c \pi_b \pi_{\text{HPT}}}\right)^{\frac{\gamma_t - 1}{\gamma_t}}\right]} \tag{4.226}$$

This expression for the optimum power split between the propeller and the jet involves all component and transmission (of power) efficiencies, as expected. However, let us assume that all efficiencies were 100% and further assume that the exhaust nozzle was perfectly expanded, i.e., $p_9 = p_0$. What does the above expression tell us about the optimum power split in a perfect turboprop engine? Let us proceed with the simplifications.

$$\alpha_{\mathrm{opt_{ideal}}} = 1 - \frac{\dfrac{\gamma_c - 1}{2} M_0^2}{\tau_{\mathrm{HPT}} \tau_\lambda - \dfrac{\tau_\lambda}{\tau_r \tau_c}} \tag{4.227}$$

From power balance between the compressor and the high-pressure turbine, we can express the following:

$$\tau_r(\tau_c - 1) = (1 + f)\tau_\lambda(1 - \tau_{\mathrm{HPT}}) \approx \tau_\lambda(1 - \tau_{\mathrm{HPT}}) \tag{4.228}$$

which simplifies to

$$\tau_\lambda \tau_{\mathrm{HPT}} \cong \tau_\lambda - \tau_r(\tau_c - 1) \tag{4.229}$$

Substitute the above equation in the optimum power split in an ideal turboprop engine, Eq. 4.227, to get

$$\alpha_{\mathrm{opt_{ideal}}} = 1 - \frac{\dfrac{\gamma_c - 1}{2} M_0^2}{\tau_\lambda - \tau_r(\tau_c - 1) - \dfrac{\tau_\lambda}{\tau_r \tau_c}} = 1 - \frac{\tau_r - 1}{\tau_\lambda - \tau_r(\tau_c - 1) - \dfrac{\tau_\lambda}{\tau_r \tau_c}} \tag{4.230}$$

At takeoff condition and low-speed climb/descent $(\tau_r \to 1)$, the optimum power split approaches 1, as expected. The propeller is the most efficient propulsor at low speeds, as it attains the highest propulsive efficiency. As flight Mach number increases, the power split term α becomes less than 1.

EXAMPLE 4.17

An advanced turboprop flies at $M_0 = 0.7$ at an altitude where $p_0 = 16$ kPa and $T_0 = -45°C$. The propeller efficiency is $\eta_{\mathrm{prop}} = 0.85$. The inlet captures airflow rate at 10 kg/s and has a total pressure recovery of $\pi_d = 0.98$. The compressor pressure ratio is $\pi_c = 30$ and its polytropic efficiency is $e_c = 0.92$. The combustor has an exit temperature $T_{t4} = 1600$ K and the fuel heating value is $Q_R = 42,000$ kJ/kg with a burner efficiency of $\eta_b = 0.99$ and the total pressure loss in the burner is $\pi_b = 0.96$. The HPT has a polytropic efficiency of $e_{t,\mathrm{HPT}} = 0.82$, and a mechanical efficiency of $\eta_{m,\mathrm{HPT}} = 0.99$. The power split

between the LPT and the engine nozzle is at $\alpha = 0.85$ and the mechanical efficiency of the LPT is $\eta_{m,\mathrm{LPT}} = 0.99$, the LPT adiabatic efficiency is $\eta_{\mathrm{LPT}} = 0.88$. A reduction gearbox is used with an efficiency of $\eta_{gb} = 0.995$. The exhaust nozzle is convergent with an adiabatic efficiency of $\eta_n = 0.95$. We will describe gas properties in the engine based only on two temperature zones (cold and hot):

Inlet and compressor sections (cold) : $\gamma_c = 1.4$,
$$c_{pc} = 1004 \text{ J/kg·K}$$

Turbines and Nozzle Sections (hot) : $\gamma_t = 1.33$,
$$c_{pt} = 1152 \text{ J/kg·K}$$

CALCULATE

(a) total pressure and temperature throughout the engine (include fuel-to-air ratio in mass/energy balance)

(b) engine core thrust in kN

(c) propeller thrust in kN

(d) power-specific fuel consumption in mg/s/kW

(e) thrust-specific fuel consumption in mg/s/N

(f) thermal and propulsive efficiencies η_{th} and η_p

(g) engine overall efficiency η_o

SOLUTION

We start with the flight parameters p_{t0}, T_{t0}, a_0, V_0. Since all our temperature calculations are expressed in absolute scale, we convert $T_0 = (-45 + 273)$ K $= 228$ K

From ambient static temperature and flight Mach number, we get $T_{t0} \cong 250.34$ K

Remember $T_t = T[1 + (\gamma - 1)M^2/2]$ at any point in the flow.

From ambient static pressure and flight Mach number, we get $p_{t0} \cong 22.194$ kPa

The speed of sound at the ambient temperature condition is $a_0 \cong 302.6$ m/s

The flight speed is $V_0 = M_0 \cdot a_0 \cong 211.8$ m/s

Now, we march through the rest of the engine, component by component, beginning with the inlet. The exit total pressure of the inlet is $p_{t2} = p_{t0} \cdot \pi_d = 21.75$ kPa. Since inlets are adiabatic, $T_{t2} = T_{t0} = 250.34$ K.

The compressor exit total pressure is the product of the compressor pressure ratio π_c and p_{t2}, i.e.,

$$p_{t3} = 30(21.75 \text{ kPa}) \cong 652.5 \text{ kPa}$$

The compressor total temperature ratio $\tau_c = \pi_c^{(\gamma_c - 1)/e_c \gamma_c}$, therefore, $T_{t3} \cong 719.9$ K. Now, we are ready for the burner. The fuel-to-air ratio follows the energy balance across the burner, namely,

$$f = \frac{h_{t4} - h_{t3}}{Q_R \eta_b - h_{t4}} \approx 0.0282 \quad \text{Given } T_{t4} = 1600\text{k}$$

The burner exit total pressure is $p_{t4} = p_{t3} \cdot \pi_b$, which gives $p_{t4} \cong 626.4$ kPa.

The power balance between the HPT and the compressor follows Eq. 4.203-a, $\eta_{mHPT}(1 + f)(h_{t4} - h_{t4.5}) = h_{t3} - h_{t2}$ with the only unknown $h_{t4.5}$. Upon solving for $h_{t4.5}$, we get $h_{t4.5} \cong 1,380.065$ kJ/kg and since $T_{t4.5} = h_{t4.5}/c_{pt}$, the HPT exit total temperature is $T_{t4.5} \cong 1198$ K.

The total pressure ratio across the HPT is related to its total temperature ratio and the HPT polytropic efficiency e_{tHPT} following Eq. 4.204, therefore $p_{t4.5} \cong 151.065$ kPa.

Since we have specified the power split between the LPT and the nozzle, we may use Eq. 4.211

$$\wp_{LPT} = \dot{m}_9 \eta_{LPT} \cdot \alpha \cdot h_{t4.5} \left[1 - \left(\frac{p_9}{p_{t4.5}} \right)^{\frac{\gamma_t - 1}{\gamma_t}} \right]$$

$$= \dot{m}_9 (h_{t4.5} - h_{t5})$$

to calculate the shaft power produced by the LPT, also note that we have the mass flow rate through the engine, $\dot{m}_9 = (1 + f)\dot{m}_0 = (1 + 0.0282)(10 \text{ kg/s}) \cong 10.282$ kg/s. The nozzle exit pressure p_9 is initially assumed to be the ambient static pressure p_0 subject to later verification.

Turboprop engines typically have a subsonic exhaust ($M_9 < 1$), so the initial assumption of $p_9 = p_0$ is often bourn out. Therefore,

$$\wp_{LPT} \cong 4.533 \text{ MW}$$

Also from Eq. 4.211, we get the LPT exit total enthalpy h_{t5} (and temperature T_{t5}). These are $h_{t5} = 939.16$ kJ/kg and $T_{t5} \cong 815.24$ K. We calculate p_{t5} from $p_{t4.5}$ and LPT total temperature ratio and polytropic efficiency e_{LPT}. Equation 4.45 relates polytropic and adiabatic efficiencies and turbine total temperature ratio. We get $e_t = 0.8533$ and then we use the efficiency exponent, i.e., $\pi_t = \tau_t^{\gamma_t/e_t(\gamma_t - 1)}$ to get $p_{t5} = 24.53$ kPa.

To calculate T_{9i}, we use isentropic pressure and temperature ratio rule between the two states t5 and 9i, to get $T_{9i} = 733.25$ K. From the definition of nozzle adiabatic efficiency and T_{9i}, we calculate the nozzle exit static temperature T_9 to be $T_9 = 737.35$ K. Now, we calculate nozzle exit velocity more accurately from $V_9^2 = 2c_{pt}(T_{t5} - T_9)$, which gives $V_9 = 423.65$ m/s.

From the following equation, we calculate propeller thrust

$$F_{prop} \cdot V_0 = \eta_{prop} \cdot \eta_{gb} \cdot \eta_{mLPT} \cdot \wp_{LPT}$$
$$F_{prop} = 17.92 \text{ kN}$$

From T_9, we calculate speed of sound and the nozzle exit Mach number to be $a_9 = 529.44$ m/s and $M_9 = 0.80$.

Since $M_9 < 1$, then $p_9 = p_0$ was a valid assumption. Based on nozzle exit Mach number M_9 and static pressure, we calculate p_{t9} to be 23.98 kPa, which gives the nozzle total pressure ratio $\pi_n = p_{t9}/p_{t5} = 0.978$.

The (net) thrust produced by the engine core is

$$(F_n)_{core} = \dot{m}_0[(1 + f)V_9 - V_0]$$
$$= 10 \text{ kg/s}[(1 + 0.0282)423.65 \text{ m/s}$$
$$- 211.8 \text{ m/s}] \approx 2.238 \text{ kN}$$

The shaft power delivered to the propeller is

$$\wp_{prop} = \eta_{gb}\eta_{mLPT}\wp_{LPT}$$
$$= (0.995)(0.99)(4.533 \text{ MW}) \approx 4.465 \text{ MW}$$

Also the engine total thrust is the sum of the propeller and core thrust, i.e.,

$$F_{total} = F_{prop} + F_{core} = (17.92 + 2.238) \text{ kN} = 20.16 \text{ kN}$$

The power specific fuel consumption is defined as $\text{PSFC} \equiv \dot{m}_f/\wp_{prop} + \wp_{core}$. The mechanical power produced by the core is $\wp_c = \dot{m}_9 V_9^2/2 - \dot{m}_0 V_0^2/2$ for a perfectly expanded nozzle. We have calculated all the parameters in the equation, therefore, $\wp_c \approx 698.37$ kW. Based on fuel-to-air ratio and air flow rate in the engine, $\dot{m}_f = 0.282$ kg/s, thus the power specific fuel consumption is $\text{PSFC} \approx 54.6$ mg/s/kW

The definition of thrust specific fuel consumption is
$$\text{TSFC} \equiv \dot{m}_f / (F_{\text{prop}} + F_{\text{core}}) \approx 13.99 \, \text{mg/s/N}$$

$$\eta_{\text{th}} \equiv \frac{\wp_{\text{prop}} + \wp_{\text{core}}}{\dot{m}_f Q_R} \approx 43.6\%$$

$$\eta_p \equiv \frac{F_{\text{total}} \cdot V_0}{\wp_{\text{prop}} + \wp_{\text{core}}} \approx 82.68\%$$

$$\eta_o = \eta_{\text{th}} \eta_p \approx 36.05\%$$

Note that the thermal efficiency is dominated by the cycle pressure ratio in a Brayton cycle and ideally is

$\eta_{\text{th-ideal}} = 1 - (PR)^{-(\gamma-1)/\gamma}$, which in this cycle, the pressure ratio is $p_{t3}/p_0 \cong 40.8$ and the ideal Brayton cycle thermal efficiency (for $\gamma = 1.33$) is $\sim 60.1\%$. Obviously, component losses contribute to a lower thermal efficiency than that predicted by an ideal cycle. On the contrary, the real advantage of turboprop is in its increased propulsive efficiency, which in this case that represents cruise, it is $\sim 83\%$.

EXAMPLE 4.18

For the turboprop engine specified in Example 4.17, vary the power split α from 0.79 to 0.97 using a spreadsheet or write a program in MATLAB and calculate and graph engine total thrust and specific fuel consumption versus the power split factor α. Also calculate the optimum power split based on the theory developed in the TP-section and compare the two results.

SOLUTION

First, the optimum power split is calculated to be

$$\alpha_{\text{opt}} = 1 - \frac{\eta_n}{(\eta_{\text{PT}} \eta_{\text{mLPT}} \eta_{\text{gb}} \eta_{\text{prop}})^2} \cdot \frac{\gamma_c - 1}{2}$$
$$\times \frac{M_0^2}{\tau_{\text{HPT}} \tau_\lambda \left[1 - \left(\dfrac{p_9/p_0}{\pi_r \pi_d \pi_c \pi_b \pi_{\text{HPT}}} \right)^{\frac{\gamma_t - 1}{\gamma_t}} \right]} \cong 0.9334$$

This problem is easily solved using a spreadsheet program. The important results are graphed. As expected, the total thrust produced by a TP engine reaches a maximum at the optimum power split α_{opt}.

Since fuel-to-air ratio remains constant (with α variation), the TSFC reaches a minimum at α_{opt}. As the power split increases, the propellers share of thrust production increases as well, as shown in the last figure. At very large power split factors (e.g., ~ 0.96 in this example), the core ram drag becomes larger than its nozzle gross thrust, thus start producing drag! This is expected, as we drain more energy from exhaust (by producing more shaft power for the prop), the nozzle exhaust velocity drops below the flight speed, i.e., $V_9 < V_0$.

The engine thermal efficiency remains nearly constant with the power split factor α. This result is also expected as the thermal efficiency is dominated by the engine pressure ratio, which remains constant with α. On the contrary, engine propulsive efficiency keeps on improving, as we reduce core exhaust velocity. This behavior is also shown graphically in the last figure.

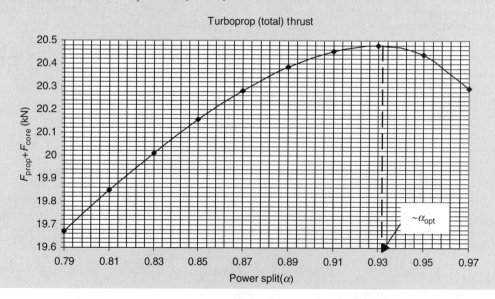

Turboprop (total) thrust

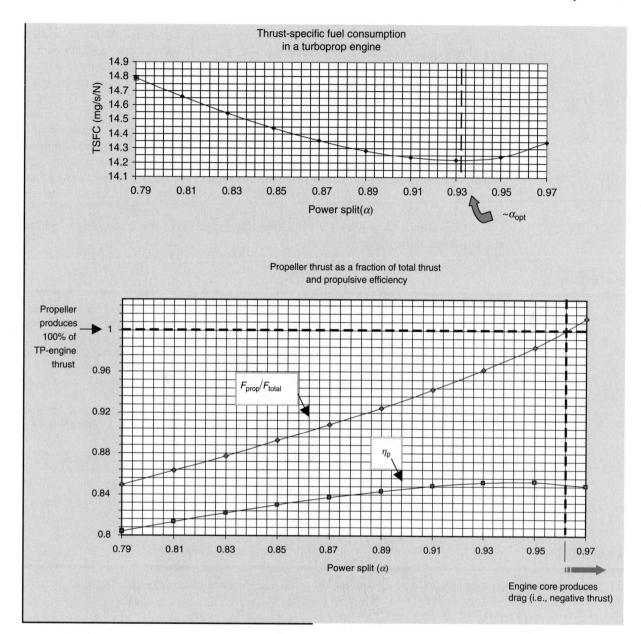

4.6   Summary

In this chapter, we learned various gas turbine engine configurations and their analysis. The performance parameters were identified to be specific thrust, specific fuel consumption, thermal, propulsive, and overall efficiencies. We had limited our approach to steady, one-dimensional flow and the "design-point" analysis. We also learned that we may analyze ramjets by setting compressor pressure ratio to one in a turbojet engine. We had essentially removed the compressor (and thus turbine) by setting its pressure ratio equal to 1. We included the effect of fluid viscosity and thermal conductivity in our cycle analysis empirically, i.e., through the introduction of component efficiency. The knowledge of component efficiency at different operating conditions is critical to their design and optimization. We always treated that knowledge, i.e., the component efficiency, as a "given" in our analysis. In reality, engine manufacturers, research laboratories, and universities continually measure component performance and sometimes report them in open literature. However, commercial engine manufacturers treat most competition-sensitive data as proprietary.

In the following five chapters, we will take on each component in an aircraft gas turbine engine individually. The nonrotating components, i.e., inlets and nozzles, are treated in a single chapter, i.e., Chapter 5. An introductory study of combustion and the gas turbine burner and afterburner configurations follow in Chapter 6. The rotating components, i.e., compressors and turbines, are treated in the turbomachinery Chapters 7–9. Finally, we allow components to be "matched" and integrated in a real engine where we explore its off-design analysis in Chapter 10.

References 1–11 provide for complimentary reading on aircraft propulsion and are recommended to the reader.

References

1. Archer, R.D. and Saarlas, M., *An Introduction of Aerospace Propulsion*, Prentice Hall, New York, 1998.
2. Cumpsty, N., *Jet Propulsion: A Simple Guide to the Aerodynamic and Thermodynamic Design and Performance of Jet Engines*, 2nd edition, Cambridge University Press, Cambridge, UK, 2003.
3. Flack, R.D. and Rycroft, M.J., *Fundamentals of Jet Propulsion with Applications*, Cambridge University Press, Cambridge, UK, 2005.
4. Heiser, W.H., Pratt, D.T., Daley, D.H., and Mehta, U.B., *Hypersonic Airbreathing Propulsion*, AIAA, Washington, DC, 1993.
5. Hesse, W.J. and Mumford, N.V.S., *Jet Propulsion for Aerospace Applications*, 2nd edition, Pittman Publishing Corporation, New York, 1964.
6. Hill, P.G. and Peterson, C.R., *Mechanics and Thermodynamics of Propulsion*, 2nd edition, Addison-Wesley, Reading, Massachusetts, 1992.
7. Kerrebrock, J.L., *Aircraft Engines and Gas Turbines*, 2nd edition, MIT Press, Cambridge, Mass, 1992.
8. Mattingly, J.D., *Elements of Gas Turbine Propulsion*, McGraw-Hill, New York, 1996.
9. Mattingly, J.D., Heiser, W.H., and Pratt, D.T., *Aircraft Engine Design*, 2nd edition, AIAA, Washington, DC, 2002.
10. Oates, G.C., *Aerothermodynamics of Gas Turbine and Rocket Propulsion*, AIAA, Washington, DC, 1988.
11. Shepherd, D.G., *Aerospace Propulsion*, American Elsevier Publication, New York, 1972.

Problems

4.1 An aircraft is flying at an altitude where the ambient static pressure is $p_0 = 25$ kPa and the flight Mach number is $M_0 = 2.5$. The total pressure at the engine face is measured to be $p_{t2} = 341.7$ kPa. Assuming the inlet flow is adiabatic and $\gamma = 1.4$, calculate

(a) the inlet total pressure recovery π_d
(b) the inlet adiabatic efficiency η_d
(c) the nondimensional entropy rise caused by the inlet $\Delta s_d / R$

4.2 A multistage axial-flow compressor has a mass flow rate of 100 kg/s and a total pressure ratio of 25. The compressor polytropic efficiency is $e_c = 0.90$. The inlet flow condition to the compressor is described by $T_{t2} = -35°C$ and $p_{t2} = 30$ kPa. Assuming the flow in the compressor is adiabatic, and constant gas properties throughout the compressor are assumed, i.e., $\gamma = 1.4$ and $c_p = 1004$ J/kg · K, calculate

(a) compressor exit total temperature T_{t3}, in K
(b) compressor adiabatic efficiency η_c
(c) compressor shaft power $\wp_c$ in MW

4.3 A gas turbine combustor has inlet condition $T_{t3} = 900$ K, $p_{t3} = 3.2$ Mpa, air mass flow rate of 100 kg/s, $\gamma_3 = 1.4$, $c_{p3} = 1004$ J/kg · K.

A hydrocarbon fuel with ideal heating value $Q_R = 42,800$ kJ/Kg is injected in the combustor at a rate of 2 kg/s. The burner efficiency is $\eta_b = 0.99$ and the total pressure at the combustor exit is 97% of the inlet total pressure, i.e., combustion causes a 3% loss in total pressure. The gas properties at the combustor exit are $\gamma_4 = 1.33$ and $c_{p4} = 1,156$ J/kg · K. Calculate

(a) fuel-to-air ratio f
(b) combustor exit temperature T_{t4} in K and p_{t4} in MPa

4.4 An uncooled gas turbine has its inlet condition the same as the exit condition of the combustor described in Problem 4.3. The turbine adiabatic efficiency is 85%. The turbine produces a shaft power to drive the compressor and other accessories at $\wp_t = 60$ MW. Assuming that the gas properties in the turbine are the same as the burner exit in Problem 4.3, calculate

    (a) turbine exit total temperature T_{t5} in K

    (b) turbine polytropic efficiency e_t

    (c) turbine exit total pressure p_{t5} in kPa

    (d) turbine shaft power $\wp_t$ based on turbine expansion ΔT_t

4.5 In a turbine nozzle blade row, hot gas mass flow rate is 100 kg/s and $h_{tg} = 1900$ kJ/kg. The nozzle blades are internally cooled with a coolant mass flow rate of 1.2 kg/s and $h_{tc} = 904$ kJ/kg as the coolant is ejected through nozzle blades trailing edge. The coolant mixes with the hot gas and causes a reduction in the mixed-out enthalpy of the gas. Calculate the mixed-out total enthalpy after the nozzle. Also for the $c_{p,\text{mixed-out}} = 1594$ J/kg $\cdot$ K, calculate the mixed out total temperature.

4.6 Consider the internally cooled turbine nozzle blade row of Problem 4.5. The hot gas total pressure at the entrance of

(a) nozzle total pressure ratio π_n

(b) nozzle area ratio A_9/A_8 for a perfectly expanded nozzle

(c) nozzle exit Mach number M_9 (perfectly expanded)

4.8 Calculate the propulsive efficiency of a turbojet engine under the following two flight conditions that represent takeoff and cruise, namely

(a) $V_0 = 100$ m/s and $V_9 = 2000$ m/s

(b) $V_0 = 750$ m/s and $V_9 = 2000$ m/s

4.9 A ramjet is in supersonic flight, as shown. The inlet pressure recovery is $\pi_d = 0.90$. The combustor burns hydrogen with $Q_R = 117,400$ kJ/kg at a combustion efficiency of $\eta_b = 0.95$. The nozzle expands the gas perfectly, but suffers from a total pressure loss of $\pi_n = 0.92$. Calculate

(a) fuel-to-air ratio f $f = \dfrac{\eta_{t4} - h_{t}}{}$

(b) nozzle exit Mach number M_9 q_R

(c) specific (net) thrust $F/\dot{m}_0$ (in N/kg/s)

(d) η_{th}, engine thermal efficiency

(e) η_p, engine propulsive efficiency

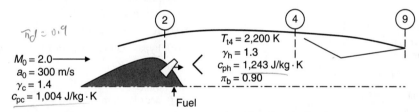

$\pi_d = 0.9$

$M_0 = 2.0 \longrightarrow$
$a_0 = 300$ m/s
$\gamma_c = 1.4$
$c_{pc} = 1,004$ J/kg $\cdot$ K

Fuel

$T_{t4} = 2,200$ K
$\gamma_h = 1.3$
$c_{ph} = 1,243$ J/kg $\cdot$ K
$\pi_b = 0.90$

■ **FIGURE P4.9**

the nozzle blade is $p_{t4} = 2.2$ MPa, $c_{pg} = 1156$ J/kg $\cdot$ K, and $\gamma_g = 1.33$. The mixed-out total pressure at the exit of the nozzle has suffered 5% loss due to both mixing and frictional losses in the blade row boundary layers. Calculate the entropy change $\Delta s/R$ across the turbine nozzle blade row.

4.7 A convergent–divergent nozzle has a pressure ratio, NPR = 12. The gas properties are $\gamma = 1.33$ and $c_p = 1156$ J/kg $\cdot$ K and remain constant in the nozzle. The nozzle adiabatic efficiency is $\eta_n = 0.94$. Calculate

4.10 A ramjet takes in 100 kg/s of air at a Mach 2 flight condition at an altitude where $p_0 = 10$ kPa and $T_0 = -25°C$. The engine throttle setting allows 3 kg/s of fuel flow rate in the combustor where a hydrocarbon fuel of 42,800 kJ/kg heating value is burned. The ramjet component efficiencies are all listed on the engine cross section. Note that the exhaust nozzle is not perfectly expanded. We intend to establish some performance parameters for this engine as well as some flow areas (i.e., physical sizes) of this engine.

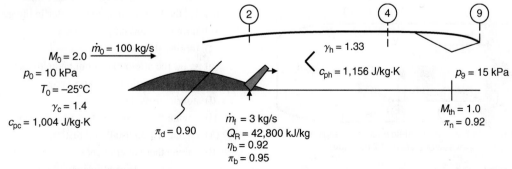

$M_0 = 2.0$ $\dot{m}_0 = 100$ kg/s
$p_0 = 10$ kPa
$T_0 = -25°C$
$\gamma_c = 1.4$
$c_{pc} = 1,004$ J/kg$\cdot$K

$\pi_d = 0.90$

$\dot{m}_f = 3$ kg/s
$Q_R = 42,800$ kJ/kg
$\eta_b = 0.92$
$\pi_b = 0.95$

$\gamma_h = 1.33$

$c_{ph} = 1,156$ J/kg$\cdot$K

$p_9 = 15$ kPa

$M_{th} = 1.0$
$\pi_n = 0.92$

■ **FIGURE P4.10**

Assuming the gas properties are split into a *cold* section and a *hot* section (perfect gas properties), namely, $\gamma_c = 1.4$ and $c_{pc} = 1.004 \, \text{kJ/kg} \cdot \text{K}$ and $\gamma_h = 1.33$ and $c_{ph} = 1.156 \, \text{kJ/kg} \cdot \text{K}$ (subscript "c" stands for "cold" and "h" for "hot"), calculate

(a) ram drag in kN

(b) the inlet capture area A_0 in m^2

(c) p_{t4} in kPa

(d) T_{t4} in K

(e) p_{t9} in kPa

(f) exit Mach number M_9

(g) exhaust velocity V_9

(h) nozzle exit area A_9 in m^2

(i) gross thrust in kN

(j) thermal efficiency[*] η_{th}

(k) propulsive efficiency[*] η_p

Note:

For exhaust nozzles that are not perfectly expanded the thermal and propulsive efficiencies are defined as

$$\eta_{th} \equiv \frac{\Delta \dot{KE}}{\dot{m}_f Q_R \eta_b} = \frac{\dot{m}_9 V_{9\,\text{eff}}^2 - \dot{m}_0 V_0^2}{2 \dot{m}_f Q_R \eta_b}$$

$$\eta_p \equiv \frac{2 F_n V_0}{\dot{m}_9 V_{9\,\text{eff}}^2 - \dot{m}_0 V_0^2}$$

where the effective exhaust velocity $V_{9\,\text{eff}}$, is defined as

$$V_{9\,\text{eff}} \equiv \frac{F_g}{\dot{m}_9}$$

4.11 A mixed-exhaust turbofan engine is described by the following design and limit parameters:

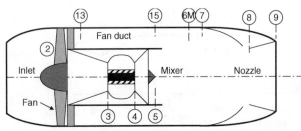

A nonafterburning mixed-flow turbofan engine
(note that stations 6*M* and 7 coincide)

■ **FIGURE P4.11**

Flight:	$M_0 = 2.2$, $p_0 = 10 \, \text{kPa}$, $T_0 = -50°\text{C}$, $R = 287 \, \text{J/kg} \cdot \text{K}$, $\gamma = 1.4$
Inlet mass flow rate and total pressure recovery:	$(1 + \alpha)\dot{m}_0 = 25 \, \text{kg/s}$, $\pi_d = 0.85$
Compressor, fan:	$\pi_c = 15$, $e_c = 0.90$, $\pi_f = 1.5$, $e_f = 0.90$
Burner:	$\pi_b = 0.95$, $\eta_b = 0.98$, $Q_R = 42,800 \, \text{kJ/kg}$, $T_{t4} = 1400°\text{C}$
Turbine:	$e_t = 0.92$, $\eta_m = 0.95$, $M_5 = 0.5$
Mixer:	$\pi_{M,f} = 0.98$
Afterburner:	None
Nozzle:	$\pi_n = 0.90$, $p_9/p_0 = 1.0$

Calculate

(a) ram drag D_R in kN

(b) p_{t2} in kPa, T_{t2} in K

(c) p_{t3} in kPa, T_{t3} in K

(d) p_{t13} in kPa, T_{t13} in K

(e) p_{t4} in kPa, T_{t5} in K

(f) fuel-to-air ratio f

(g) bypass ratio α, $\dot{m}_{core}$ in kg/s, and $\dot{m}_{fan}$ in kg/s

(h) T_{t6M} in K

(i) p_{t9} in kPa, T_{t9} in K

(j) M_9, V_9 in m/s

(k) gross thrust F_g in kN

(l) thrust-specific fuel consumption in mg/s/N

You may assume constant gas properties γ and R throughout the engine.

We may also assume that the flow in the fan duct, i.e., between stations 13 and 15, is frictionless and adiabatic.

4.12 In the afterburning turbojet engine shown, assume constant gas properties and ideal components to calculate

(a) ram drag

(b) compressor shaft power $\wp_c$

(c) fuel-to-air ratio in the primary burner

(d) τ_λ, the limit enthalpy parameter in the gas generator

(e) turbine expansion parameter τ_t

(f) turbine shaft power $\wp_t$

(g) $\tau_{\lambda AB}$, the afterburner limit enthalpy parameter

(h) fuel-to air ratio in the afterburner

(i) nozzle gross thrust

(j) engine thrust specific fuel consumption

(k) engine net uninstalled thrust

(l) engine thermal efficiency

(m) engine propulsive efficiency

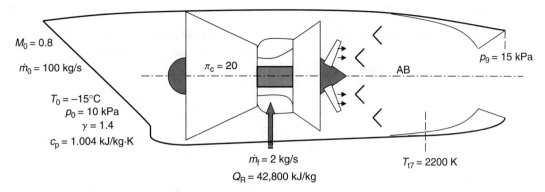

(m) engine propulsive efficiency

4.13   An ideal separate-exhaust turbofan engine has the following design parameters:

$$T_0 = -20°C, \ p_0 = 15\,\text{kPa}, \ M_0 = 0.85$$
$$\pi_c = 30$$
$$\pi_f = 1.60, \ \alpha = 6$$
$$\tau_\lambda = 6.5, \ Q_R = 42,800\,\text{kJ/kg}$$
Fan and core nozzles are convergent·

Assuming the gas is calorically perfect with $\gamma = 1.4$ and $c_p = 1.004\,\text{kJ/kg} \cdot \text{K}$, calculate

(a) compressor exit pressure p_{t3} in kPa
(b) fan exit temperature T_{t13} in K
(c) fuel-to-air ratio f
(d) turbine exit temperature T_{t5} in K
(e) fan nozzle exit Mach number M_{19}
(f) core nozzle exit Mach number M_9
(g) core nozzle exit velocity V_9 in m/s
(h) The ratio of fan-to-core thrust F_{fan}/F_{core}

4.14   In a mixed-exhaust turbofan engine, we have calculated the parameters shown on the engine diagram.

Assuming constant gas properties between the two streams and constant total pressure between the hot and cold gas streams, calculate

(a) the mixer exit total temperature T_{t6M}
(b) M_9
(c) V_9
(d) $\dfrac{F_g}{\dot{m}_5}$

4.15   In an afterburning gas turbine engine, the exhaust nozzle is equipped with a variable area throat. Calculate percent increase in nozzle throat area needed to accommodate the engine flow in the afterburning mode. With the afterburner on, the nozzle mass flow rate increases by 3%, the nozzle total temperature doubles, i.e. $(T_{t8})_{\text{AB-ON}}/(T_{t8})_{\text{AB-OFF}} = 2$, and the total pressure at the nozzle entrance is reduced by 20%. You may assume the gas properties γ and R remain constant and the nozzle throat remains choked.

4.16   For the constant-area ideal mixer shown, assuming *constant gas properties*, calculate

(a) p_5 (kPa)
(b) p_{15} (kPa)
(c) p_{6M} (kPa)

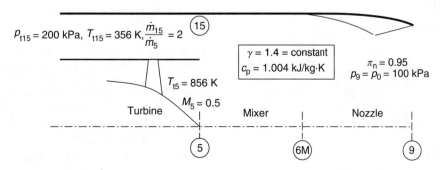

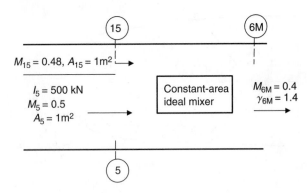

■ **FIGURE P4.16**

4.17   A turbojet engine has the following parameters at the on-design operating point:

$M_0 = 0$, $p_0 = p_{STD}$, $T_0 = T_{STD}$, $\gamma = 1.4$, $c_P = 1.004 \, \text{kJ/kg} \cdot \text{K}$
$\pi_d = 0.99$
$\pi_c = 30$, $e_c = 0.9$
$\tau_\lambda = 6.0$, $\eta_b = 0.98$, $\pi_b = 0.95$, and $Q_R = 42,800 \, \text{kJ/kg}$
$e_t = 0.90$, $\eta_m = 0.95$
$\pi_n = 0.98$ and $p_9/p_0 = 1.0$

Calculate

   (a) p_{t3} and τ_c
   (b) fuel-to-air ratio f and p_{t4}
   (c) τ_t

Now for the following off-design condition:

$M_0 = 0.85$ at 20 km U.S. standard altitude and throttle ratio, $T_{t4} = 0.9 \, T_{t4\text{-design}}$, calculate

   (d) compressor pressure ratio at the off-design operation

4.18   A turbojet engine has the following design point parameters:

$M_0 = 0$, $p_0 = 100 \, \text{kPa}$, $T_0 = 15°C$
$\pi_d = 0.98$
$\pi_c = 25$, $e_c = 0.90$
$Q_R = 42,800 \, \text{kJ/kg}$, $\pi_b = 0.95$, $\eta_b = 0.98$, $\tau_\lambda = 6.0$
$e_t = 0.85$, $\eta_m = 0.98$
$\pi_n = 0.97$, $p_9 = p_0$

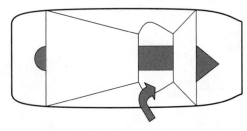

■ **FIGURE P4.18**

Calculate

   (a) fuel-to-air ratio f
   (b) turbine total temperature ratio τ_t

For the following off-design operation:

$M_0 = 0.85$, $p_0 = 10 \, \text{kPa}$, $T_0 = -15°C$
$\tau_\lambda = 6.5$

Assume $\gamma = 1.4$, $c_p = 1004 \, \text{kJ/kg} \cdot \text{K}$ and calculate

   (c) $\pi_{c\text{-Off-Design}}$ if $\tau_{t\text{-Design}} = \tau_{t\text{-Off-Design}}$

4.19   An ideal turbojet engine has the following design and limit parameters, namely,

$M_0 = 2.0$, altitude 37 kft
compressor pressure ratio π_c
maximum enthalpy ratio $\tau_\lambda = 7.0$
fuel type is hydrocarbon with $Q_R = 42,800 \, \text{kJ/kg}$
assume constant gas properties $c_p = 1.004 \, \text{kJ/kg} \cdot \text{K}$ and $\gamma = 1.4$.

   For a range of compressor pressure ratios, namely, $1 \le \pi_c \le 40$, calculate and graph (using MATLAB or a spreadsheet)

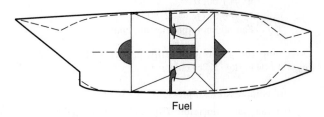

Fuel

■ **FIGURE P4.19**

   (a) engine (nondimensional) specific thrust $F_n/(\dot{m} \cdot a_0)$
   (b) thrust specific fuel consumption in mg/s/N
   (c) in order to assess the effect of gas property variations with temperature on the engine performance parameters, repeat parts (a) and (b) for the following gas properties:

Engine cold section:	$\gamma_c = 1.40$, $c_{pc} = 1.004 \, \text{kJ/kg} \cdot \text{K}$
Engine hot section:	$\gamma_t = 1.33$, $c_{pt} = 1.156 \, \text{kJ/kg} \cdot \text{K}$

   (d) To assess the effect of inlet total pressure recovery on the engine performance, calculate and graph the engine specific thrust and fuel consumption for a single compressor pressure ratio of $\pi_c = 20$, but vary π_d from 0.50 to 1.0. Use gas properties of part (c)
   (e) Now, for the following component efficiencies

$\pi_d = 0.90$, $e_c = 0.90$, $\pi_b = 0.98$, $\eta_b = 0.98$, $e_t = 0.91$,
$\eta_m = 0.99$, $\pi_n = 0.95$, $\pi_c = 20$, $\tau_\lambda = 7$, and $p_9 = p_0$,

calculate the engine performance parameters

1. specific thrust (nondimensional)
2. specific fuel consumption
3. thermal efficiency
4. propulsive efficiency

4.20   For an ideal ramjet, derive an expression for the flight Mach number in terms of the cycle limit enthalpy, τ_λ that will lead to an engine thrust of zero.

4.21   Derive an expression for an optimum Mach number that maximizes the engine-specific thrust in an ideal ramjet.

4.22   For an ideal ramjet with a perfectly expanded nozzle, show that the nozzle exit Mach number M_9 is equal to the flight Mach number M_0.

4.23   Assuming the component efficiencies in a real ramjet are
$\pi_d = 0.90$, $\pi_b = 0.95$, $\eta_b = 0.98$, $\pi_n = 0.90$ and $p_9/p_0 = 1.0$ flying at 37 kft altitude.

For the maximum enthalpy ratio $\tau_\lambda = 8.0$, the fuel heating value of 42,000 kJ/kg and a cold and hot section gas properties

Engine cold section: $\gamma_c = 1.40$, $c_{pc} = 1.004\,\text{kJ/kg}\cdot\text{K}$
Engine hot section: $\gamma_t = 1.33$, $c_{pt} = 1.156\,\text{kJ/kg}\cdot\text{K}$

calculate the optimum flight Mach number corresponding to the maximum specific thrust.

4.24   A ramjet uses a hydrocarbon fuel with $Q_R = 42,800\,\text{kJ/kg}$ flying at Mach 2 (i.e., $M_0 = 2$) in an atmosphere where $a_0 = 300\,\text{m/s}$. Its exhaust is perfectly expanded and the exhaust velocity is $V_9 = 1200\,\text{m/s}$. Assuming the inlet total pressure recovery is $\pi_d = 0.90$, the burner losses are $\pi_b = 0.98$ and $\eta_b = 0.96$, the nozzle total pressure ratio is $\pi_n = 0.98$ and $\gamma = 1.4$ and $c_p = 1.004\,\text{kJ/kg}\cdot\text{K}$ are constant throughout the engine, calculate

(a) τ_λ
(b) fuel-to-air ratio f
(c) nondimensional-specific thrust $F_n/(\dot{m}_0 a_0)$
(d) propulsive efficiency
(e) thermal efficiency

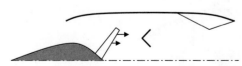

■ **FIGURE P4.24**

4.25   A large bypass ratio turbofan engine has the following design and limit parameters:
$M_0 = 0.8$, altitude = 37 kft US standard atmosphere

$\pi_d = 0.995$
$\pi_c = 40$, $e_c = 0.90$
$\alpha = 6$, $\pi_f = 1.6$, $e_f = 0.90$, $\pi_{fn} = 0.98$, fan nozzle is convergent
$\tau_\lambda = 7.0$, $Q_R = 42,800\,\text{kJ/kg}$, $\pi_b = 0.95$, $\eta_b = 0.98$
$e_t = 0.90$, $\eta_m = 0.975$
$\pi_n = 0.98$, core nozzle is convergent

Assuming the gas properties may be described by two sets of parameters, namely, cold and hot stream values, i.e.,

Engine cold section: $\gamma_c = 1.40$, $c_{pc} = 1.004\,\text{kJ/kg}\cdot\text{K}$
Engine hot section: $\gamma_t = 1.33$, $c_{pt} = 1.156\,\text{kJ/kg}\cdot\text{K}$

Calculate

(a) the ratio of compressor to fan shaft power $\wp_c/\wp_f$
(b) fuel-to-air ratio f
(c) the ratio of fan nozzle exit velocity to core nozzle exit velocity V_{19}/V_9
(d) the ratio of two gross thrusts, $F_{g,fan}/F_{g,core}$
(e) the engine thermal efficiency and compare it to an ideal Carnot cycle operating between the same temperature limits
(f) the engine propulsive efficiency and compare it to the turbojet propulsive efficiency of Problem 4.19
(g) engine thrust specific fuel consumption
(h) engine (fuel)-specific impulse I_s (in seconds)

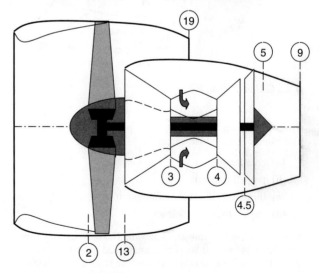

■ **FIGURE P4.25**

4.26   The flow expansion in an exhaust nozzle is shown on a T–s diagram.
Assuming $\gamma = 1.4$, $c_p = 1.004\,\text{kJ/kg}\cdot\text{K}$, calculate

(a) nozzle adiabatic efficiency η_n
(b) V_9/a_0
(c) nondimensional pressure thrust, i.e., $\dfrac{(p_9 - p_0)A_9}{\dot{m}_9 \cdot a_0}$

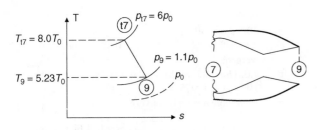

■ **FIGURE P4.26**

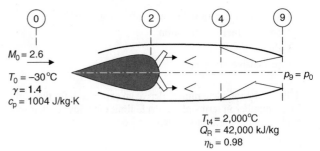

4.27 An advanced turboprop engine is flying at $M_0 = 0.75$ at 37 kft standard altitude. The component parameters are designated on the following engine drawing.

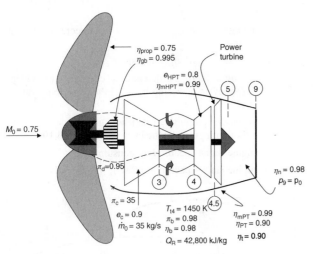

■ **FIGURE P4.27**

Assuming an optimum power split α_{opt} that leads to a maximum engine thrust, calculate

(a) core thrust
(b) propeller thrust
(c) thrust-specific fuel consumption
(d) engine overall efficiency

4.28 Consider a ramjet, as shown. The diffuser total pressure ratio is $\pi_d = 0.80$, the burner total pressure ratio is $\pi_b = 0.96$, and the nozzle total pressure ratio is $\pi_n = 0.90$.

Calculate

(a) fuel-to-air ratio f
(b) exit Mach number M_9
(c) nondimensional-specific thrust, i.e., $F_n/\dot{m}_0 a_0$

4.29 A turbojet engine flies at $V_0 = 250\,\text{m/s}$ with an exhaust velocity of $V_9 = 750\,\text{m/s}$. The fuel-to-air ratio is 2% and the actual fuel heating value is $Q_{R,\text{actual}} = 40,800\,\text{kJ/kg}$. Estimate the engine propulsive and thermal efficiencies, η_p and η_{th}.

■ **FIGURE P4.28**

Assume the nozzle is perfectly expanded.

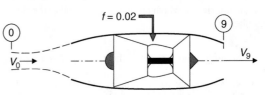

■ **FIGURE P4.29**

4.30 Consider a turboprop engine with the following parameters (note that the nozzle is convergent):

$$\dot{m}_9 = 20\,\text{kg/s}$$
$$V_0 = 220\,\text{m/s}$$
$$\eta_{prop} = 0.80$$
$$\eta_{gb} = 0.995$$
$$\eta_{mLPT} = 0.99$$

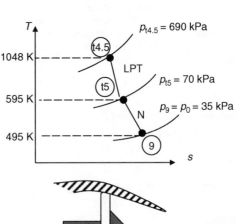

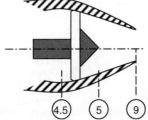

■ **FIGURE P4.30**

Calculate

- (a) propeller thrust F_{prop} (kN)
- (b) nozzle gross thrust $F_{\text{g,core}}$ (kN)
- (c) M_9
- (d) core net thrust $F_{\text{n,core}}$ (kN), assume $f = 0.02$
- (e) nozzle adiabatic efficiency η_n
- (f) power turbine adiabatic efficiency η_{PT}

Assume $\gamma = 1.4$, $c_p = 1.004\,\text{kJ/kg} \cdot \text{K}$.

4.31 A constant-area mixer operates with the inlet conditions as shown.

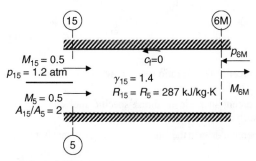

■ FIGURE P4.31

Assuming the hot stream has a total temperature of

$$T_{t5} = 4\,T_{t15}$$

Calculate

- (a) the ratio of mass flow rates $\dot{m}_{15}/\dot{m}_5$
- (b) A_{6M}/A_5
- (c) $p_{6M}(1 + \gamma M_{6M}^2)$......from impulse
- (d) T_{t6M}/T_{t5}

4.32 Consider a turboprop engine with the power turbine driving a propeller, as shown. The power turbine inlet and exit total temperatures are $T_{t4.5} = 783\,\text{K}$ and $T_{t5} = 523\,\text{K}$. The mass flow rate through the turbine is 25 kg/s. Assuming $c_p = 1,100\,\text{J/kg} \cdot \text{K}$,

Calculate

- (a) the power produced by the turbine
- (b) the power delivered to the propeller
- (c) the propeller thrust F_{prop}

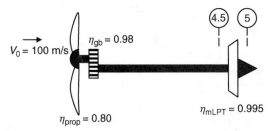

■ FIGURE P4.32

4.33 Let us study a family of turbojet engines, all with the same component parameters except the burner, as shown on the T–s diagram. The family of turbojets are at the standard sea-level condition and stationary, i.e., $M_0 = 0$. The fixed engine parameters are

$$\pi_d = 0.995$$

$$\pi_c = 20,\ e_c = 0.90,\ \gamma_c = 1.4,\ c_{pc} = 1,004\,\text{J/kg} \cdot \text{K}$$

$$\pi_b = 0.95,\ \eta_b = 0.98,\ Q_R = 42,800\,\text{kJ/kg}$$

$$\eta_m = 0.99,\ e_t = 0.85,\ \gamma_t = 1.33,\ c_{pt} = 1,146\,\text{J/kg} \cdot \text{K}$$

$$\pi_n = 0.98$$

$$p_9/p_0 = 1$$

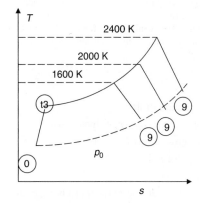

■ FIGURE P4.33

The burner exit temperature ranges from $T_{t4} = 1600\,\text{K}$ to 2400 K. Calculate and graph

- (a) nondimensional-specific gross thrust $F_g/\dot{m}_0 a_0$ versus T_{t4}
- (b) specific impulse I_s (based on fuel flow rate, in seconds) versus T_{t4}
- (c) T_9/T_0 versus T_{t4}
- (d) thermal efficiency, η_{th} versus T_{t4}

4.34 A ramjet is in flight at an altitude where $T_0 = -23°\text{C}$, $p_0 = 10\,\text{kPa}$, and the flight Mach number is M_0. Assuming $T_{t4} = 2500\,\text{K}$ and the nozzle is perfectly expanded, calculate the "optimum" flight Mach number such that ramjet specific thrust is maximized. Assume that all components are ideal, with constant γ and c_p throughout the engine and $Q_R = 42,600\,\text{kJ/kg}$. Would the fuel heating value affect the "optimum" flight Mach number?

4.35 Consider a scramjet in a Mach-6 flight. The fuel of choice for this engine is hydrogen with $Q_R = 120,000\,\text{kJ/kg}$. The inlet uses multiple oblique shocks with a total pressure

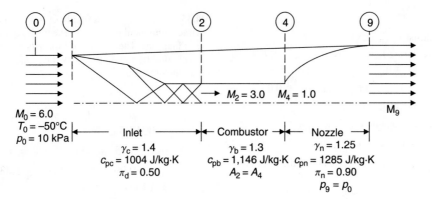

■ FIGURE P4.35

recovery of $\pi_d = 0.5$. The combustor entrance Mach number is $M_2 = 3.0$. Use Rayleigh flow approximations in the supersonic combustor to estimate the fuel-to-air ratio f for a choking exit condition, as shown, after you calculate the combustor exit total temperature T_{t4}.

Also calculate

(a) nozzle exit Mach number M_9

(b) nondimensional ram drag $D_{\text{ram}}/p_0 A_1$ (note that $A_0 = A_1$)

(c) nondimensional gross thrust $F_g/p_0 A_1$

(d) fuel specific impulse I_s in seconds

4.36 We are interested in calculating the thrust boost of an afterburning turbojet engine when the afterburner is turned on. The flight condition and engine parameters are shown.

For simplicity of calculations, we assume that gas properties γ, c_p remain constant throughout the engine.

Calculate

(a) percent thrust gained when afterburner is turned on

(b) percent increase in exhaust speed with afterburner is turned on

(c) static temperature rise at the nozzle exit when the afterburner is turned on

(d) percent fuel–air ratio increase when the afterburner is turned on

(e) percent increase in thrust specific fuel consumption with afterburner on

(f) percent drop in thermal efficiency when the afterburner is turned on

4.37 A separate-flow turbofan engine is designed with an aft-fan configuration, as shown. The fan and core engine nozzles are of convergent design.

For simplicity, you may assume constant gas properties in the engine, i.e, let γ be 1.4 and $c_p = 1,004$ J/kg $\cdot$ K.

Calculate

(a) ram drag (in kN)

(b) compressor adiabatic efficiency η_c

(c) shaft power produced by the high-pressure turbine (in MW)

(d) shaft power consumed by the fan (in MW)

(e) core net thrust (in kN)

(f) fan net thrust (in kN)

(g) thrust-specific fuel consumption (in mg/s/N)

(h) engine thermal efficiency η_{th}

(i) engine propulsive efficiency η_p

Is there an obvious advantage to an aft-fan configuration? Is there an obvious disadvantage to this design?

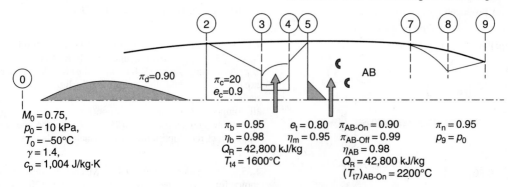

■ FIGURE P4.36

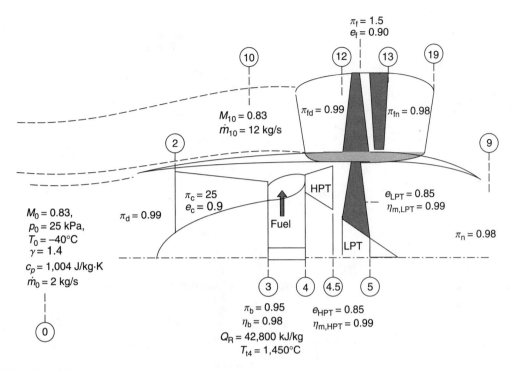

$\pi_f = 1.5$
$e_f = 0.90$

$M_{10} = 0.83$
$\dot{m}_{10} = 12$ kg/s

$\pi_{fd} = 0.99$ $\pi_{fn} = 0.98$

$\pi_c = 25$
$e_c = 0.9$ HPT

$\pi_d = 0.99$ Fuel $e_{LPT} = 0.85$
$\eta_{m,LPT} = 0.99$

$M_0 = 0.83,$
$p_0 = 25$ kPa,
$T_0 = -40°C$
$\gamma = 1.4$
$c_p = 1,004$ J/kg·K
$\dot{m}_0 = 2$ kg/s   LPT $\pi_n = 0.98$

$\pi_b = 0.95$ $e_{HPT} = 0.85$
$\eta_b = 0.98$ $\eta_{m,HPT} = 0.99$
$Q_R = 42,800$ kJ/kg
$T_{t4} = 1,450°C$

■ **FIGURE P4.37**

4.38 An advanced turboprop flies at $M_0 = 0.82$ at an altitude where $p_0 = 30$ kPa and $T_0 = -15°C$. The propeller efficiency is $\eta_{prop} = 0.85$. The inlet captures airflow rate at 50 kg/s and has a total pressure recovery of $\pi_d = 0.99$. The compressor pressure ratio is $\pi_c = 35$ and its polytropic efficiency is $e_c = 0.92$. The combustor has an exit temperature $T_{t4} = 1650$ K and the fuel heating value is $Q_R = 42,000$ kJ/kg, with a burner efficiency of $\eta_b = 0.99$ and the total pressure loss in the burner is $\pi_b = 0.96$. The HPT has a polytropic efficiency of $e_{t,HPT} = 0.80$, and a mechanical efficiency of $\eta_{m, HPT} = 0.99$. The power split between the LPT and the engine nozzle is at $\alpha = 0.75$ and the mechanical efficiency of the LPT is $\eta_{m, LPT} = 0.99$, the LPT adiabatic efficiency is $\eta_{LPT} = 0.88$. A reduction gearbox is used with an efficiency of $\eta_{gb} = 0.995$. The exhaust nozzle is convergent with an adiabatic efficiency of $\eta_n = 0.95$. We will describe gas properties in the engine based only on two temperature zones (cold and hot):

Inlet and compressor sections (cold) : $\gamma_c = 1.4$,

$c_{pc} = 1004$ J/kg · K

Turbines and nozzle sections (hot) : $\gamma_t = 1.33, c_{pt} =$

$1,152$ J/kg · K

Calculate

(a) total pressure and temperature throughout the engine (include fuel-to-air ratio in mass/energy balance)

(b) engine core thrust in kN

(c) propeller thrust in kN

(d) power-specific fuel consumption in mg/s/kW

(e) thrust-specific fuel consumption in mg/s/N

(f) thermal and propulsive efficiencies η_{th} and η_p

(g) engine overall efficiency η_o

4.39 An ideal regenerative (Brayton) cycle is shown. The cycle compression is between states 0 and 3. The compressor discharge is preheated between states 3 and 3′. The source of this thermal energy is the hot exhaust gas from the engine. The burner is responsible for the temperature rise between states 3′ and 4. The expansion in the turbine is partly between states 4 and 5 that supplies the shaft power to the compressor and partly between states 5 and 6 that produces shaft power for an external load (e.g., propeller, helicopter rotor, or electric generator). The total power production as shown in the expansion process is unaffected by the heat exchanger between states 6 and 6′. Note that the turbine exit temperature T_6 has to be higher than the compressor discharge temperature T_3 for the regenerative cycle to work. Therefore low-pressure ratio cycles can benefit from this (regenerative) concept. Also note that $T_{6'} = T_3$ and $T_{3'} = T_6$.

Show that the thermal efficiency of this cycle is

$$\eta_{th} = 1 - \frac{T_3}{T_4}$$

Calculate the thermal efficiency of a Brayton cycle with cycle pressure ratio of 10, i.e., $p_3/p_0 = 10$ and the

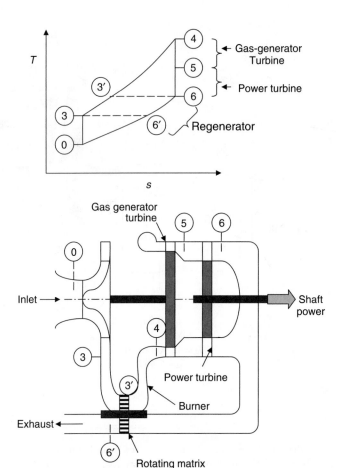

■ **FIGURE P4.39**

maximum cycle temperature ratio of $T_4/T_0 = 6.5$ with and without regeneration.

4.40 A mixed-exhaust turbofan engine with afterburner is flying at $M_0 = 2.5$, $p_0 = 25\,\text{kPa}$, and $T_0 = -35°\text{C}$. The engine inlet total pressure loss is characterized by $\pi_d = 0.85$. The fan pressure ratio is $\pi_f = 1.5$ and polytropic efficiency of the fan is $e_f = 0.90$.

The flow in the fan duct suffers 1% total pressure loss, i.e., $\pi_{fd} = 0.99$. The compressor pressure ratio and polytropic efficiency are $\pi_c = 12$ and $e_c = 0.90$, respectively. The combustor exit temperature is $T_{t4} = 1800\,\text{K}$, fuel heating value is $Q_R = 42,800\,\text{kJ/kg}$, total pressure ratio $\pi_b = 0.94$, and the burner efficiency is $\eta_b = 0.98$. The turbine polytropic efficiency is $e_t = 0.80$, its mechanical efficiency is $\eta_m = 0.95$, and the turbine exit Mach number is $M_5 = 0.5$. The constant-area mixer suffers a total pressure loss due to friction, which is characterized by $\pi_{M,f} = 0.95$. The afterburner is on with $T_{t7} = 2200\,\text{K}$, $Q_{R,AB} = 42,800\,\text{kJ/kg}$, $\pi_{AB\text{-}On} = 0.92$, and afterburner efficiency $\eta_{AB} = 0.98$. The nozzle has a total pressure ratio of $\pi_n = 0.95$ and $p_9/p_0 = 2.6$.

The gas behavior in the engine is dominated by temperature (in a thermally perfect gas), thus we consider four distinct temperature zones:

Inlet, fan, and compressor section : $\gamma_c = 1.4$,
$$c_{pc} = 1,004\,\text{J/kg} \cdot \text{K}$$

Turbine section : $\gamma_t = 1.33$, $c_{pt} = 1,152\,\text{J/kg} \cdot \text{K}$

Mixer exit : γ_{6M}, c_{p6M} (to be calculated based on mixture of gases)

Afterburner and nozzle section : $\gamma_{AB} = 1.30$,
$$c_{p,AB} = 1,241\,\text{J/kg} \cdot \text{K}$$

Calculate

(a) total pressure and temperature throughout the engine, the fan bypass ratio α, and include the contributions of fuel-to-air ratio in the primary and afterburner, f and f_{AB} and

(b) engine performance parameters, i.e., TSFC in mg/s/N, specific thrust and cycle efficiencies

CHAPTER 5

Aircraft Engine Inlets and Nozzles

Courtesy of Rolls-Royce plc

5.1   Introduction

In this chapter, we study the aerothermodynamics of aircraft engine inlets and nozzles. These two components in an aircraft engine represent "duct" flows with internal losses and hence we propose to study them in a single chapter. Despite these similarities, the inlet flowfield bears no resemblance to the exhaust flowfield. The presence of adverse pressure gradient in an inlet diffuser leads to a stalling boundary layer behavior, whereas the favorable pressure gradient in a nozzle promotes attached boundary layer flows. Another dissimilarity between these two components is in the cooling requirements of an advanced exhaust system as compared with the inlets, which remain uncooled until well into the hypersonic flight Mach numbers.

The system requirements of an aircraft intake primarily depend on the aircraft *mission specification*. In general, an aircraft intake system has to be designed to many of the following qualities, namely

1. light weight and low cost to manufacture
2. provide the engine with adequate mass flow rate at a proper Mach number at the engine face throughout the flight envelope
3. provide *spatially smooth* flow into the engine compressor, i.e., low *steady-state distortion* throughout the flight envelope
4. provide *temporally smooth* flow into the engine compressor, i.e., low *dynamic distortion* throughout the flight envelope
5. integrate well with the engine nacelle and/or fuselage, i.e., *low installation drag*
6. provide acoustic absorption of fan/engine noise, i.e., *quiet engine*

7. provide a particle separator for an engine in a vertical lift aircraft, i.e., foreign object damage (FOD) *control*

8. provide low radar signature, i.e., *low observables* for stealth requirement

9. provide favorable "*ilities*," e.g., accessibility, inspectability, reliability, maintainability, repairability

10. allow for engine *thrust growth*

By a close inspection of the above "wish" list, we note several conflicting drivers that appear in the air inlet system design. Therefore a team of engineers is tasked with seeking an "optimum compromise" for their design. The presence of conflicting requirements and seeking an optimal solution are invariably present in any commercial product development.

Our one-dimensional cycle analysis treated an aircraft engine component as a "black box," i.e., the block approach. The "block" approach linked the exit condition of a component to its inlet condition via an efficiency parameter. For example, an exit total pressure of an inlet was linked to the inlet total pressure via the inlet adiabatic efficiency parameter η_d or the total pressure recovery parameter π_d. In turn, we treated the component efficiencies as an input to our cycle analysis. Now, we propose to look inside the black box and identify design features, which affect the component efficiencies. Inevitably, we will need to examine the true multi-dimensional aspect of the flowfield in aircraft engine components.

5.2   The Flight Mach Number and Its Impact on Inlet Duct Geometry

It is customary to divide the inlet flowfield into subsonic and supersonic flow regimes. The flow in the subsonic portion of an inlet is dominated by its boundary layer behavior, whereas the supersonic portion of an inlet is dominated by the appearance of shocks and their interaction with the boundary layer on the compression ramps and the nacelle or fuselage. Hence, the emphasis is different in the two flows. In this section, we will examine the geometrical requirements for subsonic and supersonic diffusers.

From one-dimensional compressible flow studies in aerodynamics (See Anderson, 2005) we learned that

$$\frac{dA}{A} = (M_\infty^2 - 1)\frac{dV}{V} \tag{5.1}$$

Hence for a subsonic duct flow, where $M_\infty < 1$, the flow deceleration, i.e., $dV < 0$, requires an area increase, i.e., $dA > 0$, in the duct. Therefore, a cross-sectional area increase in a duct causes a subsonic flow to decelerate. On the contrary, a supersonic deceleration requires a flow area shrinkage with a turning point occurring at $M = 1$, i.e., the sonic point. This suggests a converging flow in the supersonic regime to achieve a sonic condition and then followed by a diverging duct to decelerate the flow to the desired subsonic speeds for a supersonic diffuser. An opposite behavior is observed in accelerating flows, such as those in nozzles. A subsonic flow is accelerated through a converging duct, i.e., $dA < 0$, and a supersonic acceleration is achieved in a diverging duct, i.e., $dA > 0$. Again, a supersonic nozzle needs to have a sonic "throat" before the diverging area begins for a supersonic acceleration. A schematic drawing of various duct geometries and flow regimes is shown in Fig. 5.1.

■ **FIGURE 5.1**
Geometric requirements for inlets and nozzles (* represents the sonic throat)

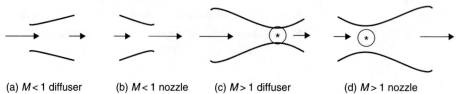

(a) *M* < 1 diffuser   (b) *M* < 1 nozzle   (c) *M* > 1 diffuser   (d) *M* > 1 nozzle

5.3   Diffusers

Now, let us introduce some preliminary nomenclature in a subsonic diffuser. A more detailed geometry of a subsonic diffusing flow is shown in Fig. 5.2, where x is the primary flow direction and y is the lateral or transverse flow direction. The symbols W and R stand for the width (in the diverging direction) of a rectangular and the radius of a conical diffuser, respectively. The axial diffuser length is called N and the diffuser wall length is depicted as L, in Fig. 5.2. We also note that the diffuser wall inclination or divergence angle is shown as ϕ_w.

The thermodynamic states of a diffuser flow are shown in Fig. 5.2 as well. Note the static pressure rise $(p_2 > p_1)$, which is accompanied by a static temperature rise in the diffuser $(T_2 > T_1)$. Also, we note a large inlet kinetic energy is shown in Fig. 5.2, $(V_1^2/2c_p)$, which diminishes to a small kinetic energy at the exit $(V_2^2/2c_p)$. The flow stagnation enthalpy remains constant in an adiabatic diffuser, which is shown in Fig. 5.2 as a constant total temperature process. Finally, we note that the total pressure at the exit of diffuser is lower than the inlet, which is the basis of an entropy rise Δs.

Now, we are ready to introduce a new performance parameter in a diffuser. This new parameter accounts for the *static* pressure rise in a diffuser, which essentially characterizes the conversion of fluid kinetic energy into the fluid static pressure. Note that in a compressible flow, the kinetic energy converts into static pressure rise as well as the internal energy of the fluid, unlike an incompressible fluid where kinetic energy is converted to static pressure only. The nondimensional pressure rise parameter in a diffuser is called the static pressure recovery coefficient C_{PR} and is defined as

$$C_{PR} \equiv \frac{p_2 - p_1}{\overline{q}_1} = \frac{\Delta p}{\rho_1 \overline{V}_1^2/2} \tag{5.2}$$

The "bar" over q_1 in the denominator represents the mass-averaged velocity to be used in the dynamic pressure calculation. This reminds us that all internal flows contain boundary layers, which make the flow nonuniform (at least within the boundary layer). We will study this new parameter and other figures-of-merit in various diffuser environments. First, let us look at an *ideal* diffuser, where the fluid is both incompressible and inviscid.

5.4   An Ideal Diffuser

Applying the Bernoulli equation between stations 1 and 2 of a diffuser for an inviscid fluid, and using the continuity equation for a one-dimensional flow, we can relate the (ideal) diffuser performance $C_{PR,ideal}$ to the diffuser area ratio A_2/A_1 as follows:

$$C_{PR_{ideal}} = \frac{q_1 - q_2}{q_1} = 1 - \left(\frac{A_1}{A_2}\right)^2 = 1 - \frac{1}{AR^2} \tag{5.3}$$

■ **FIGURE 5.2**
Definition sketch for a subsonic diffuser and a T–s diagram depicting the static and stagnation states in a diffuser

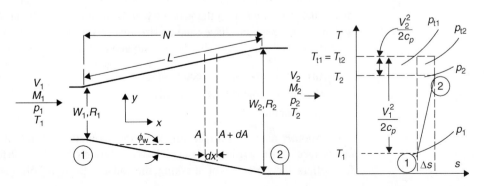

■ **FIGURE 5.3**
An ideal diffuser
pressure recovery
as a function of
diffuser area ratio

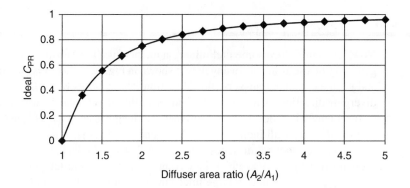

Where the *AR* depicts the diffuser exit-to-inlet area ratio. This statement shows that for a very large area ratio diffuser, the ideal static pressure recovery will approach 1. We expected this result, which states that the inlet dynamic pressure is entirely converted into static pressure rise in a duct if the exit area is infinitely large. Let us plot the ideal static pressure recovery coefficient C_{PR} versus the diffuser area ratio (Fig 5.3).

We note from the ideal diffuser pressure recovery equation 5.3 that the shape of the cross section of the diffuser does not enter the problem, i.e.,

$$C_{PR_{ideal}} \neq f(\text{geometry})_{\text{cross section}} \tag{5.4}$$

This is a consequence of our oversimplified picture of a diffusing flow in a single direction, which ignores viscosity of the fluid and thus boundary layer formation. Consequently, a major driver in diffuser performance, namely the cross-sectional shape and the shape of the diffuser centerline influencing the wall boundary layer flow is ignored in the *ideal* flow analysis. We also note that a high static pressure recovery requires a large diffuser area ratio. We will return to this parameter, i.e., the area ratio, later on in this chapter.

5.5 Real Diffusers and Their Stall Characteristics

We recognize that in a real flow environment, boundary layers are formed, and have a tendency to separate when exposed to a rising static pressure, known as an adverse (streamwise) pressure gradient. Therefore, we expect the behavior of a diffuser to be driven by the viscous region near its walls, i.e., the state of the boundary layer as in attached, separated, or transitory (unsteady). Consequently, the performance of a real diffuser should strongly depend on its inlet boundary layer condition. In addition, the geometry of the cross section as well as the centerline curvature of a diffuser both influence the cross flow tendency in the boundary layer and hence affect the three-dimensional separation characteristics of the diffuser. The formation of a skewed boundary layer is in direct response to a transverse (or lateral) pressure gradient in a diffuser duct. Therefore, we expect the geometry of the diffuser in both the streamwise area variation and the cross-sectional shape (and its streamwise variation), to be of great importance in the performance of a diffuser. The three basic geometries of interest are (See Fig. 5.4)

(a) two-dimensional rectangular
(b) axisymmetric, i.e., conical
(c) annular

Another feature of a real diffuser is the geometrical shape of its centerline. Often the engine face is hidden from an observer looking through the inlet. The feature of a hidden engine face offers the potential of masking the radar reflections off the engine face, which is

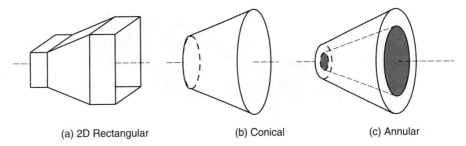

(a) 2D Rectangular	(b) Conical	(c) Annular

■ **FIGURE 5.4**   Schematic drawing of (subsonic) diffuser geometries with straight centerline

advantageous in a stealth aircraft. In addition, fighter aircraft are often designed with their engine(s) inside the fuselage to leave the wings free to carry external weapons. The so-called "buried" engine design of such aircraft requires an S-shaped subsonic diffuser duct to channel air to the engine face. From the fluid dynamics point of view, a curved duct induces a secondary flow pattern, which essentially sets up "pockets" of swirling flow at the duct exit. Often these pockets of swirling flow occur in pairs and are counterrotating. Also, since our diffuser exit is directly tied into the fan entrance duct, the pockets of swirling flow tend to locally increase or decrease (depending on their direction of rotation) the relative flow angle into the fan, which in turn can lead to cyclic loading of the fan and cause a high-cycle fatigue (HCF) problem. In severe situations, the pockets of swirling flow can produce rotating stall instability of the fan rotor. We will discuss in more detail the compressor/fan inlet flow conditions, i.e., the various types of inlet distortion that affect the stability of a compressor flow in the turbomachinery chapter. In Fig. 5.5, we show a schematic of a diffuser duct with an S-shaped (curved) centerline.

An aircraft inlet shape may be rectangular for integration and control purposes but it still has to tie in with an engine face, which is circular. Hence, the connecting diffuser duct should continually change its shape from, say, a rectangular to a circular geometry. Such ducts that change their cross-sectional shape are called "transition" ducts and pose interesting fluid mechanic problems due to their highly three-dimensional pressure pattern, i.e., the flowfield. It is interesting to note that the corners in the rectangular portion of a transition duct create, trap, and transport vortices in a complex streamwise and transverse trajectory pattern along the duct's length. Due to their prominence in modern engines, we will examine the viscous flow characteristics in transition ducts more closely in this chapter. The use of transition ducts in an aircraft engine is not limited, however, to the air inlet system. Rather, it includes circular-to-rectangular exhaust systems as well. Rectangular nozzles are useful for stealth, vector thrust, and integration purposes and will be discussed more at the end of this chapter. Figure 5.6 shows

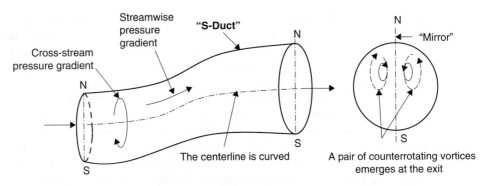

■ **FIGURE 5.5**   Schematic drawing of an S-duct with two pockets of swirling flows (known as the secondary flow pattern) generated by the bends

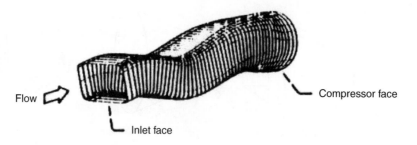

Flow ⟹

Compressor face

Inlet face

B1-B aircraft (Courtesy of USAF)

■ **FIGURE 5.6** **Curved-centerline (double-S) transition duct in a B1-B aircraft (from Anderson, 1986)**

the B1-B aircraft inlet transition duct. An isometric view of rectangular nozzle transition ducts with and without curved centerline is also shown in Fig. 5.7.

5.6   Subsonic Diffuser Performance

Subsonic diffusers exhibit four different flow regimes, or *stall* characteristics, as presented by Kline (1959) and Kline, Abbot, and Fox (1959). The flow regimes are tied to the quality of exit flow, i.e., the state of the boundary layer at the diffuser exit. Although not all different flow regimes are of interest to an aircraft inlet designer, but to present the chart and examining various features of the flow fields proves beneficial to all engineers interested in internal fluid mechanics.

Nondimensional axial length of the diffuser and the total divergence angle of the diffuser are chosen as the axes of the chart in Fig. 5.8. Let us examine this chart more closely. On the lower half of the chart where the line a–a is drawn, we note a *stable flow regime,* which is identified by the "line of first stall." Above line a–a, but below line b–b, we note that the stall phenomenon is recognized and characterized as "transitory." This behavior embodies the *unsteadiness* associated with the first appearance of stall, i.e., the transitory stall characteristics of diffuser flows. This means that a stall patch may appear and then disappear at various positions along the diffuser wall. Beyond the transitory behavior lies the fully developed stall, which is shown above line b–b and below line c–c. In this regime, a stable but separated flow is established in a diffuser. Typically, the stable separation first appears on one wall. Upon increasing the divergence angle of the diffuser, we arrive at what is known as the jet flow. In this regime, the flow emerges like a jet, with little regard for the walls of the diffuser. Two stable and large eddies appear on the walls with a central jet flow emerging in the center. Now, upon

A transition duct with a straight centerline, which is suitable for an exhaust duct in an aircraft engine, is shown on the right. The ratio of length-to-diameter for this duct is 1 and represents the "shortest" transition duct without significant flow separation.

Two transition ducts with exit-to-inlet area ratio of 1, and two different exit aspect ratios are shown on the right.

The aspect ratio at the exit plane of a rectangular nozzle is defined as the ratio of the width to height of the duct.

Converging transition ducts with area ratio of 1/2 and two different nozzle exit aspect ratio rectangular nozzles are shown on the right.

All four ducts (a–d) have a curved center-line corresponding to the exit plane being one inlet radius up from the inlet plane.

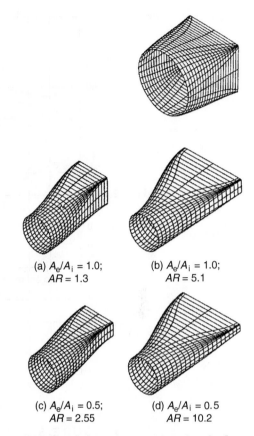

(a) $A_e/A_i = 1.0$; $AR = 1.3$ (b) $A_e/A_i = 1.0$; $AR = 5.1$

(c) $A_e/A_i = 0.5$; $AR = 2.55$ (d) $A_e/A_i = 0.5$ $AR = 10.2$

■ **FIGURE 5.7**   Isometric views of various transition ducts with and without centerline curvature. *Note*: that the duct area ratio is shown as A_e/A_i and the duct *aspect ratio* at the exit is shown as *AR* (from Farokhi, Shue, and Wu, 1989)

reducing the wall divergence angle, the flow exhibits a hysteresis behavior in that the stable separated flow from one wall may now appear on the other wall. This behavior is marked as hysteresis on the chart in Fig. 5.8. We can try to depict these flow regimes in Fig. 5.9.

Now, let us look at the hysteresis behavior in a diffuser with a fully developed stall again. In Fig. 5.9 (c), the lower wall is shown to support the stalled boundary layer flow. So the question is why the lower wall? Now, going back from (d) toward (c) in a reverse experiment that reduces the overall divergence angle of the diffuser, the upper wall may now separate and the flow on the lower wall to remain attached. Why is it that we cannot a'priori predict the wall that stalls? What else is at play that we are overlooking in our reverse experiment? The answer lies in the fine details of the flow (remember the saying "the devil is in the detail"?), namely the instability waves, i.e., the disturbances that existed in both flows and those disturbances that are generated by the diffuser wall actuation mechanism in closing/opening of the walls in our reverse experiment. Hence, our actions or external influences disturbed/created the waves that changed the overall behavior. This leads us to a new level of under-standing of the fluid behavior, namely, that a global large-scale behavior in a fluid flow must have its roots in the behavior of tiny waves, which may have found the environment in the flow ripe for interactions and growth, analogous to the environment of a mass-spring driven at its resonant frequency.

The highest static pressure recovery is achieved in a diffuser that is on the *verge* of separation, i.e., when it operates in a transitory stall mode. Although maximum static pressure recovery C_{PR} is desirable, the challenge for an aircraft inlet designer lies in the *off-design* behavior of such a diffuser, namely the possibility of a massive engine face distortion that could

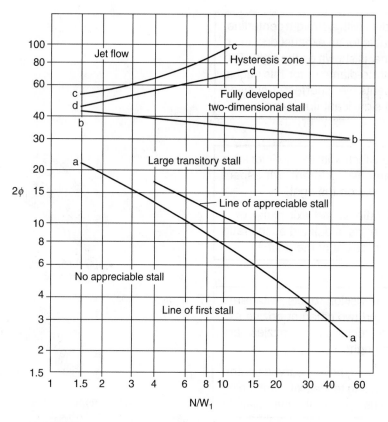

■ **FIGURE 5.8**   **Flow regimes in a two-dimensional rectangular diffuser (from Kline, Abbot, and Fox, 1959)**

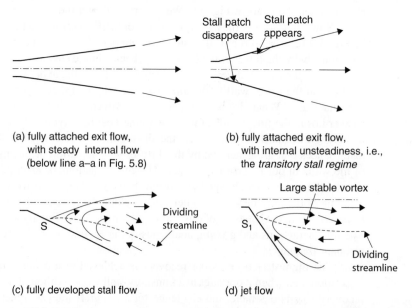

(a) fully attached exit flow, with steady internal flow (below line a–a in Fig. 5.8)

(b) fully attached exit flow, with internal unsteadiness, i.e., the *transitory stall regime*

(c) fully developed stall flow

(d) jet flow

■ **FIGURE 5.9**   **Flow pattern in four diffusers with an increasing wall divergence angle** be created due to a large-scale flow separation in the diffuser. This reminds us of an analogous situation with a pendulum. As we remember, the highest potential energy of a pendulum occurs when it is inverted, i.e., top heavy. But at that point, the slightest disturbance will create a large-scale response, i.e., toppling of the pendulum, and thus the situation with an inverted pendulum is called *meta-stable*. In a similar vein, the highest-pressure recovery diffuser is very sensitive to the inlet disturbances and thus acts in a *meta-stable* manner. However, we can invest in a flow control strategy in the air intake system in order to enhance the stability of fluid flow in diffusers. We have a variety of flow control devices/strategies at our disposal, which include boundary layer suction, blowing, vortex generator placement, and dynamic (excitation) devices such as a vibrating foil/tab that we can use in a modern diffuser.

Sovran and Klomp (1963) have developed a unified way of correlating the performance of different diffuser geometries with a straight centerline. The nondimensional parameters that describe the geometry of a diffuser are chosen to be the area ratio AR for all diffusers, and the axial length ratio N/W_1 for the rectangular diffuser, N/R_1 for the conical diffuser, and $N/\Delta R_1$ for the annular diffuser (See Fig. 5.10 for a definition sketch). These geometrical parameters, from fluid mechanics point of view, establish the (adverse) pressure gradient in a diffuser. The relative thickness of the diffuser inlet boundary layer is also recognized as an important parameter. Although similar attempts could be made with different boundary layer length scales, Sovran and Klomp chose the displacement thickness in defining an area blockage parameter B to represent the quality of the inlet flow. Blockage is defined as the ratio of *blocked-* to-geometric area, as

$$B \equiv \frac{A_\mathrm{B}}{A} \qquad (5.5)$$

where A_B is the blocked area, with the definition

$$A_\mathrm{B} \equiv A - A_\mathrm{E} = A - \int_A \frac{u}{U} dA = \int_A dA - \int_A \frac{u}{U} dA = \int_A \left(1 - \frac{u}{U}\right) dA \qquad (5.6)$$

The parameter A_E is an *effective area* accounting for the inlet flow nonuniformity. In terms of displacement thickness, the blockage is expressed as

$$B_1 = \frac{2\delta_1^*}{W_1} \qquad \text{(2D rectangular diffuser)} \qquad (5.7)$$

$$B_1 \cong \frac{2\delta_1^*}{R_1} \qquad \text{(conical diffuser)} \qquad (5.8)$$

$$B_1 \cong \frac{2\bar{\delta}_1^*}{\Delta R} \qquad \text{(annular diffuser)} \qquad \bar{\delta}_1^* \equiv \frac{\delta_\mathrm{i}^* + \delta_\mathrm{o}^*}{2} \qquad (5.9)$$

Here, we have defined the inlet displacement thickness δ_1^* in Eqs. 5.7 and 5.8 and an *average* displacement thickness over the annulus height ΔR in the annular diffuser geometry. An approximation is made in the above definitions of the inlet blockage parameter, namely, δ^{*2} terms are neglected by virtue of assuming the *inlet* boundary layer displacement thickness is reasonably thin, as compared with the channel width/radius.

The optimum performance of these rectilinear (i.e., straight centerline) diffusers, at high Reynolds numbers assuring an inlet turbulent boundary layer, and with small inlet blockage, $B < 5\%$, are correlated by Sovran and Klomp (1963), which are shown in Fig. 5.11. These charts serve a useful first step in a subsonic diffuser design. More advanced steps require the computational fluid dynamics (CFD) analysis involving Navier–Stokes equations. Also, at *transonic* Mach numbers, shock waves appear, and since their interaction with the boundary layer plays a dominant role in the diffuser performance, we cannot use these charts in the high subsonic Mach range with high degree of accuracy.

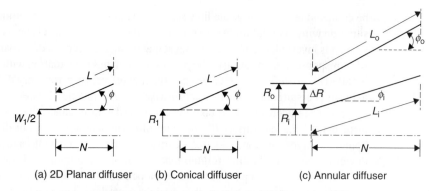

(a) 2D Planar diffuser   (b) Conical diffuser   (c) Annular diffuser

■ **FIGURE 5.10**   Definition sketch for three types of diffusers, the subscripts i and o in the annular diffuser (c) represent *inner* and *outer* walls, respectively

The hatched region on the top performance chart in Fig. 5.11 shows a low-pressure recovery zone associated with a fully developed stall behavior in 2D planar diffusers, as also shown in Fig. 5.8. Note that the hatched zone lies above the 15° half angle or above the 2ϕ of 30° in Fig. 5.8. We can use these charts in several ways, namely, for a fixed area ratio diffuser, we can seek the length ratio that optimizes the static pressure recovery of that diffuser. Conversely, the area ratio of an optimum performance diffuser can be established through these charts for a prescribed diffuser length ratio. The combination of diffuser area ratio and the axial length ratio also identifies the diffuser angle and its performance. Also note that the triangular *wedge,* created between the C_p^* and C_p^{**} lines on the 2D rectilinear diffuser performance chart of the Fig. 5.11, corresponds to a total diffuser divergence angle of $\sim 8°-10°$. This leads us to the conclusion, i.e., a rule of thumb, that an optimum wall inclination angle in a planar diffuser should lie in the neighborhood of $\phi_w \sim 4°$. This rather shallow wall inclination angle reminds us that the (natural) diffusion is a rather slow process. Consequently, rapid diffusion requires a forcing mechanism that enhances mixing of the high- and low-energy fluid in a duct. This again suggests using either vortex generators (i.e., passive control) or devise active flow control strategies to achieve enhanced mixing in a diffuser duct.

A passive method in achieving a short diffuser of a large area ratio is to install splitter plates/cones in a 2D/conical diffuser, which in essence creates a series of parallel diffusers within the overall diffuser (See Cochran and Kline, 1958). This is shown in Fig. 5.12.

5.7   Subsonic Cruise Inlet

For a subsonic cruise application, the total flow deceleration is divided into external and internal segments. The external diffusion takes place outside the inlet where the design capture streamtube itself acts like a subsonic diffuser. We depict this behavior graphically in Fig. 5.13. We have discussed this as a pre-entry drag earlier.

The inlet area at the lip, which is also known as the *highlight*, is called the inlet *capture area* and is sometimes given the symbol A_c. Since the external flow, in the absence of shocks and/or a centerbody that protrudes outside the inlet, is reversible and adiabatic, the ratio of parameters between the flight condition and the inlet face follow the isentropic rule. Namely,

$$\frac{A_0}{A_1} = \frac{M_1}{M_0}\left[\frac{1+\dfrac{\gamma-1}{2}M_0^2}{1+\dfrac{\gamma-1}{2}M_1^2}\right]^{\frac{\gamma+1}{2(\gamma-1)}}$$

$$(5.10)$$

(a) Performance chart of a 2D planar diffuser with inlet blockage of ~1.5% is shown on the right, based on the data from Kline, Abbot, and Fox (1959). Constant C_P contours are plotted. The lines of C_p^* and C_p^{**} represent the maximum diffuser static pressure recovery for a given length ratio and for a prescribed area ratio, respectively. The tick marks on the side show the diffuser half angle ϕ.

(from Sovran and Klomp, 1963)

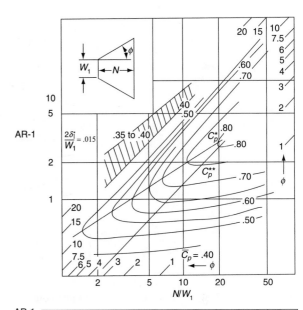

(b) Contour plots of the static pressure recovery in a conical diffuser are shown on the right, based on the data of Cockrell and Markland. The inlet blockage is $B_1 \sim 2\%$.

(from Sovran and Klomp, 1963)

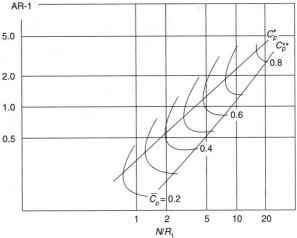

(c) Annular diffuser pressure recovery is shown on the right, based on the data of Sovran and Klomp. The inlet blockage is ~2%. The $\bar{L}$, on the abscissa, represents the average wall length of the inner and outer walls, i.e., $\bar{L} \equiv (L_i + L_o)/2$.

(from Sovran and Klomp, 1963)

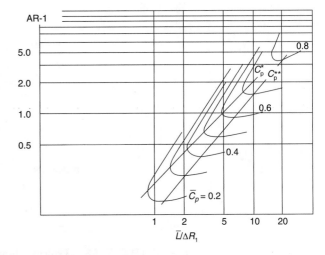

■ **FIGURE 5.11**   Diffuser performance charts (from Sovran and Klomp, 1963)

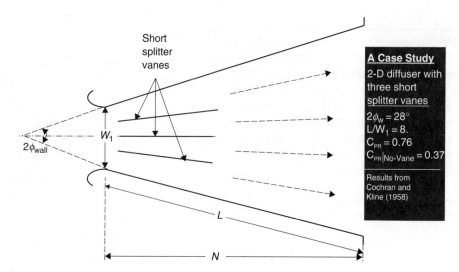

■ **FIGURE 5.12** Wide-angle diffuser is fitted with three internal splitter vanes to prevent flow separation and promote mixing

Therefore, the capture (area) ratio A_0/A_1 is a function of the desired external deceleration from the cruise Mach number M_0 to the inlet lip Mach number M_1 and is called the mass flow ratio parameter or simply MFR in abbreviation. We have graphed Eq. 5.10 for various cruise flight Mach numbers in the transonic regime in Fig. 5.14.

Now, let us address the limitations on M_1. The internal contour of a subsonic inlet lip, i.e., near the highlight, is shown in Fig. 5.15. Due to the convex shape of the inner lip contour of a subsonic inlet, as shown, the flow accelerates near the throat and thus creates a

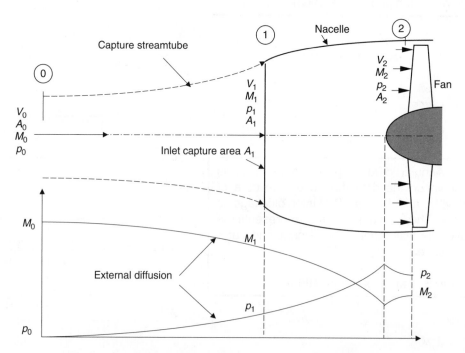

■ **FIGURE 5.13** Typical subsonic inlet at cruise condition showing external as well as internal diffusion (note that flow accelerates over the fan centerbody)

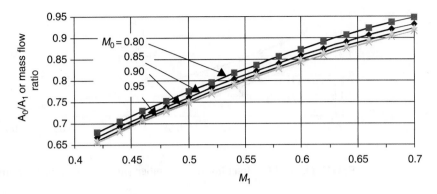

■ **FIGURE 5.14**   Capture (area) ratio and inlet face Mach number

nonuniform velocity profile, somewhat similar to the profile shown in Fig. 5.15. In order to control the "overshoot" in local speed at the throat to a value below the sonic speed, we should limit the one-dimensional throat Mach number, i.e., the mass-averaged throat Mach number, to $\overline{M}_{th} \leq 0.75$. Therefore, it is desirable to prevent a sonic bubble formation at the throat, which may terminate in a shock and possibly cause a boundary layer separation.

The one-dimensional throat Mach number of 0.75 represents an *upper bound* used for the throat sizing of a subsonic inlet. A lower value, say 0.6, is perhaps more desirable as it allows for the thrust growth potential of the engine, which invariably occurs. The average Mach number at the throat of 0.75 as compared with 0.6 represents a ~12% increase in mass flow rate and nearly the same increase in thrust. The same inlet can thus accommodate a 12% increase in thrust without a need for resizing the inlet.

Now that we have fixed the (mass) average throat Mach number to say ~0.75, applying continuity equation to the region between the highlight and the throat, and neglecting small total pressure loss due to friction on the wall, we can relate the average Mach number at highlight, M_1, to the area ratio between the highlight and the inlet throat. Continuity equation applied to an isentropic flow demands

$$\frac{A_{HL}}{A_{th}} = \frac{A_1}{A_{th}} = \frac{\overline{M}_{th}}{M_1} \left[\frac{1 + \frac{\gamma - 1}{2} M_1^2}{1 + \frac{\gamma - 1}{2} \overline{M}_{th}^2} \right]^{\frac{\gamma+1}{2(\gamma-1)}} \tag{5.11}$$

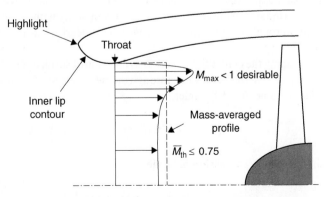

■ **FIGURE 5.15**   Flow curvature at the throat causes a nonuniform velocity profile

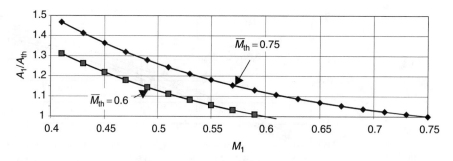

■ **FIGURE 5.16**   Inlet lip Mach number variation with the lip contraction ratio and $\overline{M}_{th}$

In Fig. 5.16, we plotted the inlet lip contraction ratio A_{HL}/A_{th} as a function of the inlet lip Mach number for an average throat Mach number of 0.75. There are at least two competing effects at play near the inlet lip, namely, the conflict between the internal and the external contours of the inlet. This is to be expected, since the flow splits into an *inner* and an *outer* flow around the inlet lip that influences the diffuser performance and the external nacelle drag characteristics, respectively. For example, a large inlet lip contraction ratio contributes to a parameter known as the lip "bluntness," which is good for low-speed, high-angle-of-attack, and side flow environments and is bad for the drag divergence characteristics of the nacelle at high speed. A low contraction ratio inlet lip offers a good high-speed characteristic for the nacelle external drag and a poor low-speed characteristic toward the engine face and flow distortion.

The high-speed performance of an air intake system is dominated by its external cowl drag behavior. The flow acceleration around the cowl external surface may exceed the local sonic speed with a potential for drag rise with shock formation and boundary layer separation. The maximum diameter of the nacelle and its axial disposition from the inlet lip (highlight) control the flow overspeed over the cowl outer surface. These parameters are denoted as A_M and l_M, respectively, for the maximum area and the length measured from the highlight to the maximum diameter. The nondimensional external cowl parameters are A_{HL}/A_M analogous to the inlet lip contraction ratio A_{HL}/A_{th}, which governed the internal performance of the inlet and l_M/d_M. Let us examine a control volume with an inlet area larger than A_M and two outlet areas, one at the capture area of the inlet and the other over the cowl maximum area, as shown in Fig. 5.17. Note that at the first appearance of sonic speed on the cowl, we reach the critical C_p (as labeled). Also note that the sonic bubble is not terminated in a shock, thus we call the inlet nacelle outer contour a "supercritical" nacelle, similar to the supercritical airfoil. Somewhere inside the bubble, the velocity peaks to V_{max} and at that location we noted the lowest C_p (as labeled). With the side of the control volume not coinciding with a stream surface, we allowed a flow *spillage* through the side at a speed of V_0, which clearly is an assumption and an acceptable one.

Let us apply the continuity equation to the control surface shown in Fig. 5.17, in order to quantify the (spillage) mass flow rate through the side in terms of the inlet mass flow rate and the geometrical parameters of the inlet.

$$\dot{m}_0 + \rho_0 V_0 (A - A_0) = \dot{m}_1 + \dot{m}_s + \rho_0 V_0 (A - A_M) \tag{5.12}$$

Upon simplification, expression 5.12 yields

$$\dot{m}_s = \rho_0 V_0 (A_M - A_0) \tag{5.13}$$

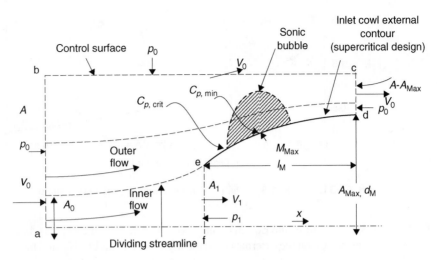

■ **FIGURE 5.17**   Control volume (a-b-c-d-e-f-a) used for the external flow analysis of subsonic inlets

Applying the (inviscid) momentum equation to the control volume in the x-direction gives

$$\dot{m}_0 V_1 + \dot{m}_s V_0 + \rho_0 V_0 (A - A_M) V_0 - \rho_0 V_0 A V_0 = -\int_{A_1}^{A_M} (p - p_0) dA - (p_1 - p_0) A_1 \quad \textbf{(5.14)}$$

$$\dot{m}_0 (V_1 - V_0) = -\int_{A_1}^{A_M} (p - p_0) dA - (p_1 - p_0) A_1 \quad \textbf{(5.15)}$$

We shall nondimensionalize the above equation by dividing through by the free-stream dynamic pressure, $q_0 = \rho_0 V_0^2/2 = \gamma p_0 M_0^2/2$ multiplied by the inlet lip area A_1 to get

$$2\frac{A_0}{A_1}\left(\frac{V_1}{V_0} - 1\right) = -\int_1^{A_M/A_1} C_p dA' - \frac{2}{\gamma M_0^2}\left(\frac{p_1}{p_0} - 1\right) \quad \textbf{(5.16)}$$

where the integrand in Eq. 5.16 is cast in terms of surface pressure coefficient C_p in the integral and the variable of integration is now changed to A', which is the nondimensional area A/A_1. The inlet lip pressure ratio in Eq. 5.16 may be written in terms of M_1 and M_0, as

$$\frac{p_1}{p_0} = \left(\frac{1 + \dfrac{\gamma - 1}{2} M_0^2}{1 + \dfrac{\gamma - 1}{2} M_1^2}\right)^{\frac{\gamma}{\gamma - 1}} \quad \textbf{(5.17)}$$

To perform the integral on the RHS of Eq. 5.16, we either input a "desired" C_p distribution, which may be a constant C_p design, on the inlet external cowl or we may express the integral in terms of a surface-mean C_p, as an area-averaged quantity, defined as

$$\int_1^{A_M/A_1} C_p dA' = \overline{C}_p \left(\frac{A_M}{A_1} - 1\right) \quad \textbf{(5.18)}$$

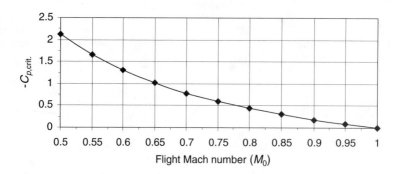

■ **FIGURE 5.18**   The pressure coefficient at the sonic point on the nacelle forebody

We may substitute Eqs. 5.17 and 5.18 into Eq. 5.16 and express A_M/A_1 in terms of the mean-surface C_p on the external cowl, the flight Mach number M_0, and the capture ratio A_0/A_1 to get

$$\frac{A_M}{A_1} = 1 + \left(2\frac{A_0}{A_1} \left(\frac{M_0}{M_1} \sqrt{\frac{1 + \frac{\gamma - 1}{2} M_0^2}{1 + \frac{\gamma - 1}{2} M_1^2}} - 1 \right) + \frac{2}{\gamma M_0^2} \left(\left(\frac{1 + \frac{\gamma - 1}{2} M_0^2}{1 + \frac{\gamma - 1}{2} M_1^2} \right)^{\frac{\gamma}{\gamma - 1}} - 1 \right) \right) / (-\overline{C}_p)$$

(5.19)

Note that the inlet Mach number M_1 in Eq. 5.19 is a function of the inlet capture ratio, as shown in Fig. 5.14. We may substitute $C_{p,\text{crit}}$ (i.e., the pressure coefficient at the sonic point) for the surface-averaged pressure coefficient, in order to plot the nacelle area ratio A_M/A_1. The critical pressure coefficient is a function of flight Mach number according to isentropic flow:

$$\overline{C}_p \approx C_{p,\text{crit}} = \frac{2}{\gamma M_0^2} \left[\left(\frac{1 + \frac{\gamma - 1}{2} M_0^2}{\frac{\gamma + 1}{2}} \right)^{\frac{\gamma}{\gamma - 1}} - 1 \right]$$

(5.20)

We have graphed Eq. 5.20 as a function of the flight Mach number in Fig. 5.18.

Now, let us incorporate the critical pressure coefficient in Eq. 5.19 and plot the external nacelle area ratio A_M/A_1 in terms of flight Mach number and the inlet capture ratio, or MFR. The result is presented in Fig. 5.19.

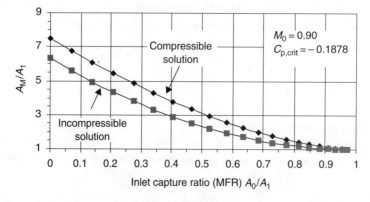

■ **FIGURE 5.19**   The minimum inlet cowl frontal area (A_M/A_1) corresponding to the sonic flow on the cowl at the cruise Mach number of 0.90

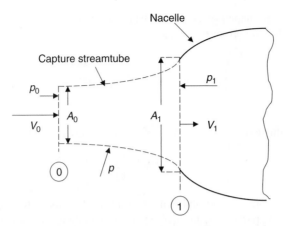

■ **FIGURE 5.20**   A control volume for additive drag derivation

We incorporated the incompressible solution in Fig. 5.19 as well to demonstrate the effect of compressibility on the cowl frontal area requirement. Once we account for the effect of density variations in the flow, the inlet cowl area ratio requirements show a higher A_M/A_1 needed than the incompressible solution, as shown in Fig. 5.19. A typical capture ratio at cruise is ~0.70, which then using Fig. 5.19 suggests a minimum cowl frontal area ratio of ~1.88 for a cruise Mach number of 0.90. This area ratio corresponds to an inlet (highlight)-to-maximum diameter ratio of ~0.73. This is somewhat lower than in the large bypass ratio turbofan engine nacelles, which have their highlight-to-maximum diameter ratio at ~0.80–0.90, primarily due to ground clearance reasons, according to Seddon and Goldsmith (1985). The higher R_1/R_M in the large bypass engines leads to a *slimline* design of the fan cowls with an attendant relief on supersonic flow and the sonic bubble formation on the cowl.

Now, let us quantify the additive drag associated with the capture streamtube that we first introduced in the engine thrust and performance parameters chapter 3. A control volume is shown in Fig. 5.20, which is composed of the capture streamtube, free stream entrance, and the inlet lip, i.e., highlight exit surfaces. The conservation of mass relates the area ratios to Mach numbers in an isentropic flow via

$$\frac{A_0}{A_1} = \frac{M_1}{M_0}\left(\frac{1 + \dfrac{\gamma - 1}{2}M_0^2}{1 + \dfrac{\gamma - 1}{2}M_1^2}\right)^{\frac{\gamma+1}{2(\gamma-1)}} \tag{5.21}$$

The conservation of momentum in the streamwise direction applied to the control volume outside the inlet may be written as

$$\dot{m}_0(V_1 - V_0) = p_0 A_0 + \int_{A_0}^{A_1} p\,dA - p_1 A_1 \tag{5.22}$$

This expression balances the momentum change in the streamwise direction by the forces that cause the change of momentum, i.e., the pressure–area terms on the RHS. By subtracting a constant pressure p_0 from all the surfaces of the control volume, we can rewrite the momentum equation as

$$\dot{m}_0(V_1 - V_0) = \int_{A_0}^{A_1} (p - p_0)\,dA - (p_1 - p_0)A_1 \tag{5.23}$$

As we recall from the chapter on engine thrust (Chapter 3), the pressure integral on the capture streamtube is called the pre-entry or additive drag, therefore, we may write it as

$$D_{\text{add}} \equiv \int_{A_0}^{A_1} (p - p_0)dA = \dot{m}_0(V_1 - V_0) + (p_1 - p_0)A_1 \qquad (5.24)$$

Now, let us cast this expression in terms of a nondimensional parameter, such as

$$\frac{D_{\text{add}}}{p_0 A_1} = \frac{\rho_0}{p_0} \frac{A_0}{A_1} V_0^2 \left(\frac{V_1}{V_0} - 1\right) + \frac{p_1}{p_0} - 1 \qquad (5.25)$$

Note the logical choice of the ambient pressure p_0 and the inlet capture area A_1 as the nondimensionalizing parameters in Eq. 5.25. These parameters do not change with the engine throttle setting, as say A_0 and p_1 would. We may introduce the flight Mach number on the RHS of Eq. 5.25, as well as the static pressure ratio in terms of flight and the inlet Mach number M_0 and M_1, respectively, using isentropic relations and the capture ratio via the continuity equation, to get

$$\frac{D_{\text{add}}}{p_0 A_1} = \gamma M_1 \left(\frac{1 + \dfrac{\gamma - 1}{2} M_0^2}{1 + \dfrac{\gamma - 1}{2} M_1^2}\right)^{\frac{\gamma+1}{2(\gamma-1)}} \left(M_1 \sqrt{\frac{1 + \dfrac{\gamma - 1}{2} M_0^2}{1 + \dfrac{\gamma - 1}{2} M_1^2}} - M_0\right)$$

$$+ \left(\frac{1 + \dfrac{\gamma - 1}{2} M_0^2}{1 + \dfrac{\gamma - 1}{2} M_1^2}\right)^{\frac{\gamma}{\gamma-1}} - 1 \qquad (5.26)$$

Note that the nondimensional additive drag is a function of the flight Mach number and the inlet lip Mach number, i.e., a function of the capture streamtube shape. A cylindrical capture streamtube with $A_0 = A_1$ would experience a vanishing pre-entry drag. Let us graph this equation for a family of inlets characterized by different inlet lip Mach number designs and flight conditions (Fig. 5.21).

Let us examine the family of additive drag curves in Fig. 5.21 more closely. Two distinct trends become apparent. First at takeoff or low-speed flight, since the inlet capture ratio is very large for this flight condition, it corresponds to a high-inlet lip Mach number, say 0.7 or higher, where

$$\left.\frac{D_{\text{add}}}{p_0 A_1}\right|_{\text{takeoff}} \sim 0.2 \qquad (5.27)$$

The second trend that we may observe from the family of nondimensional additive drag profiles, shown in Fig. 5.21, is their behavior at transonic flight condition when an engine becomes inoperative. This situation is represented by a low-inlet lip Mach number corresponding to the engine "windmilling" condition, say $M_1 \sim 0.3$ (or less), hence

$$\left.\frac{D_{\text{add}}}{p_0 A_1}\right|_{\text{windmill}} \sim 0.3 \qquad (5.28)$$

To develop an appreciation for the *magnitude* of the inlet additive drag at takeoff and the engine windmilling condition at altitude, we may graph these conditions for engines of different inlet diameters. The ambient pressure at takeoff was chosen to be the standard sea level pressure of

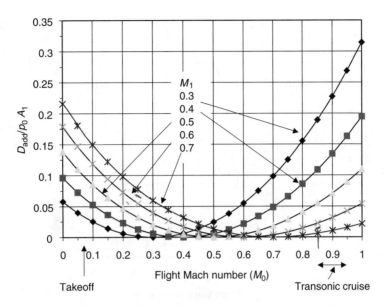

■ **FIGURE 5.21**   **Inlet additive drag profiles from takeoff to transonic cruise at different inlet lip Mach numbers M_1, $\gamma = 1.4$**

14.7 psia (or 100 kPa) and the altitude curve was chosen at 40,000 ft with an ambient pressure of 393 lb/ft^2. The result is shown in Fig. 5.22.

Although the magnitude of additive drag is very large, both at takeoff and altitude, the cowl lip suction force, on a blunt cowl, almost balances out the additive drag. In an engine out situation, however, there is a drag, a penalty due to a flow separation from the outer cowl, i.e., external nacelle drag increases due to the flow spillage from the inlet over the cowl lip. The imbalance between the additive drag and the lip suction force is called the spillage drag, as we introduced it earlier in the engine force and performance parameters Chapter 3.

$$D_{\text{spillage}} \equiv D_{\text{add}} - F_{\text{lip}} \tag{5.29}$$

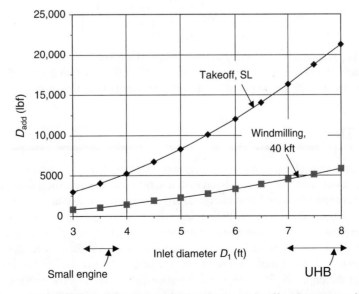

■ **FIGURE 5.22**   **Inlet additive drag at takeoff and at transonic cruise with a windmilling engine (UHB: Ultrahigh Bypass turbofan engine)**

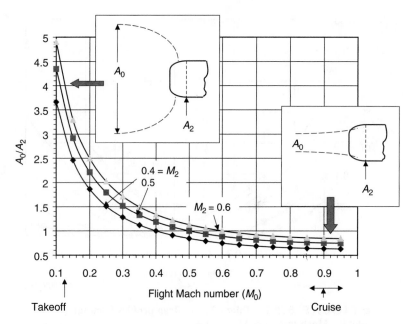

■ **FIGURE 5.23** Captured stream-to-engine face area ratio from takeoff to cruise (isentropic)

A well-designed subsonic cruise inlet attains a total pressure recovery between the flight and the engine face that is ~0.995–0.997. This is a remarkable feat achieved by the aircraft industry today. Although the total pressure recovery is reduced at takeoff and climb, as the throat Mach number is increased, we may still neglect the total pressure drop when we compare the captured stream area to the engine face area. Hence, we may link the engine face Mach number and area to the flight Mach number and the free stream area using the isentropic flow relation, namely,

$$\frac{A_0}{A_2} = \frac{M_2}{M_0} \left[\frac{1 + \frac{\gamma - 1}{2} M_0^2}{1 + \frac{\gamma - 1}{2} M_2^2} \right]^{\frac{\gamma+1}{2(\gamma-1)}}$$

(5.30)

Let us graph this equation for a typical engine face Mach number range of 0.4–0.6 and the flight Mach numbers ranging from 0.1 at takeoff to 0.95 at cruise (Fig. 5.23). The upper value corresponds to the proposed Sonic Cruiser commercial transport of Boeing.

5.8   Transition Ducts

A duct that changes its cross-sectional shape is called a transition duct. These ducts are prevalent in advanced propulsion systems, typically the transition from rectangular to round in the inlet and the round-to-rectangular cross section in the exhaust systems. The fluid mechanics of transition ducts is rich with merging boundary layers at the corners of the duct, corner vortex formation, and three-dimensional pressure field setting up fascinating swirling patterns in the cross flow, known as the *secondary flows*. Our interest in these ducts as a part of an inlet system is to be able to quantify and minimize the engine face distortion levels caused by these ducts over the entire flight envelope of the aircraft. In a rectangular exhaust system, however, the interest is in eliminating separation bubbles and minimizing corner vortices, which can *trap* the unburned fuel in the afterburner and potentially cause a burn through in the nozzle wall if

■ **FIGURE 5.24**
Transition duct flowfield visualized by near-wall velocity vectors and a cross-stream plane and on the upper right, a flow regime map of transition ducts with area ratio equal to one (from Abbott, Anderson, and Rice, 1987)

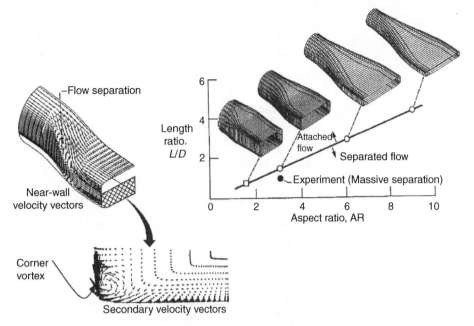

combustion takes place over the wall-bounded recirculating flows. These practical concerns for the inlets and nozzles are coupled with a design-to-weight mentality, which has been the cornerstone of aircraft system development. To minimize the length of these ducts while maintaining a healthy state of boundary layer, a designer may use a computational fluid dynamics tool, which is capable of three-dimensional viscous flow solutions. Bernhard Anderson of NASA has performed numerous computational studies of transition ducts including an extensive comparison with benchmark experimental data for validation purposes. Figure 5.24 is a composite diagram of Anderson's work (Abbott, Anderson, and Rice, 1987) with transition ducts suitable for exhaust systems (round-to-rectangular).

5.9   An Interim Summary for Subsonic Inlets

We learned that

- An inlet aerodynamic performance has an *internal* as well as an *external* component, an internal figure of merit, and an external figure of merit
- An inlet internal performance is dominated by its throat Mach number
- The one-dimensional (or average) throat Mach number should be less than 0.75
- The internal performance is governed by the *laws* of diffusers, e.g., $C_{PR}(AR, N/R_1, B)$, i.e., area ratio, length ratio, inlet blockage, and centerline shape for offset exits
- A *cubic* is proposed for the internal diffuser contour with a maximum angle of $8.7°$
- The internal performance at low speed is helped by the *lip contraction ratio*
- The captured streamtube introduces an *additive drag* with a capture ratio other than 1, which can be very large at takeoff or with a windmilling engine situation
- The capture ratio (or inlet mass flow ratio) for a subsonic cruise inlet is ~0.70
- The additive drag is nearly balanced out by the cowl lip suction force, i.e., lip *thrust*
- The uncancelled part of the additive drag by the lip suction force is called *spillage drag*, which is a strong function of lip bluntness
- Cowl lip bluntness has an internal and an external component to it

- The internal lip contour is near elliptic (i.e., a superelliptic profile)
- "Bluntness" is quantifiable
- Large capture ratios at takeoff and climb pose a potential problem for engine face distortion
- Wing and aft fuselage-mounted inlets are exposed to flow angularity, i.e., droop or toe up angles depending on their placement, i.e., the wing upwash or downwash
- Nacelle external contour sets the limitation on high-speed performance of the inlet
- Drag divergence Mach number can be raised by a *supercritical* nacelle design
- Drag reduction may be achieved through a hybrid laminar flow control on the nacelle
- Transition ducts may be needed for integration purposes, sometimes with an offset
- Secondary flow patterns are introduced in ducts with a bend or transition ducts in general, a cause of distortion

There are additional discussions that are relevant to inlets and inlet-aircraft integration:

- Inlet acoustic treatment
- Inlet (cowl) lip icing detection and anti-icing strategies
- Inlet-engine matching, steady-state, and dynamic distortion

Further readings on subsonic inlets and engine installations (references 3, 16, 18, 22, 23, 27, 29, 31, 33, 34, 45 and 47) are recommended.

5.10 Supersonic Inlets

The function of a supersonic inlet is the same as the function of a subsonic inlet, namely, to decelerate the flow to the engine face Mach number requirements (set by the throttle), typically in the range of $M_2 \sim 0.4-0.6$, efficiently, within the entire flight envelope. In practice, flow diffusion from supersonic to subsonic flow involves shocks. Therefore, the study of supersonic inlets is very much dominated by the study of shocks intersecting, interacting, reflecting, and the shock boundary layer interaction. We will also learn that shocks pose instability problems for an inlet and learn possible approaches to stabilizing them. As in subsonic inlets, we are still concerned about the external drag characteristics of the supersonic inlets. Although, there are no supersonic inlet flows that are reversible, we still propose to study them in the limit of reversible, adiabatic flow. By studying isentropic inlets, we learn about throat sizing issues, some off-design Mach numbers, and their impact on the inlet flow behavior.

5.10.1 Isentropic Convergent–Divergent Inlets

By insisting on an isentropic compression, we propose to create an isentropic compression Mach wave system that all converge to a single point, like a fan. It is theoretically possible to machine a concave surface that can decelerate a supersonic flow to a sonic flow. The concave surface is known as an isentropic compression ramp. The picture of the waves set up on an isentropic compression ramp looks like Fig. 5.25.

Now let us remember the wave and ramp angles from aerodynamics. We remember that all Mach waves make an angle, called Mach angle, with respect to the *local* flow. Therefore, the lead Mach wave makes an angle

$$\mu_0 = \sin^{-1}(1/M_0) \tag{5.31}$$

The last Mach wave, i.e., the tail wave, is normal to the sonic exit flow, i.e., $\sin^{-1}(1) = 90°$. We also remember from Prandtl–Meyer flow that the ramp angle of ν_0, which is called Prandtl–Meyer angle, corresponds to turning a *sonic flow* to achieve a supersonic Mach number M_0. Since the process of isentropic compression is reversible, it means that a flow approaching the

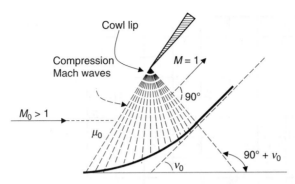

■ **FIGURE 5.25** An isentropic compression ramp decelerating a supersonic flow to $M = 1$

ramp from the top of Fig. 5.25 at sonic speed (i.e., reverse flow) will *accelerate* to achieve M_0 at the bottom of the ramp, the flow turning angle is thus the Prandtl–Meyer angle v_0 corresponding to M_0. An expression for this angle as a function of Mach number and the ratio of specific heats γ is written as Eq. 5.32, which we derived in Chapter 2.

$$v(M) \equiv \sqrt{\frac{\gamma + 1}{\gamma - 1}} \tan^{-1} \sqrt{\frac{\gamma - 1}{\gamma + 1}(M^2 - 1)} - \tan^{-1}\sqrt{M^2 - 1} \qquad \textbf{(5.32)}$$

We may decelerate the sonic exit flow from the compression ramp in a subsonic diffuser, which requires a flow area expansion, i.e., a diverging duct. Now let us incorporate such isentropic compression ramps in a duct and subsonic diffusers to create a *convergent–divergent* (C–D) duct capable of decelerating a supersonic flow to a subsonic flow reversibly and adiabatically. This duct is then called an isentropic C–D inlet. The simple geometry of it is presented in Fig. 5.26.

Let us note that the capture streamtube in Fig. 5.26 shows a capture ratio A_0/A_1 of 1, and furthermore its area ratio A_1/A_{th} is uniquely established via an isentropic A/A^* relation corresponding to M_1. For example, the inlet-to-throat area ratio is equal to 2 for an inlet Mach number of 2.2. We can read these numbers from an isentropic table for $\gamma = 1.4$. Also we note that since the flow deceleration in this inlet takes place entirely within the duct, we may call it an *internal-compression* inlet. The unique isentropic area ratio (A/A^*) at a supersonic speed creates problems at subsonic speeds. Namely, the same area ratio duct *chokes* at a subsonic Mach number too! This means that prior to reaching our target supersonic design Mach number of M_D, we have already choked the throat at a subsonic Mach number. Let us return to the example we used in this paragraph. For a supersonic design Mach number of 2.2, the C–D inlet should have its area ratio set at 2.0. By looking up the isentropic tables, in the subsonic portion of it, we note that area ratio 2.0 first chokes at Mach number ~0.3! In other words, each value of A/A^* is repeated twice in the isentropic table, once in the subsonic flow part and then the second time in the supersonic flow portion of the table. Now let us graphically represent the *dual* nature of choking in a compressible flow, i.e., the A/A^* is a *double-valued* function of Mach number. The continuity equation for an isentropic flow established A/A^* according to

$$A/A^* = (1/M)\left[\left(1 + \frac{\gamma - 1}{2}M^2\right) \middle/ \left(\frac{\gamma + 1}{2}\right)\right]^{\frac{\gamma+1}{2(\gamma-1)}} \qquad \textbf{(5.33)}$$

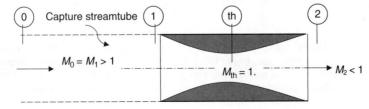

■ **FIGURE 5.26** Schematic drawing of an isentropic C–D inlet

■ **FIGURE 5.27**
Isentropic area ratio as a
function of Mach
number ($\gamma = 1.4$) shows
that a C–D duct exhibits
choking behavior at a
supersonic as well as a
subsonic Mach number

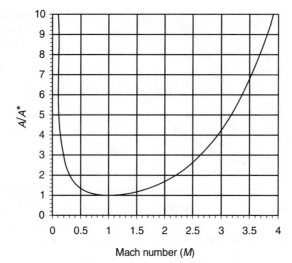

We may graph this equation for a range of Mach numbers (Fig. 5.27) to demonstrate that there are two Mach numbers, one subsonic and the other supersonic, for each A/A^*.

Now, let us examine the sequence of events that takes place outside and within the inlet from start to the design Mach number. The capture ratio for the inlet starts larger than 1, as in the subsonic inlet at takeoff. The capture ratio remains greater than 1 until the throat chokes at the corresponding subsonic A/A^*. At this moment, the capture area ratio is 1, and the throat is choked.

Various flow characteristics of an isentropic C–D inlet is shown in Fig. 5.28 as a function of flight Mach number. The first choking occurs at a unique subsonic Mach number

■ **FIGURE 5.28**
Flow characteristics of
an isentropic C–D inlet
from low speed to
supersonic conditions

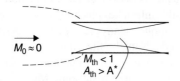

(a) At low speed the throat remains
unchoked (note the capture ratio
($A_0/A_1 > 1$)

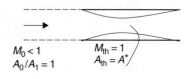

(b) At a unique subsonic Mach number
throat first chokes (note the capture
ratio $A_0/A_1 = 1$)

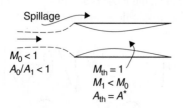

(c) At higher subsonic Mach numbers
than (b) the capture ratio falls below
1 and spillage occurs (note the
throat remains choked)

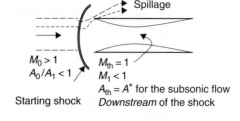

(d) At supersonic Mach numbers a
starting (bow) shock is formed in
front of the inlet causing spillage
drag while the throat remains choked

corresponding to the "bucket" graph of A/A^* shown in Fig. 5.27, where the capture ratio is exactly 1 and no spillage drag occurs. At all Mach numbers above this unique subsonic Mach number, the throat continues to remain choked and the consequence of higher flow manifests itself in a spillage over the inlet lip, as shown in parts (c) and (d) of Fig. 5.28. For supersonic flow, a bow shock is formed ahead of the inlet to allow for spillage to occur. The spillage accounts for a choked throat, which is smaller than the corresponding A/A^* at M_0. As long as the starting shock remains in front of the supersonic inlet, the spillage drag and low total pressure recovery are the consequences of this flow. This undesirable condition is called an "unstarted" inlet. We shall now examine a few methods that enable *starting* a C–D inlet.

5.10.2 Methods to Start a Supersonic Convergent–Divergent Inlet

The starting process of a supersonic C–D inlet involves swallowing the starting shock. A normal shock that is brought to the inlet lip is said to be in an incipient starting position. Any disturbance that causes the starting shock to enter the inlet will then result in the shock to move through the convergent portion of the duct and be stabilized at a location in the divergent section downstream of the throat. Normal shocks are therefore said to be unstable in a converging duct. The position of a shock in a duct is dictated by the *backpressure*, which establishes the downstream condition, or outflow condition of the shock. When the backpressure increases, a shock is pushed upstream and when the backpressure decreases, the shock moves downstream. We may use the backpressure principle to explore shock stability in converging as well as diverging ducts. First let us look at a converging duct that leads to a sonic throat. The starting shock that moves in a converging duct, by virtue of flow area contraction, will experience a decrease in its upstream Mach number M_x. Therefore the shock *weakens* as it enters a contracting duct. Consequently, the total pressure downstream of the shock in its new position is increased, as the shock is now weaker. The sonic throat, which acts as the first engine throttle, experiences an increased mass flow rate. We remember that the mass flow rate is linearly proportional to the total pressure, i.e., $\dot{m} \propto p_t$. The increase in mass flow rate downstream of the shock will reduce the shock backpressure (i.e., it creates a suction), which in turn draws the shock further in the duct and toward the throat. The shock continues moving until it has passed through the throat and is stabilized in the diverging section of the duct. The process of swallowing the starting shock, from the inlet lip to beyond the throat, occurs on the order of convective time scale, namely, $t \sim l/V$, where l is the characteristic diffuser length and V is an average flow speed in the inlet. For a diffuser of $\sim$10 ft or $\sim$3 m length and an average flow speed of $\sim$2000 ft/s or $\sim$600 m/s, the starting shock should be swallowed on the order of $\sim$10/2000 s or $\sim$5 ms. This example shows the abruptness of starting and perhaps most importantly the *unstart* process of a supersonic inlet with an internal throat.

We will address the unstart phenomenon later in this chapter. Now, let us examine the movement of a normal shock downstream of the throat, i.e., in the diverging section of a C–D inlet. Assuming the shock is in a position that matches the engine-imposed backpressure condition, any shock motion into the duct will cause M_x to increase, therefore p_t after the shock will drop and the mass flow rate is proportionately reduced. A reduced mass flow rate has the effect of an increased backpressure, which in turn pushes the shock back upstream. So, a displaced shock in the diverging section of a duct returns to its original position, which is the essence of a stable system. A similar argument can be applied to a shock displaced in the opposite direction, i.e., closer to the throat. In this case, the upstream Mach number of the shock is reduced, therefore the total pressure downstream of the shock is increased, which causes an increase in the mass flow rate and hence a reduced backpressure, which draws the shock to its original position. Again, we demonstrate shock stability in a diverging duct. We may use Fig. 5.29 to graphically depict the starting process.

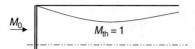

(a) The starting shock is at the lip, the so-called *incipient* starting position

(b) The shock is *unstable* in the converging section, thus it moves through the inlet

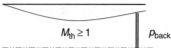

(c) The shock is swallowed and is stabilized in the diverging section according to a backpressure condition established by the engine. The inlet is started

(d) The engine backpressure is increased to place the shock closer to the throat for an increased total pressure recovery. *Best* p_{back} will place the shock at the throat for a maximum total pressure recovery

■ **FIGURE 5.29** **Starting sequence of a C–D inlet and the role of backpressure in positioning the shock**

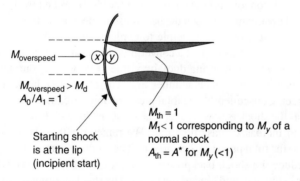

■ **FIGURE 5.30** **The starting shock is pushed toward the inlet lip through overspeeding**

We are now ready to examine different starting methods that will place the starting shock in an *incipient* position, namely, at the inlet lip. These include overspeeding, an enlarged throat, and a variable throat geometry C–D inlet.

5.10.2.1 *Overspeeding.*

For a fixed-geometry C–D inlet, a process of *overspeeding* may start the inlet. This method is feasible for low supersonic design Mach number C–D inlets only. The proposition is to offer a reduced spillage condition on the inlet so as to attract the starting shock toward the inlet lip. Eventually, for an overspeed Mach number $M_{overspeed}$, the starting shock is at the inlet lip. This condition is shown in Fig. 5.30 and is known as the incipient inlet start.

EXAMPLE 5.1

Consider an isentropic fixed-geometry C–D inlet, which is designed for $M_D = 1.5$. Calculate the overspeed Mach number that will start this inlet.

SOLUTION

We first establish the design area ratio A_1/A_{th} for a design Mach number of 1.5. From the isentropic table for $\gamma = 1.4$, we have

$$A_1/A_{th} = A_1/A^* = 1.176$$

Now, if the starting shock is at the lip, the subsonic flow at M_y will have the same A/A^* as the inlet, since the throat is still choked. Therefore we look up the isentropic table for a subsonic Mach number, which corresponds to this area ratio.

$$A_y/A^* = 1.176 \xrightarrow[M<1]{\text{Isentropic table}} M_y \approx 0.61$$

From a normal shock table, we establish a corresponding upstream Mach number M_x or $M_{overspeed}$.

$$M_y \approx 0.61 \xrightarrow{\text{Normal shock table}} M_x \approx 1.80$$

Therefore,

$$\boxed{M_{overspeed} \approx 1.80}$$

Now, let us examine an aerodynamic limitation of overspeeding. For this purpose, we assume an infinite overspeed is possible (although both structurally and propulsively impossible). From the NS-table, we have

$$M_x \sim \infty \text{ yields a } M_y \rightarrow \sim 0.378$$

Now we look up the isentropic area ratio A/A^* corresponding to $M_y = 0.378$.

$$M_y \approx 0.378 \xrightarrow{\text{Isentropic table}} A_y/A^* \approx 1.666 = A_1/A_{th}$$

For this area ratio, we find a corresponding supersonic Mach number, which then represents the design Mach number M_D

$$A_1/A^* \approx 1.666 \xrightarrow[M>1]{\text{Isentropic table}} M_1 = M_D \approx 1.99$$

(requires infinite overspeeding)

This example shows that through overspeeding, we cannot even start a C–D inlet designed for a design Mach number of $M_D = 2$. We alluded to structural limitations, infinite loads, as well as propulsive limitations, i.e., infinite thrust, earlier which prevent infinite overspeeding as a means of starting an inlet. With these limitations, we examine an enlarged throat C–D inlet that is capable of self-starting at the design Mach number. This is called Kantrowitz–Donaldson inlet.

5.10.2.2   *Kantrowitz–Donaldson Inlet.*   Consider a C–D inlet with an enlarged throat area that eliminates spillage of an isentropic C–D inlet at design Mach number. Therefore, we are interested in a C–D duct with a unique contraction ratio that places the starting shock at the inlet lip at the design Mach number. This is called a Kantrowitz–Donaldson (K–D) inlet, which has a self-starting capability (See reference 20). By virtue of having absorbed the spillage mass flow rate at the design Mach number, the throat area of a K–D inlet has to be larger than that of the corresponding isentropic inlet. Let us graph this proposition (Fig. 5.31).

We may quantify the K–D inlet contraction ratio since from the design Mach number, we can establish M_y, which is the M_1 just downstream of the shock. Also assuming an isentropic flow downstream of the shock, we may establish the required area ratio A_1/A_{th}. Once the shock is swallowed and stabilized downstream of the throat, the throat Mach number will be greater than 1. The best backpressure will place the shock at the throat. Note that a normal shock at the throat is unstable and could lead to the phenomenon of unstart, or the shock sweeping backwards through the inlet and causing tremendous spillage drag as well as a loss of engine thrust. For this reason, all supersonic inlets with an internal throat need to actively control the position of the terminal shock in the diverging section of the duct. We will address this issue

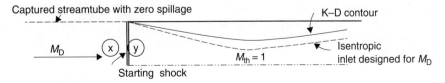

■ **FIGURE 5.31**   A K–D inlet at design Mach number (at the moment of incipient start); also note the larger throat area of a K–D inlet as compared with an isentropic inlet designed for M_D

■ **FIGURE 5.32**
The sequence of events in a Kantrowitz–Donaldson inlet

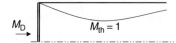

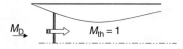

(a) At the design Mach number shock is at the inlet lip, note $A_0/A_1 = 1$, i.e., zero spillage drag

(b) The starting shock is being swallowed

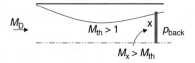

(c) Shock is in a stable position, the throat is unchoked ($M_{th} > 1$)

(d) The best backpressure places the shock at the throat

later in this chapter. Let us graphically depict the various phases of a K–D inlet flow environment from design Mach number M_D to the best backpressure (Fig. 5.32).

We now solve a problem that demonstrates the geometry and capabilities of a K–D inlet.

EXAMPLE 5.2

Calculate the contraction ratio A_1/A_{th}, and the maximum total pressure recovery of a self-starting C–D inlet designed for $M_D = 2.65$.

SOLUTION

At the design Mach number, a self-starting inlet has its starting shock at the lip (see Fig. 5.32-a), therefore

$$M_D = M_x = 2.65 \xrightarrow{\text{Normal shock table}} M_y = 0.4996 = M_1$$

$$M_1 = 0.4996\,(\sim 0.5) \xrightarrow{\text{Isentropic table}} A_1/A^*$$
$$\equiv 1.34 = A_1/A_{th}$$

$$\therefore \quad \boxed{A_1/A_{th} \equiv 1.34}$$

The best backpressure places the shock at the throat, thus we need to calculate the throat Mach number M_{th} after the inlet was started. From Fig. 5.32-c, we note that a supersonic flow at the design Mach number enters the inlet and assuming isentropic deceleration to the throat, we may arrive at the throat Mach number, via the following steps

$$M_D = 2.65 \xrightarrow{\text{Isentropic table}} A_1/A^* = 3.036$$

$$A_{th}/A^* = (A_1/A^*)/(A_1/A_{th}) = 3.036/1.34$$
$$= 2.2656 \xrightarrow{\text{Isentropic table } M > 1} M_{th} \sim 2.35$$

Now we need to place a shock at this Mach number, i.e.,

$$p_{ty}/p_{tx})_{\text{best backpressure}} \xrightarrow{\text{Normal shock table } M_x \sim 2.35} \boxed{p_{ty}/p_{tx} \sim 0.5615}$$

Note that the price of a fixed-geometry self-starting C–D inlet is a high throat Mach number and subsequent large total pressure loss. We calculated the throat Mach number, i.e., the lowest Mach number in the duct, of a started K–D inlet designed for Mach 2.65 is ~ 2.35. This performance penalty of a fixed-geometry inlet suggests the potential advantage of a variable geometry throat inlet. We shall examine such inlets next.

5.10.2.3 *Variable-Throat Isentropic C–D Inlet.* To eliminate spillage flow and attract the starting shock to the inlet lip, a variable geometry inlet opens up its throat to

■ FIGURE 5.33
Variable-geometry C–D
inlet starts by enlarging
its throat area (from A_{th}
to A'_{th})

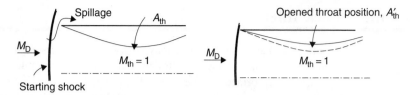

(a) Starting shock with spillage at M_D

(b) Incipient start position is achieved by
opening the throat to absorb spillage flow

accommodate a higher flow rate and thus begin the starting process. The two shock positions prior to the starting and at the incipient start condition are shown in Fig. 5.33.

After the starting shock has swept through the opened throat in the starting process, the throat Mach number becomes supersonic. The opening size of the throat and the throat Mach number are identical to those of the Kantrowitz–Donaldson inlet. However, in this case of variable geometry, we can proceed to close the inlet throat back to its original sonic condition/dimension. The attractiveness of this method of starting a supersonic inlet with an internal throat is the potential for high total pressure recovery, i.e., high performance. The price for this performance is paid through the added system weight and complexity associated with multisegmented inlet with actuators, sensors, and a controller needed to operate the inlet.

Now, let us learn, through a sample problem, the important parameters associated with a variable-geometry inlet, namely, the percent opening of the throat needed to start the inlet, the throat Mach number after the inlet has started.

EXAMPLE 5.3

A supersonic convergent–divergent inlet is to be designed for an isentropic operation (in the started mode) at $M_D = 3.3$. Calculate the inlet design contraction ratio A_1/A_{th}, the percent opening of the throat $(A'_{th} - A_{th})/A_{th}$ needed to start the inlet, and the throat Mach number in the open position M'_{th}.

SOLUTION

First, the isentropic area ratio for M_D is read from the isentropic table to be

$$M_D = 3.3 \xrightarrow{\text{Isentropic table}} A_1/A^* = A_1/A_{th} = 5.629$$

Now, at the opened throat position, the starting shock is at the lip, therefore,

$$M_D = M_x = 3.3 \xrightarrow{\text{Normal shock}} M_y = M_1$$

$$= 0.4596 \xrightarrow{\text{Isentropic table}} A_1/A'_{th} \equiv 1.425$$

Now, we can calculate the percent throat opening according to

$$\frac{A'_{th} - A_{th}}{A_{th}} \times 100 = \left(\frac{\dfrac{A'_{th}}{A_1} - \dfrac{A_{th}}{A_1}}{\dfrac{A_{th}}{A_1}} \right) \times 100$$

$$= \frac{(1/1.425) - (1/5.629)}{(1/5.629)} \cong 295\%$$

Our calculations indicate that the opened throat is roughly three times the size of the isentropic throat to start the inlet designed for Mach 3.3. We can calculate the flow Mach number at the throat, in the open position, once the shock is stabilized downstream of the throat. To get the Mach number we need A'_{th}/A^*, similar to the K–D inlet calculation, namely,

$$A'_{th}/A^* = (A'_{th}/A_1)/(A^*/A_1) = (1/1.425)/(1/5.629)$$

$$\cong 3.950 \xrightarrow{\text{Isentropic table } M>1} M'_{th} \sim 2.95$$

We finally proceed to close the throat to achieve $M_{th} = 1$ and a near isentropic flow condition.

5.11   Normal Shock Inlets

A sharp-lipped subsonic diffuser may be used in a supersonic stream without a significant aerodynamic penalty if the free stream Mach number is below ~1.6. As for all supersonic inlets, a shock is formed at or near the inlet, which decelerates the supersonic flow to subsonic. The sharp lip geometry allows for an attached shock, whereas a blunt cowl leading edge creates a bow shock with the attendant external drag penalty. This type of supersonic inlet is short, light weight, and with no movable surfaces, which is suitable for low supersonic Mach applications or low-cost weapon systems development. Let us define the geometry and salient features of a normal shock inlet in Fig. 5.34.

The position of the normal shock depends on the inlet backpressure, which is established by the engine. The best backpressure, corresponding to design Mach number, places the normal shock at the lip where the best total pressure recovery coincides with zero spillage drag. This mode of operation is called the *critical* mode. In the event of higher backpressure, i.e., when the engine mass flow rate drops, the shock stands outside the inlet and a spillage flow takes place. This is the so-called *subcritical* mode of operation. The shock is drawn into the inlet, beyond the lip, when the engine backpressure is lowered. This is the so-called *supercritical* mode of operation. In the supercritical mode, the shock Mach number is higher than M_0, hence a larger total pressure drop in the inlet results. Consequently, the corrected mass flow rate at the engine face increases, which results in an increase in axial Mach number, thereby, reducing the engine face static pressure, i.e., the inlet backpressure. Furthermore, the shock inside the duct may interact adversely with the wall boundary layer and cause separation and increase engine face distortion.

Note that a normal shock inlet has its throat at the lip and hence experiences no starting problem. Even in the sub or supercritical modes the shock is stable. The three modes of operation of a normal shock inlet are shown in Fig. 5.35. It is also customary to present the off-design performance of supersonic inlets as a plot of their total pressure recovery versus the inlet mass flow ratio. The design mass flow rate is used as the reference value to create a dimensionless mass flow ratio. A typical graph of shock total pressure recovery versus the mass flow ratio for a NS-inlet is shown in Fig. 5.36.

The total pressure recovery of a normal shock rapidly deteriorates beyond Mach ~1.6, where a ~90% total pressure recovery is achieved. The normal shock recovery as a function of Mach number is graphed in Fig. 5.37.

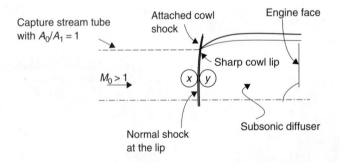

■ **FIGURE 5.34**   **Normal shock inlet is shown with the shock at the lip**

■ **FIGURE 5.35**
**Three modes of
operation of a normal
shock inlet show the
subcritical mode as an
external-performance
limiting and the
supercritical mode as the
internal performance
limiting flow conditions
with the best mode
identified as the critical
mode**

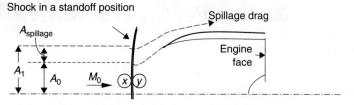

(a) Normal shock inlet in *subcritical mode* (*high* backpressure–low $\dot{m}$ and $\dot{m}_{cor}$)

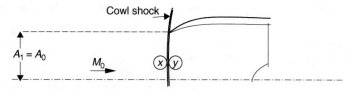

(b) Normal shock inlet in *critical mode* (*best* backpressure–$\dot{m}_{Design}$)

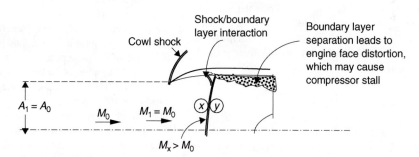

(c) Normal shock inlet in *supercritical mode* (*low* backpressure–higher $\dot{m}_{cor}$, same $\dot{m}_{Design}$)

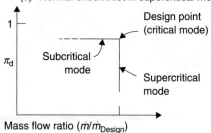

■ **FIGURE 5.36**   Typical performance plot of a normal shock inlet

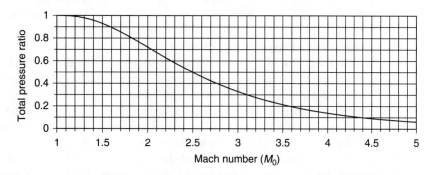

■ **FIGURE 5.37**   Normal shock recovery as a function of Mach number ($\gamma = 1.4$)

EXAMPLE 5.4

A normal shock inlet operates in a Mach 1.4 stream. Calculate

(a) inlet total pressure recovery with the shock at the lip, i.e., the best backpressure

(b) inlet total pressure recovery when the shock is inside the duct at $A_x/A_1 = 1.1$, i.e., the supercritical mode

(c) inlet total pressure recovery in subcritical mode with 10% spillage, i.e., $A_{spillage}/A_1 = 0.1$

SOLUTION

We need to establish the shock Mach number in each mode of inlet operation. In case (a), where the shock is at the lip, representing the best backpressure, the shock Mach number is M_0, therefore,

$$M_0 = 1.4 \xrightarrow{\text{Normal shock table}} \boxed{p_{t2}/p_{t0} = 0.9582}$$

To calculate the shock Mach number when the shock is inside the inlet, we note that the inlet flow may be assumed isentropic all the way to the shock. Hence, we will use A/A^* at the shock, based on upstream flow, to establish the shock Mach number M_x.

$$A_x/A^* = (A_x/A_1)(A_1/A^*)$$

The position of the shock inside the inlet is given as $A_x/A_1 = 1.1$ in the problem statement. The ratio A_1/A^* is purely a function of M_1 via isentropic flow relations, which is known to be $M_1 = M_0 = 1.4$. Therefore,

$$M_1 = 1.4 \xrightarrow{\text{Isentropic table}} A_1/A^* = 1.115$$

$$A_x/A^* = (1.1)(1.115) = 1.226 \xrightarrow{\text{Isentropic table } M > 1}$$
$$M_x \approx 1.56 \xrightarrow{\text{Normal shock table}} \boxed{p_{t2}/p_{t0} \approx 0.91}$$

Note that the shock Mach number M_x is higher than M_0 as expected (supersonic Mach number goes up in a diverging duct, i.e., with flow area increase). Also, note that the total pressure recovery is reduced in case (b), again as expected. The other penalty of the supercritical mode of operation is the potential boundary layer separation and engine face distortion, which are complex fluid phenomena and beyond the scope of the present text.

Now, let us address the case (c), where the shock stands outside the inlet and causes spillage. It is quantified as causing a 10% spillage, which is $1 - A_0/A_1$. The shock Mach number, however, is the same as M_0, hence the same total pressure recovery is expected, i.e.,

Since $M_x = M_0$ in subcritical mode, then

$$\boxed{p_{t2}/p_{t0} = 0.9582}$$

However, we note that the penalty for subcritical mode of operation is the spillage drag $D_{spillage}$.

5.12 External Compression Inlets

To achieve a higher total pressure recovery than a normal shock inlet, we need to design an inlet configuration that incorporates multiple shocks. Supersonic flow deceleration over multiple shocks is more efficient than the deceleration through a normal shock. Here, the efficiency of the supersonic diffusion process is defined by the total pressure recovery of the shock system. An external compression inlet is designed to maintain the shock system external to the inlet and hence has its throat at the cowl lip. The internal duct is hence a subsonic diffuser, which is often a transition duct with offset. Two-dimensional external compression inlets employ external compression ramps to create plane oblique shocks, whereas axisymmetric inlets employ multiple cones to create conical shocks. Single and multiple ramp external compression inlets as well as axisymmetric inlets are shown in Fig. 5.38. The position of the normal shock, which is also called a *terminal shock*, is dictated by the inlet backpressure. The optimal location for the normal shock is at the throat where the flow Mach number is the least. With the normal shock at the cowl lip, the inlet is said to operate at its critical mode, which represents the design point. Figure 5.38 shows several external compression inlets at their design point operation.

■ **FIGURE 5.38**
Shock systems of two-dimensional and axisymmetric external compression inlets with the best backpressure placing the terminal (normal) shock at the lip (throat)

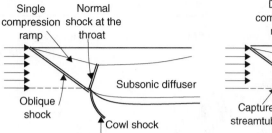

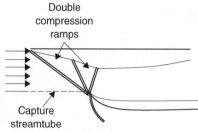

(a) A single compression ramp inlet involves one oblique shock and one normal shock

(b) A double compression ramp inlet creates two oblique shocks and one NS

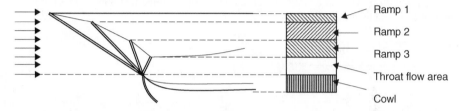

(c) The side view of a three-compression ramp inlet with three oblique shocks and one normal shock

(d) The front view of a three-ramp external compression rectangular (2D) inlet

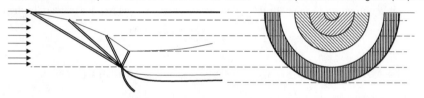

(e) A three-cone external compression inlet shows three conical shocks and one NS

(f) Front view of a three-cone inlet shows semi-circular inlet/nacelle cross sections

The total pressure recovery of the inlet shock system may be calculated via a straightforward marching technique that we learned in supersonic aerodynamics. Assuming the geometry of the external compression ramp is known (the so-called *design problem*), each ramp angle represents the flow turning θ, which combined with the information on the local Mach number M, leads to a determination of the wave angle β, by using a θ–β–M oblique shock chart (or conical shock charts). The wave angle helps establish the strength of the oblique shock, which depends on the normal component of the flow to the shock, namely, $M_n = M \sin\beta$. Since the local Mach number is only a priori known in station 0, i.e., the free stream condition, we need to start with the first oblique shock and continue marching downstream through multiple ramps and shocks. This method is best illustrated through an example.

EXAMPLE 5.5

A two-ramp external compression inlet in a supersonic flow is shown. Calculate the total pressure recovery of this inlet assuming the best backpressure has placed the normal shock on the lip. Also compare this inlet to a NS-inlet at Mach 2.0.

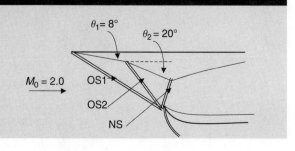

SOLUTION

The shock total pressure recovery as a function of Mach number M and the wave angle β is

$$\frac{p_{t2}}{p_{t1}} = \frac{\left(\dfrac{\dfrac{\gamma+1}{2}M^2\sin^2\beta}{1+\dfrac{\gamma-1}{2}M^2\sin^2\beta}\right)^{\frac{\gamma}{\gamma-1}}}{\left(\dfrac{2}{\gamma+1}M^2\sin^2\beta-\dfrac{\gamma-1}{\gamma+1}\right)^{\frac{1}{\gamma-1}}}$$

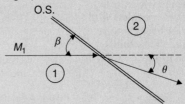

Therefore, we need to establish the wave angles and the local Mach numbers upstream of every shock. Using an oblique shock chart θ–β–M, any pair of values determines the third. For the first oblique shock (OS1) we know the Mach number and the flow turning angle, i.e.,

OBLIQUE SHOCK 1

$$M_0 = 2.0,\ \theta = 8° \quad \xrightarrow{\theta-\beta-M\text{ chart}} \quad \boxed{\beta_1 \approx 37°} \Rightarrow M_{1n} = M_0\sin\beta = 2\sin37° \equiv 1.20 \Rightarrow (p_{t2}/p_{t1})_{OS1} \cong 0.993$$

Also, from a normal shock table (or the following equation), we get M_n downstream of the first shock, i.e.,

$$M_{2n}^2 = \frac{2+(\gamma-1)M_{1n}^2}{2\gamma M_{1n}^2-(\gamma-1)}$$

$$M_{1n} \cong 1.2 \Rightarrow M_{2n} = 0.843 \Rightarrow M_2 = M_{2n}/\sin(\beta-\theta) \cong 0.843/\sin29° \cong 1.74$$

Now, we are ready for the oblique shock 2.

OBLIQUE SHOCK 2

We need the Mach number, which was calculated from above to be 1.74 and the turning angle through the second shock. Although the ramp angle is shown to be 20°, note that the flow had already turned by 8° through the first oblique shock, therefore the net turning angle through the second shock is only 12°, hence

$$M \cong 1.74,\ \theta = 12° \quad \xrightarrow{\text{Oblique shock chart}} \quad \boxed{\beta_2 \approx 48.7°} \Rightarrow M_{1n} = 1.74\sin48.7° \equiv 1.31 \Rightarrow (p_{t2}/p_{t1})_{OS2} \cong 0.978$$

$$M_{1n} = 1.31 \Rightarrow M_{2n} \cong 0.782$$

$$M_2 = \frac{M_{2n}}{\sin(\beta-\theta)} \Rightarrow M_2 \cong 0.782/\sin(36.7°) \cong 1.31$$

NORMAL SHOCK

We calculated the local Mach number downstream of the OS2 as $M = 1.31$. Since the flow is normal to the terminal shock, $\beta = 90°$, which may be substituted in the above equation for p_{t2}/p_{t1} or simply use a normal shock table.

$$M \cong 1.31,\ \beta = 90° \Rightarrow (p_{t2}/p_{t1})_{NS} \cong 0.977$$

Overall shock total pressure recovery

$$(p_{t2}/p_{t1})_{overall} \cong (p_{t2}/p_{t1})_{OS1} \cdot (p_{t2}/p_{t1})_{OS2} \cdot (p_{t2}/p_{t1})_{NS} \cong (0.993)(0.978)(0.977) \equiv 0.949$$

$$\boxed{(p_{t2}/p_{t1})_{overall} \cong 0.949}$$

COMPARISON TO A NORMAL SHOCK INLET A normal shock inlet at Mach 2, with the best back-pressure gives a dismal performance, i.e.,

$$(p_{t2}/p_{t1})_{NS} \cong 0.721$$

We see a tremendous improvement in total pressure recovery where multiple ramps are used.

5.12.1 Optimum Ramp Angles

In the previous problem, we arbitrarily chose two ramp angles of $8°$ and $12°$ net turning, respectively, for our Mach 2.0 inlet. This design recovered nearly 95% of total pressure through the shock system. A question of optimization arises as to the optimum ramp angles, which lead to a maximum pressure recovery at a given Mach number. So, let us restate the problem in general terms as follows:

Find optimum ramp angles for a multiramp external compression inlet designed for $M_0 > 0$.

This fundamental aerodynamic question was raised and answered by Oswatitsch in a 1944 German report (later translated into English in 1947 as a NACA report). The result is astonishingly simple. It states that in an optimum multishock inlet, all shocks need to be of equal strength. This speaks to the *principle of equal burden*. Hesse and Mumford (1964) in their classical propulsion textbook produced the optimum geometries for a single and double ramp external compression inlet. These are shown in Figs. 5.39 and 5.40.

5.12.2 Design and Off-Design Operation

The shock system associated with an external compression inlet determines both the internal performance of the inlet through total pressure recovery as well as the external performance through inlet external drag. The design point operation focuses the oblique shocks onto or slightly below the cowl lip and places the terminal shock exactly at the throat. These two configurations of shocks are graphed (Fig. 5.41) for discussion purposes.

Figure 5.41 raises an interesting issue about the spillage and the inlet wave drag. In part (a) the capture ratio A_0/A_1 is exactly 1, which eliminates any spillage drag. On the contrary, the static pressure rise on the cowl lip due to a strong cowl shock integrates into a wave drag. The

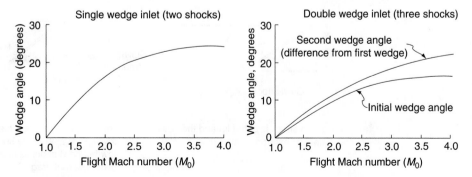

■ **FIGURE 5.39** Optimum ramp angles for a single and double-ramp external compression inlet (from Rodean, 1958.)

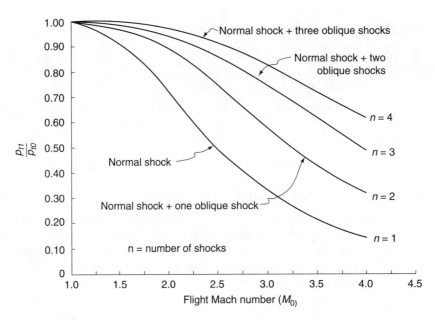

■ **FIGURE 5.40**   **Optimum total pressure recovery of external compression inlets (from Rodean, 1958.)**

part (b) configuration of the shock system intentionally misses the cowl lip and hence incurs some spillage drag. Note that the capture ratio A_0/A_1 is less than 1. But, since the flow downstream of the oblique shock has already turned by the ramp inclination angle, any additional turning due to cowl (thickness) creates a weaker cowl shock. Hence, a weaker static pressure rise on the cowl leads to a lower wave drag penalty of the inlet. Here, we learned an important lesson about *system optimization*. Simply stated, a system is more than the sum of its isolated parts! There are interaction laws between the elements of the system that inherently affect system behavior and hence optimization.

The location of the normal shock is dictated by the inlet backpressure. The cowl lip represents the throat and hence the minimum Mach number. The normal shock placement at the throat is hence considered to represent the design point operation, the so-called *critical mode*. With an increase in the backpressure, the normal shock is pushed onto the compression ramp(s) and hence it represents a *subcritical mode* of operation. With a lowered backpressure, the shock is sucked into the inlet, i.e., the subsonic diffuser, and it represents a *supercritical mode*. These are shown in Fig. 5.42.

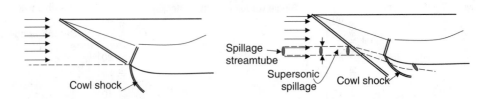

(a) The oblique shock is at the lip and the cowl shock causes external (wave) drag

(b) The oblique shock is projected below the cowl lip creating some supersonic spillage flow but cowl shock is now weaker causing lower external drag

■ **FIGURE 5.41**   **Shock location and nacelle wave drag issues**

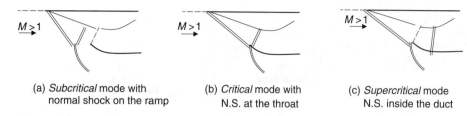

(a) *Subcritical* mode with   (b) *Critical* mode with   (c) *Supercritical* mode
    normal shock on the ramp   N.S. at the throat   N.S. inside the duct

■ **FIGURE 5.42**   Definition sketch for external compression inlet modes of operation

Separated flow

■ **FIGURE 5.43**   A *Buzz instability* cycle showing (from left) a blocked throat, reduced backpressure, shock movement toward the lip, pressure buildup, pushing the shock back out sequence

The off-design operations involve subcritical and supercritical modes. The normal shock inside the subsonic diffuser, as shown in Fig. 5.42(c), results in a lower total pressure recovery, as well as a potential boundary layer separation, engine face distortion. The lower total pressure recovery of this mode of operation is due to a stronger normal shock inside the duct than the shock at the throat. The method of calculation of the shock strength is identical to the supercritical mode of a normal shock inlet (see example problem solved). On the shock boundary layer interaction inside the duct, we may have to employ active or passive viscous flow control techniques, e.g., boundary layer suction, vortex generators, etc. We will address the flow control schemes that may be utilized to stabilize a boundary layer later in this chapter.

With an increasing backpressure in the inlet, the normal shock is pushed out of the inlet lip and onto the external compression ramps. There are three major penalties associated with this mode of operation. One, the spillage drag increases significantly. Two, the intersecting shocks outside the inlet create a vortex sheet (shear layer), which creates a distortion at the engine face. Three, the normal shock on the ramp could separate the boundary layer and cause a large amplitude oscillatory flow pulsation, with high levels of distortion at the engine. This oscillatory flow is called *Buzz* instability and is worth a closer examination. The separated boundary layer of the ramps partially blocks the throat. To supply the engine with its desired flow rate, the blocked throat experiences fluid acceleration and hence the shock backpressure drops. This will cause the shock to be attracted to the inlet lip. However, at the inlet lip position, the engine is supplied with more flow than it needs; therefore, the shock is pushed back out onto the ramps. An oscillatory back and forth flow is now ensued. A complete *Buzz instability* cycle is shown in Fig. 5.43.

It is customary to capture the overall performance of external compression inlets in a total pressure recovery graph versus mass flow ratio. A typical example is shown in Fig. 5.44.

5.13   Variable Geometry—External Compression Inlets

As noted earlier, the ramp angles for optimum total pressure recovery are a strong function of flight Mach number (see Figs. 5.39 and 5.40). In addition, the engine mass flow rate demands are different at different Mach numbers, altitudes within the flight envelope. For these reasons, the external compression ramps (or cones) need to be equipped with actuators (presently

■ **FIGURE 5.44**
Typical external compression inlet performance plot

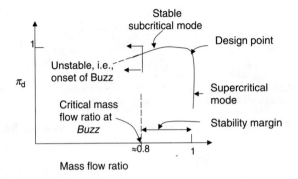

■ **FIGURE 5.45**
Variable geometry inlet shows ramp actuation at different flight phases and a bleed exit door (from Abernethy and Roberts, 1986)

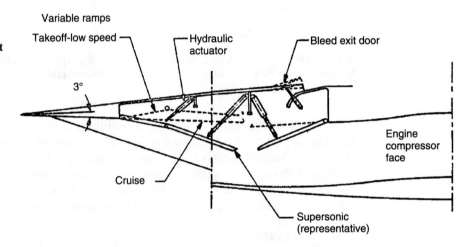

hydraulic) that control their position in various flight phases. Also the buzz instability improvement/elimination as well as engine face distortion level control demand boundary layer management techniques that require variable geometry. Let us address these and other performance optimizing techniques used in this class of supersonic inlets. Although in our discussions, we refer to ramps, as in 2D inlets, similar statements can be made for cones and axisymmetric inlets.

5.13.1 Variable Ramps

The leading ramp, for structural reasons, is a fixed-geometry solid wedge, typically 3°–5°. The following ramps may be extended into the flow or withdrawn from the flow using hydraulic actuators. The withdrawn position, which is also referred to as collapsed ramp position, allows for a larger inlet area needed for the takeoff, climb, subsonic cruise, and transonic acceleration. Figure 5.45 shows a typical 2D variable geometry external compression inlet (from Abernethy and Roberts 1986).

5.14 Mixed-Compression Inlets

To achieve the entire supersonic deceleration outside an inlet, as in the external compression inlet, is only economical at flight Mach numbers below ~2.5. The penalty comes from excessive cowl wave drag, as the cowl lip has almost the same inclination, with respect to the flow, as that of the last compression ramp of the inlet. A large flow deceleration needs a large

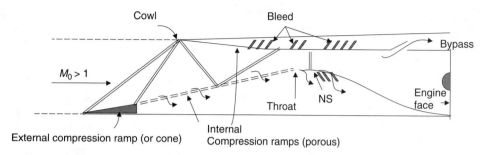

■ **FIGURE 5.46**   **A mixed compression inlet with external and internal shocks, an internal throat**

supersonic flow turning and hence the cowl lip angle may exceed the θ_{max} of the oblique shock, resulting in a bow shock formation. To alleviate this performance limitation of external compression inlets, i.e., to reduce the curvature of the cowl lip, we need to allow for internal shock reflections and obviously an internal throat. An inlet with these characteristics exhibits a mixed compression behavior and hence earns the name. A schematic drawing of a mixed compression inlet is shown in Fig. 5.46.

The operational Mach number of a mixed compression inlet extends to ∼ Mach 4.0–5.0. Since we can allow multiple ramps both outside and inside of the inlet without a concern on a cowl wave drag penalty, we may achieve a rather high total pressure recovery characteristic with this inlet. However, the presence of an internal throat indicates an inherent starting problem. The presence of external compression surfaces indicate a Buzz instability in the unstart mode. The shocks that are created and reflected from solid surfaces inside the duct alert us to the shock boundary layer interaction problems with flow separation and engine face distortion as serious by-products. The sharp cowl lips or the centerbody indicate sensitivity to misaligned flow, which is expected with takeoff, climb, landing, maneuver, and side wind. Now, let us summarize the challenge areas for mixed compression inlet design/operation:

1. Starting problem
2. *Buzz* in unstart mode (subcritical instability)
3. Shock-induced separated boundary layers and distortion
4. Unstart problem, (i.e., terminal shock stability margin)
5. Sensitivity to angle of attack, sideslip

Starting problem demands a variable throat inlet for practical purposes, as discussed earlier in the internal compression inlet section. In case of axisymmetric inlets, the centerbody assumes a translational displacement to open and close the throat area. Two-dimensional compression ramps, however, may be collapsed or deployed using a hinged flap design with hydraulic actuators. The origin of Buzz in unstart mode is the boundary layer separation off of compression ramps. Consequently, boundary layer suction and throat bleed need to be employed to overcome this instability. The separated boundary layers inside the duct, as a result of shock boundary layer interaction, also dictate a boundary layer bleed strategy through porous surfaces or slots to be incorporated in critical duct areas. The unstart problem, or providing an adequate stability margin for the normal shock in the divergent portion of the inlet, demands a shock position sensor as well as fast acting valves to alleviate the backpressure problem. A shock position sensor may be a pair of pitot tubes mounted in the desired location on the wall of the diffuser downstream of the throat. The terminal shock is to remain between these two pitot tubes. In the event of increasing backpressure, where the shock is pushed toward the throat (and potentially beyond it to the unstart position), the pitot tube reading suddenly drops

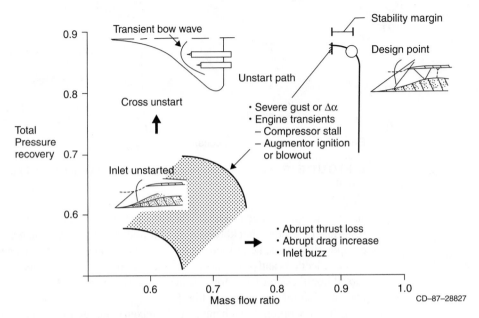

■ **FIGURE 5.47** Performance plot of a mixed-compression inlet (from Strack, 1987)

and indicates the shock crossing. Fast acting valves are to be opened to dump the excessive flow from the downstream position so as to relieve the backpressure and stabilize the shock. Since the two pitot tubes confine the position of the shock, they are also called a *shock trap*.

The sensitivity to flow angularity is caused by the sharp cowl and centerbody configuration of these inlets. Although supersonic design demands a sharp leading edge of the cowl, wing, etc., the low-speed operation is hampered by flow separation off of sharp corners/edges. Hence, articulating cowl, movable lip, cowl slot, spike bleed, and other boundary layer control techniques must be used to overcome the inherently poor low-speed characteristics of these inlets. Figure 5.47 shows the performance plot of a mixed-compression inlet (from Strack, 1987).

Schweikhard and Montoya (1974) also present valuable operational information on mixed compression inlets that is recommended for further reading.

5.15   Supersonic Inlet Types and Their Performance—A Review

Supersonic inlets invariably involve shocks to decelerate the flow to subsonic engine face Mach numbers. For subsonic throughflow engines, as in all the existing conventional engines, the deceleration to subsonic Mach number implies the placement of a normal shock somewhere in the inlet stream. The simplest supersonic diffuser, hence, utilizes a single normal shock to achieve the desired flow deceleration. This is a good and simple solution for low supersonic Mach numbers such as $M_0 \leq 1.6$. Beyond this Mach number, the normal shock deceleration becomes inefficient, as it creates excessive total pressure loss. To extend the operational Mach numbers to ~2.5, a series of oblique shocks decelerate the flow through external compression ramps or cones prior to the normal shock appearance near the cowl lip. This type of inlet was called an external compression inlet and exhibited good internal and external flow qualities for up to $M_0 \sim 2.5$. The cowl wave drag was the limiting performance parameter for the external compression inlet. We also discovered that when the normal shock is pushed out onto the ramp (i.e., the subcritical mode), the potential flow separation could cause a shock/flow oscillation that was called *Buzz instability*. To operate without buzz or simply to widen the stability margin of external compression inlets required the boundary layer removal from the compression

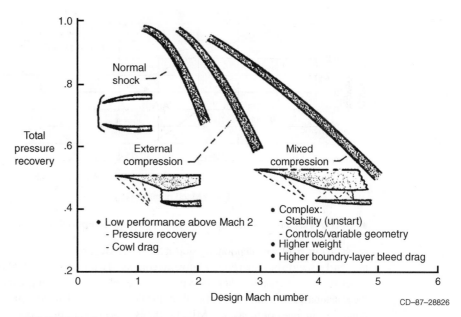

■ **FIGURE 5.48**   Performance of supersonic inlets with flight Mach number (from Strack, 1987)

ramps and incorporating a throat slot for bleed purposes. To reduce the cowl wave drag at higher supersonic Mach numbers, i.e., $M_0 > 2.5$, was to invite internal shock reflections and an internal throat into our design approach. Such inlets were called mixed-compression inlets, which produced the highest overall (internal and external) performance for $2.5 < M_0 < 5.0$. The internal throat created another form of instability, namely, the starting and unstarting instability of the inlet. Variable geometry throat offered a practical starting approach to mixed-compression inlets. To avoid unstart meant controlling, the inlet backpressure through a shock trap with a fast acting bleed valve and a bypass secondary flow control valve integrated in the subsonic diffuser. To manage the engine face distortion levels, in the face of multiple shocks inside the duct, required excessive bleeding of the boundary layer. The thrust penalty for the boundary layer bleed was the price to pay for the mixed-compression inlet. For hypersonic Mach numbers, internal compression inlets offer the lowest engine installation wave drag. In addition, flow deceleration through multiple shocks and their reflections inside a duct is preferred over fewer shock decelerations. However, the normal shock could be altogether eliminated if the engine design allowed for supersonic throughflow. A supersonic throughflow fan and a supersonic combustion ramjet (scramjet) are examples of this kind of modern thinking in hypersonic propulsion. The performance plot of the first three supersonic inlets, as a function of flight Mach number, is shown in Fig. 5.48 (from Strack, 1987).

5.16   Standards for Supersonic Inlet Recovery

Here, we introduce two standard recoveries in existence today, one according to Aircraft Industries Association (AIA) and the second one due to the Department of Defense, known as the MIL-E-5008B standard. These standards that address the shock recovery of supersonic inlets are

$$\pi_d = 1 - 0.1(M_0 - 1)^{1.5} \qquad 1 < M_0 \qquad \text{AIA-Standard} \qquad \textbf{(5.34)}$$

$$\pi_d = 1 - 0.075(M_0 - 1)^{1.35} \qquad 1 < M_0 < 5 \qquad \text{MIL-E-5008B} \qquad \textbf{(5.35)}$$

$$\pi_d = 800/(M_0^4 + 935) \qquad 5 < M_0 \qquad \text{MIL-E-5008B} \qquad \textbf{(5.36)}$$

Now, let us graph the standard recoveries and compare them to a normal shock inlet (Fig. 5.49).

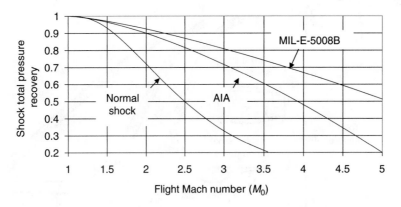

■ **FIGURE 5.49**   **Standards of inlet recovery compared to a normal shock inlet**

In Fig. 5.49, we intentionally stop at Mach 5, as we approach the hypersonic regime. The AIA standard is seen to be rather conservative, as compared with the MIL-E-5008B standard. The normal shock recovery, as noted earlier, loses its attractiveness around Mach 1.6. It is interesting to note that these standards date back to 1950s! With the development of smart structures and advanced flow control techniques of today, the MIL-E-5008B is deemed to be a conservative standard as well. These inlet recovery standards are reproduced here for reference and historical perspective.

Further readings on supersonic inlets and aircraft integration (references 19, 32 and 35) are recommended.

XB-70 (Courtesy of NASA)

5.17   Exhaust Nozzle

A modern aircraft exhaust system is tasked with the following design objectives:
- Efficient expansion of the engine gases to ambient pressure
- Low-installation drag
- Low noise
- Low-cooling requirements
- Efficient thrust reversing capability
- Efficient thrust vectoring capability
- Low observables (radar, thermal, IR signature)

- Light weight system
- Low-cost manufacturing

In this section, we address the parameters that influence the above design goals in a realistic nozzle. First, we review nozzle performance parameters/figures of merit that we learned in our one-dimensional cycle analysis, such as gross thrust, nozzle adiabatic efficiency, and nozzle total pressure ratio. Specialized topics in nozzle cooling, thrust reversing and thrust vectoring, hypersonic nozzles, and exhaust mixers are also presented.

5.18 Gross Thrust

The nozzle contribution to engine thrust is called gross thrust. It comprises two terms, namely, the momentum thrust and the pressure thrust, as we discussed in Chapters 3 and 4.

$$F_g \equiv \dot{m}_9 V_9 + (p_9 - p_0)A_9 \tag{5.37}$$

The first term on the RHS represents the momentum thrust and the second term is the pressure thrust. The gross thrust of a nozzle is maximized when it is operated as fully expanded. A nozzle is fully expanded when the static pressure in the exhaust plane and the ambient static pressure are the same, i.e., $p_9 = p_0$. As we learned earlier, all subsonic jets are perfectly expanded. In choked convergent nozzles as well as supersonic and hypersonic nozzles, the exit static pressure may be different than the ambient pressure. These are referred to as the under- or overexpanded nozzles.

5.19 Nozzle Adiabatic Efficiency

The ratio of actual kinetic energy at the nozzle exit to the ideal kinetic energy that emerges from an isentropic expansion in the nozzle is defined as the nozzle adiabatic (or isentropic) efficiency, i.e.,

$$\eta_n \equiv \frac{h_{t7} - h_9}{h_{t7} - h_{9s}} = \frac{V_9^2/2}{V_{9s}^2/2} \tag{5.38}$$

The fraction of the gas thermal energy that is not converted into directed kinetic energy at the exit plane of the nozzle is dissipated into heat in the boundary layers and across shocks. Thus, actual exhaust gas has a higher exit static temperature than the exhaust gas emerging from an isentropic nozzle, i.e., $T_9 > T_{9s}$. These behaviors are best described using a T–s diagram (see Fig. 5.50).

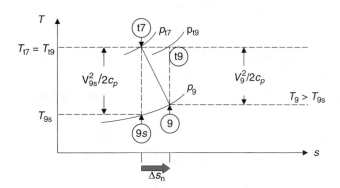

■ **FIGURE 5.50** **Adiabatic expansion process in a nozzle with loss**

5.20 Nozzle Total Pressure Ratio

We had identified nozzle total pressure ratio π_n as a figure of merit. Its definition is

$$\pi_n \equiv p_{t9}/p_{t7} \tag{5.39}$$

This parameter is a direct measure of flow irreversibility, i.e., due to friction and shock, in the nozzle. Recall that the nondimensional entropy rise in an adiabatic process is negative natural logarithm of total pressure ratio, namely,

$$\frac{\Delta s}{R} = -\ln\left(\frac{p_{t9}}{p_{t7}}\right) \tag{5.40}$$

5.21 Nozzle Pressure Ratio (NPR) and Critical Nozzle Pressure Ratio (NPR$_{crit.}$)

An important operational parameter of a nozzle is the NPR. It is defined as

$$\mathrm{NPR} \equiv \frac{p_{t7}}{p_0} \tag{5.41}$$

This parameter signifies the nozzle flow expansion potential (i.e., from the stagnation or total state at 7 to the ambient static state at 0). Since p_0 is the ambient pressure, nozzle pressure ratio is a strong function of altitude, demanding higher area expansion ratios of the nozzle with altitude (hence, a variable geometry). The choice of a convergent or a convergent–divergent nozzle can also be guided based on the magnitude of this parameter, i.e., the nozzle pressure ratio.

A critical nozzle pressure ratio leads to a sonic throat with $p_9 = p_8 = p_0$ for a convergent nozzle. It may be thought of as the minimum nozzle pressure ratio that chokes an expanding nozzle. We may write

$$p_{t8} = p_8\left(1 + \frac{\gamma - 1}{2}M_8^2\right)^{\frac{\gamma}{\gamma-1}} = p_0\left(\frac{\gamma + 1}{2}\right)^{\frac{\gamma}{\gamma-1}} \tag{5.42}$$

By introducing a total pressure loss from nozzle inlet to the throat, we may recast Eq. 5.42 as

$$\frac{p_{t7}}{p_0} = \mathrm{NPR}_{crit.} = \frac{1}{\pi_{cn}}\left(\frac{\gamma + 1}{2}\right)^{\frac{\gamma}{\gamma-1}} \tag{5.43}$$

where π_{cn} is the convergent nozzle total pressure ratio. Now, let us graph this equation for different gas temperature/composition and loss parameter in Fig. 5.51(a) (shown for a convergent nozzle).

■ **FIGURE 5.51**
Critical nozzle pressure ratio for different gas temperature/composition and internal total pressure loss parameter (a *Rule of Thumb* is that the critical nozzle pressure ratio is ~2)

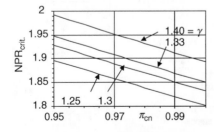

(a) Critical pressure ratio for a convergent nozzle

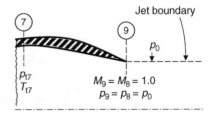

(b) Convergent nozzle operating at NPR$_{crit}$

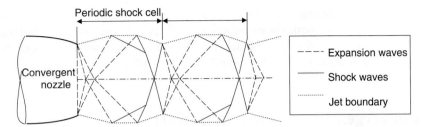

■ **FIGURE 5.52**   An underexpanded convergent nozzle with a periodic shock cell in its exhaust plume

A convergent nozzle that operates with a higher pressure ratio than the critical will still have an exit Mach number of 1, except the static pressure in the exit plane, p_9, will be higher than the ambient static pressure p_0. To relieve the high pressure in the jet at the exit plane, centered expansion waves at the nozzle exit cause a flow deflection with subsequent shock wave formations. The exhaust flowfield of a convergent nozzle that operates with $NPR > NPR_{crit}$ is schematically shown in Fig. 5.52.

5.22 Relation Between Nozzle Figures of Merit, η_n and π_n

The two figures of merit for an adiabatic nozzle flow, η_n and π_n, are related to each other, via

$$\eta_n = \frac{\left\{ NPR\left(\dfrac{p_0}{p_9}\right) \right\}^{\frac{\gamma-1}{\gamma}} - \pi_n^{-\frac{\gamma-1}{\gamma}}}{\left\{ NPR\left(\dfrac{p_0}{p_9}\right) \right\}^{\frac{\gamma-1}{\gamma}} - 1} \quad \text{or} \quad \eta_n = \frac{\{NPR\}^{\frac{\gamma-1}{\gamma}} - \pi_n^{-\frac{\gamma-1}{\gamma}}}{\{NPR\}^{\frac{\gamma-1}{\gamma}} - 1} \quad \text{for } p_9 = p_0 \quad \textbf{(5.44)}$$

Figure 5.53 is a graph of Eq. 5.44 for a perfectly expanded nozzle, $p_9 = p_0$.

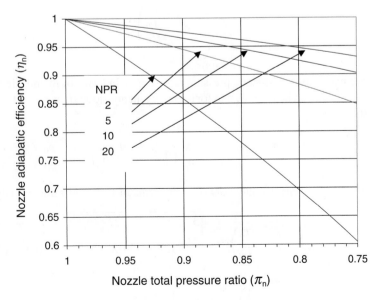

■ **FIGURE 5.53**   Exhaust nozzle figures of merit for a perfectly expanded nozzle with $\gamma = 1.30$

We note that the nozzle adiabatic efficiency improves with the nozzle pressure ratio, as expected. Furthermore, the nozzle adiabatic efficiency is diminished with a decreasing nozzle total pressure ratio, as shown in Fig. 5.53. With no loss of total pressure in the nozzle (i.e., $\pi_n = 1$), the adiabatic efficiency is 100%.

5.23   A Convergent Nozzle or a De Laval?

When do we choose a convergent–divergent nozzle with its added weight and complexity over a simple, lightweight convergent nozzle? To answer this question, we need to assess the gross thrust performance gain of a De Laval nozzle over a convergent nozzle at a given nozzle pressure ratio. Alternatively, we need to assess the gross thrust penalty of a convergent nozzle operating in an underexpanded mode. Assuming an adiabatic and reversible, i.e., isentropic, expansion process in the convergent–divergent nozzle, we can write

$$V_9 = \sqrt{2h_{t8}\left(1 - \frac{T_9}{T_{t9}}\right)} = \sqrt{2h_{t8}\left[1 - \left(\frac{p_9}{p_{t9}}\right)^{\frac{\gamma-1}{\gamma}}\right]} = \sqrt{2h_{t8}\left[1 - \left(\frac{p_0}{p_{t7}}\right)^{\frac{\gamma-1}{\gamma}}\right]} \quad (5.45)$$

The gross thrust of a perfectly expanded convergent–divergent nozzle is

$$F_{g\text{-CD}} = \dot{m}_9 V_9 \quad (5.46)$$

A convergent nozzle will produce a choked exit flow with $M_9 = 1$ for all nozzle pressure ratios beyond the critical (i.e., NPR of roughly ~2). Therefore,

$$F_{g\text{-Conv.}} = \dot{m}_9 V_9 + (p_9 - p_0)A_9 = \dot{m}_9 V_9 \left(1 + \frac{p_9(1 - p_0/p_9)A_9}{\rho_9 V_9^2 A_9}\right)$$

$$= \dot{m}_9 V_9 \left(1 + \frac{1 - \dfrac{p_0}{p_9}}{\gamma M_9^2}\right) \quad (5.47)$$

We may express the static pressure ratio p_0/p_9 in terms of the nozzle pressure ratio and exit Mach number, i.e., 1 (note that for a convergent nozzle stations 8 and 9 coincide, i.e., throat is at the exit).

$$\frac{p_0}{p_9} = \frac{p_0}{p_{t9}}\frac{p_{t9}}{p_9} = \frac{p_0}{p_{t9}}\left(1 + \frac{\gamma - 1}{2}M_9^2\right)^{\frac{\gamma}{\gamma-1}} = \frac{p_0}{p_{t7}}\left(\frac{\gamma + 1}{2}\right)^{\frac{\gamma}{\gamma-1}} \quad (5.48)$$

The exit velocity V_9 may be written in terms of total and static enthalpy according to

$$V_9 = \sqrt{2h_{t9}\left(1 - \frac{T_9}{T_{t9}}\right)} = \sqrt{2h_{t9}\left(1 - \frac{1}{1 + \dfrac{\gamma - 1}{2}M_9^2}\right)} = \sqrt{2h_{t9}\left(1 - \frac{2}{\gamma + 1}\right)}$$

$$= \sqrt{2h_{t9}\left(\frac{\gamma - 1}{\gamma + 1}\right)} \quad (5.49)$$

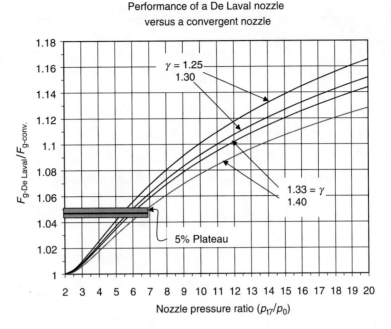

Performance of a De Laval nozzle versus a convergent nozzle

■ **FIGURE 5.54** **Thrust producing capability of an ideal convergent–divergent nozzle in comparison to a convergent nozzle**

Now, taking the ratio of the gross thrust developed by a perfectly expanded De Laval nozzle and the gross thrust produced by a convergent nozzle that is underexpanded, we have

$$
\frac{F_{g-CD}}{F_{g-Conv.}} = \sqrt{\frac{1 - NPR^{-\frac{\gamma-1}{\gamma}}}{\frac{\gamma-1}{\gamma+1}} \cdot \frac{\gamma}{\gamma + \left(1 - \left(\frac{\gamma+1}{2}\right)^{\frac{\gamma}{\gamma-1}} \cdot NPR^{-1}\right)}}
\tag{5.50}
$$

The two parameters that affect the ratio of gross thrusts in Eq. 5.50 are the nozzle pressure ratio and the gas ratio of specific heats γ. We have plotted this equation in Fig. 5.54 where we note that a ~5% gross thrust gain is achieved by a De Laval nozzle over a simple convergent nozzle at nozzle pressure ratios of ~6. In the case of a hot exhaust gas, i.e., $\gamma \sim 1.30$, we observe that the 5% gain in nozzle thrust is achieved at a lower NPR, namely, 5.5. Hence, we may conclude that for a nozzle pressure ratio >5.5, it *pays* (through a gross thrust enhancement of at least 5%) to choose a convergent–divergent nozzle over a convergent nozzle.

Our discussion of exhaust nozzle performance to this point was based on one-dimensional flow analysis. Now, we are ready to address the influence of multidimensionality of real fluid environments, such as the influence of fluid viscosity through boundary layer formation and the influence of nonaxial exit flow on nozzle gross thrust.

EXAMPLE 5.6

An exhaust nozzle has a pressure ratio of 8, i.e., $p_{t7}/p_0 = 8$. The ratio of specific heats for the gas is $\gamma = 1.3$. Calculate percent increase in gross thrust if we were to expand the gas in an ideal convergent–divergent nozzle as compared with a convergent nozzle.

SOLUTION

The gross thrust of a perfectly expanded convergent–divergent nozzle is

$$F_{g,C-D} = \dot{m}_9 V_9$$

The ideal exhaust velocity is derived in Eq. 5.45 to be

$$V_9 = \sqrt{2h_{t8}\left[1 - \left(\frac{p_0}{p_{t7}}\right)^{\frac{\gamma-1}{\gamma}}\right]} = \sqrt{2h_{t8}(0.3811)}$$

The gross thrust in a convergent nozzle is derived in Eq. 5.47 to be

$$F_{g-Conv.} = \dot{m}_9 V_9 \left(1 + \frac{1 - \dfrac{p_0}{p_9}}{\gamma M_9^2}\right)$$

$$= \dot{m}_9 V_9 [1 + (1 - p_0/p_9)/\gamma]$$

The exit Mach number $M_9 = 1.0$ and $p_9 > p_0$. We derived an expression for the static pressure ratio p_0/p_9 in Eq. 5.48 in terms of NPR to be

$$\frac{p_0}{p_9} = \frac{p_0}{p_{t7}}\left(\frac{\gamma+1}{2}\right)^{\frac{\gamma}{\gamma-1}} \approx 0.229$$

The nozzle exit velocity V_9 is $V_9 = \sqrt{2h_{t8}(\gamma-1)/(\gamma+1)} = \sqrt{2h_{t8}(0.1304)}$

Therefore, the ratio of the two gross thrusts is

$$\frac{F_{g,C-D}}{F_{g-Conv}} = \frac{V_{9C-D}}{V_{9-con}[1 + (1 - 0.229)/1.3]}$$

$$\approx \sqrt{\frac{0.381}{0.1304}\frac{1}{1.59}}$$

$$\approx 1.073 \implies \boxed{7.3\% \text{ increase in gross thrust}}$$

5.24   The Effect of Boundary Layer Formation on Nozzle Internal Performance

The presence of a boundary layer on a surface represents a displacement thickness δ^*, which acts as a *blockage* of the geometric flow area. There is also a total pressure loss associated with the boundary layer formation on the wall, namely, $p_{t8}/p_{t7} < 1$. The combined effect of these parameters on the mass flow rate is represented through a discharge coefficient C_{D8} as

$$C_{D8} \equiv \frac{\dot{m}_8}{\dot{m}_{8i}} = \frac{A_{8\,eff}}{A_{8geo}} \cong \frac{p_{t8}}{p_{t7}} \tag{5.51}$$

where the numerator is the actual mass flow rate through the nozzle throat and the denominator is the ideal mass flow rate based on geometrical throat area.

5.25   Nozzle Exit Flow Velocity Coefficient

We may define a velocity coefficient that measures the extent of viscous flow losses in the exhaust plane, i.e., let us define C_V as

$$C_V \equiv \frac{V_9}{V_{9i}} \tag{5.52}$$

where the numerator is the actual exhaust velocity and the denominator is the ideal exhaust velocity with no loss of total pressure $p_{t9} = p_{t8}$. Therefore the V_{9i} is the ideal exit velocity if only the divergent section of the nozzle were isentropic. Earlier in the definition of the nozzle adiabatic efficiency, we defined a V_{9s} that was achieved if the entire nozzle were isentropic.

A definition sketch of the three different exhaust velocities, V_9, V_{9s}, and V_{9i}, is very helpful at this point and is shown in Fig. 5.55.

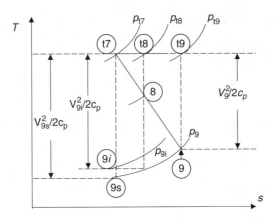

■ **FIGURE 5.55**   Definition sketch of the states of gas defined in an exhaust nozzle

After some manipulation, we may show the following expression for the velocity coefficient:

$$C_V = \sqrt{\frac{\left[1 - \left(\dfrac{p_9/p_0}{\text{NPR} \cdot \pi_n}\right)^{\frac{\gamma-1}{\gamma}}\right]}{\left[1 - \left(\dfrac{p_9/p_0}{\text{NPR} \cdot C_{D8}^2}\right)^{\frac{\gamma-1}{\gamma}}\right]}} \tag{5.53}$$

EXAMPLE 5.7

The total temperature of the gas entering a convergent–divergent nozzle is $T_{t7} = 900\,\text{K}$. The ratio of specific heats of the gas is $\gamma = 1.3$ and $c_p = 1,243.7\,\text{J/kg} \cdot \text{K}$. The nozzle pressure ratio $p_{t7}/p_0 = 8.0$, the convergent portion of the nozzle has a total pressure ratio $p_{t8}/p_{t7} = 0.98$, and the divergent section total pressure ratio is $p_{t9}/p_{t8} = 0.95$. For $p_9 = p_0$, calculate

(a) V_9 in m/s
(b) V_{9s} in m/s
(c) V_{9i} in m/s and
(d) The velocity coefficient C_V

SOLUTION

The ratio of the nozzle exit total pressure to static pressure is written in a chain rule as

$$\frac{p_{t9}}{p_9} = \frac{p_{t9}}{p_{t8}} \frac{p_{t8}}{p_{t7}} \frac{p_{t7}}{p_0} \frac{p_0}{p_9} = (0.95)(0.98)(8)(1) \approx 7.448$$

The exit Mach number M_9 is related to the above ratio, following:

$$M_9 = \sqrt{\frac{2}{\gamma-1}\left[\left(\frac{p_{t9}}{p_9}\right)^{\frac{\gamma-1}{\gamma}} - 1\right]} \approx 1.982$$

The nozzle exit static temperature is

$$T_9 = \frac{T_{t9}}{1 + (\gamma-1)M_9^2/2} = \frac{900\,\text{K}}{1 + 0.15(1.982)^2} \approx 566.2\,\text{K}$$

The speed of sound in the exit plane is

$$a_9 = \sqrt{(\gamma-1)c_p T_9} = \sqrt{0.3(1243.7)(566.2)} \approx 459.6\,\text{m/s}$$

Therefore, the nozzle exit velocity is $V_9 = M_9 a_9 = 1.982(459.6\,\text{m/s})$

$$V_9 \approx 911\,\text{m/s}$$

Now, we calculate V_{9s} from the total and static enthalpy h_{t7} and h_{9s}, respectively, according to

$$V_{9s} = \sqrt{2(h_{t7} - h_{9s})} = \sqrt{2h_{t7}\left[1 - \left(\frac{p_9}{p_{t7}}\right)^{\frac{\gamma-1}{\gamma}}\right]}$$

$$= \sqrt{2(1243.7)(900)[1 - 8^{-0.3/1.3}]} \approx 923.7\,\text{m/s}$$

To calculate the V_{9i}, we write the ideal expansion from p_{t8} to p_9, as shown in Fig. 5.55, for which we need the ratio p_{t8}/p_9

$$\frac{p_{t8}}{p_9} = \frac{p_{t8}}{p_{t7}}\frac{p_{t7}}{p_0}\frac{p_0}{p_9} = (0.98)(8)(1) \approx 7.84$$

$$V_{9i} = \sqrt{2(h_{t8} - h_{9i})} = \sqrt{2h_{t8}\left[1 - \left(\frac{p_9}{p_{t8}}\right)^{\frac{\gamma-1}{\gamma}}\right]}$$

$$= \sqrt{2(1243.7)(900)[1 - 7.84^{-0.3/1.3}]} \approx 920.2\,\text{m/s}$$

The nozzle velocity coefficient, $C_V = V_9/V_{9i} = 911/920.2 \approx 0.99$

5.26 Effect of Flow Angularity on Gross Thrust

For all nozzle geometries other than a bell-shaped nozzle, the exhaust velocity field, i.e., exit momentum, is nonaxial and hence it does not entirely contribute to an axial thrust. Let us examine a conical nozzle first for simplicity. The flow that emerges from a conical nozzle is a divergent jet flow, as shown in Fig. 5.56. Due to its radial velocity pattern, the jet flow may be seen as that of emerging from a virtual origin, somewhat like a source flow where the velocity is constant at $r = \text{constant}$. In this case at $r = R$, $V = V_9$, therefore $V = V_9$ over a spherical dome or cap.

We define an angularity loss coefficient as the ratio of axial (momentum) thrust to the momentum thrust of an equivalent bell-shaped nozzle:

$$F_g = C_A \dot{m}_9 V_9 + (p_9 - p_0)A_9 \tag{5.54}$$

where C_A is

$$C_A \equiv \frac{\int V_9 \cos\theta\, d\dot{m}}{\dot{m}_9 V_9} \tag{5.55}$$

For the element of mass flow rate $d\dot{m}$, we take the exit surface to be spherical in shape, such that the velocity vector V_9 is normal to it. The element of area on the dome is

$$dA = 2\pi R \sin\theta \cdot R\,d\theta \tag{5.56}$$

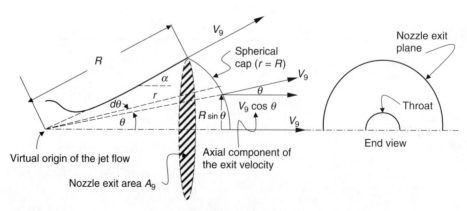

■ **FIGURE 5.56** Working sketch used to derive flow angularity loss (or divergence correction factor) for a conical nozzle of half angle α in attached exit flow

Hence, the element of mass flow rate at the nozzle exit is

$$dm = \rho_9 V_9 dA = 2\pi R^2 \rho_9 V_9 \sin\theta d\theta \qquad (5.57)$$

We integrate this equation to get the entire mass flow rate from the nozzle, namely,

$$\dot{m}_9 = \int_0^\alpha 2\pi R^2 \rho_9 V_9 \sin\theta d\theta = 2\pi R^2 \rho_9 V_9 (1 - \cos\alpha) \qquad (5.58)$$

Now, let us evaluate the axial momentum thrust, as

$$\int V_9 \cos\theta dm = \int_0^\alpha 2\pi R^2 \rho_9 V_9^2 \sin\theta\cos\theta d\theta = \pi R^2 \rho_9 V_9^2 \sin^2\alpha \qquad (5.59)$$

Substitute the axial momentum expression in the definition of the flow angularity or the divergence correction factor, to get

$$C_A = \frac{\sin^2\alpha}{2(1 - \cos\alpha)} = \frac{1 + \cos\alpha}{2} \qquad \boxed{C_{A_{conical}} = \frac{1 + \cos\alpha}{2}} \qquad (5.60)$$

EXAMPLE 5.8

Calculate and graph the divergence correction factor C_A for a conical nozzle with the exit flow angles in the range of 0–44°.

SOLUTION

We first make a spreadsheet for C_A versus α from Eq. 5.60 and then graph the resulting table. The result is shown in Fig. 5.57.

For cone angles of up to $\sim 30°$ (or half-cone angle of $\sim 15°$), we note that the divergence loss coefficient is rather insignificant with $C_A > 0.98$. The main advantage of a conical nozzle is in the low-cost manufacturing and for a modest cone angle; the thrust penalty is less than 2%. We also note that the flow angularity correction (Eq. 5.60) is purely geometrical and hence gas composition, i.e., γ, does not enter the correction process.

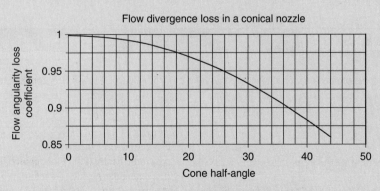

■ **FIGURE 5.57** Flow angularity loss coefficient for a conical nozzle

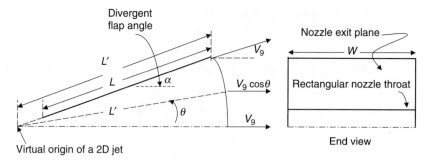

■ **FIGURE 5.58** Definition sketch for divergence loss factor in a 2D-CD nozzle

For a two-dimensional convergent–divergent nozzle, we can derive the angularity loss factor, in a straightforward manner, using the definition sketch, Fig. 5.58, as a tool.

We take the element of area at the exit to be a cylindrical surface normal to exit flow, i.e.,

$$dA = L'd\theta \cdot W \tag{5.61}$$

Therefore, the element of mass flow rate at the nozzle exit is

$$d\dot{m} = \rho_9 V_9 L'd\theta \cdot W \tag{5.62}$$

The total mass flow rate is the integral of elemental mass flow rate, i.e.,

$$\dot{m} = 2\int_0^\alpha \rho_9 V_9 L'Wd\theta = 2\rho_9 V_9 L'W\alpha \tag{5.63}$$

Hence, the correction factor C_A then becomes

$$C_A \equiv \frac{\int V_9 \cos\theta d\dot{m}}{\dot{m}_9 V_9} = \frac{2\int_0^\alpha V_9 \cos\theta \rho_9 V_9 L'Wd\theta}{2\rho_9 V_9^2 L'W\alpha} = \frac{\sin\alpha}{\alpha} \qquad \boxed{C_{A_{2D-CD}} = \frac{\sin\alpha}{\alpha}} \tag{5.64}$$

Note that the α in denominator of C_A must be in *radians* (not degrees).

EXAMPLE 5.9

Calculate and graph the divergence correction factor $C_{A,2D\text{-}CD}$ for a two-dimensional nozzle with the exit flow angles in the range of 0–44°. Then, compare the angularity correction of the conical nozzle to the two-dimensional C–D nozzle.

SOLUTION

We first make a spreadsheet for $C_{A,2D\text{-}CD}$ versus α from Eq. 5.64 and then graph the resulting table. The result is shown in Fig. 5.59.

By comparing Figs. 5.57 and 5.59, we note that a two-dimensional C–D nozzle experiences a smaller flow angularity loss than a similar conical nozzle of the same wall angle.

This result was expected, as the conical nozzle flow diverges axisymmetrically (in 3D) as opposed to a single direction in a 2D rectangular nozzle. Therefore, a conical nozzle develops less axial thrust than a 2D-C–D nozzle of the same wall divergence. Figure 5.60 shows the conical and 2D-CD nozzle performance on a same plot for comparison purposes.

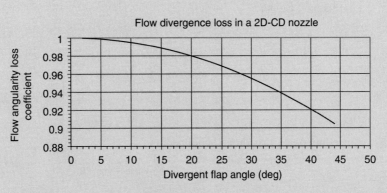

■ **FIGURE 5.59**   Flow angularity loss coefficient for a 2D-C–D nozzle

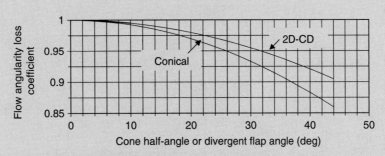

■ **FIGURE 5.60**   Divergence loss of a conical nozzle and a 2D-CD nozzle, assuming attached exit flow

5.27   Nozzle Gross Thrust Coefficient C_{fg}

We define a nozzle gross thrust coefficient C_{fg} according to

$$C_{\text{fg}} \equiv \frac{F_{\text{g-actual}}}{F_{\text{g-ideal}}} \tag{5.65}$$

where the actual gross thrust is

$$F_{\text{g-actual}} = C_{\text{A}}\dot{m}_9 V_9 + (p_9 - p_0)A_9 = \dot{m}_9 V_9 \left(C_{\text{A}} + \frac{p_9 A_9 (1 - p_0/p_9)}{\rho_9 A_9 V_9^2} \right)$$

$$= \dot{m}_9 V_9 \left[C_{\text{A}} + \frac{1 - \dfrac{p_0}{p_9}}{\gamma M_9^2} \right] \tag{5.66}$$

Now, by replacing the momentum thrust with the ideal thrust and introducing the loss coefficients, we get

$$F_{\text{g-actual}} = C_{\text{D8}} C_{\text{V}} \dot{m}_{8i} V_{9i} \left[C_{\text{A}} + \frac{1 - \dfrac{p_0}{p_9}}{\dfrac{2\gamma}{\gamma - 1} \left[\left(\dfrac{p_{t9}}{p_9} \right)^{\frac{\gamma - 1}{\gamma}} - 1 \right]} \right] \tag{5.67}$$

Since the ideal gross thrust is defined as

$$F_{\text{g-ideal}} \equiv \dot{m}_{8i} V_{9s}$$

where V_{9s} is the isentropic exit velocity of a perfectly expanded jet, which may be written as

$$V_{9s} = \sqrt{2 h_{t8} \left(1 - \frac{T_{9s}}{T_{t9}} \right)} = \sqrt{2 h_{t8} \left[1 - \left(\frac{p_0}{p_{t9}} \right)^{\frac{\gamma - 1}{\gamma}} \right]} = \sqrt{2 h_{t8} \left[1 - \left(\frac{p_0}{p_{t7}} \right)^{\frac{\gamma - 1}{\gamma}} \right]}$$

$$= \sqrt{2 h_{t8} \left[1 - \text{NPR}^{-\frac{\gamma - 1}{\gamma}} \right]} \tag{5.68}$$

Hence, the gross thrust coefficient becomes

$$C_{\text{fg}} = C_{\text{D8}} C_{\text{V}} \frac{V_{9i}}{V_{9s}} \left[C_{\text{A}} + \frac{\dfrac{\gamma - 1}{2\gamma} \left(1 - \dfrac{p_0}{p_9} \right)}{\left[\left(\pi_n \text{NPR} \dfrac{p_0}{p_9} \right)^{\frac{\gamma - 1}{\gamma}} - 1 \right]} \right] \tag{5.69}$$

$$= C_{\text{D8}} C_{\text{V}} \sqrt{\frac{1 - \left(\dfrac{p_9 / p_0}{\text{NPR} \cdot C_{\text{D8}}^2} \right)^{\frac{\gamma - 1}{\gamma}}}{1 - \text{NPR}^{-\frac{\gamma - 1}{\gamma}}}} \left[C_{\text{A}} + \frac{\dfrac{\gamma - 1}{2\gamma} \left(1 - \dfrac{p_0}{p_9} \right)}{\left(\pi_n \text{NPR} \dfrac{p_0}{p_9} \right)^{\frac{\gamma - 1}{\gamma}} - 1} \right]$$

Now, let us put this complex looking equation into perspective. An imperfect expansion in the nozzle where $p_9 \neq p_0$ leads to a pressure thrust term, which is essentially the last term in the bracket. The momentum thrust is affected by total pressure losses and flow angularity at the exit, which is the first term of the bracket. These losses are seen as C_{D8}, C_{V}, and C_{A} distributed throughout the nozzle, as shown in Fig. 5.61.

5.28 Overexpanded Nozzle Flow—Shock Losses

There is an aspect of nozzle performance, which is hampered by the presence of shocks inside the nozzle. This behavior, which is represented by a severe case of overexpansion, is not readily apparent from Eq. 5.66. An overexpanded nozzle has an area expansion ratio A_9 / A_8 in excess of that needed for a perfect expansion. This results in an exit static pressure, which falls below the ambient pressure $p_9 / p_0 < 1$. The need for an abrupt adjustment of the static pressure in the

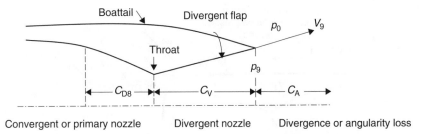

■ **FIGURE 5.61**   Convergent–divergent nozzle with the primary, secondary, and exit flow, each identified with a loss parameter, C_{D8}, C_V, and C_A, respectively

emerging supersonic flow causes a shock formation on the nozzle lip. The mild cases of overexpansion are resolved through oblique shock waves on the nozzle lip. A greater mismatch between the static pressures of the jet and ambient calls for a stronger shock, which in essence increases the wave angle with respect to stream. Eventually, a normal shock is formed at the nozzle lip. In case the normal shock at the nozzle lip is found incapable of satisfying static pressure continuity across the jet slipstream, the normal shock is then brought inside the nozzle. At this time, the exit flow is subsonic and a potential boundary layer separation needs to be investigated. Let us first examine the flow environment near the exit of an overexpanded nozzle in supersonic flight. Figure 5.62 shows the external flow around a nozzle boattail and the jet geometry dominated by wave formations. The shear layer separating the inner and outer flowfields is labeled as the jet slipstream. Due to flow curvature near the nozzle exit, local static pressure is higher than ambient pressure, i.e., local ambient static pressure near plane 9 is not p_0, which we had assumed all along in our analysis.

Note that we intentionally omitted a drawing of the wall boundary layers on the primary nozzle and divergent flaps to avoid further crowding of Fig. 5.62. This figure was produced, with simplifications, to develop an appreciation of the flow complexities in supersonic exhaust flow. Now, let us graphically examine higher levels of overexpansion in the above nozzle. With larger area ratios, assuming an attached flow, the static pressure at the exit plane drops, demanding a normal shock at the lip. Due to the presence of the wall boundary layer and a curved free shear layer (i.e., the jet slipstream), a lambda shock is formed at the lip, which is graphically depicted in Fig.5.63.

Again for simplicity, the inner boundary layer and the vortex sheet emanating from the junction of lambda shock are not graphed in Fig. 5.63. Here, we clearly see a curved shock formation, which is predominantly normal to the core flow and dominated by a lambda-shock geometry near the wall and exit shear layer. In case of a more severe overexpansion, the shock will be drawn inside the nozzle and a subsonic jet will emerge at the exit. Figure 5.64 examines this kind of overexpansion.

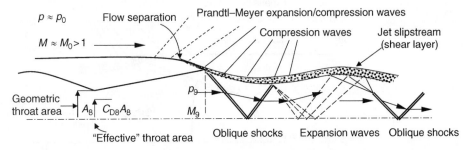

■ **FIGURE 5.62**   A cartoon of the inner and outer flow of an overexpanded nozzle in supersonic flight shows a wavy jet slipstream with a varying static pressure inside the jet

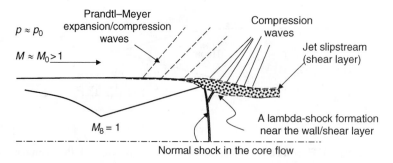

■ **FIGURE 5.63**   **Overexpanded flow from a CD-nozzle shows a lambda-shock formation at the lip**

From Fig. 5.64, we note that a separated boundary layer could alter the picture of exit flow, in particular with the flow direction as well as the effective exit flow area. How much of this complexity can we capture by simple one-dimensional models and correction factors? For example, in our gross thrust coefficient approach to account for viscous losses and nonaxial exit flow, are there any provisions for curved shocks or separated flows, such as demonstrated by some overexpanded nozzle flows? The short answer is no. The long answer starts with a may be! A plausible approach calls for identifying and bracketing off zones of invalidity from our performance charts that we derived based on simple 1D loss models and assumptions. This approach relies, in part, on the experience base of the manufacturing company and other available experimental data in the literature. For example, in separated nozzle flows, the physical meaning of the velocity coefficient C_V and the angularity loss factor C_A are completely lost, based on our definitions and simplifying assumptions. Consequently, the nozzle gross thrust coefficient C_{fg}, as described by Eq. 5.69, becomes at best inaccurate and at worst meaningless. Here, we need to exert judgment when dealing with overexpanded nozzle flows that include a shock inside the nozzle. Let us reproduce some of the criteria found in the literature for a passage shock that causes flow separation. In Kerrebrock (1992), a reference is made to Zukoski (1990) who suggests a separation point criterion in an overexpanded supersonic nozzle according to

$$p_s = p_0/(1 + M_s/2) \tag{5.70}$$

where p_s is the static pressure upstream of the shock, p_0 is the ambient static pressure, and M_s is the gas Mach number upstream of the shock. Over a wide operating range of the nozzle, i.e., NPR ranging between 2 and 50, the separation pressure ratio, Kerrebrock notes that p_s/p_0

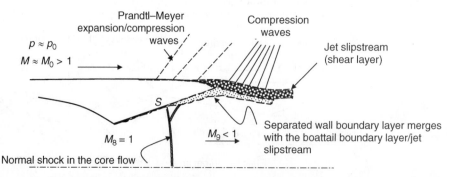

■ **FIGURE 5.64**   **Curved shock inside the nozzle creates a potential for boundary layer separation and produces a subsonic exit flow**

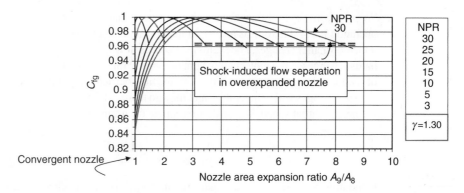

■ **FIGURE 5.65** An ideal nozzle, i.e., $p_t = $ constant, $T_t = $ constant, $C_{D8} = C_V = C_A = 1$, gross thrust coefficient with Zukoski separation criterion in overexpanded mode

ranges between 0.4 and 0.625. Mattingly (1996) recommends a $p_s/p_0 \approx 0.37$, following Summerfield, Foster, and Swan (1954) and Sutton (1992) who recommend a rough separation criterion of $p_s/p_0 \le 0.4$, immediately notes the shortcoming of this approach. In summary, several simple or more sophisticated models for shock separation criterion are introduced, which roughly identify the zones of invalidity of our *simple* nozzle theory. Figure 5.65 shows the ideal gross thrust coefficient with nozzle area ratio as a function of nozzle pressure ratio. The maximum $C_{fg} = 1$ for $p_9/p_0 = 1.0$.

Note that the nozzle overexpanded mode of operation is more prevalent for a rocket engine than an airbreathing aircraft engine. A rocket engine that operates between the ambient pressures of sea level, $\sim$14.7 psia (1 atm.), and near vacuum conditions of space, requires a large nozzle area expansion ratio (A_9/A_8) to optimize its performance. A nozzle, which is optimized for high altitude, then operates in an overexpanded mode in lower altitudes. Consequently, the thrust loss and shock interactions inside the nozzle are serious realities of rocket nozzle design. However, the airbreathing engine of an aircraft invariably operates with a smaller area expansion ratio and hence it experiences less severe over-expansion problems. In the hypersonic regime, the airbreathing and a rocket nozzle attain geometric and operational similarities and thus face the same challenges for off-design operations.

Now, we are ready to address the nozzle throat size and scheduling with afterburner operation. Also, we start the discussion on the nozzle exit area scheduling within the flight envelope and its impact on external drag penalties due to boattail and base flowfield.

5.29 Nozzle Area Scheduling, A_8 and A_9/A_8

The exhaust nozzle throat of a modern aircraft engine operates in a choked mode over a wide operating range within the flight envelope. Consequently, the nozzle throat serves as the low-pressure turbine's backpressure controller or equivalently as the engine aft *throttle station*. With an afterburner in operation, the exhaust gas temperature increases significantly with a subsequent drop of density. To accommodate a lower density gas, a choked nozzle needs to *open up* its throat to pass almost the same mass flow rate as with no afterburner operation. The one-dimensional continuity equation for a choked nozzle throat may be written as

$$\dot{m}_8 = \sqrt{\frac{\gamma_8}{R_8}} \frac{p_{t8}}{\sqrt{T_{t8}}} A_8 \left(\frac{2}{\gamma_8 + 1}\right)^{\frac{\gamma_8+1}{2(\gamma_8-1)}} \tag{5.71}$$

We note from Eq. 5.71 that the term that undergoes the most dramatic change with the afterburner operation is T_{t8} due to heat release in the combustion process. The other terms in the equation undergo smaller changes. For example, the mass flow rate increases by $\sim$2–4% the gas constant R hardly changes as the molecular weight of the new mixture has not changed much by the addition of the fuel; the total pressure drops with the afterburner, but perhaps by 2–5%; the specific heat ratio γ_8 changes with temperature and may go from $\sim$1.33 to $\sim$1.30 or 1.25. In its simplest form, the nozzle throat scheduling with afterburner for a turbojet engine may be written as

$$\frac{A_{8-AB-On}}{A_{8-AB-Off}} \approx \sqrt{\frac{T_{t8-AB-On}}{T_{t8-AB-Off}}} = \sqrt{\frac{\tau_{\lambda AB}}{\tau_t \tau_\lambda}} \tag{5.72}$$

If the aim is to be more exact with the nozzle throat area scheduling, then one must do the bookkeeping of all the terms in the continuity equation more carefully, namely,

$$\frac{A_{8-AB-On}}{A_{8-AB-Off}} \cong \frac{1+f+f_{ab}}{1+f} \sqrt{\frac{\gamma_8'}{\gamma_8} \frac{p_{t8}}{p_{t8}'}} \sqrt{\frac{\tau_{\lambda AB}}{\tau_t \tau_\lambda}} \left[\frac{\left(\frac{\gamma_8'+1}{2}\right)^{\frac{\gamma_8'+1}{2(\gamma_8'-1)}}}{\left(\frac{\gamma_8+1}{2}\right)^{\frac{\gamma_8+1}{2(\gamma_8-1)}}} \right] \tag{5.73}$$

where the primed terms are the afterburner-on quantities.

EXAMPLE 5.10

Calculate and graph the ratio of nozzle throat area with the afterburner on and off for a range of turbine expansion parameters τ_t between 0.45 and 0.65. Keep $\tau_\lambda = 6.0$ and vary $\tau_{\lambda, AB}$ from 6.5 to 9.0. Also, investigate the effect of flow losses on the throat area ratio of the nozzle with/without afterburner operation.

SOLUTION

We calculate the nozzle throat area ratio from Eq. 5.73, for a range of turbine expansion parameter τ_t and afterburner $\tau_{\lambda, AB}$ from 6.5 to 9.0. The resulting spreadsheet is graphed in Fig. 5.66.

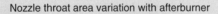

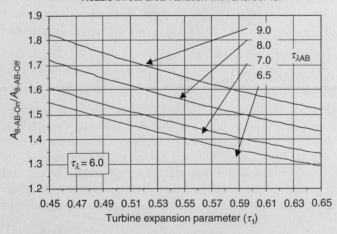

■ **FIGURE 5.66**   Ideal nozzle throat area variation for an afterburning turbojet engine

Figure 5.66 shows that the ideal nozzle throat area must be enlarged by ~30–80% in order to accommodate the afterburner flow condition (of low density, high temperature gas).

To investigate the effect of flow losses and gas property variation, we examine the case of $\tau_\lambda = 6.0$, $\tau_{\lambda, AB} = 8.0$. We assume that $\gamma = 1.33$ and $\gamma' = 1.30$ with and without afterburner operation, respectively. We also assume that the total pressure ratio with and without afterburner operation is $p'_{t8}/p_{t8} = 0.96$ where the primary burner fuel-to-air ratio is $f = 0.02$ and the afterburner fuel-to-air ratio is $f_{ab} = 0.04$. We apply these values to Eq. 5.73 and graph the result in Fig. 5.67 for a range of turbine expansion parameters.

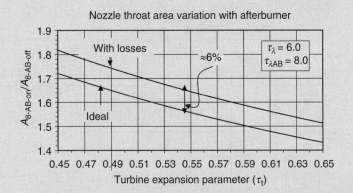

■ **FIGURE 5.67**   The impact of flow losses and gas property variation with afterburner combustion on a turbojet nozzle throat area [for $\gamma = 1.33$, $\gamma' = 1.30$, $p'_{t8}/p_{t8} = 0.96$, $f = 0.02$, $f_{ab} = 0.04$]

5.30   Nozzle Exit Area Scheduling, A_9/A_8

Unlike the choked nozzle throat that required scheduling with an afterburner operation (i.e., a single parameter control), the nozzle exit area scheduling involves a multiparameter optimization study. The reason for this added complexity is rather obvious. Exit area variation of the nozzle impacts external aerodynamics of the nacelle boattail and base flowfields, and hence external drag. Therefore, if we gain an increment of gross thrust due to a better nozzle expansion of say $p_9 \cong p_0$, and as a result incur an increment of drag rise due to nacelle boattail and base flow changes, then the nozzle area ratio A_9/A_8 needs to optimize the installed performance, namely,

$$\Delta F_g - (\Delta D_{\text{Boattail}} - \Delta D_{\text{Base}}) = \Delta F_g - \Delta D_{\text{Nacelle-Aftend}} \qquad (5.74)$$

Here, we are again reminded that the internal optimization of an airbreathing engine inlet and exhaust system is meaningless in isolation. Rather, the engine-*installed performance* should be optimized at each flight phase within the flight envelope of the aircraft. Traditionally, the nozzle exit area of a variable geometry nozzle assumed only two positions. One closed and the other open, very similar to the nozzle throat with/without afterburner. But, the limitation of only two positions for the nozzle exit area seems unnecessary today especially in light of advanced propulsion control system of modern aircraft. Here, the goal of nozzle area scheduling control is to achieve a continuous improvement of installed performance with altitude, speed, aircraft attitude, and acceleration rather than that achieved by a bang–bang (a two-position, open-closed) controller.

The contributors to the aftbody aerodynamics on an engine installation are the contributions from the boattail and base areas of the nacelle. The geometry of nozzle boattail and

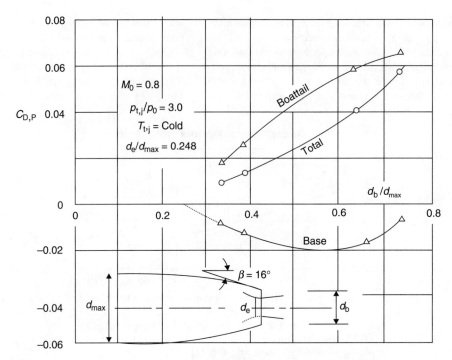

■ **FIGURE 5.68** Pressure drag contribution of the boattail and base as a function of geometry of a propulsion nacelle installation (Bowers and Tamplin, 1985)

base areas and their impact on aftbody drag are shown in Fig. 5.68 (from Bowers and Tamplin, 1985). We note that the base pressure force contribution is in the thrust direction and when integrated with the boattail drag contribution, they yield the "total" pressure drag penalty, as shown in Fig. 5.68. Also, a graphic definition of *base area* is helpful and is shown in Fig. 5.69 (from Whitford, 1987).

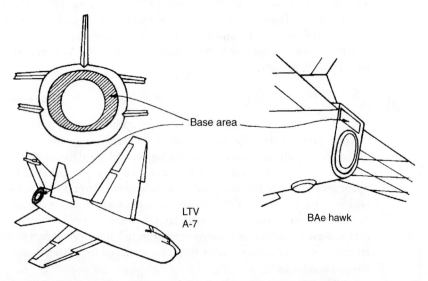

■ **FIGURE 5.69** Definition sketch of candidate nozzle/aftbody base areas (from Whitford, 1987)

Installed nozzle performance optimization

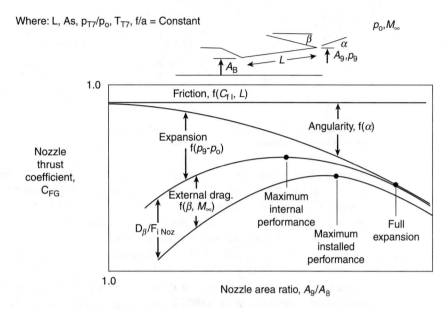

■ **FIGURE 5.70** Internal/external factors contributing to an optimized nozzle performance (from Younghans and Dusa, 1993)

An installed nozzle performance optimization that accounts for the internal as well as external losses is reported by Younghans and Dusa (1993) and reproduced here in Fig. 5.70. First, note that nozzle area ratio A_9/A_8 that is needed for a perfect expansion does not give the best-installed performance. Second, note that each loss parameter has a different area ratio for its optimization. For example, flow angularity loss vanishes for $A_9/A_8 = 1.0$. The expansion loss vanishes at full expansion ($p_9 = p_0$). The friction penalty holds in a somewhat steady manner across the nozzle area ratios. Finally, the internal performance curve is shifted down by the external drag penalties. This leads to a shift in the area ratio for optimized installed performance.

5.31 Nozzle Cooling

The operational temperatures of a modern, multitask exhaust system, with an afterburner, have required an increasing level of active nozzle cooling. In a turbofan engine, the fan stream could provide the required nozzle cooling through an engine bypass system. The supersonic inlet/engine flow match often requires an inlet excess air bypass/bleed system to be integrated into the propulsion system design. The excess air has traditionally been used for engine bay and nozzle cooling purposes. Figure 5.71 shows a schematic drawing of a flow network between an inlet and an exhaust system in a turbofan engine (from Younghans and Dusa, 1993). A typical engine airflow variation with flight Mach number and its comparison with the inlet supply are shown for a Mach 5 aircraft by Dusa (1988).

The primary goal of nozzle cooling is to operate the exhaust system components, i.e., the liner, primary and secondary nozzle walls/ramps, seals, and the actuators at a temperature that allows for reliability of operation and good material life expectancy. The material operating temperature should be well below the melting point of the material. A very useful graph from

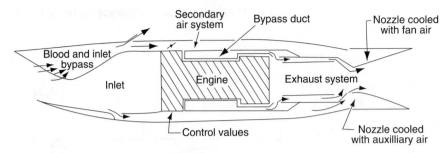

■ **FIGURE 5.71** Schematic drawing of an exhaust system cooling air sources in a turbofan engine (from Younghans and Dusa, 1993)

Wazyniak (2000) is reproduced here in Fig. 5.72 that shows the design temperature levels of jet engine materials. It is useful to examine the design versus the melting point temperature of different materials in addition to material density, as shown in Fig. 5.72. The typical uses of the materials listed are the nickel-based alloys are used in turbines, titanium in compressors and fan blades, aluminum for aircraft and nacelle skin, intermetallics and nonmetals are used in the high-temperature environment of afterburners and exhaust systems. The acceptable operating temperature of metals (e.g., stainless steel) extends to ~1700°F (or ~927°C), whereas the ceramics may operate up to ~2800°F (or ~1538°C). Carbon-carbon offers an extended operating temperature of up to ~3400°F (or ~1870°C). There are also refractory metals, such as tungsten and molybdenum that offer higher operating temperatures than stainless steel (of up to ~2880°F). However, the refractory metals have exhibited manufacturing-related problems as well as operational problems with embrittlement and oxidation.

Now, let us examine the flight total temperature, or ram temperature, as it increases with flight Mach number. Assuming an adiabatic flow of a calorically perfect gas, the inlet air total temperature follows

$$T_{t0} = T_0 \left(1 + \frac{\gamma - 1}{2} M_0^2 \right) \tag{5.75}$$

Design temperature levels of jet engine materials

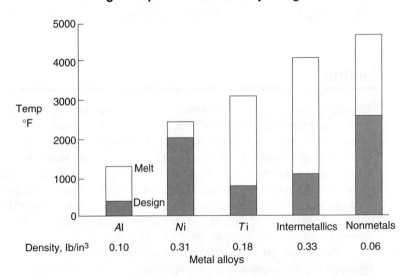

■ **FIGURE 5.72** Melting point/design temperature characteristics of various jet engine materials (from Wazyniak, 2000)

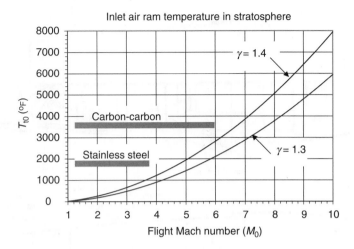

■ **FIGURE 5.73** **Stagnation temperature variation with flight Mach number in stratosphere (50 kft < alt. < 70 kft)**

Now assuming the supersonic/hypersonic flight occurs in the stratosphere (between 50 and 70 kft) where the static temperature is $\sim -60°F = $ constant, then the air ram temperature with flight Mach number varies parabolically according to

$$T_{t0} \cong 400 + 80M_0^2 \quad (\text{in } °R) \qquad \gamma = 1.4 \qquad \textbf{(5.76-a)}$$

$$T_{t0} \cong 400 + 60M_0^2 \quad (\text{in } °R) \qquad \gamma = 1.3 \qquad \textbf{(5.76-b)}$$

A graph of the inlet air ram temperature in stratosphere (for constant γ) is shown in Fig. 5.73.

We note that the ratio of specific heats γ decreases with temperature and hence the ram temperature calculation based on cold γ of 1.4 leads to an overestimation of the temperature. To put this in perspective, for a Mach 6 flight in stratosphere, T_{t0} calculated based on $\gamma = 1.4$ is 2820°F, where as $\gamma = 1.3$ yields T_{t0} of 2100°F. There is a ~720°F discrepancy between the two temperatures, although significant, it does not change the arguments about the special cooling needs of high-speed flight. Now, let us accept the goal of ~3000°F operating temperature of say carbon–carbon, the inlet air becomes too hot to be used as a coolant for flight Mach numbers $M_0 = 6^+$. Therefore, the coolant needs to be precooled in a heat exchanger, which adds weight and complexity to the engine. The other alternative is to use a cryogenic fuel that acts as a heat sink for the propulsor walls, before its injection in the combustor and reaction with air or another oxidizer, as in a rocket engine. To use a cryogenic fuel as a coolant is a common and tried approach in rocket propulsion. The method is called regenerative cooling. From Saturn-V Apollo launch vehicle (all three stages) to the Space Shuttle Main Engine, nozzles and the thrust chambers are regeneratively cooled using fuels and sometimes oxidizers. Cryogenic fuels of interest are liquid hydrogen, H_2, and methane, CH_4. Both of these fuels are available in abundance in nature in various liquid and gaseous forms. The heat sink capacity of these fuels is in part quantified by their latent heat of vaporization, namely, 446 kJ/kg for liquid hydrogen and 510 kJ/kg for liquid methane.

5.32   Thrust Reverser and Thrust Vectoring

Thrust reversers are used to decelerate the aircraft upon landing (for a shorter runway), to reduce wear on the landing gear brakes upon touchdown and landing roll, and to enhance maneuverability of a fighter aircraft in flight. The goals of thrust vectoring nozzles are to enable an aircraft to takeoff and land either vertically or use a very short runway (<500 ft) as well as to

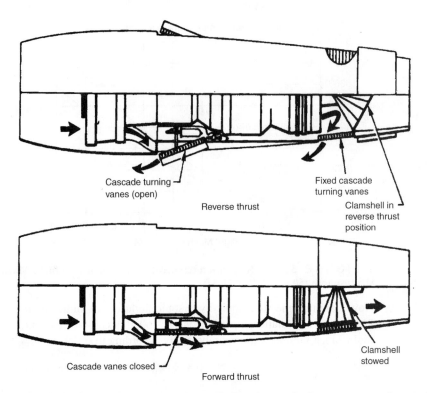

■ **FIGURE 5.74** **The use of clamshell and cascade thrust reversers in a turbofan engine (Courtesy of Pratt & Whitney)**

achieve supermaneuverability in flight. Thrust vectoring is offered as a means of reducing the size or ultimately eliminating the tail (i.e., tailless aircraft), as well as to provide the aircraft with stall recovery. Additional goals of thrust vectoring are the low-speed flight control where the conventional aerodynamic control surfaces become ineffective and the high angle of attack (the so-called high-alpha) capabilities. The latter two goals fall under the supermaneuverability category. To provide some degree of reverse thrust, we need to divert the exhaust jet partially in the flight direction. This is commonly achieved through a *block and turn* mechanism designed into exhaust systems of aircraft today. Two such concepts have been the clamshell and cascade thrust reversers. The clamshell thrust reverser is often used for nonafterburning engines to reverse the turbine exhaust gases, and the cascade type is typically used as a fan exit flow reverser in turbofan engines. Schematic drawings of the two types of thrust reverser are shown in Fig. 5.74 (from a Pratt & Whitney Aircraft, 1980).

EXAMPLE 5.11

Some details of thrust reversing and percent contributions of fan and core, in a separate- and mixed-flow turbofan engine, are shown in Fig. 5.75 (from Dietrich and Kuchar, 2000).

Compare the thrust reverser effectiveness of these two engine types.

SOLUTION

We note that in a large bypass ratio turbofan engine, the fan provides the majority of forward thrust, namely, in the example shown in Fig. 5.75, 75%. The core contributes ∼25% of the forward thrust. The blocker doors are deployed

in the fan duct and a translating nacelle aftbody exposes a cascade of turning vanes. The hot core flow continues to provide forward thrust, as no core flow thrust reverser is provisioned in the separate flow turbofan design. However, in the mixed flow case, the core flow is "dumped," as in a dump diffuser, with subsequent flow deceleration and a forward thrust reduction.

The long-duct turbofan engine provides significantly more thrust reversing capability than a separate-flow turbofan engine ($\sim$30% versus $\sim$10%). The effectiveness of a thrust reverser is defined as the fraction of forward thrust that is converted into reverse thrust. It is also customary to include the ram drag contribution to the reverse thrust in the definition of reverser effectiveness:

$$\text{Reverser effectiveness} \equiv \frac{F_{\text{Reverse}} + D_{\text{Ram}}}{F_{\text{Forward}}} \qquad (5.77)$$

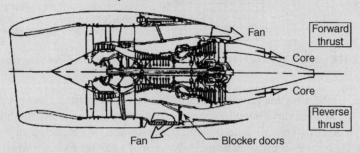

Separate-flow turbofan thrust reverser

Forward thrust

- Fan ~ 75% of total thrust
- Core ~ 25%

Reverse thrust

- Fan ~ 40% reverse thrust
- Blocker leakage ~ 5% forward thrust
- Core ~ 25% forward thrust
- Net reverser effectiveness ~ 10% reverse thrust

[40−5−25]/[100]~10%

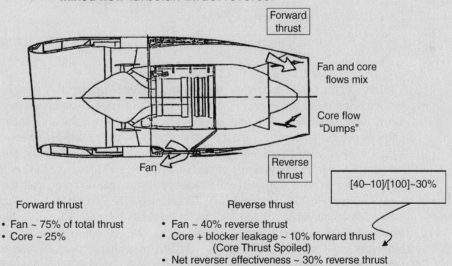

Mixed-flow turbofan thrust reverser

Forward thrust

- Fan ~ 75% of total thrust
- Core ~ 25%

Reverse thrust

- Fan ~ 40% reverse thrust
- Core + blocker leakage ~ 10% forward thrust (Core Thrust Spoiled)
- Net reverser effectiveness ~ 30% reverse thrust

[40−10]/[100]~30%

■ **FIGURE 5.75**   The thrust reverser on a separate- and mixed-flow turbofan engine with the contributions of fan and core (from Dietrich and Kuchar, 2000)

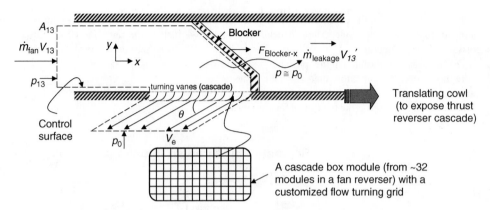

■ **FIGURE 5.76**   **A fan reverser demonstrating leakage and cascade vane angle θ**

For an elemental mass flow rate $d\dot{m}$ that attains an exit velocity from a reverser of V_e magnitude, which exits at an angle of θ with respect to the flight direction, a reverse thrust equal to $d\dot{m} \cdot V_e \cos \theta$ is created. This needs to be integrated over the exit of the turning vane cascade to evaluate the reverse thrust. A schematic diagram of a fan reverser is shown in Fig. 5.76.

$$\text{Reverse Thrust from the cascade reverser} = \int_{\dot{m}_e} V_e \cos \theta \, d\dot{m}_e \qquad (5.78)$$

The actual force exerted by the fluid on the blocker and the turning vanes, i.e., the reverse thrust force exerted on the aircraft, is established by a momentum balance of the fluid across the blocker and turning vanes, such as

$$F_{\text{Reverse}} = F_{\text{blocker}-x} + F_{\text{vanes}-x}$$

$$= \dot{m}_{\text{fan}} V_{13} - \dot{m}_{\text{leakage}} V'_{13} - (p_{13} - p_0) A_{13} + \int_{\dot{m}_e} V_e \cos \theta \, d\dot{m}_e \qquad (5.79)$$

The terms of Eq. 5.79 represent momentum and pressure–area terms, similar to the engine net thrust equation. The two positive terms on the RHS of Eq. 5.79 dominate the bulk of reverse thrust force generation. The two negative contributions are due to leakage and pressure imbalance in the fan duct, p_{13}, and the ambient pressure p_0. We may estimate these terms to get a feel for the reverser effectiveness. For example, the turning vane cascade can be assumed to operate with an inlet total pressure of the fan, i.e., p_{t13}, and an exit pressure of the ambient air, p_0. We may also assume an adiabatic efficiency of the expansion process in the turning vanes, similar to a nozzle. The vane angle may be assumed to be ~45°, and the leakage flow may be ~5% of the incoming flow. To account for the boundary layer blockage in the turning vanes, we may use a discharge coefficient C_D similar to C_{D8}. We may consider the air speed through the blocker is reduced by a factor, e.g., $V'_{13} \sim 1/2 V_{13}$ (note that a mistake in here will not be very large as the leakage mass flow rate is a small fraction of the incoming flow). For the fan exit speed V_{13}, we may estimate that based on a fan exit Mach number of ~0.5. The static pressure p_{13} may be estimated by the fan total pressure ratio and the M_{13} of ~0.5.

British Harrier, AV-8B, is the first operational fighter aircraft to achieve vertical takeoff through thrust vectoring. Harrier's four rotating nozzles turn the engine flow to a near vertical direction to achieve hover, takeoff, and landing. A three bearing swivel nozzle in Fig. 5.77 shows the mechanism for thrust vectoring in an exhaust system (from Dusa, Speir, and Rowe, 1983). The nozzle is of convergent design, hence most suitable for low nozzle pressure ratio (i.e., NPR < 6) operation of a subsonic V/STOL aircraft.

■ **FIGURE 5.77**
Three-bearing swivel nozzle (from Dusa, Speir, and Rowe, 1983)

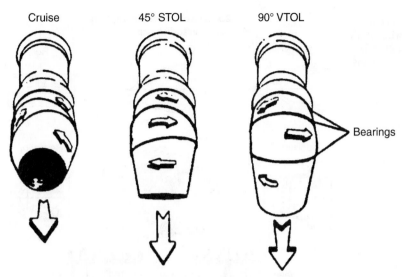

For a supersonic application, a single-expansion ramp nozzle (SERN) offers jet deflection capability through turning of the upper/lower flaps or engaging a separate deflector mechanism. The augmented-deflector exhaust nozzle (Fig. 5.78) is capable of 110° thrust vectoring angle. The deflecting flap SERN nozzle is shown in Fig. 5.79 (from Dusa, Speir, and Rowe, 1983). The two-dimensional geometry of these nozzles allows for low-drag, body-blended exhaust system integration.

Here, we note that the lower flap also serves as a reverser blocker in the STOL/SERN nozzle configuration, which is shown in Fig. 5.79. To get a feel for the cooling requirements of

■ **FIGURE 5.78**
Single expansion ramp configuration in an augmented-deflector exhaust nozzle design (ADEN) for V/STOL applications (from Dusa, Speir, and Rowe, 1983)

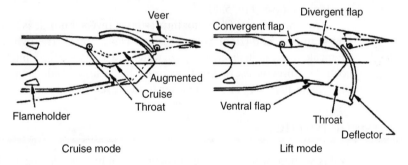

■ **FIGURE 5.79**
Single expansion ramp exhaust nozzle configuration with deflecting flaps suitable for STOL application (from Dusa, Speir, and Rowe, 1983)

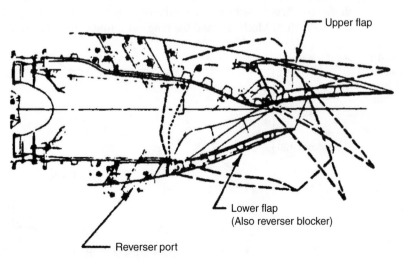

■ **FIGURE 5.80**
Schematic drawing of a multifunctional 2D-CD nozzle (from Younghans and Dusa, 1993)

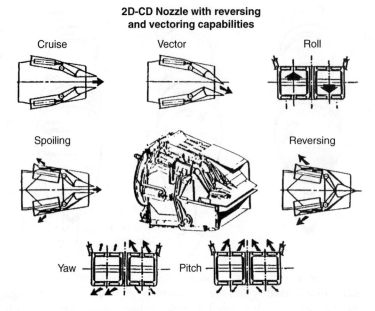

2D-CD Nozzle with reversing and vectoring capabilities

an advanced thrust-vectoring nozzle with an afterburner, we look into the experimental results of ADEN tests (at flight weight, self-cooled with the fan air on a thrust stand at NASA-Lewis). To achieve a surface temperature below the *hot streak design limit* of 2700°F, a ~5.7% cooling air was required at maximum afterburner operation of $T_{t8} = 3028°F$ and the maximum nozzle pressure ratio of 15 (from Dusa and Wooten, 1984). Now, let us examine the following twin configuration (see Fig. 5.80).

Another concept for vector thrust–thrust reverser nozzle is shown as a multiflap 2D-CD nozzle in Fig. 5.80 (from Younghans and Dusa, 1993). The twin configuration shown uses segmented, i.e., split, divergent flaps, which are used to achieve pitch and roll control. The yaw control requires the use of thrust reverser, which causes a reduction in axial thrust. Note that the convergent flaps serve as the blocker in the reverse thrust mode.

5.33 Hypersonic Nozzle

In this section, we explore the challenges of a hypersonic nozzle design. There are several design conflicts in the hypersonic exhaust system that stem from the range of flight speeds, i.e., from takeoff to Mach > 5 and the range of altitude, namely, from sea level takeoff to upper

■ **FIGURE 5.81**
Flight envelope of hypersonic vehicles with altitude in kft (from Dusa, 1988)

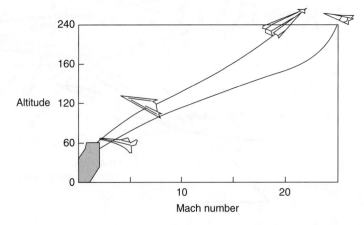

atmosphere, or 100$^+$ kft. The combination of altitude and Mach number constitutes the flight envelope of an aircraft, and in Fig. 5.81 we examine such a flight envelope (from Dusa, 1988). The shaded area in the corner of Fig. 5.81 represents the flight envelope of conventional commercial and military aircraft. We examine the representative nozzle pressure ratio for airbreathing hypersonic engines in Example 5.12.

EXAMPLE 5.12

Demonstrate that a hypersonic airbreathing propulsion system creates a nozzle pressure ratio of one or two orders of magnitude higher than that of a conventional Mach 2.5 aircraft.

SOLUTION

Let us consider a Mach 6 flight for a hypersonic cruise aircraft at high altitude, say 120 kft. For the purpose of demonstration, let us assume that the ratio of specific heats γ is constant. For an adiabatic flow of a perfect gas, the ratio of stagnation to static pressure may write as

$$\frac{p_{t0}}{p_0} = \left(1 + \frac{\gamma - 1}{2} M_0^2\right)^{\frac{\gamma}{\gamma-1}} \quad \textbf{(5.80)}$$

Assuming a Mach 6 flight and $\gamma = 1.4$, we get

$$\frac{p_{t0}}{p_0} = [1 + 0.2(36)]^{3.5} = 1580$$

Now, assuming the inlet total pressure recovery is between 40 and 50% at Mach 6, the ratio of total pressure at the combustor inlet of a ramjet (or scramjet) to ambient static pressure is

$$\frac{p_{t2}}{p_0} \approx 632 - 790$$

For a supersonic combustor with a Mach 3 inlet condition and a 30% total temperature rise, Kerrebrock (1992) shows that the combustor total pressure ratio is 37%, i.e.,

$$\frac{p_{t4}}{p_{t2}} \approx 0.37$$

Hence, the ratio of total pressure at the nozzle inlet to the ambient static pressure is

$$NPR = \frac{p_{t4}}{p_0} \approx 234 - 292$$

Note that this nozzle pressure ratio of two to three hundred was created in a Mach 6 hypersonic cruise aircraft. The NPR exponentially grows with the flight Mach number. There is no relief in the real gas effects at high temperatures since a reduced ratio of specific heats will only exacerbate the magnitude of stagnation pressure, as shown in Fig. 5.82.

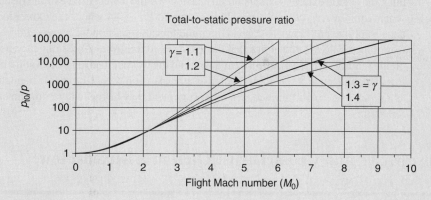

■ FIGURE 5.82   **Exponential growth of stagnation pressure with Mach number**

The nozzle area ratio as a function of nozzle pressure ratio is shown in Fig. 5.83, using a log-linear scale. We note that a hypersonic airbreathing propulsor will require a very large nozzle area ratio, whose integration into the vehicle is the challenge number one.

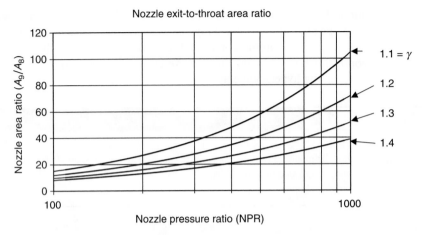

■ **FIGURE 5.83**   Nozzle exit-to-throat area ratio for a perfectly expanded nozzle, assuming a perfect gas with the ratio of specific heats γ a constant

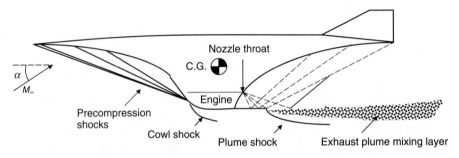

■ **FIGURE 5.84**   Schematic drawing of a generic hypersonic vehicle powered by an airbreathing engine shows a highly integrated inlet and exhaust system with the airframe

To reduce the burden of excessive weight, a hypersonic nozzle will perform an external expansion of the gas, in a geometrically similar manner as a plug nozzle. However, instead of a plug, the nozzle external expansion surface may be the lower surface of the fuselage aftbody. A schematic drawing of such a nozzle is shown in Fig. 5.84 where the expansion waves in the exhaust plume are shown as dashed lines and the compression waves as solid lines. Remember from supersonic aerodynamics that the wave reflection from a free boundary, as in the exhaust plume mixing layer, is in an *unlike* manner and the wave reflection from a solid boundary is in a *like* manner. Hence, a shock is reflected as an expansion wave from a free boundary and reflects as a shock from a solid surface. A similar statement may be made regarding an expansion wave reflections.

5.34   Exhaust Mixer and Gross Thrust Gain in a Mixed-Flow Turbofan Engine

A mixed-flow turbofan engine offers the potential of gross thrust gain and a reduction in fuel consumption as compared with an optimized separate flow turbofan engine. In this section, we examine the extent of gross thrust improvements and the nondimensional parameters that impact the performance gains.

To demonstrate the gain in the gross thrust realized by mixing a cold and a hot stream, we first treat the problem by assuming an ideal mixer. Expanding the cold, hot, or mixed gases in a

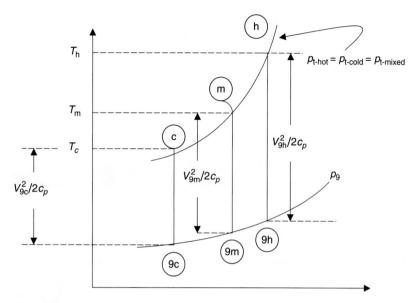

■ **FIGURE 5.85**   **An ideal expansion of a cold, hot, and a mixed stream to the same backpressure p_9**

nozzle to an exit pressure of p_9, we shall realize a conversion of the thermal energy into the kinetic energy in the exhaust nozzle. The thermodynamics of the expansion process for the three streams is shown in Fig. 5.85.

The gross thrust for a separate exhaust turbofan engine, assuming the nozzles are perfectly expanded is

$$\left.\frac{F_g}{\dot{m}_0}\right|_{\text{unmixed}} \cong V_{9h} + \alpha V_{9c} = \sqrt{2c_p T_h \left(1 - \frac{T_{9h}}{T_h}\right)} + \alpha \cdot \sqrt{2c_p T_c \left(1 - \frac{T_{9c}}{T_c}\right)} \qquad (5.81)$$

We may write the isentropic pressure ratio for the temperature ratios in Eq. 5.81 to get

$$\left.\frac{F_g}{\dot{m}_0}\right|_{\text{unmixed}} \cong \sqrt{2c_p T_h \left[1 - \left(\frac{p_9}{p_{\text{th}}}\right)^{\frac{\gamma-1}{\gamma}}\right]} + \alpha \cdot \sqrt{2c_p T_c \left[1 - \left(\frac{p_9}{p_{\text{tc}}}\right)^{\frac{\gamma-1}{\gamma}}\right]} \qquad (5.82)$$

The mixed exhaust nozzle shall have a gross thrust of

$$\left.\frac{F_g}{\dot{m}_0}\right|_{\text{mixed}} \cong (1 + \alpha)V_m = (1 + \alpha) \cdot \sqrt{2c_p T_m \left[1 - \left(\frac{p_9}{p_{\text{tm}}}\right)^{\frac{\gamma-1}{\gamma}}\right]} \qquad (5.83)$$

We note that the pressure terms in the brackets are identical hence they cancel when we express the ratio of gross thrusts. Assuming the same gas properties in the three streams, namely, we get

$$\frac{F_g|_{\text{mixed}}}{F_g|_{\text{unmixed}}} = \frac{(1 + \alpha)\sqrt{T_m}}{\sqrt{T_h} + \alpha \cdot \sqrt{T_c}} \qquad (5.84)$$

From energy balance, we know that

$$T_m = \frac{T_h + \alpha \cdot T_c}{1 + \alpha} \qquad (5.85)$$

Therefore the rise of the gross thrust as a result of mixing a cold and a hot stream may be expressed in terms of the nondimensional parameters of the respective temperature ratio T_h/T_c and the mass flow ratio, namely, the bypass ratio α. This is expressed in the following equation.

$$\frac{F_g|_{\text{mixed}}}{F_g|_{\text{unmixed}}} = \frac{\sqrt{1+\alpha} \cdot \sqrt{\dfrac{T_h}{T_c} + \alpha}}{\sqrt{\dfrac{T_h}{T_c}} + \alpha} \tag{5.86}$$

EXAMPLE 5.13

For a range of hot-to-cold temperature ratios between 1.5 and 4.5 and the turbofan bypass ratio of up to 8, calculate and graph the ratio of mixed to separate-flow turbofan engines gross thrust.

SOLUTION

We use a spreadsheet for Eq. 5.86 and graph the results in Fig. 5.86.

As expected, the gain in gross thrust due to mixing of a cold and a hot stream vanishes as $T_h/T_c \to 1$, as well as for $\alpha \to 0$. The gain peaks for a bypass ratio in the vicinity of 2, and the effect diminishes as the bypass ratio increases. The modest gain of ~2–3% is realizable in a real engine if the mixing losses in the mixer are kept low. This gross thrust increase of a mixed exhaust turbofan engine translates into a corresponding drop of 2–3% in the thrust-specific fuel

consumption of such engines compared with a separate exhaust turbofan engines. We recognize that the real process in a mixer will result in a loss of total pressure due to

1. Turbulent mixing of the cold and hot streams
2. Frictional losses on the walls of the mixer

Therefore an effective design of a mixer is crucial to realizing the performance gains promised by the mixer.

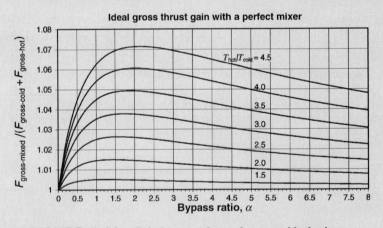

■ **FIGURE 5.86**   Gain in gross thrust due to an ideal mixer

5.35   Nozzle-Turbine (Structural) Integration

The turbine exhaust case, support struts, and a tail cone flange are shown in Fig. 5.87 (from Pratt & Whitney, 1980). The turbine exhaust cone houses the turbine bearings and provides for a smooth flow to the nozzle (or afterburner).

The exhaust cone, jet pipe, and a convergent nozzle are shown schematically in Fig. 5.88 (from Rolls-Royce, 2005). In this view of the exhaust system, the turbine rear support struts are also shown.

■ **FIGURE 5.87** Turbine exhaust's case, struts, and a tail cone flange (Courtesy of Pratt & Whitney, 1980)

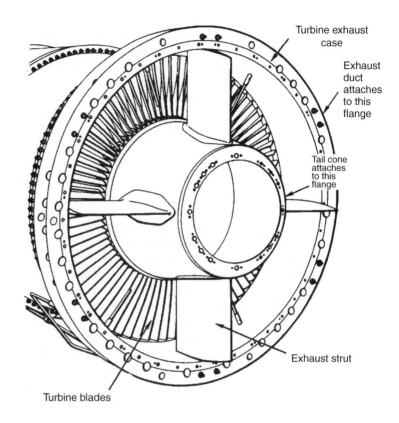

Turbine exhaust case

Exhaust duct attaches to this flange

Tail cone attaches to this flange

Exhaust strut

Turbine blades

■ **FIGURE 5.88** Drawing of exhaust cone, jet pipe, and a convergent nozzle (Courtesy of Rolls-Royce, plc., 2005)

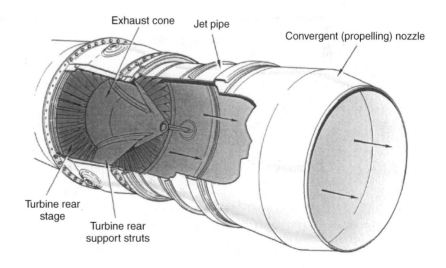

Exhaust cone

Jet pipe

Convergent (propelling) nozzle

Turbine rear stage

Turbine rear support struts

5.36 Summary of Exhaust Systems

An aircraft exhaust system is a multifunctional component whose primary mission is to convert fluid thermal energy into directed forces in any desired direction (*x*, *y*, *z*) efficiently. These forces are usually in vector thrust mode, including hover, as well as thrust reversing directions.

The system weight (and complexity) and cooling requirements are the optimizing parameters for the design and selection of exhaust systems. Advanced nozzle cooling requirements are in ~5–10% range. Airbreathing engines with afterburner require a variable geometry nozzle to open the throat with afterburner operation. The nozzle throat scheduling with the afterburner throttle setting is to decouple the afterburner/exhaust system from the gas generator. The nozzle exit area scheduling has to include the net effect of internal performance gains as well as the external boattail drag penalties. The internal performance is optimized when the nozzle is perfectly expanded. The build up of boundary layer in the convergent or primary nozzle is accounted for by a flow (discharge) coefficient at the throat, C_{D8}. The effect of total pressure loss on the exhaust velocity in the divergent section of the nozzle is accounted for by a velocity coefficient C_V. The effect of nonaxial exhaust flow appears as a reduction in axial force produced by the nozzle. The appropriate accounting of this loss is made using the flow angularity loss coefficient C_A. Lobed mixers in a mixed-stream turbofan engines improve gross thrust by a few percentage points. Off-design operation of supersonic nozzles in overexpanded mode causes shock waves to be formed and a corresponding loss of thrust. Plug nozzles have a superior overexpanded nozzle performance over the convergent–divergent nozzles, but require additional cooling and thus system complexity and weight. In hypersonic flight, the nozzle pressure ratio reaches into several hundred level, somewhat similar to a rocket engine. The required area expansion ratio for efficient expansion of the high-pressure gas demands integration with vehicle. To achieve reduced signature, an exhaust system could employ high-aspect ratio rectangular nozzles with offset to shield the turbine as well as enhanced mixing to reduce the jet temperature. The exit plane of the nozzle placed on the upper surface of the aircraft fuselage or wing provides shielding from ground radar.

Further readings on exhaust systems and installation (references 4, 7, 8, 24 and 26) are recommended.

References

1. Abbott, J.M., Anderson, B.H., and Rice, E.J., "Inlets, Ducts and Nozzles," *Aeropropulsion '87*, NASA CP-3049, 1987.

2. Abernethy, R.B. and Roberts, J.H., *In-Flight Thrust Determination and Uncertainty*, SAE Special Publication, SP-674, 1986.

3. Albers, J.A, et al., "Aerodynamic Analysis of Several High Throat Mach Number Inlets for the Quiet, Clean, Short-Haul Experimental Engine," NASA TMX-3183, 1975.

4. Aulehla, F. and Lotter, K., "Nozzle/Airframe Interference and Integration," in AGARD Lecture Series, LS-53, Paper 4, 1972.

5. Anderson, B.H., "Three-Dimensional Viscous Design Methodology of Supersonic Inlet Systems for Advanced Technology Aircraft," *Numerical Methods for Engine-Airframe Integration*, Eds. Murthy, S.N.B. and Paynter, G.C., Progress in Astronautics & Aeronautics, Vol. 102, AIAA, Washington, DC, 1986, pp. 431–480.

6. Anderson, J.D., Jr., *Fundamentals of Aerodynamics*, 4th edition, McGraw-Hill, New York, 2005.

7. Anon., Symposium on "*Developments in Aircraft Propulsion*," Haarlem, the Netherlands, 1987.

8. Berman, K. and Crimp, F.W., "Performance of Plug-Type Rocket Exhaust Nozzles," *ARS Journal*, Vol. 31, No. 1, January 1961.

9. Bowers, D.L. and Tamplin, G., "Throttle-Dependent Forces," Chapter 5 in *Thrust and Drag: Its Prediction and Verification*, Eds. Covert, E.E. et. al., Progress in Astronautics and Aeronautics, Vol. 98, AIAA, New York, 1985.

10. Cochran, D.L. and Kline, S.J., "The Use of Short Flat Vanes for Producing Efficient Wide-Angle Two-Dimensional Subsonic Diffusers," NACA TN 4309, September 1958.

11. Dietrich, D.A. and Kuchar, A.P., "High Bypass Turbofan Nacelles for Subsonic Transports," UTSI Course notes, April 2000.

12. Dusa, D.J., "Nozzles and Turbo-Ramjet Propulsion for Mach 0 to 6 Aircraft," *Short Course Notes in Hypersonic Propulsion*, NASA-Lewis, 1988.

13. Dusa, D.J., Speir, D.W., and Rowe, R.K., "Advanced Technology Exhaust Nozzle Development," AIAA Paper No. 83-1286, 1983.

14. Dusa, D.J. and Wooten, W.H., "Single Expansion Ramp Nozzle Development Status," AIAA Paper No. 84-2455, 1984.

15. Farokhi, S., Sheu, W.L., and WU, C., "On the Design of Optimum-Length Transition Ducts with Offset: A Computational Study," *Computers and Experiments in Fluid Flow*, Eds. Carlomagno, G.M. and Brebbia, C.A., Springer Verlag, Berlin 1989, pp. 215–228.

16. Hancock, J.P. and Hinson, B.L., "Inlet Development for the L-500," AIAA Paper No. 69-448, June 1969.

17. Hesse, W.J. and Mumford, N.V.S., Jr., *Jet Propulsion for Aerospace Applications*, 2nd edition, Pitman Publishing Corp., New York, 1964.

18. Holmes, B.J., "Progress in Natural Laminar Flow Research," Paper Presented at AIAA/NASA General Aviation Technology Conference, Wichita, Kansas, July 1984.

19. Johnson, C.L., "Some Development Aspects of the YF-12A Interceptor Aircraft," AIAA Paper No. 69-757, July 1969.

20. Kantrowitz, A. and Donaldson, C. du P., "Preliminary Investigation of Supersonic Diffusers," NACA ACR L5020, 1945.

21. Kerrebrock, J.L., *Aircraft Engines and Gas Turbines*, 2nd. edition, MIT Press, Cambridge, Mass, 1992.

22. Kimzey, W.F., Wehofer, S., and Covert, E.E., "Gas Turbine Engine Performance Determination," *Thrust and Drag: Its Prediction and Verification*, Ed. Covert, E.E., AIAA progress in Astronautics & Aeronautics, Vol. 98, AIAA, Washington, DC, 1985, pp. 47–119.

23. Kline, S.J., Abbott, D.E., and Fox, R.W., "Optimum Design of Straight-Walled Diffusers," *Journal of Basic Engineering*, 81, Series D, No. 3, September 1959, pp. 321–331.

24. Kozlowski, H. and Larkin, M., "Energy Efficient Engine Exhaust Mixer Model Technology Report," NASA CR-165459, June 1981.

25. Kline, S.J., "On the Nature of Stall," *Journal of Basic Engineering*, 81, series D, No. 3, September 1959, pp. 305–320.

26. Kuchar, A.P., "Variable Convergent-Divergent Exhaust Nozzle Aerodynamics," Chapter in *Aircraft Propulsion System Technology and Design*, Ed. Oates, G.C., AIAA Education Series, AIAA Inc., Washington, DC, 1989.

27. Kuechemann, D. and Webber, J., *Aerodynamics of Propulsion*, McGraw-Hill, New York, 1953.

28. Mattingly, J.D., *Elements of Gas Turbine Propulsion*, McGraw-Hill, New York, 1996.

29. Mount, J.S. and Millman, V., "Development of an Active Laminar Flow Nacelle," AIAA Paper 85-1116, 1985.

30. MIL-E-5008B, "Military Specifications-Engines, Aircraft, Turbojet, Model Specifications for," January 1959.

31. Oates, G.C., *Aerothermodynamics of Gas Turbine and Rocket Propulsion*, AIAA, Washington, DC, 1988, pp. 198–199.

32. Oswatitsch, K., "Pressure Recovery for Missiles with Reaction Propulsion at High Supersonic Speeds (The Efficiency of Shock Diffusers)," NACA TM 1140, 1947.

33. Paul, D.L. and Younghans, J.L., "Inlets & Inlet Engine Integration," Chapter 13 of *The Aerothermodynamics of Aircraft Gas Turbine Engines*, AFAPL TR-78-52, Ed. Oates, G. C., 1978.

34. Potonides, H., et al., "Design and Experimental Studies of a Type "A" V/STOL Inlet," AIAA Paper No. 78-956, 1978.

35. Powel, A.G., Welge, H.R., and Trefny, C.J., "Low-Speed Aerodynamic Test of an Axisymmetric Supersonic Inlet with Variable Cowl Slot," AIAA Paper 85-1210, 1985.

36. Pratt & Whitney Aircraft, *The Aircraft Gas Turbine Engine and Its Operation*, PWA Oper. Instr. 200, United Technologies Corporation, East Hartford, CT, June 1980.

37. Rodean, H.C., "Supersonic Inlet Analysis and Design," Ling-Temco-Vought Report No. 11174, January 1958.

38. Rolls-Royce, *The Jet Engine*, Rolls-Royce plc, Derby, England, 2005.

39. Schweikhard, W.G. and Montoya, E.J., "Research Instrumentation Requirements for Flight/Wind Tunnel Tests of the YF-12 Propulsion System and Related Flight Experience," in *Instrumentation for Airbreathing Propulsion*, AIAA Series, Progress in Astronautics and Aeronautics, Vol. 34, Eds. Fuhs, A.E. and Kingery, M., MIT Press, Cambridge, Mass. 1974.

40. Seddon, J. and Goldsmith, E.L., *Intake Aerodynamics*, AIAA, Washington, DC, 1985.

41. Sovran, G. and Klomp, E.D., "Experimentally Determined Optimum Geometries for Rectilinear Diffusers for Rectangular, Conical or Annular Cross Section," *Fluid Mechanics of Internal Flow*, Ed. Sovran, G., Elsevier Publishing Co., Amsterdam, 1963, pp. 270–319.

42. Strack, W.C., "Propulsion Challenges and Opportunities for High-Speed Transport Aircraft," in *Aeropropulsion '87*, NASA CP-10003, 1987.

43. Summerfield, M., Foster, C.R., and Swan, W.C., "Flow Separation in Over-expanded Supersonic Exhaust Nozzles," *Jet Propulsion*, Vol. 24, September–October 1954, pp. 319–321.

44. Sutton, G.P., *Rocket Propulsion Elements*, 6th edition, John Wiley & Sons, Inc., New York, 1992.

45. Wazelt, F., Ed., "Suitable Averaging Techniques in Non-Uniform Internal Flows," AGARD Advisory Report 182, June 1983.

46. Wazyniak, J., "Mission Adaptation & Technology Trends in Aero Propulsion," Short Course Notes in *Aero-Propulsion* at UTSI, 2000.

47. Whitcomb, R.T. and Clark, L.R., *An Airfoil Shape for Efficient Flight at Supercritical Mach Numbers*, NASA TMX-1109, July 1965.

48. Whitford, R., *Design for Air Combat*, Jane's Publishing, London, UK, 1987.

49. Younghans, J.L. and Dusa, D.J., "Inlets and Exhaust Systems for Multi-Mission Applications (Aero-Design and Installed Performance)," Short Course Notes, The University of Kansas, 1993.

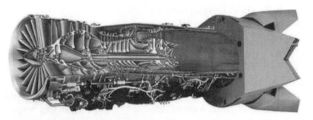

F119-PW-100 (Courtesy of Pratt & Whitney)

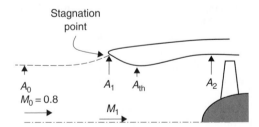

F-22 Raptor air-superiority fighter (Courtesy of USAF)

Problems

In solving the following problems, assume $\gamma = 1.4$ and $c_p = 1004 \, \text{J/kg} \cdot \text{K}$, unless otherwise stated.

5.1   In a real inlet, the total pressure loss is 10% of the flight dynamic pressure, i.e., $p_{t0} - p_{t2} = 0.1 \, q_0$ at a flight Mach number of $M_0 = 0.85$. Calculate

   (a) inlet total pressure recovery π_d

   (b) inlet adiabatic efficiency η_d.

5.2   A subsonic inlet in cruise condition is shown. Calculate

   (a) total pressure recovery π_d

   (b) area ratio A_2/A_0

   (c) static pressure ratio p_2/p_0

   (d) adiabatic efficiency η_d

   (e) entropy rise $\Delta s/R$.

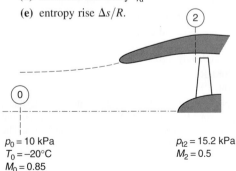

$p_0 = 10 \, \text{kPa}$
$T_0 = -20°\text{C}$
$M_0 = 0.85$

$p_{t2} = 15.2 \, \text{kPa}$
$M_2 = 0.5$

■ **FIGURE P5.2**

5.3   Consider a subsonic inlet at a flight cruise Mach number of 0.8. The captured streamtube undergoes a prediffusion external to the inlet lip, with an area ratio $A_0/A_1 = 0.92$, as shown. Calculate

■ **FIGURE P5.3**

   (a) C_p (i.e., the pressure coefficient) at the stagnation point

   (b) inlet lip Mach number M_1

   (c) lip contraction ratio A_1/A_{th} for a throat Mach number $M_{th} = 0.75$ (assume $p_{t,th}/p_{t1} = 1$)

   (d) the diffuser area ratio A_2/A_{th} if $M_2 = 0.5$ and $p_{t2}/p_{t,th} = 0.98$

   (e) the nondimensional inlet additive drag $D_{add}/p_0 A_1$.

5.4   The captured streamtube for a subsonic inlet experiences external diffusion, where flight Mach number of $M_0 = 0.85$ decelerates to M_1 of 0.65 at the inlet lip. Calculate the inlet additive drag nondimensionalized by flight static pressure p_0 and inlet area A_1.

5.5 An aircraft flies at an altitude where p_0 is 0.1915 atm and the flight total pressure p_{t0} is 1.498 atm. The engine face total pressure is measured to be, $p_{t2} = 1.348$ atm. For this inlet calculate

(a) τ_r, ram temperature ratio

(b) $\dfrac{\Delta s_d}{R}$, nondimensional entropy rise in the diffuser

(c) η_d, inlet adiabatic efficiency

5.6 A subsonic diffuser has an area ratio of $A_2/A_1 = 1.3$. The inlet Mach number to the diffuser is $M_1 = 0.72$ and the total pressure loss in the diffuser is characterized by $\Delta p_t = 0.1 q_1$. Assuming the flow in the diffuser is adiabatic, calculate

(a) the diffuser exit Mach number M_2 and

(b) diffuser static pressure recovery C_{PR}

5.7 A subsonic inlet is cruising at $M_0 = 0.85$ and the capture area ratio A_0/A_1 is 0.90 (as shown). For an altitude pressure of $p_0 = 25$ kPa, the temperature of $T_0 = -25°C$, and inlet area $A_1 = 3 \text{ m}^2$, calculate

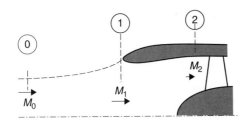

■ **FIGURE P5.7**

(a) the inlet Mach number M_1

(b) the inlet additive drag D_{add} (N)

(c) inlet mass flow rate $\dot{m}$ (kg/s)

(d) the inlet ram drag D_{ram} (kN)

(e) engine face area A_2, if $M_2 = 0.5$ (assuming $\pi_d = 0.99$)

5.8 The Mach number at the compressor face is $M_2 = 0.65$ at takeoff ($M_0 \cong 0.2$). Assuming the inlet suffers a 2% total pressure loss at takeoff, calculate the capture-to-engine face area ratio A_0/A_2.

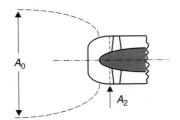

■ **FIGURE P5.8**

5.9 A subsonic aircraft cruises at $M_0 = 0.85$ and its inlet operates with a capture ratio of $A_0/A_1 = 0.7$. First, calculate the lip Mach number M_1. Second, assuming an engine becomes inoperative and the inlet lip Mach number drops to 0.3, (the so-called engine wind-milling condition), calculate the additive drag D_{add} for an inlet area of $A_1 = 4 \text{ m}^2$ and the ambient static pressure of $p_0 = 16.6$ kPa.

5.10 A subsonic inlet is flying at Mach 0.8, with an inlet capture area ratio of $A_0/A_1 = 0.7$. The inlet lip contraction ratio A_1/A_{th} is 1.15. Calculate the 1D throat Mach number and comment on the potential shock formation, near the convex surface, at the throat.

5.11 A subsonic aircraft flies at $M_0 = 0.85$ with an inlet mass flow ratio (MFR) of 0.90. Calculate the critical pressure coefficient $C_{p,crit}$ on the nacelle. Also calculate the maximum cowl (frontal) area ratio A_M/A_1 if this inlet is to experience an average surface pressure coefficient corresponding to the critical value, i.e., $\overline{C}_p \approx C_{p,crit}$.

5.12 An inlet creates a circumferential distortion at the engine face, as shown. The hub-to-tip radius ratio is $r_h/r_t = 0.5$. The spoiled sector has a 15% mass flow deficit (per unit area) as compared with a uniform flow. Assuming the static density, temperature, and pressure are uniform at the engine face and the mass flow deficit in the spoiled sector is caused by a velocity deficit (as shown), use Bernoulli equation to estimate the total pressure deficit in the spoiled sector, i.e., $[p_t - p]_{spoiled}/[p_t - p]_{uniform}$. Also, calculate

(a) $\bar{p}_{t,\,area-avg}$, area-averaged total pressure

(b) $\bar{p}_{t,\,mass-avg}$, mass-averaged total pressure

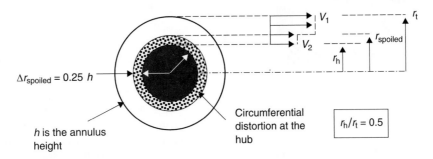

■ **FIGURE P5.12**

5.13 A subsonic inlet has a capture area ratio of $A_0/A_1 = 0.8$. Assuming the flight Mach number is 0.80 and the inlet area ratio is $A_2/A_1 = 1.25$, calculate

(a) Mach number at the inlet lip M_1

(b) diffuser exit Mach number if p_{t2}/p_{t1} is 0.95

(c) inlet static pressure ratios p_1/p_0 and p_2/p_1 for $p_{t2}/p_{t1} = 0.95$

5.14 A normal shock inlet operates in the subcritical mode, with the shock in standoff position, as shown. The bow shock is normal to the flow at the inlet centerline and weakens into an oblique shock and eventually a Mach wave away from the inlet.

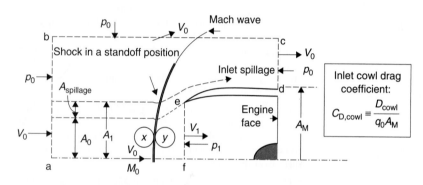

■ **FIGURE P5.14**

Apply conservations of mass and momentum to the control volume a-b-c-d-e-f-a, to approximate inlet cowl external drag force coefficient, in terms of the flight and inlet parameters that are shown, e.g., V_0, M_0, p_0, p_1, V_1, A_1, A_M, and $A_{spillage}$.

5.15 Consider a variable-geometry, convergent–divergent isentropic inlet that is designed for $M_D = 4.0$. To swallow the starting shock, i.e., to start the inlet, the throat needs to be opened. Calculate the percentage of throat area opening needed to start this inlet, $(\Delta A_{th}/A_{th}) \times 100$.

5.16 An external compression inlet is in a Mach-2 flow. The shocks are positioned according to the figure shown. Calculate

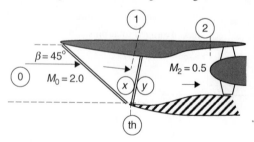

■ **FIGURE P5.16**

(a) shock total pressure ratio π_s

(b) A_2/A_{th}

You may neglect the frictional losses in the subsonic diffuser.

5.17 A normal-shock inlet is operating in a supercritical mode, as shown. Flight Mach number is $M_0 = 1.6$. The inlet capture area ratio $A_0/A_1 = 0.90$ and the diffuser area ratio $A_2/A_1 = 1.2$. Calculate

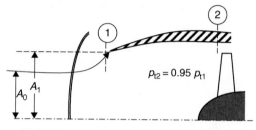

■ **FIGURE P5.17**

(a) M_1

(b) inlet total pressure recovery π_d, i.e., p_{t2}/p_{t0}

5.18 An isentropic convergent–divergent supersonic inlet is designed for $M_D = 1.6$. Calculate the inlet's

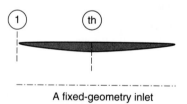

A fixed-geometry inlet

■ **FIGURE P5.18**

(a) area contraction ratio A_1/A_{th}

(b) subsonic Mach number where the throat first chokes

(c) percent spillage at $M_0 = 0.7$

(d) percent spillage at $M_0 = 1.6$ (in the unstarted mode)

(e) overspeed Mach number to start this inlet, $M_{overspeed}$

(f) throat Mach number after the inlet was started, with still $M_{overspeed}$ as the flight Mach number

5.19 Consider an isentropic fixed-geometry C–D inlet, which is designed for $M_D = 1.75$. The inlet flies at an altitude where ambient (static) pressure is 20 kPa. Calculate

(a) Overspeed Mach number that will start this inlet

(b) The flight dynamic pressure corresponding to the altitude and $M_{overspeed}$

5.20 An isentropic, convergent–divergent supersonic inlet is designed for $M_D = 3.0$. Assuming that the throat area is adjustable, calculate the percentage of the design throat area that needs to be opened to swallow the starting shock, i.e.,

$$\frac{A_{th, open} - A_{th, design}}{A_{th, design}} \times 100.$$

5.21 A supersonic C–D inlet is designed for a flight Mach number of $M_0 = 3.5$. This inlet starts by opening its throat (from A_{th} to A'_{th}). Neglecting wall frictional losses, calculate

(a) percent throat opening required

(b) throat Mach number after it starts (with the throat open at A'_{th})

5.22 A Kantrowitz–Donaldson inlet is designed for Mach 2.0. Calculate

(a) the required contraction area ratio A_1/A_{th}

(b) the inlet total pressure recovery with the best backpressure

5.23 A Kantrowitz–Donaldson inlet is designed for $M_D = 1.7$. Calculate

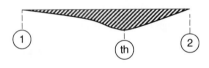

■ FIGURE P5.23

(a) the inlet contraction ratio A_1/A_{th}

(b) the throat Mach number after the inlet self started

(c) the total pressure recovery with the best backpressure.

5.24 Calculate the contraction ratio A_1/A_{th} and the maximum total pressure recovery of a self-starting C–D inlet designed for $M_D = 3.2$.

5.25 A normal-shock inlet is operating in a supercritical mode, with the shock inside the inlet. If the flight Mach number is $M_0 = 1.6$ and the shock is located at $A_s/A_t = 1.2$, calculate

(a) Mach number ahead of the shock wave, M_x

(b) percent total pressure gain if the inlet were to operate in the critical mode

5.26 A variable geometry isentropic supersonic inlet is designed for $M_D = 1.6$. Calculate

(a) percent flow spillage at $M_0 = 0.8$

(b) percent flow spillage at $M_D = 1.6$ before the inlet is started

(c) percent throat area increase needed to start this inlet

(d) throat Mach number after the shock is swallowed, M'_{th}

(e) inlet total pressure recovery with the best backpressure (with open throat)

■ FIGURE P5.26

5.27 An isentropic, fixed-geometry inlet, is designed for $M_D = 1.5$. If this inlet is to be started by overspeeding, calculate the necessary Mach number for overspeed.

5.28 A fixed-geometry, convergent–divergent, internal-compression inlet is designed for $M_D = 2.0$ and a self-starting capability. Calculate

(a) A_1/A_{th}

(b) M_{th}

(c) inlet total pressure recovery for the "best" backpressure

5.29 A normal-shock inlet is flying at a Mach number of 1.8. However, due to a nonoptimum backpressure, the normal shock is inside the duct where $A_s/A_i = 1.15$. Calculate the percent loss in the total pressure recovery due to this backpressure.

5.30 A variable-geometry supersonic convergent-divergent inlet is to be designed for an isentropic operation (in the started mode) at $M_D = 2.6$. Calculate

(a) the inlet design contraction ratio A_1/A_{th}

(b) the percent opening of the throat $(A'_{th} - A_{th})/A_{th}$ needed to start the inlet

(c) the throat Mach number in the open position, M'_{th}

5.31 A variable-geometry, internal-compression, C–D inlet is designed for $M = 2.0$. Calculate percent opening of the throat required to swallow the starting shock.

5.32 A 2D fixed geometry convergent–divergent diffuser is shown. Mach number at the entrance is $M_1 = 3.0$ and the static pressure is $p_t = 10$ kPa. A shock occurs at an area ratio $A_s/A_{th} = 1.25$ downstream of the throat, as shown. Calculate

(a) M_x

(b) p_y

(c) p_{ty}

(d) A_2/A_1 if the exit Mach number is 0.5, i.e., $M_2 = 0.5$

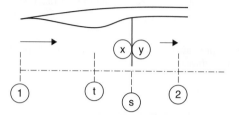

■ **FIGURE P5.32**

You may assume the throat is choked and neglect wall friction.

5.33 A normal-shock inlet operates in a Mach 1.86 stream with ambient static pressure of $p_0 = 30$ kPa. Neglecting total pressure loss in the subsonic diffuser, calculate

(a) inlet total pressure recovery with the shock at the lip, i.e., the best backpressure

(b) inlet total pressure recovery when the shock is inside the duct at $A_x/A_t = 1.2$, i.e., the supercritical mode

(c) inlet total pressure recovery in subcritical mode with 10% spillage, i.e., $A_{spillage}/A_1 = 0.1$

(d) flight dynamic pressure q_0

5.34 Consider an external compression inlet with two ramps operating in a Mach-2.5 stream of air. Calculate the total pressure recovery of the inlet shock system for the case of the best backpressure for the two ramp angles of 8° and 12°, respectively, and compare it to the normal-shock inlet at the same Mach number and with the best backpressure.

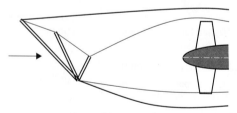

■ **FIGURE P5.34**

5.35 A supersonic flow is to be decelerated over two oblique shocks and one normal shock, similar to Problem 5.34 (i.e., $M_0 = 2.5$). The Mach number downstream of the normal shock is 0.7. Assuming the maximum total pressure recovery is obtained when the first two oblique shocks are of equal strength, calculate the necessary ramp angles.

5.36 A supersonic nozzle operates in an underexpanded mode with an area ratio $A_9/A_{th} = 2$. We know that the exhaust plume turns outward by 15°, as shown. Assuming the flow inside the nozzle is isentropic and $\gamma = 1.4$, calculate

(a) exit Mach number M_9 and

(b) the nozzle pressure ratio $p_{t,noz}/p_0$ (note that $p_{10} = p_0$)

(c) the Mach number after the expansion waves M_{10}

(d) the NPR if the nozzle was perfectly expanded

(e) the nozzle pressure ratio if a normal shock appears at the exit

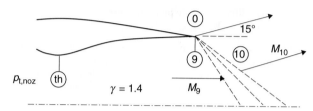

■ **FIGURE P5.36**

5.37 A convergent–divergent nozzle discharges to ambient air and has an exit-to-throat area ratio of $A_e/A_{th} = 2.4$. The total pressure at the entrance to the nozzle is $p_t = 100$ kPa. A normal shock occurs at the nozzle exit, as shown. Calculate

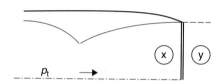

■ **FIGURE P5.37**

(a) the ambient pressure p_0 (note that $p_0 = p_y$)

(b) the exhaust temperature T_y if the temperature at the throat is known $T_{th} = 350°C$

(c) mass flow rate through the nozzle if the throat area is $A_{th} = 0.25$ m^2

Assume the medium is air with $\gamma = 1.4$, $c_p = 1.004$ kJ/kg · K.

5.38 A supersonic nozzle has an exit-to-throat area ratio of $A_e/A_{th} = 3.5$ and an upstream stagnation pressure of $p_t = 19 \times 10^5$ Pa, which remains constant with altitude. Calculate

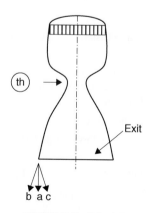

■ **FIGURE P5.38**

(a) the altitude at which the nozzle is perfectly expanded

(b) the altitude at which the exhaust plume is turned $10°$ outward

(c) the altitude at which the exhaust flow is turned $4°$ inward
Assume: $\gamma = 1.4$ and $c_p = 1,004\,\text{J/kg} \cdot \text{K}$.

5.39 A convergent–divergent nozzle operates at a pressure altitude of $p_{\text{amb}} = 1\,\text{kPa}$. The nozzle total pressure and temperature are $p_t = 101\,\text{kPa}$ and $T_t = 2000\,\text{K}$, respectively. Calculate

(a) nozzle area expansion ratio A_9/A_{th}, for perfect expansion

(b) nozzle exit Mach number M_9 for perfect expansion

(c) nozzle exit velocity V_9
Assume isentropic flow of air (with $\gamma = 1.4$ and $R = 287\,\text{J/kg} \cdot \text{K}$).

5.40 Consider an overexpanded nozzle as shown. Assuming the oblique shock at the exit makes a $40°$ angle with respect to the exit flow and the nozzle area ratio is 2.4, i.e., $A_9/A_{\text{th}} = 2.4$, calculate

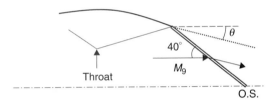

■ **FIGURE P5.40**

(a) M_9

(b) jet turning angle θ

(c) NPR, i.e., p_t/p_{amb}

5.41 A convergent–divergent nozzle has an area ratio $A_9/A_{\text{th}} = 6.79$ and a stagnation pressure of $p_t = 38.13\,\text{atm}$. First, calculate the pressure altitude (i.e., p_0) for which the nozzle is perfectly expanded. If this nozzle operated at a higher altitude, i.e., discharge to a lower backpressure atmosphere, it

will operate as an *underexpanded* nozzle with an attendant expansion fan at the exit lip. Now, if the backpressure, i.e., $p_0 = 1/2\,p_{0-\text{perfect–expansion}}$, calculate

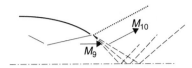

■ **FIGURE P5.41**

(a) Mach number after the (first) expansion fan, M_{10}

(b) the jet turning angle θ

(c) the tail wave angle of the first expansion fan with respect to the local flow

5.42 The exit flow from a two-dimensional C–D nozzle is shown to turn inward an angle of $15°$.

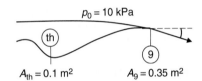

■ **FIGURE P5.42**

Calculate

(a) exit Mach number, M_9

(b) wave angle at the nozzle exit lip

(c) nozzle total pressure p_t (kPa)

5.43 A convergent nozzle experiences π_{cn} of 0.98, the gas ratio of specific heats $\gamma = 1.30$, and the gas constant is $R = 291\,\text{J/kg} \cdot \text{K}$. First, calculate the minimum nozzle pressure ratio that will choke the expanding nozzle, i.e., NPR_{crit}. This nozzle operates, however, at a higher NPR than the critical, namely, $\text{NPR} = 4.2$ and with an inlet stagnation temperature of

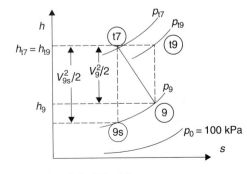

■ **FIGURE P5.43**

$T_{t7} = 939$ K. Assuming this nozzle operates in $p_0 = 100$ kPa ambient static pressure, calculate

(a) the exit static pressure and temperature p_9 and T_9, respectively

(b) the actual exit velocity V_9 in m/s

(c) nozzle adiabatic efficiency η_n

(d) the ideal exit velocity V_{9s} in m/s

(e) percent gross thrust gain, had we used a convergent–divergent nozzle with perfect expansion

(f) nozzle discharge coefficient C_{D8}

(g) draw a qualitative wave pattern in the exhaust plume

5.44   A convergent–divergent nozzle has a conical exhaust shape with the half-cone angle of $\alpha = 25°$. Calculate the divergence loss C_A for this nozzle due to nonaxial exhaust flow. Assuming the same (half) divergence angle of 25°, but in a 2D rectangular nozzle, calculate the flow angularity loss and compare it to the conical case.

5.45   The bypass ratio in a turbofan engine is $\alpha = 3.5$. We intend to mix the cold fan and the hot core flows in a mixer to enhance the engine gross thrust. Assuming the hot core temperature of the gas is $T_h = 3\,T_c$, calculate

(a) the mixed-out temperature of the gas T_m/T_c

(b) the percent increase in ideal gross thrust as a result of mixing the cold and hot streams in the mixer
You may assume a reversible adiabatic mixer flow with $\gamma_h = \gamma_c = 1.4$.

5.46   Consider a fixed-geometry supersonic nozzle with the following inlet and geometrical parameters: $T_{t7} = 2500$ K, $p_{t7} = 233.3$ kPa, $\dot{m}_8 = 100$ kg/s, $A_9/A_8 = 7.45$, $\gamma = 1.4$, and $c_p = 1004$ J/kg · K. Assuming isentropic flow in the nozzle, calculate

(a) ambient pressure p_0, if this nozzle is perfectly expanded

(b) ambient pressure if a normal shock appears at the nozzle exit

(c) nozzle throat and exit areas A_8 and A_9, respectively, in m^2

(d) the nozzle gross thrust with a NS at the exit (part b)

(e) the nozzle gross thrust in the severely overexpanded case of the N.S. being inside the nozzle at $A_s/A_8 = 4.23$. Assume the flow remains attached after the shock

(f) in reality the boundary layer in part e separates and reduces the exit flow area by the amount of "blockage".

Assuming the effective exit area is $A_{9e} = A_s$ calculate the gross thrust for this more realistic scenario.

(g) compare the gross thrust with/without separation. Why did separation help?

5.47   A convergent nozzle discharges into an open duct, as shown (i.e., the ejector concept). Through flow entrainment in the jet, air is drawn in from the upstream open end of the duct. The two flows mix in the duct and produce a mixed-out exit flow at the ambient static pressure p_0. Neglecting wall friction in the duct flow, calculate:

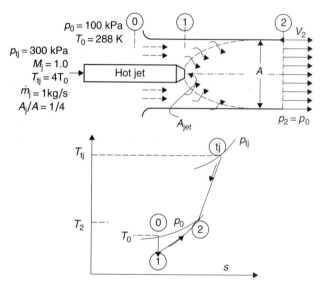

■ **FIGURE P5.47**

(a) the jet exit area A_j (in cm^2) and the ejector duct area A in cm^2

(b) the airflow drawn in from the duct inlet, i.e., the secondary flow rate $\dot{m}_s$

(c) the exit mixed-out gas velocity V_2

(d) the percent increase in mass flow rate

(e) the percent decrease in gas static temperature with the ejector

(f) gross thrust with and without the ejector

Hint: Apply conservation of mass, momentum, and energy to the control volume between 1 and 2. Flow between 0 and 1 is assumed to be isentropic.

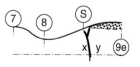 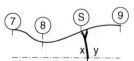 

■ **FIGURE P5.46**

5.48 The thermodynamic states of gas in an exhaust nozzle are shown on the T–s diagram.

Calculate

(a) exit Mach number M_9

(b) nozzle adiabatic efficiency η_n

Assume $\gamma = 1.30$.

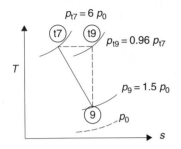

■ FIGURE P5.48

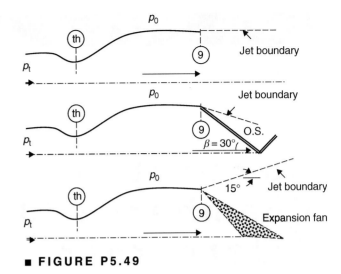

■ FIGURE P5.49

5.49 A convergent–divergent nozzle has an area ratio of $A_9/A_{th} = 5.9$. Assuming the flow is adiabatic and frictionless with $\gamma = 1.40$, calculate

(a) the nozzle pressure ratio p_t/p_0 for perfect expansion

(b) the nozzle pressure ratio p_t/p_0 if there is an oblique shock with $\beta = 30°$ at the exit lip, as shown

(c) the nozzle pressure ratio p_t/p_0 if there is an expansion wave at the lip that turns the flow 15° outward, as shown.

5.50 A scramjet flies at Mach 6 with an inlet total pressure recovery of 50%. Assuming the combustor experiences a total pressure loss of 42% (from its inlet condition), calculate the NPR, assuming $\gamma = 1.30$ and is constant.

5.51 Calculate the ratio of gross thrust produced by a convergent–divergent nozzle to the gross thrust produced by a convergent nozzle when they both operate with the same nozzle pressure ratio of NPR = 10 and $\gamma = 1.30$.

CHAPTER 6

Combustion Chambers and Afterburners

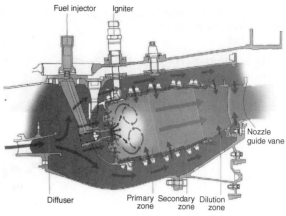

Courtesy of Rolls-Royce, plc

6.1   Introduction

To study the *mechanism of heat release* in an aircraft gas turbine engine combustion chamber, we need to revisit the thermochemistry principles that we learned in our freshman chemistry class. As engineers, we are interested in maximizing the heat release while minimizing the space requirements for the combustion chamber of an aircraft engine. Besides the *mechanism* of heat release, we are also interested in the *rate* at which chemical reactions takes place. We may think of it as one characteristic *timescale* of the problem. The question of *rate* is governed by the *chemical kinetics,* which is a subject in thermochemistry. In addition, we hope to understand the characteristic *length scales* of the problem involved in flame stability and length, which ultimately are tied to the combustion chamber geometry and sizing. In a combustion chamber of an airbreathing engine, we bring a liquid fuel and air together and an ignition source to start off the chemical reaction. We understand that the fuel has to be first vaporized before any reaction can take place between it and the oxygen in the air. We also recognize that a reaction between the oxygen in the air and the fuel must take place upon a *collision* between the two molecules. Only very energetic collisions result in the molecular dissociation of the fuel and a reaction with the oxygen. Therefore, not only the collisions are of necessity for a chemical reaction but also the *collision energy* is of importance. We may want to think of those collisions that lead to a chemical reaction as *effective collisions* in contrast to those collisions that result in a bounce back of the molecules. The question of whether collision energy is sufficient for a chemical reaction is answered by a chemical parameter known as the *activation energy* of the reaction. Now that we are relating to chemical reactions as energetic collisions between the oxygen and the fuel molecules, we may relate to the *rate of reaction* by the *frequency* of such collisions. We immediately recognize that these parameters, namely, the characteristic time and length scales in a combustion chamber depend on many parameters including the vaporization rate of the atomized fuel droplets and the turbulence intensity of the mixing process. Finally, as citizen engineers, we are committed to zero pollution, which will prove to be a daunting task in the combustion of a hydrocarbon fuel in air.

Chemical reactions take place at the molecular level and are always expressed in the following form:

$$a\,A + b\,B + \cdots \rightarrow m\,M + n\,N + \cdots \tag{6.1}$$

where a, b, m, and n are the number of *moles* of reactants A and B and the products M and N, respectively. The left-hand side of the reaction is known as the *reactants,* and the right-hand side of the reaction is known as the *products*. The number of moles of a substance is the ratio of the mass of the substance to its molecular weight, for example,

$$a \equiv \frac{m_a}{\mathrm{MW}_a} \tag{6.2}$$

with the dimensions of

$$[a] = \frac{\mathrm{kg}}{\mathrm{kg/kmol}} = \mathrm{kmol} \tag{6.3}$$

EXAMPLE 6.1

If a mixture contains 12 kg of H_2 and 8.0 kg of O_2, determine the number of moles of hydrogen and oxygen in the mixture.

$$n_{H_2} = \frac{12\,\mathrm{kg}}{2\,\mathrm{kg/kmol}} = 6.0\,\mathrm{kmol}$$

and

$$n_{O_2} = \frac{8\,\mathrm{kg}}{32\,\mathrm{kg/kmol}} = 0.25\,\mathrm{kmol}$$

Another interpretation of the mole of a substance is the amount of the substance that contains 6.023×10^{23} molecules (in a gram. mole), which is known as the Avogadro's number N_a. Avogadro's number is defined as the number of carbon atoms in 12 g of ^{12}C, which is 6.023×10^{23} molecules/gmol, therefore,

$$n = \frac{N}{N_a} \tag{6.4}$$

However, presenting the number of moles of a substance is easiest based on mass and molecular weight rather than the number of individual molecules and the Avogadro's number. In an engineering combustion problem, we typically introduce a certain *mass* of fuel and oxidizer in a reaction and not (explicitly) their number of individual molecules.

An important conservation law applicable to chemical reactions is the *law of conservation of atomic species*. It simply states that the numerical count of atoms of each species on both sides of a reaction has to be the same. The following two examples illustrate this conservation principle.

$$2H_2 + O_2 \rightarrow H_2O + OH + H$$

There are four hydrogen atoms on the reactants side as well as the product side. Two oxygen atoms appear on both sides as well. In the following reaction, methane (CH_4) is reacting with oxygen, with C, H, and O in balance.

$$CH_4 + 2O_2 \rightarrow CO_2 + 2H_2O$$

In dealing with the problems of reacting gases, such as the type experienced in a combustion chamber of a gas turbine engine, we are faced with a *mixture* of gases and hence we need to review the laws governing such mixtures. The most fundamental question regarding mixtures is whether we know the *mixture* properties, as a function of its *constituent* properties. The mixture properties range from the molecular weight of the mixture to the specific heats of the mixture, the specific enthalpy and entropy of the mixture. In addition, the constituent gases are assumed to obey the perfect gas law, which at sufficiently low densities, we understand that all gases behave as perfect.

6.2   Laws Governing Mixture of Gases

Imagine there is a volume V that contains several gases. The equilibrium mixture of gases has attained a mixture temperature T_m, which we now identify as being shared by all the constituents, i.e., all constituents have the same random kinetic energy after numerous collisions that brought them to an equilibrium state, namely,

$$T_m = T_1 = T_2 = T_3 = \cdots = T_n \tag{6.5}$$

Before we define the pressure of the mixture p_m, let us first define the *partial pressure* associated with the constituents of the mixture. A constituent of a mixture that occupies the *entire volume* of the mixture, V, and is at the mixture temperature, exerts a pressure on the vessel, which is called its partial pressure. Now, Dalton's law of additive pressures states that the mixture pressure is the sum of all its constituents' partial pressures (a direct consequence of perfect gas law), namely,

$$p_m = p_1 + p_2 + p_3 + \cdots + p_n = \sum_{i=1}^{n} p_i \tag{6.6}$$

where p_n is the partial pressure of the nth constituent.

On the question of internal energy, enthalpy and the entropy of the mixture of gases, we have the Gibbs–Dalton law that states

$$E_m = E_1 + E_2 + E_3 + \cdots + E_n = \sum_{i=1}^{n} E_i \tag{6.7}$$

$$H_m = H_1 + H_2 + H_3 + \cdots + H_n = \sum_{i=1}^{n} H_i \tag{6.8}$$

$$S_m = S_1 + S_2 + S_3 + \cdots + S_n = \sum_{i=1}^{n} S_i \tag{6.9}$$

where E is the internal energy, H the enthalpy, and S is the entropy. The constituent properties in Eqs. 6.7 through 6.9 are all based on the assumption that a constituent occupies the entire volume and is at the mixture temperature. We can also express the intrinsic variables of state, namely, the *specific* internal energy, the *specific* enthalpy, and the *specific* entropy based on the above Eqs. 6.7 through 6.9, as

$$e_m = \frac{m_1 e_1 + m_2 e_2 + m_3 e_3 + \cdots + m_n e_n}{m_m} = \left[\sum_{i=1}^{n} m_i e_i\right] \bigg/ \left[\sum_{i=1}^{n} m_i\right] \tag{6.10}$$

$$h_m = \frac{m_1 h_1 + m_2 h_2 + m_3 h_3 + \cdots + m_n h_n}{m_m} = \left[\sum_{i=1}^{n} m_i h_i\right] \bigg/ \left[\sum_{i=1}^{n} m_i\right] \tag{6.11}$$

$$s_m = \frac{m_1 s_1 + m_2 s_2 + m_3 s_3 + \cdots + m_n s_n}{m_m} = \left[\sum_{i=1}^{n} m_i s_i\right] \bigg/ \left[\sum_{i=1}^{n} m_i\right] \tag{6.12}$$

Now, the mass of the mixture m_m is obviously the sum of the individual masses m_i based on the law of conservation of mass. We can express the mass of each constituent as the product of the number of moles and the molecular weight, of that constituent, based on Eq. 6.2, as

$$m_{\mathrm{m}} = n_{\mathrm{m}}\mathrm{MW}_{\mathrm{m}} = \sum_{i=1}^{n} m_i = \sum_{i=1}^{n} n_i\mathrm{MW}_i \tag{6.13}$$

Therefore, the mixture molecular weight is now expressible in terms of the individual molecular weights and the *mole fraction* of each constituent, which we now define as χ

$$\chi_i \equiv \frac{n_i}{\sum n_i} = \frac{n_i}{n_{\mathrm{m}}} \tag{6.14}$$

and, therefore,

$$\mathrm{MW}_{\mathrm{m}} = \frac{\sum n_i \mathrm{MW}_i}{n_{\mathrm{m}}} = \sum \chi_i \mathrm{MW}_i \tag{6.15}$$

As an example, the air is composed of nitrogen, oxygen, and traces of inert gases, such as argon. Its composition is said to be nearly 78% N_2, 21% O_2, and 1% Ar, by volume. Therefore, 0.78, 0.21, and 0.01 are the *volume fractions* of the nitrogen, oxygen, and argon, respectively, in air.

From perfect gas law, written for a constituent as

$$p_i V_{\mathrm{m}} = n_i \bar{R} T_{\mathrm{m}} \tag{6.16}$$

where $\bar{R}$ is the universal gas constant with a value of

$$\bar{R} = 8.3146 \frac{\mathrm{kJ}}{\mathrm{kmol\,K}} = 1.9864 \frac{\mathrm{kcal}}{\mathrm{kmol\,K}}$$

$$\bar{R} = 1545.4 \frac{\mathrm{ft \cdot lbf}}{\mathrm{lbm \cdot mole\,°R}} = 1.9872 \frac{\mathrm{BTU}}{\mathrm{lbm \cdot mole\,°R}}$$

We can sum Eq. 6.16 for all the constituents to get

$$\sum p_i V_{\mathrm{m}} = p_{\mathrm{m}} V_{\mathrm{m}} = n_{\mathrm{m}} \bar{R} T_{\mathrm{m}} = \sum n_i \bar{R} T_{\mathrm{m}} \tag{6.17}$$

which suggests that the ratio of partial pressure of any constituent to the mixture pressure is equal to the mole fraction of that constituent, namely,

$$\frac{p_i}{p_{\mathrm{m}}} = \frac{n_i}{n_{\mathrm{m}}} \equiv \chi_i \tag{6.18}$$

Now, if we consider a mixture of gases, such as air, at a pressure, namely, p_{m}, and a temperature T_{m}, we may define a volume occupied by each constituent, V_i, such that

$$p_{\mathrm{m}} V_i = n_i \bar{R} T_{\mathrm{m}} \tag{6.19}$$

Dividing Eq. 6.19 by Eq. 6.17 yields

$$\frac{V_i}{V_{\mathrm{m}}} = \frac{n_i}{n_{\mathrm{m}}} = \chi_i \tag{6.20}$$

The above equation states that the volume fraction of a gas constituent in a mixture, as defined by Eq. 6.19 (i.e., at T_{m} and p_{m}), is equal to the constituent's mole fraction. The same conclusion about the volume fraction and mole fraction (i.e., being equal) may be reached by remembering the Law of Equal Volumes. It states that the volume occupied by one mole of any perfect gas is the same as that of any other perfect gas at a constant pressure and temperature. Consequently,

■ **FIGURE 6.1**
Definition sketch showing the volumes occupied by a constituent gas in a mixture of gases (with $V_A/V_B = n_A/n_B$)

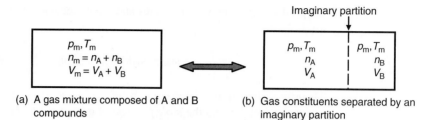

(a) A gas mixture composed of A and B compounds

(b) Gas constituents separated by an imaginary partition

when we describe the air composition as being 0.78, 0.21, and 0.01 for N_2, O_2, and Ar, respectively, by volume, we are stating the air composition on a mole fraction basis as

$$1 \text{ mole of air} = 0.78\,N_2 + 0.21\,O_2 + 0.01\,Ar \qquad (6.21)$$

with 0.78, 0.21, and 0.01 as mole fractions of the constituents N_2, O_2, and Ar. Now, we can use Eq. 6.15 to arrive at the molecular weight of air as

$$MW_{air} \cong 0.78(28) + 0.21(32) + 0.01(40) = 28.97 \approx 29 \text{ kg/kmol} \qquad (6.22)$$

Figure 6.1 shows the meaning of the volume occupied by a constituent gas in a mixture of gases. The specific heats at constant volume and pressure follow the perfect gas law:

$$\begin{aligned} de &= c_v dT \\ dh &= c_p dT \end{aligned} \qquad (6.23)$$

Now, combining these results with the mixture laws for the internal energy and enthalpy, Eqs. 6.10 and 6.11, respectively, we get

$$c_{v_m} = \frac{\sum m_i c_{vi}}{m_m} \qquad (6.24)$$

$$c_{p_m} = \frac{\sum m_i c_{pi}}{m_m} \qquad (6.25)$$

The ratio of specific heats for the mixture may now be written as

$$\gamma_m \equiv \frac{c_{p_m}}{c_{v_m}} = \frac{\sum m_i c_{pi}}{\sum m_i c_{vi}} \qquad (6.26)$$

If instead of working with masses, we prefer to work with the moles and mole fractions, we can replace each constituent mass by the product of its molecular weight and its number of moles. This approach is preferred in chemical analysis where compounds reaction is on a *molar* basis rather than mass basis. For example, let us express specific heat at constant pressure of the mixture in terms of the mole fractions instead of mass fractions, as given in Eq. 6.25,

$$c_{p_m} = \sum \frac{n_i MW_i c_{pi}}{n_m MW_m} = \frac{\sum \chi_i MW_i c_{pi}}{MW_m} \qquad (6.27)$$

6.3   Chemical Reaction and Flame Temperature

First, we define a series of terms that describe a chemical reaction, such as exothermic or endothermic. Let us assume the following chemical reaction:

$$a\,A + b\,B \longrightarrow c\,C + d\,D$$

If the reaction proceeds from the *reactants* A and B to the *products* C and D and the reaction produces heat, as a result, we shall call such reactions as *exothermic*. If the opposite occurs, i.e., if the reactants would not naturally react without an initial external heat input (stimulus), then such reactions are called *endothermic*. An example of an exothermic reaction is the combustion of hydrocarbon fuels and air. An example of an endothermic reaction is the phase transformation of water from liquid to vapor that requires a heat input. Our interest in chemical reaction lies primarily in the study of combustion and hence exothermic processes. In gas turbine engines, the reactants are typically hydrocarbon fuels and air. In a liquid propellant chemical rocket, the reactants are typically highly energetic fuels, such as hydrogen, and an oxidizer, such as oxygen. Let us consider octane (C_8H_{18}) as a typical hydrocarbon fuel in a gas turbine engine. Let us also describe air (by lumping argon into the nitrogen molecule) as being approximately composed of

$$\text{Air} \cong O_2 + \frac{79}{21} N_2 \cong O_2 + 3.76 \, N_2 \tag{6.28}$$

Now, we define a *unique* mass ratio between the fuel and oxidizer, which results in the *complete combustion* of the fuel and a *stable* product composition. This unique ratio is called the *stoichiometric* ratio and now we apply it to the combustion of octane and air, as an example.

$$C_8H_{18} + \frac{25}{2} (O_2 + 3.76 \, N_2) \rightarrow 8CO_2 + 9H_2O + \frac{25}{2} (3.76)N_2 \tag{6.29}$$

We took a mole of octane and brought in $25/2$ moles of O_2 and $(3.76)(25/2)$ moles of N_2. Note that the need for $25/2$ came from turning "C_8" of octane into "$8 \, CO_2$" and the "H_{18}" of octane into "$9 \, H_2O$," therefore $16 + 9 = 25$ oxygen atoms were needed for the complete combustion (which also means complete *oxidation*) of one mole of octane.

As noted, this *unique* proportion of fuel to oxidizer results in neither excess oxygen nor any excess fuel. Any more fuel would result in unburned fuel in the products of combustion and any more air would result in excess oxygen in the products of combustion. The nitrogen (molecule) (N_2) is treated as remaining unreacted (or *inert* in chemical terms) in the combustion process. Now, the stoichiometric fuel-to-air ratio for the combustion of octane in air is

$$f_{\text{stoich}} \equiv \frac{\text{mass}_{\text{fuel}}}{\text{mass}_{\text{oxidizer}}} = \frac{\dot{m}_f}{\dot{m}_{\text{air}}} \approx \frac{8(12) + 18(1)}{\frac{25}{2}[32 + 3.76(28)]} \cong \frac{114}{1716} \cong 0.0667 \tag{6.30}$$

Therefore, we say that the stoichiometric ratio for hydrocarbon fuels and air is about 6.7%. This is a number to remember. If we use proportionately less fuel than the stoichiometric ratio, the combustion is said to be on a *fuel lean* basis and the opposite is called a *fuel rich* combustion. A parameter that describes the fuel lean or rich condition of a combustor is called the *equivalence ratio ϕ*, which is defined as

$$\phi \equiv \frac{f}{f_{\text{stoich}}} \qquad \text{Equivalence ratio definition} \tag{6.31}$$

Consequently,

$$\phi \succ 1 \qquad \text{A fuel rich combustion} \tag{6.32}$$

$$\phi = 1 \qquad \text{Stoichiometric combustion} \tag{6.33}$$

$$\phi \prec 1 \qquad \text{A fuel lean combustion} \tag{6.34}$$

It will be demonstrated in this chapter that the highest combustion temperature is achieved very near the stoichiometric ratio (proportion of the fuel to the oxidizer). This fact, which is depicted in Fig. 6.2 should also be stored in memory as a useful concept. For many fuels, the maximum temperature, i.e., the adiabatic flame temperature, occurs at the equivalence ratio between 1 and 1.1, since the product-specific heat is reduced with a slightly fuel rich mixture ratio.

Now, assuming that we know the composition of the products in a chemical reaction, i.e., the constituents of the product and their number of moles, the fundamental question that remains is the product's temperature. How do we arrive at the product's temperature? If we treat the combustor as a black box, where the reactants enter the box at certain temperature, say T_1, and the products leave the box at another temperature, say T_2, and we can describe the heat

■ FIGURE 6.2
Flame temperature
dependence on the
equivalence ratio
showing T_{max} at $\phi \cong 1$

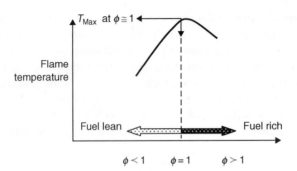

transfer through the sides of the box, i.e., the heat exchange with the surrounding, as say Q, we can use the law of conservation of energy to write:

$$Q = \sum H_{\text{Products}} - \sum H_{\text{Reactants}} \qquad (6.35)$$

where H is the *absolute* enthalpy and Q is the *external* heat interaction with the combustion chamber. We show this box schematically in Fig. 6.3, as an aid.

Remember *absolute enthalpy* from thermodynamics that was defined as the sum of heat of formation of the substance at the reference temperature (298.16 K) and what was called the *sensible enthalpy*, which raised the enthalpy of the substance from the reference temperature to the desired temperature. Absolute enthalpy H may be written as the product of mass and the specific absolute enthalpy for a substance. The mass may be expressed as the product of the number of moles and the molecular weight of the substance. Now, for a perfect gas, if we incorporate all of the above concepts, the absolute enthalpy may be written as

$$H = m \cdot h = n \cdot MW \cdot h = n \cdot MW \left[\int_{T_f}^{T} c_p dT + \Delta h_f^0 \right] \qquad (6.36)$$

where T_f is the reference temperature, Δh_f^0 is the standard heat of formation of the substance at the reference temperature per unit mass at the pressure of 1 bar. The subscript "f" refers to the reference temperature and the superscript "0" refers to the standard state of the chemical compound at the pressure of 1 bar. We define the product of the molecular weight and the specific heat at constant pressure (that appears above) as the *molar specific heat*, and give it a special symbol, namely, the c_p with a bar, i.e., let

$$\bar{c}_p \equiv MW \cdot c_p \qquad (6.37)$$

The dimensions of the molar specific heat are

$$[\bar{c}_p] = \frac{\text{kg}}{\text{kmol}} \frac{\text{kJ}}{\text{kg} \cdot \text{K}} = \frac{\text{kJ}}{\text{kmol} \cdot \text{K}} \qquad (6.38)$$

■ FIGURE 6.3
Combustion chamber is
depicted at its
boundaries, through
energy exchange

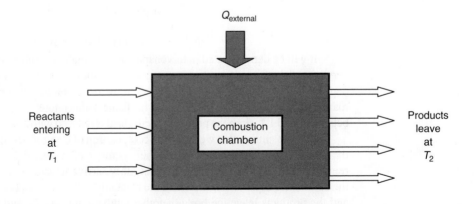

We also define the product of the molecular weight and the heat of formation per unit mass of a substance, as the *molar* heat of formation, and give it a special symbol (with a bar to identify the molar), as

$$\Delta \overline{h}_f^0 \equiv MW \cdot \Delta h_f^0 \tag{6.39}$$

The molar heat of formation is the evolved heat in forming *one mole* of the substance from its elements in their standard states at constant temperature of $T_f = 298.16\,\text{K}$ and the constant pressure of $p_f = 1\,\text{bar}$. The dimensions of the molar heat of formation at a reference temperature are

$$\left[\Delta \overline{h}_f^0\right] = \frac{\text{kg}}{\text{kmol}} \frac{\text{kJ}}{\text{kg}} = \frac{\text{kJ}}{\text{kmol}} \tag{6.40}$$

Imagine a reaction chamber, where the basic elements of a chemical compound, at their standard state, enter at a reference temperature T_f at the pressure of 1 bar and the compound is then formed in the reaction chamber. Now, in order to maintain the temperature of the compound at the reference temperature, we need to interact with the reaction chamber through heat transfer to or from the box. The heat of formation of a chemical compound is the amount of heat exchange to (positive) or from (negative) a reaction chamber, which forms the chemical compound from its constituents, and maintains a constant (known as reference) temperature. The heat of formation of all naturally occurring elements is then, by definition, zero. The reference temperature is 298.16 K, which corresponds to a standard room temperature of 25°C (77°F) and the reference pressure of 1 bar is the standard atmospheric pressure (also very close to 1 atm). This process can be schematically shown in Fig. 6.4.

The exothermic reactions from basic elements to form one mole of a chemical compound then lead to a negative heat of formation, because we need to extract heat from the reaction chamber in order to maintain a constant temperature. The opposite is true for endothermic reactions to form a compound. These concepts are best learned through a series of examples.

EXAMPLE 6.2

The standard heat of formation of the following naturally occurring elements (at $p = 1$ bar) is zero:

O_2 (g) Oxygen (molecule) in gaseous form, $(\Delta \overline{h}_f^0)_{O_2} = 0$

N_2 (g) Nitrogen (molecule) in gaseous form, $(\Delta \overline{h}_f^0)_{N_2} = 0$

H_2 (g) Hydrogen (molecule) in gaseous form, $(\Delta \overline{h}_f^0)_{H_2} = 0$

C (s) Carbon in solid form, $(\Delta \overline{h}_f^0)_{C_{(s)}} = 0$

The heat of formation of carbon dioxide can be found from the following isothermal reaction at 1 bar:

$$C_{(s)} + O_{2(g)} \xrightarrow{\;298.16\,\text{K}\;} CO_{2(g)} - 393,522\,\text{kJ/kmol}$$

■ FIGURE 6.4
Definition sketch for the *heat of formation* of (one mole of) a chemical compound in an isothermal process ($p = 1$ bar)

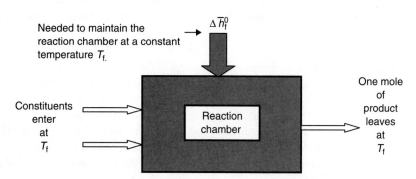

Therefore, $CO_2(g)$, carbon dioxide (gaseous), has a standard heat of formation of $\Delta \overline{h}_f^0 = -393,522\,kJ/kmol$. We note that in the previous reaction, the carbon enters the reaction as a solid and the oxygen as a gas, i.e., the standard states of carbon and oxygen, to form one mole of carbon dioxide (in gaseous form).

$$C_{(s)} + \frac{1}{2}O_{2(g)} \xrightarrow{298.16\,K} CO_{(g)} - 110,530\,kJ/kmol$$

In forming one mole of carbon monoxide in an isothermal process with $T_f = 298.16\,K$ and at a pressure of 1 bar, carbon and oxygen in their standard states have reacted according to the above. We note that a heat extraction of $110,530\,kJ/kmol$ (same as J/mol) is necessary to maintain an isothermal status for the reaction. Hence, we conclude that standard heat of formation of carbon monoxide, CO (g), is $\Delta \overline{h}_f^0 = -110,530\,kJ/kmol$. The following four reactions serve as additional examples on the standard heat of formation.

$$\frac{1}{2}O_{2(g)} \xrightarrow{298.16\,K} O + 249,170\,kJ/kmol \Rightarrow (\Delta \overline{h}_f^0)_O = +249,170\,kJ/kmol$$

$$\frac{1}{2}H_{2(g)} \xrightarrow{298.16\,K} H + 217,999\,kJ/kmol \Rightarrow (\Delta \overline{h}_f^0)_H = +217,999\,kJ/kmol$$

$$8C_{(s)} + 9H_2 \xrightarrow{298.16\,K} C_8H_{18(g)} - 208,447\,kJ/kmol \Rightarrow (\Delta \overline{h}_f^0)_{C_8H_{18(g)}} = -208,447\,kJ/kmol$$

$$8C_{(s)} + 9H_2 \xrightarrow{298.16\,K} C_8H_{18(l)} - 249,930\,kJ/kmol \Rightarrow (\Delta \overline{h}_f^0)_{C_8H_{18(l)}} = -249,930\,kJ/kmol$$

A blended jet fuel known as JP-4 has an approximate formulation, $CH_{1.93}$ and the following isothermal reaction defines its standard heat of formation in liquid form, i.e.,

$$C_{(s)} + \frac{1.93}{2}H_2 \xrightarrow{298.16\,K} CH_{1.93(l)} - 177,000\,kJ/kmol \Rightarrow (\Delta \overline{h}_f^0)_{CH_{1.93(l)}} = -177,000\,kJ/kmol$$

The energy balance equation across the burner (Eq. 6.35) may now be expressed on a molar basis, to include the complete enthalpy expression. This will account for the reactants entering the combustion chamber at other than reference temperature, namely, T_1, and the products of combustion leave the chamber at another temperature, namely, T_2, as follows:

$$Q = \sum_j [n_j \cdot \overline{c}_{pj}(T_2 - T_f)_j]_{Products} - \sum_i [n_i \cdot \overline{c}_{pi}(T_1 - T_f)_i]_{Reactants} + \Delta H_{RPf} \quad (6.41)$$

where "P" and "R" stand for products and reactants, respectively, and the last term is defined as

$$\Delta H_{RPf} = \sum_j [n_j \Delta \overline{h}_{fj}^0]_{Products} - \sum_i [n_i \Delta \overline{h}_{fi}^0]_{Reactants} \quad (6.42)$$

is the difference between the standard heats of formation of all the product constituents and the reactant constituents, at the reference temperature at the pressure of 1 bar. This quantity is also called *heat of reaction* with a symbol Q_R. Therefore, heat of reaction in a combustion process is defined as

$$Q_R \equiv \sum_P n_j \Delta \overline{h}_{fj}^0 - \sum_R n_i \Delta \overline{h}_{fi}^0 \quad (6.43)$$

We can specialize the concept of the *heat of reaction* to combustion of a fuel in pure oxygen O_2. This will be the basis for the definition of the fuel heating value. Assuming that the fuel and the gaseous oxygen, i.e.,O_2 (g), enter the combustion chamber at the reference temperature of 298.16 K and a complete combustion takes place, the heat release per unit mass of the fuel, which will return the products of combustion to the reference temperature, is called the *heating value* of the fuel. In simple terms, substitute T_f for T_1 and T_2 in Eq. 6.41, which will force the summations over the products and the reactants to zero, and note that

$$Q = Q_R = \sum_P n_j \Delta \overline{h}_{fj}^0 - \sum_R n_i \Delta \overline{h}_{fi}^0 \quad (6.44)$$

Although the negative sign on the summations of Eq. 6.43 or 6.44 shows the necessity of *removing* a quantity of heat from the combustion chamber in order to maintain an isothermal environment, it is customary to report the fuel heating value as a positive quantity. Hence, we define

$$\text{Fuel heating value} \equiv -Q_R \qquad (6.45)$$

Now, depending on the state of water in the products of combustion, i.e., as a liquid or vapor, the fuel heating value is referred to as higher heating value (HHV) or lower heating value (LHV), respectively. Since the difference between the higher and lower heating values depends on the energy spent in vaporizing the condensed water in the products of combustion, the following energy balance relates the two heating values:

$$\text{HHV} = \text{LHV} + \frac{m_{H_2O}}{m_{fuel}} h_{lg} \qquad (6.46)$$

where the last term h_{lg} is the latent heat of vaporization for water at 25°C, which is 2443 kJ/kg. Now, let us put these concepts to work and obtain a fuel heating value based on the standard heats of formation and the water/fuel mass fraction and the latent heat of vaporization of water.

EXAMPLE 6.3

Establish the lower and higher heating values of hydrogen, $H_{2(g)}$.

First consider the following chemical reaction between one mole of hydrogen and 1/2 mole of oxygen to produce one mole of water vapor:

$$H_{2(g)} + \frac{1}{2} O_{2(g)} \xrightarrow{\text{298.16K}} H_2O_{(g)}$$

We had identified the heat of reaction Q_R in a chemical reaction as Eq. 6.43, i.e.,

$$Q_R \equiv \sum_P n_j \Delta \bar{h}_{fj}^0 - \sum_R n_i \Delta \bar{h}_{fi}^0 = 1\,\text{kmol}(\Delta \bar{h}_f^0)_{H_2O(g)}$$

$$= -241,827\,\text{kJ}$$

in the above equation, we used the standard heat of formation of water in gaseous form as −241,827 kJ/kmol. Since the mass of 1 kmol of hydrogen is ∼2 kg, and the water in the product is in vapor form, we conclude that the lower heating value of hydrogen is

$$\text{LHV} \approx 241,827\,\text{kJ}/(2\,\text{kg}) = 120,913\,\text{kJ/kg}$$

And the higher heating value of hydrogen is obtained from Eq. 6.46, as

$$\text{HHV} \approx 120,913\,\text{kJ/kg} + (18/2)$$
$$\times (2,443\,\text{kJ/kg}) \approx 142,900\,\text{kJ/kg}$$

The values listed in thermochemical tables (or fuel property tables) are

$H_{2(g)}$	
LHV	HHV
120,000 kJ/kg	141,800 kJ/kg

Our result is within 1% of the above values, where our approximation of the molecular weights accounts for a portion of the difference, and the rest lies in the measurement accuracy.

Tables of thermochemical data (Table 6.1), which list the standard heat of formation of numerous compounds is found in compilations, such as JANAF or the *Handbook of Chemistry and Physics*. Here, only a partial listing is reproduced for the purposes of illustration and problem solving involving chemical reactions.

Applying Eq. 6.41, to a chemical reaction with a known product composition and a known heat exchange characteristics through the walls of the combustor, Q, will result in the calculation of the exit temperature of the products of combustion, T_2. A reference boundary condition on the combustion chamber is the adiabatic assumption of zero heat exchange with the environment, i.e., $Q \equiv 0$, which leads to a unique exit temperature value, known as the *adiabatic flame temperature* T_{af}. In addition, there are polynomial expressions that are developed for the molar specific heats of various chemical compounds, such as the sample

■ **TABLE 6.1**
Standard Heats of Formation at 298.16 K

Chemical symbol	Name	State	$\Delta \bar{h}_f^0$ (kJ/gmol) [per mole basis]	Δh_f^0 (kJ/gm) [per mass basis]
C	Carbon	Solid	0	0
C	Carbon	Gas	716.67	59.72
CO_2	Carbon dioxide	Gas	−393.522	−8.944
CO	Carbon monoxide	Gas	−110.53	−3.947
H_2	Hydrogen	Gas	0	0
H	Hydrogen atom	Gas	217.999	217.999
OH	Hydroxyl radical	Gas	39.463	2.321
H_2O	Water	Gas	−241.827	−13.435
H_2O_2	Hydrogen peroxide	Gas	−136.106	−4.003
N_2	Nitrogen	Gas	0	0
N	Nitrogen atom	Gas	472.68	33.763
NO	Nitric oxide	Gas	90.291	3.010
NO_2	Nitrogen dioxide	Gas	33.10	0.7196
N_2O	Nitrous oxide	Gas	82.05	1.8648
O_2	Oxygen	Gas	0	0
O	Oxygen atom	Gas	59.56	3.723
O_3	Ozone	Gas	34.00	0.708
CH_4	Methane	Gas	−74.873	−4.6796
CH_3OH	Methyl alcohol	Gas	−201.07	−6.2834
CH_3OH	Methyl alcohol	Liquid	−238.66	−7.4581
C_2H_5OH	Ethyl alcohol	Gas	−235.00	−5.108
C_2H_5OH	Ethyl alcohol	Liquid	−277.20	−6.026
C_3H_8	Propane	Gas	−103.90	−2.3614
C_4H_{10}	Butane	Gas	−126.148	−2.175
C_8H_{18}	Octane	Gas	−208.447	−1.8285
C_8H_{18}	Octane	Liquid	−249.93	−2.1924
$CH_{1.553}$	JP-3	Liquid		−1.11
$CH_{1.93}$	JP-4	Liquid		−1.77

Some useful conversion factors are $1 \ kcal = 4.1868 \ kJ = 3.9684 \ BTU = 3,088 \ ft \cdot lbf$ $kcal/gmol = 4186.8 \ kJ/kmol$ *and* $kJ/kmol = 1000 \ kJ/gmol$

produced in Table 6.2, which provide the remaining input to the enthalpy balance equation across the combustor, i.e., Eq. 6.41. As the polynomial expressions for specific heat are in temperature, and we do not a'priori know the temperature of the products of combustion, the process of selecting a suitable molar specific heat is iterative. Otherwise an *average* specific heat may be assumed, over the expected temperature range. Also note that in a fuel-rich combustion, $\phi \succ 1$, the heat of reaction remains the same as that of stoichiometric reaction, whereas the flame temperature will be reduced as the heat of reaction is now used to raise the temperature of additional fuel.

Now, let us proceed with two example problems that assume the molar composition of the products of combustion.

■ **TABLE 6.2**
Molar Specific Heats of Various Gases

Gases at low pressures $\bar{c}_{p0} = kJ/kmol \cdot K$ $\theta = T(K)/100$

Gas		Range (K)	Max. error (%)
N_2	$\bar{c}_{p0} = 39.060 - 512.79\theta^{-1.5} + 1072.7\theta^{-2} - 820.40\theta^{-3}$	300–3500	0.43
O_2	$\bar{c}_{p0} = 37.432 + 0.020102\theta^{1.5} - 178.57\theta^{-1.5} + 236.88\theta^{-2}$	300–3500	0.30
H_2	$\bar{c}_{p0} = 56.505 - 702.74\theta^{-0.75} + 1165.0\theta^{-1} - 560.70\theta^{-1.5}$	300–3500	0.60
CO	$\bar{c}_{p0} = 69.145 - 0.70463\theta^{0.75} - 200.77\theta^{-0.5} + 176.76\theta^{-0.75}$	300–3500	0.42
OH	$\bar{c}_{p0} = 81.564 - 59.3500\theta^{0.25} + 17.329\theta^{0.75-} - 4.2660\theta$	300–3500	0.43
HO	$\bar{c}_{p0} = 59.283 - 1.7096\theta^{0.5} - 70.613\theta^{-0.5} + 74.889\theta^{-1.5}$	300–3500	0.34
H_2O	$\bar{c}_{p0} = 143.05 - 183.54\theta^{0.25} + 82.751\theta^{0.5} - 3.6989\theta$	300–3500	0.43
CO_2	$\bar{c}_{p0} = -3.7357 + 30.529\theta^{0.5} - 4.1034\theta + 0.024\,198\theta^2$	300–3500	0.19
NO_2	$\bar{c}_{p0} = 46.045 + 216.10\theta^{-0.5} - 363.66\theta^{-0.75} + 232.550\theta^{-2}$	300–3500	0.26
CH_4	$\bar{c}_{p0} = 672.87 + 439.74\theta^{0.25} - 24.875\theta^{0.75} + 323.88\theta^{-0.5}$	300–2000	0.15
C_2H_4	$\bar{c}_{p0} = 95.395 + 123.15\theta^{0.5} - 35.641\theta^{0.75} + 182.77\theta^{-3}$	300–2000	0.07
C_2H_6	$\bar{c}_{p0} = 6.895 + 17.26\theta - 0.6402\theta^2 + 0.00728\theta^3$	300–1500	0.83
C_3H_8	$\bar{c}_{p0} = -4.042 + 30.46\theta - 1.571\theta^2 + 0.03171\theta^3$	300–1500	0.40
C_4H_{10}	$\bar{c}_{p0} = 3.954 + 37.12\theta - 1.833\theta^2 + 0.03498\,\theta^3$	300–1500	0.54

Source: Van Wylen, F.J. and Sonntag, R.E., *Fundamentals of Classical Thermodynamics*, New York: John Wiley & Sons, Inc., 1985.

EXAMPLE 6.4

Consider a chemical reaction involving H_2 and O_2 in gaseous form. The reactants enter the combustion chamber at 25°C. Assuming only water vapor exits as the composition product and the molar specific heat of water vapor is $\bar{c}_{P_{H_2O}} \cong 56.5\, kJ/kmol \cdot K$, calculate

(a) the ratio n_{H_2}/n_{O_2} of the reactants (mole ratio)
(b) the fuel-oxidizer (mass) ratio
(c) the adiabatic flame temperature T_{af}

SOLUTION

The problem is calling for a stoichiometric combustion of hydrogen and oxygen that goes to completion, hence the reaction takes on the following form:

$$H_2 + \frac{1}{2}O_2 \rightarrow H_2O \Rightarrow \boxed{\frac{n_{H_2}}{n_{O_2}} = \frac{1}{1/2} = 2}$$

The fuel-oxidizer mass ratio $= \boxed{\frac{2(1)}{1/2(32)} = \frac{1}{8} = 0.125}$

Equation 6.41 is set equal to zero for the adiabatic flame temperature calculation,

$$Q = \sum_P n_j \cdot \bar{c}_{pj}(T_2 - T_f)_j - \sum_R n_i \cdot \bar{c}_{pi}(T_1 - T_f)_i + \Delta H_{RPf} = 0$$

$$\Delta H_{RPf} = \sum_P n_j \Delta \bar{h}_{fj}^0 - \sum_R n_i \Delta \bar{h}_{fi}^0 = -241,827\, kJ$$

We also note that the reactants entrance temperature is the same as the reference temperature, hence the only contribution from the energy equation is

$$\bar{c}_{P_{H_2O}}(T_{af} - 298) = 241,827\, kJ \Rightarrow$$

$$\boxed{T_{af} = 298\, K + \frac{241,827}{56.5}\, K \cong 4,578\, K}$$

REMARKS   The adiabatic flame temperature of hydrogen–oxygen combustion is only *theoretically* 4578 K, as we have assumed a complete combustion. In an actual reaction, product dissociation takes place, which would reduce the energy release and hence not allow the adiabatic flame temperature to reach 4578 K. A fraction of water dissociates (back) into O_2, H_2, O, H, and OH. A realistic analysis would allow product dissociation to take place at an unknown molar concentration levels in a final

equilibrium state. The actual adiabatic flame temperature for the products of combustion (of H_2 and 1/2 O_2) in equilibrium is also a function of combustion pressure. We will address the methodology that allows us to calculate the equilibrium temperature and species concentration in Section 6.4.

Reactants	n_j	$\bar{h}_f^0$	Products	n_j	$\bar{h}_f^0$	$\bar{c}_{pj}$
	kmol	kJ/kmol		kmol	kJ/kmol	kJ/kmol·K
H_2	1	0	H_2O	1	−241,827	56.5
O_2	1/2	0				

EXAMPLE 6.5

One mole of methane is burned with 120% theoretical air. Assuming that the methane and air enter the combustion chamber at 25°C and the excess oxygen and nitrogen in the reaction will not dissociate, calculate

(a) the fuel-air ratio
(b) the equivalence ratio ϕ

(c) the adiabatic flame temperature

Assume:
$$\bar{c}_{p_{CO_2}} = 61.9\,\text{kJ/kmol} \cdot \text{K}, \ \bar{c}_{p_{O_2}} = 37.8\,\text{kJ/kmol} \cdot \text{K},$$
$$\bar{c}_{p_{N_2}} = 33.6\,\text{kJ/kmol} \cdot \text{K}, \ \bar{c}_{p_{H_2O}} = 52.3\,\text{kJ/kmol} \cdot \text{K},$$

SOLUTION

Theoretical air is composed of oxygen and nitrogen according to $(O_2 + 3.76\,N_2)$. First, let us write the stoichiometric reaction of methane and air, to establish a reference point:

$$CH_4 + 2(O_2 + 3.76\,N_2) \rightarrow CO_2 + 2\,H_2O + 7.52\,N_2$$
$$\text{(Stoichiometric)}$$

Now, let us implement the 120% factor for air, namely,

$$CH_4 + 2.4(O_2 + 3.76\,N_2)$$
$$\rightarrow CO_2 + 2\,H_2O + 0.4O_2 + 9.02\,N_2$$

Therefore, fuel-to-air ratio (based on mass) may be written as

$$\boxed{\frac{m_{\text{fuel}}}{m_{\text{air}}} = \frac{12 + 4}{2.4(32 + 3.76(28))} \cong 0.048}$$

For the equivalence ratio, we need the fuel-to-air ratio at stoichiometric condition as well,

$$\left(\frac{m_{\text{fuel}}}{m_{\text{air}}}\right)_{\text{Stoich.}} = \frac{12 + 4}{2(32 + 3.76(28))}$$

$$\cong 0.0583 \Rightarrow \boxed{\phi = \frac{0.048}{0.0583} \approx 0.827}$$

For the adiabatic flame temperature, let us setup a table for the "products" and "reactants," and calculate

$$\Delta H_{\text{RPf}} = \sum_P n_j \Delta \bar{h}_{fj}^0 - \sum_R n_i \Delta \bar{h}_{fi}^0 = -877,176 + 74,873$$
$$= -802,303\,\text{kJ and } \sum n_j \bar{c}_{pj} = 484.7\,\text{kJ}$$

From Eq. 6.41

$$T_2 - T_f = \frac{-\Delta H_{\text{RPF}}}{\sum n_j \bar{c}_{pj}} = \frac{802,303}{484.7}\,\text{K} \Rightarrow T_2 \cong 1,655\,\text{K},$$

REMARKS Now that we know the products' temperature, we need to go back and check the molar specific heats that were assumed at the beginning of the problem for a combustion temperature of 1655 K. We can use the polynomial expressions from Table 6.2. As expected, an iteration process ensues. Also note that our assumption of a complete combustion and no reverse reaction was flawed. We should allow for an equilibrium mixture of unburned methane, oxygen, some NO_x formation, among other products species to be present in the products of combustion. For example, if we had allowed product dissociation to occur, the final temperature would have been reduced. We clearly need more "tools" to dig deeper!

Reactants	n_j kmol	$\Delta \bar{h}_{fj}^0$ kJ/kmol	Products	n_j kmol	$\Delta \bar{h}_{fj}^0$ kJ/kmol	$\bar{c}_{pj}$ kJ/kmol · K
CH_4	1	−74,873	H_2O	2	−241,827	52.3
O_2	2.4	0	CO_2	1	−393,522	61.9
N_2	9.02	0	O_2	0.4	0	37.8
			N_2	9.02	0	33.6

6.4 Chemical Equilibrium and Chemical Composition

A chemical reaction normally continues until no more changes in molar concentration of the products occur. At that point, we say the reaction has reached an equilibrium state. We take advantage of the state of equilibrium in a chemical reaction and then try to establish the *chemical composition* of the products of combustion in that state. Take the combustion of one mole of hydrogen and one half mole of oxygen, as an example.

$$H_2 + \frac{1}{2}O_2 \leftrightarrow n_{H_2O}H_2O + n_{H_2}H_2 + n_{O_2}O_2 + n_{OH}OH + n_O O + n_H H \qquad (6.47)$$

First, note that *forward and reverse* arrows in Eq. 6.47 have replaced the forward reaction arrow of our earlier chemical reaction expressions. This simply means that a fraction of the products convert back into the reactants and as the combustion temperature increases other products may be formed and *eventually* a state of equilibrium is reached among all constituents. The word "eventually" in the previous sentence signifies a *period of time* needed to achieve the equilibrium state in a chemical reaction. Therefore, it is entirely reasonable to talk about a *rate of reaction*, and in particular the *rate of formation* of chemical species, which belong to the field of chemical kinetics. In the above equilibrium reaction of hydrogen and oxygen, we have expected/identified some water formation, some "leftover" hydrogen, some "leftover" oxygen, some hydroxyl (OH) formation, and possibly some dissociation of oxygen and hydrogen to form the atomic oxygen and hydrogen. We did not have to stop, however! Could the temperature of combustion be so high as for the oxygen and hydrogen atoms to ionize? And if so, what will the product concentrations be? These are all legitimate (and tough) questions. Now, let us stay at the level of Eq. 6.47, which identifies six constituents of unknown molar concentrations. The total number of moles of products at equilibrium is n_m, where

$$n_m = n_{H_2O} + n_{H_2} + n_{O_2} + n_{OH} + n_O + n_H \qquad (6.48)$$

The *molar concentration* of any species is simply the ratio of the number of moles of that species to the total number of moles of the product, i.e., the same as the *mole fraction* of any given species

$$\chi_i \equiv \frac{n_i}{n_m}$$

We shall apply the law of conservation of atomic species to the above reaction and conclude that

$$2n_{H_2O} + 2n_{H_2} + n_{OH} + n_H = 2 \quad \dots\dots\dots\dots\dots\dots \text{to balance H}$$

$$n_{H_2O} + 2n_{O_2} + n_{OH} + n_O = 1 \quad \dots\dots\dots\dots\dots\dots \text{to balance O}$$

Now, we have two equations and six unknowns. We need to produce four additional equations involving the six unknowns before we have a chance of solving for the six (unknown) molar concentrations.

In order to produce additional equations to assist with the unknown molar concentrations, we introduce a new law, which is called the *law of mass action*.

6.4.1 The Law of Mass Action

Consider a stoichiometric reaction of the type:

$$a\mathrm{A} + b\mathrm{B} \leftrightarrow c\mathrm{C} + d\mathrm{D}$$

where a, b, c, and d are the (stoichiometric) number of moles in the reaction. The law of mass action states that the rate of disappearance of chemical species in the reactants is proportional to the concentrations of the reactants each raised to their respective stoichiometric exponent. Applied to the forward rate of reaction r_f is proportional to the products of the concentrations of the reactants raised to their stoichiometric exponents, i.e.,

$$r_f \propto [\mathrm{A}]^a \cdot [\mathrm{B}]^b \tag{6.49}$$

The bracketed terms are molar concentrations of reactants A and B. The proportionality constant in Eq. 6.49 is called the forward reaction rate coefficient k_f, i.e.,

$$r_f = k_f [\mathrm{A}]^a \cdot [\mathrm{B}]^b \tag{6.50}$$

The law of mass action applied to a reverse reaction then relates the rate of formation of the products to the product of concentration of the products raised to their stoichiometric exponents, namely,

$$r_r = k_r [\mathrm{C}]^c \cdot [\mathrm{D}]^d \tag{6.51}$$

An equilibrium state is reached among the reactants and the product species, when the forward and reverse reaction rates are equal, namely,

$$k_f [\mathrm{A}]^a \cdot [\mathrm{B}]^b = k_r [\mathrm{C}]^c \cdot [\mathrm{D}]^d \tag{6.52}$$

Therefore the ratio of forward to reverse reaction rate coefficients in an equilibrium reaction is called the equilibrium constant K

$$K \equiv \frac{k_f}{k_r} = \frac{[\mathrm{C}]^c \cdot [\mathrm{D}]^d}{[\mathrm{A}]^a \cdot [\mathrm{B}]^b} \tag{6.53}$$

It is customary to express the concentrations of species by their partial pressures in the equilibrium mixture and the equilibrium constant is then referred to as K_P since it is based on partial pressures. Hence, in terms of the partial pressures, the law of mass action is written as

$$K_P = \frac{p_{\mathrm{C}}^c \cdot p_{\mathrm{D}}^d}{p_{\mathrm{A}}^a \cdot p_{\mathrm{B}}^b} \tag{6.54}$$

We may equivalently use the mole fractions (or molar concentrations) for species concentration in the law of mass action to arrive at an equivalent expression, namely,

$$K_n = \frac{\chi_{\mathrm{C}}^c \cdot \chi_{\mathrm{D}}^d}{\chi_{\mathrm{A}}^a \cdot \chi_{\mathrm{B}}^b} \tag{6.55}$$

The two equilibrium constants are related to each other, as we may relate the ratio of partial pressures and the mixture pressure to the mole fraction (according to Eq. 6.18), hence,

$$K_P = \frac{\chi_C^c \cdot \chi_D^d}{\chi_A^a \cdot \chi_B^b} \, p_m^{c+d-a-b} = K_n \, p_m^{c+d-a-b} \tag{6.56}$$

Among the two equilibrium constants, K_P and K_n, the former is a function of temperature alone and hence independent of pressure, whereas the latter is in general a function of both temperature and pressure, i.e.,

$$K_p = K_p(T_m) \tag{6.57}$$

$$K_n = K_n(T_m, p_m) \tag{6.58}$$

Based on a *single parameter* functional dependence of K_p, tables and charts of this equilibrium constant are produced as a function of temperature and used for product composition calculations. It is also interesting to note that the molar concentrations in a stoichiometric mixture are a function of the mixture pressure (if $c + d - a - b \neq 0$, according to Eq. 6.56), whereas the equilibrium constant K_P is only a function of the mixture temperature. The tables of values for the equilibrium constant K_P, as a function of temperature for numerous reactions of interest to combustion, are used to setup a series of auxiliary equations for the unknown molar concentrations. Some of these tables and graphical depictions of equilibrium constant K_P for many reactions are shown at the end of this section. The following example shows a simple application of this principle.

EXAMPLE 6.6

One mole of nitrogen, $N_{2(g)}$, is heated to 6250 K at the pressure of p_m. A fraction of nitrogen dissociates to nitrogen atom according to

$$xN_2 \rightarrow 2xN$$

Assuming a state of equilibrium is reached in the mixture, calculate

(a) mole fraction of N_2 at equilibrium when p_m is 1 atm
(b) mole fraction of N_2 at equilibrium when p_m is 10 atm

Assume the equilibrium constant for the reaction

$$N_2 \leftrightarrow 2N$$

is $K_p = 0.1$ atm at the temperature of 6250 K.

SOLUTION

If we start with a mole of N_2 and a fraction of it, say x mole, dissociates into nitrogen atom, then our equilibrium mixture takes on the following form:

$$N_2 \leftrightarrow (1 - x)N_2 + 2xN$$

In the above reaction, the nitrogen is in balance and we still have the unknown x to contend with. First, we need to express the mole fractions of the products in equilibrium. We start with the total number of moles of the mixture in equilibrium and then divide the species number of moles by the total, according to:

Total number of moles of the mixture at equilibrium is
$$(1 - x) + 2x = 1 + x$$

Therefore the nitrogen *mole fraction* of the mixture at equilibrium is $\dfrac{1 - x}{1 + x} = \chi_{N_2}$

The *mole fraction* of the nitrogen atom in the mixture at equilibrium is $\dfrac{2x}{1 + x} = \chi_N$

Writing an expression for the equilibrium constant of $N_2 \leftrightarrow 2N$, from its definition

$$K_p \equiv \frac{p_N^2}{p_{N_2}} = \frac{\chi_N^2}{\chi_{N_2}} \cdot p_m^{2-1}$$

$$= \left[\left(\frac{2\chi}{1+\chi} \right)^2 \right] \bigg/ \left[\frac{1-\chi}{1+\chi} \right] \cdot p_m = 0.1 \text{ atm}$$

(a) $p_m = 1$ atm

$$\frac{4x^2}{1 - x^2} = 0.1 \Rightarrow x^2 = \frac{0.1}{4.1} \Rightarrow x = 0.156$$

$$\therefore \chi_{N_2} \cong 0.73$$

(b) $p_m = 10$ atm

$$\frac{4x^2}{1 - x^2} = 0.01 \quad \Rightarrow \quad x^2 = \frac{0.01}{4.01} \quad \Rightarrow \quad x \cong 0.05$$
$$\therefore \chi_{N_2} \cong 0.90$$

REMARKS   First, note that the sum of the mole fractions in the mixture adds up to 1, as expected. Second, the fraction of nitrogen, which dissociates into nitrogen atom, x,

is suppressed by the mixture pressure, i.e., the lower mixture pressure promotes dissociation of nitrogen. Third, the mole fractions substituted in the K_p equation above were from the original reaction and not the stoichiometric reaction, $N_2 \leftrightarrow 2N$. Finally, the equilibrium constant given for the nitrogen dissociation, K_p, has a dimension of atmosphere (for this reaction) and is obtained from a table for the $N_2 \leftrightarrow 2N$ reaction at $T = 6250$ K. We will use the tables/charts for the equilibrium constants in the following sections.

6.4.2   Equilibrium Constant K_P

A first attempt in studying the molar concentrations of the products of combustion will require knowledge of the equilibrium constants (as a function of temperature) for at least the following subreactions:

- Dissociation of oxygen, nitrogen, and hydrogen into the atoms of O, N, and H, respectively
- Formation of OH, NO, NO_2, CO, O_3, NH_3 from various stoichiometric reactions
- Dissociation of water into elements, O, H, and/or OH, dissociation of CO_2

A more elaborate analysis will require additional subreactions, which go beyond the scope of our treatment. Now, let us examine some of the more useful subreactions we encounter more often in our analysis, namely,

Reaction	Equilibrium constant
1. $CO_2 + H_2 \leftrightarrow CO + H_2O$	$K_P = \dfrac{p_{CO} \cdot p_{H_2O}}{p_{CO_2} \cdot p_{H_2}} = \dfrac{\chi_{CO} \cdot \chi_{H_2O}}{\chi_{CO_2} \cdot \chi_{H_2}}$
2. $H_2O + \dfrac{1}{2}N \leftrightarrow H_2 + NO$	$K_P = \dfrac{p_{H_2} \cdot p_{NO}}{p_{H_2O} \cdot p_N^{1/2}} = \dfrac{\chi_{H_2} \cdot \chi_{NO}}{\chi_{H_2O} \cdot \chi_N^{1/2}} \cdot p_m^{1/2}$
3. $2H_2O \leftrightarrow 2H_2 + O_2$	$K_p = \dfrac{p_{H_2}^2 \cdot p_{O_2}}{p_{H_2O}^2} = \dfrac{\chi_{H_2}^2 \cdot \chi_{O_2}}{\chi_{H_2O}^2} \cdot p_m$
4. $H_2O \leftrightarrow H_2 + O$	$K_P = \dfrac{p_{H_2} \cdot p_O}{p_{H_2O}} = \dfrac{\chi_{H_2} \cdot \chi_O}{\chi_{H_2O}} \cdot p_m$
5. $\dfrac{1}{2}H_2 \leftrightarrow H$	$K_p = \dfrac{p_H}{p_{H_2}^{1/2}} = \dfrac{\chi_H}{\chi_{H_2}^{1/2}} \cdot p_m^{1/2}$
6. $H_2O \leftrightarrow \dfrac{1}{2}H_2 + OH$	$K_p = \dfrac{p_{H_2}^{1/2} \cdot p_{OH}}{p_{H_2O}} = \dfrac{\chi_{H_2}^{1/2} \cdot \chi_{OH}}{\chi_{H_2O}} \cdot p_m^{1/2}$
7. $N_2 \leftrightarrow 2N$	$K_p = \dfrac{p_N^2}{p_{N_2}} = \dfrac{\chi_N^2}{\chi_{N_2}} \cdot p_m$
8. $C + \dfrac{1}{2}O_2 \leftrightarrow CO$	$K_P = \dfrac{p_{CO}}{p_C \cdot p_{O_2}^{1/2}} = \dfrac{\chi_{CO}}{\chi_C \cdot \chi_{O_2}^{1/2}} \cdot p_m^{-1/2}$
9. $CO + \dfrac{1}{2}O_2 \leftrightarrow CO_2$	$K_P = \dfrac{p_{CO_2}}{p_{CO} \cdot p_{O_2}^{1/2}} = \dfrac{\chi_{CO_2}}{\chi_{CO} \cdot \chi_{O_2}^{1/2}} \cdot p_m^{-1/2}$

(Contiued)

Reaction	Equilibrium constant
10. $\frac{1}{2}N_2 + \frac{1}{2}O_2 \leftrightarrow NO$	$K_p = \dfrac{p_{NO}}{p_{N_2}^{1/2} \cdot p_{O_2}^{1/2}} = \dfrac{\chi_{NO}}{\chi_{N_2}^{1/2} \cdot \chi_{O_2}^{1/2}}$
11. $O_2 \leftrightarrow 2O$	$K_p = \dfrac{p_O^2}{p_{O_2}} = \dfrac{\chi_O^2}{\chi_{O_2}} \cdot p_m$
12. $CO_2 \leftrightarrow CO + \frac{1}{2}O_2$	$K_p = \dfrac{p_{CO} \cdot p_{O_2}^{1/2}}{p_{CO_2}} = \dfrac{\chi_{CO} \cdot \chi_{O_2}^{1/2}}{\chi_{CO_2}} \cdot p_m^{1/2}$

The equilibrium constants for the first 10 reactions are depicted in graphical form in Fig. 6.5 (from Hill and Peterson, 1992). The equilibrium constants for additional dissociation reactions of interest are tabulated in Table 6.3 (from Strehlow, 1984).

$$K_4 = \frac{p(H_2)\,p(O)}{p(H_2O)}$$

$$K_1 = \frac{p(CO)\,p(H_2O)}{p(CO_2)\,p(H_2)} \qquad K_5 = \frac{p(H)}{[p(H_2)]^{1/2}} \qquad K_8 = \frac{p(CO)}{p(C)[p(O_2)]^{1/2}}$$

$$K_2 = \frac{p(NO)\,p(H_2)}{[p(N)]^{1/2}\,p(H_2O)} \qquad K_6 = \frac{[p(H_2)]^{1/2}\,p(OH)}{p(H_2O)} \qquad K_9 = \frac{p(CO_2)}{p(CO)[p(O_2)]^{1/2}}$$

$$K_3 = \frac{[p(H_2)]^2\,p(O_2)}{[p(H_2O)]^2} \qquad K_7 = \frac{p(N)^2}{p(N_2)} \qquad K_{10} = \frac{p(NO)}{[p(N_2)]^{1/2}[p(O_2)]^{1/2}}$$

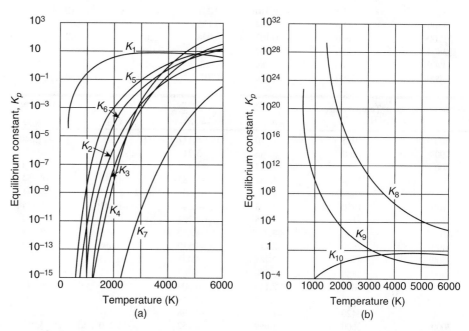

■ **FIGURE 6.5**   Equilibrium constants in terms of partial pressures, pressure in atmosphere (from Hill and Peterson, 1992)

■ **TABLE 6.3**
Logarithm to Base 10 of the Equilibrium Constant K_p

Temperature K	Temperature °R	$\frac{1}{2}O_2 \leftrightarrow O$	$\frac{1}{2}H_2 \leftrightarrow H$	$\frac{1}{2}N_2 \leftrightarrow N$	$H_2 + \frac{1}{2}O_2 \leftrightarrow H_2O$	$C + O_2 \leftrightarrow CO_2$	$C + \frac{1}{2}O_2 \leftrightarrow CO$
600	1080	−18.574	−16.336	−38.081	18.633	34.405	14.318
700	1260	−15.449	−13.599	−32.177	15.583	29.506	12.946
800	1440	−13.101	−11.539	−27.744	13.289	25.830	11.914
900	1620	−11.272	−9.934	−24.292	11.498	22.970	11.108
1000	1800	−9.807	−8.646	−21.528	10.062	20.680	10.459
1100	1980	−8.606	−7.589	−19.265	8.883	18.806	9.926
1200	2160	−7.604	−6.707	−17.377	7.899	17.243	9.479
1300	2340	−6.755	−5.958	−15.778	7.064	15.920	9.099
1400	2520	−6.027	−5.315	−14.406	6.347	14.785	8.771
1500	2700	−5.395	−4.756	−13.217	5.725	13.801	8.485
1600	2880	−4.842	−4.266	−12.175	5.180	12.940	8.234
1700	3060	−4.353	−3.833	−11.256	4.699	12.180	8.011
1800	3240	−3.918	−3.448	−10.437	4.270	11.504	7.811
1900	3420	−3.529	−3.102	−9.705	3.886	10.898	7.631
2000	3600	−3.178	−2.790	−9.046	3.540	10.353	7.469
2100	3780	−2.860	−2.508	−8.449	3.227	9.860	7.321
2200	3960	−2.571	−2.251	−7.905	2.942	9.411	7.185
2300	4140	−2.307	−2.016	−7.409	2.682	9.001	7.061
2400	4320	−2.065	−1.800	−6.954	2.443	8.625	6.946
2500	4500	−1.842	−1.601	−6.535	2.224	8.280	6.840
2600	4680	−1.636	−1.417	−6.149	2.021	7.960	6.741
2700	4860	−1.446	−1.247	−5.790	1.833	7.664	6.649
2800	5040	−1.268	−1.089	−5.457	1.658	7.388	6.563
2900	5220	−1.103	−0.941	−5.147	1.495	7.132	6.483
3000	5400	−0.949	−0.803	−4.858	1.343	6.892	6.407
3100	5580	−0.805	−0.674	−4.587	1.201	6.668	6.336
3200	5760	−0.670	−0.553	−4.332	1.067	6.458	6.269
3300	5940	−0.543	−0.439	−4.093	0.942	6.260	6.206
3400	6120	−0.423	−0.332	−3.868	0.824	6.074	6.145
3500	6300	−0.310	−0.231	−3.656	0.712	5.898	6.088
3600	6480	−0.204	−0.135	−3.455	0.607	5.732	6.034
3700	6660	−0.103	−0.044	−3.265	0.507	5.574	5.982
3800	6840	−0.007	0.042	−3.086	0.413	5.425	5.933
3900	7020	0.084	0.123	−2.915	0.323	5.283	5.886
4000	7200	0.170	0.201	−2.752	0.238	5.149	5.841

Temperature K	Temperature °R	$\frac{1}{2}O_2 + \frac{1}{2}N_2 \leftrightarrow NO$	$O_2 + \frac{1}{2}N_2 \leftrightarrow NO_2$	$\frac{1}{2}O_2 + \frac{1}{2}H_2 \leftrightarrow OH$	$\frac{1}{2}O_2 + N_2 \leftrightarrow N_2O$	$C + 2H_2 \leftrightarrow CH_4$	$C + \frac{3}{2}H_2 \leftrightarrow CH_3$
600	1080	−7.210	−6.111	−2.568	−11.040	2.001	−13.212
700	1260	−6.086	−5.714	−2.085	−10.021	0.951	−11.458
800	1440	−5.243	−5.417	−1.724	−9.253	0.146	−10.152
900	1620	−4.587	−5.185	−1.444	−8.654	−0.493	−9.145
1000	1800	−4.062	−5.000	−1.222	−8.171	−1.011	−8.344
1100	1980	−3.633	−4.848	−1.041	−7.774	−1.440	−7.693
1200	2160	−3.275	−4.721	−0.890	−7.442	−1.801	−7.153
1300	2340	−2.972	−4.612	−0.764	−7.158	−2.107	−6.698
1400	2520	−2.712	−4.519	−0.656	−6.914	−2.372	−6.309
1500	2700	−2.487	−4.438	−0.563	−6.701	−2.602	−5.974
1600	2880	−2.290	−4.367	−0.482	−6.514	−2.803	−5.681
1700	3060	−2.116	−4.304	−0.410	−6.347	−2.981	−5.423
1800	3240	−1.962	−4.248	−0.347	−6.198	−3.139	−5.195
1900	3420	−1.823	−4.198	−0.291	−6.065	−3.281	−4.991
2000	3600	−1.699	−4.152	−0.240	−5.943	−3.408	−4.808
2100	3780	−1.586	−4.111	−0.195	−5.833	−3.523	−4.642
2200	3960	−1.484	−4.074	−0.153	−5.732	−3.627	−4.492
2300	4140	−1.391	−4.040	−0.116	−5.639	−3.722	−4.355

■ **TABLE 6.3** (Continued)

Temperature K	Temperature °R	$\frac{1}{2}O_2 + \frac{1}{2}N_2$ $\leftrightarrow$ NO	$O_2 + \frac{1}{2}N_2$ $\leftrightarrow$ NO$_2$	$\frac{1}{2}O_2 + \frac{1}{2}H_2$ $\leftrightarrow$ OH	$\frac{1}{2}O_2 + N_2$ $\leftrightarrow$ N$_2$O	$C + 2H_2$ $\leftrightarrow$ CH$_4$	$C + \frac{3}{2}H_2$ $\leftrightarrow$ CH$_3$
2400	4320	−1.305	−4.008	−0.082	−5.554	−3.809	−4.230
2500	4500	−1.227	−3.979	−0.050	−5.475	−3.889	−4.115
2600	4680	−1.154	−3.953	−0.021	−5.401	−3.962	−4.009
2700	4860	−1.087	−3.928	0.005	−5.333	−4.030	−3.911
2800	5040	−1.025	−3.905	0.030	−5.270	−4.093	−3.820
2900	5220	−0.967	−3.884	0.053	−5.210	−4.152	−3.736
3000	5400	−0.913	−3.864	0.074	−5.154	−4.206	−3.659
3100	5580	−0.863	−3.846	0.094	−5.102	−4.257	−3.584
3200	5760	−0.815	−3.828	0.112	−5.052	−4.304	−3.515
3300	5940	−0.771	−3.812	0.129	−5.006	−4.349	−3.451
3400	6120	−0.729	−3.797	0.145	−4.962	−4.391	−3.391
3500	6300	−0.690	−3.783	0.160	−4.920	−4.430	−3.334
3600	6480	−0.653	−3.770	0.174	−4.881	−4.467	−3.280
3700	6660	−0.618	−3.757	0.188	−4.843	−4.503	−3.230
3800	6840	−0.585	−3.746	0.200	−4.807	−4.536	−3.182
3900	7020	−0.554	−3.734	0.212	−4.773	−4.568	−3.137
4000	7200	−0.524	−3.724	0.223	−4.741	−4.598	−3.095

Temperature K	Temperature °R	$C + H_2$ $\leftrightarrow$ CH$_2$	$C + \frac{1}{2}H_2$ $\leftrightarrow$ CH	$2C + H_2$ $\leftrightarrow$ C$_2$H$_2$	$C_2 + 2H_2$ $\leftrightarrow$ C$_2$H$_4$	$\frac{1}{2}N_2 + \frac{3}{2}H_2$ $\leftrightarrow$ NH$_3$	$\frac{1}{2}N_2 + H_2$ $\leftrightarrow$ NH$_2$	$\frac{1}{2}N_2 + \frac{1}{2}H_2$ $\leftrightarrow$ NH
600	1080	−30.678	−45.842	−16.687	−7.652	−1.377	−18.326	−31.732
700	1260	−25.898	−38.448	−13.882	−7.114	−2.023	−15.996	−27.049
800	1440	−22.319	−32.905	−11.784	−6.728	−2.518	−14.255	−23.537
900	1620	−19.540	−28.597	−10.155	−6.438	−2.910	−12.905	−20.806
1000	1800	−17.321	−25.152	−8.856	−6.213	−3.228	−11.827	−18.621
1100	1980	−15.508	−22.336	−7.795	−6.034	−3.490	−10.948	−16.834
1200	2160	−14.000	−19.991	−6.913	−5.889	−3.710	−10.216	−15.345
1300	2340	−12.726	−18.008	−6.168	−5.766	−3.897	−9.598	−14.084
1400	2520	−11.635	−16.310	−5.531	−5.664	−4.058	−9.069	−13.004
1500	2700	−10.691	−14.838	−4.979	−5.575	−4.197	−8.610	−12.068
1600	2880	−9.866	−13.551	−4.497	−5.497	−4.319	−8.210	−11.249
1700	3060	−9.139	−12.417	−4.072	−5.430	−4.426	−7.856	−10.526
1800	3240	−8.493	−11.409	−3.695	−5.369	−4.521	−7.542	−9.883
1900	3420	−7.916	−10.507	−3.358	−5.316	−4.605	−7.261	−9.308
2000	3600	−7.397	−9.696	−3.055	−5.267	−4.681	−7.009	−8.790
2100	3780	−6.929	−8.963	−2.782	−5.223	−4.749	−6.780	−8322
2200	3960	−6.503	−8.296	−2.532	−5.183	−4.810	−4.572	−7.896
2300	4140	−6.115	−7.687	−2.306	−5.146	−4.866	−6.382	−7.507
2400	4320	−5.760−	−7.130	−2.098	−5.113	−4.916	−6.208	−7.151
2500	4500	−5.433	−6.617	−1.506	−5.081	−4.963	−6.048	−6.823
2600	4680	−5.133	−6.144	−1.730	−5.052	−5.005	−5.899	−6.520
2700	4860	−4.854	−5.706	−1.566	−5.025	−5.044	−5.762	−6.240
2800	5040	−4.596	−5.300	−1.415	−5.000	−5.079	−5.635	−5.979
2900	5220	−4.356	−4.922	−1.274	−4.977	−5.112	−5.516	−5.737
3000	5400	−4.132	−4.569	−1.142	−4.955	−5.143	−5.405	−5.511
3100	5580	−3.923	−4.239	−1.019	−4.934	−5.171	−5.300	−5.299
3200	5760	−3.728	−3.930	−0.903	−4.915	−5.197	−5.203	−5.100
3300	5940	−3.544	−3.639	−0.795	−4.897	−5.221	−5.111	−4.914
3400	6120	−3.372	−3.366	−0.693	−4.880	−5.244	−5.024	−4.738
3500	6300	−3.210	−3.108	−0.597	−4.864	−5.265	−4.942	−4.572
3600	6480	−3.056	−2.865	−0.506	−4.848	−5.285	−4.865	−4.416
3700	6660	−2.912	−2.636	−0.420	−4.834	−5.304	−4.791	−4.267
3800	6840	−2.775	−2.418	−0.339	−4.821	−5.321	−4.721	−4.127
3900	7020	−2.646	−2.212	−0.262	−4.808	−5.338	−4.655	−3.994
4000	7200	−2.523	−2.016	−0.189	−4.796	−5.353	−4.592	−3.867

Let us work with these equilibrium constants to deduce equilibrium constants for other reactions that may be of interest. For example, how does the equilibrium constant of the following two Stoichiometric reactions relate to each other?

$$O_2 \leftrightarrow 2O \text{ and } \frac{1}{2}O_2 \leftrightarrow O$$

$$K_P = \frac{p_O^2}{p_{O_2}} \text{ and } K_p = \frac{p_O}{p_{O_2}^{1/2}}$$

We note that the reaction on the right depicts one half of the reaction on the left, on a molar basis, therefore the equilibrium constant on the right is the square root of the reaction on the left.

As another example, let us look at a reaction where the reactants and products are switched. The first reaction dissociates water into molecular hydrogen and oxygen, such as

$$H_2O \leftrightarrow H_2 + \frac{1}{2}O_2$$

And the second reaction is the reverse of the first, namely,

$$H_2 + \frac{1}{2}O_2 \leftrightarrow H_2O$$

The equilibrium constant for the first reaction is

$$K_{P1} = \frac{\chi_{H_2} \cdot \chi_{O_2}^{1/2}}{\chi_{H_2O}} \cdot p_m^{1/2}$$

The second reaction has

$$K_{P2} = \frac{\chi_{H_2O}}{\chi_{H_2} \cdot \chi_{O_2}^{1/2}} p_m^{-1/2}$$

as its equilibrium constant.

We note that $K_{P2} = \dfrac{1}{K_{P1}}$

After these basic manipulations, we may want to add and subtract elementary reactions to form other reactions of interest. As an example, let us consider the following (water–gas) reaction in equilibrium:

$$CO_2 + H_2 \leftrightarrow CO + H_2O \quad K_P = \frac{\chi_{CO} \cdot \chi_{H_2O}}{\chi_{CO_2} \cdot \chi_{H_2}}$$

Now, if we subtract

$$C + O_2 \leftrightarrow CO_2 \quad K_P = \frac{\chi_{CO_2}}{\chi_C \cdot \chi_{O_2}} p_m^{-1}$$

from the sum of the following two reactions

$$C + \frac{1}{2}O_2 \leftrightarrow CO \quad K_P = \frac{\chi_{CO}}{\chi_C \cdot \chi_{O_1}^{1/2}} p_m^{-1/2}$$

$$H_2 + \frac{1}{2}O_2 \leftrightarrow H_2O \quad K_P = \frac{\chi_{H_2O}}{\chi_{H_2} \cdot \chi_{O_2}^{1/2}} p_m^{-1/2}$$

we get

$$C + O_2 + H_2 - C - O_2 \leftrightarrow CO + H_2O - CO_2$$

All terms on the reactant side vanish except H_2 and we may bring the negative CO_2 from the product side to the reactant side, to get the desired reaction

$$CO_2 + H_2 \leftrightarrow CO + H_2O \quad K_P = \frac{\chi_{CO} \cdot \chi_{H_2O}}{\chi_{CO_2} \cdot \chi_{H_2}}$$

Therefore the equilibrium constant for this reaction is the product of the reactions that we added divided by the reaction that we subtracted to form the desired reaction, namely,

$$K_P = \frac{\dfrac{\chi_{CO}}{\chi_C \cdot \chi_{O_2}^{1/2}} p_m^{-1/2} \cdot \dfrac{\chi_{H_2O}}{\chi_{H_2} \cdot \chi_{O_2}^{1/2}} p_m^{-1/2}}{\dfrac{\chi_{CO_2}}{\chi_C \cdot \chi_{O_2}} p_m^{-1}} = \frac{\chi_{CO} \cdot \chi_{H_2O}}{\chi_{CO_2} \cdot \chi_{H_2}}$$

EXAMPLE 6.7

Now, let us revisit the stoichiometric combustion of hydrogen and oxygen that we started in Section 6.4. In this example, we are allowing the reaction to reach an equilibrium state, where a fraction of the fuel and oxidizer are still present, instead of the reaction going to a completion, where there is no leftover oxygen or fuel in the products of combustion.

$$H_2 + \frac{1}{2}O_2 \leftrightarrow n_{H_2O}H_2O + n_{H_2}H_2 + n_{O_2}O_2 \\ + n_{OH}OH + n_OO + n_HH \tag{6.59}$$

$$2n_{H_2O} + 2n_{H_2} + n_{OH} + n_H = 2 \ldots\ldots\ldots \text{to balance H}$$
$$n_{H_2O} + 2n_{O_2} + n_{OH} + n_O = 1 \ldots\ldots\ldots \text{to balance O}$$

From the total of six unknowns and the two equations relating the unknowns, we are still in need of four more equations that involve the unknowns. We shall use the following four stoichiometric reactions:

1. $\frac{1}{2}O_2 \leftrightarrow O \quad K_{P1} = \dfrac{\chi_O}{\chi_{O_2}^{1/2}} \cdot p_m^{1/2}$

2. $\frac{1}{2}H_2 \leftrightarrow H \quad K_{P2} = \dfrac{\chi_H}{\chi_{H_2}^{1/2}} \cdot p_m^{1/2}$

3. $\frac{1}{2}H_2 + \frac{1}{2}O_2 \leftrightarrow OH \quad K_{P3} = \dfrac{\chi_{OH}}{\chi_{H_2}^{1/2} \cdot \chi_{O_2}^{1/2}}$

4. $H_2 + \frac{1}{2}O_2 \leftrightarrow H_2O_{(g)} \quad K_{P4} = \dfrac{\chi_{H_2O}}{\chi_{H_2}^{1/2} \chi_{O_2}^{1/2}} \cdot p_m^{-1/2}$

The mole fractions in our equilibrium constants are

$$\chi_O = \frac{n_O}{n_{H_2O} + n_{H_2} + n_{O_2} + n_{OH} + n_O + n_H} \tag{6.60}$$

$$\chi_{O_2} = \frac{n_{O_2}}{n_{H_2O} + n_{H_2} + n_{O_2} + n_{OH} + n_O + n_H} \tag{6.61}$$

$$\chi_H = \frac{n_H}{n_{H_2O} + n_{H_2} + n_{O_2} + n_{OH} + n_O + n_H} \tag{6.62}$$

$$\chi_{H_2} = \frac{n_{H_2}}{n_{H_2O} + n_{H_2} + n_{O_2} + n_{OH} + n_O + n_H} \tag{6.63}$$

$$\chi_{OH} = \frac{n_{OH}}{n_{H_2O} + n_{H_2} + n_{O_2} + n_{OH} + n_O + n_H} \tag{6.64}$$

$$\chi_{H_2O} = \frac{n_{H_2O}}{n_{H_2O} + n_{H_2} + n_{O_2} + n_{OH} + n_O + n_H} \tag{6.65}$$

The denominator of the mole fractions is the number of moles of the equilibrium products. For a complete reaction of one mole of hydrogen and $1/2$ mole of oxygen we would create one mole of water. Therefore, we may assume a number slightly bigger than one (due to high temperature dissociations) as a first approximation, say 1.1, and replace the denominator by 1.1. We then check the total number of moles of the mixture against our initial guess and iterate as necessary. This method helps arrive at the unknown number of moles of the products more quickly, albeit iteratively.

In order to utilize the equilibrium constants $K_{P1}, \ldots, K_{P4}$, we need to know the mixture reaction pressure, i.e., the combustion pressure. In a gas turbine combustor, we may use the compressor discharge pressure p_{t3} as the combustion pressure. A typical value is between 25 and 50 atm (corresponding to $\pi_c \sim 25$ and 50) for a modern aircraft gas turbine engine at sea level takeoff condition. In a rocket application, the combustion chamber pressure is actually a design choice (typically ~ 1–3000 psia or ~ 70–200 atm). Although the equilibrium constants are not a function of mixture pressure, the mole fractions of the spices are, however, a function of the combustion pressure. Hence, combustion pressure affects the mole fractions of the spices in equilibrium. We also need to have an idea of the final temperature of the products of combustion, before we can use the equilibrium constants, as they are known as a function of temperature. Here, we may assume an adiabatic combustion flame temperature and proceed to solve the six equations and six unknowns for the unknown mole fractions of the products in equilibrium. Now, we need to use the calculated composition of the products of combustion to calculate the adiabatic combustion flame temperature through an enthalpy balance. If the guessed value and the calculated value matched to within a few degrees, then we stop the calculation, otherwise, we need to assume another final combustion temperature and redo the calculations. As noted earlier, determination of the species and their concentration in an equilibrium combustion product is an iterative process. An initial guess of the total number of moles of the products also helps the calculation along and requires its own iteration.

Now, let us consider the example of hydrogen–oxygen combustion at 20 atm with the combustion temperature maintained at 3500 K. Compare the estimate proposed in the above example and the exact solution.

For this problem, we need to calculate the composition and concentration of the product species at the end of equilibrium combustion. Also, since we are maintaining the final temperature, at 3500 K, we need to calculate the heat transfer needed to achieve the final temperature. An equilibrium chemical reaction analysis results in the following composition (in number of moles):

$$n_{H_2} \approx 0.1585 \quad n_{O_2} \approx 0.0500 \quad n_O \approx 0.0269$$

$$n_H \approx 0.0504 \quad n_{OH} \approx 0.1166 \quad n_{H_2O} \approx 0.7506$$

The total number of moles of the products is 1.153, which is slightly bigger than our initial guess of 1.1. The result of enthalpy balance applied to the chemical reaction in the combustor yields the necessary heat transfer to the reaction chamber to maintain the final temperature, i.e.,

$$Q \approx 5.8 \, \text{kcal/gmol} \cdot O_2$$

The adiabatic flame temperature of a stoichiometric combustion of hydrogen and oxygen yields a temperature of ~3080 K at 1 atm pressure.

The adiabatic flame temperature as a function of equivalence ratio for the Jet A fuel combustion in air at 800 K initial temperature and 25 atm initial pressure is shown in Fig. 6.6 (from Blazowski, 1985). There are several noteworthy features in Fig. 6.6 that are instructive to our study. First, the initial temperature and pressure were chosen to represent the combustion chamber inlet conditions in a modern aircraft gas turbine engine. Second, the *actual* flame temperature that allows for dissociation falls below the theoretical predictions (of a complete combustion). The difference between the actual curve and theoretical predictions starts around an equivalence ratio of ~0.4–0.5, which points to a flame temperature of ~1650–1750 K, where dissociation reactions begin. Interestingly, a combustion chamber in a modern aircraft

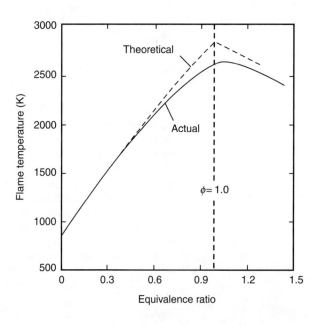

■ **FIGURE 6.6** Adiabatic flame temperature of Jet A fuel combustion in air at 800 K initial temperature and 25 atm pressure (from Blazowski, 1985)

gas turbine engine operates in the 0.4–0.5 range of the equivalence ratio, which corresponds to a fuel-to-air ratio of ~2.5–3.5%. Third, although the figure caption identifies the initial temperature at 800 K, we could have discerned this fact from the flame temperature at zero equivalence ratio on the graph, i.e., no combustion. Fourth, as noted earlier, the maximum flame temperature occurs at an equivalence ratio slightly greater than 1 (~1.1).

Also note that the burner exit (or flame) temperature is a double-valued function of equivalence ratio, or fuel-to-air ratio, i.e., we can reach a burner exit temperature from either the fuel-lean or the fuel-rich side of the stoichiometric point.

The effect of initial air temperature on the stoichiometric flame temperature of a Jet A fuel combustion in air is shown in Fig. 6.7 (a). Comparing the slope of the flame temperature rise with inlet temperature to a slope = 1.0 line drawn in Fig. 6.7 (a) for reference, we note that only ~1/2 of the inlet temperature rise appears as the rise in the final flame temperature. This behavior corresponds to higher molar concentrations of dissociated species that appear with higher temperatures. Consequently, a 100 K increase in the inlet air temperature for this reaction results in ~50 K final flame temperature rise. The effect of the combustion pressure on the stoichiometric flame temperature of Jet A fuel is shown in Fig. 6.7 (b). The increasing pressure leads to a rise in the stoichiometric flame temperature. The increase is due to a suppressed dissociation levels with pressure, as shown in Example 6.4. We observed that a lower mixture pressure promoted nitrogen dissociation and it was suppressed with an increasing mixture pressure in Example 6.4.

It is instructive to examine the effect of oxidizer, as in pure oxygen or air, on the combustion of various fuels and the impact of the oxidizer on the stoichiometric flame temperature. For this purpose, Table 6.4 is reproduced from Glassman (1987) that examines the maximum temperature at equivalence ratios near stoichiometric of various fuels reacting with air or pure oxygen. Moreover, the effect of pressure on combustion temperature is reported for methane, similar to Fig. 6.7 (b). Combustion of hydrogen in air results in a stoichiometric flame temperature of ~2400 K at 1 atm pressure. The flame temperature for the combustion of hydrogen with pure oxygen is ~3080 K, at 1 atm pressure. Air contains nitrogen (~78% by volume), which for the most part is inert and does not react with the fuel or oxidizer. Also, the fuel heating value remains the same whether the oxidizer is oxygen or air. For the same energy release (per unit mass of the fuel), there are more reactants in air that do not participate in the combustion but need to be heated to the final temperature. Consequently, the flame temperature will be lower when a fuel reacts with air as opposed to pure oxygen. This behavior is observed

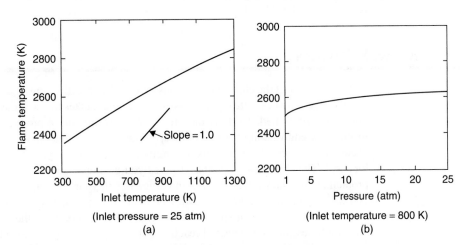

■ FIGURE 6.7   **Effect of inlet temperature and pressure on the stoichiometric flame temperature of a Jet-A fuel combustion in air (from Blazowski, 1985)**

■ **TABLE 6.4**

Maximum Flame Temperature of Fuel–Air and Fuel–Oxygen Mixtures at 25°C Temperature and 1 or 20 atm Pressure (from Glassman, 1987)

Approximate flame temperatures of various stoichiometric mixtures, critical temperature 298 K

Fuel	Oxidizer	Pressure (atm)	T(K)
Acetylene	Air	1	2600[a]
Acetylene	Oxygen	1	3410[b]
Carbon monoxide	Air	1	2400
Carbon monoxide	Oxygen	1	3220
Heptane	Air	1	2290
Heptane	Oxygen	1	3100
Hydrogen	Air	1	2400
Hydrogen	Oxygen	1	3080
Methane	Air	1	2210
Methane	Air	20	2270
Methane	Oxygen	1	3030
Methane	Oxygen	20	3460

[a] This maximum exists at $\phi = 1.3$.

[b] This maximum exists at $\phi = 1.7$.

for all (five) fuels listed in Table 6.4. Now to the effect of pressure on combustion, we note that methane reacts with air at 1 atm pressure and reaches a maximum flame temperature of 2210 K. If we increase the combustion pressure to 20 atm, the flame temperature increases to ~2270 K. This is a rather minor increase in temperature, which is caused by a decrease in the dissociation levels of the products of combustion with pressure. At this temperature, the dissociation levels were small and hence a suppressed level of dissociation did not cause a pronounced effect in final temperature. To ascertain this point, we examine the combustion of methane with pure oxygen, which reaches a flame temperature of ~3030 K at 1 atm pressure. The flame temperature rises to ~3460 K when methane reacts with pure oxygen at 20 atm pressure. At the temperature of 3030 K, the level of dissociation is much higher, hence the effect of pressure on suppressing the product dissociation is more pronounced and leads to a flame temperature of ~3460 K.

6.5   Chemical Kinetics

So far our study of the chemical reactions involved an equilibrium state where the forward and reverse reaction rates were equal. However, the time it takes to reach the equilibrium state was not explicitly studied. To study the (time) rate of formation of product species in a chemical reaction, or equivalently the rate of disappearance of reactants in a reaction, is the subject of chemical kinetics. As a propulsion engineer, why should you be interested in chemical kinetics? The short answer is that it gives you an important characteristic timescale of the problem, which may affect the size and efficiency of your combustion chamber. For example, you will need to design the combustion chamber (as in a supersonic combustion ramjet) such that the residence time of species in the chamber is longer than the characteristic chemical reaction timescale for the heat of reaction to be fully realized. Also, we learn that the reaction rate depends on the pressure of the mixture, which at high altitudes (low pressures) may become too slow to effect a combustor relight. Finally, the study of chemical kinetics helps combustion engineers meet the low-pollution design objectives of low NO_x, low CO, soot, and (visible) smoke dictated by the regulatory organizations (e.g., EPA and ICAO).

The cornerstone of chemical kinetics is the law of mass action, which for the following generic reaction

$$a\text{A} + b\text{B} \rightarrow c\text{C} + d\text{D}$$

may be written as

$$r_f = k_f [\text{A}]^a . [\text{B}]^b$$

that depicts the forward reaction rate. The reaction rate coefficient k_f is independent of reactant's concentration (note that the concentrations are already accounted for in the law of mass action itself) and in general is a function of pressure and temperature. The temperature dependency of the rate coefficient is in the form proposed by Arrhenius to be the Boltzmann factor (from the kinetic theory of gases), which is equal to the fraction of all collisions (i.e., the number of collisions per unit time and volume) with an energy greater than, say E_a, namely,

$$\text{Temperature dependence of } k_f = e^{-\frac{E_a}{RT}} \qquad (6.66)$$

where E_a is the activation energy (for the reaction, which is the minimum energy to cause a chemical reaction to occur), $\overline{R}$ is the universal gas constant, and T is the reactant's absolute temperature. The activation energy for the combustion of hydrocarbon fuel in air is in the range of 40–60 kcal/mol. The frequency of the collisions, per unit volume, contributes to the reaction rate most directly and hence the pressure dependence of the reaction rate coefficient is revealed. In general,

$$k_f = p^n f(T) e^{-\frac{E_a}{RT}} \qquad (6.67)$$

which is called the Arrhenius rule, in honor of the man who first proposed the exponential dependence, the Boltzmann factor, for the chemical reaction rate. The exponent n on the pressure term depends on the number of molecules involved in a collision to produce the chemical reaction. For example, the combustion of hydrocarbon fuels in air involves the collision of two molecules, namely, oxygen and the fuel, hence $n \approx 2$. More complex reactions involving collisions of more than two molecules in a chemical reaction lead to a higher pressure exponent n such as 3 or 4.

Now, let us examine the impact of a chemical kinetic study on a practical problem in aircraft propulsion.

6.5.1   Ignition and Relight Envelope

As noted earlier, the pressure dependence of the reaction rate coefficient may cause a relight problem at high altitude. A relight zone is plotted in Fig. 6.8 in an altitude–Mach number plot (i.e., flight envelope) for the combustion of a typical hydrocarbon fuel in air (from Henderson and Blazowski, 1989). We note that below 0.2 atm pressure, combustor relight is not possible at standard temperature, due to low reaction rate.

6.5.2   Reaction Timescale

The reaction timescale and its relation to the residence timescale of the air–fuel vapor mixture control the combustion efficiency. We require that the two timescales to be equal, to establish a minimum length requirement of the combustor. A typical variation of reaction timescale with

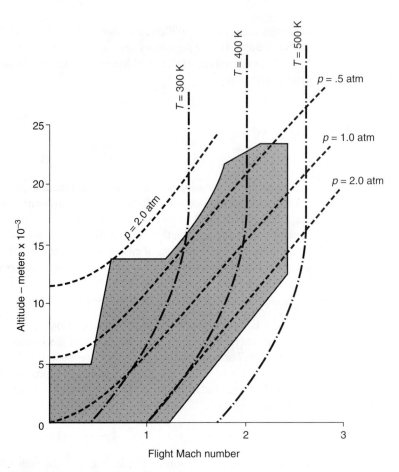

■ **FIGURE 6.8** **The typical ignition/relight envelope shows relight capability in the shaded zone (from Henderson and Blazowski, 1989)**

equivalence ratio for gasoline–air combustion is shown in Fig. 6.9 (from Kerrebrock, 1992). This shows that near stoichiometric combustion, $\phi \sim 1$, the reaction time reaches a minimum of $\sim$0.3 ms. The minimum reaction timescale is one of the indicators of the desirability of stoichiometric combustion. We shall see in the next section that the flammability limit of most fuels is a narrow band centered on the stoichiometric mixture ratio of $\phi = 1$.

The reaction timescale is inversely proportional to reaction rate (Eq. 6.67), which may be written as

$$t_{\text{reaction}} \sim k_f^{-1} \sim p^{-n}T^{-m}e^{\frac{E_a}{RT}} \tag{6.68}$$

There are no universal rules for the pressure and temperature exponents, n and m, in a gas turbine combustor, but typical value of n is $\sim$1–2 and m is $\sim$1.5–2.5 (see Zukoski, 1985). The aircraft gas turbine combustor or afterburner is designed for a residence timescale in the primary zone or the flameholder recirculation zone-mixing layer to be long compared with the reaction timescale. Hence, we are not reaction-time-limited in a conventional combustor or afterburner. However, in a scramjet combustor with a Mach 3 combustor inlet flow, we are *residence-time*-limited, hence reaction timescale becomes especially critical. Fortunately, the fuel of choice, hydrogen, offers not only a regenerative cooling option for the

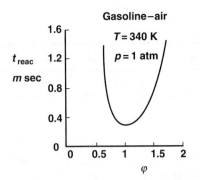

■ **FIGURE 6.9** Reaction timescale for gasoline–air combustion as a function of equivalence ratio (Mixture at 1 atm pressure and 340 K temperature) (from Kerrebrock, 1992)

engine inlet, combustor, and nozzle but also ∼1/10 of the chemical reaction timescale of hydrocarbon fuels. The pollutant formations and their dependence on the combustor characteristic timescales will be discussed in section 6.7 in this chapter.

6.5.3 Flammability Limits

A practical aspect of combustor design examines the issues of flammability limits of various fuel–air mixture ratios as a function of temperature and pressure in a quiescent environment. It is learned that a fuel–air mixture of equivalence ratio leaner than 0.6 and richer than 3.0 will not react (i.e., sustain combustion) in room temperature and pressure, as shown in Fig. 6.10 (from Blazowski, 1985). In reality, a narrow band around a stoichiometric fuel–air mixture ratio is our open window for a sustained combustion. With the increase in temperature, the flammability limit boundary widens to $0.3 < \phi < 4.0$ in the spontaneous ignition temperature range $(T > 225°C)$ for a kerosene–air mixture. The flammability limits of different fuel–air mixtures, in percent stoichiometric, are shown in Table 6.5 at the temperature of 25°C and 1 atm pressure. The most remarkable is the wide flammability limits of hydrogen. Also hydrogen, methane, and propane enjoy very large heating capacity making them suitable coolants for active cooling of the engine and airframe structure of a hypersonic aircraft.

The effect of low ambient pressure (corresponding to high altitude) on flammability limit of gasoline–air mixtures is shown in Fig. 6.11 (from Olson, Childs, and Jonash, 1955). We note

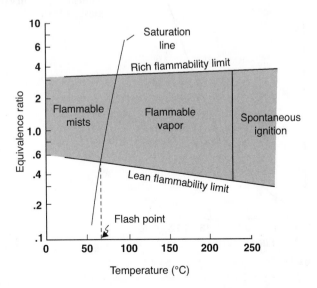

■ **FIGURE 6.10**
Flammability limits of a kerosene-type fuel in air at atmospheric pressure (from Blazowski, 1985)

■ **TABLE 6.5**
Flammability Limits of Some Fuels(25°C and 1 atm pressure)

Fuels	Jet A kerosene	Propane C_3H_8	Methane CH_4	Hydrogen H_2
Flammability limits (percent stoichiometric)	52–400	51–280	46–164	14–250

that the lean flammability limit at altitude approaches the stoichiometric ratio, i.e., $\phi \approx 1$. We also observe that the rich flammability limit becomes much wider with the increase in pressure.

As noted earlier, the flammability limits are defined for a fuel–air (gas phase) mixture as a function of temperature and pressure in a *quiescent* environment. In the presence of flow, combustion instabilities may set in to cause a flame blowout. Hence, the presence of a flow may drive an otherwise stable combustion (in the flammable zone) into an unstable reaction where the combustion efficiency goes to zero at the point of flame blowout. In order to sustain combustion in the presence of flow, i.e., to avoid blowout, we need to employ flame stabilization schemes, which are suitable for the primary combustors and the afterburners. We will address several stabilization schemes in this chapter.

A need for an external source of energy release, as provided by an electric spark discharge, is (implicitly) demonstrated in the flammability limit diagram of kerosene–air mixture in Fig. 6.10. Note that below the spontaneous ignition temperature (SIT) of the fuel–air mixture (of $\sim 225°C$), the mixture that is defined as flammable will not self-ignite. Hence, an ignition source that is an intense localized heat source is needed to initiate the reaction. The intense localized energy release will raise the temperature of a *pocket* of the vaporized fuel–air mixture to a temperature above the SIT. A flame front signifying the chemical reaction zone is then formed, which propagates in the flammable mixture at a finite speed S_T. Table 6.6 contains the spontaneous ignition temperature of some common fuels for reference (from Gouldin, 1973). A high spontaneous ignition temperature is desirable for a fuel that is to be used for its cooling capacity. For example, propane in Table 6.6 has the highest SIT at 767 K, which makes it a good candidate for cooling purposes. Since spontaneous ignition temperature of fuel–air

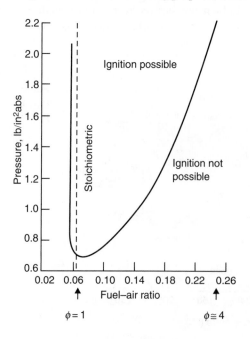

■ **FIGURE 6.11** **Flammability limits of gasoline–air mixtures as a function of combustion pressure (from Olson, Childs, and Jonash, 1955)**

■ TABLE 6.6
Spontaneous Ignition Temperatures of Common Fuels (from Gouldin, 1973)

Fuel	SIT (K)	Fuel	SIT (K)
Propane	767	Decane	481
Butane	678	Hexadecane	478
Pentane	558	Isooctane	691
Hexane	534	Kerosene (JP-8 or Jet A)	501
Heptane	496	JP-3	511
Octane	491	JP-4	515
Nonane	479	JP-5	506

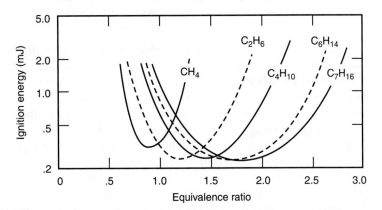

■ FIGURE 6.12 Minimum ignition energy of various vaporized fuel–air mixtures (from Lewis and von Elbe, 1961)

mixtures is a strong function of mixture pressure, the SIT information contained in Table 6.6 needs to be viewed with caution and not to be used for the "general purpose" fuel system safety calculation. The minimum external energy (stimulus) needed to initiate a reaction is a function of the equivalence ratio as well as the type of fuel for vaporized fuel–air mixtures. Figure 6.12 (from Lewis and von Elbe, 1961) shows the level of ignition energy of various vaporized fuel–air mixtures as a function of equivalence ratio. The ignition energies are in the form of "bucket" shapes with the minimum ignition energy corresponding to a particular equivalence ratio.

Note that in Fig. 6.11, the minimum ignition energy points on the "buckets" shifts to the right, i.e., toward higher equivalence ratios, with higher molecular weight fuels. Methane, CH_4, has a molecular weight of 16 and the minimum ignition energy around the stoichiometric ratio, whereas n-heptane, C_7H_{16}, has a molecular weight of 100 and the minimum ignition energy occurring around an equivalence ratio of ~2. We may express this observation in terms of the specific gravity of these fuels as well where methane's specific gravity is 0.466 (a light fuel) and n-heptane's specific gravity is 0.684 (a relatively heavier fuel).

6.5.4   Flame Speed

There are two types of flames, premixed flames and diffusion flames. When the fuel and oxidizer are in gaseous form and premixed into a (stoichiometric) flammable mixture prior to combustion initiation, the resulting flame is called a premixed flame (e.g., Bunsen burner). When the fuel and oxidizer are initially separated, as in a candle burning in air, the combustion

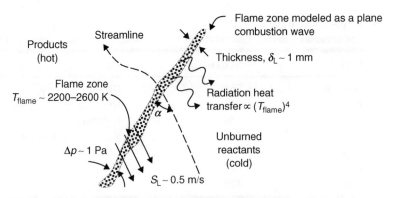

■ **FIGURE 6.13** **A schematic drawing of a thin premixed laminar flame propagating in a stoichiometric mixture of a hydrocarbon fuel and air at 1 atm pressure and $T = 25°C$.**

takes place within a diffusion flame zone where the vaporized fuel diffuses outward and meets an inwardly diffusing oxidizer. For an aircraft application where space is at a premium, a premixed flame combustor with a high level of turbulence intensity is the desirable choice. The effect of turbulence intensity is to promote mixing enhancement. It is the rapid and efficient mixing of the vaporized fuel–air mixture in a combustion chamber that is a performance limiting parameter, i.e., limiting the energy release rate, rather than the chemical reaction timescale or flame propagation speed in conventional burners. The flame zone in a premixed combustor starts propagating initially as a laminar flame and later develops into a turbulent flame. A schematic drawing of a laminar premixed flame and its typical characteristic parameters (thickness, propagation speed, temperature range, pressure drop, and wave angle, α) are shown in Fig. 6.13. The laminar flame parameters in Fig. 6.13 are extracted from Borman and Ragland (1998). S_L is the laminar flame speed, δ_L is the flame thickness, and α is the relative flow angle.

The initial mixture temperature has the most pronounced effect on laminar flame speed. The burning velocity of the laminar flame zone increases nearly fourfold when the initial mixture temperature is raised from 200 to 600 K, as reported by Dugger and Heimel (1952) for methane (CH_4), propane (C_3H_8), and ethylene (C_2H_4) combustion in air. The mixture pressure has a minor effect on the laminar flame speed as discussed by Lefebvre (1983). The experimental results of Jost (1946), shown in Fig. 6.14, demonstrates the equivalence ratio of the laminar flame speed. Moreover, it confirms the typical speed of ∼0.5 m/s for a hydrocarbon fuel, as in propane. It also shows nearly an order of magnitude increase in laminar flame speed for hydrogen, which reconfirms the status of hydrogen as an ideal fuel.

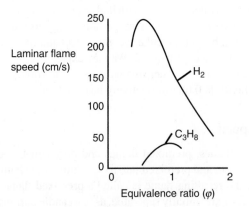

■ **FIGURE 6.14** **Laminar flame speed for propane and hydrogen as a function of equivalence ratio (from Jost, 1946)**

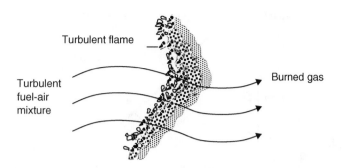

■ **FIGURE 6.15**
Graphical depiction of a
wrinkled turbulent flame
with a distribution of eddies
within the flame (adapted
from Ballal and Lefebvre,
1975).

The effect of turbulence on premixed flame speed is to enhance momentum transfer between the burning front and the unburned reactants. In addition, turbulence increases the total surface area of the flame and hence increases the heat transfer between the reaction zone and the unburned gas. To visualize the surface area of a flame front under the influence of turbulence, a *wrinkled* flame front is proposed. Under this scenario, large (energetic) turbulent eddies strike the flame front and *wrinkle* the surface, while small-scale turbulence changes the transport properties within the flame zone. Figure 6.15 shows a representative drawing of a wrinkled flame front with a range of turbulent eddy sizes within the flame (adapted from Ballal and Lefebvre, 1975). Consequently, the flame propagation speed is increased from the laminar flame speed S_L to the turbulent flame speed S_T. The first model, due to Damkohler (1947), proposes enhancing the laminar flame speed by adding the root mean square of the turbulent fluctuation speed. The increase in flame surface area due to turbulence is proportional to

$$u'_{rms} \text{ i.e.,}$$
$$S_T = S_L + u'_{rms} \tag{6.69}$$

The laminar flame speed is $S_L \sim 0.5-2\,\text{m/s}$ (depending on the initial mixture temperature), which is now enhanced by turbulent fluctuations. A low turbulence intensity environment, of say 5%, in a (fuel–air mixture) flow of $\sim 50-100\,\text{m/s}$ results in a root mean square of the turbulent fluctuation speed of $2.5-5\,\text{m/s}$. Now, in a high turbulence intensity environment, of say 15–30%, in a flow with a mean speed of $\sim 100\,\text{m/s}$, the turbulence contribution to flame propagation speed may be as high as $30\,\text{m/s}$. Although the simple model of Damkohler in Eq. 6.69 does not account for turbulence scale, the thickness of the reaction zone and other effects such as flame stretching, it serves the purpose of demonstrating the attractiveness of turbulence enhanced mixing and its impact on the reaction timescale. A review of more elaborate models is beyond the scope and intention of the present treatment. Excellent books on combustion such as those authored by Lefebvre (1983), Kuo (1986), Glassman (1987), Borman and Ragland (1998), among others are recommended for further reading in this fascinating field.

6.5.5 Flame Stability

Does a Bunsen burner or a candle in a laboratory produce a stable flame? The obvious answer is yes, but the correct answer should be a conditional yes or a maybe. Since we did not specify any source or intensity of airflow in the room, we could not rule out the possibility of flame extinction caused by the airflow. If we place a Bunsen burner or a candle in a duct (for measurement purposes) connected to a fan on one side, we may experimentally establish the wind speed in the duct that causes the candle flame to blowout. Similar experiments may be conducted in a premixed combustible mixture where a flame front is established and propagates in the mixture at the flame speed S. Now, if we set the premixed gas into motion at speed U in the opposite direction to S and equal to it in magnitude; we shall achieve a stationary flame front. Simply adding the velocity vectors demonstrates the principle. For a given mixture fuel–air

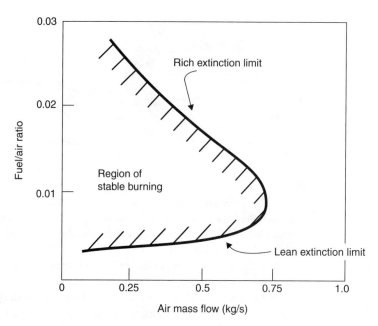

■ **FIGURE 6.16** A typical stability loop for a combustor at $p_3 = $ constant (from Lefebvre, 1983) (p_3 is combustor inlet pressure)

ratio, we may repeat the experiment for different flow speeds until we achieve an extinction of the flame (i.e., at the blowout speed). We may subsequently vary the fuel–air mixture ratio from a lean limit where extinction occurs to a rich limit at a given flow speed. The result of this experiment may be plotted in terms of the fuel–air mixture (or equivalence ratio) and the flow speed (or air mass flow rate, which is proportional to flow speed) in what is known as a stability plot or the *stability loop*, due to its shape. Figure 6.16 shows the combustion system stability loop for a constant pressure (from Lefebvre, 1983).

The lower branch of the stability loop is the lean extinction limit and the upper branch is the rich extinction limit of combustion chamber. The maximum blowout speed occurs at the maximum flame speed, which is usually near stoichiometric mixture ratio. The effect of pressure on the lean stability limit is negligible and widens the rich stability limit with the increase in pressure.

A recirculation zone in the burner (with a backward flow) is needed to stabilize a premixed flame. In studying swirling jets, we learn that when the ratio of jet angular momentum to jet axial thrust times jet diameter exceeds ~0.5, a recirculatory flow in the core of the swirling jet is established. This is known as a vortex breakdown. A schematic drawing of a swirling jet with vortex breakdown is shown in Fig. 6.17.

Flow visualization (via dye injection in water) of a bubble-type vortex breakdown is shown in Fig. 6.18 (from Sarpkaya, 1971). The complex network of vortex filaments winding in and out of the bubble and a spiraling wake formation indicates filling and emptying of the bubble. This environment, i.e., the recirculation and intense mixing, of a vortex breakdown comes close to a model of a perfectly stirred reactor that is often used in combustion studies.

To help stabilize the flame in the primary zone of a main burner, air is introduced through single or double rows of swirl vanes. Also to create a perfectly stirred reaction zone in the main burner, part of the excess air is injected in the burner through the primary air holes as radial jets. When opposing radial jets collide, a stagnation point is created with the resultant streams directed along the axis in the upstream and downstream directions. The opposing flows of the axial jet moving upstream and the spiraling air–fuel mixture create a recirculation zone with

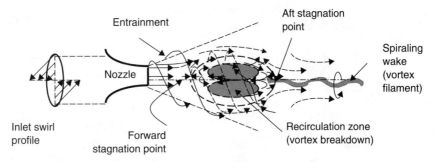

■ **FIGURE 6.17** **The flow pattern of a (bubble-type) vortex breakdown in a swirling jet**

■ **FIGURE 6.18** **Flow visualization of a bubble-type vortex breakdown (from Sarpkaya, 1971)**

intense mixing. The combustor primary zone may also be modeled as a perfectly stirred reactor. A schematic drawing of radial jets and subsequent axial jet formations are shown in Fig. 6.19.

In an afterburner, the necessary recirculation zone to stabilize the premixed flame is created by the massively separated wakes of bluff bodies. Some geometric examples of bluff bodies that create massively separated wakes are produced in Fig. 6.20.

The flameholding, i.e., creating a reversed flow environment, in a high-speed scramjet combustor may be achieved by a direct fuel injection in the supersonic air stream, or via a backward-facing step. The former involves a bow shock, and the latter involves the corner flow separation. These flowfields are of interest in high-speed propulsion and are shown in Fig. 6.21.

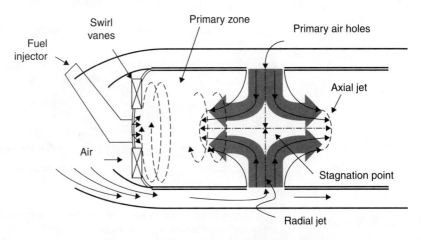

■ **FIGURE 6.19** **Collision of radial jets, stagnation point, and axial jet formation with a reverse flow in the primary zone of the combustion chamber**

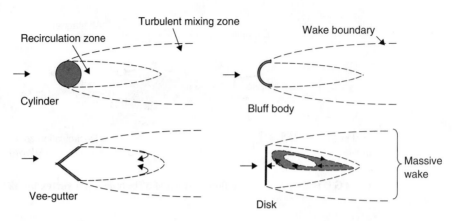

■ **FIGURE 6.20** **Bodies that create recirculation and massive wakes (suitable for subsonic flameholding)**

The primary zone of a burner and the wake of a bluff body in the afterburner are modeled as a *perfectly stirred reactor* where the reactants and products are mixed infinitely fast. Such an environment produces the maximum energy release (due to infinitely fast chemical reaction rates) for the given volume of the reactor and at a fixed reactor pressure. The theory of perfectly stirred reactor has many applications, among which the stability of premixed turbulent flames in a flow environment. The perfectly stirred reactor theory predicts

$$\frac{\dot{m}}{p^n V} = f(\phi) \tag{6.70}$$

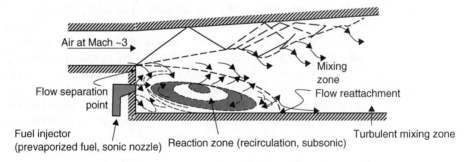

(a) Backward-facing step with base fuel injection in supersonic flow

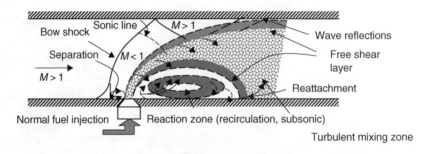

(b) Normal fuel injection in a supersonic stream creates a bow shock and a recirculation zone, which serves as the flameholder for a scramjet combustor

■ **FIGURE 6.21** **Fuel injection schemes in high-speed flow**

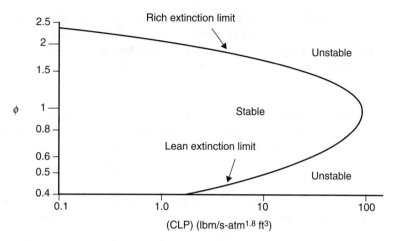

■ **FIGURE 6.22** **Combustor stability loop in terms of the loading parameter and the equivalence ratio ϕ (from Spalding, 1979)**

where the mass flow rate in the reactor is $\dot{m}$, volume of the reactor is V, pressure of the reactor is p, the exponent of pressure represents the order of reaction, which is close to 2, and ϕ is the equivalence ratio of the fuel–air mixture. The parameter on the right-hand side of Eq. 6.70 is called the combustor loading parameter (CLP). A graph of Eq. 6.70 is shown in Fig. 6.22 (from Spalding, 1979). A lean and a rich extinction limit (i.e., when flame instability causes extinction) are observed in Fig. 6.22, similar to the stability loop of Fig. 6.16. The pressure exponent "n" is taken as 1.8 in Fig. 6.22, and the dimensions of the combustor loading parameter are $\mathrm{lbm}/(\mathrm{s} \cdot \mathrm{atm}^{1.8} \cdot \mathrm{ft}^3)$ in English units.

Fundamental contributions to understanding flame stability are due to the works of Longwell, Chenevey, and Frost (1949), Haddock (1951), and Zukoski–Marble (1955). Although, we will briefly address the afterburner stability in this chapter, these references are recommended for further reading.

6.5.6 Spontaneous Ignition Delay Time

In continuing with our studies of combustion timescales and the role of chemical kinetics, we enter a new characteristic timescale, namely, the spontaneous ignition delay time. Spontaneous ignition delay time is defined as the time elapsed between the injection of a high-temperature fuel–air mixture and the appearance of a flame in the absence of an ignition source. There are two factors that contribute to an autoignition delay time. First is the rate of evaporation of fuel droplets, and the second is the chemical reaction time of the vaporized fuel and air. To promote the rate of evaporation, a liquid fuel has to be efficiently atomized to a fine spray mist. Modern aircraft engine combustors utilize spray injectors that produce an atomized fuel drop size distribution in the range of 10–400 μm. Rao and Lefebvre (1981) express the ignition delay time as

$$t_{\mathrm{i}} = t_{\mathrm{e}} + t_{\mathrm{reaction}} \tag{6.71}$$

The subscript "i" stands for ignition delay, and "e" stands for evaporation timescale in Eq. 6.71. Evaporation timescale in a stagnant mixture is proportional to the surface area of the liquid fuel droplet, as the heat transfer rate to the liquid is proportional to the surface area, i.e.,

$$t_{\mathrm{e}} \propto D^2 \quad \text{for} \quad \mathrm{Re}_{\mathrm{D}} = 0 \tag{6.72}$$

where D is the liquid fuel droplet diameter, and Re_D is the fuel droplet Reynolds number based on diameter of the droplet, turbulent fluctuation velocity u′ (rms), and the kinematic viscosity of the gas ν_g. In the presence of a highly turbulent mixture, the evaporation time is proportional to the droplet diameter to the power three halves, i.e.,

$$t_e \propto D^{1.5} \quad \text{for} \quad Re_D \gg 1 \qquad (6.73)$$

according to Ballal and Lefebvre (1980). The effect of turbulence is hence to increase the heat transfer and reduce the evaporation timescale. The reaction time is expressed in terms of the mixture temperature (in K), by Rao and Lefebvre (1981), as

$$t_{\text{reaction}} \propto e^{\frac{9160}{T_m}} \qquad (6.74)$$

From the chemical kinetic rate equation 6.67, we may also deduce the reaction timescale as the inverse of the rate constant based on initial temperature T_1, namely,

$$t_{\text{reaction}} \propto e^{\frac{E_a}{RT_1}} \qquad (6.75)$$

The ignition delay time for a range of initial temperatures between 600 and 1000 K and a range of pressures between 1 and 30 atm is shown in Fig. 6.23 (Rao and Lefebvre, 1981).

Figure 6.23 is a log–log plot. It shows an exponential drop in ignition delay time with initial temperature, as expected from Eq. 6.75. The ignition delay time is also reduced with the increase in combustion pressure. This pressure dependence becomes weaker for a high inlet temperature (1000 K) and at a high pressure (30 atm). Interestingly, the conditions of 1000 K inlet temperature and 30 atm combustion pressure are representative of a typical gas turbine engine combustor inlet conditions. Based on these, the ignition delay time is ~1–2 ms. From the reaction timescale given by Kerrebrock in Eq. 6.68, we may express the pressure dependence of the ignition delay time with an inverse relation, such as

$$t_i \sim \frac{1}{p} \qquad (6.76)$$

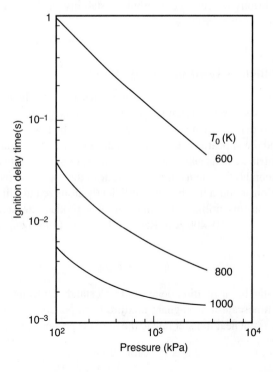

■ **FIGURE 6.23** **Effect of pressure and initial temperature on ignition delay time for a fuel–air mixture with 50 μm mean diameter fuel droplet and u′ = 0.25 m/s (from Rao and Lefebvre, 1981)**

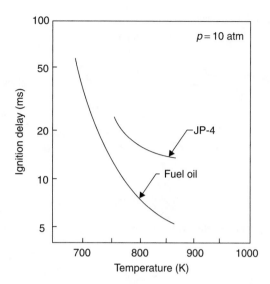

■ **FIGURE 6.24** Spontaneous ignition delay time as a function of initial temperature at 10 atm pressure (from Spadaccini, 1977)

The results of autoignition delay time studies of Spadaccini (1977) for typical hydrocarbon fuels at elevated temperatures and at the pressure of 10 atm are shown in Fig. 6.24 (a log-linear plot). We again observe an exponential drop in ignition time delay with temperature. The pressure dependence follows Eq. 6.76, or the trends shown by Rao and Lefebvre in Fig. 6.23.

6.5.7   Combustion-Generated Pollutants

The subject of pollution is an integral part of the current section, i.e., chemical kinetics. However, due to its importance, this subject is presented, in more detail, after the discussion of combustion chambers and afterburners in section 6.7.

6.6   Combustion Chamber

In this section, we explore geometric configurations of combustion chambers that are found in aircraft gas turbine engines. The flammability limits of fuel–air mixtures that we studied in the last section, required a near stoichiometric mixture ratio. To prolong the turbine life, we are forced to reduce the turbine inlet temperature to levels corresponding to an equivalence ratio of $\phi \sim 0.4-0.5$. Consequently, we have to introduce the combustor inlet air in *stages* along the combustion chamber length. The first stage admits air in a (near) stoichiometric proportion of the injected fuel flow rate. In the first stage, air and fuel mix and chemically react in what is known as the primary zone. In the second stage, air may be used to stabilize the flame of the primary zone. The subsequent stages of air are introduced along the combustor length for dilution and cooling purposes. Proper tailoring of these stages lead to a stable combustion, reduced pollutant formation as well as a combustor exit temperature profile that is reasonably uniform and devoid of hot spots. To reduce the levels of total pressure loss in a combustor, we need to decelerate the flow at the inlet to the combustion chamber. This calls for a prediffuser.

There are two combustor configurations that are in use in aircraft gas turbine engines, namely, the reversed-flow combustor and the straight through flow combustor. Figure 6.25 shows a schematic drawing of a reverse-flow combustor.

The reverse-flow combustor is best suited for application in small gas turbine engines that utilize centrifugal compressors. The flow turns nearly 180° twice and hence it suffers an added total pressure loss due to turning over a straight through flow burner. The compact design of this

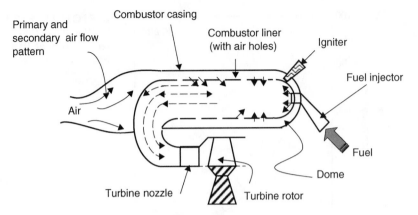

■ **FIGURE 6.25**   **Reverse-flow combustor**

burner places the turbine inlet plane near the compressor discharge plane and hence results in a shorter turbine-compressor shaft. Small airbreathing engines for drone applications as well as the early engines (e.g., Whittle engines) utilized this configuration.

A second configuration, which represents most gas turbine engines today, is the straight through flow burner. A schematic drawing of this burner is shown in Fig. 6.26.

The combustion chamber may be of a single can-type tubular design, a multican design, a can-annular design, or an annular design. These configurations are shown in Fig. 6.27 and 6.28.

The single-can and the multican combustors are heavier than the latter two types of can-annular and annular combustors. The total pressure loss in these combustors is also higher than the annular types due to their larger wetted area. However, the advantages of the can type are in their mechanical robustness and their good match with the fuel injector. Coupled with a lower development cost, the single and multican systems were widely used in the early jet engines. Presently, the lowest system weight and size belong to the winner, the annular combustor, which also offers the lowest total pressure drop. The annular combustor due to its open architecture between the fuel injectors is more susceptible to combustion and cross-coupling-related instabilities than its counterparts with their confined configurations.

It is of critical importance to combustion system designers to understand the total pressure loss mechanisms in a combustor.

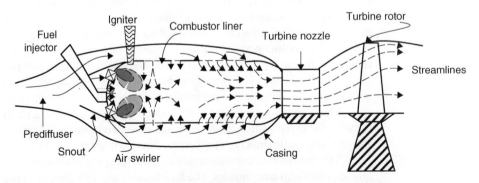

■ **FIGURE 6.26**   **Schematic drawing of a straight through flow burner**

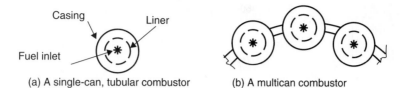

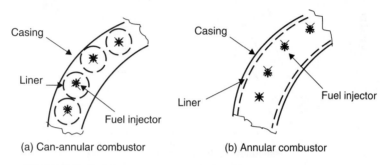

■ FIGURE 6.27 Single-can, tubular, and multican combustor configurations

■ FIGURE 6.28 Annular combustor configurations

6.6.1 Combustion Chamber Total Pressure Loss

The total pressure loss in a combustor is predominantly due to two sources. The first is due to the frictional loss in the viscous layers where we can lump in the mixing loss of the fuel and air streams prior to chemical interaction. The second source is due to heat release in an exothermic chemical reaction in the burner. The first is usually modeled as the "cold" flow loss and the second is the "hot" flow loss. We recognize, however, that the subdivision of the losses into "cold" and "hot" is artificial and the two forms of total pressure loss occur simultaneously and interact in a complex way. But throughout the ages, engineering is practiced through the art of simplification, i.e., engineers have broken down a complex problem into a series of elementary and solvable problems and then tried to predict the solution of the complex problem by devising a correlation between the elementary solutions and the observation or measurement. We intend to stay our engineering course on the topic of combustion chamber total pressure loss. There are several methods of quantifying the total pressure loss in a combustor, namely,

$$\textbf{(a)} \quad \frac{p_{t3} - p_{t4}}{p_{t3}} \text{ (Relative to the inlet total pressure)} \qquad \textbf{(6.77)}$$

$$\textbf{(b)} \quad \frac{p_{t3} - p_{t4}}{q_3} \text{ (Relative to the inlet dynamic pressure)} \qquad \textbf{(6.78)}$$

$$\textbf{(c)} \quad \frac{p_{t3} - p_{t4}}{q_r} \text{ (Relative to a reference dynamic pressure)} \qquad \textbf{(6.79)}$$

The first expression is the overall total pressure loss ratio, the second expression is a measure of the total pressure loss in terms of the inlet dynamic pressure, and the last expression uses a reference velocity for the dynamic pressure and measures the total pressure loss in terms of the

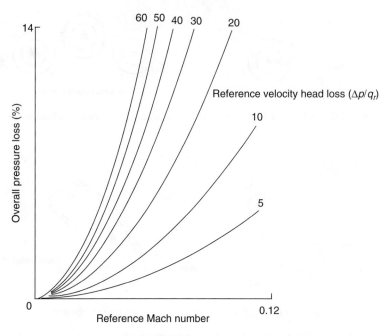

■ **FIGURE 6.29** Percent total pressure loss in a combustor (from Henderson and Blazowski, 1989)

reference dynamic pressure. The reference velocity is the mass averaged gas speed at the largest cross section of the burner, based on the combustor inlet pressure and temperature. The reference Mach number is the ratio of the reference speed to the speed of sound at the prediffuser inlet a_3. All the aerodynamic losses scale with reference dynamic pressure, as shown in Fig. 6.29 (from Henderson and Blazowski, 1989).

A parabolic dependence of the total pressure loss on the reference Mach number is shown in Fig. 6.29. Kerrebrock suggests an empirical rule for the combustor total pressure ratio, as a fraction of reference dynamic pressure. In terms of the average Mach number of the gases in the burner, M_b, Kerrebrock's empirical rule is stated as

$$\pi_b \approx 1 - \varepsilon\left(\frac{\gamma}{2}\right)M_b^2 \tag{6.80}$$

where $1 \prec \varepsilon \prec 2$.

This equation states that the total pressure loss is proportional to an "average" dynamic pressure of the gases in the combustor, with the proportionality constant ε. We note that the total pressure loss grows as the mean flow speed increases. Now, we appreciate the role of prediffuser in a combustion chamber. The total pressure loss due to combustion i.e., heat release, in the burner may be modeled by a Rayleigh flow analysis. The conservation of mass and momentum for a constant-area frictionless duct are

$$\rho_3 u_3 = \rho_4 u_4 \tag{6.81}$$

$$p_3 + \rho_3 u_3^2 = p_4 + \rho_4 u_4^2 \tag{6.82}$$

Now, instead of the compressible equation for the total pressure involving Mach number, we may use the low-speed version in an approximation, i.e., the Bernoulli equation, namely,

$$p_{t3} - p_{t4} \approx p_3 + \frac{1}{2}\rho_3 u_3^2 - p_4 - \frac{1}{2}\rho_4 u_4^2 = (p_3 + \rho_3 u_3^2) - (p_4 + \rho_4 u_4^2) - \frac{1}{2}(\rho_3 u_3^2 - \rho_4 u_4^2) \tag{6.83}$$

Canceling the first two parentheses via momentum Eq. 6.82, and using the continuity equation, we may simplify 6.83 as

$$[p_{t3} - p_{t4}]/q_3 \approx \lfloor \rho_4 u_4^2 \rfloor / \lfloor \rho_3 u_3^2 \rfloor - 1 = [\rho_3/\rho_4] - 1 \qquad (6.84)$$

Now, using the perfect gas law for the density ratio and neglecting static pressure changes between the inlet and exit of the burner in favor of the large temperature rise across the combustor, we may write the following simple expression for the total pressure loss due to heating in a constant-area burner as

$$[p_{t3} - p_{t4}]/q_3 \approx T_4/T_3 - 1 \qquad (6.85)$$

In deriving the above expression, we made a series of approximations and simplifications. Consequently, we may use Eq. 6.85 in relative terms, e.g., in describing the effect of the throttle setting (T_4) on the variation of (hot) total pressure loss in a burner. Lefebvre (1983) suggests a correlation based on the Eq. 6.85 that involves two unknown coefficients, K_1 and K_2, as

$$[(\Delta p_t)_{\text{hot}}]/q_{\text{ref}} = K_1(T_4/T_3 - K_2) \qquad (6.86)$$

The correlation coefficients K_1 and K_2 need to be established experimentally. This is another example of a simple model (Eq. 6.85) that is the basis of an engineering correlation for a complex problem. The total pressure loss due to combustion is small compared with aerodynamic losses, namely,

$$[(\Delta p_t)_{\text{hot}}]/p_{t3} \approx 0.5 - 1.0\%$$

The cold total pressure loss is in the range of $\sim$4–7%, i.e.,

$$[(\Delta p_t)_{\text{cold}}]/p_{t3} \approx 4 - 7\%$$

In an afterburner, the process of flame stabilization is achieved via flameholders, as described earlier, that creates massively separated wakes and is responsible for the bulk of the total pressure loss. We model an afterburner as a constant-area duct with a series of bluff bodies in the stream with a known drag coefficient. The diagram that describes the problem is shown in Fig. 6.30.

We may account for the wall friction in the momentum balance equation either directly or we may combine its overall drag contribution with the flameholder drag. The effect of combustion on total pressure loss is important and will be investigated later in this section.

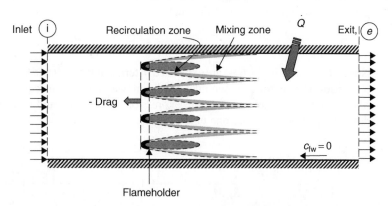

■ **FIGURE 6.30** Control volume for the estimation of total pressure loss in an afterburner

First, we model the afterburner-off condition, which is known as the "dry" analysis of the engine. The "wet" mode analysis follows the "dry" mode in this section.

From continuity equation we have

$$\rho_e u_e = \rho_i u_i \tag{6.87}$$

The fluid momentum equation in the streamwise direction gives

$$\rho_e u_e^2 - \rho_i u_i^2 = p_i - p_e - \frac{D_{\text{flameholder}}}{A} \tag{6.88}$$

We define the flameholder drag in terms of the duct cross-sectional area A as

$$D_{\text{flameholder}} \equiv C_D.q_i.A \tag{6.89}$$

Rearranging the terms of the momentum equation, we get

$$p_e \left(1 + \gamma_e M_e^2\right) = p_i \left[1 + \gamma_i M_i^2 \left(1 - \frac{C_D}{2}\right)\right] \tag{6.90}$$

Therefore the static pressure ratio is expressed in terms of the inlet and exit Mach numbers and the flameholder drag coefficient C_D as

$$\frac{p_e}{p_i} = \frac{1 + \gamma_i M_i^2 \left(1 - \dfrac{C_D}{2}\right)}{1 + \gamma_e M_e^{\,2}} \tag{6.91}$$

From the continuity equation, we express the density ratio in terms of the velocity ratio and then we replace the gas speed by the product of Mach number and the speed of sound to get

$$\frac{\rho_e}{\rho_i} = \frac{u_i}{u_e} = \frac{M_i}{M_e} \sqrt{\frac{(\gamma_i - 1)c_{pi}T_i}{(\gamma_e - 1)c_{pe}T_e}} \tag{6.92}$$

We use the energy equation for an adiabatic process, which is a statement of the conservation of total enthalpy, to get the static temperature ratio, namely,

$$\frac{c_{pe}T_e}{c_{pi}T_i} = \frac{1 + \dfrac{\gamma_i - 1}{2} M_i^2}{1 + \dfrac{\gamma_e - 1}{2} M_e^2} \tag{6.93}$$

By invoking the perfect gas law, we link the static pressure, density, and temperature ratios of the gas at the inlet and exit of the duct as

$$\frac{p_e}{p_i} = \frac{R_e}{R_i} \frac{\rho_e}{\rho_i} \frac{T_e}{T_i} \tag{6.94}$$

Now, all the ratios in Eq. 6.94 are expressed in terms of the inlet and exit Mach numbers and gas properties. Hence, for a known inlet condition, we may use this equation to establish the exit Mach number M_e. Let us substitute expressions 6.91, 6.92, and 6.93 in Eq. 6.94 to get the desired equation involving one unknown M_e,

$$\frac{1 - \gamma_i M_i^2 \left(1 - \dfrac{C_D}{2}\right)}{1 + \gamma_e M_e^2} = \frac{R_e}{R_i} \frac{M_i}{M_e} \sqrt{\frac{(\gamma_i - 1)}{(\gamma_e - 1)} \left(\frac{1 + \dfrac{\gamma_e - 1}{2} M_e^2}{1 + \dfrac{\gamma_i - 1}{2} M_i^2}\right) \frac{c_{pi}}{c_{pe}} \left(\frac{1 + \dfrac{\gamma_i - 1}{2} M_i^2}{1 + \dfrac{\gamma_e - 1}{2} M_e^2}\right)} \tag{6.95}$$

We may simplify the above equation by combining the gas properties and the last two brackets, to get

$$\frac{1 + \gamma_i M_i^2 \left(1 - \dfrac{C_D}{2}\right)}{1 + \gamma_e M_e^2} = \frac{\gamma_i M_i}{\gamma_e M_e} \sqrt{\left(\frac{\gamma_e - 1}{\gamma_i - 1}\right)\left(\frac{1 + \dfrac{\gamma_i - 1}{2} M_i^2}{1 + \dfrac{\gamma_e - 1}{2} M_e^2}\right)} \tag{6.96}$$

Fortunately, Eq. 6.96 is a quadratic equation for M_e with a closed form solution, i.e., no iteration is needed. Therefore the exit Mach number is

$$M_e^2 = \frac{2(\gamma_e - 1) - \gamma_e A^2 + \sqrt{(\gamma_e A^2 - 2\gamma_e + 2)^2 + 2(\gamma_e - 1)^2 (A^2 - 2)}}{(\gamma_e - 1)(A^2 - 2)} \tag{6.97}$$

where

$$A = \left(\frac{1 + \gamma_i M_i^2 \left(1 - \dfrac{C_D}{2}\right)}{\gamma_i M_i}\right) \sqrt{\frac{\gamma_i - 1}{1 + \dfrac{\gamma_i - 1}{2} M_i^2}} \tag{6.98}$$

With exit Mach number known, we may substitute it in Eq. 6.91 to get the static pressure ratio, and the total pressure ratio is then calculated by

$$\frac{p_{te}}{p_{ti}} = \frac{p_e}{p_i} \frac{\left(1 + \dfrac{\gamma_e - 1}{2} M_e^2\right)^{\frac{\gamma_e}{\gamma_e - 1}}}{\left(1 + \dfrac{\gamma_i - 1}{2} M_i^2\right)^{\frac{\gamma_i}{\gamma_i - 1}}} = \frac{1 + \gamma_i M_i^2 \left(1 - \dfrac{C_D}{2}\right)}{1 + \gamma_e M_e^2} \frac{\left(1 + \dfrac{\gamma_e - 1}{2} M_e^2\right)^{\frac{\gamma_e}{\gamma_e - 1}}}{\left(1 + \dfrac{\gamma_i - 1}{2} M_i^2\right)^{\frac{\gamma_i}{\gamma_i - 1}}} \tag{6.99}$$

Now we may graph the total pressure loss, due to flameholder drag, across the afterburner, as a function of inlet Mach number and gas properties at the inlet and exit.

Figure 6.31 shows the aerodynamic impact of flameholder drag on afterburner total pressure loss. The importance of low inlet Mach number to the afterburner may also be discerned from this figure. The penalty for a high inlet Mach number (of say ~0.5) is severe and thus the necessity of a prediffuser becomes evident. The presence of aerodynamic drag in the afterburner causes the exit Mach number to increase and approach a choked, i.e., $M_e = 1$, state. Frictional choking in the afterburner is to be avoided since combustion also tends to increase the Mach number, as in Rayleigh flow. A graph of the exit Mach number as a function of inlet Mach number and the flameholder drag coefficient is shown in Fig. 6.32.

The flameholder drag coefficient in our analysis was defined based on the afterburner cross-sectional area. However, the drag coefficient of bluff bodies is defined based on the maximum cross-sectional area of the bluff body. Therefore,

$$D_{afterburner} \equiv C_{D\,bluff\text{-}body} \cdot q_\infty A_{max} \cdot N = C_D \cdot q_i A_{afterburner} \tag{6.100}$$

where the A_{max} is the base area of an individual bluff body and N is the number of bluff bodies used in the flameholder arrangement, typically in the form V-gutter rings and radial V-gutter connectors. We may define a flameholder blockage area ratio as

$$\text{Blockage} \equiv B \equiv \frac{N \cdot A_{max}}{A_{afterburner}} \tag{6.101}$$

Therefore the drag coefficient used in our afterburner total pressure loss calculation is related to the individual drag coefficients via the blockage factor, i.e.,

$$C_D = B \cdot C_{D\,bluff\text{-}body} \tag{6.102}$$

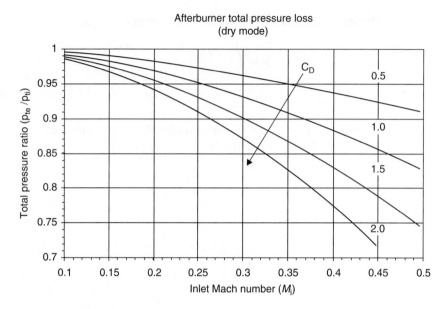

■ **FIGURE 6.31** **Afterburner total pressure loss due to flameholder drag** $(\gamma_i = 1.33,\ \gamma_e = 1.30)$

Figure 6.33 serves as a definition sketch for the afterburner duct, and the flameholder rings that contribute to the blockage parameter B. Here the radial gutters that connect the V-gutter rings to the turbine exit diffuser cone or the casing are not shown. The EJ200 of Eurojet Consortium uses radial gutters (instead of rings) for flame holding (see Fig. 6.47).

To account for higher flow velocities in the plane of the flameholder, due to flameholder blockage, the drag coefficient needs to be scaled by the dynamic pressure ratio, i.e.,

$$C_D = B \cdot \frac{\rho_{fh} V_{fh}^2}{\rho_i V_i^2} \cdot C_{D\,\text{bluff-body}} \approx B\left(\frac{V_{fh}}{V_i}\right)^2 \cdot C_{D\,\text{bluff-body}} \approx B\left(\frac{V_{fh}}{V_i}\right)^2 \tag{6.103}$$

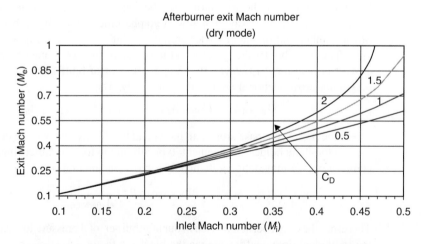

■ **FIGURE 6.32** **Frictional choking of afterburner caused by flameholder drag** $(\gamma_i = 1.33,\ \gamma_e = 1.30)$

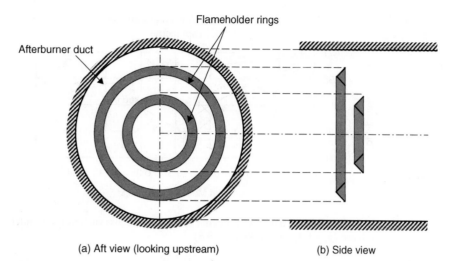

(a) Aft view (looking upstream)   **(b) Side view**

■ **FIGURE 6.33**   **Schematic drawing of the afterburner duct and the flameholder rings**

where subscript "fh" identifies the plane of the flameholder and the bluff body drag coefficient (e.g., a V-gutter) was assumed to be ~1. Note that a flat plate, in broadside, creates a drag coefficient of 2 and a cylinder in cross flow ~1.2 (both flows at a Reynolds number based on plate height or cylinder diameter of 100,000).

The analysis of an afterburner in the "wet" mode differs from the "dry" mode only in its energy balance equation, namely,

$$(1 + f_{AB})c_{pe}T_{te} - c_{pi}T_{ti} = f_{AB}Q_R\eta_{AB} \tag{6.104}$$

By neglecting the fuel-to-air ratio in the afterburner in favor of 1 on the left-hand side of the energy equation and renaming the product term on the RHS as "q," we get

$$\frac{c_{pe}T_{te}}{c_{pi}T_{ti}} \cong 1 + \frac{q}{c_{pi}T_{ti}} \tag{6.105}$$

We may relate the static enthalpy ratio to the stagnation enthalpy and Mach number as

$$\frac{c_{pe}T_e}{c_{pi}T_i} = \left(1 + \frac{q}{c_{pi}T_{ti}}\right)\frac{1 + \frac{\gamma_i - 1}{2}M_i^2}{1 + \frac{\gamma_e - 1}{2}M_e^2} \tag{6.106}$$

By replacing Eq. 6.93 with the above equation, and proceeding with the substitution of the static pressure ratio and density ratio in the perfect gas law, we obtain

$$\frac{1 + \gamma_i M_i^2\left(1 - \dfrac{C_D}{2}\right)}{1 + \gamma_e M_e^2} = \frac{\gamma_i M_i}{\gamma_e M_e}\sqrt{\left(\frac{\gamma_e - 1}{\gamma_i - 1}\right)\left(1 + \frac{q}{c_{pi}T_{ti}}\right)\left(\frac{1 + \dfrac{\gamma_i - 1}{2}M_i^2}{1 + \dfrac{\gamma_e - 1}{2}M_e^2}\right)} \tag{6.107}$$

Again, the above expression involves one unknown, i.e., M_e, and is a quadratic equation with the following closed form solution:

$$M_e^2 = \frac{2(\gamma_e - 1) - \gamma_e A^2 + \sqrt{(\gamma_e A^2 - 2\gamma_e + 2)^2 + 2(\gamma_e - 1)^2(A^2 - 2)}}{(\gamma_e - 1)(A^2 - 2)} \tag{6.108}$$

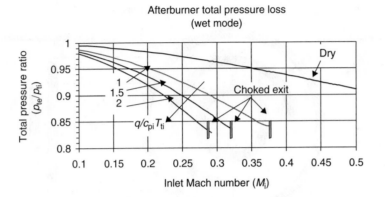

■ **FIGURE 6.34** **Afterburner total pressure loss variation with inlet Mach number and heat release** $(\gamma_i = 1.33, \ \gamma_e = 1.30, \ C_D = 0.5)$

where

$$A = \left(\frac{1 + \gamma_i M_i^2 \left(1 - \dfrac{C_D}{2}\right)}{\gamma_i M_i} \right) \sqrt{ \frac{\gamma_i - 1}{1 + \dfrac{\gamma_i - 1}{2} M_i^2} \left(\frac{1}{1 + \dfrac{q}{c_{pi} T_{ti}}} \right) } \qquad \textbf{(6.109)}$$

Now, let us graph the total pressure loss in an afterburner in the wet mode (sometimes referred to as the *reheat* mode), using the above solution for the exit Mach number. Figure 6.34 shows the afterburner total pressure loss increases with heat release in the combustor. We also note that the choked exit condition is reached at a relatively low inlet Mach number of <0.4. Higher heating levels and flameholder drag coefficients will result in earlier choking than shown in Fig. 6.34. Thus, the necessity of decelerating the flow entering an afterburner is again evident from Fig. 6.34.

We produce Fig. 6.35 to demonstrate the effect of higher flameholder drag coefficient on the afterburner wet mode total pressure loss, and the inlet Mach number range that leads to exit choking. Based on these results, an inlet's Mach number range of ∼0.2–0.25 is desirable for an afterburner.

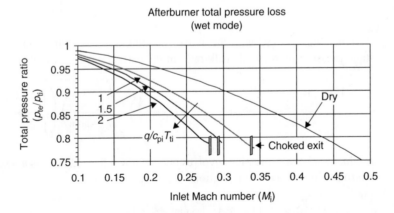

■ **FIGURE 6.35** **Afterburner total pressure loss with heat release and a higher flameholder drag** $(\gamma_i = 1.33, \ \gamma_e = 1.30, \ C_D = 1.5)$

6.6.2 Combustor Flow Pattern and Temperature Profile

A typical flow pattern in a primary burner is demonstrated in Fig. 6.36 (a) (from Tacina, 2000), which shows the burner primary zone, intermediate zone, and the dilution zone in an annular combustor. We have already discussed the primary zone in the context of flammability, flame stability, and stoichiometry. The primary air jets in opposing radial directions are also introduced to stabilize the flame and increase combustor efficiency/heat release rate. The role of intermediate zone is seen as an extension of the primary zone in achieving a complete combustion/controlling pollutant formations, and the dilution zone is seen as producing a desired turbine inlet temperature profile, as shown in Fig. 6.36 (a). Airflow distribution in a modern combustor is shown in Fig. 6.36 (b).

The temperature profile at the turbine inlet exhibits nonuniformity due to the number of fuel injectors used in the circumferential direction, the nonuniformity in dilution air cooling and mixing characteristics as well as other secondary flow patterns and instabilities that are set up in the burner. These spatial nonuniformities are described by two nondimensional parameters, the pattern factor and the profile factor.

Pattern factor

$$\text{PF} \equiv \frac{T_{t\text{-max}} - T_{t\text{-avg}}}{T_{t\text{-avg}} - T_{t\text{-in}}} \qquad (6.110)$$

where $T_{t\text{-max}}$ is the absolute maximum exit temperature in the circumferential and radial directions, $T_{t\text{-avg}}$ is the average of the exit temperature, and $T_{t\text{-in}}$ is the average inlet total temperature. Since the nonuniformity parameter in the numerator of Eq. 6.110 is the absolute

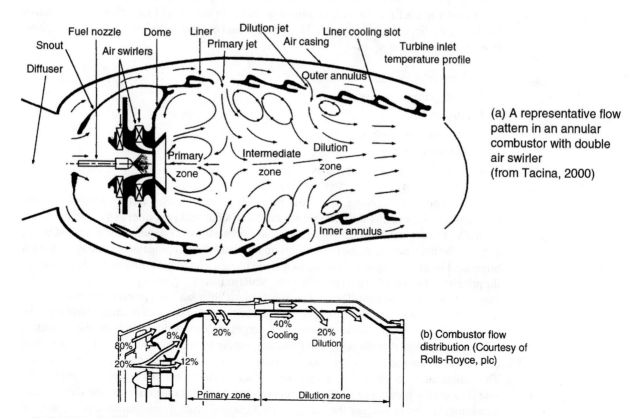

(a) A representative flow pattern in an annular combustor with double air swirler (from Tacina, 2000)

(b) Combustor flow distribution (Courtesy of Rolls-Royce, plc)

■ **FIGURE 6.36** Flow distribution and pattern in modern combustors maximum of the exit temperature (i.e., the absolute peak), this parameter is also known as the peak temperature factor. Turbine nozzle is exposed to this nonuniformity. The desired range of this parameter is between 0.15 and 0.25 in modern high-temperature rise combustors.

Profile factor

$$P_f \equiv \frac{T_{t\text{-max-avg}} - T_{t-\text{in}}}{T_{t\text{-avg}} - T_{t-\text{in}}} \tag{6.111}$$

where $T_{t\text{-max-}avg}$ is the circumferential average of the maximum temperature, i.e., the average of all the local peaks in the circumferential direction. Since the flow acceleration in the turbine nozzle reduces the temperature distortion levels, the turbine rotor that follows the nozzle is exposed to this nonuniformity. The range of this parameter is between 1.04 and 1.08.

REMARKS Flow acceleration dampens nonuniformity in the flow, whereas flow deceleration amplifies distortion. Therefore, the temperature distortion at the turbine inlet is reduced after the nozzle. By the same token, the flow distortion in the inlet diffuser or in the compressor blades is amplified. All engineers dealing with fluid flow problems should know this principle. The proof of this principle is very simple using a parallel flow model.

6.6.3 Combustor Liner and Its Cooling Methods

Combustion takes place inside a combustor liner. It consists of a dome in the primary zone, primary air holes for the primary (radial) jets, air holes for the intermediate zone (radial jets), and the cooling holes/slots in the dilution zone. The combustion temperature may exceed 2500 K (in the primary zone), whereas the liner temperature needs to be kept at ∼1200 K. The heat transfer to the liner from the combustion side is in the form of radiation and convective heating by the hot combustion gases. The conventional cooling techniques used in gas turbine engines are

1. Convective cooling
2. Impingement cooling
3. Film cooling
4. Transpiration cooling
5. Radiation cooling
6. Some combination of the above

The convective cooling method is primarily used in cooled turbine blades with the philosophy of *maximizing* the heat transfer rate from a hot wall (gas side) to the coolant inside the blade. The coolant may also impinge on a hot surface to create a stagnation point and hence enhance the heat transfer to the coolant through the wall. The impingement cooling method is often used in the leading-edge cooling of turbine blades. The film cooling method operates on the principle of *minimizing* heat transfer to the wall from the hot gases by providing a protective cool layer on the hot surface. The cool layer acts like a "blanket" that protects the surface from the hot gases. In film cooling, the coolant is injected through a series of discrete fine film holes, at a slanted angle to the flow direction, and emerges on the hot side to provide the protective cooling layer. Note that the philosophy of film cooling to minimize heat transfer to a wall is different from that of convective cooling, which maximizes the heat transfer through a wall. The transpiration cooling calls for a porous surface where the coolant emerges on the hot gas side through the pores. The cooling protection that it provides is similar to the film cooling technique, i.e., by *blanketing* the surface with a coolant layer. Transpiration cooling may be viewed as the ultimate film cooling in the limit of infinitely many and continuously distributed

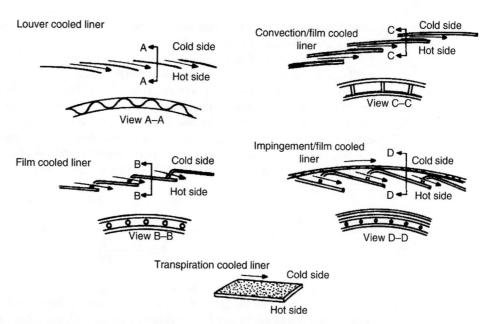

■ **FIGURE 6.37** Combustor liner cooling methods (from Henderson and Blazowski, 1989)

film holes on a surface. Finally, the radiation cooling accompanies all surfaces above the absolute zero temperature. The radiation cooling of the liner to the outer casing, however, represents only a small fraction of the overall heat transfer of the liner. We shall discuss these in the context of cooled turbine blades and shrouds in more detail in Chapter 9.

The availability of excess air at the burner inlet dominates the application of these heat transfer techniques to a combustor liner. The primary and the intermediate zone air holes are designed to provide combustion stability and maximizing the chemical reaction heat release. The remaining air, which may be about 15–30% of the combustor inlet flow, is used to provide a reliable and uniform cooling protection to the liner. A successively more sophisticated technique of liner cooling is shown in Figs. 6.37 and 6.38 (from Henderson and Blazowski, 1989). Starting from the *louver* cooling with the problems of uniform coolant distribution and

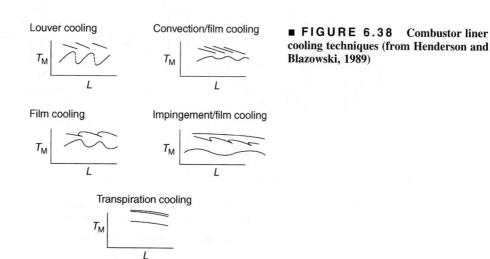

■ **FIGURE 6.38** Combustor liner cooling techniques (from Henderson and Blazowski, 1989)

control that marked the early combustor development to the most sophisticated *transpiration* cooling are schematically shown in Fig. 6.37.

The coolant mass flow requirement is diminished with cooling effectiveness. Simple film cooling, convection/film, impingement/film, and the transpiration cooling are ranked from the highest to the lowest coolant requirement, respectively. The film cooling method eliminates the problem of coolant injection uniformity and control that was shown by louver technique. The combination convection/film and impingement/film cooling techniques offer a further enhancement of cooling effectiveness of the combustor liner. The transpiration cooling through a porous liner minimizes the coolant flow requirement and offers the most uniform wall temperature distribution (as seen in Fig. 6.38). The problem of pore clogging, however, poses a challenge in transpiration cooling.

To quantify the coolant mass flow requirement, we define a cooling effectiveness parameter Φ according to

$$\Phi \equiv \frac{T_g - T_w}{T_g - T_c} \tag{6.112}$$

The cooling effectiveness is the ratio of two temperature differentials. The numerator is the difference between the hot gas T_g and the desired (average) wall temperature T_w. The denominator represents the absolute maximum potential for wall cooling. Therefore, the cooling effectiveness Φ is the fraction of the absolute maximum cooling potential that is realized in the wall cooling process. The hot gas in this application is the combustion temperature (T_g) or sometimes referred to as the flame temperature, and the coolant is the compressor discharge temperature (T_c). The desired wall temperature T_w is dictated by the wall material properties and durability requirements (e.g., 18,000-h life). The variation of cooling effectiveness with coolant flow is shown in Fig. 6.39 (data from Nealy and Reider, 1980).

In future advanced gas turbine engines, the combustor temperature may reach ~2500 K (near max. adiabatic flame temperature of hydrocarbon fuels), with a desired liner temperature of ~1200 K and a compressor discharge temperature of 900 K, the cooling effectiveness is then 0.8125. In examining Fig. 6.39, we note that a film-cooled liner may not reach this desired level of cooling effectiveness, regardless of percent coolant available. On the contrary, this level of

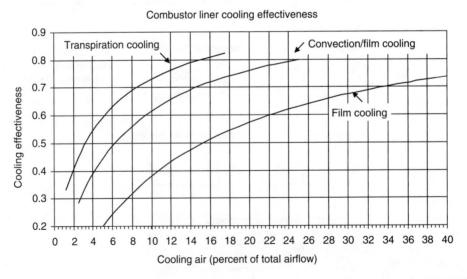

■ **FIGURE 6.39**   **Combustor liner cooling effectiveness (data from Nealy and Reider, 1980)**

cooling effectiveness represents the maximum possible limit of the transpiration cooling at close to 18% coolant availability/usage. Application of a thin layer of refractory material such as a ceramic coating lowers the heat transfer to the wall and thus allows a higher operating temperature in the combustor. To tolerate a higher heat release level in an advanced combustor, the liner material needs to tolerate higher operating temperatures and require less (to no) cooling percentage. In addition to operating at high temperatures, the liner material has to be oxidation resistant. Refractory metals such as molybdenum and tungsten suffer in this regard. A promising new liner material is the ceramic matrix composite (CMC). Development of prototype CMC combustor liner has been approached in industry by using silicon carbide fiber reinforced silicon carbide (SiC/SiC). The use of advanced composite materials as combustor liners represents a challenge in manufacturing, scaling up, life prediction, damage tolerance, and reparability characteristics.

6.6.4   Combustion Efficiency

Combustion efficiency measures the actual rate of heat release in a burner and compares it with the theoretical heat release rate possible. The theoretical heat of reaction of the fuel assumes a complete combustion with no unburned hydrocarbon fuel and no dissociation of the products of combustion. The actual heat release is affected by the quality of fuel atomization, vaporization, mixing, ignition, chemical kinetics, flame stabilization, intermediate air flow, liner cooling, and, in general, the aerodynamics of the combustor. Here a variety of timescales as in residence time, chemical reaction/reaction rate timescales, spontaneous ignition delay time that includes vaporization timescale, among other time constants enter the real combustion problem. The unburned hydrocarbon (UHC) and carbon monoxide (CO) that contribute to combustor inefficiency are also among the combustion-generated pollutants. The allowable levels of these pollutants in aircraft gas turbine engine emissions, regulated by the U.S. Environmental Protection Agency (EPA), place a 99% combustion efficiency demand on the combustor.

For a gas turbine combustor, when chemical kinetics is the limiting factor in combustor performance, Lefebvre (1959, 1966) introduces a combustor loading parameter θ, which correlates well with combustion efficiency. θ parameter is defined as

$$\theta = \frac{p_{t3}^{1.75} A_{\text{ref}} H \cdot e^{\frac{T_{t3}}{b}}}{\dot{m}_3} \tag{6.113}$$

The combustor inlet pressure, temperature, and mass flow rate are p_{t3}, T_{t3}, and $\dot{m}_3$, respectively, maximum cross-sectional area of the burner is defined as A_{ref}, the combustor height is H and b is a reaction rate parameter. The dependence of the reaction rate parameter b on the primary zone equivalence ratio ϕ is estimated by Herbert (1957) to be

$$b = 382\left(\sqrt{2} \pm \ln\frac{\phi}{1.03}\right) \quad [(+)\text{for } \phi < 1.03, \ (-)\text{for } \phi > 1.03] \tag{6.114}$$

These expressions are plotted in Fig. 6.40 (from Henderson and Blazowski, 1989). Note that dimensions of the parameters used in Fig. 6.40 (a) are in English units, as described in the definition box in the graph.

The lowest combustion efficiency is of course zero in the flameout limit. This could happen at a high altitude when the low combustor pressure in essence slows the reaction rate to a halt ($\tau_{\text{residence}} \ll \tau_{\text{reaction}}$). The graphical correlations used in the combustion efficiency (Fig. 6.40) are then used to size the combustor in conditions corresponding to high altitude (for a relight requirement) assuming a combustor efficiency of 80%.

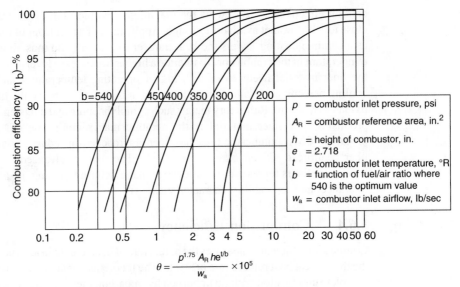

(a) A correlation of combustion efficiency

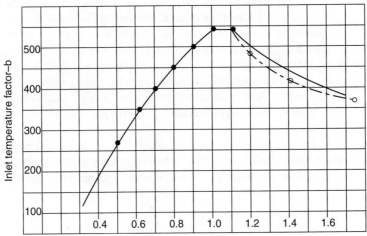

(b) Variation of reaction rate parameter b with the equivalence ratio in the primary zone

■ **FIGURE 6.40** Combustion efficiency correlation based on Lefebvre θ parameter (from Henderson and Blazowski, 1989)

6.6.5 Some Combustor Sizing and Scaling Laws

The engine size, i.e., the core (air) mass flow rate, and the compressor pressure ratio, by and large, determine the combustor inlet flow area. The flow area takes the form of an annulus, and its size is A_3. We may calculate the flow area A_3 by a one-dimensional continuity equation such as

$$A_3 = \frac{\dot{m}_3 \sqrt{T_{t3}}}{p_{t3} . M_3} \left(1 + \frac{\gamma_3 - 1}{2} M_3^2 \right)^{\frac{\gamma_3 + 1}{2(\gamma_3 - 1)}} \cdot \sqrt{\frac{R_3}{\gamma_3}} \qquad \textbf{(6.115)}$$

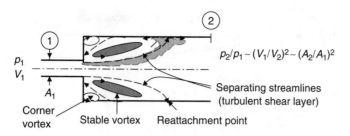

$$p_2/p_1 \sim (V_1/V_2)^2 \sim (A_2/A_1)^2$$

Separating streamlines
(turbulent shear layer)

Corner vortex   Stable vortex   Reattachment point

■ **FIGURE 6.41**   **Schematic drawing of a dump diffuser**

At design point, the engine (core) mass flow rate, the compressor discharge parameters p_{t3} and T_{t3} (which are a function of π_c and e_c), and our design choice for M_3 size the flow area A_3 according to Eq. 6.115. A typical compressor exit Mach number is $\sim$0.4–0.5. However, a combustor requires a prediffuser to decelerate the air from the compressor to improve combustion total pressure (loss) and combustion efficiency. A desired combustor inlet Mach number is $\sim$0.2. A conventional diffuser may be designed by using the design charts that were presented in the inlet and nozzle chapter. Diffusion is a slow and patient process, which normally takes place in a long duct of shallow wall divergence angles of $\sim$3–5° inclination with respect to a straight centerline. To shorten the axial length of a combustor prediffuser, one may employ a

- Diffuser with splitter vanes
- Dump diffuser
- Vortex-controlled diffuser

The subject of a wide-angle diffuser that employs splitter vanes was already presented in the inlet-nozzle chapter (references 1 and 2 should be consulted on short and hybrid diffusers). In a dump diffuser, the flow area undergoes a sudden expansion with the attendant flow separation at the lip. Figure 6.41 shows a schematic drawing of a dump diffuser.

A curved turbulent shear layer emerges from the point of separation, which reattaches in approximately 5–7 inlet channel heights, for a two-dimensional expansion. A pair of stable vortex structures is formed (in 2D) at the separation corner. In the axisymmetric case, a stable vortex ring is formed at the separation junction. The expression used in Fig. 6.41 for the static pressure ratio and velocity ratio is based on Bernoulli equation, and the last part that relates the flow speed to flow area is based on the incompressible continuity equation. The loss of total pressure between stations 1 and 2 is due to viscous/turbulent mixing and dissipation in the separated shear layer and the boundary layer formation downstream of the reattachment point. An example of an annular combustor using a dump prediffuser is shown in Fig. 6.42 (from Henderson and Blazowski, 1989).

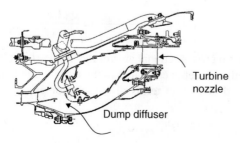

■ **FIGURE 6.42**   **F101 annular combustor with a dump (pre-) diffuser design(from Henderson and Blazowski, 1989).**

Turbine nozzle

Dump diffuser

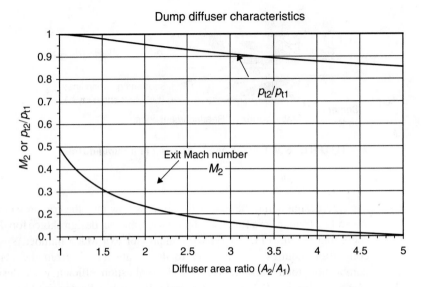

■ **FIGURE 6.43** **Total pressure ratio and exit Mach number in a dump diffuser** ($M_1 = 0.5$, $\gamma = 1.4$)

Total pressure recovery of a dump diffuser is a function of Reynolds number and the Mach number at its inlet. For Reynolds number based on the inlet diameter of $\sim$500,000 to 10^6, Barclay (1972) recommends the following correlation:

$$\frac{p_{t2}}{p_{t1}} \approx \exp\left\{-\frac{\gamma M_1^2}{2}\left[\left(1 - \frac{A_1}{A_2}\right)^2 + \left(1 - \frac{A_1}{A_2}\right)^6\right]\right\} \quad [\text{for } 1 < A_2/A_1 < 5] \qquad \textbf{(6.116)}$$

The exit Mach number M_2 is then established using a continuity equation, according to

$$\frac{p_{t2}A_2M_2}{\left(1 + \dfrac{\gamma - 1}{2}M_2^2\right)^{\frac{\gamma+1}{2(\gamma-1)}}} = \frac{p_{t1}A_1M_1}{\left(1 + \dfrac{\gamma - 1}{2}M_1^2\right)^{\frac{\gamma+1}{2(\gamma-1)}}} \qquad \textbf{(6.117)}$$

The total pressure recovery of a dump diffuser and its exit Mach number are plotted in Fig. 6.43 for an inlet Mach number of $M_1 = 0.5$ and $\gamma = 1.4$.

Figure 6.43 indicates that in order to decelerate a Mach 0.5 flow to an exit Mach number of 0.2 in a dump diffuser, we need to employ an area ratio of $\sim$2.35, and the flow in the dump diffuser will recover $\sim$0.935 of its inlet total pressure (i.e., it suffers $\sim$6.5% loss).

To establish a length for the main combustor, we relate the residence time of the fluid in the burner to the ratio of fluid mass to the flow rate in the combustor, namely,

$$\tau_{\text{resident}} \approx \frac{m_{\text{burner-gas}}}{\dot{m}} \approx \frac{\rho_{t3}A_{\text{ref}}L}{\dot{m}_3} \qquad \textbf{(6.118)}$$

where A_{ref} is the maximum cross-sectional area of the burner, L is the burner length, and the mass flow rate through the burner is estimated as the airflow rate at the combustor entrance. Now, expressing the fluid density in terms of compressor pressure ratio via an isentropic exponent, namely,

$$\rho_{t3} \sim \rho_{t2}.\pi_c^{1/\gamma}$$

We may isolate combustor length L from Eq. 6.118 and express it as

$$L \propto \frac{\dot{m}_2 . \tau_{\text{resident}}}{\rho_{t2} A_{\text{ref}} \pi_c^{1/\gamma}} = \left(\frac{\dot{m}_2}{\rho_{t2} A_2}\right) . \left(\frac{A_2}{A_{\text{ref}}}\right) \frac{\tau_{\text{resident}}}{\pi_c^{1/\gamma}} \tag{6.119}$$

In the above expression, A_2 is the engine face area and the first parenthesis is a design parameter, which is independent of the engine size. The area ratio A_2/A_{ref} may be related via continuity equation to the engine face axial Mach number (a design choice) and the combustor exit Mach number (i.e., taken at the exit of the turbine nozzle) M_4 and the total pressure and temperature ratios according to

$$\frac{A_2}{A_{\text{ref}}} \propto \frac{p_{t4}}{p_{t2}} \sqrt{\frac{T_{t2}}{T_{t4}}} \frac{\left(1 + \frac{\gamma-1}{2} M_2^2\right)^{\frac{\gamma+1}{2(\gamma-1)}}}{M_2} f(\gamma_4, R_4) \tag{6.120}$$

In Eq. 6.120, the burner exit Mach number (taken at the exit of the turbine nozzle) is set equal to 1, as the turbine nozzle remains choked over a wide range of operating conditions. We may approximate the total pressure ratio term in Eq. 6.120 by the compressor pressure ratio and express the combustor length L as

$$L \propto \frac{\tau_{\text{resident}}}{\pi_c^{1/\gamma}} \frac{\pi_c}{\sqrt{T_{t4}}} \sim \frac{\pi_c^{\frac{\gamma-1}{\gamma}}}{\sqrt{T_{t4}}} . \tau_{\text{resident}} \tag{6.121}$$

The fluid residence timescale and the chemical reaction timescale may be interchanged in a combustor

$$\tau_{\text{resident}} \cong \tau_{\text{reaction}} \tag{6.122}$$

The reaction timescale is inversely proportional to the reaction rate and attains the following general form,

$$\tau_{\text{reaction}} \propto p_{t3}^{-n} . T_{t3}^{-m} \tag{6.123}$$

Now, substituting Eq. 6.123 for the residence timescale into Eq. 6.121, we get

$$L \propto \frac{\pi_c^{\frac{\gamma-1}{\gamma}}}{\sqrt{\tau_\lambda}} \pi_c^{-n} \tau_c^{-m} \tag{6.124}$$

We may combine the exponents of the compressor pressure ratio in Eq. 6.124 as

$$L \propto \frac{\pi_c^{(1-m)(\frac{\gamma-1}{\gamma})-n}}{\sqrt{\tau_\lambda}} \tag{6.125}$$

This equation relates the combustor length L to cycle parameters π_c and τ_λ, which are independent of size. Hence, a high-pressure ratio engine occupies a smaller length combustor than a comparable size engine with a lower pressure ratio. A similar argument can be made regarding the cycle thermal loading parameter τ_λ. Table 6.6 (from Henderson and Blazowski, 1989) shows some data on contemporary combustor size, weight, and cost.

6.6.6 Afterburner

So far, we have discussed the total pressure drop in an afterburner due to flameholder drag and combustion. We also alluded to the flame stability provided by a bluff-body flameholder. In this section, we quantify the parameters governed by the fluid mechanics and the chemical reaction that afford flame stability in an afterburner. We also develop a scaling parameter that connects

■ **TABLE 6.6**
Data on combustor size, weight, and cost (Henderson and Blazowski, 1989)

Parameter	TF39	TF41	J79	JT9D	T63
Type	Annular	Cannular	Cannular	Annular	Can
Mass flow (design point)					
Airflow, lb/s	178	135	162	242	3.3
kg/s	81	61	74	110	1.5
Fuel flow, lb/h	12,850	9965	8350	16,100	235
kg/h	5829	4520	3788	7303	107
Size					
Length, in.	20.7	16.6	19.0	17.3	9.5
cm	52.6	42.2	48.3	43.9	24.1
Diameter, in.	33.3	5.3/24.1[a]	6.5/32.0[a]	38.0	5.4
cm	84.6	13.5/61.2	16.5/81.3	96.5	13.7
Weight, lb	202	64	92	217	2.2
kg	92	29	42	98	1.0
Cost, $	42,000	17,000	11,300	80,000	710

[a]Can diameter/annulus diameter.

different flameholder arrangements and duct geometries to experimental results obtained in a rectangular duct with a single flameholder. We first define a bluff body and its wake in a duct by the following length and velocity scales, as shown in Fig. 6.44.

The wake of a V-gutter of height "d" is shown to have a central recirculation zone of length "L," a wake width "W," a channel height "H," an upstream gas velocity V_1, and an accelerated gas velocity, just outside the mixing layer, V_2, known as the "edge velocity." We had defined a blockage parameter B, which for a two-dimensional duct, as shown in Fig. 6.44, is the ratio $B = d/H$. The blockage parameter B and the wake width establish the flow acceleration V_2. A continuous chemical reaction in the mixing layer will be supported only if the resident timescale of the fuel/air mixture in the mixing layer is longer than the reaction timescale. Hence, for flame stability, the following simple relationship should hold between the two characteristic timescales, namely,

$$\tau_{resident} \succ \tau_{reaction} \tag{6.126}$$

This is the basis of Zukoski–Marble (1955) characteristic time model for flame stability. The resident timescale of the unburned gas is proportional to the ratio

$$\tau_{resident} \sim \frac{L}{\bar{v}} \tag{6.127}$$

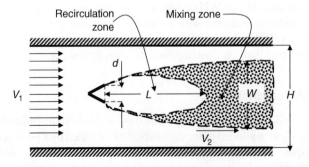

■ **FIGURE 6.44** Length and velocity scales associated with a single flameholder in a duct

where $\bar{v}$ is an average gas speed in the mixing layer. Hence, a dimensionless parameter $\bar{v}\tau/L$ emerges that could serve as a stability criterion. Experiments are conducted that establish the flame blowout condition in a rectangular duct with a single flameholder at the center of the duct with a range of chemical reaction parameters. The blowout condition parameters are labeled with a subscript "c" and by combining the proportionality constants in the unknown time constant τ_c, the average gas speed in the mixing layer is replaced by V_{2c} to produce

$$\left(\frac{\tau_c V_{2c}}{L}\right)_{\text{blowout}} = 1 \tag{6.128}$$

as the blowout stability criterion, or the *marginal* stability criterion. The measurements of L and V_{2c} then establish the critical timescale τ_c at the blowout condition. The ratio V_{2c}/L represents the fluid dynamic parameter in Eq. 6.128, whereas τ_c represents the effect of all chemical reaction parameters lumped into a single term. The timescale τ_c, which is also referred to as ignition time, is independent of the flameholder geometry and arrangement so long as the mixing layer is turbulent. It depends, however, on a number of parameters such as the fuel-to-air ratio, fuel type, inlet temperature, and the degree of vitiation of the afterburner gas. Experimental results of Zukoski (1985) over a range of two- and three-dimensional flameholders indicate that the τ_c is ~ 0.3 ms at stoichiometric fuel–air ratio. This is consistent with Fig. 6.9 (from Kerrebrock) that shows the reaction bucket has a minimum of ~ 0.3 ms near stoichiometric ratio and up to ~ 1.5 ms in fuel lean and rich limits of gasoline–air mixtures (see Fig. 6.9). A more convenient stability parameter may be defined using the upstream blowout velocity V_{1c} and the duct height H as a reference velocity and length scale,

$$\frac{\tau_c V_{1c}}{H} = \frac{V_{1c}}{V_{2c}}\frac{L}{W}\frac{W}{H}\left(\frac{\tau_c V_{2c}}{L}\right) = \frac{V_{1c}}{V_{2c}}\frac{L}{W}\frac{W}{H} \tag{6.129}$$

The length of recirculation zone to the wake width L/W is approximately 4, over a wide range of bluff-body flameholder geometries. We may also approximate the ratio of edge velocity to the upstream velocity using the continuity equation for an incompressible fluid and neglecting the entrainment in the mixing layer as

$$\frac{V_{1c}}{V_{2c}} \approx \frac{H-W}{H} = 1 - \frac{W}{H} \tag{6.130}$$

By neglecting the entrainment in the mass flow balance of Eq. 6.130; we overpredict the edge velocity V_{2c}, which makes our analysis more conservative. As a first-order approximation, we neglect the entrainment in the mass balance. Substituting Eq. 6.130 in the stability parameter of Eq. 6.129, we get

$$\frac{\tau_c V_{1c}}{H} \approx 4\frac{W}{H}\left(1 - \frac{W}{H}\right) \tag{6.131}$$

The emergence of W/H as the nondimensional parameter in the flameholder stability criterion is of great significance. This ratio depends mainly on the flameholder blockage parameter B, which may now be generalized to include different number and arrangement of flameholders in a circular duct.

Table 6.7 is reproduced from Zukoski (1985), which relates the flameholder blockage to velocity ratio for a V-gutter wedge half-angle of $15°$ and $90°$ (flat plate in broadside) in a rectangular duct.

We note that stability parameter $(W/H)(V_1/V_2)$ increases with the blockage ratio and approaches a value of 0.25 for a 50% blockage. Examining the quadratic nature of Eq. 6.131, we deduce that W/H of 0.5 maximizes the stability parameter, for a flameholder in a rectangular duct, which states that the optimum wake width is 50% of the duct height. Substituting these

■ **TABLE 6.7**
Dependence of Wake Width W, Edge Velocity V_2, and a Stability Parameter on Blockage Ratio d/H and wedge Half-Angle α (from Zukoski, 1985)

$\dfrac{d}{H}$	$\alpha = 15$ deg			$\alpha = 90$ deg		
	$\dfrac{W}{d}$	$\dfrac{V_2}{V_1}$	$\left(\dfrac{W}{H}\right)\left(\dfrac{V_1}{V_2}\right)$	$\dfrac{W}{d}$	$\dfrac{V_2}{V_1}$	$\left(\dfrac{W}{H}\right)\left(\dfrac{V_1}{V_2}\right)$
0.05	2.6	1.15	0.11	4.0	1.25	0.16
0.10	1.9	1.23	0.15	3.0	1.43	0.21
0.20	1.5	1.42	0.20	2.2	1.75	0.248
0.30	1.3	1.62	0.23	1.7	2.09	0.250
0.40	1.2	1.90	0.25	1.6	2.50	0.248
0.50	1.2	2.3	0.25	1.4	3.16	0.22

values into Eq. 6.131 yields a maximum blowout velocity in the approach stream of the flameholder, V_{1m}

$$\frac{\tau_c V_{1m}}{H} \approx 1 \tag{6.132}$$

To demonstrate the validity of this approach, Zukoski (1985) compares the result of the theoretical predictions of Table 6.7 with experimental data of Wright (1959). Figure 6.45 shows this comparison.

These results confirm the method of approach proposed by Zukoski and support the stability criterion of Eq. 6.132 for two-dimensional flameholder geometries in a rectangular duct.

For a general axisymmetric flameholder in a circular duct, the wake width is less than a two-dimensional counterpart, due to a three-dimensional relieving effect, i.e.,

$$\left(\frac{W}{d}\right)_{3D} \prec \left(\frac{W}{d}\right)_{2D} \tag{6.133}$$

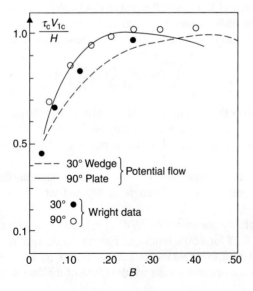

■ **FIGURE 6.45**   Comparison between theoretical stability predictions (potential flow) and the experimental data of Wright for flameholder wedge half-angles of 30° and 90° (from Zukoski, 1985)

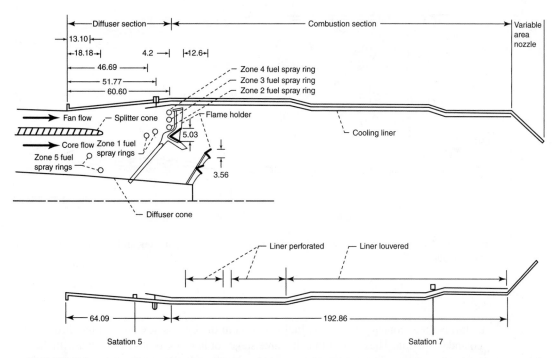

■ FIGURE 6.46   The TF30-P-3 turbofan engine with its afterburner showing the installation and scales of the flameholders, fuel injection rings and the diffuser cone(all dimensions in cm, from Zukoski, 1985)

Remember that a 3D drag coefficient is also less than a 2D drag coefficient (on a body with the same cross section), hence the wake width in 3D is less than the wake width in 2D. Consequently, the edge velocity V_2 in 3D is less than the edge velocity on a two-dimensional wake. The lower edge speed increases the residence time of the fuel–air mixture in the mixing layer, therefore the velocity in the approach stream for marginal stability is higher for a 3D flameholder, i.e.,

$$(V_{1c})_{3D} \succ (V_{1c})_{2D} \tag{6.134}$$

In an axisymmetric flameholder and circular duct, the optimum wake width is $\sim 1/3$, and the approach stream velocity for marginal stability is 50% higher in 3D than 2D, according to Zukoski (1985).

Figure 6.46 shows an afterburner with three V-gutter flameholder rings attached to the turbine exit diffuser cone via radial V-gutters (from Zukoski, 1985). There are seven fuel spray rings that are divided into five zones. The fuel spray rings are upstream of the flameholders to allow for fuel evaporation and hence reduce the ignition time in the flameholder shear layer. Zone 1 is on the shear layer of the fan and core flow. Zones 2–4 are in the fan stream and zone 5 is in the core stream. The zonal approach to fuel injection in the afterburner provides for engine thrust modulation in the reheat mode. The flow to the cooling liner is supplied by the fan stream and serves two functions. First, it protects the casing through a coolant protective layer (liner louvered) and second, the liner dampens a combustion-related acoustic instability, known as screech (liner perforated). We will examine the afterburner screech issue in more detail in section 6.9.

A longitudinal section of Eurojet Consortium, EJ200, turbofan engine afterburner, and nozzle is shown in Fig. 6.47 (from Kurzke and Riegler, 1998). Here the fuel spray in the afterburner is divided into three stages, again for thrust modulation purposes. The fuel spray bars shown represent the primary zone, and the fuel spray to the radial gutter region (in the core

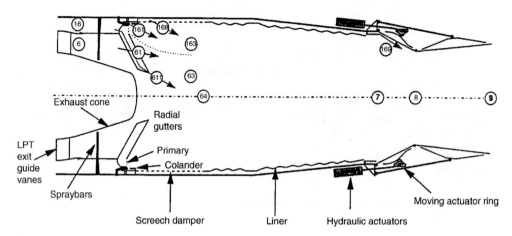

■ FIGURE 6.47 **Longitudinal section of the EJ200 afterburner and nozzle(from Kurzke and Riegler, 1998)**

stream) is the secondary zone. The fuel injection in the bypass stream is at the throat of the "colander" region. These constitute the three stages or burning zones associated with the fuel spray in this engine. The flameholder is of radial V-gutter configuration, instead of a multiring configuration, as in Fig. 6.46. The perforated screech damper and the cooling liner are also shown in Fig. 6.47.

6.7 Combustion-Generated Pollutants

Smoke, oxides of carbon, CO and CO_2, oxides of nitrogen, NO_x, and unburned hydrocarbon fuel (UHC) constitute the bulk of combustor-generated pollutants. Although oxides of sulfur, SO_x, are considered engine exhaust pollutants as well, but since sulfur content of a fuel is dictated by the aviation fuel production (or the refinery) process and is not controlled by combustion, its discussion is often absent from combustion-generated pollutants. The U.S. Environmental Protection Agency (EPA) has been charged with setting the emission standards for aircraft engines operated in the United States. There are two areas of concern with air pollution. The first deals with engine emissions near airports in landing-takeoff cycle (LTO), below 3000 feet. The second area of concern examines the effect of engine emissions at cruise altitude in the stratosphere.

6.7.1 Greenhouse Gases, CO_2 and H_2O

Carbon dioxide, CO_2, and water vapor, H_2O, constitute the products of complete combustion. The large deposition of these products in stratosphere by a fleet of commercial aircraft is feared to contribute to global warming through greenhouse effect. Both of these molecules, i.e., CO_2 and H_2O, absorb infrared radiation from earth and radiate it back toward earth. Earth's atmosphere like the glass of a greenhouse is transparent to visible light from the sun but absorbs the infrared radiation, which accounts for a warmer earth surface temperature due to a radiation absorption in its atmosphere. Now, the average temperature of the earth's surface is 298 K, whereas without the greenhouse gases in the atmosphere it would be 255 K (i.e., 43°C colder). However, a higher concentration of these greenhouse gases in the stratosphere may tilt

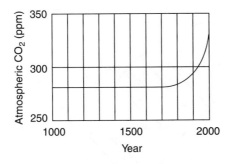

■ FIGURE 6.48   Concentration of carbon dioxide in the atmosphere over the past 1000 years based on ice core data and direct readings since 1958 (from Zumdahl, 2000)

the balance toward global warming. The impact of the additional greenhouse gases in the atmosphere is being debated. Conflicting climatic data between the Southern hemisphere and the Northern latitudes over the past century suggest a more complex dynamics than a simple infrared radiation model proposed. Figure 6.48 (from Zumdahl, 2000) shows, however, the *undeniable* rapid rise in atmospheric CO_2 concentration over the past 100 years with the increase in consumption of hydrocarbon fuels for electric power generation and transportation. At the present time, there is no emission regulation for carbon dioxide and water vapor and thus no FAA engine certification requirements. The unit of concentration in Fig. 6.48 is parts per million in molar concentration of CO_2, i.e., equal to volume fraction of CO_2 in the atmosphere.

6.7.2   Carbon Monoxide, CO, and Unburned Hydrocarbons, UHC

Carbon monoxide is formed as a result of an incomplete combustion of hydrocarbon fuels. It is a source of combustion inefficiency as well as a pollutant in the engine emissions. In combustion terms, CO is a fuel with a heating value of 2267 cal/g, which is approximately 24% of the Jet A fuel specific energy, which leaves the engine in the exhaust gases without releasing its chemical energy. Partial dissociation of carbon dioxide in the hot primary zone is also responsible for carbon monoxide formation. A fuel lean operation at low engine power setting causes a slow reaction rate in the combustor and thus carbon monoxide formation. At a low power setting, combustor inlet temperature and pressure are reduced, which directly impact the reaction rate in the combustor as we discussed earlier. In case of unburned hydrocarbons, a fraction of fuel may remain unburned due to poor atomization, vaporization, and residence time or getting trapped in the coolant film of the combustor liner. The indices of carbon monoxide and unburned hydrocarbons both drop with the engine power setting. An accumulation of aircraft engine data, which is reported in Fig.6.49 (from Henderson and Blazowski, 1989), suggests the sensitivity of CO and UHC formations to engine power setting.

It is entirely plausible to relate the combustion inefficiency, i.e., $1 - \eta_b$, particularly in the idle setting, to the emission index of CO and UHC, as they are both fuels that are present in the exhaust. Blazowski (1985) expresses the combustion inefficiency in terms of emission index EI (at idle) as

$$1 - \eta_b = [0.232(EI)_{CO} + (EI)_{UHC}] \times 10^{-3} \qquad (6.135)$$

Note that the emission index EI is the mass ratio in grams of pollutants generated per kilogram of fuel consumed. As an example, let us take emission indices of CO and UHC to be $\sim$100 and $\sim$40 g/kg, respectively, (from Fig. 6.49) to represent the engine idle power setting with combustor inlet temperature $\sim$400 K. Substitution of these emission indices in Eq. 6.135 suggests a combustor inefficiency of $\sim$6.3%. However, EPA standards on emissions applied to Eq. 6.135 demands a combustor efficiency of 99% at idle setting.

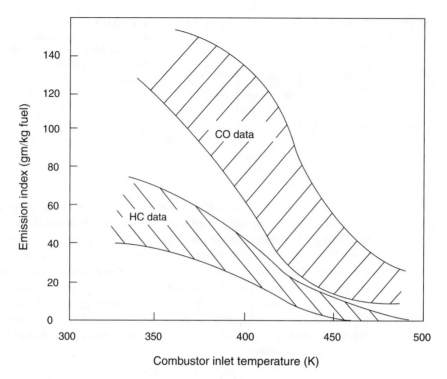

■ **FIGURE 6.49** Emission index of carbon monoxide and unburned hydrocarbons (labeled as HC Data) with engine idle setting (from Henderson and Blazowski, 1989)

6.7.3   Oxides of Nitrogen, NO and NO$_2$

Nitric oxide, NO, is formed in the high temperature region of the primary zone at temperatures above 1800 K. The chemical reaction responsible for its production is

$$N_2 + O \rightarrow NO + N \tag{6.136}$$

Since the reaction of nitrogen is with the oxygen atom in order to create nitric oxide, the regions within the flame where dissociated oxygen exists (in equilibrium) shall produce nitric oxide. This explains the high temperature requirement for NO production (in near stoichiometric mixtures). Therefore, nitric oxide is produced at high engine power settings, in direct contrast to the carbon monoxide and UHC pollutants that are generated at the low power (idle) settings. Further oxidation of NO into NO$_2$ takes place at lower temperatures of the intermediate or dilution zones of the combustor. Both types of the oxides of nitrogen are referred to as NO$_x$ in the context of engine emissions. Lipfert (1972) demonstrates the temperature sensitivity of NO$_x$ formation through an excellent correlation with numerous gas turbine engine data. His results are shown in Fig. 6.50.

We note that engine types, combustor types, and fuels had apparently no effect on the NO$_x$ correlation presented by Lipfert (see Fig. 6.50). All engines correlated with the combustor inlet temperature T_{t3}. Combustor inlet temperature is directly related to the compressor pressure ratio π_c, thus, we expect a similar dependence on the cycle pressure ratio, as shown in Fig. 6.51 (from Henderson and Blazowski, 1989).

6.7.4   Smoke

The fuel-rich regions of the primary zone in the combustor produce carbon particulates known as soot. The soot particles are then partially consumed by the high temperature gases, i.e., in the

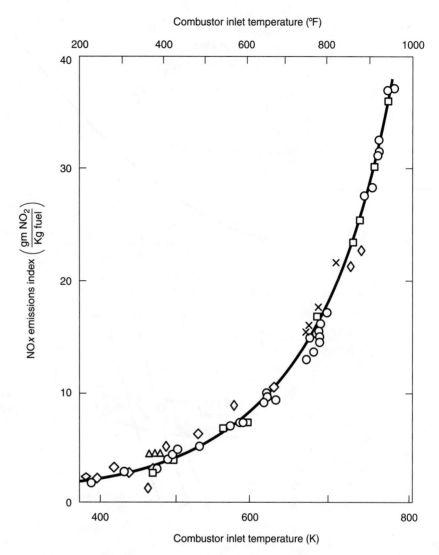

Combustor inlet temperature (°F)

NOx emissions index $\left(\dfrac{\text{gm NO}_2}{\text{Kg fuel}}\right)$

Combustor inlet temperature (K)

■ **FIGURE 6.50** Correlation of current engine NO$_x$ emissions with combustor inlet temperature (from Lipfert, 1972)

intermediate and to a lesser extent in the dilution zones of the combustor. Visible exhaust smoke corresponds to a threshold of soot concentration in the exhaust gases. Soot particles are primarily carbon (96% by weight), hydrogen, and some oxygen. An improved atomization of the fuel (using air-blast atomizers) and enhanced mixing intensity in the primary zone reduces the soot formation. The parameter that quantifies soot in the exhaust gases is called the smoke number, SN. Its measurement is based on passing a given volume of the exhaust gas through a filter for a specific time. Then compare the optical reflectance of the stained and the clean filter by using a photoelectric reflectometer. The procedure for aircraft gas turbine engines exhaust smoke measurements is detailed in a Society of Automotive Engineers (SAE) document— ARP 1179 (1970). The smoke number is defined as

$$SN \equiv 100(1 - R/R_0) \tag{6.137}$$

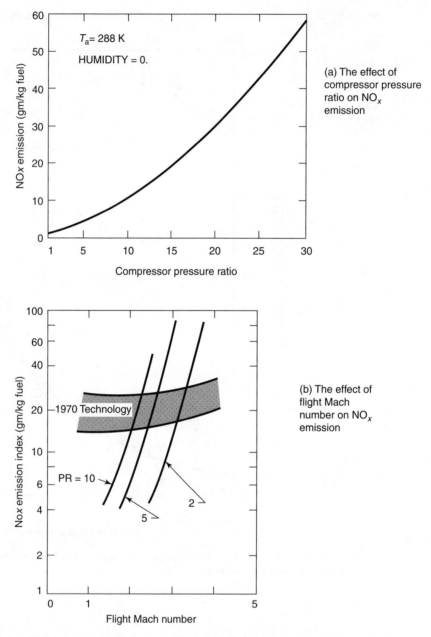

(a) The effect of compressor pressure ratio on NO_x emission

(b) The effect of flight Mach number on NO_x emission

■ **FIGURE 6.51**   **Impact of cycle pressure ratio on NO_x emission (from Henderson and Blazowski, 1989)**

where R is the absolute reflectance of the stained filter, and R_0 is the absolute reflectance of the clean filter material.

6.7.5   Engine Emission Standards

The emission standards of the U.S. EPA (1973, 1978 and 1982) and the International Civil Aviation Organization (ICAO) (1981) define the limits on CO, UHC, NO_x, and smoke production of aircraft engines near airports (Table 6.8).

■ **TABLE 6.8**
EPA and ICAO Engine Emission Standards

Pollutant	EPA (1979) Standards	ICAO (1981) Standards
CO	4.3 g/(kg · thrust · hr)	118 g/kg fuel
UHC	0.8 g/(kg · thrust · hr)	19.6 g/kg fuel
NO_x	3.0 g/(kg · thrust · hr)	$40 + 2(\pi_{00})$
Smoke	19–20	$83.6(F_{00})^{-0.274}$

In the ICAO standards, the rated cycle pressure ratio π_{00} and the rated engine thrust F_{00} (in kN) are incorporated in the regulated limits.

6.7.6   Low-Emission Combustors

To combat the problems of CO and UHC production at the low engine power settings, the concept of staged combustion is introduced. This concept breaks down a conventional combustor primary zone into a pair of separately controlled combustor stages, which may be stacked (as in parallel) or in series. The stages are referred to as a pilot and a main stage, with their separate fuel injection systems. The pilot stage serves as the burner for the low power setting at peak idle combustion efficiency. The main stage is for max power climb and cruise condition, however, operating in a leaner mixture ratio to control NO_x emissions. Complete combustion is achieved through separate fuel scheduling as well as an efficient (air-blast atomizer) fuel atomization, vaporization, and mixing in smaller volume combustors. Two concepts developed by GE-Aircraft Engines and Pratt & Whitney Aircraft are shown in Fig. 6.52.

The result of application of the double annular and Vorbix combustors to CF6–50 and JT9D-7 engines, respectively, is shown in Table 6.9 (data from Lefebvre, 1983).

The staged combustion concept is very effective in overall engine emission reduction in particular with CO and UHC, whereas NO_x continues to be a challenge. To combat NO_x, we should lower the flame temperature, which requires the combustor to operate at a lower equivalence ratio, of say $\phi \sim 0.6$. The problems of combustion stability and flameout accompany lean fuel–air mixtures. To achieve a stable lean mixture ratio to support a continuous combustion, a premixing, prevaporization approach has to be implemented. Both combustors that are shown in Fig. 6.52 employ this concept. Also the positive impact of employing air-blast atomizers is incorporated in both combustors. However, Table 6.9 indicates higher levels of smoke are generated by the double-annular and the Vorbix low-emission combustors. This reminds us of the conflict between single parameter and system optimization problems, in which different elements of the system often have opposing requirements for their optimization.

Water injection in the combustor effectively reduces the flame temperature and has demonstrated significant NO_x reduction. Figure 6.53 (from Blazowski and Henderson, 1974) shows the effect of water injection in the combustor on NO_x emission reduction. We note that the ratio of water to fuel flow is ~ 1 for a nearly 50% NO_x reduction level. This fact eliminates the potential of using water injection at the cruise altitude to lower the NO_x emissions from an aircraft gas turbine engine.

Ultralow NO_x combustors with premixed prevaporized gases are shown to operate at low equivalence ratios (below 0.6) when the combustor inlet temperature is increased. The application of this concept to automotive gas turbines has produced a very promising ultralow NO_x emission level. The burner inlet conditions of 800 K and 5.5 atm pressures in the automotive gas turbines resemble the altitude operation (or cruise operation) of an aircraft gas turbine engine. These ultralow NO_x combustor results are shown in Fig. 6.54 (from Anderson, 1974).

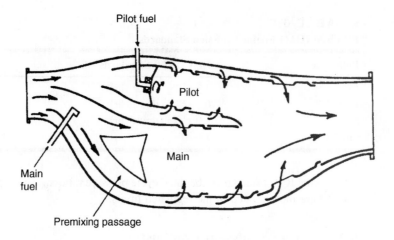

(a) GE radial two-stage (double annular) combustor for CF6-50 engine

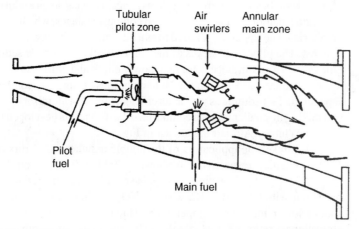

(b) Pratt & Whitney two stage series "Vorbix" combustor for JT9D-7 engine

■ **FIGURE 6.52**   Examples of staged combustion for low emissions

■ **TABLE 6.9**
Engine Emissions with Staged Combustors

	Pollutant, $g/(kg \cdot thrust \cdot hr \cdot cycle)$			
	CO	**UHC**	**NO$_x$**	**Smoke**
1979 EPA standards	4.3	0.8	3.0	20
production combustor CF6–50	10.8	4.3	7.7	13
Double-annular combustor	6.3	0.3	5.6	25
production combustor JT9D-7	10.4	4.8	6.5	4
Vorbix combustor	3.2	0.2	2.7	30

■ **FIGURE 6.53**
Water injection in the combustor reduces NO$_x$ emission(from Blazowski and Henderson, 1974)

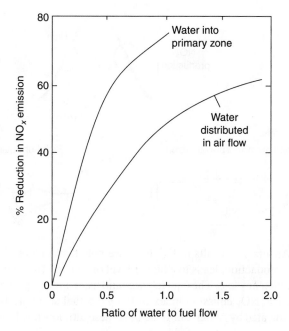

■ **FIGURE 6.54**
Effect of residence time and ϕ on NO$_x$ emission levels in ultralow NO$_x$premixed/ prevaporized burners (from Anderson, 1974)

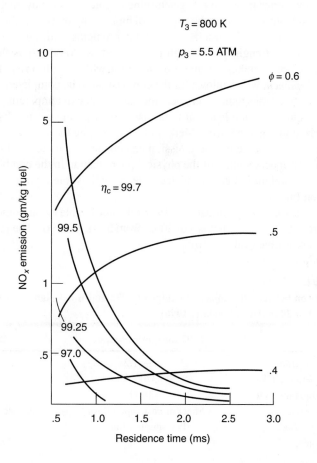

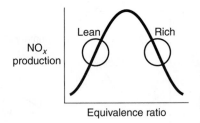

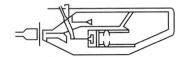

Lean premixed/prevaporized

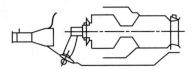

Rich burn/quick quench/lean burn

From Anderson's results in Fig. 6.54, we note that a reduced residence time, which is good for NO$_x$ production, leads to a higher level of CO and UHC emissions with an attendant combustion efficiency loss. The equivalence ratio has a large impact on NO$_x$ emissions, as seen in Fig. 6.54. The NO$_x$ emission index of 0.3 g/kg-fuel seems to be the absolute minimum possible, as indicated by Anderson. To achieve these ultralow levels of the oxides of nitrogen in the exhaust emission, an increased burner inlet temperature is required. Incorporating a solid catalytic converter in an aircraft gas turbine engine burner may enable lean combustion and ultralow NO$_x$ emission but at the expense of more complexity, weight, cost, and durability. The premixed/prevaporized burners also exhibit problems with autoignition and flashback for advanced aircraft engines with high-pressure ratios. Another possibility for low NO$_x$ combustor development takes advantage of *rich burn*, which results in a reduced flame temperature, and *quick quench*, which allows for the combustion to be completed while the temperature of the combustion gases remain very near the combustor exit temperature. This approach is called rich burn-quick quench-lean burn and has shown promising results. These approaches are summarized in Fig. 6.55, from Merkur (1996). Despite these advances, we recognize that the problem of gas turbine engine emission abatement is complex and challenging. We need to improve our understanding of the physical phenomenon in the combustor and in particular in the area of autoignition and flashback in lean combustible mixtures on the boundary of lean extinction limit.

It is instructive to compare the 1992 subsonic engine technology and the high-speed civil transport (HSCT) program goals of 2005 (from Merkur, 1996). The HSCT program is inactive at the present time (Table 6.10).

■ **TABLE 6.10**

Comparison between the subsonic engine of 1992 and the design goals for the high-speed civil transport of 2005 (from Merkur, 1996)

	1992 Subsonic engine	2005 HSCT engine
Equivalance ratio	$1.0 \rightarrow 1.2$	< 0.7 or > 1.5
Gas temperature (°F)	3700	$3400 \rightarrow 3750$
Liner temperature (°F)	< 1800	$2200 \rightarrow 2600$
Liner material	Sheet or cast superalloys	Ceramic matrix composites (CMC)
Cooling methods	Transpiration/film	Convection
Environment	oxidizing	Oxidizing or reducing
NO$_x$	36–45 g/kg f	< 5 g/kg f

6.7.7 Impact of NO on the Ozone Layer

Ozone, O_3, is toxic and a highly oxidizing agent, which is harmful to eyes, lungs, and other tissues. NO_x production in high-temperature combustion impacts the ozone in the lower as well as upper atmosphere. In this section, the chemical processes that lead to ozone creation in the lower atmosphere and to a depletion of ozone in the stratosphere are discussed.

Lower atmosphere
Combustion at high temperature (when local combustion temperature exceeds $\sim$1800 K) breaks the molecular bond of N_2 and O_2 and causes NO to be formed, i.e.,

$$N_2(g) + O_2(g) \rightarrow 2NO(g) \qquad \text{(6.138)}$$

Nitric oxide reacts with the oxygen in the atmosphere to form nitrous oxide, NO_2, according to

$$NO(g) + \frac{1}{2}O_2(g) \rightarrow NO_2 \qquad \text{(6.139)}$$

NO_2 absorbs light and decomposes to

$$NO_2(g) + h\nu(\text{radiation}) \rightarrow NO(g) + O(g) \qquad \text{(6.140)}$$

The quantity $h\nu$ in Eq. 6.140 is radiation energy with h as Plank's constant and ν as the frequency of the electromagnetic wave. Oxygen atom is a highly reactive substance and among other reactions, it reacts with oxygen to create ozone, i.e.,

$$O_2(g) + O(g) \rightarrow O_3(g) \qquad \text{(6.141)}$$

If we sum all the reactions 6.139–6.141, we get the following net result, namely,

$$\text{net } \frac{3}{2}O_2(g) \rightarrow O_3(g) \qquad \text{(6.142)}$$

Since NO facilitated the creation of ozone, without being consumed in the process, it is serving as a catalyst. In the lower atmosphere, nitric oxide, NO, helps generate harmful ozone in a catalyst role. The impact of NO on ozone changes as we enter the ozone layer in the stratosphere.

Upper atmosphere
The role of ozone in the upper atmosphere is to protect the earth against harmful radiation from the sun, namely, to absorb the highly energetic rays known as the ultraviolet radiation (with wavelength between 100 and 4000 Å).

$$O_3(g) + h\nu(\text{ultraviolet} - \text{radiation}) \rightarrow O_2(g) + O(g) \qquad \text{(6.143)}$$

The oxygen atom is highly reactive and combines with O_2 to from ozone,

$$O_2(g) + O(g) \rightarrow O_3(g) \qquad \text{(6.144)}$$

The photochemical cycle described by 6.143 and 6.144 reactions results in no net change of the ozone concentration in stratosphere. Hence, ozone serves a stable protective role in upper atmosphere. Flight of commercial supersonic aircraft with large quantities of NO emissions ($\sim$30 g/kg fuel) from their high-temperature engines, over extended cruise periods, and with a large fleet promises to deplete the ozone concentration levels. Again, NO appears to be a

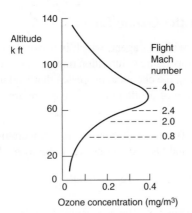

■ FIGURE 6.56 Profile of ozone concentration with altitude with a suitable cruise flight Mach number versus altitude superimposed (from Kerrebrock, 1992)

catalyst in destroying ozone in the upper atmosphere according to

$$NO(g) + O_3(g) \rightarrow NO_2(g) + O_2(g) \tag{6.145}$$

$$NO_2(g) + h\nu(\text{radiation}) \rightarrow NO(g) + O(g) \tag{6.146}$$

$$NO_2(g) + O(g) \rightarrow NO(g) + O_2(g) \tag{6.147}$$

By adding reactions 6.146 and 6.147 with *twice* the reaction in 6.145 (to get the right number of moles to cancel), we get the net photochemical reaction in the stratosphere as

$$\text{net } 2O_3(g) \rightarrow 3O_2(g) \tag{6.148}$$

Hence, ozone is depleted through a catalytic intervention of NO. The high-energy radiation absorption in stratosphere by ozone is by and large the source of warmth in this layer of earth's atmosphere. Since a rise in temperature in the stratosphere with altitude is stable with respect to vertical disturbances, the ozone concentration levels should remain stable (i.e., constant) over time. A chart of ozone concentration with altitude that also depicts the best cruise altitude for the cruise Mach number of commercial aircraft is shown in Fig. 6.56 (taken from Kerrebrock, 1992).

Now let us briefly summarize NO_x formation, combustor design parameters that impact its generation, and the ozone layer.

NO_x definition:	N_2O, NO, NO_2
Combustor reactions:	$N_2 + O \leftrightarrow NO + N$
Leading to NO_x:	$N + O_2 \leftrightarrow NO + O$
Formation:	$N + OH \leftrightarrow NO + H$
NO_x concentration:	$NO_x \propto \sqrt{p_{t3}} e^{-2400/T_{t3}} t_p$, where p_{t3} is combustion pressure, T_{t3} compressor discharge temperature, and t_p is the residence time in the primary zone.
Role of NO_x in ozone depletion:	$O_3 + h\nu \rightarrow O_2 + O$ where $h\nu$ is the UV radiation from the sun
	$O + O_2 \rightarrow O_3$
	$O + O \rightarrow O_2$
	$O + O_3 \rightarrow 2O_2$ the last two reactions tend to limit the ozone concentration in the stratosphere
	$NO + O_3 \rightarrow NO_2 + O_2$
	$NO_2 + h\nu \rightarrow NO + O$
	$NO_2 + O \rightarrow NO + O_2$ NO is maintained, O_3 destroyed

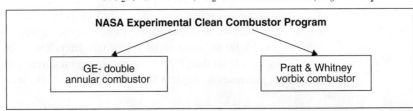

6.8   Aviation Fuels

Aircraft flight envelope, i.e., altitude Mach number, establishes the operational temperature range of the aircraft and hence its desirable fuel properties. Some of the aviation fuel properties and combustion characteristics of interest are

- Specific gravity
- Viscosity
- Vapor pressure
- Volatility
- Flashpoint
- Heating value
- Freezing point
- Thermal stability

- Initial and end boiling points
- Heat capacity
- Spontaneous ignition temperature
- Flammability limits
- Handling qualities, toxicity
- Storability
- Low fire risk
- Price/availability

Let us examine the maximum skin temperature of an aircraft in cruise and compare it to the thermal characteristics of some typical aircraft fuels. The skin temperature of an aircraft is calculated through an energy balance that accounts for aerodynamic heating, radiation cooling, regenerative wall cooling, and thermal conduction. The maximum skin temperature is, however, very near stagnation temperature of flight. Figure 6.57 shows a rapid skin temperature rise with flight Mach number, assuming the cruise is in the constant-temperature layer (i.e., tropopause, between 11 and 20 km or 36–66 kft).

At these elevated temperatures for high-speed flight, the fuel should exhibit thermal stability, which is partially addressed through its boiling point characteristics. The initial

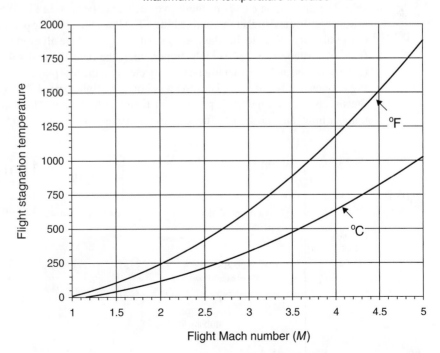

Maximum skin temperature in cruise

■ **FIGURE 6.57**   **Skin temperature rise with flight Mach number** ($\gamma = 1.4$, $T_{amb} \cong$ $-70°F$ or $-56°C$)

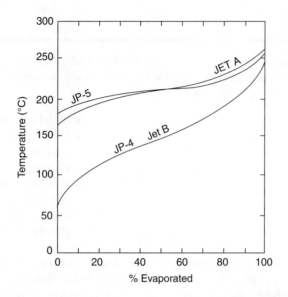

■ **FIGURE 6.58** Distillation curve for common jet fuels—A measure of fuel volatility (from Blazowski, 1985)

boiling point temperature begins when fuel vaporization starts. The end point is the temperature where all the fuel is vaporized. The range of the initial and end boiling points, known as the distillation curve, for several fuels is shown in Fig. 6.58 (from Blazowski, 1985). The military fuels have a "JP" designation, and Jet A is the most common commercial aircraft fuel. The original gas turbine fuel was kerosene, which is used as a basis of comparison and blending with other hydrocarbon jet engine fuels. The JP-4 was used primarily by the USAF and was a highly volatile fuel. A less volatile fuel, the JP-5, was a gasoline–kerosene blended fuel that was used by the USN. The lower volatility of the JP-5 made it more suitable for long-term storage in the ship tanks as well as the blended nature of the fuel allowed a wider availability on board a US ship. The distillation curves of these two fuels may be seen in Fig. 6.58, with JP-5 showing lower volatility and higher thermal stability. The low initial boiling point of the JP-4 contributes to its volatility. We note that the fuels shown in Fig. 6.58 are all evaporated at 100% when the fuels are heated to 250°C. At high temperatures, the fuels thermally decompose to form gum (coke deposits) and clog fuel filters, fuel injector, and the fuel pump. This limit corresponds to a flight Mach number of ~2.6. The obvious choice for higher speed flights is the use of cryogenic fuels such as liquid methane, propane, or liquefied hydrogen. Thermal stability, flight Mach number limit, and the relative fuel cost (tentative) are shown in Fig. 6.59 (from Strack, 1987).

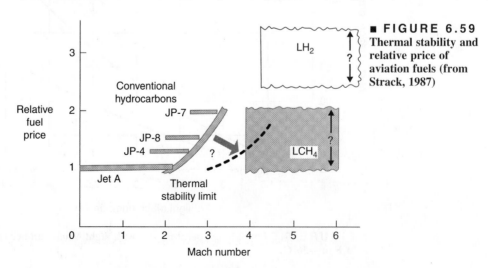

■ **FIGURE 6.59** Thermal stability and relative price of aviation fuels (from Strack, 1987)

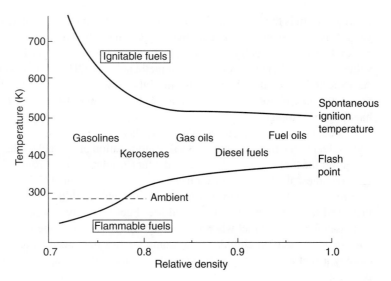

■ **FIGURE 6.60**   Ignition temperatures for petroleum fuels (from Lefebvre, 1983)

The specific gravity is a measure of relative liquid fuel density (relative to water at 4°C), which is lowest for gasoline among petroleum fuels. The flammability limits, flash point temperature, and spontaneous ignition temperature of various fuels are shown in a graph from Lefebvre (1983) in Fig. 6.60.

All hydrocarbon fuels have about the same heating value (nearly 18,600 BTU/lbm or 43.3 MJ/kg) regardless of their blends. The parameter that contributes to the heat of combustion of a fuel is the hydrogen content of the fuel. For example, the percent hydrogen content (this is the fraction of mass of the fuel molecule contributed by the hydrogen atoms) of kerosene ($\sim C_{12}H_{26}$) is ~15.3%, JP-4 ($\sim CH_{2.02}$) is at 14.5%, propane (C_3H_8) has 18.8%, methane (CH_4) has 25%, and of course pure hydrogen is at 100% hydrogen content. It is interesting to examine the heating values of these fuels and compare them to their hydrogen content. Table 6.11 shows the impact of hydrogen content on the heat of combustion (lower heating value). The higher hydrogen content in the fuel yields a higher heat of reaction, as expected.

Ragozin (1962) presents a correlation between the fuel lower heating value and the percent mass content of carbon, hydrogen, oxygen, sulfur, and water in the fuel. This useful and *approximate* correlation is

$$LHV = 0.339C + 1.03H - 0.109(O - S) - 0.025W \tag{6.149}$$

The dimensions of eq. 6.149 are MJ/kg. All parameters on the right-hand side are percent mass of carbon, hydrogen, oxygen, sulfur, and water in the fuel, respectively. The coefficient of

■ **TABLE 6.11**

Hydrogen Content and Heating Value of Common Fuels

Fuel	JP-4	Propane	Methane	Hydrogen
Hydrogen content	**14.5%**	**18.8%**	**25%**	**100%**
Lower heating value				
kcal/g	10.39	11.07	11.95	28.65
MJ/kg	43.47	46.32	49.98	119.88
BTU/lbm	18,703	19,933	21,506	51,581

hydrogen term is the largest and hence contributes the most to an increase in the fuel heating value. Also, it is interesting to note that oxygen in the fuel reduces its heating value, which is representative of all alcohol fuels (a typical alcohol fuel, ethanol, C_2H_5OH, has a lower heating value of $\sim$62% of a hydrocarbon fuel, or methanol, CH_3OH, has only $\sim$46% of heating value). As expected water dissolved in the fuel also lowers the heating value of the fuel. This parameter, i.e., dissolved water content in the fuel, may be controlled through proper fuel handling and storage.

Fuel viscosity is an important property that determines the pressure drop in the fuel lines and the fuel pump requirements, as well as atomization of the fuel by the fuel injection system. The lower the fuel viscosity the smaller the fuel droplets, hence faster vaporization rate and the ignition timescale. Poor atomization causes a higher fraction of UHC as well as higher CO and soot formation. On the contrary, a low viscosity fuel exhibits poor lubricity, which causes an increased fuel pump wear and reduced life. Fuel viscosity is a function of temperature and increases for a liquid fuel when temperature decreases and vice versa. At low temperatures corresponding to high altitudes, a subsonic aircraft may need an antifreeze fuel additive or employ a fuel heating system integrated with its fuel tank at the propulsion system design phase.

A liquid fuel in a tank always contains a certain quantity of fuel vapor, which exerts a pressure on the liquid. This is known as the vapor pressure of the liquid at the given temperature. For example, JP-4, a highly volatile fuel, has a very high vapor pressure of $\sim$0.18 atm at 38°C (or 100°F), compared with JP-5 fuel with a corresponding low vapor pressure of 0.003 atm. Although a high vapor pressure is good in a combustor and leads to better vaporization rates, but it is undesirable for its lower flash point temperature. The excessive rate of vapor production in a fuel tank or a fuel line, especially at higher temperatures of supersonic flight, could lead to an unacceptable flash point fire risk. Figure 6.60 (from Lefebvre, 1983) shows the heavier fuels have a higher flash point (therefore safer) than light fuels, such as gasoline or kerosene.

Aviation fuel is a blend of different hydrocarbons with different molecular structures and different reaction tendencies. We address two such compounds, olefins and aromatics, which are found in aviation fuels. Olefins are unsaturated hydrocarbons ($\sim C_nH_{2n}$) that are present in aviation fuel produced in the refinery process. Aromatics ($\sim C_nH_{2n-6}$) are ring compounds such as benzene (C_6H_6) that are also present in the aircraft fuel. In case of olefins, the gum formation tendency of these compounds makes them undesirable. The aromatics have lower hydrogen content than gasoline or kerosene, which reduces the heat of reaction of the fuel mixture. Aromatic content also increases the soot formation tendencies in the combustor, which again makes them undesirable. The volume fraction of fuel containing aromatic and olefinic compounds is included in the table of fuel properties. Table 6.12 from Blazowski (1985) presents important jet fuel properties for three widely used aircraft fuels, including aromatic and olefinic contents.

6.9   Combustion Instability: Screech

Acoustic waves propagate at a local speed a in a gas at rest. The acoustic waves "ride the flow," however, when the gas is in motion. The speed of propagation of a plane acoustic wave along the axis of a duct with a mean flow speed of u is $(u + a)$ in the direction of the flow and $(u - a)$ in the opposite direction. In addition, an open end of a duct or a sudden area increase/decrease in a duct causes a *reflected* wave to propagate back in the duct. These waves contribute to what is known as the *longitudinal acoustic modes* (or organ pipe modes) of a duct. We may imagine the acoustic waves propagating in the radial or in the tangential directions due to disturbances along the axis or azimuth of the duct. For example, fuel injectors in a combustor or an afterburner are placed azimuthally and flameholders in an afterburner are placed radially in the duct. Disturbances generated by these elements predominantly attain tangential and radial

■ **TABLE 6.12**
Important Jet Fuel Properties (from Blazowski, 1985)

Property	JP-4		Jet A (JP-8)		JP-5	
	Spec. req.	Typical value	Spec. req.	Typical value	Spec. req.	Typical value
Vapor pressure at 38°C (100°F), atm	0.13–0.2	0.18	—	0.007	—	0.003
Initial boiling point (°C)	—	60	—	169	—	182
End point (°C)	—	246	288	265	288	260
Flash point (°C)	—	−25	>49	52	>63	65
Aromatic content (% Vol)	<25	12	<20	16	<25	16
Olefinic content (% Vol)	<5	1	—	1	—	1
Saturates content (% Vol)	—	87	—	83	—	83
Net heat of combustion (cal/g)	>10,222	10,388	>10,222	10,333	>10,166	10,277
Specific gravity	0.751–0.802	0.758	0.755–0.830	0.810	0.788–0.845	0.818
Approximate U.S. yearly consumption (10^9 gal)		3.4		13.1		0.7

characteristics, respectively. The *radial and tangential acoustic modes* have their origin in these kinds of disturbances. Now, let us superimpose a swirl component to the mean flow and observe *spinning (or tangential) modes*. Add axial velocity and we get *helical waves*. These waves interact with each other and the sources of energy in the flow such as the mean flow or pockets of reacting gas mixtures through resonance phenomenon that leads to the growth of the disturbance wave(s) amplitude. This is similar to the dynamics of a mass-spring oscillator. The maximum energy transfer from the forcing mechanism to the driven oscillator (i.e., the mass) is found (i.e., tuned) at the resonance frequency. Nonresonant interactions are not amplified. Large-amplitude pressure waves (>10–15% of the mean) are observed in combustors at frequencies ranging from 50 to 100 Hz for longitudinal modes and as high as 5000 Hz for radial and tangential modes. The resonant phenomena in real systems do not grow without bound, however, rather they reach a limit cycle. The real systems have damping and nonlinear dynamics, which lead the oscillations to a limit cycle. Due to large amplitudes of these high-frequency disturbances, the problem of fatigue and structural failure needs to be addressed in combustors. The reflected acoustic wave from the nozzle downstream of an afterburner couples with the combustor oscillations to create a closed-loop instability wave, known as screech.

6.9.1   Screech Damper

A Helmholtz resonator is a cavity that serves as the spring to an oscillating mass, which is contained in the neck of the cavity. The damping occurs at the natural frequency of the cavity, which is a function of the geometry of the cavity and the speed of sound. An acoustic liner composed of numerous Helmholtz resonators integrated within the liner can be designed to efficiently dampen the acoustic power near the resonant frequency of the cavity. A perforated sheet covering a honeycomb substrate is often used as an acoustic liner.

6.10   Summary

Combustion is a complex and thus challenging process. It involves chemical reactions that take place at the molecular level. The process inevitably involves fuel atomization and vaporization with subsequent mixing with an oxidizer (i.e., air in airbreathing engines) leading to a chemical reaction. We learned that the laws that govern the mixture of gases are important in a chemical reaction. To simplify our analysis, we learned and applied the concept of "chemical equilibrium" to reactions in a combustor. The wealth of experimental data on equilibrium constants allowed us to use the law of mass action to calculate the equilibrium mixture of the products of combustion. Conservation of energy principle then yielded the adiabatic flame temperature. The rate of chemical reactions identified characteristic timescales, e.g., evaporation and reaction that in conjunction with the characteristic velocity scales identified a residence time, which then affected the combustor and afterburner designs.

In the combustor design section, we learned about prediffusers, flow mixing through turbulent recirculation zones that may be generated by swirl, bluff bodies, or backward-facing steps. Combustor liner cooling techniques and the use of advanced composite materials promised a lowering in the percent coolant requirement. The nonuniformity in the combustor-exit flow was characterized by two parameters, namely, a pattern factor and a profile factor. The concern with high-altitude flameout, ignition, and relight issues were presented. Combustor pollutants in the form of UHC, CO, NOx, and greenhouse gases were discussed, and the design choices that lead to an environmentally friendly combustor were identified.

We have scratched the surface of combustion and combustor design, only. The reader should follow up this material in specialized combustion textbooks, as cited throughout this chapter. In addition, references 15, 20, 21 and 35 provide background and supplementary material and are recommended for further reading.

References

1. Adkins, R. C., "A Short Diffuser with Low Pressure Loss," ASME Paper, May 1974.

2. Adkins, R. C., Matharu, D. S., and Yost, J. O., "The Hybrid Diffuser," ASME Paper 80-GT-136, 1980.

3. Anderson, D. N., "Effect of Equivalence Ratio and Dwell Time on Exhaust Emissions from an Experimental Premixing Pre-vaporizing Burner," NASA TM-X-71592, 1974.

4. Ballal, D. R. and Lefebvre, A. H., "Flame Propagation into Heterogeneous Mixtures of Fuel Droplets, Fuel Vapor and Air," *Eighteenth International Symposium on Combustion*, The Combustion Institute, Pittsburgh, 1980, pp. 321–328.

5. Ballal, D. R. and Lefebvre, A. H., "Structure and Propagation of Turbulent Flames," Proceedings of the Royal Society of London, Series A, Vol. 344, 1975 pp. 217–234.

6. Barclay, L. P., "Pressure Losses in Dump Combustors," AFAPL-TR-72–57 Air Force Aero-Propulsion Laboratory, Wright-Patterson AFB, OH, 1972.

7. Blazowski, W. S., "Fundamentals of Combustion," Chapter in *Aerothermodynamics of Aircraft Engine Components*, Ed. Oates, G. C., AIAA Education Series, AIAA Inc.Washington, DC,1985.

8. Borman, G. L. and Ragland, K. W., *Combustion Engineering*, McGraw-Hill, New York, 1998.

9. Dugger, G. L. and Heimel, S, Flame Speeds of Methane-Air, Propane-Air and Ethylene-Air Mixtures at Low Initial Temperatures, NACA TN 2624, 1952.

10. Environmental Protection Agency, Control of Air Pollution from Aircraft and Aircraft Engines, Federal Register, Vol. 38, No. 136, 1973.

11. Environmental Protection Agency, Control of Air Pollution from Aircraft and Aircraft Engines, Federal Register, Vol. 43, No. 58, 1978.

12. Environmental Protection Agency, Control of Air Pollution from Aircraft and Aircraft Engines, Federal Register, Vol. 47, No. 251, 1982.

13. Glassman, I., *Combustion*, 2nd edition, Academic Press, New York, 1987.

14. Gouldin, F. C., "Controlling Emissions from Gas Turbines—The Importance of Chemical Kinetics and Turbulent Mixing," *Combustion Science and Technology*, Vol. 7, 1973.

15. Greenhow, V. W. and Lefebvre, A. H., "Some Application of Combustion Theory to Gas Turbine Development," *Sixth International Symposium on Combustion*, Reinhold, New York, 1957, pp. 858–869.

16. Haddock, G. H., "Flame Blowoff Studies of Cylindrical Flameholders in Channeled Flow," Jet Propulsion Laboratory, Pasadena, California, Progress Report 3–24, May 1951.

17. Henderson, R. E. and Blazowski, W. S., "Turboporpulsion Combustion Technology," Chapter in *Aircraft Propulsion Systems Technology and Design*, Ed. Oates, G. C., AIAA Education Series, AIAA Inc., Washington, DC, 1989.

18. Herbert, J. D., "Theoretical Analysis of Reaction Rate Controlled Systems-Part I," AGARD Combustion Research and Reviews, Chapter 6, 1957.

19. Hill, P. G. and Peterson, C. R., *Mechanics and Thermodynamics of Propulsion*, 2nd edition, Addison-Wesely, Reading, Massachusetts, 1992.

20. Howell, J. R. and Buckius, R. O., *Fundamentals of Engineering Thermodynamics*, McGraw-Hill, New York, 1987.

21. Jones, R. E., Diehl, L. A., Petrash, D. A., and Grobman, J., "Results and Status of the NASA Aircraft Engine Emission Reduction Technology Programs," NASA TM-79009, 1978.

22. Jost, W, *Explosion and Combustion Processes in Gases*, McGraw-Hill, New York, 1946.

23. Kerrebrock, J. L., *Aircraft Engines and Gas Turbines*, 2nd edition, MIT Press, Cambridge, Mass., 1992.

24. Kuo, K. K., *Principles of Combustion*, John Wiley & Sons, Inc., New York, 1986.

25. Kurzke, J. and Riegler, C., "A Mixed-Flow Turbofan Afterburner Simulation for the Definition of Reheat Fuel Control laws," *RTO-Applied Vehicle Technology panel, Symposium on Design Principles and Methods for Aircraft gas Turbine Engines*, Toulouse, France, 1998.

26. Lefebvre, A. H., *Gas Turbine Combustion*, Hemisphere Publishing Corp., New York, 1983.

27. Lefebvre, A. H., "Theoretical Aspects of Gas Turbine Combustion Performance," College of Aeronautics Note Aero 163, Cranfield Institute of Technology, Bedford, England, 1966.

28. Lefebvre, A. H. and Halls, G. A., "Some Experiences in Combustion Scaling," AGARD Advanced Aero-Engine Testing, AGARDograph 37, Pergamon, New York, 1959 pp 177–204.

29. Lewis, B. and von Elbe, G., *Combustion Flames and Explosions of Gases*, Academic Press, New York, 1961.

30. Lipfert, F. W., "Correlations of Gas Turbine Emissions Data," ASME Paper 72-GT-60, March 1972.

31. Longwell, J. E., Chenevey, W. W., and Frost, E. E., "Flame Stabilization by Baffles in a High Velocity Gas Stream," *Third Symposium on Combustion, Flame and Explosion Phenomena*, Williams and Wilkins, Baltimore, 1949, pp. 40–44.

32. Merkur, R. A., "Propulsion System Considerations for Future Supersonic Transports—A Global Perspective," ASME Paper 96-GT-245, 1996.

33. Nealy, D. A. and Reider, S. B., "Evaluation of Laminated Porous Wall Materials for Combustor Liner Cooling," Transaction of ASME Journal of Engineering for Power, Vol. 102, No. 2, April 1980, pp. 268–276.

34. Olson, W. K., Childs, J. H., and Jonash, E. R., "The Combustion Efficiency Problem of the Turbojet at High Altitude," Transaction of ASME, Vol. 77, 1955.

35. Pinkel, B. and Karp, I. M., "A Thermodynamic Study of the Turbojet Engine," NACA Report No. 891, 1947.

36. Ragozin, N. A., *Jet Propulsion Fuels*, Pergamon, Oxford, 1962.

37. Rao, K. V. L and Lefebvre, A. H., "Spontaneous Ignition Delay Times of Hydrocarbon Fuel/Air Mixtures," *First International Combustion Specialists Symposium*, Bordeaux, France, 1981, pp. 325–330.

38. Rolls-Royce, The Jet Engine, Rolls-Royce, plc, Derby, England, 2005.

39. Sarpkaya, T., "On Stationary and Traveling Vortex Breakdown," *Journal of Fluid Mechanics*, Vol. 45, Part 3, 1971, pp. 545–559.

40. Spadaccini, L. J., "Autoignition Characteristics of Hydrocarbon Fuels at Elevated Temperatures and Pressures," *Transactions of ASME, Journal of Engineering for Power*, Vol. 99, Ser. A, No. 1, 1977, pp. 83–87 (also ASME Paper 76-GT-3, 1976).

41. Spalding, D. B., *Combustion and Mass Transfer*, Pergamon Press Inc., New York, 1979.

42. Strack, W. C., "Propulsion Challenges and Opportunities for High-Speed Transport Aircraft," Chapter in *Aeropropulsion '87*, NASA CP-3049, November 1987.

43. Strehlow, R. A., *Combustion Fundamentals*, McGraw-Hill, New York, 1984.

44. Tacina, R. R., "Combustion Technology," UTSI Short Course Notes, 2000.

45. Wright, F. H., "Bluff Body Flame Stabilization: Blockage Effects," *Combustion and Flame*, Vol. 3, 1959, p. 319.

46. Zumdahl, S. S. and Zumdahl, S. A., *Chemistry*, 5th edition, Houghton Mifflin Co., Boston, 2000.

47. Zukoski, E. E., "Afterburners," Chapter in *Aerothermodynamics of Aircraft Engine Components*, Ed. Oates, G. C., AIAA Education Series, AIAA Inc.Washington, DC, 1985.

48. Zukoski, E. E. and Marble, F. E., "The Role of Wake Transition in the Process of Flame Stabilization on Bluff Bodies," *Combustion Researches and Reviews*, 1955, Butterworths Scientific Publications, London, 1955.

49. Zukoski, E. E. and Marble, F. E., "Experiments Concerning the Mechanism of Flame Blowoff from Bluff Bodies," *Proceedings of the Gas Dynamics Symposium on Aerothermochemistry*, Northwestern University Press, Evanston, Illinois, 1955.

Problems

6.1   A mixture of gases contains 44 kg of CO_2, 112 kg of N_2, and 32 kg of O_2 at a mixture temperature of $T_m = 287$ K and the mixture pressure is $p_m = 1$ bar.

Calculate

   (a) number of moles of carbon dioxide, nitrogen, and oxygen

   (b) mole fraction of carbon dioxide

   (c) partial pressure of constituent gases, i.e., CO_2, N_2, and O_2

   (d) mixture molecular weight MW_m

   (e) volume fraction of the constituent gases

6.2   Write the chemical reaction for the combustion of hydrogen, H_2, and air, then calculate the stoichiometric fuel-to-air ratio f_{stoich}.

6.3   A gas is a mixture of 22% O_2, 33% N_2, and 45% CO_2 by volume. Calculate

(a) the mole fraction of the constituents in the mixture

(b) the mixture molecular weight MW_m

6.4   Write the chemical reaction for the complete combustion of JP-4 and air. The JP-4 has the formula $CH_{1.93}$. Also, calculate the stoichiometric fuel-to-air ratio for this blended jet fuel.

6.5   Calculate the lower and higher heating values of octane, C_8H_{18}, in the stoichiometric chemical reaction *with oxygen* at a reference temperature of 298.16 K and the pressure of 1 bar.

$$C_8H_{18(g)} + 12.5O_{2(g)} \rightarrow 8CO_{2(g)} + 9H_2O$$

6.6   Establish the higher and lower heating values of octane, C_8H_{18}, by considering the stoichiometric reaction *in air* at a reference temperature of 298.16 K and at 1 bar pressure.

$$C_8H_{18(g)} + \frac{25}{2}(O_{2(g)} + 3.76N_2)$$

$$\rightarrow 8CO_{2(g)} + 9H_2O + \frac{25}{2}(3.76)N_2$$

Compare your answer to Problem 6.5 and explain.

6.7   Adiabatic flame temperature associated with the combustion of a fuel in air depends on the temperature and pressure of air. Figure 6.6 shows the adiabatic flame temperature of Jet-A fuel in air with initial temperature of 800 K and initial pressure of 25 atm.

(a) What is the stoichiometric flame temperature of JP-A if the initial air temperature was 1000 K at 25 atm pressure?

(b) What is the stoichiometric flame temperature of Jet-A if the initial pressure of air was 1 atm and the initial temperature remained at 800 K?

6.8   Consider burning methane (CH_4) with 110% theoretical air. Calculate the equivalence ratio for this reaction. Assume that the nitrogen and the excess oxygen do not dissociate and/or chemically react to form new compounds.

6.9   One mole of octane is burned with 120% theoretical air. Assuming that the octane and air enter the combustion chamber at 25°C and the excess oxygen and nitrogen in the reaction will not dissociate, calculate

(a) the fuel–air ratio

(b) the equivalence ratio ϕ

(c) the adiabatic flame temperature T_{af}

Assume

$$\bar{c}_{p_{CO_2}} = 61.9 \text{ kJ/kmol} \cdot \text{K}, \quad \bar{c}_{p_{O_2}} = 37.8 \text{ kJ/kmol} \cdot \text{K}$$
$$\bar{c}_{p_{N_2}} = 33.6 \text{ kJ/kmol} \cdot \text{K}, \quad \bar{c}_{p_{H_2O}} = 52.3 \text{ J/kmol} \cdot \text{K}$$

6.10   One mole of oxygen, $O_{2(g)}$, is heated to 4000 K at the pressure of p_m. A fraction of oxygen dissociates to oxygen atom according to

$$xO_2 \rightarrow 2xO$$

Assuming a state of equilibrium is reached in the mixture, calculate

(a) mole fraction of O_2 at equilibrium when p_m is 1 atm.

(b) mole fraction of O_2 at equilibrium when p_m is 10 atm.

Assume the equilibrium constant for the reaction

$$O_2 \leftrightarrow 2O$$

is $K_p = 2.19$ atm at the temperature of 4000 K. Explain the effect of pressure on dissociation.

6.11   One mole of gaseous hydrogen is heated to 3000 K at the pressure of p_m. Assuming a fraction of the hydrogen dissociates upon heating into atomic hydrogen, i.e.,

$$x H_{2(g)} \leftrightarrow 2x H$$

calculate the mole fraction of hydrogen, H_2, in the equilibrium mixture for $p_m = 1$ atm and $p_m = 10$ atm. Does higher mixture pressure inhibit or promote the dissociation of hydrogen?

You may find the equilibrium constant K_p for the following stoichiometric reaction

$$\frac{1}{2}H_2 \leftrightarrow H$$

at 3000 K from Table 6.3.

6.12   Mass flow rate (combined fuel and air) into a reactor of volume 2.35 ft^3 is 100 lbm/s. The equivalence ratio is 0.6. Use the combustor loading parameter (CLP)

$$\frac{\dot{m}}{p^n V} = f(\phi)$$

and the combustor stability loop (Fig. 6.22) to establish the range of reactor pressure that leads to a stable combustion. Note that the scale is logarithmic.

6.13   Combustion involves many timescales. For example, a flammable mixture of fuel and air (under a spontaneous ignition temperature condition) experiences an ignition delay time t_i. The elements contributing to the ignition delay time are the fuel droplet evaporation time and a reaction timescales. The ignition delay results of the combustion of two hydrocarbon fuels at 10 atm of pressure are shown in Fig. 6.24. Graph an estimated ignition delay time for the same fuels but at the pressure of 30 atm by assuming pressure dependence for the reaction timescale is

(a) $1/p$

(b) $1/p^{1.5}$

For a first-order approximation, you may neglect the evaporation time dependence on the mixture pressure.

6.14 An afterburner uses flameholders with a drag coefficient of $C_D = 0.5$, based on the afterburner cross-sectional area. The inlet Mach number to the afterburner is $M_i = 0.2$, with $\gamma_i = 1.33$. The heat release due to combustion in the afterburner is $q/c_{pi}T_{ti} = 2.0$. The exhaust gas in the wet mode has a lower ratio of specific heats, namely, $\gamma_e = 1.30$. Calculate

Dry mode

    (a) exit Mach number M_e

    (b) total pressure loss as a percentage of inlet total pressure

Wet mode

    (c) exit Mach number M_e

    (d) total pressure loss as a percentage of inlet total pressure

6.15 A combustion chamber uses a prediffuser with a sudden area expansion (known as a dump diffuser) to decelerate the flow of air ($\gamma = 1.4$) before entering the combustor. Assuming the inlet Mach number to the dump diffuser is $M_1 = 0.5$, the area ratio of the dump diffuser is $A_2/A_1 = 2.0$, calculate

    (a) exit Mach number M_2

    (b) the ratio of total pressures, i.e., p_{t2}/p_{t1}

6.16 Calculate the higher and lower heating values of methyl alcohol, CH_3OH, and methane, CH_4, by considering the following reactions at the reference temperature of 298.16 K and at 1 bar pressure.

$$CH_3OH_{(g)} + \frac{3}{2}O_{2(g)} \rightarrow CO_{2(g)} + 2H_2O$$

$$CH_{4(g)} + 2O_{2(g)} \rightarrow CO_{2(g)} + 2H_2O$$

Note that the products of combustion for both reactions are the same. Explain the difference between the heating values of these two fuels that are *seemingly* so similar.

6.17 Calculate the lower and higher heating values of propane, C_3H_8, in the following chemical reaction at a reference temperature of 298.16 K and the pressure of 1 bar.

$$C_3H_{8(g)} + 5O_{2(g)} \rightarrow 3CO_{2(g)} + 4H_2O$$

6.18 Gaseous hydrogen and air enter an adiabatic combustion chamber at a temperature of 298.16 K (the reference temperature) and pressure of 1 bar.

$$H_{2(g)} + (O_2 + 3.76N_2) \rightarrow H_2O_{(g)} + \frac{1}{2}O_2 + 3.76N_2$$

Assuming the average molar specific heats of the products are

$$\bar{c}_{pO_2} = 37.8 \text{ kJ/kmol} \cdot \text{K}, \; \bar{c}_{pN_2} = 33.6 \text{ kJ/kmol} \cdot \text{K}, \; \bar{c}_{pH_2O}$$
$$= 52.3 \text{ kJ/kmol} \cdot \text{K}$$

calculate

    (a) fuel-to-air ratio

    (b) the adiabatic flame temperature T_{af}

6.19 Stoichiometric flame temperature for the combustion of methane in oxygen at 1 atm pressure and at reference temperature of 298 K is listed as 3030 K in Table 6.4. The combustion of methane in air would produce the stoichiometric flame temperature of 2210 K. Explain the difference.

6.20 An *annular combustor* utilizes a dump prediffuser, as shown. Calculate

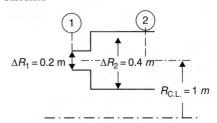

■ **FIGURE P6.20**

    (a) the diffuser area ratio A_2/A_1

    (b) the total pressure loss $\Delta p_t/p_{t1}$

    (c) exit Mach number M_2

Assume the inlet Mach number is $M_1 = 0.4$ and $\gamma = 1.4$.

6.21 Consider a coaxial dump diffuser with area ratio A_2/A_1. Assuming that the static pressure acting on the sudden expansion wall p_w is the same as the inlet static pressure p_1 and wall friction may be neglected, apply the conservation principles to show that

$$\frac{\Delta p_t}{\rho V_1^2/2} = \left(1 - \frac{A_1}{A_2}\right)^2$$

in the limit of incompressible fluid.

Assume the flow is uniform at the exit.

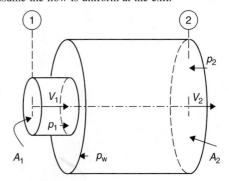

■ **FIGURE P6.21**

6.22 Take two parallel streams with initial speeds of 20 and 30 m/s, as shown.

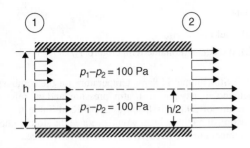

■ FIGURE P6.22

Assuming the two streams do not mix and thus remain parallel, calculate

(a) the initial distortion level

(b) the final speed of both streams

(c) the final distortion level

Here, we define the distortion level as $\Delta u/\bar{u}$ where $\bar{u}$ is the area-averaged u.

Assume the fluid density is constant at $1 \, kg/m^3$.

Why did the distortion level decrease in this case?

(d) Repeat this problem with a static pressure rise $p_1 - p_2 = -100 \, Pa$, instead of a static pressure drop. Explain the distortion level behavior.

6.23 The effect of flammability limit on gasoline–air mixture as a function of combustion pressure is presented in Fig. 6.11. First, what is the lowest pressure that combustion is possible? Second, at combustion pressure of 1.8 psia, what are the equivalence ratios corresponding to the flammability limits?

6.24 We model an afterburner as a constant-area duct with a series of bluff bodies in the stream with a known drag coefficient. Assuming the following inlet conditions for dry mode

Inlet Mach number $M_i = 0.3$
$\gamma_i = \gamma_e = 1.33$ and $c_{pi} = c_{pe}$

flameholder drag coefficient $C_D = 0.5$

Calculate

(a) static pressure ratio p_e/p_i

(b) exit Mach number M_e

(c) static temperature ratio T_e/T_i

(d) total pressure ratio p_{te}/p_{ti}

6.25 The afterburner in Problem 6.24 operates in "wet mode." Assuming the nondimensional heat release $q/c_{pi}T_{ti} = 1.0$ and $\gamma_e = 1.25$ in wet mode, calculate

(a) total pressure ratio in the wet mode

(b) exit Mach number in the wet mode

6.26 Atmospheric ozone is depleted through catalytic intervention of nitric oxide, NO. NO emissions in the upper atmosphere are estimated to be ~30 g/kg fuel from high-temperature engines, overextended cruise periods. A supersonic transport carries 50,000 kg of fuel at takeoff. Assuming 90% of the fuel is consumed during cruise calculate the amount of nitric oxide emissions for a round trip flight. Now, multiply that by 360 round trips per year to estimate the (NO) pollution (of one aircraft) per year. Finally, what is the environmental impact of a fleet of 100 aircraft?

6.27 Emission index (EI) of carbon monoxide and unburned hydrocarbons at idle power is shown in Fig. 6.49. Assuming that the combustor inlet temperature at idle power is 425 K, read an average value of the emission index for carbon monoxide and unburned hydrocarbons from the graph. Then, calculate the combustion inefficiency from

$$1 - \eta_b = [0.232(EI)_{CO} + (EI)_{UHC}] \times 10^{-3}$$

and compare the combustion efficiency η_b at idle setting to EPA standards (of 99%).

CHAPTER 7

Axial Compressor Aerodynamics

Courtesy of Rolls-Royce, plc

7.1   Introduction

In this chapter, we address compressors and turbines in an aircraft gas turbine engine. We first present the fundamental equations that are applicable to all types of turbomachinery and then follow with the flow characteristics of each machine in subsequent chapters 8 and 9.

Turbomachinery is at the heart of gas turbine engines. The role of mechanical compression of air in an engine is given to the compressor. The shaft power to drive the compressor typically is produced by expanding gases in the turbine. The machines that exchange energy with a fluid, called the working fluid, through shaft rotation are known as turbomachinery. The machines where the fluid path is predominantly along the axis of the shaft rotation are called axial-flow turbomachinery. In contrast to these machines, in radial-flow turbomachinery, the fluid path undergoes a 90° turn from the axial direction. These machines are sometimes referred to as centrifugal machines. A mixed-flow turbomachinery is a hybrid between the axial and the radial-flow machines. In aircraft gas turbine engines, the axial-flow compressors and turbines enjoy the widest application and development (Fig. 7.1). The centrifugal compressors and radial-flow turbines are used in small gas turbine engines and automotive turbocharger applications.

7.2   The Geometry

The geometry of rotating blades demands a cylindrical duct with a shaft centric configuration. This in turn leads to the choice of cylindrical coordinates for the analysis of flows in turbomachinery. The coordinates $[r, \theta, z]$ are in radial, tangential or azimuthal, and axial directions, respectively. The velocity components $[C_r, C_\theta, C_z]$ are the radial, tangential (or sometimes referred to as swirl or azimuthal velocity), and axial velocity components, respectively. These are shown in a definition sketch in Fig. 7.2.

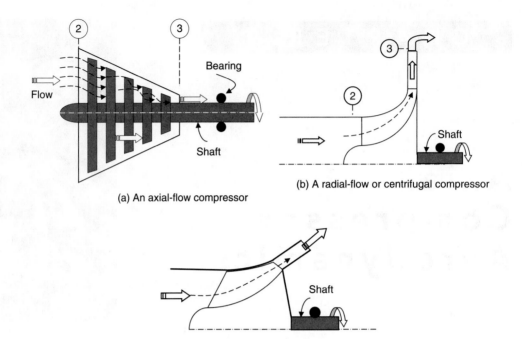

(a) An axial-flow compressor

(b) A radial-flow or centrifugal compressor

(c) A hybrid compressor

■ **FIGURE 7.1**   Schematic drawing of different types of compressors in aircraft gas turbine engines

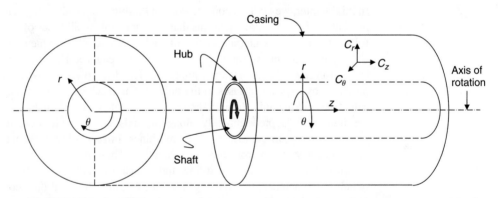

■ **FIGURE 7.2**   Definition sketch for the coordinates and the velocity components of the flow in cylindrical coordinates

7.3   Rotor and Stator Frames of Reference

In turbomachinery, energy transfer between the blades and the fluid takes place in an inherently unsteady manner. This is achieved by a set of rotating blades, called the rotor. The rotor blades are three-dimensional aerodynamic surfaces, which experience aerodynamic forces. The rotor blades are cantilevered at the hub and thus feel a root bending moment and a torque. The reaction to the blade forces and moments is exerted on the fluid, via the action–reaction principle of Newton. Stationary blades called the stator follow the rotor blades in what is known as a turbomachinery *stage*. The stator blades are three-dimensional aerodynamic surfaces as well. They are cantilevered from the casing and experience forces and

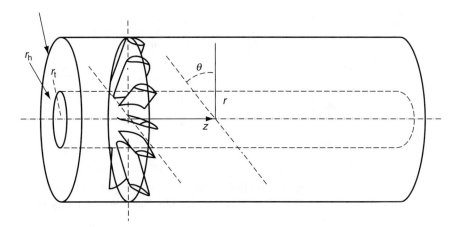

■ **FIGURE 7.3**   **Isolated rotor in a cylinder (Marble, 1964)**

moments, like the rotor. The exception is that the stator forces and moments are stationary (in the laboratory frame of reference) and thus perform no work on the fluid. The energy of the fluid is, thus, expected to remain constant in passing through the stator blades. The position of the observer is, however, important in viewing the flowfield and the energy exchange in turbomachinery. An observer fixed at the casing (or laboratory) is called an *absolute* observer. If the observer is attached to the rotor blade and spins with it, then it is called a *relative* observer. The frames of reference are then called the absolute and relative frames of reference, respectively. Consider an isolated rotor in a cylindrical duct, as shown in Fig. 7.3. An absolute observer sees the blades' aerodynamic forces are in motion, at an angular rate, i.e., the angular velocity ω of the shaft. Hence, as measured by this observer, the total enthalpy of the fluid goes up in crossing the rotor row. On the contrary, let us put ourselves in the frame of reference of a relative observer who is spinning with the rotor. According to a relative observer, the blades are not moving! An observer fixed at the rotor measures aerodynamic forces and moments of the blades, however, as the forces are stationary, there is no work done on the fluid according to this observer. Thus, the relative observer measures the same total enthalpy across the blade row.

The flowfield as seen by a relative observer attached to an isolated rotor in a cylinder is thus steady. The absolute observer on the casing, however, sees the passing of the blades and thus experiences an unsteady flowfield. As the rotor blades pass by, a periodic pressure pulse (due to blade tip) is registered at the casing, which signifies an unsteady event with a periodicity of blade passing frequency. To be able to analyze a flowfield in a steady frame of reference offers tremendous advantages in the nature and the solution of governing equations. Consequently, in analyzing the flow within a rotor blade row, we employ the relative observer stance, while the stator flows are viewed from the standpoint of an absolute observer. We need to be mindful, however, that in practice there are no isolated rotors and thus the flowfield in rotating machinery is *inherently* unsteady. We shall present some preliminary discussions on the scale and effect of unsteadiness in axial-flow compressors later in this chapter.

The velocity components as seen by observers in the two frames of reference are related. First, we note that the radial and axial velocity components are identical in the two frames, as the relative observer moves only in the θ, or the angular, direction. Therefore, the swirl or tangential velocity is the only component of the velocity vector field that is affected by the observer rotation. At a radial position r on the rotor, the relative observer rotates with a speed ωr and thus registers a tangential velocity, which is ωr less than the absolute swirl velocity, i.e.,

$$\text{Relative swirl} = \text{Absolute swirl} - \omega r \qquad \textbf{(7.1)}$$

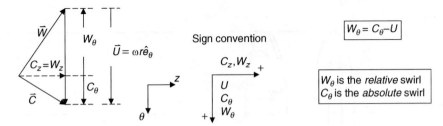

■ **FIGURE 7.4**   The velocity triangle

The fluid velocity vectors in the two frames are labeled as $\vec{C}$ and $\vec{W}$ for the absolute and relative observers, respectively. Therefore, the absolute velocity vector is described as

$$\vec{C} = C_r \hat{e}_r + C_\theta \hat{e}_\theta + C_z \hat{e}_z \tag{7.2}$$

The relative velocity vector is

$$\vec{W} = W_r \hat{e}_r + W_\theta \hat{e}_\theta + W_z \hat{e}_z \tag{7.3}$$

The rotor blade spins with an angular velocity ω, hence it describes a solid body rotation as

$$\vec{U} = \omega r \hat{e}_\theta \tag{7.4}$$

Comparing the velocity vectors as described by Eqs. 7.2–7.4, we conclude that the following vector identities hold, namely,

$$\vec{W} = \vec{C} - \vec{U} \tag{7.5-a}$$

$$\vec{C} = \vec{W} + \vec{U} \tag{7.5-b}$$

These three vectors form a triangle known as the *velocity triangle*.

Figure 7.4 shows a definition sketch of the velocity triangle and the sign convention. The positive tangential or azimuthal angle θ is in the direction of rotor rotation $\vec{U}$. The swirl velocity component is considered positive if in the direction of rotor rotation. For example, we note that the absolute swirl is pointing in the positive θ direction, therefore it is considered positive. The relative swirl velocity component W_θ is in the opposite direction, hence it is a negative quantity. The scalar relation given in Eq. 7.1 between the two swirl components is always valid.

7.4   The Euler Turbine Equation

The Euler turbine equation is called the fundamental equation of turbomachinery. Once we derive this simple yet powerful expression, its significance becomes evident. Let us consider a streamtube that enters a turbomachinery blade row. Figure 7.5 illustrates a generic streamtube with its geometry and velocity components. In general, stream surfaces undergo a radial shift when interacting with a blade row, as depicted in Fig. 7.5.

The mass flow rate in the streamtube is constant, by definition, and is labeled as $\dot{m}$. The angular momentum of the fluid in the streamtube is the moment of the tangential momentum of the fluid about the axis of rotation, namely,

$$\text{(Time rate of change of the) Fluid angular momentum} = \dot{m} r C_\theta \tag{7.6}$$

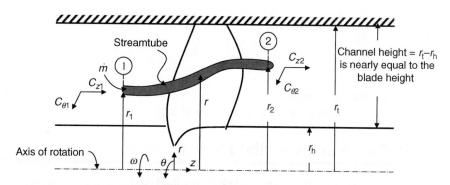

■ **FIGURE 7.5**   Streamtube interacting with a turbomachinery blade row

The change of fluid angular momentum between the exit and inlet of the streamtube is the applied torque exerted by the blade on the fluid, i.e.,

$$\dot{m}(r_2 C_{\theta 2} - r_1 C_{\theta 1}) = \tau_{\text{fluid}} \tag{7.7-a}$$

The torque is the product of blade tangential force F_θ and the moment arm r from the axis of rotation. Hence the blade torque is

$$\tau_{\text{blade}} = F_{\theta,\text{blade}} \cdot r = -\tau_{\text{fluid}} \tag{7.7-b}$$

The expression 7.7-a (or –b) is valid for the rotor as well as the stator. In case of rotor, there is an angular motion, hence the product of the angular velocity of the blade and the torque provides the power transmitted to the fluid, namely,

$$\wp = \tau_{\text{fluid}} \cdot \omega = \dot{m}\omega(r_2 C_{\theta 2} - r_1 C_{\theta 1}) = \dot{m}\omega\Delta(rC_\theta) \tag{7.8}$$

The ratio of shaft power to mass flow rate is called the specific work of the rotor, w_c for the compressor and w_t for the turbine, hence,

$$w_c \equiv \frac{\wp_c}{\dot{m}_c} = \omega\Delta(rC_\theta)_c \tag{7.9-a}$$

$$w_t \equiv \frac{\wp_t}{\dot{m}_t} = \omega\Delta(rC_\theta)_t \tag{7.9-b}$$

This is the Euler turbine equation written for the fluid interacting with a compressor or turbine rotor. As evidenced in Eq. 7.9, the rotor (specific) work appears as the change in (specific) angular momentum across a blade row times the shaft speed. The rotor and stator torques are proportional to the change in angular momentum across the rotor and stator blade rows, respectively, via Eq. 7.7.

The first law of thermodynamics applied to a steady and adiabatic process demands that the change of total enthalpy of the fluid across the blade row to be equal to the blade-specific work delivered to the fluid, namely,

$$h_{t2} - h_{t1} = \frac{\wp}{\dot{m}} = w_c \tag{7.10-a}$$

From the Euler turbine equation, the exit stagnation enthalpy in Eq. 7.10-a is related to the inlet stagnation enthalpy and the change of the angular momentum across the rotor row,

$$h_{t2} = h_{t1} + \omega\Delta(rC_\theta) \tag{7.10-b}$$

The nondimensional total enthalpy change is then

$$\frac{h_{t2}}{h_{t1}} = 1 + \frac{\omega \Delta(rC_\theta)}{h_{t1}} = \frac{T_{t2}}{T_{t1}} \tag{7.10-c}$$

Note that we assumed a calorically perfect gas in Eq. 7.10-c when we replaced the ratio of stagnation enthalpy with the ratio of total temperatures.

7.5 Axial-Flow Versus Radial-Flow Machines

In an axial-flow turbomachinery, the fluid path is predominantly along the axis of rotation. In radial-flow or centrifugal machinery, the fluid path departs from axial and attains a predominantly radial motion at the exit. As a result, the following approximation is typically made for axial-flow machines,

$$r_1 \approx r_2 \approx r \tag{7.11}$$

Therefore the Euler turbine equation may be simplified to

$$w_c \cong \omega r(\Delta C_\theta) = U(C_{\theta 2} - C_{\theta 1}) \tag{7.12}$$

The assumption of constant "r" in axial-flow turbomachinery places the stream surfaces on cylindrical surfaces. This means that stream surfaces do not undergo significant *radial deviation*. Thus, radial deviation or radial shift of stream surfaces is often assumed negligible in axial-flow machines. The simplified flowfield in an axial turbomachinery annulus may be divided into a series of $r = $ constant cylindrical cuts, as shown in Fig. 7.6.

The pitchline radius r_m is defined as the mid-radius between the hub and the casing, i.e.,

$$r_m \equiv \frac{r_h + r_t}{2} \tag{7.13}$$

Often, the pitchline radius of an axial- flow turbomachinery is assumed to represent the mean or average of the flowfield properties in the annulus and thus serves as the first line of attack in a

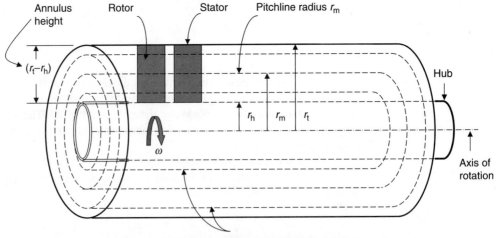

Other cylindrical cuts representing simplified stream
surfaces in a cylindrical annulus

■ **FIGURE 7.6** **A simple model of an axial compressor stage (rotor and stator) in a cylindrical annulus showing a mean or pitchline radius r_m along with other cylindrical cuts of the annulus representing simplified stream surfaces**

one-dimensional flow analysis approach. The hub and the tip radii represent the maximum deviations from the mean and thus are analyzed next. For a more accurate analysis, other cylindrical cuts are introduced in the annulus, as shown in Fig. 7.6. To further improve the accuracy of our analysis, we need to incorporate the radial disposition of the stream surfaces interacting with turbomachinery blade rows. In general, the stream surfaces undergo a radial shift interacting with a blade row and thus form *conical surfaces* in the vicinity of a blade. We need to employ a three-dimensional flow theory such as radial equilibrium theory or an actuator disc theory to approximate their exit radius r_2 of stream surfaces that have entered the blade row at the inlet radius of r_1. We will estimate the radial shift of stream surfaces using radial equilibrium theory accordingly in section 7.6.5.

A centrifugal or radial compressor receives the fluid in the axial direction near the axis of rotation and imparts large angular momentum to the fluid by expelling the fluid at a higher radius and with a large swirl kinetic energy. The radial turbines receive the fluid at a large radius and with high swirl kinetic energy and expel the fluid near the axis of rotation with a small angular momentum. In the process of absorbing the swirl kinetic energy, a large blade torque is created, which in a rotor translates into shaft power. Hence, the radial shift of stream surfaces needs to be large for an efficient centrifugal compression/expansion and thus may not be neglected from our analysis. Fortunately, unlike axial-flow turbomachinery that we have to employ three-dimensional flow theories to predict the exit radius of an incoming stream surface, in a centrifugal machine the exit radius for all stream surfaces is fixed by the impeller geometry, i.e., the impeller exit radius. We will treat centrifugal compressors in accordingly Chapter 8.

7.6 Axial-Flow Compressors and Fans

Axial-flow compressors and fans provide mechanical compression for the air stream that enters a gas turbine engine. Thermodynamically, their function is to increase the fluid pressure, efficiently. Hence, the boundary layers on the blades of compressors and fans, as well as their hub and casing, are exposed to an adverse, or rising, pressure gradient. Boundary layers exposed to an adverse pressure gradient, due to their inherently low momentum, cannot tolerate significant pressure rise. Consequently, to achieve a large pressure rise, axial-flow compressors and fans need to be staged. With this stipulation, a multistage machinery, or compression system, is born. In a stage, the rotor blade imparts angular momentum to the fluid, while the following stator blade row removes the angular momentum from the fluid. A definition sketch of a compressor stage with an inlet guide vane that imparts angular momentum to the incoming fluid is shown in Fig. 7.7. A compressor stage without an inlet guide vane is depicted in Fig. 7.8. In either case, the principle of rotor increasing the fluid angular momentum and the stator blade removing the swirl (or angular momentum) is independent of any preswirl in the incoming flow to the stage. We may introduce an inlet guide vane upstream of the rotor blades that imparts a preswirl (in the direction of the rotor motion) to the incoming stream and yet the principle of rotor–stator angular momentum interactions with the fluid remains intact.

Based on the absolute velocity field in regions 1, 2, and 3, for a generic inlet condition that may include a preswirl $C_{\theta 1}$ created by an inlet guide vane, we may write the rotor and stator torques

$$\tau_{\text{rotor}} = -\tau_{\text{fluid}} = -\dot{m} \cdot r (C_{\theta 2} - C_{\theta 1}) \tag{7.14}$$

$$\tau_{\text{stator}} = -\tau_{\text{fluid}} = -\dot{m} \cdot r (C_{\theta 3} - C_{\theta 2}) \tag{7.15}$$

■ FIGURE 7.7
Definition sketch for
station numbers and
three different planes,
r–θ, r–z, and θ–z, in a
compressor stage with an
inlet guide vane (IGV)

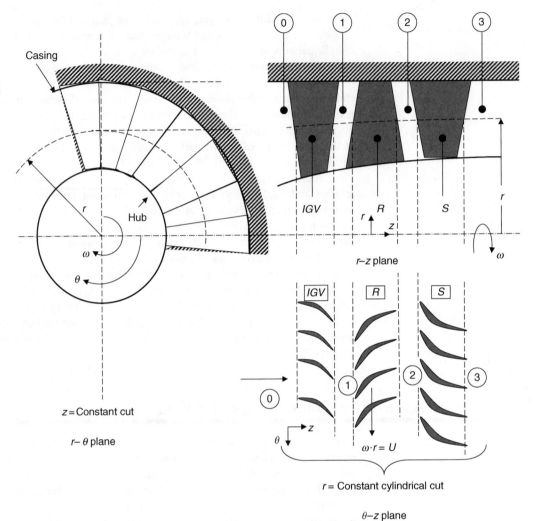

■ FIGURE 7.8
Cylindrical
(r = constant) cut of a
compressor stage with
velocity triangles
showing the rotor
imparting and the stator
removing the swirl and
angular momentum to
the fluid

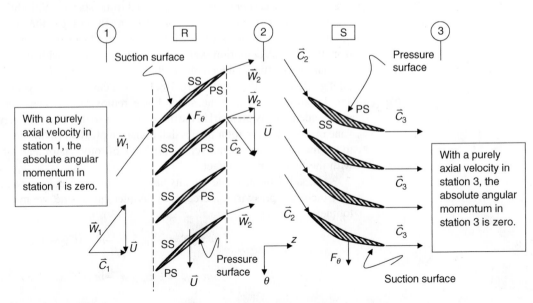

Assuming the absolute swirl and angular momentum across the stage remains the same, i.e., $C_{\theta 1} = C_{\theta 3}$ and $r_1 = r_3$ the rotor and stator torques become equal and opposite, namely,

$$\tau_{\text{rotor}} = -\tau_{\text{stator}} \tag{7.16}$$

We observe the suction and pressure surfaces of the rotor and stator blades, as shown in Fig. 7.8, and note that the blade aerodynamic forces in the θ-direction F_θ for the rotor and stator blades are in opposite directions. Since the moment of this tangential force, i.e., $r \cdot F_\theta$ from the axis of rotation, is the blade torque, we conclude that the rotor and stator torques are opposite in direction and nearly equal in magnitude.

EXAMPLE 7.1

The absolute flow at the pitchline to a compressor rotor is swirl free. The exit flow from the rotor has a positive swirl, $C_{\theta 2} = 145$ m/s. The pitchline radius is $r_m = 0.5$ m, and the rotor angular speed is $\omega = 5600$ rpm. Calculate the specific work at the pitchline and the rotor torque per unit mass flow rate.

SOLUTION

Rotor tangential speed at pitchline is

$$U_m = \omega r_m = (5600\,\text{rev/min})(2\pi\,\text{rad/rev})(\text{min}/60\,\text{s})(0.5\,\text{m})$$
$$= 293.2\,\text{m/s}$$

$$w_c \cong \omega r(\Delta C_\theta) = U(C_{\theta 2} - C_{\theta 1}) = 293.2\,\text{m/s}(145\,\text{m/s})$$
$$\approx 42.516\,\text{kJ/kg}$$
$$\tau_{r,m}/\dot{m} = r_m(C_{\theta 2} - C_{\theta 1}) = 0.5\,\text{m}(145\,\text{m/s}) = 72.5\,\text{m}^2/\text{s}$$

7.6.1 Definition of Flow Angles

The flow angles are measured with respect to the axial direction, or axis of the machine, and are labeled as α and β, which correspond to the absolute and relative flow velocity vectors $\vec{C}$ and $\vec{W}$, respectively. Figure 7.9 is a definition sketch that shows the absolute and relative flow angles in a compressor stage.

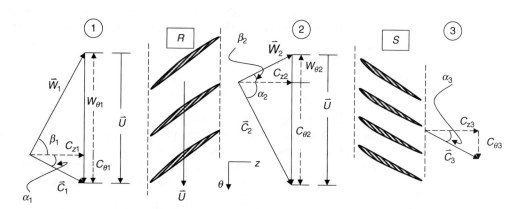

■ **FIGURE 7.9** Definition sketch for the absolute and relative flow angles in a compressor stage (a cylindrical cut of the stage, r = constant)

We may use these absolute and relative flow angles to express the velocity components in the axial and the swirl direction as

$$C_{\theta 1} = C_{z1} \cdot \tan(\alpha_1) \tag{7.17}$$

$$W_{\theta 2} = C_{z2} \cdot \tan(\beta_1) \tag{7.18}$$

$$C_{\theta 2} = C_{z2} \cdot \tan(\alpha_2) \tag{7.19}$$

$$W_{\theta 2} = C_{z2} \cdot \tan(\beta_2) \tag{7.20}$$

One method of accounting for positive and negative swirl velocities is through a convention for positive and negative flow angles. We observe that the absolute velocity vector upstream of the rotor has a swirl component in the direction of the rotor rotation. Hence, the absolute flow angle α_1 is considered positive. The opposite is true for the relative velocity vector $\vec{W}_1$, which has a swirl component in the opposite direction to the rotor rotation. Both the relative flow angle β_1 and the swirl velocity facing it $W_{\theta 1}$ are thus negative.

A turbomachinery blade row is designed to maintain an attached boundary layer, under normal operating conditions. Hence, the flow angles at the exit of the blades are primarily fixed by the blade angle at the exit plane. For example, the relative velocity vector at the exit of the rotor, $\vec{W}_2$, should nearly be tangent to the rotor suction surface at the trailing edge. Hence, β_2 is fixed by the geometry of the rotor and remains nearly constant over a wide operating range of the compressor. The same statement may be made about α_1 or α_3. These angles remain constant over a wide range of the operation of the compressor as well. Again, remember that the exit flow angle argument is made for an attached (boundary layer) flow. The other flow angles, as in β_1 or α_2, change with rotor speed U (i.e., ωr). Consequently, use is made of the nearly constant exit flow angles α_1, β_2, and α_3 in expressing performance parameters of the compressor stage and the blade row.

The axial velocity components C_{z1}, C_{z2}, or C_{z3} contribute to the mass flow rate through the machine. A common (textbook) design approach in axial-flow compressors and fans maintains a constant axial velocity throughout the stages. We shall make use of this simplifying design approach repeatedly in this chapter. Figure 7.8 shows an example of a constant axial velocity in a compressor stage. Although a simplifying design assumption, the reader needs to be aware that the goal of constant axial velocity is nearly impossible to achieve in practice. This is due to a highly three-dimensional nature of the flowfield, set up by three-dimensional pressure gradients, in a turbomachinery stage. We will address this and other topics related to three-dimensional flow in accordingly section 7.6.5.

A useful concept in turbomachinery design calls for a *repeated stage* (also referred to as *normal stage*). This implies that the velocity vectors at the exit and entrance to a stage are the same, i.e.,

$$\vec{C}_3 = \vec{C}_1 \tag{7.21-a}$$

$$\alpha_3 = \alpha_1 \tag{7.21-b}$$

The velocity triangles at the inlet and the exit of the stage shown in Fig. 7.8 have made use of the repeated stage concept. Another concept in turbomachinery calls for a *repeated row* design, which leads to flow angle implications that are noteworthy. In a repeated row design, the exit relative flow angle has the same magnitude as the absolute inlet flow angle, and the inlet relative flow angle has the same magnitude as the exit absolute flow angle, i.e.,

$$|\beta_2| = |\alpha_1| \tag{7.22-a}$$

$$|\alpha_2| = |\beta_1| \tag{7.22-b}$$

The example of the compressor stage shown in Fig. 7.9 has used the concept of repeated row. Note that a repeated row design leads to a repeated stage but the reverse is not necessarily correct. Namely, we may have a repeated stage design that does not use a repeated row concept. Figure 7.8 shows an example of a repeated stage that does not obey a repeated row design.

7.6.2   Stage Parameters

The Euler turbine equation that we derived earlier is the starting point for this section with the assumption of $r_1 \approx r_2 \approx r$,

$$\frac{T_{t2}}{T_{t1}} = 1 + \frac{U(C_{\theta 2} - C_{\theta 1})}{c_p T_{t1}}$$

We may replace the swirl velocities by the flow angles and the axial velocity components, namely,

$$\frac{T_{t2}}{T_{t1}} = 1 + \frac{U(C_{z2}\tan\alpha_2 - C_{z1}\tan\alpha_1)}{c_p T_{t1}} = 1 + \left(\frac{U^2}{c_p T_{t1}}\right)\left(\frac{C_{z1}}{U}\right)\left(\frac{C_{z2}}{C_{z1}}\tan\alpha_2 - \tan\alpha_1\right) \quad \textbf{(7.23)}$$

Expression 7.23 for the total temperature rise across the rotor (or stage) involves nondimensional groups C_{z1}/U and $c_p T_{t1}/U^2$. These groups appear throughout the turbomachinery literature and deserve a special attention. Also, the axial velocity ratio C_{z2}/C_{z1} appears that is often set equal to 1, as a first-order design assumption. Equation 7.23 is expressed in terms of the absolute flow angle at the exit of the rotor, which is not, however, a good choice, since it varies with the rotor speed. A better choice for the flow angle in plane 2, i.e., downstream of the rotor, is the relative flow angle β_2. The relative exit flow angle β_2 remains nearly unchanged as long as the flow remains attached to the blades. To express the total temperature rise across the rotor to the flow angles α_1 and β_2, we replace the absolute swirl $C_{\theta 2}$ by the relative swirl speed, namely,

$$C_{\theta 2} = U + W_{\theta 2} \quad \textbf{(7.24)}$$

Therefore,

$$\frac{T_{t2}}{T_{t1}} = 1 + \left(\frac{U^2}{c_p T_{t1}}\right)\left[1 + \frac{C_{z2}}{U}\tan\beta_2 - \frac{C_{z1}}{U}\tan\alpha_1\right] \quad \textbf{(7.25-a)}$$

If we assume a constant axial velocity design, i.e., $C_{z1} = C_{z2}$, then we get

$$\frac{T_{t2}}{T_{t1}} = 1 + \left(\frac{U^2}{c_p T_{t1}}\right)\left[1 + \left(\frac{C_z}{U}\right)(\tan\beta_2 - \tan\alpha_1)\right] \quad \textbf{(7.25-b)}$$

Here, we have expressed the stage total temperature ratio as a function of two nondimensional parameters. Note that β_2 is a negative angle and α_1 is a positive angle, according to our sign convention. Therefore, the contribution to the stage total temperature ratio falls with increasing (C_z/U) for a given wheel speed and inlet stagnation enthalpy. The ratio of axial-to-wheel speed is called the *flow coefficient* ϕ

$$\phi \equiv \frac{C_z}{U} \quad \textbf{(7.26-a)}$$

We may divide both numerator and the denominator of Eq. 7.26-a by the speed of sound in plane 1, i.e., a_1, to get the ratio of axial to blade (rotational) or tangential Mach number, namely,

$$\phi = \frac{C_z/a_1}{U/a_1} = \frac{M_z}{M_T} \quad \textbf{(7.26-b)}$$

where M_z is the axial Mach number, and M_T is the blade tangential Mach number based on U and a_1. For example, a stream surface with an axial Mach number of 0.5 that approaches a section of a rotor that is spinning at Mach 1 has a flow coefficient of 0.5. Now, let us interpret the first nondimensional group $U^2/(c_p T_{t1})$. We may divide this expression by the square of the upstream speed of sound a_1^2 to get

$$\frac{U^2/a_1^2}{c_p T_{t1}/(\gamma R T_1)} = \frac{(\gamma-1)M_T^2}{1+\left(\dfrac{\gamma-1}{2}\right)M_1^2} \tag{7.27}$$

Noting that the absolute Mach number M_1 is expressible in terms of the axial Mach number and a constant preswirl angle α_1 as

$$M_1 = \frac{M_z}{\cos \alpha_1} \tag{7.28}$$

We conclude that

$$\frac{U^2}{c_p T_{t1}} = \frac{(\gamma-1)M_T^2}{1+\left(\dfrac{\gamma-1}{2}\right)\dfrac{M_z^2}{\cos^2 \alpha_1}} \tag{7.29}$$

Based on Eqs. 7.26-b and 7.29, we may recast the stage stagnation temperature ratio in terms of blade tangential and axial Mach numbers as

$$\frac{T_{t2}}{T_{t1}} = 1 + \left[\frac{(\gamma-1)M_T^2}{1+\left(\dfrac{\gamma-1}{2}\right)\dfrac{M_z^2}{\cos^2 \alpha_1}}\right]\left[1+\left(\frac{M_z}{M_T}\right)(\tan \beta_2 - \tan \alpha_1)\right] \tag{7.30}$$

Kerrebrock (1992) offers an insightful discussion of compressor aerodynamics based on this equation. Two Mach numbers, i.e., the axial and the blade tangential Mach numbers appear to influence the stage temperature ratio in three ways, namely, via M_z, M_T, and M_z/M_T influence in Eq. 7.30. However, if we divide the first bracket on the RHS of Eq. 7.30 by the square of the blade tangential Mach number, we reduce this dependency to two parameters, namely,

$$\frac{T_{t2}}{T_{t1}} = 1 + \left[\frac{\gamma-1}{\left(\dfrac{1}{M_T^2}\right)+\left(\dfrac{\gamma-1}{2\cos^2 \alpha_1}\right)\left(\dfrac{M_z}{M_T}\right)^2}\right]\left[1+\left(\frac{M_z}{M_T}\right)(\tan \beta_2 - \tan \alpha_1)\right] \tag{7.31}$$

Consequently, two parameters govern the stage temperature ratio (or pressure ratio) and these are the blade tangential Mach number M_T and the flow coefficient or the ratio of the axial-to-tangential Mach number M_z/M_T. For given flow angles α_1 and β_2, an increase in the flow coefficient reduces the temperature rise in the stage. We may interpret an increase in the flow coefficient as an increase in the axial Mach number or the mass flow rate through the machine. As the mass flow rate increases, keeping the blade rotational Mach number the same, the blade angle of attack, or in the language of turbomachinery the *incidence angle*, decreases, hence the total temperature ratio drops. The opposite effect is observed with a decreasing mass flow rate through the machine where the incidence angle increases and thus blade work on the fluid increases to produce higher temperature or pressure rise. To visualize the effect of throughflow on rotor work, temperature and pressure rise, a rotor blade at different axial Mach numbers (or flow rate) and the same blade tangential Mach number (or wheel speed) is shown in Fig. 7.10.

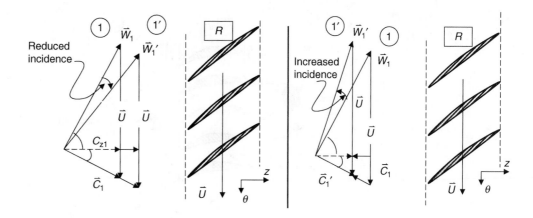

(a) The effect of *increased mass flow rate* while maintaining the same rotational speed is to reduce blade incidence angle

(b) The effect of *decreased mass flow rate*, with constant U, is to expose the blades to a higher incidence

■ **FIGURE 7.10**   Inlet velocity triangles for a compressor rotor with a changing flow rate

A reduced flow rate leads to an increased compressor temperature (or pressure) ratio. However, there is a limitation on how low the flow rate, or axial Mach number, can sink before the blades stall, for a fixed shaft rotational speed. Consequently, at reduced flow rates we could enter a blade stall flow instability, which marks the lower limit of the axial flow speed on the compressor map for a given shaft speed. The increased flow rate that leads to a reduction of the stage total temperature rise has its own limitation. The phenomenon of negative stall could be entered as the inlet Mach number is increased significantly.

The second parameter is the blade tangential Mach number M_T. A higher blade tangential Mach number increases the stage total temperature rise as deduced from Eq. 7.31. However, the limitations on the blade tangential Mach number are the appearance of strong shock waves at the tip as well as the structural limitations under centrifugal and vibratory stresses. The rotor shock losses increase (nonlinearly) with the relative tip Mach number; however, the advantage of higher work ($\propto M_T^2$) on the fluid outweigh the negatives of such a design at modest tip Mach numbers. The relative tip Mach number is defined as

$$M_{\text{tip,r}} = \sqrt{M_z^2 + \left(M_{\text{T,tip}}^2\right)} \tag{7.32}$$

where it represents the case of zero preswirl. The general case that includes a nonzero preswirl, may be written as

$$M_{\text{tip,r}} = \sqrt{M_z^2 + (M_{\text{T,tip}} - M_z \tan \alpha_1)^2} \tag{7.33}$$

The relative blade Mach numbers at the tip of operational fan blades have been supersonic for the past three decades. By necessity, these blade sections should be thin to avoid stronger bow shocks. A typical value for the relative tip Mach number is ~ 1.2 but may be designed as high as ~ 1.7. The thickness-to-chord ratio of supersonic blading may be as low as $\sim 3\%$. High strength-to-weight ratio titanium alloys represented the enabling technology that allowed the production of supersonic (tip) fans in the early 70s.

The inlet absolute flow angle to the rotor α_1 is a design choice that has the effect of reducing the rotor relative tip speed. The inlet preswirl is created by a set of inlet guide vanes, known as an IGV. To reduce the relative flow to the rotor tip, the IGV turns the flow in

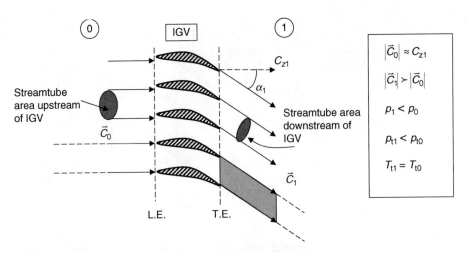

$$|\vec{C}_0| \approx C_{z1}$$

$$|\vec{C}_1| \succ |\vec{C}_0|$$

$$p_1 < p_0$$

$$p_{t1} < p_{t0}$$

$$T_{t1} = T_{t0}$$

■ **FIGURE 7.11** An inlet guide vane is seen to impart swirl to the fluid (note the shrinking stream tube area implies flow acceleration and static pressure drop)

the direction of the rotor rotation, by α_1. The flowfield entering an IGV is swirl-free and thus the function of the IGV is to impart positive swirl (or positive angular momentum) to the incoming fluid. This in effect reduces the rotor blade loading whose purpose is to impart angular momentum to the incoming fluid. An inlet guide vane and its flowfield are shown in Fig. 7.11.

We note that the blade passages in the IGV form a subsonic nozzle (i.e., contracting area) and cause flow acceleration across the blade row. The result of the flow acceleration is found in the static pressure drop due to flow acceleration, as well as a total pressure drop due to frictional losses of the blade passages. Hence, if the compressor design could avoid the use of an IGV, then certain advantages, including cost and weight savings, are gained. The advantage of operating at higher tip speeds at times outweighs the disadvantages of an IGV. The inlet guide vane may also be actuated rapidly if a quick response in thrust modulation is needed. For example, consider a lift fan, or a deflected engine exhaust flow, to support a VTOL aircraft in hover mode. The ability to modulate the jet lift (or vertical thrust) for stability and maneuver purposes may not be achieved through a spool up or spool down throttle sequence of the engine. Due to a large moment of inertia of the rotating parts in a turbomachinery, the rapid spooling is not fast enough for the control and stability purposes of a VTOL aircraft. In such applications, fast-acting IGV actuators could modulate the inlet flow to the rotor and hence the thrust/lift. An adjustable exit louver at the deflected nozzle end achieves the same goal. The inlet swirl angle α_1 may be zero or be adjustable in the range of $\pm 30°$ or more using a variable geometry IGV to relieve the relative tip Mach number, improve efficiency at compressor off-design operation, and provide for rapid thrust/lift modulation in military aircraft.

An example of a velocity triangle upstream of the rotor blade with and without the inlet guide vane is shown in Fig. 7.12. In both cases, we maintain the rotational speed and the mass flow rate, i.e., the axial velocity C_z constant.

The stage total pressure ratio is related to the stage total temperature ratio via a stage efficiency parameter η_s. Recalling from the chapter on cycle analysis,

$$\eta_s \equiv \frac{T_{t3s} - T_{t1}}{T_{t3} - T_{t1}} = \frac{\pi_s^{\frac{\gamma-1}{\gamma}} - 1}{\tau_s - 1} \tag{7.34}$$

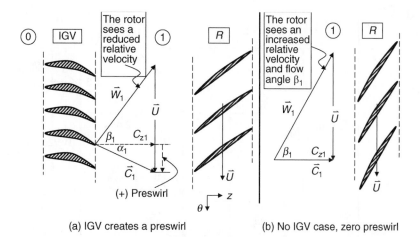

(a) IGV creates a preswirl   (b) No IGV case, zero preswirl

■ **FIGURE 7.12**   **Impact of IGV on the rotor relative flow for constant axial velocity and rotor speed**

Therefore, we may calculate the stage total pressure ratio from the velocity triangles that are used in the Euler turbine equation to establish τ_s and an efficiency parameter η_s to get

$$\pi_s = [1 + \eta_s(\tau_s - 1)]^{\frac{\gamma}{\gamma-1}} \tag{7.35}$$

Since the axial flow compressor pressure ratio per stage is small (i.e., near 1), the stage adiabatic efficiency and the polytropic efficiency are nearly equal. We recall that the polytropic efficiency was also called the "small-stage" efficiency, valid for an infinitesimal stage work. Although the approximation

$$\eta_s \cong e_c \tag{7.36}$$

is valid for low-pressure ratio axial flow compressor stages; the exact relationship between these parameters is derived in the cycle analysis chapter to be

$$\eta_s = \frac{\tau_s - 1}{\tau_s^{\frac{1}{e_c}} - 1} \tag{7.37}$$

Therefore by calculating the stage total temperature ratio from the Euler turbine equation and assuming polytropic efficiency, of say 0.90, we may calculate the stage adiabatic efficiency. Otherwise, we may calculate the stage total pressure ratio from the polytropic efficiency e_c and the stage total temperature ratio directly, via

$$\pi_s = \tau_s^{\frac{\gamma \cdot e_c}{\gamma-1}} \tag{7.38}$$

The frictional and shock losses are the main contributors to a total pressure loss in the relative frame of reference. We note that in a relative frame of reference, the blade passage is stationary and thus blade aerodynamic forces perform no work. Examining the fluid total pressure across the blade through the eyes of a relative observer, there are only losses to report. As noted, these losses stem from the viscous and turbulent dissipation of mechanical (kinetic) energy into heat as well as the flow losses associated with a shock. From the vantage point of an absolute observer, the blades are rotating and doing work on the fluid, thus increasing the fluid total temperature and pressure in the absolute frame of reference. These processes may be shown on a *T–s* diagram (see Fig. 7.13) that is instructive to review. For example, follow the static state of

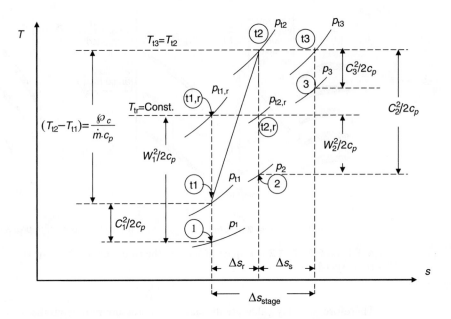

■ **FIGURE 7.13** Absolute and relative states of gas across a compressor rotor and stator

gas across the stage then follow the stagnation states in the absolute and relative frames. Explain the behavior of kinetic energy of gas as the fluid encounters the stationary and rotating blade rows, as seen by absolute and relative observers.

The *relative* total enthalpy is constant across the rotor along a stream surface (in steady flow without a radial shift in the stream surface),

$$h_{t,r} = h_1 + \frac{W_1^2}{2} = h_2 + \frac{W_2^2}{2} \tag{7.39}$$

We may replace the total kinetic energy terms in Eq. 7.39 by the sum of the kinetic energy of the component velocities, namely,

$$h_1 + W_{z1}^2/2 + W_{\theta1}^2/2 = h_2 + W_{z2}^2/2 + W_{\theta2}^2/2 \tag{7.40}$$

Let us substitute the absolute swirl minus the wheel speed $(C_\theta - U)$ for the relative swirl in Eq. 7.40, to get

$$h_1 + C_{z1}^2/2 + C_{\theta1}^2/2 - C_{\theta1} \cdot U = h_2 + C_{z2}^2/2 + C_{\theta2}^2/2 - C_{\theta2}U \tag{7.41}$$

The sum of the first three terms is the absolute total enthalpy on each side of the above equality, therefore,

$$h_{t1} - UC_{\theta1} = h_{t2} - UC_{\theta2} \tag{7.42}$$

This enthalpy constant in the rotor frame is known as "rothalpy," which may be rearranged to help us arrive at the Euler turbine equation, i.e.,

$$h_{t2} - h_{t1} = \frac{\wp}{\dot{m}} = w_c = U(C_{\theta2} - C_{\theta1}) \tag{7.43}$$

The rotor specific work on the fluid in nondimensional form is written as

$$\frac{\Delta h_t}{U^2} \equiv \psi = \frac{C_{\theta2}}{U} - \frac{C_{\theta1}}{U} = \frac{W_{\theta2} + U}{U} - \frac{C_{\theta1}}{U} = 1 + \left(\frac{C_z}{U}\right)(\tan\beta_2 - \tan\alpha_1) \tag{7.44}$$

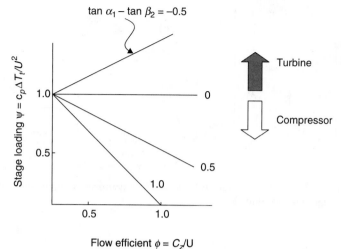

■ **FIGURE 7.14**
**Linear dependence of stage
loading and flow coefficient
parameters in ideal
turbomachinery stages
(adopted from Horlock,
1973)**

The function ψ is the nondimensional stage work parameter, which is called the *stage-loading factor or parameter*. The ratio of axial-to-wheel speed was called the flow coefficient ϕ, which forms another two-parameter family for the stage characteristics, namely,

$$\psi = 1 + \phi(\tan\beta_2 - \tan\alpha_1) \tag{7.45}$$

We note that the rotor exit relative flow angle is a negative quantity that leads to a stage-loading factor that is less than 1. In case of zero preswirl, or no inlet guide vane, the stage loading factor increases with a decreasing rotor exit flow angle β_2. In the limit of zero relative swirl at the rotor exit and no inlet guide vane, the stage loading factor approaches unity. However, a rotor relative exit flow angle of zero implies significant turning in the rotor blade passage, which may lead to flow separation. The stage-loading factor is an alternate form of expressing the stage characteristics and in essence takes the place of the rotor tangential Mach number $M_{\rm T}$, which was presented earlier. Horlock (1973) takes advantage of the linear dependence of the stage loading and the flow coefficient in Eq. 7.45 to explore the off-design behavior of ideal turbomachinery stages. We shall discuss the off-design behavior of turbomachinery later in this chapter but for now show the linear dependence of the two parameters in Fig. 7.14 (adopted from Horlock, 1973).

We define a stage *degree of reaction* °R as the fraction of static enthalpy rise across the stage that is accomplished by the rotor. Although the stator does no work on the fluid, it still acts as a diffuser that decelerates the fluid and thus causes an increase in fluid temperature, or static enthalpy. The stator takes out the swirl (kinetic energy) put in by the rotor and thus converts it to a static pressure rise. The degree of reaction measures the rotor share of the stage enthalpy rise as compared with the burden on the stator. This process is shown in an h–s diagram for the static states in a compressor stage. Remember that the static states are independent of the motion of the observer, hence they carry no subscript labels besides the station number (see Fig. 7.15).

$$°{\rm R} \equiv \frac{h_2 - h_1}{h_3 - h_1} = \frac{h_{t2} - h_{t1} - (C_2^2 - C_1^2)/2}{h_{t3} - h_{t1} - (C_3^2 - C_1^2)/2} \tag{7.46-a}$$

The stagnation enthalpy across the stator remains constant as the stator blades do no work on the fluid, hence $h_{t3} = h_{t2}$ and if we assume a repeated stage, with $C_1 = C_3$, we simplify the above expression to get

$$°{\rm R} \cong 1 - \frac{C_2^2 - C_1^2}{2(h_{t2} - h_{t1})} = 1 - \frac{C_{z2}^2 + C_{\theta2}^2 - C_{z1}^2 - C_{\theta1}^2}{2U(C_{\theta2} - C_{\theta1})} \tag{7.46-b}$$

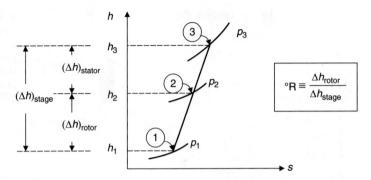

■ **FIGURE 7.15**   Static states of gas in a compressor stage and a definition sketch for the stage degree of reaction °R

Now, for a constant axial velocity across the rotor, $C_{z2} = C_{z1}$, we get a simple expression for the stage degree of reaction, as

$$°R \cong 1 - \frac{C_{\theta 2} + C_{\theta 1}}{2U} = 1 - \frac{C_{\theta,\text{mean}}}{U} \qquad \textbf{(7.46-c)}$$

where $C_{\theta \cdot \text{mean}}$ is the average swirl across the rotor. Since the flow has to fight an uphill battle with an adverse pressure gradient throughout a compressor stage, it stands to reason to expect/ design an equal burden of the static pressure rise in the rotor as that of a stator. Consequently, a 50% degree of reaction stage may be thought of as desirable. We may express the swirl velocity components across the rotor in terms of the absolute in flow and relative exit flow angles, α_1 and β_2, respectively,

$$°R = 1 - \frac{W_{\theta 2} + U + C_{\theta 1}}{2U} = \frac{1}{2} - \frac{C_z \tan \beta_2 + C_z \tan \alpha_1}{2U} = \frac{1}{2} - \left(\frac{C_z}{U}\right)\left(\frac{\tan \beta_2 + \tan \alpha_1}{2}\right) \quad \textbf{(7.47)}$$

For a 50% degree of reaction stage (at some spanwise radius r), the rotor exit flow angle has to be equal in magnitude and opposite to the inlet absolute flow angle, which is dictated by an IGV, namely,

$$\beta_2 = -\alpha_1$$

This is the condition for a repeated row design, as noted in Eq. 7.22-a and 7.22-b. Therefore a purely axial inflow (with no inlet guide vane) demands a purely axial relative outflow in order to produce a 50% degree of reaction. Again, we need to examine the net turning angle across the blade and assess the potential for flow separation. We shall introduce another parameter that will shed light on the state of the boundary layer at the blade exit and that is the diffusion factor.

On the degree of reaction, there is a body of experimental results that support the proposal that a boundary layer on a spinning blade, i.e., rotor, is more *stable* than a corresponding boundary layer on a stationary blade, i.e., stator, hence allocating a slightly higher burden of static pressure rise to the rotor. Here the word *stable* is used in the context of resistant to adverse pressure gradient or higher stalling pressure rise. Based on this, a degree of reaction of 60% may be a desirable split between the two blade rows in a compressor stage. As we shall see in the three-dimensional flow section in this chapter, the desirable degree of reaction is often compromised at different radii along a blade span to satisfy other requirements, namely, a healthy boundary layer flow.

EXAMPLE 7.2

An axial-flow compressor stage has a pitchline radius of $r_m = 0.5$ m. The rotational speed of the rotor at pitchline is $U_m = 212$ m/s. The absolute inlet flow to the rotor is described by $C_{zm} = 155$ m/s and $C_{\theta 1m} = 28$ m/s. The stage degree of reaction at pitchline is $°R_m = 0.60$, $\alpha_3 = \alpha_1$, and C_{zm} remains constant. Calculate

(a) rotor angular speed ω in rpm

(b) rotor exit swirl $C_{\theta 2m}$

(c) rotor specific work at pitchline w_{cm}

(d) relative velocity vector at the rotor exit

(e) rotor and stator torque per unit mass flow rate

(f) stage loading parameter at pitchline ψ_m

(g) flow coefficient ϕ_m.

SOLUTION

$\omega = U_m/r_m = (212 \text{ m/s})/0.5 \text{ m} = (424 \text{ rad/s})$ (rev/2π rad) (60 s/min) ≈ 4049 rpm

From Eq. 7.46-c, written at the pitchline

$$°R_m \cong 1 - \frac{C_{\theta 2m} + C_{\theta 1m}}{2U_m}$$

We isolate $C_{\theta 2m}$ to be

$$C_{\theta 2m} = 2U_m(1 - °R_m) - C_{\theta 1m}$$

$$= 2(212 \text{ m/s})(0.4) - 28 \text{ m/s} \approx 141.6 \text{ m/s}$$

Euler turbine equation describes the rotor specific work,

$$w_c \equiv \frac{\wp_c}{\dot{m}_c} = \omega \Delta(rC_\theta)_c, \quad \text{therefore,}$$

$$w_{cm} = U_m(C_{\theta 2m} - C_{\theta 1m})$$

$$= 212 \text{ m/s } (141.6 - 28)\text{m/s} \approx 24.08 \text{ kJ/kg}$$

Rotor relative swirl at the exit is

$$W_{\theta 2m} = C_{\theta 2m} - U_m = (141.6 - 212)\text{m/s} = -71 \text{ m/s}$$

Since the axial component of velocity remains constant, we can write the vector

$$\vec{W}_{2m} = 155\hat{k} - 71\hat{e}_\theta$$

Since $\alpha_3 = \alpha_1$, the rotor and stator torques are equal and opposite to eachother, i.e.,

$$\tau_{rm}/\dot{m} = r_m(C_{\theta 2m} - C_{\theta 1m}) = 0.5\text{m}(141.6 - 28)\text{m/s}$$

$$= 56.8 \text{ m}^2/\text{s} = -\tau_{sm}/\dot{m}$$

The stage loading parameter and flow coefficients are

$$\psi_m = \Delta C_\theta/U_m = (141.6 - 28)/212 \approx 0.5358$$

$$\phi_m = C_{zm}/U_m = 155/212 = 0.731$$

Another figure of merit for a compressor blade section that addresses the health of a boundary layer is, as noted earlier, the *Diffusion Factor*, or the *D*-factor. Its definitions for rotor and stator blades are, respectively

$$D_r \equiv 1 - \frac{W_2}{W_1} + \frac{|W_{\theta 2} - W_{\theta 1}|}{2\sigma_r W_1} \quad \text{(rotor } D\text{-factor)} \tag{7.48}$$

$$D_s \equiv 1 - \frac{C_3}{C_2} + \frac{|C_{\theta 3} - C_{\theta 2}|}{2\sigma_s C_2} \quad \text{(stator } D\text{-factor)} \tag{7.49}$$

where σ defines the blade solidity, i.e., the ratio of blade chord c to spacing s

$$\sigma_r \equiv \frac{c_r}{s_r} \quad \text{(rotor solidity)} \tag{7.50}$$

$$\sigma_s \equiv \frac{c_s}{s_s} \quad \text{(stator solidity)} \tag{7.51}$$

Modern compressor design utilizes a high solidity ($\sigma_m \geq 1$) blading at the pitchline radius (i.e., r_m). Since the blade spacing increases linearly with radius, the solidity of a constant chord blade also decreases linearly with the blade span, i.e.,

$$s(r) = \frac{2\pi \cdot r}{N_b} \quad \text{(blade spacing)} \tag{7.52}$$

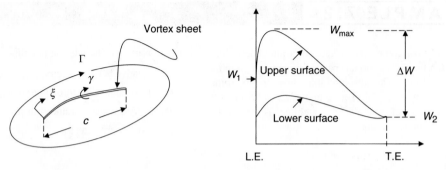

■ **FIGURE 7.16**   **Blade airfoil section is represented by a vortex sheet and a velocity distribution**

where the N_b is the number of blades in the rotor or the stator row. We can see the variation of blade row solidity with blade span by substituting Eq. 7.52 for the blade spacing, as

$$\sigma(r) = \frac{N_b \cdot c}{2\pi \cdot r} \quad \text{(blade solidity)} \qquad (7.53)$$

The hub section (r_h) has thus the highest solidity and the tip section (r_t) the lowest. The rationale for the definition of D-factor and its link to blade stall is made by Lieblein (1953, 1959, 1965). We review it here for its physical importance. First note that the definition of rotor diffusion factor is the same as the stator D-factor, except the parameters for the rotor are all in relative frame of reference. Next note that the change in swirl velocity across the rotor is the same regardless of the frame of reference, i.e., the change in absolute swirl is the same in magnitude as the change in relative swirl, namely,

$$|W_{\theta2} - W_{\theta1}| = |C_{\theta2} - C_{\theta1}| \qquad (7.54)$$

The blade circulation Γ is the integral of the vortex sheet strength γ over the chord, namely,

$$\Gamma \equiv \int_0^c \gamma(\xi)\, d\xi \approx \overline{\gamma} \cdot c \qquad (7.55)$$

Figure 7.16 shows an element of a blade section represented by a vortex sheet of local strength $\gamma(\xi)$. The local strength is equal to the local velocity jump across the blade. The average vortex sheet strength is thus the average velocity jump across the sheet, namely,

$$\overline{\gamma} \approx (\Delta W)_{\text{avg}} \approx \frac{W_{\text{max}} - W_1}{2} \qquad (7.56)$$

Therefore the average (positive) circulation around the blade section is

$$\Gamma_{\text{avg}} \approx \frac{c}{2}(W_{\text{max}} - W_1) \qquad (7.57)$$

Now, let us consider a control volume symmetrically wrapped around a blade section, as shown in Fig. 7.17.

By performing a closed line integral around the path C (see Fig. 7.17), we calculate the magnitude of blade circulation Γ as

$$\Gamma = s|W_{\theta1} - W_{\theta2}| \qquad (7.58)$$

Therefore equating the two expressions for the (magnitude of) blade circulation, we get

$$W_{\text{max}} \approx W_1 + \frac{|W_{\theta2} - W_{\theta1}|}{2\sigma} \qquad (7.59)$$

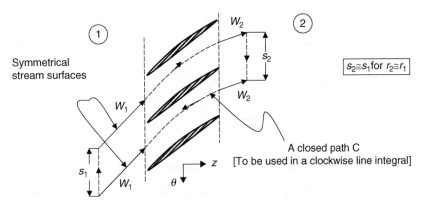

■ **FIGURE 7.17** Control volume for determining the blade circulation Γ along a stream surface

The maximum adverse pressure gradient on the blade suction surface leads to a maximum flow diffusion, which may be measured by the following parameter

$$\text{Maximum flow diffusion} \approx \frac{W_{\max} - W_2}{W_2} \approx \frac{W_{\max} - W_2}{W_1} \tag{7.60}$$

This parameter is called the diffusion factor D

$$D \equiv \frac{W_{\max} - W_2}{W_1} = \frac{W_1 + (|\Delta C_\theta|/2\sigma) - W_2}{W_1} = 1 - \frac{W_2}{W_1} + \frac{|\Delta C_\theta|}{2\sigma W_1} \tag{7.61}$$

EXAMPLE 7.3

A rotor blade row is cut at pitchline, r_m. The velocity vectors at the inlet and exit of the rotor are shown.

Assuming that $U_{1m} = U_{2m} = 200$ m/s and $C_{z1} = C_{z2} = 150$ m/s

And $\beta_2 = -35°$, calculate

(a) W_{1m} and W_{2m}

(b) D-factor D_{rm}

(c) circulation Γ_m

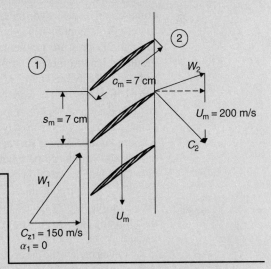

SOLUTION

$W_{1m} = [(150)^2 + (200)^2]^{1/2} = 250$ m/s

$W_{\theta2m} = C_{z2} \tan \beta_2 = (150 \,\text{m/s}) \tan(-35°)$

$W_{\theta2m} \approx -105 \,\text{m/s}$

$W_{2m} = [(150)^2 + (105)^2]^{1/2} = 183.1 \,\text{m/s}$

The D-factor for the rotor at the pitchline is

$$D_{rm} \equiv 1 - \frac{W_{2m}}{W_{1m}} + \frac{|W_{\theta2m} - W_{\theta1m}|}{2\sigma_{rm} W_{1m}}$$

The rotor solidity at the pitchline is $\sigma_{rm} = c/s = 7/7 = 1.0$

The relative tangential speed at the inlet to the rotor is equal in magnitude to the rotor speed U_m, but it has a negative value, as it points in the opposite direction to the rotor rotation

$W_{\theta1m} = -200 \,\text{m/s}$

Therefore, we have

$$D_{rm} = 1 - (183.1/250) + |(-105 + 200)|/(2.1250)$$

$$\approx 0.457$$

Since D-factor is less than ~ 0.5, the rotor boundary layer at the pitchline is expected to be attached.

Circulation around the rotor at pitchline is

$$\Gamma_m = s_m |W_{\theta 1m} - W_{\theta 2m}| = (0.07 \, \text{m})|(-200 + 105)|$$

$$\approx 6.65 \, \text{m}^2/\text{s}$$

7.6.3   Cascade Aerodynamics

A large body of experimental data supports a correlation between the blade stall and the diffusion factor. Let us first examine the attributes of a stalled flow in a compressor blade row. A separated boundary layer will create a thick wake, which may be characterized by its momentum deficit length scale θ^*. Equivalently, a wake is a region of deficit total pressure, hence in the blade frame of reference stall creates a rapid rise in the total pressure loss across the blade row. In either case, we need to examine the wake profile downstream of a blade row to quantify momentum and total pressure loss. To concentrate on the study of compressor blades, NACA developed the so-called 65 series of airfoil profiles in early 1950s. To test the aerodynamic behavior of the newly developed blade shapes, NACA (the predecessor of NASA) performed an extensive series of cascade experiments, concentrating on the impact of the 65-series compressor airfoil shapes on two-dimensional compressor loss and stall characteristics. Airfoil shapes are defined around a mean camber line, typically of circular arc or parabolic shape and a prescribed thickness distribution. A cascade is composed of a series of two-dimensional or cylindrical blades placed in a uniform flow to simulate the two-dimensional flowfield about a *spanwise section* of a compressor blade row. The goal of the investigation is to quantify the profile loss of two-dimensional blade elements as a function of the profile geometry and cascade parameters such as blade chord-to-spacing ratio, blade stagger angle, and other parameters. To construct the three-dimensional loss characteristics of compressor blades, two-dimensional cascade loss data are *stacked up* to represent the 3D picture. This approach neglects the cross interaction of the stream surfaces that is created through 3D pressure gradients. We learned in wing theory that three-dimensional pressure gradients lead to the formation of the streamwise vortices in the wake, which causes a 3D induced velocity (and induced drag) along the blade span. We shall address three-dimensional losses that are overlooked by the cascade data later in this chapter. Let us return to two-dimensional cascade studies performed at NACA. The wake profile with its momentum deficit and total pressure loss holds the key to characterizing the behavior of a compressor blade section. Figure 7.18 (a) shows a rectilinear cascade, 7.18 (b) an annular cascade, and 7.18 (c) defines the geometric parameters of the blade section and the cascade. Figure 7.19 shows periodic blade wakes downstream of a cascade where the momentum deficit and the total pressure loss are concentrated. The thickness of the wake is exaggerated in Fig. 7.19 for visualization purposes, but a thicker suction surface boundary layer than the pressure surface is intentionally graphed to show the behavior of these boundary layers that merge to form the blade wake in turbomachinery.

Let us review the cascade parameters as shown in Fig. 7.18 (c). An important geometrical cascade parameter is the solidity σ, which is defined as the ratio of chord-to-spacing. A high solidity blading is capable of a higher net turning angle than a low solidity blading. Consequently, a high solidity cascade is less susceptible to stall. Experimental evidence to support this assertion will be presented as cascade test results. As noted earlier, the modern compressor and fan design utilizes a high solidity blading ($\sigma_t \cong 1$). The angle of the mean camber line at the leading and trailing edge is used as a reference where we measure the inlet flow incidence angle i and the exit deviation angle δ^*. The incidence angle is defined as the flow angle between the tangent to the mean camber line and relative velocity vector at the inlet. In compressors, incidence angle takes the place of angle of attack in external aerodynamics. An optimum, incidence angle is defined as the incidence that causes minimum (total pressure) loss across a given cascade. A typical value of the optimum incidence angle for a subsonic flow cascade is $i_{opt} \sim 2°$. Cascade experimental results support this rule of thumb. The camber angle φ is defined as the angle formed at the intersection of the two tangents to the mean camber line at the leading and trailing edges, as shown. A large camber means a large flow turning, which may lead to flow separation. Hence, compressor blade camber is small compared with a turbine blade, which is capable of large turning due to a favorable pressure gradient. The stagger angle $\gamma°$ is sometimes referred to as the *blade-setting angle* is the blade chord angle with respect to the axial direction. The stagger angle increases with radius as the blade rotational speed (ωr) increases linearly with radius. The difference between the stagger angle and the relative inflow

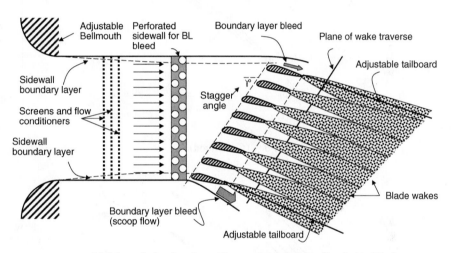

(a) Schematic drawing of a rectilinear cascade test rig with cylindrical blades

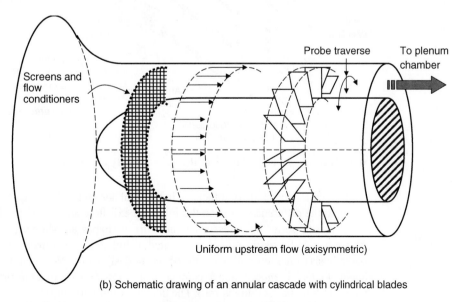

(b) Schematic drawing of an annular cascade with cylindrical blades

■ **FIGURE 7.18** **Cascade types and nomenclature**

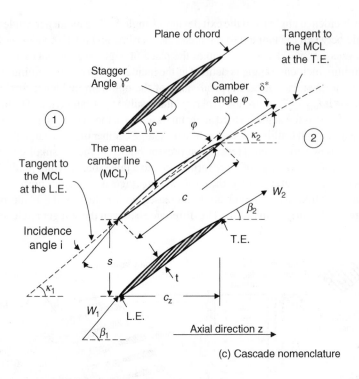

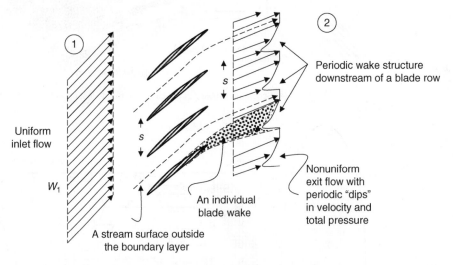

Cascade parameters

Solidity
$\sigma \equiv c/s$

Stagger angle
γ°

Camber angle
$\varphi \equiv \kappa_1 - \kappa_2$

Incidence angle
$i \equiv \beta_1 - \kappa_1$

Deviation angle
$\delta^* \equiv \beta_2 - \kappa_2$

Net turning angle
$\Delta\beta = \beta_1 - \beta_2 = \varphi + i - \delta^*$

Angle of attack
$\beta_1 - \gamma^\circ$

Thickness-to-chord ratio
t/c

(c) Cascade nomenclature

■ **FIGURE 7.18**   (Continued)

■ **FIGURE 7.19**   Survey of the wake-velocity profile downstream of a cascade

angle is called the angle of attack, as in external aerodynamics. Finally, the net turning angle refers to the difference between the inlet and exit flow angles, respectively.

The 65-series compressor cascade airfoil shapes are shown in Fig. 7.20. The design (theoretical) lift coefficient for an isolated airfoil shape (at zero angle of attack) is listed in parenthesis (times ten) following the 65-series designation. Note that the lift coefficient at zero angle of attack is entirely due to camber. The *isolated* means that the airfoil is not in a cascade configuration, i.e., solidity is zero. The thickness-to-chord ratio (in percent) comprises the last two digits of the series designation.

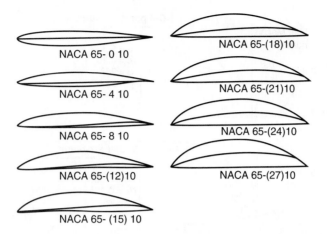

■ FIGURE 7.20 The 65-series cascade airfoil shapes with 10% thickness (Herrig, Emery, and Erwin, 1951)

For example, a NACA 65-(18)10 represents the 65-series camber shape that produces an isolated theoretical (i.e., inviscid) lift coefficient of 1.8 (at zero angle of attack) and has a 10% thickness-to-chord ratio. The cascade parameters, namely, solidity and stagger, are included in the individual loss characteristic (bucket) curves that are produced for different cascades. NACA defined a design angle of attack for the 65-series airfoils based on the smoothness of the pressure distributions on the airfoils. Figure 7.21 shows the design angle of attack for the 65-series airfoils arranged in a cascade of varying solidity. The zero solidity refers to an isolated airfoil case. Note that the combination of high solidity and high camber that leads to a high angle of attack (Fig. 7.21) does not mean that the incidence angle is very large too. The blade inlet angle κ_1 makes a large angle with respect to the blade chord for highly cambered airfoils (see Fig. 7.20 for airfoil shapes with high camber). The message from Fig. 7.21 is that higher solidity allows for larger turning, which translates into an increase in inlet flow angle (keeping the exit angle $\sim$ fixed).

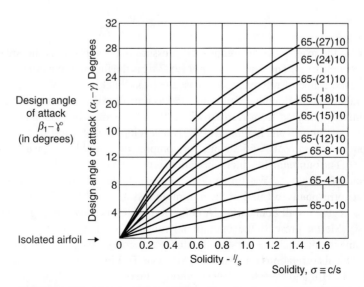

■ FIGURE 7.21 Design angle of attack, $\beta_1 - \gamma$, for the 65-series airfoils with 10% thickness as a function of solidity (from Herrig, Emery, and Erwin, 1951)

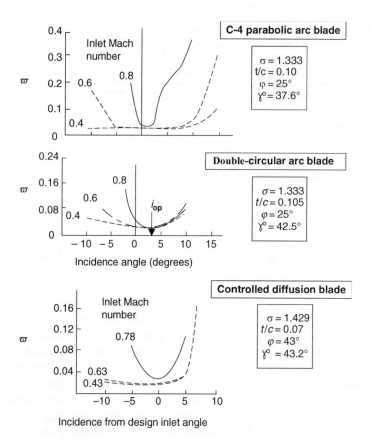

■ **FIGURE 7.22**   Variation of cascade total pressure loss parameter with inlet Mach number and flow incidence angle (adapted from Kerrebrock, 1992)

A survey of total pressure downstream of cascade reveals the presence of periodic wakes. Defining an *average* total pressure loss parameter for a cascade as

$$\varpi \equiv \frac{p_{t1} - \bar{p}_{t2}}{\rho_1 W_1^2 / 2} \tag{7.62}$$

where the average downstream total pressure may be taken as the area-average of the total pressure survey. The denominator of Eq. 7.62 is the familiar dynamic pressure $(\rho_1 W_1^2 / 2)$ in the cascade frame of reference. The cascade loss "bucket curves" are shown in Fig. 7.22 for a parabolic arc, a double-circular arc, and a controlled diffusion blading at different inlet Mach numbers (adapted from Kerrebrock, 1992). The works of Hobbs and Weingold (1984) and Hechert, Steinert and Lehmann (1985) on development of controlled diffusion airfoils and their comparison to NACA-65 airfoils for compressors should be consulted for further reading.

The minimum loss incidence angle is referred to as ϖ_{min} that corresponds to the optimum incidence angle $i_{opt.}$ We further define an acceptable operational range for the incidence angle that corresponds to 150% of the minimum loss ϖ_{min}, which Mellor referred to as the positive and negative stall boundaries. A definition sketch is shown in Fig. 7.23.

Mellor presents the operational range of each 65-series airfoils in various cascade arrangements that are very useful for preliminary design purposes. Mellor's unpublished graphical correlations (originated at MIT Gas Turbine Laboratory) were published by Horlock (1973), Hill and Peterson (1992), among others.

The correlation between the diffusion factor and the wake momentum deficit thickness θ^* is presented in Fig. 7.24 (Lieblein, 1965). The cascade data of Lieblein present 10% thick

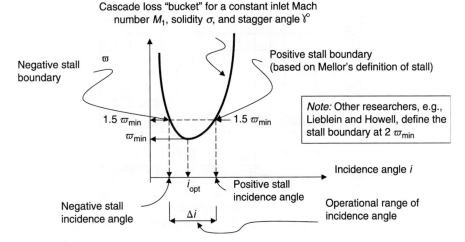

Cascade loss "bucket" for a constant inlet Mach
number M_1, solidity σ, and stagger angle $\gamma°$

■ **FIGURE 7.23**   Definition sketch for the stall boundaries based on the cascade loss curve

blades with the mean camber line shapes described by the NACA 65-series profiles and the
British C-4 parabolic arc profiles, as shown in Fig. 7.24. The Reynolds number for the cascade
tests is ~250,000. Blade stall leads to a thickening of the wake that results in a rapid increase in
profile drag and hence momentum deficit thickness. From Fig. 7.24, we note this behavior
occurs around a D-factor of 0.6. Hence, the maximum diffusion factor associated with a well-
behaved boundary layer on these classical blade profiles is $D_{\max} \sim 0.6$. A higher D-factor (of
~0.7) may be achieved in cascades of modern controlled-diffusion profiles.

How does the momentum deficit thickness θ^*/c relate to the total pressure loss parameter ϖ
in a cascade? By averaging the total pressure downstream of the cascade assuming a periodic
momentum deficit thickness with zero momentum, we can correlate these two parameters. Figure
7.25 (Lieblein, 1965) shows two total pressure loss functions and their correlation with D-factor.

To demonstrate the functional form of the correlation (see the ordinate of Fig. 7.25), we
approximate the cascade exit flow as being composed of uniform flow with periodic gaps of
zero momentum, each with θ^*/c thickness corresponding to wakes. The static pressure is
assumed to be uniform in the measuring station downstream of the cascade. Figure 7.26 shows a
definition sketch of this simplified cascade exit flow model. Although the original contribution
to this derivation is due to Lieblien–Roudebush (1956), this author has benefited from
Kerrebrock's (1992) treatment of this and other turbomachinery subjects.

The total pressure downstream of the cascade is the sum of static and dynamic pressures,
according to (incompressible) Bernoulli equation. The dynamic pressure is zero in the wake

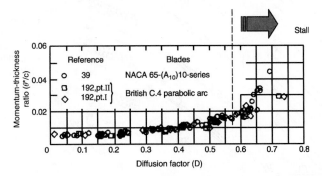

■ **FIGURE 7.24**   Correlation between the diffusion factor and the wake momentum deficit
thickness (from Lieblein 1965, reference numbers are in Lieblein)

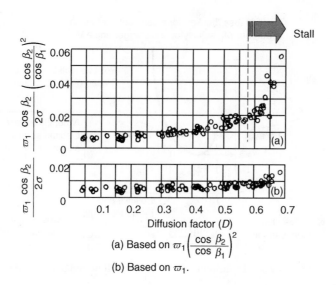

(a) Based on $\varpi_1 \left(\dfrac{\cos \beta_2}{\cos \beta_1} \right)^2$

(b) Based on ϖ_1.

■ **FIGURE 7.25**   Total pressure loss correlation with *D*-factor (from Lieblein 1965)

region of θ^* thickness. We may area-average the total pressure in the downstream region according to

$$p_{t2} \cdot (s \cdot \cos \beta_2 - \theta^*) + p_2 \cdot \theta^* = \bar{p}_{t2} \cdot s \cdot \cos \beta_2 \qquad \text{(7.63-a)}$$

$$\bar{p}_{t2} = p_{t1} - \frac{p_{t2} - p_2}{s \cdot \cos \beta_2} \cdot \theta^* = p_{t1} - \frac{\rho \cdot W_2^2}{2s \cdot \cos \beta_2} \cdot \theta^* \qquad \text{(7.63-b)}$$

Note that we have used the concept of constant total pressure in an inviscid fluid by using $p_{t1} = p_{t2}$ outside the wake region. Therefore, in the blade frame of reference, the total pressure remains conserved in the inviscid limit, assuming there are no shocks in the blade passage. In this context, we have applied the Bernoulli equation, which strictly speaking holds for an incompressible fluid. All these approximations are reasonable for our purposes here of

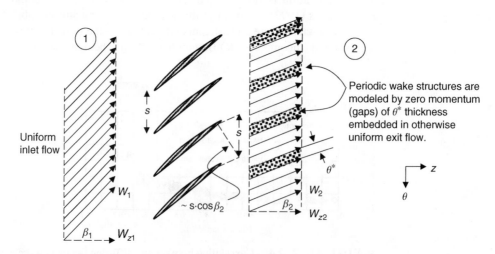

■ **FIGURE 7.26**   Definition sketch for a cascade exit flow model with periodic wakes

demonstrating subsonic cascade data correlation of total pressure loss with D-factor. Now, in terms of the total pressure loss parameter ϖ we get

$$\varpi = \frac{p_{t1} - \bar{p}_{t2}}{\rho_1 W_1^2/2} \approx \frac{\rho\left(\dfrac{W_z^2}{\cos^2 \beta_2}\right) \cdot \theta^*}{2s \cdot \cos \beta_2 \cdot \rho\left(\dfrac{W_z^2}{\cos^2 \beta_1}\right)/2} = \left(\frac{\cos \beta_1}{\cos \beta_2}\right)^2 \frac{\sigma}{\cos \beta_2}\left(\frac{\theta^*}{c}\right) \qquad \textbf{(7.64)}$$

Here we made a simple approximation of constant axial flow across the cascade. Finally, we conclude that since θ^*/c correlated with the diffusion factor, then the RHS of

$$\frac{\theta^*}{c} = \varpi \cdot \left(\frac{\cos \beta_2}{\sigma}\right)\left(\frac{\cos \beta_2}{\cos \beta_1}\right)^2 = f(D, t/c, Re_c, M, \sigma, \text{camber shape}) \qquad \textbf{(7.65)}$$

correlates with the D-factor as well, keeping other cascade parameters constant. This functional dependence of the total pressure loss parameter on D-factor is demonstrated in Fig. 7.25.

The exit flow angle deviates from the exit blade angle by what is called a deviation angle δ^* due to boundary layer buildup at the trailing edge. Cascade experiments have revealed a correlation between the deviation angle and the geometric parameters of a cascade, namely, the camber angle, the solidity, and the stagger angle. The following correlation is due to Carter (1955).

$$\delta^* = \frac{m\varphi}{\sigma^n} \qquad \text{(Carter's rule for deviation angle)} \qquad \textbf{(7.66)}$$

where φ is the camber angle, σ is the solidity, m is a function of cascade stagger angle, and n is $1/2$ for a compressor cascade and 1 for an inlet guide vane (or an accelerating passage such as turbines). The higher exponent of solidity for a turbine ($n = 1$) than the compressor ($n = 1/2$) leads to higher exit deviation angles in a compressor, as expected. The boundary layer build up in a compressor in adverse pressure gradient is greater than its counterpart in the turbine. The dependence of m on stagger angle and the profile shape is presented in Fig. 7.27, following Carter (1955).

The double-circular arc blade has a circular arc mean camber line shape where the point of maximum camber and thickness occur at mid-chord. The parabolic arc mean camber line blade that is shown in Fig. 7.27 has a maximum camber location at 40% chord. The location of the maximum camber and even the thickness affect the pressure distribution about the blade and hence the exit flow deviation angle. It is instructive to know that in turbomachinery blading the chordwise location of the maximum camber moves farther aft with an increasing inlet

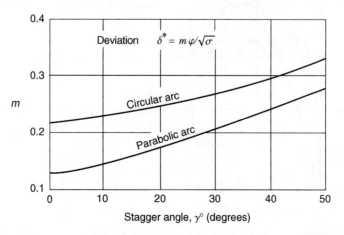

■ **FIGURE 7.27** Compressor cascade deviation angle rule following Carter (1955)

relative Mach number. For example, a multiple-circular arc blade may be used in the supersonic section of a transonic compressor with the maximum camber placed at 75% chord. We shall discuss multiple-circular arc and other profile shapes of suitable blades for supersonic applications later in this chapter.

A more general definition of the diffusion factor should involve the radial shift of stream surfaces across compressor blade rows. The constant radius, cylindrical cut approximation of the flowfield is rather restrictive. The method allows for a "pitchline," one-dimensional analysis of a compressor, but falls short of a realistic "mean" flow modeling. A general D-factor is defined as

$$D \equiv 1 - \frac{W_2}{W_1} + \frac{|r_2 W_{\theta 2} - r_1 W_{\theta 1}|}{2\sigma r_\mathrm{m} \cdot W_1} = 1 - \frac{W_2}{W_1} + \frac{|r_2 W_{\theta 2} - r_1 W_{\theta 1}|}{(r_1 + r_2)\sigma \cdot W_1} \tag{7.67}$$

Note that the change in angular momentum has replaced the change in tangential momentum in the definition of a general D-factor. We will address the general diffusion factor in the three-dimensional flow section in this chapter.

Another performance limiting parameter in a compressor in addition to the D-factor is the static pressure rise coefficient, as in a diffuser. The pressure rise coefficient and the D-factor are two different ways of measuring the same thing, namely, the tendency of the boundary layer to stall. However, the D-factor examines the stall behavior of a compressor blade suction surface, whereas the stalling pressure rise coefficient analogy with a diffuser examines the end wall stall that dominates the multistage compressor aerodynamics. The new limiting parameter is defined as

$$C_p \equiv (p_2 - p_1)/(\rho_1 W_1^2/2) \tag{7.68}$$

The static pressure rise across a blade row, in the numerator of Eq. 7.68, is independent of the observer frame of reference. The dynamic pressure at the inlet to the blade row is in the relative frame, i.e., as in a cascade parameter. From our studies of boundary layers in adverse pressure gradient we have learned that $C_{p,\mathrm{max}} \approx 0.6$. Interestingly, the numerical values of $C_{p,\mathrm{max}}$ and the limiting D-factor are both ~ 0.6, a coincidence. The nature of the compressor end wall boundary layer and its stalling characteristics is more complicated than a simple diffuser. Figure 7.28 shows

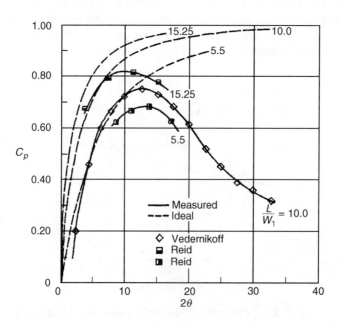

■ **FIGURE 7.28**
Static pressure rise limitation in a 2D diffuser (data from Kline, Abbott, and Fox, 1959)

the pressure rise data for a two-dimensional diffuser, where a working range of C_p within $\sim$0.4 to $\sim$0.8 is observed (from Kline et al., 1959).

The main differences between a diffuser flow and a flow in a compressor blade row are

(a) Compressor boundary layer is highly skewed due to large streamwise and radial pressure gradients, i.e., leading to (a highly) three-dimensional flow separation

(b) Inherent unsteadiness in turbomachinery due to relative motion of neighboring blade rows are absent in a diffuser

(c) Compressibility effects through the passage shock interaction with the boundary layer is absent in diffuser performance charts such as Fig. 7.28

(d) Upstream wake transport in downstream blade rows is also absent in diffusers

(e) Blade tip clearance flow is unique to turbomachinery

(f) End wall regions create secondary vortex formations as in corner and scraping vortex structures in compressors with no counterpart in a diffuser

In light of these complicating factors, to suggest a single value for the maximum (sometimes called *stalling*) pressure rise coefficient for all compressor blade sections is oversimplistic. For example, the stalling pressure rise coefficient near the end walls may be $\sim$0.48 (first stipulated by de Haller, 1953) and at the pitchline radius $\sim$0.6. A stalling pressure rise correlation for multistage axial flow compressors is successfully derived by Koch (1981). We will present Koch's correlation in the context of compressor stall margin later in this chapter. A simple application of de Haller pressure rise criterion in the incompressible limit restricts the maximum flow deceleration (W_2/W_1) in a blade row to 0.72. The shock in passage also changes the abruptness of the static pressure rise, hence the boundary layer separation. Despite these shortcomings, the concept of a stalling pressure rise in a compressor blade row is very attractive and useful. We shall examine some limiting features of stalling pressure rise coefficient in a compressor blade. First, let us use Bernoulli equation in recasting the static pressure rise coefficient C_p as

$$C_p = 1 - \frac{W_2^2}{W_1^2} - \frac{\Delta p_t}{(\rho_1 W_1^2/2)} = 1 - \frac{W_2^2}{W_1^2} - \varpi_{\text{cascade}} \qquad (7.69)$$

The cascade total pressure loss parameter $\varpi_{\min}$ is $\sim$2% for unstalled blades (see Fig. 7.21). For a maximum pressure rise coefficient of $\sim$0.6, the maximum flow deceleration (expressed in terms of W_2/W_1) is

$$W_2 \geq 0.62\, W_1$$

The implication of this (maximum) flow deceleration may be found in the turning angle limitation, namely, for a constant axial velocity W_z, we have $W_1 = W_z/\cos\beta_1$ and $W_2 = W_z/\cos\beta_2$,

$$C_p = 1 - (\cos^2\beta_1/\cos^2\beta_2) - \varpi_{\text{cascade}} \qquad (7.70)$$

We may solve Eq. 7.70 for the exit flow angle β_2 for a given cascade total pressure loss parameter, inlet flow angle, and the maximum pressure rise coefficient. The maximum turning angle $(\Delta\beta)$ associated with a limiting pressure rise coefficient is graphed in Fig. 7.29. The cases of zero total pressure loss, i.e., ideal flow, and a cascade with a total pressure loss parameter of 2, 6, and 10% are shown. The effect of total pressure loss increase is seen in Fig. 7.29 as a reduction in exit flow angle and thus an increase in flow turning in a compressor blade row. Also, from Eq. 7.70, we observe that the maximum static pressure rise coefficient $C_{p,\max}$ for a given inlet flow angle β_1 and total pressure loss occurs at the exit flow angle $\beta_2 = 0$, or axial flow direction at the exit, as expected.

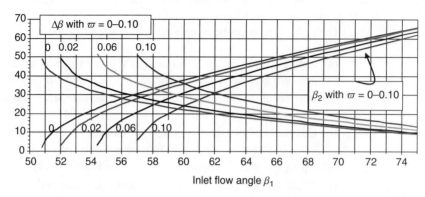

■ **FIGURE 7.29** **The exit flow and the limit turning angles (in degrees) based on a $C_{p,\max}$ of 0.6**

Another aspect of the pressure rise coefficient may be found in its relation to the relative inlet Mach number M_{1r}. From the definition of the static pressure rise coefficient, we may express the dynamic pressure in terms of the inlet static pressure and the inlet relative Mach number, as

$$C_p = \frac{p_2 - p_1}{\gamma \cdot p_1 M_{1r}^2/2} \qquad (7.71\text{-a})$$

$$\frac{p_2}{p_1} = 1 + C_p \cdot \gamma \cdot M_{1r}^2/2 \qquad (7.71\text{-b})$$

EXAMPLE 7.4

A rotor blade row at pitchline radius is shown. The rotor total pressure loss coefficient at this radius is $\varpi_{rm} = 0.03$.

(a) How much deceleration is allowed in the rotor under de Haller criterion? i.e., What is the minimum W_2?

(b) What is the static pressure rise coefficient, assuming incompressible flow and $W_{2\min}$ from de Haller criterion?

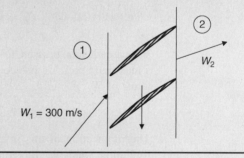

SOLUTION

de Haller criterion states that W_2/W_1 should be >0.72. Therefore, the minimum W_2 is $0.72\,W_1$ or

$$W_{2,\min} = 216\,\text{m/s}$$

The incompressible flow static pressure rise is given by Eq. 7.69

$$C_p = 1 - \frac{W_2^2}{W_1^2} - \varpi_{\text{cascade}} = 1 - (0.72)^2 - 0.03 \approx 0.4516$$

In a modern transonic fan or compressor stage, the relative Mach number to the rotor varies from hub-to-tip of $\sim$0.7 to $\sim$1.4, respectively. The variation of feasible static pressure ratio along the blade span with relative Mach number is shown in Fig. 7.30.

We observe that the static pressure ratio p_2/p_1 is nearly limited by $\sim$1.15 near the hub and may increase to $\sim$1.6 for a supersonic relative tip Mach number for a $C_{p,\max}$ of $\sim$0.45 near the

■ **FIGURE 7.30**
**Variation of static
pressure rise coefficient
C_p with inlet relative
Mach number (for
constant pressure ratio
p_2/p_1) with an arbitrarily
imposed $C_{p,\,max}$ profile**

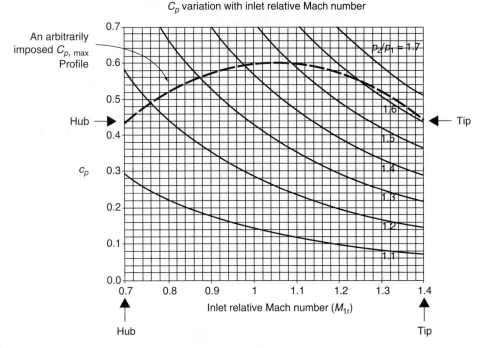

C_p variation with inlet relative Mach number

tip. The static pressure distribution downstream of a turbomachinery blade row strongly
depends on the swirl distribution and the axial velocity profile in the spanwise direction. The
most common approach in preliminary analysis of three-dimensional flows in turbomachinery
is the radial equilibrium theory that we shall discuss in this chapter.

It is useful to correlate the compressor stage efficiency, at a given radius, with the cascade
loss coefficients at the corresponding radius. Let us consider the cascade total pressure loss
coefficients of ϖ_r and ϖ_s to represent the relative total pressure loss in the rotor and the stator
blade sections, respectively, nondimensionalized with respect to the inlet dynamic pressure.

$$\varpi_r \equiv \frac{p_{t1r} - \bar{p}_{t2r}}{\rho_1 W_1^2/2} = \frac{2p_{t1r}}{\rho_1 W_1^2}\left(1 - \frac{\bar{p}_{t2r}}{p_{t1r}}\right) = \frac{2p_1\left(1 + \dfrac{\gamma-1}{2}M_{1r}^2\right)^{\frac{\gamma}{\gamma-1}}}{\rho_1 W_1^2}\left(1 - \frac{\bar{p}_{t2r}}{p_{t1r}}\right) \quad \textbf{(7.72)}$$

We may also express the ratio of static pressure to dynamic pressure in terms of the relative
Mach number to simplify the above expression as

$$\frac{\bar{p}_{t2r}}{p_{t1r}} = 1 - \varpi_r\left[\frac{\gamma M_{1r}^2/2}{\left(1 + \dfrac{\gamma-1}{2}M_{1r}^2\right)^{\frac{\gamma}{\gamma-1}}}\right] \quad \textbf{(7.73)}$$

The stagnation pressure ratio across the stator blade sections may be written in an analogous
manner to Eq. 7.73, as

$$\frac{\bar{p}_{t3}}{p_{t2}} = 1 - \varpi_s\left[\frac{\gamma M_2^2/2}{\left(1 + \dfrac{\gamma-1}{2}M_2^2\right)^{\frac{\gamma}{\gamma-1}}}\right] \quad \textbf{(7.74)}$$

The compressor stage adiabatic efficiency η_s is

$$\eta_s = \frac{\pi_s^{\frac{\gamma-1}{\gamma}} - 1}{\tau_s - 1} \tag{7.75}$$

We may use the following chain rule to express the stage total pressure ratio as

$$\pi_s \equiv \frac{p_{t3}}{p_{t1}} = \left(\frac{p_{t3}}{p_{t2}}\right)\left(\frac{p_{t2}}{p_{t2r}}\right)\left(\frac{p_{t2r}}{p_{t1r}}\right)\left(\frac{p_{t1r}}{p_{t1}}\right) \tag{7.76}$$

Equations 7.73 and 7.74 describe two of the parentheses in 7.76, and we may relate the remaining two expressions to flow Mach numbers in the absolute and relative frame, namely

$$\frac{p_{t1r}}{p_{t1}} = \left[\frac{1 + \frac{\gamma-1}{2}M_{1r}^2}{1 + \frac{\gamma-1}{2}M_1^2}\right]^{\frac{\gamma}{\gamma-1}} \tag{7.77}$$

$$\frac{p_{t2}}{p_{t2r}} = \left[\frac{1 + \frac{\gamma-1}{2}M_2^2}{1 + \frac{\gamma-1}{2}M_{2r}^2}\right]^{\frac{\gamma}{\gamma-1}} \tag{7.78}$$

From the inlet conditions, we can calculate M_1 and M_{1r} that are needed in Eq. 7.77. The absolute and relative Mach numbers in station 2, i.e., downstream of the rotor, may be calculated from the velocities C_2 and W_2 from the velocity triangles and the speed of sound a_2. The speed of sound may be related to fluid total enthalpy and kinetic energy according to

$$h_t = a^2/(\gamma - 1) + C^2/2 \tag{7.79}$$

Therefore,

$$a_2^2 = (\gamma - 1)\left[h_{t2} - \frac{C_2^2}{2}\right] = (\gamma - 1)\left\{[h_{t1} + \omega(r_2 C_{\theta2} - r_1 C_{\theta1})] - \frac{(C_{z2}^2 + C_{\theta2}^2)}{2}\right\} \tag{7.80}$$

Hence, the absolute Mach number downstream of the rotor, in terms of known quantities, is

$$M_2 = \sqrt{\frac{(C_{z2}^2 + C_{\theta2}^2)}{(\gamma - 1)[h_{t1} + \omega(r_2 C_{\theta2} - r_1 C_{\theta1}) - (C_{z2}^2 + C_{\theta2}^2)/2]}} \tag{7.81}$$

The relative Mach number M_{2r} is

$$M_{2r} = \sqrt{\frac{[C_{z2}^2 + (C_{\theta2} - \omega \cdot r_2)^2]}{(\gamma - 1)[h_{t1} + \omega(r_2 C_{\theta2} - r_1 C_{\theta1}) - (C_{z2}^2 + C_{\theta2}^2)/2]}} \tag{7.82}$$

Now, we have established all the parameters that enter the stage adiabatic efficiency Eq. 7.75 in terms of the basic velocity triangles and the cascade total pressure loss coefficients. Our analysis is valid for compressible flows in turbomachinery. However, by limiting the analysis to incompressible fluids, we may relate the stage efficiency to the stagnation pressure loss in the rotor and stator blade rows, via

$$\eta_s \approx 1 - \frac{\Delta p_{t\text{-Rotor}} + \Delta p_{t\text{-Stator}}}{\rho(h_{t2} - h_{t1})} \tag{7.83}$$

The derivation of Eq. 7.83 is straightforward and is shown in Hill and Peterson (1992). The cascade total pressure loss data directly feed into Eq. 7.83 for the stage efficiency estimation. In general, we have to supplement the cascade total pressure loss coefficient by shock losses in supersonic flow. In addition, we need to account for the effect of the shock-boundary layer interaction, which increases the blade profile losses. These effects belong to what is known as the *compressibility effects*. A host of other losses in a turbomachinery stage occur that are not usually simulated in a cascade experiment. For example, the end wall effects including the tip clearance flow, the streamwise vortices in the blades' wakes due to spanwise variation of blade circulation, as well as flow unsteadiness that result in vortex shedding in the wakes are not simulated in a typical cascade experiment. The cascade total pressure loss parameter ϖ describes the profile loss of a two-dimensional blade in a steady flow. We will use it as a foundation to construct a more elaborate blade loss model, only.

7.6.4   Aerodynamic Forces on Compressor Blades

The conservation principles that we learned and applied to wings and bodies in external aerodynamics apply to internal aerodynamics as well. There are some complicating factors in a turbomachinery flow that are absent in external aerodynamics. The most dominant feature is the presence of strong swirl in turbomachinery that is either completely absent or weak in external flows. Second, the net turning in the flow is large in turbomachinery and zero in external aerodynamics. We may characterize the flow in a turbomachinery as lacking a unique flow direction, unlike external aerodynamics. As an example, the Kutta–Joukowski theorem on lift predicts a magnitude for an ideal two-dimensional lift (on a body that creates circulation, Γ) as $\rho_\infty V_\infty \Gamma$ and a direction for the lift that is normal to V_∞. In turbomachinery blading, there are two distinct flow speeds, one upstream and the other downstream of the blade, i.e., W_1 and W_2 and not just a single V_∞. Also, on the question of the direction of this force, is it normal to W_1 or W_2? In turbomachinery, we will define a "mean" flow angle and a "mean" velocity that help us describe the blade lift and its direction. There are other complicating factors in internal flows that deal with flow distortion (e.g., upstream wake transport and interaction with downstream blades) and unsteadiness (i.e., from neighboring blade rows in relative rotation) that are inherent in a turbomachinery stage and have either no or weak counterparts in external aerodynamics. In a compressor, the flow is continually subjected to an adverse pressure gradient, as the static pressure rises along the flow direction. On the contrary, in external aerodynamics the flow is only *locally* subjected to adverse pressure gradient. Therefore, we may distinguish the two adverse pressure gradients experienced by the internal and external flows as one having a "global" and the second a "local" character, respectively. Another subtle, yet important, difference between external aerodynamics and internal flows with swirl (as in turbomachinery) deals with the stability of such flows to external disturbances. The distinguishing feature of having a mean swirl profile in a compressor leads to centrifugal instability waves that may grow and cause compressor stall. The closest external aerodynamic experience to a centrifugal instability comes from the vortex breakdown on a delta wing at high angle of attack. Kerrebrock (1977) has illuminated the behavior of instabilities in a swirling flow (in an annulus) that provides for a new understanding of disturbances in swirling flows in turbomachinery.

Figure 7.17 is reproduced here to help with the control surface definition and the application of conservation principles to the flow in a compressor blade row.

Continuity demands:

$$\rho_1 W_{z1} s_1 = \rho_2 W_{z2} s_2 \qquad (7.84)$$

Making an assumption of negligible radial shift in the stream surface from entrance to exit of the blade, we may conclude that $r_2 \approx r_1$ and therefore $s_2 \approx s_1$, hence,

$$\rho_1 W_{z1} \cong \rho_2 W_{z2} \qquad (7.85)$$

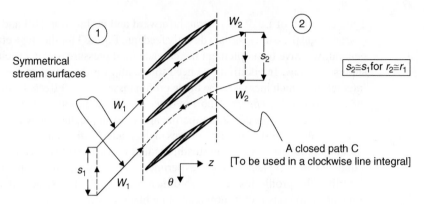

■ **FIGURE 7.17** Suitable control surface surrounding a blade along a stream surface for the application of conservation laws

The momentum equation in the axial (or z-) direction demands

$$\dot{m}(W_{z2} - W_{z1}) = (p_1 - p_2) \cdot s + F_z|_{\text{fluid}} \tag{7.86}$$

Assuming a constant axial throughflow speed, i.e., $W_{z1} = W_{z2}$, we get

$$F_z|_{\text{fluid}} = -F_z|_{\text{blade}} = (p_2 - p_1)s \tag{7.87}$$

Therefore we conclude that the axial force *acting on the blade* is in negative z-direction, i.e., it points in the upstream direction. The conservation of tangential momentum requires

$$\dot{m}(W_{\theta 1} - W_{\theta 2}) = F_\theta|_{\text{fluid}} = -F_\theta|_{\text{blade}} \tag{7.88}$$

Therefore we conclude that the tangential force on the blade is also in the negative θ-direction. These two blade forces are shown in Fig. 7.31.

We may resolve the axial and tangential blade forces into a lift and drag components but first we have to define an average flow direction through the blade row. We define a mean flow direction β_m based on the average swirl in the blade row, namely,

$$\tan \beta_m \equiv W_{\theta m}/W_z \tag{7.89}$$

where the average swirl, $W_{\theta m}$ is defined as

$$W_{\theta m} \equiv (W_{\theta 1} + W_{\theta 2})/2 \tag{7.90}$$

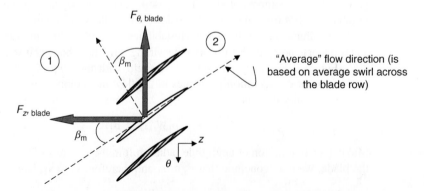

■ **FIGURE 7.31** The axial and tangential blade forces acting on a compressor blade

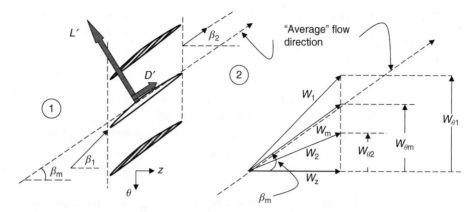

■ **FIGURE 7.32**   Blade lift and drag forces based on a definition of average flow direction β_m

The lift is the sum of the projections of the axial and tangential forces in a direction normal to the "average" flow direction defined in Fig. 7.31, i.e.,

$$L' = F_{\theta,\text{blade}} \cos \beta_m + F_{z,\text{blade}} \sin \beta_m \tag{7.91}$$

Also the drag is the sum of the projections of the two forces in the direction of "average" flow, namely,

$$D' = F_{\theta,\text{blade}} \sin \beta_m - F_{z,\text{blade}} \cos \beta_m \tag{7.92}$$

The lift and drag forces acting on a compressor blade are shown in Fig. 7.32.

Therefore the lift force (per unit span) is

$$L' = \rho \cdot W_z \cdot s(W_{\theta 1} - W_{\theta 2}) \cos \beta_m + (p_2 - p_1)s \cdot \sin \beta_m \tag{7.93}$$

Assuming an incompressible fluid and applying the Bernoulli equation to the static pressure rise term in Eq. 7.93, we get

$$L' = \rho \cdot W_z \cdot s(W_{\theta 1} - W_{\theta 2})\left(\frac{W_z}{W_m}\right) + (p_{t2} - \rho \cdot W_2^2/2 - p_{t1} + \rho \cdot W_1^2/2) \cdot s \cdot \left(\frac{W_{\theta m}}{W_m}\right) \tag{7.94-a}$$

This equation simplifies to

$$L' = \rho \cdot s(W_{\theta 1} - W_{\theta 2})\left(\frac{W_z^2}{W_m}\right) + \frac{\rho \cdot s}{2}\left(\frac{W_{\theta m}}{W_m}\right)(W_z^2 + W_{\theta 1}^2 - W_z^2 - W_{\theta 2}^2)$$
$$- \Delta p_t \cdot s \cdot \left(\frac{W_{\theta m}}{W_m}\right) \tag{7.94-b}$$

$$L' = \rho \cdot s(W_{\theta 1} - W_{\theta 2})\left(\frac{W_z^2}{W_m}\right) + \rho \cdot s\left(\frac{W_{\theta m}}{W_m}\right)\left(\frac{W_{\theta 1} + W_{\theta 2}}{2}\right)(W_{\theta 1} - W_{\theta 2})$$
$$- s \cdot \Delta p_t\left(\frac{W_{\theta m}}{W_m}\right) \tag{7.94-c}$$

Finally, the blade sectional lift force is expressed as

$$L' = \rho \cdot W_m \cdot s(W_{\theta 1} - W_{\theta 2}) - s \cdot \Delta p_t\left(\frac{W_{\theta m}}{W_m}\right) \tag{7.94-d}$$

From Eq. 7.58, we may replace the product of blade spacing and the change in swirl velocity by the blade circulation Γ to get the familiar Kutta–Joukowski theorem counterpart in turbomachinery flow, namely,

$$L' = \rho \cdot W_m \cdot \Gamma - s \cdot \Delta p_t (W_{\theta m} / W_m) \tag{7.95}$$

The first term is the familiar Kutta–Joukowski lift, and the second term is the effect of boundary layer formation and viscous losses on lift.

From Eq. 7.92, we may write the blade drag force per unit span as

$$D' = \rho \cdot W_z \cdot s (W_{\theta 1} - W_{\theta 2}) \left(\frac{W_{\theta m}}{W_m} \right) - \rho \cdot s \cdot W_{\theta m} (W_{\theta 1} - W_{\theta 2}) \left(\frac{W_z}{W_m} \right) + s \cdot \Delta p_t \cdot \left(\frac{W_z}{W_m} \right) \tag{7.96-a}$$

Upon cancellation of the first two terms, we get a simple expression, entirely based on profile losses, for the blade drag force as

$$D' = s \cdot \Delta p_t (W_z / W_m) \tag{7.96-b}$$

The lift and drag coefficients are

$$C_l \equiv \frac{L'}{\left(\dfrac{\rho \cdot W_m^2}{2} \right) \cdot c} = \frac{2\Gamma}{W_m \cdot c} - \frac{\varpi}{\sigma} \cdot \sin \beta_m \tag{7.97}$$

where the cascade total pressure loss is nondimensionalized based on the inlet relative dynamic pressure and the σ is the blade solidity.

$$C_d \equiv \frac{D'}{\left(\dfrac{\rho \cdot W_m^2}{2} \right) \cdot c} = \frac{\varpi}{\sigma} \cdot \cos \beta_m \tag{7.98}$$

As expected, the blade two-dimensional drag coefficient in the limit of low-speed flow is related to the total pressure loss in the cascade, i.e., the momentum deficit thickness in the blade wake. Furthermore, the drag coefficient is *linearly* proportional to the cascade total pressure loss measured in the wake, as expected.

We note that the wake-related loss in lift, in Eq. 7.97, is proportional to the blade drag coefficient, Eq. 7.98, therefore, the relation between the two-dimensional lift and drag coefficients may be written as

$$C_l = \frac{2\Gamma}{W_m \cdot c} - C_d \cdot \tan \beta_m \tag{7.99}$$

The first term in Eq. 7.98 is the ideal lift coefficient, following the thin airfoil theory, and the second term accounts for the loss of lift due to boundary layer/wake formation. These expressions for the lift and drag coefficients, however, lack compressibility effects, three-dimensional or spanwise effects, and the end wall effects as in a real compressor. In addition, we remember that the actual flow is unsteady and involves wake transport and vortex shedding, which are not modeled here. However, we may take additional steps in correcting for some of the cited shortcomings. From Prandtl's classical wing theory, we remember that the drag due to lift, or expressed as the induced drag coefficient C_{Di}, was related to the wing lift coefficient following

$$C_{Di} = \frac{C_L^2}{\pi \cdot e \cdot AR} \tag{7.100}$$

■ **FIGURE 7.33**
Solid wall near a vortex is
modeled by an image
vortex, which helps
satisfy the flow tangency
condition on the wall

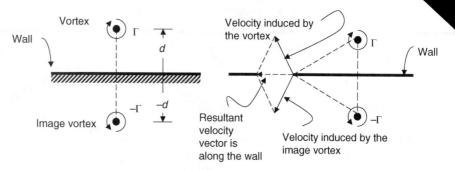

(a) a vortex and its image near a wall   (b) the induced flow at the wall

where the span efficiency factor e in the denominator represents the nonelliptic lift contribution/penalty and the term AR represents the wing aspect ratio. A distinguishing feature of the turbomachinery wake is its spiral shape rather than the flat wake of the Prandtl's lifting line theory, where Eq. 7.100 is derived. Second, the presence of solid walls acts as mirrors for the vortices (i.e., they create images), and the total induced flow has to account for the image vortices as well. A simple diagram of a vortex next to a wall and its image that renders the wall a stream surface (i.e., flow tangency condition on the wall is satisfied) is shown in Fig. 7.33. The vortex strength is shown as Γ and the image vortex strength is, thus $(-\Gamma)$ and is equally disposed on the opposite side of the wall. If we calculate the induced velocity at the wall due to these two vortices, the normal component to the wall vanishes, due to symmetry and the tangential component to the wall is doubled, again due to symmetry (see Fig. 7.33-b).

The presence of the *image vortex* with a *counter swirl* is to reduce the effective induced velocity at the blades, as compared with a vortex filament in an unbounded space. Therefore, the simple expression for the drag polar for a wing, which is the sum of the profile drag and the induced drag

$$C_D = C_{d\text{-}2D} + \frac{C_L^2}{\pi \cdot e \cdot AR} \tag{7.101}$$

should be replaced by a more complex structure in a turbomachinery blading such as

$$C_D = \underbrace{C_{d,\text{cascade}}}_{2D} + \underbrace{\kappa \cdot \left(\frac{C_L^2}{AR}\right) + C_{D,\text{End-Wall}} + C_{D,\text{shock}} + C_{D,\text{turbulent-mixing}} + C_{D,\text{rms,Vortex-shedding}}}_{3D}$$

$$\tag{7.102}$$

In Eq. 7.102, κ represents the influence factor on the induced drag due to the presence of solid walls (i.e., the annulus) and helical wakes, the end wall drag coefficient is due to viscous losses and the vortex formation (tip clearance and scraping vortex) near the wall, the drag due to shock is accounted for through $C_{D,\text{shock}}$. Turbulent mixing and the unsteady drag due to vortex shedding (the root-mean square) are reflected in the last term of Eq. 7.102. Despite its rather complete look, Eq. 7.102 is just an *empirical model*, which requires an extensive experimental database in order to be useful as a predictive tool. However, breaking down the blade loss to its basic constituent levels helps our understanding of the complex flow phenomena in turbomachinery. We will address three-dimensional losses, compressibility effects, and unsteadiness further in this chapter.

Our discussion of compressor aerodynamics has been limited to 2D flows. The behavior trend of thermodynamic and flow variables in a compressor stage with an IGV in two dimensions are reviewed in Fig. 7.34. For additional reading on cascade aerodynamics, Gostelow (1984) should be consulted.

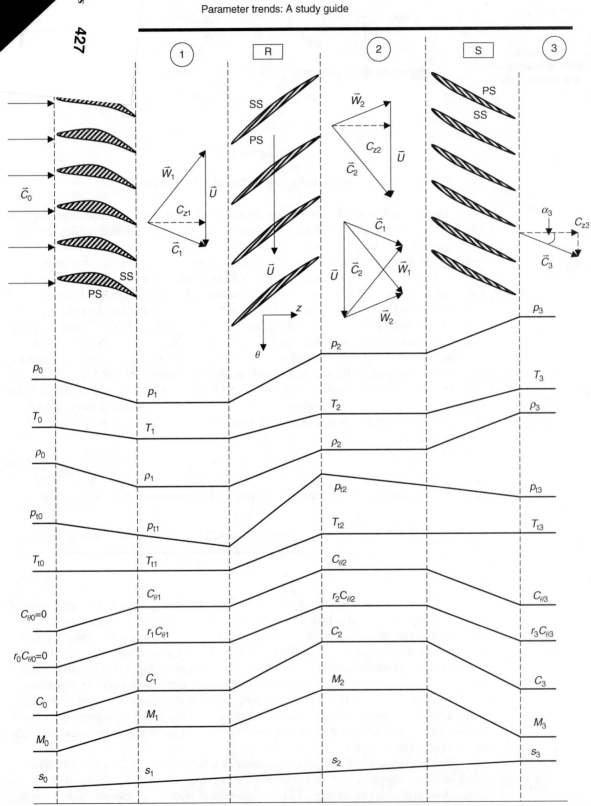

EXAMPLE 7.5

A compressor stage develops a pressure ratio of $\pi_s = 1.5$. Its stage adiabatic efficiency is $\eta_s = 0.90$. Calculate the stage total temperature ratio τ_s and compressor polytropic efficiency e_c. Assume $\gamma = 1.4$.

SOLUTION

Stage pressure and temperature ratio are related to stage adiabatic efficiency via

$$\eta_s = \frac{\pi_s^{\frac{\gamma-1}{\gamma}} - 1}{\tau_s - 1}$$

Therefore,

$$\tau_s = 1 + \frac{1}{\eta_s}\left(\pi_s^{\frac{\gamma-1}{\gamma}} - 1\right) = 1 + \left[1.5^{0.2857} - 1\right]/0.90 \approx 1.1365$$

Since $\pi_s = \tau_s^{\frac{\gamma \cdot e_c}{\gamma-1}}$, $e_c = \frac{\gamma - 1}{\gamma}\frac{\ln \pi_s}{\ln \tau_s} = 0.2857 \frac{\ln(1.5)}{\ln(1.1365)} \approx 0.905$

EXAMPLE 7.6

The profile loss parameter for a compressor blade rotor section is $\varpi = 0.05$. The rotor blade chord and spacing at this section are $c = 5.25$ cm and $s = 3.5$ cm, respectively. The relative velocity vectors up- and downstream of the rotor are shown. Calculate

(a) mean relative flow angle β_m

(b) the rotor sectional (2D) drag coefficient C_d and

(c) the rotor circulation Γ

(d) the rotor sectional (2D) lift coefficient

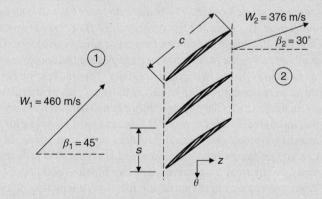

SOLUTION

The inlet swirl is $W_{\theta 1} = W_1 \cdot \sin(45°) = 325.3$ m/s
The exit swirl is $W_{\theta 2} = W_2 \cdot \sin(30°) = 188$ m/s
Therefore the average swirl across the rotor is $W_{\theta m} = (W_{\theta 1} + W_{\theta 2})/2 = 256.6$ m/s

Based on the definition of the mean flow angle β_m we have

$$\beta_m \equiv \tan^{-1}\left(\frac{W_{\theta m}}{W_z}\right)$$

The axial velocity upstream of the rotor blade is $W_{z1} = W_1 \cdot \cos(45°) = 325.4$ m/s

The axial velocity downstream of the rotor is $W_{z2} = W_2 \cdot \cos(30°) = 325.6$ m/s

Therefore, $\beta_m \approx \tan^{-1}(256.6/325.5) \approx 38.2°$

The rotor blade solidity at this radius is $\sigma = c/s = 5.25/3.5 = 1.5$

The drag coefficient is related to profile loss, solidity, and mean relative flow angle according to

$$C_d = \frac{\varpi}{\sigma} \cdot \cos \beta_m = \frac{0.05}{1.5} \cos(38.2°) \approx 0.026$$

The circulation is

$$\Gamma = s|W_{\theta 1} - W_{\theta 2}| = (3.5 \times 10^{-2}\text{ m})(325.3 - 188)\text{m/s}$$
$$\approx 4.8 \text{ m}^2/\text{s}$$

To calculate the sectional lift coefficient, we need to calculate $W_m = (W_z^2 + W_{\theta m}^2)^{0.5}$ $W_m \approx 414.5$ m/s

$$C_l = \frac{2\Gamma}{W_m \cdot c} - C_d \cdot \tan \beta_m \approx 0.4416 - 0.0206 \approx 0.4209$$

7.6.5   Three-Dimensional Flow

Our discussion of the flowfield so far has been limited to two-dimensional flows represented by the "pitchline" radius. The next level of complexity brings us to the three-dimensional flow analysis with a host of simplifying assumptions. These assumptions are necessary to render the analysis tenable. To relieve any of the assumptions requires numerical integration of the governing equations, which are coupled and nonlinear. The method of approach, in that case, is the computational fluid dynamics applied to a single or multiple stages in a compressor. Simplifying assumptions for our three-dimensional flow analysis are

1. Flow is steady, i.e., $\partial/\partial t \to 0$
2. Flow is axisymmetric, i.e., $\partial/\partial \theta \to 0$
3. Flow is adiabatic, i.e., $q_{\text{wall}} = 0$
4. Fluid is inviscid and nonheat conducting, i.e., $\mu = 0$ and $\kappa = 0$, respectively
5. Blade geometry has zero tip clearance, i.e., blades stretch between hub and casing

In addition, we may choose to limit the radial shift of the stream surfaces to within the blade rows and hence require *radial equilibrium* both upstream and downstream of the blade rows. By limiting the radial shift of the stream surfaces to within the blade rows, we are thus creating cylindrical stream surfaces outside the blade rows, i.e., $C_r = 0$ both upstream and downstream of the blade row. By assuming an axisymmetric flow, we are smearing out the blade-to-blade flow and thus lose blade periodicity for a finite number of blades. This is akin to having infinitely many blades. For the steady flow, we are placing the coordinate system with the blade row that operates in isolation in a cylindrical duct. The implication is that having a neighboring blade row in relative rotation with respect to the blade row of interest would render the flowfield unsteady. In the limit of adiabatic flow, the only mechanism for energy transfer to/ from the fluid is the mechanical shaft power. The assumption of inviscid and nonheat conducting fluid eliminates viscous and thermal boundary layer formations. In essence, the only mechanism for energy exchange and momentum transfer is the fluid pressure and not the fluid molecular viscosity and thermal conductivity. The combination of adiabatic and inviscid flow creates a fictitious isentropic environment for the flow interaction with the blades. Figure 7.35 shows cylindrical stream surfaces entering and leaving an isolated blade row in a cylindrical duct.

All parameters in this problem become a function of r only. Note that by taking a cylindrical duct downstream of the blade row, we are in essence decoupling the flow in the vicinity of the blade row from axial pressure gradients. As noted earlier, the axisymmetric flow takes out the θ-dependency, therefore the only independent parameter left is the radial position r. The velocity field in station 2, i.e., downstream of the rotor, has zero radial velocity and a

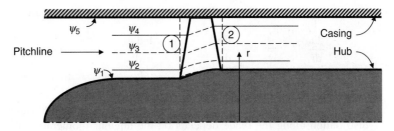

■ **FIGURE 7.35** Isolated blade row with cylindrical stream surfaces ψ_1 to ψ_5

general swirl distribution $C_\theta(r)$, as well as an axial velocity distribution $C_z(r)$ that needs to be determined from the conservation laws. It is customary to specify a desired swirl distribution downstream of a blade row and calculate the corresponding axial velocity distribution that is supported by the swirl profile. This method of approach is called the vortex design of turbomachinery blades. The pressure distribution $p(r)$ is established by imposing a radial equilibrium condition on the stream surfaces, which simplifies to a balance of pressure gradient and the centrifugal force on a swirling fluid stream, i.e.,

$$\frac{\partial p}{\partial r} = \frac{dp}{dr} = \rho \frac{C_\theta^2}{r} \tag{7.103}$$

From a known swirl distribution $C_\theta(r)$, we get the spanwise blade work distribution following the Euler turbine equation, namely,

$$w_c = h_{t2} - h_{t1} = \omega(r_2 C_{\theta 2} - r_1 C_{\theta 1}) \tag{7.104}$$

However, the only unknown in the Eq. 7.104 is the downstream radius r_2 corresponding to an upstream stream surface radius r_1. This is the crux of the problem of radial shift within the blade row, which needs to be determined. To establish the radial shift of the stream surfaces, Δr, we need to satisfy the continuity equation between the stream surfaces, from hub to tip or in our case from ψ_1 to ψ_5 upstream and downstream of the blade row. For swirl profiles that produce a constant axial velocity downstream of a blade row, the problem becomes very simple. In other cases, we need to integrate a nonuniform axial velocity and density profile in the spanwise direction to calculate the mass flow rates between the stream surfaces and thus establish the radial shift Δr.

7.6.5.1 *Blade Vortex Design.* The swirl velocity downstream of a blade at a given radius, namely pitchline, may be defined via a two-dimensional approach, as outlined in the previous sections. Anchoring the swirl profile at the pitchline, we may describe a variety of swirl profiles in the spanwise direction that could be analyzed and assessed for their practicality and utility. Some common swirl profiles assume a combination of a potential vortex, a solid body rotation, or, in general, a power profile for the swirl distribution, $C_\theta(r)$. The stagnation enthalpy in the absolute frame of reference may be written as

$$h_t \equiv h + \frac{C_z^2}{2} + \frac{C_\theta^2}{2} \tag{7.105}$$

The radial derivative of the above equation yields,

$$\frac{dh_t}{dr} = \frac{dh}{dr} + C_z \frac{dC_z}{dr} + C_\theta \frac{dC_\theta}{dr} \tag{7.106}$$

Also from Gibbs equation, we relate the enthalpy gradient to pressure and entropy gradients as

$$T \frac{ds}{dr} = \frac{dh}{dr} - \frac{1}{\rho} \frac{dp}{dr} \tag{7.107}$$

Now, let us substitute Eq. 7.107 into Eq. 7.106, to get an interim result, i.e.,

$$\underbrace{\frac{dh_t}{dr}}_{\text{blade-work-profile}} = \underbrace{T\frac{ds}{dr}}_{\text{blade-loss-profile}} + \underbrace{\frac{1}{\rho}\frac{dp}{dr}}_{\text{static-pressure-profile}} + \underbrace{C_z\frac{dC_z}{dr}}_{\text{axial-velocity-profile}} + \underbrace{C_\theta\frac{dC_\theta}{dr}}_{\text{swirl-profile}} \quad (7.108)$$

A constant-work blade will have the first term in Eq. 7.108 vanish, otherwise the work distribution along the blade depends on the angular momentum distribution upstream and downstream of the blade according to Euler turbine equation. The entropy gradient term on the RHS of Eq. 7.108 is the blade loss profile in the radial direction. This parameter has to be an input to the problem based on the accumulated loss data published in the literature or the proprietary data of the industry. It is common to ignore the radial loss profiles for the first attempt in establishing the density, velocity gradients downstream of a blade row. The static pressure gradient follows the radial equilibrium theory, namely, Eq. 7.103. The axial velocity profile is the third term on the RHS of Eq. 7.108, which is the goal of our analysis. The last term is the swirl profile that we consider an assumed function of radial position. Therefore, to establish an axial velocity profile $C_z(r)$, we need to integrate

$$\frac{1}{2}\frac{dC_z^2}{dr} = \frac{dh_t}{dr} - T\frac{ds}{dr} - \frac{C_\theta^2}{r} - C_\theta\frac{dC_\theta}{dr} \quad (7.109)$$

The last two terms in Eq. 7.109 may be combined into a single derivative involving angular momentum to get

$$\frac{1}{2}\frac{d}{dr}(C_z^2) = \frac{dh_t}{dr} - T\frac{ds}{dr} - \frac{C_\theta}{r}\frac{d(rC_\theta)}{dr} \quad (7.110)$$

The special cases of swirl profile that are considered in compressor blade design are described in the next section.

Case 1: Free-vortex design

$$rC_\theta = \text{constant} \quad (7.111)$$

This is known as the *potential vortex* or *free-vortex* design. The name association "potential vortex" is due to the similarity between this swirl profile and that of a vortex filament ($C_\theta \sim 1/r$). We also remember that the flowfield about a vortex filament is irrotational, hence a *potential* vortex. The irrotational flow around a vortex filament has zero vorticity, hence the name *free-vortex* design.

Let us consider two scenarios here. First, consider the case of zero preswirl upstream of the rotor, i.e., the case of no inlet guide vane, and the second scenario with an inlet guide vane that induces a solid-body rotation swirl profile of its own upstream of the rotor.

With no IGV, the flow upstream of the rotor is swirl free, i.e.,

$$C_{\theta 1} = 0 \quad (7.112)$$

The rotor inducing a free vortex swirl distribution in plane 2, gives

$$(rC_\theta)_2 = \text{constant} \quad (7.113)$$

The Euler turbine equation demands

$$h_{t2} - h_{t1} = \omega[(rC_\theta)_2 - (rC_\theta)_1] = \omega(rC_\theta)_2 = \text{constant} \quad (7.114)$$

The swirl profile associated with a free-vortex design blade produces a constant work along the blade span. This is known as a *constant-work* rotor. Therefore, the blade work profile and the angular momentum profile terms in Eq. 7.110 vanish, i.e.,

$$\frac{dh_t}{dr} = \frac{d(rC_\theta)}{dr} = 0 \qquad (7.115)$$

The axial velocity profile equation reduces to

$$\frac{d}{dr}(C_z^2/2) = -T\frac{ds}{dr} \qquad (7.116)$$

Here, we have a choice and that is to either specify a radial loss profile from the available data in the literature or neglect the loss term in this preliminary phase of establishing an axial velocity profile. Here, we take the second route, although we shall present and discuss radial distribution of losses associated with modern fan blades later in the chapter.

In the limit of a loss-free rotor and the free-vortex design, we conclude that the axial velocity C_z remains constant, i.e., Eq. 7.116 yields,

$$C_{z2} = \text{constant} \qquad (7.117)$$

Figure 7.36 shows the velocity pattern across the rotor.

In Fig. 7.36, we have kept the casing radius constant and have increased the hub radius to account for the density rise across the blade row. There are other choices that we could take, for example, the casing could have been tapered and the hub remained cylindrical or both hub and casing could have been tapered. Since the calculation method is identical, we shall stay with our choice of constant tip radius and an increasing hub radius. It is also customary to design turbomachinery blades based on a constant throughflow speed, i.e., axial velocity remains constant. Therefore based on continuity equation, the channel area develops inversely proportional to density ratio, assuming a uniform density profile upstream and downstream of the rotor, i.e.,

$$A_2/A_1 = \rho_1/\rho_2 \qquad (7.118\text{-a})$$

Therefore, the ratio of r_{h2} to r_t is related to the density ratio and the inlet hub-to-tip radius ratio,

$$\frac{r_{h2}}{r_t} = \sqrt{1 - \frac{\rho_1}{\rho_2}\left[1 - \left(\frac{r_{h1}}{r_t}\right)^2\right]} \qquad (7.118\text{-b})$$

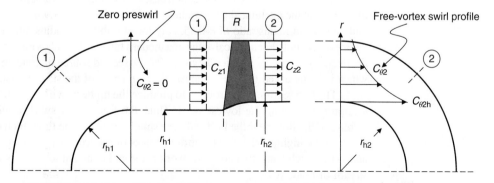

■ **FIGURE 7.36** Velocity field across a rotor blade row with zero preswirl (i.e., no IGV) and free-vortex design for the rotor blade

We tackle the density ratio problem after we examine other swirl profiles of interest. The swirl put in by the rotor needs to be taken out by the stator, therefore, we demand

$$C_{\theta 3} = 0 \tag{7.119}$$

which may be expressed in the shorthand form

$$C_\theta = \pm a/r \tag{7.120}$$

where a is a constant and the plus sign describes the rotor swirl input to the flow and the minus sign designates the stator withdrawal of the swirl put in by the rotor. The exit flow in station 3 downstream of the stator is thus swirl free just as in the flow entering the stage.

The degree of reaction for this design is rewritten from Eq. 7.46-c, with a free vortex swirl distribution substituted for rotor exit swirl as

$$°R = 1 - \frac{C_{\theta 1} + C_{\theta 2}}{2\omega \cdot r} = 1 - \frac{0 + a/r}{2\omega \cdot r} = 1 - \frac{a/\omega}{2r^2} \tag{7.121}$$

To cast this equation relative to the pitchline radius, we divide and multiply the second term in Eq. 7.121 by r_m^2 to get

$$°R(r) = 1 - \frac{(a/r_\mathrm{m})/(\omega \cdot r_\mathrm{m})}{2(r/r_\mathrm{m})^2} \tag{7.122}$$

The term a/r_m is the rotor exit swirl velocity at the pitchline radius and the second term, i.e., ωr_m, is the wheel speed at the pitchline radius. From Eq. 7.122, we conclude that the product of these two terms is related to the degree of reaction at the pitchline via

$$\left(\frac{a}{r_\mathrm{m}}\right)/(\omega \cdot r_\mathrm{m}) = 2(1 - °R_\mathrm{m}) \tag{7.123}$$

Therefore the degree of reaction for a compressor stage with no inlet guide vane and a free vortex swirl distribution produced by its rotor is

$$°R(r) = 1 - \frac{1 - °R_\mathrm{m}}{(r/r_\mathrm{m})^2} \tag{7.124}$$

We have graphed this equation for three stage designs with values of pitchline degree of reaction specified at $0.5, 0.6,$ or 0.7. We immediately note that the rotor hub is in danger of very low degrees of reaction. Since a negative degree of reaction implies that the flow in that section is not compressing (i.e., that section behaves like a turbine!), we have limited our graphical presentation in Fig. 7.37 to positive degree of reaction cases at the hub. For example, in the case of 50% degree of reaction at the pitchline the hub radius should be at $\sim 0.71\, r_\mathrm{m}$, which calls for a minimum rotor with hub-to-tip radius ratio of ~ 0.55. Any lower hub-to-tip radius ratio leads to a negative hub degree of reaction. For a typical value of a modern fan hub-tip radius ratio of ~ 0.5, we need to raise the pitchline degree or reaction beyond 50%. A second source of concern is the rotor tip region, which is being asked to bear an unusually high share of the stage pressure rise.

The hub section of the rotor is to produce the highest swirl according to Fig. 7.36. The tip section is imparting the lowest swirl. Thus, the free vortex swirl profile demands excessive turning of the flow near the hub and very small turning at the tip, which causes the free-vortex blades to be highly twisted. The attractiveness of the free-vortex design is in its simplicity of analysis, which leads to a constant-work rotor and constant axial velocity distribution downstream of blade rows.

Now, let us study the case of a compressor stage *with an IGV* that induces a preswirl of solid body rotation upstream of the rotor. The purpose of an inlet guide vane is to reduce the

■ **FIGURE 7.37**
Spanwise variation of the stage degree of reaction for a free-vortex rotor design and no inlet guide vane

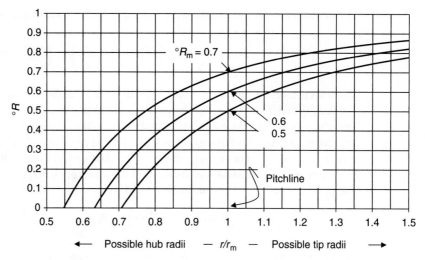

relative Mach number at the tip, hence, a free-vortex design that attains the lowest swirl velocity at the tip would make less sense than a solid body rotation type swirl distribution downstream of the IGV.

Therefore,

$$C_{\theta 1} = b \cdot r \qquad \text{(swirl profile)} \qquad (7.125)$$

where b is a constant. Combining this swirl profile with a free-vortex distribution of the rotor, we get

$$C_{\theta 2} = \frac{a}{r} + b \cdot r \qquad (7.126)$$

The rotor work distribution is

$$h_{t2} - h_{t1} = \omega[(rC_\theta)_2 - (rC_\theta)_1] \qquad (7.127\text{-a})$$

Substitute swirl profiles 7.125 and 7.126 in the rotor work input (Eq. 7.127-a) to get

$$h_{t2} - h_{t1} = \omega\left[a + br^2 - br^2\right] = \omega \cdot a = \text{constant} \qquad (7.127\text{-b})$$

We conclude that the rotor work distribution along the rotor span is constant.

Now, let us substitute the swirl profile downstream of the rotor and the constant work distribution in Eq. 7.110 to get an equation involving the axial velocity distribution,

$$\frac{1}{2}\frac{\mathrm{d}}{\mathrm{d}r}(C_z^2) = \frac{\mathrm{d}h_t}{\mathrm{d}r} - T\frac{\mathrm{d}s}{\mathrm{d}r} - \frac{C_\theta}{r}\frac{\mathrm{d}(rC_\theta)}{\mathrm{d}r} = 0 - 0 - \left(\frac{a}{r^2} + b\right)\frac{\mathrm{d}}{\mathrm{d}r}(a + br^2) \qquad (7.128)$$

This equation simplifies to

$$\frac{\mathrm{d}}{\mathrm{d}r}(C_z^2/2) = -\frac{2ab}{r} - 2b^2 r \qquad (7.129)$$

We may integrate this equation from a reference radius, e.g., the pitchline radius r_{m}, where the axial velocity is known as $C_{z\mathrm{m}}$, to any radius r to get

$$C_z^2 - C_{z\mathrm{m}}^2 = 2\left[-2ab\ln\frac{r}{r_{\mathrm{m}}} + b^2 r_{\mathrm{m}}^2\left(1 - \frac{r^2}{r_{\mathrm{m}}^2}\right)\right] \qquad (7.130)$$

or in nondimensional form the axial velocity profile downstream of the rotor is

$$\frac{C_z(r)}{C_{zm}} = \sqrt{1 + 2\left(\frac{br_m}{C_{zm}}\right)^2\left(1 - \frac{r^2}{r_m^2}\right) - \left(\frac{4ab}{C_{zm}^2}\right)\ln\left(\frac{r}{r_m}\right)} \qquad (7.131)$$

In order to calculate the downstream hub radius after the rotor, we need to integrate the product of density and axial velocity over the inlet and exit planes of the rotor, which are set equal via continuity equation. We need to ask whether the axial velocity downstream of the IGV, which is the rotor upstream, is uniform. We may use the same equation of radial equilibrium theory that we applied to the rotor and note that the IGV does no work on the fluid, hence the fluid total enthalpy remains constant. Consistent with our assumption of no radial loss in the preliminary stage of our blade row calculations, Eq. 7.110 simplifies to

$$\frac{d}{dr}(C_z^2/2) = -\frac{C_\theta}{r}\frac{d(rC_\theta)}{dr} = -b\frac{d}{dr}(br^2) = -\frac{d}{dr}(b^2r^2) \qquad (7.132)$$

Upon integration, we get a parabolic axial velocity distribution upstream of the rotor, i.e.,

$$\frac{C_z(r)}{C_{zm}} = \sqrt{1 + 2\left(\frac{br_m}{C_{zm}}\right)^2\left(1 - \frac{r^2}{r_m^2}\right)} \qquad \text{(axial velocity profile)} \qquad (7.133)$$

Let us compare the two axial velocity profiles upstream and downstream of the rotor by examining Eqs. 7.131 and 7.133. The axial velocity profile upstream of the rotor is parabolic, as noted and the profile downstream of the rotor has an extra logarithmic term, which tends to exacerbate the nonuniformity in the axial profile. This implies that the axial velocity at the tip is further reduced downstream of the rotor as compared with the upstream value. In contrast, the axial velocity near the hub shows an increase downstream of the rotor, hence an increased nonuniformity of the axial velocity profile. The swirl and axial velocity profiles across a compressor rotor that induces a free-vortex swirl distribution to an IGV flowfield with a solid-body rotation swirl profile are shown in Fig. 7.38. The choice of solid-body rotation for the inlet guide vane is consistent with the desire to reduce the rotor tip Mach number; however, combining it with a free-vortex rotor leads to a highly nonuniform axial velocity profile, which is undesirable. What happens to the spanwise distribution of the degree of reaction in this case?

To answer this question, we recall that the swirl profiles across the rotor are

$$C_{\theta 1}(r) = b \cdot r$$

and

$$C_{\theta 2}(r) = b \cdot r + \frac{a}{r}$$

■ **FIGURE 7.38**
A schematic drawing of axial and swirl profiles across a compressor rotor with an IGV that induces a solid-body rotation and the rotor is of free-vortex design

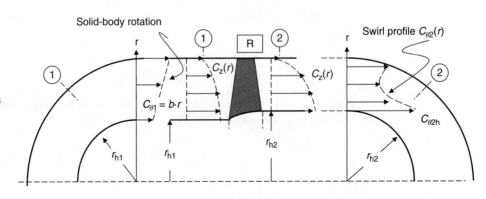

The equation for the degree of reaction is

$$\degree R = 1 - \frac{C_{\theta 1} + C_{\theta 2}}{2\omega \cdot r} = 1 - \frac{1}{2}\left[(2b/\omega) + \left(\frac{a/\omega}{r^2}\right)\right] \tag{7.134}$$

We may cast this equation in terms of the pitchline radius, similar to Eq. 7.124 as

$$\degree R = \left(1 - \frac{b}{\omega}\right) - \frac{(a/r_{\rm m})/(\omega \cdot r_{\rm m})}{2(r/r_{\rm m})^2} \tag{7.135}$$

In terms of the degree if reaction at the pitchline, $\degree R_{\rm m}$, we may write Eq. 7.135 as

$$\degree R = (1 - b/\omega) - \frac{(1 - b/\omega) - \degree R_{\rm m}}{(r/r_{\rm m})^2} \qquad (\degree R \text{ profile}) \tag{7.136}$$

EXAMPLE 7.7

Compare the degree-of-reaction profile of a compressor stage with and without an IGV for a range of hub-tip radii that result in a positive degree of reaction.

SOLUTION

We make a spreadsheet based on Eqs. 7.136 and 7.124 for a 50% degree of reaction at the pitchline radius, with varying IGV solid-body rotation swirl profile b/ω. The graph of the spreadsheet table is shown in Fig. 7.39. We note that a positive solid-body rotation (i.e., $b > 0$) causes a change in the degree of reaction in the positive direction. The case of $b = 0$ recovers the no IGV stage of Fig. 7.37.

We also note that for a special case of

$$1 - \frac{b}{\omega} = \degree R_{\rm m} \tag{7.137}$$

the reaction along the blade span becomes constant. This view is shown for the case of $b/\omega = 0.5$ in Fig. 7.39, which is plotted for a 50% degree of reaction at the pitchline.

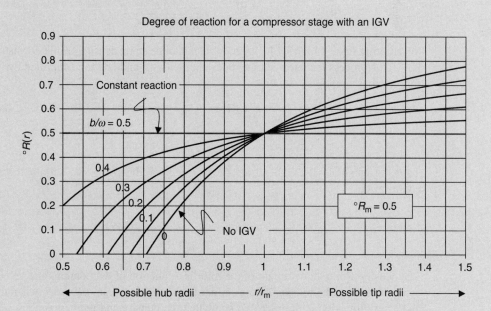

Degree of reaction for a compressor stage with an IGV

■ **FIGURE 7.39** IGV with solid-body rotation helps a free-vortex rotor in $\degree R(r)$

To get the density profile downstream of an IGV, we differentiate the perfect gas law in the radial direction and substitute the centrifugal force for the radial pressure gradient to get

$$\frac{dp}{dr} = R\frac{d}{dr}(\rho \cdot T) = \rho\frac{C_\theta^2}{r} \tag{7.138}$$

This equation may be solved for the density variation as

$$RTd\rho + R\rho \cdot dT = \rho\frac{C_\theta^2}{r}dr \tag{7.139}$$

We may isolate $d\rho/\rho$ in terms of the known functions of r, namely

$$\frac{d\rho}{\rho} = \left(\frac{1}{RT}\right)\frac{C_\theta^2}{r}dr - \frac{dT}{T} \tag{7.140}$$

Now, we need to use the conservation of total enthalpy to write an expression for the static temperature distribution downstream of the IGV, i.e.,

$$T(r) = T_{t0} - \frac{C_z^2 + C_\theta^2}{2c_p} \tag{7.141}$$

The RHS of Eq. 7.141 is a known function of the radial coordinate r via Eqs. 7.126 and 7.133. The differential of the static temperature may be written as

$$c_p dT = -d(C_z^2/2) - d(C_\theta^2/2) = 2b^2 rdr - b^2 rdr = b^2 rdr \tag{7.142}$$

The logarithmic derivative of the static temperature is the combination of Eqs. 7.141 and 7.142 as

$$\frac{dT}{T} = \frac{b^2 rdr}{c_p T_{t0} - (C_z^2 + C_\theta^2)} \tag{7.143}$$

Substituting Eq. 7.143 in 7.140, we may calculate the static density profile downstream of the IGV. This process is more accurate, but rather tedious, and is often simplified by making the following assumptions, namely,

$$C_{z1}(r) = C_{zm} = \text{constant} \tag{7.144}$$

$$\rho_1(r) = \rho_m = \text{constant} \tag{7.145}$$

Thus, to establish the hub radius downstream of the rotor, we satisfy the continuity equation via

$$\rho_m C_z \cdot \pi \cdot r_t^2\left(1 - \left(\frac{r_{h1}}{r_t}\right)^2\right) = \int_{r_{h2}}^{r_t} \rho_2(r) \cdot C_{z2}(r) \cdot 2\pi \cdot rdr \tag{7.146}$$

The density profile downstream of the rotor is a function of temperature and velocity profile using the total enthalpy expression, similar to Eq. 7.141. Again, the process is rather tedious and a simplification is in order. We know that the density and temperature are related via a polytropic exponent, namely,

$$\rho \sim T^{\frac{1}{n-1}} \tag{7.147}$$

where n is the polytropic exponent. For an isentropic flow, $n = \gamma$, and for irreversible adiabatic flows, which are encountered in turbomachinery $n < \gamma$. The density distribution is a function of the temperature distribution and the polytropic exponent, i.e.,

$$\frac{\rho}{\rho_m} = \left(\frac{T}{T_m}\right)^{\frac{1}{n-1}} \tag{7.148}$$

We may use the small stage or polytropic efficiency in a compressor, e_c, to relate pressure and density ratio to temperature ratio across a blade row as

$$\frac{p_2}{p_1} = \left(\frac{T_2}{T_1}\right)^{\frac{\gamma \cdot e_c}{\gamma-1}} \tag{7.149}$$

$$\frac{\rho_2}{\rho_1} = \left(\frac{T_2}{T_1}\right)^{\frac{1-\gamma(1-e_c)}{\gamma-1}} \tag{7.150}$$

Therefore the strategy is as follows. We first establish all parameters at the pitchline and then expand the results to other radii by using a blade vortex design choice, e.g., free vortex. Therefore, we first establish the swirl at the pitchline using a criterion based on the diffusion factor or the degree of reaction at the pitchline. Then by using the Euler turbine equation, we get the absolute total temperature at the pitchline radius, T_{t2m}. The definition of total temperature at the pitchline and the velocity components leads to the static temperature downstream of the rotor at r_m, i.e., T_{2m}. By using a value for compressor polytropic efficiency (say 0.90 or 0.91), we determine the static pressure and the density at r_m, i.e., p_{2m} and ρ_{2m}. Now, all parameters are established at the pitchline radius. We are now ready to expand our results to other radii. We choose a blade vortex design type, which fits a swirl profile along the rotor span to the pitchline swirl, $C_{\theta m}$. Therefore, the blade vortex design choice establishes the radial distribution of swirl, i.e., $C_\theta(r)$. With the knowledge of swirl downstream of the rotor, we repeat the above procedure at other radii to establish all thermodynamic and flow parameters. It is customary to repeat the procedure for an odd number of stream surfaces (5 or 7), which yield a pitchline stream surface as well. The minimum is thus three, i.e., the hub, pitchline, and the tip stream surfaces. We calculate the axial velocity distribution downstream of the rotor, using Eq. 7.110. With assumed compressor polytropic efficiency, say 0.91, we calculate the density ratio across the blade row using Eq. 7.150. Finally, we conserve the mass flow rate on the two sides of the rotor to establish the annulus area downstream of the rotor. Repeat the same process for the stator. As the process of design is iterative by nature, we need to return to our initial choices of vortex design, D-factor at the pitchline, the degree of reaction at the pitchline, or simply the blade hub-to-tip radius ratio in order to achieve an acceptable preliminary design for the stage. For the choice of blades that produce the desired flow turning at a given radius without flow separation and with an adequate margin of safety, we need to rely on cascade data presented in Fig. 7.22.

Case 2: Rotor with forced vortex design (no IGV)

The rotor induced swirl increases linearly with the radius, i.e., of solid-body rotation type, hence,

$$C_{\theta2} = b \cdot r \tag{7.151}$$

The stator removes the swirl put in by the rotor, hence the stator exit plane is swirl free, i.e., $C_{\theta1} = C_{\theta3} = 0$.

The work distribution along the rotor span is

$$w_c(r) = \omega[(rC_\theta)_2 - (rC_\theta)_1] = \omega \cdot b \cdot r^2 \tag{7.152}$$

Therefore, the rotor loading increases proportional to r^2, i.e., the rotor tip is overworked and the hub is not sufficiently loaded. The overloading of the rotor tip could result in flow separation at the tip with excessive flow turning and high diffusion. Usually, the tip region in a modern fan operates in the supersonic range, which limits the amount of flow turning and the appearance of shocks at the blade trailing edge. We shall address the issues of supersonic blade sections in a transonic compressor in a later section of this chapter. The degree of reaction for this vortex profile is

$$°R(r) = 1 - \frac{C_{\theta1} + C_{\theta2}}{2\omega \cdot r} = 1 - \frac{b}{2\omega} = \text{constant} \tag{7.153}$$

The axial velocity profile may be derived from Eq. 7.110 as

$$\frac{1}{2}\frac{d}{dr}(C_z^2) = \frac{dh_t}{dr} - T\frac{ds}{dr} - \frac{C_\theta}{r}\frac{d(rC_\theta)}{dr} = 2\omega br - 0 - b\frac{d}{dr}(br^2) = 2\omega br - 2b^2 \cdot r \tag{7.154}$$

This equation is integrated with reference to the pitchline to produce

$$\frac{C_z(r)}{C_{zm}} = \sqrt{1 + 2\left(1 - \frac{\omega}{b}\right)\left(\frac{b \cdot r_m}{C_{zm}}\right)^2\left[1 - \left(\frac{r}{r_m}\right)^2\right]} \tag{7.155}$$

For $\omega = b$, $°R = 0.5$ and the axial velocity remains uniform.

Figure 7.40 shows the spanwise distribution of the axial velocity downstream of the inlet guide vane with a solid-body rotation swirl profile. The spanwise maldistribution of axial velocity is of concern for the same reason as with the excessive tip flow turning, namely, large

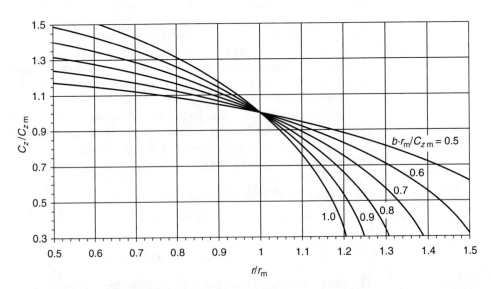

■ **FIGURE 7.40**   Axial velocity distribution downstream of an inlet guide vane, which induces a solid-body rotation swirl

flow deceleration at the tip could exceed allowable diffusion limits (either through D-factor or C_p) and lead to stall.

Case 3: A general stage vortex design (with IGV)

Let us consider a blend of the free and forced vortex swirl distributions that may be generalized by

$$C_{\theta 1} = C_{\theta 3} = -\frac{a}{r} + br^n \tag{7.156}$$

$$C_{\theta 2} = \frac{a}{r} + br^n \tag{7.157}$$

For $n = 1$, a combination of solid body rotation and a free-vortex swirl distribution are superimposed upstream and downstream of a compressor rotor. The function of stator, which is stated as to remove the swirl put in by the rotor, is seen from Eqs. 7.156 and 7.157.

The degree of reaction variation along the span may be written as

$$°R(r) = 1 - \frac{C_{\theta 1} + C_{\theta 2}}{2\omega \cdot r} = 1 - \frac{br^n}{\omega \cdot r} \qquad (°R \text{ profile}) \tag{7.158}$$

Obviously, the reaction distribution depends on the exponent n. For $n = 1$, the reaction along the stage height remains constant. This is desirable since the pressure rise along the blade becomes constant. For $n > 1$, the reaction decreases with span and for $n < 1$, the reaction increases with span. Any desired variation, including a constant, may be tailored through our choice of exponent, n of the swirl profile.

The rotor work distribution is

$$w_c(r) = \omega[(rC_\theta)_2 - (rC_\theta)_1] = \omega \cdot [(a + br^{n+1}) - (-a + br^{n+1})] = 2\omega \cdot a = \text{constant} \tag{7.159}$$

This result is independent of the exponent n in Eq. 7.156. Therefore, we create a constant work rotor by the above swirl distribution, which is desirable. The combination of constant work, which means the total enthalpy rise across the rotor is uniform, and the constant reaction (for $n = 1$), which implies a constant static enthalpy rise across the rotor, leads to a constant kinetic energy downstream of the rotor. Since swirl profile is a function of r, then the axial velocity has to attain a profile with the spanwise direction. We made this argument to show that the axial velocity may not be uniform in this case.

The axial velocity profile follows the same approach as the previous cases, namely,

$$\frac{1}{2}\frac{d}{dr}(C_z^2) = \frac{dh_t}{dr} - T\frac{ds}{dr} - \frac{C_\theta}{r}\frac{d(rC_\theta)}{dr} = 0 - 0 - \frac{\left(br^n \mp \frac{a}{r}\right)}{r}\frac{d(br^{n+1} \mp a)}{dr} \tag{7.160}$$

$$\frac{1}{2}\frac{d}{dr}(C_z^2) = -(n+1)\cdot r^{n-1} \cdot \left(b^2 r^n \mp \frac{ab}{r}\right) \tag{7.161}$$

The case of $n = 1$ is of particular interest since it leads to a constant reaction. The axial velocity profile, for $n = 1$, is

$$C_z(r) = \sqrt{C_{zm}^2 - 2b^2 r_m^2 \left(\frac{r^2}{r_m^2} - 1\right) \pm 4ab\ln\left(\frac{r}{r_m}\right)} \tag{7.162}$$

The plus and minus in Eq. 7.162 signify the upstream and downstream of the rotor, respectively. In general, the axial velocity decreases with radius with this choice of swirl profile.

7.6.5.2 *Three-Dimensional Losses.* The factors that render the flow process in a turbomachinery stage irreversible are

- End wall losses
  - □ Secondary flow losses
  - □ Tip clearance loss
  - □ Labyrinth seal and leakage flow losses
- Shock losses
  - □ Total pressure loss
  - □ Shock-boundary layer interaction
- Blade wake losses
  - □ Viscous profile drag (from cascade experiments)
  - □ Induced drag losses
  - □ Radial flow losses
- Unsteady flow losses
  - □ Upstream wake interaction
  - □ Vortex shedding in the wake
- Turbulent mixing

Although the above list seems like a formidable assortment of flow losses, but how many of them are totally decoupled from the rest? In a practical sense, the answer has to be none. Is there the possibility of duplication, i.e. double bookkeeping, in the above list? The answer is of course there is a possibility of counting a loss (or a part of it) twice or three times. Therefore, we conclude that any broken down list of factors contributing to loss is artificial at best and needs to be viewed with caution. The "list" is made only as a tool to help our understanding of complex flow phenomena in broad categories, such as

- Compressibility effects
- Viscous and turbulent dissipation
- Unsteadiness
- Three dimensionality

A transonic rotor creates a shock, which is weakened by suction surface expansion Mach waves and propagate upstream of the rotor row (see Kerrebrock, 1981). Note that since the upstream flow is subsonic in the absolute frame of reference, there is no zone of silence preventing the propagation of the rotor created waves. Figure 7.41 shows a typical flowfield.

The broad category of "end wall" losses encompasses the annulus boundary layer, corner vortex, tip clearance flow, and the seal leakage flow. The tip clearance flow is a pressure-driven phenomenon that relieves the fluid on the pressure side towards the suction surface, as shown in Fig. 7.42.

The boundary layer formation on the annulus is expected to grow with distance and in response to an adverse pressure gradient. The growth of the boundary layer with the number of stages is demonstrated by the data of Howell (1945), where the axial velocity distribution continually deforms and by the exit of the second stage there is no resemblance to a "potential core," i.e., flat top profile, that the inlet flowfield characterizes. The formation of corner vortices and the associated total pressure loss may be discerned from Howell data, as shown in Fig. 7.43. Note that the flow deflection angle near the hub and the tip of the cascade, in the boundary layer, is in excess of data outside the annulus boundary layer as seen in Fig. 7.43(b). Modern blade design accounts for this difference in the relative flow angle in the boundary layer and *bends* the blade *ends* (*end-benders*) to minimize incidence loss. A schematic drawing of an

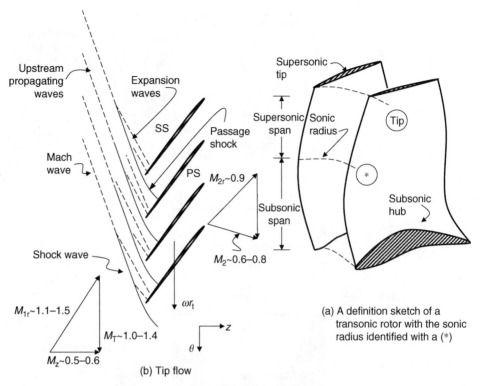

■ **FIGURE 7.41**   **Typical transonic rotor and its tip flowfield/wave pattern (bow shock and expansion waves)**

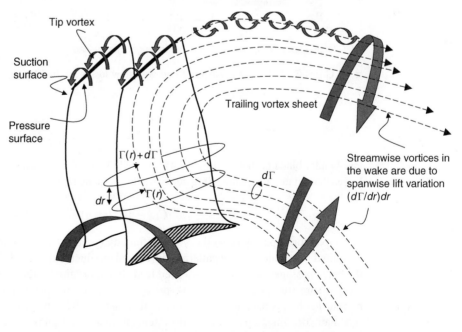

■ **FIGURE 7.42**   **Schematic drawing of tip vortex rollup and the inviscid wake composed of streamwise vortices due to spanwise lift variation on the blade**

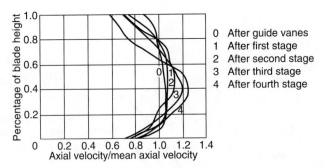

(a) The evolution of axial velocity profile with successive blade rows in a multistage compressor

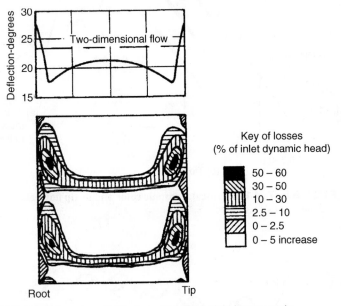

(b) Total pressure loss distribution with blade height in a cascade

■ **FIGURE 7.43** **An aspect of end wall flow losses studied by Howell (1945)**

end-bender blade is shown in Fig. 7.44. Howell (1945) proposes a simple model for end wall losses that relates an annulus drag coefficient C_{Da} to the ratio of blade spacing to blade height, i.e.,

$$C_{Da} = 0.02(s/h) \tag{7.163}$$

A comparison paper by Howell (1945) on fluid dynamics of axial compressors should also be consulted. A secondary flow pattern is developed in a duct with a bend. The flow turning within the blade passages of a compressor then leads to a migration of the boundary layer fluid from the pressure surface toward the low-pressure side, i.e., the suction surface. A flow pattern normal to the primary flow direction is then set up, which is called a secondary flow. A schematic drawing of a secondary flow generation and pattern is shown in Fig. 7.45. Pioneering work on the formulation of secondary flows is due to Hawthorne (1951) and a simplified engineering formulation is due to Squire and Winter (1951). A simple secondary flow loss

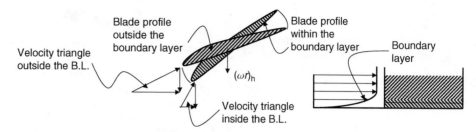

■ FIGURE 7.44 Schematic drawing of a bent tip blade to account for a larger boundary layer incidence due to a streamwise momentum deficit

model is proposed by Howell (1945) and is expressed in terms of secondary (or induced) drag coefficient C_{Ds},

$$C_{Ds} = 0.018C_L^2 \qquad (7.164)$$

The unsteadiness is an inherent mechanism for energy transfer between a rotating blade row and the fluid in turbomachinery. The upstream wake interaction with the following blade row is one source of unsteadiness (see Kerrebrock and Mikolajczek, 1970). This *wake chopping* interaction leads to unsteady lift, which, following Kelvin's circulation theorem, results in a shed vortex of opposite spin in the wake. Another source of unsteadiness to a blade row is its relative motion with respect to a downstream blade row. Although, the blade is not engaged in a viscous wake chopping activity, as in the previous case, but it is operating in an unsteady pressure field created by the downstream blade row. Operating in the unsteady pressure field is then referred to as unsteady *potential* interaction. Blade vibration in bending, twist or combined bending, and torsion is inevitable for elastic cantilevered structures. Therefore, blade vibration induces a spanwise variation of the incidence angle, hence an unsteady lift with a subsequent vortex shedding in the wake. The unsteady vortex shedding phenomenon is schematically shown in Fig. 7.46. Kotidis and Epstein (1991) have measured unsteady radial transport in the rotor wake of a transonic fan. They have related unsteady losses to the radial transport in the spanwise vortex cores as well as turbulent mixing. The individual losses and their contributions to the overall loss are depicted in Fig. 7.47, from Howell (1945). We note that at the design point

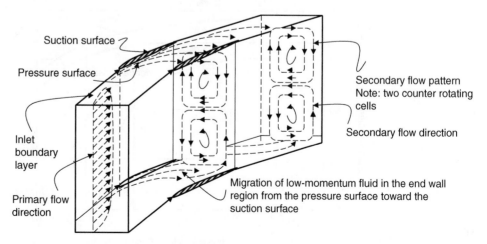

■ FIGURE 7.45 Viscous flow in a bend creates a pair of counterrotating streamwise vortices, which set up a secondary flow pattern

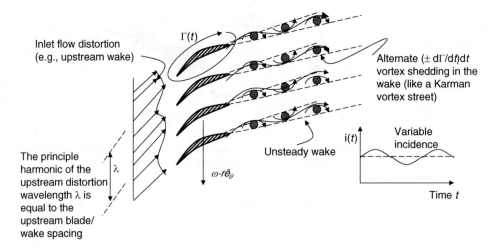

■ **FIGURE 7.46** Unsteady flow interaction with a blade row causes periodic vortex shedding in the wake

the overall losses are near minimum and in low flow, the compressor enters surge and at the high flow rates profile losses dominate due to flow separation.

7.6.5.3 *Reynolds Number Effect.*

The flow environment in a compressor, due to adverse pressure gradient, is sensitive to the state of a boundary layer. A laminar boundary layer readily separates in an adverse pressure gradient environment whereas a turbulent boundary layer may sustain a large pressure rise without separation or stall. The blade chord length c represents a suitable length scale for the Reynolds number evaluation as well as the relative velocity to the blade, W. Due to nearly flat compressor blade surfaces, we may borrow concepts from the flat plate boundary layer theory to predict the behavior of the boundary layer in adverse pressure gradient on a compressor blade. The effects of unsteadiness inherent in a turbomachinery as well as higher levels of turbulence in the core flow tend to provide for a

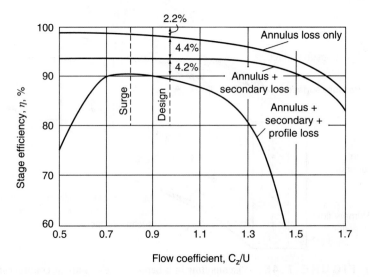

■ **FIGURE 7.47** Contributions of different losses in a compressor stage (from Howell, 1945)

higher mixing at the boundary layer and thus enhance its ability to withstand pressure rise. A more accurate approach should also include the effect of the blade curvature and its destabilizing (on a concave surface) or stabilizing (on a convex surface, such as the suction surface of a blade) effect on the boundary layer development. A simple statement can be made regarding the state of the boundary layer and its relation to compressor loss and that is to postpone/avoid flow separation, the boundary layer on a compressor blade has to be turbulent. This poses a rough minimum Reynolds number based on the blade chord of ~200,000 that serves as a rule of thumb, i.e.,

$$\mathrm{Re}_c \equiv \rho \cdot W \cdot c/\mu \geq 200{,}000$$

The Reynolds number trend on the cascade exit flow angle and total pressure loss may be seen from cascade test data of Rhoden (1956) in Fig. 7.48. We note that below $\mathrm{Re}_c \sim 100{,}000$ the cascade suffers from a rapid increase in total pressure loss and large exit flow angle deviation, which is due to the phenomenon of laminar separation. This represents a *lower critical Reynolds number.* Koch (1981) examined the stalling pressure rise capability of axial-flow compressor stages, which indicated a much-reduced sensitivity to Reynolds number below 200,000. The source of reduced sensitivity is found in higher turbulence levels and the unsteadiness inherent in a compressor stage. Koch's data normalized to Re = 130,000 is shown in Fig. 7.49.

The boundary layers in transonic compressors readily reach the turbulent regime due to a high relative velocity W. For example, Reynolds number on a fan blade operating at (relative) Mach 1.5 at the tip and 0.75 at the hub with a 1-ft (30.5 cm) chord at standard sea level conditions ranges from $\sim 5 \times 10^6$ to $\sim 10^7$. For $\mathrm{Re}_c \sim > 500{,}000$ the compressor efficiency remains nearly constant, which marks the *upper critical Reynolds number.* The blade surfaces behave as hydraulically rough and thus independent of Reynolds number for $\mathrm{Re}_c > 500{,}000$. At high altitudes where the air density is low and for a slow aircraft (e.g., a long-endurance observation platform), a suitable compressor blade should be designed with a much wider chord to combat the perils of laminar separation.

The effect of Reynolds number is expected to be minimal on secondary flow losses as the secondary flows are predominantly pressure-driven. Conventional, i.e., steady, low turbulence, cascade tests do not reproduce compressor test rig results due to higher turbulence levels that

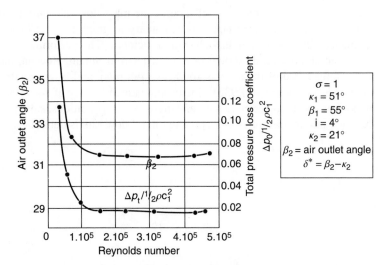

■ **FIGURE 7.48** Cascade test results for the effect of Reynolds number on exit flow angle and the total pressure loss (from Rhoden, 1956)

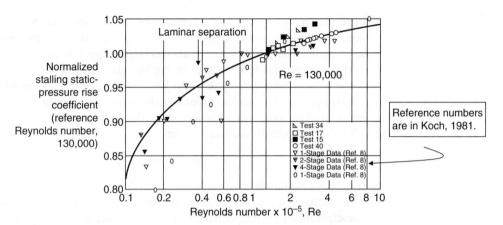

■ **FIGURE 7.49** Variation of normalized stalling pressure rise coefficient with Reynolds number in single and multistage compressor tests (from Koch, 1981)

promote mixing and the effect of unsteadiness with similar impact on the boundary layer. Finally, for the effect of Reynolds number on shock-boundary layer interaction, we may examine the work of Donaldson and Lange (1952). Figure 7.50 is a log–log graph of critical pressure rise across a shock that leads to boundary layer separation as a function of Reynolds number (from Donaldson and Lange). The two branches that appear on the graph correspond to the familiar laminar and turbulent boundary layers on a flat plate. We note that a turbulent boundary layer sustains nearly an order of magnitude (∼10 times) higher stalling pressure rise across the shock than a corresponding laminar boundary layer in a supersonic flow, as pointed out earlier in this section. Also note that the Reynolds number dependence for the stalling pressure rise in a turbulent flow $\sim Re^{-0.2}$ and for the laminar boundary layer is inversely proportional to the square root of Reynolds number, i.e., $\sim Re^{-0.5}$, which is identical to the Reynolds number behavior of momentum deficit thickness or friction drag coefficient on flat plates.

7.7 Compressor Performance Map

Compressor pressure ratio plotted against the mass flow rate through the compressor is the compressor performance map. It is customary to graph the constant rpm lines on the performance chart as well as the adiabatic efficiency. The performance map of a single-stage transonic compressor is shown in Fig. 7.51 (from Sulam, Keenan, and Flynn, 1970). The relative tip speed of the rotor is 1600 ft/s (488 m/s), and it represents a high-pressure ratio compressor stage ($\pi_s = 1.936$) with an adiabatic efficiency of 84.2% at the design point. The mass flow rate is corrected to the standard reference conditions, namely, the standard sea level pressure and temperature. The corrected mass flow rate is defined as

$$\dot{m}_c = \frac{\dot{m}\sqrt{\theta_2}}{\delta_2} \qquad (7.165)$$

where

$$\theta_2 \equiv \frac{T_{t2}}{T_{\text{ref}}} \qquad (7.166)$$

$$\delta_2 \equiv \frac{p_{t2}}{p_{\text{ref}}} \qquad (7.167)$$

Shock-boundry layer interaction

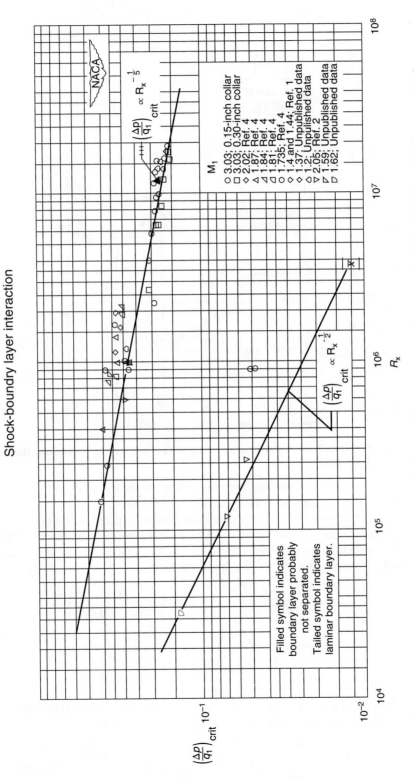

■ **FIGURE 7.50** Reynolds number dependence of the critical pressure rise across a shock that leads to boundary layer separation (from Donaldson and Lange, 1952)

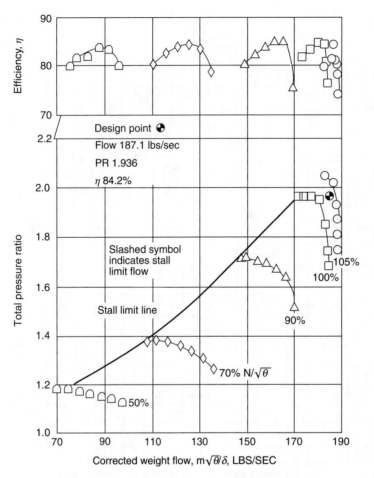

■ **FIGURE 7.51** Performance map of a transonic compressor stage (from Sulam, Keenan, and Flynn, 1970)

The reference pressure and temperature are the standard sea level conditions, namely, $p_{\text{ref}} = 1.01$ bar (or 101.33 kPa) and $T_{\text{ref}} = 288.2$ K.

The corrected mass flow rate in a compressor is a pure function of the axial Mach number operating at the standard pressure and temperature. We can easily demonstrate this by writing the continuity equation of one-dimensional flow in terms of total pressure and temperature, the axial Mach number M_z, and the flow area A, namely,

$$\dot{m} = \sqrt{\frac{\gamma}{R}} \frac{p_{\text{t}}}{\sqrt{T_{\text{t}}}} \cdot A \cdot M_z \left(\frac{1}{1 + (\gamma - 1)M_z^2/2}\right)^{\frac{\gamma+1}{2(\gamma-1)}} \tag{7.168}$$

$$\frac{\dot{m}\sqrt{T_{\text{t}}}}{p_{\text{t}}} = \sqrt{\frac{\gamma}{R}} A \cdot M_z \left(\frac{1}{1 + (\gamma - 1)M_z^2/2}\right)^{\frac{\gamma+1}{2(\gamma-1)}} \tag{7.169}$$

Now, we may multiply both sides by the standard pressure and divide by the square root of the standard temperature to get the "corrected mass flow rate,"

$$\frac{\dot{m}\sqrt{\theta}}{\delta} = \sqrt{\frac{\gamma}{R}} \frac{p_{\text{ref}}}{\sqrt{T_{\text{ref}}}} A \cdot M_z \left(\frac{1}{1 + (\gamma - 1)M_z^2/2}\right)^{\frac{\gamma+1}{2(\gamma-1)}} = f(\gamma, R, M_z) \tag{7.170}$$

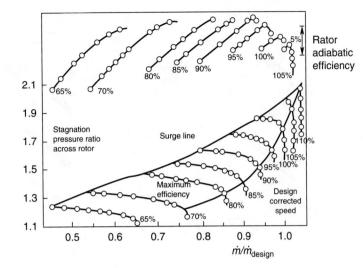

■ **FIGURE 7.52**
Performance map of a modern fan rotor (from Cumpsty, 1997)

The performance of a compressor depends on the axial and blade tangential Mach numbers M_z and M_T as discussed earlier. The corrected mass flow rate is a unique function of the axial Mach number and in addition it represents the mass flow rate into the compressor at the standard day pressure and temperature. By defining a corrected mass flow rate, we have basically taken out the effect of nonstandard atmospheric conditions of engine static testing or flight operation. The blade tangential Mach number M_T is proportional to the shaft rotational speed (or angular frequency N) divided by the local speed of sound or the square root of static temperature, namely,

$$M_T \propto \frac{N}{\sqrt{T}} \propto \frac{N}{\sqrt{\theta}} \tag{7.171}$$

The RHS of Eq. 7.171 is called the *corrected shaft speed,*

$$N_c \equiv \frac{N}{\sqrt{\theta}}$$

The corrected shaft speed is a unique function of the blade tangential Mach number, which along with the axial Mach number determines the performance of a compressor or fan, i.e.,

$$\pi_c = \pi_c(\gamma, R, M_z, M_T) = \pi_c(\gamma, R, \dot{m}_c, N_c) \tag{7.172}$$

$$\eta_c = \eta_c(\gamma, R, \dot{m}_c, N_c) \tag{7.173}$$

It is also customary to nondimensionalize the corrected mass flow rate by the "design mass flow rate," and the corrected shaft speed is nondimensionalized by the "design shaft speed" in the compressor performance map. An example of this is shown in Fig. 7.52 (from Cumpsty, 1997).

7.8   Compressor Instability—Stall and Surge

The phenomenon of stall in a compressor has its roots in aerodynamics of lifting surfaces in high angle of attack. Therefore, from aerodynamics we understand stall as flow separation from the suction or pressure surfaces of an airfoil with large positive and negative angles of attack, respectively. In a compressor operating at a constant rotational (shaft) speed when the mass flow rate drops, the axial velocity decreases, hence the incidence angle increases. We

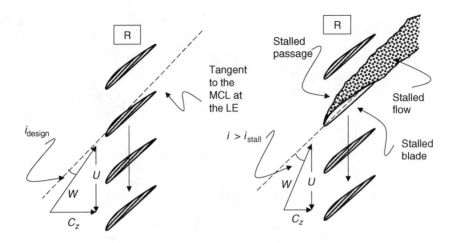

(a) A rotor operating at the design
    incidence angle

(b) The same rotor operating at a
    reduced flow and at $i > i_{stall}$

■ **FIGURE 7.53**   **A rotor blade operating at its best incidence angle and at a reduced mass flow leading to a stalling incidence angle**

remember from cascade data a parameter that was called "optimum incidence" angle, at which the blade losses were at a minimum. Large deviation from this minimum-loss incidence angle causes a rapid rise in total pressure loss, which signifies boundary layer separation. The positive and negative stall boundaries were then defined for a cascade of a given blade profile shape, solidity, stagger angle, and Mach number. Therefore, when the flow rate in a compressor drops while operating at constant shaft speed, the danger of stall lurks in the background. A stalled compressor flow is unsteady and hence offers a means of driving the naturally occurring blade vibrations into resonance. This is the mechanism for energy flow from the fluid into the blade vibration that causes *flutter*. Flutter is the *self-excited* aeroelastic instability of an aerodynamic lifting surface. In contrast, *buffet* (or *buffeting*) is a *forced vibration* of an aerodynamic surface that is in the turbulent wake of an upstream wing/ fuselage, e.g., horizontal tail buffet. Figure 7.53 shows schematic drawing of a compressor rotor with two velocity triangles, one representing its design point operation and the second corresponding to a stalled flow.

    The stalled flow in a compressor rotor may be initiated with a single blade at a certain radius. The stalled blade passage acts as a "blocker" and diverts the flow to the neighboring (as yet unstalled) blades. Following the drawing in Fig. 7.53 (b), we note that the flow diversion from the stalled passage to the one above the stalled blade causes an increase in that blades incidence angle, pushing it toward stall. Similarly, the blade below the stalled passage gets a diverted flow, which causes a reduction of the flow incidence angle to the blade below. Therefore, it moves away from stall. We note that the stalled flow in a passage is moving in the opposite direction to the blade rotation, hence it is given the name *rotating stall*. The angular speed of rotating stall propagation is $\sim 1/2$ of rotor angular speed, i.e., $\omega/2$, and in the opposite direction to the rotor rotation. Hence in the laboratory (i.e., absolute) frame of reference, a rotating stall *patch* spins with the rotor (i.e., the same direction as the rotor) but at half the speed. To get a feel for the frequency of the rotating stall cells, for example, in a large turbofan engine with a shaft speed of ~ 50 Hz, a single stall cell has half the frequency or 25 Hz. Now for a two-to-four stall cells circumferentially arranged around the compressor rotor, the stall cell frequency becomes 50–100 Hz. The rotating stall starts with a single cell and develops

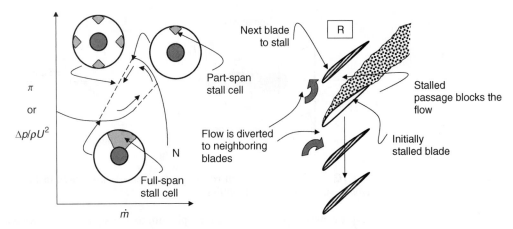

■ **FIGURE 7.54**   Schematic drawing of the mechanism of rotating stall, part and full-span stall cells and a hysteresis loop in stall recovery

into a number of cells with a partial span extent (known as the part-span stall) and may grow into a full-span stalled flow with subsequent reduction in the mass flow rate. This behavior is shown schematically in Fig. 7.54. The first appearance of stall is limited to a single cell with subsequent reduction of the flow the average pressure rise drops and the cells multiply into a periodic arrangement, as shown in Fig. 7.54. The recovery from the stalled operation in a compressor exhibits a hysteresis behavior, which accompanies systems governed by *nonlinear dynamics*.

The coupling between the compressor stall instability and the combustion chamber resonant characteristics could lead to an overall breakdown of the flow, or flow oscillation, in the compressor and combustor. The overall flow breakdown in the "system" composed of compressor and combustor is called *surge*. In a fully developed stage, surge is an axisymmetric oscillation of the flow with a characteristic timescale dictated by the plenum chamber time to empty and fill. In the initial transient stage, surge is asymmetric and thus creates transverse loads on the blades, which may rub on the casing and lead to structural damage and system breakdown. Early contributions to understanding compressor stall and surge and their distinction were made in the United States at NACA (e.g., Bullock and Finger, 1951) and Emmons, Pearson, and Grant at Harvard (1955).

A unifying approach to treat the compression system instability was first presented by Greitzer (1976). Greitzer developed a successful one-dimensional theory for the onset of surge in a compressor coupled to a plenum chamber cavity, which may serve as a model of the combustor in a gas turbine engine. The elements of the compression system model are shown in Fig. 7.55 to be composed of an inlet duct, a compressor, an exit duct connecting to a plenum chamber followed by a throttle that sets the backpressure, and the mass flow rate through the compressor.

Greitzer proposes the ratio of two characteristic timescales as the parameter that governs the dynamics of this system. One characteristic timescale is the compressor throughflow, which may be written as

$$\tau_{\text{throughflow}} \sim \frac{\rho \cdot V_{\text{c}}}{\dot{m}_{\text{c}}} \tag{7.174}$$

The second characteristic timescale is that of the plenum chamber, which is the time to charge the plenum chamber to a critical pressure rise condition for a stable compressor operation Δp_{c},

$$\tau_{\text{charge}} \sim \frac{\left(\dfrac{\Delta p_{\text{c}}}{RT}\right) \cdot V_P}{\dot{m}_{\text{c}}} \tag{7.175}$$

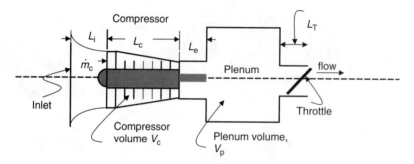

■ **FIGURE 7.55**   Compression system model that is used to study *local* (rotating stall) and *global* (surge) system instabilities

Expressing the pressure rise in the plenum as the square of the wheel speed,

$$\Delta p_c \sim \rho(\omega \cdot r)^2 \tag{7.176}$$

$$T \sim a^2 \tag{7.177}$$

and the temperature as the speed of sound squared, Greitzer has shown that the ratio of the two timescales is the square of a parameter B that dictates the fate of disturbances in a compressor, i.e.,

$$B \equiv \frac{\omega \cdot r}{2a} \sqrt{\frac{V_P}{V_c}} \sim \left(\frac{\tau_{\text{charge}}}{\tau_{\text{throughflow}}} \right)^{0.5} \tag{7.178}$$

The critical value of the B-parameter, which causes compressor instability to grow into a surge, instead of a localized rotating stall, is shown to be $\sim$0.7–0.8 by Greitzer. Experimental investigations of compression system instability have supported the proposed model. Figure 7.56 (from Greitzer, 1976) shows the computed results of a transient behavior of a compressor instability that settles into a rotating stall for low B-parameter and surge instability for a B-parameter of 0.6 and 1.58, respectively.

Current understanding of the compressor system instability has allowed strategies to be developed in its active control. Fast detection and cancellation of early oscillations are the keys to a successful active control system design. A recent review article by Paduano, Greitzer, and Epstein (2001) provides a rich exposition to the subject.

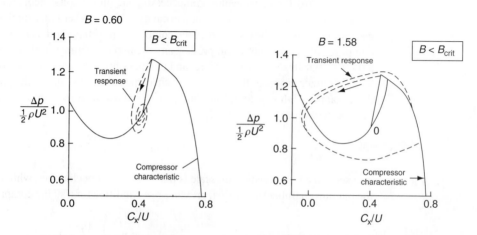

■ **FIGURE 7.56**   Transient response of a compressor to instabilities showing the appearance of a rotating stall (left) and the emergence of surge (right) (From Greitzer, 1976)

7.9   Multistage Compressors and Their Operating Line

The throttle in a gas turbine engine is the fuel flow control to the combustor. Constant throttle lines are, therefore, the lines of constant T_{t4}/T_{t2}. The path of steady-state operation of the compressor with the throttle setting is called the operating or working line in a compressor. It is often superimposed on the compressor performance map, as shown schematically in Fig. 7.57. The transient behavior of "spool up" is shown in Fig. 7.57 as well. In the context of the stall margin defined as the percent stall pressure rise divided by the corresponding percent drop in the mass flow rate, the transient engine operation deviates from the steady-state operating line and approaches the stall or surge line. The level and type of inlet flow distortion also affects the stall margin. We shall discuss inlet distortion and compressor stall later in this chapter.

Compressor performance map for the GE Energy Efficient Engine (E^3) is shown in Fig. 7.58 (from Cumpsty, 1997). Compressor test rig data as well as the engine test data are superimposed for comparison. Engine measurements are limited to the data points along the engine working line, whereas the compressor rig offers the versatility to produce the entire map. As a point of reference, E^3 program was initiated by NASA in 1973 (sparked by the oil embargo) with the goal of 12% reduction in specific fuel consumption over then current commercial aircraft engines of the day. Both GE and Pratt & Whitney developed technology demonstrator engines under this program that exceeded NASA's goal. A recent book by Garvin (1998) presents a history of aircraft engine development in the United States with a particular emphasis on GE's achievements in the commercial engine market. The lead taken by the military needs and the lag of the commercial side in aircraft engine development are presented in the context of the cold war, and the impact of business and strategic teaming/partnerships that have led to a business success. Propulsion engineers need to look at the business side of engineering to appreciate issues of cost, markets, customer service, and product support. For additional reading on the history of aircraft gas turbine engine development in the United States, the book by James St. Peter (1999) is recommended.

Let us do a simple and quick calculation of this compressor's polytropic efficiency e_c based on its design point pressure ratio and adiabatic efficiency. We estimate the compressor

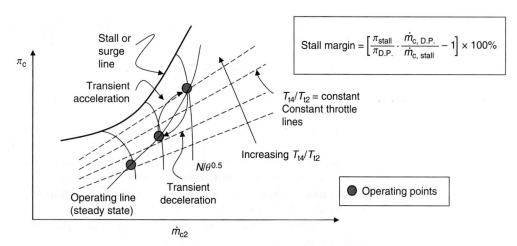

■ **FIGURE 7.57**   Constant throttle lines and a steady-state operating line are superimposed on a typical compressor performance map (D.P. is the design point)

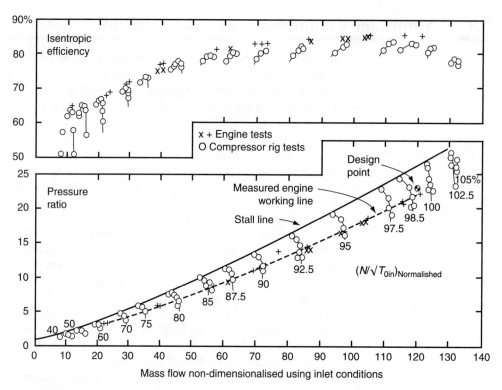

■ **FIGURE 7.58** Compressor performance map from GE E³ program with a design point pressure ratio of 23 achieved in 10 stages with variable stators in the first six stages (from Cumpsty, 1997)

adiabatic efficiency of ~0.86 from Fig. 7.58 at the design point and for the compressor pressure ratio of 23, we substitute these in the modified form of Eq. 4.22:

$$e_c = \frac{\gamma - 1}{\gamma} \left[\frac{\ln \pi_c}{\ln \left\{ 1 + \frac{1}{\eta_c} \left[\pi_c^{\frac{\gamma-1}{\gamma}} - 1 \right] \right\}} \right]$$

to get an $e_c \approx 0.907$. An average stage pressure ratio in 10 stages is

$$(\pi_s)_{\text{Avg}} \approx (\pi_c)^{\frac{1}{N}} = (23)^{\frac{1}{10}} \cong 1.368$$

This represents just an *average* of the stage pressure ratios and we recognize that the front stages operate at a higher loading (front stage pressure ratio of ~1.6) than the aft stages (pressure ratio ~1.15) due to a higher blade tangential Mach number M_T.

Now, we shall demonstrate that a throttle position T_{t4}/T_{t2} establishes the compressor pressure ratio, the mass flow rate, and the corresponding shaft speed. From the chapter on cycle analysis and off-design considerations, we argued that the turbine nozzle and the exhaust nozzle throat remain choked over a wide operating range of the engine. As we recall the turbine expansion parameter,

$$\tau_{\text{t-off-design}} = \tau_{\text{t-design}} = \text{constant}$$

The compressor-turbine power balance demands

$$\dot{m}_0(h_{t3} - h_{t2}) = \dot{m}_0(1 + f)(h_{t4} - h_{t5})$$

To simplify the equation, assume the gas is calorically perfect and neglect the small contribution of fuel-to-air ratio in favor of 1 on the RHS of the above equation, to get

$$T_{t2}(\tau_c - 1) \approx T_{t4}(1 - \tau_t)$$

Therefore the compressor temperature ratio is related to the throttle setting and a constant turbine expansion parameter τ_t via

$$\tau_c \approx 1 + (1 - \tau_t)\frac{T_{t4}}{T_{t2}} \tag{7.179}$$

Here, we see that the compressor temperature ratio increases linearly with the throttle setting. The compressor pressure ratio is related to the temperature ratio via the polytropic efficiency, therefore the compressor pressure ratio is established by the throttle setting as

$$\pi_c \approx \left[1 + (1 - \tau_t)\frac{T_{t4}}{T_{t2}}\right]^{\frac{\gamma e_c}{\gamma - 1}} \tag{7.180}$$

The continuity equation at stations 2 and 4, i.e., the compressor face and the combustor/turbine nozzle exit connects the axial Mach number at the engine face to the throttle setting according to

$$\dot{m}_2 = \sqrt{\frac{\gamma_2}{R_2}}\frac{p_{t2}}{\sqrt{T_{t2}}}A_2 \cdot M_{z2}\left(1 + \frac{\gamma_2 - 1}{2}M_{z2}^2\right)^{-\frac{(\gamma_2 + 1)}{2(\gamma_2 - 1)}}$$

$$\approx \dot{m}_4 = \sqrt{\frac{\gamma_4}{R_4}}\frac{p_{t4}}{\sqrt{T_{t4}}}A_4\left(\frac{\gamma_4 + 1}{2}\right)^{-\frac{(\gamma_4 + 1)}{2(\gamma_4 - 1)}} \tag{7.181}$$

By neglecting the combustor total pressure loss, we may replace the combustor exit total pressure p_{t4} with the compressor exit pressure p_{t3} to rewrite Eq. 7.181 as

$$M_{z2}\left(1 + \frac{\gamma - 1}{2}M_{z2}^2\right)^{-\frac{(\gamma + 1)}{2(\gamma - 1)}} \approx \text{constant}\,\frac{\pi_c}{\sqrt{T_{t4}/T_{t2}}} \tag{7.182}$$

We assumed the flow areas A_2 and A_4 remain constant in deriving Eq. 7.182 as well as all the gas property variations are included in the proportionality constant on the RHS of Eq. 7.182. The LHS of Eq. 7.182 is proportional to the corrected mass flow rate $\dot{m}_{c2}$, hence

$$\dot{m}_{c2} \propto \frac{\pi_c}{\sqrt{T_{t4}/T_{t2}}} \tag{7.183}$$

Therefore, constant throttle lines, $T_{t4}/T_{t2} = \text{constant}$, are straight lines on the compressor performance map π_c versus $\dot{m}_{c2}$, as shown schematically in Fig. 7.59. Also higher throttle constants attain higher slopes, as shown in Fig. 7.59, which get the compressor operating point closer to the compressor surge line. Figure 7.59 shows the constant throttle lines and their convergence on compressor pressure ratio of 1, corresponding to a mass flow rate of zero.

The engine face axial Mach number is established by the throttle setting via Eq. 7.182 and the compressor pressure ratio was related to the throttle setting via Eq. 7.180, therefore,

$$M_{z2}\left(1 + \frac{\gamma - 1}{2}M_{z2}^2\right)^{-\frac{(\gamma + 1)}{2(\gamma - 1)}} \approx \text{constant}\,\frac{[1 + (1 - \tau_t)(T_{t4}/T_{t2})]^{\frac{\gamma \cdot e_c}{\gamma - 1}}}{\sqrt{T_{t4}/T_{t2}}} \tag{7.184}$$

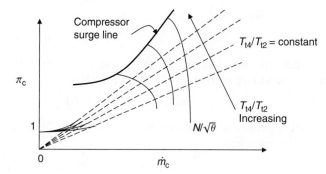

■ FIGURE 7.59
Constant throttle lines on a
compressor performance map

We may solve Eq. 7.184 for the engine face axial Mach number for a given throttle setting. We may also simplify the LHS somewhat by recognizing that the second term in the parenthesis involving axial Mach number is small compared to one and hence applying binomial expansion yields,

$$M_{z2}\left(1 - \frac{\gamma + 1}{4}M_{z2}^2\right) \approx \text{constant} \frac{[1 + (1 - \tau_t)(T_{t4}/T_{t2})]^{\frac{\gamma \cdot e_c}{\gamma - 1}}}{\sqrt{T_{t4}/T_{t2}}} \tag{7.185}$$

Now, let us demonstrate the dependence of wheel speed on the throttle setting. The compressor power is consumed by the rotor blades that experience a torque τ_r and spin at the angular rate of ω. The stator blades do not contribute to the power transfer. We need to sum overall the rotor blade rows in a multistage compressor to calculate the compressor power, namely,

$$\wp_c = \omega \cdot \sum_{j=1}^{N} \tau_{r_j} \tag{7.186}$$

where N is the number of stages. The rotor torque is the integral of angular momentum increase across the blade row and may be expressed as

$$\tau_r = \int_{r_{h2}}^{r_{t2}} r \cdot C_{\theta 2}(r) \cdot \rho_2(r) \cdot C_{z2}(r) \cdot 2\pi r \, dr - \int_{r_{h1}}^{r_{t1}} r \cdot C_{\theta 1}(r) \cdot \rho_1(r) \cdot C_{z1}(r) \cdot 2\pi r \, dr \tag{7.187}$$

The geometrical parameters are sketched in Fig. 7.60. In its simplest form the integrals are expressed as an average angular momentum at the pitchline radius r_m namely,

$$\tau_r = \dot{m}[(r_{m2}C_{\theta 2m}) - (r_{m1}C_{\theta 1m})] \tag{7.188}$$

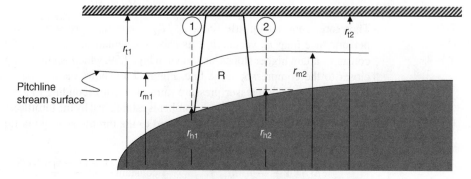

■ FIGURE 7.60   Definition sketch of the geometrical parameters

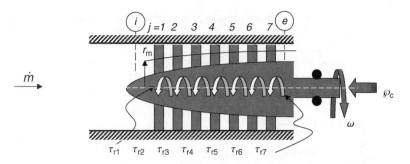

■ FIGURE 7.61 Seven-stage compressor consuming power $\wp_c$ at the shaft speed of ω and the individual rotor torques $\tau_{r1}, \tau_{r2}, \ldots, \tau_{r7}$

The rotor torque, however, is a function of the shaft speed as the absolute swirl downstream of the rotor changes with shaft speed according to

$$C_{\theta 2} = C_{z2} \tan \alpha_2 = U + W_{\theta 2} = U + C_{z2} \tan \beta_2 \qquad (7.189)$$

The rotor relative exit flow angle β_2 remains nearly constant with the shaft speed over a wide operating range of the compressor, i.e., for attached boundary layer flow to the blades. Substituting Eq. 7.189 into 7.188, we get

$$\tau_r = \dot{m}[r_{m2}(\omega \cdot r_{m2} + C_{z2} \tan \beta_2) - r_{m1} C_{z1} \tan \alpha_1] \qquad (7.190)$$

Expressing the shaft power in terms of the total enthalpy rise and the mass flow rate, we get

$$\wp_c = \dot{m}(h_{te} - h_{ti}) = \dot{m}h_{ti}(\tau_c - 1) = \omega \cdot \sum_{j=1}^{N} \tau_{rj} \qquad (7.191\text{-a})$$

Figure 7.61 shows the parameters in Eq. 7.191-a.

By substituting for the rotor torque from Eq. 7.190 to 7.191-a, for the compressor temperature ratio from Eq. 7.179, and canceling the mass flow rate, we get

$$h_{t2}(1 - \tau_t)\frac{T_{t4}}{T_{t2}} = \omega \sum_{j=1}^{N} [r_{me}(\omega \cdot r_{me} + C_{ze} \tan \beta_e) - r_{mi} C_{zi} \tan \alpha_i]_j \qquad (7.191\text{-b})$$

This is a quadratic equation in ω in terms of the throttle setting T_{t4}/T_{t2}, flow angles α_1 and β_2 that remain nearly constant, axial velocities across the rotor, and the radial position of the pitchline stream surface. In practice at a given throttle setting, the shaft finds its own speed consistent with the compressor power consumption and mass flow rate through the machine. We have thus established the interdependence of compressor pressure ratio, corrected mass flow rate, and shaft speed on the throttle setting.

The simplifications introduced in the above discussions may be removed one by one, if we are willing to iterate to find a consistent set of compressor performance parameters. For example, the choking condition in stations 4 and 8 may be relaxed in case the nozzle pressure ratio is below critical (i.e., necessary for choking). The subsonic throat Mach number may be found through application of continuity equation using a trial and error approach.

7.10   Multistage Compressor Stalling Pressure Rise and Stall Margin

Koch (1981) has developed an analogy between the stalling pressure rise capability of an axial-flow compressor stage and two-dimensional diffusers. Classical diffuser data of Reneau, Johnston, and Kline (1966) and Sovran and Klomp (1967) for a straight centerline, 2D diffuser serve as the point of analogy. The pitchline radius of the compressor stage is taken as the reference

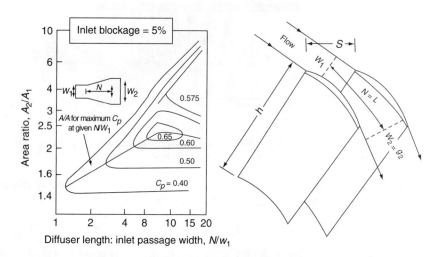

Diffuser analogy for compressor bladerow

■ **FIGURE 7.62**   Diffuser performance chart of Reneau, Johnston, and Kline (1967) and the analogy to a compressor blade row (from Wisler, 2000)

point in developing the stall analogy with the diffuser. This approach is useful in the preliminary design phase of an axial-flow compressor in estimating the maximum pressure rise potential as well as the stall margin for a given design point operation. Figure 7.62 shows the diffuser performance data of Reneau, Johnston, and Kline for an inlet boundary layer blockage of 5%. The length of the diffuser is N, which is analogous to L, as shown on the definition sketch on the right (from Wisler, 2000), to represent a diffusion length scale of a compressor blade passage at the pitchline radius.

As a first approximation, the diffusion length L is represented by the arc length of the mean camber line of the airfoil at the pitchline. Assuming a circular arc for the mean camber line, the length L is related to the camber angle φ and the chord length c via

$$L \approx c \cdot \frac{\varphi/2}{\sin(\varphi/2)} \tag{7.192}$$

Although the area ratio of a diffuser, A_2/A_1, is fixed by the geometry of the diffuser, its counterpart in a cascade depends on the blade incidence, i.e., the staggered spacing of the streamtube that enters and exits the blade row. The exit flow area of the blade channel is, however, fixed over a wide operating range of the compressor flow and therefore it serves as the reference area in the correlation development by Koch. The inlet flow area is a function of the operating point (i.e., the throttle setting) and decreases with an increasing incidence angle. Figure 7.63 shows the staggered spacing g_1 and g_2 and the flow areas at the inlet and outlet of a compressor blade row.

Here, we have adopted the terminology of Koch in representing the rotor relative velocity as V' and the absolute velocity as V. A nondimensional diffusion parameter is the ratio of the diffusion length L to the staggered spacing at the exit of the blade row, i.e., L/g_2,

$$\frac{L}{g_2} = \sigma_1 \cdot \frac{\varphi/2}{\sin(\varphi/2)} \cdot \frac{s_1}{s_2} \frac{1}{\cos \beta_2} = \sigma_1 \cdot \frac{\varphi/2}{\sin(\varphi/2)} \cdot \frac{A_1}{A_2} \cdot \frac{h_2}{h_1 \cdot \cos \beta_2} \tag{7.193}$$

where the camber angle φ is in radians and the blade inlet solidity is represented by σ_1 at the pitchline. The blade solidity, the camber angle, and the exit flow angle are all reflected in the nondimensional diffusion length ratio in a compressor blade row. The stator blade row diffusion length is calculated in a similar manner using the stator inlet solidity at the pitchline, camber angle, area ratios, and the exit absolute flow angle. The effect of higher camber angle is

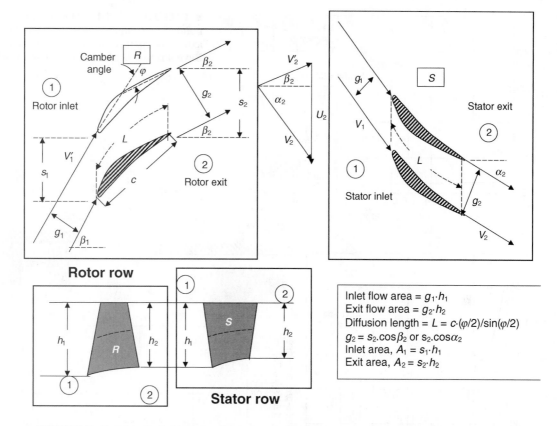

■ **FIGURE 7.63**   Definition sketch of the staggered inlet and outlet spacing and area ratio for a compressor stage (V_1' and V_2' are relative velocities W_1 and W_2)

seen as an increase in diffusion, hence static pressure rise. The effects of blade aspect ratio, tip clearance gap, and Reynolds number are accounted for through the end wall boundary layer thickness. In a diffuser, the inlet boundary layer blockage is a key parameter in the performance and stalling characteristics of diffusers and hence the end wall boundary layer thickness in a compressor stage serves the same principle.

A stage average approach is adopted in developing a correlation between the stalling pressure rise of a diffuser and a compressor stage. For example, the diffusion length ratio L/g_2 of the stage is the weighted average of the rotor and the stator values with blade row inlet dynamic head used as the weighting factor, i.e.,

$$\frac{L}{g_2}\bigg|_{\text{Stage}} = \left[\frac{(L/g_2)_{\text{Rotor}} \cdot q_1' + (L/g_2)_{\text{Stator}} \cdot q_1}{q_1' + q_1}\right] \tag{7.194}$$

where q_1' is the rotor inlet dynamic head (relative) and q_1 is the stator inlet dynamic head. Koch (1981) defines an enthalpy equivalent of the static pressure rise in a compressor stage according to

$$C_{\text{h}} \equiv \frac{c_p T_1 \left[\left(\frac{p_2}{p_1}\bigg|_{\text{Stage}}\right)^{\frac{\gamma-1}{\gamma}} - 1\right] - \frac{(U_2^2 - U_1^2)_{\text{Rotor}}}{2}}{\left[\frac{V_1'^2\big|_{\text{Rotor}} + V_1^2\big|_{\text{Stator}}}{2}\right]} \tag{7.195}$$

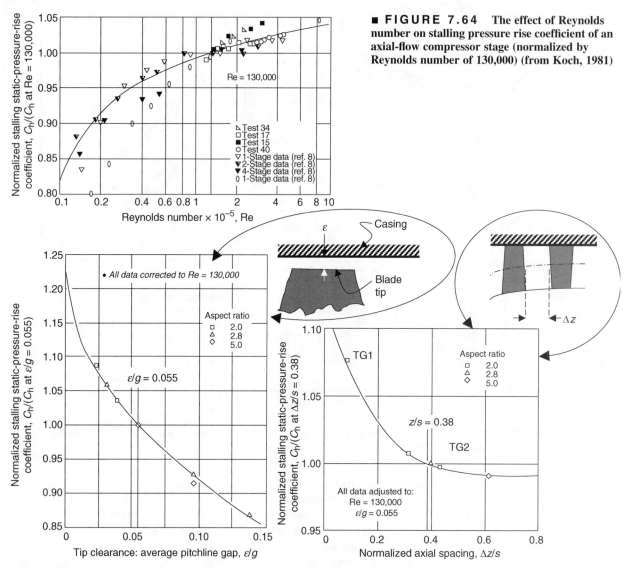

■ **FIGURE 7.64** The effect of Reynolds number on stalling pressure rise coefficient of an axial-flow compressor stage (normalized by Reynolds number of 130,000) (from Koch, 1981)

■ **FIGURE 7.65** The effect of tip clearance on the stalling pressure rise coefficient (from Koch, 1981)

■ **FIGURE 7.66** The effect of axial spacing between rotor and stator blade rows on stalling pressure rise coefficient (from Koch, 1981)

The numerator of Eq. 7.195 is the stage static enthalpy rise based on the isentropic stage temperature ratio, which is corrected for the radial shift across the rotor at the pitchline radius. To be comparable, the rotor *free work contribution* associated with the radial shift needs to be corrected for in the pressure rise comparisons of diffusers and compressor stages. The denominator is the sum of the free stream dynamic heads to the rotor and stator. There are *two* dynamic heads in the denominator corresponding to *two* static enthalpy rises across the rotor and stator in the numerator.

The contributions of Reynolds number, tip clearance, and the axial spacing between the rotor and the stator rows to the stalling pressure rise coefficient are represented in Figs. 7.64–66, respectively (from Koch, 1981). Normalizing values for Reynolds number, ratio of tip clearance to gap, and the ratio of axial distance to the blade spacing are used in presenting

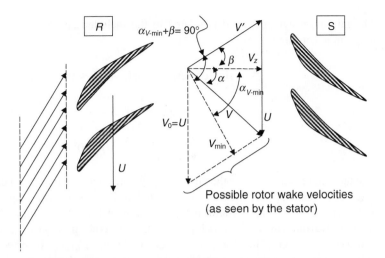

■ **FIGURE 7.67**   Velocity vector diagram upstream of the stator with possible upstream wake velocity vectors (adapted from Koch, 1981)

the stalling pressure rise coefficient. These are Reynolds number $= 130,000$, the average tip (radial) clearance-to-gap ratio $\varepsilon/g = 0.055$, and the ratio of axial clearance to the blade spacing $\Delta z/s = 0.38$. The Reynolds number is calculated based on the blade row inlet relative velocity and the chord length at the pitchline radius. The gap g and the blade spacing s represent a stage average at the pitchline radius in Figs. 7.65 and 7.66.

Note that a higher stalling pressure rise is achieved in a compressor stage with a decreasing tip clearance as well a decreasing axial spacing between the blade rows. The effect of blade stagger was significant on the stalling pressure rise capability of axial-flow compressor stages according to Koch (1981). The effect of stagger is related to a recovery potential of a total pressure deficit region (e.g., upstream wake) interacting with the stator blade row. Leroy Smith first identified this phenomenon in 1958. The velocity vector diagrams upstream of the stator are shown in Fig. 7.67. Note that the minimum wake velocity as seen by the stator occurs when the angle between the relative velocity vector V' and the wake velocity vector is 90°. In case of zero wake velocity in the relative frame, the wake as seen by the stator is equal to U and is in the tangential direction. This point is shown as $V_0 = U$ in Fig. 7.67.

The effect of low stagger (i.e., high flow coefficient) is seen in the small included angle between the relative and absolute velocity vectors $\alpha + \beta$, which will result in a low momentum wake velocity $V_{\min} \ll V$. With a low dynamic head associated with this flow, the low stagger stages seem more prone to stall than a high stagger counterpart. In a high stagger stage, i.e., a low flow coefficient case, the $\alpha + \beta > 90°$, therefore the upstream wake is entering the stator with $V_{\text{wake}} > V$, hence a higher dynamic pressure fluid is less prone to stall. The observations of $\alpha + \beta$ and 90° and its impact on the loss recovery were made by Ashby in a discussion of Smith (1958) paper. Here, we are alerted to the importance of the vector diagram in a compressor stage and its impact on its stall margin. Koch devised an *effective dynamic pressure factor* F_{ef} that accounts for the rotor wake interaction with the stator, i.e., the effect of stagger on pressure recovery. The formulation of F_{ef} is presented by Koch to be

$$F_{\text{ef}} = \frac{V_{\text{ef}}^2}{V^2} = \frac{V^2 + 2.5V_{\min}^2 + 0.5V_0^2}{4V^2} \qquad \textbf{(7.196)}$$

The minimum dynamic head is related to the included angle $(\alpha + \beta)$ of the stage vector diagram according to

$$\frac{V_{\min}^2}{V^2} = \sin^2(\alpha + \beta), \qquad \text{if}(\alpha + \beta) \leq 90° \text{ and } \beta \geq 0° \tag{7.197-a}$$

$$\frac{V_{\min}^2}{V^2} = 1.0, \qquad \text{if}(\alpha + \beta) > 90° \tag{7.197-b}$$

$$\frac{V_{\min}^2}{V^2} = \frac{V_0^2}{V^2}, \qquad \text{if}\,\beta < 0° \tag{7.197-c}$$

The procedure to account for the effects of Reynolds number, tip clearance, axial spacing, and stagger on stalling static pressure rise coefficient in a compressor stage is to first *adjust* the stalling pressure rise for Reynolds number other than 130,000 (via Fig. 7.63), the tip clearance (to gap ratio) other than 5.5% (via Fig. 7.64), and the axial spacing (to blade spacing) other than 0.38 (via Fig. 7.65). Therefore, we first calculate the $(C_h)_{\text{adj}}$. The effect of stagger and the velocity vector diagram is now introduced through the *effective dynamic head factor* F_{ef}, which is multiplied by the free stream dynamic head of the stator, i.e.,

$$(C_h)_{\text{ef}} = (C_h)_{\text{adj}} \left[\frac{V_1'^2\big|_{\text{Rotor}} + V_1^2\big|_{\text{Stator}}}{V_1'^2\big|_{\text{Rotor}} + F_{\text{ef}} \cdot V_1^2\big|_{\text{Stator}}} \right] \tag{7.198}$$

Koch's stalling effective pressure rise correlation for low- and high-speed axial-flow compressor stages is shown in Figs. 7.68 and 7.69, respectively. The maximum recovery of the two-dimensional diffuser for an inlet blockage of 9% of Sovran and Klomp (1967) marks the upper limit in Figs. 7.68 and 7.69, which serves as an approximate stalling boundary for axial-flow compressor stages. We note that the stall margin correlation for the low-speed compressor predicts typically below 0.5 of static pressure recovery at stall. The high-speed compressor stages produce a higher effective stalling static pressure rise coefficient of up to ~0.55. The case of tandem rotor compressor with an L/g_2 of 2.33 poses the highest stalling static pressure recovery that is ~23% above the diffuser stall curve, as shown in Fig. 7.69. The tandem blading

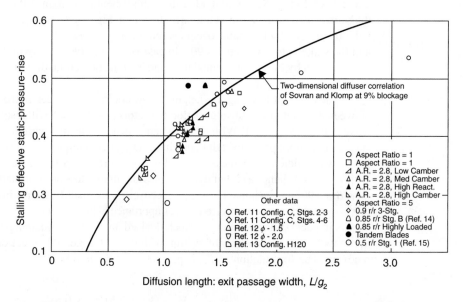

■ **FIGURE 7.68**   Correlation of effective static pressure rise coefficient at stall for low-speed compressor stages (from Koch, 1981)

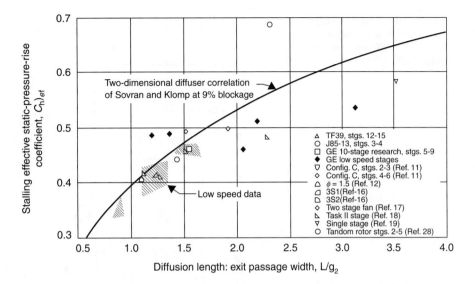

■ **FIGURE 7.69**   Correlation of effective static pressure rise coefficient at stall for high-speed compressor stages (from Koch, 1981)

serves as a means of flow control in a compressor rotor and the simple diffuser performance underpredicts the stall behavior of a compressor stage that employs flow control. An analogy of a tandem rotor compressor stage should be made with a wide-angle diffuser that employs splitter plates as a means of high static pressure recovery.

Koch demonstrates that effect of the stage reaction on the stalling pressure rise in axial-flow compressor is well predicted by the diffuser stall margin correlation of Fig. 7.68 or 7.69. The range of stage reactions from 0.39 to 1.09 (where the stator accelerates the flow, like a turbine) is shown in Fig. 7.70. It is surprising that over such a wide range of stage reaction, the stalling pressure rise in a compressor correlates with the diffuser predictions.

In a multistage high-speed compressor, the stagewise distribution of the static pressure rise at stall is compared with the stall margin correlation of Fig. 7.71 (from Koch).

Figure 7.71 shows that in a multistage compressor, there is at least one or a group of stages that rise to the predicted level of stalling pressure according to the diffuser correlation. For example, the 10-stage research compressor operating at 95% speed at stall had its 8th stage reach/exceed the stall boundary as predicted by the correlation of Fig. 7.69. The 16-stage TF39, at 97% speed, had stages 12 and 15 exceed the stall boundary and stages 13 and 14 operate within 96 percentile of the stall boundary. Note that we are not looking for a perfect match between the correlation and the stall pressure rise of every stage in a multistage compressor. We are rather interested in the trends and behavior of individual and groups of stages and their stall margin. In this context, the shaded symbols in Fig. 7.71 represent the "critical" stages, which lie above 95 percentile of the stall boundary. These stages are likely to stall first and cause a breakdown of the flow in the rest of the compressor. The stalling pressure rise data in Fig. 7.71 represent *actual* high-speed multistage compressor tests, which validate Koch's semiempirical correlation presented in this section. The stall margin correlation based on the diffuser analogy has proven to be a valuable tool in the preliminary design stage of axial-flow compressors.

We have taken the stalling effective pressure rise coefficient of Figs. 7.68 and 7.69 and graphed a family of curves with 95%, 90%, . . . , 70% of the stall correlation (see Fig. 7.72). We have thus created a chart with 5% stall margin increments as a function of the stage-averaged diffusion length ratio L/g_2. We note that increasing the diffusion length ratio L/g_2, which points to either higher solidity or lower aspect ratio blading, helps with the stall margin. Wide chord blades also operate at a higher Reynolds number, which is beneficial to boundary layer

■ **FIGURE 7.70**   **Effect of stage reaction on the ratio of the stalling pressure rise to a 2D diffuser pressure recovery (from Koch, 1981)**

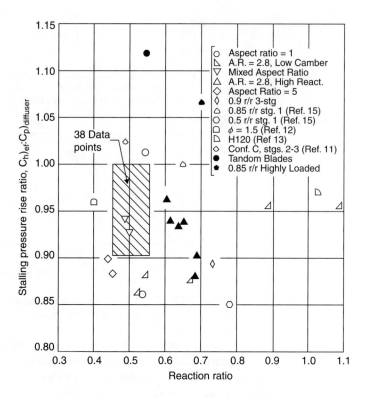

■ **FIGURE 7.71**   **Stall pressure rise, per stage, in a multistage compressor at high speed (from Koch, 1981)**

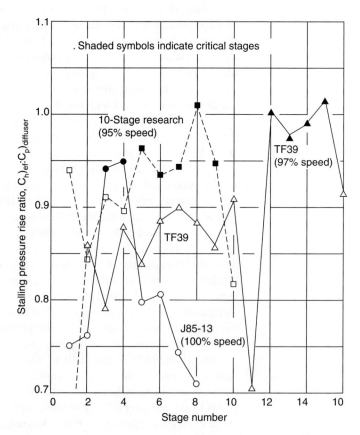

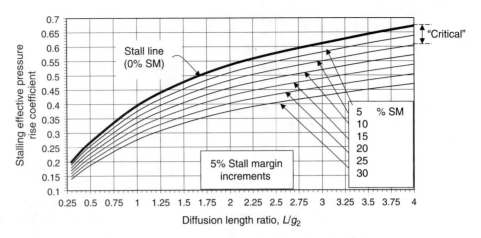

■ FIGURE 7.72 Axial-flow compressor stall margin (SM) chart with 5% stall margin increments

stability, as evidenced in Fig. 7.64. The minimum operational tip clearance of ~1% blade height and the minimum axial spacing between blade rows of ~0.25 axial chord also create higher static pressure rise in a compressor stage. Stage vector diagram affects the loss recovery ($\alpha + \beta \geq 90°$) and thus stall static pressure rise.

7.11 Multistage Compressor Starting Problem

The annulus flow area in a multistage compressor is designed based on the calculated density rise along the axis of the machine operating at the design point. The assumption of constant axial velocity, for example, implies that the channel area shrinks inversely proportional to the density rise, to conserve mass, namely,

$$\frac{A(z)}{A_1} = \frac{\rho_1}{\rho(z)} \tag{7.199}$$

This behavior is schematically shown in Fig. 7.73.

In the starting phase of a compressor, the mass flow rate is initially small, which means a higher loading for the front stages. The picture for the aft high-pressure stages is just the opposite of the front low-pressure stages. Initially, the compressor does not develop the density rise that it was designed to produce, hence with lower than design densities in the aft stages the axial velocity is

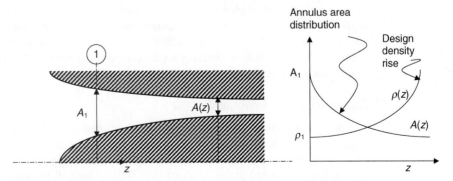

■ FIGURE 7.73 Flow annulus area and the density rise are inversely proportional in a multistage axial compressor

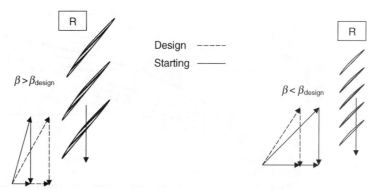

(a) Front stages see a higher incidence (loading), i.e., a lower flow coefficient ϕ

(b) Aft stages see a lower incidence (loading), i.e., a higher flow coefficient ϕ

■ **FIGURE 7.74**   Velocity triangles at design and starting phases of a high-pressure compressor

increased to satisfy the continuity equation. Higher than the design axial velocity leads to a lower loading of the aft stages. Actually the aft stages would be *windmilling* and their blades operate in a separated flow. A comparison of the velocity triangles in the starting phase and the design point will help us understand the (starting) problem in a high-pressure compressor. Figure 7.74 shows the velocity triangles for the front and aft stages of a compressor. Note that the flow coefficient is lower than the design value for the front stages, and the flow coefficient is higher for the aft stages. We remember from the compressor performance map that with a reduction of flow coefficient, the compressor can go into stall for a given shaft speed. With front stages not producing the design density ratio, the axial flow in the aft stages accelerates to conserve mass. Hence the two ends of the compressor see the opposite flow fields and are subjected to opposite flow coefficients, i.e., lower and higher than the design, respectively. However, the tendency for both front and aft stages is to stall in the starting phase of a high-pressure compressor.

The loading mismatch between the front and aft stages in the starting phase of a high-pressure axial compressor may be solved in several ways. Here we introduce three distinct approaches to the starting problem.

Proposition 1: Split shaft (or multispool shaft system)

Let the aft stages operate at a higher rotational speed than the front stages. This proposition aims at matching the relative flow angle to the rotor blades in the aft stages. Pratt & Whitney spearheaded the development of the split shaft concept in the United States in the 1950s. Today, all modern high-pressure compressors employ a two-to-three spool or shaft configuration to alleviate the problem of starting and improve compressor efficiency and stability. This method primarily attacks the aft stages, which would have been windmilling without a higher shaft speed, as in a single-shaft gas turbine engine. Rolls-Royce is the only engine company that has manufactured a production three-spool commercial engine, known as the RB 211, with several successful derivatives.

Proposition 2: Variable stators

This proposition aims at adjusting the absolute flow angle (α) through a stator variable setting to improve the mismatch between the flow and the rotor relative flow angle β. GE-Aircraft Engines spearheaded the development of high-pressure compressors with the variable stator approach in the United States also in 1950s. Today, modern high-pressure compressors use variable stators for the front several stages to help with the starting/off-design operation efficiency. This method primarily attacks the front stages. Although a variable stator benefits all stages of a compressor, its use is limited to the front few (say six) low-pressure stages as the sealing of hot gases through the variable stator seal poses an operational and maintenance problem on the high-pressure end of the compressor.

Proposition 3: Intercompressor bleed

The windmilling operation of the aft stages is the result of high axial velocities through those stages. To cut down on excessive axial velocities, we may bleed off some mass flow in the intermediate stages of the compressor, the so-called intercompressor bleed. Therefore in the starting phase, bleed ports need to be opened to help the high-pressure end of the compressor to operate properly. This proposition represents a relatively low-cost method of starting compressors that do not employ either a split shaft or the variable stators. Stationary gas turbine power plants employ intermediate bleed solution as a method of starting the compressor. This method attacks the aft as well as the front stages, as the bleed ports are located in the middle. Therefore mass withdrawal causes a reduction in the axial velocity to the aft stages as well as lowering the backpressure for the front stages, hence increasing the flow speed in the machine. The ability to *tailor* the flow through the engine to improve overall efficiency, component stability, and provide cooling to accessories and engine components all speak in favor of intercompressor bleed. In fact, all modern gas turbine engines today employ intercompressor bleed/flow control. These methods are summarized in Fig. 7.75.

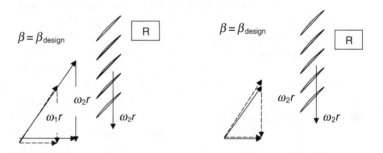

(a) Multiple shaft allows for a higher rotational speed N_2 (b) Bleed allows for a reduced axial velocity

(c) A variable stator turns the flow in the direction of the
rotor in the starting or low-speed operation

■ **FIGURE 7.75**   **Three propositions on how to start a high-pressure compressor**

7.12 The Effect of Inlet Flow Condition on Compressor Performance

Aircraft gas turbine engines operate downstream of an air intake system. The level of *distortion* that an inlet creates at the compressor face affects the performance and the stability of the compressor. First, what do we mean by *distortion*? In simple terms, distortion represents nonuniformity in the flow. The nonuniformity in total pressure as in boundary layers and wakes, the nonuniformity in temperature as in gun gas ingestion or thrust reverser flow ingestion and the nonuniformity in density, as created by hot gas ingestion are some of the different types of distortion. The common feature of all different types of distortion is found in their destabilizing impact on the compressor performance. This means that all distortions reduce the stability margin of a compressor or fan, potentially to the level of compressor stall or the engine surge. The types of distortion are

1. Total pressure distortion $p_t(r, \theta)$
2. Total temperature distortion $T_t(r, \theta)$
3. Flow angle distortion $\alpha(r, \theta)$
4. Secondary flow—swirl at the engine face $C_{\theta 1}(r, \theta)$
5. Entropy distortion $s(r, \theta)$
6. Combinations of some or all of the above distortions

The most common type of inlet distortion is the total pressure distortion that is caused by separated boundary layers in the inlet. Under normal operating conditions, the boundary layers in the inlet are well behaved and remain attached. However, if the boundary layer management system in a supersonic inlet, as in the bleed system, fails to react to an abrupt change in the flight operation (potentially due to a rapid combat maneuver), the flowfield at the engine face will contain large patches of low-energy, low-momentum flow that could cause flow separation in the front stage(s) of the fan or compressor. In describing the total pressure distortion and its impact on compressor performance, we divide the spatial extent of the spoiled flow according to its radial and circumferential extent, as shown in Fig. 7.76.

These inlet total pressure distortion patterns may be simulated by installing screens of varying porosity upstream of the compressor or fan (typically one diameter upstream) in propulsion system ground test facilities. The distortion patterns generated by screens in a test set up in ground facilities represent the "steady-state" component of the distortion and thus lack the "dynamic" or transitory nature of the distortion encountered in real flight environment. The full description of distortion requires both the steady-state and the dynamic components, as in the study of turbulent flow requiring a mean and an rms level of the fluctuation.

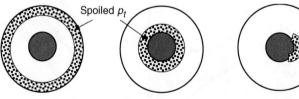

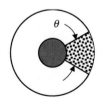

Radial tip distortion | Radial hub distortion circumferential | Circumferential hub distortion | Full-span distortion of angular extent θ

■ **FIGURE 7.76** Different types of total pressure distortion

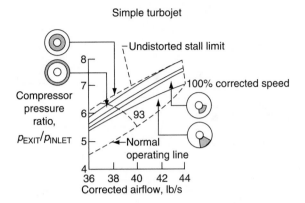

The results of NASA-Glenn 10 × 10 supersonic wind tunnel study on the response of a simple turbojet engine (J-85) to steady-state inlet total pressure distortion is shown in Fig. 7.77 (from Povodny et al., 1970).

The normal operating line, the undistorted stall boundary, and two corrected shaft speeds of 100% and 93% design are shown in dashed lines. Four solid lines correspond to the stall boundaries of the four distortion patterns simulated at the compressor face via screens. From having the least to the most impact on the stall margin deterioration, we identify the culprits as (1) radial hub, (2) radial tip, (3) circumferential hub, and (4) the full-span circumferential distortion, respectively. We also note that the full-span distortion at the 100% corrected speed operates at the stall boundary, i.e., zero-stall margin! Further research identified a critical circumferential extent of the spoiled sector that causes the maximum loss in the stall pressure ratio of a compressor is at nearly 60°, as evidenced in Fig. 7.78 (from Povodny et al., 1970).

The loss in stall pressure ratio with circumferential inlet distortion reaches to ~10% at 100% corrected speed, as shown in Fig. 7.78. At higher shaft speeds, the incidence angle in the spoiled sector is larger than at the lower shaft speed, thus the trend of higher loss of the stalling pressure ratio at higher shaft speeds becomes evident using simple velocity triangle arguments.

The temperature distortion also leads to a reduction in stall margin. In general, static temperature distortion in a flow brings about density nonuniformity, which creates a nonuniform velocity field. Consequently, it is impossible to create a static temperature distortion without creating other forms of nonuniformity, e.g., density, velocity, total pressure, in the flow. To quantify the impact of a spatial temperature distortion on engine stall behavior, NASA researchers have conducted experiments with representative data shown in Fig. 7.79 (from Povodny et al., 1970). The undistorted operating line, stall limit, and different shaft speeds are shown in dashed lines. Data points corresponding to the effect of temperature distortions of 45–120°F on the stall behavior of a variable geometry turbofan engine high-pressure compressor are plotted in solid lines. The circumferential extents of the temperature distortions were 90 and 100° in different tests. A 100°F distortion of ~90–100° circumferential extent is seen to stall the high-pressure compressor operating at its 90% corrected flow.

The inadequacy of steady-state distortion simulation in a wind tunnel is best seen in Fig. 7.80 showing F-111 flight test data of compressor stall with TF-30 engine. A steady-state distortion parameter K_D is graphed at different corrected airflows through the engine, and a band of maximum allowable distortion based on previous wind tunnel tests is shown to miss the mark of in-flight compressor stall data by a large margin (from Seddon and Goldsmith, 1985). The culprit is identified as the *dynamic distortion*.

The steady-state distortion parameter K_D expresses a weighted average of engine face total pressure distortion pattern (circumferential and radial) recorded by an engine face rake

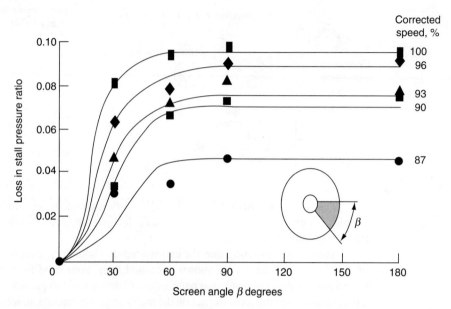

■ **FIGURE 7.78**   **Effect of the extent of circumferential spoiling on compressor performance (from Povodny et al., 1970)**

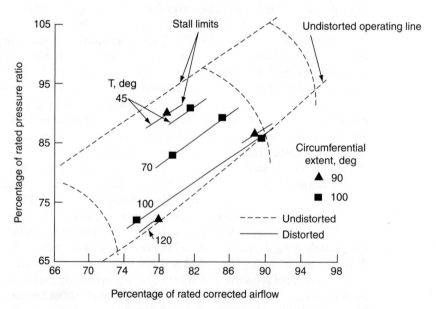

■ **FIGURE 7.79**   **Effect of spatial temperature distortions on engine stall (from Povodny et al., 1970)**

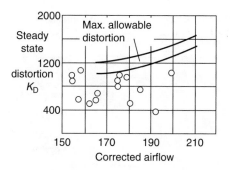

■ **FIGURE 7.80**   **In-flight measurements of engine surge (from Seddon and Goldsmith, 1985)**

system composed of several radial pitot tube measurements. For mathematical description of K_D and other distortion parameters, Hercock and Williams (1974) may be consulted. In addition, Farr and Schumacher (1974) present evaluation methods of dynamic distortion in F-15 aircraft. Schweikhand and Montoya (1974) treat research instrumentation and operational aspects in YF-12. These references are recommended for further reading.

7.13   Isometric and Cutaway Views of Axial-Flow Compressor Hardware

To meet the real world, engineers need to spend time with the real hardware. Isometric and cutaway views of the hardware provide for a three-dimensional feel but may not be a substitute for the cutaway of a real engine (or its components) in the laboratory. The real engine, albeit old and surplus, shows real fasteners, components assembly, manufacturing tolerances; some operational degradation, e.g., wear, on the rotor blade tips, shows labyrinth seals between rotating and stationary parts; it shows signs of erosion in the compressor and turbine and perhaps corrosion in the hot section. A visual inspection and a feel of the compressor and turbine blades show the turbine blades receive heavy deposits from the combustor (and its fuel additives) and give a new meaning to "surface roughness," which will not feel "hydraulically smooth."

Having pointed out some of the advantages of real hardware, let us examine some isometric and cutaway views of axial-flow compressors, designed, built, tested, and flown on aircraft in this section. Figure 7.81 is taken from a manual by Pratt & Whitney Aircraft, 1980, whereas Figs. 7.82–7.84 are taken from Rolls-Royce "The Jet Engine," 2005. Figure 7.81 shows

■ **FIGURE 7.81**
Dual-compressor rotor with two front fan stages (Courtesy of Pratt & Whitney, 1980)

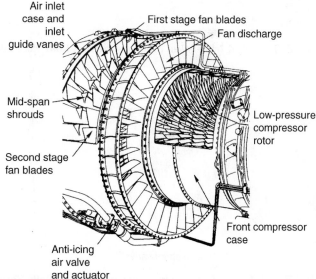

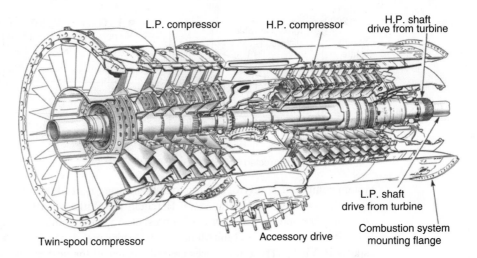

■ **FIGURE 7.82** **A twin-spool compressor (from Rolls-Royce, plc, 2005)**

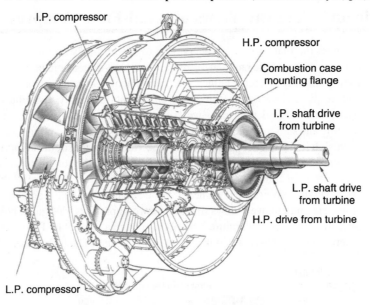

■ **FIGURE 7.83** **A triple-spool compressor (from Rolls-Roce, plc, 2005)**

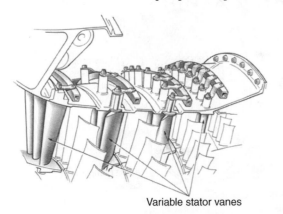

Variable stator vanes

■ **FIGURE 7.84** **Typical variable-stator vanes (from Rolls-Royce, plc, 2005)**

a two-stage fan and the low-pressure compressor rotor. Mid-span shrouds prevent the first bending mode of the first fan rotor blades. The anti-icing air valve and the actuator point to an operational need of the aircraft (i.e., flight under icing condition) and demands on the compressor air.

7.14   Compressor Design Parameters and Principles

In this section, we present design guidelines that are useful in preliminary design of axial-flow compressors. The approach to turbomachinery design in textbooks is neither unique nor exact. In general, we use the lessons learned of our predecessors and the invaluable contributions of NASA and open literature, which often come from academia and industry. One possible approach to compressor design is outlined in steps that are particularly useful for students who want to learn and practice their turbomachinery design skills.

We propose to design an axial-flow compressor for a design-point mass flow rate and pressure ratio. Here, we first review the steps and in the next section, we design an axial compressor at the pitchline. We start the design process by choosing

1. Axial Mach number at the compressor/fan face is $M_{z1} \sim 0.5$
2. Flow area at the compressor face is $A_1 = \pi r_{t1}^2 \left[1 - (r_{h1}/r_{t1})^2 \right]$ that sizes the engine and thus establishes the mass flow rate
3. Typical hub-to-tip radius ratio, r_{h1}/r_{t1} is ~ 0.4, a little less or more (may be up to ~ 0.5 or as low as ~ 0.3). The relative tip Mach number M_{r1} establishes the tip radius
4. $M_{r1} = \sqrt{M_{z1}^2 + M_{T1}^2}$

   - M_{T1}, tip tangential Mach number, i.e., $(\omega r_t)/a_1$ is in excess of sonic
   - M_{r1} is supersonic, ~ 1.2 to ~ 1.5 (remember that $M_z \sim 0.5$)

5. Thickness-to-chord ratio varies from $\sim 10\%$ in the subsonic hub region to $\sim 3\%$ in the supersonic tip region, linearly varying in between
6. The Reynolds number based on chord has to be $> 300{,}000$ (in relative frame), at altitude. This establishes the minimum chord length for turbulent boundary layer at all altitudes, the other parameter contributing to the chord length is the blade aspect ratio and bending stresses. Often chord length is several times bigger than that required for the 300,000 Reynolds number at altitude
7. Centrifugal stress at the blade root (i.e., the hub) is calculated based on the simple formula

$$\sigma_{c,\,root} \equiv \frac{F_{c,\,root}}{A_{blade \cdot root}} \tag{7.200}$$

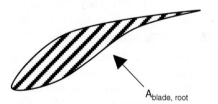

The centrifugal force is the integral of mv^2/r, which is $m\omega^2 r$, i.e.,

$$F_{c,\,root} = \int_{r_h}^{r_t} \rho_{blade} \cdot A(r)\omega^2 r \, dr \tag{7.201}$$

where ρ_{blade} is blade material density and $A(r)$ is the blade cross-sectional area as a function of span.

$$\sigma_{\text{c}} = \frac{1}{A_{\text{h}}} \int_{r_{\text{h}}}^{r_{\text{t}}} \rho_{\text{blade}} \cdot A_{\text{b}}(r) \omega^2 r \, dr \tag{7.202}$$

$$\frac{\sigma_{\text{c}}}{\rho_{\text{blade}}} = \frac{\omega^2}{A_{\text{h}}} \int_{r_{\text{h}}}^{r_{\text{t}}} A_{\text{b}}(r) r \, dr = \omega^2 \int_{r_{\text{h}}}^{r_{\text{t}}} \frac{A_{\text{b}}}{A_{\text{h}}} r \, dr \tag{7.203}$$

$$\frac{\sigma_{\text{c}}}{\rho_{\text{blade}}} = \omega^2 \int_{r_{\text{h}}}^{r_{\text{t}}} \frac{A_{\text{b}}}{A_{\text{h}}} r \, dr \tag{7.204}$$

The blade area distribution along the span, $A_{\text{b}}(r)/A_{\text{h}}$, is known as *taper* and is often approximated to be a linear function of the span. Therefore, it may be written as

$$A_{\text{b}} = A_{\text{h}} - \frac{r - r_{\text{h}}}{r_{\text{t}} - r_{\text{h}}}(A_{\text{h}} - A_{\text{t}}) \rightarrow \frac{A_{\text{b}}}{A_{\text{h}}} = 1 - \frac{r - r_{\text{h}}}{r_{\text{t}} - r_{\text{h}}}\left(1 - \frac{A_{\text{t}}}{A_{\text{h}}}\right) \tag{7.205}$$

We may substitute $A_{\text{b}}(r)/A$ in the integral and proceed to integrate; however, a customary approximation is often introduced that replaces the variable r by the pitchline radius r_{m}. The result is

$$\frac{\sigma_{\text{c}}}{\rho_{\text{blade}}} = \frac{\omega^2 A}{4\pi}\left(1 + \frac{A_{\text{t}}}{A_{\text{h}}}\right) \tag{7.206}$$

The taper ratio $A_{\text{t}}/A_{\text{r}}$ is ~ 0.8–1.0. Therefore, the ratio of centrifugal stress to the material density is related to the square of the angular speed, the taper ratio, and the flow area, $A = 2\pi r_{\text{m}}(r_{\text{t}} - r_{\text{h}})$. Equation 7.206 is the basis of the so-called AN^2 rule, where the *RHS* is related to the size (i.e., A) of the machine and the square of the angular speed (i.e., N^2), and the *LHS* is related to material property known as specific strength.

The material parameter of interest in a rotor is the *creep rupture strength*, which identifies the maximum tensile stress tolerated by the material for a given period of time at a specified operating temperature. Based on the 80% value of the allowable 0.2% creep in 1000 h for aluminum alloys and the 50% value of the allowable 0.1% creep in 1000 h for other materials, Mattingly, Heiser, and Pratt (2002) have graphed the following Figs. (7.85 and 7.86). They show the allowable stress and the allowable specific strength of different engine materials as a function of temperature.

8. The solidity at the pitchline $\sigma_{\text{m}} \sim 1$ to 2

9. The degree of reaction at the pitchline may be chosen to be $^\circ R_{\text{m}} \sim 0.5$ or 0.6 for subsonic sections and considerably higher (e.g., 0.8) for the supersonic portion of the blade. This choice specifies the $C_{\theta 2,\text{m}}$. We can use this information along with Euler turbine equation to establish the rotor-specific work at the pitchline as well as the stagnation temperature rise, i.e., we get T_{t2m}. The choice of $^\circ R$ is less critical than D-factor or de Haller criterion

10. The D-factor at the pitchline should be ≤ 0.5 or 0.55, often a quick check on the acceptable level of diffusion is made using the de Haller criterion that limits diffusion to a level corresponding to $W_2/W_1 \geq 0.72$

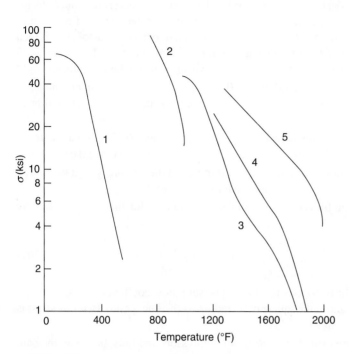

■ **FIGURE 7.85**   Allowable stress versus temperature for typical engine materials (from Mattingly, Heiser, and Pratt, 2002)

Material	No.
Aluminum alloy	1
Titanium alloy	2
Wrought nickel alloy	3
High-strength nickel alloy	4
Single-crystal superalloy	5

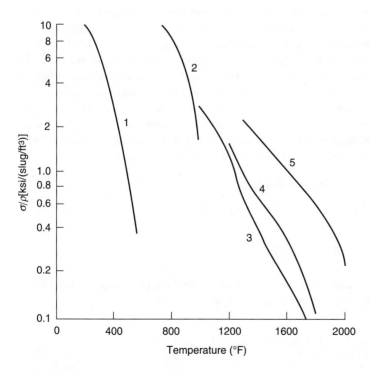

■ **FIGURE 7.86**   Allowable strength-to-weight ratio for typical engine materials (from Mattingly, Heiser, and Pratt, 2002)

11. The choice of the vortex design, i.e., $C_\theta(r)$, establishes the degree of reaction and D-factor at other radii, which we examine closely. It is acceptable for the degree of reaction to deviate from 0.5, but we do not want it to become negative in a compressor (i.e., it then behaves like a turbine!). The D-factor has to remain below the critical value of ~0.5 (or 0.55). We may iterate on the vortex design choice in order to get these parameters right

12. The polytropic efficiency is assumed to be ~0.90, i.e., e_c~0.90, which in combination with T_t ratio gives p_t ratio, i.e., we get total pressure at the pitchline at station 2, downstream the rotor

13. To get the flow parameters downstream of the stator, we preserve T_t and calculate p_{t3} based on an assumed ϖ_s of ~0.04–0.05; also the idea of repeated stage gives us the α_3

14. We assume the axial velocity remains constant throughout the compressor (at the design radius, i.e., the pitchline). This is a preliminary design choice

15. Bending stresses on a cantilevered blade, under aerodynamic loading, is estimated (by Kerrebrock, 1992) to be

$$\frac{\sigma_{\text{bending}}}{p} \approx \left(\frac{C_z}{U_t}\right)(\tau_s - 1)\left(\frac{s}{2c}\right)\left(\frac{r_t}{t_{\max}}\right)^2 \tag{7.207}$$

Bending stress is therefore proportional to gas pressure, flow coefficient at the tip, stage total temperature ratio, and to the square of tip radius-to-thickness ratio. Bending stress is also inversely proportional to solidity

16. Thermal stresses are often small in axial-flow compressors and fans; however, they are significant in the turbine section. Also, the aft stages of high-pressure compressors in modern gas turbine engines operate at high gas temperatures such as T_{gas} ~850–900 K, which require special attention to thermal strains and tip clearance, as well as total stress (i.e., the sum of centrifugal, bending, and thermal) at the hub

17. Blades as cantilevered structures attain diverse mode shapes such as first bending, second bending, first torsional, coupled first bending and torsional, as well as higher modes of vibration. Each mode shape has its own natural frequency ω_n. The effect of blade rotation is to *stiffen* the structure and thus raise these natural frequencies. The shaft and the discs also exhibit their natural vibrational mode shapes and frequencies. To avoid resonance between these frequencies and the shaft rotational speed (and its multiples), a frequency diagram, known as *Campbell diagram* (see Fig. 7.87), is generated to examine possible match points between these frequencies inherent in the compressor, i.e., in the natural modes and rotor shaft frequency. See Kerrebrock (1992), Wilson and Korakianitis (1998), and Mattingley, Heiser, and Pratt (2002) for a detailed account of engine structure.

7.14.1   Blade Design—Blade Selection

For blade design, we need to estimate the following angles:

1. Incidence angle i
2. Deviation angle δ^* $\delta^* = m\varphi/\sqrt{\sigma}$

The incidence angle is estimated at the location of minimum loss, from cascade data that is known as i_{optimum}. Typically, optimum incidence angle is ~$-5°$ to $+5°$.

Iteration loop
The deviation angle is first estimated from the simple Carter's rule:

$$\delta^* \approx \frac{\Delta\beta}{4\sqrt{\sigma}} \tag{7.208}$$

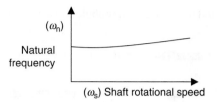

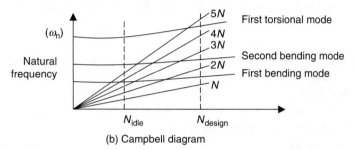

(a) Behavior of natural frequency of a vibration mode with shaft speed

(b) Campbell diagram

■ **FIGURE 7.87**   **Frequency diagram showing possible resonance condition between the dominant vibration modes and multiples of shaft frequency**

which is based on the net flow turning and the coefficient "m" in Carter's formula is taken to be 0.25.

Also, based on the inlet and exit flow angles, we may use the (NACA 65-series) cascade correlation data of Mellor (see Appendix I) to identify a suitable 65-series cascade geometry that gives an adequate stall margin. Note that the front and aft stages of a multistage compressor or fan face different stall challenges, for example, the front stages are more susceptible to positive stall and the aft stages are more susceptible to negative stall. Therefore, we may choose the design point to be farther away from the two stall boundaries for front and aft stages. Figure 7.88 is a definition sketch for the design point selection.

The optimum incidence angle is defined (by Mellor) to be at the point of lowest profile loss, although some authors prefer the optimum incidence to correspond to the angle where lift-to-drag ratio is at its maximum. Here, we adopt Mellor's definition, due to its ease of use. The cascade loss data, as shown in Fig. 7.22, can be used to arrive at the i_{opt}.

Johnsen and Bullock (1965) in NASA SP-36 outline optimum incidence estimation based on cascade data correlations for different profile types and profile thicknesses. This reference may be used for more accurate estimation of the incidence angle. From the inlet flow and an estimate of the optimum incidence angle, we calculate the leading-edge angle on the

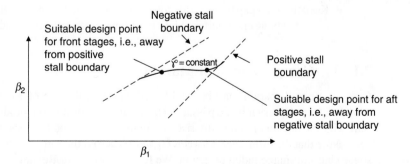

■ **FIGURE 7.88**   **Definition sketch for suitable design points for the front and aft stages of a multistage compressor or fan**

mean camber line (MCL), κ_1, and from the exit flow angle and deviation, we calculate the blade angle at the T.E.

$$i \equiv \beta_1 - \kappa_1 \qquad\Longrightarrow\qquad \kappa_1 \equiv \beta_1 - i$$

$$\delta^* \equiv \beta_2 - \kappa_2 \qquad\Longrightarrow\qquad \kappa_2 \equiv \beta_2 - \delta^* \qquad\Longrightarrow\qquad \varphi \equiv \kappa_1 - \kappa_2$$

The iteration loop on the deviation angle may be initiated at this point, since we have calculated a camber angle and we may now use the expanded version of the Carter's rule:

$$\delta^* = m\varphi/\sqrt{\sigma}$$

Although, we have introduced a possible iteration loop on the deviation angle, but in the preliminary stage of the compressor design, it is acceptable to use the simple formula of Carter for deviation.

The supersonic section is designed based on

- Double-circular arc (DCA) blades
- J-profile design
- Multiple-circular arc design (MCA)

These designs are described in Cumpsty (1989), Schobeiri (2004), and Wilson and Korakianitis (1998).

7.14.2   Compressor Annulus Design

The annulus flow area shrinks inversely proportional to the rising fluid density for a constant axial velocity, i.e.,

$$A(z)/A_1 = \rho_1/\rho(z) \quad \text{for } C_z = \text{constant(say} \sim 500 \text{ fps)}$$

We can calculate density rise per stage (at the pitchline) from the pressure and temperature rise and the use of the perfect gas law. This is an approximate method, but is adequate for the preliminary design purposes.

Common practices are

- Keep the casing, i.e., the tip, radius constant, therefore the information about the annulus area gives the hub radius
- Keep the hub radius constant and thus calculate the tip radius from the annulus area
- Keep the pitchline radius constant, which then shrinks the tip and hub equally, and we can calculate those from the annulus area
- Keep the axial gap between the rotor and stator blade rows to approximately 0.23–0.25 c_z (i.e., about one quarter axial chord length gap between blade rows)

7.14.3   Compressor Stall Margin

Compressor stall margin may be established based on the use of Koch equivalent diffuser model of the compressor blade passage. This is rather laborious, but produces good results. We start with the equivalent enthalpy rise coefficient C_h (for notation, see section 7.10).

Note that the difference between U_1 and U_2 across the rotor (in the numerator) is due to streamline shift in the radial direction. We further correct for Reynolds number, tip clearance gap, and the axial blade spacing other than the nominal values that are used in Koch's stall margin correlations. The correction factors are in Figs. 7.64–7.66. Finally, we correct the adjusted enthalpy rise coefficient for the effect of stagger and the velocity vector diagram (wake effect) to arrive at the *effective* static enthalpy rise coefficient according to

$$(C_h)_{ef} = (C_h)_{adj} \left[\frac{V'^2_1|_{Rotor} + V^2_1|_{Stator}}{V'^2_1|_{Rotor} + F_{ef} \cdot V^2_1|_{Stator}} \right]$$

The critical stalling pressure rise is shown in Fig. 7.72 as "stall line" or the case of 0% stall margin.

Guidelines on the Range of Compressor Parameters

Parameter	Range of values	Typical value
Flow coefficient ϕ	$0.3 \le \phi \le 0.9$	0.6
D-Factor	$D \le 0.6$	0.45
Axial Mach number M_z	$0.3 \le M_z \le 0.6$	0.5
Degree of reaction	$0.1 \le {}^{\circ}R \le 0.90$	0.5 (for $M < 1$)
Reynolds number based on chord	$300{,}000 \le Re_c$	500,000
Tip relative Mach number (1st Rotor)	$(M_{1r})_{tip} \le 1.7$	1.3–1.4
Stage average solidity	$1.0 \le \sigma \le 2.0$	1.4
Stage average aspect ratio	$1.0 \le AR \le 4.0$	1.0
Polytropic efficiency	$0.85 \le e_c \le 0.92$	0.90
Hub rotational speed	$\omega r_h \le 380$ m/s	300 m/s
Tip rotational speed	$\omega r_t \sim 500$ m/s	450 m/s
Loading coefficient	$0.2 \le \psi \le 0.5$	0.35
DCA blade (range)	$0.8 \le M \le 1.2$	Same
NACA-65 series (range)	$M \le 0.8$	Same
De Haller criterion	$W_2/W_1 \ge 0.72$	0.75
Blade leading-edge radius	$r_{L.E.} \sim 5$–10% of t_{max}	5% t_{max}
Compressor pressure ratio per spool	$\pi_c < 20$	up to 20
Axial gap between blade rows	$0.23\,c_z$ to $0.25\,c_z$	$0.25\,c_z$
Aspect ratio, fan	~ 2–5	~ 3
Aspect ratio, compressor	~ 1–4	~ 2
Taper ratio	~ 0.8–1.0	0.8

EXAMPLE 7.8

Design an axial-flow compressor with the following design point parameters:

1. Takeoff at ambient condition

   $T_0 = 25°C(298\text{ K}), \quad p_0 = 101\text{ kPa},$

   $\gamma = 1.4, \quad c_p = 1004\text{ J/kg} \cdot K, \quad M_0 = 0$

2. Overall compressor pressure ratio is 20

3. Polytropic efficiency is assumed constant along the compressor, $e_c = 0.90$

4. Mass flow rate is 100 kg/s

Design choices

 No IGV
 Repeated stage
 Constant axial velocity C_z

Calculate

Compressor
 annulus geometry at the inlet and exit
 number of stages and stage pressure ratio

First Stage
 Reynolds number based on chord
 blade section design at the pitchline

SOLUTION

The design axial Mach number at the compressor face is assumed to be

$$M_{z1} = 0.5$$

The static temperature at the engine face is

$$T_1 = T_{t1}/[1 + (\gamma - 1)M_1^2/2]$$
$$= 298 \, \text{K}[1 + 0.2(0.5)^2] \cong 283.8 \, \text{K}$$

$$a_1 = \sqrt{(\gamma - 1)c_{p1}T_1} = 337.6 \, \text{m/s}$$

Therefore, the axial velocity is $C_{z1} \approx 168.8$ m/s.

The static pressure at the engine face is

$$p_1 = p_{t1}/[1 + (\gamma - 1)M_1^2/2]^{\gamma/(\gamma-1)}$$
$$= 101 \, \text{kPa}/[1 + 0.2(0.5)^2]^{3.5} \cong 85.14 \, \text{kPa}$$

The static density at engine face is

$$\rho_1 = p_1/RT_1$$

The gas constant $R = (\gamma - 1)c_p/\gamma$, which gives $R = 286.86$ J/kg · K and the static density at the engine face is

$$\rho_1 \approx 1.046 \, \text{kg/m}^3$$

The continuity equation for steady, uniform flow is

$$\dot{m} = \rho AV = \rho_1 A_1 C_{z1}$$

We calculate the flow area at the engine face

$$A_1 = (100 \, \text{kg/s})/(1.046 \, \text{kg/m}^3)/(168.8 \, \text{m/s})$$
$$\approx 0.56644 \, \text{m}^2$$

The flow area at the engine face may be written in terms of the tip radius and hub-to-tip radius ratio, according to

$$A_1 = \pi r_{t1}^2 \left[1 - \left(\frac{r_{h1}}{r_{t1}}\right)^2\right]$$

Here, we enter our next assumption, i.e., we choose the hub-to-tip radius ratio at the first stage to be

$$\frac{r_{h1}}{r_{t1}} = 0.5$$

This assumption along with the area A_1 gives the tip radius r_{t1} from the above equation, i.e., we calculate tip radius at the engine face to be

$$r_{t1} = \sqrt{A_1/\pi \left[1 - \left(\frac{r_{h1}}{r_{t1}}\right)^2\right]} \cong 0.49 \, \text{m}$$

Obviously, our hub radius is half the tip radius by design choice, i.e.,

$$r_{h1} \cong 0.245 \, \text{m}$$

Now, let us calculate the compressor discharge condition. The exit total pressure is 20 times the inlet total pressure, therefore,

$$p_{t,\text{exit}} = 20(101 \, \text{kPa}) = 2{,}020 \, \text{kPa}$$

The compressor temperature and pressure ratio are related via the polytropic efficiency following

$$\tau_c = \pi_c^{(\gamma-1)/e_c\gamma}$$

Therefore, $\tau_c \approx 2.5884$, which gives the discharge total temperature to be

$$T_{t,\text{exit}} = 771.3 \, \text{K}$$

We assume the compressor discharge is swirl free and since we had also assumed the axial velocity to be constant along the compressor, we know that

$$C_{\text{exit}} = C_{z,\text{exit}} = \text{Constant} = 168.8 \, \text{m/s}$$

We calculate the static temperature of the gas at compressor discharge, T_{exit}, according to

$$T_{\text{exit}} = T_{t,\text{exit}} - C_{\text{exit}}^2/2c_{p,\text{exit}}$$

Assuming a calorically perfect gas $c_p = 1004$ J/kg · K = constant and, therefore,

$$T_{\text{exit}} = 757.14 \, \text{K}$$

The speed of sound at the exit is

$$a_{\text{exit}} = \sqrt{(\gamma - 1)c_p T_{\text{exit}}} \cong 551.4 \, \text{m/s}$$

Thus, the exit Mach number at the compressor discharge is

$$M_{\text{exit}} = 168.8/551.4 \approx 0.306$$

Note that our choice of constant axial velocity led to a reduction of flow Mach number at the compressor discharge, since the gas temperature and thus speed of sound increases along the compressor, which cause the Mach number to drop. This is a good situation for the burner, since we have to decelerate the compressor discharge gas to around Mach 0.2 for efficient combustion. The prediffuser in the burner has to decelerate the gas from Mach ~0.3 to ~0.2.

From M_{exit} and $p_{t,\text{exit}}$ we get p_{exit}

$$p_{\text{exit}} = p_{t,\text{exit}}/[1 + (\gamma - 1)M_{\text{exit}}^2/2]^{\gamma/(\gamma-1)}$$
$$= 2020 \, \text{kPa}/[1 + 0.2(0.306)^2]^{3.5} \cong 1{,}892.9 \, \text{kPa}$$

We calculate the exit density ρ_{exit} from the exit temperature and pressure, from perfect gas law

$$\rho_{\text{exit}} \approx 1892.9 \, \text{kPa}/[(286.86 \, \text{J/kg} \cdot \text{K})(757.14 \, \text{K})]$$

$$\rho_{\text{exit}} \approx 8.715 \, \text{kg/m}^3$$

From continuity equation, we get the exit flow area,

$$A_{exit} = \frac{\dot{m}}{\rho_{exit} C_z} = \frac{100 \text{ kg/s}}{8.715 \text{ kg/m}^3 (168.8 \text{ m/s})} \cong 0.06797 \text{ m}^2$$

Assuming a constant tip radius r_t = constant, we calculate the hub radius at the compressor exit to be

$$r_{h, exit} = \sqrt{r_t^2 - A_{exit}/\pi}$$
$$= \sqrt{(0.49)^2 - (0.06797)/3.14159265} \approx 0.4677 \text{ m}$$

Note that the channel height (or blade height) at the compressor exit is 0.02258 m $\sim$ 2.2 cm. If we had shrunk the casing, i.e., the tip radius, instead of keeping it constant, the blade height at the exit would increase.

(b) the constant pitchline radius $r_m = (r_h + r_t)/2 = 0.3677$ m, which gives h_{blade} of $\sim$2.94 cm

(c) the constant hub radius, gives an exit channel/blade height of $h_{blade} \sim 4.07$ cm

We note that the increase in blade height between cases (a) and (b) is small, whereas the blade height nearly doubles in case (c) as compared to (a).

Pitchline calculations
Pitchline radius at the compressor face is at $r_{m1} = (r_{h1} + r_{t1})/2 \cong 0.3677$ m
The shaft rpm is selected to give $U_{tip} = 450$ m/s, which is very desirable for transonic compressor pressure ratio. The choice of shaft rpm and the blade geometry will be tested against the centrifugal stress calculations at the blade root.

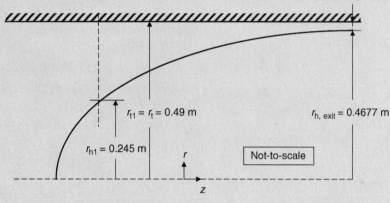

Channel height at the exit

$r_{t1} = r_t = 0.49$ m

$r_{h, exit} = 0.4677$ m

$r_{h1} = 0.245$ m

Not-to-scale

Here, we may do a quick trade study of the effect of outer casing shrinkage on the blade height in the last stage. The annulus area may be written in terms of the pitchline radius and the channel height as

$$A = 2\pi r_m(r_t - r_h)$$

For the exit flow area of 0.06797 m^2, we have the product of pitchline radius and the blade height as

$$r_m \cdot h_{blade} \cong 0.010818 \text{ m}^2$$
$$h_{blade} \cong 0.010818 \text{ m}^2/r_m$$

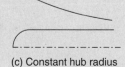

Therefore, the blade or channel height is inversely proportional to the pitchline radius. Here, we examine three common choices:

(a) the constant tip radius, which resulted in h_{blade} of $\sim$2.2 cm

But, for now we choose U_{tip} and then establish shaft rpm.

$U_{tip} = \omega r_t$, which gives

$$\omega = 450 \text{ m/s}/0.49 \text{ m} = 918.4 \text{ rad/s} \cong 8,770 \text{ rpm}$$

Therefore, the rotor speed at pitchline is

$$U_m = U_{tip}(r_m/r_t) = (450 \text{ m/s})(0.3677/0.49)$$
$$\approx 337.5 \text{ m/s}.$$

The flow angles at the inlet to the rotor blade are

$$\alpha_1 = 0 \qquad \text{no IGV(design choice)}$$

$$\beta_1 = \tan^{-1}(337.5/168.8) \approx 63.43°$$

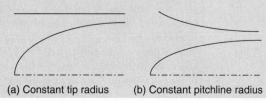

(a) Constant tip radius   (b) Constant pitchline radius   (c) Constant hub radius

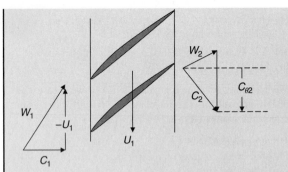

The relative velocity is $W_{m1} \approx 377.4$ m/s, which also gives the relative inlet Mach number at the pitchline of

$$M_{1r,m} = 377.4/337.6 = 1.12 \qquad \text{(pitchline is supersonic)}$$

In order to establish the flow angles at the rotor exit, we may choose the degree of reaction at the pitchline to be 0.5, i.e.,

$$^{\circ}R_m = 0.5 \qquad 1^{st} \text{ Design choice on } ^{\circ}R_m$$

This choice immediately gives $C_{\theta 2m}$, since

$$^{\circ}R_m = 1 - \frac{C_{\theta 1,m} + C_{\theta 2,m}}{2U}$$

The choice of degree of reaction of $1/2$ will make $C_{\theta 2,m} = U_m$ and thus we are asking the rotor relative exit flow to be purely axial, i.e., $\beta_{2m} = 0$.

The problem with this choice is that the net flow turning at the pitchline is $\sim 63°$, which means the blade needs too much camber and thus the excessive diffusion will cause boundary layer to separate. This should be evident from our calculation of the D-factor. Let us see if that fact is borne by our D-factor calculation.

$$D_{r,m} = 1 - \frac{W_{2m}}{W_{1m}} + \frac{|\Delta C_\theta|}{2\sigma_m W_{1m}}$$

$$W_{2m} = C_z = 168.8 \text{ m/s}$$

$$W_{1m} = \left[(168.8)^2 + (337.5)^2\right]^{1/2} = 377.4 \text{ m/s}$$

$$\Delta C_\theta = 337.5 \text{ m/s}$$

Let the solidity at the pitchline be $\sigma_m = 1$ (design choice), therefore we calculate the D-factor at the pitchline to be

$$D_{r,m} = 1.053 \qquad \text{(Unacceptable)}$$

As we suspected, the amount of turning is not tolerable by the boundary layer and it will stall. The D-factors above 0.55 indicate that the boundary layer is on the verge of stall. We are also guided by de Haller criterion that limits the amount of diffusion in the blade, i.e.,

De Haller criterion is $\dfrac{W_2}{W_1} \geq 0.72$

Using de Haller criterion, we arrive at the exit relative velocity at the pitchline of

$$W_{2m} = 271.7 \text{ m/s}$$

Therefore, $\beta_2 = \cos^{-1}(168.8/271.7) \cong 51.6°$

The relative swirl at the rotor exit is

$$W_{\theta 2m} = C_{z2} \tan \beta_{2m} \cong 212.9 \text{ m/s}$$

The loading parameter at the pitchline is

$$\psi_m = 124.6/337.5 \approx 0.369$$

The absolute swirl is $C_{\theta 2m} \approx (337.5 - 212.9)$ m/s = 124.6 m/s,

The absolute flow angle at the rotor exit at the pitchline radius is

$$\alpha_{2m} = \tan^{-1}(124.6/168.8) \approx 36.4°$$

The absolute velocity at the rotor exit is $C_{2m} = (C_z^2 + C_{\theta 2m}^2)^{1/2} = 209.8$ m/s

The stator thus has to turn the flow $36.4°$ and decelerate the flow to the axial velocity of 168.8 m/s if we are to achieve a repeated stage design. Where does this deceleration fall with respect to the de Haller criterion?

$$C_3/C_2 = 168.8/209.8 = 0.805$$

This clearly meets the De Haller criterion. Now, let us supplement our knowledge of rotor and stator diffusion by calculating their D-factors.

$$D_{r,m} = 1 - \frac{W_{2m}}{W_{1m}} + \frac{|\Delta C_\theta|}{2\sigma_{r,m} W_{1m}} \cong 1 - \frac{271.7}{377.4}$$
$$+ \frac{124.6}{2(1)(377.4)} \cong 0.445$$

$$D_{s,m} = 1 - \frac{C_{3m}}{C_{2m}} + \frac{|\Delta C_\theta|}{2\sigma_{s,m} C_{2m}} \cong 1 - \frac{168.8}{209.8}$$
$$+ \frac{124.6}{2(1.5)(209.8)} \cong 0.433$$

In the rotor calculation, we chose the pitchline solidity of $\sigma_{r,m} = 1$, whereas for the stator, we chose a pitchline solidity of $\sigma_{s,m} = 1.25$ (see the stator blade design section where this choice becomes clear). These are the kind of design choices that we can make in early stages of compressor preliminary design. We may revisit them at any point in our design process.

The pitchline degree of reaction is $^{\circ}R_m = 1 - (C_{\theta 1m} + C_{\theta 2m})/2U_m = 1 - 124.6/2(337.5) = 0.815$

Rotor blade design at pitchline

We choose a double-circular arc blade for the supersonic section at the pitchline. The relative flow in supersonic blade sections is tangent to the upper surface. Also, we have to choose a blade thickness-to-chord ratio at the pitchline. The minimum t/c is at the tip, and for structural reasons it is limited to 3%. The hub, which operates in the subsonic relative flow, may be given a t/c of 10%. If we choose these t/cs as our design choices, and then impose a linear variation of the t/c along span, we arrive at the pitchline t/c of 6.5%.

Before we embark on the blade section design, we may establish the minimum chord length that is required for the turbulent boundary layer formation on the blade. The criterion of $Re_c > 300,000$ is clearly a useful guideline, but we have to satisfy that at all altitudes. In the low-density flight associated with the highest flight altitude in the flight envelope, we calculate the required chord length.

At 16-km altitude; the air density is ~ 0.136 times the sea level density. Therefore, at the engine face we will have

$$\rho_1 \approx (0.136)(1.046\,\text{kg/m}^3) \cong 0.142\,\text{kg/m}^3$$

Also the kinematic viscosity ν at 16-km altitude is ~ 5.849 times the kinematic viscosity of air at sea level.

$$Re_c = \frac{W_{1m} \cdot c_m}{\nu_1} = \frac{377.4(\text{m/s}) \cdot c_m}{8.54 \times 10^{-5}\,\text{m}^2/\text{s}}$$
$$\geq 300,000 \quad \Longrightarrow \quad c_m > 6.8\,\text{cm}$$

We may expand the chord length beyond this minimum for the following reasons:

- blade structural calculations (demanding a wider chord)
- blade aspect ratio (pointing to the advantages of a wide-chord design)
- blade vibrational modes (from Campbell diagram)

The relative inlet flow is aligned with the upper surface, therefore, the incidence angle may be written as

$$i \approx \frac{\theta_{\text{L.E.}}}{2} \approx \left(\frac{t_{\max}}{c}\right) \quad \Longrightarrow \quad i \approx 3.7°$$

The deviation angle may be estimated form the Carter's basic rule and then supplemented by $2°$, since the effect of shock-boundary layer interaction is to thicken the boundary layer on the suction surface, thus, causing a higher deviation angle, i.e.,

$$\delta^* \approx \frac{\Delta\beta}{4\sqrt{\sigma}} + 2° \quad \Longrightarrow \quad \delta^* \approx 5°$$

The blade leading edge angle $\kappa_{1,m} = \beta_{1,m} - i$
$$\Longrightarrow \quad \kappa_{1,m} \approx 59.7°$$

The blade trailing-edge angle $\kappa_{2,m} = \beta_{2,m} - \delta^*$
$$\Longrightarrow \quad \kappa_{2,m} \approx 46.6°$$

Therefore, the camber angle is $\varphi_m = \kappa_{1,m} - \kappa_{2,m}$
$$\Longrightarrow \quad \varphi_m \approx 13.1°$$

The stagger angle, for a double-circular arc blade, is related to the inlet flow angle, camber, and incidence angle according to

$$\beta_1 = \gamma° + \frac{\varphi}{2} + i \quad \Longrightarrow \quad \gamma° \approx 60.6°$$

The number of rotor blades may now be calculated since we have chosen the solidity at the pitchline and selected the chord length. The blade spacing at the pitchline is

$$s_{r1,m} = \frac{2\pi r_{m1}}{N_{R1}}$$

We also chose the rotor solidity of 1 at the pitchline, which gives

$$\sigma_{r1,m} = \frac{c_{r1,m}}{s_{r1,m}} = 1.0$$

Therefore, the number of rotor blades is $N_{R1} = 2\pi r_{m1}/c_{rm,1}$

For a rotor chord length of 7 cm (or 0.07 m) and the pitchline radius of 0.3677 m, we get the number of rotor blades in the first stage to be

$$N_{R1} = 33$$

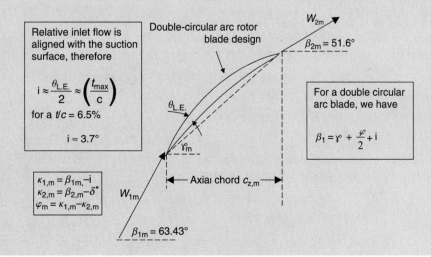

Relative inlet flow is aligned with the suction surface, therefore

$$i \approx \frac{\theta_{\text{L.E.}}}{2} \approx \left(\frac{t_{\max}}{c}\right)$$

for a $t/c = 6.5\%$

$$i \approx 3.7°$$

$$\kappa_{1,m} = \beta_{1m} - i$$
$$\kappa_{2,m} = \beta_{2,m} - \delta^*$$
$$\varphi_m = \kappa_{1,m} - \kappa_{2,m}$$

W_{1m}

$\beta_{1m} = 63.43°$

Double-circular arc rotor blade design

W_{2m}

$\beta_{2m} = 51.6°$

$\theta_{\text{L.E.}}$

γ_m

Axial chord $c_{z,m}$

For a double circular arc blade, we have

$$\beta_1 = \gamma° + \frac{\varphi}{2} + i$$

Stator blade design at pitchline

The design of the stator blade at the pitchline follows the flow angles and Mach numbers in the absolute frame, i.e., we know that

$$\alpha_2 = 36.4°$$

$$\alpha_3 = 0° \quad \text{(repeated stage)}$$

To calculate the Mach number at the entrance to the stator, we need to calculate the static temperature and speed of sound at the pitchline radius downstream of the rotor.

From Euler turbine equation $T_{t2,m} = T_{t1} + U_m(C_{\theta 2} - C_{\theta 1})/c_p$

We made the assumption of no radial shift across the rotor and a calorically perfect gas. These assumptions or approximations are reasonable at the preliminary stage of design.

$$T_{t1} = 298\,\text{K},\ U_m = 337.5\,\text{m/s},\ C_{\theta 1} = 0,$$
$$C_{\theta 2} = 124.6\,\text{m/s},\ c_p = 1{,}004\,\text{J/kg·K}$$

Substitute these values in the Euler equation to get $T_{t2,m} = 339.9$ K

Since the static temperature and total temperature are related via $T = T_t - C^2/2c_p$, we calculate the static temperature upstream of the stator at the pitchline radius

$$T_{2,m} \approx 318\,\text{K with a corresponding}$$
$$a_{2,m} = 357.3\,\text{m/s}$$

$$M_{2,m} = 209.8/357.3 \approx 0.587$$

$$\text{(subsonic stator flow)}$$

$$C_{2,m} = 209.8\,\text{m/s}$$

The stator blade at the pitchline is a subsonic section. Therefore, we may choose NACA-65 series cascade profiles from the data of Mellor (see Appendix I for a complete description) that

(a) best match the inlet and exit flow angles

(b) provide for a reasonable "stall margin"

We choose the **65-(18)10** cascade with a solidity of $\sigma = 1.25$, where the intersection of the inlet and exit flow angles yield (note that the inlet and exit angles to the stator are α_2 and α_3, respectively)

$$\gamma° \approx 15°$$

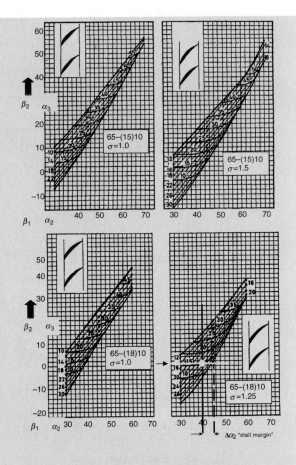

Angle of attack, $\alpha_2 - \gamma° \approx 22°$ [which is consistent with our $\alpha_2 = 36.7°$ and the stagger of 15°)

We note a tolerance to inlet flow angle variation of ~6.5° margin to the positive stall boundary. We may use this range of inlet flow angle variation to arrive at an approximate "stall margin" for the stator blade section at the pitchline in the first stage. A more accurate approach to stall margin calculation is based on Koch's method that we have detailed in this chapter.

Stage pressure ratio and number of stages

The total temperature rise across the first stage, at the pitchline, is

$$\Delta T_t = 41.9\,\text{K}$$

For a repeated stage and assuming the blade rotational speed U remains nearly constant at the pitchline of subsequent stages, we conclude that every stage achieves $\Delta T_t = 41.9$ K. The total stagnation temperature rise for the compressor was calculated to be

$$\Delta T_{t,\text{compressor}} = 771.3\,\text{K} - 298\,\text{K} = 473.3\,\text{K}$$

(assuming a calorically-perfect gas)

If each stage produces 41.9 K total temperature rise, we need 11.29 stages to achieve the needed total temperature rise for the compressor (at the pitchline). Based on the approximations that we have made, e.g., calorically perfect gas and representing the 3D behavior of the compressor by its pitchline, we choose 12 stages for the compressor.

$$N_{\text{stages}} = 12$$

We calculate the stage pressure ratio π_s for each stage, and graph it in the following figure.

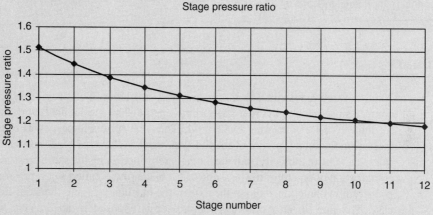

Stage pressure ratio

Note that the compressor pressure ratio at the pitchline, after 12 stages, will be 22.5 (which is >20, the compressor design pressure ratio). The pitchline, however, does not represent the "average" of the stage pressure ratio. The main reason is that the pitchline is away from the end walls where losses are predominant. Therefore, we expect lower total pressure ratio per stage as that represented by our pitchline calculation. Another note is that the polytropic efficiency e_c should not be treated as constant along a multistage compressor. As compression progresses, the end wall losses grow and thus the slope of ideal-to-actual work (for small stage) drops. The polytropic efficiency may be as high as 0.90 (or even 0.92) for early stages, and as low as 0.85 in the aft stages.

Rotor blade stress calculations
The rotor blade is subject to centrifugal as well as bending, vibrational, and thermal stresses. The dominant stress in a rotor, however, is the centrifugal stress σ_c.

$$\sigma_c \equiv \frac{F_c}{A_{\text{hub}}}$$

$$F_c = \int_{r_h}^{r_t} \rho_{\text{blade}} \cdot A_b(r)\omega^2 r\,dr$$

$$\sigma_c = \frac{1}{A_h} \int_{r_h}^{r_t} \rho_{\text{blade}} \cdot A_b(r)\omega^2 r\,dr$$

$$\frac{\sigma_c}{\rho_{\text{blade}}} = \frac{\omega^2}{A_h} \int_{r_h}^{r_t} A_b(r)r\,dr = \omega^2 \int_{r_h}^{r_t} \frac{A_b}{A_h} r\,dr$$

$$\frac{\sigma_c}{\rho_{\text{blade}}} = \omega^2 \int_{r_h}^{r_t} \frac{A_b}{A_h} r\,dr$$

The blade area distribution along the span, $A_b(r)/A_h$, is known as *taper* and is often approximated to be a linear function of the span. Therefore, it may be written as

$$A_b = A_h - \frac{r - r_h}{r_t - r_h}(A_h - A_t)$$

$$\frac{A_b}{A_h} = 1 - \frac{r - r_h}{r_t - r_h}\left(1 - \frac{A_t}{A_h}\right)$$

We may substitute $A_b(r)/A$ in the integral and proceed to integrate; however, a customary approximation is often introduced that replaces the variable r by the pitchline radius r_m. The result is

$$\frac{\sigma_c}{\rho_{\text{blade}}} = \frac{\omega^2 A}{4\pi}\left(1 + \frac{A_t}{A_h}\right)$$

Therefore, the ratio of centrifugal stress to the material density is related to the square of the angular speed, the taper ratio, and the flow area $A = 2\pi r_m(r_t - r_h)$.

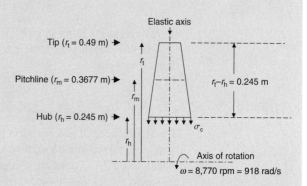

Titanium alloy is suitable for compressor/fan rotor blades, due to its low density, high-strength characteristics, which makes it a high strength-to-weight ratio, or high specific-strength material of choice for rotor blades in a compressor or fan. The density of titanium alloy is

$$\rho_{titanium} = 4,680 \, \text{kg/m}^3$$

The material parameter of interest in a rotor is the *creep rupture strength*, which identifies the maximum tensile stress tolerated by the material for a given period of time at a specified operating temperature. Based on the 80% value of the allowable 0.2% creep in 1000 h, for aluminum alloys and the 50% value of the allowable 0.1% creep in 1000 h for other materials, Mattingly, Heiser, and Pratt (2002) have graphed figures 7.85 and 7.86. They show the allowable specific strength of different engine materials as a function of temperature.

The allowable strength-to-weight ratio for the titanium alloy is noted to be $\sim$9 ksi/slug/ft^3, which is equivalent to

Allowable creep   120,560 m^2/s^2 (or $\sim$120 kPa/kg/m^3)
rupture strength
Required strength   68,326 m^2/s^2 (or $\sim$68 kPa/kg/m^3)

7.15   Summary

Axial-flow compressors are highly evolved machines that provide efficient mechanical compression (of the gas) in a gas turbine engine. The mechanical work is delivered to the medium by a set of rotating blade rows. The rotor blades impart swirl to the flow, thereby, increasing the total pressure of the fluid. In between the rotor blade rows are the stator blade rows that remove the swirl from the fluid, thereby increasing the static pressure of the fluid. The combination of one rotor row and one stator blade row is called a compressor stage. Since the primary flow direction is along the axis of the machine, staging the axial-flow compressor is easy. The flow in an axial-flow compressor is exposed to an *adverse pressure gradient* environment, i.e., a climbing pressure hill that tends to stall the boundary layer. Therefore, the phenomenon of *stall* is experienced in a compressor. The stalled compressor flow is inherently unsteady and thus may cause a system-wide instability (between the compressor and combustor) known as *surge*.

In analyzing the flow in the rotor, it is most convenient to use the rotating frame of reference that spins with the rotor, which is known as the *relative frame of reference*. The flow in stator blade row is best analyzed by a stationary observer (i.e., a frame of reference that is fixed with respect to the casing) known as the *absolute frame of reference*. The velocity vectors in the two frames of reference form a triangle, described by

$$\vec{C} - \omega r \hat{e}_\theta = \vec{W}$$

The *Euler turbine equation* is known as the fundamental equation in turbomachinery that relates the change of angular momentum $\Delta (rC_\theta)$ across a rotor blade row to the rotor specific work via

$$w_c \equiv \frac{\wp_c}{\dot{m}} = \omega\Delta(rC_\theta) \tag{7.9-a}$$

If we express the jump in swirl across a rotor in terms of the (rotor) inlet absolute and exit relative flow angles (that remain constant over a wide operating range of the compressor), we demonstrate that

$$\tau_c - 1 = \left(\frac{U^2}{c_p T_{t1}}\right)\left[1 + \frac{C_{z2}}{U}\tan\beta_2 - \frac{C_{z1}}{U}\tan\alpha_1\right] \tag{7.25-a}$$

The significance of this equation is the appearance of U^2 in front of the bracket on the RHS, which indicates that the total temperature rise across a rotor is proportional to the square of the wheel speed U^2. Therefore a high blade Mach number is desirable, which is exactly the argument for the development of *transonic compressors*. A rule of thumb is that the blade tangential Mach number at the tip is slightly supersonic, $M_T \sim 1.2$–1.3. The structural limitations currently limit the blade tip Mach number to ~ 1.5 in conventional (subsonic throughflow) compressors. A high strength-to-weight ratio material, such as titanium, is desirable for fan blade construction.

Since we have to limit the level and extent of the flow diffusion in a compressor blade row, we define a *diffusion factor D* for the rotor and stator blade rows according to

$$D_r \equiv 1 - \frac{W_2}{W_1} + \frac{|W_{\theta 2} - W_{\theta 1}|}{2\sigma_r W_1} \qquad \text{(rotor } D\text{-Factor)}$$

$$D_s \equiv 1 - \frac{C_3}{C_2} + \frac{|C_{\theta 3} - C_{\theta 2}|}{2\sigma_s C_2} \qquad \text{(stator } D\text{-Factor)}$$

The experience shows that diffusion factor is to be limited to ~ 0.6 about the mid-span of a blade row and to ~ 0.4 near the hub or the tip. The lower diffusion imposed on the hub and the tip is due to a complex hub boundary layer/corner vortex formation and the tip clearance flows that dominate the two ends of a blade row. We may view the diffusion factor as a blade-row-specific figure of merit. We define a stage-based figure of merit as well, which is called the *degree of reaction* $°R$,

$$°R \equiv \frac{h_2 - h_1}{h_3 - h_1}$$

Degree of reaction is the ratio of the rise of the static enthalpy across the rotor to that of the stage. In essence, it speaks to the rotor's share of the static pressure rise to that of the stator. Therefore, a 50% degree of reaction stage equally divides the burden of the pressure rise to the rotor and the stator. Although 50% sounds desirable for equal burden, but since the rotor blades spin, their boundary layers are more stable and thus can withstand higher static pressure rise than the stationary stator blades. Consequently, a 60% degree of reaction may be more optimal than the 50%.

Cascade aerodynamics provides a rich background for a two-dimensional (subsonic) blade design. The minimum-loss incidence angle i_{opt} and the cascade (total pressure) loss bucket are used for the preliminary 2D design of compressor blade sections as well as identifying the *positive and negative stall boundaries*. The limited supersonic cascade data make the preliminary design of the supersonic sections of transonic fans less grounded. The general rules are based in gas dynamics, which indicate the selection of the thinnest possible sections in order to minimize the wave drag. The thinnest (structurally feasible) section is $\sim 3\%$ thick and the subsonic root is often $\sim 10\%$ thick.

Three-dimensional design of blades, in the preliminary stage, is achieved by the so-called *vortex design* approach. Here we introduce a catalog of swirl profiles, such as free-vortex, solid-body rotation, and others that will be anchored at the pitchline and then give us a swirl profile along the span. We applied radial equilibrium theory to calculate an axial velocity profile. Three-dimensional losses are due to secondary flows, tip clearance, and the blade junction corner vortex. The unsteady flow losses are related to upstream wake chopping (by the downstream blade row) and the subsequent vortex shedding in the wake. Losses due to compressibility are minimized by thin profile designs that form the sections of swept blades.

A single-spool compressor is of limited capability in high-pressure compressor applications. The primary reason for that is the differing rotational requirement of the front versus the aft stages. This is the crux of the classical *starting problem* for a high-pressure ratio compressor. Multistage compressors are often driven by different shafts in order to spin the high-pressure stages faster than the low-pressure compressor. Variable stators in the front stages are always adjustable to help with the starting problem as well as the off-design operation of the compressor. A fraction of airflow in the compressor is bled at the intermediate and exit sections to cool the HPT, the casing, and the exhaust nozzle. The cooling fraction is ~10–15% in modern engines.

As an upper bound, multispool, high-pressure ratio compressors can achieve a pressure ratio of ~45–50. The exit temperature of ~900 K is the limit of current materials for an uncooled compressor, which limit the compressor pressure ratio. The advances in computational fluid dynamics, parallel processing, and computer memory have elevated the design of compressors to be based on flow physics and with less reliance on empiricism and cascade data. The result has been the appearance of high-efficiency unconventional transonic blades and stages with *forward and aft swept blades* and *lean*. The advances in integrated manufacturing technology that involve super plastic forming and diffusion bonding (SPF/DB) are used for the manufacturing of modern (composite) wide-chord fan blades. New materials and manufacturing technology, e.g., BLISK, offer weight savings in compressors with the subsequent improvement on the engine thrust-to- weight ratio.

Dixon [7], Cheng et al. [2], Prince, Wisler and Hilvers [41], Whitcomb and Clark [53] are recommended for additional reading.

References

1. Bullock, R.O. and Finger, H.B., "Compressor Surge Investigated by NACA," *SAE Journal*, Vol. 59, September 1951, pp. 42–45.

2. Cheng, P., Prell, M.E., Greitzer, E.M., and Tan, C.S., "Effects of Compressor Hub Treatment on Stator Stall and Pressure Rise," *AIAA Journal*, Vol. 21, No. 7, July 1984.

3. Carter, A.D.S., "The Axial Compressor," in *Gas Turbine Principles and Practice*, Ed., Cox, H.R., Newnes Ltd., London, UK, 1955.

4. Cumpsty, N., *Compressor Aerodynamics*, Cambridge University Press, UK, 1989.

5. Cumpsty, N., *Jet Propulsion*, Cambridge University Press, Cambridge, UK, 1997.

6. De Haller, P., "Das Verhalten von Tragfluegelgittern in Axialverdichtern und im Windkanal," *Brenstoff und Waermekraft*, Vol. 5, 1953.

7. Dixon, S.L., *Fluid Mechanics, Thermodynamics of Turbomachinery*, 2nd edition, Pergamon Press, Oxford, UK, 1975.

8. Donaldson, C. Du P., and Lange, R.H., "Study of Pressure Rise Across Shock Waves Required to Separate Laminar and Turbulent Boundary Layers," NACA Technical Note 2770, 1952.

9. Emmons, H.W., Pearson, C.E., and Grant, H.P., "Compressor Surge and Stall Propagation," *Transactions of the ASME*, Vol. 79, May 1955, pp. 455–467.

10. Farr, A.P. and Schumacher, G.A., "System for Evaluation of F-15 Inlet Dynamic Distortion," Paper in *Instrumentation for Airbreathing Propulsion*, Progress in Astronautics and Aeronautics, Vol. 34, Eds. Fuhs, A.E. and Kingery, M., MIT Press, Cambridge, Mass, 1974.

11. Garvin, R.V., *Starting Something Big: The Commercial Emergence of GE Aircraft Engines*, AIAA, Inc., Reston, Virginia, 1998.

12. Greitzer, E.M., "Surge and Rotating Stall in Axial Flow Compressors," *ASME Journal of Engineering for Power*, Vol. 98, No. 2, 1976, p. 190.

13. Gostelow, J.P., *Cascade Aerodynamics*, Pergamon Press, Oxford, UK, 1984.

14. Hawthorne, W.R., "Secondary Circulation in Fluid Flow," *Proceedings of Royal Society*, London, Vol. 206, 374, 1951.

15. Hechert, H., Steinert, W., and Lehmann, K., "Comparison of Controlled Diffusion Airfoils with Conventional NACA-65 Airfoils Developed for Stator Blade Application in a Multistage Axial Compressor," *Transactions of the ASME, Journal of Engineering for Gas Turbines and Power*, Vol. 107, April 1985, pp. 494–498.

16. Hercock, R.G. and Williams, D.D., "Distortion-Induced Engine Instability: Aerodynamic Response," AGARD, LS72-Paper No. 3, 1974.

17. Herrig, L.J., Emery, J.C., and Erwin, J.R., "Systematic Two-Dimensional Cascade Tests of NACA 65-Series Compressor Blades at Low Speeds," NACA-RM L51G31, 1951.

18. Hill, P.G. and Peterson, C.R., "Mechanics and Thermodynamics of Propulsion," 2nd edition, Addison-Wesely, Reading, Massachusetts, 1992.

19. Hobbs, D.E. and Weingold, H.D., "Development of Controlled Diffusion Airfoils for Multistage Compressor Applications," *Transactions of ASME, Journal of Engineering for Gas Turbines and Power*, Vol. 106, April 1984, pp. 271–278.

20. Horlock, J.H., *Axial Flow Compressors*, Krieger Publishing Company, Huntington, NY, 1973.

21. Howell, A.R., "Design of Axial Compressors", *Proceedings of Institution of Mechanical Engineers*, London, Vol. 153, 1945.

22. Howell, A.R., "Fluid Dynamics of Axial Compressors", *Proceedings of Institution of Mechanical Engineers*, London, Vol. 153, 1945.

23. Johnsen, I.A. and Bullock, R.O.,"Aerodynamic Design of Axial-Flow Compressors," NASA SP- 36, Washington, D.C., 1965.

24. Kerrebrock, J.L., "Flow in Transonic Compressors," *AIAA Journal*, Vol. 19, No. 1, 1981, pp. 4–19.

25. Kerrebrock, J.L., *Aircraft Engines and Gas Turbines,* 2nd edition, MIT Press, Cambridge, Mass., 1992.

26. Kerrebrock, J.L., "Small Disturbances in Turbomachine Annuli with Swirl," *AIAA Journal*, Vol. 15, June 1977, pp. 794–803.

27. Kerrebrock, J.L., Mikolajczak, A.A., "Intra-Stator Transport of Rotor Wakes and Its Effect on Compressor Performance," *ASME Journal of Engineering for Power*, October 1970, p. 359.

28. Kline, S.J., Abbott, D.E., and Fox, R.W., "Optimum Design of Straight-Walled Diffusers," *Journal of Basic Engineering*, Vol. 81, serried, D, No. 3, September 1959, pp. 321–331.

29. Koch, C.C., "Stalling Pressure Rise Capability of Axial Flow Compressor Stages," *Transactions of the ASME, Journal of Engineering for Power*, Vol. 103, October 1981, pp. 645–656.

30. Kotidis, P.A. and Epstein, A.H., "Unsteady Radial Transport in a Transonic Compressor Stage," *Transactions of ASME, Journal of Turbomachinery*, Vol. 113, April 1991, pp. 207–218.

31. Lieblein, S., Schwenk, F.D., and Broderick, R.L., "Diffusion Factor for Estimating Losses and Limiting Blade loadings in Axial-Flow Compressor Blade Element," NACA RM E53D01, June 1953.

32. Lieblein, S.,"Loss and Stall Analysis of Compressor Cascades," *Transactions of the ASME, Journal of Basic Engineering*, September 1959, pp. 387–400.

33. Lieblein, S., "Experimental Flow in Two-Dimensional Cascades," in *Aerodynamic Design of Axial Flow Compressors*, NASA SP-36, 1965.

34. Lieblein, S. and Roudebush, W.H., "Theoretical Loss Correlation for Low-Speed Two-Dimensional Cascade Flow," NACA TN 3662, 1956.

35. Marble, F.E., Three-Dimensional Flow in Turbomachines, in *Aerodynamics of Turbines and Compressors* Vol. X, Ed. Hawthorne, W.R., Princeton Series on High Speed Aerodynamics and Jet Propulsion,Princeton University Press, Princeton, N.J., 1964.

36. Mattingly, J.D., Heiser, W.H., and Pratt, D.T., *Aircraft Engine Design*, 2nd edition, AIAA Education Series, AIAA, Reston, VA, 2002.

37. Oyama, A., Liou, M.-S., and Obayashi, S., "Transonic Axial-Flow Blade Shape Optimization using Evolutionary Algorithm and Three-Dimensional Navier-Stokes Solver," AIAA Paper Number 2002–5642, 2002.

38. Paduano, J.D., Greitzer, E.M., and Epstein, A.H., "Compression System Stability and Active Control,"

Annual Review of Fluid Mechanics, Vol. 33, 2001, pp. 491–517.

39. Povodny, J.H. et al. "Effects of Engine Inlet Disturbances on Engine Stall Performance," Paper in NASA-SP 259, Aircraft Propulsion, 1970, pp. 313–351.

40. Pratt and Whitney, *Aircraft Gas Turbine Engines and Its Operation*, P&W Operations Manual 200, 1980.

41. Prince, D.C., Jr., Wisler, D.C., and Hilvers, D.E., "Study of Casing Treatment Stall Margin Improvement Phenomena," ASME Paper No. 75-GT-60, 1975.

42. Reneau, L.R., Johnston, J.P. and Kline, S.J., "Performance and Design of Straight, Two-Dimensional Diffusers," *Journal of Basic Engineering*, ASME Transactions, Series D, Vol. 89, 1967, pp. 141–150.

43. Rolls-Royce, *The Jet Engine*, Rolls-Royce plc, Derby, England, 2005.

44. Rhoden, H.G., "Effects of Reynolds Number on the Flow of Air through a Cascade of Compressor Blades," ARC R&M 2919, 1956.

45. Schobeiri, M.T., *Turbomachinery Flow Physics and Dynamic Performance*, Springer Verlag, Berlin, 2004.

46. Schweikhard, W.G., Montoya, E.J., "Research Instrumentation Requirements for Flight Wind Tunnel Tests of the YF-12 Propulsion System and Related Flight Experience," Paper in *Instrumentation for Airbreathing Propulsion*, Progress in Astronautics and Aeronautics, Vol. 34, Eds. Fuhs, A.E. and Kingery, M., MIT Press, Cambridge, Mass, 1974.

47. Seddon, J. and Goldsmith, E.L., *Intake Aerodynamics*, American Institute of Aeronautics and Astronautics, Inc., Washington, D.C., 1985, pp. 292–319.

48. Smith, L.H., "Recovery Ratio—A Measure of the Loss Recovery Potential of Compressor Stages," *Transactions of the ASME*, Vol. 80, No. 3, April 1958, pp. 517–524.

49. Sovran, G. and Klomp, E.D., "Experimentally Determined Optimum Geometries for Rectilinear Diffusers with Rectangular, Conical or Annular Cross Section," *Fluid Mechanics of Internal Flow*, Elsevier Publishing, Amsterdam, The Netherlands, 1967.

50. Squire, H.B. and Winter, K.G.,"The Secondary Flow in a Cascade of Airfoils in a Non-Uniform Stream," *Journal of Aeronautical Sciences*, Vol.18, No. 271,1951.

51. St. Peter, J., *The History of Aircraft gas Turbine Engine Development in the United States*, International Gas Turbine Institute, Atlanta,1999.

52. Sulam, D.H., Keenan, M.J., and Flynn, J.T., "Data and Performance of a Multiple Circular Arc Rotor," *Single-Stage Evaluation of a Highly-Loaded High-Mach-Number Compressor Stages*, Vol.II, NASA CR-72694, 1970.

53. Whitcomb, R.T. and Clark, L.R., "An Airfoil Shape for Efficient Flight at Supercritical Mach Numbers," NASA TMX-1109, July 1965.

54. Wisler, D.C.,"Advanced Compressor and Fan Systems," *UTSI Short course notes on Aero-Propulsion Systems,* April 2000.

Problems

7.1 The absolute flow at the pitchline to a compressor rotor has a coswirl with $C_{\theta 1} = 78$ m/s. The exit flow from the rotor has a positive swirl, $C_{\theta 2} = 172$ m/s. The pitchline radius is at $r_m = 0.6$ m and the rotor angular speed is $\omega = 5220$ rpm. Calculate the specific work at the pitchline and the rotor torque per unit mass flow rate.

7.2 An axial-flow compressor stage has a pithline radius of $r_m = 0.6$ m. The rotational speed of the rotor at pitchline is $U_m = 256$ m/s. The absolute inlet flow to the rotor is described by $C_{zm} = 155$ m/s and $C_{\theta 1m} = 28$ m/s. Assuming that the stage degree of reaction at pitchline is $^\circ R_m = 0.50$, $\alpha_3 = \alpha_1$, and C_{zm} remains constant, calculate

(a) rotor angular speed ω in rpm
(b) rotor exit swirl $C_{\theta 2m}$
(c) rotor specific work at pitchline, w_{cm}
(d) relative velocity vector at the rotor exit
(e) rotor and stator torques per unit mass flow rate
(f) stage loading parameter at pitchline, ψ_m
(g) flow coefficient φ_m

7.3 A rotor blade row is cut at pitchline, r_m. The velocity vectors at the inlet and exit of the rotor are shown.

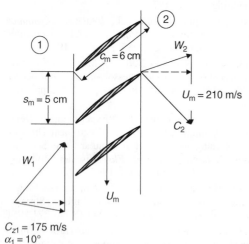

■ **FIGURE P7.3**

Assuming that $U_{m1} = U_{m2} = 210$ m/s and $C_{z1} = C_{z2} = 175$ m/s, $\rho_1 = 1$ kg/m^3 $\beta_2 = -25^\circ$, and $\varpi_r = 0.03$, calculate

(a) $W_{\theta 1}$ and $W_{\theta 2}$
(b) W_{1m} and W_{2m}
(c) D-factor D_{rm}
(d) Circulation Γ_m
(e) rotor lift at pitchline per unit span
(f) lift coefficient at pitchline

(g) rotor-specific work at r_m
(h) loading coefficient ψ_m
(i) degree of reaction $^\circ R_m$

7.4 A rotor blade row at the hub radius is shown. The rotor total pressure loss coefficient at this radius is $\varpi_{rm} = 0.04$.

(a) How much deceleration is allowed in the rotor under de Haller criterion? i.e., What is the minimum W_2?
(b) What is the static pressure rise coefficient, assuming incompressible flow and W_{2min} from de Haller criterion?
(c) Compare the C_p in part (b) to the Arbitrary C_{pmax} shown in Fig. 7.30 at the hub.

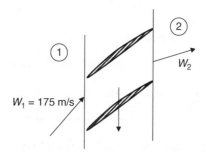

■ **FIGURE P7.4**

7.5 A compressor stage develops a pressure ratio of $\pi_s = 1.6$. Its polytropic efficiency is $e_c = 0.90$. Calculate the stage total temperature ratio τ_s and compressor stage adiabatic efficiency η_s. Assume $\gamma = 1.4$.

7.6 An axial-flow compressor stage is shown at the pitchline. Assuming

$\alpha_1 = \alpha_3$
$\beta_2 = -30^\circ$ $p_{t1} = 10^5$ Pa
$C_{z1} \cong C_{z2} \cong C_{z3}$ $T_{t1} = 290$ K
$r_1 \cong r_2 \cong r_3$ $M_1 = 0.5$
$\varpi_r = 0.03$ $\alpha_1 = 30^\circ$
$\varpi_s = 0.02$ $\gamma = 1.4$
Calculate $c_p = 1{,}004$ J/kg·K

(a) w_c (in kJ/kg)
(b) T_{t3}/T_{t1}
(c) M_{2r}
(d) p_{t2}/p_{t1}
(e) p_{t3}/p_{t2}
(f) η_s
(g) $^\circ R_m$

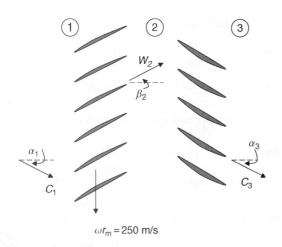

■ **FIGURE P7.6**

7.7 The flow at the entrance to an axial-flow compressor rotor has zero preswirl and an axial velocity of 175 m/s. The shaft angular speed is 5000 rpm. If at a radius of 0.5 m, the rotor exit flow has zero relative swirl, calculate at this radius

(a) rotor specific work w_c in kJ/kg

(b) degree of reaction $^\circ R$

7.8 The absolute flow angle at the inlet of a stator blade in a compressor is $\alpha_2 = 45^\circ$, as shown. The absolute total pressure and temperature in station 2 are $p_{t2} = 150$ kPa and $T_{t2} = 300$ K, respectively. The total pressure loss coefficient for this section of the the stator blade is $\varpi_s = 0.02$. Assuming the axial velocity remains constant and gas properties are $\gamma = 1.4$ and $c_p = 1.004$ kJ/kg · K, calculate

(a) entrance Mach number M_2

(b) exit total pressure p_{t3}

(c) exit Mach number M_3

(d) stator torque for a mass flow rate of $\dot{m} = 100$ kg/s

(e) static pressure rise, $\Delta p = p_3 - p_2$

(f) static temperature rise $\Delta T = T_3 - T_2$

(g) entropy rise $(s_3 - s_2)/R$

Note that the radius of this cut (section) is at $r \cong 0.5$ m from the axis of rotation, as shown in the diagram.

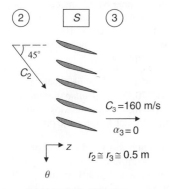

■ **FIGURE P7.8**

7.9 An axial-flow compressor with four stages is shown. Assuming a repeated stage design, with constant throughflow speed, $C_z = 150$ m/s, and 50% degree of reaction at the pitchline with zero preswirl, calculate

(a) rotor specific work at the pitchline (kJ/kg)

(b) stage pressure ratio for an $\eta_s = 0.90$

(c) compressor pressure ratio π_c

(d) shaft power (in MW) for a mass flow rate of $\dot{m}_0 = 100$ kg/s

(e) D-factor for the first rotor at the pitchline for $\sigma_r = 2.0$

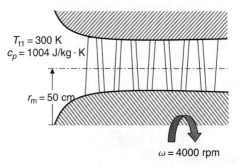

■ **FIGURE P7.9**

7.10 An axial-flow compressor stage is designed on the principle of constant through flow speed. The flow at the entrance to the rotor has 100 m/s of positive swirl and 180 m/s of axial velocity. Assuming we are at the pitchline radius $r_m = 0.5$ m, where the rotor rotational speed is $U_m = 230$ m/s, the degree of reaction $^\circ R_m = 0.5$, the radial shift in the streamtube is negligible, i.e., $r_{1m} \approx r_{2m} \approx r_{3m}$ and also assuming a repeated stage design principle is implemented, calculate

(a) α_{1m} and β_{1m}

(b) α_{2m} and β_{2m}

(c) rotor specific work at the pitchline w_{cm} in kJ/kg

(d) stator torque at the pitchline per unit mass flow rate, $\tau_S/\dot{m}$

7.11 In an axial-flow compressor test rig with no inlet guide vanes, a 1-m diameter fan rotor blade spins with a sonic tip speed, i.e., $U_{tip}/a_1 = 1.0$. If the speed of sound in the laboratory is $a_0 = 300$ m/s, and the axial velocity to the fan is $w_1 = 150$ m/s, calculate the fan rotational speed ω in rpm.

$$R = 287 \text{ J/kg} \cdot \text{K and } \gamma = 1.4$$

7.12 An axial-flow compressor rotor has an angular velocity of $\omega = 5000$ rpm. The flow entering the compressor rotor has zero preswirl and an axial velocity of $C_{z1} = 150$ m/s. Assuming the axial velocity is constant throughout the stage, and the rotor specific work at the radius $r = 0.5$ m is $w_c = 62$ kJ/kg ($\gamma = 1.4$ and $R = 287$ J/kg · K) calculate

(a) stage degree of reaction, $^\circ R$, at this radius

(b) total pressure ratio across the rotor, p_{t2}/p_{t1}, at this radius, assuming a polytropic efficiency of 90% and $T_1 = 20$ °C.

7.13 An axial-flow compressor rotor at the pitchline has a radius of $r_m = 0.35$ m. The shaft rotational speed is $\omega = 5000$ rpm. The inlet flow to the rotor has zero pre-swirl and the axial velocity is $C_{z1} = C_{z2} = C_{z3} = 175$ m/s. The rotor has a 50% degree of reaction at the pitchline. The stage adiabatic efficiency is nearly equal to the polytropic efficiency $\eta_s \cong e_c = 0.92$. Assuming the inlet total temperature is $T_{t1} = 288$ K and $c_p = 1.004$ kJ/kg·K, calculate

(a) rotor specific work at $r = r_m$
(b) stage loading ψ at $r = r_m$
(c) flow coefficient at $r = r_m$
(d) rotor relative Mach number at the pitchline, $M_{1r,m}$
(e) stage total pressure ratio at the pitchline

7.14 For the multistage compressor, as shown, calculate

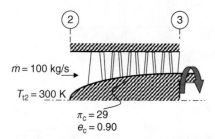

$\dot{m} = 100$ kg/s

$T_{t2} = 300$ K

$\pi_c = 29$
$e_c = 0.90$

■ **FIGURE P7.14**

(a) compressor adiabatic efficiency η_c
(b) shaft power $\wp_c$
(c) exit total temperature T_{t3}
(d) average π_s for ten stages, i.e., $N = 10$
Assume $\gamma = 1.4$ and $c_p = 1004$ J/kg · K.

7.15 A compressor test rig operates in a laboratory where $\theta_2 = 0.95$ and $\delta_2 = 0.95$. The mass flow rate is measured at the compressor face to be $\dot{m} = 100$ kg/s and the shaft power is measured to be $\wp_c = 100$ MW. Assuming compressor polytropic efficiency is $e_c = 0.90$, calculate

(a) compressor (total) pressure ratio π_c
(b) compressor corrected mass flow rate $\dot{m}_{c2}$
(c) compressor adiabatic efficiency η_c

7.16 A compressor adiabatic efficiency is measured to be $\eta_c = 0.85$ for a compressor total pressure ratio of $\pi_c = 20$. What is the "small-stage" efficiency for this compressor?

7.17 A compressor has a polytropic efficiency of $e_c = 0.92$ and a pressure ratio, $\pi_c = 25$ for an inlet condition of $T_{t1} = 520$ °R and $c_p = 0.24$ BTU/lbm · °R, calculate

(a) exit total temperature T_{t2}
(b) compressor adiabatic efficiency η_c
(c) compressor specific work w_c
(d) shaft power $\wp_s$ for a 100 lbm/s flow rate

7.18 A multistage compressor develops a total pressure ratio $\pi_c = 25$, and is designed with eight identical (i.e., "repeated") stages. The compressor polytropic efficiency is $e_c = 0.92$. Calculate

(a) *average* stage total pressure ratio π_s
(b) stage adiabatic efficiency η_s
(c) compressor total temperature ratio τ_c

7.19 Plot compressor adiabatic efficiency η_c versus π_c ranging from 1.0 to 50, for the following polytropic efficiencies $e_c = 0.95, 0.90$, and 0.85 as the running parameter.

7.20 A rotor blade row has a solidity of 1.0 at its pitchline. The absolute flow enters the rotor with no preswirl at $C_{z1m} = 500$ fps, and the rotor rotational speed is $U_m = 1200$ fps (at the pitchline). If the rotor exit flow angle at r_m is $\beta_{2m} = -30°$, calculate

(a) exit swirl velocities $W_{\theta2m}$ and $C_{\theta2m}$
(b) blade torque (per unit mass flow rate) at the pitchline assuming $r_m = 1.0$ ft
(c) the nondimensional total temperature rise across the rotor $\Delta T_t/T_{t1}$
(d) exit speed of sound a_2 assuming inlet speed of sound is $a_1 = 1100$ fps
(e) exit absolute Mach number M_{2m}
(f) inlet static pressure p_1 for an inlet total pressure of $P_{t1} = 14.7$ psia
(g) exit total pressure, assuming rotor adiabatic efficiency is 0.9
(h) rotor static pressure rise, $C_{Pm} = \Delta p_m/(1/2 \ \rho W_{1m}^2)$
(i) shaft rotational speed ω (rpm)
(j) rotor torque at r_m per unit mass flow rate
(k) the axial force on the blades at r_m, assuming the mean chord $c_m = 4$ in.
(l) tangential force on the blade at r_m
(m) sectional lift-to-drag ratio, L'/D'
(n) rotor specific work at r_m

7.21 Calculate the Reynolds number based on chord at the pitchline for the rotor blade described in Problem 7.20, assuming fluid coefficient of viscosity is $\mu = 1.8 \times 10^{-5}$ kg/m·s. Compare the Reynolds number that you calculate to the upper critical Reynolds number in a compressor.

7.22 Calculate the circulation at the pitchline for the rotor blade row described in Problem 7.20. Also calculate the fraction of "ideal" lift that was destroyed by the total pressure losses in the blade row. Is there any indication of shock losses at the pitchline?

7.23 A rotor blade row has a hub-to-tip radius ratio of 0.5, solidity at the pitchline of 1.0, the axial velocity is 160 m/s, and zero preswirl. The mean section has a design diffusion factor of $D_m = 0.5$. Calculate and plot where appropriate

(a) exit swirl at the pitchline assuming the shaft rpm of 6000 and $r_m = 1.0$ ft (0.3 m)

(b) downstream swirl distribution $C_{\theta 2}$ (r) assuming a free-vortex design rotor

(c) the radial distribution of degree of reaction $°R$ along the blade span

(d) radial distribution of diffusion factor $D_r(r)$.

7.24 An axial-flow compressor stage is downstream of an IGV that turns the flow $15°$ in the direction of the rotor rotation, as shown. The axial velocity component remains constant throughout the stage at $C_z = 150$ m/s. The rotor rotational speed is $\omega = 3000$ rpm and the pitchline radius is $r_m = 0.5$ m. The rotor relative exit flow angle is $\beta_2 = -15°$. The static temperature and pressure of air upstream of the rotor are $T_1 = 20°C$ and $p_1 = 10^5$ Pa, respectively. Assuming $c_p = 1.004$ kJ/kg·K and $\gamma = 1.4$, calculate

(a) relative Mach number to the rotor, M_{1r}

(b) absolute total temperature T_{t1}

(c) rotor specific work w_c in kJ/kg

(d) total temperature downstream of the rotor, T_{t2}

(e) relative Mach number downstream of the rotor, M_{2r}

(f) stage total pressure ratio at the pitchline radius for $e_c = 0.92$

(g) stage degree of reaction $°R_m$ at the pitchline

7.25 A compressor stage with an inlet guide vane is shown at its pitchline radius $r_m = 0.5$ m. The rotor angular speed is $\omega = 4000$ rpm. The axial velocity is constant throughout at $C_z = 150$ m/s and the IGV imparts a preswirl of 75 m/s in the direction of rotor rotation, as shown. Assuming the inlet flow to IGV has $p_{t0} = 100$ kPa and $T_{t0} = 25°C$, calculate

(a) T_0, M_0, p_0

Assuming the IGV has a total pressure loss coefficient of $\varpi_{IGV} = 0.02$, calculate

(b) p_{t1}, T_{t1}, T_1, M_1, p_1, M_{1r} and p_{t1r}

Knowing that the compressor stage has a degree of reaction of $°R = 0.5$ at the pitchline, calculate

(c) $C_{\theta 2}$, T_{t2}, T_2, M_2, M_{2r}

For a rotor total pressure loss coefficient of $\varpi_r = 0.03$ at the pitchline, calculate

(d) p_{t2}

Assume: $\gamma = 1.4$ and $c_p = 1004$ J/kg · K throughout the stage.

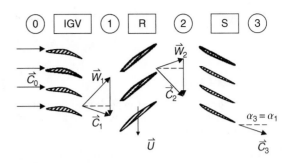

■ **FIGURE P7.25**

7.26 Apply Euler turbine equation to a streamtube that enters a compressor rotor blade row at $r = 2.5$ ft with zero preswirl and exits the row at $r = 2.7$ ft and attains 1000 fps of swirl velocity in the absolute frame. Assume that the shaft rotational speed is 5000 rpm.

7.27 Assume that we can analyze a 3D compressor rotor by stacking up its 2D flowfield, what is known as the "strip theory." Now, let us consider one such section. The inlet flow approaches the rotor blade at $\beta_1 = -45°$. The relative exit flow angle is $\beta_2 = -30°$ for a net $15°$ turning. The solidity of the rotor at this section is 1.5. Identify the most suitable NACA 65-series profile for this section that could produce the largest positive stall tolerance. Estimate the stagger angle that the blades need to be set at this section. To solve this problem, you have to use the cascade data. What is the safe operating range of the incidence angle (or equivalently the inlet flow angles), in degrees, for this section?

7.28 A simple method to establish the annulus geometry in a multistage compressor is to assume a constant throughflow (i.e., axial) speed C_z. Calculate

(a) the density ratio ρ_3/ρ_2

(b) the exit-to-inlet area ratio A_3/A_2

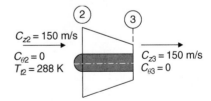

■ **FIGURE P7.28**

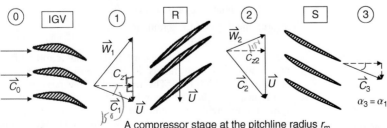

A compressor stage at the pitchline radius r_m

■ **FIGURE P7.24**

for

$\pi_c = 35$

$e_c = 0.90$

$C_z = $ constant

$\gamma = 1.4$, $c_p = 1{,}004$ J/kg · K

Also, if the hub radius is constant and for a mass flow rate of 100 kg/s and the inlet total pressure of $p_{t2} = 100$ kPa, calculate the tip-to-tip radius ratio $(r_3/r_2)_{tip}$

7.29 Consider a compressor stage with no inlet guide vane. The hub-to-tip radius ratio for the rotor is $r_{h1}/r_{t1} = 0.5$ and the mass flow rate (of air) in the compressor is 100 kg/s, at the standard sea level condition, $p_{t1} = 100$ kPa and $T_{t1} = 288$ K. For an axial Mach number of $M_z = 0.5$, calculate

(a) the rotor hub and casing radii r_{h1} and r_{t1} (in meters)
To achieve a relative tip Mach number of $(M_{1r})_{tip} = 1.4$, calculate

(b) the rotor rotational speed ω (in radians per second)
For a design pitchline degree of reaction of $°R_m = 0.5$, and constant axial velocity, calculate

(c) the rotor exit swirl at the pitchline radius $C_{\theta 2m}$ (in m/s)

(d) the total temperature T_{t2m} (in K) using Euler equation

(e) the total pressure p_{t2m} (in kPa), assuming $\varpi_{rm} = 0.005$

(f) the fluid density ρ_{2m} in kg/m³
Now, assume that we chose a free-vortex design for the rotor.

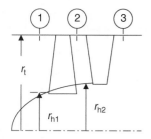

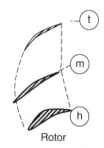

Rotor

■ FIGURE P7.29

Calculate

(g) the degree of reaction for the rotor hub and tip $°R_h$ and $°R_t$

(h) the diffusion factor at the pitchline D_m

(i) the hub radius in station 2, r_{h2}, from continuity equation.

7.30 A compressor rotor velocity triangles at the pitchline are shown. The compressor hub-to-tip radius ratio is $r_h/r_t = 0.4$. The rotor solidity at the pitchline is $\sigma_m = 1.0$. Assuming constant axial velocity across the rotor at pitchline, calculate

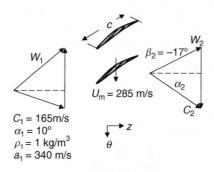

$C_1 = 165$ m/s
$\alpha_1 = 10°$
$\rho_1 = 1$ kg/m³
$a_1 = 340$ m/s

■ FIGURE P7.30

(a) solidity at the tip for $c_t = c_m$

(b) solidity at the hub for $c_h = c_m$

(c) degree of reaction $°R_m$

(d) diffusion factor D_m

(e) de Haller criterion

(f) rotor loading coefficient ψ_m

(g) flow coefficient φ_m

(h) Assuming that the rotor has a free-vortex design, calculate

(i) diffusion factor at the tip D_t

(j) the degree of reaction at the hub, $°R_h$

(k) the axial velocity distribution downstream of the rotor (assuming radial equilibrium)

7.31 The flow coefficient to a rotor at pitchline is $\varphi_m = 0.8$, its loading coefficient is $\psi_m = 1.0$. The inlet flow to the rotor has zero swirl in the absolute frame of reference. Assuming axial velocity $C_{zm} = $ constant across the rotor, calculate

(a) the relative inlet flow angle β_{1m}

(b) the relative exit flow angle β_{2m}

(c) the degree of reaction $°R_m$

7.32 The end wall boundary layers and a rotor blade are shown. The hub radius is $r_h = 0.5$ m and the shaft speed is $\omega = 4000$ rpm. The rotor inlet flow has zero pre-swirl, i.e., $\alpha_1 = 0$. The rotor blade is twisted to meet the flow at zero incidence within the boundary layer. Calculate the blade stagger angle at the hub and its change within the boundary layer if $\delta_{bl,h} = 10$ cm and it has a 1/7th power law profile. $C_z(y)/C_{zm} = (y/\delta)^{1/7}$ and $C_{zm} = 150\,\mathrm{m/s}$. [Assume blade stagger angle is $\approx$ relative flow angle, β_1]

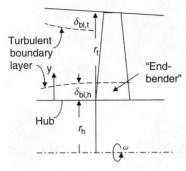

■ FIGURE P7.32

7.33 A multistage compressor has 12 repeated stages. Each stage produce a constant total temperature rise of $\Delta T_t = 25°C$. The inlet total temperature is $T_{t1} = 288$ K, and stage adiabatic efficiency is $\eta_s = 0.90$ and is assumed constant for the 12 stages. Calculate and graph the stage total pressure ratio for all 12 stages. What is the compressor overall total pressure ratio?

7.34 A compressor stage has 37 rotor blades and 41 stator blades. The shaft rotational speed is 5000 rpm. Calculate

   (a) the rotor blade passing frequency as seen by the stator blades

   (b) the stator blade passing frequency as seen by the rotor blades

7.35 The absolute flow to a compressor rotor has a coswirl with $\alpha_1 = 15°$. The exit flow from the rotor has an absolute flow angle $\alpha_2 = 35°$. The pitchline radius is at $r_m = 0.6$ m and the rotor angular speed is $\omega = 5220$ rpm. Assuming the axial velocity is $C_{zm} = 150$ m/s and is constant across the rotor, calculate

   (a) the specific work at the pitchline

   (b) the rotor torque per unit mass flow rate

   (c) the degree of reaction

7.36 A rotor blade row has a hub-to-tip radius ratio of 0.4, solidity at the pitchline of 1.2, the axial Mach number of 0.6, and zero preswirl. The mean section has a design diffusion factor of $D_m = 0.5$. Assuming $a_1 = 330$ m/s, calculate and plot where appropriate

   (a) exit swirl at the pitchline assuming the shaft rpm of 5000 and $r_m = 0.4$ m

   (b) downstream swirl distribution $C_{\theta2}(r)$, assuming a solid-body rotation vortex design rotor

   (c) the radial distribution of degree of reaction $°R(r)$, along the blade span

   (d) radial distribution of diffusion factor $D_r(r)$.

7.37 Bending stresses on a cantilevered blade, under aerodynamic loading, is estimated (by Kerrebrock, 1992) to be

$$\frac{\sigma_{bending}}{p} \approx \left(\frac{C_z}{U_t}\right)(\tau_s - 1)\left(\frac{s}{2c}\right)\left(\frac{r_t}{t_{max}}\right)^2$$

Estimate the ratio of bending stress to inlet pressure $(\sigma_{bending}/p)$ in a rotor with axial velocity 165 m/s, the rotor tip radius is $r_t = 0.75$ m, the mean solidity of 1.5, angular speed is $\omega = 4000$ rpm. The maximum blade thickness is $t_{max} = 1$ cm. The stage total pressure ratio is 1.6 and the stage adiabatic efficiency is 0.88.

7.38 Centrifugal stress is proportional to AN^2 as we discussed in the compressor design section. It follows

$$\frac{\sigma_c}{\rho_{blade}} = \frac{\omega^2 A}{4\pi}\left(1 + \frac{A_t}{A_h}\right)$$

for a linear taper ratio. For a titanium rotor blade of $r_h = 0.4$ m, $r_t = 0.8$ m, and taper ratio $A_t/A_h = 0.8$, calculate the acceptable shaft speed ω if the allowable strength-to-weight ratio is ~ 9 ksi/slug/ft^3, which is equivalent to

allowable creep rupture strength	120,560 m^2/s^2 (or $\sim$120 kPa/kg/m^3)
required strength	68,326 m^2/s^2 (or $\sim$68 kPa/kg/m^3)

7.39 A rotor section has de Haller criterion $W_2/W_1 = 0.75$. Assuming the axial velocity remains constant across the rotor and the velocity triangles are as shown, calculate the corresponding D-factor and degree of reaction for the rotor section.

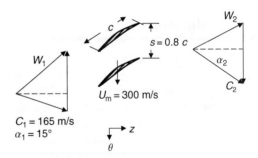

■ FIGURE P7.39

7.40 A portion of a compressor map surrounding the design point is shown.

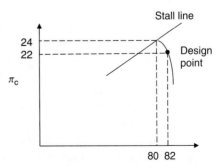

■ FIGURE P7.40

Calculate this compressor's stall margin.

CHAPTER 8

Centrifugal Compressor Aerodynamics

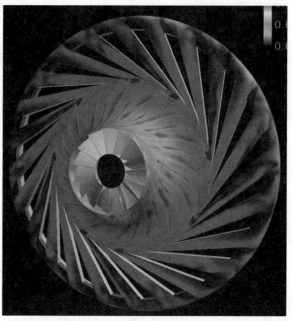

CFD results shown in a centrifugal compressor (Courtesy of NASA)

8.1   Introduction

Centrifugal compressors belong to the general category of turbomachines. The flow may enter a centrifugal compressor in the axial direction. The rotor, which is known as the impeller, imparts energy to the fluid by the rotation of its aerodynamic surfaces (i.e., blades or impeller vanes) that are highly curved and twisted. The initial curvature and twist in the impeller, known as the inducer section, has the function of meeting the incoming flow at its relative flow angle. The second function of the inducer is to turn the relative flow toward the axial direction before it begins its journey in the radial direction. The third function of the inducer is to increase the fluid static pressure in the passage by decelerating the gas. The inducer exit flow is then further decelerated radially outward, by virtue of centrifugal force acting on the fluid in the spinning impeller (vaned) passages. Since the pressure rise in this type of configuration is primarily produced by centrifugal compression/force, this kind of turbomachinery is known as centrifugal compressor. In contrast to axial-flow compressors where the flow deviation in the radial direction is negligible, the principle of operation of the centrifugal compressor is based on large radial shift between the inlet and exit of the impeller (see Fig. 8.1). Since the impeller exit radius is by design much larger then the impeller inlet radius, the centrifugal compressors have a lower mass flow rate per frontal area than the axial-flow compressors. A low mass flow per frontal area is of critical concern to aircraft propulsion system designers, since it translates into an increased drag count. However, in low-speed applications where the drag penalty is less significant, centrifugal compressors offer the advantage of high-pressure ratio per stage and robustness of construction that is less prone to structural failure. There are other applications for small gas generators on board aircraft that are suitable for centrifugal compressor application. For example, APUs (auxiliary power unit) are used to start the engines and provide power to aircraft, which are entirely embedded in the fuselage, therefore there are no concerns

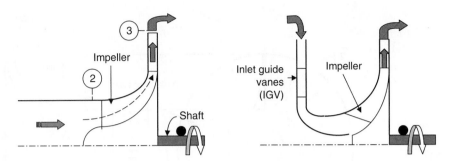

■ **FIGURE 8.1**   Schematic drawing of a first (left) and a second stage (right) centrifugal compressor

of external drag penalties for their use. Industrial and automotive turbocharging applications also use centrifugal compressors.

The radially pumped flow from the impeller first needs to be decelerated through a radial and vaned diffuser, and then it needs to turn back toward the axis of rotation for either the next compressor stage or the combustion chamber. Since any ducting that involves turns is a source of (total pressure) loss, multistage centrifugal compressors are considered cumbersome (and with lower efficiency) as compared with axial-flow compressors. However, the total pressure ratio of a single-stage centrifugal compressor may be as high as 10–12, whereas an axial-flow compressor stage produces a high of $\sim$1.6–2.0 in advanced transonic fan stages.

8.2   Centrifugal Compressors

A centrifugal compressor is a robust mechanical compression system that pumps the gas from primarily an axial inlet condition to a radial exit direction. The elements of a centrifugal compressor rotor, known as the impeller, are shown in Fig. 8.2.

The inlet section of an impeller is called the inducer, which is turned in the direction of impeller rotation to meet the flow at a small incidence angle. The radial displacement of the fluid in the inducer is small but the flow turning is rather large. The inducer turns the fluid from the inlet relative flow direction β_1 to the axial direction. The static pressure rise in the inducer is due to the conversion of relative swirl kinetic energy in a curved diffuser. The outer portion of the impeller is responsible for turning the flow in the radial direction and accelerating the fluid in the tangential direction. The torque acting on the fluid causes the angular momentum to change across the rotor blade following Euler turbine equation:

$$\tau_{\text{fluid}} = -\tau_{\text{blade}} = \dot{m}(r_2 C_{\theta 2} - r_1 C_{\theta 1})$$

As in the axial-flow compressors, we note that the torque acting on the fluid is positive across the rotor (as $r_2 C_{\theta 2} > r_1 C_{\theta 1}$) and thus the rotor torque being equal and opposite is negative. Then multiplying the rotor torque with the angular speed of the shaft, ω, we get a negative shaft power that is delivered to the rotor. This is consistent with the thermodynamic convention employed in the first law that considers the work done *on the surrounding* as positive. Therefore the work done *by the surrounding* (as needed by compressors) is then negative. The torque acting on the fluid crossing the stator, as in the axial-flow compressors is negative. Therefore the stator vanes experience a positive torque. There will not be any contribution to the fluid power by the stator blades as they are stationary and thus perform no work on the medium. The shaft power delivered to the fluid is converted into total enthalpy rise according to

$$\wp = \dot{m}\omega(r_2 C_{\theta 2} - r_1 C_{\theta 1}) = \dot{m}(h_{t2} - h_{t1}) \tag{8.1}$$

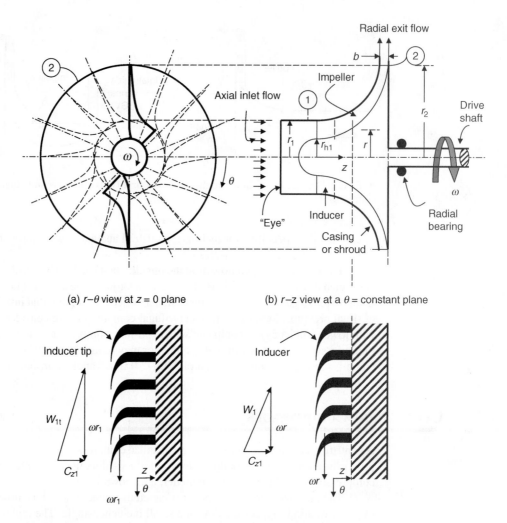

(a) $r-\theta$ view at $z = 0$ plane (b) $r-z$ view at a θ = constant plane

(c) $\theta-z$ view at the inducer tip, r_1 = constant plane (d) $\theta-z$ view at an r = constant ($r < r_1$) plane

■ **FIGURE 8.2** Definition sketch of a centrifugal compressor impeller with an inducer

The rotor total temperature ratio is deduced from Eq. 8.1, i.e.,

$$\frac{T_{t2}}{T_{t1}} = 1 + \frac{\omega \cdot r_2}{c_p T_{t1}}\left[C_{\theta 2} - \left(\frac{r_1}{r_2}\right)C_{\theta 1}\right] \tag{8.2}$$

In the above equation, the impeller radius ratio (r_1/r_2) is small, which makes the second term in the bracket negligible compared with the first term. In addition, either due to a lack of inlet guide vanes, which results in $C_{\theta 1}$ to be identically zero, or the fact that $C_{\theta 2} \gg C_{\theta 1}$ in a centrifugal compressor, we often neglect the second term in the bracket in Eq. 8.2 in favor of the first term, i.e.,

$$\frac{T_{t2}}{T_{t1}} \cong 1 + \frac{\omega \cdot r_2 \cdot C_{\theta 2}}{c_p T_{t1}} = 1 + \frac{U_2^2}{c_p T_{t1}}\left(\frac{C_{\theta 2}}{U_2}\right) \tag{8.3}$$

However, note that we do not *have to* neglect the second term in Eq. 8.2, as the inlet swirl $C_{\theta 1}$ may be treated as a known input to the problem.

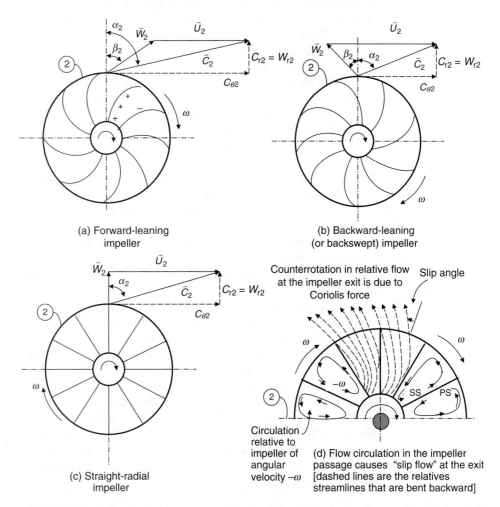

■ **FIGURE 8.3**   Definition sketch of the impeller blade shapes, flow angles, ideal velocity triangles at the impeller exit, and the phenomenon of slip

There are three types of impeller geometries:

- Radial impeller
- Forward-leaning impeller
- Backward-leaning (or backswept) impeller

A schematic drawing of impeller geometries and their corresponding velocity triangles are shown in Fig. 8.3.

The radial impeller offers radial passages to accelerate the fluid in the rotor and ideally attain a radial exit flow. The forward-leaning impeller geometry accelerates the fluid in a spiral passage leaning in the direction of the rotor rotation. The backswept impeller is composed of spiral passages that direct the relative flow in the opposite direction to rotor rotation. The geometry of the impeller dictates the exit flow angle β_2 and thus strongly impacts the magnitude of the exit swirl velocity $C_{\theta2}$. The three types of impeller geometry are shown in Fig. 8.3, with the flow angles at the impeller exit are now measured with respect to the radial direction, as shown. The velocity triangles at the impeller exit show that the absolute swirl is increased when the impeller geometry is forward leaning (with $\beta_2 < 0$). For the case of the straight radial impeller, the exit absolute swirl should theoretically be the same as the wheel speed U_2. The backward-leaning or backswept impeller geometry turns the relative flow in the opposite direction to the wheel rotation (with $\beta_2 > 0$), therefore the absolute swirl in the exit plane is reduced. All these arguments are the results of simple interpretation of the velocity triangles (see Fig. 8.3).

Euler turbine equation suggests that the forward-leaning blades produce the highest total enthalpy rise and hence absorb the maximum shaft power. The impeller exit Mach number is the highest of the three designs and thus limits the static pressure recovery and efficiency of the radial diffuser. Also, the blades are under severe structural loads in bending due to their curvature. Similarly, the blades of a backward-leaning impeller are under high structural loading and the shaft power absorption is reduced. The exit Mach number of the backward-leaning impeller is the lowest of the three designs and will be explored as an alternative to radial impeller design. The lower exit Mach number of the backward-leaning impeller is attractive in achieving higher efficiency diffusion in the radial diffuser. The optimum geometry from structural standpoint seems to be the straight radial impeller with good power absorption, acceptable stability characteristics, and a low level of structural loading (in bending) due to purely radial design. However, the pressure rise in the compressor increases with the wheel speed, hence a backward-leaning impeller design could operate at higher U_T once the structural design issues are resolved.

In the rotor frame of reference, the flow in the impeller is primarily in the radial direction, which bends in the opposite direction to the wheel rotation due to a Coriolis force, i.e., $\propto 2\omega W_r$. The presence of tangential pressure gradients in addition to the Coriolis effect combine to create a countercirculation in impeller blade passages toward the trailing edge. The counter-rotation in the impeller exit flow causes a reduction in the absolute swirl, therefore a reduction in the power absorption and thus leads to a reduced pressure ratio. This phenomenon is known as *slip* in centrifugal compressors. An example of the slip in a straight radial impeller is shown in part (d) of Fig. 8.3. The other two examples of impeller configurations, i.e., the forward- and the backward-leaning types, are subject to the phenomenon of slip in their respective passages. We define a slip factor for radial impellers as the ratio of absolute swirl to the exit wheel speed, which represents the *ideal* swirl had the fluid attained the same tangential velocity as the disk, namely,

$$\varepsilon \equiv \frac{C_{\theta 2}}{U_2} \quad \text{(slip factor)} \tag{8.4}$$

Hence, the total temperature rise across the straight radial impeller is related to the wheel speed and the slip factor ε following Euler turbine equation, i.e.,

$$T_{t2} = T_{t1} + \frac{\varepsilon \cdot U_2^2}{c_p} \tag{8.5-a}$$

From Eq. 8.5-a, the impeller loading coefficient $\Delta h_t / U_2^2$ gives a simple expression for the radial impeller, which is independent of mass flow rate, assuming constant slip factor, i.e.,

$$\Delta h_t / U_2^2 = \varepsilon \tag{8.5-b}$$

In general, the slip factor is defined as

$$\varepsilon \equiv \frac{(C_{\theta 2})_{\text{actual}}}{(C_{\theta 2})_{\text{ideal}}} \quad \text{(slip factor)} \tag{8.6}$$

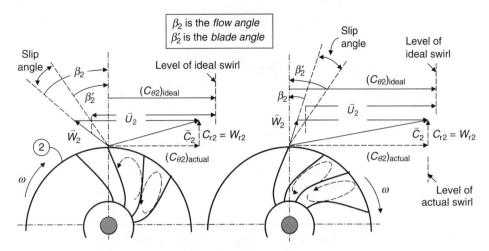

(a) Backswept impeller with β_2 and $\beta_2' > 0$ (b) forward-leaning impeller with β_2 and $\beta_2' < 0$

■ **FIGURE 8.4**   **Definition sketch for the slip flow in a backswept and forward-leaning impeller**

where the ideal exit swirl is calculated based on the *impeller blade exit angle* β_2', and the actual exit swirl is based on the *actual exit flow angle* β_2. Figure 8.4 shows a definition sketch of these angles for the backward and forward-leaning impellers.

    If we adopt the sign convention that the impeller exit angle is measured from the vane position to the radial direction and is positive in the direction of the rotor rotation, we note that a backswept impeller has a positive β_2' and a forward-leaning impeller has a negative vane exit angle β_2'. With a similar sign convention for the exit flow angle, we may write the ideal and actual exit swirl velocities in terms of the impeller exit angle β_2', flow angle β_2 as

$$C_{\theta2,\text{ideal}} = U_2 - C_{r2} \cdot \tan \beta_2' \qquad \textbf{(8.7-a)}$$

$$C_{\theta2,\text{actual}} = U_2 - C_{r2} \cdot \tan \beta_2 \qquad \textbf{(8.7-b)}$$

From Eq. 8.7-a, we note that a backswept impeller, with a positive exit vane angle, reduces the absolute (ideal) swirl, whereas a forward-leaning impeller with a negative vane angle increases the exit absolute swirl. Similar argument can be made regarding the exit swirl velocity in the actual flow as described in Eq. 8.7-b. We define the slip factor as the ratio of actual to the ideal exit swirl as

$$\varepsilon \equiv \frac{C_{\theta2,\text{actual}}}{C_{\theta2,\text{ideal}}} = \frac{U_2 - C_{r2} \cdot \tan \beta_2}{U_2 - C_{r2} \cdot \tan \beta_2'} \qquad \textbf{(8.8)}$$

The impeller specific work and the loading are written as

$$\Delta h_t = U_2 C_{\theta2} = U_2(U_2 - C_{r2} \tan \beta_2) = U_2^2\left[1 - \frac{C_{r2}}{U_2} \tan \beta_2\right] \qquad \textbf{(8.9-a)}$$

$$\frac{\Delta h_t}{U_2^2} = 1 - \left(\frac{C_{r2}}{U_2}\right) \tan \beta_2 \qquad \textbf{(8.9-b)}$$

    We may graph Eq. 8.9-b, i.e., the variation of the nondimensional work, or the impeller-loading coefficient $\Delta h_t/U_2^2$, with flow coefficient C_{r2}/U_2 (which is similar to C_z/U in axial-flow

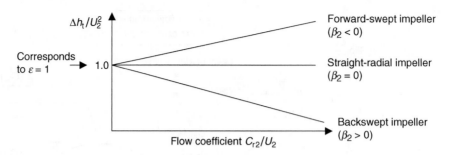

■ **FIGURE 8.5**   **Centrifugal compressor impeller loading variation with flow coefficient or mass flow rate**

compressors), for different impellers. Figure 8.5 shows the impeller loading variation with flow coefficient for the three impeller types (with a constant slip factor ε).

There are several correlations between the slip factor ε and the number of the impeller vanes N in radial impellers, which are proposed by Stodola (1927), Busemann (1928), and Stanitz–Ellis (1949). Stanitz–Ellis model, which is similar to Stodola's, is considered a reasonable approximation for the slip factor in straight-radial impellers (and a first-order approximation for other impeller types),

$$\varepsilon = 1 - \frac{1.98}{N} \tag{8.10}$$

A 20-bladed impeller, for example, will experience a slip factor of ~0.9 in its energy transfer to the fluid. For radial impellers, the slip factor (as expressed in Eq. 8.10) is independent of the mass flow rate. However, slip factor in reality is a function of the mass flow rate for all three configurations, i.e., radial, forward- and backward-leaning impellers. Increasing the number of the impeller blades reduces the effect of slip, which in turn increases the weight. To achieve the effect of higher blade count without investing heavily in the system weight increase, a compromise solution is found by incorporating splitter vanes in the outer half of an impeller disk. The presence of splitter vanes acts as an inhibitor of counterswirl produced in the outer portion of an impeller disk and possible reduction of flow separation in the impeller. Figure 8.6 shows the splitter vanes concept in a centrifugal turbomachinery. A parallel argument in the creation of slip is worthy of discussion. The flow at the exit of the inducer is turned parallel to the axis and is starting its journey in the radial direction. Therefore, the rotor flow is purely radial at the exit of the inducer. This resembles a source flow, which is irrotational. For the flow to remain irrotational (in the absence of external forces, i.e., in rotor frame of reference),

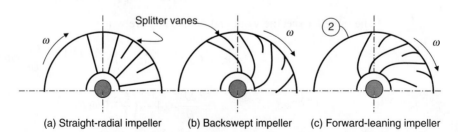

■ **FIGURE 8.6**   **Schematic drawing of splitter vanes on the impeller disk of a centrifugal compressor**

counter eddies of the same angular speed need to be set up in the impeller blade passages. The counter eddies thus lead to an exit flow that is rotated in the opposite direction to the wheel rotation, by an angle known as the slip angle.

The stage total temperature ratio is the same as the impeller total temperature ratio, since the stationary blades are adiabatic and do no work on the fluid, therefore, we may write Eq. 8.3 as

$$\tau_{stage} = \frac{T_{t2}}{T_{t1}} = 1 + \varepsilon \frac{U_2^2}{\left(\dfrac{\gamma R T_{t1}}{\gamma - 1}\right)} = 1 + (\gamma - 1)\varepsilon \frac{U_2^2}{a_{t1}^2} \qquad (8.11)$$

The stagnation speed of sound at the inlet condition is a_{t1} in Eq. 8.11. The ratio of wheel speed at the impeller exit to the stagnation speed of sound at the inlet is a nondimensional parameter that is called Mach index Π_M in the centrifugal compressor literature. In terms of the Mach index, we may write the stage total temperature ratio for a straight-radial impeller with zero preswirl as

$$\tau_s = 1 + (\gamma - 1)\varepsilon \cdot \Pi_M^2 \qquad (8.12)$$

Upon closer examination of this parameter, i.e., Mach index, we note that it is directly proportional to the impeller tangential Mach number at the exit and inversely proportional to the axial Mach number at the inlet of the compressor. This relation is shown in Eq. 8.13.

$$\Pi_M^2 \equiv \frac{U_2^2}{\gamma \cdot R \cdot T_{t1}} = \frac{U_2^2}{\gamma \cdot R \cdot T_1 \left(1 + \dfrac{\gamma - 1}{2} M_1^2\right)} = \frac{M_{T2}^2}{1 + \dfrac{\gamma - 1}{2} M_1^2} \qquad (8.13)$$

where $M_{T2} \equiv \dfrac{U_2}{a_1}$.

EXAMPLE 8.1

Use a spreadsheet to calculate and graph the ratio of Mach index to the impeller tip tangential Mach number (from Eq. 8.13) for a range of inlet Mach numbers from 0.2 to 0.6.

SOLUTION

■ FIGURE 8.7
Comparison between the Mach index and impeller tip Mach number M_{T2}

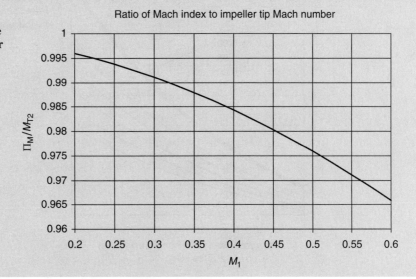

The ratio of the Mach index and the impeller tip tangential Mach number M_{T2} is

$$\frac{\Pi_M}{M_{T2}} = \left(1 + \frac{\gamma - 1}{2} M_1^2\right)^{-0.5} \quad (8.14)$$

The result of spreadsheet calculation of Eq. 8.14 is graphed in Fig. 8.7.

At inlet axial Mach number of ~0.45, the Mach index is ~98% of the rotor tip tangential Mach number. The impeller tip tangential Mach number is thus the dominant parameter that determines the stage temperature ratio and pressure ratio, similar to the axial-flow compressor.

The impeller total pressure ratio is related to the impeller total temperature ratio via polytropic efficiency according to

$$\frac{p_{t2}}{p_{t1}} = \left(\frac{T_{t2}}{T_{t1}}\right)^{\frac{\gamma \cdot e_c}{\gamma - 1}} \cong \left[1 + (\gamma - 1)\varepsilon \cdot M_{T2}^2\right]^{\frac{\gamma \cdot e_c}{\gamma - 1}} \quad (8.15)$$

where we replaced the Mach index with the tip tangential Mach number. Now, let us graph this equation for different tip tangential Mach numbers. Figure 8.8 shows the functional dependence of pressure ratio and tip Mach number. We note an exponential growth of the impeller total pressure ratio with the tangential Mach number at the impeller tip. There are two limitations to the exponential growth performance of centrifugal compressors with tip tangential Mach number depicted in Fig. 8.8. One limiting factor is due to excessive centrifugal stresses and the structural loads, and the second limiting factor is due to a drop in efficiency of diffusion process in the radial diffuser at high inlet supersonic Mach numbers. To tackle the structural problem, a high strength-to-weight ratio material is desirable for the rotor application. Titanium alloys are thus suitable, and the lighter, lower cost aluminum is still considered a material of choice for small centrifugal compressor applications. The impeller rim speed of ~1700 ft/s (or ~520 m/s) represents the state-of-the art in centrifugal compressors. The corresponding tangential rim Mach number defined as the ratio of rim speed to the inlet speed of sound a_1 at standard sea level inlet condition, i.e., a_1 of 1100 ft/s or 340 m/s, is limited to $M_{T2} \sim 1.55$. From the three possible impeller geometries, the forward-leaning design would create a higher exit Mach number than

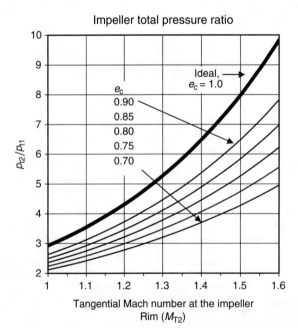

■ FIGURE 8.8 Variation of impeller pressure ratio with tip tangential Mach number (based on the inlet speed of sound a_1) for a straight-radial impeller with slip factor $\varepsilon = 0.9$

the rim tangential Mach number, based on the velocity triangle depicted in Fig. 8.3(a). Therefore the inlet Mach number to the diffuser will be even higher than 1.55, based on the inlet speed of sound (a_1). The straight-radial impeller would roughly achieve the same diffuser inlet Mach number as the impeller tangential rim Mach number. Note that the radial flow Mach number is small compared with a supersonic tip. To reduce the diffuser inlet Mach number to lower than 1.55 (based on a_1), we may lean the impeller blades backward, known as backsweep. The main disadvantage of curved impellers, i.e., both the forward- and backward-leaning types, is in their higher (bending) stress levels than the radial vanes. The curved impellers experience large bending stresses due to the centrifugal force, whereas the radial impellers theoretically experience no bending stresses due to centrifugal loads. The higher cost of manufacturing attributed to curved impellers is considered their second disadvantage as compared with straight-radial impellers. Despite higher manufacturing costs and higher stresses, the trend is to maximize the wheel speed for efficient centrifugal compression while reducing the diffuser entrance Mach number by backward-leaning the impeller blades. The backswept impellers of $\sim 45°$ exit sweep represent the state of the art in high performance centrifugal compressors.

So far we have referenced impeller exit velocity components to the inlet speed of sound. The behavior of actual flow in the diffuser depends, however, on the Mach number based on the local speed of sound at the impeller exit, a_2. The Euler turbine equation applied to an impeller with zero inlet swirl (or preswirl) is the starting point of our calculation of T_2, or a_2, i.e.,

$$\frac{T_{t2}}{T_{t1}} - 1 \cong \frac{U_2^2}{c_p T_{t1}} \left(\frac{C_{\theta 2}}{U_2} \right)$$

From the definition of slip factor, Eq. 8.8, we will substitute for the ratio of exit absolute swirl to the wheel speed in the above equation to get

$$\frac{T_{t2}}{T_{t1}} - 1 \cong (\gamma - 1) \Pi_M^2 \cdot \varepsilon \cdot \left[1 - \left(\frac{C_{r2}}{U_2} \right) \tan \beta_2' \right] \tag{8.16}$$

Now, the total temperature ratio may be related to the static temperatures and the fluid kinetic energy following the definition of total enthalpy, i.e.,

$$\frac{T_{t2}}{T_{t1}} = \frac{T_2 + C_2^2 / 2 c_p}{T_1 + C_1^2 / 2 c_p} \tag{8.17}$$

The contribution of kinetic energy at the entrance to the centrifugal compressor may be neglected in favor of the static temperature and thus Eq. 8.17 may be approximated as

$$\frac{T_{t2}}{T_{t1}} \cong \frac{T_2 + C_2^2 / 2 c_p}{T_1} = \frac{T_2}{T_1} + \frac{C_2^2}{2 c_p T_1} \tag{8.18}$$

The RHS of Eq. 8.18 may be simplified if we replace the absolute kinetic energy with the swirl kinetic energy, which essentially neglects small contribution of the exit radial kinetic energy. We may solve for the static temperature ratio as

$$\frac{T_2}{T_1} \cong \frac{T_{t2}}{T_{t1}} - \frac{C_{\theta 2}^2}{2 c_p T_1} = \frac{T_{t2}}{T_{t1}} - \frac{\gamma - 1}{2} \Pi_M^2 \left(\frac{C_{\theta 2}}{U_2} \right)^2 \tag{8.19}$$

By substituting for the total temperature ratio from Eq. 8.16 and the ratio of exit swirl to the wheel speed from Eq. 8.8, we get the impeller static temperature ratio as

$$\frac{T_2}{T_1} - 1 \cong (\gamma - 1) \Pi_M^2 \cdot \varepsilon \cdot \left[1 - \left(\frac{C_{r2}}{U_2} \right) \tan \beta_2' \right] \left\{ 1 - \frac{1}{2} \cdot \varepsilon \cdot \left[1 - \left(\frac{C_{r2}}{U_2} \right) \tan \beta_2' \right] \right\} \tag{8.20}$$

The parameters on the RHS of Eq. 8.20 are all nondimensional design parameters. In its simplest form, where we set radial velocity nearly equal to zero for curved impellers or in the case of straight-radial blades, $\beta'_2 = 0$, and neglect slip, i.e., $\varepsilon = 1$, the static temperature ratio reduces to

$$\frac{T_2}{T_1} - 1 \approx \left(\frac{\gamma - 1}{2}\right)\Pi_M^2 = \frac{1}{2}\left(\frac{T_{t2}}{T_{t1}} - 1\right) \tag{8.21}$$

Since the total temperature through the stator remains constant, the impeller static temperature raise is half of the stage total temperature rise, which for low speed at the exit of the diffuser becomes one half of the stage static temperature rise ($T_3/T_1 - 1$). Therefore under these conditions, the rotor and the stator produce the same static temperature rise, or what was called a 50% degree of reaction. The ratio of the speed of sound at the impeller exit to the inlet speed of sound is the square root of the static temperature ratio as

$$\frac{a_2}{a_1} \approx \sqrt{1 + \left(\frac{\gamma - 1}{2}\right)\Pi_M^2} \tag{8.22}$$

From Eq. 8.14, we may substitute for the Mach index in terms of the wheel tangential Mach number M_{T2} and the inlet Mach number M_1 to write the ratio of the speeds of sound across the impeller as

$$\frac{a_2}{a_1} \approx \sqrt{1 + \frac{\left(\frac{\gamma - 1}{2}\right)M_{T2}^2}{1 + \left(\frac{\gamma - 1}{2}\right)M_1^2}} \tag{8.23}$$

We may now graph this expression to examine the rise of speed of sound across the impeller. Figure 8.9 shows the ratio of speed of sound across the impeller.

In the limiting case of $M_{T2} \sim 1.55$, the ratio of speed sounds across the impeller a_2/a_1 is ~ 1.2. Therefore all the local Mach numbers at the impeller exit that we had referenced to the inlet speed of sound are $\sim 20\%$ smaller when we use local speed of sound a_2 instead of a_1. Now, let us recap, through an example, the impeller exit Mach numbers as viewed by a radial diffuser that follows the impeller.

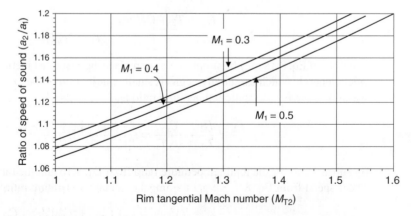

■ **FIGURE 8.9** **The ratio of speed of sound across the impeller for three inlet Mach numbers of $M_1 = 0.3$, 0.4, and 0.5 and zero preswirl ($\gamma = 1.4$)**

EXAMPLE 8.2

Consider a centrifugal compressor impeller with the following inlet and design parameters:

Known/design inlet conditions

- Inlet absolute stagnation temperature and pressure are 288 K and 101×10^5 Pa, respectively, i.e., our design point operation is at the standard sea level static condition
- $C_1 = C_{z1} = 150 \, \text{m/s}$, this gives an axial inlet Mach number of ~0.5, no preswirl, which is our design choice
- Hub-to-tip radius ratio is $r_{h1}/r_1 = 0.5$, this too is a design choice
- Limit the relative Mach number at the eye radius to 1.0, to avoid shock losses in the inducer
- Compressor adiabatic efficiency $\eta_c = 0.72$ estimated based on similar machines
- Compressor mass flow rate is $\dot{m} = 5 \, \text{kg/s}$
- Gas properties are $c_p = 1004 \, \text{J/kg} \cdot \text{K}$ and $\gamma = 1.4$

Known/design exit conditions

- $C_{r2} = C_{z1} = 150 \, \text{m/s}$, this could be our design choice, which sizes the impeller exit area

- Impeller rim speed is $U_2 = 520 \, \text{m/s}$, we chose the maximum rim speed in this example
- Impeller exit vane angle $\beta_2' = 20°$, i.e., we chose a backsweep design to reduce M_2
- Allow for a 10° slip angle, which makes the exit relative flow angle $\beta_2 = 30°$.
- Assume equal total pressure loss parameter ϖ in the impeller and the radial diffuser

Calculate

(a) the absolute Mach number entering the radial diffuser, M_2, and its components, i.e., radial and tangential M_{r2} and $M_{\theta2}$

(b) the relative Mach number at the impeller exit, M_{2r}

(c) the compressor pressure ratio π_c and its polytropic efficiency e_c

(d) the inlet flow area and the "eye" radius r_1

(e) the shaft angular speed ω

(f) the impeller exit radius r_2

(g) the impeller exit width b

(h) the shaft power

SOLUTION

We first draw the impeller exit velocity triangle as an aid in calculating the velocity components.

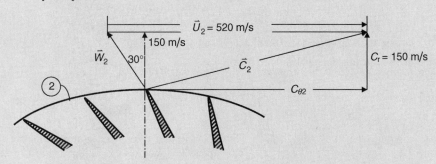

The static temperature at the impeller inlet is

$$T_1 = T_{t1} - C_1^2/2c_p = 288 \, \text{K} - (150)^2/[2(1004)]$$

$$\text{K} = 276.8 \, \text{K} \qquad \Longrightarrow \qquad \boxed{T_1 = 267.8 \, \text{K}}$$

$$a_1 = \sqrt{(\gamma - 1)c_p T_1} \qquad \Longrightarrow \qquad a_1 = [0.4(1004)(276.8)]^{1/2}$$

$$= 333.4 \, \text{m/s} \qquad \Longrightarrow \qquad \boxed{a_1 = 333.4 \, \text{m/s}}$$

The absolute Mach number in the inlet is $M_1 = 150/333.4 \cong 0.45$ $\qquad \Longrightarrow \qquad \boxed{M_1 \cong 0.45}$

The absolute exit swirl velocity is the difference between the rim speed and the relative swirl velocity $W_{\theta2}$, which from the velocity triangle, it is

$$C_{\theta2} = 520(\text{m/s}) - 150.\tan30°(\text{m/s})$$

$$= 433.4 \, \text{m/s} \qquad \Longrightarrow \qquad \boxed{C_{\theta2} = 433.4 \, \text{m/s}}$$

Therefore, the absolute velocity at the exit has a magnitude of

$$C_2 = [(433.4^2) + (150)^2]^{1/2} \, \text{m/s}$$

$$= 458.6 \, \text{m/s} \implies \boxed{C_2 = 458.6 \, \text{m/s}}$$

We need the speed of sound at the exit, a_2, in order to establish the absolute exit Mach numbers. We may use the approximate form of the ratio of speeds of sound as expressed in Eq. 8.22, i.e.,

$$\frac{a_2}{a_1} \approx \sqrt{1 + \frac{\left(\frac{\gamma - 1}{2}\right) M_{T2}^2}{1 + \left(\frac{\gamma - 1}{2}\right) M_1^2}}$$

The impeller tip tangential Mach number is $M_{T2} = U_2/a_1 = 520/333.4 = 1.560$
The inlet absolute Mach number is $M_1 = C_1/a_1 = 150/333.4 = 0.450$

$$\frac{a_2}{a_1} \approx \sqrt{1 + \frac{(0.2)(1.560)^2}{1 + (0.2)(0.45)^2}}$$

$$\cong 1.212 \implies \boxed{a_2 \cong 1.212(333.4) \, \text{m/s} = 404.1 \, \text{m/s}}$$

We may decide to use the more accurate Eq. 8.16 to calculate the exit temperature and the speed of sound, i.e.,

$$\frac{T_{t2}}{T_{t1}} = \frac{T_2 + C_2^2/2c_p}{T_1 + C_1^2/2c_p}$$

We will use Euler turbine equation to calculate the total temperature ratio across the impeller as

$$\frac{T_{t2}}{T_{t1}} \cong 1 + \frac{\omega \cdot r_2 \cdot C_{\theta2}}{c_p T_{t1}} = 1 + \frac{U_2^2}{c_p T_{t1}}\left(\frac{C_{\theta2}}{U_2}\right)$$

Substituting in the above equation for total temperature ratio, we get

$$\frac{T_{t2}}{T_{t1}} = 1 + \frac{(520)^2}{(1004)(288)}\left(\frac{433.4}{520}\right)$$

$$\cong 1.7794 \implies \boxed{T_{t2} \cong 512.5 \, \text{K}}$$

The static temperature is related to total temperature via

$$T_2 = T_{t2} - C_2^2/2c_p \cong 512.47 \, \text{K} - (458.6)^2/2(1004)$$

$$\text{K} \cong 407.7 \, \text{K} \implies \boxed{T_2 \cong 407.7 \, \text{K}}$$

The speed of sound at the impeller exit is

$$a_2 = \sqrt{\gamma \cdot RT_2} = \sqrt{(\gamma - 1)c_p T_2}$$

$$= \sqrt{(0.4)(1004)(407.7)}$$

$$\cong 404.7 \, \text{m/s} \implies \boxed{a_2 = 404.7 \, \text{m/s}}$$

The difference between the approximate and the more exact formulation of the speed of sound at the impeller exit, as we calculated, is negligible (404.1 m/s versus 404.7 m/s).

Now we can calculate the impeller exit Mach numbers, based on the local speed of sound, namely,

$$M_{\theta2} \equiv C_{\theta2}/a_2 = 433.4/404.7 \cong 1.071$$

$$M_{r2} \equiv C_r/a_2 = 150/404.7 \cong 0.371 \implies \boxed{\begin{aligned} M_{\theta2} &= 1.071 \\ M_{r2} &= 0.371 \\ M_2 &= 1.133 \end{aligned}}$$

$$M_2 \equiv C_2/a_2 = 458.6/404.7 \cong 1.133$$

The impeller exit Mach number is W_2/a_2 and $W_2 = 150 \, \text{m/s}/\cos 30° = 173.2 \, \text{m/s}$

Therefore, the impeller exit relative Mach number is

$$173.2/404.7 \cong 0.428 \quad \boxed{M_{2r} \cong 0.428}$$

Note that our design choices have led to a slightly supersonic inlet condition to the radial diffuser that follows the impeller. To calculate the compressor total pressure ratio, we use the adiabatic efficiency η_c and the compressor total temperature ratio T_{t2}/T_{t1} as we derived earlier

$$\pi_c = [1 + \eta_c(\tau_c - 1)]^{\frac{\gamma}{\gamma-1}} = [1 + 0.72(1.7794 - 1)]^{3.5}$$

$$\cong 4.75 \implies \boxed{\pi_c = p_{t2}/p_{t1} = 4.75}$$

Compressor polytropic and adiabatic efficiencies are related via

$$\eta_c = \left[\frac{\pi_c^{\frac{\gamma-1}{\gamma}} - 1}{\pi_c^{\frac{\gamma-1}{\gamma \cdot e_c}} - 1}\right]$$

Solving the above equation for e_c, we get

$$e_c = \frac{\left(\frac{\gamma - 1}{\gamma}\right)\ln(\pi_c)}{\ln\left[1 + \frac{\pi_c^{\frac{\gamma-1}{\gamma}} - 1}{\eta_c}\right]} = \frac{0.2857 \ln(4.75)}{\ln\left[1 + \frac{(4.75)^{0.2857} - 1}{0.72}\right]}$$

$$\cong 0.773 \implies \boxed{e_c = 0.773}$$

Now let us size the inlet flow area and the eye radius r_1 for the mass flow rate of 5 kg/s, under standard sea level static condition at the inlet. First, we need to calculate the fluid density at the inlet, ρ_1. We can calculate the density from the perfect gas law, which needs a gas constant R, or from the speed of sound, which requires γ. We calculate the static pressure at the inlet, i.e.,

$$p_1 = \frac{p_{t1}}{\left[1 + \frac{\gamma - 1}{2} M_1^2\right]^{\frac{\gamma}{\gamma-1}}}$$

$$= \frac{1.01 \times 10^5 \, \text{Pa}}{[1 + 0.2(0.45)^2]^{3.5}} \cong 8.7897 \times 10^4 \, \text{Pa}$$

$$a_1^2 = \frac{\gamma \cdot p_1}{\rho_1}$$

$$\rho_1 = (1.4)(8.7897 \times 10^4)/(333.4)^2 \, \text{kg/m}^3 \cong 1.107 \, \text{kg/m}^3$$

$$\dot{m} = \rho_1 A_1 C_{z1}$$

$$A_1 = (5 \, \text{kg/s})/[(1.107 \, \text{kg/m}^3)(150 \, \text{m/s})] = 0.03011 \, \text{m}^2$$
$$= 301.1 \, \text{cm}^2$$

$$A_1 = \pi(r_1^2 - r_{h1}^2) = \pi \cdot r_1^2 \left[1 - \left(\frac{r_{h1}}{r_1} \right)^2 \right]$$

$$r_1 = [(301.1 \, \text{cm}^2)/\pi(1 - 0.25)]^{1/2}$$
$$\cong 11.30 \, \text{cm or } r_1 \sim 4.466 \, \text{inches}$$

$$\boxed{r_e = r_1 \cong 11.30 \, \text{cm}}$$

From the inlet relative Mach number of 1.0 at the eye radius r_e, we establish the angular speed of the shaft ω,

$$C_{z1}^2 + (\omega \cdot r_1)^2 = a_1^2 \cdot M_{r1}^2 = (333.4)^2 (1.0)^2$$
$$= 111,156 \, \text{m}^2/\text{s}^2$$

Therefore,

$$(\omega r_1)^2 = [111,156 - (150)^2] \text{m}^2/\text{s}^2$$
$$= 88,656 \, \text{m}^2/\text{s}^2, \text{ or } \omega r_1 \cong 297.7 \, \text{m/s},$$
$$\text{or } \omega \cong 2,635 \, \text{rad/s}$$
$$\text{or } \omega \cong 25,160 \, \text{rpm}$$

$$\boxed{\omega \cong 2,635 \, \text{rad/s} \cong 25,160 \, \text{rpm}}$$

The impeller exit radius r_2 is established by the rim speed of 520 m/s specified in the problem according to

$$r_2 = U_2/\omega = (520 \, \text{m/s})/(2635 \, \text{rad/s})$$

$$\boxed{r_2 \cong 19.73 \, \text{cm}}$$

This gives a radius ratio of ~ 0.572 for the eye to exit, i.e., $r_e/r_2 \sim 0.572$. We shall examine the effect of this parameter on the inducer diffusion factor in the next section. The impeller exit width (or blade axial span at the exit) b is established from continuity equation and our design choice for the exit radial velocity $C_{r2} = W_{r2} = 150 \, \text{m/s}$, i.e.,

$$A_2 = \frac{\dot{m}}{\rho_2 \cdot C_{r2}} = 2\pi \cdot r_2 \cdot b$$

In this preliminary stage, note that we made a first order approximation in the flow area by not accounting for the blade thickness multiplied by the number of blades that reduces the exit flow area. There are two unknowns in the

above equation. One is the blade axial span at the exit, b, and the other is the gas density at the impeller exit, ρ_2. To calculate the gas density at the impeller exit, we first proceed to calculate the static pressure at the impeller exit p_2. Then by using a perfect gas law at station 2, we calculate the density of the gas, ρ_2. The exit static pressure p_2 is calculated from our assumption of equal total pressure loss parameter ϖ across the rotor and the diffuser. From the definition of $\varpi_{\text{imp.}}$, we have

$$\varpi_{\text{imp.}} \equiv \frac{p_{t1r} - p_{t2r}}{q_{1r}}$$

$$= \frac{p_1 \left[1 + \frac{\gamma - 1}{2} M_{r1}^2 \right]^{\frac{\gamma}{\gamma - 1}} - p_2 \left[1 + \frac{\gamma - 1}{2} M_{r2}^2 \right]^{\frac{\gamma}{\gamma - 1}}}{\frac{\gamma}{2} p_1 M_{1r}^2}$$

Also the diffuser total pressure loss parameter is

$$\varpi_{\text{dif.}} \equiv \frac{p_{t2} - p_{t3}}{q_2}$$

$$= \frac{p_2 \left[1 + \frac{\gamma - 1}{2} M_2^2 \right]^{\frac{\gamma}{\gamma - 1}} - \pi_s \cdot p_1 \left[1 + \frac{\gamma - 1}{2} M_1^2 \right]^{\frac{\gamma}{\gamma - 1}}}{\frac{\gamma}{2} p_2 M_2^2}$$

We replaced the exit total pressure p_{t3} by the product of the stage total pressure ratio π_s and the inlet absolute total pressure. By setting the above two total pressure loss parameters equal to each other, we calculate the impeller exit static pressure p_2, namely,

$$p_2/p_1 \cong 1.795 \text{ or } p_2 \cong 1.578 \times 10^5 \, \text{Pa}$$

$$\boxed{p_2 \cong 1.578 \times 10^5 \, \text{Pa}}$$

Therefore, the impeller exit density is $\rho_2 = \gamma p_2/(a_2)^2$, which is

$$\rho_2 = (1.4)(1.578 \times 10^5 \, \text{Pa})/(404.7 \, \text{m/s})^2 \cong 1.3487 \, \text{kg/m}^3$$

$$\boxed{\rho_2 \cong 1.3487 \, \text{kg/m}^3}$$

The impeller exit width b is now estimated from the continuity equation,

$$b \approx \frac{\dot{m}}{2\pi \cdot r_2 \rho_2 \cdot C_{r2}}$$

$$= \frac{5 \, \text{kg/s}}{2\pi \cdot (0.1973 \, \text{m})(1.3487 \, \text{kg/m}^3)(150 \, \text{m/s})}$$

$$\cong 0.01994 \, \text{m} \qquad \boxed{b \cong 2.0 \, \text{cm}}$$

The shaft power is the product of the mass flow rate and the total enthalpy rise across the rotor,

$$\wp_s = \dot{m} c_p (T_{t2} - T_{t1}) = 5 \, \text{kg/s}(1004 \, \text{J/kg} \cdot \text{K})(512.5 - 288)$$

$$\text{K} \cong 1,127 \, \text{kW} \qquad \boxed{\wp_s \cong 1127 \, \text{kW}}$$

8.3   Radial Diffuser

The flow that leaves the impeller of a centrifugal compressor enters a radial diffuser. As the absolute flow at the impeller exit has swirl, the streamlines in the radial diffuser are spirals in shape. A definition sketch of a radial diffuser with its spiral streamlines and the station numbers is shown in Fig. 8.10.

The discharge of the centrifugal impeller enters a vaneless radial diffuser at first, which may be followed by a vaned diffuser. The function of the radial diffuser, as in all diffusers, is to decelerate the flow and convert the kinetic energy to static pressure rise.

The diffuser in a centrifugal compressor lies between parallel walls with a spiraling flow that moves outward in the radial direction. Due to the parallel sidewalls, the diffuser area increases monotonically with radius according to

$$\frac{A(r)}{A_2} = \frac{r}{r_2} \tag{8.24}$$

The swirl component of the flow in the radial diffuser decays as a result of

1. The conservation of angular momentum reduces the swirl inversely proportional to an increasing radius in the radial diffuser, and
2. The external torque on the fluid applied by the wall (due to fluid viscosity) is in the opposite direction to the fluid angular momentum, thus it acts to retard the fluid swirl as it moves outward in spirals.

The conservation of angular momentum, in the absence of external torque, requires

$$rC_\theta = r_2 C_{\theta 2} \tag{8.25-a}$$

$$\frac{C_\theta(r)}{C_{\theta 2}} = \frac{r_2}{r} \tag{8.25-b}$$

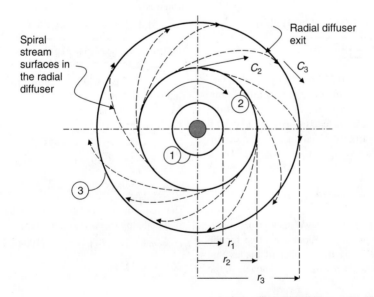

■ **FIGURE 8.10**   Definition sketch of the flow in the radial diffuser

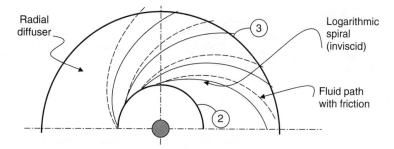

■ **FIGURE 8.11** Schematic drawing of the fluid path in a radial diffuser with and without the effect of wall friction

The fluid path whose angular momentum is conserved results in a *logarithmic spiral*, as schematically shown in Fig. 8.11. The effect of external retarding torque is

$$\tau_{\text{ext}} = \dot{m}(r_3 C_{\theta 3} - r_2 C_{\theta 2})$$ **(8.26-a)**

Therefore, in the presence of a retarding torque (due to friction), the angular momentum is reduced and swirl diminish even faster than a logarithmic spiral, namely,

$$\frac{r C_\theta}{r_2 C_{\theta 2}} = 1 - \frac{\tau_{\text{wall}}}{\dot{m}(r_2 C_{\theta 2})}$$ **(8.26-b)**

where the (external) wall torque is written as negative, which applies a retarding influence on the fluid angular momentum. The fluid path with friction *unwinds* faster than the inviscid solution. A schematic drawing of the two stream patterns in a radial diffuser with/without friction is shown in Fig. 8.11.

The inlet Mach number M_2 to the radial diffuser may be calculated using the velocity triangle at the impeller exit.

$$C_2 = \sqrt{C_{r2}^2 + C_{\theta 2}^2} = \sqrt{C_{r2}^2 + (U_T - C_{r2} \tan \beta_2)^2}$$ **(8.27)**

The speed of sound at the impeller exit is based on T_2, for which we derived a simple expression,

$$\frac{T_2}{T_1} \approx 1 + \left(\frac{\gamma - 1}{2}\right) M_T^2$$

Therefore the speed of sound at the impeller exit is related to the speed of sound at the inlet according to

$$\frac{a_2}{a_1} \approx \sqrt{1 + \left(\frac{\gamma - 1}{2}\right) M_T^2}$$ **(8.28)**

The combination of the exit speed to the local speed of sound is the Mach number, therefore,

$$M_2 \approx \frac{\sqrt{M_{z1}^2 + (M_T - M_{z1} \tan \beta_2)^2}}{\sqrt{1 + \left(\frac{\gamma - 1}{2}\right) M_T^2}}$$ **(8.29)**

In expression 8.29, we approximated the radial velocity at the impeller exit by the inlet axial velocity C_{z1}. We may use this as a design choice to approximate the blade axial span at the impeller exit b.

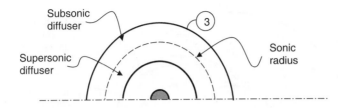

■ **FIGURE 8.12**
Radial diffuser with supersonic inlet flow and a sonic radius

The effect of backward-leaning design (i.e., $\beta_2 > 0$) on the impeller exit Mach number is seen from Eq. 8.28 to be reducing the exit velocity, thus the Mach number. A reduced impeller exit Mach number then relieves the requirements on the radial diffuser and improve its performance. Despite the choice of backward-leaning designs in reducing the exit velocity, our desire to achieve higher pressure ratios has pushed the exit Mach number of the impeller, i.e., the diffuser inlet Mach number into supersonic regime. The radial diffuser is then divided into a supersonic diffuser followed by a subsonic diffuser. The dividing line between the two parts is the invisible sonic circle, as shown in Fig. 8.12. The subsonic diffuser may be bladed (vanes) to improve the pressure recovery.

For the conversion of the entire kinetic energy to the fluid static enthalpy rise, we get

$$h_3 - h_2 = \frac{C_2^2}{2} \approx \frac{U_T^2}{2} \tag{8.30}$$

Therefore, the static temperature ratio across the compressor stage is related to the square of the impeller exit Mach number, according to

$$\frac{T_3}{T_1} = \frac{T_2}{T_1} + \left(\frac{\gamma - 1}{2}\right) M_T^2 \tag{8.31}$$

Now, substituting for the static temperature ratio for the rotor from Eq. 8.31, we get

$$\frac{T_3}{T_1} \approx 1 + (\gamma - 1) M_T^2 \tag{8.32}$$

The static pressure ratio in the limit of reversible and adiabatic flow for the stage is

$$\frac{p_3}{p_1} \approx \left[1 + (\gamma - 1) M_T^2 \right]^{\frac{\gamma}{\gamma - 1}} \tag{8.33}$$

In this model, we note that the static temperature rise across the diffuser is the same as the rotor, which produces the other half of the static temperature rise. In the context of the degree of reaction, the stage that we defined is a 50% degree of reaction type.

The placement of cambered vanes in radial diffusers similar to the splitter plates in axial diffusers significantly improves the static pressure recovery of radial diffusers. The vane leading edge needs to be aligned with the local relative flow in order to avoid lip separation. A schematic drawing of cambered vanes or wedges in a centrifugal compressor diffuser is shown in Fig. 8.13.

The guiding hands of the vanes' sidewalls assist the flow in remaining attached to diffuser walls. This is another example of three-dimensional flow separation being delayed by taking one spatial degree of freedom away from the flow, i.e., turning a 3D flow into an effective 2D flow. The diffuser throat blockage, as presented earlier in the inlet chapter, dominates the performance of the cambered vane diffusers. Pratt & Whitney has introduced "pipe" diffusers, which have produced a superior performance to the cambered vane or channel diffusers (see Kenny, 1984).

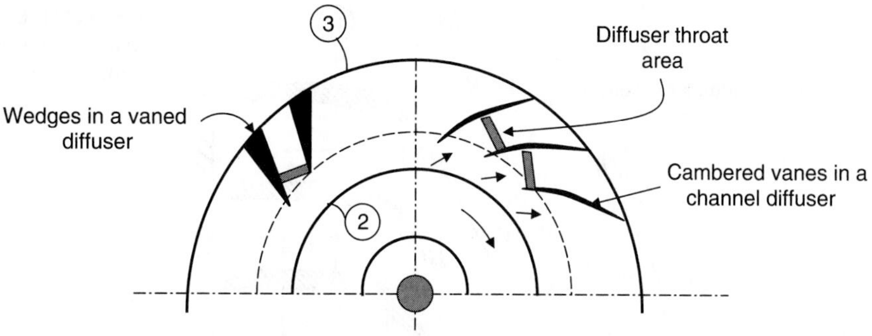

■ **FIGURE 8.13**   Schematic drawing of cambered vanes or wedges in a radial diffuser

EXAMPLE 8.3

For an impeller tip Mach number of 1.2, and equal static pressure rise across the rotor and diffuser, calculate

• the static pressure ratio across the rotor and diffuser, p_3/p_1

• the static pressure ratio across the diffuser, p_3/p_2

• estimate diffuser static pressure rise (assume $M_2 \approx M_T$)

SOLUTION

In the limit of reversible adiabatic flow, we may use Eq. 8.33 to get

$$\frac{p_3}{p_1} \approx [1 + 0.4(1.2)^2]^{3.5} \cong 4.9$$

The static pressure ratio across the diffuser is square root of the stage pressure ratio, namely,

$$\frac{p_3}{p_2} \approx \sqrt{4.9} \approx 2.2$$

The static pressure rise coefficient for the diffuser is

$$C_p \equiv \frac{p_3 - p_2}{q_2} = \frac{p_3 - p_2}{\gamma \cdot p_2 M_2^2/2} = \frac{2}{\gamma \cdot M_2^2}\left(\frac{p_3}{p_2} - 1\right) \quad \textbf{(8.34)}$$

The Mach number at the impeller exit is

$$M_2^2 = M_{r2}^2 + M_{\theta 2}^2 \quad \textbf{(8.35)}$$

We need to analyze the velocity triangle at the impeller tip to evaluate these terms, i.e., the radial and tangential Mach

numbers. For the sake of discussion, let us take M_2 nearly the same as M_T, which forces the static pressure rise coefficient in the radial diffuser to be

$$C_p \approx \frac{2}{(1.4)(1.2)^2}(2.2 - 1) \approx 1.19$$

From our earlier diffuser studies we know that this level of static pressure rise is not tolerable in a practical diffuser. In the axial-flow compressor Chapter 7, we set a maximum value of $C_{p,\text{max}}$ of ~ 0.6, which in this case is nearly twice the maximum value. We conclude that the diffuser section needs to be more lightly loaded than the rotor if the boundary layer is to remain attached in the diffuser. This is not surprising that in centrifugal compressors the rotor tip Mach number plays such a dominant role in establishing the performance of the machine. We experienced a similar behavior in axial-flow machinery.

8.4   Inducer

The inducer is the part of the impeller that meets the flow at the compressor inlet at the relative flow angle β_1. Its primary function is to turn the flow toward the axial direction, as shown in Fig. 8.13, thereby increasing its static pressure. In this context, the function of inducer is very similar to the rotor blades in an axial-flow compressor. There are some differences though that we need to consider. One is that the rotor blades in an axial-flow compressor do not need to turn the relative

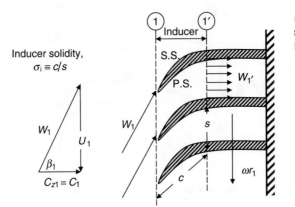

flow completely in the axial direction, whereas the inducer has to turn the flow completely in the axial direction. This causes a higher loading, i.e., turning demand, on the inducer. The second difference is subtler than the first. The difference arises from the trailing edge of the blades. Namely, the inducer trailing edge is not free, i.e., it is connected to the rest of the impeller. The axial-flow compressor rotor trailing edge is free and thus obeys the Kutta condition, which influences the trailing edge flow and the wake pattern. For example, the axial-flow compressor rotor immediately adjusts to the incoming flow disturbances by vortex shedding in the wake or adjusting the wake vortex strength in the spanwise direction. Despite these differences, we still proceed to analyze the inducer blades using the same tools we developed in the axial-flow compressors.

Let us identify station $1'$ as the exit of the inducer, as shown in Fig. 8.14. We may define the diffusion factor for the inducer as

$$D_{\text{inducer}} \equiv 1 - \frac{W_{1'}}{W_1} + \frac{|\Delta C_\theta|}{2\sigma_i W_1} \quad (\text{inducer } D-\text{factor}) \tag{8.36}$$

The change of swirl across the inducer is approximately the wheel speed U_1, since the inlet flow to the inducer is swirl free (in the absolute frame) and the exit flow is nearly axial in the relative frame. Now, if we assume that the axial velocity in the inducer remains constant, i.e.,

$$C_{z1} \cong W_{1'} \tag{8.37}$$

We may write the diffusion factor in terms of axial flow Mach number at the inlet and the impeller tip Mach number M_T, according to

$$D_{\text{inducer}} \cong 1 - \frac{C_{z1}}{\sqrt{C_{z1}^2 + U_1^2}} + \frac{U_1}{2\sigma_i \sqrt{C_{z1}^2 + U_1^2}} \tag{8.38}$$

$$D_{\text{inducer}} \cong 1 - \frac{1}{\sqrt{1 + \left(\dfrac{U_1}{U_2}\right)^2 \left(\dfrac{U_2}{C_{z1}}\right)^2}} + \frac{U_1/C_{z1}}{2\sigma_i \sqrt{1 + \left(\dfrac{U_1}{U_2}\right)^2 \left(\dfrac{U_2}{C_{z1}}\right)^2}} \tag{8.39-a}$$

$$D_{\text{inducer}} \cong 1 - \frac{1}{\sqrt{1 + \left(\dfrac{r_1}{r_2}\right)^2 \left(\dfrac{M_T}{M_{z1}}\right)^2}} + \frac{(M_T/M_{z1})(r_1/r_2)}{2\sigma_i \sqrt{1 + \left(\dfrac{r_1}{r_2}\right)^2 \left(\dfrac{M_T}{M_{z1}}\right)^2}} \tag{8.39-b}$$

Here, we note that the geometric parameters such as the impeller radius ratio and the inducer solidity are tied together with the choices of the tangential and axial flow Mach numbers to arrive at a reasonable inducer diffusion factor $D_{inducer}$, i.e., $D_{inducer} < 0.6$. The axial Mach number varies in the range of $M_{z1} \sim 0.4$–0.6 and the impeller tip Mach number $M_T \sim 1$–1.5, the radius ratio, and the solidity may be plotted as a function of inducer diffusion factor.

EXAMPLE 8.4

Calculate and graph the inducer D-factor for solidity of one and over a range of impeller tip Mach numbers and radius ratios.

SOLUTION

A spreadsheet is produced based on Eq. 8.39-b for solidity of 1. The eye-to-tip radius ratio varied from 0.1 to 0.5, and the tip-to-axial Mach number ratio varied from 2 to 4.0. The graph of the tabulated spreadsheet is shown in Fig. 8.15. The parameter M_T/M_{z1} is chosen as a running parameter in Fig. 8.15, as it determines the compressor pressure ratio. We observe that for a given solidity, e.g., $\sigma_i = 1$, and a maximum diffusion factor, of say 0.6, a high-pressure ratio compressor impeller requires a low eye-to-tip radius ratio (~ 0.22). But since the mass flow rate through the machine is proportional to the inlet area, for a high-pressure ratio

impeller, the mass flow per unit frontal area (i.e., A_2 or A_3) drops. This inverse relationship between the radius ratio and the pressure ratio limits the practical designs of centrifugal compressors to a compromise between these conflicting requirements.

The advances in transonic rotor design in axial-flow compressors have prompted maximum relative Mach numbers approaching the inducer tip to be as high as ~ 1.5. The supersonic relative tip Mach number creates shocks, which demand extensive inducer geometric tailoring via viscous CFD analysis to minimize shock boundary layer interaction losses.

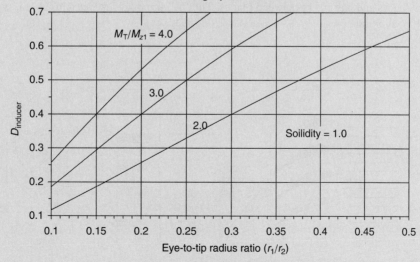

■ **FIGURE 8.15**   Inducer design parameters (for an inducer solidity of one)

EXAMPLE 8.5

In Example 8.1, we calculated an eye-to-tip radius ratio of 0.57, an axial Mach number of $M_{z1} \sim 0.45$ and $M_T \sim 1.56$. Calculate the inducer diffusion factor in terms of its solidity. Is there any solidity that allows acceptable diffusion in the inducer?

SOLUTION

Upon substitution of the Mach numbers and radius ratio in Eq. 8.39-b that relates the diffusion factor to the impeller solidity, we get

$$D_i \cong 0.79 + \frac{0.446}{\sigma_i}$$

As suspected, the diffusion is too large and there is no choice of solidity (σ_i) that can decrease the inducer diffusion factor to below 0.79. Therefore, we are prompted to change our design parameters in order to reduce the inducer (diffusion) loading.

8.5   Inlet Guide Vanes (IGVs) and Inducer-Less Impellers

By incorporating inlet guide vanes in the incoming flow, we may impart a preswirl to the fluid, which is in the direction of impeller rotation. Therefore, it is possible to eliminate the need for an inducer if the relative flow to the impeller is already in the axial direction. Figure 8.16 shows the IGV exit flow and the impeller inlet relative flow, which is purely in the axial direction. In part (c) of Fig. 8.16, we are comparing an impeller with inducer to an inducer-less impeller in the meridional plane. The presence of inducer offers a larger throat area than the inducer-less impeller. This geometrical influence causes the centrifugal compressor with an inducer-less impeller reach choking condition before a geometrically comparable centrifugal compressor with an inducer section. This may be seen from the data of Eckardt (1977). Figure 8.17 shows Eckardt's comparison of two centrifugal compressors with the same exit radius r_2 and width b. The "A" compressor has inducer and backsweep of 40°, whereas the "B" impeller is inducer-less and has 30° backsweep blades. The "A" impeller reaches a higher mass flow rate before choking.

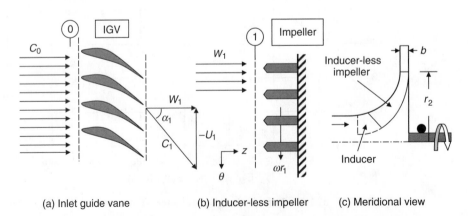

(a) Inlet guide vane   (b) Inducer-less impeller   (c) Meridional view

■ **FIGURE 8.16   IGV-coupled inducer-less impeller and geometrical comparison of two impellers**

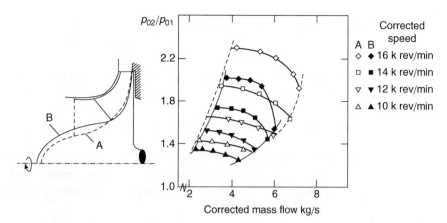

■ **FIGURE 8.17** Performance maps of two impellers (A) with and (B) without inducer (from Eckhardt, 1977)

8.6 Impeller Exit Flow and Blockage Effects

We have introduced the phenomenon of slip at the impeller exit. The turning of the relative flow in the opposite direction to that of the rotor starts inside the impeller passage. We have graphically depicted the relative streamlines in the impeller in Fig. 8.3(d). The accumulation of the flow on the pressure side results in a jet-like behavior and the sparse low-momentum flow on the suction side of the impeller vanes gives an appearance of the wake (or deficit momentum) flow. This behavior at the exit of the centrifugal compressor impellers is known as the *jet-wake flow*, first described by Dean and Senoo (1960). Figure 8.18 shows the jet-wake exit flow behavior in a radial impeller. The first observation is that the impeller exit flow is highly nonuniform. Therefore the one-dimensional approximations that we often make need to be modified to reflect the nonuniform flow behavior. The second implication is that the boundary layer on the suction surface near the impeller exit is likely separated, as schematically shown in Fig. 8.18. The combined effects of jet-wake velocity profile and separated flow gives rise to a high level of *blockage* at the exit.

Through the concept of blockage, we define an effective flow area $A_{2\text{eff}}$. The definition of blockage from diffuser studies in Chapter 5 is repeated here, where effective and geometrical areas are applied to the impeller exit:

$$B_2 \equiv 1 - \frac{A_{2\text{eff}}}{A_2}$$

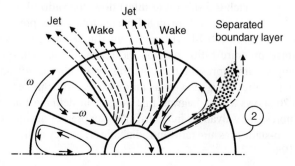

■ **FIGURE 8.18** The exit flow pattern from an impeller shows a "jet-wake" profile

where A_2 is the geometric flow area that may be written as

$$A_2 = (2\pi r_2 b - Ntb)\cos\beta'_2 \tag{8.40}$$

In Eq. 8.40, N is the number of the impeller vanes, t is thickness of the blades at the exit, b is the axial span of vanes at 2, and β'_2 is the blade sweep angle at the exit (measured from the radial direction). The cosine term in Eq. 8.40 gives a projection of the exit area that is in the radial direction. The conclusions from this section are

(a) One-dimensional flow analysis is totally inadequate and gives erroneous results

(b) Jet-wake profile and flow separation have to be accounted for by a suitable averaging technique and an estimate of blockage B_2, respectively. In either case, more experimental data are needed.

8.7   Efficiency and Performance

Centrifugal compressor efficiency has always lagged its counterpart, the axial-flow compressor, for the same pressure ratio. The flow path is rather torturous as it twists from one plane to another in a centrifugal compressor, thus massive secondary flow and separation losses become imminent. In addition, for high-pressure ratio centrifugal compressors, the flow in the diffuser becomes supersonic and its efficient diffusion rather complex. In a historical perspective presented by Kenny (1984), different loss sources, future design trends, and limitations of centrifugal compressors are discussed. Although it is very difficult, if not impossible, to separate losses in turbomachinery into their components without overlapping, which is double/triple accounting, it is still instructive to expose general loss sources in turbomachinery. In a centrifugal compressor, the losses are attributed to

1. Tip clearance loss in an unshrouded impeller
2. Secondary flow losses due to flow turning
3. Disk friction losses
4. Losses due to compressibility effects in the inducer due to shocks and flow separation
5. Jet-wake mixing flow losses in the diffuser
6. Diffuser compressibility losses due to high absolute Mach numbers
7. Flow unsteadiness and vortex shedding in the impeller wake

Also a comparison of two centrifugal compressors designed for an 8:1 total-to-static pressure ratio and their efficiencies at off-design are shown in Fig. 8.19 (also from Kenny). Note that the efficiencies are listed for total exit-to-static inlet state of the gas, as well as the pressure ratio. For a Mach 0.5 (absolute inlet) flow, the ratio of total-to-static pressure is $[1 + (0.2)(0.5)^2]^{3.5} \cong 1.186$, therefore an 8:1 design total-static pressure ratio means a total-to-total pressure ratio of $\pi_s \cong 6.74$. Although centrifugal compressors are less efficient; they can produce stage pressure ratios outside the reach of the axial-flow compressors. The centrifugal compressors represent the compressor of choice for small aircraft engines, APUs, industrial applications, as well as automotive turbochargers.

Figure 8.20 shows a T–s diagram that is used in defining total-to-static efficiency of centrifugal compressors. Advances in centrifugal compressor design are demonstrated by an increase in total-to-static pressure ratio and adiabatic efficiency as shown in Figs. 8.21 and 8.22 (from Kenny, 1984).

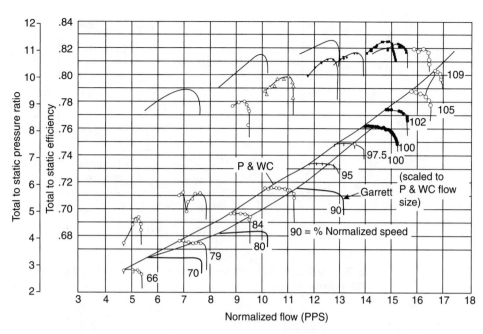

■ **FIGURE 8.19**   The performance map for two centrifugal compressors with an 8:1 total-static design pressure ratio (from Kenny, 1984)

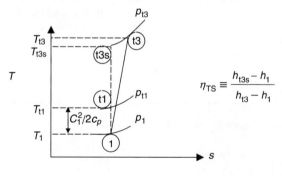

■ **FIGURE 8.20**   *T–s* diagram used in defining the total-to-static efficiency of a centrifugal compressor

$$\eta_{TS} \equiv \frac{h_{t3s} - h_1}{h_{t3} - h_1}$$

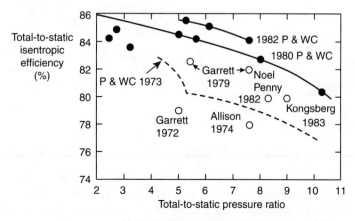

Note: All efficiencies are at 11% surge margin

■ **FIGURE 8.21**   Advances in centrifugal compressor performance (From Kenny, 1984)

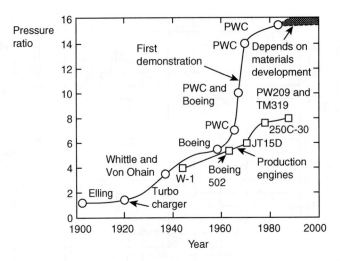

■ **FIGURE 8.22**   **Historic perspective on centrifugal compressor (total to static) pressure ratio (From Kenny, 1984)**

EXAMPLE 8.6

The inlet flow to a centrifugal compressor is described by

$$C_1 = 200\,\text{m/s}$$
$$p_{t1} = 101\,\text{kPa}$$
$$T_{t1} = 288\,\text{K}$$

The compressor pressure ratio at the design point is $p_{t3}/p_{t1} = 6.8$ with a total-to-total adiabatic efficiency of $\eta_{TT} = 0.80$. Calculate the compressor total-to-static efficiency. Assume $\gamma = 1.4$ and $c_p = 1004\,\text{J/Kg} \cdot \text{K}$.

SOLUTION

From the definition of total–total adiabatic efficiency, we calculate T_{t3} following

$$T_{t3} = T_{t1}\left\{1 + \frac{1}{\eta_{TT}}\left[\left(\frac{p_{t3}}{p_{t1}}\right)^{\frac{\gamma-1}{\gamma}} - 1\right]\right\} \approx 550.5\,\text{K}$$

Also, T_{t2s} is related to T_{t1} and compressor pressure ratio isentropically, i.e.,

$$T_{t2s} = T_{t1}\left(\frac{p_{t3}}{p_{t1}}\right)^{\frac{\gamma-1}{\gamma}} \approx 498\,\text{K}$$

The static temperature of the inlet gas is calculated from

$$T_1 = T_{t1} - \frac{C_1^2}{2c_p} \approx 288\,\text{K} - (200)^2/2/1004 \approx 268.1\,\text{K}$$

The definition of total-to-static efficiency is $\eta_{TS} \equiv \dfrac{h_{t3s} - h_1}{h_{t3} - h_1}$ and we have calculated all the parameters in this equation, therefore, $\eta_{TS} \approx 0.814$

8.8   Summary

Centrifugal compressors achieve high compression per stage that is predominantly produced by centrifugal force with a large exit-to-inlet impeller radius ratio. The flow area through the inlet is thus a small fraction of the frontal area of the machine. Therefore the mass flow rate per frontal area is lower in centrifugal than the axial-flow compressors.

The pressure ratio per stage is high, however, with lower adiabatic efficiency than the axial counterpart. The flow path from the inlet to the exit twists and turns, thus secondary flow losses are significant. The phenomenon of slip creates a jet-wake nonuniform velocity profile at the impeller exit. The geometry of the impeller blade passages is characterized by a large wetted perimeter and a small flow area, which makes for an equivalently small hydraulic diameter with large frictional losses. The radial flow from the impeller is decelerated in radial and vaned diffusers. The diffuser exit flow is turned 90° for a follow up combustor or even 180° degrees for the next centrifugal compressor stage. Therefore, staging centrifugal compressors involve turnaround ducts, which make them more cumbersome than axial compressors.

The robust construction of the centrifugal compressors and their high pressure ratio per stage capability make them ideal for small gas turbine engines, APUs, automotive turbochargers, and other industrial uses. The inducer-less and shrouded impellers are used in industrial applications, whereas the unshrouded impellers with inducers that offer lower weight and higher mass flow capability are suitable for aerospace applications.

Classical literature on turbomachinery is found in Stodola's text (1924, 1927) as well as Traupel's book (1977). Taylor (1964), Marble (1964), Kerrebrock (1992), Cumpsty (2004) and Dixon (1975) have treated compressors in detail and are recommended for further reading. Historical perspectives presented by Garvin (1998) and St. Peter (1999) compliment the subject. Hill and Peterson's book (1992) as well as Paduano, Greitzer and Epstein's review paper (2001) on compression system stability are recommended to the reader.

References

1. Busemann, A., "Das Foerderhuehenverhaeltnis radialer Kreiselpumpen mit Logarithmisch-spiraligen Schaufeln," *Zeitschrift angewandete Mathematik und Mechanik*, Vol. 8, No. 5, 1928.

2. Cumpsty, N.A., *Compressor Aerodynamics*, Krieger Publishing Co., Malabar, Florida, 2004.

3. Dean, R.C. and Senoo, Y., "Rotating Wakes in Vaneless Diffusers,: Transactions of ASME," *Journal of Basic Engineering*, Vol. 82, 1960, pp. 563–574.

4. Dixon, S.L., *Fluid Mechanics, Thermodynamics of Turbomachinery*, 2nd , edition, Pergamon Press, Oxford, UK, 1975.

5. Eckhardt, D., Vergleichende Stroemungsuntersuchungen an drei Radiaverdichter-Laufraden mit konventionellen Messferfahren, Forschungsbericht Verbrennungskraftmaschinen, Vorhaben 182, Vol. 237, 1977.

6. Garvin, R.V., *Starting Something Big: The Commercial Emergence of GE Aircraft Engines*, AIAA, Inc., Reston, Virginia, 1998.

7. Hill, P.G. and Peterson, C.R., *Mechanics and Thermodynamics of Propulsion*, 2nd edition, Addison-Wesely, Reading, Massachusetts, 1992.

8. Kenny, D.P., "The History and Future of the Centrifugal Compressor in Aviation Gas Turbines," SAE Paper No. 841635, 1984.

9. Kerrebrock, J.L., *Aircraft Engines and Gas Turbines*, 2nd edition, MIT Press, Cambridge, Mass., 1992.

10. Marble, F.E., "Three-Dimensional Flow in Turbomachines," in *Aerodynamics of Turbines and Compressors*, Vol. X, Ed. Hawthorne, W.R., Princeton Series on High Speed Aerodynamics and Jet Propulsion, Princeton University Press, Princeton, N.J., 1964.

11. Paduano, J.D., Greitzer, E.M., and Epstein, A.H., "Compression System Stability and Active Control," *Annual Review of Fluid Mechanics*, Vol. 33, 2001, pp. 491–517.

12. Stanitz, J.D. and Ellis, G.O., "Two-Dimensional Compressible Flow in Centrifugal Compressors with Straight Blades," NACA Tech. Note 1932, 1949.

13. St. Peter, J., *The History of Aircraft gas Turbine Engine Development in the United States*, International Gas Turbine Institute, Atlanta, 1999.

14. Stodola, A., *Dampf- und Gasturbinen*, Springer Verlag, 1924.

15. Stodola, A., *Steam and Gas Turbines*, Vos. 1 and 2, McGraw-Hill, New York, 1927.

16. Taylor, E.S., "The Centrifugal Compressor", Chapter in *Aerodynamics of Turbines and Compressors*, Ed. Hawthorne, W.R., Princeton Series in High-Speed Aerodynamics and Jet Propulsion, Vol. X, Princeton University Press, Princeton, NJ, 1964.

17. Traupel, W., *Thermische Turbomaschinen*, 3rd edition, Springer Verlag, 1977.

Problems

8.1 A centrifugal compressor has 22 radial impellers with an (impeller) exit radius of $r_2 = 0.25$ m. Assuming the mass flow rate through the compressor is $\dot{m} = 10$ kg/s, the shaft speed is $\omega = 10,000$ rpm, and the inlet flow is swirl free, calculate

(a) the shaft power $\wp_s$ in kW

(b) the total temperature rise in the compressor (assume $\gamma = 1.4$ and $c_p = 1004$ J/kg · K)

8.2 Size the exit radius of a centrifugal compressor impeller, r_2, that is to reach a tangential Mach number of $M_T = 1.5$. The shaft rotational speed is 25,000 rpm and the inlet flow condition is characterized by

$$p_1 = 100 \text{ kPa}$$
$$T_1 = 288 \text{ K}$$
$$M_1 = 0.5$$
$$\gamma = 1.4 \text{ and } c_p = 1,004 \text{ J/Kg · K}$$

8.3 We are interested in a parametric study of the diffusion factor for an inducer section of an impeller. We know from Eq. 8.39-b that

$$D_{inducer} \cong 1 - \frac{1}{\sqrt{1 + \left(\dfrac{r_1}{r_2}\right)^2 \left(\dfrac{M_{T2}}{M_{z1}}\right)^2}}$$
$$+ \frac{(M_{T2}/M_{z1})(r_1/r_2)}{2\sigma_i \sqrt{1 + \left(\dfrac{r_1}{r_2}\right)^2 \left(\dfrac{M_{T2}}{M_{z1}}\right)^2}}$$

There are three nondimensional groups that appear in the above equation.

(a) impeller radius ratio r_1/r_2

(b) ratio of impeller tip Mach umber to the inlet axial Mach number M_{T2}/M_{z1}

(c) inducer solidity σ_i

Graph $D_{inducer}$ versus one of the three nondimensional groups while keeping the other two groups constant.

8.4 A centrifugal compressor has no inlet guide vanes and thus zero preswirl. The axial Mach number is $M_1 = 0.5$ and is uniform at the impeller face. The inlet condition is entirely known, $p_1 = 100$ kPa, $T_1 = 288$ K, and the geometry of the inlet, $r_{1h} = 10$ cm and $r_{1t} = 25$ cm. The exit radius is $r_2 = 0.35$ cm and shaft rotational speed is 10,000 rpm. The centrifugal compressor has 20 radial impeller blades.

Assuming the compressor polytropic efficiency $e_c = 0.85$, calculate

(a) mass flow rate $\dot{m}_1$ (in kg/s)

(b) absolute swirl velocity at the impeller exit, $C_{\theta 2}$ in m/s

(c) compressor specific work, at the pitchline, w_c in kJ/kg

(d) compressor shaft power in kW

(e) compressor pressure ratio π_s

(f) exit (absolute) Mach number at 2, M_2, assuming $C_{r2} = C_{z1}$

(g) swirl velocity at the exit of the radial vaneless diffuser

(assuming zero frictional losses in the diffuser) and radius of 45 cm

8.5 A centrifugal compressor has 25 impeller blades of radial design. The rotor exit diameter is $r_2 = 0.4$ m, and its rotational speed is $\omega = 8000$ rpm. Assuming the inlet to rotor flow is purely axial and the air mass flow rate is $\dot{m} = 25$ kg/s through the compressor, calculate

(a) compressor shaft power $\wp_c$ in MW

(b) (time rate of change of) angular momentum at the rotor exit

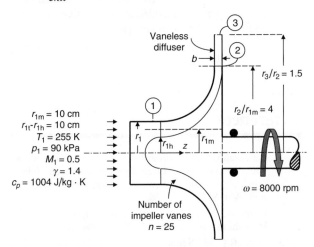

■ **FIGURE P8.5**

8.6 The inlet flow to a radial (vaneless) diffuser has a swirl component of $C_{\theta 2} = 200$ m/s and a radial component of $C_{r2} = 100$ m/s. The impeller exit radius $r_2 = 25$ cm and the flow static pressure and temperature are

$$p_2 = 300 \text{ kPa}$$
$$T_2 = 425 \text{ K}$$

Assuming inviscid flow in the radial diffuser, with $\gamma = 1.4$ and $R = 287$ J/Kg · K, calculate

(a) $C_r(r)$

(b) $C_\theta(r)$

8.7 A centrifugal compressor discharge pressure and temperature are measured to be

$$p_{t3} = 303 \text{ kPa}$$
$$T_{t3} = 408 \text{ K}$$

The inlet flow is purely axial with $C_{z1} = 100\,\text{m/s}$ and the inlet total pressure and temperature are

$$p_{t1} = 101\,\text{kPa}$$
$$T_{t1} = 288\,\text{K}$$

Calculate

(a) total-to-static pressure ratio p_{t3}/p_1
(b) total-to-total pressure ratio
(c) total-to-static adiabatic efficiency η_{TS}
(d) total-to-total adiabatic efficiency η_{TT}
(e) polytropic efficiency e_c for total-static and total-total processes

8.8 To eliminate the inducer from a centrifugal compressor impeller, we need to induce a swirl profile in the incoming flow (known as preswirl) that is in the direction of the rotor and its relative flow to the impeller is purely axial. This task is done through an IGV.

For an inducer-less impeller of

$$r_{1h} = 5\,\text{cm}$$
$$r_{1t} = 10\,\text{cm}$$
$$\omega = 11{,}500\,\text{rpm}$$
$$C_{z1} = 100\,\text{m/s}$$

Calculate

(a) swirl profile $C_{\theta1}(r)$
(b) absolute flow angle $\alpha_1(r)$
(c) the IGV twist from hub-to-tip, assuming a constant deviation angle with span

8.9 The inlet static pressure to an impeller is $p_1 = 100\,\text{kPa}$ and its exit static pressure is $p_2 = 200\,\text{kPa}$ with inlet relative dynamic pressure $q_{1r} = 140\,\text{kPa}$. Treating the impeller as a diffuser, assuming $\gamma = 1.4$, calculate

(a) inlet relative Mach number M_{1r}
(b) static pressure recovery coefficient C_{PR}

8.10 A radial diffuser has an exit-to-inlet radius ratio r_3/r_2 of 4. Assume the flow is incompressible and the fluid is inviscid. Calculate

(a) the diffuser area ratio A_3/A_2
(b) exit-to-inlet radial velocity ratio C_{r3}/C_{r2}
(c) exit-to-inlet swirl ratio $C_{\theta3}/C_{\theta2}$
(d) static pressure rise coefficient or static pressure recovery C_{PR}

8.11 A centrifugal compressor has a total pressure ratio of $p_{t3}/p_{t1} = 8$. Calculate and graph this compressor's total-to-static pressure ratio p_{t3}/p_1 for a range of inlet Mach numbers M_1 from 0.25 to 0.75 in steps of 0.05. Assume $\gamma = 1.4$.

8.12 The total-to-total compressor adiabatic efficiency for a centrifugal compressor is $\eta_{TT} = 0.78$. The compressor total

pressure ratio is $p_{t3}/p_{t1} = 7$. Assuming the inlet Mach number is $M_1 = 0.4$ and $\gamma = 1.4$, calculate

(a) total-to-static pressure ratio
(b) total temperature ratio T_{t3}/T_{t1}
(c) total-to-static adiabatic efficiency η_{TS}.

8.13 A backswept impeller has a tip speed of $U_2 = 366\,\text{m/s}$ and the impeller blade exit angle is $\beta'_2 = 30°$. The radial velocity at the impeller exit is $C_{r2} = 125\,\text{m/s}$. Stagnation speed of sound at the impeller inlet is $a_{t1} = 340\,\text{m/s}$, assuming the inlet flow is purely axial, and the slip angle at the impeller exit is $6°$, i.e., $\beta_2 - \beta'_2 = 6°$, calculate

(a) slip factor ε
(b) Mach index Π_M
(c) Impeller specific work w_c in kJ/kg
(d) stage total temperature ratio τ_{stage}
(e) stage total pressure ratio if $e_c = 0.89$

Assume $\gamma = 1.4$ and $c_p = 1004\,\text{J/Kg·K}$.

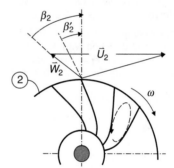

■ **FIGURE P8.13**

8.14 A forward-swept impeller has a tip speed of $U_2 = 400\,\text{m/s}$ and the impeller blade exit angle is $\beta'_2 = -30°$. The radial velocity at the impeller exit is $C_{r2} = 100\,\text{m/s}$. Stagnation speed of sound at the impeller inlet is $a_{t1} = 330\,\text{m/s}$, assuming the inlet flow is purely axial and the slip angle at the impeller exit is $-6°$, i.e., $\beta'_2 - \beta_2 = -6°$, calculate

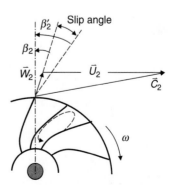

■ **FIGURE P8.14**

(a) slip factor ε

(b) Mach index Π_M

(c) impeller specific work w_c in kJ/kg

(d) stage total temperature ratio τ_{stage}

(e) stage total pressure ratio if $e_c = 0.89$.

8.15   A 19-bladed radial impeller has an exit radius $r_2 = 25$ cm. The impeller width at the exit is $b = 1$ cm. We wish to calculate and compare some impeller exit flow areas with successive levels of detail.

Case-1
Assume the individual trailing edge thickness of impeller blades is zero, i.e., $t = 0$, and calculate the geometric flow area.

Case-2
Assume the individual trailing edge thickness is $t = 2$ mm (per blade), and calculate the geometric flow area at the impeller exit.

Case-3
Assume the impeller exit blockage is 10% and now calculate the effective flow area.

Assuming the same slip factor between the three cases, relate the average radial velocity C_{r2} in cases 1, 2, and 3.

8.16   Air enters a centrifugal compressor at $p_{t1} = 101$ kPa, $T_{t1} = 288$ K, and purely in an axial direction with $M_z = 0.4$. Assuming the impeller radius at the exit is $r_2 = 25$ cm and the impeller is of radial design with slip factor $\varepsilon = 0.9$ and $M_{T2} = 1.5$, calculate

(a) U_2 in m/s

(b) specific work w_c in kJ/kg

(c) impeller exit total temperature T_{t2} in K

(d) shaft rotational speed ω in rpm

8.17   The flow pattern at the exit of a radial centrifugal compressor impeller shows a jet-wake behavior, as shown in Fig. 8.18, which is reproduced here with additional descriptions. In order to estimate the blockage, we assume that the wake region has zero momentum and is of angular extent $\theta_{2,\,wake}$ (per passage) at the impeller exit, as shown. The number of impeller blades is N, and the blade thickness at the exit is t (note the blade sweep angle is zero).

Calculate

(a) the geometric flow area at 2 in terms of r_2, t, N, and b

(b) the effective flow area in terms of $\theta_{2,\,wake}$

(c) the blockage B_2

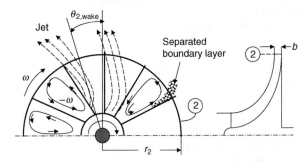

■ **FIGURE P8.17**

CHAPTER 9

Aerothermo-dynamics of Gas Turbines

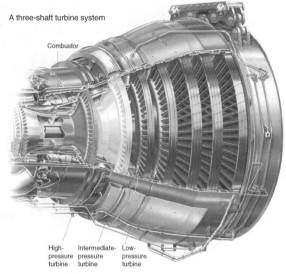

A three-shaft turbine system

Combustor

High-pressure turbine

Intermediate-pressure turbine

Low-pressure turbine

Courtesy of Rolls-Royce, plc

9.1   Introduction

Gas turbines produce shaft power for the compressor in the gas generator as well as other external loads, such as the fan in a turbofan, propeller in a turboprop, helicopter rotor in a turboshaft, and electric generator. The high pressure and temperature gas from the combustor enters the gas turbine first through a stationary blade row, known as the nozzle. The first nozzle is thus exposed to the highest gas temperatures in the engine ($T_{t4} \sim 1750-2000$ K). However, the stationary nature of gas turbine nozzle saves it from an additional centrifugal stress that rotor blades face. Since the gas path temperature is 400–800 K higher than the blade service temperature, all modern gas turbines are cooled. The coolant, which is bled from compressor accounts for $\sim$10–15% of the airflow rate in the gas generator. A reduction of $\sim$3% turbine efficiency per 1% cooling flow is attributed to cooling losses. In this chapter, we will introduce aerodynamics of turbine blades, optimal nozzle exit tangential Mach number, optimum solidity, turbine losses, and cooling. We conclude the chapter with axial-flow turbine design and practices and an example. Figure 9.1 shows a schematic drawing of a gas turbine.

9.2   Axial-Flow Turbines

The flowfield in a turbine is dominated by favorable pressure gradients unlike the compressor flow. The boundary layers in a turbine are thus less susceptible to stall than their counterparts in a compressor. The notion that a turbine is a "mirror image" of a compressor is therefore misleading. The evidence of many compressor stages ($\sim$5 in modern gas turbine engines of the same mass flow rate) being driven by a single turbine stage is a compelling reason for the inherent differences in the fluid mechanics of the two devices. The challenges in a turbine are primarily in the cooling techniques and their effectiveness; tip clearance control, blade life, and material characteristics suitable for a high-temperature environment and corrosion. In the

527

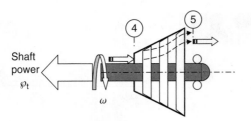

■ **FIGURE 9.1** Schematic drawing of an axial-flow turbine

compressor, the flow stability (rotating stall and surge) and efficiency are highly critical to the design.

The aerodynamics of turbine stages is governed by the conservation principles of mass, momentum, and energy. The conservation of angular momentum is expressed by the Euler turbine equation, which we derived in the early part of Chapter 7. It states that the time-rate-of-change of the fluid angular momentum is balanced by a net external torque that acts on the fluid, namely,

$$\tau_{\text{fluid}} = \dot{m}(r_2 C_{\theta 2} - r_1 C_{\theta 1}) \tag{9.1}$$

Let us show a schematic drawing of a turbine stage, assign station numbers, and show the velocity triangles in the stage. A turbine stage begins with a stationary blade row, known as the nozzle, followed by a rotating blade row, known as the rotor or sometimes bucket. The combustor exit flow is expanded, i.e., accelerated, in a turbine nozzle, which converts the fluid thermal energy into the fluid kinetic energy. The inlet flow to the nozzle is typically swirl-free and thus the nozzle imparts swirl to the flow. This behavior is opposite to a compressor stator, which removes the swirl from the flow. The nozzle flow is choked over a wide operating range of the engine, and the exit of the nozzle may also operate in the supersonic regime. The rotor blade row follows the nozzle and exchanges energy with the fluid. The rotor thus removes the swirl put in by the nozzle, again an opposite behavior to the compressor rotor. An examination of velocity triangles in a turbine stage (in Fig. 9.2) reveals the swirl deposition and extraction of the nozzle and the rotor rows in a turbine, respectively.

The sign convention on the swirl velocity component is the same as the compressors, namely, the swirl is considered positive in the direction of the rotor rotation. For example, the inlet swirl upstream of the turbine nozzle is shown in the positive direction in Fig. 9.2.

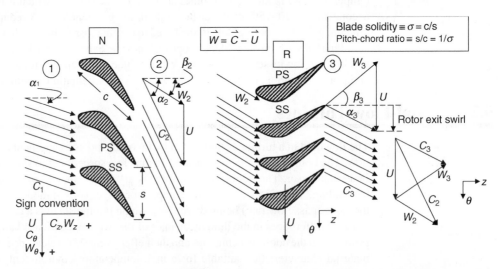

■ **FIGURE 9.2** Schematic drawing of a turbine stage with velocity triangles, sign convention flow angle definitions, and blade solidity

The absolute and relative swirls downstream of the nozzle are both in the positive direction, as also shown in Fig. 9.2. The relative swirl downstream of the rotor blade has a negative swirl velocity, whereas the absolute swirl is shown in the positive direction.

The power exchange between the rotor and the fluid may be observed in Fig. 9.2 as the rotor rotation is in the same direction as the aerodynamic force acting on it in the tangential direction. We may express the rotor shaft power in thermodynamic terms as

$$\wp_{\text{rotor}} = -\wp_{\text{fluid}} = \dot{m}(h_{t2} - h_{t3}) \tag{9.2}$$

We may express the rotor power in mechanical terms via the Euler turbine equation as

$$\wp_{\text{rotor}} = -\wp_{\text{fluid}} = \dot{m}\omega[(r_2 C_{\theta 2}) - (r_3 C_{\theta 3})] \tag{9.3}$$

Since the stator or nozzle puts in swirl, i.e., angular momentum, in the flow and the rotor or bucket removes the angular momentum in a turbine, the bracket on the RHS of Eq. 9.3 is positive. The implication is that the turbine rotor produces power and the power of the fluid is thus drained or diminished. These are all consistent with our expectations. The turbine nozzle imparts a positive torque on the fluid and thus feels an equal and opposite torque in reaction. We may express the nozzle torque as

$$\tau_n = -\tau_{\text{fluid}} = \dot{m}[(r_1 C_{\theta 1}) - (r_2 C_{\theta 2})] \tag{9.4}$$

From the magnitude of the angular momentum up- and downstream of the nozzle, we note that the fluid torque is positive and the nozzle torque is negative. The rotor toque is

$$\tau_r = -\tau_{\text{fluid}} = \dot{m}[(r_2 C_{\theta 2}) - (r_3 C_{\theta 3})] \tag{9.5}$$

Since the bracket on the RHS of Eq. 9.5 is positive, as the fluid angular momentum is higher upstream of the rotor than downstream, we may conclude that the torque acting on the fluid is negative and the blade torque is in the positive direction. In case of repeated stages, we have the rotor exit angular momentum equal to the nozzle entrance angular momentum, i.e.,

$$r_3 C_{\theta 3} = r_1 C_{\theta 1} \tag{9.6}$$

Then the rotor torque becomes equal and opposite the nozzle torque. Consequently, a turbine stage, as a whole, does not impart a net torque on the engine structure.

In axial-flow turbines, as in axial-flow compressors, the radial shift of the stream surfaces is small and usually neglected in a first approximation. A second approximation is made about the constant axial velocity through the stage or the density–velocity ratio across blade rows in a stage. The choice of the degree of reaction in a turbine ranges from zero to upward of 50%. A zero degree of reaction turbine stage, known as the impulse stage, produces the entire static enthalpy drop of the stage across the nozzle. The rotor produces no static enthalpy drop. The degree of reaction in a turbine is defined as the ratio of static enthalpy drop in the rotor to that of the stage, namely,

$$^{\circ}R \equiv \frac{h_2 - h_3}{h_1 - h_3} \tag{9.7}$$

Let us rewrite Eq. 9.7 in terms of the stagnation states and the fluid kinetic energy,

$$^{\circ}R = \frac{h_{t2} - h_{t3} - C_2^2/2 + C_3^2/2}{h_{t1} - h_{t3} - C_1^2/2 + C_3^2/2}$$

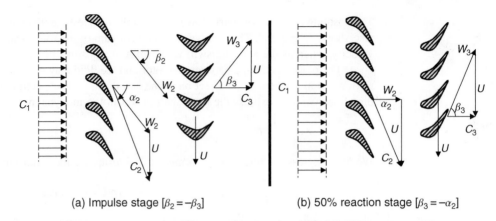

(a) Impulse stage [$\beta_2 = -\beta_3$] (b) 50% reaction stage [$\beta_3 = -\alpha_2$]

■ **FIGURE 9.3**   An Impulse and a 50% reaction stage with swirl-free exit flow

Assuming the inlet and exit kinetic energies of the stage are equal, the above expression is simplified to

$$°R = 1 - \frac{C_2^2 - C_3^2}{2(h_{t2} - h_{t3})} = 1 - \frac{C_{\theta2}^2 - C_{\theta3}^2}{2U(C_{\theta2} - C_{\theta3})} = 1 - \frac{C_{\theta2} + C_{\theta3}}{2U} = 1 - \frac{C_{\theta\,\text{mean}}}{U} \qquad (9.8)$$

where we have further assumed that the axial velocity remains constant and the radial shift of stream surfaces is small. The form of the degree of reaction, as expressed in Eq. 9.8, is identical to the formulation of the compressor degree of reaction. It is the nondimensional average swirl velocity ($C_{\theta\text{mean}}/U$) in the rotor that establishes the degree of reaction at a given radius. Thus, the 50% degree of reaction demands the average swirl in the rotor to be one half of the rotational speed, which suggests a symmetrical velocity triangle across the rotor and nozzle (stator). This behavior is identical to the compressor. The zero degree of reaction, known as the impulse turbine, demands the average swirl in the rotor to be the same as the rotor speed. The velocity triangles and the blade row geometry for an axial exit flow are shown in Fig. 9.3.

For a 50% degree of reaction turbine stage and a swirl-free exit flow condition, i.e., $C_{\theta3} = 0$, we have

$$C_{\theta m} = \frac{U}{2} = \frac{C_{\theta2}}{2} \quad \text{or} \quad C_{\theta2} = U \qquad (9.9)$$

The rotor specific work, via Euler turbine equation, is

$$w_t \equiv \frac{\wp_t}{\dot{m}_t} \cong U\Delta C_\theta = U^2 \qquad (9.10)$$

The zero degree of reaction of an impulse turbine stage requires

$$C_{\theta m} = U = \frac{C_{\theta2}}{2} \quad \text{or} \quad C_{\theta2} = 2U \qquad (9.11)$$

The rotor specific work for an impulse turbine is, therefore,

$$w_t \equiv \frac{\wp_t}{\dot{m}_t} \cong U\Delta C_\theta = 2U^2 \qquad (9.12)$$

Comparing the specific work of the rotors from Eqs. 9.10 and 9.12, we note that the impulse design produces twice as much shaft power per unit mass flow rate than the 50% reaction

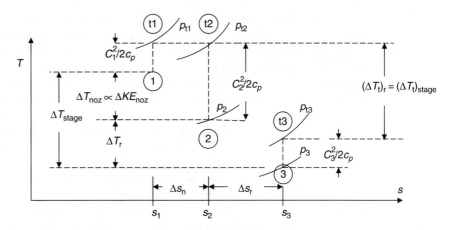

■ **FIGURE 9.4** *T–s* diagram of thermodynamic states of gas in a reaction turbine stage

turbine for the same rotor speed *U*. Although the impulse turbine designs have the potential of producing large specific works, they are less efficient than the reaction type turbines. The primary reason is in the rotor where the relative flow speed remains constant and hence the viscous flow losses are larger than reaction turbines where the flow continuously accelerates. It is also instructive to show the thermodynamic states of the gas in both types of turbine stages on a *T–s* diagram. Figure 9.4 shows the static and total (or stagnation) states of the fluid in a reaction turbine stage, i.e., $^{\circ}R > 0$.

We note that the static temperature drop occurred across the nozzle as well as the rotor. Therefore, the turbine stage that is shown in Fig. 9.4 is a reaction turbine. The stage work may also be noted in Fig. 9.4 as proportional to the stage total temperature drop. The nozzle suffers some total pressure loss $p_{t1} - p_{t2}$ due to viscous effects in the boundary layer and possible shock losses. The rotor loss may be observed in the entropy rise Δs_r in Fig. 9.4. The thermodynamic states of gas in an impulse turbine stage, for comparison purposes, are shown in Fig. 9.5.

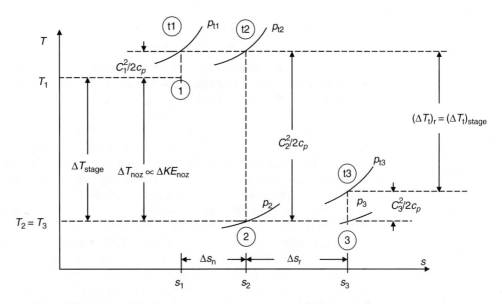

■ **FIGURE 9.5** Thermodynamic states of gas in an impulse turbine stage ($^{\circ}R = 0$) show a large power production capability, $\propto (\Delta T_t)_{\text{stage}}$

Similar observations to the reaction turbine may also be made on the impulse stage, as shown in Fig. 9.5. One exception is that the static temperature change across the rotor is zero in the impulse turbine. Consequently, the static temperature drop ΔT_{stage} occurs solely across the nozzle. Also, we note a large expansion across the nozzle, as evidenced by a large ΔT_{noz}, as compared with the reaction type turbine. The entropy rise in an impulse stage is intentionally depicted as larger than the corresponding entropy rise in the reaction turbine stage to signify higher losses. We also note that the rotor specific work, which is proportional to the total temperature drop across the rotor, is depicted roughly twice as much (in Fig. 9.5) as the reaction turbine stage of Fig. 9.5 to reflect Eqs. 9.10 and 9.12. In some texts (e.g., Dixon, 1975), an impulse stage is defined as having zero static pressure drop across the rotor. In this case, then a zero-reaction turbine, which has its static enthalpy constant across the rotor, is not strictly speaking an impulse stage. In a zero-reaction turbine, there is still some static pressure drop due to frictional losses within the blade. Thus, an impulse turbine would have a slightly negative degree of reaction. In this book, the words impulse and zero reaction are interchangeably used, albeit, more loosely than the strict classical definition.

The turbine specific work could be expressed in terms of the nozzle exit flow angle α_2 and the condition of zero swirl at the rotor exit, as

$$w_t = U(C_{\theta 2} - C_{\theta 3}) \approx UC_{\theta 2} = UC_{z2} \tan \alpha_2 \tag{9.13}$$

For a constant axial flow velocity C_z, we note that the rotor work per unit mass flow rate increases linearly with the increase in wheel speed U and (nonlinearly with) the flow turning angle α_2 through the tangent of α_2 term in Eq. 9.13. For practical reasons, the flow angle at the nozzle exit is limited to $\sim 70°$, otherwise flow losses become excessive, i.e.,

$$\boxed{\alpha_{2,\text{max}} \approx 70°} \tag{9.14}$$

The wheel speed U is a quadratic contributor to the blade centrifugal stresses σ_c, hence, it is limited by the structural stresses and the blade life requirements. To gain further insight, we may cast the above equation in terms of the blade tangential and the flow Mach numbers downstream of the nozzle as

$$w_t = UC_2 \sin \alpha_2 = a_2^2 M_T M_2 \cdot \sin \alpha_2 = a_{t2}^2 \cdot M_T M_2 \sin \alpha_2 \bigg/ \left[1 + \left(\frac{\gamma - 1}{2} \right) M_2^2 \right] \tag{9.15}$$

Since the stagnation temperature does not change across the nozzle, the stagnation speed of sound at 2 is the same as 1, hence, proportional to the combustor exit total temperature T_{t4}, a cycle parameter. We may conclude from Eq. 9.15 that to maximize the rotor specific work, we need to have high inlet stagnation temperatures, T_{t1} (or T_{t4} from our cycle analysis notation), a high tangential Mach number of the blade, M_T, a high swirl Mach number at the nozzle exit, $M_2 \sin \alpha_2$. The blade tangential Mach number M_T has a maximum at the rotor blade tip radius $M_T(r_t)$. The high gas temperatures in the high-pressure turbine (HPT) demand internal cooling of the rotor blades. A cooled rotor blade leading edge needs to be blunt to allow for sufficient surface area to cool, as it represents the stagnation point of the blade with the attendant high heat flux. A blunt leading edge is best suited for a subsonic relative flow, which then limits the maximum M_T parameter in the HPT. In the low-pressure turbine, the gas temperature is significantly reduced, hence internal cooling of the blades may not be required at all. Therefore, the rotor tip in the LPT may operate with a sharp leading edge, which is then suitable for a supersonic operation. In addition, the speed of sound for the first rotor is about $\sim \sqrt{1/\tau_t}$ higher than the speed of sound in the last rotor. This means that for the same tip speeds ωr_t, the blade tangential Mach number is higher in the LPT than the HPT, i.e., inversely proportional to the local speed of sound. These arguments are made to show the conflicting requirements of internal

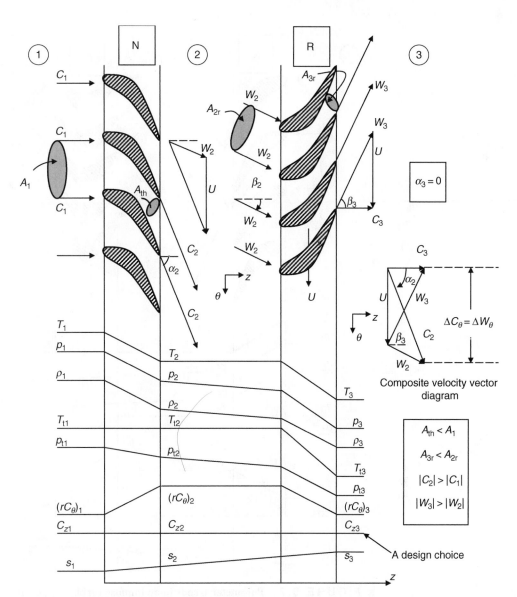

■ **FIGURE 9.6**   Parameter trends in a reaction turbine stage with zero exit swirl and designed for constant axial velocity (in an uncooled stage)

cooling (due to high gas temperatures) and centrifugal stresses with the desire to operate the turbine rotors with a supersonic tip to maximize the rotor work output. Let us summarize the parameter trends in reaction and impulse turbine stages graphically in Fig. 9.6 and 9.7 respectively.

We have examined parameter trends in turbine stages with *zero exit swirl* in Fig. 9.6 and 9.7. Let us provide a rationale for that. The swirl in the exit flow does not contribute to the thrust production, which is the function of our engine, and even increases the total pressure loss in the nozzle. Consequently, it is desirable to design a turbine stage with zero exit swirl. Also, we have depicted the streamtubes upstream and downstream of the turbine blades with corresponding flow areas A_1 and A_2 for the nozzle and A_{2r} and A_{3r} in the rotor frame of reference for the rotor. In a reaction turbine, the streamtube flow areas continually shrink, as flow continually accelerates in the blade rows. In the impulse stage, the relative flow in the rotor attains a constant speed and

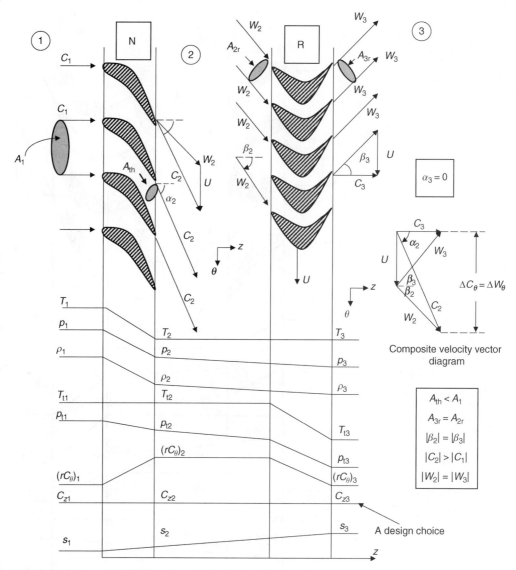

■ **FIGURE 9.7** Parameter trends in an impulse turbine stage with zero exit swirl and designed for constant axial velocity (in an uncooled stage)

thus the relative streamtube does not shrink. Thus, the static temperature remains constant and the static pressure remains unchanged in the limit of inviscid flow and drops with frictional losses in the boundary layers. The relative flow downstream of the rotor in an impulse stage is the mirror image of the relative flow upstream of the rotor (see Fig. 9.7). Also note that the choice of constant axial velocity C_z, as shown in Figs. 9.6 and 9.7, is depicted as a design choice, similar to a compressor. The annulus sizing and the geometry are directly affected by our choice of the axial velocity and the mass flow rate through continuity equation. With a constant axial velocity and a decrease in gas density in the flow direction, the annulus flow area needs to grow inversely proportional to the density drop. In the case of a cooled turbine, the coolant mass flow rate needs to be added to the hot gas flow rate in establishing the annulus flow area, via continuity equation. Now, let us examine the design parameters that affect the mass flow rate in a turbine stage, the power production, and the turbine efficiency. Three turbine annuli are shown in Fig. 9.8 and the station numbers at the entrance and exit correspond to our cycle

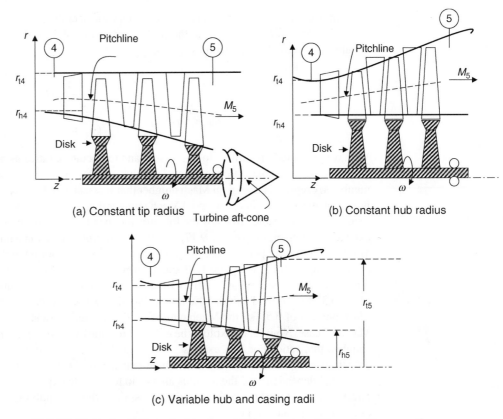

■ **FIGURE 9.8** Schematic drawing of the geometry of three turbine annuli

analysis notation, stations 4 and 5, respectively. In this example, we show three types of annuli, one with a constant tip radius, i.e., the casing is of cylindrical shape, and a variable hub radius for the annulus geometry, two, the case of constant hub radius Fig. 9.8(b), and three, with a varying hub and casing radii Fig. 9.8(c). In a mixed-stream turbofan engine, the choice of constant tip radius for the outer casing of the turbine annulus leads to a better integration of the cold and hot streams in a forced mixer. The second advantage of constant outer radius is in lower centrifugal stresses of the rotor blades, simply by the virtue of a smaller moment arm. The third advantage is in engine frontal area and weight. Integration with the turbine aft-cone in aircraft gas turbine engines will also benefit from a tapered hub radius, as shown schematically in Fig. 9.8(a). In some applications, e.g., stationary gas turbine power plants, the choice of constant hub radius is considered advantageous as it integrates with an annular exhaust diffuser as well as reduced manufacturing costs of the turbine rotor disks (i.e., all with the same diameter). The varying hub and tip case shown schematically in Fig. 9.8(c) offers a constant radius pitchline advantage, but at an increased manufacturing costs and potentially an increased weight over the first two designs. In general, a turbine annulus with an increasing tip radius adds weight to the rotor disks and turbine overall weight.

Our one-dimensional continuity equation written between stations 4 and 5 yields

$$\dot{m}_4 = \sqrt{\frac{\gamma_4}{R_4}}\frac{p_{t4}}{\sqrt{T_{t4}}}A_4M_4\left(1+\left(\frac{\gamma_4-1}{2}\right)M_4^2\right)^{-\frac{\gamma_4+1}{2(\gamma_4-1)}}=\dot{m}_5$$

$$=\sqrt{\frac{\gamma_5}{R_5}}\frac{p_{t5}}{\sqrt{T_{t5}}}A_5M_5\left(1+\left(\frac{\gamma_5-1}{2}\right)M_5^2\right)^{-\frac{\gamma_5+1}{2(\gamma_5-1)}} \qquad (9.16)$$

Since the exit of the nozzle remains choked over a wide operating range of the engine, it is more convenient to use $M_4 = 1$ and A_4 as the choked throat area of the turbine nozzle. Therefore, the annulus area ratio A_5/A_4 becomes

$$\frac{A_5}{A_4} = \sqrt{\frac{\gamma_4/\gamma_5}{R_4/R_5}}\,\frac{1}{M_5}\,\frac{\sqrt{\tau_t}}{\pi_t}\,\frac{\left(1 + \left(\dfrac{\gamma_5 - 1}{2}\right)M_5^2\right)^{\frac{\gamma_5+1}{2(\gamma_5-1)}}}{(2/(\gamma_4 + 1))^{\frac{\gamma_4+1}{2(\gamma_4-1)}}} \tag{9.17}$$

We have presented the turbine total pressure and temperature ratios as π_t and τ_t, respectively, which we used as a convention in the cycle analysis chapter. For a zero exit swirl, the flow Mach number M_5 represents the axial Mach number that we treat as a design choice. To minimize the flow area, we need to choose an axial Mach number that is not too low, e.g., $M_5 = M_{z5} = 0.50$ seems to be reasonable. To show the functional dependence of the area ratio on the choice of axial Mach number, we plot Eq. 9.17, keeping the turbine expansion parameters π_t and τ_t fixed as well as the gas properties. The normalized area ratio as a function of the axial Mach number is shown in Fig. 9.9, for a $\gamma_5 = 1.33$ and a normalizing axial Mach number of 0.5. We note a rapid rise in the required area ratio with a decrease in axial Mach number. For example, at $M_5 = 0.3$, the required exit is 50% larger than the exit area corresponding to a design axial Mach number of 0.5. For the Mach number of 0.7, the exit area shrinks to 81.5% of that of a Mach 0.5 case. Now, let us explore the effect of turbine expansion parameters π_t and τ_t on the annulus area ratio. For this exercise, we keep the axial Mach number fixed and the gas properties constant, as we vary the turbine expansion parameters. First, let us simplify the functional dependence of the annulus area ratio to a single expansion parameter, e.g., τ_t. We relate the pressure and temperature ratios across the turbine via polytropic efficiency e_t as in the cycle analysis chapter, i.e.,

$$\pi_t = \tau_t^{\frac{\gamma}{(\gamma-1)e_t}}$$

Therefore, the functional dependence of the turbine annulus area ratio and the turbine expansion parameter τ_t is

$$A_5/A_4 \sim (1/\tau_t)^{3.978} \tag{9.18}$$

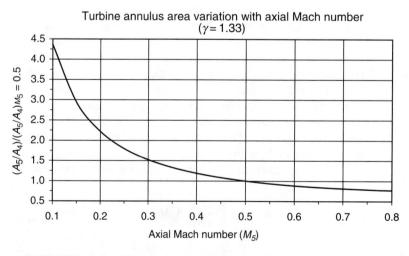

■ **FIGURE 9.9** The effect of axial Mach number on the annulus area of a turbine (normalized for $M_5 = 0.5$, $\tau_t = $ constant)

Turbine annulus area ratio
($\gamma = 1.33$, $e_t = 0.90$)

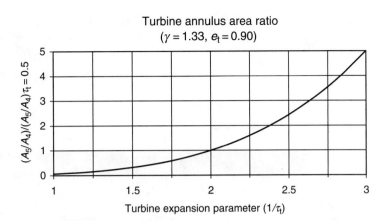

■ **FIGURE 9.10**   **Turbine annulus area as a function of the turbine expansion parameter** $(1/\tau_t)$ **(normalized for** $\tau_t = 0.5$, $M_5 = $ **constant)**

Where we chose a $\gamma_5 = 1.33$ and $e_t = 0.9$. A graph of this function normalized with respect to a turbine expansion parameter, $(1/\tau_t)$ of 2.0, is shown in Fig. 9.10. We observe that the annulus area ratio is a strong function of the turbine expansion parameter, as expected. The exponential growth in the flow area is in response to the exponential decay in the gas density with the temperature ratio, while maintaining the constant axial Mach number.

EXAMPLE 9.1

The combustor discharge into a turbine nozzle has a total temperature of 1800 K and axial Mach number of 0.55, as shown.

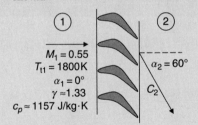

Assuming that the nozzle is uncooled, the axial velocity remains constant across the nozzle, and the absolute flow angle at the nozzle exit is $\alpha_2 = 60°$, calculate

(a) inlet velocity C_1 in m/s

(b) the exit absolute Mach number M_2

(c) nozzle torque per unit mass flow rate for $r_1 \approx r_2 = 0.35$ m

SOLUTION

We use total temperature and Mach number at the nozzle inlet to calculate the static temperature T_1 according to

$$T_1 = \frac{T_{t1}}{1 + (\gamma - 1)M_1^2/2} \approx 1714 \, \text{K}$$

The speed of sound is

$$a_1 = \sqrt{(\gamma - 1)c_p T_1} \approx 809 \, \text{m/s}$$

Therefore, the absolute gas velocity into the turbine nozzle has a magnitude of

$$C_1 = M_1 a_1 \approx 445 \, \text{m/s}$$

Downstream of the nozzle, we know the axial component of the velocity and the flow angle. Therefore, we calculate C_2 from

$$C_2 = C_{z2}/\cos\alpha_2 \approx 445/\cos(60°) \, \text{m/s} \approx 889.2 \, \text{m/s}$$

To calculate speed of sound downstream of the nozzle, we need to know the static temperature of the gas, T_2. We know the total temperature of the gas T_{t2} since (for a calorically perfect gas) is the same as T_{t1} for an uncooled turbine, therefore, we may use the following equation to calculate T_2

$$T_2 = T_{t2} - C_2^2/2c_p = 1800 \, \text{K} - (889.2)^2/2/1157$$

$$\approx 1458.3 \, \text{K}$$

The speed of sound, a_2 is $a_2 = \sqrt{(\gamma - 1)c_p T_2} \approx 746.2 \text{ m/s}$

Therefore, $M_2 = C_2/a_2$ ➡ $M_2 \approx 1.19$

The nozzle torque is equal to the change of angular momentum across the nozzle according to Eq. 9.4, i.e.,

$$\tau_n/\dot{m} = r_1 C_{\theta 1} - r_2 C_{\theta 2}$$

The nozzle exit swirl is related to axial velocity C_{z2} and the swirl angle α_2, $C_{\theta 2} = C_{z2} \tan \alpha_2 \approx 445 \tan(60°) \approx 769.8 \text{ m/s}$
Therefore, the nozzle torque per unit mass flow rate is

$$\tau_n/\dot{m} = r_1 C_{\theta 1} - r_2 C_{\theta 2} \approx -269.4 \text{ m}^2/\text{s}$$

EXAMPLE 9.2

The turbine nozzle in Example 9.1 produced a slightly supersonic exit flow with $M_2 \approx 1.19$. If we turn the flow to a reduced α_2 level, we may reduce the exit Mach number M_2. Calculate the nozzle exit flow angle α_2 if we wish the exit Mach number to be $M_2 \approx 1.0$, i.e., choked.

SOLUTION

In this case we know M_2, which we use with T_{t2} to calculate T_2 from

$$T_2 = T_{t2}/[1 + (\gamma - 1)M_2^2/2)] = 1800 \text{ K}/[1 + 0.33/2]$$

$$\approx 1545 \text{ K}$$

Now, we calculate the speed of sound, a_2 to be ➡ $a_2 \approx 768.1 \text{ m/s}$

Therefore, $C_2 = M_2 a_2 = a_2$
The exit flow angle is related to the axial velocity according to

$$\cos \alpha_2 = \frac{C_{z2}}{C_2} = \frac{445}{768.1} \approx 0.5794 \implies \alpha_2 \approx 54.6°$$

EXAMPLE 9.3

A turbine stage at pitchline has the following velocity vectors, as shown.

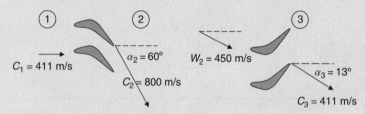

Calculate

(a) the axial velocities up- and downstream of the rotor

(b) the rotor velocity U_m

(c) the degree of reaction at this radius.

SOLUTION

From the flow condition downstream of the nozzle, we may calculate C_{z2} to be

$$C_{z2} = C_2 \cos(60°) = 400 \text{ m/s}$$

From the flow condition downstream of the rotor, we may calculate C_{z3} to be

$$C_{z3} = C_3 \cos(13°) \approx 400 \text{ m/s}$$

The rotor velocity $U_m = C_{\theta 2m} - W_{\theta 2m}$

$$C_{\theta 2m} = 400 \tan 60° \approx 693 \text{ m/s and}$$

$$W_{\theta 2m} = [(450)^2 - (400)^2]^{0.5} \approx 206 \text{ m/s}$$

Therefore, $U_m \approx 693 - 206 \approx 487 \text{ m/s}$

Also $C_{\theta 3} = C_3 \sin(13°) \approx 92.4 \text{ m/s}$

For the degree of reaction, since $C_1 = C_3$, we use Eq. 9.8 to get

$$°R_m = 1 - \frac{C_{\theta 2m} + C_{\theta 3m}}{2U_m}$$

$$= 1 - (693 + 92.4)/2/487 \approx 0.194$$

9.2.1 Optimal Nozzle Exit Swirl Mach Number $M_{\theta 2}$

The effect of the degree of reaction on the mass flow rate and optimal swirl in a turbine stage, for a given blade tangential Mach number M_T, is considered in this section. For simplicity of the expressions, we assume that the exit flow from the rotor is swirlfree, i.e., $C_{\theta 3} = 0$. The degree of reaction may be successively written as

$$°R = 1 - \frac{C_{\theta m}}{U} = 1 - \frac{C_{\theta 2}}{2U} = 1 - \frac{C_2 \sin \alpha_2}{2U} = 1 - \frac{M_2 \sin \alpha_2}{2M_T} \qquad (9.19)$$

This reduces to

$$M_2 \sin \alpha_2 = M_{\theta 2} = 2M_T(1 - °R) \qquad (9.20)$$

Assuming the total temperature, total pressure, and flow area remain constant across a turbine nozzle, the continuity demands (keeping gas properties constant)

$$M_1 \left[1 + \left(\frac{\gamma - 1}{2}\right)M_1^2\right]^{-\frac{\gamma+1}{2(\gamma-1)}} = M_2 \cos \alpha_2 \left[1 + \left(\frac{\gamma - 1}{2}\right)M_2^2\right]^{-\frac{\gamma+1}{2(\gamma-1)}} \qquad (9.21)$$

Therefore,

$$\cos \alpha_2 = \frac{M_1}{M_2}\left[\frac{1 + (\gamma - 1)M_2^2/2}{1 + (\gamma - 1)M_1^2/2}\right]^{\frac{\gamma+1}{2(\gamma-1)}} \qquad (9.22)$$

Let us graph nozzle exit flow angle α_2 from Eq. 9.22 for a given nozzle inlet Mach number M_1. Figure 9.11 shows the nozzle exit flow angle for inlet Mach numbers of 0.2–0.7

■ FIGURE 9.11
Nozzle exit flow variation with inlet Mach number

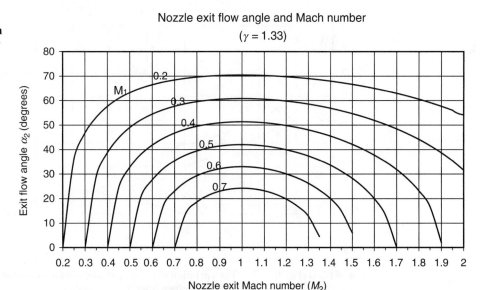

Nozzle exit flow angle and Mach number
($\gamma = 1.33$)

in increments of 0.1, as a function of the exit Mach number M_2. The nozzle exit flow angle α_2 reaches a maximum for the exit Mach number of 1.

Although we can discern the maximum nozzle exit flow angle from Fig. 9.11, it is the product of $M_2 \sin \alpha_2$ that contributes to the specific work output of the stage. Hence, to maximize the nozzle exit swirl Mach number $M_{\theta 2}$ for a given inlet Mach number M_1, we may rewrite Eq. 9.22 as

$$(M_2 \sin \alpha_2)^2 = M_2^2 - M_1^2 \left[\frac{1 + (\gamma - 1)M_2^2/2}{1 + (\gamma - 1)M_1^2/2} \right]^{\frac{\gamma+1}{\gamma-1}} \tag{9.23}$$

Now, let us find the exit Mach number M_2, which maximizes the above function, i.e., the square of the swirl Mach number. We differentiate the RHS of Eq. 9.23 with respect to M_2 and set it equal to zero to find

$$M_2 = \sqrt{\frac{2}{\gamma - 1} \left\{ \left[\left(\frac{2}{\gamma + 1} \right) \frac{(1 + (\gamma - 1)M_1^2/2)^{\frac{\gamma+1}{\gamma-1}}}{M_1^2} \right]^{\frac{\gamma-1}{2}} - 1 \right\}} \tag{9.24}$$

Figure 9.12 shows nozzle flow parameters for a turbine stage that produces maximum work per unit mass flow rate, as a function of turbine inlet Mach number and the stage degree of reaction. The maximum specific work is produced at the condition of maximum nozzle exit swirl, assuming the rotor exit is swirl-free. The degree of reaction, from Eq. 9.20 is also plotted in Fig. 9.12 for the conditions of optimal swirl and a blade tangential Mach number of $M_T = 0.5$. The ratio of specific heat was assumed to be $\gamma = 1.33$ to represent gas in the turbine environment.

Under these optimal conditions, a choice of the stage degree of reaction establishes a turbine inlet Mach number, M_1. An impulse turbine stage, with $^\circ R = 0$, demands a nozzle inlet Mach number of ~0.4 and a 50% reaction stage corresponds to an inlet Mach number ~0.65. We thus observe that with the increase in reaction, the mass flow rate per unit area increases and vice versa. An impulse stage is thus capable of producing maximum specific work, but at a

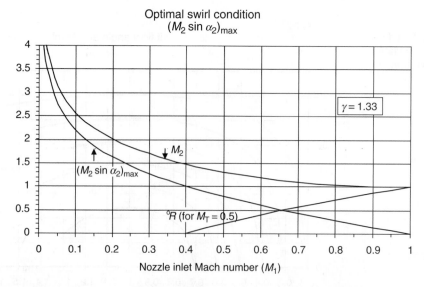

■ **FIGURE 9.12** **The variation of nozzle parameters for maximum swirl Mach number and the stage degree of reaction for a given inlet Mach number**

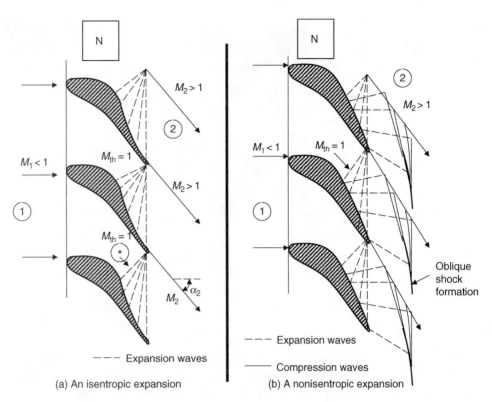

(a) An isentropic expansion   (b) A nonisentropic expansion

■ **FIGURE 9.13**   Aerodynamic shapes of turbine nozzle blades with supersonic exit flow

reduced mass flow density. The optimal exit Mach number M_2 that maximizes the swirl and stage work, then varies between 1.2 and 1.5 for a stage reaction that varies between 50% and 0, respectively. A supersonic exit flow from the nozzle, i.e., $M_2 > 1$, demands a convergent–divergent blade passage geometry, with an ideal (i.e., isentropic) area ratio of A/A^* in the range of 1.03–1.18 (for a $\gamma = 1.33$). Since the nozzle area ratios are small, the nozzle throat may be placed at the exit plane and the suction surface of the blade downstream of the throat to serve as the supersonic expansion ramp, as in a plug nozzle. By necessity, the portion of the suction surface downstream of the throat then has to be concave to cancel the expansion waves emanating from the blade trailing edge. The external aerodynamic shape of the nozzle with an isentropic expansion is depicted in Fig. 9.13(a) (note the curvature reversal on the suction surface from convex to concave). In the case of a continuously convex suction surface, the expansion waves are reflected from the blade and the wake to form oblique shocks, as schematically shown in Fig. 9.13(b). At off-design points, an isentropic expansion nozzle will also create wave reflections and shock formations. In addition, the curvature reversal from convex to concave on the suction surface causes an increased heat transfer to the blade (due to Goertler instability), thus the on-design advantage of an isentropic expansion nozzle is completely wiped out by its disadvantages at off-design.

In the HPT, the blades are internally cooled and their trailing edge often serves as the orifice for the coolant ejection. The resultant coolant jet is then mixed with the merging blade boundary layers downstream of the trailing edge. This gives us a picture of a thick jet-wake flow under rapid mixing conditions introduced by large shear and the combustor-generated turbulence levels that are high. Consequently, the supersonic flow in the exit is confronted with a thick and growing shear layer, which then causes oblique shock waves to form at the trailing edge, as depicted in Fig. 9.14. These waves are in response to wake

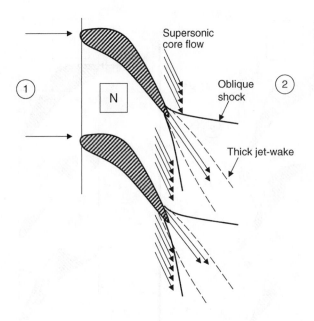

■ **FIGURE 9.14** Nozzle trailing-edge flow with coolant ejection and "fishtail" shock formation

formation and are in addition to the ones shown in Fig. 9.15. A turbine nozzle flow is visualized using schlieren optics in Fig. 9.15 (from Gostelow 1984). The expansion waves and their reflections as well as shock formations and the wakes are visible (note, there is no trailing-edge blowing).

9.2.2 Turbine Blade Losses

We may categorize the losses in a turbine blade row as having their origins in

1. Profile loss, which is established based on 2D cascade studies
2. Secondary flow losses, which may be estimated using secondary flow theory of Hawthorne or the approximate version by Squire and Winter
3. Annulus losses, which includes tip clearance loss, (seal) leakage loss, casing boundary layer loss, corner vortex loss

■ **FIGURE 9.15** Schlieren photograph of a turbine nozzle cascade showing expansion waves and shock formations downstream of an uncooled turbine nozzle (from Gostelow, 1984)

4. Coolant-related losses, including mixing loss, induced separation loss

5. Shock and shock-boundary layer interaction losses in transonic stages of LPT

6. Unsteady flow loss causing shock oscillation and vortex shedding in the wake

Again, a reminder is in order that these losses are not all independent of each other rather they are, in most part, coupled. It is impossible to separate these losses, i.e., decouple them from each other, with high degree of certainty. We may concentrate on any one aspect of these losses and proceed to study the isolated effect of that loss contributor, e.g., 2D profile loss. However, the overall loss is by no means the linear sum of the elementary losses that we studied or considered.

9.2.2.1   *Blade Profile Loss.*

Ainley (1950) has assembled and presented cascade test data for rotor and nozzle blades in a turbine with limited compressibility and thickness-to-chord ratio variations. Figure 9.16 shows the profile loss of a reaction rotor blade with blade solidity (inverse of pitch-chord ratio) and the exit flow angle β_3.

The loss "buckets" identify an optimum solidity for a given exit flow angle. The higher turning blades (e.g., $\beta_3 = -70°$) require a higher solidity (lower pitch-chord ratio) than the rotor blades with lower turning, as expected. The blade thickness (20%) is high, which is typical of HPTs and is needed for internal cooling and for stress considerations. The Reynolds number based on chord of ~200,000 seems to be transitional on a flat plate, but remember that the turbine blades are highly cambered and thus flat plate transition criterion does not directly apply. Figure 9.17 shows the profile loss coefficient of an impulse rotor at zero incidence and the same flow conditions as the reaction rotor of Fig. 9.16. Here, we note that the impulse blade profile losses are significantly higher than their counterparts in a reaction rotor (compare the levels of Figs. 9.16 and 9.17). We expected the extra losses in an impulse rotor because the flow remains nearly at constant speed within the rotor as opposed to a continuously accelerating flow in a reaction rotor. Here again an optimum solidity is observed as a function of the exit flow angle. We will address the issue of optimum solidity in a turbine blade row in the next section.

Nozzle profile loss coefficient is presented in Fig. 9.18 (from Ainley and Mathieson, 1951). The levels are comparable to the reaction rotor of Fig. 9.16. We observe that the level

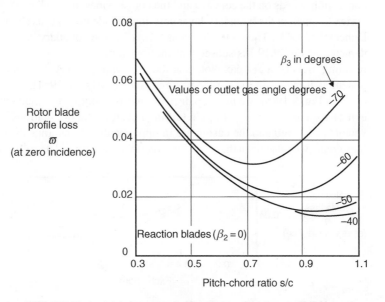

■ FIGURE 9.16   **Profile loss coefficient for reaction (rotor) blades at zero incidence $Re_c = 2 \times 10^5$, $M < 0.6$, $t/c = 0.20$ (adapted from Ainley, 1950)**

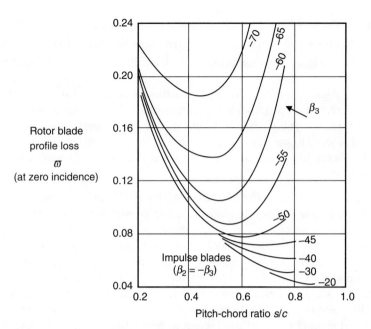

■ **FIGURE 9.17**   Profile loss coefficient for impulse blades at zero incidence Re$_c$ = 2 × 10^5, M < 0.6, t/c = 0.20 (adapted from Ainley, 1950)

of loss increases as the exit flow angle increases, as expected. The 70° exit angle is considered the upper limit of acceptable loss, as the 80° exit flow nozzle represents nearly doubling the profile loss of say 60° exit flow angle nozzle. Here too, we note an optimum solidity represented by the loss buckets.

9.2.2.2   *Secondary Flow Losses.*   As we discussed earlier in the axial-flow compressor section, the secondary flows are set up in a curved channel. The mechanism for the secondary flow generation is the creation of streamwise vorticity in a bend as a result of different convection speeds on the concave and the convex sides of the channel. Take, for example, the vortex filament in the boundary layer entering a blade passage, as shown in Fig. 9.19(a) (from Duncombe, 1964). The vortex filament "A_1A_1" in the boundary layer is normal to the flow as depicted in Fig. 9.19. Its subsequent motion in the passage is frozen at four time intervals resulting in A_2A_2, A_3A_3, etc. Note that the normal filament A_1A_1 develops a streamwise component as it moves in the curved passage (see Hawthrone 1951).

In Fig. 9.19(b), we note a pair of counterrotating vortex filaments are generated at the exit of the curved passage (e.g., a turbine nozzle). The orientations of the normal vorticity vectors at the hub and the casing are in opposite directions, thus the streamwise vortex that they create will also be in the opposite directions, as shown. The right-hand rule applied to the

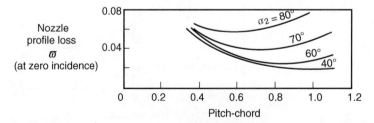

■ **FIGURE 9.18**   Nozzle profile loss coefficient at zero incidence Re$_c$ = 2 × 10^5, M < 0.6, t/c = 0.20 (adapted from Ainley and Mathieson, 1951)

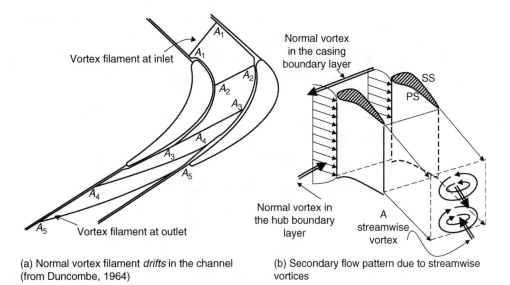

(a) Normal vortex filament *drifts* in the channel
(from Duncombe, 1964)

(b) Secondary flow pattern due to streamwise
vortices

■ **FIGURE 9.19** A normal vortex filament at the entrance to a curved channel develops a
streamwise component and thus creates secondary flow

streamwise vortices identifies the direction of the secondary stream pattern, i.e., the circulatory flow in the clockwise or counterclockwise direction. To calculate the loss associated with the secondary flow, we need to integrate the kinetic energy trapped in the induced (secondary) flow field. The procedure and the associated uncertainties in the strength of the normal vorticity make this approach cumbersome. There are empirical models in the literature that are useful in preliminary design. One such model is of Dunham and Came (1970), which relates the secondary flow loss to the blade aspect ratio (h/c), exit flow angle β_2, the blade inlet angle κ_1, blade pitch-chord ratio, lift coefficient C_L, and the mean flow angle β_m. The secondary loss correlation is

$$Y_s \equiv \frac{\bar{p}_{t1,rel} - \bar{p}_{t2,rel}}{\bar{p}_{t1,rel} - p_1} = 0.0334 \left(\frac{c}{h}\right) \left(\frac{\cos \beta_2}{\cos \kappa_1}\right) \left(\frac{C_L}{s/c}\right)^2 \left(\frac{\cos^2 \beta_2}{\cos^3 \beta_m}\right) \tag{9.25}$$

The total pressure terms in the secondary flow loss coefficient are the mass average of the inlet and exit flowfields, representing a mean spanwise average quantity at the inlet and exit planes. Note that the correlation in Eq. 9.25 scales with C_L^2/AR, which is proportional to induced drag from finite wing theory of Prandtl. The product of the last two brackets in Eq. 9.25 is known as Ainley loading parameter, which was initially proposed by Ainley. In reality, separation of the secondary flow from end wall losses is impractical. The two are so intertwined, i.e., strongly coupled, that any dissection results in a mutilated picture of the flowfield. However, we can observe trends and obtain rough estimates of the secondary loss from models such as Dunham and Came (1970). A graph of secondary loss parameter in Ainley's model in Fig. 9.20 (from Ainley and Mathieson, 1951) confirms the trend and also indicates a large data scatter.

Separation of 2D profile loss from overall loss in a turbine nozzle is shown in Fig. 9.21 (from Johnston, 1953). From this figure, we discern an optimum (mean) spacing that results in a minimum overall loss in three-dimensional nozzle blades. Also note that for the mean outlet angle of 52.8°, for a pitch/chord ratio of ∼1.22, the overall loss is nearly twice the minimum. The minimum loss occurs at an average pitch/chord ratio of ∼0.9 with a mean outlet angle of ∼60°.

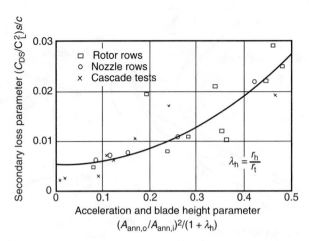

■ **FIGURE 9.20**   Correlation of secondary loss with flow acceleration and blade height (from Ainley and Mathieson, 1951)

9.2.2.3   Annulus Losses.

The growth of annulus boundary layer and its interactions with blade tip clearance flow at the tip, the corner vortex at the junction of the blade, and the hub (for rotor) and casing (for the nozzle) and any seal leakage flow combine to form this category of loss. It is a smorgasbord of *all the remaining* losses that dominate the annulus region. We may examine the tip clearance loss portion of the turbine annulus loss in more detail.

Turbine Rotor Tip Clearance Loss.   A free-tip rotor experiences an over-the-tip flow from the pressure side toward the suction surface. With the blade tip operating very close to the casing to minimize tip leakage flow, the boundary layer on the annulus wall is scraped, rolled, and formed into a vortex, known as the scraping vortex. Schematic drawing of the scraping vortex in both a turbine and a compressor rotor is shown in Fig. 9.22 (from Farokhi, 1986, 1988). The formation of passage scraping vortex is seen to modify the tip loading to reduce the pressure difference at the rotor tip in a turbine, therefore reduce the tip leakage loss. The counterpart of

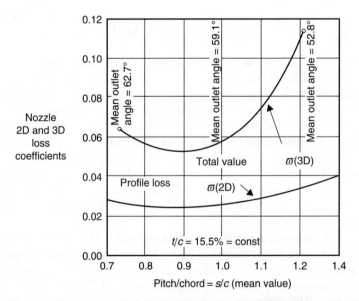

■ **FIGURE 9.21**   Profile and overall loss coefficients for a turbine nozzle (profile loss was estimated from Ainley, 1950) (adapted from Johnston, 1953)

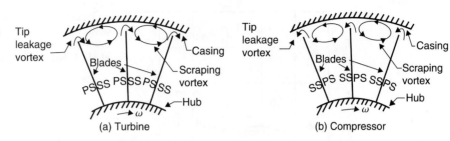

■ **FIGURE 9.22**
Schematic drawing of the tip vortex and the passage scraping vortex interaction in turbomachinery free-tip rotors (from Farokhi, 1986, 1988)

this effect is observed in a compressor where the scraping vortex circulation leads to an increased tip clearance flow due to higher tip loading.

The geometrical parameters of a turbine rotor are shown in Fig. 9.23 to include the hub, tip and pitchline radii, and the tip clearance gap ε. The nondimensional tip clearance may be expressed as a fraction of the blade height h or the local chord length c_{tip} and is referred to as *relative clearance height*.

There are several other nondimensional parameters that are of interest in the study of tip clearance loss and directly impact/measure its magnitude. These parameters are

- A_ε/A_0, which is the ratio of tip clearance area to the zero-clearance annulus flow area downstream of the turbine rotor,
- $\dot{m}_\varepsilon/\dot{m}_0$, which is the fraction of the mass flow rate in the tip clearance region to the zero-clearance annulus mass flow rate,
- $\Delta\eta/\eta_0$, which is the ratio of the turbine efficiency loss due to tip clearance to the zero-clearance turbine efficiency,
- $\Delta\wp_t/\wp_{t0}$, which is the fraction of turbine power loss due to tip clearance to the turbine power with zero tip clearance.

We note that the above parameters are not independent of each other. For example, the clearance area ratio and the clearance mass flow ratio are proportional to each other, or the loss of power in the turbine due to clearance must be proportional to the fraction of the mass flow rate. Also, the loss of efficiency and the loss of power are not independent. The earliest attempt in understanding/modeling the tip clearance loss in a turbine is due to Anderhub (1912) who performed experimental measurements in a steam turbine. Stodola (1924, 1927) used Anderhub's data to present a tip clearance loss efficiency model,

$$\frac{\Delta\eta}{\eta_0} = (1 - \alpha^*) \cdot 2 \cdot \frac{A_\varepsilon}{A_0} \quad \text{(Stodola's model)} \tag{9.26}$$

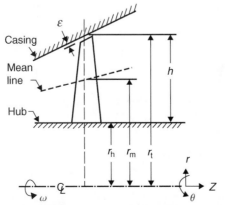

■ **FIGURE 9.23** **Geometry of a free-tip turbine rotor**

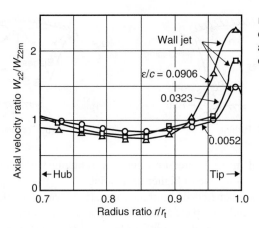

■ **FIGURE 9.24** Axial velocity distribution at the rotor exit in a single-stage axial-flow turbine (at design point) (Bammert et al., 1968)

where α^* is the energy recovery factor for the tip clearance flow and the factor 2 in Stodola's model is based on Anderhub's observations that the axial velocity of the gas emerging from the tip clearance region is nearly twice as large as the axial velocity in the core flow. In effect, the tip clearance flow emerges as a wall jet with no loss of energy, especially in light of no power interaction with the rotor. The experimental evidence of the wall jet emergence in the rotor tip clearance region is seen in Fig. 9.24 (from Bammert, Klaeukens, and Hartmann, 1968).

The tip clearance is presented as a fraction of the tip chord length in Fig. 9.24. The axial velocity in the tip clearance region has exceeded twice the core value for a clearance-to-chord ratio of $\sim$9%. Interestingly, the emergence of the wall jet due to tip clearance in the turbine exit flow if coupled with an axial exhaust diffuser (typical of stationary gas turbine power plants) is rather beneficial in the exhaust diffuser performance. Farokhi (1986, 1987) has investigated the effect of the rotor tip clearance flow on the axial exhaust diffuser and the turbine performance in a trade-off study. The analysis shows that turbine power loss due to rotor tip clearance is partially or completely offset by the enhanced axial diffuser pressure recovery due to the presence of a highly energetic wall jet.

Traupel (1958) proposes a tip clearance loss that is proportional to the clearance area ratio with the proportionality constant a function of the net flow turning (i.e., $\Delta\beta$) in the blade tip region. Bammert, Klaeukens, and Hartmann (1968, 1987) proposes an efficiency loss model that is proportional to the clearance mass flow ratio. To derive an expression for the axial velocities in the tip clearance and the core, Bammert assumes the conservation of angular momentum in the clearance flow and equal expansion of the tip and the core flow. These two assumptions are clearly plausible and the velocity triangles inside the core and the blade tip region are shown in Fig. 9.25, following Bammert, Klaeukens, and Hartmann (1968).

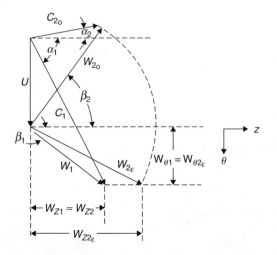

■ **FIGURE 9.25** Velocity triangles in the clearance gap and the core

Equal expansion causes the relative velocity magnitude to be preserved, as shown by the circular arc in Fig. 9.25, and the conservation of angular momentum has kept the relative swirl in the upstream flow unchanged in the clearance flow downstream of the rotor, i.e., $W_{\theta\varepsilon} = W_{\theta 1}$. We can deduce the mass flow ratio from Fig. 9.25, to be

$$\frac{\dot{m}_\varepsilon}{\dot{m}_0} = \frac{A_\varepsilon}{A_0}\left[1 + \tan^2\beta_2 - \left(\frac{W_{z1}}{W_{z2}}\tan\beta_1\right)^2\right]^{1/2} \tag{9.27}$$

Lakshminarayana (1970) derives a semiempirical expression for the decrease in turbomachinery stage efficiency due to clearance, which is based on the induced flowfield of a potential vortex on the blades and the radial motion of the fluid in the blade boundary layer near the tip. Lakshminarayana's expression for $\Delta\eta$ is

$$\Delta\eta = \frac{0.7\tau \cdot \psi}{\cos\beta_m}\left[1 + 10\left(\frac{\phi \cdot \tau \cdot A}{\psi \cdot \cos\beta_m}\right)^{1/2}\right] \quad \text{(Lakshminarayana's model)} \tag{9.28}$$

Lakshminarayana's model is derived for both types of turbomachinery, namely, compressors and turbines. However, as it lacks the effect of tip loading in compressors and unloading in turbines due to relative wall motion, as shown schematically in Fig. 9.24, it cannot enjoy a general applicability. The effect of blade tip configurations (other than flat) on tip clearance flow is not accounted for in Lakshminarayana's model.

The fraction of turbine power loss due to rotor tip clearance is proportional to the fraction of tip leakage flow, i.e.,

$$\frac{\Delta\wp_t}{\wp_{t0}} \propto \frac{\dot{m}_\varepsilon}{\dot{m}_0} \tag{9.29}$$

The zero-clearance turbine power is defined as

$$\wp_{t0} \equiv \eta_0 \cdot \wp_{t,\text{ideal}} \tag{9.30}$$

The turbine power output with tip clearance is

$$\wp_{t\varepsilon} = \eta_\varepsilon \cdot \wp_{t,\text{ideal}} \tag{9.31}$$

Now, defining incremental efficiency and power due to clearance as

$$\Delta\wp_t \equiv \wp_{t0} - \wp_{t\varepsilon} \tag{9.32}$$

$$\Delta\eta \equiv \eta_0 - \eta_\varepsilon \tag{9.33}$$

We get

$$\frac{\Delta\wp_t}{\wp_t} = \frac{\Delta\eta}{\eta_0} \propto \frac{\dot{m}_\varepsilon}{\dot{m}_0} \tag{9.34}$$

We expect the proportionality factor that emerges in Eq. 9.34 between the incremental efficiency loss and the clearance mass fraction to include the following effects:

- The effect of tip shape
- Relative wall motion
- Clearance gap Reynolds number
- Clearance-to-blade height ratio $\tau \equiv \varepsilon/h$

Farokhi (1988) suggests that the proportionality function in Eq. 9.34 is the blade tip discharge coefficient C_D where the discharge coefficient is in general a function of

$$C_D = C_D \left(\frac{\varepsilon}{h}, \frac{\varepsilon}{c}, \text{Re}_\varepsilon, \phi_{\text{tip}}, \psi_{\text{tip}}, \frac{\delta_w^*}{h}, \frac{\theta_w}{h} \right) \tag{9.35}$$

Therefore, the turbine efficiency loss due to rotor tip clearance is (from Farokhi, 1988)

$$\frac{\Delta\eta}{\eta_0} = C_D \frac{A_\varepsilon}{A_0} \left[1 + \tan^2 \beta_2 - \left(\frac{W_{z1}}{W_{z2}} \tan \beta_1 \right)^2 \right]^{1/2} \tag{9.36}$$

We may simplify the above equation by assuming the axial velocity remains nearly constant across the rotor, i.e., $W_{z1} \cong W_{z2}$, and by recasting it in terms of the flow coefficient and the stage loading factor, and the mean flow angle in the rotor as β_m,

$$\frac{\Delta\eta}{\eta_0} \cong C_D \frac{A_\varepsilon}{A_0} \left(1 - \frac{\psi}{\phi} \tan \beta_m \right)^{1/2} \tag{9.37}$$

The area ratio term may also be simplified as

$$\frac{A_\varepsilon}{A_0} \approx \left(\frac{r_t}{r_m} \right) \left(\frac{\varepsilon}{h} \right) = \left(\frac{r_t}{r_m} \right) \cdot \tau \tag{9.38}$$

when we neglect the $(h\tau^2/2r_m)$. Therefore the decrease in turbine stage efficiency is expressed as

$$\Delta\eta \cong \eta_0 C_D \cdot \tau \cdot \left(\frac{r_t}{r_m} \right) \left(1 - \frac{\psi}{\phi} \tan \beta_m \right)^{1/2} \tag{9.39}$$

Note that the mean flow angle β_m is either a negative angle (for reaction stages) or zero for an impulse stage. Therefore the square root term in Eq. 9.39 is always positive. The functional dependence of efficiency loss and the clearance height fraction is seen from the model to be linear. This behavior is shown in Fig. 9.26 for a constant stage loading, flow coefficient, and pitchline radius ratio (from Farokhi, 1988).

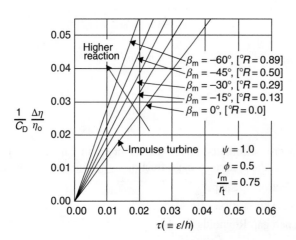

■ **FIGURE 9.26** Variation of turbine efficiency loss parameter with tip clearance (from Farokhi, 1988)

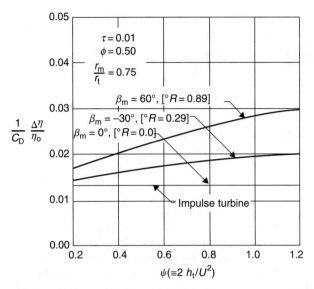

■ **FIGURE 9.27**   Turbine efficiency loss due to tip clearance increases with the stage loading for reaction turbines and remains unchanged for an impulse turbine stage (for $\tau = 0.01$, $\phi = 0.50$, and $r_m/r_t = 0.75$) (from Farokhi, 1988)

The behavior of turbine efficiency loss with stage loading and with flow coefficient is examined in Figs. 9.27 and 9.28 (from Farokhi, 1988).

In a series of cascade experiments conducted in a water table, Booth, Dodge, and Hepworth (1982) measured leakage discharge coefficients over blade tips of various shapes. The baseline blade tip shape selected was that of a Garrett low-aspect ratio turbine. The variety of tip shapes included squealer, winglet, flat, knife, and grooved modifications to the baseline. These tip configurations are shown in Fig. 9.29 (from Booth, Dodge, and Hepworth, 1982).

Relative tip discharge coefficients for advanced blade tip configurations are presented in Fig. 9.30 (from Booth, Dodge, and Hepworth, 1982).

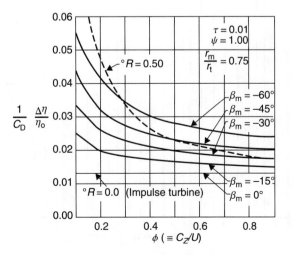

■ **FIGURE 9.28**   Turbine efficiency loss due to tip clearance decreases with the flow coefficient for reaction turbines and remains unchanged for an impulse turbine stage (for $\tau = 0.01$, $\psi = 1.0$, and $r_m/r_t = 0.75$) (from Farokhi, 1988)

■ **FIGURE 9.29**
Turbine rotor blade tip configurations studied by Booth, Dodge, and Hepworth (1982)

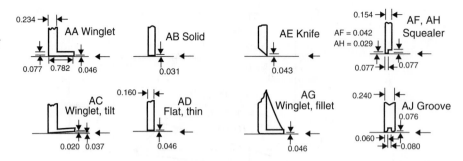

The flat tip discharge coefficient is reported by Booth, Dodge, and Hepworth to be ~0.81–0.83, which may be used to estimate the discharge coefficients for tip configurations shown in Fig. 9.30. The use of squealer and knife tip geometries has a potential of reducing the tip leakage flow by ~30%. As the experiments of Booth, Dodge, and Hepworth did not include the effect of relative wall motion on tip unloading, their results need to be modified for the relative wall motion effect. The relative wall motion causes the scraping passage vortex, which in a turbine causes the tip unloading (as seen in Fig. 9.22) or equivalently, a reduced tip clearance flow. Graham (1986) has verified the reduction of tip loading by virtue of rotation in a turbine blade row. Graham measured the static pressure distribution around the tip- and mid-span profiles of turbine blades in an experimental rig with relative wall motion. Figure 9.31 shows the ratio of tip- to the mid-span loading at different tip clearance fractions.

The stationary end wall case in Fig. 9.31 serves as the basis to judge the effect of relative wall motion on tip loading. The experimental data range on tip clearance starts at ~1/2% clearance-to-span ratio and stretches to ~2.5%. It is clearly evident that the relative wall motion causes tip unloading and at higher engine equivalent speeds (2 × EES), the tip unloading is exacerbated. Since the tip clearance flow is predominantly a pressure-driven phenomenon and

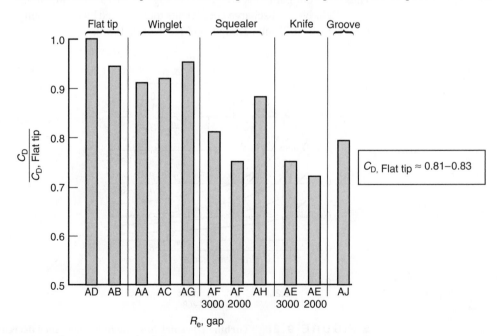

■ **FIGURE 9.30**   Relative rotor tip discharge coefficient (from Booth, Dodge, and Hepworth, 1982)

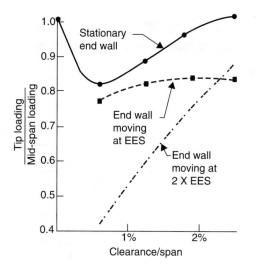

the effect of blade rotation is reducing the pressure difference or loading at the tip section, Farokhi (1988) proposes the following simple model for the tip discharge coefficient that accounts for blade rotation,

$$C_{D\omega} \sim C_{D,\omega=0}\left(\frac{\Delta p_\omega}{\Delta p_{\omega=0}}\right)^{1/2} \qquad \textbf{(9.40)}$$

The subscript ω stands for the rotation in Eq. 9.40. The basis for the model is the Bernoulli equation, which sets velocity proportional to the square root of static pressure drop through an orifice. In light of Graham's data on tip unloading, we note that the magnitude of the pressure unloading term in Eq. 9.40 is between 5 and 15%. This leads to a reduction in tip discharge coefficient by ~4–10% over the measurements of Booth, Dodge, and Hepworth, for example, that did not include relative wall motion.

EXAMPLE 9.4

A turbine stage is characterized by a loading coefficient ψ of 1.0 and a flow coefficient of ϕ of 0.5. The rotor is unshrouded and has a tip clearance gap-to-blade height ratio of 2%. The ratio of the tip to pitchline radius is $r_t/r_m = 1.25$. The rotor blade tip has a knife configuration with a tip discharge coefficient of 0.5. The mean relative flow angle in the turbine rotor is $\beta_m = -20°$. Calculate the loss of turbine efficiency due to the tip clearance.

SOLUTION

According to Eq. 9.39, the loss of turbine efficiency is

$$\Delta\eta \cong \eta_0 C_D \cdot \tau \cdot \left(\frac{r_t}{r_m}\right)\left(1 - \frac{\psi}{\phi}\tan\beta_m\right)^{1/2} \approx 0.013\,\eta_0$$

9.2.3   Optimum Solidity

Although the flow in turbines is characterized as having favorable pressure gradient, the flow acceleration on the suction surface of a turbine blade has to face an adverse pressure gradient

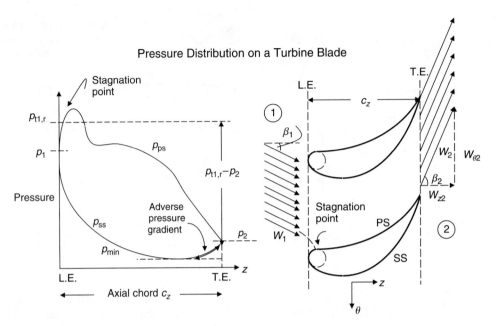

■ **FIGURE 9.32** **A typical pressure distribution on a turbine blade**

toward the trailing edge. This behavior is typical of all lifting surfaces whether a compressor or a turbine blade. A typical pressure distribution on a turbine blade is shown in Fig. 9.32 where the flow angles and velocity vectors are relative to the blade.

The blade loading is the integral of the pressure difference on the pressure and suction surfaces, which produces the blade tangential force F_θ.

$$F_\theta = c_z \int_0^1 (p_p - p_s)d(z/c_z) \tag{9.41}$$

Also from the momentum balance across the blade row in the tangential direction, we get the blade tangential force as

$$F_\theta = \rho_2 s \cdot W_{z2}(W_{\theta 1} - W_{\theta 2}) \tag{9.42}$$

We note from this equation that the blade tangential force increases with blade spacing s for a given velocity vector diagram, i.e., swirl distribution and through flow. With an increase in tangential blade force the integral of the pressure jump across the blade increases, according to Eq. 9.41. With higher loading, the adverse pressure gradient on the suction surface may cause the boundary layer separation and a higher blade profile loss. On the contrary, keeping the blade spacing (relative to chord) small to reduce the blade loading will increase the wetted area and thus the frictional losses of the blade row. Hence, we expect that optimum blade solidity exists, which produces the lowest profile and wetted surface losses for a blade row. Zweifel (1945) has addressed the issue of optimum solidity in turbomachinery by introducing a loading parameter and a criterion that is widely accepted. Here we present Zweifel's methodology.

Equating the two expressions for the blade tangential force (per unit span) and isolating axial solidity c_z/s yields

$$\sigma_z \equiv \frac{c_z}{s} = \frac{\rho_2 W_{z2}(W_{\theta 1} - W_{\theta 2})}{\int_0^1 (p_p - p_s)d(z/c_z)} \tag{9.43}$$

Zweifel (1945) has defined a blade loading parameter ψ_z as the ratio of the blade tangential force per unit span and axial chord to the difference of inlet stagnation pressure and the exit static pressure, i.e.,

$$\psi_Z \equiv \frac{\int_0^1 (p_p - p_s)d(z/c_z)}{p_{t1,r} - p_2}$$

(9.44)

Now, in terms of the Zweifel loading parameter, we may cast Eq. 9.43 into the following form,

$$\sigma_z \equiv \frac{\rho_2 W_{z2}(W_{\theta 1} - W_{\theta 2})}{(p_{t1,r} - p_2)\psi_Z}$$

(9.45)

The axial and tangential velocities may be replaced by

$$W_{z2} = W_2 \cos \beta_2$$

$$W_{\theta 2} = W_2 \sin \beta_2$$

We may substitute these expressions in Eq. 9.45 and with some minor manipulations we get

$$\sigma_z \equiv \left(\frac{\rho_2 W_2^2/2}{p_{t1,r} - p_2} \right) \frac{\left(\dfrac{W_{\theta 1}}{W_{\theta 2}} - 1 \right) \sin(2\beta_2)}{\psi_Z}$$

(9.46)

If we ignore the total pressure loss between the inlet and blade exit, we may cast the pressure difference term in Eq. 9.46 as

$$p_{t1,r} - p_2 \approx p_{t2,r} - p_2 = p_2 \left[\left(1 + \frac{\gamma - 1}{2} M_{2r}^2 \right)^{\frac{\gamma}{\gamma - 1}} - 1 \right]$$

(9.47)

The dynamic pressure term in Eq. 9.46 may be written in terms of static pressure and local Mach number as

$$\rho_2 W_2^2/2 = \gamma \cdot p_2 M_{2r}^2/2$$

(9.48)

Therefore, the axial solidity may be expressed as

$$\sigma_z \equiv \left(\frac{(\gamma/2)M_{2r}^2}{\left(1 + \dfrac{\gamma - 1}{2} M_{2r}^2 \right)^{\frac{\gamma}{\gamma - 1}} - 1} \right) \frac{\left(\dfrac{W_{\theta 1}}{W_{\theta 2}} - 1 \right) \sin(2\beta_2)}{\psi_Z}$$

(9.49)

In the incompressible limit (i.e., $M_{2r} \to 0$), the above expression simplifies to

$$\sigma_z \equiv \frac{\left(\dfrac{W_{\theta 1}}{W_{\theta 2}} - 1 \right) \sin(2\beta_2)}{\psi_Z}$$

(9.50)

Zweifel recommended a value of 0.8 (to $\sim$1) for the loading parameter ψ_Z. Hence, based on the exit flow angle and swirl ratio across the blade row, an optimum solidity is obtained from Eq. 9.49 or 9.50. Equation 9.50 has a maximum for β_2 of 45° for a blade row with a fixed swirl

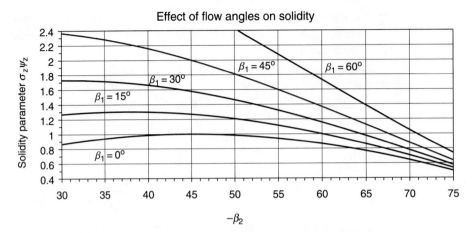

■ **FIGURE 9.33**   **Effect of flow angles in a turbine blade row on the solidity parameter** $\sigma_z \cdot \psi_z$

ratio. To express the blade solidity purely in terms of the flow angles, we write the swirl velocity components in terms of the axial flow and the tangent of the flow angle as

$$W_{\theta 1} = W_{z1} \tan \beta_1$$
$$W_{\theta 2} = W_{z2} \tan \beta_2$$

By substituting these expressions in Eq. 9.50 and assuming constant axial velocity, we get

$$\sigma_z \psi_Z = \left(\frac{\tan \beta_1}{\tan \beta_2} - 1 \right) \sin(2\beta_2) \tag{9.51}$$

After some minor trigonometric manipulations, we express Eq. 9.51 in terms of flow turning angle in the blade row as

$$\sigma_z \psi_Z = \frac{2 \cos \beta_2}{\cos \beta_1} \sin(\beta_1 - \beta_2) \tag{9.52}$$

Figure 9.33 is a plot of the axial solidity parameter with turbine flow angles. In the region of interest in a turbine, i.e., for the exit flow angles between $-45°$ and $-75°$, we observe that the solidity parameter decreases with the decrease in exit flow angle for a given inlet flow angle. In order to interpret this result, we take the Zweifel loading parameter as a constant, namely, ~ 0.8. Therefore, a decrease in exit flow angle, for a fixed inlet flow angle, increases the stage reaction, which tends to reduce the blade loading. However, to maintain the blade loading fixed, the solidity has to be reduced, which inversely affects the blade loading. Also, for a given exit flow angle $-\beta_2$, the increase in the inlet flow angle causes a lower stage reaction (i.e., moving toward impulse stage) and thus an increase in blade loading. To maintain the loading as described by a constant value of Zweifel parameter, the blade solidity has to increase, which has a tendency to reduce the blade loading. The variation of stage reaction with the flow angles is depicted in Fig. 9.34 for a turbine stage with zero exit swirl.

A summary of the results of the optimum pitch-chord ratio for reaction and impulse turbines as a function of blade relative exit flow angle is shown in Fig. 9.35 (after Duncombe, 1964). The two criteria for optimum solidity, i.e., Zweifel's and Howell's, show a very limited range of agreement with the test data, as depicted in Fig. 9.35.

Effect of flow angles on degree of reaction
(stage with zero exit swirl)

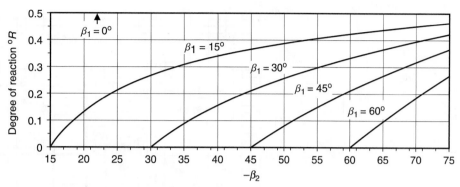

■ **FIGURE 9.34**   The effect of flow angles on a turbine stage with zero exit swirl (β_1 and β_2 are the rotor relative inlet and exit flow angles, respectively)

9.2.4   Turbine Cooling

The entrance flow to a turbine nozzle is the combustor exit flow with its spatially nonuniform characteristics, such as the hot spots. The average combustor exit temperature in modern gas turbine engines is ~1750–2000 K and the temperature nonuniformity or "hot spots" may expose the nozzle to an additional temperature of 100–200 K. This is clearly beyond the temperature limits of metals, including the nickel-based alloys used in turbines. Figure 9.36 shows a turbine temperature map with three distinct temperature ranges. Turbine *gas-path temperature* is the hot gas temperature, which is typical of modern gas turbine combustor exit

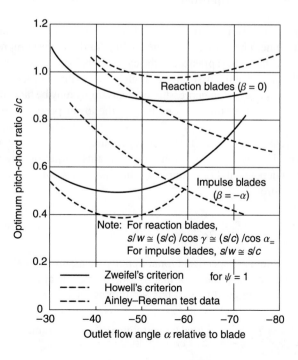

■ **FIGURE 9.35**   Optimum pitch-chord ratio (the inverse of solidity adapted from Duncombe, 1964)

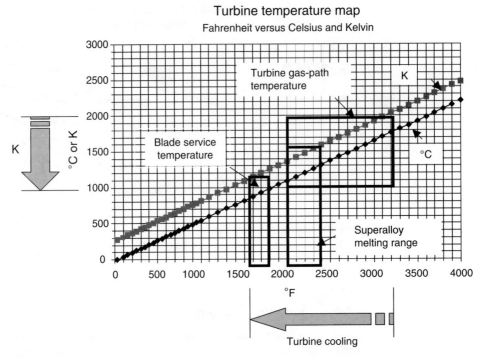

■ **FIGURE 9.36** Turbine temperature map shows the extent of turbine cooling in a modern GT-engine

conditions. The superalloy *melting range* identifies the melting temperature for (Nickel-based) superalloys that are used in turbine manufacturing. The *blade service temperature* represents the blade surface operating temperature, which is about 80% of the melting temperature of the superalloy.

Therefore the need arises to cool the first and possibly the second and the third stages of a high-pressure turbine. The source of turbine cooling fluid is in the compressor. The compressor discharge achieves the highest pressure in the cycle, therefore it is best suited to cool the nozzle, the rotor, and the casing of the first stage of high-pressure turbine. As the gas pressure drops in subsequent turbine stages, a lower pressure coolant supply may be tapped for cooling purposes. Intercompressor bleed may thus be used to cool the lower pressure regions (i.e., blades and casing) in a turbine.

The gas is characterized by its static temperature T_g, which is independent of the motion of the observer. Then by virtue of its motion, it attains a total or stagnation temperature T_t, which is observer dependent. For example, the turbine nozzle, which is stationary, is exposed to the total temperature

$$T_t = T_g + \frac{C^2}{2c_p} \quad \text{total temperature in the absolute frame} \tag{9.53}$$

where C is the absolute gas speed and for simplicity we assumed a calorically perfect gas with constant c_p. On the contrary, the rotor that follows the nozzle attributes a relative gas speed to the incoming flow, therefore the total temperature as measured by a rotor is

$$T_{t,r} = T_g + \frac{W^2}{2c_p} \quad \text{total temperature in the rotor frame of reference} \tag{9.54}$$

The kinetic energy contribution to the gas total temperature, i.e., the second term on the RHS of Eqs. 9.53 and 9.54, may be significant and thus affects the cooling requirements of nozzle and rotor blades. The coolant temperature T_c represents another temperature that we work with in a turbine-cooling problem. Unless we cool the coolant, T_c is the compressor discharge (static) temperature. The wall temperature T_w represents the actual wall temperature that (along with the state of stress) determines the material life. The temperature of the wall on the hot gas side is T_{wg} and the wall temperature on the coolant side is T_{wc}. In case the wall is insulated, the flow of a thermally conducting fluid over the wall creates a wall temperature, which is known as the adiabatic wall temperature T_{aw}.

In summary, we deal with several temperatures in a cooled turbine, these are

1. T_g, which is the gas static temperature
2. T_{aw} is the adiabatic wall temperature, i.e., the temperature of the insulated wall
3. T_t, the total gas temperature in the absolute frame, e.g., in the nozzle or casing frame of reference
4. $T_{t,r}$, the total gas temperature as seen by the turbine rotor
5. T_{wg} is the wall temperature on the hot gas side (i.e., actual or desired surface temperature)
6. T_{wc} is the wall temperature on the coolant side
7. T_c is the coolant temperature

The adiabatic wall temperature in gas turbine is very nearly equal to the total or stagnation temperature of the gas. For a flat plate, the adiabatic wall temperature is

$$T_{aw} = T_g + r\frac{C^2}{2c_p} \quad \text{for a stator} \tag{9.55}$$

$$T_{aw,r} = T_g + r\frac{W^2}{2c_p} \quad \text{for a rotor} \tag{9.56}$$

The coefficient r in the above equations is known as the *recovery factor*. The recovery factor is a function of gas property, known as Prandtl number and the state of the boundary layer, following

$$r = \sqrt{Pr} \quad \text{for (flat plate) laminar boundary layers} \tag{9.57}$$

$$r = \sqrt[3]{Pr} \quad \text{for a flat plate turbulent boundary layer} \tag{9.58}$$

where Pr is the gas Prandtl number, which is a fluid property, defined as

$$Pr \equiv \frac{\mu c_p}{k} \tag{9.59}$$

where μ is the fluid viscosity, c_p is the specific heat at constant pressure, and κ is the fluid thermal conductivity. The Prandtl number for air is $\sim$0.71 (for a wide range of temperatures), therefore the recovery factor r is either $\sim$0.8426 for a laminar boundary layer or $\sim$0.8921 for a turbulent boundary layer. The turbine boundary layers are predominantly turbulent, as a result of high free-stream turbulence intensity that is produced in the combustor ($\sim$10–20%). Therefore, the adiabatic wall temperature and the gas total temperature are very nearly equal. The implication is that unless we cool the blade (or casing), the surface metal temperature reaches nearly that of the stagnation temperature of the gas (in the appropriate frame of reference, e.g., the nozzle will see in Eq. 9.55 and the rotor will see Eq. 9.56). The static

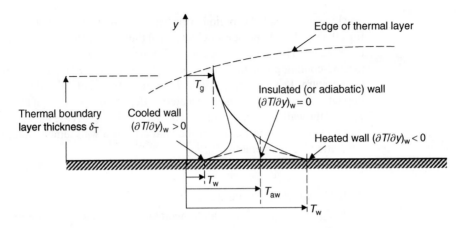

■ **FIGURE 9.37** Static temperature distribution $T(y)$ within the thermal boundary layer subjected to different wall boundary conditions (i.e., cooled, adiabatic, or heated)

temperature distribution in a thermal boundary layer varies according to the wall boundary condition, as shown in Fig. 9.37. The edge of the thermal layer is depicted by δ_T thickness, analogous to the viscous boundary layer thickness δ.

EXAMPLE 9.5

The total gas temperature at the exit of a turbine nozzle is $T_{tg} = 1700$ K and the local gas Mach number is $M_2 = 1.0$. The gas ratio of specific heats is $\gamma = 1.33$, $c_p = 1157\,\text{J/kg}\cdot\text{K}$, and Prandtl number is Pr = 0.71. First calculate

(a) the gas static temperature T_g

(b) the adiabatic wall temperature T_{aw} on the nozzle for a turbulent boundary layer

(c) the adiabatic wall temperature on the nozzle for a laminar boundary layer

Assuming that the gas speed relative to the rotor is $W_2 = 455$ m/s, calculate

(d) the total temperature of the gas on the rotor, i.e., as measured in the rotor frame of reference, $T_{t,r}$.

SOLUTION

The static temperature is related to the total temperature and Mach number $T = T_t/[1 + (\gamma - 1)M^2/2]$, therefore,

$$T_g = 1700\,\text{K}/[1 + 0.165] \approx 1459.2\,\text{K}$$

The local speed of sound is $a_2 = [(\gamma - 1)c_p T_2]^{0.5} \approx 746.4$ m/s

Therefore, the gas speed is the product of local Mach number and the speed of sound, i.e., $C_2 = a_2 = 746.4$ m/s.

The adiabatic wall temperature is given by Eq. 9.55 (for a stator)

$$T_{aw} = T_g + r\frac{C^2}{2c_p} \quad \text{where } r = \sqrt{\text{Pr}} \text{ for laminar and } r = \sqrt[3]{\text{Pr}} \text{ for turbulent BL.}$$

$$T_{aw} = 1459.2\,\text{K} + (0.71)^{1/3}(746.4)^2/1157 \approx 1889\,\text{K}$$

(for turbulent boundary layer)

$$T_{aw} = 1459.2\,\text{K} + (0.71)^{1/2}(746.4)^2/1157 \approx 1865\,\text{K}$$

(for laminar boundary layer)

In the rotor frame of reference, the contribution of gas kinetic energy is based on W_2, therefore,

$$T_{tr} = T_g + W_2^2/2cp = 1459.2\,\text{K} + (455)^2/2/1157$$

$$\approx 1549\,\text{K}$$

The rotor feels a lower stagnation gas temperature than the stator by about 150 K.

The temperature distribution in the thermal layer is created as a result of fluid thermal conductivity and is governed by the Fourier's law of heat conduction, i.e.,

$$q_y = \frac{\dot{Q}}{A} = -k\frac{\partial T}{\partial y} \tag{9.60}$$

where q_y is the heat transfer per unit area (i.e., the heat flux), due to conduction, in the y-direction, k is the thermal conductivity of the fluid (or solid), and the last term is the temperature gradient in the y-direction. The negative sign indicates the flow of heat is in the opposite direction to that of the temperature gradient, which forces heat to flow from hot to cold. The dimensions of thermal conductivity are discerned from Eq. 9.60 to be $W/m \cdot K$.

In the case of adiabatic wall, heat transfer rate is zero, therefore the temperature gradient at the insulated wall becomes zero, i.e.,

$$\left.\frac{\partial T}{\partial y}\right|_{y=0} = 0 \quad \text{for an insulated, or adiabatic, wall} \tag{9.61}$$

The flow of heat from the wall to the fluid makes q_y positive (for the fluid), thus the temperature gradient negative at the wall

$$\left.\frac{\partial T}{\partial y}\right|_{y=0} \prec 0 \quad \text{for a heated wall} \tag{9.62}$$

The cooled wall case will pose a negative heat transfer condition to the fluid, thus the temperature gradient will attain a positive slope at the wall.

$$\left.\frac{\partial T}{\partial y}\right|_{y=0} \succ 0 \quad \text{for a cooled wall} \tag{9.63}$$

All these boundary conditions on the slope of static temperature distribution at the wall are shown in Fig. 9.37.

The cooling of a surface may be achieved in different ways, for example,

1. We may coat the surface with a low-thermal conductivity layer, i.e., a protective layer like a ceramic coating, which reduces the heat transfer through a surface
2. Radiation cooling, which is natural to all surfaces above absolute zero temperature
3. We may increase the heat transfer through the wall by internal cooling, i.e., convective cooling or impingement cooling
4. We may decrease heat transfer by *blanketing* the wall with a layer of coolant, as in film cooling
5. We may *blanket* the entire wall by transpiration cooling through a porous skin
6. Combination of the above methods

The first two methods are *passive* thermal protection system and surface radiation, whereas methods 3–6 involve *active* cooling of the surface. The radiation cooling in a gas turbine is only a small contributor to the total heat transfer and thus often neglected. The active cooling methods are divided into *internal* versus *external* cooling schemes.

9.2.4.1   *Convective Cooling.*   Convective heat transfer is governed by the Newton's Law of Cooling, according to

$$q \equiv \frac{\dot{Q}_c}{A} = h\Delta T \tag{9.64}$$

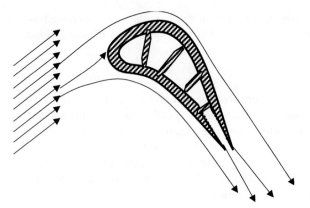

■ **FIGURE 9.38** An internally cooled turbine blade is shown with separate cooling channels and trailing-edge ejection

where h is the coefficient of heat transfer and the temperature difference ΔT is the *driving* term for heat transfer. This means that heat transfer is proportional to area as well as the temperature difference. For example, $(T_{aw} - T_{wg})$ drives the heat transfer to the wall on the gas side of a cooled turbine blade. Similarly, $(T_{wc} - T_c)$ is the driver for heat transfer on the coolant side of the same blade. The dimensions of the heat transfer coefficient h are (as seen from Eq. 9.64) $W/m^2 K$. A convectively cooled turbine blade is shown in Fig. 9.38. The internal passage is divided into a nose (or leading edge) section, three central sections, and one trailing-edge passage. The arrangement of these internal cooling passages meters the amount of coolant to the desired section of the blade. For example, the stagnation point heating is very critical in turbine blade heat transfer and cooling therefore the leading-edge region has its own separate nose channel. The options with the nose channel include impingement cooling that we will discuss in the next section.

The static temperature distribution on the hot gas side, across the turbine wall and in the coolant side, is shown on an internally cooled turbine blade in Fig. 9.39.

The steady-state heat transfer rate from the hot gas to the coolant suggests a balance

$$\dot{Q}_c = A_g h_g (T_{aw} - T_{wg}) = A_w k_w \frac{T_{wg} - T_{wc}}{t_w} = A_c h_c (T_{wc} - T_c) \qquad (9.65)$$

To see a correlation between the film coefficients h_g and/or h_c and the flow Reynolds number, Prandtl number, etc., we may write the heat transfer rate from the hot gas using Stanton number according to

$$\dot{Q}_w = A_g \cdot \text{St} \cdot \rho_g u_g c_{pg} (T_{aw} - T_{wg}) \qquad (9.66)$$

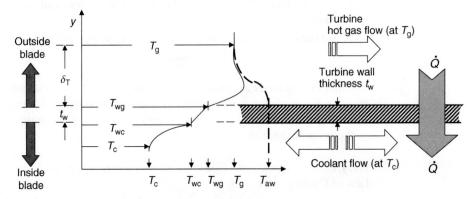

■ **FIGURE 9.39** Static temperature distribution on a turbine blade with internal cooling (the dashed line shows the temperature distribution for an uncooled/insulated wall)

where ρ_g is the gas density, u_g is the gas speed, c_{pg} is the specific heat at constant pressure on the gas side, and St is the Stanton number and is nondimensional. Therefore by equating the wall heat transfer rate and the convective heat transfer to the wall in Eqs. 9.65 and 9.66, we note

$$h_g = \rho_g u_g c_{pg} \cdot \text{St} \tag{9.67}$$

The simplest expression for Reynolds analogy between the skin friction and heat transfer is derived for a gas of Prandtl number unity, i.e., for $\text{Pr} = 1$, we can show that

$$\text{St} = \frac{c_f}{2} \tag{9.68}$$

where c_f is the local skin friction coefficient. We remember that local skin friction coefficient is a function of Reynolds number, and it jumps through the transition from laminar to turbulent state of the boundary layer. In boundary layer theory we learned that

$$c_f \propto \text{Re}^{-1/2} \quad \text{(Laminar)} \tag{9.69}$$

and

$$c_f \propto \text{Re}^{-1/5} \quad \text{(Turbulent)} \tag{9.70}$$

Therefore, the higher heat transfer rates for turbulent flow suggest that we should "stir up" the coolant flow inside the cooling passages as much as possible. The practical approach has been through introducing a staggered array of pins in the cooling channel (or ribs normal to the coolant flow direction), as shown in Fig. 9.40. A limitation to the extent of flow resistance that we introduce inside a cooling channel is the loss of total pressure, which is a "commodity" of value for film-cooled blades.

Let us substitute Eq. 9.67 into 9.65 to express the heat transfer rate to the blade in terms of gas mass flow rate, i.e.,

$$\dot{Q} = A_g \rho_g u_g c_{pg} \cdot \text{St}_g \cdot (T_{aw} - T_{wg}) = \dot{m}_g c_{pg} \cdot \text{St}_g \cdot (T_{aw} - T_{wg}) \tag{9.71}$$

The equivalent form of heat transfer on the coolant side may be written as

$$\dot{Q} = A_c \rho_c u_c c_{pc} \cdot \text{St}_c \cdot (T_{wc} - T_c) = \dot{m}_c c_{pc} \cdot \text{St}_c \cdot (T_{wc} - T_c) \tag{9.72}$$

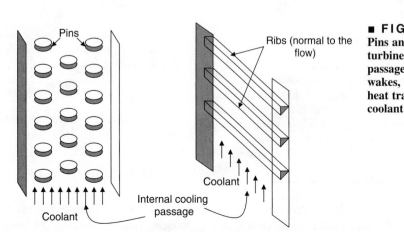

■ **FIGURE 9.40**
Pins and ribs inside turbine blade cooling passages create turbulent wakes, which enhance heat transfer to the coolant

Setting the heat transfer rate on the gas side and the coolant side equal to each other and expressing the ratio of the coolant-to-gas flow rate as ε, we get an expression for the coolant (mass) fraction, namely,

$$\varepsilon \equiv \frac{\dot{m}_c}{\dot{m}_g} = \frac{c_{pc}}{c_{pg}} \frac{St_c}{St_g} \left(\frac{T_{wc} - T_c}{T_{aw} - T_{wg}} \right) \tag{9.73}$$

From the heat conduction equation through the wall, we express

$$T_{wc} = T_{wg} - \frac{\dot{Q} \cdot t_w}{A_w k_w} = T_{wg} - \left(\frac{t_w}{k_w} \right) \left(\frac{A_g}{A_w} \right) \rho_g u_g c_{pg} St_g (T_{aw} - T_{wg}) \tag{9.74}$$

We note that the gas flow area is proportional to the blade passage spacing s and the wall area of the blade is proportional to twice the blade chord $2c$, therefore the area ratio of the gas to blade is nearly the inverse of twice the solidity, $1/2\sigma$, i.e.,

$$T_{wc} = T_{wg} - \left(\frac{t_w}{k_w} \right) \frac{1}{2\sigma} \rho_g u_g c_{pg} St_g (T_{aw} - T_{wg}) \tag{9.75}$$

We may also cast the gas mass density, i.e., $\rho_g u_g$, in terms of the pressure, temperature, and Mach number of the gas following continuity equation

$$\rho_g u_g = \frac{\dot{m}_g}{A_g} = \sqrt{\frac{\gamma_g}{R_g}} \frac{p_{tg}}{\sqrt{T_{tg}}} M_g \left(1 + \frac{\gamma - 1}{2} M_g^2 \right)^{-\frac{\gamma+1}{2(\gamma-1)}} \tag{9.76}$$

We recognize that the gas total pressure (in the high-pressure turbine) is $\sim p_{t4} \sim p_{t3} = \sim p_{t2} \cdot \pi_c$. The gas total temperature is $\sim T_{t4}$. The average gas Mach number is ~ 0.75. Therefore, we may substitute these parameters from our cycle analysis to estimate the coolant mass fraction in a high-pressure turbine. The following example highlights the method for estimating the coolant fraction.

EXAMPLE 9.6

Let us consider a gas turbine engine with the following cycle parameters at the sea level static condition:

$T_0 = 288\,K$
$p_0 = 100\,kPa$
$\gamma = 1.4,\ c_{pc} = 1004.5\,J/kg \cdot K$
$\pi_c = 25,\ e_c = 0.90$
$T_{t4} = 2000\,K$
$\gamma_g = 1.33,\ c_{pg} = 1188\,J/kg \cdot K$

$T_{t3} \approx 800\,K$
$p_{t3} \approx 2.5\,MPa$

Therefore, assuming the gas-side Stanton number is $St_g \approx 0.005$, the adiabatic wall temperature is nearly the gas stagnation temperature, i.e., $T_{aw} = 2000\,K$, the gas total pressure is nearly the compressor discharge total pressure, $p_{tg} \approx 2.5\,MPa$, and the desired gas-side wall temperature is $T_{wg} = 1200\,K$, the thickness of the internally cooled wall is ~ 2 mm, the blade mean solidity in the HPT is ~ 2, the

thermal conductivity of nickel-based alloy is $k_w = 14.9\,W/m \cdot K$, we get

$$T_{wc} \cong T_{wg} - 330 = 870\,K$$

$$\varepsilon = \left(\frac{1.0045}{1.188} \right) \frac{St_c}{St_g} \left(\frac{870 - 800}{2000 - 1200} \right) \cong 0.074 \frac{St_c}{St_g}$$

For a cooling passage to the gas-side Stanton number ratio of $St_c/St_g \sim 1/2$, we estimate the cooling fraction to be nearly 3.7% for the first blade row in a HPT, i.e.,

$$\varepsilon \approx \frac{0.074}{2} = 0.037$$

The overall cooling requirement may be three to four times this level or nearly ~ 10–14% in a modern gas turbine engine.

Now, let us address some of the assumptions and uncertainties that entered our coolant fraction estimation. First, the cooling passage modeling with the roughness elements and turbulators (as in pins and ribs), which is also rotating (as in a rotor), is a complex matter and very difficult. It does not lend itself to a simple "rough pipe" modeling approach. Therefore, our assumed ratio of the coolant-to-gas side Stanton number of half is purely an estimate. In our heat flow balance from the gas to the coolant, we neglected the heat transfer (through conduction) to the disk as well as radiation heat transfer. This may impact the heat flux by upward of 5%. Also, the Stanton number is in general a function of the Prandtl number, Mach number, surface roughness, and rotation parameter, i.e.,

$$\mathrm{St} = f(\mathrm{Re}_x, \mathrm{Pr}, M, \text{roughness, rotation}) \tag{9.77}$$

which we did not include in our estimation. Most convective heat transfer correlations have their origin in flat plate studies and thus are in the form of

$$\mathrm{St} = \text{constant} \, \mathrm{Pr}^{-n} \cdot \mathrm{Re}_x^{-m} \tag{9.78}$$

An example is the Eckert–Livingood (1953) correlation (subscript "g" stands for the gas) for a turbulent boundary layer on a flat plate with constant wall temperature

$$\mathrm{St_g} = 0.0296 \, \mathrm{Pr_g}^{-2/3} \mathrm{Re}_x^{-1/5} \tag{9.79-a}$$

and for a laminar boundary layer on a flat plate with constant wall temperature

$$\mathrm{St_g} = 0.332 \, \mathrm{Pr_g}^{-2/3} \mathrm{Re}_x^{-1/2} \tag{9.79-b}$$

The blade leading edge, or stagnation point heating, is modeled as a cylinder in cross flow. We will discuss a suitable correlation for leading-edge heating/cooling in the next section. In addition, blade rotation has a strong influence on the turbine blade heat transfer and thus deserves special attention. There are specialized heat transfer texts and reference books that should be consulted for accurate correlations of heat transfer functions of the type expressed in Eq. 9.77. Eckert (1971), Rohsenow, Hartnett, and Cho (1998), and Incropera–DeWitt (2001) are among the recommended references.

9.2.4.2 *Impingement Cooling.*

Impingement Cooling. The turbine blade stagnation point, near the leading edge, represents the highest heat flux area of the blade. A typical heat flux distribution on a turbine blade (for a given free-stream turbulence intensity) is presented in Fig. 9.41. "s" is a natural coordinate measuring the surface length from the leading edge on the suction and pressure surfaces. Due to longer length of the suction surface, as compared with the pressure surface, and a different location of the transition point on the two sides of the blade, the heat flux graph looks lob-sided.

An important observation (from Fig. 9.41) is that the highest heat flux occurs at the leading edge, or the stagnation point heating in a turbine blade is the most critical. The second message is the rapid rise of heat transfer due to boundary layer transition from laminar to turbulent. The third observation is the *curvature switch* from convex to concave on the suction and pressure surfaces, respectively, thus affecting the transition point on the blade. In the theory of curved viscous flows, a convex curvature has a stabilizing effect on the flow, whereas a concave curvature has a destabilizing effect. The concave curvature case leads to the appearance of streamwise Goertler vortices that cause an enhanced mixing of the flow at the surface. The free-stream turbulence intensity Tu also enhances the heat transfer to a surface in two ways; (1) it promotes earlier transition and (2) it enhances mixing at the surface. An accepted correlation for leading-edge heat transfer finds its roots in a cylinder in cross flow problem (see Colladay, 1975), which is

$$h_{\mathrm{g,le}} = a \left[1.14 \frac{k_\mathrm{g}}{D} \mathrm{Re}_\mathrm{D}^{1/2} \mathrm{Pr}^{0.4} \left(1 - \left| \frac{\varPhi}{90} \right|^3 \right) \right] \qquad -80^\circ < \varPhi < 80^\circ \tag{9.80}$$

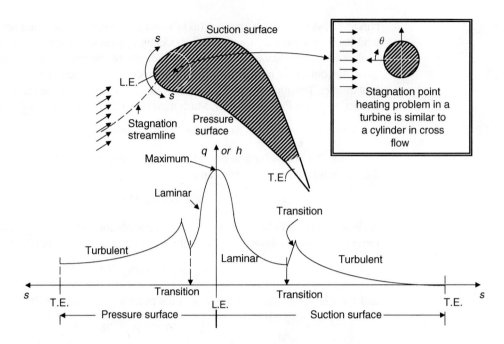

■ **FIGURE 9.41**   Schematic drawing of heat flux q (or heat transfer coefficient h) distribution on an uncooled turbine blade (for a free-stream turbulence intensity)

where

   a augmentation factor (from 1.2 to 1.8, based on free-stream turbulence)

   D diameter of leading-edge circle

   Φ angular distance from the leading-edge stagnation point, in degrees

To effectively cool the leading edge of a turbine blade, the internal cooling passage at the nose "showers" the leading edge with the coolant through a series of holes, as shown in Fig. 9.42. Since, the angle of impact between the coolant and the surface is nearly normal, hence, the name "impingement" is attributed to this type of cooling. Some of the coolant that enters the

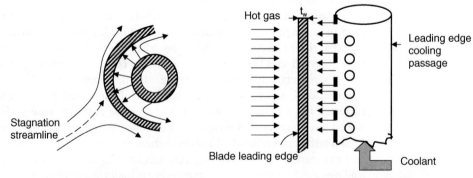

(a) The top view of the nose cooling passage with radial holes for the coolant to "impinge" on the hot surface

(b) The side view of the impingement cooling along the turbine leading edge with the holes in the radial direction

■ **FIGURE 9.42**   Schematic drawing of an impingement cooling scheme suitable for the leading edge of a turbine blade

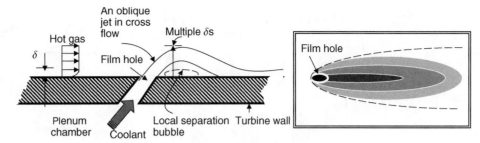

(a) Side view of a film hole used in film cooling

(b) Top view of the same film hole
the contours are for constant heat flux

■ **FIGURE 9.43** Coolant ejection from a film hole on a turbine blade

leading-edge channel may discharge through the blade tip or it may be confined to within the blade. The example shown in Fig. 9.42 (b) has sealed off the nose channel exit, thus the entire coolant is used in the impingement cooling of the blade leading edge. To study heat transfer correlations with impingement cooling, Kercher and Tabakoff (1970) may be consulted.

9.2.4.3 *Film Cooling.*

The most critical areas of a turbine blade may be film cooled through a row of film-cooling holes. The coolant is ejected through a hole at an angle with respect to the flow, which in turn bend and cover a portion of the surface with a "blanket" of coolant.

Figure 9.43 (a) shows a slanted jet emerging from a surface at an angle. Note the scale of the gas boundary layer thickness δ as depicted in Fig. 9.43 (a), and compare it to the penetration of the coolant jet in the hot gas flow. The coolant jet penetrates the hot gas free stream (i.e., inviscid core) and is deflected by the external forces in the free stream. The penetration of the slanted film columns in the free stream and the associated local flow separation immediately downstream of the film hole causes the profile drag of turbine blades to increase. Therefore, the film cooling of turbine blades is more disruptive to the external aerodynamics of the blades, as compared with internal cooling scheme. The contours of constant heat flux are shown in Fig. 9.43 (b).

The cooling effect of the ejected jet that emerges from the film hole covers only a small region in the immediate vicinity of the ejection hole. For this reason, practical film cooling in gas turbines involves numerous film holes in one or multiple (staggered) rows to cover a significant portion of a surface. Figure 9.45 shows a staggered array of two rows of film holes with *typical* length scales, i.e., hole diameter D and spacing (or pitch P) that are noted on the graph.

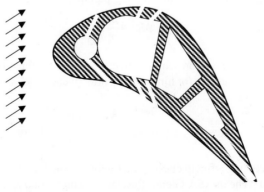

■ **FIGURE 9.44** Schematic drawing of a film-cooled turbine blade (with six film holes)

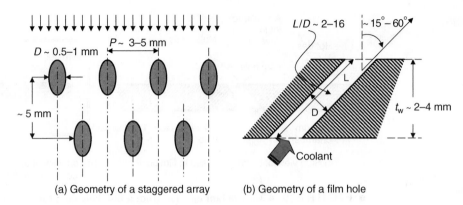

(a) Geometry of a staggered array   (b) Geometry of a film hole

■ **FIGURE 9.45** Definition sketch used in film cooling and some *typical* scales

The diameter of the film holes range ~ 0.5–1 mm. Although it is possible to reduce the film-hole diameter to below 0.5 mm using advanced manufacturing techniques (e.g., using electron beam), in practice such small holes are prone to clogging, especially in the gas turbine environment. The products of combustion include particulates and by-products that *cling* to surface and clog the film holes. It is intuitively expected that the nondimensional pitch–diameter ratio of the film hole P/D to be an important parameter in heat transfer, thus film-cooling effectiveness. Also, the length–diameter ratio L/D for the film hole is important in the lateral spread of the film, and the size of the local separation bubble immediately downstream of the film hole. To add to the complexity of the film cooling, we note that the shape of the coolant plenum chamber also impacts the coolant jet velocity distribution and its mixing with the free stream and thus important to the film-cooling effectiveness. A specialized reference on the fundamentals of film cooling and gas turbine heat transfer is the von Karman Lecture Series (VKI-LS 1982).

A film-cooling effectiveness parameter η_f may be defined as

$$\eta_f \equiv \frac{T_g - T_{\text{aw-f}}}{T_g - T_c} \tag{9.81}$$

where the only new temperature in the equation, i.e., $T_{\text{aw-f}}$, is the adiabatic wall temperature in the presence of the film and excluding other cooling effects. Note that the true adiabatic wall temperature with film cooling, $T_{\text{aw-f}}$, is a very difficult parameter to measure. Therefore, it is possible (and preferable) to define a film-cooling effectiveness parameter that utilizes a different and more easily measured temperature in the experiment, e.g., the actual wall temperature, in the presence of film cooling. The Stanton number for a film-cooled blade involves an additional "blowing parameter" M_b with typical values for low and high blowing rates are 0.5 and 1.0, respectively. The blowing parameter is defined as

$$M_b \equiv \frac{\rho_c u_c}{\rho_g u_g} \tag{9.82}$$

Considering all of the arguments presented above, we expect the functional form of the Stanton number for a film-cooled surface (or film-cooling effectiveness) to be represented by (at least) the following parameters:

$$\text{St} = f\left(\text{Re}_x, \text{Pr}, M, M_b, \frac{P}{D}, \frac{L}{D}, \text{Tu}, \text{roughness}, \text{rotation}, \text{plenum} - \text{geometry}\right) \tag{9.83}$$

Research on film-cooling effectiveness is actively pursued in the laboratory and in the computational field. The NASA-Glenn Research Center conducts film-cooling research

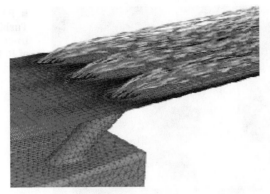

■ **FIGURE 9.46** **A computational research in film cooling (2005) (From Fluent Inc., www.fluent.com)**

■ **FIGURE 9.47**
Definition sketch for the transpiration cooling scheme

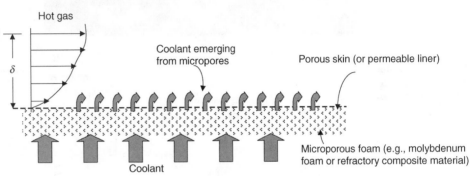

in-house, works with universities as collaborators as well as industry. Their Web site www. nasa.gov/centers/glenn/home/index.html should be used as a resource for the latest research in aircraft gas turbine engines. Figure 9.46 shows a 3D computational mesh used in film-cooling research (from Fluent Inc., 2005).

9.2.4.4   Transpiration Cooling. The coolant may emerge from very small pores ($\sim$10–100 μm) of a porous surface and thus be embedded entirely within the viscous boundary layer of the gas turbine blade. This is analogous to human perspiration as a means of cooling and is known as *transpiration cooling*. The technique involves pumping a coolant through microporous foam, which is bonded to a porous outer skin, as shown in Fig. 9.47.

The appeal of transpiration cooling is in its effectiveness with minimal coolant mass flux requirement (see Wang, Messner, and Stetter, 2004). The disadvantage of the scheme is in its impracticality of keeping the micropores unclogged in a gas turbine environment. Other material characteristics such as oxidation resistance, material life, and manufacturing costs all impact the practicality of transpiration cooling for an aircraft gas turbine engine. From fluid mechanics point of view, the static pressure drop across the porous foam is large (per unit mass flux); hence, the pressurized coolant requirement is more stringent for a transpiration-cooled surface as compared with film cooling. Figure 9.48 shows different cooling schemes from Rolls-Royce.

9.3   Turbine Performance Map

As demonstrated in the earlier part of this chapter, the performance of a turbomachinery stage is fully determined by two parameters; (1) the axial Mach number or equivalently the corrected mass flow rate and (2) the tangential blade Mach number, or equivalently the corrected shaft speed. The turbine performance map is thus a graph of $1/\pi_t$ versus the corrected mass flow rate

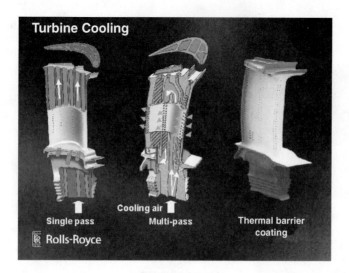

■ **FIGURE 9.48**
Turbine cooling schemes
from Rolls-Royce (courtesy
of Rolls-Royce, plc)

■ **FIGURE 9.49**
**Typical turbine
performance map**

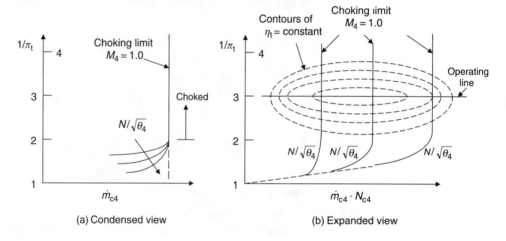

(a) Condensed view   (b) Expanded view

$\dot{m}_{c4}$ for the corrected shaft speeds N_{c4}. Typical turbine performance maps are shown in Fig. 9.49. In part (a), the choking limit (i.e., $M_4 = 1.0$) is approached with the increase in corrected shaft speed. In part (b), however, we may graph the product of the corrected mass flow rate and the corrected shaft speed in order to (graphically) separate the individual choking limits. In addition, the contours of constant turbine adiabatic efficiency are superimposed (dashed lines).

9.4   The Effect of Cooling on Turbine Efficiency

The impact of cooling on turbine efficiency may be attributed to the following effects:

1. The coolant mass flow rate does not participate in turbine power production, therefore, per 1% cooling fraction there is ~1% loss in power production (due to loss of working fluid)

2. The coolant injection in the hot gas stream causes mixing losses of two streams as well as an increase in profile drag loss on the blades (for disruption of flow on the blades, as in film holes)

3. The coolant suffers a total pressure drop inside the cooling passages due to friction, turbulators, pins, and other roughness elements inside the cooled turbine blades. The reduced total pressure on the part of the coolant then causes the stage total pressure ratio π_t to drop

4. The heat transfer between the hot gas and the coolant causes an entropy rise for the mixed-out gas

The turbine efficiency may be defined as the ratio of actual turbine work per total airflow (that includes the coolant fraction) and the ideal turbine work, achieved isentropically, across the actual turbine expansion, $(p_{t5}/p_{t4})_{\text{actual}}$.

The actual turbine work (per unit mass flow) for the two streams is the sum of individual streams reaching the same exit total temperature T_{t5}

$$w_{t,\text{actual}} = (1 - \varepsilon)c_{pt}(T_{t4} - T_{t5}) + \varepsilon c_{pc}(T_{tc} - T_{t5}) \tag{9.84}$$

The ideal (i.e., isentropic) turbine work for the two streams expanding through the actual pressure ratio is

$$w_{t,\text{ideal}} = (1 - \varepsilon)c_{pt}T_{t4}\left[1 - (p_{t5}/p_{t4})^{(\gamma_t-1)/\gamma_t}\right] + \varepsilon c_{pc}T_{tc}\left[1 - (p_{t5}/p_{t4})^{(\gamma_t-1)/\gamma_t}\right] \tag{9.85}$$

Therefore, the cooled turbine efficiency may be written as

$$\eta_t = \frac{(1 - \varepsilon)c_{pt}(T_{t4} - T_{t5}) + \varepsilon c_{pc}(T_{tc} - T_{t5})}{[(1 - \varepsilon)c_{pt}T_{t4} + \varepsilon c_{pc}T_{tc}][1 - (p_{t5}/p_{t4})^{(\gamma_t-1)/\gamma_t}]} \tag{9.86}$$

Kerrebrock (1992) shows that turbine efficiency in Eq. 9.86 may be approximated by

$$\eta_t \approx 1 - \left(\frac{\tau_t}{1 - \tau_t}\right)\left[2\sigma\text{St}\left(\frac{T_{t4}}{T_{tc}} - 1\right)\left(1 - \frac{T_w}{T_{t4}}\right) + \left(\frac{\gamma}{\gamma - 1}\right)\varepsilon\left(\frac{\Delta p_f}{p_{t4}}\right)\right] \tag{9.87}$$

Where σ is the blade solidity, Stanton number is St, and Δp_f is the total pressure loss due to friction inside the blade cooling passages. Kerrebrock estimates the loss of turbine efficiency to be ~2.7% per percent of cooling flow for a typical gas turbine, based on Eq. 9.87. Accounting for other sources of loss as in kinetic energy loss in film-cooled blades, Kerrebrock estimates an additional 1/2% to be added to the 2.7% to get an estimated 3.2% turbine efficiency loss per percent of cooling flow. More experimental data and validation are needed to cover a wide array of internal cooling configurations and internal/external loss estimation. Figure 9.50 shows the evolution of turbine blade cooling (courtesy of Rolls-Royce plc, 2005).

■ **FIGURE 9.50**
The evolution of turbine blade cooling (courtesy of Rolls-Royce plc, 2005)

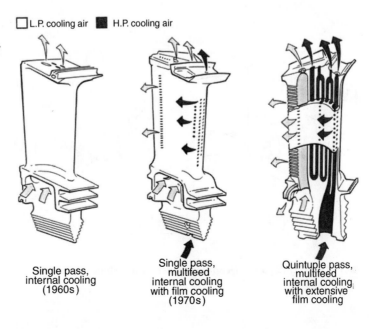

☐ L.P. cooling air   ■ H.P. cooling air

Single pass, internal cooling (1960s)

Single pass, multifeed internal cooling with film cooling (1970s)

Quintuple pass, multifeed internal cooling with extensive film cooling

9.5 Turbine Blade Profile Design

A definition sketch of a turbine cascade is shown in Fig. 9.51. The basic parameters are the same as the compressor cascade. For example, the net flow turning is the difference between the inlet and exit flow angles (β_1 and β_2), or the camber angle (φ) is defined as the sum of the angles of the tangent to the mean camber line at the leading and trailing edge (κ_1 and κ_2). Also note that the deviation angle is the flow angle beyond the tangent to the mean camber line at the trailing edge (δ^*). The blade setting or stagger angle is defined the same way as in a compressor cascade γ°. The blade chord and spacing are the same as c and s, respectively. Now, let us examine some of the distinguishing features in a turbine cascade, such as an inlet induced flow angle $\Delta\theta_{ind}$, or throat opening o, or the suction surface curvature e, downstream of the throat, and finally the trailing-edge thickness $t_{t.e.}$.

9.5.1 Angles

In order to construct a suitable turbine blade profile, we need to estimate the inlet and exit blade angles and their relation to the actual incidence and deviation angles. The incidence angle in a turbine cascade accounts for the flow curvature near the leading edge, called the *induced turning*, $\Delta\theta_{ind}$ and is called *actual incidence* i_{ac}. The correlation between the induced angle, inlet flow angle, and blade solidity is (from Wilson and Korakianitis, 1998)

$$\Delta\theta_{ind} = 14\left(1 - \frac{\beta_1}{70^\circ}\right) + 9(1.8 - \sigma) \qquad 0 \le \beta_1 \le 70^\circ \tag{9.88}$$

The actual incidence and flow angles are corrected by the induced angle according to

$$i_{ac} = i + \Delta\theta_{ind} \tag{9.89}$$

$$\beta_{ac} = \beta_1 + \Delta\theta_{ind} \tag{9.90}$$

The flow and turbine blade angles at the leading edge follow the same relation as the compressor, i.e.,

$$i = \beta_1 - \kappa_1 \tag{9.91}$$

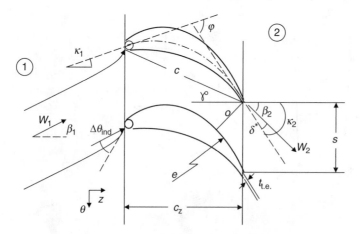

■ **FIGURE 9.51** Definition sketch of a turbine cascade

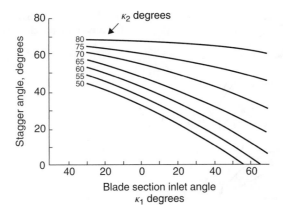

■ **FIGURE 9.52** Turbine blades stagger in relation to blade leading and trailing-edge angles (from Kacker and Okapuu, 1981)

The deviation angle is also important to the turbine profile design, as it *adds* to the blade camber, and if it is underpredicted, the exit swirl will be less than the design value and thus blade torque and in case of rotor, power, will be less than expected. Carter's rule for deviation angle in a turbine, although not the most accurate, is adequate for the preliminary design purposes,

$$\delta^* = \frac{m\Delta\varphi}{\sigma} \approx \frac{\Delta\varphi}{8\sigma} \approx \frac{\Delta\beta}{8\sigma} \qquad \textbf{(9.92)}$$

Stagger, or blade setting, angle is critical to the *smoothness* of turbine flow passage (area distribution) design. The simple approximation equates the stagger to the mean flow angle in a blade row, i.e.,

$$\gamma \approx \beta_m \approx \tan^{-1}(W_{\theta m}/C_z) \quad \text{(rotor)} \qquad \textbf{(9.93-a)}$$

$$\gamma \approx \alpha_m \approx \tan^{-1}(C_{\theta m}/C_z) \quad \text{(nozzle)} \qquad \textbf{(9.94-b)}$$

A more accurate determination of the stagger angle, based on the blade leading and trailing-edge angles, κ_1 and κ_2 is shown in Fig. 9.52 (from Kacker and Okapuu, 1981).

9.5.2 Other Blade Geometrical Parameters

Conventional turbine blade passages have their throat at the exit, as shown in Fig. 9.51. It is desirable to expand the flow beyond the throat on the suction surface, i.e., provide a convex curvature beyond the throat. This geometrical feature is advantageous to favorable pressure gradient and thus smaller deviation angle. The nondimensional radius of curvature s/e characterizes the convex curvature. The upper value for the convex curvature parameter s/e is 0.75 with typical range corresponding to $0.25 \leq s/e \leq 0.625$. Also note that the pressure surface at the trailing edge assumes a concave curvature of radius $\sim(e + o)$.

In a turbine, the blade leading-edge radius $r_{l.e.}$ is critical to effective cooling and thus blade life. The value of nondimensional leading-edge radius $r_{l.e.}/s$ is between 0.05 and 0.10.

The trailing-edge thickness $t_{t.e.}$, in a turbine is finite. The main reasons are structural integrity as well as trailing-edge coolant slots. The trailing-edge thickness adversely impacts the flow blockage and blade profile losses. Thus, we wish to minimize the trailing-edge thickness consistent with the blade structural and cooling requirements. The typical nondimensional values of $t_{t.e.}/c$ fall between 0.015 and 0.05.

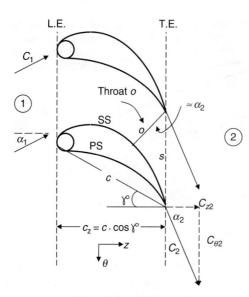

■ **FIGURE 9.53** Definition sketch used for a turbine nozzle throat sizing

9.5.3 Throat Sizing

The throat sizing in a turbine nozzle (or rotor) is very important both for choked and unchoked nozzles. The geometry that is shown in the following definition sketch (Fig. 9.53) is used to relate the throat width or opening o to the blade spacing s.

The throat opening o is related to the spacing and cosine of the exit flow angle α_2 in nozzle and β_2 in rotor, following

$$o \approx s \cdot \cos \alpha_2 \tag{9.94}$$

This approximation is acceptable for the subsonic exit flow, however, for the supersonic exit Mach numbers (but below 1.3), we correct the throat opening by the inverse of A/A^* corresponding to the supersonic exit Mach number, i.e.,

$$\frac{o}{s} \approx \frac{\cos \alpha_2}{(A/A^*)_{M_2}} \tag{9.95}$$

The design exit Mach number for the first turbine nozzle should slightly exceed 1, i.e., $M_2 > 1$, and is commonly taken to be ~ 1.1. The exit Mach numbers from the subsequent blades (in relative-to-blade frame of reference) in a turbine should remain below 1, i.e., unchoked. For example, the design Mach number at the first rotor exit M_{3r} is chosen to be as high as 0.90, but never 1 or above. Also, all subsequent blades in a multistage turbine, on the same spool, remain unchoked. For multispool gas turbines, the first nozzle, on all spools, is choked and its design exit Mach number is ~ 1.1.

9.5.4 Throat Reynolds Number Re_o

The throat Reynolds number should preferably be in the range of 10^5 to 10^6. Experimental data demonstrate a strong correlation between blade profile loss and the throat Reynolds number. The definition of throat Reynolds number uses the relative exit flow velocity from the blade and the static conditions at the throat, i.e.,

$$Re_o \equiv \rho_{th} C_2 o_n / \mu_{th} \quad \text{(Nozzle)} \qquad 10^5 \leq Re_o \leq 10^6 \tag{9.96-a}$$

$$Re_o \equiv \rho_{th} W_3 o_r / \mu_{th} \quad \text{(Rotor)} \qquad 10^5 \leq Re_o \leq 10^6 \tag{9.96-b}$$

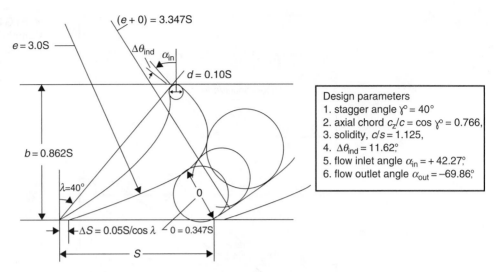

■ **FIGURE 9.54**   Example of a turbine blade profile design (from Wilson and Korakianitis, 1998)

The throat opening o is given a subscript n and r in the previous definitions, to signify the nozzle and rotor throat openings, respectively.

9.5.5 Turbine Blade Profile Design

We have identified some definite structure for the turbine profile at and beyond the throat. For example, we have the throat opening o/s related to exit flow angle and Mach number, or we have a range for the trailing-edge thickness, also a curvature on the suction side and a curvature on the pressure side, all near the trailing edge. At the leading edge, we have some design guidelines for the leading-edge radius, and some correlations for the induced flow turning, besides the flow angles at the inlet and exit. The stagger angle is also estimated using Eq. 9.93 or Fig. 9.51.

Once the trailing-edge passage beyond the throat is constructed and the leading-edge radius (or a range of radii) is chosen, the trial-and-error phase of curve fitting to the upper and lower surfaces begins. The goal is to produce a flow passage that *smoothly and uniformly* contracts to the throat section. Therefore, beyond the trailing-edge construction of the turbine blade profile, the rest of the approach deals with flow passage design (i.e., with a smooth area contraction).

Wilson and Korakianitis (1998) constructed the following turbine profile (Fig. 9.54) based on the input shown in the box and the methodology of this section. Schobeiri (2004) also provides details in the construction of turbine profiles and is recommended for further reading.

9.5.6 Blade Vibration and Campbell Diagram

Campbell diagram is of interest because it shows possible matches between blade vibrational mode frequency and multiples of shaft rotational speed. The multiples of shaft rotational speeds are caused by the struts and blades (wakes) in neighboring rows and they serve as the source of *excitation*. In essence, the blade passing frequency, which is the product of the number of blades times the shaft frequency, is the source of excitation for the blades in the next/previous row. Vibration frequency in kilohertz for the first two bending and the first two torsional modes is shown in Fig. 9.55 (from Wilson and Korakianitis, 1998) to vary with rotor shaft speed (in rpm), due to the so-called stiffening effect that rotation has on a structure. The

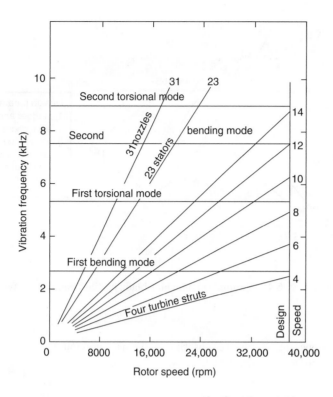

Campbell frequency diagram for a turbine blade (from Wilson and Korakianitis, 1998)

design shaft speed is also identified on the chart (to be ∼37,000 rpm). The straight lines corresponding to multiple shaft speeds are drawn. The first or fundamental bending mode, known as the *first-flap* mode, has a natural frequency that lies between the fourth and sixth multiples of shaft rpm at design speed. Since the fifth multiple of shaft speed lies halfway between the fourth and sixth, we note that the first bending mode is below the fifth multiple of shaft speed at design rpm. Closer examination of Fig. 9.54 also indicates that 4, 23, and 31 multiples of shaft speeds have a special significance in this turbine (rotor) blade row. These correspond to the number of struts and stator or nozzle blades that serve as the excitation source for the rotor through blade passing frequency of wakes and mutual interference effects of their rotating pressure fields.

The structural design of blades should clearly indicate a resonance-free operating condition at the design speed, idle speed, and other operational speeds where significant time is spent. However, it is impossible to avoid all resonant frequencies as we speed up to the design, or other operational shaft speeds. Therefore, spool-up speed/acceleration, or the spool-down speed/deceleration are important to the cyclic loads and fatigue life of blades, struts, disks, and other engine components.

9.5.7   Turbine Blade and Disk Material Selection and Design Criteria

Turbine blade and disk materials and the year of development are shown in Fig. 9.56 (from Wilson and Korakianitis, 1998). These, by necessity, are high-temperature materials. Some have high thermal conductivity, as in nickel-based alloys for the blades and disks thermal stress alleviation, and some are low thermal conductivity materials, such as ceramics, that reduce heat transfer to the blades. All materials, especially for turbine blades, use a thermal

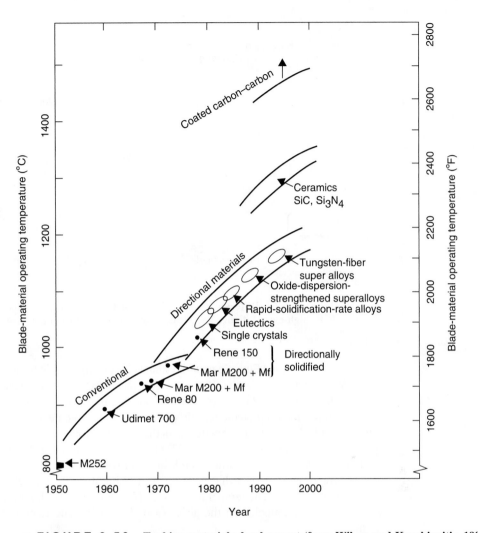

■ FIGURE 9.56   Turbine materials development (from Wilson and Korakianitis, 1998)

protection coating to reduce the surface operating temperature and thus in effect increase component life.

There are four clusters of materials that are labeled in Fig. 9.56. The *conventional* alloys exhibit the lowest operating temperature capability, whereas *directional materials* that include single crystals, rapid solidification rate alloys, oxide-dispersion-strengthened superalloys, and Tungsten–fiber superalloys achieve high temperature capability. The ceramics as in silicon carbide offer an additional tolerance to high temperature. The *coated carbon–carbon* composite material offers the highest temperature capability but the issues of cost, damage tolerance, inspectability, and reliability hamper its use in large operational gas turbine engines. The turbine design example at the end of this chapter uses a design blade surface temperature of 1200 K, which based on Fig. 9.56 implies the use of directionally solidified material.

Different parts of turbine blade and disk are subject to different mechanical design criteria, as shown in Fig. 9.57 from Wilson and Korakianitis (1998). Mechanical designs of turbine components address low- and high-cycle fatigue, oxidation/corrosion, and creep rupture problems.

The number of cycles to failure for conventionally cast and directionally solidified material shows the fatigue strength of Rene 80, directionally solidified Rene 150, and

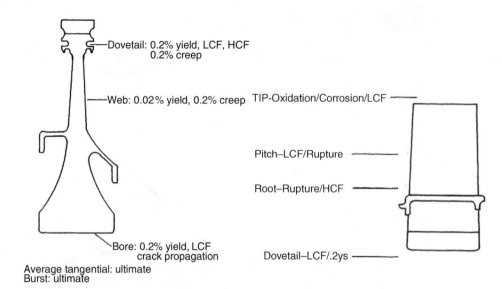

■ **FIGURE 9.57** **Different design criteria for turbine blade and disk (LCF: low-cycle fatigue HCF: high-cycle fatigue, YS: yield strength) (from Wilson and Korakianitis, 1998)**

directionally solidified Eutectic material (that are typically used in turbine blades) in Fig. 9.58 (from Wilson and Korakianitis, 1998). We note that directional solidification improves fatigue strength by a factor of 2 or 3 over conventionally cast Rene 80 as shown in Fig. 9.58.

Another material characteristic of interest to turbine designers is the *creep rupture strength*. It is the maximum tensile stress that material tolerates without failure over a time period at a given temperature. The 0.2% creep design criterion is listed for the turbine disk web, blade slot in the hub, blade pitchline, and root. This 0.2% creep rupture strength is plotted for three materials that are used in turbine disks in Fig. 9.59 (from Wilson and Korakianitis, 1998). The temperature (in Kelvin) and time (in hours) are combined in the Larson–Miller parameter on the abscissa of Fig. 9.59. The Larson–Miller parameter is defined as

$$T(\text{K}) \cdot [\text{C} + \log t(\text{hours})] \tag{9.97}$$

Where C is constant for a material (in this case, 25).

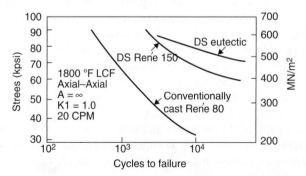

■ **FIGURE 9.58** **Fatigue strength of turbine blade material (from Wilson and Korakianitis, 1998) (DS: directionally solidified)**

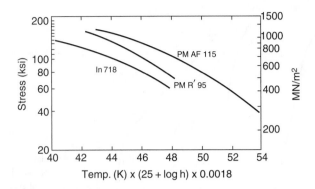

■ **FIGURE 9.59** 0.2% *Creep strength of turbine disk alloys (from Wilson and Korakianitis, 1998)*

The treatment of the turbine material selection and design criteria in this section has been by necessity very brief. We have not even scratched the surface of the vast and specialized field of gas turbine high-temperature materials and mechanical design. The reader is to refer to specialized texts and references on the subject.

9.6   Stresses in Turbine Blades and Disks and Useful Life Estimation

Turbine blades and disks are subjected to centrifugal stresses due to shaft rotation as well as thermal stresses due to temperature differentials in the material due to cooling, gas bending stresses due to gas loads, and vibratory stresses due to cyclic loading and blade vibration. The centrifugal stresses take on the same form as the one developed in the compressor section.

The dominant stress in a rotor and disk is the centrifugal stress σ_c. At the blade hub, the ratio of the centrifugal force F_c to the blade area at the hub A_h is the centrifugal stress, σ_c,

$$\sigma_c \equiv \frac{F_c}{A_{hub}} \tag{9.98}$$

$$F_c = \int_{r_h}^{r_t} \rho_{blade} \cdot A_b(r)\omega^2 r dr \tag{9.99}$$

$$\sigma_c = \frac{1}{A_h} \int_{r_h}^{r_t} \rho_{blade} \cdot A_b(r)\omega^2 r dr \tag{9.100}$$

$$\frac{\sigma_c}{\rho_{blade}} = \frac{\omega^2}{A_h} \int_{r_h}^{r_t} A_b(r) r dr = \omega^2 \int_{r_h}^{r_t} \frac{A_b}{A_h} r dr \tag{9.101}$$

$$\frac{\sigma_c}{\rho_{blade}} = \omega^2 \int_{r_h}^{r_t} \frac{A_b}{A_h} r dr \tag{9.102}$$

The blade area distribution along the span $A_b(r)/A_h$ is known as *taper* and is often approximated to be a linear function of the span. Therefore, it may be written as

$$A_b = A_h - \frac{r - r_h}{r_t - r_h}(A_h - A_t) \implies \frac{A_b}{A_h} = 1 - \frac{r - r_h}{r_t - r_h}\left(1 - \frac{A_t}{A_h}\right) \tag{9.103}$$

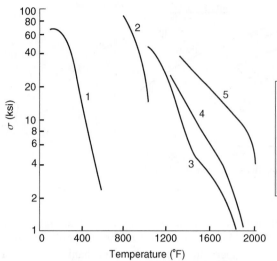

Material	No.
Aluminum alloy	1
Titanium alloy	2
Wrought nickel alloy	3
High-strength nickel alloy	4
Single-crystal superalloy	5

■ **FIGURE 9.60** **Allowable stress versus temperature for typical engine materials (from Mattingly, Heiser, and Pratt, 2002)**

We may substitute $A_b(r)/A$ in the integral and proceed to integrate; however, a customary approximation is often introduced that replaces the variable r by the pitchline radius r_m. The result is

$$\frac{\sigma_c}{\rho_{blade}} = \frac{\omega^2 A}{4\pi}\left(1 + \frac{A_t}{A_h}\right) \tag{9.104}$$

Therefore, the ratio of centrifugal stress to the material density is related to the square of the angular speed, the taper ratio, and the flow area, $A = 2\pi r_m(r_t - r_h)$. This equation is the basis of the so-called AN^2 rule, i.e., where A is the flow area and N is the shaft angular speed, often expressed in the customary unit of rpm. The right-hand side of Eq. 9.104 incorporates turbomachinery size (throughflow area) and the impact of angular speed, whereas the left-hand side of Eq. 9.104 is a material property known as the (tensile) specific strength.

The material parameter of interest in a rotor is the *creep rupture strength*, which identifies the maximum tensile stress tolerated by the material for a given period of time at a specified operating temperature. Based on the 80% value of the allowable 0.2% creep in 1000 h, for aluminum alloys and the 50% value of the allowable 0.1% creep in 1000 h for other materials, Mattingly, Heiser, and Pratt [28] have graphed Figs. 9.60 and 9.61. They show the allowable stress and the allowable specific strength of different engine materials as a function of temperature. Note that the unit of stress in the following two figures is ksi, which is 1000 lb per square inch (i.e., 1000 psi) and the temperatures are expressed in degree Fahrenheit.

Thermal stresses are calculated based on thermal strains that are set up in a material with differential temperature ΔT from

$$\varepsilon_t = \alpha \Delta T \tag{9.105}$$

where ε_t is thermal strain (i.e., elongation per unit length) and α is the coefficient of (linear) thermal expansion, which is a material property. The linear stress–strain relationship demands

$$\sigma_t = E\varepsilon_t = \alpha E \Delta T \tag{9.106}$$

where E is the modulus of elasticity. The thermal stresses in a disk of constant thickness with no center hole and with the radius r_h is a simple model of a turbine disk. If the disk has a

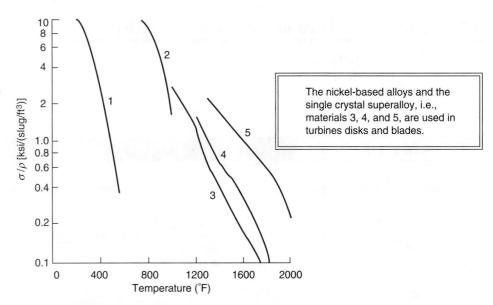

The nickel-based alloys and the single crystal superalloy, i.e., materials 3, 4, and 5, are used in turbines disks and blades.

linear temperature distribution in the radial direction, the thermal stresses in radial and tangential directions are shown (by Mattingly, Heiser, and Pratt, 2002) to be

$$\sigma_{tr} = \frac{\alpha E \Delta T}{3}\left(1 - \frac{r}{r_h}\right) \tag{9.107-a}$$

$$\sigma_{t\theta} = \frac{\alpha E \Delta T}{3}\left(1 - 2\frac{r}{r_h}\right) \tag{9.107-b}$$

The maximum of both stresses occur at the center of the disk, i.e., $r = 0$, and for typical values of coefficient of thermal expansion of nickel-based alloys, $\alpha \sim 10.2 \times 10^{-6}$ in/in.°F at 1400 °F (that corresponds to gas turbine temperatures), as well as the modulus of elasticity E for nickel alloys (at 1400 °F) is $\sim 20.5 \times 10^6$ psi, which for a 100 °F temperature differential gives thermal stresses in the radial and tangential directions of about ~ 6970 psi (equivalent to 1011 kPa). This example illustrates the need for special attention to the thermal stresses in turbine disks and material property that is of utmost interest is the thermal conductivity. The transient operations of the gas turbine that expose the turbine disk to high temperatures at the rim while the center of the disk is still at a low temperature pose the highest levels of thermal stress. Nickel alloys have the highest levels of thermal conductivity, in metals, at high temperatures, which are suitable for use in turbine disks and blades.

The total stress in the material, σ_{total}, which is the sum of the centrifugal, bending (i.e., gas loads), thermal and vibratory stresses, and the material operating temperature combine to estimate material (useful) life. There are stress–temperature–material life curves, for a variety of materials, in gas turbine industrial practice. Figure 9.62 shows an example of a family of stress–temperature–life curves for any given material, e.g., nickel alloys.

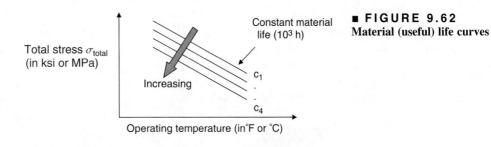

■ **FIGURE 9.62**
Material (useful) life curves

9.7   Axial-Flow Turbine Design and Practices

In this section, we apply some of the concepts that we learned in the chapter to the preliminary design of a cooled gas turbine. The approach to turbine design is as varied and diverse as the textbooks written on the subject. Therefore, there is no *unique* approach to turbine design, rather the author's preference in applying a set of guidelines to turbine design.

EXAMPLE 9.7

We intend to design an axial-flow turbine to drive a particular compressor (or other external loads, such as an electric generator). Therefore the mass flow rate (at the design point) through the machine is established by the compressor–combustor in the gas turbine engine. Also, the shaft connecting the compressor to the turbine, on a given spool, rotates at the same speed. In addition to these two constraints, we have the third input to the turbine design problem, which is the combustor exit temperature T_{t4}. The fourth input is the combustor exit (total) pressure, which is dictated by the compressor, i.e., cycle, pressure ratio. Therefore, at a minimum, we have the following problem statement:

Design an axial-flow turbine for a gas generator with

$$\dot{m}_1 = 93\,\text{kg/s} \quad \Longleftarrow \quad \text{Turbine inlet mass flow rate}$$
$$\omega = 8770\,\text{rpm} \quad \Longleftarrow \quad \text{Shaft rpm}$$
$$T_{t1} = 2000\,\text{K}$$
$$\quad \Longleftarrow \quad \text{Turbine inlet (total) temperature}$$
$$p_{t1} = 1960\,\text{kPa} \qquad \text{and pressure}$$

$$r_{m1} = 0.37\,\text{m} \quad \Longleftarrow \quad \text{Pitchline radius at the compressor inlet}$$

In addition, there is a specification that 10 kg/s of air was bled off at the compressor exit to provide turbine cooling. The coolant information is thus

$$\dot{m}_c = 10\,\text{kg/s}$$
$$T_{tc} = 771\,\text{K} \quad \Longleftarrow \quad \begin{array}{l}\text{Coolant characterization}\\ \text{(bleed at the compressor exit)}\end{array}$$
$$p_{tc} = 2020\,\text{kPa}$$

The design ambient pressure and temperature and flight Mach number are also known, i.e., we have

$$M_0 = 0$$
$$p_0 = 101\,\text{kPa} \quad \Longleftarrow \quad \text{Design altitude and Mach number}$$
$$T_0 = 25°\text{C}$$

Finally, we have (or assume) gas characteristics (γ, c_p) at each station in the turbine. For the starters, we choose

$$c_{pc} = 1004.5\,\text{J/kg} \cdot \text{K}, \gamma_c = 1.40$$
$$\quad \Longleftarrow \quad \begin{array}{l}\text{Gas}\\ \text{properties}\end{array}$$
$$c_{pt} = 1243.67\,\text{J/kg} \cdot \text{K}, \gamma_t = 1.30$$

SOLUTION

Preliminary design of a cooled turbine
We start with the turbine shaft power that is consumed by the compressor, bearings, and other external loads. Let us assume that turbine provides shaft power only to the compressor.

$$\wp_t = \frac{\wp_c}{\eta_m}$$

The role of mechanical efficiency η_m term in the power balance equation is to account for power loss in bearings and other external sources, e.g., electric generator, requiring turbine shaft power. Here we assume that the mechanical efficiency is ~0.995. Also from the compressor side, we

calculate the shaft power absorbed by the compressor (based on cycle station numbers),

$$\wp_c \cong \dot{m}_0 c_{pc}(T_{t3} - T_{t2})$$

The air mass flow rate in the compressor is ~100 kg/s with an inlet and exit total temperatures of 298 and 771 K, respectively (i.e., for a compressor pressure ratio of 20 and polytropic efficiency of $e_c = 0.90$). The shaft power consumed by the compressor is

$$\wp_c = 47.489\,\text{MW}$$

Therefore, the turbine shaft power required is

$$\wp_t \cong 47.728\,\text{MW}$$

The turbine power is related to the hot gas and coolant expansion to the same exit temperature $T_{t5\text{-cooled}}$ as discussed in this chapter (and based on cycle station numbers), we had

$$\wp_{t,actual} = \dot{m}_0[(1 - \varepsilon)c_{pt}(T_{t4}\text{-}T_{t5\text{-cooled}})$$
$$+ \varepsilon c_{pc}(T_{tc} - T_{t5\text{-cooled}})]$$

The compressor airflow rate $\dot{m}_0$ and ε (as coolant mass fraction) are known to be 100 kg/s and $\sim 10\%$ (i.e., 0.01), respectively. We solve the turbine power equation for the unknown, $T_{t5\text{-cooled}}$, to get

$$T_{t5\text{-cooled}} \approx 1507\,\text{K}$$

The total temperature drop across the turbine is $\Delta T_t = 493\,\text{K}$. The question that we ask is how many stages of turbine expansion will it take to create a 493 K total temperature drop?

The answer depends on our stage degree of reaction design choice. We have seen that an impulse turbine produces (theoretically) twice as much power per stage than a 50% reaction turbine (for the case swirl-free turbine exit flow). But we also noted that efficiency of the impulse turbine is lower than the reaction type. In addition, note that the desire to reach a zero exit swirl condition at the turbine exit is one of the goals that we will try to match at the turbine exit but it may be nonzero for intermediate stages. In this sense, the turbine stages will not be repeated. The only repetition from one stage to another seems to be in the constant axial velocity assumption that we introduce in our design strategy. Also, our calculations are limited to the turbine pitchline radius.

First turbine nozzle pitchline design (C_z – constant design)
The maximum turning angle α_2 appears at the nozzle exit Mach number of ~ 1. Since the first turbine nozzle has to operate in a choked state (over a wide operating range), we choose the first nozzle exit Mach number slightly supersonic, i.e.,

$$M_2 = 1.1 \quad \text{(design choice for the 1st turbine nozzle)}$$

From the stagnation temperature $T_{t1} = 2000\,\text{K}$ and the exit Mach number $M_2 = 1.1$, we calculate the static temperature at the nozzle exit,

$$T_2 = 2000\,\text{K}/[1 + 0.15(1.1)^2] \approx 1693\,\text{K}$$

The speed of sound in station 2 is

$$a_2 = 794.7\,\text{m/s}$$

Therefore the absolute gas speed at the nozzle exit is

$$C_2 = 874.2\,\text{m/s}$$

For the first nozzle exit angle of $\sim 60°$, we calculate the nozzle exit swirl and axial velocity components,

For $\alpha_2 \approx 60°$, $C_{\theta 2} \approx 757.1\,\text{m/s}$ and $C_{z2} \approx 437.1\,\text{m/s}$

Since we adopted a constant axial velocity design approach, the nozzle inlet velocity is assumed to be purely in the axial direction as well as being equal to the nozzle exit axial velocity, i.e.,

$$C_1 = C_{z1} = C_{z2} = 437.1\,\text{m/s}$$

Now, we may calculate the nozzle inlet static temperature T_1 and speed of sound a_1

$$T_1 = T_{t1} - C_1^2/2c_{pt} = 2000\,\text{K} - (437.1)^2/2/1243.67$$
$$= 1923.2\,\text{K} \quad \Longrightarrow \quad a_1 = 847.1\,\text{m/s}$$

Therefore, the gas Mach number at the nozzle entrance is: $M_1 = M_{z1} = 0.516$

First turbine nozzle optimal solidity
Zweifel related optimal (axial) solidity to a loading parameter ψ_z and relative flow angles β_1 and β_2 according to

$$\sigma_z \psi_Z = \frac{2\cos\beta_2}{\cos\beta_1}\sin(\beta_1 - \beta_2)$$

The graph of the optimal solidity parameter $\sigma_z \cdot \psi_Z$ is shown in Fig. 9.33. In the first nozzle, the inlet flow is axial, i.e., $\alpha_1 = 0$, and the exit flow angle was assumed to be $\alpha_2 \sim 60°$.

From Fig. 9.33, we read an optimum solidity parameter of $\sigma_z \cdot \psi_Z \approx 0.85$. Therefore, following the recommended loading parameter of $\psi_Z \approx 1.0$, we get the axial solidity of the first nozzle to be

$$\sigma_z \approx 0.85 \quad \text{(optimum axial solidity according to Zweifel)}$$

The actual solidity is $\sigma = \sigma_z/\cos\gamma°$, where $\gamma°$ is the stagger angle. A first approximation to stagger angle in a nozzle is the average flow angle α_m (based on the mean swirl) and in the rotor is the mean relative flow angle. The mean swirl across the nozzle is $C_{\theta m}$ and the mean flow angle is $\alpha_{2m} \approx \tan^{-1}(C_{\theta m}/C_z) \approx 40.9°$. Therefore, the stagger $\gamma° \approx 40.9°$ and the actual optimum solidity is $\sigma = \sigma_z/\cos\gamma° \quad \Longrightarrow \quad \sigma_n \approx 1.1$

Although, we estimated optimum nozzle solidity from Zweifel's method in this instance, we will not hesitate to change it to meet other blade row constraints, as in cooling requirements. The lack of overwhelming experimental support for the optimum solidity as formulated by Zweifel makes this choice an obvious one to change in subsequent turbine design iterations.

First turbine nozzle deviation angle
The turbine blades, by the virtue of their predominantly favorable pressure gradient, have lower deviation angles than compressor blades (of comparable camber). Carter's rule for deviation angle in turbines (with subsonic exit flow) may be written as

$$\delta^* = \frac{m\Delta\varphi}{\sigma} \approx \frac{\Delta\varphi}{8\sigma} \approx \frac{\Delta\beta}{8\sigma} \quad \text{(subsonic exit)}$$

However, for sonic or supersonic exit flows, the deviation angle becomes negligibly small since the flow continues to accelerate beyond the throat and thus boundary layers remain very thin,

$$\delta^* \approx 0 \quad \text{(supersonic exit)}$$

Since, the first nozzle exit Mach number was chosen to be 1.1 we neglect the deviation angle at the first nozzle exit.

First turbine nozzle throat sizing
The throat sizing in a turbine nozzle (or rotor) is very important both for choked and unchoked nozzles. The geometry that is shown in the definition sketch (Fig. 9.51) is used to relate the throat width or opening o to the blade spacing s.

The throat opening o is related to the spacing and cosine of the exit flow angle α_2 following

$$o \approx s \cdot \cos \alpha_2$$

This approximation is acceptable for the subsonic exit flow; however, for the supersonic exit Mach numbers (but below 1.3), we correct the throat opening by the inverse of A/A^* corresponding to the supersonic exit Mach number, i.e.,

$$\frac{o}{s} \approx \frac{\cos \alpha_2}{(A/A^*)_{M_2}}$$

The size of correction is small (for low supersonic exit Mach numbers) and to demonstrate we use the current example for $\alpha_2 = 60°$ and $M_2 = 1.1$, for which the A/A^* is ~1.008, i.e.,

$$\frac{o}{s} \approx \cos \alpha_2 = 0.5 \quad M_2 \leq 1$$

$$\frac{o}{s} \approx \frac{\cos \alpha_2}{(A/A^*)_{M_2}} \approx 0.496 \quad M_2 = 1.1$$

First turbine nozzle cooling
We had related the cooling mass fraction in a blade row to the gas and wall temperatures and Stanton numbers on the coolant and gas side following the steady-state heat transfer equation:

$$\varepsilon \equiv \frac{\dot{m}_c}{\dot{m}_g} = \frac{c_{pc}}{c_{pg}} \frac{\mathrm{St}_c}{\mathrm{St}_g} \left(\frac{T_{wc} - T_c}{T_{aw} - T_{wg}} \right) \quad \textbf{(9.73)}$$

The wall temperature on the coolant side was related to the wall temperature on the gas side, wall thickness, blade solidity, gas mass density, and thermal conductivity of the wall according to the Fourier heat conduction and steady-state heat transfer from the gas through the solid and to the coolant:

$$T_{wc} = T_{wg} - \left(\frac{t_w}{k_w} \right) \frac{1}{2\sigma} \rho_g u_g c_{pg} \mathrm{St}_g (T_{aw} - T_{wg}) \quad \textbf{(9.75)}$$

Note that the blade solidity σ has entered blade cooling problem through Eq. 9.75. Finally, the gas mass density is

related to *average flow passage Mach number*, total pressure, and temperature following continuity equation:

$$\rho_g u_g = \frac{\dot{m}_g}{A_g}$$

$$= \sqrt{\frac{\gamma_g}{R_g}} \frac{p_{tg}}{\sqrt{T_{tg}}} M_g \left(1 + \frac{\gamma - 1}{2} M_g^2 \right)^{-\frac{\gamma+1}{2(\gamma-1)}} \quad \textbf{(9.76)}$$

The average flow passage Mach number is simply taken to be the mean of the inlet and exit Mach numbers, i.e.,

$$M_g \approx (M_1 + M_2)/2$$

Therefore, if we approximate gas mass density in the nozzle based on

$$p_{tg} \approx 1960 \, \text{kPa}$$
$$T_{tg} \approx 2000 \, \text{K}$$
$$M_g \approx (M_2 + M_1)/2 = 0.808$$

we get the mass density in the nozzle to be $\rho_g u_g \approx 1665.9 \, \text{kg/m}^2\text{s}$

Now, we calculate the wall temperature on the coolant side, T_{wc} using the estimated gas mass density and the following parameters:

$T_{wg} = 1200 \, \text{K}$ (desired gas wall temperature, a design choice that impacts life)

$t_w = 1.5 \, \text{mm}$ (1st design choice on the wall thickness)

$k_w = 14.9 \, \text{W/mK}$ (wall thermal conductivity, nickel-based alloys used in turbines)

$\mathrm{St}_g \approx 0.005$ (gas side Stanton number, an estimate based on C_f)

$T_{aw} \approx T_{tg} = 2000 \, \text{K}$ (adiabatic wall temperature ≈ the gas total temperature)

We calculate the wall temperature on the coolant side to be

$$T_{wc} \approx 820.8 \, \text{K}.$$

We may now estimate, using Eq. 9.73, the coolant mass fraction that is needed in the first nozzle to maintain the gas-side surface temperature on the nozzle at the desired level (of 1200 K). In addition, we have approximated the ratio of coolant-to-gas side Stanton numbers, $\mathrm{St}_c/\mathrm{St}_g$, to be ~0.5 (suggested by Kerrebrock, 1992) which gives

$$\varepsilon_{N1} \approx 0.0252 \quad \text{(nozzle-1 coolant mass fraction)}$$

Here, we have estimated the first nozzle internal cooling fraction of 2.52%, which leaves another 7.48% internal cooling for the first rotor, the casing and the second nozzle and the rotor, if needed. In case the amount of coolant fraction (of 10%) is insufficient (for internal cooling purposes), we have to go back and bleed a higher coolant fraction from the compressor.

First turbine rotor pitchline design (C_z—constant design)
The first rotor rotational speed at the pitchline is

$$U_m = \omega r_m = (8770\,\text{rpm})(2\pi\,\text{rad/rev})(\text{min}/60\,\text{s})(0.37\,\text{m})$$

$$\approx 339.8\,\text{m/s}$$

Here, we have used the pitchline radius from the compressor (inlet) side as a first estimate of pitchline radius in the turbine. However, we can change the pitchline radius in the turbine to accomplish our design goals in power production and stage-by-stage load variations. Therefore, the pitchline radius in the turbine is clearly a design variable, but as a first approximation, the compressor inlet value may be used to start off the turbine design iteration cycle.

The rotor (relative) exit Mach number M_{3r} impacts the power production in the rotor as well as the turbine performance and is a design choice. The performance improves with the increase in relative exit Mach number. For starters, we choose $M_{3r} \sim 0.8$ but we will be willing to change M_{3r} up to 0.9 and even go below 0.8, if necessary. Note that the first turbine rotor is to remain unchoked. The relative Mach number is written in terms of axial and tangential velocities and speed of sound as

$$M_{3r} = \frac{1}{a_3}\left[C_{z3}^2 + W_{\theta3}^2\right]^{1/2} = 0.80$$

(rotor relative exit Mach number design choice)

And from the conservation of energy in the relative frame, we write

$$\frac{a_3^2}{\gamma_t - 1} + \frac{W_3^2}{2} = \frac{a_2^2}{\gamma_t - 1} + \frac{W_2^2}{2}$$

We may combine the above two equations and use constant axial velocity to write

$$W_{\theta3}^2 = \frac{M_{3r}^2[a_2^2 + (\gamma_t - 1)W_{\theta2}^2/2] - C_z^2}{1 + (\gamma_t - 1)M_{3r}^2/2}$$

The relative swirl downstream of the first rotor is in opposite direction to the rotor rotation, therefore it is negative. We take the minus solution of the above equation, i.e.,

$$W_{\theta3} = -\sqrt{\frac{M_{3r}^2[a_2^2 + (\gamma_t - 1)W_{\theta2}^2/2] - C_z^2}{1 + (\gamma_t - 1)M_{3r}^2/2}} \implies$$

$$W_{\theta3} = -458\,\text{m/s}$$

From relative exit swirl, we calculate the absolute exit swirl $C_{\theta3} = -118.2\,\text{m/s}$ and thus the degree of reaction

$$^{\circ}R_m = 1 - \frac{C_{\theta m}}{U_m} = 1 - \frac{C_{\theta2} + C_{\theta3}}{2U_m} \implies {}^{\circ}R_m \approx 0.06$$

Based on the velocity components across the rotor, we have all the flow angles at the rotor exit, i.e., α_3 and β_3. Therefore

the first rotor velocity components, flow angles, and Mach number at the pitchline are

$$W_{\theta2} = C_{\theta2} - U_m = 417.3\,\text{m/s} \implies \beta_2 \approx +43.7^{\circ}$$

$$W_{\theta3} = -458\,\text{m/s} \implies \beta_3 \approx -46.3^{\circ}$$

$$C_{\theta3} = W_{\theta3} + U_m = -118.2\,\text{m/s} \implies \alpha_3 \approx -15.1^{\circ}$$

$$W_2 = [C_{z2}^2 + W_{\theta2}^2]^{0.5} = 604.2\,\text{m/s} \implies M_{2r} = 604.2/794.7$$
$$= 0.760$$

The total pressure and temperature of the gas depends on the motion of the observer. Therefore, the rotor *sees* p_{t2r} and T_{t2r} and not the absolute p_{t2} and T_{t2}. In order to calculate the relative total pressure and temperature, we first determine the static pressure and temperature in station 2 and then use relative Mach number M_{2r} or relative kinetic energy to calculate the relative total pressure and temperature as seen by the rotor. The static temperature of the gas in station 2 is

$$T_2 = T_{t2} - C_2^2/2c_{pt} = 2000 - (874.2)^2/2/1243.67$$
$$= 1692.8\,\text{K}$$

Therefore the relative total temperature is $\implies$

$$T_{t2r} = T_2 + W_2^2/2c_{pt} \approx 1839.6\,\text{K}$$

Note that the rotor *sees* a stagnation temperature of 1839.6 K instead of 2000 K that the nozzle feels. The rotor operates in a ~160 K cooler gas, by virtue of its rotation!

To calculate the static pressure at nozzle exit plane, we need to estimate a total pressure loss factor that the flow in the nozzle suffers due to profile loss, secondary flow loss, other (than secondary flow) three-dimensional losses, as well as shock losses in a supersonic nozzle. Using Fig. 9.21, we estimate the total pressure loss factor in the nozzle to be $\varpi_n \sim 0.06$. Here again, we may assign a higher loss coefficient to the blade row in our design based on the relevant cascade loss data available. From the definition of total pressure loss parameter,

$$\varpi_n \equiv \frac{p_{t1} - p_{t2}}{p_{t1} - p_1}$$

We first calculate the static pressure p_1 from p_{t1} and M_1 (0.516) that we had calculated earlier,

$$p_1 \approx 1654\,\text{kPa}$$

Therefore the total pressure at the nozzle exit is $p_{t2} \approx 1941.6\,\text{kPa}$

The static pressure p_2 may be calculated from p_{t2} and $M_2 = 1.1$ to be

$$p_2 \approx 942.5\,\text{kPa}$$

Since we calculated M_{2r} to be 0.760, we may now calculate p_{t2r}

$$p_{t2r} \approx 1351.5 \, \text{kPa}$$

To complete our first stage calculations of pressure and temperature, we should calculate the rotor exit conditions as well. The total temperature T_{t3} may be calculated from the Euler turbine equation and the known velocity triangles, i.e.,

$$T_{t3} = 1760.9 \, \text{K}$$

The static temperature T_3 is calculated from $T_3 = T_{t3} - C_3^2/2c_{pt}$ to be $T_3 = 1678.4 \, \text{K}$

The corresponding speed of sound is $a_3 = 791.3 \, \text{m/s}$ The absolute Mach number at the exit of the rotor is $M_3 = 0.572$

If we assume a relative total pressure loss coefficient in the rotor $\varpi_r \sim 0.08$ (note that we chose a higher loss in the rotor than the nozzle to account for tip clearance flow in the rotor), we can calculate the rotor exit relative total pressure p_{t3r}

$$p_{t3r} = 1318.8 \, \text{kPa}$$

Since we had chosen an M_{3r} as a design choice (we took 0.8), we can calculate p_3

$$p_3 = 886.4 \, \text{kPa}$$

From the optimum solidity approach of Zweifel we estimate the axial solidity parameter for the first turbine rotor to be

$$\sigma_z \psi_z \approx 1.8$$

Based on the recommended value of $\psi_z \approx 1.0$, we get the optimum axial solidity of the first rotor, $\sigma_{zr} = 1.8$. The stagger angle at the pitchline is estimated to be at the mean swirl angle in the rotor, i.e.,

$$\beta_m \approx \tan^{-1}(W_{\theta m}/C_z) = \tan^{-1}\{[(-458 + 417.3)/2]/437.1\} \approx -2.7°$$

Therefore the optimum rotor solidity is calculated to be $1.8/\cos(2.7°) \approx 1.8$

In the spreadsheet calculation, we started with this value of solidity, but we had to reduce the rotor solidity in order to lower the coolant fraction in the blade row. The value of first rotor solidity that we accepted was $\sigma_{r1} \approx 1.1$.

Rotor deviation angle
Using Carter's rule for a turbine, we estimate the first rotor deviation angle δ^*

$$\delta^* \approx \frac{\Delta \beta}{8\sigma_r} \approx \frac{90°}{8(1.1)} \approx 10.2°$$

Note that the relatively large deviation angle of $\sim 10°$ is caused by a very large flow turning in the rotor of $\sim 90°$. With a reduced camber, or net flow turning angle, the deviation angle drops.

First turbine rotor blade cooling
In turbine rotor cooling calculations, we encounter the heat transfer in *relative frame of reference*. Consequently, the stagnation temperature becomes *as observed by the rotor*. We calculated the relative frame stagnation pressure and temperature to be 1351.5 kPa and 1839.6 K, respectively. We proceed to calculate the parameters in the coolant mass fraction equation to the rotor:

$$\varepsilon \equiv \frac{\dot{m}_c}{\dot{m}_g} = \frac{c_{pc}}{c_{pg}} \frac{\text{St}_c}{\text{St}_g} \left(\frac{T_{wc} - T_c}{T_{aw} - T_{wg}} \right)$$

The coolant wall temperature is estimated from the relative frame of reference parameters in equation

$$T_{wc} = T_{wg} - \left(\frac{t_w}{k_w} \right) \frac{1}{2\sigma} \rho_g u_g c_{pg} \text{St}_g (T_{aw} - T_{wg})$$

We choose

$t_w = 2.0 \, \text{mm}$ (a slightly thicker wall chosen for the rotor due to centrifugal stresses)

$k_w = 14.9 \, \text{W/m·K}$ (same high thermal conductivity material as the nozzle)

$\sigma_r = 1.1$ (we changed this value in the spreadsheet from that indicated by Zweifel)

$T_{aw} \approx T_{t2r} = 1839.6 \, \text{K}$ (first approximation for adiabatic wall temperature)

$T_{wg} = 1200 \, \text{K}$ (desired wall temperature—design choice that impacts blade life)

$\text{St}_g = 0.0065$ (higher than nozzle due to higher turbulence level in the rotor passage)

For the mass density in the rotor blade row, we use the same continuity equation, but in relative frame of reference variables,

$$(\rho_g u_g)_r = \frac{\dot{m}_g}{A_g} = \sqrt{\frac{\gamma_g}{R_g}} \frac{p_{tg,r}}{\sqrt{T_{tg,r}}} M_{g,r} \left(1 + \frac{\gamma - 1}{2} M_{g,r}^2 \right)^{-\frac{\gamma+1}{2(\gamma-1)}}$$

The gas Mach number in the rotor varies from inlet $M_{2r} = 0.760$ to $M_{3r} = 0.80$, therefore we get an average gas Mach number of ~ 0.78 in the rotor frame. Now, let us substitute these numbers in the mass density equation to get

$$(\rho_g u_g)_r \approx 1183.6 \, \text{kg/m}^2\text{s}$$

The wall temperature on the coolant side is calculated to be $T_{wc} \approx 826.6 \, \text{K}$

The coolant mass fraction for the first rotor is estimated to be $\varepsilon_{R1} \approx 0.035$

The first stage of the turbine has consumed (2.5% in the nozzle plus 3.5% in the rotor) 6% of coolant, leaving $\sim 4\%$ for the casing and the second stage, if it required cooling.

First stage loading parameter and shaft power

$$w_t = U_m(C_{\theta 2} - C_{\theta 3}) = 297,407 \, \text{J/kg}$$
$$\psi = w_t/U_m^2 = 2.57$$

The shaft power that the first stage (rotor) produces is

$$\wp_{\text{Stage-1}} = \dot{m}_{\text{Stage-1}} \cdot (w_t)_{\text{Stage-1}} = 93 \, \text{kg/s} \, (297,407 \, \text{J/kg})$$

$$\approx 27.66 \, \text{MW}$$

By comparing the power that is produced by the first stage (rotor) to that of the entire turbine, i.e., 47.7 MW, we note that the second stage is needed and it needs to produce ~ 20 MW of power. Also, since the gas temperature at the exit of stage 1 exceeds the desirable gas-side wall temperature, the second stage has to be cooled as well.

Second turbine nozzle pitchline design (Cz—constant design)
To begin the second stage nozzle design, we start by listing the inlet conditions to the second stage, i.e.,

$$T_{t3} = 1760.8 \, \text{K}$$
$$p_{t3} = 1091.1 \, \text{kPa}$$
$$C_3 = 452.8 \, \text{m/s}$$
$$\alpha_3 = -15.1°$$

Now, we are ready to choose some of the second stage design variables, e.g., either the nozzle exit Mach number M_4 or the nozzle exit flow angle α_4, here we chose the exit flow angle and calculated the exit Mach number based on the constant axial velocity design principle, i.e.,

$\alpha_4 = 55°$ (design choice, which can be adjusted up to 70° or down to $\sim 40°$) leads to the nozzle exit Mach number (that has to be subsonic, or unchoked), $M_4 = 0.887$

Here, we have an additional choice about the pitchline radius for the second stage, i.e., r_{m3} does not have to be the same as r_{m1}. We started our calculations (in the spreadsheet) with the same value and then increased it to produce more shaft power. Finally, we arrived at the pitchline radius for the second stage:

$r_{m3} = 0.47$ m (an increase of 10 cm of radial shift for the pitchline in stage 2 over stage 1)

With these choices, we calculate the second nozzle exit flow conditions, and with an assumed total pressure loss parameter $\varpi_{n2} \sim 0.06$, we calculate the total pressure as well,

$$U_3 = 431.6 \, \text{m/s}$$
$$W_{\theta 4} = 89.3 \, \text{m/s}$$
$$\beta_4 = 11.5°$$
$$W_4 = 446.1 \, \text{m/s}$$
$$M_{4r} = 0.582$$
$$p_{t4} = 1078.8 \, \text{kPa}$$
$$p_{t4r} = 824.6 \, \text{kPa}$$

Our choice of the second nozzle solidity $\sigma_{n2} = 0.8$ yields a deviation angle of $\delta^* = 10.2°$. For the blade internal cooling calculation, we select

$$St_g = 0.0065$$
$$t_w = 2.2 \, \text{mm}$$

As in the first stage, we assume the ratio of Stanton number of the coolant and gas to be $St_c/St_g = 0.5$ to calculate

$$\rho_g u_g = 951.1 \, \text{kg/m}^2\text{s}$$
$$T_{wc} = 802 \, \text{K}$$

Finally, the second nozzle coolant fraction, $\varepsilon_{N2} = 0.022$ The throat sizing of the second nozzle gives $(o/s)_{N2} = 0.643$

Second turbine rotor pitchline design (Cz—constant design)
The second rotor exit Mach number is a design choice for which we have chosen

$$M_{5r} = 0.75 \quad \text{(a design choice, always less than 1)}$$

In addition, we assumed the solidity of second rotor, wall thickness for the cooled second rotor, Stanton number on the gas side, and the ratio of coolant to gas Stanton numbers according to

$$\sigma_{r2} = 0.44$$
$$t_w = 2 \, \text{mm}$$
$$St_g = 0.0065$$
$$St_c/St_g = 0.5$$

We calculate the second rotor parameters, such as $T_{t4r} = 1655$ K, $\rho_g u_g = 723.1 \, \text{kg/m}^2$ and $T_{wc} = 794.3$ K The second rotor coolant mass fraction is $\varepsilon_{R2} = 0.0207$ The second rotor exit flow conditions are

$$C_5 = 437.5 \, \text{m/s}$$
$$C_{\theta 5} = 19.9 \, \text{m/s}$$
$$\alpha_5 = 2.6° \quad \text{(nearly swirl-free exit, as we aimed)}$$
$$M_5 = 0.582 \quad \text{(a reasonable exit Mach number)}$$

The second stage degree of reaction is

$$°R = 0.373$$

Second stage loading parameter and shaft power

$$w_t = U_m(C_{\theta 4} - C_{\theta 5}) = 216,248 \, \text{J/kg}$$

$$\psi = w_t/U_m^2 = 1.161$$

Shaft power for the second stage (rotor),

$$\wp_{\text{Stage2}} = \dot{m}_{\text{Stage2}} \cdot (w_t)_{\text{Stage2}} = 93 \, \text{kg/s} (216,248 \, \text{J/kg})$$

$$\approx 20.11 \, \text{MW}$$

Impact of cooling on turbine exit temperature and pressure
It is important to note that we treated the blade-cooling problem completely *internally* to achieve a desired blade surface temperature T_{wg}. By this technique, we estimated the coolant fraction for internal blade cooling purposes. We did not inject the coolant in the gas stream per blade row, as it is commonly done in a real cooled gas turbine blade. The coolant is always discharged through film holes, trailing-edge slots, and other discharge orifices.

Therefore, there are at least three effects that we ignored by our internal cooling approach. First, the mass flow rate through the next blade row increases by the amount of coolant injected in the gas path. Second, the temperature of the gas is not reduced due to the injection of coolant per blade row. Finally, the total pressure of the gas is not reduced due to turbulent mixing losses of the hot and cold gas streams. There are other important details that accompany mixing of two streams with relative flow angle with respect to each other, as in coolant flow emerging from film holes. For example, the flow angle associated with the *mixed-out state* is not exactly the same as the gas path angle that we calculate in the absence of cooling. There are two approaches that we can take to remedy the cooling problem. The first approach will correct the turbine exit total temperature and pressure for all the coolant suddenly *dumped* at the turbine exit. The second approach will do a stepwise calculation per blade row as the coolant is injected in the gas path and achieves a mixed-out state with the hot gas. The level of detail in the second approach is slightly above the scope of the present book, but definitely feasible and straightforward. Here we adopt the first approach for simplicity.

A corrective action on the temperature and pressure may be taken by first performing an energy balance between the turbine discharge total temperature where coolant was not introduced in the gas path and the coolant flow at the coolant temperature T_{tc}. The conservation of energy yields

$$T_{t5-\text{cooled}} = \frac{\dot{m}_g c_{pt} T_{t5} + \dot{m}_c c_{pc} T_{tc}}{(\dot{m}_g + \dot{m}_c)\bar{c}_p}$$

where the mean specific heat at constant pressure for the mixture is

$$\bar{c}_p = \frac{\dot{m}_g c_{pt} + \dot{m}_c c_{pc}}{\dot{m}_g + \dot{m}_c}$$

First, the average specific heat is calculated to be $\bar{c}_p \approx 1220.4 \, \text{J/kg} \cdot \text{K}$

The mixed-out total temperature is then $T_{t5\text{-cooled}} \approx 1521.8 \, \text{K}$

Our power balance at the beginning of the turbine design had indicated that the exit total temperature of the cooled turbine will be $T_{t5\text{-cooled}} \approx 1507 \, \text{K}$

As expected, these estimations are based on the same principles and are thus very close, i.e., within 1% of each other.

To estimate the turbine total pressure at the exit, we approximate the adiabatic efficiency of a cooled turbine and then use the efficiency and temperature ratio to get the turbine pressure ratio. The adiabatic efficiency of a cooled turbine was estimated to suffer $\sim$3% per 1% coolant mass fraction. Therefore, a 10% coolant mass fraction causes the turbine adiabatic efficiency to be $\eta_t \sim 70\%$. In terms of the adiabatic efficiency, we may express the turbine total pressure ratio according to

$$\pi_t = \left[1 - \frac{1}{\eta_t}(1 - \tau_t) \right]^{\frac{\gamma_t}{\gamma_t - 1}}$$

We may substitute for $\tau_t = 1507/2000 = 0.7537$ and 0.7 for the adiabatic efficiency of our cooled turbine to get the turbine pressure ratio and exit total pressure of the cooled turbine:

$$\pi_t = 0.1527 \quad \Longrightarrow \quad p_{t5\text{-cooled}} \approx 299.2 \, \text{kPa}$$

Interim summary

In this section, we recapitulate the outcomes of our design so far in meeting the required shaft power, the target cooling mass fraction (albeit for internal cooling), and the condition of zero exit swirl that we wish to impose on the turbine exit flow.

Total shaft power

Total power produced by the internally cooled turbine is the sum of the two stages, i.e., 27.66 MW and 20.11 MW, which is 47.77 MW. The required power was calculated to be 47.75 MW. There is a match.

Cooling mass fraction

The sum of the coolant fractions for the first two stages is 0.025, 0.035, 0.022, and 0.021, which gives 0.1033. We started with a 10% coolant mass fraction, which we have totally consumed in four internally cooled blade rows. However, the casing needs to be cooled as well as the exhaust nozzle. Therefore we need to demand a higher level of compressor bleed for turbine and nozzle cooling purposes. But for the present (turbine preliminary design) purposes, we have achieved our internal cooling target.

Turbine exit swirl

The goal of zero turbine exit swirl is nearly achieved. We ended up with a 2.6° of swirl angle at the exit of the second stage rotor, which in terms of absolute swirl velocity is $\sim$20 m/s.

Turbine annulus sizing

The nozzle inlet Mach number $M_1 \approx 0.516$ and the mass flow rate will be used to size the nozzle inlet flow area (A_1) around a pitchline radius of 0.37 m,

$$\dot{m}_1 = \sqrt{\frac{\gamma_t}{R_t}} \frac{p_{t1}}{\sqrt{T_{t1}}} A_1 M_{z1} \left(1 + \left(\frac{\gamma_t - 1}{2} \right) M_{z1}^2 \right)^{-\frac{\gamma_t + 1}{2(\gamma_t - 1)}}$$

$$\Longrightarrow \quad A_1 = 0.071 \, \text{m}^2$$

We calculate the first nozzle blade height h_1, i.e., ($r_{t1} - r_{h1}$) from the flow area and the mean radius r_m, i.e.,

$$r_{t1} - r_{h1} = \frac{A_1}{2\pi r_m} \approx 3.054 \, \text{cm} \quad \Longrightarrow \quad h_1 \approx 3.054 \, \text{cm}$$

Now, we can calculate the turbine exit flow area A_5 from the one-dimensional continuity equation based on the cooled turbine exit total pressure and temperature

$$\dot{m}_5 = \dot{m}_4 + \dot{m}_c$$

$$= \sqrt{\frac{\gamma_t}{R_t}} \frac{p_{t5-\text{cooled}}}{\sqrt{T_{t5-\text{cooled}}}} A_5 M_{z5} \left(1 + \left(\frac{\gamma_t - 1}{2}\right) M_{z5}^2 \right)^{-\frac{\gamma_t+1}{2(\gamma_t-1)}}$$

We calculate the turbine exit flow area, A_5 ➡

$$A_5 \approx 0.4988 \text{ m}^2$$

The blade/channel height at the turbine exit, h_5, assuming the pitchline radius is $r_{m5} = r_{m4} = 0.47$ m is

$$r_{t5} - r_{h5} = \frac{A_5}{2\pi r_{m5}} \approx 16.89 \text{ cm} \quad \Longrightarrow \quad h_5 \approx 16.89 \text{ cm}$$

Figure 9.63 is a schematic drawing of a two-stage turbine with its corresponding station numbers.

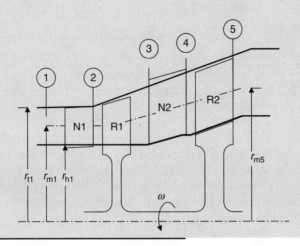

■ **FIGURE 9.63**
Definition sketch of a two-stage turbine and its station numbers

9.8 Gas Turbine Design Summary

We used some commonly accepted design practices, e.g., constant axial velocity, to design suitable velocity triangles at the pitchline. In the process, we encountered some *hard* and *soft* design constraints. Some examples of hard and soft design constraints are listed below:

Hard design constraints		Soft design constraints	
$M_2 > 1$	(choked first nozzle)	$\sigma \neq \sigma_{\text{opt}}$	(solidity other than the optimum)
$M_{3r} < 1$	(unchoked rotor exit flow)	$r_m \neq$ constant	(pitchline radius is variable)
$M_{5r} < 1$	(unchoked rotor exit flow)	$\alpha_{\text{exit}} \neq 0$	(turbine exit swirl not zero)
T_{wg}	(maximum wall temperature)	$t_{\text{wall}} = 2$ mm	(wall thickness other than nominal)
$^\circ\text{R} > 0$	(positive degree of reaction)	$0 \leq {}^\circ\text{R} \leq 1$	(wide-ranging choice of $^\circ$R)
		$T_{\text{wg}} = 1200$ K	(a function of material selection)
		$\psi > 2$	(higher loading than 2 is acceptable)
		$C_z \neq$ constant	(acceptable)

The design process requires iteration. To facilitate successive calculations, a spreadsheet was developed. It is important to recognize the approximations that were used in the analysis. For example, we used the same T_c in our internal cooling calculations for the nozzle *and* the rotor. The coolant temperature that we used in the nozzle was the compressor discharge temperature. However, when the coolant is injected in the hub of a rotating blade row, as in the rotor, the *relative* stagnation temperature is different than the absolute total temperature. But, we did not distinguish between the two. In practice, the coolant is injected at an angle in the direction of rotor rotation with the resultant relative stream in the axial direction. The schematic drawing of a cooled turbine rotor blade and the coolant

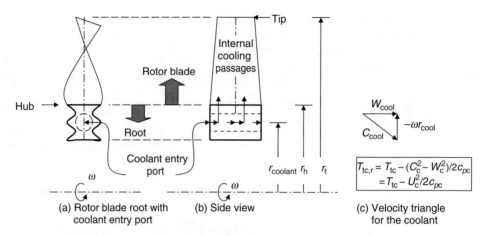

■ **FIGURE 9.64**   Definition sketch for coolant entry at the rotor blade root and coolant total temperature in the rotor frame of reference

velocity triangle is shown in Fig. 9.64 The correction for the coolant relative total temperature is shown to be $\sim U_c^2/2c_{pc}$. Therefore, the coolant entering the blade root *feels cooler* to the rotor than the coolant in the nozzle. Here, the emphasis in the summary is not on the extent of correction to the stagnation temperature, rather on the awareness of the two frames of reference.

9.9   Summary

Modern gas turbines truly represent the most technologically challenging component in an aircraft engine. Critical technologies in the development of gas turbines are (single crystal) material, internal/external cooling, thermal protection coating, aerodynamics, active tip clearance control, and manufacturing.

Degree of reaction in a turbine influences power production, efficiency, and stage mass flow density (i.e., mass flow rate per unit area). For example, a 50% reaction turbine stage produces (for a swirl-free rotor exit flow)

$$w_t \equiv \frac{\wp_t}{\dot{m}_t} \cong U\Delta C_\theta = U^2$$

And an impulse turbine (with 0% reaction) stage produces (for a swirl-free rotor exit flow)

$$w_t \equiv \frac{\wp_t}{\dot{m}_t} \cong U\Delta C_\theta = 2U^2$$

The choice of nozzle optimal exit swirl Mach number coupled with axial flow from the rotor demonstrated that the choice of degree of reaction establishes a turbine inlet Mach number M_1. An impulse turbine stage, with $^\circ R = 0$, demands a nozzle inlet Mach number of ~ 0.4 and a 50% reaction stage corresponds to an inlet Mach number ~ 0.65. With the increase in reaction, the mass flow rate per unit area increases and vice versa. Although an impulse stage is capable of producing maximum specific work, it also reduces the mass flow density. The optimal nozzle exit Mach number M_2 that maximizes the swirl and stage work varies between 1.2 and 1.5 for a stage reaction that varies between 50% and 0, respectively. The practical maximum exit flow angle in a nozzle is $\alpha_{2,\max} \approx 70^\circ$.

Turbine losses are attributed to the following sources:

1. Profile loss, which is established based on 2D cascade studies
2. Secondary flow losses, which may be estimated using secondary flow theory of Hawthorne or the approximate version by Squire and Winter
3. Annulus losses, which includes tip clearance loss, (seal) leakage loss, casing boundary layer loss, corner vortex loss
4. Coolant-related losses, including mixing loss, induced separation loss, and internal passage coolant total pressure loss
5. Shock and shock-boundary layer interaction losses in transonic stages of LPT
6. Unsteady flow loss causing shock oscillation and vortex shedding in the wake

In examining blade profile losses, we have noted that the impulse blade profile losses are significantly higher than their counterparts in a reaction rotor. This contributes to lower stage efficiency in an impulse type versus a reaction turbine. Tip clearance is a major contributor to turbine losses, and consequently the active control of tip clearance is an adopted/practiced strategy. The other option is to shroud the turbine rotor blades (at the tip) to eliminate the tip clearance loss. The added weight and centrifugal stresses in shrouded rotors are the penalty paid for such a relief.

We introduced the concept of optimal solidity in turbine aerodynamics. However, the two criteria for optimum solidity, i.e., Zweifel's and Howell's, showed a very limited range of agreement with the test data.

Turbine cooling relies on internal and external cooling schemes as well as thermal protection coating. Leading-edge impingement and internal convective cooling are practiced internally. Either film or transpiration cooling achieves the external cooling. The role of the thermal protection coating is to reduce the heat transfer to the blade by coating the blades with a very low thermal conductivity layer (as in silicon-based paint, or ceramic coating).

Glassman (1973), Hill and Peterson (1992), Marble (1964) and Schobeiri (2004) are recommended for additional reading on gas turbines. VKI Lecture series (1982) and Glass, Dilley and Kelly (1999) provide valuable discussions on advanced turbine cooling. The contributions of Garvin (1998) and St. Peter (1999) on the history of gas turbine development in the United States are also recommended.

References

1. Ainley, D.G., "An Approximate Method for Estimation of the Design Point Efficiency of Axial Flow Turbines," British Aeronautical Research Council Current Paper 30, 1950.

2. Ainley, D.G. and Mathieson, G.C.R. "A Method for Performance Estimation for Axial Flow Turbines," Aeronautical Research Council, Research and Memorandum No. 2974, 1951.

3. Anderhub, W., "Untersuchungen ueber die Dampfstroemung im radialen Schaufelspalts bei Ueberdruckturbinen," Doctoral Dissertation, Swiss Federal Institute of Technology, Zurich, Switzerland, 1912.

4. Bammert, K., Klaeukens, H., and Hartmann, D., "Der Einfluss des radialen Schaufelspalts auf den Wirkungs- grad mehrstufige Turbinen," *VDI-Zeitschrift*, Vol. 110, No. 10, April 1968, pp. 390–395.

5. Booth, T.C., Dodge, P.R., and Hepworth, H.K., "Rotor Tip Leakage: Part I-Basic Methodology," *Transactions of ASME, Journal of Engineering for power*, Vol. 104, January 1982, pp. 154–161.

6. Colladay, R.S., "Turbine Cooling," A Chapter in *Turbine Design and Applications*, Glassman, A.J. (editor), NASA SP-290, Washington, DC, 1975.

7. Dixon, S.L., *Fluid Mechanics, Thermodynamics of Turbomachinery*, 2nd, edition. Pergamon Press, Oxford, UK, 1975.

8. Duncombe, E., "Aerodynamic Design of Axial Flow Turbines," Section H, in *Aerodynamics of Turbines and*

Compressors, Vol. X, Hawthorne, W. R. (editor), Princeton Series on High Speed Aerodynamics and Jet Propulsion, Princeton University Press, Princeton, N.J., 1964.

9. Dunham, J. and Came, P., "Improvements to the Ainley-Mathieson Method of Turbine Performance Prediction," Transactions of the ASME, Series A, Vol. 92, 1970.

10. Eckert, E.R.G., *Analysis of Heat and Mass Transfer* McGraw-Hill, New York, 1971.

11. Eckert, E.R.G. and Livingood, N.B.,"Comparison of the Effectiveness of Convection-, Transpiration-, and Film-Cooling Methods with Air as Coolant," NACA Report 1182, 1953.

12. Farokhi, S., "The Effect of Rotor Tip Clearance Flow on the Axial Exhaust Diffuser and the Turbine Performance," *Proceedings of the First International Power Conference in Beijing-China,* October 1986, pp. 348–353.

13. Farokhi, S., "A Trade-Off Study of Rotor Tip Clearance Flow in a Turbine/Exhaust Diffuser System," ASME Paper No. 87-GT-229, June 1987.

14. Farokhi, S., "Analysis of Rotor Tip-Clearance Loss in Axial Flow Turbines,"*AIAA Journal of Propulsion and Power*, Vol. 4, No. 5., September-October, 1988 pp. 452–457

15. Garvin, R.V., *Starting Something Big: The Commercial Emergence of GE Aircraft Engines,* AIAA, Inc., Reston, Virginia, 1998.

16. Glass, D.E., Dilley, A.D., and Kelly, H.N., "Numerical Analysis of Convection/Transpiration Cooling," NASA/TM-1999-209828, 1999.

17. Glassman, A.J. (editor), *Turbine Design and Application,* 3 volumes, NASA SP-290, 1973.

18. Graham, J.A.H., "Investigation of Tip Clearance Cascade in a Water Analogy Rig," *Transactions of ASME, Journal of Engineering for Gas Turbines and Power*, Vol. 108, January 1986, pp. 38–46.

19. Gostelow, J.P., *Cascade Aerodynamics,* Pergamon Press, Oxford, UK, 1984.

20. Hawthorne, W.R., *"Secondary Circulation in Fluid Flow,"* *Proceedings of Royal Society,* London, Vol. 206, 374, 1951.

21. Hill, P.G. and Peterson, C.R., *Mechanics and Thermodynamics of Propulsion,* 2nd edition, Addison-Wesely, Reading, Massachusetts, 1992.

22. Incropera, F.P. and DeWitt, D.P., *Fundamentals of Heat and Mass Transfer,* 5th edition, Wiley, New York, 2001.

23. Johnston, I.H., "An Analysis of Air Flow Through the Nozzle Blades of a Single Stage Turbine," ARC Current Paper 131, 1953.

24. Kacker, S.C. and Okapuu, U,"A mean-line prediction method for axial-flow-turbine efficiency," Paper no. 81-GT-58, ASME, New York, 1981.

25. Kercher, D.M. and Tabakoff, W., "Heat Transfer by a Square Array of Round Air Jets Impinging Perpendicular to a Flat Surface Including the Effect of Spent Air,"*ASME Journal of Engineering for Power*, Vol. 92, No. 1, January, 1970, pp. 73–82.

26. Kerrebrock, J.L., *Aircraft Engines and Gas Turbines*, 2nd Edition , MIT Press, Cambridge, Mass. 1992.

27. Lakshminarayana, B., "Methods of Predicting the Tip Clearance Effects in Axial Flow Turbomachinery," *Transactions of ASME, Journal of Basic Engineering*, Vol. 92, September, 1970 pp. 467–482.

28. Mattingly, J.D., Heiser, W.H., and Pratt, D. T. *Aircraft Engine Design*, 2nd Edition, AIAA Education Series, AIAA, Reston, VA, 2002.

29. Marble, F.E., "Three-Dimensional Flow in Turbomachines," in, *Aerodynamics of Turbines and Compressors*, Vol. X, Hawthorne W.R. (editor), Princeton Series on High Speed Aerodynamics and Jet Propulsion, Princeton University Press, Princeton, NJ, 1964.

30. Rohsenow, W.M., Hartnett, J.P., and Cho, Y.I., *Handbook of Heat Transfer,* 3rd edition, McGraw-Hill, New York, 1998.

31. Rolls-Royce, *The Jet Engine*, Rolls-Royce plc, Derby, England, 2005.

32. Schobeiri, M., *Turbomachinery Flow Physics and Dynamic Performance.* Springer Verlag, Berlin, 2004.

33. St.Peter, J., *The History of Aircraft gas Turbine Engine Development in the United States*, International Gas Turbine Institute, Atlanta, 1999.

34. Stodola, A., *Dampf- und Gasturbinen*, Springer Verlag, 1924.

35. Stodola, A., *Steam and Gas Turbines*, Vols. 1 and 2, McGraw-Hill, New York, 1927.

36. Traupel, W., *Thermische Turbomaschinen*, Vol. 1, Springer Verlag, 1958, pp. 295–296.

37. VKI-Lecture Series, "Film Cooling and Turbine Blade Heat Transfer," VKI-LS 82–02, 1982.

38. Wang, J.H., Messner, J., and Stetter, H., "An Experimental Investigation on Transpiration Cooling Part II: Comparison of Cooling Methods and Media,*"* *International Journal of Rotating Machinery*, Vol. 10, No. 5, 2004, pp. 355–363.

39. Wilson, D.G. and Korakianitis, T. *The Design of High-Efficiency Turbomachinery and Gas Turbines,* 2nd edition, Prentice Hall, New York, 1998.

40. Zweifel, O., "The Spacing of Turbomachine Blading, Especially with Large Angular Deflection,"*Brown Boveri Review*, Vol. 32, No. 12, December 1945, pp. 436–444.

Problems

9.1 The combustor discharge into a turbine nozzle has a total temperature of 1850 K and inlet Mach number of 0.50, as shown.

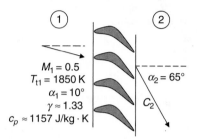

■ **FIGURE P9.1**

Assuming that the nozzle is uncooled, the axial velocity remains constant across the nozzle and the absolute flow angle at the nozzle exit is $\alpha_2 = 65°$, calculate

(a) inlet velocity C_1 in m/s

(b) the exit absolute Mach number M_2 and

(c) nozzle torque per unit mass flow rate for $r_1 \approx r_2 = 0.40$ m

9.2 Calculate the nozzle exit flow angle α_2 in Problem 9.1, if we wish the exit Mach number to be $M_2 = 1.0$, i.e., choked.

9.3 A turbine stage at pitchline has the following velocity vectors, as shown.

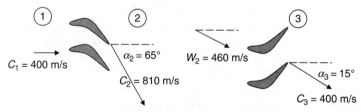

■ **FIGURE P9.3**

Calculate

(a) the axial velocities up- and downstreams of the rotor

(b) relative flow angle, β_2 in degrees

(c) the rotor velocity U_m

(d) the degree of reaction at this radius

(e) rotor specific work, w_m in kJ/kg

9.4 A turbine stage is designed with a constant axial velocity of 250 m/s and zero exit swirl. For a rotor rotational speed U_m at the pitchline of 600 m/s.

Calculate

(a) the nozzle exit flow angle, α_2 in degrees for $°R_m = 0.50$

(b) the nozzle exit flow angle, α_2 in degrees for $°R_m = 0.0$

(c) the rotor specific work at the pitchline radius, for $°R_m = 0.5$ and $°R_m = 0.0$

9.5 An axial-flow turbine nozzle turns the flow from an axial direction in the inlet to an exit flow angle of $\alpha_2 = 70°$. The rotor wheel speed is $U = 400$ m/s at the pitchline. The rotor is of impulse design and the exit flow from the rotor has zero swirl, i.e., $\alpha_3 = 0$. Calculate

(a) the rotor-specific work

(b) the stage loading at the pitchline

9.6 An axial-flow turbine stage at the pitchline is shown.

The flow entering and exiting the turbine stage is axial, i.e., $\alpha_1 = \alpha_3 = 0$

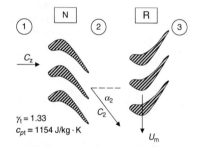

■ **FIGURE P9.6**

The nozzle exit flow is $\alpha_2 = 65°$. The shaft speed is $\omega = 5500$ rpm and the pitchline radius is $r_m = 50$ cm. Assuming $C_z = 250$ m/s = constant.

Calculate

(a) turbine-specific work w_t (kJ/kg)

(b) β_3 (degrees)

(c) $°R_m$

9.7 The combustor discharge total temperature and pressure are $T_{t1} = 2000$ K and $p_{t1} = 2$ MPa, respectively, with $\gamma_t = 1.30$ and $c_{pt} = 1244$ J/kg · K. The flow speed is 400 m/s and is in the axial direction. Calculate

(a) the combustor exit static temperature T_1 in K

(b) the Mach number M_1

(c) the combustor exit static pressure p_1 (in kPa)

9.8 The inlet flow condition to a turbine nozzle is characterized by $T_{t1} = 1800$ K and $p_{t1} = 2.4$ MPa, $M_1 = 0.5$, $\alpha_1 = 5°$, with $\gamma_t = 1.30$ and $c_{pt} = 1244$ J/kg · K. Assuming the nozzle is designed for constant axial velocity, i.e., $C_z =$ constant, calculate the nozzle exit flow angle α_2 that will produce the exit Mach number of $M_2 = 1.1$.

9.9   For the flow condition across a nozzle as shown, calculate

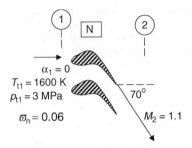

■ **FIGURE P9.9**

(a) T_2 (in K)

(b) C_{z2} (in m/s)

(c) $C_{\theta 2}$ (in m/s)

(d) M_1

(e) p_{t2} (in MPa)

(f) p_2 (in kPa)

Assume $C_z =$ constant, $\gamma_t = 1.30$, and $c_{pt} = 1244$ J/kg·K.

9.10   The velocity triangles across a turbine rotor are shown. The axial velocity remains constant across the rotor. If the total temperature at the rotor inlet is 1850 K with $\gamma_t = 1.30$ and $c_{pt} = 1244$ J/kg·K, calculate

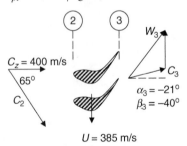

■ **FIGURE P9.10**

(a) inlet absolute Mach number M_2

(b) inlet relative Mach number M_{2r}

(c) the degree of reaction $^\circ R$

(d) rotor specific work w_t in kJ/kg

(e) exit static temperature $T_3(K)$

(f) exit relative Mach number M_{3r}.

9.11   A rotor is designed for constant axial velocity. The velocity triangles are as shown. The rotor total pressure loss coefficient is known to be 0.08 with $\gamma_t = 1.30$ and $R_t = 287$ J/kg·K.

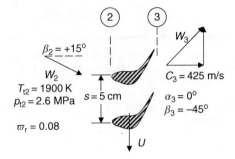

■ **FIGURE P9.11**

Calculate

(a) rotor speed U in m/s

(b) rotor specific work w_t in kJ/kg

(c) stage degree of reaction $^\circ R$

(d) rotor circulation Γ (in m²/s)

(e) inlet absolute Mach number M_2

(f) inlet gas static density ρ_2 (in kg/m³)

(g) exit relative Mach number M_{3r}

(h) exit total pressure p_{t3} (in MPa)

(i) exit static density ρ_3 (in kg/m³).

9.12   Coolant air is bled from a compressor exit at $T_{tc} = 800$ K with $c_{pc} = 1004$ J/kg·K. The coolant is given a (positive) preswirl before it enters the rotor blade root in the direction of the rotor rotation. Assuming the port of coolant entry into the rotor is at $r_c = 42$ cm with rotor angular speed $\omega = 12000$ rpm, and the coolant enters the rotor blade root axially, as shown in Fig. 9.64, calculate the coolant relative total temperature as it enters the rotor blade.

9.13   A multistage turbine is to be designed with constant axial velocity. The total temperature ratio across the entire turbine is $\tau_t = 0.72$ and the turbine polytropic efficiency is $e_t = 0.85$ with $\gamma_t = 1.30$ and $c_{pt} = 1244$ J/kg·K. Calculate

(a) The turbine total pressure ratio π_t

(b) the turbine adiabatic efficiency η_t

(c) turbine area ratio, A_5/A_4, for choked inlet, i.e., $T_{t4} = 1675$ K and $C_z = 445$ m/s and swirl-free exit flow

Assume γ and R remain constant across the turbine.

9.14   A 50% degree of reaction turbine stage is shown. The nozzle turns the flow 65° and the rotor exit flow is swirl-free. Assuming axial velocity remains constant throughout the stage, calculate

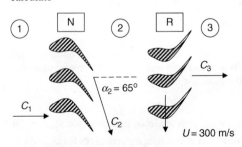

■ **FIGURE P9.14**

(a) axial velocity C_z m/s

(b) rotor specific work w_t kJ/kg

(c) stage loading ψ

(d) flow coefficient φ

(e) static temperature drop $\Delta T = T_1 - T_2$

(f) static temperature drop $\Delta T = T_2 - T_3$

Assume $c_{pt} = 1.156$ kJ/kg · K.

9.15 A simple method to establish the annulus geometry in a multistage turbine is to assume a constant throughflow (i.e., axial) speed C_z. Calculate

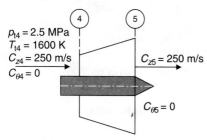

$p_{t4} = 2.5$ MPa
$T_{t4} = 1600$ K
$C_{z4} = 250$ m/s
$C_{\theta4} = 0$

$C_{z5} = 250$ m/s

$C_{\theta5} = 0$

■ **FIGURE P9.15**

(a) the density ratio ρ_5/ρ_4

(b) the exit-to-inlet area ratio A_5/A_4

for

$1/\pi_t = 3.5$

$e_t = 0.80$

$C_z = $ const.

$\gamma_t = 1.33, c_{pt} = 1156$ J/kg · K

Also, if the hub radius is constant, calculate the tip-to-tip radius ratio $(r_5/r_4)_{tip}$ for $(r_h/r_t)_4 = 0.80$

9.16 A turbine stage has an inlet total temperature T_{t1} of 2000 K. If the rotational speed of the rotor at the pitchline is 200 m/s and the axial velocity component is $C_z = 250$ m/s, calculate

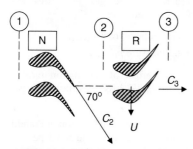

■ **FIGURE P9.16**

(a) the stagnation temperature as seen by the rotor, $T_{t2,r}$

(b) the gas static temperature upstream of the rotor, T_2

(c) the adiabatic wall temperature for the rotor, $T_{aw,r}$ assuming a turbulent boundary layer

Assume $\gamma = 1.33, c_p = 1156$ J/kg · K.

9.17 A turbine stage is characterized by a loading coefficient, ψ of 0.8, and a flow coefficient of ϕ of 0.6. The rotor is unshrouded and has a tip clearance gap-to-blade height ratio of 2.5%. The ratio of the tip to pitchline radius is $r_t/r_m = 1.50$. The rotor blade tip has a knife configuration with a tip discharge coefficient of 0.60. The mean relative flow angle in the turbine rotor is $\beta_m = -30°$. Calculate the loss of turbine efficiency due to the tip clearance.

9.18 The relative flow angles in a turbine rotor are $\beta_2 = 25°$ and $\beta_3 = -45°$.

Calculate

(a) optimum axial solidity parameter

(b) the axial solidity σ_z based on the recommended value of the loading coefficient ψ_z

9.19 The thermal boundary layer associated with the flow of a hot gas over an insulated flat plate is shown.

Assume that the gas Prandtl number is Pr = 0.71, the ratio of specific heats is $\gamma = 1.33$, the gas Mach number is $M_g = 0.75$.

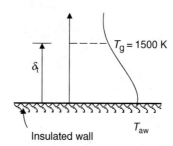

$T_g = 1500$ K

δ_t

T_{aw}

Insulated wall

■ **FIGURE P9.19**

Calculate

(a) the adiabatic wall temperature T_{aw} (in K)

(b) the gas stagnation temperature T_{tg} (in K)

First assume that the viscous boundary layer is laminar and then solve the problem for a turbulent boundary layer.

9.20 Consider the flow of high-temperature, ($M_g = 1.0, \gamma_g = 1.33, c_{pg} = 1156$ J/kg · K) high Mach number gas over a flat wall, ($T_g = 1420$ K, $p_g = 12$ kpa, $Pr_g = 0.70$) as shown. We intend to internally cool the wall to achieve a wall temperature of $T_{wg} = 1200$ K.

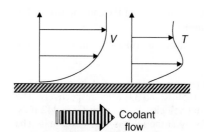

V

T

Coolant
flow

■ **FIGURE P9.20**

Assuming the gas-side Stanton number is $St_g = 0.005$, calculate

(a) the gas stagnation temperature T_{tg} (K)

(b) the adiabatic wall temperature T_{aw} (K) for a turbulent boundary layer

(c) the gas-side film coefficient h_g (W/m^2 K)

(d) the heat flux to the wall q_w (kW/m^2)

For a wall thickness of $t_w = 3$ mm, and a thermal conductivity, $k_w = 14.9$ W/m · K, calculate

(e) the wall temperature on the coolant side T_{wc} (K)

9.21 The leading edge on a turbine nozzle is to be internally cooled using an impingement cooling technique, as shown. The leading-edge diameter is 8 mm. Calculate the heat transfer coefficient h_g at the leading edge, assuming an augmentation factor $a = 1.5$ due to a high-intensity turbulent flow in the turbine. (use Eq. 9.80 for a cylinder in cross flow).

$M_g = 0.5$,
$\gamma_g = 1.33$,
$c_{pg} = 1156$ J/kg · K
$k_g = 0.082$ W/m · K
$T_g = 1500$ K
$p_g = 1.0$ Mpa
$Pr_g = 0.70$
$\mu_g = 4.9 \times 10^{-5}$ kg/m · s

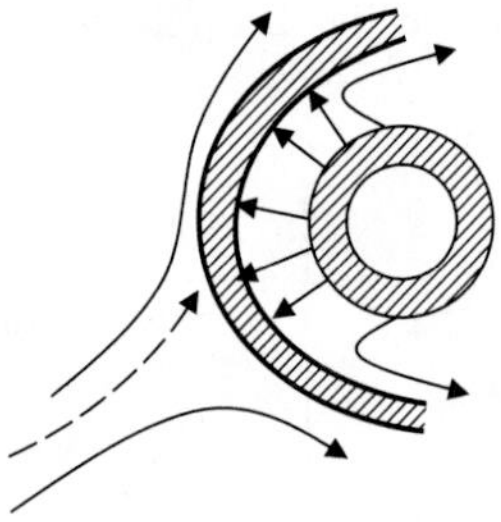

■ **FIGURE P9.21**

9.22 The exit flow angle in a turbine nozzle at the pitchline is 70°. The blade spacing is $s = 6$ cm. The total pressure and temperature at the nozzle throat are 2.6 MPa and 1700 K respectively. Assuming the nozzle throat is choked calculate the mass flow rate (per unit span) through a single nozzle blade passage. The gas properties may be assumed to be $\gamma_t = 1.30$ and $c_{pt} = 1244$ J/kg · K.

9.23 A turbine blade row is cooled with a coolant mass fraction of 3% and the coolant inlet temperature of $T_{tc} = 775$ K. The coolant is ejected at the blade trailing edge at a temperature of 825 K as it mixes with the hot gas at the temperature of 1825 K. Assuming gas properties for the coolant and hot gas are $\gamma_c = 1.40$, $c_{pc} = 1004$ J/kg · K and $\gamma_t = 1.30$ and $c_{pt} = 1244$ J/kg · K and the coolant flow rate is 3 kg/s, calculate

(a) the rate of heat transfer to the coolant from the blade (in kW)

(b) the mixed-out specific heat (of the coolant and the hot gas after mixing)

(c) the mixed-out temperature (of the coolant and the hot gas after mixing)

9.24 The total temperature and pressure at the entrance to a rotor are known to be 1850 K and 3.0 MPa, respectively. The rotor is of constant axial velocity design with $C_z = 442$ m/s. The rotor relative velocity at the inlet is $W_2 = 562$ m/s. The rotor relative exit Mach number is to be $M_{3r} = 0.90$. Assuming $\alpha_2 = 60°$, $\gamma = 1.33$ and $c_p = 1156$ J/kg · K calculate

(a) the relative inlet Mach number M_{2r}

(b) the relative flow angle at the exit β_3 (degrees)

(c) the absolute exit flow angle α_3 (degrees)

(d) stage degree of reaction $°R$

9.25 A turbine rotor has a tip clearance height of 1% compared with blade height. The loading coefficient is $\psi = 2.5$ and the flow coefficient is $\varphi = 0.6$. The turbine efficiency for zero tip clearance is $\eta_0 = 0.85$. The tip-to-mean radius ratio is $r_t/r_m = 1.25$, and the mean flow angle in the rotor is $\beta_m = -15°$. Using the following tip clearance efficiency loss model,

$$\Delta\eta \cong \eta_0 C_D . \tau . \left(\frac{r_t}{r_m}\right)\left(1 - \frac{\psi}{\phi}\tan\beta_m\right)^{1/2}$$

calculate the turbine efficiency loss due to 1% tip clearance for the following rotor tip shapes

(a) *knife* with gap Reynolds number of 2000

(b) *groove*

Assuming the rotor with a flat-tip has a discharge coefficient of $C_{D,flat-tip} = 0.83$.

9.26 The free-stream gas temperature is $T_\infty = 1600$ K and the free-stream gas speed is $V_\infty = 850$ m/s. The gas properties are $\gamma_t = 1.30$, $c_{pt} = 1244$ J/kg · K and Prandtl number Pr $= 0.73$. Consider the flow of this gas over a flat plate at a high Reynolds number corresponding to turbulent flow. Calculate

(a) the gas total temperature $°T_{t\infty}$ (in K) in the freestream

(b) the adiabatic wall temperature T_{aw} (in K)

(c) percent error if we assume adiabatic wall temperature is the same as total temperature of the gas

9.27 A gas turbine is to provide a shaft power of 50 MW to a compressor with mass flow rate of 100 kg/s. The compressor bleeds 10 kg/s at its exit for turbine internal cooling purposes. The combustor exit mass flow rate is 93 kg/s, which accounts for 3 kg/s of fuel flow rate in the combustor. The turbine inlet temperature is $T_{t4} = 1850$ K and $p_{t4} = 2.0$ MPa with $\gamma_t = 1.30$, $c_{pt} = 1244$ J/kg · K Assuming the coolant total temperature is $T_{tc} = 785$ K and the gas properties are $\gamma_c = 1.40$, $c_{pc} = 1004$ J/kg · K, calculate

(a) the turbine exit total temperature $T_{t5\text{-cooled}}$

(b) the turbine exit total pressure $p_{t5\text{-cooled}}$.

You may assume that the effect of cooling on turbine adiabatic efficiency is about ~2.8% loss (of efficiency) per 1% cooling.

9.28 The Stanton number is in general a function of Prandtl and Reynolds numbers, among other nondimensional parameters such as Mach number, roughness, curvature, free-stream turbulence intensity, etc. Eckert–Livingood model for a flat plate with constant wall temperature, excluding all other effects except Prandtl number and the Reynolds number, is

$$St_g = 0.0296 Pr_g^{-2/3} Re_x^{-1/5}$$

for a turbulent boundary layer. The Prandtl number for the gas is 0.704 and remains constant along the plate. Make a spreadsheet calculation of Stanton number St_g with respect to Reynolds number in the range of $200,000 \leq Re_x \leq 500,000$. Graph the Stanton number versus Reynolds number. Also, calculate the wall-averaged Stanton number.

9.29 The nozzle throat opening o is to be sized for a throat Reynolds number of $Re_o = 500,000$. The Throat is choked with $T_{t\text{-throat}} = 2000$ K, $p_{t\text{-throat}} = 2.1$ MPa, and the gas properties at the throat are $\gamma_t = 1.30$, $c_{pt} = 1244$ J/kg · K, $\mu_{throat} = 6.5 \times 10^{-5}$ kg/m · s. Calculate the nozzle throat opening o in centimeters.

9.30 A turbine rotor has a hub and tip radii

$$r_h = 35 \text{ cm}$$
$$r_t = 50 \text{ cm}$$

The rotor angular speed is $\omega = 10,000$ rpm. The rotor blade taper ratio A_t/A_h is 0.75. Estimate the ratio of blade centrifugal stress at the hub to blade material density, σ_c/ρ_{blade}. If this rotor operates at 1200°F, what are the suitable materials for this rotor? (*Hint*: Use Fig. 9.61 as a guide)

CHAPTER 10

Courtesy of Rolls-Royce, plc

Aircraft Engine Component Matching and Off-Design Analysis

10.1   Introduction

How does an existing engine behave in an off-design operating condition? For an existing engine, we know its geometry and design-point parameters, such as compressor pressure ratio or turbine entry temperature. All the parameters that we used in the cycle analysis were in effect "design-point" parameters. For example, take compressor pressure ratio; it may be designed into a number of stages, each with a certain design pressure ratio. Although we cannot change the number of stages after the engine is manufactured, we may operate the compressor at a different shaft speed than its design speed, or, equivalently, we may operate the engine at a different throttle setting than its design. In general, the independent parameters that may be set and create an off-design flight condition for an airbreathing jet engine are

- Altitude of flight, which may include nonstandard ambient pressure and temperature, e.g., arctic flight or seasonal effects
- Flight Mach number, from takeoff, climb, cruise, dash, approach, landing
- Throttle setting, i.e., turbine entry temperature
- Afterburner setting, e.g., -on, -off, or partial setting
- Nozzle area ratio setting, e.g., to optimize installed thrust

We will first consider individual component matching in a jet engine. Then we will examine the entire system from the gas generator to the whole aircraft engine at on- and off-designs.

10.2   Engine (Steady-State) Component Matching

What are the physical laws that need to be satisfied by engine components that interact with each other regardless of their mode of operation? In steady-state operation, the physical mass flow rate (e.g., in kg/s) through a component should be delivered in its entirety to the next component(s). For example, the air mass flow rate through the inlet must be equal to the fan mass flow rate, unless some air was bypassed in the inlet before getting to the fan, in an inlet flow stability scheme. This satisfies the law of conservation of mass.

The next conservation law that applies to the components that interact is the conservation of energy. For example, the low-pressure spool connecting the low-pressure turbine to the low-pressure compressor/fan via a shaft must deliver turbine shaft power output (e.g., in MW) to the low-pressure compressor allowing for possible power extraction for electric generator and frictional losses (i.e., energy dissipation) in the bearings supporting the rotating machinery. In addition, the compressor and turbine stages on the same shaft are constrained by the same physical shaft (angular) speed, i.e., $N_{LPC} = N_{LPT}$ and $N_{HPC} = N_{HPT}$ for a two-spool machine.

There is also the law of conservation of atomic species. We studied this topic in Chapter 6 for a mixture of gases that were chemically reacting. In this chapter, we allow for a variable gas composition through a component, but we will not do an equilibrium chemical analysis of gas mixtures, except for the combustor. For example, the chemical reaction in a combustor produces a certain gas composition at the turbine entry. The flow expansion in the turbine, i.e., the temperature drop, however, causes the gas composition to change. Applying the equilibrium gas chemistry principles will establish the gas composition at every operating condition. To simplify the process, i.e., to avoid running the equilibrium gas chemistry codes (see, for example, Gordon and McBride, 1994) for every component, we choose to specify "appropriate" gas constants, γ, R or γ, and c_p, per component entrance and exit condition (or cold and hot). In cycle analysis, we used three levels of gas constants, (1) inlet and compressor (or cold), (2) turbine (or hot section), and (3) the afterburner and nozzle (or very hot section). Although the principle variable in establishing gas constants is the gas temperature, but the fraction of hydrocarbon fuel content in the gas also impacts these variables.

A remark about the law of conservation of momentum is also in order, as it appears that it is omitted from our component matching discussions. In reality, we still use the law of conservation of momentum to calculate the force magnitude and direction that each component produces in the propulsion system (remember the section on fluid impulse in Chapter 2). However, in the steady-state component matching case, it is the mass and energy that dictate the interaction.

Finally, a brief review of engine corrected parameters is in order, as the component performance maps are often presented in these coordinates. There are five component and engine corrected parameters. These are (1) Corrected mass flow rate, (2) Corrected shaft speed, (3) Corrected fuel flow rate, (4) Corrected thrust, and (5) Corrected thrust-specific fuel consumption.

10.2.1   Engine Corrected Parameters

We studied in Chapters 7–9 that the corrected mass flow rate and corrected shaft speed are the parameters that fully describe the behavior of a compressor or turbine. These parameters are related to the axial and tangential Mach numbers in turbomachines, respectively. We had defined reference pressure and temperature, $p_{ref} = 101.33\,kPa$ and $T_{ref} = 288.2\,K$, respectively. Based on these reference conditions, we defined a δ and a θ parameter, for station i, according to

$$\delta_i \equiv p_{ti}/p_{ref} \tag{10.1}$$

$$\theta_i \equiv T_{ti}/T_{ref} \tag{10.2}$$

The corrected mass flow rate at station i is the mass flow rate through station i at the reference pressure and temperature:

$$\dot{m}_{ci} \equiv \frac{\dot{m}_i \sqrt{\theta_i}}{\delta_i} = \sqrt{\frac{\gamma_i}{R_i}} \frac{p_{ref}}{\sqrt{T_{ref}}} A_i M_i \left(1 + \frac{\gamma_i - 1}{2} M_i^2 \right)^{-\frac{\gamma_i + 1}{2(\gamma_i - 1)}} \tag{10.3}$$

The corrected shaft speed is

$$N_{ci} \equiv \frac{N}{\sqrt{\theta_i}} \tag{10.4}$$

We may define a corrected fuel flow rate according to

$$\dot{m}_{fc} \equiv \frac{\dot{m}_f}{\delta_2 \sqrt{\theta_2}} \tag{10.5}$$

Corrected thrust and thrust-specific fuel consumption, are

$$F_c \equiv F/\delta_0 \tag{10.6}$$

$$\text{TSFC}_c \equiv \frac{\text{TSFC}}{\sqrt{\theta_0}} \tag{10.7-a}$$

By taking the ratio of Eqs. 10.5 and 10.6, we relate three corrected parameters according to

$$\text{TSFC}_c = \pi_d \frac{\dot{m}_{fc}}{F_c} \tag{10.7-b}$$

10.2.2 Inlet-Compressor Matching

A simple schematic drawing of an inlet coupled to a fan (or compressor) is shown as a guide to discussion in Fig. 10.1. The fixed areas are the inlet lip A_1, engine face area A_2, and compressor

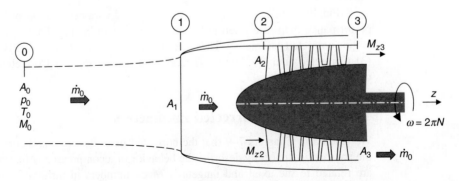

■ **FIGURE 10.1** Schematic representation of the inlet-compressor matching

exit area A_3. The variable is the free stream capture area A_0, which is a strong function of flight condition, e.g., takeoff, climb, or cruise.

The ambient static pressure and temperature p_0 and T_0 and flight Mach number are independent variables as well. We may establish flight total pressure and temperature p_{t0} and T_{t0} using the static values and flight Mach number M_0. We will use the flight Mach number and total gas properties in the cycle component matching and off-design analysis.

The physical mass flow rate through an area A is tied to axial Mach number M_z, local p_t, and T_t. The expression for the mass flow rate in terms of these variables is

$$\dot{m} = \sqrt{\frac{\gamma}{R}} \frac{p_t}{\sqrt{T_t}} \cdot A \cdot M_z \left(1 + \frac{\gamma - 1}{2} M_z^2\right)^{\frac{-(\gamma+1)}{2(\gamma-1)}} \tag{10.8}$$

The engine face (physical) mass flow rate (e.g., in kg/s or lbm/s) is

$$\dot{m}_2 = \sqrt{\frac{\gamma}{R}} \frac{p_{t2}}{\sqrt{T_{t2}}} \cdot A_2 \cdot M_{z2} \left(1 + \frac{\gamma - 1}{2} M_{z2}^2\right)^{\frac{-(\gamma+1)}{2(\gamma-1)}} \tag{10.9}$$

The inlet mass flow rate is

$$\dot{m}_0 = \sqrt{\frac{\gamma}{R}} \frac{p_{t0}}{\sqrt{T_{t0}}} \cdot A_0 \cdot M_0 \left(1 + \frac{\gamma - 1}{2} M_0^2\right)^{\frac{-(\gamma+1)}{2(\gamma-1)}} \tag{10.10}$$

The law of conservation of mass demands that the inlet flow rate and the airflow rate into the engine to be equal (assuming there is no inlet bleed), i.e.,

$$\sqrt{\frac{\gamma}{R}} \frac{p_{t2}}{\sqrt{T_{t2}}} \cdot A_2 \cdot M_{z2} \left(1 + \frac{\gamma - 1}{2} M_{z2}^2\right)^{\frac{-(\gamma+1)}{2(\gamma-1)}} = \sqrt{\frac{\gamma}{R}} \frac{p_{t0}}{\sqrt{T_{t0}}} \cdot A_0 \cdot M_0 \left(1 + \frac{\gamma - 1}{2} M_0^2\right)^{\frac{-(\gamma+1)}{2(\gamma-1)}}$$
$$\tag{10.11}$$

The conservation of energy, along with the assumption of the adiabatic flow in the inlet, identifies a constant total enthalpy flow or equivalently constant total temperature, if we assume constant c_p. Therefore, eliminating the terms that remain unchanged between the flight condition and engine face in Eq. 10.11, we get

$$\pi_d \frac{A_2}{A_0} \cdot M_{z2} \left(1 + \frac{\gamma - 1}{2} M_{z2}^2\right)^{\frac{-(\gamma+1)}{2(\gamma-1)}} = M_0 \left(1 + \frac{\gamma - 1}{2} M_0^2\right)^{\frac{-(\gamma+1)}{2(\gamma-1)}} \tag{10.12}$$

The RHS in Eq. 10.12 is completely known for a given flight Mach number M_0. The LHS contains three unknowns, (1) inlet total pressure recovery π_d, (2) the free stream capture area A_0 (or the ratio A_2/A_0), and (3) the axial Mach number at the engine face, M_{z2}. A typical inlet performance map contains the inlet recovery variation with flight Mach number, i.e., $\pi_d = f(M_0)$, or we may use the AIA or MIL-E 5008-B standards that we presented in Chapter 5. Therefore, the mass and energy balance for the inlet/compressor results in two unknowns A_0 and M_{z2} for any flight Mach number M_0.

The main area of concern in inlet-compressor matching is, however, tied to the inlet distortion problem. We have addressed both the inlet steady state and dynamic distortion effects on compressor performance extensively in Chapter 7. The ensued unsteadiness and instability in a compressor/fan, i.e., stall and surge, find their roots often in inlet distortion.

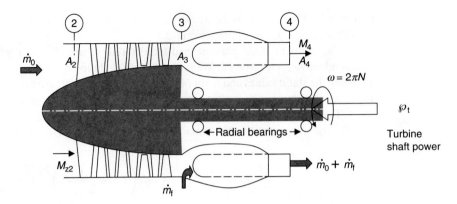

■ **FIGURE 10.2** Compressor–combustor matching with their respective mass flow rates, areas, and Mach numbers

10.2.3 Compressor–Combustor Matching

In a steady-state operation, the compressor (air) mass flow rate combines with the fuel flow rate in the combustor to form the exit mass flow rate at the entrance to the turbine. Figure 10.2 shows the combination of the compressor and burner and the three stations that connects them.

The mass flow rate at the burner exit is

$$\dot{m}_4 = \dot{m}_3 + \dot{m}_f = (1 + f)\dot{m}_0 = (1 + f)\dot{m}_2 \tag{10.13}$$

The unknowns on the RHS of Eq. 10.13 are the fuel-to-air ratio f and the axial Mach number at the engine face, M_{z2}. The fuel-to-air ratio is set by the throttle, which is an independent parameter. Therefore, we consider the throttle setting as known in an off-design operating condition. We can also write the burner exit mass flow rate in terms of total pressure p_{t4} and temperature T_{t4}, flow area A_4, and the local Mach number M_4, namely,

$$\dot{m}_4 = \sqrt{\frac{\gamma_t}{R_t}} \frac{p_{t4}}{\sqrt{T_{t4}}} \cdot A_4 \cdot M_4 \left(1 + \frac{\gamma_t - 1}{2} M_4^2\right)^{\frac{-(\gamma_t+1)}{2(\gamma_t-1)}} \tag{10.14}$$

Now, let us examine the RHS of Eq. 10.14. If we take the station 4 as the exit of the first turbine nozzle, then over a wide operating condition, it remains choked, i.e., $M_4 = 1$. The combustor exit total pressure $p_{t4} = \pi_b \cdot p_{t3}$, which for an assumed combustor total pressure loss π_b is the compressor discharge total pressure p_{t3}, which may be written as

$$p_{t4} = \pi_b p_{t3} = \pi_b \pi_c p_{t2} = \pi_d \pi_b \pi_c p_{t0} \tag{10.15}$$

The combustor exit total temperature T_{t4} is throttle setting dependent and is thus treated as known. In light of Eq. 10.15, let us recast Eq. 10.13 in the following form:

$$\sqrt{\frac{\gamma_t}{R_t}} \frac{\pi_b \pi_c}{\sqrt{T_{t4}/T_{t2}}} \cdot A_4 \left(\frac{\gamma_t + 1}{2}\right)^{\frac{-(\gamma_t+1)}{2(\gamma_t-1)}} = (1 + f)\sqrt{\frac{\gamma_c}{R_c}} A_2 \cdot M_{z2} \left(1 + \frac{\gamma_c - 1}{2} M_{z2}^2\right)^{\frac{-(\gamma_c+1)}{2(\gamma_c-1)}} \tag{10.16}$$

The RHS of Eq. 10.16 is a function of axial flow Mach number at the engine face, M_{z2} and throttle setting. Also we note that the RHS of Eq. 10.16 is proportional to the corrected mass flow rate at the compressor face. Therefore, we conclude that for a given compressor corrected mass flow rate and engine throttle setting, T_{t4}/T_{t2} and f, we can establish a unique compressor pressure ratio π_c. This is consistent with our earlier studies of turbomachinery as depicted in Fig. 10.3 (reproduced from Fig. 7.57) where a compressor performance map shows the throttle

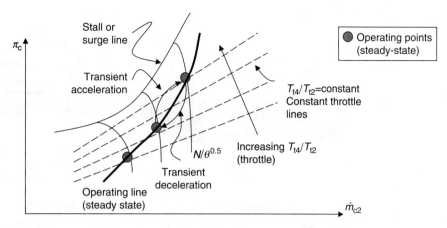

■ FIGURE 10.3   Operating line (or compressor–combustor matching) on a compressor map

lines ($T_{t4}/T_{t2} =$ constant, or nominal operating line) and the engine corrected speed lines N_c. The path of steady-state engine operation is shown as the engine "operating line" in Fig. 10.3. An alternative view of Eq. 10.16 is that for a given compressor pressure ratio and the throttle setting, there is a unique corrected compressor mass flow rate, which is a pure function of the axial Mach number into the engine, M_{z2}. We may use Eq. 10.16 with the design choices for π_c, M_{z2}, T_{t4}/T_{t2}, and f (and an assumed π_b) to calculate A_4/A_2. This area ratio remains constant (for the fixed-area turbines) at off-design.

The energy balance across the combustor reveals the fuel-to-air ratio f as

$$f = \frac{h_{t4} - h_{t3}}{Q_R \eta_b - h_{t4}} \tag{10.17-a}$$

The unknowns on the RHS of Eq. 10.17-a are h_{t3} and η_b, assuming that the throttle setting has specified T_{t4} in the off-design condition. Once we establish a compressor pressure ratio at the off-design operation, h_{t3} becomes known. We may use either a burner efficiency correlation for η_b (e.g., from Fig. 6.40) or an assumed value to proceed with the calculation of the new fuel-to-air ratio f at off-design. In terms of the constant-throttle-line parameter T_{t4}/T_{t2}, as shown in Fig. 10.3, and compressor temperature ratio, we can rewrite Eq. 10.17-a by dividing its numerator and denominator by the total enthalpy at the engine face, h_{t2}, to get

$$f Q_R \eta_b / h_{t2} \approx \frac{c_{pt}}{c_{pc}} \left(\frac{T_{t4}}{T_{t2}} \right) - \tau_c \tag{10.17-b}$$

Finally, the compressor power is matched to the shaft power produced by the turbine when we account for any external (electric) power extraction and bearing frictional losses. The physical shaft rotational speed N is the same for the compressor–turbine stages on the same spool (i.e., shaft). We will address these matching requirements in the compressor–turbine matching section.

10.2.4   Combustor–Turbine Matching

The steady-state combustor–turbine interaction is dominated by (1) choked turbine nozzle (i.e., $M_4 = 1$, which acts as the first throttle station in the gas turbine engine), (2) turbine cooling requirements, (3) hot spots in turbine entry, and (4) loss of turbine efficiency due to cooling. The latter three issues relate to either combustor design or turbine cooling and its impact on efficiency. Our goal in turbine off-design analysis is primarily to calculate turbine expansion

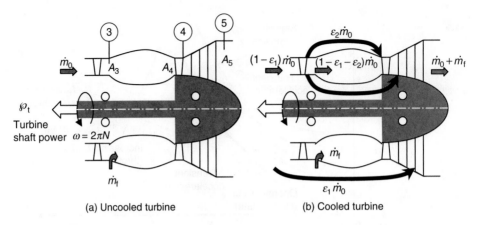

(a) Uncooled turbine (b) Cooled turbine

■ **FIGURE 10.4** Combustor–turbine matching

parameter (i.e., its power production capability). The effect of cooling on turbine aerodynamics is often modeled as an equivalent uncooled turbine operating at a lower adiabatic efficiency, e.g., $\eta_t = 0.75-0.85$. Figure 10.4 shows the schematic drawing of a combustor–turbine with part (a) showing an uncooled turbine and part (b) showing a cooled turbine. The two epsilons ε_1 and ε_2 in Fig. 10.4(b) depict air-cooling mass flow fractions that are bled from the compressor at an intermediate pressure point and at the compressor exit, respectively.

The physical mass flow rates at stations 4 and 5 are equal, if the turbine is uncooled. Therefore, by writing the corrected mass flow rates in both stations as

$$\dot{m}_{c4} = \dot{m}_4 \sqrt{\theta_4}/\delta_4 \tag{10.18-a}$$

$$\dot{m}_{c5} = \dot{m}_5 \sqrt{\theta_5}/\delta_5 \tag{10.18-b}$$

And taking the ratio of the above two expressions, we relate the corrected mass flow rates at turbine entrance and turbine exit according to

$$\frac{\dot{m}_{c4}}{\dot{m}_{c5}} = \frac{\pi_t}{\sqrt{\tau_t}} \tag{10.19}$$

The RHS of Eq. 10.19 relates to turbine expansion, and it is this parameter that controls the axial Mach numbers, i.e., mass flow rate, through the turbine. We will encounter this parameter again in this chapter. In the cooled turbine case, we need to account for the mass flow rate of the coolant, which is injected between stations 4 and 5 in the turbine. Therefore,

$$\dot{m}_4 = (1 + f - \varepsilon_1 - \varepsilon_2)\dot{m}_0 \tag{10.20-a}$$

$$\dot{m}_5 = (1 + f)\dot{m}_0 \tag{10.20-b}$$

The corrected mass flow rates, in the cooled turbine case, are related by cooling fractions according to

$$\frac{\dot{m}_{c4}}{\dot{m}_{c5}} = \left(\frac{1 + f - \varepsilon_1 - \varepsilon_2}{1 + f}\right) \frac{\pi_t}{\sqrt{\tau_t}} = \left(1 - \frac{\varepsilon_1 + \varepsilon_2}{1 + f}\right) \frac{\pi_t}{\sqrt{\tau_t}} \tag{10.21}$$

One last comment about τ_t is that in a cooled turbine case, T_{t5} is reduced as a result of coolant injection in the hot gas stream. Therefore τ_t (cooled turbine) is reduced compared with τ_t in an uncooled turbine. Another impact of cooling the turbine is on π_t, which is also reduced as the

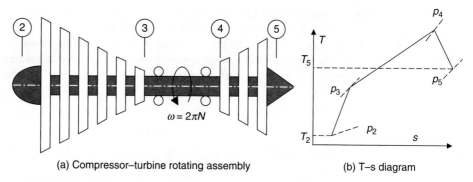

(a) Compressor–turbine rotating assembly   (b) T–s diagram

■ **FIGURE 10.5**   **Schematic drawing of a compressor–turbine rotating assembly on the same shaft**

turbulent mixing of the coolant injection into the hot gas causes a total pressure drop in the turbine.

10.2.5   Compressor–Turbine Matching and Gas Generator Pumping Characteristics

The law of conservation of energy is applied to a compressor that is driven by a turbine on the same shaft. The schematic drawing of a compressor–turbine rotating assembly is shown in Fig. 10.5. We may account for power dissipation in bearings through mechanical efficiency η_m, which results in the following form of law of conservation of energy:

$$\wp_\mathrm{c} = \dot{m}_0(h_{t3} - h_{t2}) = \eta_\mathrm{m}\wp_\mathrm{t} = \eta_\mathrm{m}(\dot{m}_0 + \dot{m}_\mathrm{f})(h_{t4} - h_{t5}) \tag{10.22-a}$$

Dividing Eq. 10.22-a by the air mass flow rate and the total enthalpy at the engine face (which is the same as the flight total enthalpy), we get

$$\tau_\mathrm{c} - 1 = \eta_\mathrm{m}(1 + f)\frac{c_{pt}}{c_{pc}}\frac{T_{t4}}{T_{t2}}(1 - \tau_\mathrm{t}) \tag{10.22-b}$$

We shall demonstrate that the turbine expansion parameter τ_t remains constant over a wide operating range in the engine off-design condition; therefore, we may conclude from Eq. 10.22-b that the compression parameter $(\tau_\mathrm{c} - 1)$ varies linearly with the throttle parameter T_{t4}/T_{t2} in off-design. Ignoring the variation of $\eta_\mathrm{m}(1 + f)$ in on- and off-designs, we get

$$\frac{\tau_\mathrm{c,O-D} - 1}{\tau_\mathrm{c,D} - 1} \cong \frac{(T_{t4}/T_{t2})_\mathrm{O-D}}{(T_{t4}/T_{t2})_\mathrm{D}} \tag{10.23}$$

From Eq. 10.23, we arrive at the off-design compressor pressure ratio based on either compressor polytropic or the adiabatic efficiency according to

$$\pi_\mathrm{c} = \tau_\mathrm{c}^{e_c\gamma_c/(\gamma_c-1)}$$

$$\pi_\mathrm{c} = [1 + \eta_\mathrm{c}(\tau_\mathrm{c} - 1)]^{\gamma_c/(\gamma_c-1)}$$

In terms of the throttle ratio T_{t4}/T_{t2} and adiabatic efficiency, the compressor off-design pressure ratio is

$$\pi_\mathrm{c,O-D} \cong \left[1 + \eta_\mathrm{c}(\tau_\mathrm{c,D} - 1)\frac{(T_{t4}/T_{t2})_\mathrm{O-D}}{(T_{t4}/T_{t2})_\mathrm{D}}\right]^{\gamma_c/(\gamma_c-1)} \tag{10.24}$$

Since compressor and turbine performance maps are always shown in terms of the corrected parameters, i.e., corrected mass flow rate and the corrected shaft speed, we express the off-design performance of the gas generator in corrected terms as well.

The common shaft between the compressor and turbine has the same angular speed $\omega = 2\pi N$. Therefore, the corrected shaft speeds of the compressor and turbine, which are defined as

$$N_{c2} \equiv N/\sqrt{\theta_2} \tag{10.25-a}$$

$$N_{c4} \equiv N/\sqrt{\theta_4} \tag{10.25-b}$$

are related to each other by the throttle parameter, i.e.,

$$N_{c2} \equiv N_{c4}\sqrt{T_{t4}/T_{t2}} \tag{10.25-c}$$

Since we had demonstrated in Chapter 7 that compressor (specific) work is proportional to the square of shaft speed, i.e.,

$$w_c \propto T_{t2}(\tau_c - 1) \propto N^2 \tag{10.26}$$

Divide Eq. 10.26 by T_{t2} to arrive at the corrected shaft speed, i.e.,

$$\tau_c - 1 \propto N_{c2}^2 \tag{10.27}$$

Now, by comparing Eq. 10.27 with Eq. 10.25-c, we conclude that for $\tau_t = $ constant, the corrected shaft speed N_{c4} remains constant, and N_{c2} is proportional to the square root of the throttle parameter T_{t4}/T_{t2}, i.e.,

$$N_{c4,O-D} \approx N_{c4,D} \approx \text{constant} \tag{10.28-a}$$

$$\boxed{\frac{N_{c2,O-D}}{N_{c2,D}} \approx \frac{\sqrt{(T_{t4}/T_{t2})_{O-D}}}{\sqrt{(T_{t4}/T_{t2})_{D}}}} \tag{10.28-b}$$

Again, the corrected mass flow rates in compressors and turbines are

$$\dot{m}_{c2} = \dot{m}_2\sqrt{\theta_2}/\delta_2 \tag{10.29-a}$$

$$\dot{m}_{c4} = \dot{m}_4\sqrt{\theta_4}/\delta_4 \tag{10.29-b}$$

The physical mass flow rates in compressors and turbines are different only by the amount of fuel that we consume in the burner, therefore Eqs. 10.29-a and −b yield

$$(1 + f)\dot{m}_{c2}\delta_2/\sqrt{\theta_2} = \dot{m}_{c4}\delta_4/\sqrt{\theta_4}$$

Since the flow in station 4 is choked (over a wide operating range), the corrected mass flow rate is constant at 4 and we thus conclude that

$$\dot{m}_{c2} = \frac{\dot{m}_{c4}}{1 + f}\frac{p_{t4}}{p_{t2}}\bigg/\sqrt{\frac{T_{t4}}{T_{t2}}} = \frac{\dot{m}_{c4}\pi_b\pi_c}{(1 + f)\sqrt{T_{t4}/T_{t2}}} \tag{10.30-a}$$

$$\boxed{\frac{\dot{m}_{c2,O-D}}{\dot{m}_{c2,D}} \cong \frac{\pi_{c,O-D}}{\pi_{c,D}}\frac{\sqrt{(T_{t4}/T_{t2})_D}}{\sqrt{(T_{t4}/T_{t2})_{O-D}}}} \tag{10.30-b}$$

Figure 10.6 shows typical compressor and turbine performance maps in corrected parameters, with constant adiabatic efficiency contours and the respective operating lines.

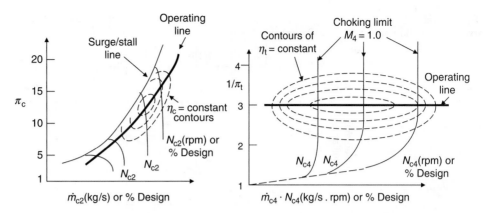

■ **FIGURE 10.6**   **Typical compressor and turbine performance maps used in component matching studies**

10.2.5.1   *Gas Generator Pumping Characteristics.*

In the previous sections, we studied the component matching at on- and off-designs for an "on-paper" aircraft engine with a chosen set of design parameters. However, in the case of an existing gas generator where component performance maps are available, we can study all its possible operating conditions by establishing its so-called *pumping characteristics*. They are (1) the corrected airflow rate $\dot{m}_{c2}$, (2) the pressure ratio of the gas generator, p_{t5}/p_{t2}, (3) the temperature ratio of the gas generator, T_{t5}/T_{t2}, and (4) the fuel flow parameter, $fQ_R\eta_b/c_pT_{t2}$ or the corrected fuel flow rate $\dot{m}_{fc}$. The pumping characteristics are often expressed as a function of percent corrected shaft speed $\%N_{c2}$ and the throttle parameter T_{t4}/T_{t2}.

From Eq. 10.30-a, we have

$$\dot{m}_{c2} = \frac{\dot{m}_{c4}}{1+f}\frac{p_{t4}}{p_{t2}}\Bigg/\sqrt{\frac{T_{t4}}{T_{t2}}} = \frac{\dot{m}_{c4}\pi_b\pi_c}{(1+f)\sqrt{T_{t4}/T_{t2}}} \tag{10.31}$$

$$\pi_c = \frac{(1+f)}{\pi_b}\frac{A_2}{A_4}\frac{\dot{m}_{c2}/A_2}{\dot{m}_{c4}/A_4}\sqrt{\frac{T_{t4}}{T_{t2}}} \tag{10.32}$$

There are two corrected mass flow rates per unit area in Eq. 10.32. These are pure functions of axial Mach number and gas constants. Since the turbine nozzle remains choked at design and over a wide operating range of the engine, its corrected mass flow rate is

$$\dot{m}_{c4}/A_4 = \sqrt{\frac{\gamma_t}{R_t}}\frac{p_{ref}}{\sqrt{T_{ref}}}\left(\frac{\gamma_t+1}{2}\right)^{\frac{-(\gamma_t+1)}{2(\gamma_t-1)}}$$

$$= f(\gamma_t, R_t)\quad [\approx 237(\text{kg/s})/\text{m}^2 \text{ for } \gamma_t = 1.33,\ R_t = 286.8\,\text{J/kg}\cdot\text{K}]$$

The corrected mass flow rate at the engine face, per unit area, is

$$\dot{m}_{c2}/A_2 = \sqrt{\frac{\gamma_c}{R_c}}\frac{p_{ref}}{\sqrt{T_{ref}}}M_{z2}\left(1+\frac{\gamma_c-1}{2}M_{z2}^2\right)^{\frac{-(\gamma_c+1)}{2(\gamma_c-1)}} = f(M_{z2}, \gamma_c, R_c) \tag{10.33}$$

Figure 10.7 (a) shows a graph of Eq. 10.33. A general graph of the corrected mass flow rate per unit area is shown in Fig. 10.7(b). The typical compressor and turbine entry (Mach number) range are identified.

■ **FIGURE 10.7**
Corrected mass flow rate per unit area A

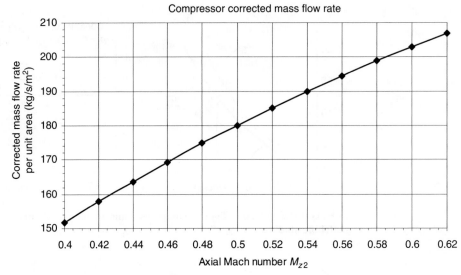

(a) Corrected flow rate at the compressor face (per unit area A_2) for $\gamma = 1.4$

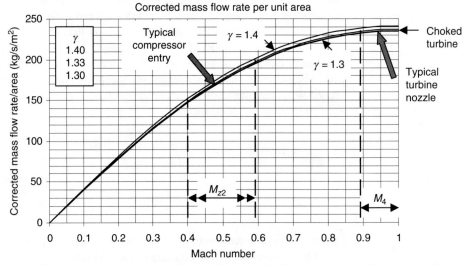

(b) General variation of corrected mass flow rate with Mach number and different γ

As stated earlier, from the engine design point, we calculate A_2/A_4 from Eq. 10.25 to be

$$\frac{A_4}{A_2} = \frac{(1 + f)}{\pi_b \dot{m}_{c4}/A_4} \frac{\dot{m}_{c2}/A_2}{\pi_c} \sqrt{\frac{T_{t4}}{T_{t2}}} \qquad (10.34)$$

The RHS of Eq. 10.34 involves known parameters at the design point, for example, we may have chosen the following design values:

$$M_{z2} = 0.5,$$

$$\pi_c = 25, \ e_c = 0.90, \ \gamma_c = 1.4, \ c_{pc} = 1004 \, \text{J/kg} \cdot \text{K}$$

$$T_{t4}/T_{t2} = 6$$

$$M_4 = 1.0, \text{ therefore, } \dot{m}_{c4}/A_4 \cong 237 \, \text{kg/s/m}^2$$

$$Q_R = 42,000 \, \text{kJ/kg}, \ \eta_b = 0.995, \ \gamma_t = 1.33, \ c_{pt} = 1156 \, \text{J/kg} \cdot \text{K}$$

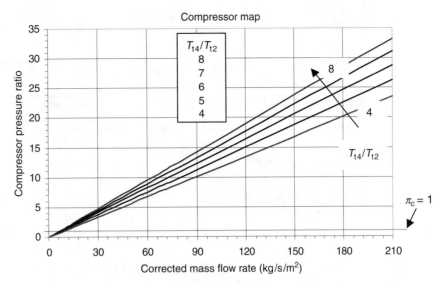

■ **FIGURE 10.8** Compressor map (note that the graph is valid for $\pi_c > 1$)

The fuel-to-air ratio is

$$f = \frac{c_{pt}T_{t4} - c_{pc}T_{t3}}{Q_R\eta_b - c_{pt}T_{t4}} = \frac{(c_{pt}/c_{pc})T_{t4}/T_{t2} - \tau_c}{Q_R\eta_b/c_{pc}T_{t2} - (c_{pt}/c_{pc})T_{t4}/T_{t2}} \approx 0.030$$

If we substitute these values and $\pi_b \approx 0.97$, in Eq. 10.34, we get the area ratio $A_4/A_2 \approx 0.08$. Now, we have the constants in Eq. 10.34, i.e.,

$$\pi_c = \frac{(1+f)}{\pi_b}\frac{A_2}{A_4}\frac{\dot{m}_{c2}/A_2}{\dot{m}_{c4}/A_4}\sqrt{\frac{T_{t4}}{T_{t2}}} \approx 0.0560\sqrt{\frac{T_{t4}}{T_{t2}}}(\dot{m}_{c2}/A_2) \tag{10.35}$$

Let us graph Eq. 10.35 for the compressor pressure ratio versus the corrected mass flow rate for different throttle parameters. Figure 10.8 shows the compressor map based on Eq. 10.35.

An exploded view of the compressor map at low-pressure ratios is shown in Fig. 10.9. The power balance between the compressor and turbine yields an equation for τ_t according to

$$c_{pc}(T_{t3} - T_{t2}) = \eta_m(1+f)c_{pt}(T_{t4} - T_{t5}) \tag{10.36-a}$$

$$c_{pc}(T_{t3}/T_{t2} - 1) = \eta_m(1+f)c_{pt}T_{t4}/T_{t2}(1 - T_{t5}/T_{t4}) \tag{10.36-b}$$

$$\tau_t = 1 - \left(\frac{c_{pc}}{c_{pt}}\right)\frac{\tau_c - 1}{\eta_m(1+f)(T_{t4}/T_{t2})} \tag{10.36-c}$$

Also for the gas generator pumping characteristics, T_{t5}/T_{t2}, we write

$$\frac{T_{t5}}{T_{t2}} = \tau_t\frac{T_{t4}}{T_{t2}} = \frac{T_{t4}}{T_{t2}} - \left(\frac{c_{pc}}{c_{pt}}\right)\frac{\tau_c - 1}{\eta_m(1+f)} \implies \boxed{\frac{T_{t5}}{T_{t2}} = \frac{T_{t4}}{T_{t2}} - \left(\frac{c_{pc}}{c_{pt}}\right)\frac{\tau_c - 1}{\eta_m(1+f)}} \tag{10.37}$$

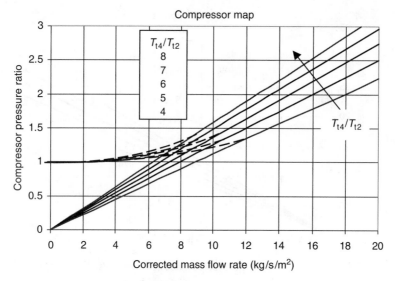

■ **FIGURE 10.9** An exploded view of compressor map at low-pressure ratio

Another pumping characteristic is p_{t5}/p_{t2}, which is the product of component pressure ratios according to

$$\frac{p_{t5}}{p_{t2}} = \pi_t \pi_b \pi_c \qquad (10.38)$$

The turbine pressure ratio π_t is related to turbine temperature ratio τ_t and adiabatic efficiency η_t according to

$$\pi_t = [1 - (1 - \tau_t)/\eta_t]^{\gamma_t/(\gamma_t - 1)} \implies \boxed{\frac{p_{t5}}{p_{t2}} = \pi_b \pi_c [1 - (1 - \tau_t)/\eta_t]^{\gamma_t/(\gamma_t - 1)}} \qquad (10.39)$$

The fuel flow parameter in the gas generator is

$$\frac{f Q_R \eta_b}{\bar{c}_p T_{t2}} \approx \frac{T_{t4}}{T_{t2}} - \frac{T_{t3}}{T_{t2}} \implies \boxed{\frac{f Q_R \eta_b}{\bar{c}_p T_{t2}} \approx \frac{T_{t4}}{T_{t2}} - \tau_c} \qquad (10.40)$$

The $\bar{c}_p$ in Eq. 10.40 is the "average" of the cold and hot c_p. The turbine corrected shaft speed is related to the compressor corrected shaft speed via Eq. 10.25-c. It is rewritten and boxed for the gas generator matching.

$$N_{c2} \equiv N_{c4} \sqrt{T_{t4}/T_{t2}} \implies \boxed{N_{c4} = N_{c2}/\sqrt{T_{t4}/T_{t2}}} \qquad (10.41)$$

An example illustrates the calculation technique for gas generator pumping characteristics over a range of corrected shaft speeds for a constant throttle parameter T_{t4}/T_{t2}.

EXAMPLE 10.1

A compressor performance map is shown. The nominal operating line (which is a straight line) corresponds to $T_{t4}/T_{t2} = 6.0$. A turbine performance map is also shown.

Calculate and graph the gas generator pumping characteristics as a function of $N_{c2}/N_{c2,D}$ for $N_{c2,D} = 14,000$ rpm.

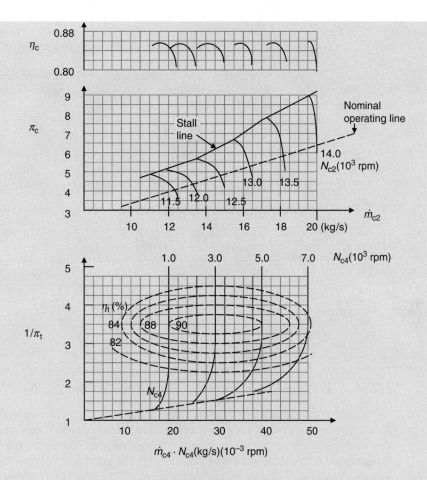

SOLUTION

First, we read compressor pressure ratio, corrected mass flow rate, and the adiabatic efficiency as a function of corrected shaft speed from the compressor performance map (on the nominal operating line). The compressor map data are summarized in the following table:

N_{c2} (1,000 rpm)	p_{t3}/p_{t2}(or π_c)	$\dot{m}_{c2}$(kg/s)	η_c
14.0	6.50	20.0	0.82
13.5	5.88	18.1	0.84
13.0	5.32	16.4	0.83
12.5	4.81	14.8	0.83
12.0	4.36	13.4	0.83
11.5	4.00	12.2	0.84

We may now calculate T_{t3}/T_{t2} or τ_c at each corrected shaft speed, following the definition of compressor adiabatic efficiency, i.e.,

$$\tau_c = 1 + \frac{1}{\eta_c}\left[\pi_c^{(\gamma_c-1)/\gamma_c} - 1\right]$$

using a spreadsheet program. Also, the fuel flow parameter from Eq. 10.40

$$\frac{fQ_R\eta_b}{\bar{c}_p T_{t2}} \approx \frac{T_{t4}}{T_{t2}} - \frac{T_{t3}}{T_{t2}} = 6 - \tau_c$$

is readily calculated for each N_{c2}. From Eq. 10.36-c

$$\tau_t = 1 - \left(\frac{c_{pc}}{c_{pt}}\right)\frac{\tau_c - 1}{\eta_m(1 + f)(T_{t4}/T_{t2})}$$

we calculate τ_t (we estimate $f \approx 0.03$, $\eta_m \approx 0.995$, $c_{pc} = 1004\,\text{J/kg} \cdot \text{K}$ and $c_{pt} = 1156\,\text{J/kg} \cdot \text{K}$).
 From Eq. 10.41,

$$N_{c4} = N_{c2}/\sqrt{T_{t4}/T_{t2}},$$

we calculate N_{c4} (since T_{t4}/T_{t2} is known to be 6.0).
 From Eq. 10.31

$$\dot{m}_{c2} = \frac{\dot{m}_{c4}\pi_b\pi_c}{(1 + f)\sqrt{T_{t4}/T_{t2}}}$$

we calculate the turbine corrected mass flow rate $\dot{m}_{c4}$ (we estimate $\pi_b \approx 0.95$).

From the product of $\dot{m}_{c4} N_{c4}$ and N_{c4}, we refer to the turbine performance map to estimate the turbine adiabatic efficiency η_t. From turbine adiabatic efficiency and temperature ratio, we use $\pi_t = [1 - (1 - \tau_t)/\eta_t]^{\gamma_t/(\gamma_t - 1)}$ to calculate turbine pressure ratio p_{t5}/p_{t4} (we use $\gamma_t = 1.33$).

We are now ready to calculate the pumping characteristics p_{t5}/p_{t2} and T_{t5}/T_{t2}. Finally, we use the table of data

that we generated (in the spreadsheet) to graph the gas generator pumping characteristics, as shown in the following graphs. The gas generator pumping characteristics at other values of throttle parameter T_{t4}/T_{t2} will be needed to complete the operating range of the gas generator.

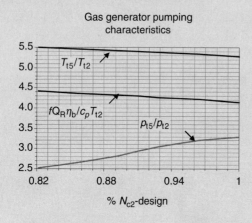

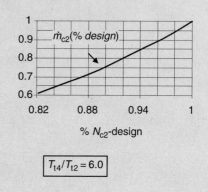

10.2.6 Turbine–Afterburner–(Variable-Geometry) Nozzle Matching

As we are marching along the components of a gas turbine engine, the turbine may be followed by an afterburner, as in fighter/military aircraft, or a variable-geometry exhaust nozzle. It is thus convenient to analyze all three components in this section. What is the role of afterburner in matching with the turbine and the exhaust nozzle? The afterburner-off mode, which is referred to as the "dry" mode, is just a duct with total pressure loss associated with the viscous effects in an adiabatic flow (i.e., the afterburner-off case). The afterburner-on mode, which is referred to as the "wet" mode, impacts the mass flow rate in the exhaust nozzle in two major ways: (1) the hot gas in the afterburner has a much reduced density and (2) the mass flow rate in the nozzle is increased by the amount of fuel burned in the afterburner. There is also an additional total pressure loss due to combustion in the afterburner that we discussed in Chapter 6. The choked exhaust nozzle throat adjusts itself with afterburner operation, i.e., it opens, to accommodate the extra mass flow rate at a reduced density with higher p_t-losses, but it still remains choked. Figure 10.10 shows the turbine matched to an afterburner and an exhaust nozzle with afterburner-off (top portion) and -on mode (bottom portion).

■ **FIGURE 10.10**
Schematic drawing of turbine–afterburner–nozzle matching in "dry" and "wet" modes

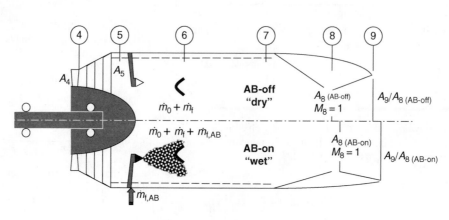

The exhaust nozzle throat, as indicated in Fig. 10.10, remains choked in both dry and wet modes. In the dry mode, the total temperature at the turbine exit remains constant throughout the afterburner and the nozzle. The law of conservation of mass for stations 4 and 8 reveal

$$
\dot{m}_4 = \sqrt{\frac{\gamma_4}{R_4}} \frac{p_{t4}}{\sqrt{T_{t4}}} \cdot A_4 \cdot M_{z4} \left(1 + \frac{\gamma_4 - 1}{2} M_{z4}^2\right)^{\frac{-(\gamma_4+1)}{2(\gamma_4-1)}} = \dot{m}_8
$$

$$
\dot{m}_8 = \sqrt{\frac{\gamma_8}{R_8}} \frac{p_{t8}}{\sqrt{T_{t8}}} \cdot A_{8,\text{dry}} \cdot M_{z8} \left(1 + \frac{\gamma_8 - 1}{2} M_{z8}^2\right)^{\frac{-(\gamma_8+1)}{2(\gamma_8-1)}}
$$

(10.42)

Since both stations 4 and 8 are choked, $M_4 = M_8 = 1$, we may rewrite Eq. 10.42 as

$$
\frac{p_{t8}/p_{t4}}{\sqrt{T_{t8}/T_{t4}}} = \sqrt{\frac{\gamma_4/\gamma_8}{R_4/R_8}} \frac{A_4}{A_{8\cdot\text{dry}}} \left(\frac{2}{\gamma_4 + 1}\right)^{\frac{\gamma_4+1}{2(\gamma_4-1)}} \bigg/ \left(\frac{2}{\gamma_8 + 1}\right)^{\frac{\gamma_8+1}{2(\gamma_8-1)}}
$$

$$
= f(\gamma_4, \gamma_8, R_4/R_8, A_4/A_{8,\text{dry}})
$$

(10.43)

The stagnation pressure at the nozzle throat is $(p_{t5})\pi_{\text{AB}-\text{off}}$ and $T_{t8} = T_{t5}$, therefore, Eq. 10.43 yields

$$
\frac{\pi_t}{\sqrt{\tau_t}} = f(\pi_{\text{AB,dry}}, \gamma_4, R_4, \gamma_8, R_8, A_4/A_{8,\text{dry}})
$$

(10.44)

The RHS of Eq. 10.44 remains nearly constant, between on- and off-design operations of the engine. This result is very important in engine off-design analysis, i.e.,

$$
\boxed{\pi_t/\sqrt{\tau_t} \approx \text{constant (AB-off)}}
$$

(10.45)

With the afterburner-on or wet mode, the continuity equation written for the two choked stations 4 and 8 gives

$$
\frac{p_{t8}/p_{t4}}{\sqrt{T_{t8}/T_{t4}}} = \sqrt{\frac{\gamma_4/\gamma_8}{R_4/R_8}} \left(\frac{1+f}{1+f+f_{\text{AB}}}\right) \frac{A_4}{A_{8,\text{wet}}} \left(\frac{2}{\gamma_4 + 1}\right)^{\frac{\gamma_4+1}{2(\gamma_4-1)}} \bigg/ \left(\frac{2}{\gamma_8 + 1}\right)^{\frac{\gamma_8+1}{2(\gamma_8-1)}}
$$

$$
= f(\gamma_4, \gamma_8, R_4/R_8, A_4/A_{8,\text{wet}})
$$

(10.46)

$$
\frac{\pi_{\text{AB,wet}}}{\sqrt{\tau_{\text{AB,wet}}}} \frac{\pi_t}{\sqrt{\tau_t}} \cong f(\gamma_4, R_4, \gamma_8, R_8, A_4/A_{8,\text{wet}}, f_{\text{AB}})
$$

(10.47)

The nozzle throat in the AB-on mode is actuated to pass through the afterburner flow with sonic condition without affecting the backpressure of the turbine. In essence, the exhaust nozzle throat is just a valve. Therefore, since $A_{8(\text{AB-on})}$ is inversely proportional to $\pi_{\text{AB,wet}}$ and directly proportional to the square root of $\tau_{\text{AB,wet}}$ and is sized to accept the additional mass flow rate due to the afterburner (i.e., f_{AB}), we conclude that

$$
\boxed{\pi_t/\sqrt{\tau_t} \approx \text{constant (AB-on)}}
$$

(10.48)

Note that the two constants on the RHS of Eqs. 10.45 and 10.48 are only different as the gas properties vary with afterburner operation, otherwise, turbine expansion remains unaffected with/without afterburner operation in on- and off-design modes when the gas generator is coupled to a variable-area nozzle.

Theoretically, the nozzle exit area A_9 variation is an independent parameter, which is intended to *optimize the installed thrust* with afterburner in operation. We will thus treat the nozzle area ratio A_9/A_8 as a given or prescribed parameter in our off-design analysis. The ideal value of the nozzle area ratio A_9/A_8 corresponds to the perfect expansion, i.e., $p_9 = p_0$, which maximizes the nozzle gross thrust. In practice however, there is an $A_{9,\max}$ that is dictated by the nacelle aft-end geometry/envelope. In selection/scheduling of A_9/A_8 with flight Mach number, altitude, and afterburner operation, the A_8 is dictated by the afterburner operation and the $A_{9,\text{Max}}$ by the engine envelope.

10.2.6.1 *Fixed-Geometry Convergent Nozzle Matching.*

In fixed-geometry convergent nozzles, the turbine is no longer decoupled from the exhaust nozzle. The turbine backpressure is continually affected by the (mass flow rate in the) nozzle, which in turn sets the turbine shaft power delivered to compressor and thus the compressor pressure ratio. The compressor pressure ratio and the corrected mass flow rate then establish the shaft speed. The only independent variable is thus the fuel flow rate (or throttle setting), which finds a consistent shaft speed with the gas generator pumping characteristics and the mass flow rate through a constant-area convergent nozzle. From continuity equation written between station 8 and 2, we have

$$\frac{A_8}{A_2} = \frac{(1+f)}{\dot{m}_{c8}/A_8} \frac{\dot{m}_{c2}/A_2}{(p_{t5}/p_{t2})} \sqrt{\frac{T_{t5}}{T_{t2}}} \tag{10.49}$$

Equation 10.49 contains, for the most part, gas generator pumping characteristics. A corrected shaft speed N_{c2} gives us p_{t5}/p_{t2}, T_{t5}/T_{t2} and $\dot{m}_{c2}$ for a throttle setting T_{t4}/T_{t2}. Also, the nozzle corrected mass flow rate per unit area is either the maximum value corresponding to a choked exit (i.e., $M_8 = 1.0$), or it may be calculated based on the nozzle pressure ratio p_{t7}/p_0. In either case, there is a unique relationship between the pumping characteristics variables in Eq. 10.49 that is dictated by $A_8/A_2 = $ constant, which corresponds to a certain throttle setting. Therefore, we may not specify both the corrected shaft speed and the throttle parameter simultaneously in a gas turbine engine with a fixed-area convergent exhaust nozzle. In general, a given throttle setting T_{t4}/T_{t2} finds a corrected shaft speed N_{c2} that provides for the mass flow and the (shaft) power balance between all the components in the engine.

10.3 Engine Off-Design Analysis

We may now apply the principles learned in steady-state component matching to engine off-design analysis. We first start with a simple (i.e., nonafterburning) turbojet engine. The known set of design parameters is

1.	Design flight condition	M_0, p_0, T_0
2.	Design compressor pressure ratio	π_c
3.	Ideal heating value of the fuel	Q_R
4.	Design turbine inlet temperature	T_{t4} (or expressed as τ_λ)
5.	Design component efficiencies	$\pi_d, e_c, \pi_b, \eta_b, e_t, \eta_m, \pi_n$
6.	Design nozzle exit pressure or area ratio	p_9/p_0 or A_9/A_8
7.	Gas properties at design point	$\gamma_c, c_{pc}, \gamma_t, c_{pt}$
8.	Design corrected mass flow rate	$\dot{m}_{c2}$
9.	Design corrected shaft speed	N_{c2}
10.	Design axial Mach number at 2	M_{z2}

Then, we subject the turbojet engine to off-design conditions, namely,

1.	Off-design flight conditions (specified)	M_0, p_0, T_0
2.	Off-design turbine inlet temperature (specified)	T_{t4} or τ_λ

■ FIGURE 10.11
Schematic diagram of a turbojet engine, its station numbers, and design parameters

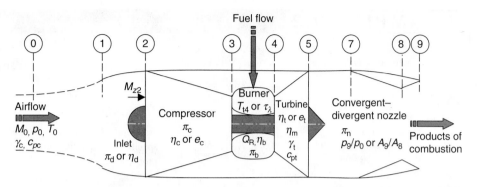

If we assume that the component efficiencies remain the same (between on- and off-designs), the only cycle parameter that needs to be established in off-design condition is the compressor pressure ratio π_c, which in turn establishes the off-design corrected mass flow rate and corrected shaft speed. We may also choose to specify component efficiencies at off-design based on the component performance map or empirical data. To aid with the analysis, a schematic drawing of a simple turbojet engine and its station numbers is shown in Fig. 10.11. Note that the combination of the axial Mach number and the corrected mass flow rate at any station determines the flow area in that station. The corrected mass flow rate at the compressor face and the design Mach number M_{z2} sizes the fan/compressor face area A_2.

10.3.1 Off-Design Analysis of a Turbojet Engine

To simplify the task of engine off-design analysis, we make two assumptions:

1. Turbine inlet is choked at both design and off-design operations, i.e., $M_4 = 1.0$
2. Exhaust nozzle is choked at both design and off-design operations, i.e., $M_8 = 1.0$

Assuming the same gas properties between the on- and off-design conditions as well as fixed areas A_4 and A_8, we had shown (in Eq. 10.44) that

$$\frac{\pi_t}{\sqrt{\tau_t}} \approx \text{constant} \tag{10.50}$$

But, we also remember that π_t and τ_t are related via turbine polytropic efficiency e_t according to

$$\pi_t = \tau_t^{\frac{\gamma_t}{e_t(\gamma_t-1)}}$$

Therefore, if we assume that the turbine polytropic efficiency remains constant between the on- and off-design operations, we may conclude that the turbine expansion parameter τ_t remains constant, i.e.,

$$\tau_t \approx \text{constant} \tag{10.51}$$

Based on this conclusion, the key to solving for the unknown compressor pressure ratio at off-design is

1. establishing τ_t from the design-mode operation
2. use power balance between the compressor and known turbine expansion parameter τ_t to calculate the off-design compressor pressure ratio
3. calculate the new corrected mass flow rate and shaft speed at off-design

The best way to demonstrate the approach to engine off-design analysis is to solve an example problem.

EXAMPLE 10.2

Assume a turbojet engine has the following design-point parameters:

$M_0 = 0$, $p_0 = 0.1\,\text{MPa}$, $T_0 = 15°C$
$\pi_d = 0.98$
$\pi_c = 25$, $e_c = 0.90$
$Q_R = 42,800\,\text{kJ/kg}$, $\pi_b = 0.98$, $\eta_b = 0.99$, $T_{t4} = 1500°C$
$e_t = 0.85$, $\eta_m = 0.995$
$\dot{m}_{c2} = 73\,\text{kg/s}$
$N_{c2}(\text{rpm}) = 6000$
$M_{z2} = 0.6$
$\pi_n = 0.97$, $p_9/p_0 = 1.0$

If this engine is operating in the following off-design condition

$M_0 = 0.8$, $p_0 = 33\,\text{kPa}$, $T_0 = -15°C$
$T_{t4} = 1375°C$
$\pi_d = 0.995$
$p_9/p_0 = 1.0$

assuming all other component efficiencies (except π_d that is specified) remain the same (as design) at off-design and gas properties are $\gamma_c = \gamma_t = 1.4$ and $c_{pc} = c_{pt} = 1004\,\text{J/kg} \cdot \text{K}$, calculate

(a) $\pi_{c\text{-O-D}}$

(b) $\dot{m}_{c2,\text{O-D}}\,(\text{in kg/s})$

(c) $N_{c2,\text{O-D}}\,(\text{in rpm})$

(d) $M_{z2,\text{O-D}}$

SOLUTION

Power balance between the compressor and turbine at design point yields

$$T_{t2}(\tau_c - 1) = (1 + f)\eta_m T_{t4}(1 - \tau_t)$$

Therefore, we may isolate the turbine expansion ratio τ_t to get

$$\tau_t = 1 - \frac{1}{(1+f)\eta_m}\frac{T_{t2}}{T_{t4}}(\tau_c - 1) = 1 - \frac{1}{(1+f)\eta_m}\frac{\tau_r}{\tau_\lambda}(\tau_c - 1)$$

The same turbine expansion τ_t between the on- and off-design operations (as derived earlier), as well as assuming nearly constant $(1 + f)\eta_m$ between the on- and off-design modes, suggest that

$$\left.\frac{\tau_r}{\tau_\lambda}(\tau_c - 1)\right|_{\text{off-design}} \approx \left.\frac{\tau_r}{\tau_\lambda}(\tau_c - 1)\right|_{\text{design}}$$

The off-design compressor temperature ratio is

$$\tau_{c,\text{O-D}} \approx 1 + \frac{(\tau_r/\tau_\lambda)_D}{(\tau_r/\tau_\lambda)_{\text{O-D}}}(\tau_c - 1)_D$$

Now, let us calculate the parameters and substitute in the above equation

$$(\tau_r/\tau_\lambda)_D = \left(\frac{T_{t2}}{T_{t4}}\right)_D = \left(\frac{T_0}{T_{t4}}\right)_D = \frac{273 + 15}{273 + 1500} \cong 0.16244$$

$$(\tau_c)_D = (\pi_c)_D^{\frac{\gamma-1}{e_c\gamma}} = (25)^{\frac{0.4}{(0.9)(1.4)}} \cong 2.7784$$

$$(\tau_r/\tau_\lambda)_{\text{O-D}} = \left(\frac{T_{t2}}{T_{t4}}\right)_{\text{O-D}} = \left(\frac{T_0[1 + (\gamma - 1)M_0^2/2]}{T_{t4}}\right)_{\text{O-D}} = \frac{(273 - 15)[1 + 0.2(0.64)]}{273 + 1375} \cong 0.1766$$

Therefore,

$$\tau_{c,\text{O-D}} \approx 1 + \frac{0.1624}{0.1766}(2.7784 - 1) \cong 2.6354$$

Hence, the compressor pressure ratio at off-design is

$$\pi_{c,O-D} = \left[(\tau_c)_{O-D} \right]^{\frac{\gamma \cdot e_c}{\gamma - 1}} = (2.6354)^{0.9(3.5)} \cong 21.2 \implies \boxed{\begin{array}{c} \text{Off-design} \\ \hline \pi_c \approx 21.2 \end{array}}$$

With the compressor pressure ratio calculated at off-design, we may proceed to calculate engine performance parameters and figures of merit, such as specific thrust, specific fuel consumption, and thermal and propulsive efficiencies. The corrected mass flow rate at off-design is related to the design value by Eq. 10.31.

$$\frac{\dot{m}_{c2,O-D}}{\dot{m}_{c2,D}} \cong \frac{\pi_{c,O-D}}{\pi_{c,D}} \frac{\sqrt{(T_{t4}/T_{t2})_D}}{\sqrt{(T_{t4}/T_{t2})_{O-D}}} = \frac{21.2}{25} \sqrt{\frac{0.1766}{0.16244}} \approx 0.8842$$

$$\dot{m}_{c2,O-D} \approx 0.8842(73 \text{ kg/s}) \approx 64.5 \text{ kg/s} \implies \boxed{\begin{array}{c} \text{Off-design} \\ \hline \dot{m}_{c2} \approx 64.5 \text{ kg/s} \end{array}}$$

The off-design corrected shaft speed is related to the design value via Eq. 10.28-b,

$$\frac{N_{c2,O-D}}{N_{c2,D}} = \frac{\sqrt{(T_{t4}/T_{t2})_{O-D}}}{\sqrt{(T_{t4}/T_{t2})_D}} = \sqrt{\frac{0.16244}{0.1766}} \approx 0.959$$

$$N_{c2,O-D}(\text{rpm}) \approx 0.9591(6000) \approx 5754 \implies \boxed{\begin{array}{c} \text{Off-design} \\ \hline N_{c2} \approx 5754 \text{ rpm} \end{array}}$$

In addition to the corrected mass flow rate at off-design, we can calculate the physical mass flow rate at off-design as well. For that purpose, Eq. 8.10 may be written as

$$\dot{m}_2 = \dot{m}_{c2} \delta_2 / \sqrt{\theta_2}$$

$$\delta_2 \equiv p_{t2}/p_{\text{ref}} \text{ where } p_{\text{ref}} = 101.33 \text{ kPa}$$

$$p_{t0} = p_0(1 + 0.2M_0^2)^{3.5} = 33 \text{ kPa}[1 + 0.2(0.64)]^{3.5} \approx 50.3 \text{ kPa}$$

$$p_{t2} = \pi_d \, p_{t0} = (0.995)(50.3 \text{ kPa}) = 50.05 \text{ kPa, therefore } \delta_2 = 50.05 \text{ kPa}/101.33 \text{ kPa} = 0.4939$$

$$\theta_2 \equiv T_{t2}/T_{\text{ref}} \text{ where } T_{\text{ref}} = 288.2 \text{ K}$$

$$T_{t2} = T_0(1 + 0.2 M_0^2) = 258 \text{ K}[1 + 0.2(0.64)] \approx 291 \text{ K, therefore } \theta_2 = 291 \text{ K}/288.2 \text{ K} \approx 1.01$$

We may substitute these values in the mass flow rate equation, to get

$$\dot{m}_2 = \dot{m}_{c2} \delta_2 / \sqrt{\theta_2} = 64.5 \text{ kg/s} (0.4939)/\sqrt{1.01} \approx 31.7 \text{ kg/s} \implies \boxed{\begin{array}{c} \text{Off-design} \\ \hline \dot{m}_2 \approx 31.7 \text{ kg/s} \end{array}}$$

Now, we can calculate the off-design axial Mach number at the engine face, M_{z2}.
  From the definition of corrected mass flow rate

$$\dot{m}_{c2} = \frac{\dot{m}_2 \sqrt{\theta_2}}{\delta_2} = \sqrt{\frac{\gamma_2}{R_2}} \frac{p_{\text{ref}}}{\sqrt{T_{\text{ref}}}} \cdot A_2 \cdot M_{z2} \left(1 + \frac{\gamma_2 - 1}{2} M_{z2}^2 \right)^{\frac{-(\gamma_2+1)}{2(\gamma_2-1)}} = f(\gamma_2, R_2, M_{z2})$$

we conclude that

$$\frac{\dot{m}_{c2,D}}{\dot{m}_{c2,O-D}} = \frac{M_{c2,D}}{M_{c2,O-D}} \left[\frac{1 + (\gamma_2 - 1)M_{z2,D}^2/2}{1 + (\gamma_2 - 1)M_{z2,D}^2/2} \right]^{\frac{-(\gamma_2+1)}{2(\gamma_2-1)}} = \frac{73}{64.5} = \frac{0.6}{M_{z2,O-D}} \left[\frac{1 + 0.2M_{z2,O-D}^2}{1 + 0.2(0.6)^2} \right]^3$$

Solving for the only unknown in the above equation yields

$$M_{z2,O-D} = 0.4975 \implies \boxed{\begin{array}{c} \text{Off-design} \\ \hline M_{z2} \approx 0.4975 \end{array}}$$

Note that although the physical mass flow rate dropped by more than half, the axial Mach number at the engine face dropped from 0.6 to ~0.5.

10.3.2   Off-Design Analysis of an Afterburning Turbojet Engine

As we learned in the cycle analysis in Chapter 4, the design parameters for an afterburning turbojet engine are

1. M_0, p_0, T_0, γ_c, and c_{pc}
2. π_d
3. π_c and e_c
4. $Q_R, \pi_b, \eta_b, T_{t4}$
5. $e_t, \eta_m, \gamma_t, c_{pt}$
6. $Q_{RAB}, \pi_{AB}, \eta_{AB}, T_{t7}, \gamma_{AB}, c_{pAB}$
7. p_9/p_0
8. $\dot{m}_{c2}, N_{c2}, M_{z2}$

Assuming that the first turbine nozzle and the throat of the exhaust nozzle are choked at design point, i.e., $M_4 = M_8 = 1.0$, and remain choked for off-design operation, we have shown in the previous section that

$\pi_t/\sqrt{\tau_t} \approx$ constant for both AB-on and AB-off modes of operation at on- and off-designs.

We may also conclude that τ_t remains constant as π_t is related to τ_t by the polytropic efficiency of the turbine, e_t and γ_t. Again, the strategy is to calculate the turbine expansion parameter τ_t at design point and use it to calculate the compressor pressure ratio at off-design operation. The approach is identical to the simple turbojet that we studied in the previous section.

At off-design, the following flight, throttle, and efficiency parameters are specified or assumed:

1. M_0, p_0, and T_0, γ_c, c_{pc}
2. π_d
3. e_c
4. $Q_R, \pi_b, \eta_b, T_{t4}$
5. $e_t, \eta_m, \gamma_t, c_{pt}$
6. $Q_{RAB}, \pi_{AB}, \eta_{AB}, T_{t7}, \gamma_{AB}, c_{pAB}$
7. p_9/p_0

Comparing the two lists, we note that the only missing parameter at off-design is the compressor pressure ratio π_{c-O-D}, which in turn determines the corrected mass flow rate and the shaft speed at off-design. The enabling concept that bridges the two modes (i.e., on- and off-designs) is

$$\tau_t \approx \text{constant}$$

Now, let us solve an example problem.

EXAMPLE 10.3

An afterburning turbojet engine has the following design-point parameters:

$M_0 = 0$, $p_0 = 101.33 \, \text{kPa}$, $T_0 = 288.2 \, \text{K}$,

$\gamma_c = 1.4$, $c_{pc} = 1004 \, \text{J/kg} \cdot \text{K}$

$\pi_d = 0.95$

$\pi_c = 20,\ e_c = 0.90$

$\dot{m}_{c2} = 33\,\text{kg/s}$

$N_{c2} = 7120\,\text{rpm}$

$M_{z2} = 0.6$

$Q_R = 42{,}800\,\text{kJ/kg},\ \pi_b = 0.98,\ \eta_b = 0.97,\ T_{t4} = 1850\,\text{K}$

$\gamma_t = 1.33\ c_{pt} = 1156\,\text{J/kg}\cdot\text{K}$

$e_t = 0.80,\ \eta_m = 0.995$

$Q_{R,AB} = 42{,}800\,\text{kJ/kg},\ \pi_{AB} = 0.95,\ \eta_{AB} = 0.98,$

$T_{t7} = 2450\,\text{K}$

$\gamma_{AB} = 1.3,\ c_{pc} = 1243\,\text{J/kg}\cdot\text{K}$

$\pi_n = 0.93,\ p_9/p_0 = 1.0$

We wish to calculate the engine off-design performance characteristics that correspond to the supersonic flight

condition of the aircraft at high altitude. The off-design conditions are

$M_0 = 2.0,\ p_0 = 20\,\text{kPa},\ T_0 = 223\,\text{K},\ \gamma_c = 1.4,$

$c_{pc} = 1004\,\text{J/kg}\cdot\text{K}$

$\pi_d = 0.80$

$e_c = 0.90$

$Q_R = 42{,}800\,\text{kJ/kg},\ \pi_b = 0.98,\ \eta_b = 0.97,$

$T_{t4} = 1850\,\text{K}$

$\gamma_t = 1.33\ c_{pt} = 1156\,\text{J/kg}\cdot\text{K}$

$e_t = 0.80,\ \eta_m = 0.995$

$Q_{R,AB} = 42{,}800\,\text{kJ/kg},\ \pi_{AB} = 0.95,\ \eta_{AB} = 0.98,$

$T_{t7} = 2450\,\text{K}$

$\gamma_{AB} = 1.3,\ c_{pc} = 1243\,\text{J/kg}\cdot\text{K}$

$\pi_n = 0.93,\ p_9/p_0 = 1.0$

SOLUTION

DESIGN-POINT ANALYSIS   Our first goal is to calculate the turbine expansion parameter τ_t at the design condition. The power balance between the compressor and turbine gives

$$\tau_t = 1 - \frac{1}{(1+f)\eta_m}\frac{c_{pc}T_{t2}}{c_{pt}T_{t4}}(\tau_c - 1)$$

We start calculating the parameters in the above equation.

Since the design point corresponds to takeoff or $M_0 = 0$ condition, $T_{t2} = T_0 = 288.2\,\text{K}$
Also, we know that the compressor pressure and temperature ratios are related via

$$\tau_c = \pi_c^{(\gamma_c - 1)/\gamma_c e_c} = 20^{0.4/1.4/0.9} \approx 2.5884$$

The compressor discharge temperature is $T_{t3} = 2.5884(288.2\,\text{K}) = 745.97\,\text{K}$

The last unknown in the above equation for τ_t is the fuel-to-air ratio f

$$f = \frac{c_{pt}T_{t4} - c_{pc}T_{t3}}{Q_R\eta_b - c_{pt}T_{t4}} = \frac{1156(1850) - 1004(745.97)}{42800(1000)(0.97) - 1156(1850)} \approx 0.0360$$

Now, we have calculated all the terms in τ_t, therefore,

$$\tau_t = 1 - \frac{1004(288.2)(2.5884 - 1)}{1.036(0.995)(1156)(1850)} \approx 0.7915 \quad\Longrightarrow\quad \boxed{\tau_t \approx 0.7915 \text{ at on- and off-design}}$$

OFF-DESIGN ANALYSIS   With $\tau_t = 0.7915$ at off-design, we are now ready to calculate the new compressor pressure ratio. Again, the power balance between the compressor and turbine holds the key, i.e.,

$$\tau_{c,O-D} = 1 + (1+f)\eta_m\frac{c_{pt}T_{t4}}{c_{pc}T_{t2}}\bigg|_{O-D}(1 - \tau_t)$$

The total temperature at the engine face, T_{t2}, is equal to the flight total temperature T_{t0}, which is

$$T_{t2} = T_{t0} = (223\,\text{K})[1 + 0.2(2.0)^2] \approx 401.4\,\text{K}$$

We may first use the same fuel-to-air ratio in off-design as design in the above equation to calculate $\tau_{c,O-D}$ and after we calculate the off-design T_{t3}, we can recalculate the fuel-to-air ratio in the burner. We may repeat the calculation loop, until we arrive at the off-design $\tau_{c,O-D}$ to within 10^{-4} (i.e., $1/100$ of 1%) accuracy. In this example, we will check the accuracy of using $(1+f)_D$ and $(1+f)_{O-D}$ on $\tau_{c,O-D}$.

$$\tau_{c,O-D} = 1 + (1.036)(0.995)\frac{1156(1850)}{1004(401.4)}\bigg|_{O-D}(1 - 0.7915) \approx 2.1405 \quad\Longrightarrow\quad \boxed{\begin{array}{c}\text{Off-design}\\ \pi_c \approx 10.994\end{array}}$$

$$\pi_{c,O-D} = (\tau_{c,O-D})^{\gamma_c e_c / (\gamma_c - 1)} = 2.1405^{3.5(0.9)} \approx 10.994$$

With the calculated compressor pressure ratio at off-design, we may proceed to calculate all other parameters of interest, namely,

$$\frac{\dot{m}_{c2,O-D}}{\dot{m}_{c2,D}} \cong \frac{\pi_{c,O-D}}{\pi_{c,D}} \frac{\sqrt{(T_{t4}/T_{t2})_D}}{\sqrt{(T_{t4}/T_{t2})_{O-D}}} = \frac{10.994}{20} \sqrt{\frac{1850/288.2}{1850/401.4}} \approx 0.6487 \implies \boxed{\begin{array}{l} \text{Off-design} \\ \dot{m}_{c2} \approx 21.41 \, \text{kg/s} \end{array}}$$

$$\dot{m}_{c2,O-D} = 0.6487(33 \, \text{kg/s}) \approx 21.41 \, \text{kg/s}$$

$$\frac{N_{c2,O-D}}{N_{c2,D}} = \frac{\sqrt{(T_{t4}/T_{t2})_{O-D}}}{\sqrt{(T_{t4}/T_{t2})_D}} = \frac{\sqrt{1850/401.4}}{\sqrt{1850/288.2}} \approx 0.8473 \implies \boxed{\begin{array}{l} \text{Off-design} \\ N_{c2} \approx 6,033 \, \text{rpm} \end{array}}$$

$$N_{c2,O-D} \approx 0.8473(7,120 \, \text{rpm}) \approx 6,033 \, \text{rpm}$$

$$\delta_2 \equiv p_{t2}/p_{\text{ref}} \text{ where } p_{\text{ref}} = 101.33 \, \text{kPa}$$

$$p_{t0} = p_0(1 + 0.2 M_0^2)^{3.5} = 20 \, \text{kPa}[1 + 0.2(2.0)^2]^{3.5} \approx 156.5 \, \text{kPa}$$

$$p_{t2} = \pi_d \, p_{t0} = (0.8)(156.5 \, \text{kPa}) = 125.2 \, \text{kPa}, \text{ therefore } \delta_2 = 125.2 \, \text{kPa}/101.33 \, \text{kPa} = 1.2356$$

$$\theta_2 \equiv T_{t2}/T_{\text{ref}} \text{ where } T_{\text{ref}} = 288.2 \, \text{K}$$

$$\theta_2 = 401.4/288.2 \approx 1.3928$$

$$\dot{m}_2 = \dot{m}_{c2}\delta_2/\sqrt{\theta_2} = 21.41 \, \text{kg/s}(1.2356)/\sqrt{1.3928} \approx 22.41 \, \text{kg/s} \implies \boxed{\begin{array}{l} \text{Off-design} \\ \dot{m}_2 \approx 22.41 \, \text{kg/s} \end{array}}$$

We calculate compressor discharge temperature $T_{t3} \approx 859.2 \, \text{K}$, and with $T_{t4} = 1850 \, \text{K}$ specified, we calculate the fuel-to-air ratio at off-design, in the combustor to be

$$f_{O-D} = 0.03305$$

We had used $f = 0.0360$ in calculating $\tau_{c,O-D}$. What is the impact of this inaccuracy (in f) on τ_c? We note from the following compressor–turbine power balance equation:

$$\tau_{c,O-D} = 1 + (1 + f)\eta_m \frac{c_{pt}T_{t4}}{c_{pc}T_{t2}}\bigg|_{O-D} (1 - \tau_t)$$

that $(\tau_{c,O-D} - 1)$ is proportional to $(1 + f)$, therefore,

$$\frac{(\tau_{c,O-D} - 1)_{\text{new}}}{(\tau_{c,O-D} - 1)_{\text{old}}} = \frac{(1 + f_{O-D})_{\text{new}}}{(1 + f_{O-D})_{\text{old}}} = \frac{1.03305}{1.0360} \approx 0.99715$$

If we substitute for the old $\tau_{c,O-D}$, we calculate the new $\tau_{c,O-D}$ from above, which is within 1/10 of 1% of the old value. For higher accuracy, we can repeat the off-design calculations with the new value of $\tau_{c,O-D}$ and proceed to calculate a new value for the fuel-to-air ratio f. However, in this example, we are satisfied with the level of accuracy as it falls well within all the other approximations that we have introduced in the problem, namely, a'priori specified gas properties and component efficiencies in off-design.

We continue to march through the engine (at off-design conditions) and calculate

$$p_{t5} = 413.7 \, \text{kPa}, \quad p_{t7} = 393.04 \, \text{kPa}, \quad f_{AB} \approx 0.0367, \quad p_{t9} = 365.52 \, \text{kPa}, \quad M_9 = 2.524, \quad T_9 = 1253 \, \text{K},$$

$$V_9 = 1725 \, \text{m/s}. \text{ The nondimensional specific thrust and specific fuel consumptions are}$$

$$\frac{F_n}{\dot{m}_0 a_0} \approx 4.166 \text{ and TSFC} = 55.94 \, \text{mg/s/N}$$

The cycle thermal efficiency, the engine propulsive efficiency, and the overall efficiency are

$$\eta_{\text{th}} = 0.4822, \quad \eta_p = 0.5283, \text{ and } \eta_o = 0.2547$$

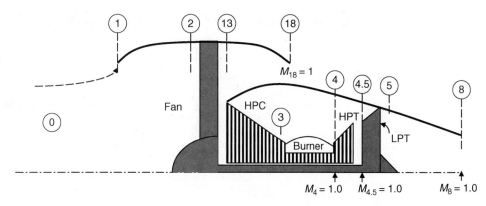

■ **FIGURE 10.12**
Two-spool separate-flow
turbofan engine with
choked convergent nozzles

10.3.3 Off-Design Analysis of a Separate-Flow Turbofan (Two-Spool) Engine

We will approach the off-design analysis problem of separate-flow turbofan engines the same way we did in the turbojet problem, i.e., we assume the first nozzle choking stations in the turbine as well as the exhaust nozzle throat in off-design engine operation. Then we set the mass flow rates between the choked stations equal to each other and establish the constants of gas turbine engine operation.

Figure 10.12 shows a definition sketch of a two-spool separate-flow turbofan engine. The fan is driven by the low-pressure turbine (LPT), and the high-pressure compressor (HPC) is driven by the high-pressure turbine (HPT). Again, we assume that the first turbine nozzles of the HPT and LPT are choked at design point and remain choked in off-design operation, i.e.,

$$M_4 = M_{4.5} = 1.0 \tag{10.52-a}$$

The second assumption deals with the exhaust nozzle throats, i.e.,

$$M_8 = M_{18} = 1.0 \tag{10.52-b}$$

The known design parameters for this turbofan engine are

1. Design flight condition $\qquad\qquad M_0, p_0, T_0$
2. Design compressor and fan pressure
   ratios $\qquad\qquad\qquad\qquad\qquad \pi_c, \pi_f$
3. Design bypass ratio $\qquad\qquad\qquad \alpha$
4. Ideal heating value of the fuel $\qquad Q_R$
5. Design turbine inlet temperature $\quad T_{t4}$ (or expressed as τ_λ)
6. Design component efficiencies $\qquad \pi_d, e_c, e_f, \pi_b, \eta_b, e_t, \eta_m, \pi_n, \pi_{nf}, p_8/p_0, p_{18}/p_0$
7. Gas properties at design point $\qquad \gamma_c, c_{pc}, \gamma_t, c_{pt}$
8. Design choked stations $\qquad\qquad M_4 = M_{4.5} = M_8 = M_{18} = 1.0$

In off-design condition, we may have a different throttle setting and fly at a different altitude Mach number, i.e., we subject the turbofan engine to

1. Off-design flight conditions (specified) $\quad M_0, p_0, T_0$
2. Off-design turbine inlet temperature (specified) $\quad T_{t4}$ or τ_λ

And off-design inlet losses.

The primary unknowns, in the off-design operation, are three cycle parameters, namely,

1. $\pi_{c,O-D}$
2. $\pi_{f,O-D}$
3. α_{O-D}

The mass flow rates between the stations 4.5 and the nozzle throat are nearly the same, therefore,

$$\dot{m}_{4.5} = \sqrt{\frac{\gamma_{4.5}}{R_{4.5}}} \frac{p_{t4.5}}{\sqrt{T_{t4.5}}} A_{4.5} \left(\frac{2}{\gamma_{4.5}+1}\right)^{\frac{\gamma_{4.5}+1}{2(\gamma_{4.5}-1)}} \cong \dot{m}_8 = \sqrt{\frac{\gamma_8}{R_8}} \frac{p_{t8}}{\sqrt{T_{t8}}} A_8 \left(\frac{2}{\gamma_8+1}\right)^{\frac{\gamma_8+1}{2(\gamma_8-1)}}$$

Using similar arguments as in the turbojet section, e.g., constant areas, constant gas properties, etc., we conclude that,

$$\boxed{\pi_{tL}/\sqrt{\tau_{tL}} \approx \text{constant}} \tag{10.53-a}$$

or assuming the same polytropic efficiency that relates pressure and temperature ratios in the turbine, we conclude that the first constant of operation in our separate-flow turbofan engine is

$$\tau_{tL} \approx \text{constant} \tag{10.53-b}$$

Now let us consider the mass flow rates between two other stations, namely, station 4 and 4.5, i.e.,

$$\dot{m}_4 = \sqrt{\frac{\gamma_4}{R_4}} \frac{p_{t4}}{\sqrt{T_{t4}}} A_4 \left(\frac{2}{\gamma_4+1}\right)^{\frac{\gamma_4+1}{2(\gamma_4-1)}} \cong \dot{m}_{4.5} = \sqrt{\frac{\gamma_{4.5}}{R_{4.5}}} \frac{p_{t4.5}}{\sqrt{T_{t4.5}}} A_{4.5} \left(\frac{2}{\gamma_{4.5}+1}\right)^{\frac{\gamma_{4.5}+1}{2(\gamma_{4.5}-1)}}$$

For fixed area turbines, A_4 and $A_{4.5}$ remain constant, i.e., they are not adjustable, and assuming the gas properties do not appreciably change between the on- and off-design conditions, we may conclude that the second constant of engine operation is

$$\boxed{\pi_{tH}/\sqrt{\tau_{tH}} \approx \text{constant}} \tag{10.54-a}$$

and similarly, assuming constant e_{tH} between on- and off-designs, we conclude that

$$\tau_{tH} \approx \text{constant} \tag{10.54-b}$$

Note that the combination of Eqs. 10.53-b and 10.54-b results in the constant overall expansion across the turbine, i.e.,

$$\tau_t = \tau_{tH} \cdot \tau_{tL} \approx \text{constant}$$

The fan nozzle is assumed to remain choked in off-design as well as on-design operation, which states that its mass flow rate is

$$\dot{m}_{18} = \sqrt{\frac{\gamma_{18}}{R_{18}}} \frac{p_{t18}}{\sqrt{T_{t18}}} A_{18} \left(\frac{2}{\gamma_{18}+1}\right)^{\frac{\gamma_{18}+1}{2(\gamma_{18}-1)}} = \alpha \cdot \dot{m}_0 \tag{10.55}$$

The main nozzle mass flow rate may be written as

$$\dot{m}_8 = \sqrt{\frac{\gamma_8}{R_8}} \frac{p_{t8}}{\sqrt{T_{t8}}} A_8 \left(\frac{2}{\gamma_8+1}\right)^{\frac{\gamma_8+1}{2(\gamma_8-1)}} = (1+f)\dot{m}_0 \tag{10.56}$$

Taking the ratio of the two mass flow rates in 10.55 and 10.56 and assuming the gas properties and the flow nozzle throat areas do not change between the on- and off-designs, we get

$$\frac{\alpha}{1+f} = \text{constant} \frac{p_{t18}/\sqrt{T_{t18}}}{p_{t8}/\sqrt{T_{t8}}} = \text{constant} \frac{p_{t13}/\sqrt{T_{t13}}}{p_{t5}/\sqrt{T_{t5}}}$$

We may divide the numerator and denominator by $p_{t2}/(T_{t2})^{1/2}$ to get

$$\alpha \approx \text{constant} \frac{\left(\dfrac{p_{t13}}{p_{t2}}\right)\Big/\sqrt{\dfrac{T_{t13}}{T_{t2}}}}{\left(\dfrac{p_{t5}}{p_{t4}}\dfrac{p_{t4}}{p_{t3}} \cdot \dfrac{p_{t3}}{p_{t13}}\dfrac{p_{t13}}{p_{t2}}\right)\Big/\sqrt{\dfrac{T_{t5}}{T_{t4}}\dfrac{T_{t4}}{T_0}\dfrac{T_0}{T_{t2}}}} \approx \text{constant} \frac{\sqrt{\tau_\lambda/\tau_r} \cdot \tau_f}{\pi_{cH}} \tag{10.57}$$

Note that all the constants were combined in the coefficient in front of the RHS, e.g., turbine pressure ratio or temperature ratio, which do not change were lumped in the constant of the above equation. Since the turbine expansion parameters τ_t and π_t remain constant in off-design, we may further simplify Eq. 10.57 to get

$$\boxed{\alpha\pi_{cH}\sqrt{\frac{\tau_r\tau_f}{\tau_\lambda}} \approx \text{constant}} \tag{10.58}$$

The last equation, i.e., Eq. 10.58 constitutes the third constant of turbofan operation, within the approximations that we introduced. We use these three constants of operation between the on- and off-design modes to establish the three unknowns at off-design, namely, the fan and the compressor pressure ratio as well as the bypass ratio.

Off-design analysis
The power balance between the HPT and HPC yields

$$T_{t13}(\tau_{cH} - 1) = \eta_{mH}(1 + f)T_{t4}(1 - \tau_{tH})$$

We may divide both sides by T_{t2} to get the nondimensional expression

$$\tau_f(\tau_{cH} - 1) = \eta_{mH}(1 + f)(1 - \tau_{tH})\frac{\tau_\lambda}{\tau_r} \approx \text{constant}\frac{\tau_\lambda}{\tau_r} \tag{10.59}$$

Note that this has only two unknowns and they are τ_{cH} and τ_f since we specify the throttle setting at off-design, T_{t4} (or τ_λ) as well as flight Mach number produces τ_r information. We may cast Eq. 10.59 as

$$\frac{\tau_r\tau_f}{\tau_\lambda}(\tau_{cH} - 1) \approx \text{constant} = C_1 \tag{10.60}$$

The power balance between the fan and the LPT yields

$$(1 + \alpha)T_{t2}(\tau_f - 1) = \eta_{mL}(1 + f)T_{t4.5}(1 - \tau_{tL}) = \eta_{mL}(1 + f)T_{t4}\tau_{tH}(1 - \tau_{tL}) \tag{10.61}$$

We note that RHS of the above equation is completely known, either from the design point calculations as in τ_{tH} and τ_{tL} or from the off-design throttle setting T_{t4}. The LHS of Eq. 10.61 contains two unknowns, namely, the off-design bypass ratio and the off-design fan temperature ratio. The Eq. 10.61 simplifies to

$$(1 + \alpha)(\tau_f - 1)\frac{\tau_r}{\tau_\lambda} \approx \text{constant} = C_2 \tag{10.62}$$

Now, we use the third constant, i.e., Eq. 10.58, to establish a third equation involving the three unknowns, α and τ_f, and π_{cH} at off-design.

$$\alpha \cdot \pi_{cH}\sqrt{\frac{\tau_r \cdot \tau_f}{\tau_\lambda}} \approx \text{constant} = C_3 \tag{10.63}$$

Is there a closed form solution for the three unknowns in terms of the three constants C_1, C_2, and C_3? The answer is no, since these equations are coupled and involve unknowns with the following exponent:

$$\pi_{cH} = \tau_{cH}^{\frac{\gamma \cdot e_{cH}}{\gamma - 1}} \tag{10.64}$$

Off-design solution strategy

From the on-design cycle analysis, calculate the three constants, C_1, C_2, and C_3. Then solve Eqs. 10.60, 10.62, and 10.63 iteratively to arrive at the unknowns α, τ_{cH}, and τ_f. We may eliminate the bypass ratio α from Eqs. 10.62 and 10.63, and after some minor simplification, we get

$$\left(\frac{\tau_r}{\tau_\lambda} + \frac{C_3}{\pi_{cH}}\sqrt{\frac{\tau_r}{\tau_\lambda\tau_f}}\right)(\tau_f - 1) = C_2 \tag{10.65}$$

From Eq. 10.60, we isolate τ_f, to get

$$\tau_f = \frac{C_1\tau_\lambda}{\tau_r(\tau_{cH} - 1)} \tag{10.66}$$

We introduce expression 10.66 into Eq. 10.65 and replace pressure ratio by the temperature ratio via Eq. 10.64 will result in a one equation, one unknown expression,

$$\left(\frac{\tau_r}{\tau_\lambda} + \frac{C_3}{(\tau_{cH})^{\frac{\gamma \cdot e_{cH}}{\gamma - 1}}}\sqrt{\frac{\tau_r}{\tau_\lambda\left[\frac{C_1\tau_\lambda}{\tau_r(\tau_{cH} - 1)}\right]}}\right)\left[\frac{C_1\tau_\lambda}{\tau_r(\tau_{cH} - 1)} - 1\right] = C_2 \tag{10.67}$$

Finally, we need to iterate for τ_{cH} in Eq. 10.67. Then introducing it in Eq. 10.66 yields τ_f and substituting them in Eq. 10.63, we get the off-design bypass ratio α.

To demonstrate the methodology, we solve an example.

EXAMPLE 10.4

Consider a separate-flow turbofan engine (as in Fig. 10.12) with the following design-point parameters:

1. $M_0 = 0$, $p_0 = 0.1\,\text{MPa}$, $T_0 = 15°C$
2. $\pi_d = 0.98$
3. $\pi_f = 2.0$, $e_f = 0.90$
4. $\alpha = 6.0$
5. $\pi_{cH} = 15$, $e_{cH} = 0.90$
6. $T_{t4} = 1700°C$, $Q_R = 42{,}800\,\text{kJ/kg}$, $\eta_b = 0.99$, $\pi_b = 0.95$
7. $e_{tH} = 0.85$, $\eta_{mH} = 0.995$
8. $e_{tL} = 0.89$, $\eta_{mL} = 0.995$
9. $\pi_n = \pi_{nf} = 0.98$, $p_8 = p_{18} = p_0$
10. $\gamma_c = 1.4$, $c_{pc} = 1004\,\text{J/kg} \cdot \text{K}$

11. $\gamma_t = 1.33$, $c_{pt} = 1146\,\text{J/kg} \cdot \text{K}$
12. $M_4 = M_{4.5} = M_8 = M_{18} = 1.0$

The off-design operation of this engine is represented by a cruise altitude flight such as

$$M_0 = 0.85,\ p_0 = 10\,\text{kPa},\ T_0 = -15°C$$
$$T_{t4} = 1500°C$$
$$\pi_d = 0.995$$

All other efficiencies and gas properties remain constant.

Calculate the following parameters at off-design condition:

(a) Fan pressure ratio π_f
(b) High-pressure compressor pressure ratio π_{cH}
(c) Bypass ratio α

SOLUTION

We first calculate the design values for $\tau_r, \tau_f, \tau_{cH}$, and τ_λ

$$\tau_r = 1 + 0.2(0) = 1$$

$$\tau_f = \pi_f^{0.2857/0.90} = (2)^{0.31746} = 1.2461$$

$$\tau_{cH} = \pi_{cH}^{0.2857/0.90} = (15)^{0.31746} = 2.3624$$

$$\tau_\lambda = \frac{c_{pt}T_{t4}}{c_{pc}T_0} = \left(\frac{1146}{1004}\right)\left(\frac{1700 + 273}{15 + 273}\right) \cong 7.82$$

Now, we can calculate the first constant C_1

$$C_1 = \frac{\tau_r \tau_f}{\tau_\lambda}(\tau_{cH} - 1) = \frac{1.2461}{7.82}(1.3624) \cong 0.2171$$

Constants C_2 and C_3 are

$$C_2 = (1 + \alpha)(\tau_f - 1)\frac{\tau_r}{\tau_\lambda} = 7(0.2461)(1/7.82) \cong 0.2203$$

$$C_3 = \alpha \cdot \pi_{cH}\sqrt{\frac{\tau_r \cdot \tau_f}{\tau_\lambda}} = 6(15)\sqrt{\frac{1.2461}{7.82}} \cong 35.927$$

Let us calculate τ_r and τ_λ for the off-design operation

$$\tau_{r,O-D} = 1 + 0.2(0.85)^2 = 1.1445$$

$$\tau_{\lambda,O-D} = \frac{c_{pt}T_{t4}}{c_{pc}T_0} = \left(\frac{1146}{1004}\right)\left(\frac{1500 + 273}{273 - 15}\right) \cong 7.8440$$

Now, let us substitute all the parameters in Eq. 10.67

$$\left(\frac{\tau_r}{\tau_\lambda} + \frac{C_3}{(\tau_{cH})^{\frac{\gamma \cdot e_{cH}}{\gamma - 1}}}\sqrt{\frac{\tau_r}{\tau_\lambda\left[\frac{C_1\tau_\lambda}{\tau_r(\tau_{cH} - 1)}\right]}}\right)\left[\frac{C_1\tau_\lambda}{\tau_r(\tau_{cH} - 1)} - 1\right] = C_2$$

$$\left(\frac{1.1445}{7.844} + \frac{35.927}{\tau_{cH}^{3.15}}\sqrt{\frac{1.1445}{7.844\left[\frac{0.2171(7.844)}{1.1445(\tau_{cH} - 1)}\right]}}\right)\left[\frac{0.2171(7.844)}{1.1445(\tau_{cH} - 1)}\right] = 0.2203$$

The solution to this equation is found using an Excel spreadsheet to be

$$\tau_{cH} \approx 2.1184$$

Therefore the high-pressure compressor pressure ratio at off design is

$$\pi_{cH} = \tau_{cH}^{\gamma e/(\gamma - 1)} = (2.1184)^{3.15} \approx 10.64$$

> Off-design
> $\pi_{cH} \approx 10.64$

The fan pressure ratio at off–design is calculated from

$$\tau_f = \frac{C_1\tau_\lambda}{\tau_r(\tau_{cH} - 1)} = 1.3304$$

> Off-design
> $\pi_f \approx 2.458$

$$\pi_f = \tau_f^{3.15} = 1.3304^{3.15} \approx 2.458$$

The off-design bypass ratio is calculated from

$$\alpha \cdot \pi_{cH}\sqrt{\frac{\tau_r \cdot \tau_f}{\tau_\lambda}} \approx \text{constant} = C_3$$

> Off-design
> $\alpha \approx 7.66$

$$\alpha \approx 7.66$$

Based on these off-design parameters, we may proceed to calculate engine off-design performance such as thrust-specific fuel consumption or specific thrust and thermal and propulsive efficiencies.

10.4 Unchoked Nozzles and Other Off-Design Iteration Strategies

Let us reexamine the assumptions that we made in component matching and engine off-design analysis, in particular, choked flow condition at stations: 4, 4.5, and 8, i.e., the HPT entrance, the LPT entrance, and the exhaust nozzle throat. These choked stations simplified our solution methodology, as the corrected mass flow rate in those stations remained fixed. However, what if those stations were not choked, i.e., what if $M_4 \neq 1.0$, $M_{4.5} \neq 1$, and $M_8 \neq 1$, in some off-design operations? How do we know our assumption was correct?

10.4.1 Unchoked Exhaust Nozzle

We may start our off-design analysis with the assumption of choked stations at 4, 4.5, and 8, as described earlier. With these assumptions, we may calculate the missing off-design cycle parameters, namely, the compressor and fan pressure ratios and bypass ratio (if a turbofan engine). Then, we can march through the engine and calculate all the total pressures and temperatures in the engine including p_{t9} if a convergent–divergent (C–D) nozzle or p_{t8} if the nozzle was convergent. Now, we apply the nozzle-choking criterion, or what we called critical nozzle pressure ratio (NPR)$_{\text{crit}}$, for a choked throat in a C–D nozzle we must have

$$\frac{p_{t9}}{p_0} \geq \left(\frac{\gamma_9 + 1}{2}\right)^{\gamma_9/(\gamma_9 - 1)} \tag{10.68-a}$$

For a choked throat in a convergent nozzle we must have

$$\frac{p_{t8}}{p_0} \geq \left(\frac{\gamma_8 + 1}{2}\right)^{\gamma_8/(\gamma_8 - 1)} \tag{10.68-b}$$

With these conditions met, we had made correct assumptions (i.e., $M_8 = 1.0$). Otherwise, the nozzle throat Mach number is less than 1 and its value is

$$M_8 = \sqrt{\frac{2}{\gamma_8 - 1}\left[\left(\frac{p_{t8}}{p_0}\right)^{(\gamma_8 - 1)/\gamma_8} - 1\right]} \tag{10.69}$$

With a reduced Mach number at the throat, the corrected mass flow rate through the nozzle throat drops.

$$\dot{m}_{c8} = \frac{\dot{m}_8 \sqrt{\theta_8}}{\delta_8} = \sqrt{\frac{\gamma_8}{R_8}} \frac{p_{\text{ref}}}{\sqrt{T_{\text{ref}}}} \cdot A_8 \cdot M_8 \left(1 + \frac{\gamma_8 - 1}{2} M_8^2\right)^{\frac{-(\gamma_8 + 1)}{2(\gamma_8 - 1)}} \tag{10.70}$$

If we write the corrected mass flow rate at the turbine inlet as

$$\dot{m}_{c4} = \frac{\dot{m}_4 \sqrt{\theta_4}}{\delta_4} = \sqrt{\frac{\gamma_4}{R_4}} \frac{p_{\text{ref}}}{\sqrt{T_{\text{ref}}}} \cdot A_4 \cdot M_4 \left(1 + \frac{\gamma_4 - 1}{2} M_4^2\right)^{\frac{-(\gamma_4 + 1)}{2(\gamma_4 - 1)}} \tag{10.71}$$

And take the ratio of the two corrected mass flow rates while ignoring variation of gas constants, we get

$$\frac{\dot{m}_{c4}}{\dot{m}_{c8}} = \frac{p_{t8}/p_{t4}}{\sqrt{T_{t8}/T_{t4}}} = \frac{\pi_t}{\sqrt{\tau_t}} \quad \text{(AB-off)} \tag{10.72}$$

Note that although the corrected mass flow rate at 8 has dropped (to below critical), it may not have caused the turbine nozzle to unchoke. Therefore, we continue with the assumption that turbine nozzle is choked, i.e., $M_4 = 1.0$ or $\dot{m}_{c4} = $ constant, and our new turbine expansion equation becomes

$$\boxed{\frac{\pi_t}{\sqrt{\tau_t}} = \frac{\text{constant}}{\dot{m}_{c8}}} \tag{10.73}$$

Compare this equation with Eq. 10.45, when we had made the assumption of choked flow at 8, i.e.,

$$\boxed{\pi_t/\sqrt{\tau_t} \approx \text{constant}} \tag{10.45}$$

Equation 10.73 indicates that turbine expansion is no longer constant when the throat unchokes, rather it is inversely proportional to the nozzle corrected mass flow rate at the nozzle throat. A reduction in the corrected mass flow rate at the nozzle throat then causes the turbine pressure ratio π_t or backpressure p_{t5} to increase, which in turn reduces the turbine shaft power, which causes the compressor pressure ratio and mass flow rate to drop, in a domino effect.

In summary, the iteration strategy for an unchoked exhaust nozzle throat is

1. Calculate the nozzle throat Mach number from Eq. 10.69 (using p_{t8}/p_0 from round 1)
2. Calculate the corrected mass flow rate at 8 from Eq. 10.70
3. Calculate a new turbine expansion parameter from Eq. 10.73
4. Calculate the new cycle pressure ratio, bypass ratio, fan pressure ratio, etc.
5. Calculate the new p_{t8}/p_0 from the new cycle parameters
6. Calculate the new nozzle throat Mach number from Eq. 10.69 (from round 2)
7. Compare the two throat Mach numbers (from round 1 and 2)
8. Repeat the process until the two nozzle throat Mach numbers are within say $\sim 1\%$

10.4.2 Unchoked Turbine Nozzle

In certain low mass flow rate conditions, it is possible for the turbine nozzle to operate in an unchoked mode. Typically, an inverse turbine pressure ratio $(1/\pi_t)$ of ~ 2 is the boundary between choked and unchoked turbine operation at low mass flow rates. The following rule of thumb is thus of interest:

Rule of thumb: For $\quad 1/\pi_t \approx 2$ or less $\quad$ turbine unchokes, i.e., $M_4 < 1$

When the turbine unchokes (mainly due to low mass flow rate conditions), its corrected mass flow rate at station 4 drops (to below critical corresponding to Mach 1 condition). The new turbine expansion parameter, from Eq. 10.72 then follows the more general rule of

$$\frac{\pi_t}{\sqrt{\tau_t}} = \frac{\dot{m}_{c4}}{\dot{m}_{c8}} \tag{10.74}$$

Here we have fixed the exhaust nozzle corrected flow $\dot{m}_{c8}$ as in the previous section, but we now have to iterate on the turbine corrected flow $\dot{m}_{c4}$. The strategy here is that for a given exhaust nozzle corrected mass flow rate $\dot{m}_{c8}$ that we calculate from Eq. 10.70, we assume a new (and lower) $\dot{m}_{c4}$, and then using Eq. 10.74, we calculate a new turbine expansion parameter. The new cycle parameters at off-design are then calculated, which lead to a test of M_8 as in step #7 in the iteration strategy of section 10.4.1.

10.4.3 Turbine Efficiency at Off-Design

So far we have considered the turbine pressure and temperature ratios that are related via constant polytropic efficiency e_t according to

$$\pi_t = \tau_t^{\frac{\gamma_t}{e_t(\gamma_t - 1)}} \tag{10.75}$$

Alternatively, we have turbine adiabatic efficiency η_t that connects the turbine pressure and temperature ratios following

$$\tau_t = 1 - \eta_t\left(1 - \pi_t^{(\gamma_t - 1)/\gamma_t}\right) \tag{10.76}$$

In reality, turbine efficiency, either e_t or η_t, changes with operating condition and the turbine expansion parameter that was kept constant between the on- and off-design was

$$\frac{\pi_{t,O-D}}{\sqrt{\tau_{t,O-D}}} = \frac{\pi_{t,D}}{\sqrt{\tau_{t,D}}} = C_1 \tag{10.77}$$

From our design analysis, we determine the RHS of the above equation (constant C_1), then we have two equations and two unknowns, for τ_t and π_t, if we know (or can estimate) the off-design efficiency η_t. For example, we have to numerically solve the following equation for τ_t.

$$\tau_t = 1 - \eta_t \left[1 - \left(C_1 \tau_t^{1/2} \right)^{(\gamma_t - 1)/\gamma_t} \right] \tag{10.78}$$

10.4.4 Variable Gas Properties

Up to this point in our analysis, we have prescribed gas properties at both design and off-design conditions. In the off-design analysis, the compressor pressure ratio and therefore its exit temperature T_{t3} are both unknown. Since for a thermally perfect gas we have

$$\gamma = \gamma(T)$$
$$c_p = c_p(T)$$

we need to recalculate all gas properties based on our cycle temperatures and in fact we have to include the effect of fuel-to-air ratio on the gas constants after the burner. Based on the recalculated gas properties, we have to repeat the cycle analysis and continue the loop until certain level of accuracy (in, for example, τ_t) is achieved. Every one of these steps represents a higher level of refinement in engine performance simulation. The complexities of the iteration loops and gas modeling have prompted the creation of computer codes. One example is the NNEP, which stands for Navy/NASA Engine Program, which also interfaces with CEC (complex chemical equilibrium composition).

10.5 Summary

The purpose of this chapter was to integrate our individual component studies into a complete propulsion system. We asked and answered the question of how an aircraft gas turbine engine (that was designed for certain operating conditions) behaved in off-design. We started with a review of the five corrected parameters in an engine:

1. $\dot{m}_{ci} \equiv \dfrac{\dot{m}_i \sqrt{\theta_i}}{\delta_i} = \sqrt{\dfrac{\gamma_i}{R_i}} \dfrac{p_{ref}}{\sqrt{T_{ref}}} A_i M_i \left(1 + \dfrac{\gamma_i - 1}{2} M_i^2 \right)^{-\frac{\gamma_i + 1}{2(\gamma_i - 1)}}$

2. $N_{ci} \equiv \dfrac{N}{\sqrt{\theta_i}}$

3. $\dot{m}_{fc} \equiv \dfrac{\dot{m}_f}{\delta_2 \sqrt{\theta_2}}$

4. $F_c \equiv F / \delta_0$

5. $\text{TSFC}_c \equiv \dfrac{\text{TSFC}}{\sqrt{\theta_0}}$

We learned about the individual component interaction in a steady-state mode of operation. The conservation principles of mass and energy provided the link, i.e., the match, between the components. In case the compressor and turbine performance maps were available, we used them to calculate the gas generator pumping characteristics for a given throttle setting. Pumping characteristics are

1. the corrected air/flow rate $\dot{m}_{c2}$
2. the pressure ratio p_{t5}/p_{t2}
3. the temperature ratio T_{t5}/T_{t2}
4. the fuel flow parameter $fQ_R\eta_b/c_pT_{t2}$ or the corrected fuel flow rate $\dot{m}_{fc}$

If the component performance maps were not available, we relied on the persistence of certain choking stations in the engine (as in stations 4 and 8 in a single-spool engine and 4, 4.5, and 8 in a two-spool engine) in on- and off-designs to calculate the off-design performance of the engine. The highest level of simplifications, as in constant component efficiencies, gas properties, negligible fuel-to-air ratio variation, etc., resulted in a constant turbine expansion parameter in a *single-spool turbojet* engine

$$\boxed{\pi_t/\sqrt{\tau_t} = \text{constant}}$$

between the on- and off-designs. We used the same principles in a *two-spool turbofan* engine that resulted in three constants between the on- and off-design turbine expansion parameters and one involving bypass ratio

$$\boxed{\pi_{tL}/\sqrt{\tau_{tL}} \approx \text{constant}}$$

$$\boxed{\pi_{tH}/\sqrt{\tau_{tH}} \approx \text{constant}}$$

$$\boxed{\alpha\pi_{cH}\sqrt{\frac{\tau_r\tau_f}{\tau_\lambda}} \approx \text{constant}}$$

In case the exhaust nozzle throat was unchoked, the corrected mass flow rate at the nozzle throat was reduced and the turbine expansion parameter τ_t was no longer constant, i.e.,

$$\boxed{\frac{\pi_t}{\sqrt{\tau_t}} = \frac{\text{constant}}{\dot{m}_{c8}}}$$

An iteration strategy to arrive at a consistent operating condition of the engine in off-design was outlined in Section 10.4.1. Other iteration strategies were presented for unchoked turbine and variable gas properties.

The most critical and perhaps interesting part of the engine component matching is in the study of engine transients and unsteady interactions. For example, stall and surge are unsteady behavior of the compression–combustor system. The phenomenon of inlet unstart and upstream propagation of "hammer shock" finds its root in supersonic mixed-compression inlet instability. External compression inlets may initiate "buzz" instability and subsequent compressor stall. The acceleration and deceleration paths (i.e., spool up and down) in an engine are the transients that they too may cause compressor instability. In general, the question that we need to ask is what happens if we disturb the steady state? Would the oscillations in the system decay or grow? What is the impact of the rate of change that we introduce in the dynamic system, as in the rate of fuel addition, spool up or spool down, or rapid actuation of (stator) blades? Kerrebrock's paper (1977) on small disturbance theory examines the growth/decay characteristics of pressure, entropy, and vorticity perturbations in swirling flows in turbomachinery. Schobeiri (2005) has treated the time-dependent, dynamic performance of turbomachinery and gas turbine systems extensively in his recent book.

Finally, there are many aircraft engine simulation codes, for example, NNEP that provide for an accurate estimation of the engine performance characteristics over a wide operating range of the engine/aircraft. Also, a major new initiative on propulsion system simulation is NASA-Glenn's "Numerical Propulsion System Simulation" (NPSS) Program (see Lytle, 1999) that promises to bring high fidelity to fully three-dimensional transient simulation of complex aircraft engine configurations.

The principles that we learned in this chapter help us to (1) produce engine off-design performance for preliminary design purposes, (2) use compressor and turbine performance maps to calculate the gas generator pumping characteristics, and (3) understand and interpret the results of engine simulation codes, if they are available and used.

Additional references (e.g., 1-5, 7-10 and 13-17) on gas turbine and aircraft propulsion compliment the subject of this chapter and are recommended for further reading.

References

1. Archer, R.D. and Saarlas, M., *An Introduction of Aerospace Propulsion*, Prentice Hall, New York, 1998.

2. Bathie, W., *Fundamentals of Gas Turbines*, 2nd edition, John Wiley & Sons, Inc. New York, 1995.

3. Cumpsty, N., *Jet Propulsion: A Simple Guide to the Aerodynamic and Thermodynamic Design and Performance of Jet Engines*, 2nd edition, Cambridge University Press, Cambridge, UK, 2003.

4. Flack, R.D., *Fundamentals of Jet Propulsion with Applications*, Cambridge University Press, Cambridge, UK, 2005.

5. Gordon, S. and McBride, B.J., "Computer Program for Computation of Complex Chemical Equilibrium Compositions, Rocket Performance, Incident and Reflected Shocks, and Chapman-Jouguet Detonations," NASA SP-273, 1976.

6. Gordon, S. and McBride, B.J., "Computer Program for Calculation of Complex Chemical Equilibrium Compositions and Applications I. Ananlysis", NASA RP-1311, 1994.

7. Heiser, W.H., Pratt, D.T., Daley, D.H., and Mehta, U.B., *Hypersonic Airbreathing Propulsion*, AIAA, Washington, DC, 1993.

8. Hesse, W.J. and Mumford, N.V.S., *Jet Propulsion for Aerospace Applications*, 2nd edition, Pittman Publishing Corporation, New York, 1964.

9. Hill, P.G. and Peterson, C.R., *Mechanics and Thermodynamics of Propulsion*, 2nd edition, Addison-Wesley, Reading, Massachusetts, 1992.

10. Kerrebrock, J.L., *Aircraft Engines and Gas Turbines*, 2nd edition, MIT Press, Cambridge, Mass, 1992.

11. Kerrebrock, J.L., "Small Disturbances in Turbomachine Annuli with Swirl,"*AIAA Journal*, Vol. 15, June 1977, pp. 794–803.

12. Lytle, J.K., "The Numerical Propulsion System Simulation: A Multidisciplinary Design System for Aerospace Vehicles," NASA TM-1999-209194.

13. Mattingly, J.D., *Elements of Gas Turbine Propulsion*, McGraw-Hill, New York, 1996.

14. Mattingly, J.D., Heiser, W.H., and Pratt, D.T., *Aircraft Engine Design*, 2nd edition, AIAA, Washington, DC, 2002.

15. Oates, G.C., *Aerothermodynamics of Gas Turbine and Rocket Propulsion*, AIAA, Washington, DC, 1988.

16. Schobeiri, M.T., *Turbomachinery Performance and Flow Physics*, Springer Verlag, New York, 2005.

17. Shepherd, D.G., *Aerospace Propulsion*, American Elsevier Publication, New York, 1972.

Problems

10.1 In a turbojet engine, the compressor face total pressure and temperature are 112 kPa and 268 K, respectively. The shaft speed is 6400 rpm. The air mass flow rate is 125 kg/s and the fuel mass flow rate is 2.5 kg/s. The fuel heating value is 42000 kJ/kg and the engine produces 145 kN of thrust. Express the following engine corrected parameters:

(a) the corrected (air) mass flow rate $\dot{m}_{c2}$ in kg/s assuming
$p_{t2} = 0.99 \, p_{t0}$

(b) the corrected shaft speed N_{c2} in rpm

(c) the corrected fuel flow rate, $\dot{m}_{fc}$ in kg/s

(d) the corrected thrust F_c in kN

(e) the corrected thrust-specific fuel consumption $TSFC_c$ in mg/s/N.

Note: $p_{ref} = 101.33 \, \text{kPa}$ and $T_{ref} = 288.2 \, \text{K}$

10.2 The corrected mass flow rate at the engine face is $\dot{m}_{c2} = 180 \, \text{kg/s}$. Calculate the axial Mach number M_{z2} at the engine face for $A_2 = 1 \, \text{m}^2$. Also calculate the capture area A_0 for a flight Mach number of $M_0 = 0.85$ and assume an inlet total pressure recovery $\pi_d = 0.995$. Assume $\gamma_c = 1.4$. $R_c = 287 \, \text{J/kg} \cdot \text{K}$, $p_0 = 30 \, \text{kPa}$ and $T_0 = 250 \, \text{K}$.

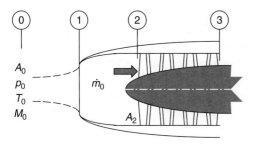

■ **FIGURE P10.2**

10.3 A compressor has an air mass flow rate of 180 kg/s. The fuel flow rate in the burner is 6 kg/s. The flow areas are $A_2 = 1\,\text{m}^2$, $A_3 = 0.14\,\text{m}^2$. Compressor total pressure ratio is $\pi_c = 10$ and $e_c = 0.9$. The fuel heating value is $Q_R = 42000\,\text{kJ/kg}$, burner efficiency and total pressure ratio are $\eta_b = 0.99$ and $\pi_b = 0.95$, respectively.

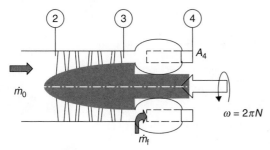

■ **FIGURE P10.3**

Assuming $M_{z2} = 0.5$, $T_{t2} = 300\,\text{K}$ and $M_{z4} = 1.0$, Calculate:

(a) T_{t4}/T_{t2}

(b) p_{t4}/p_{t2}

(c) A_4 in m^2

(d) the physical and corrected mass flow rates at 4, i.e., $\dot{m}_4$ and $\dot{m}_{c4}$ Gas properties are $\gamma_c = 1.4$, $c_{pc} = 1004\,\text{J/kg}\cdot\text{K}$, $\gamma_t = 1.33$, $c_{pt} = 1156\,\text{J/kg}\cdot\text{K}$

10.4 A gas turbine operates with a choked nozzle $M_4 = 1.0$ with $\gamma_t = 1.33$, and $c_{pt} = 1156\,\text{J/kg}\cdot\text{K}$. Turbine expansion parameter $\tau_t = 0.80$ and turbine adiabatic efficiency is $\eta_t = 0.86$. The burner total pressure ratio is $\pi_b = 0.95$ and the burner total temperature ratio is $\tau_b = 1.80$.

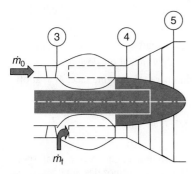

■ **FIGURE P10.4**

Calculate

(a) the ratio of corrected mass flow rates $\dot{m}_{c4}/\dot{m}_{c5}$

(b) the ratio of corrected mass flow rates $\dot{m}_{c4}/\dot{m}_{c3}$, assuming $f = 0.03$.

10.5 A multistage compressor is connected to a multistage turbine on the same shaft. The shaft speed is $N = 8000\,\text{rpm}$. The throttle parameter is $T_{t4}/T_{t2} = 6.0$. The compressor inlet flow has a $p_{t2} = p_{\text{ref}}$ and $T_{t2} = T_{\text{ref}}$. Compressor discharge temperature is $T_{t3} = 872\,\text{K}$. The engine corrected mass flow rate is $\dot{m}_{c2} = 360\,\text{kg/s}$.

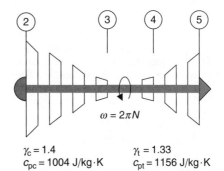

■ **FIGURE P10.5**

Calculate

(a) the corrected shaft speed N_{c2} (rpm)

(b) the corrected shaft speed N_{c4} (rpm)

(c) the compressor pressure ratio π_c assuming $e_c = 0.90$

(d) the compressor shaft power in MW

(e) the fuel-to-air ratio assuming $\pi_b = 0.94$, $\eta_b = 0.99$, and $Q_R = 42000\,\text{kJ/kg}$

(f) turbine expansion parameters T_{t5}/T_{t4} and p_{t5}/p_{t4} for $\eta_t = 0.85$ and $\eta_m = 0.99$

(g) gas generator pumping characteristics p_{t5}/p_{t2} and T_{t5}/T_{t2}

10.6 A high-pressure ratio compressor performance map is shown. The nominal operating line corresponds to $T_{t4}/T_{t2} = 6.5$. Assuming a constant turbine adiabatic efficiency of $\eta_t = 0.88$ at on- and off-designs, calculate and plot the pumping characteristics of the gas generator similar to Example 10.1. The design corrected mass flow rate is 180 kg/s, with $\pi_{c,\text{design}} = 27.5$.

Assume

$$\gamma_c = 1.4, \; c_{pc} = 1004\,\text{J/kg}\cdot\text{K}$$
$$\gamma_t = 1.4, \; c_{pt} = 1156\,\text{J/kg}\cdot\text{K}$$
$$f \approx 0.03$$
$$\eta_m = 0.995$$
$$\pi_b = 0.95$$

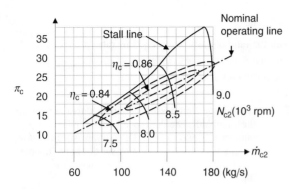

■ **FIGURE P10.6**

10.7 An afterburner on- or off-mode should not affect the turbine back pressure. This requirement is often met by a variable throat convergent–divergent exhaust nozzle. The "dry" mode is characterized by a lower total pressure loss and an adiabatic flow:

$$\pi_{AB-dry} = 0.96, \quad \tau_{AB-dry} = T_{t7}/T_{t5} = 1.0,$$
$$\gamma_{AB-dry} = 1.33, \quad c_{pAB-dry} = 1156 \, J/kg \cdot K$$

The "wet" mode is characterized by a higher total pressure loss and chemical energy release

$$\pi_{AB-wet} = 0.90, \quad \tau_{AB-wet} = T_{t7}/T_{t5} = 2.0,$$
$$\gamma_{AB-wet} = 1.30, \quad c_{pAB-wet} = 1243 \, J/kg \cdot K,$$
$$Q_{R, AB} = 42{,}000 \, kJ/kg, \quad \eta_{AB} = 0.95$$

The turbine entry temperature (TET) is $T_{t4} = 1760 \, K$ and $p_{t4} = 2.0 \, MPa$ (in both dry and wet modes) and the turbine expansion parameter $\tau_t = 0.80$ and $\eta_t = 0.85$. The gas properties in the turbine are $\gamma_t = 1.33$ and $c_{pt} = 1156 \, J/kg \cdot K$. The corrected mass flow rate at turbine entry is $\dot{m}_{c4} = 80 \, kg/s$ and turbine nozzle is choked, $M_4 = 1.0$.

The exhaust nozzle in dry and wet modes is choked, i.e., $M_8 = 1.0$. The total pressure ratio in the convergent (part of the) nozzle is $p_{t8}/p_{t7} = 0.98$ for dry and 0.95 for wet mode. The nozzle divergent section has a total pressure ratio of $p_{t9}/p_{t8} = 0.99$ for dry and 0.95 for wet operation.

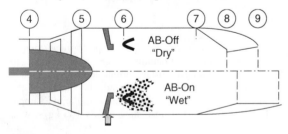

■ **FIGURE P10.7**

Calculate

(a) $A_4(m^2)$

(b) $A_5(m^2)$ for $M_5 = 0.5$

(c) f_{AB}

(d) $A_g(m^2)$ "dry"

(e) $A_g(m^2)$ "wet"

(f) A_9/A_8 "dry" for $p_9 = p_0 = 100 \, kPa$

(g) A_9/A_8 "wet" for $p_9 = p_0 = 100 \, kPa$

(h) nozzle gross thrust (kN) "dry"

(i) nozzle gross thrust (kN) "wet"

10.8 A turbojet engine has the following design-point parameters:

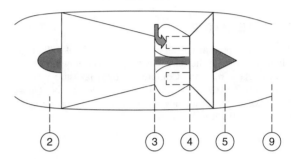

■ **FIGURE P10.8**

$$M_0 = 0, \quad p_0 = 101.33 \, kPa, \quad T_0 = 15.2°C$$
$$\pi_d = 0.98$$
$$\pi_c = 25, \quad e_c = 0.90$$
$$M_4 = 1.0$$
$$Q_R = 42{,}800 \, kJ/kg, \quad \pi_b = 0.95, \quad \eta_b = 0.98, \quad \tau_\lambda = 6.0$$
$$e_t = 0.85, \quad \eta_m = 0.98$$
$$\pi_n = 0.97, \quad p_9 = p_0$$
$$\dot{m}_{c2} = 80 \, kg/s, \quad M_{z2} = 0.50$$

Calculate

(a) fuel-to-air ratio f

(b) turbine total temperature ratio τ_t

For the following off-design operation

$$M_0 = 0.85, \quad p_0 = 20 \, kPa, \quad T_0 = -15°C$$
$$\tau_\lambda = 6.5$$
$$e_t = 0.80, \quad \eta_m = 0.98, \quad \pi_n = 0.97, \quad p_9 = p_0$$

Assume a calorically perfect gas with $\gamma = 1.4$ and $c_p = 1004 \, kJ/kg \cdot K$ constant throughout the engine, and calculate

(c) $\pi_{c-off-design}$

(d) the ratio of corrected shaft speeds $N_{c2, O-D}/N_{c2, D}$

(e) the corrected mass flow rate at off-design (kg/s)

(f) the axial Mach number at the engine face, M_{z2}, at off-design

(g) thrust-specific fuel consumption at design and off-design in mg/s/N

10.9 A separate-flow turbofan engine has a dual spool configuration, as shown. Fan and core nozzles are convergent and choked.

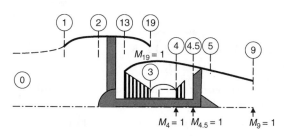

■ **FIGURE P10.9**

The design parameters for this engine are:

1. $M_0 = 0$, $p_0 = 101.33\,\text{kPa}$, $T_0 = 15.2°\text{C}$
2. $\pi_\text{d} = 0.98$
3. $\pi_\text{f} = 1.8$, $e_\text{f} = 0.90$
4. $\alpha = 5.0$
5. $\pi_\text{cH} = 14$, $e_\text{cH} = 0.90$
6. $T_\text{t4} = 1600°\text{C}$, $Q_R = 42{,}800\,\text{kJ/kg}$, $\eta_\text{b} = 0.99$, $\pi_\text{b} = 0.95$
7. $e_\text{tH} = 0.85$, $\eta_\text{mH} = 0.995$
8. $e_\text{tL} = 0.89$, $\eta_\text{mL} = 0.995$
9. $\pi_\text{n} = \pi_\text{nf} = 0.98$, $p_8 = p_{18} = p_0$
10. $\gamma_\text{c} = 1.4$, $c_{pc} = 1004\,\text{J/kg} \cdot \text{K}$
11. $\gamma_\text{t} = 1.33$, $c_{pt} = 1146\,\text{J/kg} \cdot \text{K}$
12. $M_4 = M_{4.5} = M_9 = M_{19} = 1.0$

An off-design operating condition is described by

(a) $M_0 = 0.90$, $p_0 = 20\,\text{kPa}$, $T_0 = -20°\text{C}$
(b) $T_\text{t4} = 1300°\text{C}$

Assuming all component efficiencies remain constant, calculate

(a) fan pressure ratio π_f
(b) hip-pressure compressor pressure ratio π_cH
(c) the bypass ratio α

10.10   A turbojet engine has the following design parameters (which is at takeoff):

$M_0 = 0$

$p_0 = 101.33\,\text{kPa}$, $T_0 = 15.2°\text{C}$, $\gamma_\text{c} = 1.4$,

$c_{pc} = 1004\,\text{J/kg} \cdot \text{K}$

$\pi_\text{d} = 0.95$

$\pi_\text{c} = 30$, $e_\text{c} = 0.90$

$Q_R = 42{,}600\,\text{kJ/kg}$, $\pi_\text{b} = 0.95$, $\eta_\text{b} = 0.98$, $T_\text{t4} = 1700°\text{C}$

$\eta_\text{m} = 0.98$, $e_\text{t} = 0.85$, $\gamma_\text{t} = 1.33$, $c_{pt} = 1156\,\text{J/kg} \cdot \text{K}$

$\pi_\text{n} = 0.90$

$p_9 = p_0$

This engine powers an aircraft that cruises at $M_0 = 0.80$ at an altitude where $T_0 = -35°\text{C}$, $p_0 = 20\,\text{kPa}$. The turbine entry temperature at cruise is $T_\text{t4} = 1500°\text{C}$. Assume that the engine has the same component efficiencies at cruise and takeoff, and the nozzle is perfectly expanded at cruise, as well.

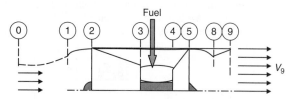

■ **FIGURE P10.10**

Calculate

(a) the exhaust velocity V_9 (in m/s) at the design point, i.e., at takeoff
(b) the thermal efficiency η_th at the design point
(c) thrust-specific fuel consumption at the design point
(d) the compressor pressure ratio at cruise
(e) the exhaust velocity V_9 (in m/s) at cruise
(f) the thermal efficiency η_th at cruise
(g) the propulsive efficiency η_p at cruise
(h) the thrust-specific fuel consumption at cruise

10.11   In a gas generator, the compressor and burner performance maps are shown. The turbine adiabatic efficiency is assumed nearly constant at $\eta_\text{t} = 0.85$. The nominal operating

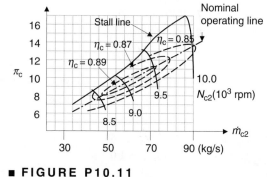

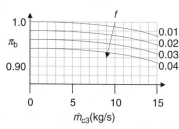

■ **FIGURE P10.11**

line on the compressor performance map represents the $T_{t4}/T_{t2} = 7.0$ throttle line.

The design corrected shaft speed is $N_{c2} = 10,000$ rpm and the compressor pressure ratio at design is $\pi_{c,D} = 13.5$ (note that the corrected mass flow rate at the compressor face is 89 kg/s at design).

Assuming

$$\gamma_c = 1.4, \; c_{pc} = 1004 \, \text{J/kg} \cdot \text{K}$$
$$\gamma_t = 1.33, \; c_{pt} = 1156 \, \text{J/kg} \cdot \text{K}$$
$$f \approx 0.03$$
$$\eta_m = 0.995$$

Calculate and graph the gas generator pumping characteristics, as percent corrected shaft speed N_{c2} (% design).

10.12 The gas generator in Problem 10.11 is to be matched to a variable-throat area convergent nozzle, as shown.

The nozzle performance map is also shown here as a graph of π_n versus the corrected mass flow rate (at the engine face). First, by matching the mass flow rate at station 8 (or 9) to that of 2, show that the corrected mass flow rates at 8 (or 9) and 2 are related to pumping characteristics according to

$$\dot{m}_{c9} = (1 + f) \frac{\sqrt{T_{t5}/T_{t2}}}{\pi_n(p_{t5}/p_{t2})} \dot{m}_{c2}$$

Assuming the nozzle throat remains choked, i.e., $M_8 = 1.0$, graph A_8 variation (or percent $A_8/A_{8,D}$ variation) as a function of the percent-design corrected shaft speed N_{c2} (% design).

9. $\pi_n = \pi_{nf} = 0.98$, $p_8 = p_{18} = p_0$
10. $\gamma_c = 1.4$, $c_{pc} = 1004 \, \text{J/kg} \cdot \text{K}$
11. $\gamma_t = 1.33$, $c_{pt} = 1146 \, \text{J/kg} \cdot \text{K}$
12. $M_4 = M_{4.5} = M_8 = M_{18} = 1.0$

The off-design operation of this engine is represented by a cruise altitude flight such as

$$M_0 = 0.85, \; p_0 = 10 \, \text{kPa}, \; T_0 = -15°\text{C}$$
$$T_{t4} = 1500°\text{C}$$
$$\pi_d = 0.995$$

Assuming all other efficiencies and gas properties remaining constant; calculate the following parameters at off-design condition:

(a) fan pressure ratio π_f

(b) high-pressure compressor pressure ratio π_{cH}

(c) bypass ratio α

10.14 A turbojet engine has the following design-point parameters:

1. $M_0 = 0$, $p_0 = 0.1 \, \text{MPa}$, $T_0 = 15°\text{C}$
2. $\pi_d = 0.98$
3. $\pi_c = 15$, $e_c = 0.90$
4. $Q_R = 42,800 \, \text{kJ/kg}$, $\pi_b = 0.97$, $\eta_b = 0.98$, $T_{t4} = 1485°\text{C}$

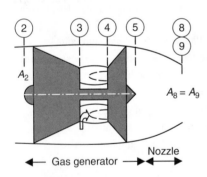
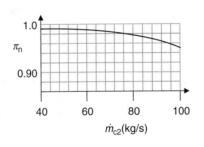

■ **FIGURE P10.12**

10.13 A separate-flow turbofan engine has the following design-point parameters:

1. $M_0 = 0$, $p_0 = 0.1 \, \text{MPa}$, $T_0 = 15°\text{C}$
2. $\pi_d = 0.98$
3. $\pi_f = 1.65$, $e_f = 0.90$
4. $\alpha = 7.0$
5. $\pi_{cH} = 20$, $e_{cH} = 0.90$
6. $T_{t4} = 1650°\text{C}$, $Q_R = 42,800 \, \text{kJ/kg}$, $\eta_b = 0.99$, $\pi_b = 0.95$
7. $e_{tH} = 0.85$, $\eta_{mH} = 0.995$
8. $e_{tL} = 0.89$, $\eta_{mL} = 0.995$

5. $e_t = 0.80$, $\eta_m = 0.995$
6. $\dot{m}_{c2} = 24 \, \text{kg/s}$
7. $N_{c2}(\text{rpm}) = 6,000$
8. $M_{z2} = 0.6$
9. $\pi_n = 0.97$, $p_9/p_0 = 1.0$

The off-design flight condition is described by

$$M_0 = 2.0, \; p_0 = 18 \, \text{kPa}, \; T_0 = -15°\text{C}$$
$$T_{t4} = 1475°\text{C}$$
$$\pi_d = 0.88$$
$$p_9/p_0 = 1.0$$

Assuming all other component efficiencies (except π_d that is specified) remain the same (as design) at off-design and gas properties are $\gamma_c = \gamma_t = 1.4$ and $c_{pc} = c_{pt} = 1004\,\text{J/kg} \cdot \text{K}$, calculate

(a) π_{c-O-D}

(b) $\dot{m}_{c2,O-D}$ (in kg/s)

(c) $N_{c2,O-D}$ (in rpm)

(d) $M_{z2,O-D}$

10.15 An afterburning turbojet engine's design-point parameters are

1. $M_0 = 0$, $p_0 = 101.33\,\text{kPa}$, $T_0 = 288.2\,\text{K}$, $\gamma_c = 1.4$, $c_{pc} = 1004\,\text{J/kg} \cdot \text{K}$

2. $\pi_d = 0.95$

3. $\pi_c = 18$, $e_c = 0.90$

4. $\dot{m}_{c2} = 67\,\text{kg/s}$

5. $N_{c2} = 7120\,\text{rpm}$

6. $M_{z2} = 0.5$

7. $Q_R = 42,800\,\text{kJ/kg}$, $\pi_b = 0.98$, $\eta_b = 0.97$, $T_{t4} = 1773\,\text{K}$

8. $\gamma_t = 1.33$, $c_{pt} = 1156\,\text{J/kg} \cdot \text{K}$

9. $e_t = 0.80$, $\eta_m = 0.995$

10. $Q_{R,AB} = 42,800\,\text{kJ/kg}$, $\pi_{AB} = 0.95$, $\eta_{AB} = 0.98$, $T_{t7} = 2250\,\text{K}$

11. $\gamma_{AB} = 1.3$, $c_{pc} = 1243\,\text{J/kg} \cdot \text{K}$

12. $\pi_n = 0.90$, $p_9/p_0 = 1.0$

The off-design conditions correspond to supersonic flight at high altitude

$M_0 = 2.5$, $p_0 = 15\,\text{kPa}$, $T_0 = 223\,\text{K}$, $\gamma_c = 1.4$,
$c_{pc} = 1004\,\text{J/kg} \cdot \text{K}$
$\pi_d = 0.82$
$e_c = 0.90$
$Q_R = 42,800\,\text{kJ/kg}$, $\pi_b = 0.98$, $\eta_b = 0.97$, $T_{t4} = 1850\,\text{K}$
$\gamma_t = 1.33$ $c_{pt} = 1156\,\text{J/kg} \cdot \text{K}$
$e_t = 0.80$, $\eta_m = 0.995$
$Q_{R,AB} = 42,800\,\text{kJ/kg}$, $\pi_{AB} = 0.95$, $\eta_{AB} = 0.98$,
$T_{t7} = 2450\,\text{K}$
$\gamma_{AB} = 1.3$, $c_{pc} = 1243\,\text{J/kg} \cdot \text{K}$
$\pi_n = 0.88$, $p_9/p_0 = 1.0$

Calculate

(a) compressor pressure ratio at off-design

(b) the corrected and physical mass flow rates at the compressor face at off-design in kg/s

(c) the fuel-to-air ratio at off-design

(d) the exhaust speed at off-design in m/s

(e) thrust specific fuel consumption in mg/s/N at on- and off-design

CHAPTER 11

Chemical Rocket and Hypersonic Propulsion

Courtesy of NASA (Constellation Program)

11.1   Introduction

A chemical rocket is a self-contained jet engine that carries both fuel and oxidizer that is needed for combustion. The fuel and oxidizer are jointly called *propellant*. Since the rocket carries its own oxidizer, it does not need an air inlet to provide oxygen for combustion, as in a gas turbine engine. It is this independence of rocket operation on atmospheric air that makes it a good candidate for space propulsion. The relative simplicity and manufacturing cost of a rocket engine as compared with an airbreathing jet engine have made rockets the propulsor of choice for low-cost weapons and aerospace vehicle boost applications. For example, Space Shuttle uses two solid rocket boosters (known as SRBs) to help achieve liftoff and vehicle acceleration to orbit. Space Shuttle also uses liquid propellant rocket engines as its main engine, dubbed SSME or Space Shuttle Main Engine. Depending on the form of propellant that is used in combustion, a chemical rocket may be called a *liquid propellant*, a *solid propellant*, a *gaseous propellant* rocket or if the fuel is in solid form and the oxidizer is in liquid (or gaseous) form, the rocket is called a *hybrid*. For a liquid propellant rocket, the fuel and oxidizer are stored in separate tanks in liquid form. The solid propellant rocket has the fuel and oxidizer in solid form, packed or cast in the rocket case, in the form of a paste. Since rockets operate on the principle of jet propulsion, which is achieved when a fluid is ejected through a nozzle at finite speed, the fluid and the form of acceleration opens new categories of rocket propulsion devices. For example, if the fluid is made up of ions (of an atom, say Cesium) that are accelerated in an electric field (in lieu of a nozzle), the rocket achieves a thrust and this class of rocket propulsion belongs to electric or ion propulsion. For the case of fluid (like hydrogen) that is heated by a nuclear reactor before ejection through a nozzle, the category of engine is called a nuclear rocket propulsion device. For atmospheric applications that require stringent emissions standards, only the liquid and solid propellant rockets are suitable. These belong to the general category of *chemical rocket propulsion*. Forward (1995) presents intriguing advanced propulsion concepts (other than pure rockets) that are thought provoking and thus recommended for reading.

The main drawback of rocket engines operating in Earth's atmosphere is their high propellant consumption rate per unit thrust produced as compared to the fuel consumption rate in airbreathing engines. The figure of merit that compares rockets to airbreathing engines is the specific impulse I_s (in seconds), that we defined in Chapter 3, repeated here for convenience:

$$I_s \equiv \frac{F}{\dot{m}_p g_0} \quad (\text{s}) \tag{3.32}$$

$$\text{where} \quad \dot{m}_p \equiv \dot{m}_f + \dot{m}_o \tag{3.33}$$

The numerator in specific impulse is the net (uninstalled) thrust produced, which for an airbreathing engine, we derived to be (Eq. 3.14)

$$F_n)_{\text{uninstalled}} = (\dot{m}_0 + \dot{m}_f)V_9 - \dot{m}_0 V_0 + (p_9 - p_0)A_9 \tag{3.14}$$

Since, rockets do not suffer from any ram drag, they produce thrust according to (Eq. 3.30):

$$F_{\text{rocket}} = (\dot{m}_0 + \dot{m}_f)V_9 + (p_9 - p_0)A_9 \tag{3.30}$$

The denominator of specific impulse is the propellant weight flow rate for a rocket (i.e., the sum of oxidizer and fuel weight flow rates) and only the fuel weight flow rate for an airbreathing engine. The use of a more energetic fuel like hydrogen leads to directly higher specific impulse, since a smaller, i.e., nearly two and half times lower, consumption of hydrogen is needed to achieve the same temperature in the combustor. Now, if we graph representative values for the specific impulse of typical airbreathing and rocket engines (see Fig. 11.1, also Kerrebrock, 1992), for a hydrocarbon and hydrogen fuel, as a function of flight Mach number, we note the following important features:

- rocket-specific impulse is independent of vehicle speed, as it does not suffer from any ram drag penalty, nor stagnation temperature rise of air limits its combustion process
- the specific impulse for a chemical rocket is about 250–450 s

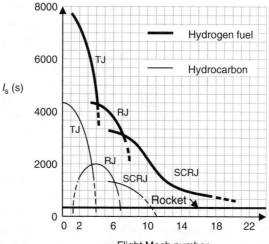

■ **FIGURE 11.1** Approximate variation of specific impulse with flight Mach number for different airbreathing engines (TJ: turbojet, RJ: ramjet, and SCRJ: scramjet) and a typical chemical rocket

- the specific impulse for airbreathing engines is shifted upward when hydrogen is used as fuel in comparison to hydrocarbon fuels
- a gas turbine engine achieves takeoff thrust and the best low-speed performance, as depicted by a specific impulse of between ~4000 and 8000 s for a turbojet and (not shown) higher than 10,000 s for a turbofan
- the specific impulse of gas turbine engines drop with flight Mach number
- the rapid fall of a gas turbine engine performance with forward speed is broken up if we switch to the ramjet family of airbreathing engines, for example, near Mach 3, a conventional ramjet starts to outperform a turbojet
- the rapid fall of a conventional ramjet engine performance with flight Mach number is halted if we switch to the supersonic combustion ramjet (or scramjet)
- the flight Mach number of ~6 is where scramjet outperforms other airbreathing engines
- deterioration of scramjet performance with forward speed is inevitable as the flight stagnation temperature increases and components losses mount
- although the upper operational Mach number of scramjets is unknown, it is suspected that for high flight Mach numbers, e.g., Mach ~10–15, scramjet performance may begin to fall below a chemical rocket.

11.2   From Takeoff to Earth Orbit

A vehicle requires a flight Mach number of ~25 to achieve a circular low earth orbit (LEO). The proof of that is very simple. On a circular orbit of radius r (which for LEO is ≥ 100 km above the Earth of ~6000 km radius), a vehicle of mass m experiences a centrifugal force of magnitude mV^2/r. The gravitational pull on the vehicle toward the Earth is *roughly* $\sim mg_0$. These two forces are in balance if the vehicle is to maintain its circular orbit, therefore,

$$V \approx \sqrt{rg_0} \approx \sqrt{6000\,\text{km}(9.8\,\text{m/s}^2)} \approx 7668\,\text{m/s}$$

Which for a speed of sound of ~300 m/s (at high altitude), we get a flight (or orbital) Mach number of ~25.5. The purpose of this simple demonstration was to get a *rough order of magnitude* for the Mach number required at LEO to maintain a circular orbit. We recognize that we made several approximations, for example, in the gravitational force on the vehicle when we applied g_0 on the Earth's surface, instead of a reduced gravitational acceleration that corresponds to an altitude of 100 km, or when we did not account for any atmospheric drag at 100 km, or when we used an "average" Earth radius (of 6000 km) as in a homogeneous spherical earth.

To reach LEO and maintain a circular orbit, we need to fly about Mach 25. For single-stage to orbit aircraft (what is known as SSTO), the engine(s) have to produce takeoff thrust, maintain climb rate, and acceleration until the vehicle has achieved Mach 25. If we wish to use the most efficient means of reaching Mach 25 from takeoff, we have to employ a variety of engines, as depicted by the specific impulse chart of Fig. 11.1. At takeoff, we could use a turbofan engine and gradually reduce its bypass ratio (as in a variable-bypass TF engine) with flight Mach number until it operates as a turbojet. Then, we should be able to shut down the gas generator all together near Mach 3 and switch over to a conventional (or subsonic combustion) ramjet for up to ~Mach 6. The scramjet is to take over beyond Mach 6 and accelerate the vehicle through Mach 10–15, depending on the hydrocarbon or hydrogen fuel, respectively. At these Mach numbers, we are still too slow to maintain a circular orbit at LEO. Therefore, chemical rocket engines have to be fired for the last leg of our launch, which should take the vehicle to Mach 25 and low-Earth orbit. The main challenge for such a *combination propulsion system* (CPS) is complex system integration into vehicle and the mechanical complexities involved in transition from one set or class of engines to another.

11.3   Chemical Rockets

Various components of a chemical rocket are shown in a schematic drawing in Fig. 11.2.

The example shown in Fig. 11.2 uses liquid fuel and oxidizer as propellant. The fuel and oxidizer are stored in insulated tanks in liquid state to reduce the volumetric requirements of the propellant. The fuel and oxidizer pumps provide the propellant feed system into the combustion chamber. The pumps are driven by a turbine (not shown in the schematic drawing), therefore the name *turbopump* system. The propellant is fed into the combustion chamber through an injector plate. The pressure in the combustion chamber is ~20% lower than the feed pressure to avoid pressure oscillation instability known as chugging. The combustion temperature is governed by the heat of reaction of the fuel and oxidizer. The products of combustion are then accelerated through an exhaust nozzle. The thrust may be controlled (to the point of shutting down and restarting) by the fuel and oxidizer pumps. The guidance, navigation, control, and payload are shown in the nose chamber. The fins provide aerodynamic control and stability of the rocket.

Elements of a solid rocket motor are shown in Fig. 11.3. The propellant is in solid form and is called the grain or charge. Propellant is bonded to the rocket motor case, which typically has an internal burning grain cavity, as shown. The igniter initiates the burning of the solid grain. Once burning is initiated and hot gases are expelled through an exhaust nozzle, the thrust of the solid motor is established by the burning grain surface area and the propellant-burning rate. Since solid rocket motors have no pumps or valves to control the amount of propellant engaged in the combustion process, the thrust cannot be fully controlled. We can, however, terminate the operation of a solid rocket motor by opening thrust termination ports to cancel the thrust, as shown.

A hybrid rocket typically uses a solid fuel and a liquid oxidizer, as shown in Fig. 11.4. The oxidizer flow is regulated by the pressurized tank and a valve and injects the stream of oxidizer through a solid fuel port. The chemical reaction may be initiated by an igniter, similar to a solid propellant rocket motor. The hot combustion gases are expelled through the exhaust nozzle to produce thrust.

11.4   Chemical Rocket Applications

Chemical rockets are capable of producing large thrust and operate at any altitude, including space. They also lend themselves to *staging* for increased flight efficiency. These diverse capabilities in addition to the low cost of manufacturing put rockets on many flight vehicle applications, including as strap-on boosters. Some of the main examples are listed here.

■ **FIGURE 11.2**
Schematic diagram of a (simplified) chemical rocket with liquid hydrogen as fuel and liquid oxygen (LO$_2$) as the oxidizer (with turbopump feed system)

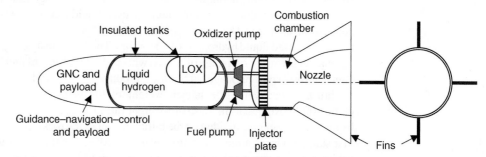

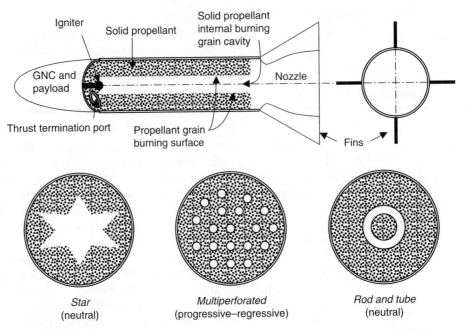

■ **FIGURE 11.3** **Schematic drawing of a solid-propellant chemical rocket with internal burning grain (grain designs referred to neutral, progressive, and regressive describe thrust-time behavior)**

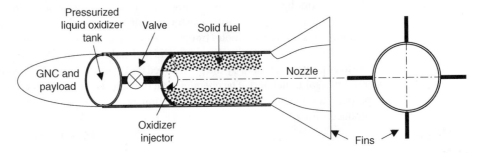

■ **FIGURE 11.4** **Schematic drawing of a hybrid rocket showing solid fuel and gaseous oxidizer propellant**

11.4.1 Launch Engines

Saturn V was a three-stage liquid propellant rocket that was successfully used to launch Apollo spacecraft toward the moon. In a multistage rocket, each stage is self-contained and operates autonomously. All three stages of Saturn V used liquid oxygen (LO_2) as their oxidizer in the combustion chamber. The first stage used RP-1 (a hydrocarbon fuel), whereas the second and third stages used liquid hydrogen (LH_2) for fuel. The low density of hydrogen makes it less than optimal for use in the first stage of launch where a large volume of fuel is required to produce the launch thrust. The three stages of Saturn V produced 7,570,000 lb, 1,125,000 lb, and 230,000 lb of thrust, respectively. The launch weight of the vehicle that included the payload was 6,262,000 lb. Note that the thrust-to-weight ratio for Saturn at launch is the thrust produced by its first stage engine divided by the initial weight of the rocket, i.e., 7570/6262 = 1.209. The first stage produces a modest thrust level of ~20% over the initial weight (to accelerate the vehicle), i.e., it imparts $0.2\,g_0$ initial acceleration to the vehicle.

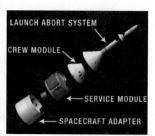

■ **FIGURE 11.5** Artist's concept of Orion Crew Exploration Vehicle (CEV) approaching the International Space Station, Ares I launch vehicle, and components of CEV (Courtesy of NASA)

Another example is the Space Shuttle Main Engine that uses LO_2/LH_2 as propellant. There are three SSMEs and each produce 375,000 lb at sea level and 470,000 lb at altitude. The SSME starts at launch and operates for ~8.5 min. The use of liquid hydrogen in SSME for launch has required an external (propellant) tank that is jettisoned after the liquid hydrogen and oxygen are consumed. The external tank (ET) is the only expendable component of the Space Shuttle.

The new generation of launch and human exploration vehicles, post Space Shuttle, are Ares I launch vehicle and the Orion Crew Exploration Vehicle (CEV). Orion's first flight with astronauts onboard is planned for 2014 to the International Space Station. Orion's first flight to the moon is planned for 2020. Figure 11.5 shows artist's concepts and models of the new family of human space exploration launch and crew exploration vehicle (courtesy of NASA).

11.4.2 Boost Engines

Boosters are capable of providing thrust, often as a simple add-on or strap-on propulsion system. For example, there are two solid rocket boosters on the Space Shuttle launch vehicle. Each is capable of producing 2,700,000 lb of thrust. The solid rocket boosters start at launch and operate for 123 s. The empty boosters are then jettisoned and recovered at sea for reuse.

11.4.3 Space Maneuver Engines

Space Shuttle is equipped with two orbital maneuver systems, which are responsible for the orbiter reentry and orbit correction maneuvers.

11.4.4 Attitude Control Rockets

Space Shuttle has 38 primary thrusters and 6 vernier thrusters that are used to impart small velocity corrections to the orbiter. The duration of operation of these reaction control thrusters (known as *pulses*) is between a few milliseconds and a few seconds.

11.5 New Parameters in Rocket Propulsion

Rocket propulsion, as a self-contained system, introduces new parameters to our propulsion vocabulary. For example, thrust-time behavior, as in neutral, progressive or regressive, or total impulse I_t, which is defined as the integral of the thrust over time, has no counterparts in airbreathing engines. The total impulse is a parameter that is directly measured in rockets on thrust stand and used to estimate the engine-specific impulse I_s. The total impulse is defined as

$$I_t \equiv \int_0^{t_b} F(t)dt \qquad (11.1)$$

The time at burnout is t_b. The weight of the propellant is

$$w_p = g_0 \int_0^{t_b} \dot{m}_p dt \qquad (11.2)$$

and the specific impulse is the ratio of total impulse to propellant weight, namely,

$$I_s \equiv \frac{\int_0^{t_b} F(t)dt}{g_0 \int_0^{t_b} \dot{m}_p dt} = \frac{I_t}{m_p g_o} \qquad (11.3)$$

Figure 11.6 shows three different thrust-time behaviors for solid propellant rockets where the shape of the propellant grain dictates the burning surface area (variation with time) and thus thrust production. The initial (ignition delay) transient to produce thrust and the burnout transient are nearly parabolic in shape and are also shown in Fig. 11.6.

The flight performance of rockets strongly depends on vehicle design through the use of stages, and the efficiency of components. For example, the rocket initial mass is m_0, and the final mass, i.e., after propellant burnout, is m_f. The ratio of final-to-initial mass is an important parameter, and a figure-of-merit in rocket design, which is known as the *mass ratio*, MR,

$$MR \equiv \frac{m_f}{m_0} \qquad (11.4)$$

Propellant mass m_p is the difference between the initial and final masses, for a given stage, i.e.,

$$m_p = m_0 - m_f \qquad (11.5)$$

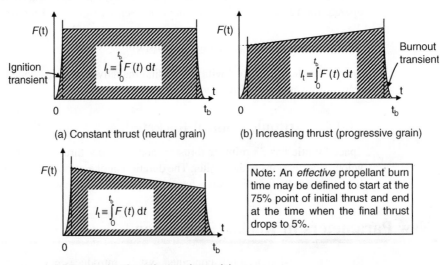

(a) Constant thrust (neutral grain)

(b) Increasing thrust (progressive grain)

(c) Decreasing thrust (regressive grain)

Note: An *effective* propellant burn time may be defined to start at the 75% point of initial thrust and end at the time when the final thrust drops to 5%.

■ **FIGURE 11.6** Total impulse for three different grain designs in solid rocket motors

The ratio of propellant-to-initial mass is known as the *propellant mass fraction ζ,*

$$\zeta \equiv \frac{m_{\mathrm{p}}}{m_0} = 1 - \mathrm{MR} \tag{11.6}$$

A large value of propellant mass fraction is desirable as it indicates an efficient use of *system mass* on the rocket that is directly responsible for thrust production.

Finally, an effective exhaust velocity c is defined as, in effect, the mass-averaged nozzle gross thrust, namely,

$$c \equiv F/\dot{m}_{\mathrm{p}} \tag{11.7}$$

Using Eq. 3.32, at the beginning of this chapter, we relate the effective exhaust velocity c to the specific impulse,

$$c = \dot{m}_{\mathrm{p}} g_0 I_{\mathrm{s}}/\dot{m}_{\mathrm{p}} = g_0 I_{\mathrm{s}} \tag{11.8}$$

We may also relate the effective exhaust velocity to the actual exhaust velocity and pressure thrust through

$$c = V_2 + \frac{(p_2 - p_0)A_2}{\dot{m}_{\mathrm{p}}} \tag{11.9}$$

Station numbers in a rocket engine are defined in Fig. 11.7. The station c refers to the combustion chamber, station 1 is at the inlet to the convergent–divergent nozzle, station th is at the nozzle throat, and station 2 is at the nozzle exit plane. The ambient state is identified by station 0. For most applications, there is no distinction between stations c and 1, but in reality the mixture of gases in the combustion chamber are chemically reacting, whereas the station 1 assumes the combustion is complete and the mixture has reached a chemical equilibrium state. Also, we usually associate a near stagnant gas state in the combustion chamber with $V_{\mathrm{c}} \approx 0$, but we attribute certain finite gas speed V_1 at the nozzle inlet that follows from the area ratio A_1/A_{th} based on the choked throat $M_{\mathrm{th}} = 1.0$ condition. Note parameter trends inside a rocket engine (in Fig. 11.7).

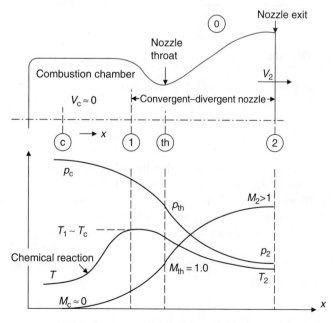

■ **FIGURE 11.7**   **Rocket engine (thrust chamber) stations and corresponding static pressure, temperature, and Mach number variations**

EXAMPLE 11.1

Space Shuttle Main Engine (SSME) produces 375,000 lb of thrust at sea level and 470,000 lb at altitude. Assuming constant momentum thrust with altitude, estimate the diameter D_2 of the SSME nozzle exit area A_2.

SOLUTION

The difference between the sea level and altitude thrust is primarily caused by the difference between the pressure thrusts, i.e.,

$$F_{vacuum} - F_{sea\,level} \equiv (p_{sl} - 0)A_2$$

Therefore, $A_2 \approx 95000\,\text{lb}/(14.7\,\text{lb/in}^2)/(144\,\text{in}^2/\text{ft}^2) \approx 44.88\,\text{ft}^2$

Diameter $D_2 = (4A_2/\pi)^{1/2} \approx 7.56\,\text{ft or } 90.7\,\text{in}.$
   The actual (SSME nozzle) exit diameter is listed at 94 in. There is a small discrepancy (i.e., ~3.5%) due to our simplified approach, e.g., assuming identical momentum thrust at sea level and altitude or neglecting pressure at altitude.

EXAMPLE 11.2

A rocket engine has a propellant mass flow rate of 1000 kg/s and the specific impulse is $I_s = 340$ s. Calculate

(a)   rocket thrust F

(b)   effective exhaust velocity c

SOLUTION

From the definition of specific impulse, we get

$$F = \dot{m}_p g_0 I_s = (1000\,\text{kg/s})(9.8\,\text{m/s}^2)(340\,\text{s}) \approx 3.332 \times 10^6\,\text{N}$$

The effective exhaust velocity is the ratio of thrust to propellant mass flow rate, $c = F/\dot{m}_p \approx 3332\,\text{m/s}$

11.6   Thrust Coefficient, C_F

It is customary to express the rocket thrust as a product of chamber pressure p_c, the throat area A_{th}, and a coefficient C_F, known as the *thrust coefficient*, i.e.,

$$F \equiv C_F p_c A_{th} \tag{11.10}$$

In this section, we will demonstrate that thrust coefficient is a function of the nozzle area expansion ratio A_2/A_{th}, the pressure ratio p_2/p_c, and the ratio of specific heats γ. Therefore, the thrust coefficient is explicitly independent of combustion temperature (thus propellant combination) and is treated as a nozzle-related parameter. The ratio of specific heats γ is the only link between thrust coefficient and the propellant combination or combustion temperature.

   The mass flow rate through the nozzle may be written at the sonic throat, through the use of continuity equation:

$$\dot{m}_p = \sqrt{\frac{\gamma}{R}} \frac{p_c}{\sqrt{T_c}} A_{th} \left(\frac{2}{\gamma + 1} \right)^{\frac{\gamma + 1}{2(\gamma - 1)}} \tag{11.11}$$

The exhaust velocity V_2 is written based on the isentropic expansion through the nozzle from the stagnation pressure of p_c to the exit static pressure p_2 through these simple steps:

$$T_c = T_2 + V_2^2/2c_p$$

$$V_2 = \sqrt{2c_p(T_c - T_2)} = \sqrt{2c_p T_c \left[1 - \left(\frac{p_2}{p_c}\right)^{\frac{\gamma-1}{\gamma}}\right]} \qquad \textbf{(11.12)}$$

Since the momentum thrust is the product of mass flow rate and the exhaust velocity (Eqs. 11.11 and 11.12), we may write the following expression for the thrust coefficient

$$C_F = \frac{\dot{m}_p V_2 + (p_2 - p_0)A_2}{p_c A_{th}} \qquad \textbf{(11.13-a)}$$

$$C_F = \sqrt{\left(\frac{2\gamma^2}{\gamma - 1}\right)\left(\frac{2}{\gamma + 1}\right)^{\frac{\gamma+1}{\gamma-1}}\left[1 - \left(\frac{p_2}{p_c}\right)^{\frac{\gamma-1}{\gamma}}\right]} + \frac{p_2 - p_0}{p_c}\left(\frac{A_2}{A_{th}}\right) \qquad \textbf{(11.13-b)}$$

A perfectly expanded nozzle meets the condition of static pressure match $p_2 = p_0$, therefore the maximum thrust coefficient is a function of p_2/p_c and the ratio of specific heat γ. We may graph the optimum thrust coefficient, i.e., C_F for a perfectly expanded nozzle with isentropic expansion using an Excel spreadsheet calculation and graph of the function

$$C_{F,\,opt} = \sqrt{\left(\frac{2\gamma^2}{\gamma - 1}\right)\left(\frac{2}{\gamma + 1}\right)^{\frac{\gamma+1}{\gamma-1}}\left[1 - \left(\frac{p_2}{p_c}\right)^{\frac{\gamma-1}{\gamma}}\right]} \qquad \textbf{(11.13-c)}$$

in Fig. 11.8 (with abscissa in logarithmic scale).

Before leaving the subject of thrust coefficient, we recognize the bracket term under the square root in Eqs. 11.13-b or -c as the ideal thermal efficiency of a Brayton cycle operating between the two pressures p_c and p_2, i.e.,

$$\eta_{th} = 1 - \left(\frac{p_2}{p_c}\right)^{\frac{\gamma-1}{\gamma}} \qquad \textbf{(11.14)}$$

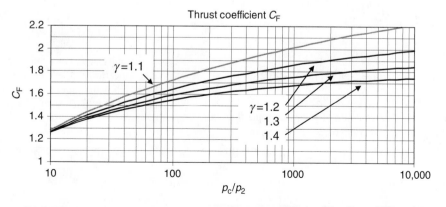

■ **FIGURE 11.8** Variation of optimum thrust coefficient for different gas ratio of specific heats (note that p_c/p_2 is the familiar *nozzle pressure ratio* (NPR) for a perfectly expanded nozzle)

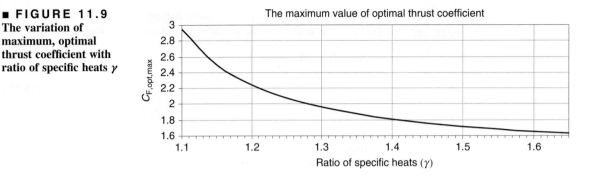

■ **FIGURE 11.9**
The variation of maximum, optimal thrust coefficient with ratio of specific heats γ

For very large chamber pressures p_c and very large area ratio nozzles where the exit static pressure p_2 is low, the thermal efficiency approaches 1, which then points to the maximum attainable thrust coefficient $C_{F,opt,Max}$, i.e.,

$$C_{F,\,opt,\,Max} = \sqrt{\left(\frac{2\gamma^2}{\gamma-1}\right)\left(\frac{2}{\gamma+1}\right)^{\frac{\gamma+1}{\gamma-1}}} \qquad \textbf{(11.15)}$$

The maximum value of thrust coefficient is a pure function of γ and is shown in Fig. 11.9.

EXAMPLE 11.3

A rocket engine has a chamber pressure of $p_c = 200$ atm and the throat area is $A_{th} = 0.25\,\mathrm{m}^2$. Assuming that the nozzle is perfectly expanded with the gas ratio of specific heats $\gamma = 1.3$ and the ambient pressure of $p_0 = 1$ atm, calculate

(a) optimum thrust coefficient $C_{F,opt}$
(b) thrust F in N or lbf
(c) nozzle exit Mach number M_2
(d) nozzle area expansion ratio A_2/A_{th}

SOLUTION

We substitute for p_c/p_2 of 200/1 and $\gamma = 1.3$ in Eq. 11.13-c, to get

$$C_{F,opt.} = 1.650$$

Therefore, thrust is the product of thrust coefficient, chamber pressure, and throat area. The chamber pressure is first converted to N/m² (or Pa),

$$p_c = 200\,\mathrm{atm} = 200(101\,\mathrm{kPa}) = 20.2\,\mathrm{MPa}$$

$$F = 8.332 \times 10^6\,\mathrm{N} \quad (\mathrm{or}\ 1.873 \times 10^6\,\mathrm{lbf})$$

The nozzle exit Mach number for an isentropic flow in a nozzle is related to the ratio of p_c/p_2, according to

$$p_c/p_2 \approx p_{t2}/p_2 = \left[1 + (\gamma-1)M_2^2/2\right]^{\gamma/(\gamma-1)}$$

Therefore, the exit Mach number is

$$M_2 = \sqrt{\frac{2}{\gamma-1}\left[\left(\frac{p_c}{p_2}\right)^{\frac{\gamma-1}{\gamma}}-1\right]} \approx 4.0$$

From continuity equation written between the exit and the throat, taking throat Mach number as unity, we express A_2/A^*, which is the same as A_2/A_{th} in terms of exit Mach number M_2 and a function of γ, according to

$$\frac{A_2}{A_{th}} = \frac{1}{M_2}\left[\frac{2}{\gamma+1}\left(1+\frac{\gamma-1}{2}M_2^2\right)\right]^{\frac{\gamma+1}{2(\gamma-1)}} \approx 15.89$$

11.7   Characteristic Velocity, c^*

A velocity parameter that is linked to the combustion chamber is called characteristic velocity c^*. We expect this velocity to be related to the speed of sound in the combustion chamber, as the characteristic speed. But, first its definition

$$c^* \equiv \frac{p_c A_{th}}{\dot{m}_p} \tag{11.16}$$

In terms of specific impulse and thrust coefficient, or effective exhaust speed and thrust coefficient, the characteristic velocity may be written as

$$c^* = I_s g_0 / C_F = c / C_F \tag{11.17}$$

Returning to the definition of c^* in 11.16, we may express the propellant mass flow rate for the sonic flow at the throat and gas total pressure and temperature, to get

$$c^* = \frac{\sqrt{\gamma R T_c}}{\gamma \left[\dfrac{2}{\gamma + 1}\right]^{\frac{\gamma+1}{2(\gamma-1)}}} \tag{11.18}$$

As expected, the numerator in 11.18 is the speed of sound in the combustion chamber. The propellant combination of fuel and oxidizer and their proportions establish the equivalence ratio, the products of combustion, and the mixture temperature for a combustion pressure p_c, as we studied in Chapter 6. From Eq. 11.16, we note that the experimental estimation of characteristic velocity is rather straightforward, since the chamber pressure and propellant flow rate are easily measured and the throat area is a geometric parameter. The Table 11.1 shows some liquid propellant combinations and the characteristic speed c^*, also known as *chamber c^** (from Sutton and Biblarz, 2001). The mixture ratio is the *oxidizer-to-fuel* mass ratio. The chamber pressure for the ideal performance values in Table 11.1 is taken to be 1000 psia (this corresponds to ~6.87 MPa). The 1000 psia chamber pressure is often taken as a reference value in rocket propulsion. RP-1 is a hydrocarbon fuel [$CH_{1.953}$] similar to kerosene.

■ **TABLE 11.1**

Theoretical Performance of Liquid Propellant Combinations (Data from Sutton and Biblarz, 2001)

Oxidizer	Fuel	Mixture ratio (Mass)	Chamber temperature (K)	Chamber c^* (m/s)
Oxygen	Methane	3.2	3526	1835
		3.0	3526	1853
	Hydrazine	0.74	3285	1871
		0.90	3404	1892
	Hydrogen	3.40	2959	2428
		4.02	2999	2432
	RP-1	2.24	3571	1774
		2.56	3677	1800
Fluorine	Hydrazine	1.83	4553	2128
		2.30	4713	2208
	Hydrogen	4.54	3080	2534
		7.60	3900	2549

■ **TABLE 11.2**
Theoretical Performance of Solid Propellant Combinations (Data from Sutton and Biblarz, 2001)

Oxidizer	Fuel	Chamber temperature (K)	Chamber c^* (m/s)
Ammonium nitrate	11% binder and 7% additive	1282	1209
Ammonium Perchlorate 78–66%	18% organic polymer binder and 4–20% Al	2816	1590
Ammonium Perchlorate 84–68%	12% polymer binder and 4–20% Al	3371	1577

A few examples of solid propellant combinations and the corresponding chamber c^* values are shown in Table 11.2 (from Sutton and Biblarz, 2001). Again, the reference chamber pressure is taken to be 1000 psia. The composition of solid propellant is characterized by the amount of binder (i.e., glue or epoxy), additives, and aluminum, which enhance combustion chamber temperature and chamber c^*.

EXAMPLE 11.4

A liquid propellant rocket uses hydrogen fuel and oxygen for combustion. The oxidizer-to-fuel mass ratio is 4.02. The ratio of specific heats of the combustion gas is $\gamma = 1.26$ and the chamber pressure is 1000 psia. Use Eq. 11.18 to estimate combustion gas constant R as well as the molecular weight of the mixture, MW.

SOLUTION

From Table 11.1, we read $T_c = 2999$ K and $c^* = 2432$ m/s corresponding to the liquid propellant described above.

$$c^* = \frac{\sqrt{\gamma R T_c}}{\gamma \left[\dfrac{2}{\gamma+1}\right]^{\frac{\gamma+1}{2(\gamma-1)}}} = 2432\,\text{m/s} \implies$$

$$\sqrt{\gamma R T_c} = 1801.5\,\text{m/s}$$

By substituting for γ and T_c, we estimate $R \approx 858.9$ J/kg · K and since the universal gas constant is $\bar{R} = 8.3146$ kJ/kmol · K,

$$MW = \frac{\bar{R}}{R} \approx 9.68\,\frac{\text{kg}}{\text{kmol}}$$

EXAMPLE 11.5

Consider a fuel-rich combustion of hydrogen and oxygen in a liquid propellant rocket according to

$$4H_2 + O_2 \longrightarrow 2H_2O + 2H_2$$

Calculate

(a) the oxidizer-to-fuel mixture ratio

(b) the molecular weight of the mixture of gases in the products of combustion

SOLUTION

The oxidizer-to-fuel mass ratio is (32/8), which is 4, very close to the mixture ratio in Example 11.4.

The mixture molecular weight is

$$MW = \frac{2(18) + 2(2)}{4} = 10 \frac{kg}{kmol}$$

This simple calculation of chemical reaction and the mixture of gases in the combustion products support our earlier calculations of molecular weight in Example 11.4 where we found that the average molecular weight of products of combustion was ~9.68 kg/kmol.

11.8   Flight Performance

Consider a rocket of instantaneous mass $m(t)$ flying at the instantaneous speed V. The flight path angle with respect to horizon is θ, where vertical is the direction of the gravitational force, as shown in Fig. 11.10. The atmospheric drag D acts along the vehicle axis, as shown.

We may relate the net forces along the vehicle axis to the vehicle instantaneous acceleration according to

$$m\frac{dV}{dt} = F - D - mg\sin\theta \tag{11.19}$$

Case 1 Zero gravitational force and atmospheric drag
This case is of interest as it produces a simple closed-form solution, since

$$m\frac{dV}{dt} = F = \dot{m}_p c \tag{11.20}$$

We recognize that propellant flow rate is the rate at which the vehicle loses mass, therefore substituting

$$\dot{m}_p = -\frac{dm}{dt}$$

in Eq. 11.20, we have

$$m\frac{dV}{dt} = \dot{m}_p c = -\frac{dm}{dt}c \tag{11.21-a}$$

which separates to the following form that is suitable for integration

$$dV = -c\frac{dm}{m} \tag{11.21-b}$$

The vehicle attains an incremental speed change ΔV as natural logarithm of inverse mass ratio, and proportional to the nozzle effective exhaust velocity c, i.e.,

$$\Delta V = -\bar{c}\ln\left(\frac{m_f}{m_0}\right) = \bar{c}\ln\left(\frac{1}{MR}\right) \tag{11.22}$$

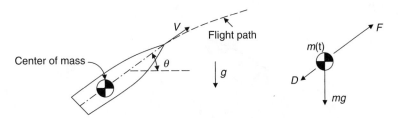

■ **FIGURE 11.10**   Definition sketch of a rocket in flight and the forces that act on the vehicle

This is known as the *fundamental equation of rocketry*. For a rocket designed to have a final mass of 5% of its initial mass, the inverse mass ratio is 20 and the natural logarithm of 20 is 2.9957. Therefore, the rocket may be gaining speed of $\sim 3c$, and as shown in Eq. 11.8, the effective exhaust velocity is $\sim g_0 I_s$. For an average specific impulse I_s of ~ 400 s and the (earth) gravitational acceleration of $\sim 10 \, \text{m/s}^2$, we have an effective exhaust speed of ~ 4000 m/s. Thus, the rocket terminal speed, without accounting for gravity and drag, is $\sim 3c$, which is nearly 12,000 m/s. However, we need to correct for the effects of gravitational force and aerodynamic drag that we neglected in case-1. Both of these effects cause a reduction in the terminal speed or the attained incremental speed ΔV.

Before we embark on case 2, let us show the fundamental equation of rocketry derived from a stationary observer frame of reference outside the rocket. The rocket speed V is measured relative to the stationary observer. The effective exhaust speed c is relative to the rocket, which translates into $(c - V)$ in the stationary observer frame of reference. At time t, the vehicle mass is m and the instantaneous speed V. At an incremental time later, i.e., $t + dt$, the vehicle has a reduced mass $(m - dm)$ and has attained an incremental speed change $(V + dV)$. During the same dt time period, the propellant of mass dm is ejected through the nozzle backward and has attained a speed $(c - V)$ with respect to the stationary observer. The external forces acting on the rocket for the observer are only the aerodynamic drag and the gravitational force. The thrust produced by the rocket motor is an internal force to the vehicle and thus the observer will not record that as external force acting on the vehicle. The graphical depiction of the two snapshots of the rocket is shown in Fig. 11.11.

The change of momentum along the rocket axis for the vehicle and propellant *slug* between times t and $t + dt$ (with positive taken to be in the flight direction) is

$$[(m - dm)(V + dV) - (c - V)dm] - mV \cong mdV - cdm \quad \textbf{(11.23)}$$

We neglected $(dm)(dV)$ in Eq. 11.23 as a second order effect. The change of momentum along the vehicle axis is equal to the impulse of external forces (along the vehicle axis and in the flight direction) in time dt, namely,

$$(-D - mg \sin \theta)dt \quad \textbf{(11.24)}$$

Set Eq. 11.23 equal to 11.24 and divide by dt to get

$$m\frac{dV}{dt} - c\frac{dm}{dt} = -D - mg \sin \theta \quad \textbf{(11.25)}$$

We recognize the second term on the LHS of Eq. 11.25 as the rocket thrust force, which then reproduces Eq. 11.19, i.e.,

$$m\frac{dV}{dt} = F - D - mg \sin \theta$$

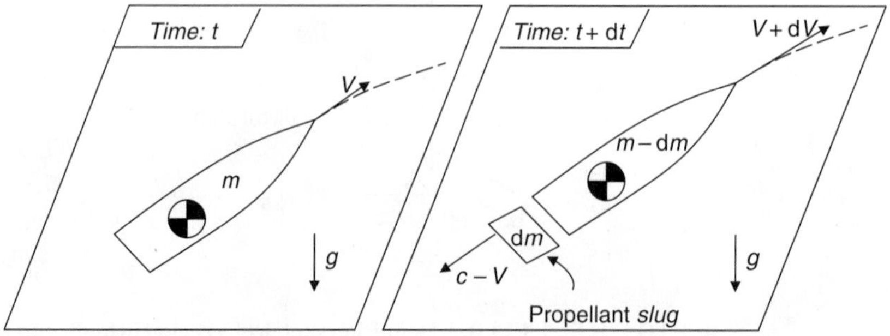

■ FIGURE 11.11
The rocket as observed by a stationary frame of reference before and after a propellant *slug* is ejected from the nozzle

■ **FIGURE 11.12**
Velocity increment for a
rocket in zero gravity, zero
drag environment (with
$g_0 = 9.8\,\text{m/s}^2$)

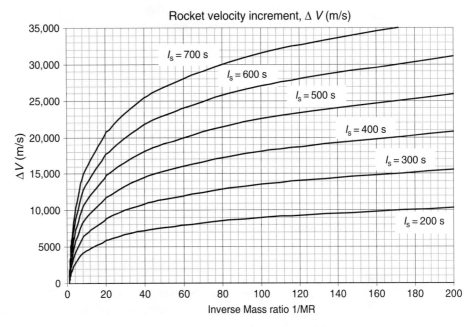

The fundamental equation of rocketry, Eq. 11.22 is expressed in terms of specific impulse and earth's surface gravitational acceleration as

$$\Delta V = g_0 \bar{I}_s \ln\left(\frac{1}{\text{MR}}\right) \qquad (11.26)$$

The graph of velocity increment for different specific impulse and vehicle mass ratio is shown in Fig. 11.12. We note that as mass ratio is reduced (or inverse of mass ratio is increased), velocity increment grows logarithmically. The impact of specific impulse on velocity increment may also be discerned from Fig. 11.12. For a mass ratio of 5%, the velocity increment varies between 6000–20,000 m/s, when the specific impulse varies between 200 and 700 s. The mass ratio of 5% indicates a rocket vehicle design that has 95% of its initial mass occupied by its propellant and only 5% devoted to structure, skin, payload, navigation, and control. This value, i.e., MR of 0.05, represents the best of a single-stage rocket design today. It is, however, possible to improve the rocket performance in attaining a higher terminal velocity by discarding of empty propellant tanks. This introduces the concept of a multistage rocket. Each rocket stage is completely self-contained, i.e., they have their own motor, propellant tanks, and feed system.

EXAMPLE 11.6

The specific impulse of a chemical rocket is 400 s. The rocket is in a zero gravity vacuum flight. If we could improve the propellant fraction ζ from 0.90 to 0.95, calculate the percent improvement in ΔV that can be achieved.

SOLUTION

Since the propellant mass fraction is $\zeta = 1 - \text{MR}$, therefore, the mass ratio changes from 0.1 to 0.05 as ζ varies between 0.9 and 0.95. We substitute, 0.1 and 0.05 in Eq. 11.24, to get

$\Delta V = (9.8\,\text{m/s}^2)(400\,\text{s})\ln(1/0.10) = 9026\,\text{m/s}$ for $\zeta = 0.90$
$\Delta V = (9.8\,\text{m/s}^2)(400\,\text{s})\ln(1/0.05) = 11{,}743\,\text{m/s}$ for $\zeta = 0.95$

We may realize a 30% increase in ΔV, or equivalently ~2700 m/s increase in the terminal velocity of the rocket.

Case 2 Rocket performance including the effect of gravity

From Eq. 11.19, we keep the gravity term on the RHS as we neglect the aerodynamic drag,

$$m\frac{dV}{dt} = F - mg\sin\theta = -c\frac{dm}{dt} - mg\sin\theta \tag{11.27}$$

Divide both sides by m/dt, to get

$$dV = -c\frac{dm}{m} - g\sin\theta\, dt \tag{11.28}$$

If we integrate Eq. 11.28 and use the time-averaged value of $g \cdot \sin\theta$ (denoted by subscript, m), we get

$$\Delta V = -\bar{c}\ln(\text{MR}) - (g\sin\theta)_m t_b \tag{11.29}$$

Gravitational acceleration g is governed by Newton's law of gravitation, i.e.,

$$F_g = G\frac{Mm}{r^2} \tag{11.30}$$

It is customary to combine the product of universal gravitational constant G, the mass of the planet M, and inverse of planet radius squared, into a single term, called the gravitational acceleration on the planet's surface, i.e.,

$$g_0 \equiv \frac{Gm}{R^2} \tag{11.31}$$

Therefore, gravitational acceleration g as a function of altitude h above a planet of radius R is related to the acceleration on the surface of the planet by

$$g = \frac{g_0}{(1 + h/R)^2} \tag{11.32}$$

The mean radius of earth is 6000 km and the gravitational acceleration on earth's surface is 9.8 m/s^2, therefore, at 100 km altitude, where low-earth orbit starts, gravitational acceleration is ≈ 9.48 m/s^2 or $\sim$97% of its sea level value. Figure 11.13 shows the variation of gravitational acceleration with altitude for up to 300 km. We note that since $h \ll R$ (in this graph), the variation is nearly linear, i.e., if we apply binomial expansion to Eq. 11.32 for $h/R \ll 1$, we get

$$g \approx g_0(1 - 2h/R) \quad \text{for } h \ll R \tag{11.33}$$

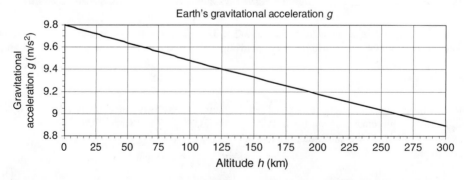

■ **FIGURE 11.13**   **Variation of earth's gravitational acceleration with altitude h (in km)**

EXAMPLE 11.7

A rocket is vertically launched and operates for 30 s and has a mass ratio of 0.10. The (mean) rocket-specific impulse is 420 s. Assuming the average gravitational acceleration over the burn period is 9.65 m/s^2, calculate the terminal velocity of the rocket with and without gravitational effects. Neglect the effect of aerodynamic drag in both cases.

SOLUTION

We use Eq. 11.29 with effective exhaust velocity replaced with the product of specific impulse and gravitational acceleration on earth's surface

$$\Delta V = -g_0 \bar{I}_s \ln(\text{MR}) - (g \sin \theta)_m t_b$$

The vertical launch gives $\theta = 90°$ and the average gravitational acceleration is also given to be 9.65 m/s^2, therefore,

$$\Delta V = -(9.8)(420)\ln(0.1)\,\text{m/s} \approx 9477\ \text{m/s (without gravity)}$$
$$\Delta V = -(9.8)(420)\ln(0.1)\,\text{m/s} - 9.65(30)\ \text{m/s} \approx 9188\ \text{m/s}$$
$$\text{(with gravity)}$$

Note that the gravitational effect/deceleration caused ~3% reduction in terminal speed.

Case 3 Rocket flight performance including the effects of gravity and aerodynamic drag
The motion of the rocket in a gravitational field including atmospheric drag was derived in Eq. 11.19, to be

$$m\frac{dV}{dt} = F - D - mg \sin \theta$$

The thrust is again represented by $-(dm/dt) \cdot c$ and the rocket drag is proportional to the flight dynamic pressure and the frontal (cross-sectional) area of the rocket A_f, with the proportionality constant called drag coefficient C_D according to

$$D = C_D \left(\frac{\rho V^2}{2}\right) A_f \tag{11.34}$$

The variation of drag coefficient with flight Mach number may be modeled based on wind tunnel data, computational fluid dynamics codes, or design charts that are developed by industry (proprietary). In general, the presence and the shape of the exhaust plume affect the pressure distribution on the rocket and thus vehicle drag. Therefore, it is customary to depict the baseline drag coefficient in the absence of the exhaust jet. Also, a winged rocket experiences aerodynamic lift and thus we need to model the lift coefficient C_L as a function of Mach number and angle of attack. Some typical aerodynamic data, including lift variation with flight Mach number and angle of attack α are shown in Fig. 11.14 for V-2 rocket, without the jet and exhaust plume effect (from Sutton and Biblarz, 2001). In Fig. 11.14, we note that the drag coefficient exhibits a transonic drag rise behavior as well as a *plateau* region at high supersonic Mach numbers. This behavior is indeed known as the *hypersonic Mach independence principle*.

The equation of motion of the rocket is rewritten as

$$m\frac{dV}{dt} = -c\frac{dm}{dt} - C_D\rho V^2 A_f/2 - mg \sin \theta \tag{11.35-a}$$

And by dividing both sides by m/dt, we get

$$dV = -c\frac{dm}{m} - \frac{C_D \rho V^2 A_f}{2m} dt - g \sin \theta\, dt \tag{11.35-b}$$

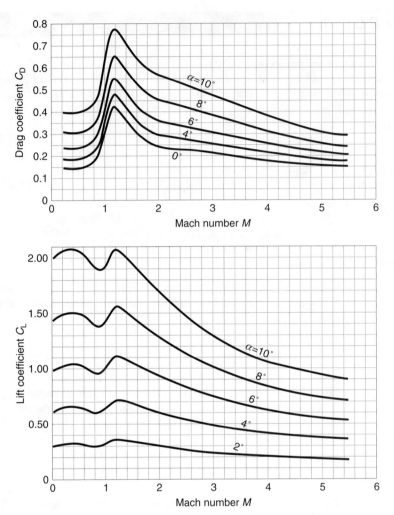

We may integrate Eq. 11.35-b, to get

$$\int_{V_0}^{V_f} dV = -\int_{m_0}^{m_f} c \frac{dm}{m} - \int_{0}^{t_b} \frac{C_D \rho V^2 A_f}{2m} dt - \int_{0}^{t_b} g \sin \theta \, dt \tag{11.35-c}$$

$$\Delta V = -\bar{c} \ln(\text{MR}) - (g \sin \theta)_m t_b - \int_{0}^{t_b} \frac{C_D \rho V^2 A_f}{2m} dt \tag{11.35-d}$$

We may write the rocket instantaneous mass m in terms of initial mass, propellant mass, and burn time, as

$$m(t) = m_0 - \frac{m_p}{t_b} t = m_0 \left[1 - \varsigma \left(\frac{t}{t_b} \right) \right] \tag{11.36}$$

where ς is the propellant mass fraction. The last integral in Eq. 11.35-d may now be expressed as

$$\int_{0}^{t_b} \frac{C_D \rho V^2 A_f}{2m} dt = \frac{A_f}{2m_0} \int_{0}^{t_b} \frac{C_D \rho V^2}{(1 - \varsigma \, t/t_b)} dt \tag{11.37}$$

Note that the integrand in Eq. 11.37 is an *implicit* function of time, since time is related to altitude and density is a function of altitude; also time implies rocket speed V, which is related to the Mach number at the altitude and drag coefficient is a function of flight Mach number. Therefore, we may proceed in time, in a stepwise fashion, from zero to the burnout time t_b. The governing equation is thus

$$\Delta V = -\bar{c}\ln(MR) - (g\sin\theta)_m t_b - \frac{A_f}{2m_0}\int_0^{t_b}\frac{C_D\rho V^2}{1-\varsigma t/t_b}\,dt \tag{11.38}$$

Note that the flight path angle, in this nonlifting case, denoted by θ is changing with time. Therefore rocket undergoes *gravity turn* as evidenced by the trio of forces shown in Fig. 11.10. Hence at each time step, we need to calculate the new flight path angle θ. The incremental change of flight path direction is dictated by the component of gravitational force *normal* to the rocket axis, namely,

$$m\frac{dv}{dt} = -mg\cos\theta \tag{11.39}$$

which integrates to

$$\Delta v = -(g\cos\theta)_m t \tag{11.40}$$

The rocket moves to a new position along its axis Δx and normal to its axis Δy according to

$$\Delta x = \overline{V}\Delta t \tag{11.41-a}$$

And

$$\Delta y = \bar{v}\Delta t \tag{11.41-b}$$

Figure 11.15 shows the velocities along the vehicle axis and the turning speed due to the component of gravitational force normal to the axis. Incremental step Δx is taken along vehicle axis and the step Δy is normal to that direction.

   An interesting case in hypersonic propulsion involves integrating rockets and airbreathing propulsion in an aircraft. Under those circumstances, the airbreathing engines rather than the rocket primarily drive the choice of the flight trajectory. The structural loads limit the upper bounds of acceptable flight dynamic pressure, and the lower bounds of the dynamic pressure are limited by the mass flow rate per unit area at a given altitude (i.e., thrust requirement). Hence, the flight trajectory for an airbreathing hypersonic aircraft falls within a constant dynamic pressure band, as shown in Fig. 11.16 (from Heiser, et al., 1994). Mach number versus altitude for three constant dynamic pressures (an upper, a lower, and a mean) is shown in Fig. 11.16.

   Assuming a constant dynamic pressure trajectory, we may integrate Eq. 11.38 to get

$$\Delta V = -\bar{c}\ln(MR) - (g\sin\theta)_m t_b + \left(\frac{A_f}{m_0}\right)\left(\frac{\rho V^2}{2}\right)\left(\frac{t_b}{\varsigma}\right)\overline{C}_D\ln(1-\varsigma) \tag{11.42-a}$$

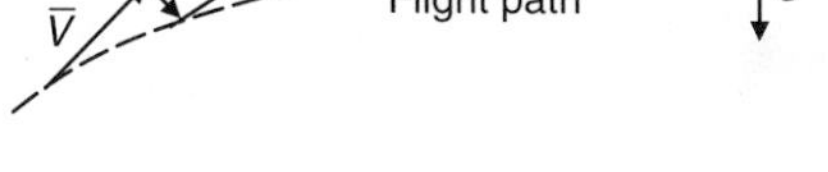

■ **FIGURE 11.15**   **Definition sketch of velocity components and the rocket flight path in a *gravity turn***

■ **FIGURE 11.16**
Standard day geometric
altitude flight trajectory for
constant dynamic pressure
(from Heiser et al., 1994)

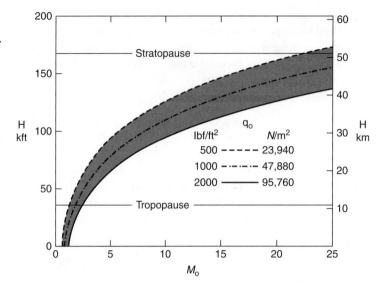

where the drag coefficient is replaced by a suitable average $\overline{C}_D$ (shown with a bar). By combining the first and last terms on the RHS of Eq. 11.42-a, we obtain an interesting form for terminal velocity, namely,

$$\Delta V = -\ln(1 - \varsigma)\left[g_0\overline{I}_s - \left(\frac{A_f}{m_0}\right)\left(\frac{\rho V^2}{2}\right)\left(\frac{t_b}{\varsigma}\right)\overline{C}_D\right] - (g\sin\theta)_m t_b \qquad \textbf{(11.42-b)}$$

The bracket on the RHS of Eq. 11.42-b contains the contribution of thrust and the penalty of aerodynamic drag. We note that a low dynamic pressure trajectory reduces the drag penalty as well as a high initial mass per unit frontal area, i.e., m_0/A_f. This requirement tends to produce long slender vehicles for low frontal cross-sectional area and large mass. Also, low drag coefficient C_D is desirable in achieving high terminal speeds, which implies a slender conical nose or tangent ogive to reduce the vehicle drag coefficient.

EXAMPLE 11.8

A rocket plane has a propellant mass fraction of $\varsigma = 0.80$ and a mean-specific impulse of 345 s. The flight trajectory is described by a constant dynamic pressure of $q_0 = 100\,\text{kPa}$. The mean drag coefficient is approximated to be 0.3, the vehicle initial mass is $m_0 = 500,000\,\text{kg}$, and the vehicle frontal cross-sectional area is $A_f = 20\,\text{m}^2$. For a burn time of 60 s, calculate the rocket terminal speed while neglecting gravitational effect.

SOLUTION

The contribution of rocket thrust to terminal speed is

$$(\Delta V)_{\text{thrust}} = -\ln(1 - \varsigma)g_0\overline{I}_s$$

$$= -(9.8\,\text{m/s}^2)(345\,\text{s})\ln(0.2) \approx 5442\,\text{m/s}$$

The penalty of drag on terminal speed is

$$(\Delta V)_{\text{drag}} = \ln(1 - \varsigma)\left(\frac{A_f}{m_0}\right)\left(\frac{\rho V^2}{2}\right)\left(\frac{t_b}{\varsigma}\right)\overline{C}_D$$

The initial rocket mass per unit cross-sectional area is, $m_0/A_f = 25,000\,\text{kg/m}^2$. The penalty for vehicle drag on terminal speed is thus

$$(\Delta V)_{\text{drag}} \approx -580\,\text{m/s}$$

Therefore, the rocket vehicle will attain a terminal speed (excluding gravitational effect) of

$$\Delta V = (\Delta V)_{\text{thrust}} + (\Delta V)_{\text{drag}} \approx 4862\,\text{m/s}$$

11.9   Multistage Rockets

A multiple-stage rocket has the capability to improve the flight performance of a single-stage rocket. There have been many forms of staging used in rocket launch systems to date. An example of various staging concepts is shown in Fig. 11.17 (from Sutton and Biblarz, 2001).

Strap-on boosters that jettison after their propellant is depleted are examples of staging that are widely used in launch systems. For example, the space shuttle has two solid rocket boosters that separate after their propellant is fully consumed (i.e., in 123 s after launch). Also, the external tank that carries liquid hydrogen and liquid oxygen for launch is separated after ~8.5 min. Figure 11.18 (a) shows the space shuttle orbiter with solid rocket boosters (SRB) and the external tank (ET). The internal compartments of the external tank are shown in Fig. 11.18 (b) from Isakowitz (1995).

The terminal speed of a multistage rocket (with n stages) is the sum of the terminal speeds of its individual stages, namely,

$$\Delta V = \sum_{i=1}^{n} \Delta V_i \tag{11.43}$$

In the absence of gravity and atmospheric drag, the individual stage incremental speed is

$$\Delta V_i = -c_i \ln(MR)_i \tag{11.44}$$

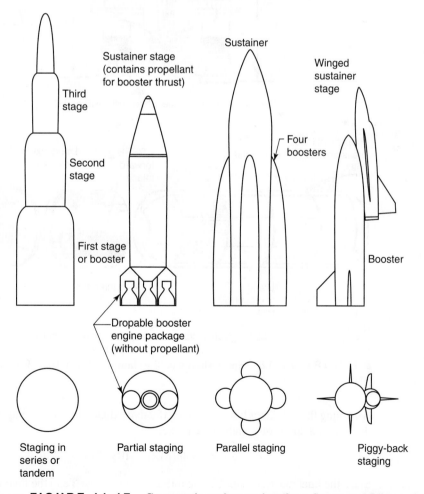

■ FIGURE 11.17   **Concepts in rocket staging (from Sutton and Biblarz, 2001)**

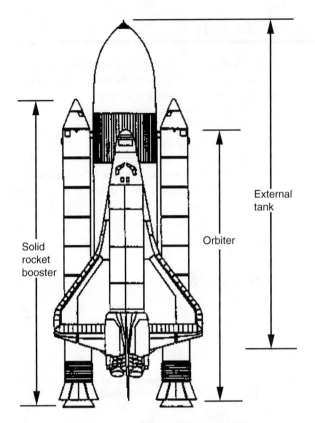

(a) Space shuttle orbiter with SRB and ET

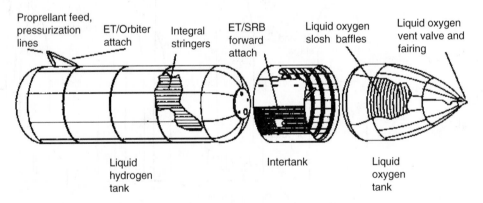

(b) The internal compartments of the external tank (ET)

■ **FIGURE 11.18**   **Space shuttle orbiter launch system, from Isakowitz (1995)**

Assuming that all stages have the same effective exhaust speed, for simplicity, we may express the terminal speed of a multistage rocket by

$$\Delta V = -c \ln(\mathrm{MR}_1 \cdot \mathrm{MR}_2 \cdot \mathrm{MR}_3 \cdots \mathrm{MR}_n) \qquad (11.45)$$

Since the final mass of stage 1 is the initial mass of stage 2 and the final mass of stage 2 is the initial mass of stage 3, the product of n stage mass ratios in Eq. 11.45 simplifies to the payload or

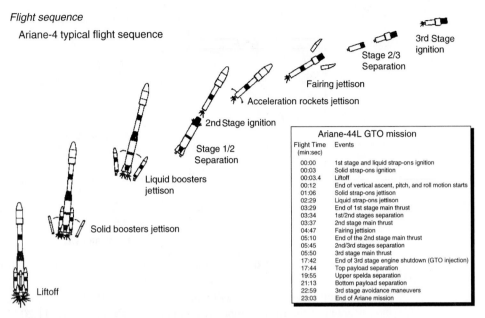

Flight sequence

Ariane-4 typical flight sequence

3rd Stage ignition

Stage 2/3 Separation

Fairing jettison

Acceleration rockets jettison

2nd Stage ignition

Stage 1/2 Separation

Liquid boosters jettison

Solid boosters jettison

Liftoff

Ariane-44L GTO mission

Flight Time (min:sec)	Events
00:00	1st stage and liquid strap-ons ignition
00:03	Solid strap-ons ignition
00:03.4	Liftoff
00:12	End of vertical ascent, pitch, and roll motion starts
01:06	Solid strap-ons jettison
02:29	Liquid strap-ons jettison
03:29	End of 1st stage main thrust
03:34	1st/2nd stages separation
03:37	2nd stage main thrust
04:47	Fairing jettison
05:10	End of the 2nd stage main thrust
05:45	2nd/3rd stages separation
05:50	3rd stage main thrust
17:42	End of 3rd stage engine shutdown (GTO injection)
17:44	Top payload separation
19:55	Upper spelda separation
21:13	Bottom payload separation
22:59	3rd stage avoidance maneuvers
23:03	End of Ariane mission

■ **FIGURE 11.19**   **Typical flight sequence of Ariane-4 (from Isakowitz, 1995)**

final mass (of the nth stage) divided by the initial rocket (or launch) mass. Here, we have called the final mass of the last stage as the *payload* mass m_{L}. Therefore,

$$\Delta V = -c \ln(m_{\text{L}}/m_0) \tag{11.46}$$

As noted earlier, the ratio of final-to-initial mass for a single-stage rocket is at best 0.05, or its inverse mass ratio is 20. For a multistage rocket, however, the ratio of payload to initial mass may be ~0.01 or even lower. This gives an inverse mass ratio in excess of 100. Therefore, the flight performance shown in Fig. 11.12 where the inverse mass ratio varies from 1 to 300 should be viewed in the context of possible single or multistage rocket vehicles.

It is instructive to view the flight sequence of a multistage rocket. Figure 11.19 shows the flight sequence of Ariane-4 with table to the right listing the flight times (from Isakowitz, 1995).

11.10 Propulsive and Overall Efficiencies

The fraction of the thrust power and the residual kinetic power in the jet that is converted to thrust power is called the propulsive efficiency of a rocket. The residual kinetic power in the jet is the kinetic power as observed by a stationary observer, i.e.,

$$\text{Residual kinetic power in the jet} \equiv \dot{m}_{\text{p}}(c - V)^2/2 \tag{11.47}$$

Therefore, the propulsive efficiency is

$$\eta_{\text{p}} \equiv \frac{F \cdot V}{F \cdot V + \dot{m}_{\text{p}}(c - V)^2/2} \tag{11.48}$$

Replacing thrust by the product of propellant flow rate and the effective exhaust speed and simplifying, we get

$$\eta_{\text{p}} \equiv \frac{c \cdot V}{c \cdot V + (c - V)^2/2} = \frac{2cV}{c^2 + V^2} = \frac{2(V/c)}{1 + (V/c)^2} \tag{11.49}$$

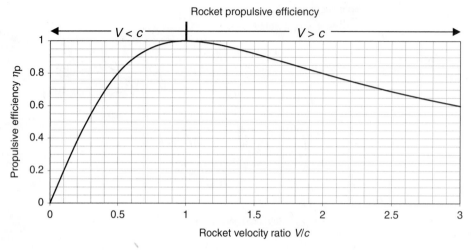

Rocket propulsive efficiency

■ **FIGURE 11.20**   Rocket propulsive efficiency η_p stretches from $V/c < 1$ to $V/c > 1$

The propulsive efficiency will reach its peak (of 100%) when the ratio V/c approaches unity. A graph of the rocket propulsive efficiency is shown in Fig. 11.20.

A distinction between a rocket and an airbreathing engine is that a rocket may operate in $V/c > 1$, whereas an airbreathing engine has to operate with an exhaust velocity that is higher than the flight speed, since there is a ram drag penalty for the captured airstream. As a result, the range of $V/c > 1$ is unique to rockets.

Similar to an airbreathing engine, the overall efficiency of a chemical rocket engine is defined by the fraction of the chemical power invested in the combustor that is converted to the vehicle thrust power, i.e.,

$$\eta_o \equiv \frac{F \cdot V}{\dot{m}_p Q_R + \dot{m}_p v_{inj}^2 / 2} \approx \frac{c \cdot V}{Q_R} \tag{11.50}$$

The second term in the denominator of Eq. 11.50 is the kinetic power of the propellant at the injector plate, i.e., inlet to the thrust chamber. Often times, the contribution of the injected propellant kinetic power is small and thus neglected as compared with the chemical power of combustion.

EXAMPLE 11.9

A liquid propellant rocket uses hydrogen and oxygen as propellant. The heat of reaction for the combustion is $Q_R = 13.3$ MJ/kg. The specific impulse is 421 s and the flight speed is 2000 m/s. Neglecting the propellant kinetic power at the injector plate, calculate

(a) effective exhaust speed c in m/s
(b) propulsive efficiency
(c) overall efficiency

SOLUTION

Effective exhaust speed is $c = g_0 I_s$, therefore, $c \approx 4126$ m/s
Propulsive efficiency is $\eta_p = 2(2000/4126)/[1 + (2000/4126)^2] \approx 0.785$

Overall efficiency $\eta_o \approx c \cdot V / Q_R \approx 0.620$

11.11   Chemical Rocket Combustion Chamber

The liquid propellant rocket thrust chamber is composed of an injector plate, a combustion chamber, and an exhaust nozzle. The injector plate serves the function of an atomizer and mixer of the fuel and oxidizer streams. Since the thrust chamber needs to be cooled, the choice of propellant often calls for cryogenic propellants such as hydrogen and oxygen. To minimize storage volume, propellant is stored in liquid form in insulated tanks. The fuel and oxidizer latent heats of vaporization and subsequent heating of the gaseous form of the propellant provide cooling of the thrust chamber walls. Vaporizing and preheating of the propellant is actually advantageous to combustion efficiency and allows the products of combustion to attain higher temperatures in the combustor. Solid propellant combustion chamber houses the propellant and uses liners, insulators, and inhibitors to protect the walls from excessive heating. The period of operation of solid motors is thus limited by the amount of propellant that is carried inside the combustion chamber, which typically lasts from a few seconds to about 2 min for boosters.

11.11.1   Liquid Propellant Combustion Chambers

The design of injector plate is critical to stable combustion of the liquid propellant and high combustion efficiency. The stability of combustion is achieved through a large static pressure drop across the injector plate. The conventional design rule calls for 20% drop across the injector plate, i.e., the chamber pressure p_c is 20% less than the feed pressure of propellant upstream of the injector plate. This relatively large static pressure drop decouples small pressure isolations in the combustor from the propellant pressure feed oscillations. This instability is called chugging and is the easiest to be avoided. To achieve high efficiency in chemical reaction of the fuel and oxidizer streams, the injector plate design plays a critical role. The precombustion processes of atomization, vaporization, and mixing need to be efficiently accomplished through the injector plate. There are several design approaches to the injector plate, namely, fuel–oxidizer, or unlike stream, impingement type, fuel–fuel, and oxidizer–oxidizer, or like stream, impingement type, parallel stream, or showerhead, nozzle type, and premix chamber swirl nozzle type. These injector plate design types are schematically shown in Fig. 11.21.

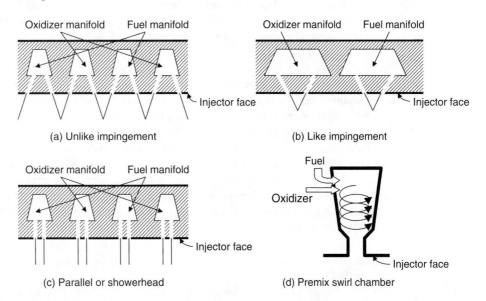

■ **FIGURE 11.21**   Schematic drawing of some injector plate design concepts

Precision in manufacturing of injector plates ensures proper mixture ratio of fuel and oxidizer that is brought together at a desired distance from the injector plate and at proper angle with respect to combustion chamber axis. Any misalignment in the orifice geometry and angle will cause inefficient mixing, poor combustion efficiency, and possibly chemical reaction that takes place at the wall with subsequent structural failure due to inadequate cooling. On the contrary, proper injector plate design will reduce the amount of backsplash and heat transfer to the thrust chamber walls and produce stable combustion with high efficiency. In sizing the orifice areas for the fuel and oxidizer, we use the Bernoulli equation corrected for viscous fluid effects, according to,

$$v_f = C_{df}\sqrt{\frac{2(\Delta p)_f}{\rho_f}} \tag{11.51-a}$$

$$v_o = C_{do}\sqrt{\frac{2(\Delta p)_o}{\rho_o}} \tag{11.51-b}$$

where C_d is the discharge coefficient, which is a function of the orifice geometry and Reynolds number. The mass flow rate per orifice opening follows the continuity equation, i.e.,

$$\dot{m}_f = \rho_f v_f A_f = C_{df} A_f \sqrt{2\Delta p_f \cdot \rho_f} \tag{11.52-a}$$

$$\dot{m}_o = \rho_o v_o A_o = C_{do} A_o \sqrt{2\Delta p_o \cdot \rho_o} \tag{11.52-b}$$

The ratio of the oxidizer-to-fuel mass flow rates is known as the *mixture ratio r*

$$r \equiv \frac{\dot{m}_o}{\dot{m}_f} = \left(\frac{C_{do}}{C_{df}}\right)\left(\frac{A_o}{A_f}\right)\sqrt{\left(\frac{\Delta p_o}{\Delta p_f}\right)\left(\frac{\rho_o}{\rho_f}\right)} \tag{11.53}$$

From our studies of combustion (Chapter 6), we remember that for stoichiometric mixture ratio, the products of combustion reach nearly their maximum temperature. However, for a hydrogen (fuel) rich mixture ratio, $r < r_{stoich}$, the products of combustion will have a lower mixture molecular weight and thus higher specific impulse. As a result, mixture ratio is often less than the stoichiometric value. Figure 11.22 (from Sutton and Biblarz, 2001) shows the result of thermochemical calculations for oxygen and a hydrocarbon fuel, known as RP-1, in a combustion chamber of 1000 psia pressure and perfect expansion to sea level ambient condition in the nozzle.

Note that there are two sets of nozzle calculations shown in Fig. 11.22. The *frozen equilibrium* calculation assumes constant combustion product composition in the nozzle, i.e., the equilibrium mole fractions of the products of combustion remain unchanged as the result of flow expansion (i.e., cooling) in the nozzle. Frozen equilibrium calculations are simple but not very accurate in predicting rocket thrust chamber performance. The *shifting equilibrium* calculation, however, allows for chemical reaction in the nozzle as a result of flow expansion, thus it represents a more accurate description of the rocket thrust chamber performance. In either case, we note that the specific thrust has reached maximum at either the mixture ratio of ~2.3 for frozen equilibrium or 2.5 for the shifting equilibrium calculations. Based on the chemical formula for RP-1, i.e., $CH_{1.953}$, stoichiometric oxidizer-to-fuel ratio is 3.413. Thus, the choice of the mixture ratio r is based on maximum specific thrust I_s rather than maximum chamber temperature T_c. Also note that a fuel-rich rocket nozzle exit flow is advantageous to a ram-rocket configuration since the excess fuel may be used as the fuel in the ramjet/scramjet combustor.

Figure 11.23 (from Sutton and Biblarz, 2001) shows injector discharge coefficients for different orifice geometries that are suitable for rocket propulsion applications.

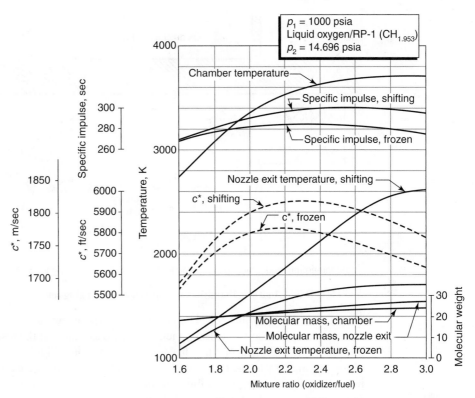

■ **FIGURE 11.22**   Thermochemical calculations for oxygen-RP-1 combustion with $p_c = 1000$ psia, $p_2 = p_0 = 14.7$ psia (from Sutton and Biblarz, 2001)

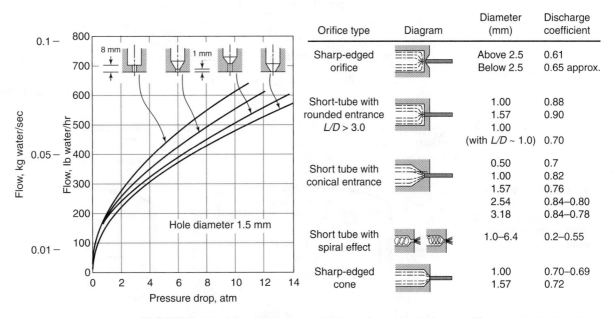

■ **FIGURE 11.23**   Discharge coefficients for typical injector orifice geometries, based on water tests (from Sutton and Biblarz, 2001)

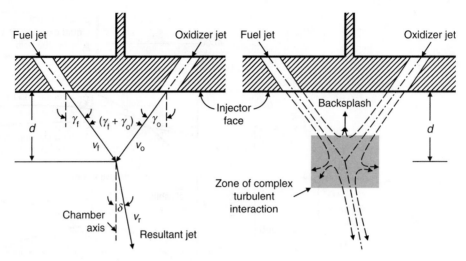

(a) Momentum model for two impinging jets   (b) "Real" impingement jets with backsplash

■ **FIGURE 11.24** **Definition sketch of two impinging jets**

For preliminary injector plate calculation purposes, we use the law of conservation of momentum applied to two colliding jets. Here, we assume that the resultant jet has the same mass flow rate as the sum of individual jets' mass flow rates. Figure 11.24 is a definition sketch of two impinging jets, their angles γ_f and γ_o, and collision distance d from the wall. The *real* impingement jets, however, exhibit some backsplash as well as complex turbulent interactions, as schematically indicated in Fig. 11.24 (b).

The resultant jet is assumed to emerge at angle δ with respect to chamber axis, as shown. The momentum balance in the chamber axis direction is

$$\dot{m}_f v_f \cos \gamma_f + \dot{m}_o v_o \cos \gamma_o = (\dot{m}_f + \dot{m}_o) v_r \cos \delta \tag{11.54}$$

The momentum balance normal to the chamber axis direction is

$$\dot{m}_f v_f \sin \gamma_f - \dot{m}_o v_o \sin \gamma_o = (\dot{m}_f + \dot{m}_o) v_r \sin \delta \tag{11.55}$$

The ratio of the two momentum equations will isolate the resultant jet flow angle δ

$$\tan \delta = \frac{\dot{m}_f v_f \sin \gamma_f - \dot{m}_o v_o \sin \gamma_o}{\dot{m}_f v_f \cos \gamma_f + \dot{m}_o v_o \cos \gamma_o} \tag{11.56}$$

The resultant jet may be designed to emerge in the axial direction near the center of the combustion chamber and toward the center near the chamber walls. The schematic drawing of the resultant propellant stream is shown in Fig. 11.25 (a). The chamber wall is regeneratively

■ **FIGURE 11.25**
Schematic drawing of possible propellant injection pattern in the combustion chamber

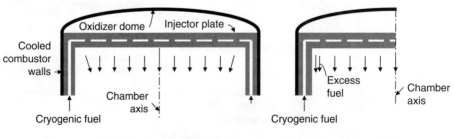

(a) Directing the resultant stream away from the wall

(b) Directing excess fuel along the wall/film cooling

cooled by cryogenic propellant. Since rocket combustion often involves a fuel-rich mixture of the propellant, we may inject the extra fuel adjacent to the combustor wall as a means of thermal protection for the wall (see Fig. 11.25 (b))

The total included angle at the impingement point is the sum of the two angles $\gamma_o + \gamma_f$. The total impingement angle impacts the extent of backsplash and potential chemical reaction and heating at the injector plate.

EXAMPLE 11.10

A liquid propellant rocket injector plate uses an unlike impingement design. The fuel and oxidizer orifice discharge coefficients are $C_{df} = 0.82$ and $C_{do} = 0.65$ respectively.

The static pressure drop across the injector plate for both oxidizer and fuel jets is the same, $\Delta p_f = \Delta p_o = 200 \, \text{kPa}$. The fuel and oxidizer densities are $\rho_f = 85 \, \text{kg/m}^3$ and $\rho_o = 1350 \, \text{kg/m}^3$. The mixture ratio is $r = 2.50$. Calculate

(a) oxidizer-to-fuel orifice area ratio A_o/A_f

(b) oxidizer and fuel jet velocities v_o and v_f

(c) graph injection angle γ_f versus γ_o for axial resultant stream $\delta = 0$ and different mixture ratios, i.e., for $r = 2.5, 3.5,$ and 4.5

SOLUTION

The orifice area ratio follows Eq. 11.53

$$r \equiv \frac{\dot{m}_o}{\dot{m}_f} = \left(\frac{C_{do}}{C_{df}}\right)\left(\frac{A_o}{A_f}\right)\sqrt{\left(\frac{\Delta p_o}{\Delta p_f}\right)\left(\frac{\rho_o}{\rho_f}\right)} \implies$$

$$A_o/A_f \approx 0.791$$

From Eq. 11.51 we get the fuel and oxidizer orifice discharge speed:

$$v_f = C_{df}\sqrt{2(\Delta p)_f/\rho_f} \implies v_f \approx 56.2 \, \text{m/s}$$

$$v_o = C_{do}\sqrt{2(\Delta p)_o/\rho_o} \implies v_o \approx 11.2 \, \text{m/s}$$

The momentum balance normal to the axis must be zero to get a purely axial resultant flow, i.e.,

$$\dot{m}_f v_f \sin\gamma_f - \dot{m}_o v_o \sin\gamma_o = 0$$

Therefore, the flow angles are related to each other and the mixture ratio according to

$$\sin\gamma_f = r\left(\frac{v_o}{v_f}\right)\sin\gamma_o$$

$$= r \cdot (11.2/56.2)\sin\gamma_o \approx 0.199 \cdot r \cdot \sin\gamma_o$$

The graph of this function is produced using Excel spreadsheet program.

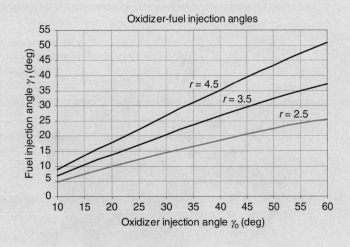

Oxidizer-fuel injection angles

11.11.1.1 *Some Design Guidelines for Injector Plate.* Backsplash for impingement-type injectors, as noted earlier, can bring the propellant back toward the injector plate, where it may lead to chemical reaction and heat release with subsequent injector plate structural failure. To control backsplash, we place design limitations on impingement distance d and the total included (impingement) angle $(\gamma_o + \gamma_f)$, namely,

$$d \sim 5 \text{ to } 7(d_o + d_f)/2 \tag{11.57}$$

$$(\gamma_o + \gamma_f) < 60° \tag{11.58}$$

To control chugging instability, we build in a large pressure drop across the injector plate, i.e., $(\Delta p)_{inj} \approx 0.2 \, p_c$. This is a *ballpark* rule as the actual pressure drop depends on the injector design, e.g., unlike impingement, like-impingement, etc. Orifice diameters are to be small ($\sim$0.4–2.4 mm) for good atomization. Smaller orifice diameters are prone to plugging. The orifice diameter ratio $d_o/d_f \sim 1.22$ is recommended for unlike impingement. NASA publications (e.g., NASA SP-8089 and NASA SP-8120) on injectors and rocket nozzles are sources for additional reading that are highly recommended.

11.11.1.2 *Combustion Instabilities.* There are three main types of instabilities associated with chemical rockets. These are categorized based on their frequency range. The first type, which is the low frequency disturbance, is known as *chugging*. Its frequency range is between 10 and 200 Hz. The source is instability is the coupling between the pressure feed oscillations and the combustion chamber pressure oscillations. This form of longitudinal instability appears as pogo instability on the vehicle's motion. The *fix* for this instability is in designing large static pressure drop ($\sim$20% p_c) across the injector plate as well as stiffening the coupling between the components of propellant feed system and the thrust chamber. The second mid-range instability is known as acoustical, buzzing, or entropy instability. The frequency range for this instability is 200–1000 Hz. The source of this intermediate frequency instability is the mechanical vibration of the propulsion system and possible coupling between the injector plate spray pattern and the turbulent eddies in the combustor. Acoustic coupling between the combustion and the injector plate propellant spray pattern is also deemed responsible for this instability. The third instability is of high frequency, i.e., higher than 1000 Hz, known as the *screeching or screaming* instability. This type has its roots in the energy release pattern in the combustion chamber and resonant behavior with the combustor cavity. Injector face baffles, as shown in Fig. 11.26 (from Sutton and Biblarz, 2001), have proven effective in control of screaming instability in liquid propellant rockets.

■ **FIGURE 11.26**
Main injector assembly of Space Shuttle Main Engine showing baffle with five outer compartments (from Sutton and Biblarz, 2001)

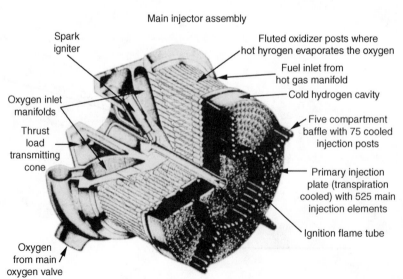

Main injector assembly

Spark igniter

Fluted oxidizer posts where hot hyrogen evaporates the oxygen

Fuel inlet from hot gas manifold

Cold hydrogen cavity

Oxygen inlet manifolds

Five compartment baffle with 75 cooled injection posts

Thrust load transmitting cone

Primary injection plate (transpiration cooled) with 525 main injection elements

Ignition flame tube

Oxygen from main oxygen valve

11.11.2 Solid Propellant Combustion Chambers

Solid propellants embody oxidizer, fuel, and a binder in solid or rubbery form. Solid propellants are categorized based on their composition. They are known as double-base (DB), composite, and composite modified double-base (CMDB). The double-base solid propellant is a *homogeneous* mixture of oxidizer, fuel, binder, and some additives (e.g., metal oxides and metal particles). The composite propellant is the *heterogeneous* mixture of fuel, oxidizer, binder, and additives. The composite modified double-base propellant is the combination of the first two types. As noted, the additives are often metal oxides and metal particles that enhance burning rates. Examples of solid propellants and their characteristics are shown in Table 11.3 (from Sutton and Biblarz, 2001). The department of defense (DOD) gives solid propellants (referred to as "explosives") a hazard classification of 1.1 or 1.3, which regulates their manufacturing, handling, storage, labeling, and shipping (see Explosive Hazard Classification Procedure, 1989).

Solid propellant burning rate r is defined as the speed of chemical reaction, i.e., combustion, that progresses normal to a propellant grain surface. Therefore, the gaseous mass flow rate that emerges from a burning solid propellant grain surface is proportional to the surface burning area, the speed of propellant surface recession, and the grain density according to continuity equation

$$\dot{m} = \rho_p r A_b \qquad (11.59)$$

where ρ_p is the (solid) propellant grain density, in kg/m^3 (or $slugs/ft^3$), r is the surface recession or burning rate in cm/s (or in/s), and A_b is the surface area of solid propellant grain that is burning. An empirical formula that relates the burning rate to combustion chamber conditions, for an end-burning grain, is

$$r = a\, p_c^n \qquad (11.60)$$

where p_c is the combustion chamber pressure, a and n are indices of combustion that are constant for different propellants. The constant a is a function of the propellant grain temperature and n is the pressure exponent, thus burning rate dependence on combustion chamber pressure. We can demonstrate that for stable operation, n has to be less than 1. Note that the mass flow rate through the nozzle is proportional to chamber pressure p_c, and if the chamber pressure drops the mass flow rate through the nozzle decreases. With a decrease in chamber pressure, the burning rate drops to a lesser extent if $n < 1$ and thus cause the pressure to build up in the combustion chamber. However, if $n > 1$, the burning rate drops faster than the flow rate in the nozzle, thus leading to a further drop in the chamber pressure. This is clearly an unstable situation. Therefore, based on stability arguments, we conclude that the pressure exponent n in solid propellant grain burning rate has to be less than 1. Figure 11.27 shows the burning rate dependence on chamber pressure and initial grain temperature for several solid propellants. We note that the burning rate varies between ~0.04 and 3 in/s or ~0.1 and 7.5 cm/s. Also, the linear variation in the log–log plot supports the empirical formula 11.60. The indices a and n may be determined from the data in Fig. 11.27. Note that there is a "Plateau DB" propellant that shows a constant burning rate with chamber pressure, thus it must have its pressure exponent, $n = 0$.

Assuming a steady-state flow, we may equate the gasification of the solid propellant to the mass flow rate through a choked nozzle throat to get the area ratio A_b/A_{th}

$$\dot{m} = \rho_p r A_b = p_1 A_{th} \sqrt{\frac{\gamma}{RT_1}\left[\frac{2}{\gamma+1}\right]^{\frac{\gamma+1}{\gamma-1}}} \qquad (11.61)$$

■ **TABLE 11.3**

Characteristics of Some Solid Propellants (from Sutton and Biblarz, 2001)

Propellant type[a]	I_s Range (sec)[b]	Flame Temperature[e]		Density or spec. gravity[e]		Metal content (wt %)	Burning rate[c,e] (in./sec)	Pressure exponent[e] n	Hazard classification	Stress (psi)/ Strain (%)		Processing method
		(°F)	(K)	(lb/in³)	(sp. gr.)					−60°F	+150°F	
DB	220–230	4100	2550	0.058	1.61	0	0.05–1.2	0.30	1.1	4600/2	490/60	Extruded
DB/AP/Al	260–265	6500	3880	0.065	1.80	20–21	0.2–1.0	0.40	1.3	2750/5	120/50	Extruded
DB/AP-HMX/Al	265–270	6700	4000	0.065	1.80	20	0.2–1.2	0.49	1.1	2375/3	50/33	Solvent cast
PVC/AP/Al	260–265	5600	3380	0.064	1.78	21	0.3–0.9	0.35	1.3	369/150	38/220	Cast or extruded
PU/AP/Al	260–265	5700	3440	0.064	1.78	16–20	0.2–0.9	0.15	1.3	1170/6	75/33	Cast
PBAN/AP/Al	260–263	5800	3500	0.064	1.78	16	0.25–1.0	0.33	1.3	520/16 (at −10°F)	71/28	Cast
CTPB/AP/Al	260–265	5700	3440	0.064	1.78	15–17	0.25–2.0	0.40	1.3	325/26	88/75	Cast
HTPB/AP/Al	260–265	5700	3440	0.067	1.86	4–17	0.25–3.0	0.40	1.3	910/50	90/33	Cast
PBAA/AP/Al	260–265	5700	3440	0.064	1.78	14	0.25–1.3	0.35	1.3	500/13	41/31	Cast
AN/Polymer	180–190	2300	1550	0.053	1.47	0	0.06–0.5	0.60	1.3	200/5	NA	Cast

[a]Al, aluminum; AN, ammonium nitrate; AP, ammonium perchlorate; CTPB, carboxy-terminated polybutadiene; DB, double-base; HMX, cyclotetramethylene tetranitramine; HTPB, hydroxyl-terminatd polybutadiene; PBAA, polybutadiene–acrylic acid polymer; PBAN, polybutadiene–acrylic acid–acrylonitrile terpolymer; PU, polyurethane; PVC, polyvinyl chloride.

[b]At 1000 psia expanding to 14.7 psia, ideal or theoretical value at reference conditions.

[c]At 1000 psia.

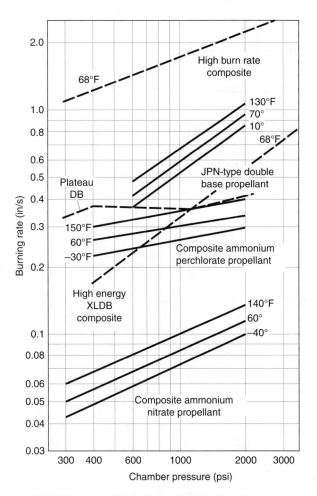

■ **FIGURE 11.27**   Dependence of burning rate of typical solid rocket propellants on chamber pressure and initial grain temperature (from Sutton and Biblarz, 2001)

Therefore, the ratio of burning-to-throat area is expressed as

$$\frac{A_b}{A_{th}} = \frac{p_1}{\rho_p \, r} \sqrt{\frac{\gamma}{RT_1}\left[\frac{2}{\gamma+1}\right]^{\frac{\gamma+1}{\gamma-1}}} = \frac{p_1^{1-n}}{\rho_p \, a\sqrt{RT_1}} \sqrt{\gamma\left[\frac{2}{\gamma+1}\right]^{\frac{\gamma+1}{\gamma-1}}} \tag{11.62}$$

Note that the RHS of Eq. 11.62 is composed of three groups; the first is the combustor pressure dependence, the second is the combustion temperature dependence (in the denominator), and the last term is a function of gas property γ. For a given propellant, we note that chamber pressure is related to the area ratio A_b/A_{th} following

$$p_1 \propto \left(\frac{A_b}{A_{th}}\right)^{\frac{1}{1-n}} \tag{11.63}$$

Since n is less than 1, the exponent of Eq. 11.63 shows the amplification of the chamber pressure with a change in burning surface area increase. Therefore, a sudden appearance of a crack in the grain has the potential of creating a large chamber pressure rise, which may lead to a structural failure. This parameter, i.e., the burning-to-throat area ratio, is an *internal ballistic parameter*

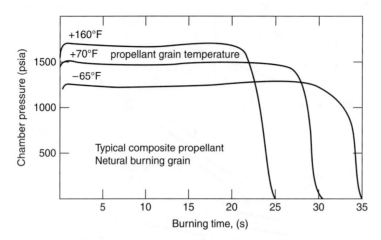

■ **FIGURE 11.28** Thrust-time behavior for different initial grain temperatures for a neutral solid propellant rocket motor conserves the total impulse (from Sutton and Biblarz, 2001)

to solid rocket motors and is given a symbol K. The temperature sensitivity of the grain burning rate for a constant chamber pressure and the temperature sensitivity of chamber pressure for a constant K are known as σ_p and π_K, respectively. These are defined as

$$\sigma_p \equiv \left(\frac{\delta \ln r}{\delta T}\right)_p = \frac{1}{r}\left(\frac{\delta r}{\delta T}\right)_p \tag{11.64}$$

$$\pi_K \equiv \left(\frac{\delta \ln p}{\delta T}\right)_K = \frac{1}{p_1}\left(\frac{\delta p}{\delta T}\right)_K \tag{11.65-a}$$

σ_p represents percent change in burning rate r per degree change in initial grain temperature T while keeping the chamber pressure constant. π_K represents the percent change in chamber pressure per degree change in initial grain temperature for the same burning to throat area ratio K. These initial grain temperature sensitivities form the remaining *internal ballistic parameters* for a solid rocket motor. According to Kubota (1984), typical values for σ_p range from 0.002 to 0.008/K (or per °C) and the range of π_K is related to σ_p via pressure index n according to

$$\pi_K = \frac{\sigma_p}{1-n} \tag{11.65-b}$$

Since σ_p is the parameter that sizes the burning rate enhancement due to initial temperature of grain and n leads to the growth of chamber pressure, a solid propellant of low σ_p and n is desirable. Both burning rate and chamber pressure increase with initial grain temperature, therefore, the propellant is consumed at a faster rate and thus the burn time will be reduced. Since thrust is proportional to propellant mass flow rate, it is, therefore, inversely proportional to the burn time for a given propellant mass. But the burn time is reduced inversely proportional to the mass flow rate. Therefore, the integral of thrust over time, i.e., the total impulse, remains constant. Figure 11.28 shows the effect of initial grain temperature on the thrust-time behavior of a solid rocket motor (from Sutton and Biblarz, 2001).

EXAMPLE 11.11

A solid rocket motor has a design chamber pressure of 7 MPa, an end-burning grain with $n = 0.5$ and $r = 5$ cm/s at the design chamber pressure, and design grain temperature of 15°C. The temperature sensitivity of the burning rate is

$\sigma_p = 0.002$ per °C and chamber pressure sensitivity to initial grain temperature is $\pi_K = 0.004$ per °C. The nominal effective burn time for the rocket is 60 s, i.e., at design conditions.   Calculate

(a) the new chamber pressure and burning rate when the initial grain temperature is 45°C

(b) the corresponding reduction in burn time Δt_b in seconds

SOLUTION

The change in initial grain temperature is $\Delta T = 30$°C, therefore the new chamber pressure is

$$p_c = 7\,\text{MPa}[1 + 0.004(30)] \approx 7.84\,\text{MPa}$$

We first calculate the burning rate at the new chamber pressure of 7.84 MPa while maintaining the same exponent n and coefficient a according to

$$r = a\,p_1^n$$

$$r = (5\,\text{cm/s})\left(\frac{7.84}{7}\right)^{0.5} \approx 5.29\,\text{cm/s}$$

Now, we are ready to correct for the effect of the grain temperature on burning rate, which keeps the chamber pressure constant, i.e., through σ_p,

$$r = 5.29\,\text{cm/s}[1 + 0.002(30)] \approx 5.61\,\text{cm/s}$$

The length of the end-burning grain that was designed to be burned at 5 cm/s for 60 s is approximately

$$L \approx 5\,\text{cm/s}(60\,\text{s}) = 3\,\text{m}$$

The time to burn 3 m of end-burning grain at 5.61 cm/s is therefore

$$t_b \approx 300\,\text{cm}/5.61\,\text{cm/s} \approx 53.5\,\text{s}$$

We may also note that thrust is proportional to chamber pressure, therefore the burn time is inversely proportional to the chamber pressure (for a constant total impulse), i.e.,

$$t_b \approx 60\,\text{s}(7\,\text{MPa}/7.84\,\text{MPa}) \approx 53.6\,\text{s}$$

Therefore, a reduction in burn time of $\Delta t_b \approx 6.5\text{s}$ is caused by a higher grain temperature.

Solid propellant grain designs that incorporate a port area A_p experience an enhanced burning rate, known as *erosive burning*. For erosive-burning grains, heat transfer to the grain is increased due to the flow of hot combustion gases over the grain surface. The empirical rule for the burning rate thus includes the effect of gas speed. For example, we may express the erosive burning grain rate as

$$r = r_0(1 + kv) \tag{11.66}$$

where r_0 is the reference burning rate for an end burning grain and the parenthesis represents the erosive burning *augmentation factor* due to gas speed v, and the coefficient k is an empirical parameter. Figure 11.29 shows a definition sketch for a solid rocket grain that experiences erosive burning. The port area A_p varies along the length of the grain, as the enhanced burning rate consumes the propellant at a faster rate toward the nozzle throat, as shown.

Figure 11.30 shows the erosive burning augmentation ratio, r/r_0, from Razdan and Kuo (1984). The linear relationship between the augmented burning rate and the freestream velocity is clearly demonstrated in Fig. 11.30. We may extract the proportionality factor k from the

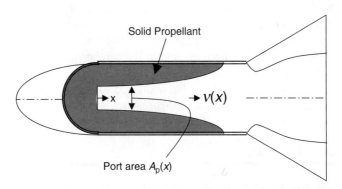

■ **FIGURE 11.29**   Schematic drawing of an erosive burning solid propellant grain

■ **FIGURE 11.30**
Erosive burning
augmentation ratio (from
Razdan and Kuo, 1984)

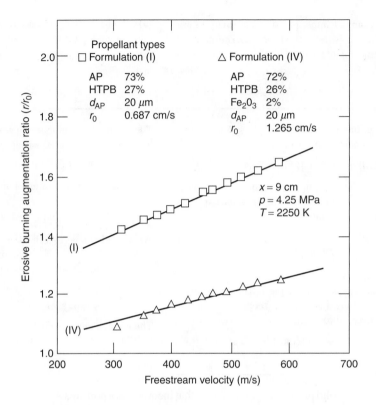

experimental data of Fig. 11.30. This task is left to a problem at the end of this chapter. Also note that erosive burning rate may be enhanced by nearly 50% or more (depending on the propellant type and gas speed) as compared with an end burning grain. This will clearly impact the burn time and rocket flight performance as well as pose special design issues for the casing thermal protection.

There are other factors that contribute to burning rate enhancement besides erosive burning. For example, longitudinal and lateral vehicle accelerations that are encountered in rapid maneuvers or for stability purposes can amplify the rate of gasification in a solid rocket motor. Spin-stabilized platforms and antimissile rockets experience large lateral and longitudinal accelerations. These longitudinal and lateral loads may also cause stress buildup and crack development in the grain. Cracks are structural failures that will impact the performance of the rocket. Due to random, i.e., unpredictable, nature of crack initiation and development with subsequent burning along the crack and with a potential compromise of the integrity of the liner, inhibitor, or the insulator, the mechanical aspect of grain design, processing, and installation are of critical importance to solid rocket motors.

Solid propellant motors encounter combustion instabilities that often involve pressure oscillations in the combustion chamber that are due to coupling between the combustion gas, the combustor cavity, and the flow field in the combustor. The treatment of these complex interactions and the phenomenological behavior of the instabilities are beyond the scope of the present text. Price (1984), however, has addressed solid propellant combustion instabilities in detail, which is recommended for further reading.

11.12 Thrust Chamber Cooling

In this section, we examine the cooling requirements and challenges in liquid and solid propellant rocket thrust chambers.

11.12.1 Liquid Propellant Thrust Chambers

Liquid propellant rockets usually operate for extended periods, i.e., typically for several minutes, and thus expose the combustion chamber walls to intense heating. The combustion products reach (near stoichiometric level) temperatures of ~3000–4700 K depending on the propellant combination (see Table 11.1). All modes of heat transfer, i.e., convection, conduction, and radiation exist in a rocket combustor and contribute to the combustor wall heating. The products of combustion then expand in a convergent–divergent nozzle to produce thrust. Although expansion of the combustion gases in the nozzle convert their thermal energy into kinetic energy, extended sections of the nozzle downstream of the throat are still exposed to excessive gas temperatures and thus need to be cooled. Fortunately, the liquid propellant onboard offers a cooling capacity that is tapped for the so-called *regenerative cooling* purposes of the thrust chamber walls. The cooling capacity is due to the phase change, i.e., from liquid to vapor, known as latent heat of vaporization. Figure 11.31 shows a definition sketch of the regenerative cooling scheme in a liquid propellant rocket. Note that we switched the chamber designation of temperature, pressure, etc. from using the subscript "c" (for chamber) to "g" for (hot) gas. Since we are cooling the walls of the thrust chamber, we need the subscript "c" for the coolant.

As we discussed heat transfer in Chapter 9 for turbine cooling, the steady-state heat transfer from the hot combustion gases to the coolant takes place convectively over the combustor inner wall, across a hot gas film, radiatively due to the volume of hot combustion gases to the wall, conductively through the combustor wall, and convectively over a coolant film on the coolant channel side. The net one-dimensional heat flux to the thrust chamber wall is the sum of the convection as well as the radiative heat transfer according to

$$\dot{q}_w = \dot{q}_c + \dot{q}_r \tag{11.67}$$

The radiative heat transfer from the volume of hot gas to the solid wall in the combustor is governed by

$$\dot{q}_r = \varepsilon_g \sigma T_g^4 \tag{11.68}$$

where ε_g is the emissivity of the gas (equal to 1 for blackbody radiation and less than 1 for graybody radiation), σ is the Stefan–Boltzmann constant, which is $5.67 \times 10^{-8} \, \text{W/m}^2\text{K}^4$ and T_g is the absolute gas temperature. Although Stefan–Boltzmann constant is very small (due to

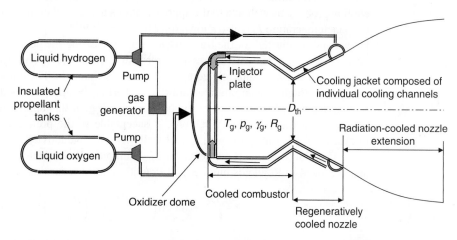

■ **FIGURE 11.31**   Definition sketch of a regeneratively cooled rocket thrust chamber [subscript "g" stands for (hot) gas]

10^{-8} term), the temperature contribution to radiation heat transfer grows exponentially with temperature, i.e., T_g^4. Therefore according to Sutton and Biblarz (2001), radiation accounts for 3–40% of the combustion chamber heat flux for the gas temperature range of 1900–3900 K. This is a reminder that we may not neglect radiative heat transfer unless the gas temperature is low, namely, ~700 K. The presence of solid particles in the rocket combustion chamber increases the emissivity ε_g in Eq. 11.68. The upper value of 1, i.e. blackbody radiation, may be chosen for high concentration of solid particles in the gas stream. The convective heat transfer is governed by the Newton's law of cooling according to

$$\dot{q}_c = h_g(T_{aw} - T_{wg}) \tag{11.69}$$

where h_g is the gas-side film coefficient, T_{aw} is the adiabatic wall temperature (on the gas side), and T_{wg} is the gas-side wall temperature. The Fourier's law of heat conduction governs the heat conduction through the wall,

$$\dot{q}_{cond} = k_w \left(\frac{T_{wg} - T_{wc}}{t_w} \right) \tag{11.70}$$

where k_w is the thermal conductivity of the wall material (a property of the wall), T_{wc} is the wall temperature on the coolant side, and t_w is the wall thickness (a design parameter). Note that the minus in Fourier heat conduction law is absorbed in the parenthesis, i.e., the heat transfer is from the combustor toward the coolant side. Now, combining these sources of heat flow in a steady-state (one-dimensional) rocket thrust chamber-cooling problem, we can write

$$\dot{q}_w = \underbrace{h_g(T_{aw} - T_{wg}) + \varepsilon_g \sigma T_g^4}_{\text{Combustion chamber}} = \underbrace{k_w \left(\frac{T_{wg} - T_{wc}}{t_w} \right)}_{\text{Wall}} = \underbrace{h_c(T_{wc} - T_c)}_{\text{Coolant}} \tag{11.71}$$

We may eliminate the gas and coolant sidewall temperatures, T_{wg} and T_{wc}, from the heat flux equation to get

$$\dot{q}_w = \frac{T_{aw} - T_c + (\varepsilon_g \sigma T_g^4 / h_g)}{(1/h_g) + (t_w/k_w) + (1/h_c)} \tag{11.72}$$

Let us examine the parameters in Eq. 11.72. The stagnation temperature of the gas often approximates the adiabatic wall temperature T_{aw} as we encountered in turbine cooling section. This approximation, i.e., $T_{aw} \approx T_{tg}$, is appropriate in applying to the rocket thrust chamber cooling problem as well due to the approximations that are inherent in heat transfer solutions. Coolant temperature T_c is also known since the stored liquid temperature in the insulated tank is known and is adjusted for the transmission line connecting the pump and the thrust chamber heat transfer. Approximate expressions for film coefficients h_g and h_c are also known from the heat transfer literature related to the long tubes/channels. The emissivity of gas ε_g in the presence of (high concentration of) solid particles may be approximated as 1, i.e., treating the particle-laden gas as a perfect blackbody radiator. Again this approximation, in the context of other heat transfer approximations, is acceptable in the preliminary design phase. The thickness of the combustor wall t_w and wall thermal conductivity k_w are design choices.

The film coefficients on the gas side h_g is empirically estimated for heat transfer in tubes and is related to Reynolds number and Prandtl number according to

$$\frac{h_g D_g}{k_g} = 0.026 \left(\frac{\rho_g v_g D_g}{\mu_g} \right)^{0.8} \left(\frac{\mu_g c_{pg}}{k_g} \right)^{0.4} \tag{11.73}$$

where D_g is the (local) diameter of the thrust chamber, k_g is the gas thermal conductivity, $\rho_g v_g$ is the average gas mass flow rate per unit area in the combustion chamber, μ_g is the gas coefficient of viscosity, and c_{pg} is the specific heat of the gas at constant pressure. The left-hand side of Eq. 11.73 is the Nusselt number; the parentheses on the RHS are Reynolds and Prandtl numbers, respectively. Since the fluid density and properties μ_g, k_g, and c_{pg} are functions of gas temperature, the question of "which gas temperature" in the combustion chamber should we use to estimate these parameters arise. There is no unique answer to this question; however, for engineering approximations an "average film" temperature T_f is used as an acceptable reference temperature, i.e.,

$$T_f \equiv \frac{T_g + T_{wg}}{2} \tag{11.74}$$

The coolant side film coefficient (for the coolant in the liquid state) is related to the Reynolds and Prandtl numbers on the coolant side following

$$\frac{h_c D_c}{k_c} = 0.023 \left(\frac{\rho_c v_c D_c}{\mu_c}\right)^{0.8} \left(\frac{\mu_c \bar{c}_c}{k_c}\right)^{0.33} \tag{11.75}$$

where $\bar{c}_c$ is the coolant average specific heat and D_c is the coolant passage hydraulic diameter. We may express the film coefficient on the coolant side in terms of on coolant mass flux, $\rho_c v_c$, and specific heat:

$$\frac{h_c}{(\rho_c v_c)\bar{c}_c} = 0.023 \left(\frac{\rho_c v_c D_c}{\mu_c}\right)^{-0.2} \left(\frac{\mu_c \bar{c}_c}{k_c}\right)^{-0.67} \tag{11.76}$$

This expression clearly shows that for a given coolant Reynolds number and Prandtl number, the coolant film coefficient h_c increases with coolant mass flux and the coolant specific heat. We may thus tailor the cross section of the coolant passage in critical areas such as the nozzle throat to increase the local mass flux, $\rho_c v_c$. Also, we may choose a liquid fuel with high specific heat. For example, hydrogen has a very high specific heat compared with other rocket fuels and is thus a coolant of choice for regeneratively cooling rocket thrust chambers. Table 11.4 from Sutton and Biblarz (2001) show some physical properties of some liquid propellants.

In general, we have to integrate the steady-state heat transfer Eq. 11.71 along the thrust chamber (axis). We may divide the thrust chamber into cylindrical sections of small axial length Δx where all the heat transfer parameters are assumed constant. By further assuming axisymmetric condition per cylindrical sections, we can march along the axis from the combustion chamber out toward the nozzle exit. The overall heat transfer to an axisymmetric thrust chamber is the integral of the heat flux along the wall, which is thus the sum of the segments according to

$$\dot{Q} = \int \dot{q}\, dA \cong \sum_{i=1}^{N} 2\pi r_i \dot{q}_i \Delta x_i \tag{11.76}$$

The overall heat transfer causes the temperature of the coolant to rise according to

$$\dot{Q} = \dot{m}_c \bar{c}_c (T_2 - T_1)_c \tag{11.77}$$

where T_1 represents the entrance temperature of the coolant into the jacket, and T_2 is the bulk coolant temperature as it (leaves the cooling jacket and) enters the injector. As a design choice, we may limit the bulk coolant exit temperature to below the boiling point, i.e., $T_2 < T_{\text{boiling-point}}$. Table 11.5 lists the heat transfer characteristics of some liquid propellants (from Sutton and Biblarz, 2001). The pressure in the cooling jacket has a strong influence on the boiling point temperature of the fuel. For example, hydrazine boils at 387 K for a pressure of 0.101 MPa (i.e., 1 atm), whereas it boils at 588 K (for an increase of 201 K) if the pressure is raised to 6.89 MPa. The cooling jacket pressure is to be provided by a turbopump (or a pressure) feed system and the

■ **TABLE 11.4**
Physical Properties of Some Liquid Propellants (from Sutton and Biblarz, 2001)

Propellant	Liquid fluorine	Hydrazine	Liquid hydrogen	Methane	Monomethyl-hydrazine	Nitric Acid[a] (99% pure)	Nitrogen tetroxide	Liquid oxygen	Rocket fuel RP-1	Unsymmetrical dimethyl-hydrazine (UDMH)	Water
Chemical formula	F_2	N_2H_4	H_2	CH_4	CH_3NHNH_2	HNO_3	N_2O_4	O_2	Hydrocarbon $CH_{1.97}$	$(CH_3)_2NNH_2$	H_2O
Molecular mass	38.0	32.05	2.016	16.03	46.072	63.016	92.016	32.00	~175	60.10	18.02
Melting or freezing point (K)	53.54	274.69	14.0	90.5	220.7	231.6	261.95	54.4	225	216	273.15
Boiling point (K)	85.02	386.66	20.4	111.6	360.6	355.7	294.3	90.0	460–540	336	373.15
Heat of vaporization (kJ/kg)	166.26[b]	44.7[b]	446	510[b]	875	480	413[b]	213	246[b]	542 (298 K)	2253[b]
Specific heat (kcal/kg-K)	0.368 (85 K); 0.357 (69.3 K)	0.736 (293 K); 0.758 (338 K)	1.75[b] (20.4 K); —	0.835[b]	0.698 (293 K); 0.735 (393 K)	0.042 (311 K); 0.163 (373 K)	0.374 (290 K); 0.447 (360 K)	0.4 (65 K)	0.45 (298 K)	0.672 (298 K); 0.71 (340 K)	1.008 (273.15 K)
Specific gravity[c]	1.636 (66 K); 1.440 (93 K)	1.005 (293 K); 0.952 (350 K)	0.071 (20.4 K); 0.076 (14 K)	0.424 (111.5 K)	0.8788 (293 K); 0.857 (311 K)	1.549 (273.15 K); 1.476 (313.15 K)	1.447 (293 K); 1.38 (322 K)	1.14 (90.4 K); 1.23 (77.6 K)	0.58 (422 K); 0.807 (289 K)	0.856 (228 K); 0.784 (244 K)	1.002 (373.15 K); 1.00 (293.4 K)
Viscosity (centipoise)	0.305 (77.6 K); 0.397 (70 K)	0.97 (298 K); 0.913 (330 K)	0.024 (14.3 K); 0.013 (20.4 K)	0.12 (111.6 K); 0.22 (90.5 K)	0.855 (293 K); 0.40 (344 K)	1.45 (273 K)	0.47 (293 K); 0.33 (315 K)	0.87 (53.7 K); 0.19 (90.4 K)	0.75 (289 K); 0.21 (366 K)	4.4; 0.48 (220 K); 0.48 (300 K)	0.284 (373.15 K); 1.000 (277 K)
Vapor pressure (MPa)	0.0087 (100 K); 0.00012 (66.5 K)	0.0014 (293 K); 0.016 (340 K)	0.2026 (23 K); 0.87 (30 K)	0.033 (100 K); 0.101 (117 K)	0.0073 (300 K); 0.638 (428 K)	0.0027 (273.15 K); 0.605 (343 K)	0.01014 (293 K); 0.2013 (328 K)	0.0052 (88.7 K)	0.002 (344 K); 0.023 (422 K)	0.0384 (289 K); 0.1093 (339 K)	0.00689 (312 K); 0.03447 (345 K)

[a] Red fuming nitric acid (RFNA) has 5 to 20% dissolved NO_2 with an average molecular weight of about 60, and a density and vapor pressure somewhat higher than those of pure nitric acid.

[b] At boiling point.

[c] Reference for specific gravity ratio: 10^3 kg/m^3 or 62.42 lbm/ft^3.

■ **TABLE 11.5**

Heat Transfer Characteristics of Some Liquid Propellants (from Sutton and Biblarz, 2001)

Liquid coolant	Boiling characteristics		Critical Temp. (K)	Critical Pressure (MPa)	Nucleate boiling characteristics			
	Pressure (MPa)	Boiling Temp. (K)			Temp. (K)	Pressure (MPa)	Velocity (m/sec)	q_{max} (MW/m^2)
Hydrazine	0.101	387	652	14.7	322.2	4.13	10	22.1
	0.689	455					20	29.4
	3.45	540			405.6	4.13	10	14.2
	6.89	588					20	21.2
Kerosene	0.101	490	678	2.0	297.2	0.689	1	2.4
	0.689	603					8.5	6.4
	1.38	651			297.2	1.38	1	2.3
	1.38	651					8.5	6.2
Nitrogen tetroxide	0.101	294	431	10.1	288.9	4.13	20	11.4
	0.689	342			322.2			9.3
	4.13	394			366.7			6.2
Unsymmetrical dimethyl hydrazine	0.101	336	522	6.06	300	2.07	10	4.9
	1.01	400					20	7.2
	3.45	489			300	5.52	10	4.7

higher pressure ratio will then require multistage pumps and very heavy turbopump feed system.

Also note that the critical heat flux, i.e., the maximum heat transfer rate per unit area, in the thrust chamber theoretically occurs at the throat where the flow passage area is at a minimum. Experimental evidence of the peak heat flux *near* the throat is well documented in the literature, e.g., Witte and Harper (1962). To lower the wall temperature on the gas side, we have to increase the coolant side film coefficient h_c, which according to Eq. 11.76 behaves as

$$h_c \propto (\rho_c v_c)^{0.8} / D_c^{0.2} \tag{11.78}$$

The design requirement on T_{wg} at the throat dictates the following characteristics on the coolant passage side:

■ High mass flux $\rho_c v_c$ or since the liquid density is nearly constant, high coolant speed v_c
■ Small diameter D_c cooling tubes

The coolant speed at the throat may be as high as $\sim$15–20 m/s. The hydraulic diameter of the cooling tubes near the throat scales in millimeters. The most effective throat cooling in regeneratively cooled thrust chambers combines these two effects. Therefore, the cooling jacket is composed of small-diameter tubes with a tapered cross-sectional area at the throat with relatively high coolant speed.

In very small liquid propellant thrust chambers, where either insufficient coolant flow is available or insufficient (throat) surface area is available for cooling, ablative materials are used. Also, in cases where the motor is pulsed as in vernier rockets, passive cooling technique, such as ablative cooling, is used. An ablative material directly transitions from the solid-to-gaseous phase while absorbing latent heat of vaporization in the process. Additionally, the emergence (and the flow) of the (ablative) gas on the surface acts as a cooling layer that protects the surface.

11.12.2 Cooling of Solid Propellant Thrust Chambers

Solid propellant rockets may use several (nonregenerative) methods to protect the thrust chamber walls (or motorcase) against the heat loads in the combustor. The purpose of an *inhibitor* is to inhibit grain burning where the designer wishes to protect the thrust chamber walls (or as a means of controlling thrust-time or vehicle dynamic behavior). The purpose of an insulating layer, i.e., the one with low thermal conductivity, next to the wall is also to protect the wall from excessive heating. In addition to these two techniques, solid propellant grain designer may place a cooler burning grain in the outer shell next to the wall to protect the motorcase. However, this is an expensive manufacturing proposition seldom used. Since ablative materials have been extensively used for thermal protection with success during the Apollo era, and graphite has emerged as a viable high temperature material, designers use ablative materials and graphite as throat inserts (or for motorcase insulation) in solid propellant rockets today with confidence. The use of graphite offers the added advantage of erosion resistance (as compared with ablative materials that decompose), thus maintaining the throat (cross-sectional flow) area nearly constant as a function of burn time. Detailed engineering design considerations in solid rocket motors, including practical thermal management approaches are presented by Heister (1995), which is suggested for further reading.

EXAMPLE 11.12

A liquid propellant rocket combustor is regeneratively cooled and is characterized by

The gas static and stagnation temperatures are nearly equal, with $T_g \approx T_{tg} = 2750$ K

The coolant bulk temperature is 300 K

The wall thickness is $t_w = 2$ mm

The thermal conductivity of the wall is $k_w = 43$ W/m · °C

The gas side film coefficient $h_g = 657$ W/m²K

The coolant side film coefficient $h_c = 26,000$ W/m²K

The emissivity of the hot gas is given by $\varepsilon_g = 0.05$

Calculate

(a) the heat flux due to radiation in kW/m²

(b) the total heat flux in kW/m²

(c) the convection heat flux in kW/m²

(d) wall temperature on the gas side T_{wg} in K

(e) wall temperature on the coolant side T_{wc} in K

SOLUTION

The radiation heat flux is governed by

$$\dot{q}_r = \varepsilon_g \sigma T_g^4$$
$$= 0.05(5.67 \times 10^{-8} \text{ W/m}^2\text{K}^4)(2750)^4 \text{ K}^4$$
$$\approx 162.1 \text{ kW/m}^2$$

We will use Eq. 11.72 for the total heat flux and approximate adiabatic wall temperature by the gas stagnation temperature, to get

$$\dot{q}_w \approx \frac{T_{tg} - T_c + (\varepsilon_g \sigma T_g^4/h_g)}{(1/h_g) + (t_w/k_w) + (1/h_c)}$$
$$= [2750 - 300 + 162.1(10^3)/657]/[(1/657)$$
$$+ (0.002/43) + (1/26000)] \approx 1.678 \text{ MW/m}^2$$

The heat flux due to convection is the difference between the total and radiative heat flux, i.e.,

$$\dot{q}_c = \dot{q}_w - \dot{q}_r = (1.678 - 10^3) - 162.1) \text{ kW/m}^2$$
$$= 1.516 \text{ MW/m}^2$$

The wall temperature on the gas side is calculated from Eq. 11.69 to be

$$\dot{q}_c = h_g(T_{aw} - T_{wg}) \quad \Longrightarrow \quad T_{wg} \approx 443 \text{ K}$$

The liquid-side wall temperature is

$$\dot{q}_w = h_c(T_{wc} - T_c) \quad \Longrightarrow \quad T_{wc} \approx 364.5 \text{ K}$$

11.13   Combustor Volume and Shape

Chemical rocket combustion chambers are cylindrical in shape, characterized by a length L_1, diameter D_1, an area contraction ratio A_1/A_{th}, (toward the nozzle throat), and a contraction length L_c (or a contraction angle θ_c). Figure 11.32 shows a definition sketch of a combustor volume. The combustor volume V_c includes the convergent section to the throat.

Chamber diameter D_1 is primarily governed by the vehicle size or thrust magnitude. In flight performance section, we desired a low cross-sectional area for the vehicle (through m_0/A), i.e., our goal is to minimize D_1, consistent with thrust production. The area contraction A_1/A_{th} sets the gas speed in the chamber. Remember that the throat is sonic and area ratio (A/A^*) establishes the upstream Mach number M_1. The desire is toward large area ratio, yet limited by minimum A_1. The contraction angle θ_c provides for a smooth flow at the throat with high discharge coefficient. The desire is to have a large angle consistent with high efficiency, which translates to about 30° or less (typically in the range of 20°–30°). The chamber length L_1 is the minimum length needed for a complete combustion in liquid propellant rockets. Therefore, chamber length is directly proportional to the product of mean speed of the propellant in the combustor and the propellant *residence time*, i.e.,

$$L_1 \approx v_1 \cdot t_{res} \tag{11.79}$$

Residence time depends on the evaporation, mixing, and reaction timescales, which is a strong function of the propellant combination. Residence time is also a function of the injector plate design through the atomization scale. Therefore, chamber length is affected by the injector design and the propellant choice. Residence or stay time range between 1 and 40 ms for different propellants. It is customary to define a characteristic length scale L^* based on the combustor (total) volume that includes the convergent section to the throat area, namely,

$$L^* \equiv \frac{V_c}{A_{th}} \tag{11.80}$$

Huzel and Huang (1992) present historical numbers for L^* for different propellants. As these numbers are not based on fundamental physics, they should be used for comparative purposes only. Table 11.6 shows the historical data from Huzel and Huang (1992).

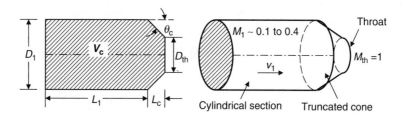

■ **FIGURE 11.32**   **Definition sketch of a rocket combustion chamber geometry**

11.14   Rocket Nozzles

The aerodynamic principles of exhaust nozzles and their design considerations are presented in Chapter 5. The range of operation of the rocket nozzles, however, is vastly different than airbreathing engines. A rocket may operate in ambient conditions ranging from sea level to vacuum in space. Therefore, the nozzle pressure ratio (NPR) in a rocket may range from $\sim$50 at

■ **TABLE 11.6**
Combustor Characteristic Length L^* (from Huzel and Huang, 1992)

Propellants	L^*(m), low	L^*(m), high
Liquid fluorine/hydrazine	0.61	0.71
Liquid fluorine/gaseous H_2	0.56	0.66
Liquid fluorine/liquid H_2	0.64	0.76
Nitric acid/hydrazine	0.76	0.89
N_2O_4/hydrazine	0.60	0.89
Liquid O_2/ammonia	0.76	1.02
Liquid O_2/gaseous H_2	0.56	0.71
Liquid O_2/liquid H_2	0.76	1.02
Liquid O_2/RP-1	1.02	1.27
H_2O_2/RP-1 (including catalyst)	1.52	1.78

sea level to infinity (assuming ambient pressure at orbit is zero) in space. Consequently, chemical rockets with such large pressure ratios require very large expansion area ratios, i.e., $A_2/A_{th} \gg 1$, in order to maximize their thrust production. Area expansion ratios of 100 or more are designed for rocket nozzles for space applications. It is a challenge to design a variable-area rocket nozzle due to mechanical complexity and weight penalty. A successful example is shown in Fig. 11.33 where the RL-10B-2 rocket engine uses a two-piece extendable nozzle skirt.

Possible rocket nozzle configurations are shown in Table 11.6 (from Sutton and Biblarz, 2001). The nozzles are divided into two categories (1) the ones with and (2) the ones without a centerbody. The second category of exhaust nozzles, without a centerbody, is the conventional conical, bell-shaped, and partial-bell configurations. All nozzles with an exit flow component that is not purely in the axial direction suffer the so-called (momentum) angularity loss. We have addressed and quantified this loss in Chapter 5. For example, the conical nozzle that offers a simple geometry for manufacturing; it introduces an angularity loss factor to the momentum thrust that is given by Eq. 5.62 and is repeated here for convenience

$$C_{A_{\text{Conical}}} = \frac{1 + \cos \alpha}{2}$$

where α is the half-cone angle. The two-dimensional nozzles with exit flow angle α, suffered the following momentum angularity loss, as derived in Chapter 5.

$$C_{A_{2D-CD}} = \frac{\sin \alpha}{\alpha}$$

A conical nozzle with $15°$ half-cone angle creates an angularity momentum loss of only 1.7%. However, the low-altitude, i.e., overexpanded, operation of conical nozzles involves shock formations (inside the nozzle) that cause flow separation, as shown in Table 11.6. The bell-shaped nozzles, either full bell or partial bell are contoured and are more complex to manufacture, but by virtue of smaller exit angularity, the nozzle momentum thrust efficiency can be better than a conical nozzle of the same area ratio, or length. We note that the sea-level performance of bell-shaped nozzles (as shown in Table 11.6) tends to create a full exit flow with oblique shocks hanging at the lip. The category of nozzles that use a centerbody, as in a plug (or aerospike) nozzle or the expansion-deflection nozzle, tends to produce better off-design performance, both at the low and high altitudes, than the ones without centerbody. We

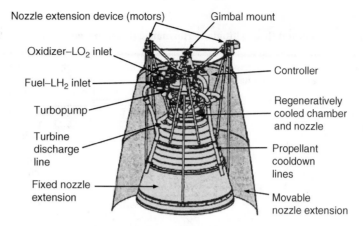

Nozzle extension device (motors)
Gimbal mount
Oxidizer–LO$_2$ inlet
Controller
Fuel–LH$_2$ inlet
Regeneratively cooled chamber and nozzle
Turbopump
Turbine discharge line
Propellant cooldown lines
Fixed nozzle extension
Movable nozzle extension

(a) Half section of nozzle extension in stowed position

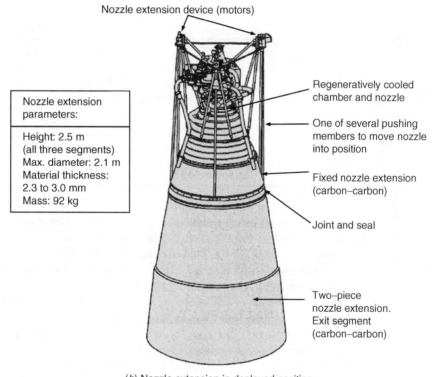

Nozzle extension device (motors)

Nozzle extension parameters:

Height: 2.5 m
(all three segments)
Max. diameter: 2.1 m
Material thickness:
2.3 to 3.0 mm
Mass: 92 kg

Regeneratively cooled chamber and nozzle

One of several pushing members to move nozzle into position

Fixed nozzle extension (carbon–carbon)

Joint and seal

Two–piece nozzle extension. Exit segment (carbon–carbon)

(b) Nozzle extension in deployed position

■ **FIGURE 11.33** **The RL-10B-2 rocket engine with extendable nozzle skirt (courtesy of Pratt & Whitney)**

may note the flow distribution at the nozzle exit, in Table 11.6, shows a fully attached flow, i.e., without any shock-induced separations. The off-design performance of aerospike and other nozzles with a centerbody is thus superior to conical and bell-shaped nozzles, but the system penalty is paid by the centerbody needs to be cooled. More details of the flow pattern for a linear (truncated) aerospike are shown in Fig. 11.34. This concept is used in XRS-2200 aerospike linear rocket engine that uses 20 individual thrust cells or modules and two concave

■ **TABLE 11.6**
Rocket Nozzle Configurations and their Altitude Performance (from Sutton and Biblarz, 2001)

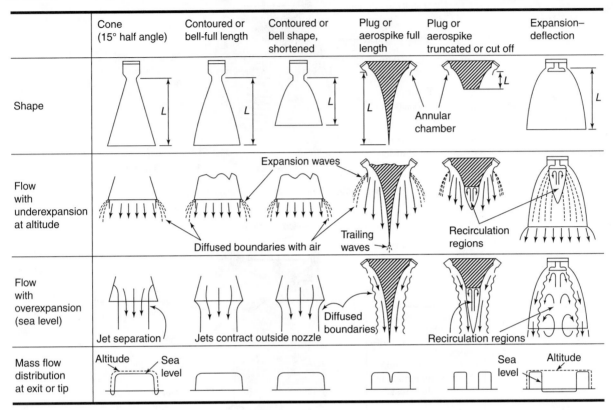

regeneratively (fuel) cooled (external) expansion ramps. The truncated base is porous where the gas generator flow (from the turbo pump system) fills the base and provides for a continuous and smooth expansion. The pressure distribution on the ramp shows the presence of (periodic) compression and expansion waves. The local compression (or shock) waves are followed by the reflected expansion waves from the jet boundary (or shear layer) and thus do not cause boundary layer separation. However, the presence of the compression waves on the ramp will cause an increase in the wall heat transfer and thus a more conservative fuel-cooling approach needs to be employed. The two-dimensional aspect of the linear aerospike offers better integration potential with a winged aircraft than the axisymmetric configuration. In addition, the action of (20) individual thrusters may be used for stability and control purposes of the aircraft.

The radius of curvature upstream and downstream of the nozzle throat, the initial and exit angles of the bell nozzles, and the angularity loss parameter of conical and bell-shaped nozzles are presented and compared in Fig. 11.35 (adapted from Huzel and Huang, 1992). These curves that relate the geometry and performance of various nozzles may be used in preliminary design trade studies. The nozzle length comparisons in Fig. 11.35 are suitable for weight estimation studies.

11.14.1 Multiphase Flow in Rocket Nozzles

In our analysis so far, we have treated the nozzle fluid to be in the gaseous state. There are at least three sources for liquid or solid particle presence in a rocket nozzle flow. The first is the

Linear aerospike flow pattern and pressure distribution

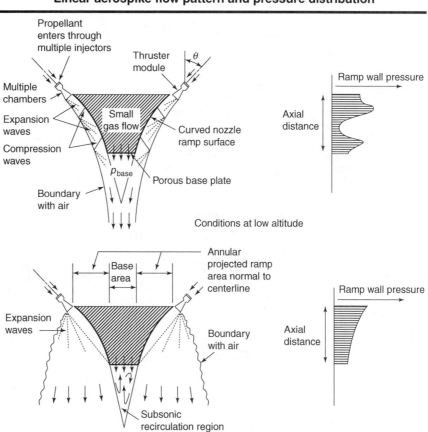

■ **FIGURE 11.34**   Flow pattern and pressure distribution in a linear aerospike nozzle at low and high altitudes (from Sutton and Biblarz, 2001)

propellant itself where solid particles are intentionally embedded or dispersed to increase the rocket performance through higher specific impulse. The second source is the combustion, which may lead to soot and particulate formation in the products of combustion. The third source is the appearance of liquid droplets, i.e., condensation, in the exhaust stream due to static temperature drop in the gases. The very large area expansion ratios bring about very large temperature and pressure expansion ratios, which result in the condensation of a fraction of the exhaust stream. It is the impact of these solid or liquid fractions (present in the nozzle gases) on rocket performance that we address in this section.

The thermal and dynamic interaction of the two phases in a fluid, say solid and gas, pose complex fluid mechanics and heat transfer problems. However, in four limiting cases, we may arrive at closed-form solutions and shed light on the effects of multiphase flow on rocket performance.

These four limiting cases are

1. Solid particles reach the same temperature as the gas at the nozzle exit, the so-called thermal equilibrium case

2. Solid particles are fully accelerated and reach the same speed as the gas at the nozzle exit, i.e., they reach momentum equilibrium

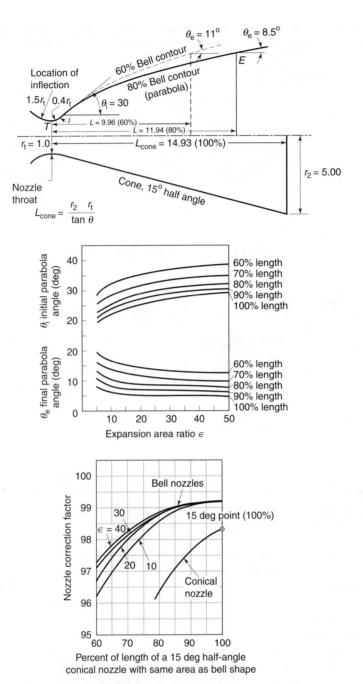

■ **FIGURE 11.35**   **Conical, bell and partial bell nozzle design, and performance comparison (adapted from Huzel and Huang, 1992)**

3. Solid particles maintain their temperatures, i.e., the case of zero thermal interaction or heat transfer between the two phases

4. Solid particles are not accelerated by the gas and thus maintain their negligible momentum at the nozzle exit

Note that an underlying assumption in these four limiting cases is that the solid fraction present at the nozzle inlet remains unchanged, i.e., *frozen*, in the nozzle and appear in whole at the

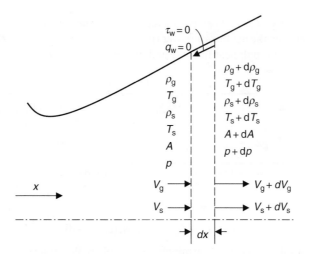

■ FIGURE 11.36 Definition sketch of a slab in a two-phase flow in a nozzle

nozzle exit. The four cases noted above address only the fate of these particles in heat and momentum transfer with the gas. To derive the fundamental equations of two-phase flow, we return to a *slab* concept where we apply the (one-dimensional) conservation principles to a slab of fluid. The fluid and flow parameters on the two sides of the slab are related to each other via a linear relationship using the Taylor series expansion of a continuous function. Figure 11.36 shows the definition sketch of a *slab* in a two-phase fluid flow with zero friction and heat transfer (i.e., reversible and adiabatic condition) at the wall.

By stipulating a constant mass flow rate for the gas and the solid phases in the fluid, we get

$$\dot{m}_g = \rho_g V_g A = (\rho_g + d\rho_g)(V_g + dV_g)(A + dA) \tag{11.81}$$

$$\dot{m}_s = \rho_s V_s A = (\rho_s + d\rho_s)(V_s + dV_s)(A + dA) \tag{11.82}$$

Note that the density of the solid is the mass of the solid contained in a unit volume of the gas–solid mixture and is not to be confused by the density of the solid itself. The gas density is the mass of the gas per unit volume of the mixture, as expected. By conserving the mass flow rate of solid and gas individually, through Eqs. 11.81 and 11.82, we conserve the collective mass flow rate in the nozzle. The conservation of momentum applied to the slab gives

$$\underbrace{\dot{m}_g(V_g + dV_g) + \dot{m}_s(V_s + dV_s)}_{\text{"Momentum"}_{\text{out}}} - \underbrace{(\dot{m}_g V_g - \dot{m}_s V_s)}_{\text{"Momentum"}_{\text{in}}}$$

$$= \underbrace{pA - (p + dp)(A + dA) - (p + dp/2)dA}_{\text{Net pressure force in the flow direction}} \tag{11.83}$$

Note that the last term on the RHS of the Eq. 11.83 is the pressure force exerted by the sidewall on the fluid (in the *x*-direction). Also the "momentum" in quotation mark is used as the shorthand for "time rate of change of momentum." Momentum equation simplifies to

$$\dot{m}_g dV_g + \dot{m}_s dV_s = -A dp \tag{11.84}$$

We may divide both sides of Eq. 11.84 by the flow area A to get

$$\rho_g V_g dV_g + \rho_s V_s dV_s = -dp \tag{11.85}$$

The steady energy equation written for a (calorically) perfect gas and solid phases in a flow is the balance between the net flux of power (in fluid and solid phases) and the rate of the external energy exchange through heat transfer and mechanical shaft power, i.e.,

$$
\overbrace{\begin{aligned}&\left[\dot{m}_g[c_{pg}(T_g + dT_g) + (V_g + dV_g)^2/2] + \dot{m}_s[c_s(T_s + dT_s) + (V_s + dV_s)^2/2\right] - \\ &\underbrace{\left[\dot{m}_g[c_{pg}T_g + V_g^2/2] + \dot{m}_s[c_sT_s + V_s^2/2]\right]}_{\text{Fluid and solid power in}} = \underbrace{\dot{Q}_w - \wp_s \equiv 0}_{\substack{\text{Net power exchange} \\ \text{w/surrounding}}}\end{aligned}}^{\text{Fluid and solid power out}}
\tag{11.86}
$$

The energy equation simplifies to

$$
\dot{m}_g(c_{pg}dT_g + V_g dV_g) + \dot{m}_s(c_s dT_s + V_s dV_s) = 0
\tag{11.87}
$$

The fraction of solid particles in the mixture of two-phase flow may be given a symbol X defined as

$$
X \equiv \frac{\dot{m}_s}{\dot{m}_g + \dot{m}_s} = \frac{\rho_s V_s}{\rho_g V_g + \rho_s V_s}
\tag{11.88}
$$

In terms of this parameter, i.e., solid mass fraction X that constitutes a constant in our problem based on our frozen-phase assumption, we may write the energy equation as

$$
(1 - X)(c_{pg}dT_g + V_g dV_g) + X(c_s dT_s + V_s dV_s) = 0
\tag{11.89}
$$

We replace the solid particle density–velocity term $\rho_s V_s$ by

$$
\rho_s V_s = \frac{X}{1 - X} \rho_g V_g
$$

in the momentum equation to get

$$
V_g dV_g = -\frac{dp}{\rho_g} - \frac{X}{1 - X} V_g dV_s
\tag{11.90}
$$

If we divide Eq. 11.89 by $(1 - X)$ and substitute Eq. 11.90 for $\rho_g dV_g$, we get

$$
c_{pg}dT_g + \frac{X}{1 - X} c_s dT_s + \frac{X}{1 - X}(V_s - V_g)dV_s = \frac{dp}{\rho_g}
\tag{11.91}
$$

Case 1

The solid and gas are in thermal equilibrium, i.e., $dT_s = dT_g = dT$, therefore, Eq. 11.91 reduces to

$$
\left(c_{pg} + \frac{X}{1 - X} c_s\right)dT + \frac{X}{1 - X}(V_s - V_g)dV_s = \frac{dp}{\rho_g}
\tag{11.92}
$$

We may consider two scenarios for the solid particles momentum behavior in the nozzle flow. First, if the solid particles are accelerated by the gas to achieve the same speed as the gas, we get $V_s = V_g$, therefore, Eq. 11.92 simplifies to

$$
\left(c_{pg} + \frac{X}{1 - X} c_s\right)dT = \frac{dp}{\rho_g}
\tag{11.93-a}
$$

This equation readily integrates for a perfect gas where $\rho_g = p/R_g T$ to the following form:

$$
\frac{p_2}{p_c} = \left(\frac{T_2}{T_c}\right)^{\left(c_{pg} + \frac{X}{1-X}c_s\right)/R_g} \quad \text{or} \quad \frac{T_2}{T_c} = \left(\frac{P_2}{P_c}\right)^{R_g/\left(c_{pg} + \frac{X}{1-X}c_s\right)}
\tag{11.93-b}
$$

Also Eq. 11.89 for $V_s = V_g = V$ and $T_s = T_g = T$ gives

$$[(1 - X)c_{pg} + Xc_s]dT + VdV = 0 \tag{11.94}$$

which may be integrated from the chamber condition (with $V_c \approx 0$) to the nozzle exit, V_2

$$V_2 = \sqrt{2[(1 - X)c_{pg} + Xc_s]T_c(1 - T_2/T_c)} \tag{11.95}$$

In terms of pressure ratio, nozzle exit velocity is

$$V_2 = \sqrt{2\left[(1 - X)c_{pg} + Xc_s\right]T_c\left[1 - \left(\frac{p_2}{p_c}\right)^{R_g/\left(c_{pg} + \frac{X}{1-X}c_s\right)}\right]} \tag{11.96}$$

Since solid particles have attained the same speed as the gas, the ratio of specific impulse with and without solid particulate flow is the ratio of the two exhaust speeds, namely,

$$\frac{I_s(X)}{I_s(0)} = \sqrt{(1 - X + Xc_s/c_{pg})\left[\frac{1 - \left(\frac{p_2}{p_c}\right)^{R_g/\left(c_{pg} + \frac{X}{1-X}c_s\right)}}{1 - \left(\frac{p_2}{p_c}\right)^{R_g/c_{pg}}}\right]} \tag{11.97}$$

For a gas molecular weight of 20.1 kg/kmol, the ratio of specific heats $\gamma = 1.26$, and specific heat of the solid (aluminum) particles $= 903\,J/kg \cdot K$, we calculate and graph Eq. 11.97 in Fig. 11.37.

Figure 11.37 shows a drop in rocket performance with increasing fraction of the solid flow. The impact of cycle pressure ratio is small, as expected. The presence of solid particles in the propellant is, however, to enhance the burning rate and combustion temperature T_c, which was assumed constant in this calculation. The effect of higher burning rate and combustion temperature due to solids should, therefore, outweigh the penalty of the two-phase flow in the nozzle, as depicted in Fig. 11.37.

The second scenario for the particle momentum/acceleration is to assume the solid particles are not accelerated by the gas, i.e., $dV_s = 0$ while their temperature has reached that of the gas. The temperature–pressure relation remains the same as Eq. 11.93-b. However, the energy equation gives the variation of the gas speed with temperature drop of the mixture, i.e.,

$$[(1 - X)c_{pg} + Xc_s]dT + (1 - X)V_gdV_g = 0 \tag{11.98-a}$$

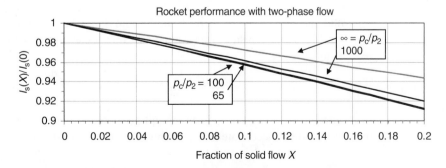

■ **FIGURE 11.37** The effect of solid particles on rocket performance (assuming $T_s = T_g, V_s = V_g$ and constant T_c)

The solution of the above equation is

$$V_{g2} = \sqrt{2\left[c_{pg} + \frac{X}{1-X}c_s\right]T_c\left[1 - \left(\frac{p_2}{p_c}\right)^{R_g/\left(c_{pg} + \frac{X}{1-X}c_s\right)}\right]} \qquad \textbf{(11.98-b)}$$

The thrust of the two streams of gas and solid, one moving at speed V_{g2} and the second moving at speed V_s, is

$$F = \dot{m}_g V_{g2} + \dot{m}_s V_s + (p_2 - p_0)A_2 \qquad \textbf{(11.99)}$$

Therefore, the stream-thrust-averaged speed at the nozzle exit (for a perfectly expanded nozzle) is

$$\overline{V}_2 = \frac{\dot{m}_g V_{g2} + \dot{m}_s V_s}{\dot{m}_g + \dot{m}_s} = (1-X)V_{g2} + XV_{s2} \qquad \textbf{(11.100-a)}$$

Since the second term on the RHS of Eq. 11.100-a is the product of two small numbers (remember that the solid was assumed to be unaccelerated and X is the small solid fraction), therefore, the thrust-averaged exit velocity is

$$\overline{V}_2 \approx (1-X)V_{g2} \qquad \textbf{(11.100-b)}$$

and the specific thrust is expressed in terms of the thrust-averaged exit velocity as

$$I_s = \overline{V}_2/g_0 \approx (1-X)V_{g2}/g_0 \qquad \textbf{(11.101)}$$

The ratio of specific impulse with/without solid particles that are in thermal equilibrium with the gas, but are not accelerated by the gas, is

$$\frac{I_s(X)}{I_s(0)} = (1-X)\sqrt{\left(1 + \frac{X}{1-X}\frac{c_s}{c_{pg}}\right)\left[\frac{1 - \left(\frac{p_2}{p_c}\right)^{R_g/\left(c_{pg} + \frac{X}{1-X}c_s\right)}}{1 - \left(\frac{p_2}{p_c}\right)^{R_g/c_{pg}}}\right]} \qquad \textbf{(11.102)}$$

The graph of Eq. 11.102 superimposed on the rocket performance chart of Fig. 11.37 is shown in Fig. 11.38. Note that the specific impulse is further degraded when the solid particles are not accelerated by the gas (i.e., the case of small particle drag).

Case 2

The solid and gas are not in thermal equilibrium, i.e., when the solid temperature remains constant, $dT_s = 0$ as a result of negligible heat transfer with the gas. In this case, Eq. 11.91 reduces to

$$c_{pg}dT_g + \frac{X}{1-X}(V_s - V_g)dV_s = \frac{dp}{\rho_g} \qquad \textbf{(11.103)}$$

In addition to $T_s = $ constant, either for full acceleration where $V_s = V_g$ or when the solid particle remains unaccelerated, i.e., when $dV_s = 0$, Eq. 11.103 gives

$$c_{pg}dT_g = \frac{dp}{\rho_g} \quad \text{or} \quad \frac{T_2}{T_c} = \left(\frac{p_2}{p_c}\right)^{c_{pg}/R_g} \qquad \textbf{(11.104)}$$

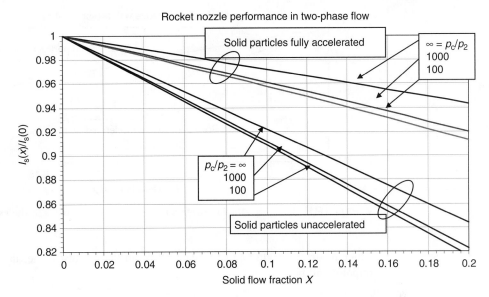

■ **FIGURE 11.38** Variation of specific impulse with solid flow fraction in the gaseous exhaust nozzle (assuming $T_s = T_g$, and constant T_c)

The energy Eq. 11.89 reduces to the following form for $V_s = V_g$, i.e., full acceleration case,

$$(1 - X)c_{pg}dT_g + V_g dV_g = 0 \quad \text{or} \quad V_{g2} = \sqrt{2(1 - X)c_{pg}T_c\left[1 - \left(\frac{p_2}{p_c}\right)^{c_{pg}/R_g}\right]} \quad \textbf{(11.105)}$$

Therefore, the specific impulse ratio is

$$\frac{I_s(X)}{I_s(0)} = \sqrt{1 - X} \quad \textbf{(11.106)}$$

The energy Eq. 11.89 for the case of constant V_s, i.e., unaccelerated solid particles in the two-phase flow reduces to

$$c_{pg}dT_g + V_g dV_g = 0 \quad \textbf{(11.107)}$$

which has a solution following

$$V_{g2} = \sqrt{2c_{pg}T_c\left[1 - \left(\frac{p_2}{p_c}\right)^{c_{pg}/R_g}\right]} \quad \textbf{(11.108)}$$

Although the gas phase seems to be unaffected by the presence of the solid particles, the thrust is affected since the two phases of the flow have different momentums at the nozzle exit. The thrust averaged exit velocity is

$$\overline{V}_2 = \frac{\dot{m}_g V_{g2} + \dot{m}_s V_s}{\dot{m}_g + \dot{m}_s} = (1 - X)V_{g2} + XV_s \quad \textbf{(11.109)}$$

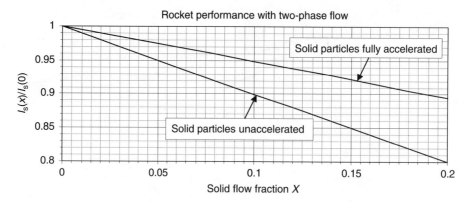

■ **FIGURE 11.39**   **Rocket performance with solid particles in the exhaust stream (assumed T_s = constant)**

which again for negligible solid velocity V_s in the exhaust plane, the ratio of the specific impulse will be

$$\frac{I_s(X)}{I_s(0)} \approx 1 - X \tag{11.110}$$

Now, let us graph Eqs. 11.106 and 11.110 in Fig. 11.39. The specific impulse diminishes with solid flow fraction, as in the previous cases. The nozzle (or cycle) pressure ratio has no effect on specific impulse ratio if the solid particles are not heated to the gas temperature, i.e., when they are not in thermal equilibrium with the gas. The unaccelerated solid particles cause a larger reduction in specific impulse than the fully accelerated ones. The reason is that the fully accelerated solid particles contribute to momentum, whereas the unaccelerated particles with negligible exit momentum cause a reduction in thrust. Therefore, the solid particle drag causes the particles to accelerate and contribute to thrust. The case studies in this section are useful only in establishing a range for the impact that multiphase flow makes on rocket performance. In reality, multiphase flow is more complex and accurate simulation needs computational approach.

EXAMPLE 11.13

A solid propellant rocket motor uses a composite propellant with 16% aluminum. The same propellant with 18% aluminum enhances the combustion temperature by 5.7%. Assuming in both cases that the solid particles are fully accelerated (i.e., $V_s = V_g$) in the nozzle and are in thermal equilibrium with the gas (i.e., $T_s = T_g$), calculate the ratio of specific impulse in the two cases. Aluminum specific heat is $c_s = 903\,\text{J/kg} \cdot \text{K}$ and the specific heat at constant pressure for the gas is $c_{pg} = 2006\,\text{J/kg} \cdot \text{K}$.

SOLUTION

Equation 11.96 that is repeated here is derived precisely for the two-phase flow case at hand, namely, $V_s = V_g$ and $T_s = T_g$. In the limit of large pressure ratio (p_c/p_2) and small variations in X, the pressure bracket will not contribute to the ratio of specific impulse significantly, hence, the exhaust speed may be approximated by

$$V_2 = \sqrt{2\big[(1-X)c_{pg} + Xc_s\big]T_c\left[1 - \left(\frac{p_2}{p_c}\right)^{R_g/\left(c_{pg} + \frac{X}{1-X}c_s\right)}\right]}$$

$$\approx \sqrt{2[(1-X)c_{pg} + Xc_s]T_c}$$

Based on this approximation, we get the ratio of specific impulse in the two cases to be

$$\frac{I_s(X = 0.18)}{I_s(X = 0.16)} \approx \sqrt{\frac{[(1 - 0.18)2006 + 0.18(903)]}{[(1 - 0.16)2006 + 0.16(903)]}}(1.057)$$

$$\approx 1.022$$

This gives an improvement of ~2.2% in specific impulse because the higher combustion temperature of 5.7% more than compensated for the loss due to higher levels of solid flow rate (i.e., 18% vs. 16%) in the nozzle. The higher specific impulse means higher effective exhaust velocity and thrust.

11.14.2 Flow Expansion in Rocket Nozzles

The products of combustion do not in reality stop reacting once they enter the nozzle. Actually, the conversion of thermal to kinetic energy in the nozzle causes a *cooling down effect* of the gases as well as the pressure drop, which impact the reaction rates of some of the products of combustion. Therefore, chemical reaction takes place continually from the combustor through the nozzle. Sometimes the nozzle reverses some of the reactions that took place in the combustor. For example, dissociation in the combustor, which is a direct result of high combustion temperature, may reverse, through the recombination process, to some extent in the nozzle when the temperature of the mixture drops. In addition, chemical reactions between the constituents are often considered to be in equilibrium. However, that too is an assumption that leads to major simplifications in the analysis and may be questioned. Whether chemical equilibrium is reached in a mixture depends on the chemical kinetics and the reaction rates in the mixture. In a rocket expansion process that the timescales are very short (in milliseconds) due to high (convection) speeds, chemical equilibrium may be questioned legitimately in such environments. Computational approaches to complex chemical reactions are fortunately available, e.g., Gordon and McBride, 1996, that may be used for an accurate analysis of chemically reacting mixtures of gases in different thermal environments.

However, the choice of *frozen (concentration of) species* of gas mixture in the nozzle (known as *frozen chemistry* model) is often made for preliminary rocket design calculations. As evidenced from the computational results of Olson (1962), shown in Fig. 11.40 (or Fig. 11.22 from Sutton and Biblarz, 2001), the *corrected* specific impulse for hydrogen–oxygen and JP4–oxygen both indicate to be higher with chemical reactions in the nozzle. Also, the corrected experimental results in Olson show that the equilibrium model gives an overestimation of the actual performance. This important result points to the frozen chemistry

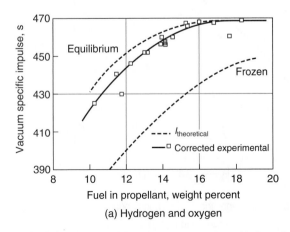

(a) Hydrogen and oxygen

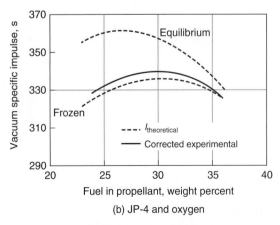

(b) JP-4 and oxygen

■ **FIGURE 11.40**   Comparison of specific impulse between experimental and computational models (from Olson, 1962)

model as a *conservative* estimation of the rocket performance. A conservative estimation is precisely what is needed in a preliminary design stage.

11.14.3 Thrust Vectoring Nozzles

The conventional gimbals with hydraulic or electromechanical actuators may be used to swivel the entire rocket thrust chamber. The range of control authority using this approach is $\sim \pm 7°$. Injection of a fluid, i.e., either a propellant or an inert gas, in the nozzle, known as the secondary injection, causes an oblique shock formation and thus secondary injection thrust vector control (SITVC). The range of control authority using secondary injection is $\sim \pm 5°$. The use of rocket clusters in the vehicle design opens the opportunity to differential operation of the opposite rockets in the cluster to achieve control. The use of jet deflectors, as in jet tabs or jet vanes, offers a limited-duration thrust vector control (in short burn-time rockets) since they are exposed to the hot exhaust gases. The control authority offered in this scheme is $\sim \pm 10°$. Humble et al. (1995) present a practical engineering approach to rocket thrust vector control and is recommended for further reading.

A modern approach to (an old) nozzle design has produced linear aerospike configuration that uses many (e.g., 20) rocket engine cells. Figures 11.41 and 11.42 show aerospike nozzle thrust coefficient variation with nozzle pressure ratio, and RS-2200 is a linear aerospike rocket engine on a test stand (from Boeing). The linear aerospike nozzles have superior off-design performance over a bell nozzle as well as having a vector control capability.

11.15 High-Speed Airbreathing Engines

Ramjets offer the highest (fuel) specific impulse at flight Mach numbers above ~ 4, as depicted in Fig. 11.1. However, they produce no static thrust. The analysis of conventional, i.e., subsonic combustion, ramjets is the same as other airbreathing engines such as turbojets. A typical flow path in a conventional ramjet engine is shown in Fig. 11.43.

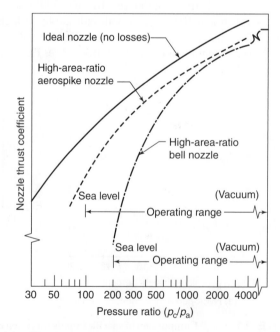

■ **FIGURE 11.41**   **Nozzle thrust coefficient for aerospike (from Huzel and Huang 1992)**

■ **FIGURE 11.42**
RS-2200 Linear aerospike rocket engine test (Courtesy of Boeing)

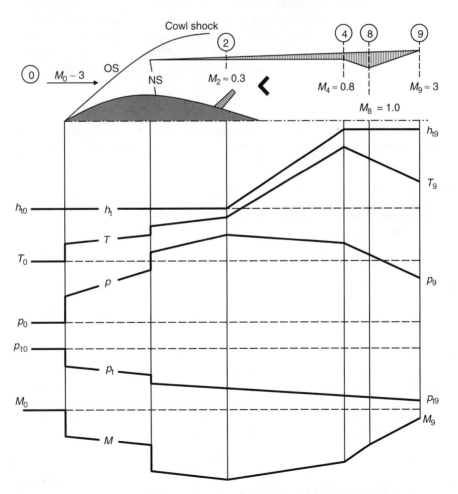

■ **FIGURE 11.43**   Flow parameter trends in a typical conventional ramjet

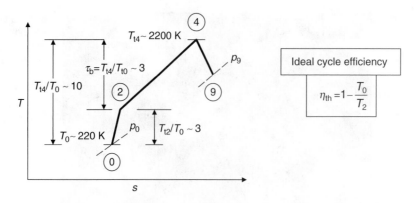

■ **FIGURE 11.44**   **Conventional ramjet cycle and its ideal thermal efficiency**

The *T–s* diagram of the ramjet cycle is shown in Fig. 11.44. The ram compression in supersonic–hypersonic flow includes shock compression that causes an increase in gas temperature and a reduction in the Mach number. In the conventional ramjet, there is a normal shock in the inlet that transitions the flow into subsonic regime. The typical combustor Mach numbers in a conventional ramjet is $\sim$0.2–0.3. The combustor exit temperature T_{t4} may reach stoichiometric levels of $\sim$2000–2500 K.

The equations that are used in the calculation of conventional ramjet performance are listed in Tables 11.7 and 11.8. The equations in these tables are listed sequentially; therefore, they are useful in computer-based calculations/simulation of ramjets. These are the same equations as in turbojets that we derived in Chapter 4, except we have set $\pi_c = 1$ to simulate a ramjet.

■ **TABLE 11.7**

Summary of Ramjet Equations in Terms of Dimensional Parameters (For calorically-perfect gas)

Given M_0, p_0, T_0 (or altitude), γ, and R
  η_d or π_d
  π_d, η_d, Q_R, and T_{t4} or τ_λ
  π_n or η_n and p_9

Calculate $f, F_n/\dot{m}_0$, TSFC, I_s, η_{th}, η_p, and η_o

$T_{t0} = T_0\left(1 + \dfrac{\gamma - 1}{2}M_0^2\right) = T_{t2}$	get flight total temperature
$p_{t0} = p_0\left(1 + \dfrac{\gamma - 1}{2}M_0^2\right)^{\frac{\gamma}{\gamma - 1}}$	get flight total pressure
$V_0 \equiv M_0 \cdot a_0 = M_0\sqrt{\gamma R T_0}$	get flight velocity
$\dfrac{p_{t2}}{p_0} = \left\{1 + \eta_d\dfrac{\gamma - 1}{2}M_0^2\right\}^{\frac{\gamma}{\gamma - 1}}$ or $p_{t2} = \pi_d \cdot p_{t0}$	get total pressure at diffuser exit
$p_{t4} = p_{t2} \cdot \pi_b$	get burner exit total pressure
$f = \dfrac{h_{t4} - h_{t0}}{Q_R\eta_b - h_{t4}} = \dfrac{T_{t4} - T_{t0}}{\dfrac{Q_R\eta_b}{c_p} - T_{t4}}$	get fuel-to-air ratio in the burner
$\pi_n = \left\{\left(\dfrac{p_{t4}}{p_9}\right)^{\frac{\gamma - 1}{\gamma}} - \eta_n\left[\left(\dfrac{p_{t4}}{p_9}\right)^{\frac{\gamma - 1}{\gamma}} - 1\right]\right\}^{\frac{-\gamma}{\gamma - 1}}$	get nozzle total pressure ratio

■ **TABLE 11.7**
(Continued)

$$p_{t9} = p_{t4} \cdot \pi_n$$
get total pressure at nozzle exit

$$M_9 = \sqrt{\frac{2}{\gamma - 1}\left[\left(\frac{p_{t9}}{p_9}\right)^{\frac{\gamma - 1}{\gamma}} - 1\right]}$$
get nozzle exit Mach number

$$T_9 = \frac{T_{t9}}{1 + \frac{\gamma - 1}{2}M_9^2}$$
get nozzle exit static temperature

$$a_9 = \sqrt{\gamma R T_9}$$
get speed of sound at nozzle exit

$$V_9 = a_9 \cdot M_9$$
get nozzle exit velocity

$$\frac{F_n}{\dot{m}_0} = (1 + f)V_9\left(1 + \frac{1}{\gamma M_9^2}\left(1 - \frac{p_0}{p_9}\right)\right) - V_0$$
get specific thrust

$$\text{TSFC} \equiv \frac{\dot{m}_f}{F_n} = \frac{f}{F_n/\dot{m}_0}$$
get thrust-specific fuel consumption

$$I_s = 1/(g_0 \, \text{TSFC})$$
get (fuel)-specific impulse

$$\eta_p = \frac{2(F_n/\dot{m}_0)V_0}{(1 + f)V_9^2 - V_0^2}$$
get propulsive efficiency

$$\eta_{th} = \frac{(1 + f)V_9^2 - V_0^2}{2fQ_R\eta_b}$$
get cycle thermal efficiency

$$\eta_o = \eta_p \cdot \eta_{th} = \frac{(F_n/\dot{m}_0)V_0}{fQ_R\eta_b} = \frac{V_0/Q_R\eta_b}{\text{TSFC}}$$
get overall efficiency

■ **TABLE 11.8**
Summary of Ramjet Equations in Terms of Nondimensional Parameters (For calorically perfect gas)

Given M_0, p_0, T_0 (or altitude), γ, and R
η_d or π_d
π_d, η_d, Q_R, and T_{t4} or τ_λ
π_n or η_n and p_9

Calculate $f, F_n/\dot{m}_0, \text{TSFC}, I_s, \eta_{th}, \eta_p$, and η_o

$$\tau_r = 1 + \frac{\gamma - 1}{2}M_0^2$$
get ram temperature ratio

$$\pi_r = \left(1 + \frac{\gamma - 1}{2}M_0^2\right)^{\frac{\gamma}{\gamma - 1}} = \tau_r^{\frac{\gamma}{\gamma - 1}}$$
get ram pressure ratio

$$\pi_d = \frac{p_{t2}}{p_{t0}} = \left(\frac{1 + \eta_d \frac{\gamma - 1}{2}M_0^2}{1 + \frac{\gamma - 1}{2}M_0^2}\right)^{\frac{\gamma}{\gamma - 1}}$$
get inlet total pressure ratio

$$f = \frac{\tau_\lambda - \tau_r}{\frac{Q_R\eta_b}{c_pT_0} - \tau_\lambda}$$
get fuel-to-air ratio

(Continued)

■ **TABLE 11.8**
(Continued)

$$\pi_n = \left\{ \left(\pi_b \pi_d \pi_r \frac{p_0}{p_9} \right)^{\frac{\gamma-1}{\gamma}} - \eta_n \left[\left(\pi_b \pi_d \pi_r \frac{p_0}{p_9} \right)^{\frac{\gamma-1}{\gamma}} - 1 \right] \right\}^{\frac{-\gamma}{\gamma-1}}$$ get nozzle pressure ratio

$$M_9 = \sqrt{ \frac{2}{\gamma-1} \left[\left(\pi_n \pi_b \pi_d \pi_r \frac{p_0}{p_9} \right)^{\frac{\gamma-1}{\gamma}} - 1 \right] }$$ get nozzle exit Mach number

$$V_9 = M_9 \sqrt{ \gamma R T_0 \frac{\tau_\lambda}{1 + \frac{\gamma-1}{2} M_9^2} } = a_0 M_9 \sqrt{ \frac{\tau_\lambda}{1 + \frac{\gamma-1}{2} M_9^2} }$$ get exhaust velocity

$$\frac{F_n}{\dot{m}_0} = (1 + f) V_9 \left(1 + \frac{1}{\gamma M_9^2} \left(1 - \frac{p_0}{p_9} \right) \right) - V_0$$ get specific thrust

$$\text{TSFC} \equiv \frac{\dot{m}_f}{F_n} = \frac{f}{F_n / \dot{m}_0}$$ get thrust-specific fuel consumption

$$I_s = 1/(g_0 \cdot \text{TSFC})$$ get fuel-specific impulse

$$\eta_p = \frac{2(F_n/\dot{m}_0) V_0}{(1 + f) V_9^2 - V_0^2}$$ get propulsive efficiency η_p

$$\eta_{th} = \frac{(1 + f) V_9^2 - V_0^2}{2 f Q_R \eta_b}$$ get thermal efficiency η_{th}

$$\eta_o = \eta_p \cdot \eta_{th} = \frac{(F_n/\dot{m}_0) V_0}{f Q_R \eta_b} = \frac{V_0/Q_R \eta_b}{\text{TSFC}}$$ get overall efficiency η_o

EXAMPLE 11.13

A conventional ramjet has a maximum temperature $T_{t4} = 2000$ K that is set constant for all flight Mach numbers. The inlet total pressure recovery π_d varies with flight Mach number according to

$$\pi_d = 1 - 0.075(M_0 - 1)^{1.35}$$

Ramjet burns a hydrocarbon fuel with $Q_R = 42800$ kJ/kg, and combustor efficiency and total pressure ratio are $\eta_b = 0.98$ and $\pi_b = 0.95$, respectively. The nozzle is perfectly expanded and has a total pressure ratio $\pi_n = 0.95$. Assuming a calorically perfect gas with $\gamma = 1.4$ and $R = 287$ J/kg · K, use a spreadsheet to calculate the ramjet (fuel)-specific impulse, propulsive, thermal, and overall efficiencies over a range of flight Mach number starting at $M_0 = 1.4$ up until ramjet ceases to produce any thrust. Altitude pressure and temperature are 10 kPa and 240 K, respectively.

SOLUTION

A spreadsheet based on sequential equations in Table 11.7 is constructed. Graphs of specific impulse and efficiencies are generated in the spreadsheet program. There are many interesting trends/features that we observe. For example, specific impulse increases at first, peaks around Mach 3.5, then drops to zero as flight Mach number reaches ~5.7. This behavior of specific impulse is first attributed to the improvements of cycle efficiency (as calculated and graphed) and later drops as flight stagnation temperature

increase would allow less fuel to be burned in the combustor (to reach T_{max} in the cycle). Consequently, thrust production capability diminishes with increase in flight speed. Eventually, we reach a point that flight stagnation temperature and cycle maximum temperature are equal. Before we reach this flight Mach number, exhaust speed V_9 falls to the levels of flight speed V_0, which creates a zero net thrust. We may also calculate and graph burner total temperature ratio τ_b as well as fuel-to-air ratio in the burner to observe the effect of flight Mach number on these parameters. The fuel-to-air ratio and $(\tau_b - 1)$ both approach zero, as flight Mach number approaches 6.

The propulsive efficiency of a ramjet continually improves with flight Mach number, since nozzle exit velocity drops. We recall that $\eta_p \approx 2/(1 + V_9/V_0)$, which approaches 1 as V_9/V_0 approaches 1.

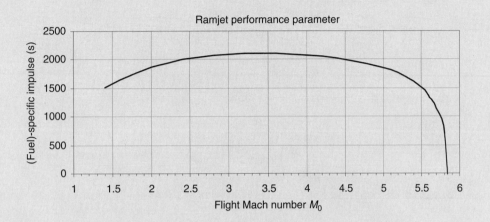

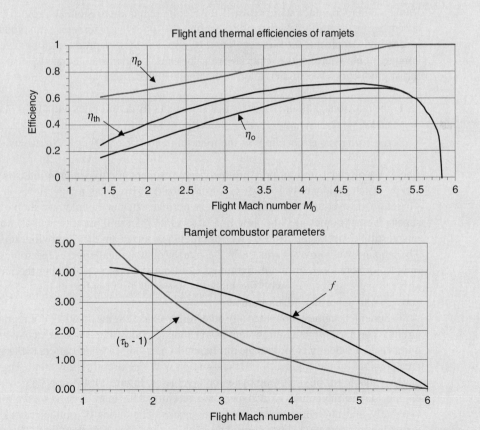

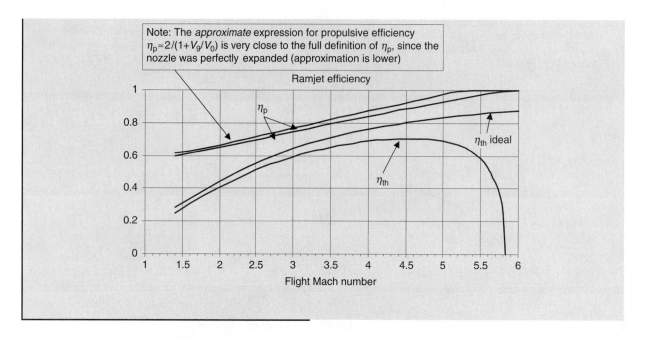

Note: The *approximate* expression for propulsive efficiency $\eta_p \approx 2/(1+V_9/V_0)$ is very close to the full definition of η_p, since the nozzle was perfectly expanded (approximation is lower)

11.15.1 Supersonic Combustion Ramjet

Ramjets cannot produce static thrust. They require ram compression, i.e., some forward flight speed, before these simple airbreathing engines (or "ducts with burners") can produce thrust. It is seemingly ironic that the same engines (i.e., ramjets) *cease* to produce thrust at very high ram compressions! There are two aspects of ram compression that, at high speeds, are detrimental to thrust production. We studied them in Example 11.13. These are

1. the inlet total pressure recovery that exponentially deteriorates with flight Mach number, and

2. the rising gas temperature in the inlet that cuts back $(\Delta T_t)_{burner}$ to eventually zero

The worst offender in total pressure recovery of supersonic/hypersonic inlets is the normal shock, which incidentally is also responsible for the rising gas temperatures in the burner. Therefore, if we could only do away with the normal shock in the inlet, we should be getting a super efficient ramjet. But the flow in a supersonic inlet without a (terminal) normal shock would still be supersonic! How do we burn fuel in an airstream that is moving supersonically? The short answer is *not very efficiently*! Despite the obvious challenges, supersonic combustion ramjets, or scramjets, are born. Scramjets hold out the promise of being the most efficient airbreathing engines, i.e., with the highest (fuel)-specific impulse, at flight Mach numbers above ~6 to suborbital Mach numbers.

Scramjet engines, more than any other airbreathing engine, need to be integrated with the aircraft. The need for integration stem from a long forebody that is needed at hypersonic Mach numbers to efficiently compress the air. In addition, a long forebody offers the largest capture area possible for the engine, A_0. Aft integration with the aircraft allows for a large area ratio nozzle suitable for high altitude hypersonic vehicles. Figure 11.45 shows a generic scramjet engine that is integrated in a hypersonic aircraft. The inlet achieves compression both externally and internally through oblique shocks. The series of oblique shock reflections inside the inlet create a shock train, known as the *isolator*. The flow that emerges from the isolator is supersonic, say Mach 3.0, as it enters the combustor. Achieving efficient combustion at supersonic speeds is a challenge.

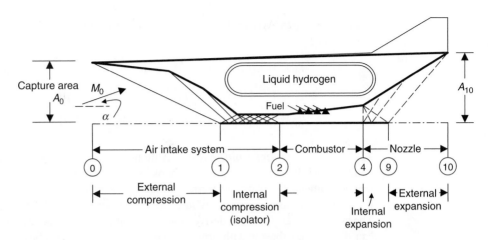

■ FIGURE 11.45   Typical arrangement of a generic scramjet engine on a hypersonic aircraft

Fuel injection, atomization, vaporization, mixing, and chemical reaction timescales should, by necessity, be short. If air is moving at Mach 3 where speed of sound is ~333 m/s, it traverses 1 m in ~1 ms. This is just an indication of the convective or residence timescale in the combustor. Hydrogen offers these qualities. For example, hydrogen offers ~1/10 of the chemical reaction timescale of hydrocarbon fuels, as discussed in Chapter 6. Additionally, hydrogen is in a cryogenic state as a liquid; therefore, it offers aircraft and engine structure regenerative cooling opportunities.

By accepting high Mach numbers inside the scramjet engine, the static temperature of the gas will be lower throughout the engine and thus the burner is allowed to release heat in the combustor that is needed for propulsion. The h–s diagram of the scramjet cycle is shown in Fig. 11.46.

11.15.1.1 Inlet Analysis.   Scramjet inlets are integrated with the aircraft forebody, involving several external and internal oblique shocks. In addition, the vehicle's angle of attack, α, impacts the inlet recovery through oblique shock waves' angles. For a given forebody shape, we may use shock-expansion theory (discussed in Chapter 2) to calculate the waves' orientations and the associated total pressure recovery. In the absence of the detailed aircraft forebody geometry, or in the preliminary phase, we may assume an inlet recovery or resort to

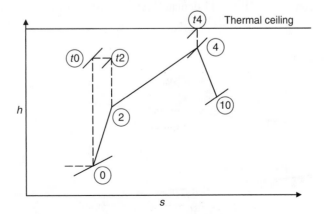

■ FIGURE 11.46   Static and stagnation states of gas in a scramjet engine

standards such as AIA or MIL-E-5008B. We listed these in Chapter 5, repeated here for convenience.

$$\pi_d = 1 - 0.1(M_0 - 1)^{1.5} \qquad 1 < M_0 \qquad \text{AIA- Standard} \tag{5.34}$$

$$\pi_d = 1 - 0.075(M_0 - 1)^{1.35} \quad 1 < M_0 < 5 \quad \text{MIL-E5008B} \tag{5.35}$$

$$\pi_d = 800/(M_0^4 + 935) \qquad 5 < M_0 \qquad \text{MIL-E-5008B} \tag{5.36}$$

11.15.1.2 *Scramjet Combustor.*

Although the detail design and analysis of supersonic combustion is beyond the scope of this book, we can still present global models and approaches that are amenable to closed-form solution. Supersonic branch of Rayleigh flow, i.e., a frictionless, constant-area duct with heating is a possible model for a scramjet combustor. The fact that the analysis was derived for a constant-area duct is not a major limitation, since we can divide a combustor with area change into a series of combustors with constant areas. Rayleigh flow has been detailed in Chapter 2 and we will not review it here. However, we may offer to analyze a variable-area combustor that maintains a constant pressure. From cycle analysis viewpoint, constant-pressure combustion is advantageous to cycle efficiency and thus we consider it here further. Figure 11.47 shows a frictionless duct with heat exchange, but with constant static pressure. We assume the gas is calorically perfect and the flow is steady and one-dimensional. The contribution of fuel mass flow rate to the gas in the duct is small; therefore, combustion is treated as heat transfer through the wall. The problem statement specifies the inlet condition and heat transfer rate (or fuel flow rate or fuel-to-air ratio) and assumes a constant static pressure in a duct with a variable area. The purpose of the analysis is to calculate the duct exit flow condition as well as the area A_4.

We apply conservation principles to the slab of fluid, as shown. The continuity requires

$$\rho A V = (\rho + d\rho)(V + dV)(A + dA) \quad \Longrightarrow \quad \frac{dV}{V} + \frac{d\rho}{\rho} + \frac{dA}{A} = 0 \tag{11.111}$$

The balance of x-momentum gives

$$\dot{m}(V + dV) - \dot{m}V = pA - p(A + dA) + p\,dA \equiv 0 \quad \Longrightarrow \quad dV = 0 \tag{11.112}$$

Equation 11.112 signifies a constant-velocity flow, i.e.,

$$V_4 = V_2 \tag{11.113}$$

Therefore, continuity Eq. 11.111 demands $\rho A = $ constant, or area is inversely proportional to density.

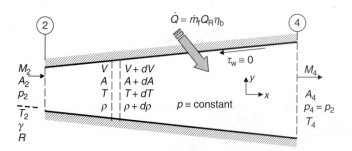

■ **FIGURE 11.47**   A frictionless, constant-pressure combustor

The energy balance written for the duct gives

$$h_{t4} - h_{t2} = \frac{\dot{Q}}{\dot{m}} = q = f Q_R \eta_b \qquad \textbf{(11.114-a)}$$

Since with constant velocity, the kinetic energy (per unit mass) remains constant, then

$$h_{T4} - h_{t2} = h_4 - h_2 \qquad \textbf{(11.114-b)}$$

$$T_4 = T_2 + q/c_p = T_2 + f Q_R \eta_b / c_p \qquad \textbf{(11.115)}$$

The RHS of Eq. 11.115 is known (as an input to the problem), which then establishes the exit static temperature T_4. Since static pressure is constant, we can calculate the exit density ρ_4. Also, the area ratio follows the inverse of the density ratio, i.e.,

$$A_4 = A_2(\rho_2/\rho_4) \qquad \textbf{(11.116)}$$

which establishes the exit flow area. The exit static temperature calculates the exit speed of sound and Mach number (since $V = $ constant).

We may calculate a critical heat flux q^* (or critical fuel-to-air ratio f^*) that will choke the duct at its exit. Equation 11.115 is written as

$$\frac{T_4}{T_2} = 1 + \frac{q}{c_p T_2} \qquad \textbf{(11.117)}$$

Also, Mach number ratio M_4/M_2 is related to the ratio of speeds of sound following

$$\frac{M_4}{M_2} = \frac{V_4 a_2}{V_2 a_4} = \frac{a_2}{a_4} = \sqrt{\frac{T_2}{T_4}} = \frac{1}{\sqrt{1 + \dfrac{q}{c_p T_2}}} \qquad \textbf{(11.118)}$$

For a choked exit, $M_4 = 1$; therefore, the critical heat flux q^* in terms of duct inlet flow conditions is expressed as

$$\frac{q^*}{c_p T_2} = M_2^2 - 1 \qquad \textbf{(11.118-a)}$$

In terms of fuel-to-air ratio, we may express q^* as $f^* Q_R \eta_b$ to get

$$f^* \approx \frac{c_p T_2}{Q_R \eta_b}(M_2^2 - 1) \qquad \textbf{(11.118-b)}$$

With calculated exit Mach number, we may arrive at the total pressure at the exit, following

$$p_{t4} = p_2\left(1 + \frac{\gamma - 1}{2}M_4^2\right)^{\frac{\gamma}{\gamma-1}}$$

An input to this analysis is the burner efficiency η_b. However, combustion efficiency in supersonic streams is not as well established or understood as the conventional low-speed burners. Mixing efficiency in supersonic shear layers, without chemical reaction, forms the foundation of scramjet combustion efficiency. A supersonic stream mixes with a lower speed stream (in our case the fuel) along a shear layer by vortex formations and supersonic wave interactions. These interactions require space along the shear layer or flow direction. Consequently, burner efficiency is a function of the combustor length and continually grows with distance along the scramjet combustor. Burrows and Kurkov (1973) may be consulted for some data and analysis related to supersonic combustion of hydrogen in vitiated air. Heiser et al. (1994) should be consulted for detailed discussion of hypersonic airbreathing propulsion.

11.15.1.3 Scramjet Nozzle. We have discussed hypersonic nozzles, to some extent, in Chapter 5. The large area ratios needed for high-speed, high-altitude flight prompted the use of aircraft underbody as the expansion ramp for the scramjet nozzle (shown in Fig. 11.45). For perfect expansion, we may approximate the area ratio requirements, A_{10}/A_9, by one-dimensional gas dynamic equations. Severe overexpansion at lower altitudes causes a shock to appear on the aft underbody of the vehicle. The complicating factors in the analysis of hypersonic airbreathing nozzles are similar to those in the rocket nozzles, i.e., continuing chemical reaction in the nozzle, flow separation, cooling, and flow unsteadiness, among others. Simple nozzle design based on the method of characteristics is a classical approach that students in gas dynamics learn. However, developing robust, high-fidelity computational fluid dynamics codes for direct numerical simulation of viscous reacting flows is the most powerful tool that researchers and industry are undertaking.

11.16 Rocket-Based Airbreathing Propulsion

Integrating rockets with ramjets has long been the solution of overcoming ramjets lack of takeoff capability. A ram-rocket may be a configuration similar to the one shown in Fig. 11.48.

A fuel-rich solid propellant rocket provides the takeoff thrust. The air intake is sealed off until forward speed of the aircraft can produce the needed ram compression for the ramjet combustor. The rocket motor serves as the gas generator for the ramjet, i.e., combustion gases from the fuel-rich solid propellant in the rocket motor mix with the air and the mixture is combusted in the ramjet burner to produce thrust. Variations of this scheme, such as a separate ramjet fuel, may be used in a gas generator ram-rocket. There is no new theory to be presented here, at least at the preliminary level, i.e., we have developed the tools in this book to analyze both components individually and combined. A ram-rocket configuration where the rocket provides takeoff thrust is shown in Fig. 11.49. The ramjet fuel is injected in the airstream for sustained thrust, as shown.

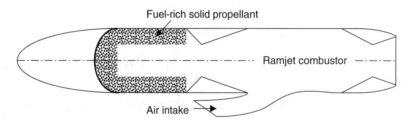

■ **FIGURE 11.48** **Schematic drawing of a gas generator ram-rocket**

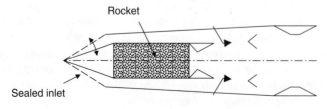

■ **FIGURE 11.49** **A ram-rocket configuration**

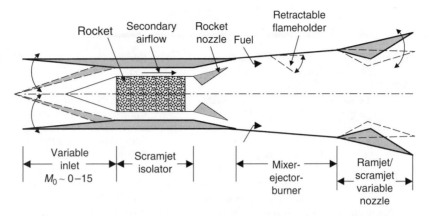

■ **FIGURE 11.50** Concept in RBCC propulsion

A rocket placed in a duct will draw air in and through mixing of the cold and hot gases will enhance its thrust level. This is the ejector principle. Since the propulsion is based on both the rocket and assisted by the secondary air mixing through an inlet, this may be referred to as an *airbreathing rocket*. The duct that follows the rocket nozzle is then the mixer-ejector, which may also serve as the combustor of a conventional or scramjet engine for combined operation. Now, we have clearly passed the airbreathing and rocket propulsion boundaries and entered into the realm of rocket-based combined cycle (RBCC) propulsion (see Fig. 11.50). This is the future of hypersonic flight from takeoff to orbit. It clearly is an exciting time to be a propulsion engineer/researcher.

11.17   Summary

High-speed atmospheric flight benefits from airbreathing engines. The oxygen in the air is the *infinite* oxidizer supply that combustion engines need. Requirements of takeoff to earth orbit necessitates multitude of airbreathing and rocket engines. Interestingly, the synergy of rocket-based combined cycle brings about the promise of efficient, cost-effective, and reliable hypersonic propulsion.

The unifying figure of merit in propulsion systems is the specific impulse. Although chemical rockets' specific impulse is in the range of 200–500 s and airbreathing engines an order of magnitude higher (i.e., $\sim$2000–10,000 s), chemical rockets produce thrust that is independent of vehicle speed. The airbreathing engines lose their advantage as flight Mach number increases into hypersonic regime. Chemical rockets come with liquid, solid, hybrid, or gaseous propellants. Liquid propellant rockets require a feed system that involves either a pressurized feed system with regulators and valves or a gas generator-driven turbopump system. These systems are heavy and add to the propulsion mass fraction and cost. In contrast, solid propellant rockets have no feed system and are, thus, lightweight and low cost and often are used as strap-on boosters. The controllability of thrust in a liquid rocket engine is the clear advantage that it has over a solid rocket motor. The ejector concept of airbreathing rockets combined with the dual-mode subsonic/supersonic combustion ramjets forms the foundation of a promising new engine that is suitable for continuous operation from runway to orbit.

References

1. Burrows, M.C. and Kurkov, A.P., "Analytical and Experimental Study of Supersonic Combustion of Hydrogen in a Vitiated Airstream," NASA TMX-2628, 1973.

2. "Explosive Hazard Classification Procedures," DOD, U.S. Army Technical Bulletin TB 700-2, 1989.

3. Forward, R.L., "Advanced Propulsion Systems," Chapter 11 in *Space Propulsion Analysis and Design*, Humble, R.W., Henry, G.N., and Larson, W.J. (editors), McGraw-Hill, New York, 1995.

4. Gordon, S. and McBride, S., "Computer Program for Calculation of Complex Chemical Equilibrium Compositions and Applications, Vol. 1: Analysis," October 1994 and "Vol. 2: User Manual and Program Description," NASA Reference Publication 1311, 1996.

5. Heiser, H.W., Pratt, D.T., Daley, D.H., and Mehta, U.B., *Hypersonic Airbreathing Propulsion*, AIAA, Inc., Reston, Virginia, 1994.

6. Heister, S., "Solid Rocket Motors," Chapter 6 in *Space Propulsion Analysis and Design*, Humble, R.W., Henry, G.N., and Larson, W.J. (editors), McGraw-Hill, New York, 1995.

7. Hill, P.G. and Peterson, C.R., *Mechanics and Thermodynamics of Propulsion*, 2nd edition, Addison-Wesley, Reading, Massachusetts, 1992.

8. Humble, R.W., Lewis, D., Bissell, W., and Sakheim, R., "Liquid Rocket Propulsion Systems," Chapter 5 in *Space Propulsion Analysis and Design*, Humble, R.W., Henry, G.N., and Larson, W.J. (editors), McGraw-Hill, New York, 1995.

9. Huzel, D. K. and Huang, D.H., *Design of Liquid Propellant Rocket Engines*, AIAA, Inc., Reston, Virginia, 1992.

10. Isakowitz, S.J., *International Reference Guide to Space Launch Systems*, 2nd edition, AIAA, Inc., Reston, Virginia, 1995.

11. Kerrebrock, J.L., Gas *Turbines and Aircraft Engines*, 2nd edition, MIT Press, Cambridge, Massachusetts, 1992.

12. Kubota, N., "Survey of Rocket Propellants and their Combustion Characteristics," Chapter 1 in *Fundamentals of Solid Propellant Combustion*, Kuo, K.K. and Summerfield, M. (editors), AIAA Progress Series in Astronautics and Aeronautics, Vol. 90, AIAA Inc., Reston, Virginia, 1984.

13. NASA SP-8089, "Liquid Rocket Engine Injectors," March 1976.

14. NASA SP-8120, "Liquid Rocket Engine Nozzles," July 1976.

15. Olson, W.T., "Recombination and Condensation Processes in High Area Ratio Nozzles," *Journal of American Rocket Society*, Vol. 32, No. 5, May 1962, pp.672–680.

16. Price, E.W., "Experimental Observations of Combustion Instability," Chapter 13 in *Fundamentals of Solid Propellant Combustion*, Kuo, K.K. and Summerfield, M. (editors), AIAA Progress Series in Astronautics and Aeronautics, Vol. 90, AIAA Inc., Reston, Virginia, 1984.

17. Razdan, M.K. and Kuo, K.K., "Erosive Burning of Solid Propellants," Chapter 10, in *Fundamentals of Solid Propellant Combustion*. Kuo, K.K. and Summerfield, M. (editors), AIAA Progress Series in Astronautics and Aeronautics, Vol. 90, AIAA Inc., Reston, Virginia, 1984.

18. Sutton, G.P. and Biblarz, O., *Rocket Propulsion Elements*, 7th edition , John Wiley & Sons, Inc. New York, 2001.

19. Witte, A.B. and Harper, E.Y., "Experimental Investigation and Empirical Correlation of Local Heat Transfer Rates in Rocket-Engine Thrust Chambers," Technical Report Number 32-244 Jet Propulsion Laboratory, California Institute of Technology, Pasadena, California, March 1962.

Problems

11.1   A booster stage produces 190,400 lb of thrust at sea level and 242,500 lb at altitude. Assuming the momentum thrust remains nearly constant with altitude, estimate the nozzle exit diameter D_2 for the booster.

11.2   RD-170 is a Russian launch vehicle booster. It has a nozzle throat diameter of 235.5 mm and exit diameter of 1430 mm. Propellant flow rate is 2393 kg/s and the nozzle exit static pressure is 7300 Pa. RD-170 develops a vacuum-specific impulse of 337 s. Calculate

(a) nozzle area ratio, A_2/A_{th}

(b) effective exhaust speed c (in m/s)

(c) vacuum thrust (in MN)

(d) pressure thrust in vacuum (in kN)

11.3   A rocket engine has a propellant mass flow rate of 1000 kg/s and an effective exhaust speed of $c = 3500$ m/s. Calculate

(a) rocket thrust F in kN

(b) specific impulse I_s in seconds

11.4   A rocket engine has a chamber pressure of $p_c = 1000$ psia and the throat area is $A_{th} = 1.5$ ft^2. Assuming that the nozzle is perfectly expanded with the gas ratio of specific heats $\gamma = 1.2$ and the ambient pressure of $p_0 = 14.7$ psia, calculate

(a) optimum thrust coefficient $C_{F,opt}$

(b) thrust F in lbf

(c) nozzle exit Mach number M_2

(d) nozzle area expansion ratio A_2/A_{th}

11.5 Consider a fuel-rich combustion of the hydrocarbon fuel known as RP-1 and oxygen in a liquid propellant rocket according to

$$2CH_{1.953} + n_{O_2}O_2 \longrightarrow CO_2 + (1.953/2)H_2O + CH_{1.953}$$

Calculate

(a) number of moles of oxygen n_{O_2}

(b) the mixture ratio r, i.e., the oxidizer-to-fuel ratio

(c) the molecular weight of the mixture of gases in the products of combustion

11.6 The average specific impulse of a chemical rocket is 360 s. The rocket is in a zero gravity vacuum flight. Calculate and graph vehicle terminal speed ΔV for the propellant fraction ζ, that ranges between 0.80 and 0.95.

11.7 A rocket is vertically launched and operates for 60 s and has a mass ratio of 0.05. The (mean) rocket specific impulse is 375 s. Assuming the average gravitational acceleration over the burn period is 9.70 m/s^2, calculate the terminal velocity of the rocket with and without gravitational effects. Neglect the effect of aerodynamic drag in both cases.

11.8 A rocket has a mass ratio of MR = 0.10 and a mean specific impulse of 365 s. The flight trajectory is described by a constant dynamic pressure of $q_0 = 50$ kPa. The mean drag coefficient is approximated to be 0.25, the vehicle initial mass is $m_0 = 100,000$ kg, and the vehicle (maximum) frontal cross-sectional area A_f is 5 m^2. For a burn time of 100 s, calculate the rocket terminal speed while neglecting gravitational effect.

11.9 In comparing the flight performance of a single-stage with a two-stage rocket, let us consider the two rockets have the same initial mass m_0, the same payload mass m_L, and the same overall structural mass m_s. The structural mass fraction ε, which is defined as the ratio of stage structural mass to the initial stage mass, is also assumed to be the same for the single-stage rocket and each of the two stages of the two-stage rocket. For the effective exhaust speed of 3500 m/s be constant for the single-stage and each stage of the two-stage rocket, calculate the terminal velocity for the two rockets in zero gravity and vacuum flight, for

$$m_0 = 100,000 \text{ kg}$$
$$m_L = 500 \text{ kg}$$
$$m_s = 10,000 \text{ kg}$$
$$\varepsilon_{single\text{-}stage} = 0.1$$
$$\varepsilon_{stage\text{-}1} = \varepsilon_{stage\text{-}2} = 0.1$$

11.10 A liquid propellant rocket uses a hydrocarbon fuel and oxygen as propellant. The heat of reaction for the combustion is $Q_R = 18.7$ MJ/kg. The specific impulse is 335 s and the flight speed is 2500 m/s. Neglecting the propellant kinetic power at the injector plate, calculate

(a) effective exhaust speed c in m/s

(b) propulsive efficiency η_p

(c) overall efficiency η_o

11.11 An injector plate uses an unlike impingement design. The fuel and oxidizer orifice discharge coefficients are $C_{df} = 0.80$ and $C_{do} = 0.75$. The static pressure drop across the injector plate for both oxidizer and fuel jets is the same, $\Delta p_f = \Delta p_o = 180$ kPa. The fuel and oxidizer densities are $\rho_f = 325$ kg/m^3 and $\rho_o = 1200$ kg/m^3 and the oxidizer-fuel mass ratio is $r = 3.0$. Calculate

(a) oxidizer-to-fuel orifice area ratio A_o/A_f

(b) oxidizer and fuel jet velocities v_o and v_f

11.12 A solid rocket motor has a design chamber pressure of 10 MPa, an end-burning grain with $n = 0.4$ and $r = 3$ cm/s at the design chamber pressure and design grain temperature of 15°C. The temperature sensitivity of the burning rate is $\sigma_p = 0.002/°$C, and chamber pressure sensitivity to initial grain temperature is $\pi_K = 0.005/°$C. The nominal effective burn time for the rocket is 120 s, i.e., at design conditions. Calculate

(a) the new chamber pressure and burning rate when the initial grain temperature is 75°C

(b) the corresponding reduction in burn time Δt_b in seconds

11.13 Extract the erosive burning parameter k from the data of Fig. 11.30 for the two solid propellants shown.

11.14 A regeneratively cooled rocket thrust chamber has its maximum heat flux of 15 MW/m^2 near its throat. The hot gas stagnation temperature is 3000 K and the local gas Mach number is assumed to be ~1.0. The gas mean molecular weight is MW = 23 kg/kmol and the ratio of specific heats is $\gamma = 1.24$. Calculate

(a) gas static temperature T_g (in K)

(b) gas speed near the throat (in m/s)

(c) gas-side film coefficient h_g for $T_{wg} \sim 1000$ K

11.15 A rocket combustion chamber is designed for a chamber pressure of $p_c = 50$ MPa. The combustion gas has a ratio of specific heats $\gamma = 1.25$. If this rocket is to operate between sea level and 200,000 ft altitude, calculate the range of area ratios in the nozzle that will lead to perfect expansion at all altitudes. Assume isentropic flow in the nozzle.

11.16 The propellant flow rate in a chemical nozzle is 10,000 kg/s, the nozzle exhaust speed is 2200 m/s, and the nozzle exit pressure is $p_2 = 0.01$ atm. Assuming the nozzle exit diameter is $D_2 = 2$ m, calculate

(a) the pressure thrust (in MN) at sea level

(b) the effective exhaust speed c (in m/s) at sea level

11.17 A solid propellant rocket motor uses a composite propellant with 16% aluminum. The same propellant with 18% aluminum enhances the combustion temperature by 5.7%. Assuming in both cases that the solid particles are fully accelerated (i.e., $V_s = V_g$) in the nozzle but the solid temperature remains constant (i.e., T_s = constant), calculate the ratio of specific impulse in the two cases. Aluminum specific heat is $c_s = 903$ J/kg · K, and the specific heat at constant pressure for the gas is $c_{pg} = 2006$ J/kg · K.

11.18 The coefficient of linear thermal expansion for a solid propellant grain is $1.5 \times 10^{-4}/{}^\circ$C. Calculate the change of length ΔL for a 1 m long propellant grain that experiences a temperature change from -30°C to $+70^\circ$C.

11.19 A ramjet has a maximum temperature $T_{t4} = 2500$ K. The inlet total pressure recovery π_d varies with flight Mach number according to

$$\pi_d = 1 - 0.075(M_0 - 1)^{1.35}$$

Ramjet burns hydrogen fuel with $Q_R = 120,000$ kJ/kg and combustor efficiency and total pressure ratio are $\eta_b = 0.99$ and $\pi_b = 0.95$, respectively. The nozzle is perfectly expanded and has a total pressure ratio $\pi_n = 0.90$. Assuming a calorically perfect gas with $\gamma = 1.4$ and $R = 287$ J/kg · K, use a spreadsheet to calculate the ramjet (fuel)-specific impulse, propulsive, thermal, and overall efficiencies over a range of flight Mach

number starting at $M_0 = 3$ up until ramjet ceases to produce any thrust. Altitude pressure and temperature are 15 kPa and 250 K, respectively.

11.20 Consider a scramjet in a Mach 6 flight. The fuel for this engine is hydrogen with $Q_R = 120,000$ kJ/kg. The inlet uses multiple oblique shocks with a total pressure recovery following MIL-E-5008B standards for $M_0 > 5$, i.e.,

$$\pi_d = 800/(M_0^4 + 935)$$

The combustor entrance Mach number is $M_2 = 2.6$. Use frictionless, constant-pressure heating, i.e., $C_f = 0$ and $p_4 = p_2$, to simulate the combustor with combustor exit Mach number $M_4 = 1.0$. All component parameters and gas constants are shown in the schematic drawing below.
Calculate

(a) Inlet static temperature ratio T_2/T_0

(b) combustor exit temperature T_4 (in K)

(c) combustor static pressure ratio p_4/p_2

(d) fuel-to-air ratio f

(e) nozzle exit Mach number M_{10}

(f) nondimensional ram drag D_{ram}/p_0A_1 (note that $A_0 = A_1$)

(g) nondimensional gross thrust F_g/p_0A_1

(h) fuel-specific impulse I_s in seconds

(i) combustor area ratio A_4/A_2

(j) nozzle area ratio A_{10}/A_4

(k) thermal efficiency

(l) propulsive efficiency

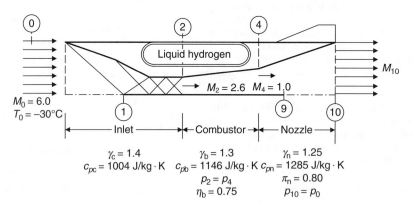

■ **FIGURE P11.20**

A P P E N D I C E S

U.S. Standard Atmosphere

U.S. Standard Atmosphere U.S. Units

In this table **alt** is altitude in thousands of feet.
sigma is density divided by sea-level density.
delta is pressure divided by sea-level pressure.
theta is temperature divided by sea-level temperature.
temp is temperature in degrees Rankine.
press is pressure in pounds per square foot.
dens is density in slugs per cubic foot.
a is the speed of sound in feet per second.
visc is viscosity in 10^{-6} slugs per foot-second.
k.visc is kinematic viscosity in 10^{-6} square feet per second.
ratio is speed of sound divided by kinematic viscosity times 10^6.

alt	sigma	delta	theta	temp	press	dens	a	visc	k.visc	ratio
−1	1.0296	1.0367	1.0069	522.2	2193.8	0.0024472	1120.3	0.376	1.54E−4	7.30
0	1.0000	1.0000	1.0000	518.7	2116.2	0.0023769	1116.5	0.374	1.57E−4	7.10
1	0.9711	0.9644	0.9931	515.1	2040.9	0.0023081	1112.6	0.372	1.61E−4	6.91
2	0.9428	0.9298	0.9863	511.5	1967.7	0.0022409	1108.7	0.370	1.65E−4	6.72
3	0.9151	0.8963	0.9794	508.0	1896.7	0.0021752	1104.9	0.368	1.69E−4	6.54
4	0.8881	0.8637	0.9725	504.4	1827.7	0.0021109	1101.0	0.366	1.73E−4	6.36
5	0.8617	0.8321	0.9656	500.8	1760.9	0.0020482	1097.1	0.364	1.78E−4	6.18
6	0.8359	0.8014	0.9588	497.3	1696.0	0.0019869	1093.2	0.362	1.82E−4	6.01
7	0.8107	0.7717	0.9519	493.7	1633.1	0.0019270	1089.3	0.360	1.87E−4	5.84
8	0.7861	0.7429	0.9450	490.2	1572.1	0.0018685	1085.3	0.358	1.91E−4	5.67
9	0.7621	0.7149	0.9381	486.6	1512.9	0.0018113	1081.4	0.355	1.96E−4	5.51
10	0.7386	0.6878	0.9313	483.0	1455.6	0.0017555	1077.4	0.353	2.01E−4	5.35
11	0.7157	0.6616	0.9244	479.5	1400.1	0.0017011	1073.4	0.351	2.07E−4	5.20
12	0.6933	0.6362	0.9175	475.9	1346.2	0.0016480	1069.4	0.349	2.12E−4	5.05
13	0.6715	0.6115	0.9107	472.3	1294.1	0.0015961	1065.4	0.347	2.18E−4	4.90
14	0.6502	0.5877	0.9038	468.8	1243.6	0.0015455	1061.4	0.345	2.23E−4	4.75
15	0.6295	0.5646	0.8969	465.2	1194.8	0.0014962	1057.4	0.343	2.29E−4	4.61
16	0.6092	0.5422	0.8901	461.7	1147.5	0.0014480	1053.3	0.341	2.35E−4	4.47
17	0.5895	0.5206	0.8832	458.1	1101.7	0.0014011	1049.2	0.339	2.42E−4	4.34
18	0.5702	0.4997	0.8763	454.5	1057.5	0.0013553	1045.1	0.337	2.48E−4	4.21
19	0.5514	0.4795	0.8695	451.0	1014.7	0.0013107	1041.0	0.335	2.55E−4	4.08
20	0.5332	0.4599	0.8626	447.4	973.3	0.0012673	1036.9	0.332	2.62E−4	3.95
21	0.5153	0.4410	0.8558	443.9	933.3	0.0012249	1032.8	0.330	2.70E−4	3.83
22	0.4980	0.4227	0.8489	440.3	894.6	0.0011836	1028.6	0.328	2.77E−4	3.71
23	0.4811	0.4051	0.8420	436.7	857.2	0.0011435	1024.5	0.326	2.85E−4	3.59
24	0.4646	0.3880	0.8352	433.2	821.2	0.0011043	1020.3	0.324	2.93E−4	3.48
25	0.4486	0.3716	0.8283	429.6	786.3	0.0010663	1016.1	0.322	3.02E−4	3.37
26	0.4330	0.3557	0.8215	426.1	752.7	0.0010292	1011.9	0.319	3.10E−4	3.26
27	0.4178	0.3404	0.8146	422.5	720.3	0.0009931	1007.7	0.317	3.19E−4	3.15
28	0.4031	0.3256	0.8077	419.0	689.0	0.0009580	1003.4	0.315	3.29E−4	3.05
29	0.3887	0.3113	0.8009	415.4	658.8	0.0009239	999.1	0.313	3.39E−4	2.95
30	0.3747	0.2975	0.7940	411.8	629.7	0.0008907	994.8	0.311	3.49E−4	2.85
31	0.3611	0.2843	0.7872	408.3	601.6	0.0008584	990.5	0.308	3.59E−4	2.76
32	0.3480	0.2715	0.7803	404.7	574.6	0.0008270	986.2	0.306	3.70E−4	2.66
33	0.3351	0.2592	0.7735	401.2	548.5	0.0007966	981.9	0.304	3.82E−4	2.57
34	0.3227	0.2474	0.7666	397.6	523.5	0.0007670	977.5	0.302	3.93E−4	2.48
35	0.3106	0.2360	0.7598	394.1	499.3	0.0007382	973.1	0.300	4.06E−4	2.40
36	0.2988	0.2250	0.7529	390.5	476.1	0.0007103	968.7	0.297	4.19E−4	2.31
37	0.2852	0.2145	0.7519	390.0	453.9	0.0006780	968.1	0.297	4.38E−4	2.21
38	0.2719	0.2044	0.7519	390.0	432.6	0.0006463	968.1	0.297	4.59E−4	2.11
39	0.2592	0.1949	0.7519	390.0	412.4	0.0006161	968.1	0.297	4.82E−4	2.01
40	0.2471	0.1858	0.7519	390.0	393.1	0.0005873	968.1	0.297	5.06E−4	1.91
41	0.2355	0.1771	0.7519	390.0	374.7	0.0005598	968.1	0.297	5.30E−4	1.83
42	0.2245	0.1688	0.7519	390.0	357.2	0.0005336	968.1	0.297	5.56E−4	1.74
43	0.2140	0.1609	0.7519	390.0	340.5	0.0005087	968.1	0.297	5.84E−4	1.66
44	0.2040	0.1534	0.7519	390.0	324.6	0.0004849	968.1	0.297	6.12E−4	1.58
45	0.1945	0.1462	0.7519	390.0	309.4	0.0004623	968.1	0.297	6.42E−4	1.51

(Continued)

alt	sigma	delta	theta	temp	press	dens	a	visc	k.visc	ratio
46	0.1854	0.1394	0.7519	390.0	295.0	0.0004407	968.1	0.297	6.74E−4	1.44
47	0.1767	0.1329	0.7519	390.0	281.2	0.0004201	968.1	0.297	7.07E−4	1.37
48	0.1685	0.1267	0.7519	390.0	268.1	0.0004005	968.1	0.297	7.41E−4	1.31
49	0.1606	0.1208	0.7519	390.0	255.5	0.0003817	968.1	0.297	7.78E−4	1.24
50	0.1531	0.1151	0.7519	390.0	243.6	0.0003639	968.1	0.297	8.16E−4	1.19
51	0.1460	0.1097	0.7519	390.0	232.2	0.0003469	968.1	0.297	8.56E−4	1.13
52	0.1391	0.1046	0.7519	390.0	221.4	0.0003307	968.1	0.297	8.98E−4	1.08
53	0.1326	0.0997	0.7519	390.0	211.0	0.0003153	968.1	0.297	9.42E−4	1.03
54	0.1264	0.0951	0.7519	390.0	201.2	0.0003006	968.1	0.297	9.88E−4	0.98
55	0.1205	0.0906	0.7519	390.0	191.8	0.0002865	968.1	0.297	1.04E−3	0.93
56	0.1149	0.0864	0.7519	390.0	182.8	0.0002731	968.1	0.297	1.09E−3	0.89
57	0.1096	0.0824	0.7519	390.0	174.3	0.0002604	968.1	0.297	1.14E−3	0.85
58	0.1044	0.0785	0.7519	390.0	166.2	0.0002482	968.1	0.297	1.20E−3	0.81
59	0.0996	0.0749	0.7519	390.0	158.4	0.0002367	968.1	0.297	1.25E−3	0.77
60	0.0949	0.0714	0.7519	390.0	151.0	0.0002256	968.1	0.297	1.32E−3	0.74
61	0.0905	0.0680	0.7519	390.0	144.0	0.0002151	968.1	0.297	1.38E−3	0.70
62	0.0863	0.0649	0.7519	390.0	137.3	0.0002050	968.1	0.297	1.45E−3	0.67
63	0.0822	0.0618	0.7519	390.0	130.9	0.0001955	968.1	0.297	1.52E−3	0.64
64	0.0784	0.0590	0.7519	390.0	124.8	0.0001864	968.1	0.297	1.59E−3	0.61
65	0.0747	0.0562	0.7519	390.0	118.9	0.0001777	968.1	0.297	1.67E−3	0.58

U.S. Standard Atmosphere SI Units

In this table **alt** is altitude in kilometers.
sigma is density divided by sea-level density.
delta is pressure divided by sea-level pressure.
theta is temperature divided by sea-level temperature.
temp is temperature in Kelvins.
press is pressure in Newtons per square meter (or Pa).
dens is density in kilograms per cubic meter.
a is the speed of sound in meters per second.
visc is viscosity in 10^{-6} kilograms per meter-second.
k.visc is kinematic viscosity in square meters per second.

alt km	sigma	delta	theta	temp K	press Pa	dens kg/m^3	a m/s	visc kg/m-s	k.visc m^2/s
−2	1.2067E+0	1.2611E+0	1.0451	301.2	1.278E+5	1.478E+0	347.9	18.51	1.25E−5
0	1.0000E+0	1.0000E+0	1.0000	288.1	1.013E+5	1.225E+0	340.3	17.89	1.46E−5
2	8.2168E−1	7.8462E−1	0.9549	275.2	7.950E+4	1.007E+0	332.5	17.26	1.71E−5
4	6.6885E−1	6.0854E−1	0.9098	262.2	6.166E+4	8.193E−1	324.6	16.61	2.03E−5
6	5.3887E−1	4.6600E−1	0.8648	249.2	4.722E+4	6.601E−1	316.5	15.95	2.42E−5
8	4.2921E−1	3.5185E−1	0.8198	236.2	3.565E+4	5.258E−1	308.1	15.27	2.90E−5
10	3.3756E−1	2.6153E−1	0.7748	223.3	2.650E+4	4.135E−1	299.5	14.58	3.53E−5
12	2.5464E−1	1.9146E−1	0.7519	216.6	1.940E+4	3.119E−1	295.1	14.22	4.56E−5
14	1.8600E−1	1.3985E−1	0.7519	216.6	1.417E+4	2.279E−1	295.1	14.22	6.24E−5
16	1.3589E−1	1.0217E−1	0.7519	216.6	1.035E+4	1.665E−1	295.1	14.22	8.54E−5
18	9.9302E−2	7.4662E−2	0.7519	216.6	7.565E+3	1.216E−1	295.1	14.22	1.17E−4
20	7.2578E−2	5.4569E−2	0.7519	216.6	5.529E+3	8.891E−2	295.1	14.22	1.60E−4
22	5.2660E−2	3.9945E−2	0.7585	218.6	4.047E+3	6.451E−2	296.4	14.32	2.22E−4
24	3.8316E−2	2.9328E−2	0.7654	220.6	2.972E+3	4.694E−2	297.7	14.43	3.07E−4
26	2.7964E−2	2.1597E−2	0.7723	222.5	2.188E+3	3.426E−2	299.1	14.54	4.24E−4
28	2.0470E−2	1.5950E−2	0.7792	224.5	1.616E+3	2.508E−2	300.4	14.65	5.84E−4
30	1.5028E−2	1.1813E−2	0.7861	226.5	1.197E+3	1.841E−2	301.7	14.75	8.01E−4
32	1.1065E−2	8.7740E−3	0.7930	228.5	8.890E+2	1.355E−2	303.0	14.86	1.10E−3
34	8.0709E−3	6.5470E−3	0.8112	233.7	6.634E+2	9.887E−3	306.5	15.14	1.53E−3
36	5.9245E−3	4.9198E−3	0.8304	239.3	4.985E+2	7.257E−3	310.1	15.43	2.13E−3
38	4.3806E−3	3.7218E−3	0.8496	244.8	3.771E+2	5.366E−3	313.7	15.72	2.93E−3
40	3.2615E−3	2.8337E−3	0.8688	250.4	2.871E+2	3.995E−3	317.2	16.01	4.01E−3
42	2.4445E−3	2.1708E−3	0.8880	255.9	2.200E+2	2.995E−3	320.7	16.29	5.44E−3
44	1.8438E−3	1.6727E−3	0.9072	261.4	1.695E+2	2.259E−3	324.1	16.57	7.34E−3
46	1.3992E−3	1.2961E−3	0.9263	266.9	1.313E+2	1.714E−3	327.5	16.85	9.83E−3
48	1.0748E−3	1.0095E−3	0.9393	270.6	1.023E+2	1.317E−3	329.8	17.04	1.29E−2
50	8.3819E−4	7.8728E−4	0.9393	270.6	7.977E+1	1.027E−3	329.8	17.04	1.66E−2
52	6.5759E−4	6.1395E−4	0.9336	269.0	6.221E+1	8.055E−4	328.8	16.96	2.10E−2
54	5.2158E−4	4.7700E−4	0.9145	263.5	4.833E+1	6.389E−4	325.4	16.68	2.61E−2
56	4.1175E−4	3.6869E−4	0.8954	258.0	3.736E+1	5.044E−4	322.0	16.40	3.25E−2
58	3.2344E−4	2.8344E−4	0.8763	252.5	2.872E+1	3.962E−4	318.6	16.12	4.07E−2
60	2.5276E−4	2.1668E−4	0.8573	247.0	2.196E+1	3.096E−4	315.1	15.84	5.11E−2
62	1.9647E−4	1.6468E−4	0.8382	241.5	1.669E+1	2.407E−4	311.5	15.55	6.46E−2
64	1.5185E−4	1.2439E−4	0.8191	236.0	1.260E+1	1.860E−4	308.0	15.26	8.20E−2

(Continued)

alt km	sigma	delta	theta	temp K	press Pa	dens kg/m^3	a m/s	visc kg/m-s	k.visc m^2/s
66	1.1668E−4	9.3354E−5	0.8001	230.5	9.459E+0	1.429E−4	304.4	14.97	1.05E−1
68	8.9101E−5	6.9593E−5	0.7811	225.1	7.051E+0	1.091E−4	300.7	14.67	1.34E−1
70	6.7601E−5	5.1515E−5	0.7620	219.6	5.220E+0	8.281E−5	297.1	14.38	1.74E−1
72	5.0905E−5	3.7852E−5	0.7436	214.3	3.835E+0	6.236E−5	293.4	14.08	2.26E−1
74	3.7856E−5	2.7635E−5	0.7300	210.3	2.800E+0	4.637E−5	290.7	13.87	2.99E−1
76	2.8001E−5	2.0061E−5	0.7164	206.4	2.033E+0	3.430E−5	288.0	13.65	3.98E−1
78	2.0597E−5	1.4477E−5	0.7029	202.5	1.467E+0	2.523E−5	285.3	13.43	5.32E−1
80	1.5063E−5	1.0384E−5	0.6893	198.6	1.052E+0	1.845E−5	282.5	13.21	7.16E−1
82	1.0950E−5	7.4002E−6	0.6758	194.7	7.498E−1	1.341E−5	279.7	12.98	9.68E−1
84	7.9106E−6	5.2391E−6	0.6623	190.8	5.308E−1	9.690E−6	276.9	12.76	1.32E+0
86	5.6777E−6	3.6835E−6	0.6488	186.9	3.732E−1	6.955E−6	274.1	12.53	1.80E+0

APPENDIX B

Isentropic Table

sentropic tables are produced, for calorically perfect gas (i.e., constant γ), based on the following equations:

$$\frac{T_t}{T} = 1 + \left(\frac{\gamma - 1}{2}\right)M^2 \tag{B.1}$$

$$\frac{p_t}{p} = \left(\frac{T_t}{T}\right)^{\frac{\gamma}{\gamma-1}} = \left[1 + \left(\frac{\gamma - 1}{2}\right)M^2\right]^{\frac{\gamma}{\gamma-1}} \tag{B.2}$$

$$\frac{\rho_t}{\rho} = \left(\frac{p_t}{p}\right)^{\frac{1}{\gamma}} = \left[1 + \left(\frac{\gamma - 1}{2}\right)M^2\right]^{\frac{1}{\gamma-1}} \tag{B.3}$$

$$\frac{A}{A^*} = \frac{1}{M}\left(\frac{1 + \frac{\gamma-1}{2}M^2}{\frac{\gamma+1}{2}}\right)^{\frac{\gamma+1}{2(\gamma-1)}} \tag{B.4}$$

$$\frac{T}{T^*} = \frac{\frac{\gamma+1}{2}}{1 + \left(\frac{\gamma-1}{2}\right)M^2} \tag{B.5}$$

$$\frac{p}{p^*} = \left[\frac{\frac{\gamma+1}{2}}{1 + \left(\frac{\gamma-1}{2}\right)M^2}\right]^{\frac{\gamma}{\gamma-1}} \tag{B.6}$$

$$\frac{\rho}{\rho^*} = \frac{p}{p^*}\bigg/\frac{T}{T^*} = \left[\frac{\frac{\gamma+1}{2}}{1 + \left(\frac{\gamma-1}{2}\right)M^2}\right]^{\frac{1}{\gamma-1}} \tag{B.7}$$

We derived these equations in Chapter 2.

Isentropic Table $\gamma = 1.4$

M	T_t/T	p_t/p	ρ_t/ρ	A/A^*	T/T^*	p/p^*	ρ/ρ^*
0.02	1.0001E+00	1.0003E+00	1.0002E+00	2.8942E+01	1.1999E+00	1.8924E+00	1.5771E+00
0.04	1.0003E+00	1.0011E+00	1.0008E+00	1.4481E+01	1.1996E+00	1.8908E+00	1.5762E+00
0.06	1.0007E+00	1.0025E+00	1.0018E+00	9.6658E+00	1.1991E+00	1.8882E+00	1.5746E+00
0.08	1.0013E+00	1.0045E+00	1.0032E+00	7.2615E+00	1.1985E+00	1.8845E+00	1.5724E+00
0.10	1.0020E+00	1.0070E+00	1.0050E+00	5.8218E+00	1.1976E+00	1.8797E+00	1.5696E+00
0.12	1.0029E+00	1.0101E+00	1.0072E+00	4.8643E+00	1.1966E+00	1.8740E+00	1.5661E+00
0.14	1.0039E+00	1.0138E+00	1.0098E+00	4.1823E+00	1.1953E+00	1.8672E+00	1.5621E+00
0.16	1.0051E+00	1.0180E+00	1.0128E+00	3.6727E+00	1.1939E+00	1.8594E+00	1.5574E+00
0.18	1.0065E+00	1.0229E+00	1.0163E+00	3.2779E+00	1.1923E+00	1.8506E+00	1.5522E+00
0.20	1.0080E+00	1.0283E+00	1.0201E+00	2.9635E+00	1.1905E+00	1.8409E+00	1.5463E+00
0.22	1.0097E+00	1.0343E+00	1.0244E+00	2.7076E+00	1.1885E+00	1.8302E+00	1.5399E+00
0.24	1.0115E+00	1.0409E+00	1.0290E+00	2.4955E+00	1.1863E+00	1.8185E+00	1.5329E+00
0.26	1.0135E+00	1.0481E+00	1.0341E+00	2.3173E+00	1.1840E+00	1.8060E+00	1.5254E+00
0.28	1.0157E+00	1.0560E+00	1.0397E+00	2.1655E+00	1.1815E+00	1.7926E+00	1.5173E+00
0.30	1.0180E+00	1.0644E+00	1.0456E+00	2.0350E+00	1.1788E+00	1.7783E+00	1.5086E+00
0.32	1.0205E+00	1.0735E+00	1.0520E+00	1.9218E+00	1.1759E+00	1.7633E+00	1.4995E+00
0.34	1.0231E+00	1.0833E+00	1.0588E+00	1.8229E+00	1.1729E+00	1.7474E+00	1.4898E+00
0.36	1.0259E+00	1.0937E+00	1.0661E+00	1.7358E+00	1.1697E+00	1.7308E+00	1.4797E+00
0.38	1.0289E+00	1.1048E+00	1.0738E+00	1.6587E+00	1.1663E+00	1.7134E+00	1.4691E+00
0.40	1.0320E+00	1.1166E+00	1.0819E+00	1.5901E+00	1.1628E+00	1.6953E+00	1.4580E+00
0.42	1.0353E+00	1.1290E+00	1.0905E+00	1.5289E+00	1.1591E+00	1.6766E+00	1.4465E+00
0.44	1.0387E+00	1.1422E+00	1.0996E+00	1.4740E+00	1.1553E+00	1.6573E+00	1.4345E+00
0.46	1.0423E+00	1.1561E+00	1.1092E+00	1.4246E+00	1.1513E+00	1.6373E+00	1.4222E+00
0.48	1.0461E+00	1.1708E+00	1.1192E+00	1.3801E+00	1.1471E+00	1.6168E+00	1.4094E+00
0.50	1.0500E+00	1.1862E+00	1.1297E+00	1.3398E+00	1.1429E+00	1.5958E+00	1.3963E+00
0.52	1.0541E+00	1.2024E+00	1.1407E+00	1.3034E+00	1.1384E+00	1.5743E+00	1.3828E+00
0.54	1.0583E+00	1.2194E+00	1.1522E+00	1.2703E+00	1.1339E+00	1.5523E+00	1.3690E+00
0.56	1.0627E+00	1.2373E+00	1.1643E+00	1.2403E+00	1.1292E+00	1.5299E+00	1.3549E+00
0.58	1.0673E+00	1.2560E+00	1.1768E+00	1.2130E+00	1.1244E+00	1.5072E+00	1.3405E+00
0.60	1.0720E+00	1.2755E+00	1.1898E+00	1.1882E+00	1.1194E+00	1.4841E+00	1.3258E+00
0.62	1.0769E+00	1.2959E+00	1.2034E+00	1.1656E+00	1.1143E+00	1.4607E+00	1.3108E+00
0.64	1.0819E+00	1.3173E+00	1.2176E+00	1.1451E+00	1.1091E+00	1.4370E+00	1.2956E+00
0.66	1.0871E+00	1.3396E+00	1.2322E+00	1.1265E+00	1.1038E+00	1.4131E+00	1.2801E+00
0.68	1.0925E+00	1.3628E+00	1.2475E+00	1.1096E+00	1.0984E+00	1.3890E+00	1.2645E+00
0.70	1.0980E+00	1.3871E+00	1.2633E+00	1.0944E+00	1.0929E+00	1.3647E+00	1.2487E+00
0.72	1.1037E+00	1.4124E+00	1.2797E+00	1.0806E+00	1.0873E+00	1.3402E+00	1.2327E+00
0.74	1.1095E+00	1.4387E+00	1.2967E+00	1.0681E+00	1.0815E+00	1.3157E+00	1.2165E+00
0.76	1.1155E+00	1.4661E+00	1.3143E+00	1.0570E+00	1.0757E+00	1.2911E+00	1.2002E+00

M	T_t/T	p_t/p	ρ_t/ρ	A/A^*	T/T^*	p/p^*	ρ/ρ^*
0.78	1.1217E+00	1.4947E+00	1.3325E+00	1.0470E+00	1.0698E+00	1.2665E+00	1.1838E+00
0.80	1.1280E+00	1.5243E+00	1.3514E+00	1.0382E+00	1.0638E+00	1.2418E+00	1.1673E+00
0.82	1.1345E+00	1.5552E+00	1.3709E+00	1.0305E+00	1.0578E+00	1.2172E+00	1.1507E+00
0.84	1.1411E+00	1.5873E+00	1.3910E+00	1.0237E+00	1.0516E+00	1.1925E+00	1.1340E+00
0.86	1.1479E+00	1.6207E+00	1.4118E+00	1.0179E+00	1.0454E+00	1.1680E+00	1.1173E+00
0.88	1.1549E+00	1.6553E+00	1.4333E+00	1.0129E+00	1.0391E+00	1.1436E+00	1.1006E+00
0.90	1.1620E+00	1.6913E+00	1.4555E+00	1.0089E+00	1.0327E+00	1.1192E+00	1.0838E+00
0.92	1.1693E+00	1.7287E+00	1.4784E+00	1.0056E+00	1.0263E+00	1.0950E+00	1.0670E+00
0.94	1.1767E+00	1.7675E+00	1.5020E+00	1.0031E+00	1.0198E+00	1.0710E+00	1.0502E+00
0.96	1.1843E+00	1.8078E+00	1.5264E+00	1.0014E+00	1.0132E+00	1.0471E+00	1.0334E+00
0.98	1.1921E+00	1.8496E+00	1.5515E+00	1.0003E+00	1.0066E+00	1.0234E+00	1.0167E+00
1.00	1.2000E+00	1.8929E+00	1.5774E+00	1.0000E+00	1.0000E+00	1.0000E+00	1.0000E+00
1.02	1.2081E+00	1.9379E+00	1.6041E+00	1.0003E+00	9.9331E−01	9.7679E−01	9.8336E−01
1.04	1.2163E+00	1.9846E+00	1.6316E+00	1.0013E+00	9.8658E−01	9.5382E−01	9.6679E−01
1.06	1.2247E+00	2.0330E+00	1.6599E+00	1.0029E+00	9.7982E−01	9.3112E−01	9.5030E−01
1.08	1.2333E+00	2.0831E+00	1.6891E+00	1.0051E+00	9.7302E−01	9.0870E−01	9.3390E−01
1.10	1.2420E+00	2.1351E+00	1.7191E+00	1.0079E+00	9.6618E−01	8.8656E−01	9.1759E−01
1.12	1.2509E+00	2.1890E+00	1.7500E+00	1.0113E+00	9.5932E−01	8.6473E−01	9.0139E−01
1.14	1.2599E+00	2.2449E+00	1.7818E+00	1.0153E+00	9.5244E−01	8.4321E−01	8.8531E−01
1.16	1.2691E+00	2.3028E+00	1.8145E+00	1.0198E+00	9.4554E−01	8.2201E−01	8.6935E−01
1.18	1.2785E+00	2.3628E+00	1.8481E+00	1.0248E+00	9.3861E−01	8.0113E−01	8.5353E−01
1.20	1.2880E+00	2.4250E+00	1.8827E+00	1.0304E+00	9.3168E−01	7.8060E−01	8.3784E−01
1.22	1.2977E+00	2.4894E+00	1.9183E+00	1.0366E+00	9.2473E−01	7.6041E−01	8.2231E−01
1.24	1.3075E+00	2.5560E+00	1.9549E+00	1.0432E+00	9.1777E−01	7.4057E−01	8.0692E−01
1.26	1.3175E+00	2.6251E+00	1.9925E+00	1.0504E+00	9.1080E−01	7.2108E−01	7.9170E−01
1.28	1.3277E+00	2.6967E+00	2.0311E+00	1.0581E+00	9.0383E−01	7.0195E−01	7.7664E−01
1.30	1.3380E+00	2.7707E+00	2.0708E+00	1.0663E+00	8.9686E−01	6.8318E−01	7.6175E−01
1.32	1.3485E+00	2.8474E+00	2.1116E+00	1.0750E+00	8.8989E−01	6.6478E−01	7.4704E−01
1.34	1.3591E+00	2.9269E+00	2.1535E+00	1.0842E+00	8.8292E−01	6.4674E−01	7.3250E−01
1.36	1.3699E+00	3.0091E+00	2.1965E+00	1.0940E+00	8.7596E−01	6.2907E−01	7.1815E−01
1.38	1.3809E+00	3.0942E+00	2.2407E+00	1.1042E+00	8.6901E−01	6.1177E−01	7.0398E−01
1.40	1.3920E+00	3.1823E+00	2.2861E+00	1.1149E+00	8.6207E−01	5.9484E−01	6.9001E−01
1.42	1.4033E+00	3.2734E+00	2.3327E+00	1.1261E+00	8.5514E−01	5.7827E−01	6.7623E−01
1.44	1.4147E+00	3.3678E+00	2.3805E+00	1.1379E+00	8.4822E−01	5.6207E−01	6.6264E−01
1.46	1.4263E+00	3.4654E+00	2.4296E+00	1.1501E+00	8.4133E−01	5.4623E−01	6.4925E−01
1.48	1.4381E+00	3.5665E+00	2.4800E+00	1.1629E+00	8.3445E−01	5.3075E−01	6.3606E−01
1.50	1.4500E+00	3.6710E+00	2.5317E+00	1.1762E+00	8.2759E−01	5.1564E−01	6.2306E−01
1.52	1.4621E+00	3.7792E+00	2.5848E+00	1.1899E+00	8.2075E−01	5.0088E−01	6.1027E−01
1.54	1.4743E+00	3.8911E+00	2.6392E+00	1.2042E+00	8.1393E−01	4.8648E−01	5.9769E−01
1.56	1.4867E+00	4.0068E+00	2.6951E+00	1.2190E+00	8.0715E−01	4.7242E−01	5.8530E−01

(Continued)

M	T_t/T	p_t/p	ρ_t/ρ	A/A^*	T/T^*	p/p^*	ρ/ρ^*
1.58	1.4993E+00	4.1266E+00	2.7524E+00	1.2344E+00	8.0038E−01	4.5872E−01	5.7312E−01
1.60	1.5120E+00	4.2504E+00	2.8111E+00	1.2502E+00	7.9365E−01	4.4535E−01	5.6114E−01
1.62	1.5249E+00	4.3785E+00	2.8714E+00	1.2666E+00	7.8695E−01	4.3232E−01	5.4937E−01
1.64	1.5379E+00	4.5110E+00	2.9332E+00	1.2835E+00	7.8027E−01	4.1963E−01	5.3780E−01
1.66	1.5511E+00	4.6479E+00	2.9965E+00	1.3010E+00	7.7363E−01	4.0726E−01	5.2643E−01
1.68	1.5645E+00	4.7896E+00	3.0614E+00	1.3190E+00	7.6703E−01	3.9522E−01	5.1526E−01
1.70	1.5780E+00	4.9360E+00	3.1280E+00	1.3376E+00	7.6046E−01	3.8350E−01	5.0430E−01
1.72	1.5917E+00	5.0874E+00	3.1962E+00	1.3567E+00	7.5392E−01	3.7208E−01	4.9353E−01
1.74	1.6055E+00	5.2439E+00	3.2662E+00	1.3764E+00	7.4742E−01	3.6098E−01	4.8296E−01
1.76	1.6195E+00	5.4057E+00	3.3378E+00	1.3967E+00	7.4096E−01	3.5017E−01	4.7259E−01
1.78	1.6337E+00	5.5729E+00	3.4113E+00	1.4175E+00	7.3454E−01	3.3966E−01	4.6242E−01
1.80	1.6480E+00	5.7458E+00	3.4865E+00	1.4390E+00	7.2816E−01	3.2945E−01	4.5244E−01
1.82	1.6625E+00	5.9244E+00	3.5636E+00	1.4610E+00	7.2181E−01	3.1951E−01	4.4265E−01
1.84	1.6771E+00	6.1091E+00	3.6426E+00	1.4836E+00	7.1551E−01	3.0986E−01	4.3305E−01
1.86	1.6919E+00	6.2998E+00	3.7235E+00	1.5069E+00	7.0925E−01	3.0047E−01	4.2365E−01
1.88	1.7069E+00	6.4970E+00	3.8063E+00	1.5307E+00	7.0304E−01	2.9136E−01	4.1442E−01
1.90	1.7220E+00	6.7006E+00	3.8912E+00	1.5552E+00	6.9686E−01	2.8250E−01	4.0539E−01
1.92	1.7373E+00	6.9111E+00	3.9781E+00	1.5804E+00	6.9073E−01	2.7390E−01	3.9653E−01
1.94	1.7527E+00	7.1284E+00	4.0671E+00	1.6062E+00	6.8465E−01	2.6555E−01	3.8786E−01
1.96	1.7683E+00	7.3530E+00	4.1582E+00	1.6326E+00	6.7861E−01	2.5744E−01	3.7936E−01
1.98	1.7841E+00	7.5849E+00	4.2514E+00	1.6597E+00	6.7262E−01	2.4957E−01	3.7104E−01
2.00	1.8000E+00	7.8244E+00	4.3469E+00	1.6875E+00	6.6667E−01	2.4192E−01	3.6289E−01
2.02	1.8161E+00	8.0718E+00	4.4446E+00	1.7159E+00	6.6076E−01	2.3451E−01	3.5491E−01
2.04	1.8323E+00	8.3273E+00	4.5447E+00	1.7451E+00	6.5491E−01	2.2732E−01	3.4710E−01
2.06	1.8487E+00	8.5911E+00	4.6471E+00	1.7750E+00	6.4910E−01	2.2034E−01	3.3945E−01
2.08	1.8653E+00	8.8635E+00	4.7518E+00	1.8056E+00	6.4334E−01	2.1357E−01	3.3197E−01
2.10	1.8820E+00	9.1447E+00	4.8590E+00	1.8369E+00	6.3762E−01	2.0700E−01	3.2464E−01
2.12	1.8989E+00	9.4350E+00	4.9687E+00	1.8690E+00	6.3195E−01	2.0063E−01	3.1747E−01
2.14	1.9159E+00	9.7347E+00	5.0809E+00	1.9018E+00	6.2633E−01	1.9445E−01	3.1046E−01
2.16	1.9331E+00	1.0044E+01	5.1957E+00	1.9354E+00	6.2076E−01	1.8846E−01	3.0360E−01
2.18	1.9505E+00	1.0363E+01	5.3132E+00	1.9698E+00	6.1523E−01	1.8266E−01	2.9689E−01
2.20	1.9680E+00	1.0693E+01	5.4333E+00	2.0050E+00	6.0976E−01	1.7703E−01	2.9033E−01
2.22	1.9857E+00	1.1033E+01	5.5561E+00	2.0409E+00	6.0433E−01	1.7157E−01	2.8391E−01
2.24	2.0035E+00	1.1384E+01	5.6818E+00	2.0777E+00	5.9895E−01	1.6629E−01	2.7763E−01
2.26	2.0215E+00	1.1746E+01	5.8103E+00	2.1153E+00	5.9361E−01	1.6116E−01	2.7149E−01
2.28	2.0397E+00	1.2119E+01	5.9416E+00	2.1538E+00	5.8833E−01	1.5620E−01	2.6549E−01
2.30	2.0580E+00	1.2504E+01	6.0759E+00	2.1931E+00	5.8309E−01	1.5138E−01	2.5962E−01
2.32	2.0765E+00	1.2902E+01	6.2133E+00	2.2333E+00	5.7790E−01	1.4672E−01	2.5388E−01
2.34	2.0951E+00	1.3312E+01	6.3536E+00	2.2744E+00	5.7276E−01	1.4220E−01	2.4827E−01
2.36	2.1139E+00	1.3734E+01	6.4971E+00	2.3164E+00	5.6767E−01	1.3782E−01	2.4279E−01

M	T_t/T	p_t/p	ρ_t/ρ	A/A^*	T/T^*	p/p^*	ρ/ρ^*
2.38	2.1329E+00	1.4170E+01	6.6438E+00	2.3592E+00	5.6262E−01	1.3358E−01	2.3743E−01
2.40	2.1520E+00	1.4620E+01	6.7937E+00	2.4031E+00	5.5762E−01	1.2948E−01	2.3219E−01
2.42	2.1713E+00	1.5084E+01	6.9469E+00	2.4478E+00	5.5267E−01	1.2550E−01	2.2707E−01
2.44	2.1907E+00	1.5562E+01	7.1034E+00	2.4936E+00	5.4777E−01	1.2164E−01	2.2207E−01
2.46	2.2103E+00	1.6054E+01	7.2634E+00	2.5403E+00	5.4291E−01	1.1791E−01	2.1718E−01
2.48	2.2301E+00	1.6562E+01	7.4268E+00	2.5880E+00	5.3810E−01	1.1429E−01	2.1240E−01
2.50	2.2500E+00	1.7086E+01	7.5938E+00	2.6367E+00	5.3333E−01	1.1079E−01	2.0773E−01
2.52	2.2701E+00	1.7626E+01	7.7643E+00	2.6864E+00	5.2862E−01	1.0740E−01	2.0317E−01
2.54	2.2903E+00	1.8182E+01	7.9385E+00	2.7372E+00	5.2394E−01	1.0411E−01	1.9871E−01
2.56	2.3107E+00	1.8755E+01	8.1165E+00	2.7890E+00	5.1932E−01	1.0093E−01	1.9435E−01
2.58	2.3313E+00	1.9346E+01	8.2982E+00	2.8419E+00	5.1474E−01	9.7848E−02	1.9009E−01
2.60	2.3520E+00	1.9954E+01	8.4839E+00	2.8959E+00	5.1020E−01	9.4865E−02	1.8593E−01
2.62	2.3729E+00	2.0581E+01	8.6734E+00	2.9511E+00	5.0571E−01	9.1975E−02	1.8187E−01
2.64	2.3939E+00	2.1227E+01	8.8669E+00	3.0073E+00	5.0127E−01	8.9177E−02	1.7790E−01
2.66	2.4151E+00	2.1892E+01	9.0646E+00	3.0647E+00	4.9687E−01	8.6467E−02	1.7402E−01
2.68	2.4365E+00	2.2577E+01	9.2663E+00	3.1232E+00	4.9251E−01	8.3843E−02	1.7023E−01
2.70	2.4580E+00	2.3283E+01	9.4723E+00	3.1830E+00	4.8820E−01	8.1301E−02	1.6653E−01
2.72	2.4797E+00	2.4010E+01	9.6825E+00	3.2439E+00	4.8393E−01	7.8841E−02	1.6292E−01
2.74	2.5015E+00	2.4758E+01	9.8971E+00	3.3061E+00	4.7971E−01	7.6458E−02	1.5938E−01
2.76	2.5235E+00	2.5528E+01	1.0116E+01	3.3695E+00	4.7553E−01	7.4150E−02	1.5593E−01
2.78	2.5457E+00	2.6322E+01	1.0340E+01	3.4341E+00	4.7139E−01	7.1915E−02	1.5256E−01
2.80	2.5680E+00	2.7138E+01	1.0568E+01	3.5001E+00	4.6729E−01	6.9751E−02	1.4927E−01
2.82	2.5905E+00	2.7979E+01	1.0801E+01	3.5673E+00	4.6323E−01	6.7656E−02	1.4605E−01
2.84	2.6131E+00	2.8844E+01	1.1038E+01	3.6359E+00	4.5922E−01	6.5626E−02	1.4291E−01
2.86	2.6359E+00	2.9735E+01	1.1281E+01	3.7058E+00	4.5525E−01	6.3661E−02	1.3984E−01
2.88	2.6589E+00	3.0651E+01	1.1528E+01	3.7771E+00	4.5132E−01	6.1757E−02	1.3684E−01
2.90	2.6820E+00	3.1594E+01	1.1780E+01	3.8497E+00	4.4743E−01	5.9914E−02	1.3391E−01
2.92	2.7053E+00	3.2564E+01	1.2037E+01	3.9238E+00	4.4358E−01	5.8129E−02	1.3105E−01
2.94	2.7287E+00	3.3563E+01	1.2300E+01	3.9993E+00	4.3977E−01	5.6400E−02	1.2825E−01
2.96	2.7523E+00	3.4590E+01	1.2567E+01	4.0762E+00	4.3600E−01	5.4725E−02	1.2552E−01
2.98	2.7761E+00	3.5646E+01	1.2840E+01	4.1546E+00	4.3226E−01	5.3103E−02	1.2285E−01
3.00	2.8000E+00	3.6733E+01	1.3119E+01	4.2345E+00	4.2857E−01	5.1533E−02	1.2024E−01
3.05	2.8605E+00	3.9586E+01	1.3839E+01	4.4410E+00	4.1951E−01	4.7818E−02	1.1399E−01
3.10	2.9220E+00	4.2646E+01	1.4595E+01	4.6573E+00	4.1068E−01	4.4387E−02	1.0808E−01
3.15	2.9845E+00	4.5925E+01	1.5388E+01	4.8838E+00	4.0208E−01	4.1218E−02	1.0251E−01
3.20	3.0480E+00	4.9437E+01	1.6219E+01	5.1209E+00	3.9370E−01	3.8290E−02	9.7256E−02
3.25	3.1125E+00	5.3196E+01	1.7091E+01	5.3690E+00	3.8554E−01	3.5584E−02	9.2295E−02
3.30	3.1780E+00	5.7219E+01	1.8005E+01	5.6286E+00	3.7760E−01	3.3082E−02	8.7613E−02
3.35	3.2445E+00	6.1520E+01	1.8961E+01	5.9000E+00	3.6986E−01	3.0769E−02	8.3192E−02

(Continued)

M	T_t/T	p_t/p	ρ_t/ρ	A/A^*	T/T^*	p/p^*	ρ/ρ^*
3.40	$3.3120E+00$	$6.6117E+01$	$1.9963E+01$	$6.1836E+00$	$3.6232E-01$	$2.8630E-02$	$7.9018E-02$
3.45	$3.3805E+00$	$7.1029E+01$	$2.1011E+01$	$6.4800E+00$	$3.5498E-01$	$2.6650E-02$	$7.5076E-02$
3.50	$3.4500E+00$	$7.6272E+01$	$2.2108E+01$	$6.7895E+00$	$3.4783E-01$	$2.4818E-02$	$7.1352E-02$
3.55	$3.5205E+00$	$8.1868E+01$	$2.3255E+01$	$7.1127E+00$	$3.4086E-01$	$2.3122E-02$	$6.7833E-02$
3.60	$3.5920E+00$	$8.7837E+01$	$2.4453E+01$	$7.4500E+00$	$3.3408E-01$	$2.1550E-02$	$6.4508E-02$
3.65	$3.6645E+00$	$9.4200E+01$	$2.5706E+01$	$7.8019E+00$	$3.2747E-01$	$2.0095E-02$	$6.1364E-02$
3.70	$3.7380E+00$	$1.0098E+02$	$2.7015E+01$	$8.1690E+00$	$3.2103E-01$	$1.8745E-02$	$5.8392E-02$
3.75	$3.8125E+00$	$1.0820E+02$	$2.8381E+01$	$8.5516E+00$	$3.1475E-01$	$1.7494E-02$	$5.5581E-02$
3.80	$3.8880E+00$	$1.1589E+02$	$2.9807E+01$	$8.9505E+00$	$3.0864E-01$	$1.6334E-02$	$5.2922E-02$
3.85	$3.9645E+00$	$1.2407E+02$	$3.1295E+01$	$9.3660E+00$	$3.0269E-01$	$1.5257E-02$	$5.0406E-02$
3.90	$4.0420E+00$	$1.3277E+02$	$3.2847E+01$	$9.7989E+00$	$2.9688E-01$	$1.4258E-02$	$4.8024E-02$
3.95	$4.1205E+00$	$1.4201E+02$	$3.4465E+01$	$1.0250E+01$	$2.9123E-01$	$1.3329E-02$	$4.5770E-02$
4.00	$4.2000E+00$	$1.5184E+02$	$3.6151E+01$	$1.0719E+01$	$2.8571E-01$	$1.2467E-02$	$4.3634E-02$
4.05	$4.2805E+00$	$1.6227E+02$	$3.7908E+01$	$1.1207E+01$	$2.8034E-01$	$1.1666E-02$	$4.1612E-02$
4.10	$4.3620E+00$	$1.7334E+02$	$3.9739E+01$	$1.1715E+01$	$2.7510E-01$	$1.0920E-02$	$3.9695E-02$
4.15	$4.4445E+00$	$1.8509E+02$	$4.1644E+01$	$1.2243E+01$	$2.7000E-01$	$1.0227E-02$	$3.7879E-02$
4.20	$4.5280E+00$	$1.9755E+02$	$4.3628E+01$	$1.2791E+01$	$2.6502E-01$	$9.5821E-03$	$3.6157E-02$
4.25	$4.6125E+00$	$2.1075E+02$	$4.5692E+01$	$1.3362E+01$	$2.6016E-01$	$8.9817E-03$	$3.4523E-02$
4.30	$4.6980E+00$	$2.2475E+02$	$4.7839E+01$	$1.3955E+01$	$2.5543E-01$	$8.4225E-03$	$3.2974E-02$
4.35	$4.7845E+00$	$2.3957E+02$	$5.0072E+01$	$1.4570E+01$	$2.5081E-01$	$7.9014E-03$	$3.1504E-02$
4.40	$4.8720E+00$	$2.5526E+02$	$5.2392E+01$	$1.5210E+01$	$2.4631E-01$	$7.4158E-03$	$3.0108E-02$
4.45	$4.9605E+00$	$2.7186E+02$	$5.4804E+01$	$1.5873E+01$	$2.4191E-01$	$6.9630E-03$	$2.8783E-02$
4.50	$5.0500E+00$	$2.8941E+02$	$5.7310E+01$	$1.6562E+01$	$2.3762E-01$	$6.5406E-03$	$2.7525E-02$
4.55	$5.1405E+00$	$3.0798E+02$	$5.9912E+01$	$1.7276E+01$	$2.3344E-01$	$6.1463E-03$	$2.6329E-02$
4.60	$5.2320E+00$	$3.2759E+02$	$6.2614E+01$	$1.8018E+01$	$2.2936E-01$	$5.7783E-03$	$2.5193E-02$
4.65	$5.3245E+00$	$3.4832E+02$	$6.5418E+01$	$1.8786E+01$	$2.2537E-01$	$5.4345E-03$	$2.4113E-02$
4.70	$5.4180E+00$	$3.7020E+02$	$6.8328E+01$	$1.9583E+01$	$2.2148E-01$	$5.1133E-03$	$2.3086E-02$
4.75	$5.5125E+00$	$3.9330E+02$	$7.1346E+01$	$2.0408E+01$	$2.1769E-01$	$4.8130E-03$	$2.2110E-02$
4.80	$5.6080E+00$	$4.1766E+02$	$7.4477E+01$	$2.1263E+01$	$2.1398E-01$	$4.5322E-03$	$2.1180E-02$
4.85	$5.7045E+00$	$4.4337E+02$	$7.7722E+01$	$2.2149E+01$	$2.1036E-01$	$4.2695E-03$	$2.0296E-02$
4.90	$5.8020E+00$	$4.7046E+02$	$8.1086E+01$	$2.3067E+01$	$2.0683E-01$	$4.0236E-03$	$1.9454E-02$
4.95	$5.9005E+00$	$4.9901E+02$	$8.4571E+01$	$2.4017E+01$	$2.0337E-01$	$3.7934E-03$	$1.8652E-02$
5.00	$6.0000E+00$	$5.2909E+02$	$8.8182E+01$	$2.5000E+01$	$2.0000E-01$	$3.5777E-03$	$1.7889E-02$
5.10	$6.2020E+00$	$5.9410E+02$	$9.5792E+01$	$2.7069E+01$	$1.9349E-01$	$3.1862E-03$	$1.6467E-02$
5.20	$6.4080E+00$	$6.6608E+02$	$1.0395E+02$	$2.9283E+01$	$1.8727E-01$	$2.8419E-03$	$1.5176E-02$
5.30	$6.6180E+00$	$7.4566E+02$	$1.1267E+02$	$3.1649E+01$	$1.8132E-01$	$2.5386E-03$	$1.4000E-02$
5.40	$6.8320E+00$	$8.3352E+02$	$1.2200E+02$	$3.4174E+01$	$1.7564E-01$	$2.2710E-03$	$1.2930E-02$
5.50	$7.0500E+00$	$9.3038E+02$	$1.3197E+02$	$3.6869E+01$	$1.7021E-01$	$2.0346E-03$	$1.1953E-02$
5.60	$7.2720E+00$	$1.0370E+03$	$1.4260E+02$	$3.9740E+01$	$1.6502E-01$	$1.8253E-03$	$1.1062E-02$
5.70	$7.4980E+00$	$1.1543E+03$	$1.5394E+02$	$4.2797E+01$	$1.6004E-01$	$1.6399E-03$	$1.0247E-02$

M	T_t/T	p_t/p	ρ_t/ρ	A/A^*	T/T^*	p/p^*	ρ/ρ^*
5.80	7.7280E+00	1.2830E+03	1.6602E+02	4.6049E+01	1.5528E−01	1.4754E−03	9.5014E−03
5.90	7.9620E+00	1.4242E+03	1.7888E+02	4.9507E+01	1.5072E−01	1.3291E−03	8.8186E−03
6.00	8.2000E+00	1.5789E+03	1.9255E+02	5.3179E+01	1.4634E−01	1.1989E−03	8.1925E−03
6.10	8.4420E+00	1.7481E+03	2.0707E+02	5.7076E+01	1.4215E−01	1.0829E−03	7.6180E−03
6.20	8.6880E+00	1.9329E+03	2.2248E+02	6.1209E+01	1.3812E−01	9.7930E−04	7.0901E−03
6.30	8.9380E+00	2.1347E+03	2.3884E+02	6.5589E+01	1.3426E−01	8.8673E−04	6.6047E−03
6.40	9.1920E+00	2.3547E+03	2.5617E+02	7.0227E+01	1.3055E−01	8.0389E−04	6.1578E−03
6.50	9.4500E+00	2.5942E+03	2.7452E+02	7.5133E+01	1.2698E−01	7.2966E−04	5.7461E−03
6.60	9.7120E+00	2.8548E+03	2.9395E+02	8.0322E+01	1.2356E−01	6.6306E−04	5.3664E−03
6.70	9.9780E+00	3.1380E+03	3.1449E+02	8.5804E+01	1.2026E−01	6.0323E−04	5.0158E−03
6.80	1.0248E+01	3.4454E+03	3.3620E+02	9.1592E+01	1.1710E−01	5.4941E−04	4.6920E−03
6.90	1.0522E+01	3.7787E+03	3.5913E+02	9.7701E+01	1.1405E−01	5.0095E−04	4.3925E−03
7.00	1.0800E+01	4.1398E+03	3.8332E+02	1.0414E+02	1.1111E−01	4.5725E−04	4.1152E−03
7.10	1.1082E+01	4.5307E+03	4.0883E+02	1.1093E+02	1.0828E−01	4.1780E−04	3.8584E−03
7.20	1.1368E+01	4.9533E+03	4.3572E+02	1.1808E+02	1.0556E−01	3.8216E−04	3.6203E−03
7.30	1.1658E+01	5.4098E+03	4.6405E+02	1.2560E+02	1.0293E−01	3.4991E−04	3.3993E−03
7.40	1.1952E+01	5.9026E+03	4.9386E+02	1.3352E+02	1.0040E−01	3.2070E−04	3.1941E−03
7.50	1.2250E+01	6.4339E+03	5.2522E+02	1.4184E+02	9.7959E−02	2.9421E−04	3.0034E−03
7.60	1.2552E+01	7.0064E+03	5.5819E+02	1.5058E+02	9.5602E−02	2.7017E−04	2.8260E−03
7.70	1.2858E+01	7.6227E+03	5.9283E+02	1.5976E+02	9.3327E−02	2.4833E−04	2.6608E−03
7.80	1.3168E+01	8.2855E+03	6.2922E+02	1.6940E+02	9.1130E−02	2.2846E−04	2.5070E−03
7.90	1.3482E+01	8.9979E+03	6.6740E+02	1.7951E+02	8.9008E−02	2.1038E−04	2.3636E−03
8.00	1.3800E+01	9.7629E+03	7.0745E+02	1.9011E+02	8.6957E−02	1.9389E−04	2.2297E−03
8.10	1.4122E+01	1.0584E+04	7.4945E+02	2.0121E+02	8.4974E−02	1.7885E−04	2.1048E−03
8.20	1.4448E+01	1.1464E+04	7.9345E+02	2.1284E+02	8.3056E−02	1.6512E−04	1.9881E−03
8.30	1.4778E+01	1.2407E+04	8.3954E+02	2.2502E+02	8.1202E−02	1.5257E−04	1.8789E−03
8.40	1.5112E+01	1.3416E+04	8.8778E+02	2.3776E+02	7.9407E−02	1.4109E−04	1.7768E−03
8.50	1.5450E+01	1.4496E+04	9.3826E+02	2.5108E+02	7.7670E−02	1.3058E−04	1.6812E−03
8.60	1.5792E+01	1.5651E+04	9.9104E+02	2.6501E+02	7.5988E−02	1.2095E−04	1.5917E−03
8.70	1.6138E+01	1.6884E+04	1.0462E+03	2.7956E+02	7.4359E−02	1.1211E−04	1.5077E−03
8.80	1.6488E+01	1.8201E+04	1.1039E+03	2.9476E+02	7.2780E−02	1.0400E−04	1.4290E−03
8.90	1.6842E+01	1.9605E+04	1.1641E+03	3.1063E+02	7.1250E−02	9.6551E−05	1.3551E−03
9.00	1.7200E+01	2.1103E+04	1.2269E+03	3.2719E+02	6.9767E−02	8.9698E−05	1.2857E−03
9.10	1.7562E+01	2.2699E+04	1.2925E+03	3.4445E+02	6.8329E−02	8.3392E−05	1.2204E−03
9.20	1.7928E+01	2.4398E+04	1.3609E+03	3.6246E+02	6.6934E−02	7.7584E−05	1.1591E−03
9.30	1.8298E+01	2.6207E+04	1.4322E+03	3.8122E+02	6.5581E−02	7.2231E−05	1.1014E−03
9.40	1.8672E+01	2.8130E+04	1.5065E+03	4.0077E+02	6.4267E−02	6.7292E−05	1.0471E−03
9.50	1.9050E+01	3.0174E+04	1.5839E+03	4.2113E+02	6.2992E−02	6.2734E−05	9.9590E−04
9.60	1.9432E+01	3.2345E+04	1.6645E+03	4.4232E+02	6.1754E−02	5.8523E−05	9.4767E−04
9.70	1.9818E+01	3.4650E+04	1.7484E+03	4.6436E+02	6.0551E−02	5.4629E−05	9.0220E−04

(Continued)

M	T_t/T	p_t/p	ρ_t/ρ	A/A^*	T/T^*	p/p^*	ρ/ρ^*
9.80	2.0208E+01	3.7096E+04	1.8357E+03	4.8730E+02	5.9382E−02	5.1027E−05	8.5930E−04
9.90	2.0602E+01	3.9690E+04	1.9265E+03	5.1114E+02	5.8247E−02	4.7693E−05	8.1880E−04
10.00	2.1000E+01	4.2439E+04	2.0209E+03	5.3593E+02	5.7143E−02	4.4603E−05	7.8056E−04
10.10	2.1402E+01	4.5351E+04	2.1190E+03	5.6168E+02	5.6070E−02	4.1739E−05	7.4442E−04
10.20	2.1808E+01	4.8435E+04	2.2210E+03	5.8843E+02	5.5026E−02	3.9082E−05	7.1025E−04
10.30	2.2218E+01	5.1697E+04	2.3268E+03	6.1621E+02	5.4010E−02	3.6616E−05	6.7794E−04
10.40	2.2632E+01	5.5148E+04	2.4367E+03	6.4504E+02	5.3022E−02	3.4324E−05	6.4736E−04
10.50	2.3050E+01	5.8796E+04	2.5508E+03	6.7495E+02	5.2061E−02	3.2195E−05	6.1841E−04
10.60	2.3472E+01	6.2651E+04	2.6692E+03	7.0599E+02	5.1125E−02	3.0214E−05	5.9099E−04
10.70	2.3898E+01	6.6721E+04	2.7919E+03	7.3816E+02	5.0213E−02	2.8371E−05	5.6500E−04
10.80	2.4328E+01	7.1019E+04	2.9192E+03	7.7152E+02	4.9326E−02	2.6654E−05	5.4036E−04
10.90	2.4762E+01	7.5553E+04	3.0512E+03	8.0609E+02	4.8461E−02	2.5054E−05	5.1700E−04
11.00	2.5200E+01	8.0334E+04	3.1879E+03	8.4190E+02	4.7619E−02	2.3563E−05	4.9483E−04
11.10	2.5642E+01	8.5375E+04	3.3295E+03	8.7899E+02	4.6798E−02	2.2172E−05	4.7378E−04
11.20	2.6088E+01	9.0687E+04	3.4762E+03	9.1739E+02	4.5998E−02	2.0873E−05	4.5379E−04
11.30	2.6538E+01	9.6281E+04	3.6280E+03	9.5714E+02	4.5218E−02	1.9661E−05	4.3479E−04
11.40	2.6992E+01	1.0217E+05	3.7852E+03	9.9828E+02	4.4458E−02	1.8527E−05	4.1674E−04
11.50	2.7450E+01	1.0837E+05	3.9478E+03	1.0408E+03	4.3716E−02	1.7468E−05	3.9957E−04
11.60	2.7912E+01	1.1489E+05	4.1160E+03	1.0848E+03	4.2992E−02	1.6477E−05	3.8324E−04
11.70	2.8378E+01	1.2174E+05	4.2900E+03	1.1303E+03	4.2286E−02	1.5549E−05	3.6770E−04
11.80	2.8848E+01	1.2895E+05	4.4698E+03	1.1774E+03	4.1597E−02	1.4680E−05	3.5291E−04
11.90	2.9322E+01	1.3651E+05	4.6557E+03	1.2260E+03	4.0925E−02	1.3866E−05	3.3882E−04
12.00	2.9800E+01	1.4446E+05	4.8478E+03	1.2762E+03	4.0268E−02	1.3103E−05	3.2540E−04
12.10	3.0282E+01	1.5281E+05	5.0462E+03	1.3281E+03	3.9628E−02	1.2388E−05	3.1260E−04
12.20	3.0768E+01	1.6157E+05	5.2511E+03	1.3816E+03	3.9002E−02	1.1716E−05	3.0040E−04
12.30	3.1258E+01	1.7075E+05	5.4626E+03	1.4369E+03	3.8390E−02	1.1086E−05	2.8877E−04
12.40	3.1752E+01	1.8038E+05	5.6810E+03	1.4940E+03	3.7793E−02	1.0494E−05	2.7767E−04
12.50	3.2250E+01	1.9048E+05	5.9064E+03	1.5529E+03	3.7209E−02	9.9376E−06	2.6707E−04
12.60	3.2752E+01	2.0106E+05	6.1390E+03	1.6136E+03	3.6639E−02	9.4146E−06	2.5696E−04
12.70	3.3258E+01	2.1215E+05	6.3788E+03	1.6762E+03	3.6082E−02	8.9227E−06	2.4729E−04
12.80	3.3768E+01	2.2375E+05	6.6262E+03	1.7408E+03	3.5537E−02	8.4599E−06	2.3806E−04
12.90	3.4282E+01	2.3590E+05	6.8812E+03	1.8074E+03	3.5004E−02	8.0242E−06	2.2924E−04
13.00	3.4800E+01	2.4862E+05	7.1441E+03	1.8761E+03	3.4483E−02	7.6139E−06	2.2080E−04
13.10	3.5322E+01	2.6191E+05	7.4150E+03	1.9468E+03	3.3973E−02	7.2273E−06	2.1274E−04
13.20	3.5848E+01	2.7582E+05	7.6942E+03	2.0196E+03	3.3475E−02	6.8629E−06	2.0502E−04
13.30	3.6378E+01	2.9036E+05	7.9817E+03	2.0947E+03	3.2987E−02	6.5193E−06	1.9763E−04
13.40	3.6912E+01	3.0555E+05	8.2779E+03	2.1719E+03	3.2510E−02	6.1951E−06	1.9056E−04
13.50	3.7450E+01	3.2143E+05	8.5828E+03	2.2515E+03	3.2043E−02	5.8892E−06	1.8379E−04
13.60	3.7992E+01	3.3800E+05	8.8967E+03	2.3334E+03	3.1586E−02	5.6003E−06	1.7731E−04
13.70	3.8538E+01	3.5531E+05	9.2198E+03	2.4177E+03	3.1138E−02	5.3275E−06	1.7109E−04

M	T_t/T	p_t/p	ρ_t/ρ	A/A^*	T/T^*	p/p^*	ρ/ρ^*
13.80	3.9088E+01	3.7338E+05	9.5523E+03	2.5044E+03	3.0700E−02	5.0697E−06	1.6514E−04
13.90	3.9642E+01	3.9223E+05	9.8944E+03	2.5936E+03	3.0271E−02	4.8260E−06	1.5943E−04
14.00	4.0200E+01	4.1190E+05	1.0246E+04	2.6854E+03	2.9851E−02	4.5956E−06	1.5395E−04
14.10	4.0762E+01	4.3241E+05	1.0608E+04	2.7797E+03	2.9439E−02	4.3776E−06	1.4870E−04
14.20	4.1328E+01	4.5379E+05	1.0980E+04	2.8767E+03	2.9036E−02	4.1714E−06	1.4366E−04
14.30	4.1898E+01	4.7608E+05	1.1363E+04	2.9764E+03	2.8641E−02	3.9761E−06	1.3883E−04
14.40	4.2472E+01	4.9930E+05	1.1756E+04	3.0789E+03	2.8254E−02	3.7912E−06	1.3418E−04
14.50	4.3050E+01	5.2349E+05	1.2160E+04	3.1842E+03	2.7875E−02	3.6160E−06	1.2972E−04
14.60	4.3632E+01	5.4868E+05	1.2575E+04	3.2924E+03	2.7503E−02	3.4500E−06	1.2544E−04
14.70	4.4218E+01	5.7491E+05	1.3002E+04	3.4035E+03	2.7138E−02	3.2926E−06	1.2133E−04
14.80	4.4808E+01	6.0221E+05	1.3440E+04	3.5177E+03	2.6781E−02	3.1433E−06	1.1737E−04
14.90	4.5402E+01	6.3061E+05	1.3890E+04	3.6349E+03	2.6431E−02	3.0017E−06	1.1357E−04
15.00	4.6000E+01	6.6016E+05	1.4351E+04	3.7552E+03	2.6087E−02	2.8674E−06	1.0992E−04
16.00	5.2200E+01	1.0277E+06	1.9687E+04	5.1445E+03	2.2989E−02	1.8420E−06	8.0127E−05
17.00	5.8800E+01	1.5589E+06	2.6512E+04	6.9204E+03	2.0408E−02	1.2143E−06	5.9499E−05
18.00	6.5800E+01	2.3110E+06	3.5121E+04	9.1592E+03	1.8237E−02	8.1911E−07	4.4915E−05
19.00	7.3200E+01	3.3557E+06	4.5843E+04	1.1946E+04	1.6393E−02	5.6409E−07	3.4409E−05
20.00	8.1000E+01	4.7830E+06	5.9049E+04	1.5377E+04	1.4815E−02	3.9576E−07	2.6714E−05
21.00	8.9200E+01	6.7031E+06	7.5147E+04	1.9558E+04	1.3453E−02	2.8240E−07	2.0991E−05
22.00	9.7800E+01	9.2509E+06	9.4590E+04	2.4606E+04	1.2270E−02	2.0462E−07	1.6677E−05
23.00	1.0680E+02	1.2589E+07	1.1788E+05	3.0650E+04	1.1236E−02	1.5036E−07	1.3382E−05
24.00	1.1620E+02	1.6913E+07	1.4555E+05	3.7832E+04	1.0327E−02	1.1192E−07	1.0838E−05
25.00	1.2600E+02	2.2454E+07	1.7821E+05	4.6304E+04	9.5238E−03	8.4302E−08	8.8517E−06
26.00	1.3620E+02	2.9486E+07	2.1649E+05	5.6235E+04	8.8106E−03	6.4197E−08	7.2864E−06
27.00	1.4680E+02	3.8330E+07	2.6110E+05	6.7806E+04	8.1744E−03	4.9385E−08	6.0414E−06
28.00	1.5780E+02	4.9360E+07	3.1280E+05	8.1211E+04	7.6046E−03	3.8350E−08	5.0430E−06
29.00	1.6920E+02	6.3009E+07	3.7239E+05	9.6662E+04	7.0922E−03	3.0042E−08	4.2360E−06
30.00	1.8100E+02	7.9777E+07	4.4075E+05	1.1438E+05	6.6298E−03	2.3728E−08	3.5790E−06
31.00	1.9320E+02	1.0024E+08	5.1882E+05	1.3462E+05	6.2112E−03	1.8885E−08	3.0404E−06
32.00	2.0580E+02	1.2504E+08	6.0759E+05	1.5763E+05	5.8309E−03	1.5138E−08	2.5962E−06
33.00	2.1880E+02	1.5494E+08	7.0814E+05	1.8369E+05	5.4845E−03	1.2217E−08	2.2276E−06
34.00	2.3220E+02	1.9077E+08	8.2159E+05	2.1309E+05	5.1680E−03	9.9224E−09	1.9200E−06
35.00	2.4600E+02	2.3349E+08	9.4916E+05	2.4614E+05	4.8780E−03	8.1070E−09	1.6619E−06
40.00	3.2100E+02	5.9261E+08	1.8461E+06	4.7853E+05	3.7383E−03	3.1942E−09	8.5446E−07
45.00	4.0600E+02	1.3485E+09	3.3214E+06	8.6063E+05	2.9557E−03	1.4038E−09	4.7494E−07
50.00	5.0100E+02	2.8147E+09	5.6182E+06	1.4554E+06	2.3952E−03	6.7252E−10	2.8078E−07
100.00	2.0010E+03	3.5840E+11	1.7911E+08	4.6365E+07	5.9970E−04	5.2816E−12	8.8072E−09
1000.00	2.0000E+05	3.5778E+18	1.7889E+13	4.6296E+12	6.0000E−06	5.2908E−19	8.8181E−14

Isentropic Table γ = 1.3

M	T_t/T	p_t/p	ρ_t/ρ	A/A^*	T/T^*	p/p^*	ρ/ρ^*
0.02	1.0001E+00	1.0003E+00	1.0002E+00	2.9268E+01	1.1499E+00	1.8319E+00	1.5931E+00
0.04	1.0002E+00	1.0010E+00	1.0008E+00	1.4644E+01	1.1497E+00	1.8305E+00	1.5921E+00
0.06	1.0005E+00	1.0023E+00	1.0018E+00	9.7740E+00	1.1494E+00	1.8281E+00	1.5905E+00
0.08	1.0010E+00	1.0042E+00	1.0032E+00	7.3423E+00	1.1489E+00	1.8248E+00	1.5883E+00
0.10	1.0015E+00	1.0065E+00	1.0050E+00	5.8860E+00	1.1483E+00	1.8206E+00	1.5855E+00
0.12	1.0022E+00	1.0094E+00	1.0072E+00	4.9174E+00	1.1475E+00	1.8154E+00	1.5820E+00
0.14	1.0029E+00	1.0128E+00	1.0098E+00	4.2275E+00	1.1466E+00	1.8093E+00	1.5779E+00
0.16	1.0038E+00	1.0167E+00	1.0129E+00	3.7118E+00	1.1456E+00	1.8022E+00	1.5732E+00
0.18	1.0049E+00	1.0212E+00	1.0163E+00	3.3123E+00	1.1444E+00	1.7943E+00	1.5679E+00
0.20	1.0060E+00	1.0263E+00	1.0201E+00	2.9940E+00	1.1431E+00	1.7855E+00	1.5619E+00
0.22	1.0073E+00	1.0318E+00	1.0244E+00	2.7349E+00	1.1417E+00	1.7759E+00	1.5554E+00
0.24	1.0086E+00	1.0380E+00	1.0291E+00	2.5202E+00	1.1401E+00	1.7654E+00	1.5484E+00
0.26	1.0101E+00	1.0447E+00	1.0342E+00	2.3396E+00	1.1385E+00	1.7540E+00	1.5407E+00
0.28	1.0118E+00	1.0520E+00	1.0397E+00	2.1859E+00	1.1366E+00	1.7419E+00	1.5325E+00
0.30	1.0135E+00	1.0598E+00	1.0457E+00	2.0537E+00	1.1347E+00	1.7290E+00	1.5237E+00
0.32	1.0154E+00	1.0683E+00	1.0521E+00	1.9389E+00	1.1326E+00	1.7153E+00	1.5145E+00
0.34	1.0173E+00	1.0773E+00	1.0590E+00	1.8385E+00	1.1304E+00	1.7009E+00	1.5047E+00
0.36	1.0194E+00	1.0870E+00	1.0663E+00	1.7502E+00	1.1281E+00	1.6857E+00	1.4943E+00
0.38	1.0217E+00	1.0973E+00	1.0740E+00	1.6719E+00	1.1256E+00	1.6699E+00	1.4836E+00
0.40	1.0240E+00	1.1082E+00	1.0823E+00	1.6023E+00	1.1230E+00	1.6534E+00	1.4723E+00
0.42	1.0265E+00	1.1198E+00	1.0910E+00	1.5401E+00	1.1204E+00	1.6363E+00	1.4606E+00
0.44	1.0290E+00	1.1321E+00	1.1001E+00	1.4843E+00	1.1175E+00	1.6186E+00	1.4484E+00
0.46	1.0317E+00	1.1450E+00	1.1098E+00	1.4341E+00	1.1146E+00	1.6004E+00	1.4358E+00
0.48	1.0346E+00	1.1586E+00	1.1199E+00	1.3888E+00	1.1116E+00	1.5815E+00	1.4228E+00
0.50	1.0375E+00	1.1730E+00	1.1306E+00	1.3479E+00	1.1084E+00	1.5622E+00	1.4094E+00
0.52	1.0406E+00	1.1880E+00	1.1417E+00	1.3107E+00	1.1052E+00	1.5424E+00	1.3956E+00
0.54	1.0437E+00	1.2038E+00	1.1534E+00	1.2770E+00	1.1018E+00	1.5221E+00	1.3815E+00
0.56	1.0470E+00	1.2204E+00	1.1656E+00	1.2464E+00	1.0983E+00	1.5015E+00	1.3670E+00
0.58	1.0505E+00	1.2378E+00	1.1783E+00	1.2186E+00	1.0948E+00	1.4804E+00	1.3523E+00
0.60	1.0540E+00	1.2560E+00	1.1916E+00	1.1932E+00	1.0911E+00	1.4590E+00	1.3372E+00
0.62	1.0577E+00	1.2750E+00	1.2055E+00	1.1702E+00	1.0873E+00	1.4372E+00	1.3218E+00
0.64	1.0614E+00	1.2948E+00	1.2199E+00	1.1492E+00	1.0834E+00	1.4152E+00	1.3062E+00
0.66	1.0653E+00	1.3156E+00	1.2349E+00	1.1302E+00	1.0795E+00	1.3929E+00	1.2903E+00
0.68	1.0694E+00	1.3372E+00	1.2505E+00	1.1129E+00	1.0754E+00	1.3703E+00	1.2742E+00
0.70	1.0735E+00	1.3598E+00	1.2667E+00	1.0972E+00	1.0713E+00	1.3476E+00	1.2579E+00
0.72	1.0778E+00	1.3833E+00	1.2835E+00	1.0831E+00	1.0670E+00	1.3246E+00	1.2414E+00
0.74	1.0821E+00	1.4079E+00	1.3010E+00	1.0703E+00	1.0627E+00	1.3016E+00	1.2247E+00
0.76	1.0866E+00	1.4334E+00	1.3191E+00	1.0589E+00	1.0583E+00	1.2784E+00	1.2079E+00

M	T_t/T	p_t/p	ρ_t/ρ	A/A^*	T/T^*	p/p^*	ρ/ρ^*
0.78	1.0913E+00	1.4600E+00	1.3379E+00	1.0486E+00	1.0538E+00	1.2551E+00	1.1910E+00
0.80	1.0960E+00	1.4877E+00	1.3574E+00	1.0395E+00	1.0493E+00	1.2317E+00	1.1739E+00
0.82	1.1009E+00	1.5165E+00	1.3775E+00	1.0315E+00	1.0446E+00	1.2083E+00	1.1567E+00
0.84	1.1058E+00	1.5464E+00	1.3984E+00	1.0245E+00	1.0399E+00	1.1849E+00	1.1394E+00
0.86	1.1109E+00	1.5776E+00	1.4200E+00	1.0185E+00	1.0352E+00	1.1615E+00	1.1221E+00
0.88	1.1162E+00	1.6100E+00	1.4424E+00	1.0134E+00	1.0303E+00	1.1382E+00	1.1047E+00
0.90	1.1215E+00	1.6436E+00	1.4655E+00	1.0092E+00	1.0254E+00	1.1149E+00	1.0872E+00
0.92	1.1270E+00	1.6786E+00	1.4895E+00	1.0058E+00	1.0204E+00	1.0917E+00	1.0698E+00
0.94	1.1325E+00	1.7149E+00	1.5142E+00	1.0032E+00	1.0154E+00	1.0685E+00	1.0523E+00
0.96	1.1382E+00	1.7526E+00	1.5397E+00	1.0014E+00	1.0103E+00	1.0455E+00	1.0349E+00
0.98	1.1441E+00	1.7917E+00	1.5661E+00	1.0004E+00	1.0052E+00	1.0227E+00	1.0174E+00
1.00	1.1500E+00	1.8324E+00	1.5934E+00	1.0000E+00	1.0000E+00	1.0000E+00	1.0000E+00
1.02	1.1561E+00	1.8746E+00	1.6216E+00	1.0003E+00	9.9476E−01	9.7748E−01	9.8263E−01
1.04	1.1622E+00	1.9184E+00	1.6506E+00	1.0014E+00	9.8947E−01	9.5516E−01	9.6533E−01
1.06	1.1685E+00	1.9639E+00	1.6806E+00	1.0030E+00	9.8413E−01	9.3304E−01	9.4809E−01
1.08	1.1750E+00	2.0111E+00	1.7116E+00	1.0054E+00	9.7876E−01	9.1115E−01	9.3093E−01
1.10	1.1815E+00	2.0600E+00	1.7436E+00	1.0083E+00	9.7334E−01	8.8950E−01	9.1386E−01
1.12	1.1882E+00	2.1108E+00	1.7766E+00	1.0119E+00	9.6788E−01	8.6809E−01	8.9690E−01
1.14	1.1949E+00	2.1635E+00	1.8106E+00	1.0160E+00	9.6239E−01	8.4695E−01	8.8005E−01
1.16	1.2018E+00	2.2182E+00	1.8457E+00	1.0208E+00	9.5687E−01	8.2608E−01	8.6332E−01
1.18	1.2089E+00	2.2749E+00	1.8818E+00	1.0262E+00	9.5131E−01	8.0549E−01	8.4672E−01
1.20	1.2160E+00	2.3337E+00	1.9192E+00	1.0321E+00	9.4572E−01	7.8520E−01	8.3026E−01
1.22	1.2233E+00	2.3947E+00	1.9576E+00	1.0386E+00	9.4011E−01	7.6520E−01	8.1395E−01
1.24	1.2306E+00	2.4579E+00	1.9973E+00	1.0457E+00	9.3447E−01	7.4551E−01	7.9779E−01
1.26	1.2381E+00	2.5235E+00	2.0381E+00	1.0533E+00	9.2881E−01	7.2614E−01	7.8180E−01
1.28	1.2458E+00	2.5915E+00	2.0802E+00	1.0616E+00	9.2313E−01	7.0709E−01	7.6597E−01
1.30	1.2535E+00	2.6620E+00	2.1236E+00	1.0703E+00	9.1743E−01	6.8836E−01	7.5032E−01
1.32	1.2614E+00	2.7351E+00	2.1683E+00	1.0797E+00	9.1171E−01	6.6997E−01	7.3485E−01
1.34	1.2693E+00	2.8108E+00	2.2144E+00	1.0896E+00	9.0598E−01	6.5191E−01	7.1956E−01
1.36	1.2774E+00	2.8894E+00	2.2619E+00	1.1001E+00	9.0024E−01	6.3418E−01	7.0446E−01
1.38	1.2857E+00	2.9708E+00	2.3107E+00	1.1111E+00	8.9448E−01	6.1680E−01	6.8956E−01
1.40	1.2940E+00	3.0552E+00	2.3611E+00	1.1227E+00	8.8872E−01	5.9976E−01	6.7486E−01
1.42	1.3025E+00	3.1427E+00	2.4129E+00	1.1349E+00	8.8294E−01	5.8306E−01	6.6036E−01
1.44	1.3110E+00	3.2334E+00	2.4663E+00	1.1477E+00	8.7717E−01	5.6670E−01	6.4606E−01
1.46	1.3197E+00	3.3275E+00	2.5213E+00	1.1610E+00	8.7138E−01	5.5069E−01	6.3198E−01
1.48	1.3286E+00	3.4249E+00	2.5779E+00	1.1750E+00	8.6560E−01	5.3502E−01	6.1810E−01
1.50	1.3375E+00	3.5259E+00	2.6362E+00	1.1895E+00	8.5981E−01	5.1970E−01	6.0443E−01
1.52	1.3466E+00	3.6306E+00	2.6962E+00	1.2046E+00	8.5403E−01	5.0471E−01	5.9098E−01
1.54	1.3557E+00	3.7390E+00	2.7579E+00	1.2203E+00	8.4825E−01	4.9007E−01	5.7775E−01
1.56	1.3650E+00	3.8515E+00	2.8215E+00	1.2367E+00	8.4247E−01	4.7577E−01	5.6473E−01

(Continued)

M	T_t/T	p_t/p	ρ_t/ρ	A/A^*	T/T^*	p/p^*	ρ/ρ^*
1.58	1.3745E+00	3.9680E+00	2.8869E+00	1.2536E+00	8.3669E−01	4.6180E−01	5.5194E−01
1.60	1.3840E+00	4.0887E+00	2.9543E+00	1.2712E+00	8.3092E−01	4.4816E−01	5.3936E−01
1.62	1.3937E+00	4.2138E+00	3.0236E+00	1.2895E+00	8.2517E−01	4.3486E−01	5.2699E−01
1.64	1.4034E+00	4.3435E+00	3.0949E+00	1.3083E+00	8.1942E−01	4.2188E−01	5.1485E−01
1.66	1.4133E+00	4.4778E+00	3.1682E+00	1.3279E+00	8.1368E−01	4.0922E−01	5.0293E−01
1.68	1.4234E+00	4.6170E+00	3.2437E+00	1.3481E+00	8.0795E−01	3.9688E−01	4.9122E−01
1.70	1.4335E+00	4.7612E+00	3.3214E+00	1.3690E+00	8.0223E−01	3.8486E−01	4.7974E−01
1.72	1.4438E+00	4.9107E+00	3.4013E+00	1.3906E+00	7.9653E−01	3.7315E−01	4.6847E−01
1.74	1.4541E+00	5.0655E+00	3.4835E+00	1.4128E+00	7.9085E−01	3.6174E−01	4.5741E−01
1.76	1.4646E+00	5.2259E+00	3.5680E+00	1.4359E+00	7.8518E−01	3.5064E−01	4.4657E−01
1.78	1.4753E+00	5.3921E+00	3.6550E+00	1.4596E+00	7.7952E−01	3.3983E−01	4.3595E−01
1.80	1.4860E+00	5.5643E+00	3.7445E+00	1.4841E+00	7.7389E−01	3.2931E−01	4.2553E−01
1.82	1.4969E+00	5.7427E+00	3.8365E+00	1.5093E+00	7.6827E−01	3.1908E−01	4.1533E−01
1.84	1.5078E+00	5.9274E+00	3.9311E+00	1.5353E+00	7.6268E−01	3.0914E−01	4.0533E−01
1.86	1.5189E+00	6.1189E+00	4.0284E+00	1.5621E+00	7.5711E−01	2.9947E−01	3.9554E−01
1.88	1.5302E+00	6.3171E+00	4.1284E+00	1.5897E+00	7.5156E−01	2.9007E−01	3.8596E−01
1.90	1.5415E+00	6.5225E+00	4.2313E+00	1.6181E+00	7.4603E−01	2.8093E−01	3.7658E−01
1.92	1.5530E+00	6.7353E+00	4.3371E+00	1.6474E+00	7.4052E−01	2.7206E−01	3.6739E−01
1.94	1.5645E+00	6.9556E+00	4.4458E+00	1.6775E+00	7.3504E−01	2.6344E−01	3.5841E−01
1.96	1.5762E+00	7.1838E+00	4.5576E+00	1.7085E+00	7.2958E−01	2.5507E−01	3.4961E−01
1.98	1.5881E+00	7.4202E+00	4.6725E+00	1.7404E+00	7.2415E−01	2.4695E−01	3.4102E−01
2.00	1.6000E+00	7.6650E+00	4.7906E+00	1.7732E+00	7.1875E−01	2.3906E−01	3.3261E−01
2.02	1.6121E+00	7.9185E+00	4.9121E+00	1.8069E+00	7.1337E−01	2.3140E−01	3.2438E−01
2.04	1.6242E+00	8.1811E+00	5.0369E+00	1.8415E+00	7.0802E−01	2.2398E−01	3.1635E−01
2.06	1.6365E+00	8.4529E+00	5.1651E+00	1.8772E+00	7.0270E−01	2.1677E−01	3.0849E−01
2.08	1.6490E+00	8.7345E+00	5.2970E+00	1.9138E+00	6.9741E−01	2.0979E−01	3.0081E−01
2.10	1.6615E+00	9.0260E+00	5.4324E+00	1.9514E+00	6.9215E−01	2.0301E−01	2.9331E−01
2.12	1.6742E+00	9.3278E+00	5.5716E+00	1.9901E+00	6.8691E−01	1.9644E−01	2.8598E−01
2.14	1.6869E+00	9.6403E+00	5.7147E+00	2.0298E+00	6.8171E−01	1.9008E−01	2.7883E−01
2.16	1.6998E+00	9.9638E+00	5.8616E+00	2.0706E+00	6.7653E−01	1.8390E−01	2.7183E−01
2.18	1.7129E+00	1.0299E+01	6.0126E+00	2.1125E+00	6.7139E−01	1.7792E−01	2.6501E−01
2.20	1.7260E+00	1.0646E+01	6.1678E+00	2.1555E+00	6.6628E−01	1.7213E−01	2.5834E−01
2.22	1.7393E+00	1.1005E+01	6.3271E+00	2.1997E+00	6.6120E−01	1.6651E−01	2.5184E−01
2.24	1.7526E+00	1.1376E+01	6.4908E+00	2.2450E+00	6.5615E−01	1.6107E−01	2.4548E−01
2.26	1.7661E+00	1.1761E+01	6.6590E+00	2.2916E+00	6.5114E−01	1.5581E−01	2.3928E−01
2.28	1.7798E+00	1.2159E+01	6.8317E+00	2.3394E+00	6.4615E−01	1.5070E−01	2.3324E−01
2.30	1.7935E+00	1.2571E+01	7.0091E+00	2.3884E+00	6.4120E−01	1.4576E−01	2.2733E−01
2.32	1.8074E+00	1.2997E+01	7.1913E+00	2.4388E+00	6.3629E−01	1.4098E−01	2.2157E−01
2.34	1.8213E+00	1.3439E+01	7.3784E+00	2.4904E+00	6.3140E−01	1.3635E−01	2.1595E−01
2.36	1.8354E+00	1.3895E+01	7.5705E+00	2.5434E+00	6.2655E−01	1.3187E−01	2.1047E−01

M	T_t/T	p_t/p	ρ_t/ρ	A/A^*	T/T^*	p/p^*	ρ/ρ^*
2.38	1.8497E+00	1.4368E+01	7.7678E+00	2.5977E+00	6.2174E−01	1.2753E−01	2.0513E−01
2.40	1.8640E+00	1.4857E+01	7.9703E+00	2.6535E+00	6.1695E−01	1.2334E−01	1.9992E−01
2.42	1.8785E+00	1.5363E+01	8.1783E+00	2.7107E+00	6.1220E−01	1.1928E−01	1.9483E−01
2.44	1.8930E+00	1.5886E+01	8.3918E+00	2.7693E+00	6.0749E−01	1.1535E−01	1.8987E−01
2.46	1.9077E+00	1.6428E+01	8.6110E+00	2.8295E+00	6.0281E−01	1.1154E−01	1.8504E−01
2.48	1.9226E+00	1.6988E+01	8.8360E+00	2.8912E+00	5.9816E−01	1.0786E−01	1.8033E−01
2.50	1.9375E+00	1.7567E+01	9.0670E+00	2.9544E+00	5.9355E−01	1.0431E−01	1.7574E−01
2.52	1.9526E+00	1.8167E+01	9.3040E+00	3.0193E+00	5.8897E−01	1.0086E−01	1.7126E−01
2.54	1.9677E+00	1.8787E+01	9.5473E+00	3.0857E+00	5.8443E−01	9.7536E−02	1.6689E−01
2.56	1.9830E+00	1.9428E+01	9.7970E+00	3.1539E+00	5.7992E−01	9.4317E−02	1.6264E−01
2.58	1.9985E+00	2.0091E+01	1.0053E+01	3.2238E+00	5.7544E−01	9.1203E−02	1.5850E−01
2.60	2.0140E+00	2.0777E+01	1.0316E+01	3.2954E+00	5.7100E−01	8.8193E−02	1.5446E−01
2.62	2.0297E+00	2.1486E+01	1.0586E+01	3.3688E+00	5.6660E−01	8.5282E−02	1.5052E−01
2.64	2.0454E+00	2.2219E+01	1.0863E+01	3.4440E+00	5.6223E−01	8.2467E−02	1.4668E−01
2.66	2.0613E+00	2.2978E+01	1.1147E+01	3.5211E+00	5.5789E−01	7.9746E−02	1.4295E−01
2.68	2.0774E+00	2.3761E+01	1.1438E+01	3.6001E+00	5.5359E−01	7.7115E−02	1.3930E−01
2.70	2.0935E+00	2.4572E+01	1.1737E+01	3.6810E+00	5.4932E−01	7.4572E−02	1.3576E−01
2.72	2.1098E+00	2.5410E+01	1.2044E+01	3.7639E+00	5.4509E−01	7.2113E−02	1.3230E−01
2.74	2.1261E+00	2.6276E+01	1.2358E+01	3.8489E+00	5.4089E−01	6.9736E−02	1.2893E−01
2.76	2.1426E+00	2.7171E+01	1.2681E+01	3.9359E+00	5.3672E−01	6.7439E−02	1.2565E−01
2.78	2.1593E+00	2.8096E+01	1.3012E+01	4.0251E+00	5.3259E−01	6.5218E−02	1.2246E−01
2.80	2.1760E+00	2.9052E+01	1.3351E+01	4.1164E+00	5.2849E−01	6.3072E−02	1.1935E−01
2.82	2.1929E+00	3.0040E+01	1.3699E+01	4.2099E+00	5.2443E−01	6.0997E−02	1.1631E−01
2.84	2.2098E+00	3.1061E+01	1.4056E+01	4.3057E+00	5.2040E−01	5.8992E−02	1.1336E−01
2.86	2.2269E+00	3.2116E+01	1.4422E+01	4.4038E+00	5.1640E−01	5.7054E−02	1.1049E−01
2.88	2.2442E+00	3.3206E+01	1.4797E+01	4.5043E+00	5.1244E−01	5.5181E−02	1.0769E−01
2.90	2.2615E+00	3.4333E+01	1.5181E+01	4.6072E+00	5.0851E−01	5.3371E−02	1.0496E−01
2.92	2.2790E+00	3.5496E+01	1.5576E+01	4.7125E+00	5.0462E−01	5.1622E−02	1.0230E−01
2.94	2.2965E+00	3.6698E+01	1.5980E+01	4.8204E+00	5.0075E−01	4.9931E−02	9.9714E−02
2.96	2.3142E+00	3.7939E+01	1.6394E+01	4.9308E+00	4.9692E−01	4.8297E−02	9.7194E−02
2.98	2.3321E+00	3.9222E+01	1.6818E+01	5.0439E+00	4.9313E−01	4.6718E−02	9.4741E−02
3.00	2.3500E+00	4.0546E+01	1.7254E+01	5.1596E+00	4.8936E−01	4.5192E−02	9.2351E−02
3.05	2.3954E+00	4.4049E+01	1.8389E+01	5.4611E+00	4.8009E−01	4.1598E−02	8.6648E−02
3.10	2.4415E+00	4.7845E+01	1.9596E+01	5.7806E+00	4.7102E−01	3.8298E−02	8.1311E−02
3.15	2.4884E+00	5.1954E+01	2.0879E+01	6.1190E+00	4.6215E−01	3.5269E−02	7.6316E−02
3.20	2.5360E+00	5.6403E+01	2.2241E+01	6.4774E+00	4.5347E−01	3.2487E−02	7.1643E−02
3.25	2.5844E+00	6.1215E+01	2.3687E+01	6.8569E+00	4.4498E−01	2.9933E−02	6.7270E−02
3.30	2.6335E+00	6.6420E+01	2.5221E+01	7.2584E+00	4.3668E−01	2.7588E−02	6.3177E−02
3.35	2.6834E+00	7.2045E+01	2.6849E+01	7.6833E+00	4.2856E−01	2.5433E−02	5.9347E−02
3.40	2.7340E+00	7.8123E+01	2.8575E+01	8.1326E+00	4.2063E−01	2.3455E−02	5.5763E−02

(Continued)

M	T_t/T	p_t/p	ρ_t/ρ	A/A^*	T/T^*	p/p^*	ρ/ρ^*
3.45	2.7854E+00	8.4687E+01	3.0404E+01	8.6076E+00	4.1287E−01	2.1637E−02	5.2407E−02
3.50	2.8375E+00	9.1771E+01	3.2342E+01	9.1096E+00	4.0529E−01	1.9967E−02	4.9267E−02
3.55	2.8904E+00	9.9415E+01	3.4395E+01	9.6399E+00	3.9787E−01	1.8431E−02	4.6326E−02
3.60	2.9440E+00	1.0766E+02	3.6569E+01	1.0200E+01	3.9063E−01	1.7020E−02	4.3573E−02
3.65	2.9984E+00	1.1654E+02	3.8869E+01	1.0791E+01	3.8354E−01	1.5722E−02	4.0994E−02
3.70	3.0535E+00	1.2612E+02	4.1303E+01	1.1416E+01	3.7662E−01	1.4529E−02	3.8579E−02
3.75	3.1094E+00	1.3643E+02	4.3876E+01	1.2074E+01	3.6985E−01	1.3431E−02	3.6316E−02
3.80	3.1660E+00	1.4752E+02	4.6596E+01	1.2769E+01	3.6323E−01	1.2421E−02	3.4196E−02
3.85	3.2234E+00	1.5946E+02	4.9471E+01	1.3501E+01	3.5677E−01	1.1491E−02	3.2209E−02
3.90	3.2815E+00	1.7230E+02	5.2508E+01	1.4273E+01	3.5045E−01	1.0634E−02	3.0346E−02
3.95	3.3404E+00	1.8611E+02	5.5714E+01	1.5087E+01	3.4427E−01	9.8457E−03	2.8599E−02
4.00	3.4000E+00	2.0094E+02	5.9099E+01	1.5944E+01	3.3824E−01	9.1191E−03	2.6962E−02
4.05	3.4604E+00	2.1686E+02	6.2670E+01	1.6846E+01	3.3233E−01	8.4494E−03	2.5425E−02
4.10	3.5215E+00	2.3396E+02	6.6437E+01	1.7796E+01	3.2657E−01	7.8320E−03	2.3984E−02
4.15	3.5834E+00	2.5230E+02	7.0408E+01	1.8795E+01	3.2093E−01	7.2626E−03	2.2631E−02
4.20	3.6460E+00	2.7197E+02	7.4594E+01	1.9847E+01	3.1541E−01	6.7373E−03	2.1361E−02
4.25	3.7094E+00	2.9306E+02	7.9004E+01	2.0953E+01	3.1003E−01	6.2525E−03	2.0169E−02
4.30	3.7735E+00	3.1565E+02	8.3649E+01	2.2115E+01	3.0476E−01	5.8050E−03	1.9049E−02
4.35	3.8384E+00	3.3985E+02	8.8540E+01	2.3337E+01	2.9961E−01	5.3916E−03	1.7996E−02
4.40	3.9040E+00	3.6575E+02	9.3687E+01	2.4621E+01	2.9457E−01	5.0098E−03	1.7008E−02
4.45	3.9704E+00	3.9347E+02	9.9103E+01	2.5970E+01	2.8965E−01	4.6568E−03	1.6078E−02
4.50	4.0375E+00	4.2312E+02	1.0480E+02	2.7386E+01	2.8483E−01	4.3305E−03	1.5204E−02
4.55	4.1054E+00	4.5482E+02	1.1079E+02	2.8872E+01	2.8012E−01	4.0287E−03	1.4383E−02
4.60	4.1740E+00	4.8870E+02	1.1708E+02	3.0432E+01	2.7552E−01	3.7494E−03	1.3609E−02
4.65	4.2434E+00	5.2488E+02	1.2369E+02	3.2069E+01	2.7101E−01	3.4909E−03	1.2882E−02
4.70	4.3135E+00	5.6352E+02	1.3064E+02	3.3785E+01	2.6660E−01	3.2516E−03	1.2197E−02
4.75	4.3844E+00	6.0475E+02	1.3793E+02	3.5584E+01	2.6230E−01	3.0299E−03	1.1552E−02
4.80	4.4560E+00	6.4875E+02	1.4559E+02	3.7470E+01	2.5808E−01	2.8244E−03	1.0944E−02
4.85	4.5284E+00	6.9566E+02	1.5362E+02	3.9447E+01	2.5395E−01	2.6340E−03	1.0372E−02
4.90	4.6015E+00	7.4566E+02	1.6205E+02	4.1517E+01	2.4992E−01	2.4573E−03	9.8329E−03
4.95	4.6754E+00	7.9894E+02	1.7088E+02	4.3685E+01	2.4597E−01	2.2934E−03	9.3245E−03
5.00	4.7500E+00	8.5569E+02	1.8015E+02	4.5954E+01	2.4211E−01	2.1413E−03	8.8451E−03
5.10	4.9015E+00	9.8040E+02	2.0002E+02	5.0815E+01	2.3462E−01	1.8690E−03	7.9662E−03
5.20	5.0560E+00	1.1215E+03	2.2182E+02	5.6134E+01	2.2745E−01	1.6338E−03	7.1833E−03
5.30	5.2135E+00	1.2810E+03	2.4570E+02	6.1947E+01	2.2058E−01	1.4304E−03	6.4851E−03
5.40	5.3740E+00	1.4608E+03	2.7183E+02	6.8294E+01	2.1399E−01	1.2543E−03	5.8617E−03
5.50	5.5375E+00	1.6634E+03	3.0039E+02	7.5216E+01	2.0767E−01	1.1015E−03	5.3044E−03
5.60	5.7040E+00	1.8913E+03	3.3157E+02	8.2756E+01	2.0161E−01	9.6883E−04	4.8056E−03
5.70	5.8735E+00	2.1471E+03	3.6556E+02	9.0963E+01	1.9579E−01	8.5337E−04	4.3587E−03
5.80	6.0460E+00	2.4341E+03	4.0259E+02	9.9885E+01	1.9021E−01	7.5277E−04	3.9578E−03

M	T_t/T	p_t/p	ρ_t/ρ	A/A^*	T/T^*	p/p^*	ρ/ρ^*
5.90	6.2215E+00	2.7554E+03	4.4288E+02	1.0958E+02	1.8484E−01	6.6499E−04	3.5978E−03
6.00	6.4000E+00	3.1147E+03	4.8668E+02	1.2009E+02	1.7969E−01	5.8827E−04	3.2740E−03
6.10	6.5815E+00	3.5160E+03	5.3422E+02	1.3149E+02	1.7473E−01	5.2114E−04	2.9826E−03
6.20	6.7660E+00	3.9635E+03	5.8579E+02	1.4383E+02	1.6997E−01	4.6230E−04	2.7201E−03
6.30	6.9535E+00	4.4619E+03	6.4168E+02	1.5718E+02	1.6538E−01	4.1066E−04	2.4832E−03
6.40	7.1440E+00	5.0163E+03	7.0217E+02	1.7162E+02	1.6097E−01	3.6527E−04	2.2692E−03
6.50	7.3375E+00	5.6322E+03	7.6759E+02	1.8721E+02	1.5673E−01	3.2533E−04	2.0758E−03
6.60	7.5340E+00	6.3156E+03	8.3828E+02	2.0403E+02	1.5264E−01	2.9012E−04	1.9008E−03
6.70	7.7335E+00	7.0729E+03	9.1458E+02	2.2216E+02	1.4870E−01	2.5906E−04	1.7422E−03
6.80	7.9360E+00	7.9112E+03	9.9688E+02	2.4169E+02	1.4491E−01	2.3161E−04	1.5984E−03
6.90	8.1415E+00	8.8380E+03	1.0856E+03	2.6271E+02	1.4125E−01	2.0732E−04	1.4678E−03
7.00	8.3500E+00	9.8615E+03	1.1810E+03	2.8532E+02	1.3772E−01	1.8580E−04	1.3492E−03
7.10	8.5615E+00	1.0991E+04	1.2837E+03	3.0961E+02	1.3432E−01	1.6672E−04	1.2412E−03
7.20	8.7760E+00	1.2235E+04	1.3941E+03	3.3569E+02	1.3104E−01	1.4976E−04	1.1430E−03
7.30	8.9935E+00	1.3604E+04	1.5126E+03	3.6366E+02	1.2787E−01	1.3469E−04	1.0534E−03
7.40	9.2140E+00	1.5109E+04	1.6398E+03	3.9365E+02	1.2481E−01	1.2127E−04	9.7169E−04
7.50	9.4375E+00	1.6763E+04	1.7762E+03	4.2578E+02	1.2185E−01	1.0931E−04	8.9708E−04
7.60	9.6640E+00	1.8577E+04	1.9223E+03	4.6017E+02	1.1900E−01	9.8631E−05	8.2890E−04
7.70	9.8935E+00	2.0566E+04	2.0787E+03	4.9695E+02	1.1624E−01	8.9093E−05	7.6652E−04
7.80	1.0126E+01	2.2744E+04	2.2461E+03	5.3626E+02	1.1357E−01	8.0562E−05	7.0941E−04
7.90	1.0362E+01	2.5126E+04	2.4250E+03	5.7825E+02	1.1099E−01	7.2923E−05	6.5708E−04
8.00	1.0600E+01	2.7731E+04	2.6161E+03	6.2308E+02	1.0849E−01	6.6075E−05	6.0908E−04
8.10	1.0842E+01	3.0574E+04	2.8201E+03	6.7089E+02	1.0607E−01	5.9929E−05	5.6502E−04
8.20	1.1086E+01	3.3676E+04	3.0377E+03	7.2185E+02	1.0373E−01	5.4409E−05	5.2454E−04
8.30	1.1334E+01	3.7057E+04	3.2697E+03	7.7614E+02	1.0147E−01	4.9444E−05	4.8732E−04
8.40	1.1584E+01	4.0740E+04	3.5169E+03	8.3394E+02	9.9275E−02	4.4975E−05	4.5307E−04
8.50	1.1838E+01	4.4746E+04	3.7800E+03	8.9544E+02	9.7149E−02	4.0948E−05	4.2153E−04
8.60	1.2094E+01	4.9102E+04	4.0600E+03	9.6082E+02	9.5088E−02	3.7316E−05	3.9246E−04
8.70	1.2354E+01	5.3834E+04	4.3578E+03	1.0303E+03	9.3091E−02	3.4036E−05	3.6565E−04
8.80	1.2616E+01	5.8969E+04	4.6741E+03	1.1041E+03	9.1154E−02	3.1072E−05	3.4090E−04
8.90	1.2882E+01	6.4538E+04	5.0102E+03	1.1824E+03	8.9275E−02	2.8390E−05	3.1803E−04
9.00	1.3150E+01	7.0573E+04	5.3668E+03	1.2655E+03	8.7452E−02	2.5963E−05	2.9690E−04
9.10	1.3422E+01	7.7108E+04	5.7451E+03	1.3536E+03	8.5683E−02	2.3762E−05	2.7735E−04
9.20	1.3696E+01	8.4179E+04	6.1462E+03	1.4469E+03	8.3966E−02	2.1766E−05	2.5925E−04
9.30	1.3973E+01	9.1823E+04	6.5712E+03	1.5458E+03	8.2299E−02	1.9954E−05	2.4248E−04
9.40	1.4254E+01	1.0008E+05	7.0213E+03	1.6504E+03	8.0679E−02	1.8308E−05	2.2694E−04
9.50	1.4537E+01	1.0900E+05	7.4977E+03	1.7611E+03	7.9106E−02	1.6810E−05	2.1252E−04
9.60	1.4824E+01	1.1862E+05	8.0016E+03	1.8781E+03	7.7577E−02	1.5447E−05	1.9913E−04
9.70	1.5113E+01	1.2899E+05	8.5345E+03	2.0018E+03	7.6091E−02	1.4205E−05	1.8670E−04
9.80	1.5406E+01	1.4016E+05	9.0976E+03	2.1324E+03	7.4646E−02	1.3073E−05	1.7514E−04

(Continued)

M	T_t/T	p_t/p	ρ_t/ρ	A/A^*	T/T^*	p/p^*	ρ/ρ^*
9.90	1.5701E+01	1.5218E+05	9.6924E+03	2.2703E+03	7.3241E−02	1.2040E−05	1.6440E−04
10.00	1.6000E+01	1.6513E+05	1.0320E+04	2.4159E+03	7.1875E−02	1.1096E−05	1.5439E−04
10.10	1.6301E+01	1.7904E+05	1.0983E+04	2.5694E+03	7.0546E−02	1.0234E−05	1.4508E−04
10.20	1.6606E+01	1.9399E+05	1.1682E+04	2.7313E+03	6.9252E−02	9.4452E−06	1.3640E−04
10.30	1.6913E+01	2.1004E+05	1.2419E+04	2.9019E+03	6.7993E−02	8.7233E−06	1.2831E−04
10.40	1.7224E+01	2.2727E+05	1.3195E+04	3.0815E+03	6.6767E−02	8.0621E−06	1.2076E−04
10.50	1.7537E+01	2.4575E+05	1.4013E+04	3.2707E+03	6.5574E−02	7.4559E−06	1.1371E−04
10.60	1.7854E+01	2.6555E+05	1.4873E+04	3.4698E+03	6.4411E−02	6.8999E−06	1.0713E−04
10.70	1.8173E+01	2.8676E+05	1.5779E+04	3.6792E+03	6.3279E−02	6.3894E−06	1.0098E−04
10.80	1.8496E+01	3.0948E+05	1.6732E+04	3.8993E+03	6.2176E−02	5.9205E−06	9.5230E−05
10.90	1.8821E+01	3.3378E+05	1.7734E+04	4.1308E+03	6.1100E−02	5.4894E−06	8.9850E−05
11.00	1.9150E+01	3.5977E+05	1.8787E+04	4.3739E+03	6.0052E−02	5.0929E−06	8.4815E−05
11.10	1.9481E+01	3.8754E+05	1.9893E+04	4.6293E+03	5.9030E−02	4.7279E−06	8.0099E−05
11.20	1.9816E+01	4.1721E+05	2.1054E+04	4.8973E+03	5.8034E−02	4.3916E−06	7.5680E−05
11.30	2.0153E+01	4.4889E+05	2.2274E+04	5.1786E+03	5.7062E−02	4.0817E−06	7.1537E−05
11.40	2.0494E+01	4.8269E+05	2.3553E+04	5.4737E+03	5.6114E−02	3.7959E−06	6.7652E−05
11.50	2.0837E+01	5.1874E+05	2.4895E+04	5.7831E+03	5.5189E−02	3.5321E−06	6.4005E−05
11.60	2.1184E+01	5.5717E+05	2.6302E+04	6.1074E+03	5.4286E−02	3.2885E−06	6.0582E−05
11.70	2.1533E+01	5.9812E+05	2.7776E+04	6.4472E+03	5.3405E−02	3.0634E−06	5.7366E−05
11.80	2.1886E+01	6.4172E+05	2.9321E+04	6.8031E+03	5.2545E−02	2.8552E−06	5.4343E−05
11.90	2.2241E+01	6.8812E+05	3.0939E+04	7.1757E+03	5.1705E−02	2.6627E−06	5.1502E−05
12.00	2.2600E+01	7.3749E+05	3.2632E+04	7.5657E+03	5.0885E−02	2.4844E−06	4.8829E−05
12.10	2.2961E+01	7.8999E+05	3.4405E+04	7.9738E+03	5.0084E−02	2.3193E−06	4.6313E−05
12.20	2.3326E+01	8.4579E+05	3.6260E+04	8.4006E+03	4.9301E−02	2.1663E−06	4.3944E−05
12.30	2.3693E+01	9.0507E+05	3.8199E+04	8.8468E+03	4.8537E−02	2.0244E−06	4.1713E−05
12.40	2.4064E+01	9.6801E+05	4.0227E+04	9.3133E+03	4.7789E−02	1.8928E−06	3.9611E−05
12.50	2.4437E+01	1.0348E+06	4.2346E+04	9.8006E+03	4.7059E−02	1.7706E−06	3.7628E−05
12.60	2.4814E+01	1.1057E+06	4.4560E+04	1.0310E+04	4.6345E−02	1.6571E−06	3.5759E−05
12.70	2.5193E+01	1.1809E+06	4.6872E+04	1.0841E+04	4.5647E−02	1.5516E−06	3.3995E−05
12.80	2.5576E+01	1.2606E+06	4.9287E+04	1.1396E+04	4.4964E−02	1.4535E−06	3.2329E−05
12.90	2.5961E+01	1.3450E+06	5.1807E+04	1.1975E+04	4.4296E−02	1.3623E−06	3.0757E−05
13.00	2.6350E+01	1.4344E+06	5.4436E+04	1.2579E+04	4.3643E−02	1.2774E−06	2.9271E−05
13.10	2.6741E+01	1.5291E+06	5.7179E+04	1.3210E+04	4.3004E−02	1.1983E−06	2.7867E−05
13.20	2.7136E+01	1.6292E+06	6.0040E+04	1.3866E+04	4.2379E−02	1.1246E−06	2.6539E−05
13.30	2.7533E+01	1.7352E+06	6.3022E+04	1.4551E+04	4.1767E−02	1.0559E−06	2.5283E−05
13.40	2.7934E+01	1.8473E+06	6.6130E+04	1.5265E+04	4.1168E−02	9.9186E−07	2.4095E−05
13.50	2.8337E+01	1.9657E+06	6.9368E+04	1.6008E+04	4.0582E−02	9.3210E−07	2.2970E−05
13.60	2.8744E+01	2.0908E+06	7.2740E+04	1.6782E+04	4.0008E−02	8.7631E−07	2.1905E−05
13.70	2.9153E+01	2.2230E+06	7.6252E+04	1.7588E+04	3.9446E−02	8.2421E−07	2.0896E−05
13.80	2.9566E+01	2.3626E+06	7.9908E+04	1.8426E+04	3.8896E−02	7.7552E−07	1.9940E−05

M	T_t/T	p_t/p	ρ_t/ρ	A/A^*	T/T^*	p/p^*	ρ/ρ^*
13.90	2.9981E+01	2.5099E+06	8.3713E+04	1.9299E+04	3.8357E−02	7.3001E−07	1.9034E−05
14.00	3.0400E+01	2.6652E+06	8.7672E+04	2.0207E+04	3.7829E−02	6.8745E−07	1.8174E−05
14.10	3.0821E+01	2.8291E+06	9.1790E+04	2.1151E+04	3.7312E−02	6.4763E−07	1.7359E−05
14.20	3.1246E+01	3.0019E+06	9.6072E+04	2.2133E+04	3.6805E−02	6.1036E−07	1.6585E−05
14.30	3.1673E+01	3.1839E+06	1.0052E+05	2.3153E+04	3.6308E−02	5.7546E−07	1.5851E−05
14.40	3.2104E+01	3.3758E+06	1.0515E+05	2.4213E+04	3.5821E−02	5.4276E−07	1.5153E−05
14.50	3.2537E+01	3.5778E+06	1.0996E+05	2.5315E+04	3.5344E−02	5.1211E−07	1.4491E−05
14.60	3.2974E+01	3.7905E+06	1.1495E+05	2.6459E+04	3.4876E−02	4.8338E−07	1.3861E−05
14.70	3.3413E+01	4.0143E+06	1.2014E+05	2.7648E+04	3.4417E−02	4.5642E−07	1.3263E−05
14.80	3.3856E+01	4.2498E+06	1.2553E+05	2.8881E+04	3.3967E−02	4.3113E−07	1.2694E−05
14.90	3.4301E+01	4.4975E+06	1.3112E+05	3.0162E+04	3.3526E−02	4.0739E−07	1.2152E−05
15.00	3.4750E+01	4.7579E+06	1.3692E+05	3.1490E+04	3.3094E−02	3.8509E−07	1.1637E−05
16.00	3.9400E+01	8.1991E+06	2.0810E+05	4.7777E+04	2.9188E−02	2.2347E−07	7.6570E−06
17.00	4.4350E+01	1.3693E+07	3.0874E+05	7.0781E+04	2.5930E−02	1.3381E−07	5.1610E−06
18.00	4.9600E+01	2.2235E+07	4.4828E+05	1.0265E+05	2.3185E−02	8.2403E−08	3.5545E−06
19.00	5.5150E+01	3.5208E+07	6.3840E+05	1.4603E+05	2.0852E−02	5.2039E−08	2.4959E−06
20.00	6.1000E+01	5.4497E+07	8.9339E+05	2.0417E+05	1.8852E−02	3.3620E−08	1.7835E−06
21.00	6.7150E+01	8.2630E+07	1.2305E+06	2.8101E+05	1.7126E−02	2.2173E−08	1.2949E−06
22.00	7.3600E+01	1.2295E+08	1.6706E+06	3.8125E+05	1.5625E−02	1.4901E−08	9.5381E−07
23.00	8.0350E+01	1.7984E+08	2.2382E+06	5.1048E+05	1.4312E−02	1.0188E−08	7.1193E−07
24.00	8.7400E+01	2.5891E+08	2.9624E+06	6.7532E+05	1.3158E−02	7.0764E−09	5.3788E−07
25.00	9.4750E+01	3.6738E+08	3.8773E+06	8.8350E+05	1.2137E−02	4.9872E−09	4.1095E−07
26.00	1.0240E+02	5.1432E+08	5.0226E+06	1.1440E+06	1.1230E−02	3.5623E−09	3.1724E−07
27.00	1.1035E+02	7.1112E+08	6.4443E+06	1.4673E+06	1.0421E−02	2.5764E−09	2.4726E−07
28.00	1.1860E+02	9.7192E+08	8.1950E+06	1.8653E+06	9.6965E−03	1.8851E−09	1.9444E−07
29.00	1.2715E+02	1.3141E+09	1.0335E+07	2.3518E+06	9.0444E−03	1.3942E−09	1.5417E−07
30.00	1.3600E+02	1.7590E+09	1.2934E+07	2.9424E+06	8.4559E−03	1.0416E−09	1.2320E−07
31.00	1.4515E+02	2.3324E+09	1.6069E+07	3.6548E+06	7.9228E−03	7.8552E−10	9.9161E−08
32.00	1.5460E+02	3.0655E+09	1.9829E+07	4.5089E+06	7.4386E−03	5.9766E−10	8.0359E−08
33.00	1.6435E+02	3.9957E+09	2.4312E+07	5.5274E+06	6.9973E−03	4.5852E−10	6.5539E−08
34.00	1.7440E+02	5.1677E+09	2.9631E+07	6.7355E+06	6.5940E−03	3.5454E−10	5.3775E−08
35.00	1.8475E+02	6.6342E+09	3.5909E+07	8.1612E+06	6.2246E−03	2.7616E−10	4.4373E−08
40.00	2.4100E+02	2.0989E+10	8.7092E+07	1.9781E+07	4.7718E−03	8.7289E−11	1.8296E−08
45.00	3.0475E+02	5.8033E+10	1.9043E+08	4.3233E+07	3.7736E−03	3.1570E−11	8.3675E−09
50.00	3.7600E+02	1.4423E+11	3.8360E+08	8.7061E+07	3.0585E−03	1.2702E−11	4.1538E−09
100.00	1.5010E+03	5.8105E+13	3.8711E+10	8.7770E+09	7.6616E−04	3.1530E−14	4.1162E−11
1000.00	1.5000E+05	2.6889E+22	1.7926E+17	4.0630E+16	7.6666E−06	6.8124E−23	8.8889E−18

Normal Shock Table

$\mathbf{N}$ormal shock relations that are used in calculating the jump conditions, for a calorically perfect gas (i.e., $\gamma = $ constant), are summarized below:

$$M_2^2 = \frac{2 + (\gamma - 1)M_1^2}{2\gamma M_1^2 - (\gamma - 1)} \tag{C.1}$$

$$\frac{\rho_2}{\rho_1} = \frac{(\gamma + 1)M_1^2}{2 + (\gamma - 1)M_1^2} \tag{C.2}$$

$$\frac{p_2}{p_1} = 1 + \frac{2\gamma}{\gamma + 1}(M_1^2 - 1) \tag{C.3}$$

$$\frac{T_2}{T_1} = \left[1 + \frac{2\gamma}{\gamma + 1}(M_1^2 - 1)\right]\left[\frac{2 + (\gamma - 1)M_1^2}{(\gamma + 1)M_1^2}\right] \tag{C.4}$$

$$\frac{p_{t2}}{p_{t1}} = \left[1 + \frac{2\gamma}{\gamma + 1}(M_1^2 - 1)\right]\left[\frac{1 + \frac{\gamma - 1}{2}\left(\frac{2 + (\gamma - 1)M_1^2}{2\gamma M_1^2 - (\gamma - 1)}\right)}{1 + \frac{\gamma - 1}{2}M_1^2}\right]^{\frac{\gamma}{\gamma - 1}} \tag{C.5}$$

$$\Delta s/R = -\ln(p_{t2}/p_{t1}) \tag{C.6}$$

$$\frac{p_{t2}}{p_1} = \frac{p_{t2}}{p_{t1}}\frac{p_{t1}}{p_1} = \left[1 + \frac{2\gamma}{\gamma + 1}(M_1^2 - 1)\right]\left[+\frac{\gamma - 1}{2}\left(\frac{2 + (\gamma - 1)M_1^2}{2\gamma M_1^2 - (\gamma - 1)}\right)\right]^{\frac{\gamma}{\gamma - 1}} \tag{C.7}$$

We have derived these expressions in Chapter 2.

Normal Shock Table $\gamma = 1.4$

M_1	M_2	T_2/T_1	p_2/p_1	ρ_2/ρ_1	p_{t2}/p_{t1}	p_{t2}/p_1	$\Delta s/R$
1	1	1	1	1	1	1.8929	0
1.02	0.9805	1.0132E+00	1.0471E+00	1.0334E+00	9.9999E−01	1.9379E+00	9.9554E−06
1.04	0.9620	1.0263E+00	1.0952E+00	1.0671E+00	9.9992E−01	1.9844E+00	7.6696E−05
1.06	0.9444	1.0393E+00	1.1442E+00	1.1009E+00	9.9975E−01	2.0325E+00	2.4928E−04
1.08	0.9277	1.0522E+00	1.1941E+00	1.1349E+00	9.9943E−01	2.0819E+00	5.6942E−04
1.1	0.9118	1.0649E+00	1.2450E+00	1.1691E+00	9.9893E−01	2.1328E+00	1.0725E−03
1.12	0.8966	1.0776E+00	1.2968E+00	1.2034E+00	9.9821E−01	2.1851E+00	1.7885E−03
1.14	0.8820	1.0903E+00	1.3495E+00	1.2378E+00	9.9726E−01	2.2388E+00	2.7426E−03
1.16	0.8682	1.1029E+00	1.4032E+00	1.2723E+00	9.9605E−01	2.2937E+00	3.9559E−03
1.18	0.8549	1.1154E+00	1.4578E+00	1.3069E+00	9.9457E−01	2.3500E+00	5.4461E−03
1.2	0.8422	1.1280E+00	1.5133E+00	1.3416E+00	9.9280E−01	2.4075E+00	7.2276E−03
1.22	0.8300	1.1405E+00	1.5698E+00	1.3764E+00	9.9073E−01	2.4663E+00	9.3122E−03
1.24	0.8183	1.1531E+00	1.6272E+00	1.4112E+00	9.8836E−01	2.5263E+00	1.1709E−02
1.26	0.8071	1.1657E+00	1.6855E+00	1.4460E+00	9.8568E−01	2.5875E+00	1.4427E−02
1.28	0.7963	1.1783E+00	1.7448E+00	1.4808E+00	9.8268E−01	2.6500E+00	1.7469E−02
1.3	0.7860	1.1909E+00	1.8050E+00	1.5157E+00	9.7937E−01	2.7136E+00	2.0842E−02
1.32	0.7760	1.2035E+00	1.8661E+00	1.5505E+00	9.7575E−01	2.7784E+00	2.4547E−02
1.34	0.7664	1.2162E+00	1.9282E+00	1.5854E+00	9.7182E−01	2.8444E+00	2.8585E−02
1.36	0.7572	1.2290E+00	1.9912E+00	1.6202E+00	9.6758E−01	2.9115E+00	3.2958E−02
1.38	0.7483	1.2418E+00	2.0551E+00	1.6549E+00	9.6304E−01	2.9798E+00	3.7665E−02
1.4	0.7397	1.2547E+00	2.1200E+00	1.6897E+00	9.5819E−01	3.0492E+00	4.2704E−02
1.42	0.7314	1.2676E+00	2.1858E+00	1.7243E+00	9.5306E−01	3.1198E+00	4.8074E−02
1.44	0.7235	1.2807E+00	2.2525E+00	1.7589E+00	9.4765E−01	3.1915E+00	5.3772E−02
1.46	0.7157	1.2938E+00	2.3202E+00	1.7934E+00	9.4196E−01	3.2643E+00	5.9795E−02
1.48	0.7083	1.3069E+00	2.3888E+00	1.8278E+00	9.3600E−01	3.3382E+00	6.6139E−02
1.5	0.7011	1.3202E+00	2.4583E+00	1.8621E+00	9.2979E−01	3.4133E+00	7.2800E−02
1.52	0.6941	1.3336E+00	2.5288E+00	1.8963E+00	9.2332E−01	3.4894E+00	7.9775E−02
1.54	0.6874	1.3470E+00	2.6002E+00	1.9303E+00	9.1662E−01	3.5667E+00	8.7057E−02
1.56	0.6809	1.3606E+00	2.6725E+00	1.9643E+00	9.0970E−01	3.6450E+00	9.4644E−02
1.58	0.6746	1.3742E+00	2.7458E+00	1.9981E+00	9.0255E−01	3.7244E+00	1.0253E−01
1.6	0.6684	1.3880E+00	2.8200E+00	2.0317E+00	8.9520E−01	3.8050E+00	1.1071E−01
1.62	0.6625	1.4018E+00	2.8951E+00	2.0653E+00	8.8765E−01	3.8866E+00	1.1917E−01
1.64	0.6568	1.4158E+00	2.9712E+00	2.0986E+00	8.7992E−01	3.9693E+00	1.2792E−01
1.66	0.6512	1.4299E+00	3.0482E+00	2.1318E+00	8.7201E−01	4.0531E+00	1.3695E−01
1.68	0.6458	1.4440E+00	3.1261E+00	2.1649E+00	8.6394E−01	4.1379E+00	1.4625E−01
1.7	0.6405	1.4583E+00	3.2050E+00	2.1977E+00	8.5572E−01	4.2238E+00	1.5581E−01
1.72	0.6355	1.4727E+00	3.2848E+00	2.2304E+00	8.4736E−01	4.3108E+00	1.6563E−01
1.74	0.6305	1.4873E+00	3.3655E+00	2.2629E+00	8.3886E−01	4.3989E+00	1.7571E−01

(Continued)

M_1	M_2	T_2/T_1	p_2/p_1	ρ_2/ρ_1	p_{t2}/p_{t1}	p_{t2}/p_1	$\Delta s/R$
1.76	0.6257	1.5019E+00	3.4472E+00	2.2952E+00	8.3024E−01	4.4880E+00	1.8604E−01
1.78	0.6210	1.5167E+00	3.5298E+00	2.3273E+00	8.2151E−01	4.5782E+00	1.9661E−01
1.8	0.6165	1.5316E+00	3.6133E+00	2.3592E+00	8.1268E−01	4.6695E+00	2.0741E−01
1.82	0.6121	1.5466E+00	3.6978E+00	2.3909E+00	8.0376E−01	4.7618E+00	2.1845E−01
1.84	0.6078	1.5617E+00	3.7832E+00	2.4224E+00	7.9476E−01	4.8552E+00	2.2971E−01
1.86	0.6036	1.5770E+00	3.8695E+00	2.4537E+00	7.8569E−01	4.9497E+00	2.4120E−01
1.88	0.5996	1.5924E+00	3.9568E+00	2.4848E+00	7.7655E−01	5.0452E+00	2.5290E−01
1.9	0.5956	1.6079E+00	4.0450E+00	2.5157E+00	7.6736E−01	5.1418E+00	2.6480E−01
1.92	0.5918	1.6236E+00	4.1341E+00	2.5463E+00	7.5812E−01	5.2394E+00	2.7691E−01
1.94	0.5880	1.6394E+00	4.2242E+00	2.5767E+00	7.4884E−01	5.3381E+00	2.8922E−01
1.96	0.5844	1.6553E+00	4.3152E+00	2.6069E+00	7.3954E−01	5.4378E+00	3.0173E−01
1.98	0.5808	1.6713E+00	4.4071E+00	2.6369E+00	7.3021E−01	5.5386E+00	3.1442E−01
2	0.5774	1.6875E+00	4.5000E+00	2.6667E+00	7.2087E−01	5.6404E+00	3.2729E−01
2.02	0.5740	1.7038E+00	4.5938E+00	2.6962E+00	7.1153E−01	5.7433E+00	3.4034E−01
2.04	0.5707	1.7203E+00	4.6885E+00	2.7255E+00	7.0218E−01	5.8473E+00	3.5357E−01
2.06	0.5675	1.7369E+00	4.7842E+00	2.7545E+00	6.9284E−01	5.9523E+00	3.6696E−01
2.08	0.5643	1.7536E+00	4.8808E+00	2.7833E+00	6.8351E−01	6.0583E+00	3.8051E−01
2.1	0.5613	1.7705E+00	4.9783E+00	2.8119E+00	6.7420E−01	6.1654E+00	3.9422E−01
2.12	0.5583	1.7875E+00	5.0768E+00	2.8402E+00	6.6492E−01	6.2735E+00	4.0809E−01
2.14	0.5554	1.8046E+00	5.1762E+00	2.8683E+00	6.5567E−01	6.3827E+00	4.2210E−01
2.16	0.5525	1.8219E+00	5.2765E+00	2.8962E+00	6.4645E−01	6.4929E+00	4.3626E−01
2.18	0.5498	1.8393E+00	5.3778E+00	2.9238E+00	6.3727E−01	6.6042E+00	4.5056E−01
2.2	0.5471	1.8569E+00	5.4800E+00	2.9512E+00	6.2814E−01	6.7165E+00	4.6500E−01
2.22	0.5444	1.8746E+00	5.5831E+00	2.9784E+00	6.1905E−01	6.8298E+00	4.7956E−01
2.24	0.5418	1.8924E+00	5.6872E+00	3.0053E+00	6.1002E−01	6.9442E+00	4.9426E−01
2.26	0.5393	1.9104E+00	5.7922E+00	3.0319E+00	6.0105E−01	7.0597E+00	5.0908E−01
2.28	0.5368	1.9285E+00	5.8981E+00	3.0584E+00	5.9214E−01	7.1762E+00	5.2401E−01
2.3	0.5344	1.9468E+00	6.0050E+00	3.0845E+00	5.8329E−01	7.2937E+00	5.3906E−01
2.32	0.5321	1.9652E+00	6.1128E+00	3.1105E+00	5.7452E−01	7.4122E+00	5.5423E−01
2.34	0.5297	1.9838E+00	6.2215E+00	3.1362E+00	5.6581E−01	7.5319E+00	5.6950E−01
2.36	0.5275	2.0025E+00	6.3312E+00	3.1617E+00	5.5718E−01	7.6525E+00	5.8487E−01
2.38	0.5253	2.0213E+00	6.4418E+00	3.1869E+00	5.4862E−01	7.7742E+00	6.0035E−01
2.4	0.5231	2.0403E+00	6.5533E+00	3.2119E+00	5.4014E−01	7.8969E+00	6.1592E−01
2.42	0.5210	2.0595E+00	6.6658E+00	3.2367E+00	5.3175E−01	8.0207E+00	6.3159E−01
2.44	0.5189	2.0788E+00	6.7792E+00	3.2612E+00	5.2344E−01	8.1455E+00	6.4734E−01
2.46	0.5169	2.0982E+00	6.8935E+00	3.2855E+00	5.1521E−01	8.2713E+00	6.6318E−01
2.48	0.5149	2.1178E+00	7.0088E+00	3.3095E+00	5.0707E−01	8.3982E+00	6.7911E−01
2.5	0.5130	2.1375E+00	7.1250E+00	3.3333E+00	4.9901E−01	8.5261E+00	6.9512E−01
2.52	0.5111	2.1574E+00	7.2421E+00	3.3569E+00	4.9105E−01	8.6551E+00	7.1120E−01
2.54	0.5092	2.1774E+00	7.3602E+00	3.3803E+00	4.8318E−01	8.7851E+00	7.2736E−01

M_1	M_2	T_2/T_1	p_2/p_1	ρ_2/ρ_1	p_{t2}/p_{t1}	p_{t2}/p_1	$\Delta s/R$
2.56	0.5074	2.1976E+00	7.4792E+00	3.4034E+00	4.7540E−01	8.9161E+00	7.4360E−01
2.58	0.5056	2.2179E+00	7.5991E+00	3.4263E+00	4.6772E−01	9.0482E+00	7.5990E−01
2.6	0.5039	2.2383E+00	7.7200E+00	3.4490E+00	4.6012E−01	9.1813E+00	7.7626E−01
2.62	0.5022	2.2590E+00	7.8418E+00	3.4714E+00	4.5263E−01	9.3155E+00	7.9269E−01
2.64	0.5005	2.2797E+00	7.9645E+00	3.4937E+00	4.4522E−01	9.4506E+00	8.0918E−01
2.66	0.4988	2.3006E+00	8.0882E+00	3.5157E+00	4.3792E−01	9.5869E+00	8.2573E−01
2.68	0.4972	2.3217E+00	8.2128E+00	3.5374E+00	4.3071E−01	9.7241E+00	8.4233E−01
2.7	0.4956	2.3429E+00	8.3383E+00	3.5590E+00	4.2359E−01	9.8624E+00	8.5899E−01
2.72	0.4941	2.3642E+00	8.4648E+00	3.5803E+00	4.1657E−01	1.0002E+01	8.7570E−01
2.74	0.4926	2.3858E+00	8.5922E+00	3.6015E+00	4.0965E−01	1.0142E+01	8.9245E−01
2.76	0.4911	2.4074E+00	8.7205E+00	3.6224E+00	4.0283E−01	1.0283E+01	9.0925E−01
2.78	0.4896	2.4292E+00	8.8498E+00	3.6431E+00	3.9610E−01	1.0426E+01	9.2610E−01
2.8	0.4882	2.4512E+00	8.9800E+00	3.6636E+00	3.8946E−01	1.0569E+01	9.4298E−01
2.82	0.4868	2.4733E+00	9.1111E+00	3.6838E+00	3.8293E−01	1.0714E+01	9.5991E−01
2.84	0.4854	2.4955E+00	9.2432E+00	3.7039E+00	3.7649E−01	1.0859E+01	9.7687E−01
2.86	0.4840	2.5179E+00	9.3762E+00	3.7238E+00	3.7014E−01	1.1006E+01	9.9387E−01
2.88	0.4827	2.5405E+00	9.5101E+00	3.7434E+00	3.6389E−01	1.1154E+01	1.0109E+00
2.9	0.4814	2.5632E+00	9.6450E+00	3.7629E+00	3.5773E−01	1.1302E+01	1.0280E+00
2.92	0.4801	2.5861E+00	9.7808E+00	3.7821E+00	3.5167E−01	1.1452E+01	1.0451E+00
2.94	0.4788	2.6091E+00	9.9175E+00	3.8012E+00	3.4570E−01	1.1603E+01	1.0622E+00
2.96	0.4776	2.6322E+00	1.0055E+01	3.8200E+00	3.3982E−01	1.1754E+01	1.0793E+00
2.98	0.4764	2.6555E+00	1.0194E+01	3.8387E+00	3.3404E−01	1.1907E+01	1.0965E+00
3	0.4752	2.6790E+00	1.0333E+01	3.8571E+00	3.2834E−01	1.2061E+01	1.1137E+00
3.02	0.4740	2.7026E+00	1.0474E+01	3.8754E+00	3.2274E−01	1.2216E+01	1.1309E+00
3.04	0.4729	2.7264E+00	1.0615E+01	3.8935E+00	3.1723E−01	1.2372E+01	1.1481E+00
3.06	0.4717	2.7503E+00	1.0758E+01	3.9114E+00	3.1180E−01	1.2529E+01	1.1654E+00
3.08	0.4706	2.7744E+00	1.0901E+01	3.9291E+00	3.0646E−01	1.2687E+01	1.1827E+00
3.1	0.4695	2.7986E+00	1.1045E+01	3.9466E+00	3.0121E−01	1.2846E+01	1.1999E+00
3.12	0.4685	2.8230E+00	1.1190E+01	3.9639E+00	2.9605E−01	1.3006E+01	1.2172E+00
3.14	0.4674	2.8475E+00	1.1336E+01	3.9811E+00	2.9097E−01	1.3167E+01	1.2345E+00
3.16	0.4664	2.8722E+00	1.1483E+01	3.9981E+00	2.8597E−01	1.3329E+01	1.2519E+00
3.18	0.4654	2.8970E+00	1.1631E+01	4.0149E+00	2.8106E−01	1.3492E+01	1.2692E+00
3.2	0.4643	2.9220E+00	1.1780E+01	4.0315E+00	2.7623E−01	1.3656E+01	1.2865E+00
3.22	0.4634	2.9471E+00	1.1930E+01	4.0479E+00	2.7148E−01	1.3821E+01	1.3039E+00
3.24	0.4624	2.9724E+00	1.2081E+01	4.0642E+00	2.6681E−01	1.3987E+01	1.3212E+00
3.26	0.4614	2.9979E+00	1.2232E+01	4.0803E+00	2.6222E−01	1.4155E+01	1.3386E+00
3.28	0.4605	3.0234E+00	1.2385E+01	4.0963E+00	2.5771E−01	1.4323E+01	1.3559E+00
3.3	0.4596	3.0492E+00	1.2538E+01	4.1120E+00	2.5328E−01	1.4492E+01	1.3733E+00
3.32	0.4587	3.0751E+00	1.2693E+01	4.1276E+00	2.4892E−01	1.4662E+01	1.3906E+00
3.34	0.4578	3.1011E+00	1.2848E+01	4.1431E+00	2.4463E−01	1.4834E+01	1.4080E+00

(Continued)

M_1	M_2	T_2/T_1	p_2/p_1	ρ_2/ρ_1	p_{t2}/p_{t1}	p_{t2}/p_1	$\Delta s/R$
3.36	0.4569	3.1273E+00	1.3005E+01	4.1583E+00	2.4043E−01	1.5006E+01	1.4253E+00
3.38	0.4560	3.1537E+00	1.3162E+01	4.1734E+00	2.3629E−01	1.5180E+01	1.4427E+00
3.4	0.4552	3.1802E+00	1.3320E+01	4.1884E+00	2.3223E−01	1.5354E+01	1.4600E+00
3.42	0.4544	3.2069E+00	1.3479E+01	4.2032E+00	2.2823E−01	1.5530E+01	1.4774E+00
3.44	0.4535	3.2337E+00	1.3639E+01	4.2179E+00	2.2431E−01	1.5706E+01	1.4947E+00
3.46	0.4527	3.2607E+00	1.3800E+01	4.2323E+00	2.2045E−01	1.5884E+01	1.5121E+00
3.48	0.4519	3.2878E+00	1.3962E+01	4.2467E+00	2.1667E−01	1.6062E+01	1.5294E+00
3.5	0.4512	3.3151E+00	1.4125E+01	4.2609E+00	2.1295E−01	1.6242E+01	1.5467E+00
3.52	0.4504	3.3425E+00	1.4289E+01	4.2749E+00	2.0929E−01	1.6423E+01	1.5640E+00
3.54	0.4496	3.3701E+00	1.4454E+01	4.2888E+00	2.0570E−01	1.6604E+01	1.5813E+00
3.56	0.4489	3.3978E+00	1.4619E+01	4.3026E+00	2.0218E−01	1.6787E+01	1.5986E+00
3.58	0.4481	3.4257E+00	1.4786E+01	4.3162E+00	1.9871E−01	1.6971E+01	1.6159E+00
3.6	0.4474	3.4537E+00	1.4953E+01	4.3296E+00	1.9531E−01	1.7156E+01	1.6332E+00
3.62	0.4467	3.4819E+00	1.5122E+01	4.3429E+00	1.9197E−01	1.7341E+01	1.6504E+00
3.64	0.4460	3.5103E+00	1.5291E+01	4.3561E+00	1.8869E−01	1.7528E+01	1.6677E+00
3.66	0.4453	3.5388E+00	1.5462E+01	4.3692E+00	1.8547E−01	1.7716E+01	1.6849E+00
3.68	0.4446	3.5674E+00	1.5633E+01	4.3821E+00	1.8230E−01	1.7905E+01	1.7021E+00
3.7	0.4439	3.5962E+00	1.5805E+01	4.3949E+00	1.7919E−01	1.8095E+01	1.7193E+00
3.72	0.4433	3.6252E+00	1.5978E+01	4.4075E+00	1.7614E−01	1.8286E+01	1.7365E+00
3.74	0.4426	3.6543E+00	1.6152E+01	4.4200E+00	1.7314E−01	1.8478E+01	1.7536E+00
3.76	0.4420	3.6836E+00	1.6327E+01	4.4324E+00	1.7020E−01	1.8671E+01	1.7708E+00
3.78	0.4414	3.7130E+00	1.6503E+01	4.4447E+00	1.6731E−01	1.8865E+01	1.7879E+00
3.8	0.4407	3.7426E+00	1.6680E+01	4.4568E+00	1.6447E−01	1.9060E+01	1.8050E+00
3.82	0.4401	3.7723E+00	1.6858E+01	4.4688E+00	1.6168E−01	1.9256E+01	1.8221E+00
3.84	0.4395	3.8022E+00	1.7037E+01	4.4807E+00	1.5895E−01	1.9454E+01	1.8392E+00
3.86	0.4389	3.8323E+00	1.7216E+01	4.4924E+00	1.5626E−01	1.9652E+01	1.8562E+00
3.88	0.4383	3.8625E+00	1.7397E+01	4.5041E+00	1.5362E−01	1.9851E+01	1.8733E+00
3.9	0.4377	3.8928E+00	1.7578E+01	4.5156E+00	1.5103E−01	2.0051E+01	1.8903E+00
3.92	0.4372	3.9233E+00	1.7761E+01	4.5270E+00	1.4848E−01	2.0253E+01	1.9073E+00
3.94	0.4366	3.9540E+00	1.7944E+01	4.5383E+00	1.4598E−01	2.0455E+01	1.9243E+00
3.96	0.4360	3.9848E+00	1.8129E+01	4.5494E+00	1.4353E−01	2.0658E+01	1.9412E+00
3.98	0.4355	4.0158E+00	1.8314E+01	4.5605E+00	1.4112E−01	2.0863E+01	1.9581E+00
4	0.4350	4.0469E+00	1.8500E+01	4.5714E+00	1.3876E−01	2.1068E+01	1.9750E+00
4.1	0.4324	4.2048E+00	1.9445E+01	4.6245E+00	1.2756E−01	2.2111E+01	2.0592E+00
4.2	0.4299	4.3666E+00	2.0413E+01	4.6749E+00	1.1733E−01	2.3179E+01	2.1427E+00
4.3	0.4277	4.5322E+00	2.1405E+01	4.7229E+00	1.0800E−01	2.4273E+01	2.2256E+00
4.4	0.4255	4.7017E+00	2.2420E+01	4.7685E+00	9.9481E−02	2.5393E+01	2.3078E+00
4.5	0.4236	4.8751E+00	2.3458E+01	4.8119E+00	9.1698E−02	2.6539E+01	2.3893E+00
4.6	0.4217	5.0523E+00	2.4520E+01	4.8532E+00	8.4587E−02	2.7710E+01	2.4700E+00
4.7	0.4199	5.2334E+00	2.5605E+01	4.8926E+00	7.8086E−02	2.8907E+01	2.5499E+00

M_1	M_2	T_2/T_1	p_2/p_1	ρ_2/ρ_1	p_{t2}/p_{t1}	p_{t2}/p_1	$\Delta s/R$
4.8	0.4183	5.4184E+00	2.6713E+01	4.9301E+00	7.2140E−02	3.0130E+01	2.6291E+00
4.9	0.4167	5.6073E+00	2.7845E+01	4.9659E+00	6.6699E−02	3.1379E+01	2.7076E+00
5	0.4152	5.8000E+00	2.9000E+01	5.0000E+00	6.1716E−02	3.2653E+01	2.7852E+00
5.1	0.4138	5.9966E+00	3.0178E+01	5.0326E+00	5.7151E−02	3.3954E+01	2.8621E+00
5.2	0.4125	6.1971E+00	3.1380E+01	5.0637E+00	5.2966E−02	3.5280E+01	2.9381E+00
5.3	0.4113	6.4014E+00	3.2605E+01	5.0934E+00	4.9126E−02	3.6632E+01	3.0134E+00
5.4	0.4101	6.6097E+00	3.3853E+01	5.1218E+00	4.5601E−02	3.8009E+01	3.0878E+00
5.5	0.4090	6.8218E+00	3.5125E+01	5.1489E+00	4.2361E−02	3.9412E+01	3.1615E+00
5.6	0.4079	7.0378E+00	3.6420E+01	5.1749E+00	3.9383E−02	4.0841E+01	3.2344E+00
5.7	0.4069	7.2577E+00	3.7738E+01	5.1998E+00	3.6643E−02	4.2296E+01	3.3065E+00
5.8	0.4059	7.4814E+00	3.9080E+01	5.2236E+00	3.4120E−02	4.3777E+01	3.3779E+00
5.9	0.4050	7.7091E+00	4.0445E+01	5.2464E+00	3.1795E−02	4.5283E+01	3.4484E+00
6	0.4042	7.9406E+00	4.1833E+01	5.2683E+00	2.9651E−02	4.6815E+01	3.5183E+00
6.1	0.4033	8.1760E+00	4.3245E+01	5.2893E+00	2.7672E−02	4.8373E+01	3.5873E+00
6.2	0.4025	8.4153E+00	4.4680E+01	5.3094E+00	2.5845E−02	4.9957E+01	3.6556E+00
6.3	0.4018	8.6584E+00	4.6138E+01	5.3287E+00	2.4156E−02	5.1566E+01	3.7232E+00
6.4	0.4011	8.9055E+00	4.7620E+01	5.3473E+00	2.2594E−02	5.3201E+01	3.7901E+00
6.5	0.4004	9.1564E+00	4.9125E+01	5.3651E+00	2.1148E−02	5.4862E+01	3.8562E+00
6.6	0.3997	9.4113E+00	5.0653E+01	5.3822E+00	1.9808E−02	5.6549E+01	3.9217E+00
6.7	0.3991	9.6700E+00	5.2205E+01	5.3987E+00	1.8566E−02	5.8261E+01	3.9864E+00
6.8	0.3985	9.9326E+00	5.3780E+01	5.4145E+00	1.7414E−02	5.9999E+01	4.0505E+00
6.9	0.3979	1.0199E+01	5.5378E+01	5.4298E+00	1.6345E−02	6.1763E+01	4.1138E+00
7	0.3974	1.0469E+01	5.7000E+01	5.4444E+00	1.5351E−02	6.3553E+01	4.1765E+00
7.1	0.3968	1.0744E+01	5.8645E+01	5.4586E+00	1.4428E−02	6.5368E+01	4.2386E+00
7.2	0.3963	1.1022E+01	6.0313E+01	5.4722E+00	1.3569E−02	6.7209E+01	4.3000E+00
7.3	0.3958	1.1304E+01	6.2005E+01	5.4853E+00	1.2769E−02	6.9076E+01	4.3608E+00
7.4	0.3954	1.1590E+01	6.3720E+01	5.4980E+00	1.2023E−02	7.0969E+01	4.4209E+00
7.5	0.3949	1.1879E+01	6.5458E+01	5.5102E+00	1.1329E−02	7.2887E+01	4.4804E+00
7.6	0.3945	1.2173E+01	6.7220E+01	5.5220E+00	1.0680E−02	7.4831E+01	4.5393E+00
7.7	0.3941	1.2471E+01	6.9005E+01	5.5334E+00	1.0075E−02	7.6801E+01	4.5977E+00
7.8	0.3937	1.2772E+01	7.0813E+01	5.5443E+00	9.5102E−03	7.8797E+01	4.6554E+00
7.9	0.3933	1.3078E+01	7.2645E+01	5.5550E+00	8.9819E−03	8.0818E+01	4.7125E+00
8	0.3929	1.3387E+01	7.4500E+01	5.5652E+00	8.4878E−03	8.2865E+01	4.7691E+00
8.1	0.3925	1.3700E+01	7.6378E+01	5.5751E+00	8.0254E−03	8.4938E+01	4.8251E+00
8.2	0.3922	1.4017E+01	7.8280E+01	5.5847E+00	7.5924E−03	8.7037E+01	4.8806E+00
8.3	0.3918	1.4338E+01	8.0205E+01	5.5940E+00	7.1866E−03	8.9162E+01	4.9355E+00
8.4	0.3915	1.4662E+01	8.2153E+01	5.6030E+00	6.8061E−03	9.1312E+01	4.9899E+00
8.5	0.3912	1.4991E+01	8.4125E+01	5.6117E+00	6.4492E−03	9.3488E+01	5.0438E+00
8.6	0.3909	1.5324E+01	8.6120E+01	5.6201E+00	6.1141E−03	9.5689E+01	5.0972E+00

(Continued)

M_1	M_2	T_2/T_1	p_2/p_1	ρ_2/ρ_1	p_{t2}/p_{t1}	p_{t2}/p_1	$\Delta s/R$
8.7	0.3906	1.5660E+01	8.8138E+01	5.6282E+00	5.7994E−03	9.7917E+01	5.1500E+00
8.8	0.3903	1.6000E+01	9.0180E+01	5.6361E+00	5.5036E−03	1.0017E+02	5.2023E+00
8.9	0.3901	1.6345E+01	9.2245E+01	5.6437E+00	5.2255E−03	1.0245E+02	5.2542E+00
9	0.3898	1.6693E+01	9.4333E+01	5.6512E+00	4.9639E−03	1.0475E+02	5.3056E+00
9.1	0.3895	1.7045E+01	9.6445E+01	5.6584E+00	4.7175E−03	1.0708E+02	5.3565E+00
9.2	0.3893	1.7401E+01	9.8580E+01	5.6653E+00	4.4856E−03	1.0944E+02	5.4069E+00
9.3	0.3891	1.7760E+01	1.0074E+02	5.6721E+00	4.2669E−03	1.1182E+02	5.4569E+00
9.4	0.3888	1.8124E+01	1.0292E+02	5.6787E+00	4.0608E−03	1.1423E+02	5.5064E+00
9.5	0.3886	1.8492E+01	1.0513E+02	5.6850E+00	3.8664E−03	1.1666E+02	5.5554E+00
9.6	0.3884	1.8863E+01	1.0735E+02	5.6912E+00	3.6828E−03	1.1912E+02	5.6041E+00
9.7	0.3882	1.9238E+01	1.0961E+02	5.6972E+00	3.5095E−03	1.2161E+02	5.6523E+00
9.8	0.3880	1.9617E+01	1.1188E+02	5.7031E+00	3.3458E−03	1.2412E+02	5.7000E+00
9.9	0.3878	2.0001E+01	1.1418E+02	5.7088E+00	3.1911E−03	1.2665E+02	5.7474E+00
10	0.3876	2.0388E+01	1.1650E+02	5.7143E+00	3.0448E−03	1.2922E+02	5.7943E+00
11	0.3859	2.4471E+01	1.4100E+02	5.7619E+00	1.9451E−03	1.5626E+02	6.2425E+00
12	0.3847	2.8943E+01	1.6783E+02	5.7987E+00	1.2866E−03	1.8587E+02	6.6557E+00
13	0.3837	3.3805E+01	1.9700E+02	5.8276E+00	8.7709E−04	2.1806E+02	7.0389E+00
14	0.3829	3.9055E+01	2.2850E+02	5.8507E+00	6.1380E−04	2.5282E+02	7.3958E+00
15	0.3823	4.4694E+01	2.6233E+02	5.8696E+00	4.3953E−04	2.9016E+02	7.7298E+00
16	0.3817	5.0722E+01	2.9850E+02	5.8851E+00	3.2119E−04	3.3008E+02	8.0435E+00
17	0.3813	5.7138E+01	3.3700E+02	5.8980E+00	2.3899E−04	3.7257E+02	8.3391E+00
18	0.3810	6.3944E+01	3.7783E+02	5.9088E+00	1.8072E−04	4.1763E+02	8.6186E+00
19	0.3806	7.1139E+01	4.2100E+02	5.9180E+00	1.3865E−04	4.6527E+02	8.8836E+00
20	0.3804	7.8722E+01	4.6650E+02	5.9259E+00	1.0777E−04	5.1548E+02	9.1355E+00
21	0.3802	8.6694E+01	5.1433E+02	5.9327E+00	8.4778E−05	5.6827E+02	9.3755E+00
22	0.3800	9.5055E+01	5.6450E+02	5.9387E+00	6.7414E−05	6.2364E+02	9.6047E+00
23	0.3798	1.0381E+02	6.1700E+02	5.9438E+00	5.4140E−05	6.8158E+02	9.8239E+00
24	0.3796	1.1294E+02	6.7183E+02	5.9484E+00	4.3877E−05	7.4209E+02	1.0034E+01
25	0.3795	1.2247E+02	7.2900E+02	5.9524E+00	3.5859E−05	8.0519E+02	1.0236E+01
30	0.3790	1.7594E+02	1.0498E+03	5.9669E+00	1.4531E−05	1.1593E+03	1.1139E+01
35	0.3788	2.3914E+02	1.4290E+03	5.9756E+00	6.7571E−06	1.5777E+03	1.1905E+01
40	0.3786	3.1206E+02	1.8665E+03	5.9813E+00	3.4771E−06	2.0606E+03	1.2569E+01
45	0.3784	3.9469E+02	2.3623E+03	5.9852E+00	1.9339E−06	2.6078E+03	1.3156E+01
50	0.3784	4.8706E+02	2.9165E+03	5.9880E+00	1.1438E−06	3.2194E+03	1.3681E+01

Normal Shock Table $\gamma = 1.3$

M_1	M_2	T_2/T_1	p_2/p_1	ρ_2/ρ_1	p_{t2}/p_{t1}	p_{t2}/p_1	$\Delta s/R$
1	1	1	1	1	1	1.8324	0
1.02	0.9805	1.0104E+00	1.0457E+00	1.0349E+00	9.9999E−01	1.8746E+00	9.7482E−06
1.04	0.9619	1.0206E+00	1.0922E+00	1.0702E+00	9.9992E−01	1.9183E+00	7.7203E−05
1.06	0.9442	1.0307E+00	1.1397E+00	1.1058E+00	9.9975E−01	1.9634E+00	2.5262E−04
1.08	0.9273	1.0407E+00	1.1881E+00	1.1416E+00	9.9942E−01	2.0099E+00	5.7916E−04
1.1	0.9112	1.0506E+00	1.2374E+00	1.1777E+00	9.9891E−01	2.0578E+00	1.0938E−03
1.12	0.8958	1.0605E+00	1.2876E+00	1.2141E+00	9.9817E−01	2.1070E+00	1.8282E−03
1.14	0.8810	1.0703E+00	1.3387E+00	1.2507E+00	9.9719E−01	2.1575E+00	2.8094E−03
1.16	0.8669	1.0801E+00	1.3907E+00	1.2876E+00	9.9595E−01	2.2092E+00	4.0603E−03
1.18	0.8533	1.0898E+00	1.4436E+00	1.3246E+00	9.9442E−01	2.2622E+00	5.6003E−03
1.2	0.8403	1.0995E+00	1.4974E+00	1.3618E+00	9.9258E−01	2.3164E+00	7.4458E−03
1.22	0.8278	1.1092E+00	1.5521E+00	1.3993E+00	9.9044E−01	2.3718E+00	9.6103E−03
1.24	0.8159	1.1189E+00	1.6077E+00	1.4368E+00	9.8797E−01	2.4283E+00	1.2105E−02
1.26	0.8043	1.1286E+00	1.6642E+00	1.4746E+00	9.8517E−01	2.4861E+00	1.4939E−02
1.28	0.7932	1.1383E+00	1.7217E+00	1.5125E+00	9.8204E−01	2.5450E+00	1.8120E−02
1.3	0.7825	1.1480E+00	1.7800E+00	1.5505E+00	9.7858E−01	2.6050E+00	2.1653E−02
1.32	0.7722	1.1578E+00	1.8392E+00	1.5886E+00	9.7478E−01	2.6661E+00	2.5542E−02
1.34	0.7623	1.1676E+00	1.8994E+00	1.6268E+00	9.7065E−01	2.7283E+00	2.9791E−02
1.36	0.7527	1.1774E+00	1.9604E+00	1.6651E+00	9.6618E−01	2.7917E+00	3.4402E−02
1.38	0.7435	1.1872E+00	2.0224E+00	1.7035E+00	9.6139E−01	2.8561E+00	3.9374E−02
1.4	0.7346	1.1971E+00	2.0852E+00	1.7419E+00	9.5628E−01	2.9217E+00	4.4709E−02
1.42	0.7260	1.2070E+00	2.1490E+00	1.7804E+00	9.5084E−01	2.9883E+00	5.0405E−02
1.44	0.7176	1.2170E+00	2.2136E+00	1.8189E+00	9.4510E−01	3.0560E+00	5.6461E−02
1.46	0.7096	1.2271E+00	2.2792E+00	1.8574E+00	9.3906E−01	3.1247E+00	6.2874E−02
1.48	0.7018	1.2372E+00	2.3457E+00	1.8960E+00	9.3273E−01	3.1945E+00	6.9644E−02
1.5	0.6942	1.2473E+00	2.4130E+00	1.9346E+00	9.2611E−01	3.2654E+00	7.6765E−02
1.52	0.6869	1.2575E+00	2.4813E+00	1.9731E+00	9.1921E−01	3.3373E+00	8.4236E−02
1.54	0.6799	1.2678E+00	2.5505E+00	2.0117E+00	9.1206E−01	3.4103E+00	9.2052E−02
1.56	0.6730	1.2782E+00	2.6206E+00	2.0502E+00	9.0465E−01	3.4843E+00	1.0021E−01
1.58	0.6664	1.2886E+00	2.6916E+00	2.0887E+00	8.9700E−01	3.5593E+00	1.0870E−01
1.6	0.6599	1.2991E+00	2.7635E+00	2.1272E+00	8.8911E−01	3.6354E+00	1.1753E−01
1.62	0.6537	1.3097E+00	2.8363E+00	2.1656E+00	8.8101E−01	3.7125E+00	1.2668E−01
1.64	0.6476	1.3204E+00	2.9100E+00	2.2039E+00	8.7270E−01	3.7906E+00	1.3616E−01
1.66	0.6417	1.3311E+00	2.9846E+00	2.2422E+00	8.6420E−01	3.8697E+00	1.4595E−01
1.68	0.6360	1.3419E+00	3.0601E+00	2.2804E+00	8.5551E−01	3.9499E+00	1.5606E−01
1.7	0.6304	1.3529E+00	3.1365E+00	2.3185E+00	8.4665E−01	4.0311E+00	1.6647E−01
1.72	0.6250	1.3638E+00	3.2138E+00	2.3565E+00	8.3762E−01	4.1133E+00	1.7719E−01
1.74	0.6198	1.3749E+00	3.2921E+00	2.3944E+00	8.2845E−01	4.1965E+00	1.8820E−01

(Continued)

M_1	M_2	T_2/T_1	p_2/p_1	ρ_2/ρ_1	p_{t2}/p_{t1}	p_{t2}/p_1	$\Delta s/R$
1.76	0.6146	1.3861E+00	3.3712E+00	2.4322E+00	8.1914E−01	4.2808E+00	1.9950E−01
1.78	0.6097	1.3974E+00	3.4512E+00	2.4698E+00	8.0970E−01	4.3660E+00	2.1109E−01
1.8	0.6048	1.4087E+00	3.5322E+00	2.5074E+00	8.0015E−01	4.4523E+00	2.2296E−01
1.82	0.6001	1.4201E+00	3.6140E+00	2.5448E+00	7.9049E−01	4.5396E+00	2.3510E−01
1.84	0.5955	1.4317E+00	3.6968E+00	2.5821E+00	7.8074E−01	4.6278E+00	2.4752E−01
1.86	0.5911	1.4433E+00	3.7804E+00	2.6193E+00	7.7090E−01	4.7171E+00	2.6019E−01
1.88	0.5867	1.4550E+00	3.8650E+00	2.6563E+00	7.6100E−01	4.8074E+00	2.7313E−01
1.9	0.5825	1.4668E+00	3.9504E+00	2.6932E+00	7.5103E−01	4.8987E+00	2.8631E−01
1.92	0.5784	1.4788E+00	4.0368E+00	2.7299E+00	7.4101E−01	4.9909E+00	2.9975E−01
1.94	0.5744	1.4908E+00	4.1241E+00	2.7664E+00	7.3094E−01	5.0842E+00	3.1342E−01
1.96	0.5704	1.5029E+00	4.2122E+00	2.8028E+00	7.2084E−01	5.1785E+00	3.2733E−01
1.98	0.5666	1.5151E+00	4.3013E+00	2.8390E+00	7.1072E−01	5.2738E+00	3.4148E−01
2	0.5629	1.5274E+00	4.3913E+00	2.8750E+00	7.0058E−01	5.3700E+00	3.5585E−01
2.02	0.5592	1.5398E+00	4.4822E+00	2.9108E+00	6.9043E−01	5.4673E+00	3.7043E−01
2.04	0.5557	1.5523E+00	4.5740E+00	2.9465E+00	6.8029E−01	5.5656E+00	3.8524E−01
2.06	0.5522	1.5650E+00	4.6667E+00	2.9820E+00	6.7015E−01	5.6648E+00	4.0025E−01
2.08	0.5488	1.5777E+00	4.7603E+00	3.0173E+00	6.6003E−01	5.7651E+00	4.1548E−01
2.1	0.5455	1.5905E+00	4.8548E+00	3.0524E+00	6.4993E−01	5.8663E+00	4.3090E−01
2.12	0.5423	1.6034E+00	4.9502E+00	3.0873E+00	6.3985E−01	5.9685E+00	4.4651E−01
2.14	0.5391	1.6165E+00	5.0465E+00	3.1219E+00	6.2982E−01	6.0718E+00	4.6232E−01
2.16	0.5361	1.6296E+00	5.1437E+00	3.1564E+00	6.1983E−01	6.1760E+00	4.7831E−01
2.18	0.5331	1.6428E+00	5.2418E+00	3.1907E+00	6.0988E−01	6.2812E+00	4.9449E−01
2.2	0.5301	1.6562E+00	5.3409E+00	3.2248E+00	5.9999E−01	6.3873E+00	5.1084E−01
2.22	0.5272	1.6696E+00	5.4408E+00	3.2587E+00	5.9016E−01	6.4945E+00	5.2736E−01
2.24	0.5244	1.6832E+00	5.5416E+00	3.2923E+00	5.8039E−01	6.6027E+00	5.4406E−01
2.26	0.5217	1.6969E+00	5.6434E+00	3.3257E+00	5.7069E−01	6.7118E+00	5.6091E−01
2.28	0.5190	1.7106E+00	5.7460E+00	3.3590E+00	5.6106E−01	6.8220E+00	5.7792E−01
2.3	0.5163	1.7245E+00	5.8496E+00	3.3920E+00	5.5151E−01	6.9331E+00	5.9509E−01
2.32	0.5138	1.7385E+00	5.9540E+00	3.4248E+00	5.4204E−01	7.0452E+00	6.1241E−01
2.34	0.5112	1.7526E+00	6.0594E+00	3.4573E+00	5.3266E−01	7.1583E+00	6.2987E−01
2.36	0.5088	1.7668E+00	6.1656E+00	3.4896E+00	5.2336E−01	7.2724E+00	6.4748E−01
2.38	0.5064	1.7812E+00	6.2728E+00	3.5218E+00	5.1416E−01	7.3874E+00	6.6523E−01
2.4	0.5040	1.7956E+00	6.3809E+00	3.5536E+00	5.0505E−01	7.5035E+00	6.8310E−01
2.42	0.5017	1.8101E+00	6.4898E+00	3.5853E+00	4.9603E−01	7.6205E+00	7.0111E−01
2.44	0.4994	1.8248E+00	6.5997E+00	3.6167E+00	4.8712E−01	7.7385E+00	7.1925E−01
2.46	0.4972	1.8395E+00	6.7105E+00	3.6479E+00	4.7830E−01	7.8575E+00	7.3751E−01
2.48	0.4950	1.8544E+00	6.8222E+00	3.6789E+00	4.6959E−01	7.9775E+00	7.5589E−01
2.5	0.4929	1.8694E+00	6.9348E+00	3.7097E+00	4.6099E−01	8.0985E+00	7.7438E−01
2.52	0.4908	1.8845E+00	7.0483E+00	3.7402E+00	4.5249E−01	8.2204E+00	7.9299E−01
2.54	0.4888	1.8997E+00	7.1627E+00	3.7705E+00	4.4410E−01	8.3433E+00	8.1171E−01

M_1	M_2	T_2/T_1	p_2/p_1	ρ_2/ρ_1	p_{t2}/p_{t1}	p_{t2}/p_1	$\Delta s/R$
2.56	0.4868	1.9150E+00	7.2780E+00	3.8005E+00	4.3582E−01	8.4672E+00	8.3053E−01
2.58	0.4848	1.9304E+00	7.3942E+00	3.8304E+00	4.2765E−01	8.5921E+00	8.4945E−01
2.6	0.4829	1.9459E+00	7.5113E+00	3.8600E+00	4.1959E−01	8.7180E+00	8.6847E−01
2.62	0.4810	1.9616E+00	7.6293E+00	3.8894E+00	4.1165E−01	8.8449E+00	8.8759E−01
2.64	0.4791	1.9774E+00	7.7482E+00	3.9185E+00	4.0381E−01	8.9727E+00	9.0680E−01
2.66	0.4773	1.9932E+00	7.8681E+00	3.9474E+00	3.9609E−01	9.1015E+00	9.2610E−01
2.68	0.4755	2.0092E+00	7.9888E+00	3.9761E+00	3.8849E−01	9.2313E+00	9.4549E−01
2.7	0.4738	2.0253E+00	8.1104E+00	4.0045E+00	3.8100E−01	9.3621E+00	9.6496E−01
2.72	0.4721	2.0415E+00	8.2330E+00	4.0328E+00	3.7362E−01	9.4938E+00	9.8451E−01
2.74	0.4704	2.0578E+00	8.3564E+00	4.0608E+00	3.6636E−01	9.6266E+00	1.0041E+00
2.76	0.4687	2.0743E+00	8.4808E+00	4.0885E+00	3.5921E−01	9.7603E+00	1.0238E+00
2.78	0.4671	2.0908E+00	8.6060E+00	4.1161E+00	3.5218E−01	9.8950E+00	1.0436E+00
2.8	0.4655	2.1075E+00	8.7322E+00	4.1434E+00	3.4526E−01	1.0031E+01	1.0635E+00
2.82	0.4639	2.1243E+00	8.8592E+00	4.1705E+00	3.3845E−01	1.0167E+01	1.0834E+00
2.84	0.4624	2.1412E+00	8.9872E+00	4.1973E+00	3.3175E−01	1.0305E+01	1.1034E+00
2.86	0.4609	2.1582E+00	9.1161E+00	4.2240E+00	3.2517E−01	1.0444E+01	1.1234E+00
2.88	0.4594	2.1753E+00	9.2458E+00	4.2504E+00	3.1870E−01	1.0583E+01	1.1435E+00
2.9	0.4580	2.1925E+00	9.3765E+00	4.2766E+00	3.1234E−01	1.0724E+01	1.1637E+00
2.92	0.4565	2.2099E+00	9.5081E+00	4.3026E+00	3.0609E−01	1.0865E+01	1.1839E+00
2.94	0.4551	2.2273E+00	9.6406E+00	4.3283E+00	2.9995E−01	1.1008E+01	1.2041E+00
2.96	0.4538	2.2449E+00	9.7740E+00	4.3538E+00	2.9392E−01	1.1151E+01	1.2245E+00
2.98	0.4524	2.2626E+00	9.9083E+00	4.3792E+00	2.8799E−01	1.1296E+01	1.2448E+00
3	0.4511	2.2804E+00	1.0043E+01	4.4043E+00	2.8217E−01	1.1441E+01	1.2652E+00
3.02	0.4498	2.2983E+00	1.0180E+01	4.4291E+00	2.7646E−01	1.1588E+01	1.2857E+00
3.04	0.4485	2.3164E+00	1.0317E+01	4.4538E+00	2.7085E−01	1.1735E+01	1.3062E+00
3.06	0.4472	2.3345E+00	1.0455E+01	4.4783E+00	2.6534E−01	1.1883E+01	1.3267E+00
3.08	0.4460	2.3528E+00	1.0593E+01	4.5025E+00	2.5994E−01	1.2033E+01	1.3473E+00
3.1	0.4448	2.3711E+00	1.0733E+01	4.5265E+00	2.5463E−01	1.2183E+01	1.3679E+00
3.12	0.4436	2.3896E+00	1.0874E+01	4.5503E+00	2.4943E−01	1.2334E+01	1.3886E+00
3.14	0.4424	2.4082E+00	1.1015E+01	4.5739E+00	2.4432E−01	1.2487E+01	1.4093E+00
3.16	0.4412	2.4270E+00	1.1158E+01	4.5973E+00	2.3932E−01	1.2640E+01	1.4300E+00
3.18	0.4401	2.4458E+00	1.1301E+01	4.6205E+00	2.3440E−01	1.2794E+01	1.4507E+00
3.2	0.4389	2.4648E+00	1.1445E+01	4.6435E+00	2.2958E−01	1.2950E+01	1.4715E+00
3.22	0.4378	2.4838E+00	1.1590E+01	4.6663E+00	2.2486E−01	1.3106E+01	1.4923E+00
3.24	0.4368	2.5030E+00	1.1736E+01	4.6889E+00	2.2023E−01	1.3263E+01	1.5131E+00
3.26	0.4357	2.5223E+00	1.1883E+01	4.7113E+00	2.1568E−01	1.3421E+01	1.5339E+00
3.28	0.4346	2.5417E+00	1.2031E+01	4.7335E+00	2.1123E−01	1.3580E+01	1.5548E+00
3.3	0.4336	2.5613E+00	1.2180E+01	4.7555E+00	2.0687E−01	1.3740E+01	1.5757E+00
3.32	0.4326	2.5809E+00	1.2330E+01	4.7772E+00	2.0259E−01	1.3901E+01	1.5966E+00

(Continued)

M_1	M_2	T_2/T_1	p_2/p_1	ρ_2/ρ_1	p_{t2}/p_{t1}	p_{t2}/p_1	$\Delta s/R$
3.34	0.4316	2.6007E+00	1.2480E+01	4.7988E+00	1.9839E−01	1.4064E+01	1.6175E+00
3.36	0.4306	2.6206E+00	1.2632E+01	4.8202E+00	1.9428E−01	1.4227E+01	1.6384E+00
3.38	0.4296	2.6406E+00	1.2784E+01	4.8415E+00	1.9026E−01	1.4391E+01	1.6594E+00
3.4	0.4287	2.6607E+00	1.2937E+01	4.8625E+00	1.8631E−01	1.4556E+01	1.6803E+00
3.42	0.4277	2.6809E+00	1.3092E+01	4.8833E+00	1.8245E−01	1.4722E+01	1.7013E+00
3.44	0.4268	2.7012E+00	1.3247E+01	4.9039E+00	1.7866E−01	1.4889E+01	1.7223E+00
3.46	0.4259	2.7217E+00	1.3403E+01	4.9244E+00	1.7495E−01	1.5057E+01	1.7433E+00
3.48	0.4250	2.7423E+00	1.3560E+01	4.9447E+00	1.7131E−01	1.5225E+01	1.7643E+00
3.5	0.4241	2.7630E+00	1.3717E+01	4.9648E+00	1.6775E−01	1.5395E+01	1.7853E+00
3.52	0.4232	2.7838E+00	1.3876E+01	4.9847E+00	1.6427E−01	1.5566E+01	1.8063E+00
3.54	0.4224	2.8047E+00	1.4036E+01	5.0044E+00	1.6085E−01	1.5738E+01	1.8273E+00
3.56	0.4215	2.8257E+00	1.4196E+01	5.0239E+00	1.5751E−01	1.5911E+01	1.8483E+00
3.58	0.4207	2.8469E+00	1.4358E+01	5.0433E+00	1.5423E−01	1.6085E+01	1.8693E+00
3.6	0.4199	2.8681E+00	1.4520E+01	5.0625E+00	1.5102E−01	1.6259E+01	1.8903E+00
3.62	0.4191	2.8895E+00	1.4683E+01	5.0815E+00	1.4788E−01	1.6435E+01	1.9113E+00
3.64	0.4183	2.9110E+00	1.4847E+01	5.1004E+00	1.4481E−01	1.6612E+01	1.9324E+00
3.66	0.4175	2.9327E+00	1.5012E+01	5.1190E+00	1.4180E−01	1.6790E+01	1.9534E+00
3.68	0.4167	2.9544E+00	1.5178E+01	5.1375E+00	1.3885E−01	1.6968E+01	1.9744E+00
3.7	0.4160	2.9763E+00	1.5345E+01	5.1559E+00	1.3596E−01	1.7148E+01	1.9954E+00
3.72	0.4152	2.9982E+00	1.5513E+01	5.1741E+00	1.3314E−01	1.7328E+01	2.0164E+00
3.74	0.4145	3.0203E+00	1.5682E+01	5.1921E+00	1.3037E−01	1.7510E+01	2.0374E+00
3.76	0.4138	3.0425E+00	1.5851E+01	5.2099E+00	1.2766E−01	1.7693E+01	2.0583E+00
3.78	0.4130	3.0648E+00	1.6022E+01	5.2276E+00	1.2501E−01	1.7876E+01	2.0793E+00
3.8	0.4123	3.0873E+00	1.6193E+01	5.2451E+00	1.2242E−01	1.8061E+01	2.1003E+00
3.82	0.4116	3.1098E+00	1.6365E+01	5.2625E+00	1.1988E−01	1.8246E+01	2.1212E+00
3.84	0.4109	3.1325E+00	1.6539E+01	5.2797E+00	1.1740E−01	1.8433E+01	2.1422E+00
3.86	0.4103	3.1553E+00	1.6713E+01	5.2967E+00	1.1496E−01	1.8620E+01	2.1631E+00
3.88	0.4096	3.1782E+00	1.6888E+01	5.3136E+00	1.1258E−01	1.8809E+01	2.1841E+00
3.9	0.4089	3.2012E+00	1.7063E+01	5.3303E+00	1.1025E−01	1.8998E+01	2.2050E+00
3.92	0.4083	3.2243E+00	1.7240E+01	5.3469E+00	1.0797E−01	1.9188E+01	2.2259E+00
3.94	0.4076	3.2476E+00	1.7418E+01	5.3634E+00	1.0574E−01	1.9380E+01	2.2467E+00
3.96	0.4070	3.2710E+00	1.7597E+01	5.3796E+00	1.0356E−01	1.9572E+01	2.2676E+00
3.98	0.4064	3.2944E+00	1.7776E+01	5.3958E+00	1.0142E−01	1.9765E+01	2.2885E+00
4	0.4058	3.3181E+00	1.7957E+01	5.4118E+00	9.9330E−02	1.9960E+01	2.3093E+00
4.1	0.4028	3.4378E+00	1.8872E+01	5.4896E+00	8.9525E−02	2.0946E+01	2.4132E+00
4.2	0.4000	3.5605E+00	1.9810E+01	5.5639E+00	8.0727E−02	2.1956E+01	2.5167E+00
4.3	0.3975	3.6862E+00	2.0771E+01	5.6350E+00	7.2834E−02	2.2991E+01	2.6196E+00
4.4	0.3950	3.8147E+00	2.1755E+01	5.7029E+00	6.5752E−02	2.4050E+01	2.7219E+00
4.5	0.3927	3.9462E+00	2.2761E+01	5.7678E+00	5.9398E−02	2.5134E+01	2.8235E+00
4.6	0.3906	4.0806E+00	2.3790E+01	5.8299E+00	5.3695E−02	2.6241E+01	2.9244E+00
4.7	0.3886	4.2180E+00	2.4841E+01	5.8893E+00	4.8574E−02	2.7374E+01	3.0247E+00

M_1	M_2	T_2/T_1	p_2/p_1	ρ_2/ρ_1	p_{t2}/p_{t1}	p_{t2}/p_1	$\Delta s/R$
4.8	0.3867	4.3583E+00	2.5915E+01	5.9461E+00	4.3976E−02	2.8530E+01	3.1241E+00
4.9	0.3849	4.5015E+00	2.7011E+01	6.0005E+00	3.9844E−02	2.9711E+01	3.2228E+00
5	0.3832	4.6476E+00	2.8130E+01	6.0526E+00	3.6129E−02	3.0917E+01	3.3207E+00
5.1	0.3816	4.7967E+00	2.9272E+01	6.1025E+00	3.2788E−02	3.2147E+01	3.4177E+00
5.2	0.3801	4.9488E+00	3.0437E+01	6.1503E+00	2.9780E−02	3.3401E+01	3.5139E+00
5.3	0.3786	5.1038E+00	3.1623E+01	6.1961E+00	2.7072E−02	3.4679E+01	3.6093E+00
5.4	0.3773	5.2617E+00	3.2833E+01	6.2400E+00	2.4630E−02	3.5982E+01	3.7038E+00
5.5	0.3760	5.4225E+00	3.4065E+01	6.2822E+00	2.2428E−02	3.7309E+01	3.7974E+00
5.6	0.3747	5.5863E+00	3.5320E+01	6.3226E+00	2.0441E−02	3.8661E+01	3.8902E+00
5.7	0.3736	5.7531E+00	3.6597E+01	6.3614E+00	1.8646E−02	4.0037E+01	3.9821E+00
5.8	0.3725	5.9228E+00	3.7897E+01	6.3986E+00	1.7023E−02	4.1437E+01	4.0732E+00
5.9	0.3714	6.0954E+00	3.9220E+01	6.4344E+00	1.5555E−02	4.2862E+01	4.1634E+00
6	0.3704	6.2710E+00	4.0565E+01	6.4688E+00	1.4226E−02	4.4311E+01	4.2527E+00
6.1	0.3694	6.4495E+00	4.1933E+01	6.5018E+00	1.3021E−02	4.5785E+01	4.3412E+00
6.2	0.3685	6.6309E+00	4.3323E+01	6.5336E+00	1.1929E−02	4.7282E+01	4.4288E+00
6.3	0.3676	6.8153E+00	4.4737E+01	6.5641E+00	1.0937E−02	4.8804E+01	4.5156E+00
6.4	0.3668	7.0027E+00	4.6172E+01	6.5935E+00	1.0037E−02	5.0351E+01	4.6015E+00
6.5	0.3660	7.1930E+00	4.7630E+01	6.6218E+00	9.2182E−03	5.1922E+01	4.6866E+00
6.6	0.3652	7.3862E+00	4.9111E+01	6.6491E+00	8.4733E−03	5.3517E+01	4.7708E+00
6.7	0.3645	7.5824E+00	5.0615E+01	6.6753E+00	7.7950E−03	5.5137E+01	4.8543E+00
6.8	0.3638	7.7815E+00	5.2141E+01	6.7006E+00	7.1768E−03	5.6781E+01	4.9369E+00
6.9	0.3631	7.9836E+00	5.3690E+01	6.7250E+00	6.6129E−03	5.8449E+01	5.0187E+00
7	0.3625	8.1886E+00	5.5261E+01	6.7485E+00	6.0982E−03	6.0142E+01	5.0998E+00
7.1	0.3619	8.3966E+00	5.6855E+01	6.7712E+00	5.6280E−03	6.1859E+01	5.1800E+00
7.2	0.3613	8.6075E+00	5.8471E+01	6.7931E+00	5.1981E−03	6.3600E+01	5.2595E+00
7.3	0.3607	8.8214E+00	6.0110E+01	6.8142E+00	4.8047E−03	6.5366E+01	5.3382E+00
7.4	0.3601	9.0382E+00	6.1772E+01	6.8346E+00	4.4444E−03	6.7156E+01	5.4161E+00
7.5	0.3596	9.2579E+00	6.3457E+01	6.8543E+00	4.1142E−03	6.8970E+01	5.4933E+00
7.6	0.3591	9.4806E+00	6.5163E+01	6.8733E+00	3.8114E−03	7.0809E+01	5.5698E+00
7.7	0.3586	9.7063E+00	6.6893E+01	6.8917E+00	3.5334E−03	7.2672E+01	5.6455E+00
7.8	0.3581	9.9348E+00	6.8645E+01	6.9095E+00	3.2780E−03	7.4560E+01	5.7205E+00
7.9	0.3577	1.0166E+01	7.0420E+01	6.9267E+00	3.0433E−03	7.6472E+01	5.7948E+00
8	0.3573	1.0401E+01	7.2217E+01	6.9434E+00	2.8273E−03	7.8408E+01	5.8684E+00
8.1	0.3568	1.0638E+01	7.4037E+01	6.9595E+00	2.6285E−03	8.0369E+01	5.9414E+00
8.2	0.3564	1.0879E+01	7.5880E+01	6.9751E+00	2.4453E−03	8.2354E+01	6.0136E+00
8.3	0.3560	1.1122E+01	7.7745E+01	6.9902E+00	2.2764E−03	8.4363E+01	6.0852E+00
8.4	0.3557	1.1368E+01	7.9633E+01	7.0048E+00	2.1205E−03	8.6397E+01	6.1561E+00
8.5	0.3553	1.1618E+01	8.1543E+01	7.0190E+00	1.9767E−03	8.8455E+01	6.2263E+00
8.6	0.3549	1.1870E+01	8.3477E+01	7.0327E+00	1.8437E−03	9.0537E+01	6.2960E+00

(Continued)

M_1	M_2	T_2/T_1	p_2/p_1	ρ_2/ρ_1	p_{t2}/p_{t1}	p_{t2}/p_1	$\Delta s/R$
8.7	0.3546	1.2125E+01	8.5432E+01	7.0461E+00	1.7208E−03	9.2644E+01	6.3650E+00
8.8	0.3543	1.2383E+01	8.7410E+01	7.0590E+00	1.6071E−03	9.4775E+01	6.4333E+00
8.9	0.3539	1.2644E+01	8.9411E+01	7.0715E+00	1.5018E−03	9.6931E+01	6.5011E+00
9	0.3536	1.2908E+01	9.1435E+01	7.0837E+00	1.4043E−03	9.9111E+01	6.5682E+00
9.1	0.3533	1.3175E+01	9.3481E+01	7.0954E+00	1.3138E−03	1.0132E+02	6.6348E+00
9.2	0.3530	1.3445E+01	9.5550E+01	7.1069E+00	1.2300E−03	1.0354E+02	6.7008E+00
9.3	0.3528	1.3717E+01	9.7641E+01	7.1180E+00	1.1521E−03	1.0580E+02	6.7662E+00
9.4	0.3525	1.3993E+01	9.9755E+01	7.1288E+00	1.0798E−03	1.0807E+02	6.8310E+00
9.5	0.3522	1.4272E+01	1.0189E+02	7.1393E+00	1.0126E−03	1.1038E+02	6.8953E+00
9.6	0.3520	1.4554E+01	1.0405E+02	7.1495E+00	9.5006E−04	1.1270E+02	6.9590E+00
9.7	0.3517	1.4838E+01	1.0623E+02	7.1594E+00	8.9190E−04	1.1505E+02	7.0222E+00
9.8	0.3515	1.5126E+01	1.0844E+02	7.1690E+00	8.3775E−04	1.1743E+02	7.0848E+00
9.9	0.3513	1.5416E+01	1.1066E+02	7.1784E+00	7.8730E−04	1.1983E+02	7.1469E+00
10	0.3510	1.5710E+01	1.1291E+02	7.1875E+00	7.4028E−04	1.2225E+02	7.2085E+00
11	0.3491	1.8806E+01	1.3665E+02	7.2663E+00	4.1085E−04	1.4782E+02	7.7973E+00
12	0.3476	2.2198E+01	1.6265E+02	7.3274E+00	2.3840E−04	1.7583E+02	8.3416E+00
13	0.3464	2.5884E+01	1.9091E+02	7.3757E+00	1.4380E−04	2.0628E+02	8.8471E+00
14	0.3455	2.9865E+01	2.2143E+02	7.4145E+00	8.9725E−05	2.3916E+02	9.3188E+00
15	0.3448	3.4141E+01	2.5422E+02	7.4460E+00	5.7683E−05	2.7448E+02	9.7606E+00
16	0.3442	3.8712E+01	2.8926E+02	7.4721E+00	3.8077E−05	3.1223E+02	1.0176E+01
17	0.3436	4.3578E+01	3.2657E+02	7.4938E+00	2.5735E−05	3.5242E+02	1.0568E+01
18	0.3432	4.8739E+01	3.6613E+02	7.5121E+00	1.7765E−05	3.9505E+02	1.0938E+01
19	0.3429	5.4194E+01	4.0796E+02	7.5277E+00	1.2499E−05	4.4011E+02	1.1290E+01
20	0.3426	5.9945E+01	4.5204E+02	7.5410E+00	8.9464E−06	4.8761E+02	1.1624E+01
21	0.3423	6.5990E+01	4.9839E+02	7.5525E+00	6.5046E−06	5.3754E+02	1.1943E+01
22	0.3421	7.2331E+01	5.4700E+02	7.5625E+00	4.7972E−06	5.8991E+02	1.2247E+01
23	0.3419	7.8966E+01	5.9787E+02	7.5713E+00	3.5846E−06	6.4471E+02	1.2539E+01
24	0.3417	8.5896E+01	6.5100E+02	7.5789E+00	2.7108E−06	7.0196E+02	1.2818E+01
25	0.3415	9.3121E+01	7.0639E+02	7.5858E+00	2.0729E−06	7.6163E+02	1.3087E+01
30	0.3410	1.3367E+02	1.0173E+03	7.6103E+00	6.2331E−07	1.0966E+03	1.4288E+01
35	0.3406	1.8159E+02	1.3847E+03	7.6252E+00	2.2492E−07	1.4924E+03	1.5308E+01
40	0.3404	2.3688E+02	1.8086E+03	7.6349E+00	9.2847E−08	1.9491E+03	1.6192E+01
45	0.3403	2.9955E+02	2.2890E+03	7.6415E+00	4.2498E−08	2.4667E+03	1.6974E+01
50	0.3401	3.6959E+02	2.8260E+03	7.6463E+00	2.1110E−08	3.0452E+03	1.7674E+01

Rayleigh Flow

The parameters corresponding to a Rayleigh line, for a calorically perfect gas (i.e., $\gamma = $ constant), are summarized below:

$$\frac{p}{p^*} = \frac{\gamma + 1}{1 + \gamma M^2} \tag{D.1}$$

$$\frac{T}{T^*} = M^2 \left(\frac{\gamma + 1}{1 + \gamma M^2} \right)^2 \tag{D.2}$$

$$\frac{\rho}{\rho^*} = \frac{1}{M^2} \left(\frac{1 + \gamma M^2}{\gamma + 1} \right) \tag{D.3}$$

$$\frac{T_t}{T_t^*} = M^2 \left(\frac{\gamma + 1}{1 + \gamma M^2} \right)^2 \left(\frac{1 + \dfrac{\gamma - 1}{2} M^2}{\dfrac{\gamma + 1}{2}} \right) \tag{D.4}$$

$$\frac{p_t}{p_t^*} = \left(\frac{\gamma + 1}{1 + \gamma M^2} \right) \left(\frac{1 + \dfrac{\gamma - 1}{2} M^2}{\dfrac{\gamma + 1}{2}} \right)^{\frac{\gamma}{\gamma - 1}} \tag{D.5}$$

$$\frac{\Delta s}{R} = \left(\frac{\gamma}{\gamma - 1} \right) \ln \left(\frac{T}{T^*} \right) - \ln \left(\frac{p}{p^*} \right) \tag{D.6}$$

We have derived these expressions in Chapter 2.

Rayleigh Flow $\gamma = 1.4$

M	p/p^*	T/T^*	ρ/ρ^*	T_t/T_t^*	p_t/p_t^*	$\Delta s/R$
0.10	2.3669E+00	5.6020E−02	4.2250E+01	4.6777E−02	1.2591E+00	1.0949E+01
0.12	2.3526E+00	7.9698E−02	2.9519E+01	6.6606E−02	1.2554E+00	9.7088E+00
0.14	2.3359E+00	1.0695E−01	2.1842E+01	8.9471E−02	1.2510E+00	8.6724E+00
0.16	2.3170E+00	1.3743E−01	1.6859E+01	1.1511E−01	1.2461E+00	7.7865E+00
0.18	2.2959E+00	1.7078E−01	1.3443E+01	1.4324E−01	1.2406E+00	7.0169E+00
0.20	2.2727E+00	2.0661E−01	1.1000E+01	1.7355E−01	1.2346E+00	6.3402E+00
0.22	2.2477E+00	2.4452E−01	9.1921E+00	2.0574E−01	1.2281E+00	5.7395E+00
0.24	2.2209E+00	2.8411E−01	7.8171E+00	2.3948E−01	1.2213E+00	5.2023E+00
0.26	2.1925E+00	3.2496E−01	6.7470E+00	2.7446E−01	1.2140E+00	4.7193E+00
0.28	2.1626E+00	3.6667E−01	5.8980E+00	3.1035E−01	1.2064E+00	4.2828E+00
0.30	2.1314E+00	4.0887E−01	5.2130E+00	3.4686E−01	1.1985E+00	3.8870E+00
0.32	2.0991E+00	4.5119E−01	4.6523E+00	3.8369E−01	1.1904E+00	3.5271E+00
0.34	2.0657E+00	4.9327E−01	4.1877E+00	4.2056E−01	1.1822E+00	3.1989E+00
0.36	2.0314E+00	5.3482E−01	3.7984E+00	4.5723E−01	1.1737E+00	2.8992E+00
0.38	1.9964E+00	5.7553E−01	3.4688E+00	4.9346E−01	1.1652E+00	2.6250E+00
0.40	1.9608E+00	6.1515E−01	3.1875E+00	5.2903E−01	1.1566E+00	2.3740E+00
0.42	1.9247E+00	6.5346E−01	2.9454E+00	5.6376E−01	1.1480E+00	2.1439E+00
0.44	1.8882E+00	6.9025E−01	2.7355E+00	5.9748E−01	1.1394E+00	1.9331E+00
0.46	1.8515E+00	7.2538E−01	2.5525E+00	6.3007E−01	1.1308E+00	1.7397E+00
0.48	1.8147E+00	7.5871E−01	2.3918E+00	6.6139E−01	1.1224E+00	1.5624E+00
0.50	1.7778E+00	7.9012E−01	2.2500E+00	6.9136E−01	1.1141E+00	1.3998E+00
0.52	1.7409E+00	8.1955E−01	2.1243E+00	7.1990E−01	1.1059E+00	1.2509E+00
0.54	1.7043E+00	8.4695E−01	2.0122E+00	7.4695E−01	1.0979E+00	1.1145E+00
0.56	1.6678E+00	8.7227E−01	1.9120E+00	7.7249E−01	1.0901E+00	9.8977E−01
0.58	1.6316E+00	8.9552E−01	1.8219E+00	7.9648E−01	1.0826E+00	8.7577E−01
0.60	1.5957E+00	9.1670E−01	1.7407E+00	8.1892E−01	1.0753E+00	7.7174E−01
0.62	1.5603E+00	9.3584E−01	1.6673E+00	8.3983E−01	1.0682E+00	6.7696E−01
0.64	1.5253E+00	9.5298E−01	1.6006E+00	8.5920E−01	1.0615E+00	5.9078E−01
0.66	1.4908E+00	9.6816E−01	1.5399E+00	8.7708E−01	1.0550E+00	5.1260E−01
0.68	1.4569E+00	9.8144E−01	1.4844E+00	8.9350E−01	1.0489E+00	4.4187E−01
0.70	1.4235E+00	9.9290E−01	1.4337E+00	9.0850E−01	1.0431E+00	3.7807E−01
0.72	1.3907E+00	1.0026E+00	1.3871E+00	9.2212E−01	1.0376E+00	3.2072E−01
0.74	1.3585E+00	1.0106E+00	1.3442E+00	9.3442E−01	1.0325E+00	2.6940E−01
0.76	1.3270E+00	1.0171E+00	1.3047E+00	9.4546E−01	1.0278E+00	2.2370E−01
0.78	1.2961E+00	1.0220E+00	1.2682E+00	9.5528E−01	1.0234E+00	1.8324E−01
0.80	1.2658E+00	1.0255E+00	1.2344E+00	9.6395E−01	1.0193E+00	1.4767E−01
0.82	1.2362E+00	1.0276E+00	1.2030E+00	9.7152E−01	1.0157E+00	1.1668E−01
0.84	1.2073E+00	1.0285E+00	1.1738E+00	9.7807E−01	1.0124E+00	8.9953E−02
0.86	1.1791E+00	1.0283E+00	1.1467E+00	9.8363E−01	1.0095E+00	6.7220E−02

M	p/p^*	T/T^*	ρ/ρ^*	T_t/T_t^*	p_t/p_t^*	$\Delta s/R$
0.88	1.1515E+00	1.0269E+00	1.1214E+00	9.8828E−01	1.0070E+00	4.8217E−02
0.90	1.1246E+00	1.0245E+00	1.0977E+00	9.9207E−01	1.0049E+00	3.2700E−02
0.92	1.0984E+00	1.0212E+00	1.0756E+00	9.9506E−01	1.0031E+00	2.0444E−02
0.94	1.0728E+00	1.0170E+00	1.0549E+00	9.9729E−01	1.0017E+00	1.1237E−02
0.96	1.0479E+00	1.0121E+00	1.0354E+00	9.9883E−01	1.0008E+00	4.8812E−03
0.98	1.0236E+00	1.0064E+00	1.0172E+00	9.9971E−01	1.0002E+00	1.1930E−03
1.00	1.0000E+00	1.0000E+00	1.0000E+00	1.0000E+00	1.0000E+00	0.0000E+00
1.02	9.7698E−01	9.9304E−01	9.8382E−01	9.9973E−01	1.0002E+00	1.1412E−03
1.04	9.5456E−01	9.8554E−01	9.6857E−01	9.9895E−01	1.0008E+00	4.4660E−03
1.06	9.3275E−01	9.7755E−01	9.5417E−01	9.9769E−01	1.0017E+00	9.8337E−03
1.08	9.1152E−01	9.6913E−01	9.4056E−01	9.9601E−01	1.0031E+00	1.7112E−02
1.10	8.9087E−01	9.6031E−01	9.2769E−01	9.9392E−01	1.0049E+00	2.6179E−02
1.12	8.7078E−01	9.5115E−01	9.1550E−01	9.9148E−01	1.0070E+00	3.6917E−02
1.14	8.5123E−01	9.4169E−01	9.0394E−01	9.8871E−01	1.0095E+00	4.9219E−02
1.16	8.3222E−01	9.3196E−01	8.9298E−01	9.8564E−01	1.0124E+00	6.2984E−02
1.18	8.1374E−01	9.2200E−01	8.8258E−01	9.8230E−01	1.0157E+00	7.8116E−02
1.20	7.9576E−01	9.1185E−01	8.7269E−01	9.7872E−01	1.0194E+00	9.4525E−02
1.22	7.7827E−01	9.0153E−01	8.6328E−01	9.7492E−01	1.0235E+00	1.1213E−01
1.24	7.6127E−01	8.9108E−01	8.5432E−01	9.7092E−01	1.0279E+00	1.3085E−01
1.26	7.4473E−01	8.8052E−01	8.4578E−01	9.6675E−01	1.0328E+00	1.5061E−01
1.28	7.2865E−01	8.6988E−01	8.3765E−01	9.6243E−01	1.0380E+00	1.7135E−01
1.30	7.1301E−01	8.5917E−01	8.2988E−01	9.5798E−01	1.0437E+00	1.9299E−01
1.32	6.9780E−01	8.4843E−01	8.2247E−01	9.5341E−01	1.0497E+00	2.1548E−01
1.34	6.8301E−01	8.3766E−01	8.1538E−01	9.4873E−01	1.0561E+00	2.3876E−01
1.36	6.6863E−01	8.2689E−01	8.0861E−01	9.4398E−01	1.0629E+00	2.6277E−01
1.38	6.5464E−01	8.1613E−01	8.0212E−01	9.3914E−01	1.0701E+00	2.8747E−01
1.40	6.4103E−01	8.0539E−01	7.9592E−01	9.3425E−01	1.0777E+00	3.1281E−01
1.42	6.2779E−01	7.9469E−01	7.8997E−01	9.2931E−01	1.0856E+00	3.3874E−01
1.44	6.1491E−01	7.8405E−01	7.8427E−01	9.2434E−01	1.0940E+00	3.6522E−01
1.46	6.0237E−01	7.7346E−01	7.7880E−01	9.1933E−01	1.1028E+00	3.9221E−01
1.48	5.9018E−01	7.6294E−01	7.7356E−01	9.1431E−01	1.1120E+00	4.1968E−01
1.50	5.7831E−01	7.5250E−01	7.6852E−01	9.0928E−01	1.1215E+00	4.4758E−01
1.52	5.6676E−01	7.4215E−01	7.6368E−01	9.0424E−01	1.1315E+00	4.7589E−01
1.54	5.5552E−01	7.3189E−01	7.5902E−01	8.9920E−01	1.1419E+00	5.0458E−01
1.56	5.4458E−01	7.2173E−01	7.5455E−01	8.9418E−01	1.1527E+00	5.3361E−01
1.58	5.3393E−01	7.1168E−01	7.5024E−01	8.8917E−01	1.1640E+00	5.6295E−01
1.60	5.2356E−01	7.0174E−01	7.4609E−01	8.8419E−01	1.1756E+00	5.9259E−01
1.62	5.1346E−01	6.9190E−01	7.4210E−01	8.7922E−01	1.1877E+00	6.2250E−01
1.64	5.0363E−01	6.8219E−01	7.3825E−01	8.7429E−01	1.2002E+00	6.5265E−01

(Continued)

M	p/p^*	T/T^*	ρ/ρ^*	T_t/T_t^*	p_t/p_t^*	$\Delta s/R$
1.66	4.9405$E-$01	6.7259$E-$01	7.3454$E-$01	8.6939$E-$01	1.2131$E+$00	6.8303$E-$01
1.68	4.8472$E-$01	6.6312$E-$01	7.3096$E-$01	8.6453$E-$01	1.2264$E+$00	7.1360$E-$01
1.70	4.7562$E-$01	6.5377$E-$01	7.2751$E-$01	8.5971$E-$01	1.2402$E+$00	7.4436$E-$01
1.72	4.6677$E-$01	6.4455$E-$01	7.2418$E-$01	8.5493$E-$01	1.2545$E+$00	7.7529$E-$01
1.74	4.5813$E-$01	6.3545$E-$01	7.2096$E-$01	8.5019$E-$01	1.2692$E+$00	8.0636$E-$01
1.76	4.4972$E-$01	6.2649$E-$01	7.1785$E-$01	8.4551$E-$01	1.2843$E+$00	8.3757$E-$01
1.78	4.4152$E-$01	6.1765$E-$01	7.1484$E-$01	8.4087$E-$01	1.2999$E+$00	8.6889$E-$01
1.80	4.3353$E-$01	6.0894$E-$01	7.1193$E-$01	8.3628$E-$01	1.3159$E+$00	9.0031$E-$01
1.82	4.2573$E-$01	6.0036$E-$01	7.0912$E-$01	8.3174$E-$01	1.3324$E+$00	9.3183$E-$01
1.84	4.1813$E-$01	5.9191$E-$01	7.0640$E-$01	8.2726$E-$01	1.3494$E+$00	9.6342$E-$01
1.86	4.1072$E-$01	5.8359$E-$01	7.0377$E-$01	8.2283$E-$01	1.3669$E+$00	9.9507$E-$01
1.88	4.0349$E-$01	5.7540$E-$01	7.0122$E-$01	8.1845$E-$01	1.3849$E+$00	1.0268$E+$00
1.90	3.9643$E-$01	5.6734$E-$01	6.9875$E-$01	8.1414$E-$01	1.4033$E+$00	1.0585$E+$00
1.92	3.8955$E-$01	5.5941$E-$01	6.9636$E-$01	8.0987$E-$01	1.4222$E+$00	1.0903$E+$00
1.94	3.8283$E-$01	5.5160$E-$01	6.9404$E-$01	8.0567$E-$01	1.4417$E+$00	1.1221$E+$00
1.96	3.7628$E-$01	5.4392$E-$01	6.9180$E-$01	8.0152$E-$01	1.4616$E+$00	1.1539$E+$00
1.98	3.6988$E-$01	5.3636$E-$01	6.8962$E-$01	7.9742$E-$01	1.4821$E+$00	1.1858$E+$00
2.00	3.6364$E-$01	5.2893$E-$01	6.8750$E-$01	7.9339$E-$01	1.5031$E+$00	1.2176$E+$00
2.05	3.4866$E-$01	5.1087$E-$01	6.8248$E-$01	7.8355$E-$01	1.5579$E+$00	1.2971$E+$00
2.10	3.3454$E-$01	4.9356$E-$01	6.7782$E-$01	7.7406$E-$01	1.6162$E+$00	1.3764$E+$00
2.15	3.2122$E-$01	4.7696$E-$01	6.7347$E-$01	7.6493$E-$01	1.6780$E+$00	1.4555$E+$00
2.20	3.0864$E-$01	4.6106$E-$01	6.6942$E-$01	7.5613$E-$01	1.7434$E+$00	1.5342$E+$00
2.25	2.9675$E-$01	4.4582$E-$01	6.6564$E-$01	7.4768$E-$01	1.8128$E+$00	1.6126$E+$00
2.30	2.8551$E-$01	4.3122$E-$01	6.6210$E-$01	7.3954$E-$01	1.8860$E+$00	1.6905$E+$00
2.35	2.7487$E-$01	4.1723$E-$01	6.5878$E-$01	7.3173$E-$01	1.9634$E+$00	1.7679$E+$00
2.40	2.6478$E-$01	4.0384$E-$01	6.5567$E-$01	7.2421$E-$01	2.0451$E+$00	1.8448$E+$00
2.45	2.5522$E-$01	3.9100$E-$01	6.5275$E-$01	7.1699$E-$01	2.1311$E+$00	1.9211$E+$00
2.50	2.4615$E-$01	3.7870$E-$01	6.5000$E-$01	7.1006$E-$01	2.2218$E+$00	1.9968$E+$00
2.55	2.3754$E-$01	3.6691$E-$01	6.4741$E-$01	7.0340$E-$01	2.3173$E+$00	2.0718$E+$00
2.60	2.2936$E-$01	3.5561$E-$01	6.4497$E-$01	6.9700$E-$01	2.4177$E+$00	2.1463$E+$00
2.65	2.2158$E-$01	3.4478$E-$01	6.4267$E-$01	6.9084$E-$01	2.5233$E+$00	2.2200$E+$00
2.70	2.1417$E-$01	3.3439$E-$01	6.4049$E-$01	6.8494$E-$01	2.6343$E+$00	2.2931$E+$00
2.75	2.0712$E-$01	3.2442$E-$01	6.3843$E-$01	6.7926$E-$01	2.7508$E+$00	2.3655$E+$00
2.80	2.0040$E-$01	3.1486$E-$01	6.3648$E-$01	6.7380$E-$01	2.8731$E+$00	2.4373$E+$00
2.85	1.9399$E-$01	3.0568$E-$01	6.3463$E-$01	6.6855$E-$01	3.0014$E+$00	2.5083$E+$00
2.90	1.8788$E-$01	2.9687$E-$01	6.3288$E-$01	6.6350$E-$01	3.1359$E+$00	2.5787$E+$00
2.95	1.8205$E-$01	2.8841$E-$01	6.3121$E-$01	6.5865$E-$01	3.2768$E+$00	2.6483$E+$00
3.00	1.7647$E-$01	2.8028$E-$01	6.2963$E-$01	6.5398$E-$01	3.4245$E+$00	2.7173$E+$00
3.10	1.6604$E-$01	2.6495$E-$01	6.2669$E-$01	6.4516$E-$01	3.7408$E+$00	2.8532$E+$00
3.20	1.5649$E-$01	2.5078$E-$01	6.2402$E-$01	6.3699$E-$01	4.0871$E+$00	2.9863$E+$00

M	p/p^*	T/T^*	ρ/ρ^*	T_t/T_t^*	p_t/p_t^*	$\Delta s/R$
3.30	1.4773E−01	2.3766E−01	6.2159E−01	6.2940E−01	4.4655E+00	3.1168E+00
3.40	1.3966E−01	2.2549E−01	6.1938E−01	6.2236E−01	4.8783E+00	3.2446E+00
3.50	1.3223E−01	2.1419E−01	6.1735E−01	6.1580E−01	5.3280E+00	3.3699E+00
3.60	1.2537E−01	2.0369E−01	6.1548E−01	6.0970E−01	5.8173E+00	3.4926E+00
3.70	1.1901E−01	1.9390E−01	6.1377E−01	6.0401E−01	6.3488E+00	3.6128E+00
3.80	1.1312E−01	1.8478E−01	6.1219E−01	5.9870E−01	6.9256E+00	3.7307E+00
3.90	1.0765E−01	1.7627E−01	6.1073E−01	5.9373E−01	7.5505E+00	3.8463E+00
4.00	1.0256E−01	1.6831E−01	6.0938E−01	5.8909E−01	8.2268E+00	3.9595E+00
4.10	9.7823E−02	1.6086E−01	6.0812E−01	5.8473E−01	8.9579E+00	4.0706E+00
4.20	9.3400E−02	1.5388E−01	6.0695E−01	5.8065E−01	9.7473E+00	4.1796E+00
4.30	8.9266E−02	1.4734E−01	6.0587E−01	5.7682E−01	1.0599E+01	4.2865E+00
4.40	8.5397E−02	1.4119E−01	6.0486E−01	5.7322E−01	1.1516E+01	4.3914E+00
4.50	8.1772E−02	1.3540E−01	6.0391E−01	5.6982E−01	1.2502E+01	4.4944E+00
4.60	7.8370E−02	1.2996E−01	6.0302E−01	5.6663E−01	1.3563E+01	4.5955E+00
4.70	7.5174E−02	1.2483E−01	6.0220E−01	5.6362E−01	1.4702E+01	4.6948E+00
4.80	7.2167E−02	1.2000E−01	6.0142E−01	5.6078E−01	1.5923E+01	4.7923E+00
4.90	6.9336E−02	1.1543E−01	6.0069E−01	5.5809E−01	1.7232E+01	4.8881E+00
5.00	6.6667E−02	1.1111E−01	6.0000E−01	5.5556E−01	1.8634E+01	4.9822E+00
5.20	6.1767E−02	1.0316E−01	5.9874E−01	5.5088E−01	2.1734E+01	5.1658E+00
5.40	5.7383E−02	9.6019E−02	5.9762E−01	5.4667E−01	2.5268E+01	5.3432E+00
5.60	5.3447E−02	8.9584E−02	5.9662E−01	5.4288E−01	2.9281E+01	5.5150E+00
5.80	4.9900E−02	8.3765E−02	5.9572E−01	5.3944E−01	3.3822E+01	5.6814E+00
6.00	4.6693E−02	7.8487E−02	5.9491E−01	5.3633E−01	3.8946E+01	5.8427E+00
6.50	3.9900E−02	6.7263E−02	5.9320E−01	5.2970E−01	5.4683E+01	6.2256E+00
7.00	3.4483E−02	5.8264E−02	5.9184E−01	5.2438E−01	7.5414E+01	6.5824E+00
7.50	3.0094E−02	5.0943E−02	5.9074E−01	5.2004E−01	1.0229E+02	6.9162E+00
8.00	2.6490E−02	4.4910E−02	5.8984E−01	5.1647E−01	1.3662E+02	7.2298E+00
8.50	2.3495E−02	3.9883E−02	5.8910E−01	5.1349E−01	1.7992E+02	7.5254E+00
9.00	2.0979E−02	3.5650E−02	5.8848E−01	5.1098E−01	2.3388E+02	7.8048E+00
9.50	1.8846E−02	3.2053E−02	5.8795E−01	5.0885E−01	3.0041E+02	8.0698E+00
10.00	1.7021E−02	2.8972E−02	5.8750E−01	5.0702E−01	3.8161E+02	8.3217E+00
15.00	7.5949E−03	1.2979E−02	5.8519E−01	4.9752E−01	2.6488E+03	1.0325E+01

Rayleigh Flow $\gamma = 1.3$

M	p/p^*	T/T^*	ρ/ρ^*	T_t/T_t^*	p_t/p_t^*	$\Delta s/R$
0.10	2.2705E+00	5.1551E−02	4.4043E+01	4.4894E−02	1.2471E+00	1.3669E+01
0.12	2.2577E+00	7.3402E−02	3.0758E+01	6.3966E−02	1.2437E+00	1.2132E+01
0.14	2.2429E+00	9.8596E−02	2.2748E+01	8.5987E−02	1.2397E+00	1.0847E+01
0.16	2.2259E+00	1.2684E−01	1.7549E+01	1.1072E−01	1.2351E+00	9.7477E+00
0.18	2.2070E+00	1.5782E−01	1.3984E+01	1.3790E−01	1.2300E+00	8.7922E+00
0.20	2.1863E+00	1.9120E−01	1.1435E+01	1.6726E−01	1.2245E+00	7.9514E+00
0.22	2.1639E+00	2.2662E−01	9.5483E+00	1.9849E−01	1.2185E+00	7.2046E+00
0.24	2.1398E+00	2.6373E−01	8.1135E+00	2.3131E−01	1.2121E+00	6.5363E+00
0.26	2.1142E+00	3.0216E−01	6.9969E+00	2.6541E−01	1.2053E+00	5.9347E+00
0.28	2.0873E+00	3.4156E−01	6.1109E+00	3.0050E−01	1.1983E+00	5.3908E+00
0.30	2.0591E+00	3.8159E−01	5.3961E+00	3.3629E−01	1.1909E+00	4.8971E+00
0.32	2.0298E+00	4.2189E−01	4.8111E+00	3.7250E−01	1.1834E+00	4.4476E+00
0.34	1.9995E+00	4.6217E−01	4.3263E+00	4.0886E−01	1.1756E+00	4.0374E+00
0.36	1.9684E+00	5.0213E−01	3.9200E+00	4.4512E−01	1.1677E+00	3.6624E+00
0.38	1.9365E+00	5.4150E−01	3.5762E+00	4.8106E−01	1.1596E+00	3.3190E+00
0.40	1.9040E+00	5.8002E−01	3.2826E+00	5.1647E−01	1.1515E+00	3.0043E+00
0.42	1.8710E+00	6.1748E−01	3.0300E+00	5.5115E−01	1.1434E+00	2.7156E+00
0.44	1.8375E+00	6.5369E−01	2.8110E+00	5.8494E−01	1.1352E+00	2.4506E+00
0.46	1.8038E+00	6.8849E−01	2.6200E+00	6.1769E−01	1.1271E+00	2.2073E+00
0.48	1.7699E+00	7.2173E−01	2.4523E+00	6.4928E−01	1.1191E+00	1.9840E+00
0.50	1.7358E+00	7.5329E−01	2.3043E+00	6.7960E−01	1.1111E+00	1.7791E+00
0.52	1.7018E+00	7.8310E−01	2.1731E+00	7.0858E−01	1.1033E+00	1.5911E+00
0.54	1.6678E+00	8.1108E−01	2.0562E+00	7.3614E−01	1.0957E+00	1.4188E+00
0.56	1.6339E+00	8.3719E−01	1.9516E+00	7.6224E−01	1.0882E+00	1.2610E+00
0.58	1.6002E+00	8.6140E−01	1.8577E+00	7.8684E−01	1.0809E+00	1.1166E+00
0.60	1.5668E+00	8.8370E−01	1.7729E+00	8.0993E−01	1.0739E+00	9.8476E−01
0.62	1.5336E+00	9.0410E−01	1.6963E+00	8.3151E−01	1.0671E+00	8.6447E−01
0.64	1.5008E+00	9.2263E−01	1.6267E+00	8.5158E−01	1.0605E+00	7.5499E−01
0.66	1.4684E+00	9.3930E−01	1.5633E+00	8.7015E−01	1.0543E+00	6.5555E−01
0.68	1.4365E+00	9.5417E−01	1.5055E+00	8.8726E−01	1.0483E+00	5.6550E−01
0.70	1.4050E+00	9.6728E−01	1.4525E+00	9.0294E−01	1.0426E+00	4.8418E−01
0.72	1.3740E+00	9.7870E−01	1.4039E+00	9.1722E−01	1.0373E+00	4.1102E−01
0.74	1.3436E+00	9.8849E−01	1.3592E+00	9.3016E−01	1.0323E+00	3.4548E−01
0.76	1.3136E+00	9.9671E−01	1.3180E+00	9.4180E−01	1.0276E+00	2.8706E−01
0.78	1.2843E+00	1.0034E+00	1.2799E+00	9.5219E−01	1.0233E+00	2.3529E−01
0.80	1.2555E+00	1.0088E+00	1.2446E+00	9.6139E−01	1.0193E+00	1.8974E−01
0.82	1.2272E+00	1.0127E+00	1.2118E+00	9.6944E−01	1.0157E+00	1.5001E−01
0.84	1.1996E+00	1.0154E+00	1.1814E+00	9.7642E−01	1.0124E+00	1.1572E−01

M	p/p^*	T/T^*	ρ/ρ^*	T_t/T_t^*	p_t/p_t^*	$\Delta s/R$
0.86	1.1726E+00	1.0169E+00	1.1531E+00	9.8238E−01	1.0095E+00	8.6524E−02
0.88	1.1461E+00	1.0173E+00	1.1267E+00	9.8736E−01	1.0070E+00	6.2098E−02
0.90	1.1203E+00	1.0166E+00	1.1020E+00	9.9143E−01	1.0049E+00	4.2138E−02
0.92	1.0951E+00	1.0150E+00	1.0789E+00	9.9465E−01	1.0031E+00	2.6359E−02
0.94	1.0704E+00	1.0124E+00	1.0573E+00	9.9707E−01	1.0018E+00	1.4496E−02
0.96	1.0464E+00	1.0090E+00	1.0370E+00	9.9873E−01	1.0008E+00	6.3002E−03
0.98	1.0229E+00	1.0049E+00	1.0179E+00	9.9969E−01	1.0002E+00	1.5407E−03
1.00	1.0000E+00	1.0000E+00	1.0000E+00	1.0000E+00	1.0000E+00	0.0000E+00
1.02	9.7768E−01	9.9446E−01	9.8312E−01	9.9971E−01	1.0002E+00	1.4748E−03
1.04	9.5591E−01	9.8833E−01	9.6720E−01	9.9885E−01	1.0008E+00	5.7748E−03
1.06	9.3470E−01	9.8165E−01	9.5217E−01	9.9748E−01	1.0018E+00	1.2722E−02
1.08	9.1403E−01	9.7448E−01	9.3797E−01	9.9563E−01	1.0032E+00	2.2148E−02
1.10	8.9390E−01	9.6686E−01	9.2454E−01	9.9334E−01	1.0049E+00	3.3897E−02
1.12	8.7429E−01	9.5883E−01	9.1182E−01	9.9065E−01	1.0071E+00	4.7821E−02
1.14	8.5518E−01	9.5045E−01	8.9977E−01	9.8759E−01	1.0097E+00	6.3784E−02
1.16	8.3658E−01	9.4175E−01	8.8833E−01	9.8420E−01	1.0127E+00	8.1654E−02
1.18	8.1847E−01	9.3276E−01	8.7747E−01	9.8050E−01	1.0161E+00	1.0131E−01
1.20	8.0084E−01	9.2353E−01	8.6715E−01	9.7653E−01	1.0199E+00	1.2264E−01
1.22	7.8367E−01	9.1408E−01	8.5733E−01	9.7231E−01	1.0241E+00	1.4553E−01
1.24	7.6695E−01	9.0444E−01	8.4798E−01	9.6786E−01	1.0288E+00	1.6989E−01
1.26	7.5068E−01	8.9465E−01	8.3908E−01	9.6322E−01	1.0338E+00	1.9562E−01
1.28	7.3484E−01	8.8473E−01	8.3059E−01	9.5840E−01	1.0392E+00	2.2263E−01
1.30	7.1942E−01	8.7470E−01	8.2249E−01	9.5342E−01	1.0451E+00	2.5083E−01
1.32	7.0442E−01	8.6458E−01	8.1475E−01	9.4830E−01	1.0514E+00	2.8016E−01
1.34	6.8980E−01	8.5440E−01	8.0736E−01	9.4306E−01	1.0581E+00	3.1052E−01
1.36	6.7558E−01	8.4417E−01	8.0029E−01	9.3772E−01	1.0653E+00	3.4186E−01
1.38	6.6173E−01	8.3392E−01	7.9352E−01	9.3229E−01	1.0728E+00	3.7411E−01
1.40	6.4825E−01	8.2365E−01	7.8705E−01	9.2679E−01	1.0809E+00	4.0721E−01
1.42	6.3513E−01	8.1339E−01	7.8084E−01	9.2122E−01	1.0893E+00	4.4109E−01
1.44	6.2235E−01	8.0314E−01	7.7489E−01	9.1561E−01	1.0982E+00	4.7571E−01
1.46	6.0990E−01	7.9292E−01	7.6919E−01	9.0996E−01	1.1075E+00	5.1101E−01
1.48	5.9779E−01	7.8274E−01	7.6371E−01	9.0428E−01	1.1173E+00	5.4694E−01
1.50	5.8599E−01	7.7261E−01	7.5845E−01	8.9858E−01	1.1275E+00	5.8347E−01
1.52	5.7449E−01	7.6253E−01	7.5340E−01	8.9287E−01	1.1382E+00	6.2053E−01
1.54	5.6330E−01	7.5253E−01	7.4855E−01	8.8716E−01	1.1494E+00	6.5810E−01
1.56	5.5240E−01	7.4259E−01	7.4388E−01	8.8145E−01	1.1611E+00	6.9614E−01
1.58	5.4177E−01	7.3274E−01	7.3938E−01	8.7576E−01	1.1732E+00	7.3460E−01
1.60	5.3142E−01	7.2297E−01	7.3505E−01	8.7008E−01	1.1858E+00	7.7346E−01
1.62	5.2134E−01	7.1330E−01	7.3089E−01	8.6443E−01	1.1989E+00	8.1269E−01
1.64	5.1151E−01	7.0372E−01	7.2687E−01	8.5880E−01	1.2125E+00	8.5224E−01

(Continued)

M	p/p^*	T/T^*	ρ/ρ^*	T_t/T_t^*	p_t/p_t^*	$\Delta s/R$
1.66	5.0193E−01	6.9424E−01	7.2300E−01	8.5321E−01	1.2266E+00	8.9211E−01
1.68	4.9260E−01	6.8486E−01	7.1926E−01	8.4766E−01	1.2412E+00	9.3225E−01
1.70	4.8350E−01	6.7560E−01	7.1566E−01	8.4215E−01	1.2563E+00	9.7264E−01
1.72	4.7463E−01	6.6644E−01	7.1218E−01	8.3668E−01	1.2719E+00	1.0133E+00
1.74	4.6598E−01	6.5739E−01	7.0882E−01	8.3125E−01	1.2881E+00	1.0541E+00
1.76	4.5754E−01	6.4846E−01	7.0558E−01	8.2588E−01	1.3049E+00	1.0951E+00
1.78	4.4931E−01	6.3964E−01	7.0244E−01	8.2056E−01	1.3222E+00	1.1363E+00
1.80	4.4129E−01	6.3095E−01	6.9941E−01	8.1529E−01	1.3400E+00	1.1776E+00
1.82	4.3346E−01	6.2236E−01	6.9648E−01	8.1008E−01	1.3584E+00	1.2190E+00
1.84	4.2582E−01	6.1390E−01	6.9364E−01	8.0492E−01	1.3775E+00	1.2606E+00
1.86	4.1837E−01	6.0556E−01	6.9089E−01	7.9983E−01	1.3971E+00	1.3022E+00
1.88	4.1110E−01	5.9733E−01	6.8823E−01	7.9479E−01	1.4173E+00	1.3440E+00
1.90	4.0400E−01	5.8922E−01	6.8566E−01	7.8982E−01	1.4381E+00	1.3858E+00
1.92	3.9708E−01	5.8124E−01	6.8316E−01	7.8490E−01	1.4595E+00	1.4276E+00
1.94	3.9031E−01	5.7337E−01	6.8074E−01	7.8005E−01	1.4816E+00	1.4695E+00
1.96	3.8371E−01	5.6562E−01	6.7839E−01	7.7526E−01	1.5043E+00	1.5114E+00
1.98	3.7726E−01	5.5798E−01	6.7612E−01	7.7053E−01	1.5277E+00	1.5533E+00
2.00	3.7097E−01	5.5047E−01	6.7391E−01	7.6587E−01	1.5518E+00	1.5953E+00
2.05	3.5586E−01	5.3218E−01	6.6868E−01	7.5449E−01	1.6150E+00	1.7001E+00
2.10	3.4160E−01	5.1461E−01	6.6381E−01	7.4350E−01	1.6826E+00	1.8047E+00
2.15	3.2814E−01	4.9773E−01	6.5928E−01	7.3290E−01	1.7551E+00	1.9091E+00
2.20	3.1541E−01	4.8151E−01	6.5505E−01	7.2269E−01	1.8324E+00	2.0130E+00
2.25	3.0338E−01	4.6595E−01	6.5110E−01	7.1285E−01	1.9150E+00	2.1165E+00
2.30	2.9199E−01	4.5101E−01	6.4741E−01	7.0339E−01	2.0031E+00	2.2194E+00
2.35	2.8120E−01	4.3668E−01	6.4395E−01	6.9428E−01	2.0970E+00	2.3217E+00
2.40	2.7097E−01	4.2293E−01	6.4070E−01	6.8551E−01	2.1970E+00	2.4233E+00
2.45	2.6127E−01	4.0973E−01	6.3765E−01	6.7709E−01	2.3033E+00	2.5242E+00
2.50	2.5205E−01	3.9707E−01	6.3478E−01	6.6898E−01	2.4164E+00	2.6243E+00
2.55	2.4330E−01	3.8492E−01	6.3208E−01	6.6119E−01	2.5367E+00	2.7236E+00
2.60	2.3498E−01	3.7326E−01	6.2953E−01	6.5370E−01	2.6644E+00	2.8221E+00
2.65	2.2707E−01	3.6207E−01	6.2713E−01	6.4649E−01	2.7999E+00	2.9198E+00
2.70	2.1953E−01	3.5133E−01	6.2486E−01	6.3956E−01	2.9438E+00	3.0165E+00
2.75	2.1235E−01	3.4101E−01	6.2271E−01	6.3290E−01	3.0964E+00	3.1125E+00
2.80	2.0550E−01	3.3110E−01	6.2067E−01	6.2649E−01	3.2582E+00	3.2075E+00
2.85	1.9897E−01	3.2158E−01	6.1875E−01	6.2033E−01	3.4296E+00	3.3016E+00
2.90	1.9274E−01	3.1243E−01	6.1692E−01	6.1440E−01	3.6113E+00	3.3949E+00
2.95	1.8679E−01	3.0364E−01	6.1518E−01	6.0869E−01	3.8037E+00	3.4872E+00
3.00	1.8110E−01	2.9518E−01	6.1353E−01	6.0320E−01	4.0073E+00	3.5786E+00
3.10	1.7046E−01	2.7923E−01	6.1046E−01	5.9282E−01	4.4507E+00	3.7588E+00
3.20	1.6070E−01	2.6446E−01	6.0768E−01	5.8319E−01	4.9466E+00	3.9354E+00

M	p/p^*	T/T^*	ρ/ρ^*	T_t/T_t^*	p_t/p_t^*	$\Delta s/R$
3.30	1.5175E−01	2.5076E−01	6.0514E−01	5.7424E−01	5.5003E+00	4.1085E+00
3.40	1.4350E−01	2.3804E−01	6.0283E−01	5.6592E−01	6.1179E+00	4.2782E+00
3.50	1.3589E−01	2.2622E−01	6.0071E−01	5.5818E−01	6.8059E+00	4.4444E+00
3.60	1.2887E−01	2.1522E−01	5.9877E−01	5.5096E−01	7.5712E+00	4.6074E+00
3.70	1.2236E−01	2.0497E−01	5.9698E−01	5.4423E−01	8.4216E+00	4.7671E+00
3.80	1.1633E−01	1.9540E−01	5.9533E−01	5.3794E−01	9.3652E+00	4.9237E+00
3.90	1.1072E−01	1.8646E−01	5.9380E−01	5.3206E−01	1.0411E+01	5.0772E+00
4.00	1.0550E−01	1.7810E−01	5.9239E−01	5.2656E−01	1.1569E+01	5.2277E+00
4.10	1.0064E−01	1.7027E−01	5.9108E−01	5.2139E−01	1.2850E+01	5.3754E+00
4.20	9.6106E−02	1.6293E−01	5.8986E−01	5.1655E−01	1.4264E+01	5.5202E+00
4.30	9.1864E−02	1.5604E−01	5.8873E−01	5.1201E−01	1.5824E+01	5.6624E+00
4.40	8.7894E−02	1.4956E−01	5.8768E−01	5.0773E−01	1.7544E+01	5.8018E+00
4.50	8.4172E−02	1.4347E−01	5.8669E−01	5.0370E−01	1.9436E+01	5.9388E+00
4.60	8.0679E−02	1.3773E−01	5.8576E−01	4.9991E−01	2.1517E+01	6.0732E+00
4.70	7.7397E−02	1.3232E−01	5.8490E−01	4.9633E−01	2.3802E+01	6.2053E+00
4.80	7.4309E−02	1.2722E−01	5.8409E−01	4.9296E−01	2.6308E+01	6.3350E+00
4.90	7.1400E−02	1.2240E−01	5.8333E−01	4.8976E−01	2.9055E+01	6.4624E+00
5.00	6.8657E−02	1.1784E−01	5.8261E−01	4.8675E−01	3.2061E+01	6.5877E+00
5.20	6.3620E−02	1.0945E−01	5.8130E−01	4.8118E−01	3.8939E+01	6.8319E+00
5.40	5.9114E−02	1.0190E−01	5.8013E−01	4.7617E−01	4.7127E+01	7.0680E+00
5.60	5.5066E−02	9.5092E−02	5.7908E−01	4.7166E−01	5.6835E+01	7.2966E+00
5.80	5.1417E−02	8.8935E−02	5.7814E−01	4.6757E−01	6.8300E+01	7.5181E+00
6.00	4.8117E−02	8.3349E−02	5.7729E−01	4.6386E−01	8.1790E+01	7.7329E+00
6.50	4.1127E−02	7.1461E−02	5.7551E−01	4.5595E−01	1.2641E+02	8.2427E+00
7.00	3.5549E−02	6.1922E−02	5.7409E−01	4.4961E−01	1.9131E+02	8.7179E+00
7.50	3.1029E−02	5.4156E−02	5.7295E−01	4.4443E−01	2.8385E+02	9.1625E+00
8.00	2.7316E−02	4.7754E−02	5.7201E−01	4.4017E−01	4.1338E+02	9.5803E+00
8.50	2.4230E−02	4.2416E−02	5.7124E−01	4.3661E−01	5.9167E+02	9.9740E+00
9.00	2.1637E−02	3.7921E−02	5.7059E−01	4.3361E−01	8.3332E+02	1.0346E+01
9.50	1.9438E−02	3.4100E−02	5.7003E−01	4.3106E−01	1.1562E+03	1.0699E+01
10.00	1.7557E−02	3.0826E−02	5.6957E−01	4.2888E−01	1.5822E+03	1.1035E+01
15.00	7.8365E−03	1.3817E−02	5.6715E−01	4.1752E−01	2.0348E+04	1.3706E+01

APPENDIX E

Fanno Flow

The equations used to tabulate Fanno flow functions, for a calorically perfect gas, are summarized below:

$$4C_f \frac{L^*}{D_h} = \frac{1 - M^2}{\gamma M^2} + \frac{\gamma + 1}{2\gamma} \ln\left[\frac{(\gamma + 1)M^2}{2\left(1 + \frac{\gamma - 1}{2}M^2\right)}\right] \tag{E.1}$$

$$\frac{T}{T^*} = \frac{\gamma + 1}{2\left(1 + \frac{\gamma - 1}{2}M^2\right)} \tag{E.2}$$

$$\frac{\rho}{\rho^*} = \frac{1}{M}\sqrt{\frac{2\left(1 + \frac{\gamma - 1}{2}M^2\right)}{\gamma + 1}} \tag{E.3}$$

$$\frac{p}{p^*} = \frac{1}{M}\sqrt{\frac{\gamma + 1}{2\left(1 + \frac{\gamma - 1}{2}M^2\right)}} \tag{E.4}$$

$$\frac{I}{I^*} = \frac{p(1 + \gamma M^2)}{p^*(1 + \gamma)} = \frac{1}{M}\sqrt{\frac{\gamma + 1}{2[1 + (\gamma - 1)M^2/2]}}\left(\frac{1 + \gamma M^2}{\gamma + 1}\right) \tag{E.5}$$

$$\frac{p_t}{p_t^*} = \frac{1}{M}\left(\frac{1 + \frac{\gamma - 1}{2}M^2}{\frac{\gamma + 1}{2}}\right)^{\frac{\gamma + 1}{2(\gamma - 1)}} \tag{E.6}$$

$$\frac{\Delta s}{R} = \left(\frac{\gamma}{\gamma - 1}\right) \ln\left(\frac{T}{T^*}\right) - \ln\left(\frac{p}{p^*}\right) \tag{E.7}$$

We derived these expressions in Chapter 2.

Fanno Flow $\gamma = 1.4$

M	$4C_fL^*/D$	T/T^*	p/p^*	ρ/ρ^*	p_t/p_t^*	I/I^*	$\Delta s/R$
0.10	6.6922E+01	1.1976E+00	1.0944E+01	9.1378E+00	5.8218E+00	4.6236E+00	1.7616E+00
0.12	4.5408E+01	1.1966E+00	9.1156E+00	7.6182E+00	4.8643E+00	3.8747E+00	1.5819E+00
0.14	3.2511E+01	1.1953E+00	7.8093E+00	6.5333E+00	4.1824E+00	3.3432E+00	1.4309E+00
0.16	2.4198E+01	1.1939E+00	6.8291E+00	5.7200E+00	3.6727E+00	2.9474E+00	1.3009E+00
0.18	1.8543E+01	1.1923E+00	6.0662E+00	5.0879E+00	3.2779E+00	2.6422E+00	1.1872E+00
0.20	1.4533E+01	1.1905E+00	5.4554E+00	4.5826E+00	2.9635E+00	2.4004E+00	1.0864E+00
0.22	1.1596E+01	1.1885E+00	4.9554E+00	4.1694E+00	2.7076E+00	2.2046E+00	9.9606E−01
0.24	9.3866E+00	1.1863E+00	4.5383E+00	3.8255E+00	2.4956E+00	2.0434E+00	9.1451E−01
0.26	7.6877E+00	1.1840E+00	4.1851E+00	3.5347E+00	2.3173E+00	1.9088E+00	8.4040E−01
0.28	6.3573E+00	1.1815E+00	3.8820E+00	3.2857E+00	2.1656E+00	1.7950E+00	7.7268E−01
0.30	5.2993E+00	1.1788E+00	3.6191E+00	3.0702E+00	2.0351E+00	1.6979E+00	7.1053E−01
0.32	4.4468E+00	1.1759E+00	3.3887E+00	2.8818E+00	1.9219E+00	1.6144E+00	6.5329E−01
0.34	3.7520E+00	1.1729E+00	3.1853E+00	2.7158E+00	1.8229E+00	1.5420E+00	6.0042E−01
0.36	3.1802E+00	1.1697E+00	3.0042E+00	2.5684E+00	1.7358E+00	1.4789E+00	5.5146E−01
0.38	2.7055E+00	1.1663E+00	2.8420E+00	2.4367E+00	1.6587E+00	1.4236E+00	5.0603E−01
0.40	2.3086E+00	1.1628E+00	2.6958E+00	2.3184E+00	1.5901E+00	1.3749E+00	4.6382E−01
0.42	1.9744E+00	1.1591E+00	2.5634E+00	2.2115E+00	1.5289E+00	1.3318E+00	4.2455E−01
0.44	1.6916E+00	1.1553E+00	2.4428E+00	2.1145E+00	1.4740E+00	1.2937E+00	3.8798E−01
0.46	1.4510E+00	1.1513E+00	2.3326E+00	2.0261E+00	1.4246E+00	1.2598E+00	3.5391E−01
0.48	1.2454E+00	1.1471E+00	2.2313E+00	1.9451E+00	1.3801E+00	1.2296E+00	3.2215E−01
0.50	1.0691E+00	1.1429E+00	2.1381E+00	1.8708E+00	1.3398E+00	1.2027E+00	2.9255E−01
0.52	9.1747E−01	1.1384E+00	2.0519E+00	1.8024E+00	1.3034E+00	1.1786E+00	2.6497E−01
0.54	7.8667E−01	1.1339E+00	1.9719E+00	1.7391E+00	1.2703E+00	1.1571E+00	2.3927E−01
0.56	6.7362E−01	1.1292E+00	1.8975E+00	1.6805E+00	1.2403E+00	1.1378E+00	2.1535E−01
0.58	5.7572E−01	1.1244E+00	1.8282E+00	1.6260E+00	1.2130E+00	1.1205E+00	1.9310E−01
0.60	4.9086E−01	1.1194E+00	1.7634E+00	1.5753E+00	1.1882E+00	1.1050E+00	1.7244E−01
0.62	4.1723E−01	1.1143E+00	1.7026E+00	1.5279E+00	1.1656E+00	1.0912E+00	1.5328E−01
0.64	3.5333E−01	1.1091E+00	1.6456E+00	1.4836E+00	1.1451E+00	1.0788E+00	1.3553E−01
0.66	2.9788E−01	1.1038E+00	1.5919E+00	1.4421E+00	1.1265E+00	1.0678E+00	1.1915E−01
0.68	2.4980E−01	1.0984E+00	1.5413E+00	1.4032E+00	1.1097E+00	1.0579E+00	1.0405E−01
0.70	2.0817E−01	1.0929E+00	1.4935E+00	1.3665E+00	1.0944E+00	1.0492E+00	9.0181E−02
0.72	1.7217E−01	1.0873E+00	1.4482E+00	1.3320E+00	1.0806E+00	1.0414E+00	7.7490E−02
0.74	1.4114E−01	1.0815E+00	1.4054E+00	1.2994E+00	1.0681E+00	1.0345E+00	6.5923E−02

(Continued)

M	$4C_fL^*/D$	T/T^*	p/p^*	ρ/ρ^*	p_t/p_t^*	I/I^*	$\Delta s/R$
0.76	1.1449E−01	1.0757E+00	1.3647E+00	1.2686E+00	1.0570E+00	1.0284E+00	5.5434E−02
0.78	9.1691E−02	1.0698E+00	1.3261E+00	1.2395E+00	1.0471E+00	1.0231E+00	4.5979E−02
0.80	7.2306E−02	1.0638E+00	1.2893E+00	1.2119E+00	1.0382E+00	1.0185E+00	3.7517E−02
0.82	5.5946E−02	1.0578E+00	1.2542E+00	1.1858E+00	1.0305E+00	1.0145E+00	3.0009E−02
0.84	4.2269E−02	1.0516E+00	1.2208E+00	1.1609E+00	1.0237E+00	1.0112E+00	2.3419E−02
0.86	3.0976E−02	1.0454E+00	1.1889E+00	1.1373E+00	1.0179E+00	1.0083E+00	1.7713E−02
0.88	2.1804E−02	1.0391E+00	1.1583E+00	1.1148E+00	1.0129E+00	1.0059E+00	1.2858E−02
0.90	1.4520E−02	1.0327E+00	1.1291E+00	1.0934E+00	1.0089E+00	1.0040E+00	8.8238E−03
0.92	8.9194E−03	1.0263E+00	1.1011E+00	1.0730E+00	1.0056E+00	1.0025E+00	5.5815E−03
0.94	4.8199E−03	1.0198E+00	1.0743E+00	1.0535E+00	1.0031E+00	1.0014E+00	3.1035E−03
0.96	2.0601E−03	1.0132E+00	1.0485E+00	1.0348E+00	1.0014E+00	1.0006E+00	1.3636E−03
0.98	4.9614E−04	1.0066E+00	1.0238E+00	1.0170E+00	1.0003E+00	1.0001E+00	3.3708E−04
1.00	0.0000E+00	1.0000E+00	1.0000E+00	1.0000E+00	1.0000E+00	1.0000E+00	0.0000E+00
1.02	4.5728E−04	9.9331E−01	9.7711E−01	9.8369E−01	1.0003E+00	1.0001E+00	3.2967E−04
1.04	1.7657E−03	9.8658E−01	9.5507E−01	9.6805E−01	1.0013E+00	1.0005E+00	1.3043E−03
1.06	3.8337E−03	9.7982E−01	9.3383E−01	9.5306E−01	1.0029E+00	1.0012E+00	2.9032E−03
1.08	6.5792E−03	9.7302E−01	9.1335E−01	9.3868E−01	1.0051E+00	1.0020E+00	5.1061E−03
1.10	9.9283E−03	9.6618E−01	8.9359E−01	9.2486E−01	1.0079E+00	1.0031E+00	7.8941E−03
1.12	1.3815E−02	9.5932E−01	8.7451E−01	9.1159E−01	1.0113E+00	1.0043E+00	1.1249E−02
1.14	1.8179E−02	9.5244E−01	8.5608E−01	8.9883E−01	1.0153E+00	1.0057E+00	1.5152E−02
1.16	2.2967E−02	9.4554E−01	8.3826E−01	8.8655E−01	1.0198E+00	1.0073E+00	1.9587E−02
1.18	2.8130E−02	9.3861E−01	8.2103E−01	8.7473E−01	1.0248E+00	1.0090E+00	2.4537E−02
1.20	3.3625E−02	9.3168E−01	8.0436E−01	8.6335E−01	1.0304E+00	1.0108E+00	2.9986E−02
1.22	3.9412E−02	9.2473E−01	7.8822E−01	8.5238E−01	1.0366E+00	1.0128E+00	3.5919E−02
1.24	4.5456E−02	9.1777E−01	7.7258E−01	8.4181E−01	1.0432E+00	1.0149E+00	4.2321E−02
1.26	5.1723E−02	9.1080E−01	7.5743E−01	8.3161E−01	1.0504E+00	1.0170E+00	4.9177E−02
1.28	5.8185E−02	9.0383E−01	7.4274E−01	8.2176E−01	1.0581E+00	1.0193E+00	5.6474E−02
1.30	6.4814E−02	8.9686E−01	7.2848E−01	8.1226E−01	1.0663E+00	1.0217E+00	6.4199E−02
1.32	7.1588E−02	8.8989E−01	7.1465E−01	8.0308E−01	1.0750E+00	1.0241E+00	7.2338E−02
1.34	7.8484E−02	8.8292E−01	7.0122E−01	7.9421E−01	1.0842E+00	1.0267E+00	8.0878E−02
1.36	8.5482E−02	8.7596E−01	6.8818E−01	7.8563E−01	1.0940E+00	1.0292E+00	8.9808E−02
1.38	9.2565E−02	8.6901E−01	6.7551E−01	7.7734E−01	1.1042E+00	1.0319E+00	9.9115E−02
1.40	9.9716E−02	8.6207E−01	6.6320E−01	7.6931E−01	1.1149E+00	1.0346E+00	1.0879E−01
1.42	1.0692E−01	8.5514E−01	6.5122E−01	7.6154E−01	1.1262E+00	1.0373E+00	1.1882E−01
1.44	1.1416E−01	8.4822E−01	6.3958E−01	7.5402E−01	1.1379E+00	1.0401E+00	1.2919E−01
1.46	1.2144E−01	8.4133E−01	6.2825E−01	7.4673E−01	1.1501E+00	1.0430E+00	1.3989E−01
1.48	1.2873E−01	8.3445E−01	6.1722E−01	7.3967E−01	1.1629E+00	1.0458E+00	1.5092E−01
1.50	1.3602E−01	8.2759E−01	6.0648E−01	7.3283E−01	1.1762E+00	1.0487E+00	1.6226E−01
1.52	1.4332E−01	8.2075E−01	5.9602E−01	7.2619E−01	1.1899E+00	1.0516E+00	1.7391E−01
1.54	1.5060E−01	8.1393E−01	5.8583E−01	7.1975E−01	1.2042E+00	1.0546E+00	1.8584E−01

M	$4C_fL^*/D$	T/T^*	p/p^*	ρ/ρ^*	p_t/p_t^*	I/I^*	$\Delta s/R$
1.56	$1.5787E{-}01$	$8.0715E{-}01$	$5.7591E{-}01$	$7.1351E{-}01$	$1.2190E{+}00$	$1.0575E{+}00$	$1.9807E{-}01$
1.58	$1.6511E{-}01$	$8.0038E{-}01$	$5.6623E{-}01$	$7.0745E{-}01$	$1.2344E{+}00$	$1.0605E{+}00$	$2.1057E{-}01$
1.60	$1.7233E{-}01$	$7.9365E{-}01$	$5.5679E{-}01$	$7.0156E{-}01$	$1.2502E{+}00$	$1.0635E{+}00$	$2.2333E{-}01$
1.62	$1.7950E{-}01$	$7.8695E{-}01$	$5.4759E{-}01$	$6.9584E{-}01$	$1.2666E{+}00$	$1.0665E{+}00$	$2.3636E{-}01$
1.64	$1.8664E{-}01$	$7.8027E{-}01$	$5.3862E{-}01$	$6.9029E{-}01$	$1.2836E{+}00$	$1.0695E{+}00$	$2.4963E{-}01$
1.66	$1.9373E{-}01$	$7.7363E{-}01$	$5.2986E{-}01$	$6.8489E{-}01$	$1.3010E{+}00$	$1.0725E{+}00$	$2.6315E{-}01$
1.68	$2.0078E{-}01$	$7.6703E{-}01$	$5.2131E{-}01$	$6.7965E{-}01$	$1.3190E{+}00$	$1.0755E{+}00$	$2.7690E{-}01$
1.70	$2.0777E{-}01$	$7.6046E{-}01$	$5.1297E{-}01$	$6.7455E{-}01$	$1.3376E{+}00$	$1.0785E{+}00$	$2.9088E{-}01$
1.72	$2.1471E{-}01$	$7.5392E{-}01$	$5.0482E{-}01$	$6.6959E{-}01$	$1.3567E{+}00$	$1.0815E{+}00$	$3.0508E{-}01$
1.74	$2.2158E{-}01$	$7.4742E{-}01$	$4.9686E{-}01$	$6.6476E{-}01$	$1.3764E{+}00$	$1.0845E{+}00$	$3.1949E{-}01$
1.76	$2.2840E{-}01$	$7.4096E{-}01$	$4.8909E{-}01$	$6.6007E{-}01$	$1.3967E{+}00$	$1.0875E{+}00$	$3.3411E{-}01$
1.78	$2.3516E{-}01$	$7.3454E{-}01$	$4.8149E{-}01$	$6.5550E{-}01$	$1.4175E{+}00$	$1.0905E{+}00$	$3.4893E{-}01$
1.80	$2.4185E{-}01$	$7.2816E{-}01$	$4.7407E{-}01$	$6.5105E{-}01$	$1.4390E{+}00$	$1.0935E{+}00$	$3.6394E{-}01$
1.82	$2.4847E{-}01$	$7.2181E{-}01$	$4.6681E{-}01$	$6.4672E{-}01$	$1.4610E{+}00$	$1.0965E{+}00$	$3.7913E{-}01$
1.84	$2.5503E{-}01$	$7.1551E{-}01$	$4.5972E{-}01$	$6.4250E{-}01$	$1.4836E{+}00$	$1.0995E{+}00$	$3.9450E{-}01$
1.86	$2.6152E{-}01$	$7.0925E{-}01$	$4.5278E{-}01$	$6.3839E{-}01$	$1.5069E{+}00$	$1.1024E{+}00$	$4.1005E{-}01$
1.88	$2.6794E{-}01$	$7.0304E{-}01$	$4.4600E{-}01$	$6.3439E{-}01$	$1.5308E{+}00$	$1.1054E{+}00$	$4.2576E{-}01$
1.90	$2.7429E{-}01$	$6.9686E{-}01$	$4.3936E{-}01$	$6.3048E{-}01$	$1.5553E{+}00$	$1.1083E{+}00$	$4.4164E{-}01$
1.92	$2.8057E{-}01$	$6.9073E{-}01$	$4.3287E{-}01$	$6.2668E{-}01$	$1.5804E{+}00$	$1.1112E{+}00$	$4.5767E{-}01$
1.94	$2.8677E{-}01$	$6.8465E{-}01$	$4.2651E{-}01$	$6.2297E{-}01$	$1.6062E{+}00$	$1.1141E{+}00$	$4.7385E{-}01$
1.96	$2.9291E{-}01$	$6.7861E{-}01$	$4.2029E{-}01$	$6.1935E{-}01$	$1.6326E{+}00$	$1.1170E{+}00$	$4.9018E{-}01$
1.98	$2.9897E{-}01$	$6.7262E{-}01$	$4.1421E{-}01$	$6.1582E{-}01$	$1.6597E{+}00$	$1.1198E{+}00$	$5.0665E{-}01$
2.00	$3.0495E{-}01$	$6.6667E{-}01$	$4.0825E{-}01$	$6.1237E{-}01$	$1.6875E{+}00$	$1.1227E{+}00$	$5.2325E{-}01$
2.05	$3.1961E{-}01$	$6.5200E{-}01$	$3.9388E{-}01$	$6.0412E{-}01$	$1.7600E{+}00$	$1.1297E{+}00$	$5.6531E{-}01$
2.10	$3.3381E{-}01$	$6.3762E{-}01$	$3.8024E{-}01$	$5.9635E{-}01$	$1.8369E{+}00$	$1.1366E{+}00$	$6.0810E{-}01$
2.15	$3.4756E{-}01$	$6.2354E{-}01$	$3.6728E{-}01$	$5.8902E{-}01$	$1.9185E{+}00$	$1.1434E{+}00$	$6.5157E{-}01$
2.20	$3.6086E{-}01$	$6.0976E{-}01$	$3.5494E{-}01$	$5.8210E{-}01$	$2.0050E{+}00$	$1.1500E{+}00$	$6.9563E{-}01$
2.25	$3.7374E{-}01$	$5.9627E{-}01$	$3.4319E{-}01$	$5.7557E{-}01$	$2.0964E{+}00$	$1.1565E{+}00$	$7.4024E{-}01$
2.30	$3.8618E{-}01$	$5.8309E{-}01$	$3.3200E{-}01$	$5.6938E{-}01$	$2.1931E{+}00$	$1.1628E{+}00$	$7.8533E{-}01$
2.35	$3.9821E{-}01$	$5.7021E{-}01$	$3.2133E{-}01$	$5.6353E{-}01$	$2.2953E{+}00$	$1.1690E{+}00$	$8.3085E{-}01$
2.40	$4.0984E{-}01$	$5.5762E{-}01$	$3.1114E{-}01$	$5.5798E{-}01$	$2.4031E{+}00$	$1.1751E{+}00$	$8.7676E{-}01$
2.45	$4.2107E{-}01$	$5.4533E{-}01$	$3.0141E{-}01$	$5.5272E{-}01$	$2.5168E{+}00$	$1.1810E{+}00$	$9.2300E{-}01$
2.50	$4.3193E{-}01$	$5.3333E{-}01$	$2.9212E{-}01$	$5.4772E{-}01$	$2.6367E{+}00$	$1.1867E{+}00$	$9.6954E{-}01$
2.55	$4.4241E{-}01$	$5.2163E{-}01$	$2.8323E{-}01$	$5.4298E{-}01$	$2.7630E{+}00$	$1.1923E{+}00$	$1.0163E{+}00$
2.60	$4.5253E{-}01$	$5.1020E{-}01$	$2.7473E{-}01$	$5.3846E{-}01$	$2.8960E{+}00$	$1.1978E{+}00$	$1.0633E{+}00$
2.65	$4.6232E{-}01$	$4.9906E{-}01$	$2.6658E{-}01$	$5.3417E{-}01$	$3.0359E{+}00$	$1.2031E{+}00$	$1.1105E{+}00$
2.70	$4.7176E{-}01$	$4.8820E{-}01$	$2.5878E{-}01$	$5.3007E{-}01$	$3.1830E{+}00$	$1.2083E{+}00$	$1.1578E{+}00$
2.75	$4.8089E{-}01$	$4.7761E{-}01$	$2.5131E{-}01$	$5.2617E{-}01$	$3.3377E{+}00$	$1.2133E{+}00$	$1.2053E{+}00$
2.80	$4.8971E{-}01$	$4.6729E{-}01$	$2.4414E{-}01$	$5.2245E{-}01$	$3.5001E{+}00$	$1.2182E{+}00$	$1.2528E{+}00$

(Continued)

M	$4C_fL^*/D$	T/T^*	p/p^*	ρ/ρ^*	p_t/p_t^*	I/I^*	$\Delta s/R$
2.85	4.9823E−01	4.5723E−01	2.3726E−01	5.1890E−01	3.6707E+00	1.2230E+00	1.3004E+00
2.90	5.0646E−01	4.4743E−01	2.3066E−01	5.1551E−01	3.8498E+00	1.2277E+00	1.3480E+00
2.95	5.1441E−01	4.3788E−01	2.2431E−01	5.1227E−01	4.0376E+00	1.2322E+00	1.3957E+00
3.00	5.2210E−01	4.2857E−01	2.1822E−01	5.0918E−01	4.2346E+00	1.2366E+00	1.4433E+00
3.10	5.3672E−01	4.1068E−01	2.0672E−01	5.0337E−01	4.6573E+00	1.2450E+00	1.5384E+00
3.20	5.5038E−01	3.9370E−01	1.9608E−01	4.9804E−01	5.1210E+00	1.2530E+00	1.6333E+00
3.30	5.6317E−01	3.7760E−01	1.8621E−01	4.9314E−01	5.6286E+00	1.2605E+00	1.7279E+00
3.40	5.7515E−01	3.6232E−01	1.7704E−01	4.8862E−01	6.1837E+00	1.2676E+00	1.8219E+00
3.50	5.8637E−01	3.4783E−01	1.6851E−01	4.8445E−01	6.7896E+00	1.2743E+00	1.9154E+00
3.60	5.9689E−01	3.3408E−01	1.6055E−01	4.8059E−01	7.4501E+00	1.2807E+00	2.0082E+00
3.70	6.0677E−01	3.2103E−01	1.5313E−01	4.7701E−01	8.1691E+00	1.2867E+00	2.1004E+00
3.80	6.1605E−01	3.0864E−01	1.4620E−01	4.7368E−01	8.9506E+00	1.2924E+00	2.1917E+00
3.90	6.2478E−01	2.9688E−01	1.3971E−01	4.7059E−01	9.7990E+00	1.2978E+00	2.2823E+00
4.00	6.3300E−01	2.8571E−01	1.3363E−01	4.6771E−01	1.0719E+01	1.3029E+00	2.3720E+00
4.10	6.4074E−01	2.7510E−01	1.2793E−01	4.6502E−01	1.1715E+01	1.3077E+00	2.4608E+00
4.20	6.4804E−01	2.6502E−01	1.2257E−01	4.6250E−01	1.2792E+01	1.3123E+00	2.5488E+00
4.30	6.5492E−01	2.5543E−01	1.1753E−01	4.6015E−01	1.3955E+01	1.3167E+00	2.6358E+00
4.40	6.6142E−01	2.4631E−01	1.1279E−01	4.5794E−01	1.5210E+01	1.3208E+00	2.7219E+00
4.50	6.6757E−01	2.3762E−01	1.0833E−01	4.5587E−01	1.6562E+01	1.3247E+00	2.8071E+00
4.60	6.7338E−01	2.2936E−01	1.0411E−01	4.5393E−01	1.8018E+01	1.3285E+00	2.8914E+00
4.70	6.7888E−01	2.2148E−01	1.0013E−01	4.5210E−01	1.9583E+01	1.3320E+00	2.9747E+00
4.80	6.8410E−01	2.1398E−01	9.6371E−02	4.5037E−01	2.1264E+01	1.3354E+00	3.0570E+00
4.90	6.8904E−01	2.0683E−01	9.2812E−02	4.4875E−01	2.3067E+01	1.3386E+00	3.1384E+00
5.00	6.9373E−01	2.0000E−01	8.9443E−02	4.4721E−01	2.5000E+01	1.3416E+00	3.2189E+00
5.20	7.0242E−01	1.8727E−01	8.3220E−02	4.4439E−01	2.9283E+01	1.3473E+00	3.3770E+00
5.40	7.1028E−01	1.7564E−01	7.7611E−02	4.4186E−01	3.4175E+01	1.3525E+00	3.5315E+00
5.60	7.1741E−01	1.6502E−01	7.2540E−02	4.3959E−01	3.9740E+01	1.3572E+00	3.6824E+00
5.80	7.2389E−01	1.5528E−01	6.7941E−02	4.3754E−01	4.6050E+01	1.3615E+00	3.8297E+00
6.00	7.2980E−01	1.4634E−01	6.3758E−02	4.3568E−01	5.3180E+01	1.3655E+00	3.9737E+00
6.50	7.4247E−01	1.2698E−01	5.4823E−02	4.3173E−01	7.5134E+01	1.3740E+00	4.3193E+00
7.00	7.5273E−01	1.1111E−01	4.7619E−02	4.2857E−01	1.0414E+02	1.3810E+00	4.6458E+00
7.50	7.6114E−01	9.7959E−02	4.1731E−02	4.2601E−01	1.4184E+02	1.3867E+00	4.9547E+00
8.00	7.6812E−01	8.6957E−02	3.6860E−02	4.2390E−01	1.9011E+02	1.3915E+00	5.2476E+00
8.50	7.7397E−01	7.7670E−02	3.2787E−02	4.2214E−01	2.5109E+02	1.3955E+00	5.5258E+00
9.00	7.7891E−01	6.9767E−02	2.9348E−02	4.2066E−01	3.2719E+02	1.3989E+00	5.7905E+00
9.50	7.8313E−01	6.2992E−02	2.6419E−02	4.1940E−01	4.2113E+02	1.4019E+00	6.0429E+00
10.00	7.8676E−01	5.7143E−02	2.3905E−02	4.1833E−01	5.3594E+02	1.4044E+00	6.2840E+00
15.00	8.0577E−01	2.6087E−02	1.0768E−02	4.1276E−01	3.7552E+03	1.4177E+00	8.2309E+00

Fanno Flow $\gamma = 1.3$

M	$4C_f L^*/D$	T/T^*	p/p^*	ρ/ρ^*	p_t/p_t^*	I/I^*	$\Delta s/R$
0.10	7.2202E+01	1.1483E+00	1.0716E+01	9.3320E+00	5.8860E+00	4.7196E+00	1.7726E+00
0.12	4.9020E+01	1.1475E+00	8.9269E+00	7.7793E+00	4.9174E+00	3.9539E+00	1.5928E+00
0.14	3.5120E+01	1.1466E+00	7.6486E+00	6.6705E+00	4.2275E+00	3.4102E+00	1.4417E+00
0.16	2.6157E+01	1.1456E+00	6.6895E+00	5.8393E+00	3.7118E+00	3.0053E+00	1.3116E+00
0.18	2.0058E+01	1.1444E+00	5.9432E+00	5.1932E+00	3.3123E+00	2.6929E+00	1.1977E+00
0.20	1.5732E+01	1.1431E+00	5.3459E+00	4.6765E+00	2.9940E+00	2.4452E+00	1.0967E+00
0.22	1.2562E+01	1.1417E+00	4.8569E+00	4.2540E+00	2.7349E+00	2.2445E+00	1.0061E+00
0.24	1.0177E+01	1.1401E+00	4.4491E+00	3.9022E+00	2.5202E+00	2.0792E+00	9.2438E-01
0.26	8.3414E+00	1.1385E+00	4.1038E+00	3.6047E+00	2.3396E+00	1.9411E+00	8.5004E-01
0.28	6.9035E+00	1.1366E+00	3.8076E+00	3.3499E+00	2.1859E+00	1.8242E+00	7.8207E-01
0.30	5.7595E+00	1.1347E+00	3.5507E+00	3.1293E+00	2.0537E+00	1.7244E+00	7.1966E-01
0.32	4.8370E+00	1.1326E+00	3.3257E+00	2.9364E+00	1.9389E+00	1.6385E+00	6.6215E-01
0.34	4.0848E+00	1.1304E+00	3.1271E+00	2.7663E+00	1.8385E+00	1.5639E+00	6.0900E-01
0.36	3.4653E+00	1.1281E+00	2.9503E+00	2.6153E+00	1.7502E+00	1.4989E+00	5.5974E-01
0.38	2.9507E+00	1.1256E+00	2.7920E+00	2.4804E+00	1.6719E+00	1.4418E+00	5.1401E-01
0.40	2.5200E+00	1.1230E+00	2.6493E+00	2.3591E+00	1.6023E+00	1.3915E+00	4.7149E-01
0.42	2.1572E+00	1.1204E+00	2.5202E+00	2.2494E+00	1.5401E+00	1.3470E+00	4.3190E-01
0.44	1.8499E+00	1.1175E+00	2.4026E+00	2.1499E+00	1.4843E+00	1.3075E+00	3.9500E-01
0.46	1.5882E+00	1.1146E+00	2.2951E+00	2.0591E+00	1.4341E+00	1.2724E+00	3.6059E-01
0.48	1.3645E+00	1.1116E+00	2.1965E+00	1.9760E+00	1.3888E+00	1.2410E+00	3.2849E-01
0.50	1.1724E+00	1.1084E+00	2.1056E+00	1.8997E+00	1.3479E+00	1.2130E+00	2.9855E-01
0.52	1.0071E+00	1.1052E+00	2.0217E+00	1.8293E+00	1.3107E+00	1.1880E+00	2.7062E-01
0.54	8.6433E-01	1.1018E+00	1.9438E+00	1.7642E+00	1.2770E+00	1.1655E+00	2.4457E-01
0.56	7.4082E-01	1.0983E+00	1.8715E+00	1.7039E+00	1.2464E+00	1.1454E+00	2.2030E-01
0.58	6.3377E-01	1.0948E+00	1.8040E+00	1.6478E+00	1.2186E+00	1.1273E+00	1.9771E-01
0.60	5.4087E-01	1.0911E+00	1.7409E+00	1.5956E+00	1.1932E+00	1.1112E+00	1.7670E-01
0.62	4.6018E-01	1.0873E+00	1.6818E+00	1.5468E+00	1.1702E+00	1.0966E+00	1.5720E-01
0.64	3.9008E-01	1.0834E+00	1.6264E+00	1.5011E+00	1.1492E+00	1.0836E+00	1.3913E-01
0.66	3.2919E-01	1.0795E+00	1.5742E+00	1.4583E+00	1.1302E+00	1.0720E+00	1.2241E-01
0.68	2.7632E-01	1.0754E+00	1.5250E+00	1.4181E+00	1.1129E+00	1.0616E+00	1.0700E-01
0.70	2.3049E-01	1.0713E+00	1.4786E+00	1.3802E+00	1.0972E+00	1.0524E+00	9.2820E-02
0.72	1.9082E-01	1.0670E+00	1.4347E+00	1.3446E+00	1.0831E+00	1.0441E+00	7.9830E-02
0.74	1.5658E-01	1.0627E+00	1.3931E+00	1.3109E+00	1.0703E+00	1.0369E+00	6.7977E-02
0.76	1.2713E-01	1.0583E+00	1.3536E+00	1.2790E+00	1.0589E+00	1.0304E+00	5.7215E-02
0.78	1.0192E-01	1.0538E+00	1.3161E+00	1.2489E+00	1.0486E+00	1.0248E+00	4.7501E-02
0.80	8.0447E-02	1.0493E+00	1.2804E+00	1.2203E+00	1.0395E+00	1.0199E+00	3.8796E-02
0.82	6.2305E-02	1.0446E+00	1.2464E+00	1.1932E+00	1.0315E+00	1.0156E+00	3.1063E-02
0.84	4.7118E-02	1.0399E+00	1.2140E+00	1.1674E+00	1.0246E+00	1.0120E+00	2.4266E-02

(Continued)

M	$4C_fL^*/D$	T/T^*	p/p^*	ρ/ρ^*	p_t/p_t^*	I/I^*	$\Delta s/R$
0.86	3.4562E−02	1.0352E+00	1.1831E+00	1.1429E+00	1.0185E+00	1.0089E+00	1.8372E−02
0.88	2.4350E−02	1.0303E+00	1.1535E+00	1.1195E+00	1.0134E+00	1.0064E+00	1.3350E−02
0.90	1.6230E−02	1.0254E+00	1.1251E+00	1.0973E+00	1.0092E+00	1.0043E+00	9.1719E−03
0.92	9.9783E−03	1.0204E+00	1.0980E+00	1.0760E+00	1.0058E+00	1.0027E+00	5.8086E−03
0.94	5.3962E−03	1.0154E+00	1.0720E+00	1.0557E+00	1.0032E+00	1.0015E+00	3.2342E−03
0.96	2.3076E−03	1.0103E+00	1.0470E+00	1.0363E+00	1.0014E+00	1.0006E+00	1.4236E−03
0.98	5.5556E−04	1.0052E+00	1.0231E+00	1.0178E+00	1.0004E+00	1.0002E+00	3.5312E−04
1.00	0.0000E+00	1.0000E+00	1.0000E+00	1.0000E+00	1.0000E+00	1.0000E+00	0.0000E+00
1.02	5.1576E−04	9.9476E−01	9.7782E−01	9.8297E−01	1.0003E+00	1.0001E+00	3.4258E−04
1.04	1.9908E−03	9.8947E−01	9.5646E−01	9.6664E−01	1.0014E+00	1.0006E+00	1.3602E−03
1.06	4.3247E−03	9.8413E−01	9.3588E−01	9.5097E−01	1.0030E+00	1.0013E+00	3.0329E−03
1.08	7.4272E−03	9.7876E−01	9.1604E−01	9.3592E−01	1.0054E+00	1.0022E+00	5.3418E−03
1.10	1.1217E−02	9.7334E−01	8.9689E−01	9.2146E−01	1.0083E+00	1.0033E+00	8.2685E−03
1.12	1.5622E−02	9.6788E−01	8.7840E−01	9.0755E−01	1.0119E+00	1.0047E+00	1.1796E−02
1.14	2.0574E−02	9.6239E−01	8.6054E−01	8.9417E−01	1.0160E+00	1.0063E+00	1.5906E−02
1.16	2.6017E−02	9.5687E−01	8.4327E−01	8.8129E−01	1.0208E+00	1.0080E+00	2.0584E−02
1.18	3.1894E−02	9.5131E−01	8.2657E−01	8.6887E−01	1.0262E+00	1.0099E+00	2.5813E−02
1.20	3.8159E−02	9.4572E−01	8.1040E−01	8.5691E−01	1.0321E+00	1.0119E+00	3.1578E−02
1.22	4.4766E−02	9.4011E−01	7.9475E−01	8.4538E−01	1.0386E+00	1.0141E+00	3.7866E−02
1.24	5.1677E−02	9.3447E−01	7.7958E−01	8.3425E−01	1.0457E+00	1.0165E+00	4.4660E−02
1.26	5.8854E−02	9.2881E−01	7.6488E−01	8.2350E−01	1.0533E+00	1.0189E+00	5.1949E−02
1.28	6.6266E−02	9.2313E−01	7.5062E−01	8.1313E−01	1.0616E+00	1.0215E+00	5.9718E−02
1.30	7.3882E−02	9.1743E−01	7.3679E−01	8.0310E−01	1.0703E+00	1.0241E+00	6.7955E−02
1.32	8.1675E−02	9.1171E−01	7.2336E−01	7.9341E−01	1.0797E+00	1.0269E+00	7.6647E−02
1.34	8.9622E−02	9.0598E−01	7.1032E−01	7.8403E−01	1.0896E+00	1.0297E+00	8.5782E−02
1.36	9.7699E−02	9.0024E−01	6.9765E−01	7.7497E−01	1.1001E+00	1.0327E+00	9.5349E−02
1.38	1.0589E−01	8.9448E−01	6.8534E−01	7.6619E−01	1.1111E+00	1.0357E+00	1.0534E−01
1.40	1.1416E−01	8.8872E−01	6.7337E−01	7.5769E−01	1.1227E+00	1.0387E+00	1.1573E−01
1.42	1.2252E−01	8.8294E−01	6.6173E−01	7.4945E−01	1.1349E+00	1.0419E+00	1.2652E−01
1.44	1.3093E−01	8.7717E−01	6.5040E−01	7.4148E−01	1.1477E+00	1.0451E+00	1.3771E−01
1.46	1.3939E−01	8.7138E−01	6.3937E−01	7.3374E−01	1.1610E+00	1.0483E+00	1.4926E−01
1.48	1.4788E−01	8.6560E−01	6.2863E−01	7.2624E−01	1.1750E+00	1.0516E+00	1.6119E−01
1.50	1.5639E−01	8.5981E−01	6.1817E−01	7.1896E−01	1.1895E+00	1.0549E+00	1.7347E−01
1.52	1.6492E−01	8.5403E−01	6.0798E−01	7.1190E−01	1.2046E+00	1.0583E+00	1.8610E−01
1.54	1.7344E−01	8.4825E−01	5.9805E−01	7.0505E−01	1.2203E+00	1.0617E+00	1.9907E−01
1.56	1.8196E−01	8.4247E−01	5.8837E−01	6.9839E−01	1.2367E+00	1.0651E+00	2.1237E−01
1.58	1.9047E−01	8.3669E−01	5.7893E−01	6.9193E−01	1.2536E+00	1.0686E+00	2.2600E−01
1.60	1.9895E−01	8.3092E−01	5.6972E−01	6.8564E−01	1.2712E+00	1.0721E+00	2.3993E−01
1.62	2.0740E−01	8.2517E−01	5.6073E−01	6.7954E−01	1.2895E+00	1.0756E+00	2.5417E−01
1.64	2.1581E−01	8.1942E−01	5.5196E−01	6.7360E−01	1.3083E+00	1.0791E+00	2.6870E−01

M	$4C_fL^*/D$	T/T^*	p/p^*	ρ/ρ^*	p_t/p_t^*	I/I^*	$\Delta s/R$
1.66	2.2419E−01	8.1368E−01	5.4340E−01	6.6783E−01	1.3279E+00	1.0826E+00	2.8352E−01
1.68	2.3252E−01	8.0795E−01	5.3504E−01	6.6222E−01	1.3481E+00	1.0861E+00	2.9863E−01
1.70	2.4081E−01	8.0223E−01	5.2687E−01	6.5675E−01	1.3690E+00	1.0897E+00	3.1400E−01
1.72	2.4904E−01	7.9653E−01	5.1889E−01	6.5143E−01	1.3906E+00	1.0933E+00	3.2964E−01
1.74	2.5721E−01	7.9085E−01	5.1109E−01	6.4626E−01	1.4129E+00	1.0968E+00	3.4554E−01
1.76	2.6532E−01	7.8518E−01	5.0347E−01	6.4122E−01	1.4359E+00	1.1004E+00	3.6169E−01
1.78	2.7338E−01	7.7952E−01	4.9601E−01	6.3631E−01	1.4596E+00	1.1039E+00	3.7808E−01
1.80	2.8136E−01	7.7389E−01	4.8873E−01	6.3152E−01	1.4841E+00	1.1075E+00	3.9471E−01
1.82	2.8928E−01	7.6827E−01	4.8160E−01	6.2686E−01	1.5093E+00	1.1111E+00	4.1157E−01
1.84	2.9713E−01	7.6268E−01	4.7463E−01	6.2232E−01	1.5353E+00	1.1146E+00	4.2866E−01
1.86	3.0491E−01	7.5711E−01	4.6781E−01	6.1789E−01	1.5621E+00	1.1182E+00	4.4596E−01
1.88	3.1262E−01	7.5156E−01	4.6113E−01	6.1357E−01	1.5897E+00	1.1217E+00	4.6347E−01
1.90	3.2025E−01	7.4603E−01	4.5459E−01	6.0935E−01	1.6181E+00	1.1252E+00	4.8119E−01
1.92	3.2781E−01	7.4052E−01	4.4820E−01	6.0524E−01	1.6474E+00	1.1287E+00	4.9911E−01
1.94	3.3529E−01	7.3504E−01	4.4193E−01	6.0123E−01	1.6775E+00	1.1322E+00	5.1722E−01
1.96	3.4269E−01	7.2958E−01	4.3579E−01	5.9732E−01	1.7085E+00	1.1357E+00	5.3552E−01
1.98	3.5002E−01	7.2415E−01	4.2978E−01	5.9350E−01	1.7404E+00	1.1392E+00	5.5401E−01
2.00	3.5727E−01	7.1875E−01	4.2390E−01	5.8977E−01	1.7732E+00	1.1427E+00	5.7267E−01
2.05	3.7506E−01	7.0536E−01	4.0969E−01	5.8082E−01	1.8592E+00	1.1513E+00	6.2006E−01
2.10	3.9235E−01	6.9215E−01	3.9617E−01	5.7238E−01	1.9514E+00	1.1597E+00	6.6845E−01
2.15	4.0915E−01	6.7912E−01	3.8330E−01	5.6440E−01	2.0501E+00	1.1681E+00	7.1776E−01
2.20	4.2546E−01	6.6628E−01	3.7103E−01	5.5686E−01	2.1555E+00	1.1763E+00	7.6791E−01
2.25	4.4130E−01	6.5364E−01	3.5932E−01	5.4973E−01	2.2682E+00	1.1844E+00	8.1885E−01
2.30	4.5665E−01	6.4120E−01	3.4815E−01	5.4297E−01	2.3884E+00	1.1923E+00	8.7050E−01
2.35	4.7154E−01	6.2897E−01	3.3748E−01	5.3656E−01	2.5167E+00	1.2001E+00	9.2282E−01
2.40	4.8598E−01	6.1695E−01	3.2728E−01	5.3047E−01	2.6535E+00	1.2078E+00	9.7573E−01
2.45	4.9997E−01	6.0514E−01	3.1751E−01	5.2469E−01	2.7992E+00	1.2153E+00	1.0292E+00
2.50	5.1352E−01	5.9355E−01	3.0817E−01	5.1920E−01	2.9544E+00	1.2226E+00	1.0831E+00
2.55	5.2665E−01	5.8217E−01	2.9922E−01	5.1397E−01	3.1196E+00	1.2298E+00	1.1375E+00
2.60	5.3937E−01	5.7100E−01	2.9063E−01	5.0899E−01	3.2954E+00	1.2368E+00	1.1924E+00
2.65	5.5169E−01	5.6005E−01	2.8240E−01	5.0424E−01	3.4823E+00	1.2437E+00	1.2475E+00
2.70	5.6362E−01	5.4932E−01	2.7450E−01	4.9972E−01	3.6810E+00	1.2504E+00	1.3030E+00
2.75	5.7518E−01	5.3880E−01	2.6692E−01	4.9540E−01	3.8921E+00	1.2570E+00	1.3588E+00
2.80	5.8637E−01	5.2849E−01	2.5963E−01	4.9127E−01	4.1164E+00	1.2634E+00	1.4148E+00
2.85	5.9721E−01	5.1840E−01	2.5263E−01	4.8733E−01	4.3545E+00	1.2697E+00	1.4710E+00
2.90	6.0771E−01	5.0851E−01	2.4590E−01	4.8356E−01	4.6072E+00	1.2758E+00	1.5274E+00
2.95	6.1788E−01	4.9883E−01	2.3942E−01	4.7995E−01	4.8753E+00	1.2817E+00	1.5840E+00
3.00	6.2774E−01	4.8936E−01	2.3318E−01	4.7650E−01	5.1596E+00	1.2876E+00	1.6407E+00
3.10	6.4653E−01	4.7102E−01	2.2139E−01	4.7002E−01	5.7806E+00	1.2988E+00	1.7543E+00

(Continued)

M	$4C_fL^*/D$	T/T^*	p/p^*	ρ/ρ^*	p_t/p_t^*	I/I^*	$\Delta s/R$
3.20	6.6419E−01	4.5347E−01	2.1044E−01	4.6406E−01	6.4774E+00	1.3095E+00	1.8681E+00
3.30	6.8077E−01	4.3668E−01	2.0025E−01	4.5857E−01	7.2584E+00	1.3196E+00	1.9819E+00
3.40	6.9636E−01	4.2063E−01	1.9075E−01	4.5349E−01	8.1326E+00	1.3293E+00	2.0956E+00
3.50	7.1103E−01	4.0529E−01	1.8189E−01	4.4880E−01	9.1096E+00	1.3385E+00	2.2091E+00
3.60	7.2484E−01	3.9063E−01	1.7361E−01	4.4444E−01	1.0200E+01	1.3472E+00	2.3221E+00
3.70	7.3784E−01	3.7662E−01	1.6586E−01	4.4040E−01	1.1416E+01	1.3555E+00	2.4347E+00
3.80	7.5010E−01	3.6323E−01	1.5860E−01	4.3664E−01	1.2769E+01	1.3634E+00	2.5467E+00
3.90	7.6166E−01	3.5045E−01	1.5179E−01	4.3313E−01	1.4273E+01	1.3709E+00	2.6581E+00
4.00	7.7258E−01	3.3824E−01	1.4539E−01	4.2986E−01	1.5944E+01	1.3781E+00	2.7687E+00
4.10	7.8289E−01	3.2657E−01	1.3938E−01	4.2681E−01	1.7796E+01	1.3849E+00	2.8786E+00
4.20	7.9263E−01	3.1541E−01	1.3372E−01	4.2395E−01	1.9847E+01	1.3914E+00	2.9877E+00
4.30	8.0185E−01	3.0476E−01	1.2838E−01	4.2126E−01	2.2115E+01	1.3975E+00	3.0959E+00
4.40	8.1058E−01	2.9457E−01	1.2335E−01	4.1875E−01	2.4621E+01	1.4034E+00	3.2032E+00
4.50	8.1885E−01	2.8483E−01	1.1860E−01	4.1638E−01	2.7386E+01	1.4090E+00	3.3096E+00
4.60	8.2669E−01	2.7552E−01	1.1411E−01	4.1416E−01	3.0432E+01	1.4143E+00	3.4151E+00
4.70	8.3413E−01	2.6660E−01	1.0986E−01	4.1207E−01	3.3785E+01	1.4194E+00	3.5196E+00
4.80	8.4119E−01	2.5808E−01	1.0584E−01	4.1009E−01	3.7470E+01	1.4243E+00	3.6231E+00
4.90	8.4790E−01	2.4992E−01	1.0202E−01	4.0823E−01	4.1517E+01	1.4289E+00	3.7257E+00
5.00	8.5427E−01	2.4211E−01	9.8408E−02	4.0647E−01	4.5954E+01	1.4333E+00	3.8272E+00
5.20	8.6611E−01	2.2745E−01	9.1715E−02	4.0323E−01	5.6134E+01	1.4416E+00	4.0273E+00
5.40	8.7686E−01	2.1399E−01	8.5666E−02	4.0032E−01	6.8294E+01	1.4492E+00	4.2234E+00
5.60	8.8663E−01	2.0161E−01	8.0181E−02	3.9770E−01	8.2756E+01	1.4561E+00	4.4154E+00
5.80	8.9554E−01	1.9021E−01	7.5195E−02	3.9533E−01	9.9885E+01	1.4624E+00	4.6035E+00
6.00	9.0369E−01	1.7969E−01	7.0649E−02	3.9318E−01	1.2009E+02	1.4683E+00	4.7877E+00
6.50	9.2121E−01	1.5673E−01	6.0906E−02	3.8861E−01	1.8721E+02	1.4809E+00	5.2317E+00
7.00	9.3547E−01	1.3772E−01	5.3016E−02	3.8494E−01	2.8532E+02	1.4914E+00	5.6530E+00
7.50	9.4721E−01	1.2185E−01	4.6544E−02	3.8196E−01	4.2578E+02	1.5000E+00	6.0533E+00
8.00	9.5698E−01	1.0849E−01	4.1172E−02	3.7950E−01	6.2308E+02	1.5073E+00	6.4340E+00
8.50	9.6519E−01	9.7149E−02	3.6669E−02	3.7745E−01	8.9544E+02	1.5134E+00	6.7966E+00
9.00	9.7215E−01	8.7452E−02	3.2858E−02	3.7573E−01	1.2655E+03	1.5186E+00	7.1425E+00
9.50	9.7809E−01	7.9106E−02	2.9606E−02	3.7426E−01	1.7611E+03	1.5231E+00	7.4729E+00
10.00	9.8322E−01	7.1875E−02	2.6810E−02	3.7300E−01	2.4159E+03	1.5270E+00	7.7890E+00
15.00	1.0102E+00	3.3094E−02	1.2128E−02	3.6647E−01	3.1490E+04	1.5476E+00	1.0356E+01

Prandtl–Meyer Function and Mach Angle

In sonic and supersonic flows, Mach waves are formed. The local Mach wave angle is called a Mach angle μ, and it is related to the local Mach number via

$$\sin \mu = \frac{1}{M} \text{ or } \mu = \sin^{-1}\left(\frac{1}{M}\right) \tag{F.1}$$

Each Mach wave causes an infinitesimal turn of the flow and a corresponding infinitesimal change in fluid pressure, temperature, density, and Mach number. Through infinitely many such waves, i.e., Mach waves, flow can turn a finite angle and attain a finite change in pressure, temperature, density, and Mach number. Prandtl–Meyer function $\nu(M)$ defines an angle that a sonic flow has to turn in order to achieve a supersonic Mach number M.

$$\nu(M) = \sqrt{\frac{\gamma + 1}{\gamma - 1}} \tan^{-1} \sqrt{\frac{\gamma - 1}{\gamma + 1}(M^2 - 1)} - \tan^{-1} \sqrt{M^2 - 1} \tag{F.2}$$

A graphical representation of the Prandtl–Meyer function and the Mach angle is shown.

M	ν (deg)	μ (deg)	M	ν (deg)	μ (deg)
1.0	0.00	90.00	5.8	83.54	9.93
1.1	1.34	65.38	5.9	84.26	9.76
1.2	3.56	56.44	6	84.96	9.59
1.3	6.17	50.28	6.2	86.30	9.28
1.4	8.99	45.58	6.4	87.56	8.99
1.5	11.91	41.81	6.6	88.76	8.71
1.6	14.86	38.68	6.8	89.90	8.46
1.7	17.81	36.03	7	90.97	8.21
1.8	20.73	33.75	7.5	93.44	7.66
1.9	23.59	31.76	8	95.63	7.18
2.0	26.38	30.00	8.5	97.57	6.76
2.1	29.10	28.44	9	99.32	6.38
2.2	31.73	27.04	9.5	100.89	6.04
2.3	34.28	25.77	10	102.32	5.74
2.4	36.75	24.62	10.5	103.61	5.47
2.5	39.12	23.58	11	104.80	5.22
2.6	41.42	22.62	11.5	105.88	4.99
2.7	43.62	21.74	12	106.88	4.78
2.8	45.75	20.92	12.5	107.80	4.59
2.9	47.79	20.17	13	108.65	4.41
3.0	49.76	19.47	13.5	109.44	4.25
3.1	51.65	18.82	14	110.18	4.10
3.2	53.47	18.21	14.5	110.87	3.95
3.3	55.22	17.64	15	111.51	3.82
3.4	56.91	17.10	15.5	112.11	3.70
3.5	58.53	16.60	16	112.68	3.58
3.6	60.09	16.13	16.5	113.21	3.47
3.7	61.60	15.68	17	113.71	3.37
3.8	63.05	15.26	17.5	114.18	3.28
3.9	64.44	14.86	18	114.63	3.18
4.0	65.79	14.48	18.5	115.05	3.10
4.1	67.08	14.12	19	115.45	3.02
4.2	68.33	13.77	20	116.2	2.87
4.3	69.54	13.45	21	116.9	2.73
4.4	70.71	13.14	22	117.5	2.61
4.5	71.83	12.84	23	118.0	2.49
4.6	72.92	12.56	24	118.6	2.39
4.7	73.97	12.28	25	119.0	2.29
4.8	74.99	12.02	26	119.5	2.20
4.9	75.97	11.78	27	119.9	2.12
5	76.92	11.54	28	120.2	2.05
5.1	77.84	11.31	29	120.6	1.98
5.2	78.73	11.09	30	120.9	1.91
5.3	79.60	10.88	31	121.2	1.85
5.4	80.43	10.67	32	121.5	1.79
5.5	81.25	10.48	33	121.8	1.74
5.6	82.03	10.29	34	122.0	1.69
5.7	82.80	10.10	35	122.3	1.64

M	ν (deg)	μ (deg)	M	ν (deg)	μ (deg)
36	122.5	1.59	60	125.7	0.95
37	122.7	1.55	70	126.4	0.82
38	122.9	1.51	80	126.9	0.72
39	123.1	1.47	90	127.3	0.64
40	123.3	1.43	100	127.6	0.57
50	124.7	1.15			

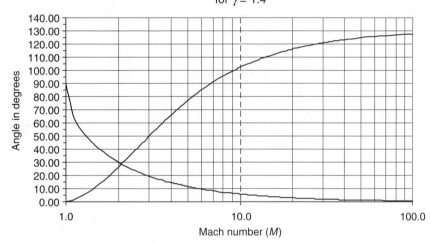

Prandtl–Meyer function and Mach angle
for $\gamma = 1.4$

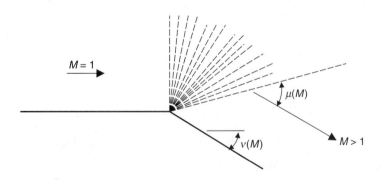

Oblique Shock
Charts

Oblique Shock Charts $\gamma = 1.4$

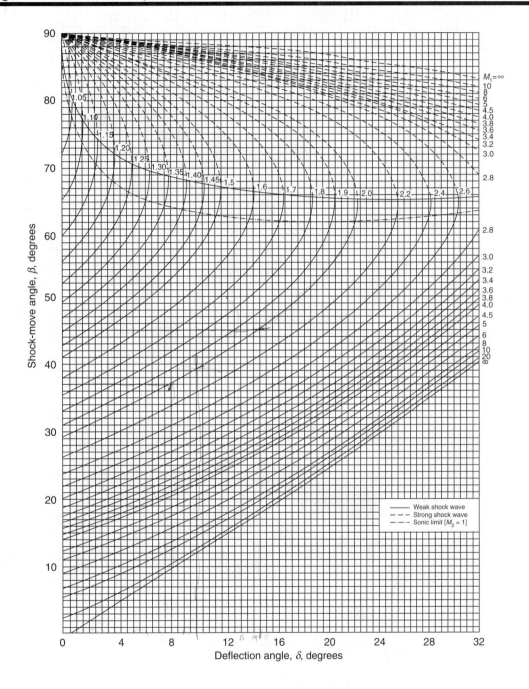

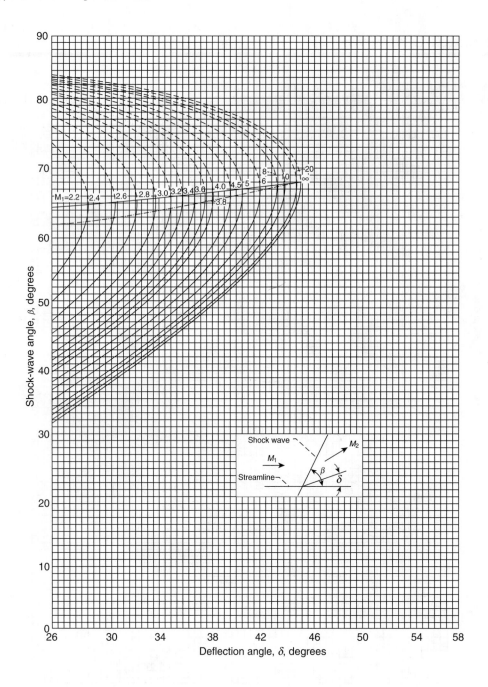

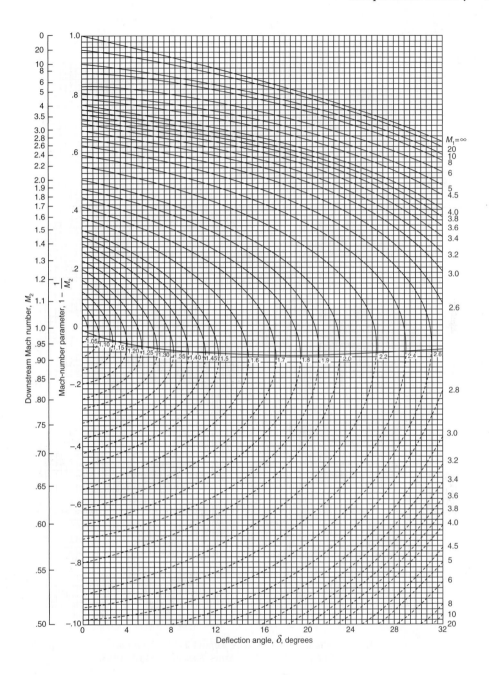

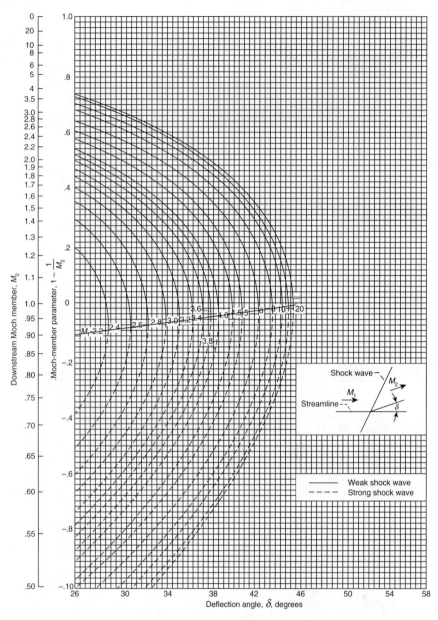

The oblique shock charts are from NACA Report 1135, *Equations, Tables, and Charts for Compressible Flow*, Ames Research Staff, 1953.

APPENDIX H

Conical Shock Charts

Conical Shock Charts $\gamma = 1.4$

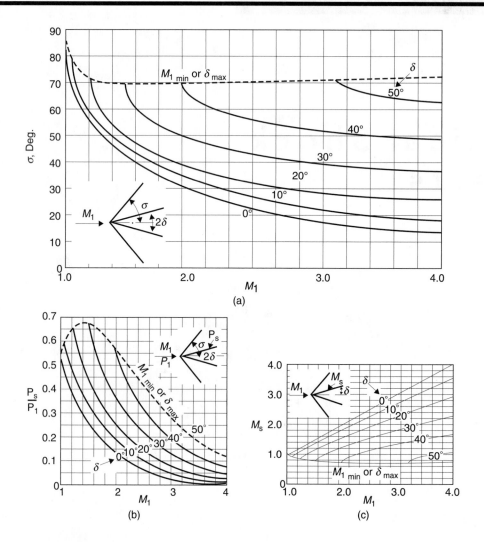

(a) Shock angle versus approach Mach number, with cone angle as parameter.

(b) Ratio of surface pressure to free stream stagnation pressure versus free stream Mach number.

(c) Surface Mach number versus free stream Mach number, with cone angle as parameter.

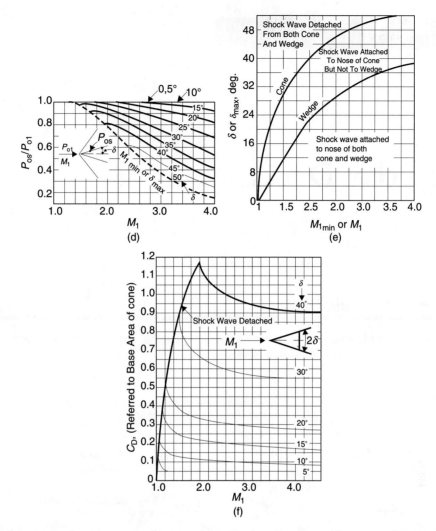

(**d**) Ratio of surface stagnation pressure to free stream stagnation pressure, with cone angle as parameter.

(**e**) Regions of shock attachment and detachment for cone and wedge.

(**f**) Pressure drag coefficient based on projected frontal area.

From: A. H. Shapiro, *The Dynamics and Thermodynamics of Compressible Fluid Flow*,
Vol. II, The Ronald Press Company, 1954.

APPENDIX I

Cascade Data

NACA-65 Series Cascade Data

Cascade total pressure loss data are plotted in the shape of a "bucket" in terms of incidence angle. A definition sketch is shown.

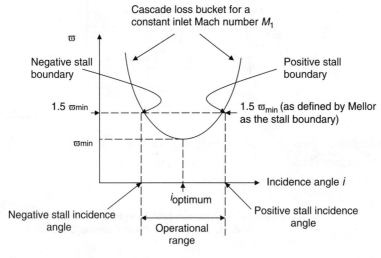

Definition sketch for the cascade stall boundaries, according to Mellor

Mellor presents the operational range of each 65-series airfoils in various cascade arrangements that are very useful for preliminary design purposes. Mellor's unpublished graphical correlations (originated at MIT's Gas Turbine Laboratory) were published by Horlock (1973), Hill and Peterson (1992), among others. In this appendix, Mellor cascade data presentation is graphed according to the following definition sketch.

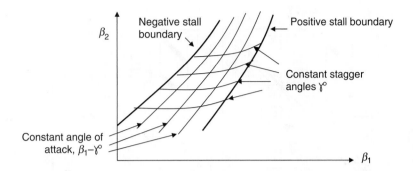

Definition sketch for the 65-series cascade data as presented by Mellor

Cascade Data

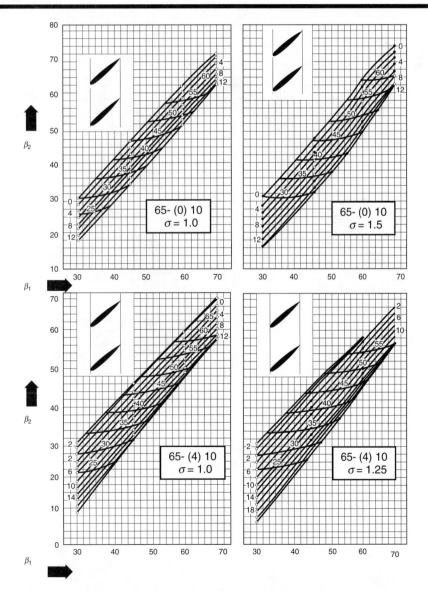

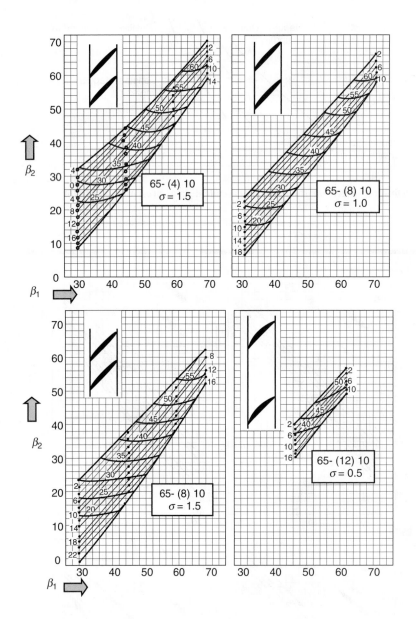

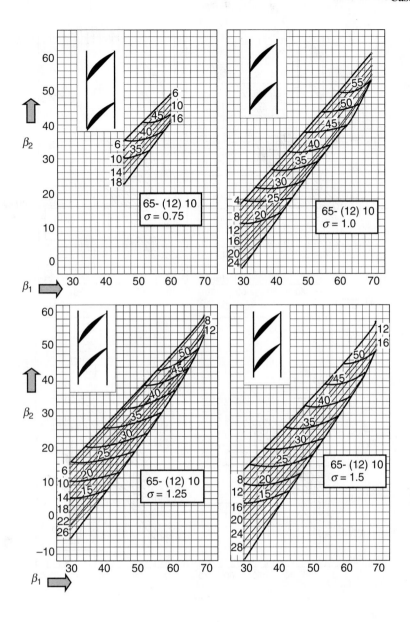

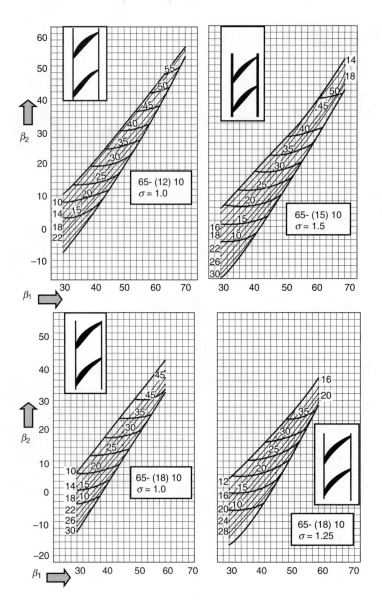

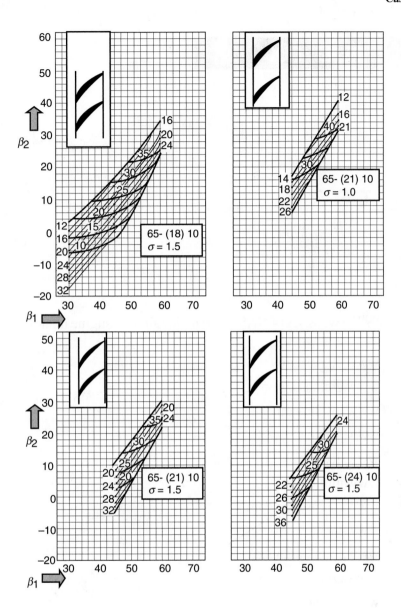

APPENDIX J

Websites

The websites of NASA and major aircraft engine manufacturers are listed for reference.

www.hq.nasa.gov

www.grc.nasa.gov

www.pw.utc.com

www.pwc.ca

www.ge.com

www.mtu.de

www.rolls-royce.com

www.snecma.com

www.enginealliance.com

www.cfm56.com

www.geae.com

www.V2500.com

INDEX